Deep-Water Contourite Systems:
Modern Drifts and Ancient Series, Seismic and Sedimentary Characteristics

Geological Society Special Publications
Society Book Editors
A. J. FLEET (CHIEF EDITOR)
P. DOYLE
F. J. GREGORY
J. S. GRIFFITHS
A. J. HARTLEY
R. E. HOLDSWORTH
A. C. MORTON
N. S. ROBINS
M. S. STOKER
J. P. TURNER

Society Publication reviewing procedures

The Society makes every effort to ensure that the scientific and production quality of its books matches that of its journals. Since 1997, all book proposals have been refereed by specialist reviewers as well as by the Society's Books Editorial Committee. If the referees identify weaknesses in the proposal, these must be addressed before the proposal is accepted.

Once the book is accepted, the Society has a team of Book Editors (listed above) who ensure that the volume editors follow strict guidelines on refereeing and quality control. We insist that individual papers can only be accepted after satisfactory review by two independent referees. The questions on the review forms are similar to those for *Journal of the Geological Society*. The referees' forms and comments must be available to the Society's Book Editors on request.

Although many of the books result from meetings, the editors are expected to commission papers that were not presented at the meeting to ensure that the book provides a balanced coverage of the subject. Being accepted for presentation at the meeting does not guarantee inclusion in the book.

Geological Society Publications are included in the ISI Index of Scientific Book Contents, but they do not have an impact factor, the latter being applicable only to journals.

More information about submitting a proposal and producing Society Publications can be found on the Society's web site: www.geolsoc.org.uk.

GEOLOGICAL SOCIETY MEMOIR NO. 22

Deep-Water Contourite Systems: Modern Drifts and Ancient Series, Seismic and Sedimentary Characteristics

EDITED BY

D. A. V. STOW
University of Southampton, UK

C. J. PUDSEY
British Antarctic Survey, UK

J. A. HOWE
Scottish Association for Marine Science, UK

J.-C. FAUGÈRES
University of Bordeaux, France

A. R. VIANA
PETROBRAS E&P, Brazil

2002
Published by
The Geological Society
London

THE GEOLOGICAL SOCIETY

The Geological Society of London (GSL) was founded in 1807. It is the oldest national geological society in the world and the largest in Europe. It was incorporated under Royal Charter in 1825 and is Registered Charity 210161.

The Society is the UK national learned and professional society for geology with a worldwide Fellowship (FGS) of 9000. The Society has the power to confer Chartered status on suitably qualified Fellows, and about 2000 of the Fellowship carry the title (CGeol). Chartered Geologists may also obtain the equivalent European title, European Geologist (EurGeol). One fifth of the Society's fellowship resides outside the UK. To find out more about the Society, log on to www.geolsoc.org.uk.

The Geological Society Publishing House (Bath, UK) produces the Society's international journals and books, and acts as European distributor for selected publications of the American Association of Petroleum Geologists (AAPG), the American Geological Institute (AGI), the Indonesian Petroleum Association (IPA), the Geological Society of America (GSA), the Society for Sedimentary Geology (SEPM) and the Geologists' Association (GA). Joint marketing agreements ensure that GSL Fellows may purchase these societies' publications at a discount. The Society's online bookshop (accessible from www.geolsoc.org.uk) offers secure book purchasing with your credit or debit card.

To find out about joining the Society and benefiting from substantial discounts on publications of GSL and other societies world-wide, consult www.geolsoc.org.uk, or contact the Fellowship Department at: The Geological Society, Burlington House, Piccadilly, London W1J 0BG: Tel. +44 (0)20 7434 9944; Fax +44 (0)20 7439 8975; Email: enquiries@geolsoc.org.uk.

Published by The Geological Society from:
The Geological Society Publishing House
Unit 7, Brassmill Enterprise Centre
Brassmill Lane
Bath BA1 3JN, UK

(*Orders*: Tel. +44 (0)1225 445046
Fax +44 (0)1225 442836)
Online bookshop: http: //bookshop.geolsoc.org.uk

The publishers make no representation, express or implied, with regard to the accuracy of the information contained in this book and cannot accept any legal responsibility for any errors or omissions that may be made.

© The Geological Society of London 2002. All rights reserved. No reproduction, copy or transmission of this publication may be made without written permission. No paragraph of this publication may be reproduced, copied or transmitted save with the provisions of the Copyright Licensing Agency, 90 Tottenham Court Road, London W1P 9HE. Users registered with the Copyright Clearance Center, 27 Congress Street, Salem, MA 01970, USA: the item-fee code for this publication is 0305–8719/02/$15.00.

British Library Cataloguing in Publication Data
A catalogue record for this book is available from the British Library.

ISBN 1-86239-092-4
ISSN 0435-4052

Typeset by Type Study, Scarborough, UK
Printed by Cambrian, Aberyswyth, UK.

Distributors

USA
AAPG Bookstore
PO Box 979
Tulsa
OK 74101–0979
USA
Orders: Tel. + 1 918 584–2555
Fax +1 918 560–2652
E-mail *bookstore@aapg.org*

India
Affiliated East-West Press PVT Ltd
G-1/16 Ansari Road, Daryaganj,
New Delhi 110 002
India
Orders: Tel. +91 11 327–9113
Fax +91 11 326–0538
E-mail *affiliat@nda.vsnl.net.in*

Japan
Kanda Book Trading Co.
Cityhouse Tama 204
Tsurumaki 1–3-10
Tama-shi
Tokyo 206–0034
Japan
Orders: Tel. +81 (0)423 57–7650
Fax +81 (0)423 57–7651

Contents

Preface	vi
McCave, I. N. Charles Davis Hollister, 1936–1999: A personal scientific appreciation of the father of 'contourites'	1
Stow, D. A. V., Faugères, J.-C., Howe, J. A., Pudsey, C. J. & Viana, A. R. Bottom currents, contourites and deep-sea sediment drifts: current state-of-the-art	7
McCave, I. N., Chandler, R. C., Swift, S. A. & Tucholke, B. E. Contourites of the Nova Scotian continental rise and the HEBBLE area	21
Tucholke, B. E. The Greater Antilles Outer Ridge: development of a distal sedimentary drift by deposition of fine-grained contourites	39
Laberg, J. S., Vorren, T. O. & Knutsen, S.-M. The Lofoten Drift, Norwegian Sea,	57
Howe, J. A., Stoker, M. S., Stow, D. A. V. & Akhurst, M. C. Sediment drifts and contourite sedimentation in the northeastern Rockall Trough and Faroe–Shetland Channel, North Atlantic Ocean	65
Akhurst, M. C., Stow, D. A. V. & Stoker, M. S. Late Quaternary glacigenic contourite, debris flow and turbidite process interaction in the Faroe–Shetland Channel, NW European continental margin	73
Knutz, P. C., Jones, E. J. W., Howe, J. A., van Weering, T. J. C. & Stow, D. A. V. Wave-form sheeted contourite drift on the Barra Fan, NW UK continental margin	85
Stow, D. A. V., Armishaw, J. E. & Holmes, R. Holocene contourite sand sheet on the Barra Fan slope, NW Hebridean margin	99
Sivkov, V., Gorbatskiy, V., Kuleshov, A. & Zhurov, Y. Muddy contourites in the Baltic Sea: an example of a shallow-water contourite system	121
Stow, D. A. V., Faugères, J.-C., Gonthier, E., Cremer, M., Llave, E., Hernández-Molina, F. J., Somoza, L. & Díaz-del-Río, V. Faro–Albufeira drift complex, northern Gulf of Cadiz	137
Ercilla, G., Baraza, J., Alonso, B., Estrada, F., Casas, D. & Farrán, M. The Ceuta Drift, Alboran Sea, southwestern Mediterranean	155
Reeder, M. S., Rothwell, G. & Stow, D. A. V. The Sicilian gateway: anatomy of the deep-water connection between East and West Mediterranean basins	171
Roveri, M. Sediment drifts of the Corsica Channel, northern Tyrrhenian Sea	191
Faugères, J.-C., Zaragosi, S., Mézerais, M. L. & Massé, L. The Vema contourite fan in the south Brazilian basin	209
Faugères, J.-C., Lima, A. F., Massé, L. & Zaragosi, S. The Columbia Channel–Levee system: a fan drift in the southern Brazil Basin	223
Gomes, P. O. & Viana, A. R. Contour currents, sediment drifts and abyssal erosion on the northeastern continental margin off Brazil	239
Viana, A. R., Hercos, C. M., de Almeida jr., W., Magalhães, J. L. C. & de Andrade, S. B. Evidence of bottom current influence on the Neogene to Quaternary sedimentation along the northern Campos slope, SW Atlantic Margin	249
Viana, A. R., de Almeida jr., W. & de Almeida, C. W. Upper slope sands: late Quaternary shallow-water sandy contourites of Campos Basin, SW Atlantic Margin	261
Uenzelmann-Neben, G. Contourites on the Agulhas Plateau, SW Indian Ocean: indications for the evolution of currents since Palaeogene times	271
Pudsey, C. J. The Weddell Sea: contourites and hemipelagites at the northern margin of the Weddell Gyre	289
Michels, K. H., Kuhn, G., Hillenbrand, C.-D., Diekmann, B., Fütterer, D. K., Grobe, H. & Uenzelmann-Neben, G. The southern Weddell Sea: combined contourite–turbidite sedimentation at the southeastern margin of the Weddell Gyre	305
Pudsey, C. J. & Howe, J. A. Mixed biosiliceous–terrigenous sedimentation under the Antarctic Circumpolar Current, Scotia Sea	325
Cunningham, A. P., Howe, J. A. & Barker, P. F. Contourite sedimentation in the Falkland Trough, western South Atlantic	337
Rebesco, M., Pudsey, C. J., Canals, M., Camerlenghi, A., Barker, P. F., Estrada, F. & Giorgetti, A. Sediment drifts and deep-sea channel systems, Antarctic Peninsula Pacific Margin	353
Escutia, C., Nelson, C. H., Acton, G. D., Eittreim, S. L., Cooper, A. K., Warnke, D. A. & Jaramillo, J. M. Current controlled deposition on the Wilkes Land continental rise, Antarctica	373
Carter, L. & McCave, I. N. Eastern New Zealand drifts, Miocene–Recent	385
Stow, D. A. V., Ogawa, Y., Lee, I. T. & Mitsuzawa, K. Neogene contourites, Miura–Boso forearc basin, SE Japan	409
Ito, M. Kuroshio Current-influenced sandy contourites from the Plio-Pleistocene Kazusa forearc basin, Boso Peninsula, Japan	421
Luo, S., Gao, Z., He, Y. & Stow, D. A. V. Ordovician carbonate contourite drifts in Hunan and Gansu Provinces, China	433
Stow, D. A. V., Kahler, G. & Reeder, M. Fossil contourites: type example from an Oligocene palaeoslope system, Cyprus	443
Index	457

It is recommended that reference to all or part of this book should be made in one of the following ways:

Stow, D. A. V., Pudsey, C. J., Howe, J. A., Faugères, J.-C. & Viana, A. R. (eds). 2002. *Deep-Water Contourite Systems: Modern Drifts and Ancient Series, Seismic and Sedimentary Characteristics.* Geological Society, London, Memoirs, **22**.

Akhurst, M. C., Stow, D. A. V. & Stoker, M. S. 2002. Late Quaternary glacigenic contourite, debris flow and turbidite process interaction in the Faroe–Shetland Channel, NW European continental margin. *In*: Stow, D. A. V., Pudsey, C. J., Howe, J. A., Faugères, J.-C. & Viana, A. R. (eds) *Deep-Water Contourite Systems: Modern Drifts and Ancient Series, Seismic and Sedimentary Characteristics.* Geological Society, London, Memoirs, **22**, 73–84.

Preface

Contourites are an extremely significant but still relatively little known group of sediments that represent a widespread component of deep ocean basins and their margins. They are coming to play an increasingly critical role in paleoceanographic studies, in that they hold an important key to the decoding of paleocirculation records encapsulated in oceanic drifts and corresponding hiatuses. These are closely linked to past climatic change. Furthermore, they are an important part of the spectrum of deposits that confront the oil industry as exploration moves into progressively greater water depths. Thick units of sandy contourites together with bottom-current reworked sandy turbidites are potentially important as hydrocarbon reservoirs where suitably buried in association with source rocks. The nature and effects of bottom currents on margin stability, the wear and tear of submarine cables, and on subsea engineering projects still need to be carefully evaluated.

However, because they are complex deposits with very subtle characteristics that are not easily recognized and decoded, they have been surrounded by controversy since they were first recognised in the early 1960s. Their occurrence and recognition in ancient series now exposed on land or in deeply buried successions of oil exploration boreholes has become particularly contentious in recent years.

In order to foster greater international dialogue on these issues, to help resolve some of the more controversial aspects, and to stimulate more focused research, an *International Geological Correlation Programme* initiative (Project 432) on *Bottom Currents, Contourites and Paleocirculation* was launched in 1998. One of the aims of this project was to publish a compendium of examples of contourite systems, both modern and ancient, and this volume is the result. Our task as editors has been facilitated by a series of IGCP432 workshop meetings, some of which have resulted in separate publications – *Deep-Water Sedimentary Systems: New Models for the 21st Century* (Stow & Mayall, eds., 2000, Marine & Petroleum Geology, v17); *Recognition and Interpretation of Deep-Water Sediment Waves* (Wynn & Stow, eds., 2002, Marine Geology); and *Seismic Expression of Contourites and Related Deposits* (Rebesco & Stow, eds., 2003, Marine Geophysical Researches, in press).

This Memoir includes 30 papers involving over 75 key scientists from around the world. Following an introductory state-of-the-art paper by the editors, there are 25 separate case studies on modern drifts and four on ancient contourite series. As far as possible, we have tried to ensure that each of these contributions highlights the specific geological and oceanographic setting, bathymetry, physiographic and stratigraphic context, seismic attributes and sedimentary characteristics of the contourite system in question. Case studies range from some of the well-documented North Atlantic drifts to those much less known from the Mediterranean, from important syntheses of the Gulf of Cadiz and Vema Channel Gateway, to completely new data on South Atlantic, Pacific and Antarctic margin systems. The four papers on ancient series – from Japan, China and Cyprus – serve to emphasize the complex nature and subtle characteristics of contourites, which make their identification such a scientific challenge.

This volume is dedicated to the memory of Charley Hollister (1936–1999), one of the founding fathers and pioneers of contourite research. Charley was an enthusiastic supporter of IGCP 432, and had agreed to write the *Preface* for this Memoir before his untimely death in 1999.

Charles Davis Hollister, 1936–1999
A personal scientific appreciation of the father of 'contourites'

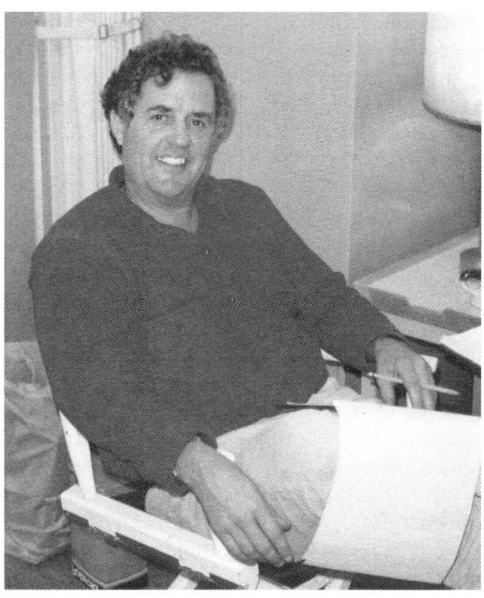

Charley Hollister on board R. V. Knorr cruise KN103 during HEBBLE in 1983.

The untimely death of Charley Hollister in a climbing accident robbed the contourite community of the principal originator and proponent of the concept of deep-ocean current-controlled sedimentation. He possessed boundless enthusiasm, a strong belief that science should be and was fun, a wonderful ability to get disparate groups of people to work together and a fine capacity to see connections between varied strands of data. The latter ability in particular allowed him as a graduate student at Lamont Geological Observatory in the 1960s to put together physical oceanography from Georg Wust, early seismic profiler results from the Ewings, deep-sea photographs, stratigraphic and sedimentological data from cores with his supervisor Bruce Heezen, and develop the notion of sedimentation controlled by deep geostrophic flows.

Charley was born into a landed Californian family whose fortunes were founded on cattle ranching. Eschewing the role of sedate conformity and leadership normally reserved for elder sons, his boyhood was, by his own account (thoroughly confirmed by his family), happily irresponsible. His schooling was chequered and his undergraduate academic performance at Oregon State University undistinguished. However, although his best developed abilities were as a marksman and mountaineer, late in the day the spark of what he really wanted to do struck. Doc Ewing, director of Lamont was persuaded that Charley's enthusiasm once channelled, could lead to a successful career as a marine geologist and admitted him to the graduate programme in 1960.

In the late 1950s ideas about the deep circulation developed rapidly with Wust's (1955) recalculation of geostrophic velocities, Swallow & Worthington's (1957) demonstration of high flow speed in the Western Boundary Undercurrent, and Stommel's (1958) proposal of a scheme for the deep sea circulation. These new ideas were quickly picked up by Charley's supervisor Bruce Heezen who wrote in 1959 '... it would appear that the deep ripples and bottom scour must be the work of currents related to the general circulation of the ocean ...'. However, a strong emphasis was then placed on the continental rise and abyssal plains displaying features originating in turbidity current flow. Hollister assembled the photographs showing current bedforms corresponding in location to Wust's predicted strong abyssal flows in the South Atlantic and convinced Heezen of the importance of current reworking of the continental rise. This became Hollister's topic for a PhD dissertation and he turned to examination of further evidence in photographs and cores for bottom currents, finding a positive result at several other deep locations where Wust had predicted rapid flows, particularly the western North Atlantic (Heezen & Hollister 1964). Wust's presence at Lamont through this period (1960–1964) acted as a great stimulus to the task. This work culminated in the proposal that the deep western boundary current was responsible for most of the deposition on, and form of, the eastern United States' continental rise (Heezen et al. 1966). The importance of turbidites was explicitly denied, '... recognisable turbidites constitute a small proportion of the glacial and postglacial sediments of the continental rise' (Heezen & Hollister 1964). Those statements underplay the clear recognition by the authors that they were dealing with terrigenous sediments which had come downslope, probably in turbidity currents. The argument really lay in the degree of completeness to which the turbidites were reworked and deposited as contourites.

There was by now a large collection of cores from the continental margin, and the development in 1961 of seismic reflection profiling by John Ewing at Lamont had yielded several hundred thousand miles of data by 1964 (Ewing & Ewing 1964). In these data were the striking profiles of the Blake Outer Ridge showing it to be 2 km of deep-sea sediments at a location where turbidity current emplacement was impossible. There was thus abundant data on the sediment characteristics and distribution of contourites as is made clear by figures 1 and 3 of Heezen et al. (1966). This paper was rapidly followed by others from Heezen's group consolidating their position that sediment transport in contour currents was responsible for much of the sedimentary topography of the American Basin, and by Charley's PhD thesis (Hollister 1967) a significant part of which is in Hollister & Heezen (1972). It was in the latter paper that the term 'contourite' was first defined, though it had already been in conversational use for a while.

A new source of data added by Hollister (1967; Hollister & Heezen 1972) was the 12 kHz echogram character of the bottom.

From: Stow, D. A. V., Pudsey, C. J., Howe, J. A., Faugères, J.-C. & Viana, A. R. (eds)
Deep-Water Contourite Systems: Modern Drifts and Ancient Series, Seismic and Sedimentary Characteristics.
Geological Society, London, Memoirs, **22**, 1–5. 0435-4052/02/$15.00 © The Geological Society of London 2002.

This new development of 'echo-character mapping' by Hollister was greatly expanded in the following decade using the now ubiquitous 3.5 kHz profiler. The 12 kHz system revealed contour-parallel variation on the continental rise with poorest reflectivity [hyperbolae and indistinct ('mushy') echoes] in the location of supposed maximum bottom current. The variation was suggested to be due to variations in microtopography. A superb book of photographs with linking narrative by Bruce & Charley *The Face of the Deep* (1971) showing this microtopography was a further product of their collaboration. It remains the best published collection of deep-sea photographs, but sadly is long out of print.

Many questions relating to sedimentation and the genesis of echo character could be answered only by a closer look. Charley was quick to realise the potential offered by the Scripps deep-tow system (Spiess & Mudie 1970; Spiess & Tyce 1973). The deep-tow gave side-scan sonar and 4 kHz reflection profiles of a quality comparable to that used by workers on continental shelves where these techniques were commonplace. In addition, twin 35 mm cameras allowed stereographic resolution of the dimensions of smaller features. The principal elements of microtopography surveyed were ripples, sand dunes and furrows as well as mud waves which were also studied via surface ship 3.5 kHz records. Most records produced great surprises from ripples in mud to barchan dunes of sand to extensive fields of linear furrows (Hollister *et al.* 1974, 1976; Lonsdale & Spiess 1977). The furrows appeared to be part erosional, part depositional and pointed to the possibility of helical secondary circulation in the bottom mixed layer proposed by Charley's student Roger Flood (1978). Furrows were also found to be responsible for hyperbolic echoes mapped previously. This closer look also involved him in physical oceanographic measurements in relation to large scale sediment bodies and the deep Western Boundary Undercurrent (WBUC). Geologists often had to conduct the physical oceanographic investigation of the flow setting in relation to bedform development, a fact with which he used to upbraid his P.O. colleagues at WHOI who at that time did not set current meters deeper than 4000 m, even where depth was 5000 m.

Staff at Lamont, particularly Maurice Ewing, were instrumental in establishing the Deep Sea Drilling Project with Leg 1 in 1968. Although a recent PhD, Charley sailed on Leg 11 in 1970 as Co-Chief Scientist with John Ewing. Everything was new then, and they found the cherts producing the 'Horizon A' seismic reflector, stacks of contourites and the Mid-Cretaceous black mudstones. At the end-of-leg press conference in New York Charley made the point about carbon content by setting a piece alight. He always knew how to get attention. The first attempt at reconstruction of the palaeocirculation of the Atlantic in a series of palaeogeographic maps from Jurassic to recent in Berggren & Hollister (1974) was a seminal product of his Atlantic drilling work. DSDP Leg 35 to the inhospitable Bellinghausen Sea of the SE Pacific in 1974 was his last drilling leg, for other things had started to occupy him.

Standard piston cores of 6 cm internal diameter (i.d.) do not provide much material to work with and sidewall deformation generally distorts sedimentary structures. So early in his career at Woods Hole (which he joined in 1967 after an interlude climbing the high peaks of Antarctica) Charley conceived of the Giant Piston Corer. This involved a very large weight and pipe that held 10 cm i.d. liner (compared with the 6 cm then current, giving three times the volume of material). Gradually, in an experimental process that involved destroying much ships equipment; cables, sheaves etc., and leaving many GPC's in the sea bed as a result of these failures, the system was brought to an operational state. The cores obtained were of stunning quality and those from the Central Pacific (GPC-3), Bermuda Rise (GPC-5) and Bahama Outer Ridge (GPC-9) are still being sampled, and at the hands of Lloyd Keigwin and Ed Boyle paved the way for modern high-resolution palaeoceanography. Today's IMAGES programme using a 7.5 ton corehead and 11 cm i.d. liner, but with modern Kevlar cable, owes much to Charley's foresight.

This development was rather expensive, particularly in view of the amount of hardware destroyed or left on the sea bed, so it was fortunate that he had stumbled on a new funding source which had not previously had dealings with oceanographers. A chance meeting in Washington in the early seventies led to the investigation of what he liked to call 'the least valuable real estate on earth', the mid-gyre deep-sea bed, for disposal of high-level radioactive waste. He became aware that the Washington agencies in need of answers to problems of waste disposal had substantial budgets. It was also very clear to him that investigation of this problem would be very expensive indeed! The project was to examine the feasibility of, and problems associated with, subsurface disposal of 'hot' nuclear waste. If waste canisters or projectiles were to be emplaced below the sea bed, feasibility would partly depend on the geotechnical properties of the bed. Soil mechanical tests needed larger samples than were available from small diameter piston cores. Charley had the imagination to see that this was a heaven-sent opportunity to mount a multi-faceted approach to problems of the deep-sea environment while also funding his equipment developments.

This project, run through the Department of Energy's Sandia Laboratories funded a large amount of work on seismic profiling, sedimentology, biology, micropalaeontology, stratigraphy, mineralogy, modelling, and science policy studies of the deep sea bed. The work culminated in an authoritative paper in *Science* in 1981. Charley was, when among scientists, a disinterested observer concerned mainly with maximizing research income for important investigations of the deep sea environment in connection with a very significant societal problem. There were times when he appeared to be advocating deep sea disposal and this made him the target for the more politically motivated commentators in environmental organizations. Much good science came out of the Sandia project and Hollister recognized at an early stage that environmental policy studies had to be conducted alongside the scientific work. Technical feasibility and safety might be demonstrated, but projects could, and frequently still do, fall on politically motivated campaigns of scientific misinformation

The final major science project led by Charley Hollister was HEBBLE, the High Energy Benthic Boundary Layer Experiment funded by the US Office of Naval Research. In 1974 I organized the NATO Conference on the Benthic Boundary Layer at the skiing resort of Les Arcs at which Charley was an invited keynote speaker (McCave 1976). He saw that his interests in current-controlled sedimentation and bedforms could be given full expression under the banner of boundary layer research and found a ready ear in Tom Pyle, program manager at the ONR. (He was also delighted to find that his interest in mountains and skiing could so easily be fitted under 'oceanography', that he took our lesson of Les Arcs and repeated it many times for HEBBLE at Keystone, Colorado). After a few exploratory meetings and probing possible experimental areas we assembled an executive committee and fixed on the lower Nova Scotian rise as the experimental site. It was probably not accidental that this was both part of Charley's original thesis area, and that it lay under the main path of Soviet submarines (and aircraft who often inspected us) down the US east coast to Cuba.

In seeking the simplest type of boundary layer for HEBBLE it was decided to try and avoid areas of furrows and longitudinal ripples with possible helical circulations, and mud waves which might have associated standing internal waves (Flood 1978), but nevertheless investigate an area where dynamic processes of erosion and deposition were well developed. As with his previous projects, HEBBLE assembled geologists, geochemists, biologists, engineers and physical oceanographers focussing on flow-bed interactions considered very broadly. The results came out in two special issues of *Marine Geology*, and many other publications, totalling over a hundred papers. The ideas of deep-sea storms, turbulence in a planetary boundary layer, nepheloid layers, bioturbation, benthic habitats, sediment sorting, and many other matters bearing on the formation of contourites were probed more deeply and revealingly than ever before in the field.

Early on in HEBBLE Charley moved from being a Senior Scientist in the Department of Geology and Geophysics to taking charge of the institution's graduate program as Dean of Education and an Associate Director of WHOI. Although he sailed on most of the cruises and took an active interest in steering all the aspects of HEBBLE, and particularly in stereophotogrammetry of the sea bed, he was increasingly occupied in matters of educational policy and science policy more generally. The end of HEBBLE in 1987 really saw the end of Charley's active involvement in experimental science, and in 1989 after ten years as Dean of the Graduate Program, during which he had come to understand very clearly that the education program was in dire need of endowed funding, he took his final position at Woods Hole as Vice-President of the Corporation and Associate Director for External Affairs in charge of fund raising and communications. In this, as in his marine geological activities, the combination of enthusiasm and a deep love of the subject, coupled with a vital objective, the survival and enhancement of the quality of the Woods Hole Oceanographic Institution, proved most successful and WHOI has been enriched in many ways by his activities.

Those of us who were his principal collaborators and friends, evident from his publication record given below, have felt a deep sense of loss. The wider world of those concerned with the interactions of currents and the deep-sea bed, will I hope, through these brief remarks, realise the substantial contribution made to the field by Charley Hollister, to whose memory this book is dedicated. He has the best claim to be regarded as the father of the concept of contourite sediments.

I. Nicholas McCave
Cambridge, 1999

References (other than in Hollister bibliography below)

EWING, M. & EWING, J. 1964. Distribution of oceanic sediments. *In*: YOSHIDA, K. (ed.) *Studies on Oceanography*. Tokyo University Press, Tokyo, 525–537.

FLOOD, R. D. 1978. *Studies of deep-sea sedimentary microtopography in the North Atlantic Ocean*. PhD Thesis, WHOI, Woods Hole, Mass.

HEEZEN, B. C. 1959. Dynamic processes of abyssal sedimentation; erosion, transportation and redeposition on the deep-sea floor. *Geophysical Journal of the Royal Astronomical Society*, **2**, 142–163.

LONSDALE, P. & SPIESS, F. N. 1977. Abyssal bedforms explored with a deeply towed instrument package. *Marine Geology*, **23**, 57–75.

MCCAVE, I. N. 1976. *The Benthic Boundary Layer*. Plenum, New York.

SPIESS, F. N. & MUDIE, J. D. 1970. Small-scale topographic and magnetic features. *In*: MAXWELL, A. E. (ed.) *The Sea*. John Wiley, New York, **4A**, 205–250.

SPIESS, F. N. & TYCE, R. C. 1973. *Marine Physical Laboratory deep-tow instrumentation system*. Scripps Institute of Oceanography Ref., 73–74.

STOMMEL, H. 1958. The abyssal circulation. *Deep-Sea Research*, **5**, 80–82.

SWALLOW, J. C. & WORTHINGTON, L. V. 1957. Measurements of deep currents in the western North Atlantic. *Nature*, **179**, 1183.

WÜST, G. 1955. Stromgeschwindigkeiten im Tiefen- und Bodenwasser des Atlantischen Ozeans auf Grund dynamischer Berechnung der Meteor-Profile der Deutschen Atlantischen Expedition 1925/27. *In*: *Papers in Marine Biology and Oceanography*. Bigelow Volume, supplement to Deep-Sea Research, **3**, 373–397.

Scientific Publications of C. D. Hollister

HEEZEN, B. C. & **HOLLISTER, C. D.** 1964. Deep sea current evidence from abyssal sediments. *Marine Geology*, **1**, 141–174.

HEEZEN, B. C. & **HOLLISTER, C. D.** 1964. Turbidity currents and glaciation. *In*: NAIRN, A. E. M. (ed.) *Problems in Paleoclimatology*. Interscience, London/New York, 99–108.

HOLLISTER, C. D. & HEEZEN, B. C. 1964. Modern graywacke-type sands. *Science*, **145**, 1573–1574.

HEEZEN, B. C., **HOLLISTER, C. D.** & RUDDIMAN, W. F. 1966. Shaping of the continental rise by deep geostrophic contour currents. *Science*, **152**, 502–508.

HOLLISTER, C. D. & HEEZEN, B. C. 1966. Ocean bottom currents. *In*: FAIRBRIDGE, R. W. (ed.) *The Encyclopedia of Oceanography*. Dowden, Hutchinson & Ross, Stroudsburg, Pa, 576–583.

HOLLISTER, C. D. 1967. *Sediment distribution and deep circulation in the western North Atlantic*. PhD thesis, Columbia University, New York.

HOLLISTER, C. D. & HEEZEN, B. C. 1967. The floor of the Bellingshausen Sea. *In*: HERSEY, J. B. (ed.) *Deep–Sea Photography*. Johns Hopkins University Press, Baltimore, 177–189.

HEEZEN, B. C. & **HOLLISTER, C. D.** 1967. Physiography and bottom currents in the Bellingshausen Sea. *Antarctic Journal of the US*, **2**, 184–185.

SCHNEIDER, E. D., FOX, P. J., **HOLLISTER, C. D.**, NEEDHAM, D. & HEEZEN, B. C. 1967. Further evidence of contour currents in the western north Atlantic. *Earth and Planetary Science Letters*, **2**, 351–359.

HOLLISTER, C. D. & ELDER, R. 1969. Contour currents in the Weddell Sea. *Deep-Sea Research*, **16**, 99–101.

HEEZEN, B. C., JOHNSON, G. L. & **HOLLISTER, C. D.** 1969. The Northwest Atlantic Mid-Ocean Canyon. *Canadian Journal of Earth Sciences*, **6**, 1441–1453.

HEEZEN, B. C. & **HOLLISTER, C. D.** 1971. *The Face of the Deep*. Oxford University Press, New York.

SOUTHARD, J. B., YOUNG, R. A. & **HOLLISTER, C. D.** 1971. Experimental erosion of calcareous ooze. *Journal of Geophysical Research*, **76**, 5903–5909.

HOLLISTER, C. D. & HEEZEN, B. C. 1972. Geologic effects of ocean bottom currents. *In*: A. L. Gordon (ed.) *Studies in Physical Oceanography – A Tribute to George Wust on his 80th Birthday*. Gordon & Breach, New York, **2**, 37–66.

HOLLISTER, C. D., EWING, J. I. et al. 1972. *Initial Reports of the Deep Sea Drilling Project*, **11**. National Science Foundation, US Government Printing Office, Washington, D.C.

HOLLISTER, C. D. & EWING, J. I. 1972. Lithology of sediments from the western North Atlantic: Leg XI, Deep Sea Drilling Project. *In*: **HOLLISTER, C. D.**, EWING, J. I. et al. (eds) *Initial Reports of the Deep Sea Drilling Project*, **11**, 901–949.

HOLLISTER, C. D. & EWING, J. I. 1972. Regional aspects of deep sea drilling in the western North Atlantic. *In*: **HOLLISTER, C. D.** EWING, J. I. et al. (eds) *Initial Reports of the Deep Sea Drilling Project*, **11**, 951–973.

SILVA, A. J. & **HOLLISTER, C. D.** 1973. Geotechnical properties of ocean sediments recovered with a giant piston corer: 1. Gulf of Maine. *Journal of Geophysical Research*, **78**, 3597–3616.

BOUMA, A. H. & **HOLLISTER, C. D.** 1973. Deep ocean basin sedimentation. *In*: MIDDLETON, G. V. & BOUMA, A. H. (eds) *Turbidites and Deep-Water Sedimentation*. S.E.P.M. Pacific Section, 79–118.

HOLLISTER, C. D., SILVA, A. J. & DRISCOLL, A. H. 1973. A giant piston corer. *Ocean Engineering*, **2**, 159–168.

MACDONALD, K. C. & **HOLLISTER, C. D.** 1973. Near bottom thermocline in the Samoan Passage, west equatorial Pacific. *Nature*, **243**, 461–462.

HOLLISTER, C. D. 1973. Continental shelf and slope of the United States: Texture of surface sediments from New Jersey to southern Florida. *US Geological Survey, Prof. Paper*, 529-M.

TUCHOLKE, B. E. & **HOLLISTER, C. D.** 1973. Late Wisconsin glaciation off southeastern New England: New evidence from the marine environment. *Geological Society of American Bulletin*, **84**, 3279–3296.

TUCHOLKE, B. E., OLDALE, R. N. & **HOLLISTER, C. D.** 1973. Map showing echo sounding survey (3.5 kHz) of Massachusetts and Cape Cod Bays, western Gulf of Maine. *US Geological Survey, Miscellaneous Geologic Investigations*, **Map I-716**.

TUCHOLKE, B. E., WRIGHT, W. R. & **HOLLISTER, C. D.** 1973. Abyssal circulation over the Greater Antilles Outer Ridge. *Deep-Sea Research*, **20**, 973–995.

HOLLISTER, C. D., JOHNSON, D. A. & LONSDALE, P. F. 1974. Current controlled abyssal sedimentation: Samoan Passage, equatorial West Pacific. *Journal of Geology*, **82**, 275–299.

FLOOD, R. D., **HOLLISTER, C. D.**, JOHNSON, D. A., LONSDALE, P. F. &

SOUTHARD, J. B. 1974. Abyssal furrows and hyperbolic echo traces on the Bahama Outer Ridge. *Geology*, **2**, 395–400.

FLOOD, R. D. & **HOLLISTER, C. D.** 1974. Current-controlled topography on the continental margin off the eastern United States. *In*: BURKE, C. A. & DRAKE, C. L. (eds) *The Geology of Continental Margins*. Springer-Verlag, New York, 197–205.

BERGGREN, W. A. & **HOLLISTER, C. D.** 1974. Paleogeography, paleobiogeography and the history of circulation in the Atlantic Ocean. *In*: HAY, W. W. (ed.) *Studies in Paleo-Oceanography*. Society of Economic Paleontologists and Mineralogists, Special Publication, **20**, 126–186.

DRISCOLL, A. H. & **HOLLISTER, C. D.** 1974. The W.H.O.I. giant piston core: State of the art. *Marine Technology Society Tenth Annual Conference Proceedings*, 663–676.

YOUNG, R. A. & **HOLLISTER, C. D.** 1974. Quaternary sedimentation on the Northwest African continental rise. *Journal of Geology*, **82**, 675–689.

BISHOP, W. P. & **HOLLISTER, C. D.** 1974. Seabed disposal – where to look. *Nuclear Technology*, **24**, 425–443.

HOLLISTER, C. D. & CRADDOCK, C. 1974. Deep Sea Drilling Project, Leg 35: Bellingshausen Sea. *Antarctic Journal of the US*, **9**, 154–155.

HOLLISTER, C. D., HEEZEN, B. C. & NAFE, K. E. 1975. Animal traces on the deep-sea floor. *In*: FREY, R. W. (ed.) *The Study of Trace Fossils*. Springer-Verlag, New York, 493–510.

HOLLISTER, C. D., CRADDOCK, C. *et al.* 1976. *Initial Reports of the Deep Sea Drilling Project*, **35**. National Science Foundation, US Government Printing Office, Washington, D.C.

HOLLISTER, C. D. & CRADDOCK, C. 1976. Introduction, principal results – Leg 35, Deep Sea Drilling Project. *In*: **HOLLISTER, C. D.** & CRADDOCK, C. (eds) *Initial Reports of the Deep Sea Drilling Project*. **35**.

PETERS, C. S. & **HOLLISTER, C. D.** 1976. Heavy mineral characteristics and dispersal patterns from DSDP Leg 35, Southeast Pacific Basin. *In*: **HOLLISTER, C. D.** & CRADDOCK, C. (eds) *Initial Reports of the Deep Sea Drilling Project*. **35**, 291–300.

TUCHOLKE, B. E., **HOLLISTER, C. D.**, WEAVER, F. M. & VENNUM, W. R. 1976. Continental rise and abyssal plain sedimentation in the Southeast Pacific Basin, Leg 35 Deep Sea Drilling Project. *In*: **HOLLISTER, C. D.** & CRADDOCK, C. (eds) *Initial Reports of the Deep Sea Drilling Project*. **35**, 359–400.

SILVA, A. J., **HOLLISTER, C. D.**, LAINE, E. P. & BEVERLY, B. E. 1976. Geotechnical properties of deep sea sediments: Bermuda Rise. *Marine Geotechnology*, **1**, 195–232.

HOLLISTER, C. D., SOUTHARD, J. B., FLOOD, R. D. & LONSDALE, P. F. 1976. Flow phenomena in the benthic boundary layer and bed forms beneath deep-current systems. *In*: MCCAVE, I. N. (ed.) *The Benthic Boundary Layer*. Plenum Press, New York, 183–204.

BERGGREN, W. A. & **HOLLISTER, C. D.** 1977. Plate tectonics and paleocirculation–commotion in the ocean. *Tectonophysics*, **38**, 11–48.

FROSCH, R. A., **HOLLISTER, C. D.** & DEESE, D. A. 1978. Radioactive waste disposal in the oceans. *The Ocean Yearbook*, **1**, 340–348.

HOLLISTER, C. D., GLENN, M. F. & LONSDALE, P. F. 1978. Morphology of seamounts in the western Pacific and Philippine Basin from multibeam sonar data. *Earth and Planetary Science Letters*, **41**, 405–418.

SILVA, A. J. & **HOLLISTER, C. D.** 1979. Geotechnical properties of ocean sediments recovered with giant piston corer: Blake Bahama Outer Ridge. *Marine Geology*, **29**, 1–22.

LONSDALE, P. F. & **HOLLISTER, C. D.** 1979. A near-bottom traverse of Rockall Trough: hydrographic and geologic inferences. *Oceanologica Acta*, **2**, 91–105.

FLOOD, R. D., **HOLLISTER, C. D.** & LONSDALE, P. F. 1979. Disruption of the Feni sediment drift by debris flows from Rockall Bank. *Marine Geology*, **32**, 311–334.

CORLISS, B. H. & **HOLLISTER, C. D.** 1979. Cenozoic sedimentation in the central North Pacific. *Nature*, **282**, 707–709.

LONSDALE, P. F. & **HOLLISTER, C. D.** 1979. Cut-offs at an abyssal meander. *Geology*, **7**, 597–601.

HOLLISTER, C. D., CORLISS, B. H. & ANDERSON, D. R. 1980. Submarine geologic disposal of nuclear wastes. *In*: *Underground Disposal of Radioactive Wastes*, **Vol. I**, IAEA-SM-243/99, IAEA, Vienna, 131–139.

FLOOD, R. D. & **HOLLISTER, C. D.** 1980. Submersible studies of deep-sea furrows and transverse ripples in cohesive sediments. *Marine Geology*, **36**, M1–M9.

GARDNER, W. D., GLOVER, L. K. & **HOLLISTER, C. D.** 1980. Canyons off Northwest Puerto Rico: Studies of their origin and maintenance with the nuclear research submarine NR-1. *Marine Geology*, **37**, 41–70.

SHOR, A. N., LONSDALE, P. F. & **HOLLISTER, C. D.** 1980. Charlie-Gibbs Fracture Zone: Bottom water transport and its geological effects. *Deep-Sea Research*, **27A**, 325–345.

MCCAVE, I. N., LONSDALE, P. F., **HOLLISTER, C. D.** & GARDNER, W. D. 1980. Sediment transport over the Hatton and Gardar contourite drifts. *Journal of Sedimentary Petrology*, **50**, 1049–1062.

BOWEN, V. T. & **HOLLISTER, C. D.** 1981. Pre- and post-dumping investigations for inauguration of new low level radioactive waste dump sites. *Radioactive Waste Management*, **1**, 235–269.

HOLLISTER, C. D. 1981. Chapter 5: Sub-seabed disposal of high-level nuclear wastes. *In*: JACKSON, T. C. (ed.), *Nuclear Waste Management: The Ocean Alternative*. Pergamon Press, New York.

LONSDALE, P. F., **HOLLISTER, C. D.** & MAYER, L. 1981. Erosion and deposition in interplain channels of the Maury Channel system, Northeast Atlantic. *Oceanologica Acta*, **4**, 185–201.

LAINE, E. P. & **HOLLISTER, C. D.** 1981. Geologic effects of the Gulf Stream system on the northern Bermuda Rise. *Marine Geology*, **39**, 277–310.

HOLLISTER, C. D., ANDERSON, D. R. & HEATH, G. R. 1981. Sub-sea bed disposal of nuclear wastes. *Science*, **213**, 1321–1335.

CORLISS, B. H. & **HOLLISTER, C. D.** 1982. A paleoenvironmental model for Cenozoic sedimentation in the central North Pacific. *In*: SCRUTTON, R. A. & TALWANI, M. (eds) *The Ocean Floor*. John Wiley, Chichester, 277–304.

HINGA, K. R., HEATH, G. R., ANDERSON, D. R. & **HOLLISTER, C. D.** 1982. Disposal of high-level radioactive wastes by burial in the sea floor. *Environmental Science and Technology*, **16**, 28A–37A.

MCCAVE, I. N., **HOLLISTER, C. D.**, LAINE, E. P., LONSDALE, P. F. & RICHARDSON, M. J. 1982. Erosion and deposition on the eastern margin of the Bermuda Rise in the Late Quaternary. *Deep-Sea Research*, **29**, 535–561.

NOWELL, A. R. M., **HOLLISTER, C. D.** & JUMARS, P. A. 1982. High Energy Benthic Boundary Layer Experiment: HEBBLE. *EOS, Transactions American Geophysical Union*, **63**(31), 594–595.

MCCAVE, I. N., **HOLLISTER, C. D.**, DEMASTER, D. J., NITTROUER, C. A., SILVA, A. & YINGST, J. Y. 1984. Analysis of a longitudinal ripple from the Nova Scotian Continental Rise. *Marine Geology*, **58**, 275–286.

HOLLISTER, C. D. & MCCAVE, I. N. 1984. Sedimentation under sea storms. *Nature*, **309**, 220–225.

TUCKHOLKE, B. E. & **HOLLISTER, C. D.** 1984. Current zonation and bedforms on Nova Scotian continental rise determined from bottom photography. *In*: SMITH, P. F. (ed.) *Underwater Photography, Scientific and Engineering Applications*. Van Nostrand Reinhold, New York, 137–140.

GARDNER, W. D., SOUTHARD, J. D. & **HOLLISTER, C. D.** 1985. Sedimentation, resuspension and chemistry of particles in the Northwest Atlantic. *Marine Geology*, **65**, 199–242.

NOWELL, A. R. M. & **HOLLISTER, C. D.** (eds) 1985. *Deep Ocean Sediment Transport, Preliminary Results of the High Energy Benthic Boundary Layer Experiment*. Elsevier, New York.

NOWELL, A. R. M. & **HOLLISTER, C. D.** 1985. The objectives and rationale of HEBBLE. *Marine Geology*, **66**, 1–12.

MCCAVE, I. N. & **HOLLISTER, C. D.** 1985. Sedimentation under deep-sea current systems: Pre-HEBBLE ideas. *Marine Geology*, **66**, 13–24.

TUCHOLKE, B. E., **HOLLISTER, C. D.**, BISCAYE, P. & GARDNER, W. 1985. Abyssal current character determined from sediment bedforms on the Nova Scotian Continental Rise. *Marine Geology*, **66**, 43–58.

SWIFT, S. A., **HOLLISTER, C. D.** & CHANDLER, R. S. 1985. Close-up stereo photographs of abyssal bedforms on the Nova Scotian Continental Rise. *Marine Geology*, **66**, 303–322.

NOWELL, A. R. M., MCCAVE, I. N. & **HOLLISTER, C. D.** 1985. Contributions of HEBBLE to understanding marine sedimentation. *Marine Geology*, **66**, 397–409.

RICHARDSON, M. J. & **HOLLISTER, C. D.** 1987. Compositional changes in particulate matter on the Iceland Rise, through the water column, and at the seafloor. *Journal of Marine Research*, **45**, 175–200.

NOWELL, A. R. M. & **HOLLISTER, C. D.** 1988. Graduate students in oceanography: recruitment, success and career prospects. *EOS, Transactions American Geophysical Union*, **69**, 834–843.

HOLLISTER, C. D. & NOWELL, A. R. M. 1991. Prologue: Abyssal storms as a global geologic process. *Marine Geology*, **99**, 275–280.

HOLLISTER, C. D. & NOWELL, A. R. M. 1991. HEBBLE epilogue. *Marine Geology*, **99**, 445–460.

HOLLISTER, C. D. 1992. Potential use of the deep seafloor for waste disposal. *In*: HSU, K. & THIEDE, J. (eds) *Use and Misuse of the Seafloor*. John Wiley, Chichester, 127–130.

HOLLISTER, C. D. 1993. The concept of deep-sea contourites, *Sedimentary Geology*, **82**, 1–7.

VARTANOV, R. & HOLLISTER, C. D. 1997. Nuclear legacy of the Cold War. *Marine Policy*, **21**, 1–15.

Popular Articles

HEEZEN, B. C. & HOLLISTER, C. D. 1971. The deep, deep sea floor. *Natural History*, **80**(5), 30–33.

BERGGREN, W. A. & HOLLISTER, C. D. 1974. Currents of time. *Oceanus*, **17**(2), 28–33.

HOLLISTER, C. D. 1977. The seabed option. *Oceanus*, **20**(1), 18–25.

HOLLISTER, C. D., FLOOD, R. D. & MCCAVE, I. N. 1978. Plastering and decorating in the North Atlantic. *Oceanus*, **21**(1), 5–13.

HOLLISTER, C. D. 1983. In pursuit of oceanography and a better life for all. *Oceanus*, **26**(2), 10–16.

HOLLISTER, C. D., NOWELL, A. R. M. & JUMARS, P. A. 1984. The dynamic abyss. *Scientific American*, **250**(3), 42–53.

HOLLISTER, C. D. 1990. Options for waste: Space, land or sea? *Oceanus*, **33**(2), 13–17.

NOWELL, A. R. M. & HOLLISTER, C. D. 1990. Undergraduate and graduate education in oceanography. *Oceanus*, **33**(3), 31–35.

HOLLISTER, C. D. & NADIS, S. 1998. Burial of radioactive waste under the seabed. *Scientific American*, **278**(1), 2–7.

Bottom currents, contourites and deep-sea sediment drifts: current state-of-the-art

DORRIK A. V. STOW[1], JEAN-CLAUDE FAUGÈRES[2], JOHN A. HOWE[3], CAROL J. PUDSEY[4]
& ADRIANO R. VIANA[5]

[1]*Southampton Oceanography Centre, University of Southampton, Waterfront Campus, Southampton SO14 3ZH, UK (e-mail: davs@soc.soton.ac.uk)*
[2]*Department of Oceanography, University of Bordeaux I, Talence 33405, France*
[3]*Scottish Association for Marine Science, University of the Highlands and Islands, Oban PA34 4AD, Scotland*
[4]*British Antarctic Survey, High Cross, Madingley Rise, Cambridge CB3 0ET, UK*
[5]*Petrobras Av. Elias Agostinho, 665 Macae, Rio de Janeiro, Brazil*

Abstract: This paper provides both an introduction to and summary for the Atlas of Contourite Systems that has been compiled as part of the International Geological Correlation Project – IGCP 432. Following the seminal works of George Wust on the physical oceanography of bottom currents, and Charley Hollister on contourite sediments, a series of significant advances have been made over the past few decades. While accepting that ideas and terms must remain flexible as our knowledge base continues to increase, we present a consensus view on terminology and definitions of bottom currents, contourites and drifts. Both thermohaline and wind-driven circulation, influenced by Coriolis Force and molded by topography, contribute to the oceanic system of bottom currents. These semi-permanent currents show significant variability in time and space, marked by periodic benthic storm events in areas of high surface kinetic energy.

Six different drift types are recognized in the ocean basins and margins at depths greater than about 300 m: (i) contourite sheet drifts; (ii) elongate mounded drifts; (iii) channel related drifts; (iv) confined drifts; (v) infill drifts; and (vi) modified drift-turbidite systems. In addition to this overall geometry, their chief seismic characteristics include: a uniform reflector pattern that reflects long-term stability, drift-wide erosional discontinuities caused by periodic changes in bottom current regime, and stacked broadly lenticular seismic depositional units showing oblique to downcurrent migration. At a smaller scale, a variety of seismic facies can be recognized that are here related to bottom current intensity. A model for seismic facies cyclicity (alternating transparent/reflector zones) is further elaborated, and linked to bottom current/climate change. Both erosional features and depositional bedforms are diagnostic of bottom current systems and velocities.

Many different contourite facies are now known to exist, encompassing all compositional types. We propose here a C1–5 notation for the standard contourite facies sequence, which can be interpreted in terms of fluctuation in bottom current velocity and/or sediment supply. Several proxies can be utilized to decode contourite successions in terms of current fluctuation. Gravel lag and shale chip contourites, as well as erosional discontinuities are indicative of still greater velocities. There are a small but growing number of land-based examples of fossil contourites, based on careful analysis using the recommended three-stage approach to interpretation. Debate still surrounds the recognition and interpretation of bottom current reworked turbidites.

Contourites are an extremely important but still relatively little known group of sediments, that have been surrounded by controversy since they were first recognized in the early 1960s. They are one of the keys to our further understanding of bottom-water circulation and the ocean-climate link, and play an increasingly critical role in palaeoceanographic studies. They are an important part of the spectrum of deposits that confront the oil industry as exploration moves into progressively greater water depths.

In order to foster greater international dialogue on these issues, as well as to stimulate more focussed research, an International Geological Correlation Programme initiative (Project 432) was launched in 1998. One of the early aims of this project was to publish a compendium of examples of contourite systems, both modern and ancient, and this volume is the result. This introductory paper provides an opportunity for the volume editors to summarise current knowledge and understanding of these systems, to highlight key areas of study, to document a consensus view on terminology, and to lay some pointers towards the direction of future research. This overview derives from many sources, including: our individual research experience and publications; discussions with many other scientists, facilitated by various IGCP432 workshop meetings; informal research reports in the IGCP432 Newsletter series; and a recent synthesis written by the senior author for the *Encyclopedia of the Oceans* (Stow 2001, in part based on Stow 1994). In such a concise synthesis, it has not always been possible to fully reference the many ideas and data that are included. Instead we provide below a summary of some of the key publications used. We also refer, as appropriate, to the 35 other contributions in this volume.

Historical perspective

At about the same time that marine geologists first began to recognize the significance of turbidity (density) currents in the erosion and deposition of sediments, the German physical oceanographer George Wust initially proposed that bottom currents driven by thermohaline circulation might be sufficiently strong to influence sediment flux in the deep ocean basins. But at that time, in 1936, his contention was loudly decried by other physical oceanographers and thus went largely unheard by geologists. It was not until the 1960s, following pioneering work by the American team of Bruce Heezen and Charlie Hollister, that the concept was forced centre-stage in marine science, with combined geological and oceanographic evidence that was irrefutable.

In their seminal paper of 1966, Heezen *et al.* demonstrated the very significant effects of contour following bottom currents (also known as contour currents) in shaping sedimentation on the deep continental rise off eastern North America. The deposits of these semi-permanent alongslope currents soon became known as contourites, clearly distinguishing them from the deposits of downslope event processes known as turbidites. The ensuing decade saw a profusion of research on contourites and bottom currents in and beneath the present-day oceans, and the demarcation of slope-parallel, elongate, mounded sediment bodies made up largely of contourites that became known as drifts. Their early identification in ancient rocks exposed on land, however, proved mostly inaccurate, as this was based on comparisons with the North American rise sediments. Subsequent work has demonstrated a much more complex interbedding of fine-grained turbidites, bottom current reworked turbidites, and contourites on the eastern North American rise (Stow 1979; Hollister 1993).

Other significant stepping stones that have helped to direct contourite research through the 1980s and 1990s are highlighted below, with example references only. A more accurate view was developed of contourite sediments from coring of drift deposits, and standard facies models were developed (Stow 1979, 1982; Faugères et al. 1984; Gonthier et al. 1984). The direct link between bottom current strength and nature of the contourite facies, especially grain size, was demonstrated (Ellwood & Ledbetter 1977; Stow et al. 1986). This has been taken forward through the work by Nick McCave and associates (Robinson & McCave 1994; McCave et al. 1995) in decoding the often very subtle signatures captured in contourites in terms of variation in deep-sea palaeo-circulation. Discrimination was made between contourites and other deep-sea facies, such as turbidites deposited by catastrophic downslope flows and hemipelagites that result from continuous vertical settling in the open ocean (Stow & Lovell 1979; Stow & Tabrez 1998). Much progress has been made on the types and distribution of sediment drifts (McCave et al. 1988; Faugères & Stow 1993; Howe et al. 1994; Stoker et al. 1998a), as well as on their seismic characteristics (Faugères et al. 1999).

For the most part, physical oceanographers have worked independently of geologists on the nature and variability of bottom currents, so that much integration is still required between these disciplines. Important contributions that to some extent bridge this divide have come from the HEBBLE project on the Nova Scotian Rise (Nowell & Hollister 1985; McCave et al. 1988), extensive work around the Antarctic margin (Pudsey et al. 1988; Gilbert et al. 1998), and recent work along the Brazilian continental margin (Viana et al. 1998a, b). The international deep-sea drilling programme in its various guises (DSDP, IPOD, ODP) has contributed enormously to contourite research; the palaeoceanographic context and study of oceanic gateways remain primary targets at present (see review in Stow et al. 1998). Although much effort also has been made to correctly identify fossil contourites in ancient series on land, much confusion and controversy still abounds (Stow et al. 1998a; Shanmugam 2000). Their recognition in oil company boreholes, therefore, and the contribution of contourite facies to hydrocarbon reservoir intervals remains a target for further study (Shanmugam et al. 1993, 1995).

Several edited volumes of papers dealing in part or wholly with contourite systems have been published or are in press at the time of writing, emphasising the current level of interest and research. These include Nowell & Hollister 1985; McCave et al. 1988; Stow & Faugères 1993, 1998; Gao et al. 1998; Mienert 1998; Stoker et al. 1998b; Maldonado & Nelson 1999; Stow & Mayall 2000; Wynn & Stow in press; Rebesco & Stow in press; as well as the present publication.

Consensus on terminology

While recognizing the need to retain a certain fluidity in our use of terms to allow for developments in understanding, most workers would currently agree on the following broad definitions and usage. As far as possible, contributions to this volume conform with this view.

Bottom currents is the generally accepted term for those currents that operate in deep-water and that are part either of the normal thermohaline or of the major wind-driven circulation pattern of the oceans and their marginal seas. In general they are semi-permanent in nature with a net flow alongslope. In detail, they are extremely variable in direction and velocity, typically exhibiting giant eddies, and local downslope or oblique-to-slope flow, especially at the exit of narrow gateways. They do not hug rigidly to the contours, although the term *contour current* is still widely used synonymous with bottom current. Other types of current that operate in deep water include: internal tidal and internal wave-related currents, downwelling slope currents, upwelling slope currents, and clear-water up and down canyon currents. It is not necessarily possible to distinguish the seismic features or sediment facies that result from these currents from those of bottom currents. Where a distinction can be made (especially in modern or sub-recent) systems, then a modifying term should be applied.

Contourites are the sediments deposited by or significantly affected by the action of bottom currents. A wide range of contourite facies can be recognized from muddy to gravel-lag facies, and of all different compositions depending on the sediment supply system. Because of process interaction and process continuums in the deep sea, many transitional facies can occur. Where distinction can be made, then terms such as *bottom current moulded hemipelagite* or *bottom current reworked turbidite* should be used. The term *fossil contourite* is now widely applied to ancient contourites exposed on land, as well as to those interpreted as contourites in deep boreholes.

There has always been a problem surrounding the appropriate water depth at which to call a bottom current deposit, a contourite, recognizing that many different currents will act upon the seafloor everywhere from the shoreline outwards, as well as in lacustrine settings. It makes little sense to be too rigid in defining an upper depth limit for contourites *sensu stricto*, as the precise water depth at which sub-recent and older contourites accumulated is generally unknown. This is even more true of fossil contourites. We therefore favour retaining, as a guideline, a water depth of around 300 m as the upper limit for contourite deposition. Clearly, if a mid to upper slope current system straddles this depth limit, perhaps on a seasonal basis, then the whole deposit should be given the same name. A qualifying term, such as *shallow water contourite* is recommended in this case, and the same should be used for drift deposits in relatively shallow gateways.

Sediment drift is a general term for a sediment accumulation, of no definitive or unique geometry, that has experienced some current control on deposition. It is not restricted to bottom current deposits. *Contourite drift* is the specific term for a sediment drift that has been formed principally (though not necessarily exclusively) by bottom currents. Various types of contourite drifts are recognized (see below), including mixed drift systems for those where there has been significant process interaction, e.g. bottom current modified turbidite drift, etc.

The terms *mound, lobe, fan, channel, levee* are all also associated closely with turbidite and associated downslope systems. When applying these terms to contourite systems it is therefore important to modify with the contourite prefix.

Bottom currents

At the present day, deep-ocean bottom water is formed by the cooling and sinking of surface water at high latitudes, followed by the deep slow thermohaline circulation of these polar water masses throughout the world's ocean (Figs 1 and 2). Antarctic Bottom Water (AABW) is the coldest, densest and hence deepest water in the oceans, forming close to and beneath floating ice shelves around Antarctica, with localized areas of major generation such as the Weddell Sea. Once formed at the surface, partly by cooling and partly as freezing seawater leaves behind water of greater salinity, AABW rapidly descends the continental slope, circulates eastwards around the continent and then flows northwards through deep-ocean gateways into the Pacific, Atlantic and

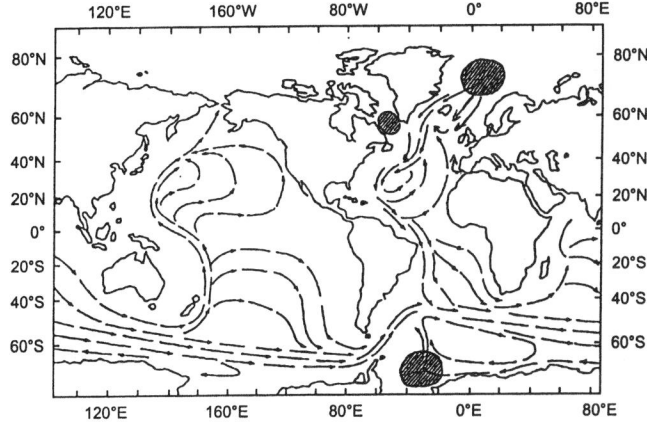

Fig. 1. Global pattern of abyssal circulation. Shaded areas are regions of production of bottom waters (after Stow *et al.* 1996).

Indian Oceans (**Pudsey *et al.***, **Pudsey & Howe**, **Carter & McCave**). Here it is further compartmentalized by topographic barriers, such as the mid-ocean ridges and aseismic ridge systems, either circulating within the sub-basin or escaping to an adjacent sub-basin where the gateway sill is sufficiently deep.

Arctic Bottom Water (ABW) forms in the vicinity of the subpolar surface water gyre in the Norwegian and Greenland Seas and then overflows intermittently to the south through narrow gateways across the Scotland–Iceland–Geenland topographic barrier, into the Rockall, Iceland and Greenland basins (**Howe *et al.*** and **Akhurst *et al.***). It mixes with cold deep Labrador Sea Water as it flows south along the Greenland–North American continental margin, and with recirculated North Atlantic Bottom Water in the Rockall Basin. Above these bottom waters, the ocean basins are further stratified into water masses of different temperature, salinity and density characteristics, each of which has its own circulatory pattern and shows slow mixing with adjacent water masses.

Bottom waters generally move very slowly throughout the ocean basins, at velocities no greater than 1–2 cm s^{-1}. However, they are significantly affected as they flow by the Coriolis Force, which results from the Earth's spin, and by basin or gateway topography. The Coriolis effect is to constrain water masses against the continental slopes on the western margins of basins, where they become restricted and intensified forming distinct Western Boundary Undercurrents that commonly attain velocities of 10–20 cm s^{-1}. These velocities may exceed 100 cm s^{-1} where the flow is particularly restricted or the slope especially steep. Through narrow passages or gateways on the deep seafloor, flow velocities in excess of 200 cm s^{-1} have been recorded.

Many bottom currents, therefore, are a semi-permanent part of the thermohaline circulation pattern, and sufficiently competent in parts to erode, transport and deposit sediment, especially clay, silt and fine sand grades and, more rarely, coarser sands and gravels. The currents are also highly variable in velocity, direction and, therefore, in their precise location at any one time. Mean flow velocity generally decreases from the core to the margins of the current, where large eddies peel of and move at high angles or in a reverse direction to the main flow. Tidal, seasonal and less regular periodicities have been recorded during long-term measurements (i.e. > six-month duration), and complete flow reversals are also commonplace.

Other bottom currents that are distinct from purely thermohaline circulation are the major wind-driven systems (Fig. 3). In some cases, these act throughout much of the water column, still registering significant flow at 4000 m depth. This is especially true of those currents that flow along western margins of ocean basins and are intensified by the Coriolis Force, such as the Gulf Stream and Kuroshio Current. The Circumpolar Antarctic Current is also well known for its effect on the deep slope and rise around the Antarctic continent (**Rebesco *et al.*** and **Escutia & Nelson**). Eddy kinetic energy associated with sea-surface topographic variations in regions of surface current instability and meandering can be transmitted through the water column, and so result in marked variation in kinetic energy at the seafloor (Fig. 4). This in turn has been shown to result in the alternation of short (days to weeks) episodes of high bottom current velocity known as benthic storms, and longer periods (weeks to months) of lower velocity. Benthic storms can result in the erosion and resuspension of large volumes of sediment, that becomes incorporated into the bottom nepheloid layer (**McCave & Tucholke**), although we know little more about their cumulative, long-term effects on the nature of contourites.

Deep and intermediate depth water is also formed from relatively warm surface waters that are subject to excessive evaporation at low latitudes, and hence to an increase in relative density. This process is most effective in semi-enclosed marginal seas and basins. The Mediterranean Sea is currently the principal source of warm, highly saline, intermediate water, formed principally in the eastern Mediterranean or Levantine Sea (**Reeder *et al.***). The bottom water so formed flows through the Sicily gateway (between Sicily and North Africa), through the western Mediterranean, out

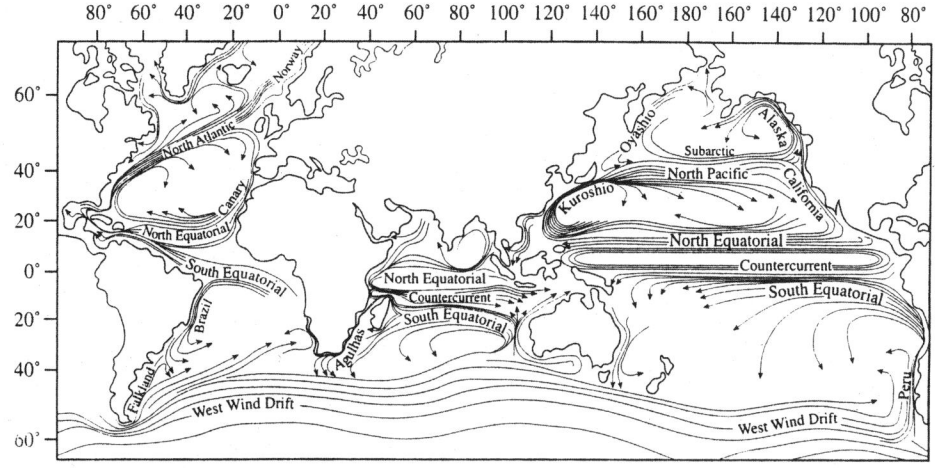

Fig. 2. Global pattern of surface circulation, showing some of the principal wind-driven currents that act as bottom currents along continental margins (after Charnock 1996).

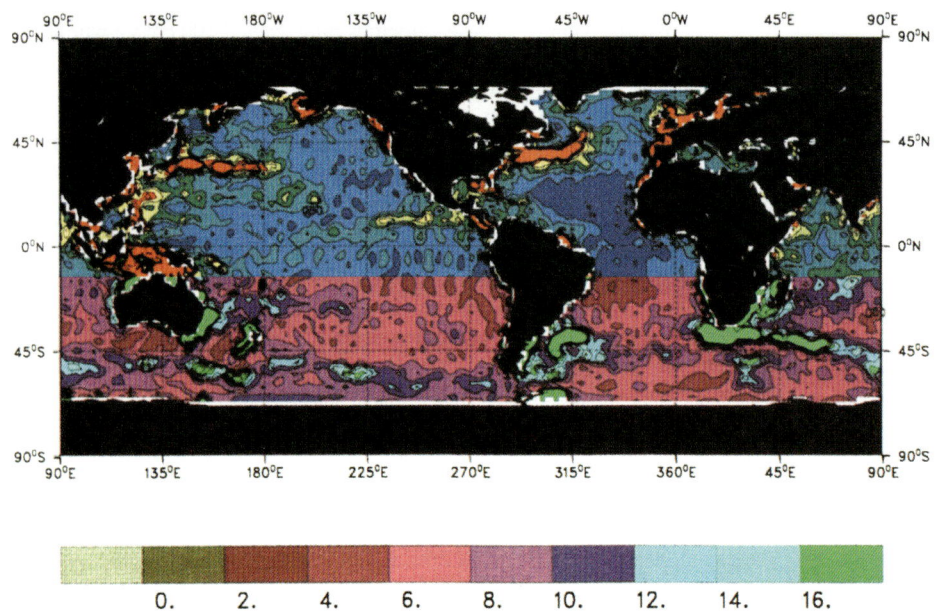

Fig. 3. Statistics of eddy energy derived from variability of sea-surface slope (obtained from the Topex/Poseidon satellite). Highest eddy energy is indicated by the yellow-red end of the spectrum. This can be transmitted through the water column to the seafloor (after Richards & Gould 1996).

through the Straits of Gibraltar and then northwards along the Iberian and north European margin (**Stow** *et al.* and **Faugères** *et al.*). At different periods of Earth history warm saline bottom waters will have been equally or more important than cold water masses in the global thermohaline circulation.

Contourite drifts

The recognition of contourite drifts in seismic profiles, both at the surface and within section, has been a fast evolving field, with ever more examples and types of drift being added to the data base. At present, contourite accumulations can be grouped into six main classes on the basis of their overall morphology and setting: (i) contourite sheet drifts; (ii) elongate mounded drifts; (iii) channel-related drifts; (iv) confined drifts; (v) infill drifts; and (vi) modified drift-turbidite systems (Table 1, Fig. 4). It is important to note, however, that these distinctive morphologies are simply type members within a continuous spectrum, so that hybrid types also occur. They are also found at all depths within the oceans, including all deep water (> 2000 m) and mid-water (300–2000 m) settings. Those current controlled sediment bodies that occur in shallower water (50–300 m) on the outer shelf or uppermost slope are not considered contourite drifts *sensu stricto*, but may be referred to as shallow water drifts (**Viana** *et al.*). The occurrence and geometry of these different types is controlled principally by five interrelated factors: the morphological context or topography; the current velocity and variability, at both a short-period and longer timescale; the amount and type of sediment available; the length of time over which the bottom current processes have operated; and modification by interaction with downslope processes and their deposits.

Contourite sheet drifts

These form extensive very low-relief accumulations, either as part of the fill of basin plains or plastered against the continental margin. They comprise a layer of more or less constant thickness (up to a few hundreds of metres) that covers a large area, but that demonstrates a very slight decrease in thickness towards its margins, i.e. having a very broad low-mounded geometry. The internal seismofacies is typically one of low amplitude, discontinuous reflectors or, in some parts, is more or less transparent.

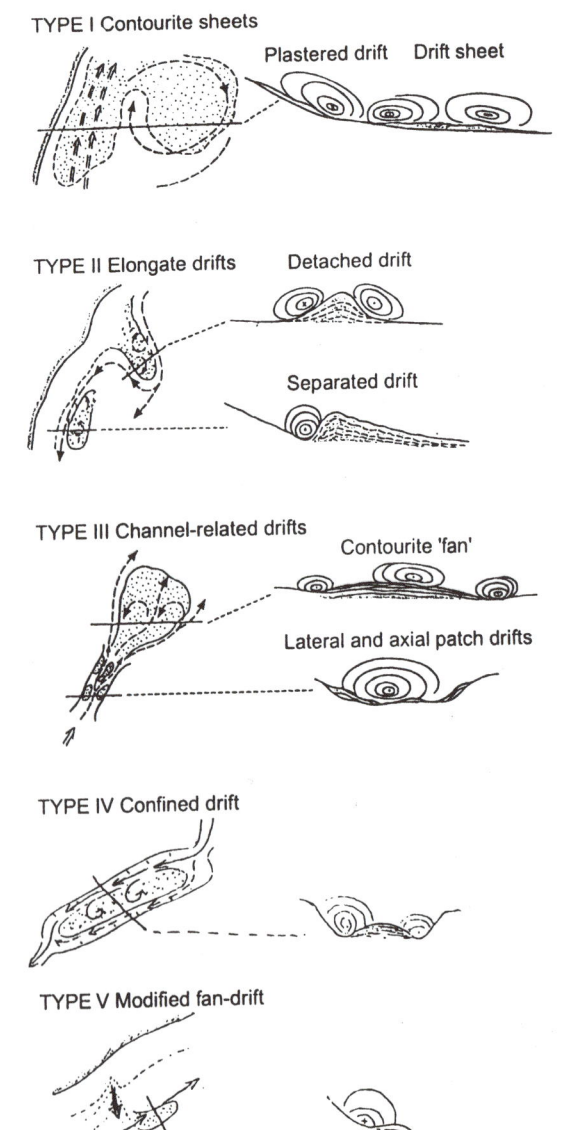

Fig. 4. Contourite drift models (modified from Faugères *et al.* 1999; Stow 2001).

Table 1. *Drift morphology, classification and dimensions. Modified from Faugères et al. (1999) and Stow (2001)*

Drift type	Subdivisions	Size	Examples
Contourite sheet drift	a) abyssal sheet	10^5–10^6 km^2	a) Argentine basin; Gloria Drift
	b) slope (plastered sheet)	10^3–10^4 km^2	b) Gulf of Cadiz; Campos margin
	c) slope (patch) sheets	$< 10^3$ km^2	
Elongated mounded drift	a) detached drift	10^3–10^5 km^2	a) Eirek drift; Blake drift
	b) separated drift	10^3–10^4 km^2	b) Feni drift; Faro drift
Channel-related drift	a) patch-drift	10–10^3 km^2	a) NE Rockall trough
	b) contourite-fan	10^3–10^5 km^2	b) Vema Channel exit
Confined drift		10^3–10^5 km^2	Sumba drift; E Chatham rise
Modified drift – turbidite systems	a) extended turbidite bodies	10^3–10^4 km^2	a) Columbia levee S Brazil Basin; Hikurangi fandrift
	b) sculptured turbidite bodies	10^3–10^4 km^2	b) SE Weddell Sea
	c) intercalated turbidite-contourite bodies	can be very extensive	c) Hatteras rise

They may be covered by large fields of sediment waves, as in the case of the South Brazilian and Argentinian basins where they are also capped in the central region by giant elongate bifurcated drifts.

The different hydrological and morphological contexts define either abyssal sheets or slope sheets (also known as plastered drifts). The former carpet the floors of abyssal plains and other deep water basins including those of the South Atlantic and the central Rockall trough in the NE Atlantic. The basin margin relief partially traps the bottom currents and determines a very complex gyratory circulation. Slope sheets occur near the foot of slopes where outwelling or downwelling bottom currents exist, such as in the Gulf of Cadiz as a result of the deep Mediterranean Sea Water outwelling at an intermediate water level into the Atlantic, or around the Antarctic margins (e.g. the Weddell Sea slope) as a result of the formation and downwelling of cold AABW. They are also found plastered against the slope at any level, particularly where gentle relief and smooth topography favours a broad non-focussed bottom current, such as along the Hebrides margin and Nova Scotian margin.

Abyssal sheet drifts typically comprise fine-grained contourite facies, including silts and muds, biogenic-rich pelagic material, or manganiferous red clay, interbedded with other basin plain facies. Accumulation rates are generally low, around 2–4 cm ka^{-1}. Slope sheets are more varied in grain size, composition and rates of accumulation. Thick sandy contourites have been recovered from base-of-slope sheets in the Gulf of Cadiz, and rates of over 20 cm ka^{-1} are found in sandy-muddy contourite sheets on the Hebridean slope. Case studies in this volume that include sheet drifts are given by **McCave & Tucholke**, **Akhurst et al.**, **Knutz et al.**, **Stow et al.** (Barra) and **Reeder et al.**

Elongate mounded drifts

This type of contourite accumulation is distinctly mounded and elongate in shape, with dimensions variable from a few tens of km to over 1000 km long, length to width ratios from 2:1 to 10:1, and thicknesses up to 2 km. They may occur anywhere from the outer shelf/upper slope, such as those east of New Zealand to the abyssal plains, depending on the depth at which the bottom current flows. They are very common throughout the North Atlantic, but occur also in all the other ocean basins and some marginal seas. One or both lateral margins are generally flanked by distinct moats along which the flow axis occurs and which experience intermittent erosion and non-deposition. Elongate drifts associated with channels or confined basins are classified separately.

Both the elongation trend and direction of progradation are dependent upon an interaction between the local topography, the current system and intensity, and the Coriolis Force. Elongation is generally parallel or subparallel to the margin, with both detached and separated types recognized, but progradation can lead to parts of the drift being elongated almost perpendicular to the margin. Internal seismic character reflects the individual style of progradation, typically with lenticular, convex-upward depositional units overlying a major erosional discontinuity. Fields of migrating sediment waves are common.

Sedimentation rates depend very much on the amount and supply of material to the bottom currents. On average, rates are greater than for sheet drifts, being between 2 and 10 cm ka^{-1}, but may range from < 2 cm ka^{-1} for open ocean pelagic biogenic-rich drifts, to > 60 cm ka^{-1} for some marginal drifts (e.g. along the Hebridean margin). The sediment type also varies according to input, including biogenic, volcaniclastic and terrigenous types. Grain size varies from muddy to sandy as a result of long-term fluctuations in bottom current strength. Many of the case studies presented in this volume include elongate mounded drifts: **Tucholke**, **Laberg et al.**, **Howe et al.**, **Stow et al.** (Faro), **Ercilla et al.**, **Gomes & Viana**, **Uenzelmann-Neben**, **Michels et al.** and **Pudsey & Howe**.

Channel-related drifts

This type of contourite deposit is related to deep channels, passageways or gateways through which the bottom circulation is constrained so that flow velocities are markedly increased (e.g. Vema Channel, Kane Gap, Samoan Passage, Almirante Passage, Faroe–Shetland Channel etc.). Gateways are very important narrow conduits that cut across the sills between ocean basins and thereby allow the exchange of deep and intermediate water masses. In addition to significant erosion and scouring of the passage floor, irregular discontinuous sediment bodies are deposited on the floor and flanks of the channel, as axial and lateral patch drifts, and at the downcurrent exit of the channel, as a contourite fan.

Patch drifts are typically small (a few tens of square kilometres in area, 10–150 m thick) and either irregular in shape or elongate in the direction of flow. They can be reflector-free or with a more chaotic seismic facies, and may have either a sheet or mounded geometry. Contourite fans are much larger cone-shaped deposits, up to 100 km or more in width and radius and 300 m in thickness (e.g. the Vema contourite fan). Channel floor deposits include patches of coarse grained (sand and gravel) lag contourites, mud-clast contourites and associated hiatuses that result from substrate erosion, as well as patch drifts of finer grained muddy and silty contourites where current velocities are locally reduced.

Manganiferous mud contourites and nodules are also typical in places. Accumulation rates range from very low, due to non-deposition and erosion, to as much as 10 cm ka^{-1} in some patch drifts and contourite fans. The case studies in this volume from channel-related settings include those by: **Akhurst *et al.*, Reeder *et al.*, Roveri, Faugères *et al.*** (Vema and Columbia papers) and **Carter & McCave**.

Confined drifts

Relatively few examples are currently known of drifts confined within small basins. These typically occur in tectonically active areas, such as the Sumba drift in the Sumba forearc basin of the Indonesian arc system, the Meiji drift in the Aleutian trench and an unnamed drift in the Falkland Trough. Apart from their topographic confinement, the gross seismic character appears similar to mounded elongate drifts having distinct moats along both margins. Sediment type and grain size depend very much on the nature of input to the bottom current system, but are inadequately known to allow generalisation at this stage. The two case studies from this volume that include confined drifts are those by **Reeder *et al.*** and **Cunningham *et al.***

Infill drifts

This drift type is the most recent addition to our end-member models and so also lacks many well documented examples. They are mostly moderate relief, variable-shape, small scale features that are formed as the local infill of topographic depressions, which have developed beneath the flow pathway of a bottom current system. Typically they occur as infills and partial infills at the head of a slump scar, or at the margins and toe region of a large slump/slide mass. They were first recognized as such from the Hebridean slope on the NW UK continental margin, but are probably very widespread beneath other slope centred bottom currents. Drift infill of a channel of unknown origin has been noted on the Faro-Albufeira drift complex in the Gulf of Cadiz (**Stow *et al.*** and **Faugères *et al.***). The seismic geometry is one that has clearly moulded to fill progressively, with downcurrent prograding reflectors, the topographic depression or irregularity involved. We have no data on the sediment fill facies or rates. Some of the Campos margin contourites fit into this category (**Viana *et al.***).

Modified drift-turbidite systems

The interaction of downslope and alongslope processes and deposits at all scales is the normal condition on the margins as well as within the central parts of present ocean basins. Interaction with slow pelagic and hemipelagic accumulation is also the norm, but these deposits do not substantially affect the drift type or morphology. Over a relatively long timescale, there has been an alternation of periods during which either downslope or alongslope processes have dominated as a result of variations in climate, sea-level and bottom circulation coupled with basin morphology and margin topography. This has been particularly true since the late Eocene onset of the current period of intense thermohaline circulation, and with the marked alternation of depositional style reflecting glacial-interglacial episodes during the past 2 Ma.

At the scale of the drift deposit, this interaction can have different expressions as exemplified in the following examples:

(a) Nova Scotian Margin: regular interbedding of thin muddy contourite sheets deposited during interglacial periods and fine grained turbidites dominant during glacials; marked asymmetry of channel levees on the Laurentian Fan, with the larger levees and extended tail in the direction of the dominant bottom current flow.
(b) Cape Hatteras Margin: complex imbrication of downslope and alongslope deposits on the lower continental rise, that has been referred to as a *companion drift-fan*.
(c) The Chatham–Kermadec Margin: the deep western boundary current in this region scours and erodes the Bounty Fan south of the Chatham Rise and directly incorporates fine grained material from turbidity currents that have travelled down the Hikurangi Channel. This material, together with hemipelagic, is swept north from the downstream end of the turbidity current channel to form a *fan-drift* deposit.
(d) West Antarctic Peninsula Margin: eight large sediment mounds, elongated perpendicular to the margin and separated by turbidity current channels, have an asymmetry that indicates construction by entrainment of the suspended load of down-channel turbidity currents within the ambient southwesterly directed bottom currents and their deposition downcurrent.
(e) Hebridean Margin: complex pattern of intercalation of downslope (slides, debrites and turbidites), alongslope contourites and glaciomarine hemipelagites in both time and space; the alongslope distribution of these mixed facies types by the northward-directed slope current has led to the term *composite slope-front fan* for the Barra Fan.

Several of the case studies in this volume describe either modified drift-turbidite systems of the sort outlined above, or simply systems with close interbedding of turbidites and contourites. They include papers by **Knutz *et al.*, Reeder *et al.*, Faugères *et al.*** (Columbia), **Viana *et al.*, Rebesco *et al.*** and **Escutia & Nelson**.

Seismic characteristics

In many cases, the first means of identification of contourite drifts will be on seismic profiles. However, with growing recognition of the widespread occurrence of drifts in the deep sea, their variety of types, scales and depositional settings, as well as their similarity to features typical of other deep-sea facies (such as turbidite lobes, levees and fans), it has become necessary to erect a set of seismic criteria that will help distinguish drifts from other similar bodies. This becomes much more difficult, of course, where close interbedding of different facies has occurred as in the mixed drift systems noted above. The current set of seismic criteria for enabling positive identification, slightly modified from the three-scale approach developed by Faugères *et al.* (1999), is summarized below. The key attributes are shown schematically in Figure 5, and abundantly illustrated in many of the case studies presented in this volume.

Large scale (i.e. drift scale)

Contourite drifts form as an integral part of the depositional environment in which they occur, in some cases as an isolated sediment body but more usually as a complex of drift types and other deep-water architectural elements. The large-scale features of these accumulations reflect long-lasting (temporally) stable conditions in the bottom current regime and/or oceanographic setting.

(1) *Drift geometry*. The variety of drift geometries now known to exist have been described in some detail in the previous section. Those with a more distinctly mounded rather than low-relief sheet-like geometry are most easily identified. This is especially true where the sediment body occurs beneath an existing bottom

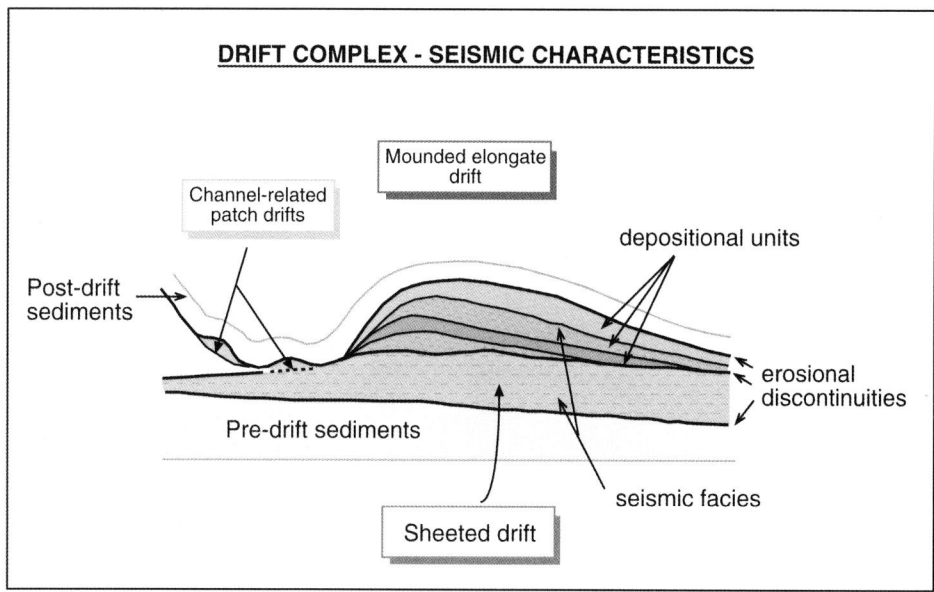

Fig. 5. Schematic model of principal seismic characteristics of contourite drifts.

current system and is clearly isolated from other possible sediment sources (such as downslope supply routes).

(2) *Drift elongation.* An overall downcurrent elongation direction is typical of most drifts.

(3) *Erosional discontinuities.* There are typically widespread discontinuities both at the base and within the drift, extending across the accumulation as a whole. These are commonly marked by continuous high-amplitude reflector, that may also underline a change in seismofacies. Some of these unconformities will be on a subregional scale, beyond the confines of the drift, while others (such as the basal horizon) will even link into oceanwide discontinuities. These reflect periodic changes in bottom current conditions.

(4) *Uniform reflector pattern.* Drifts are commonly represented by extensive, sub-parallel, moderate- to low-amplitude reflectors, with mainly gradational changes typical between seismic facies, in addition to the erosional discontinuities noted above. These reflect the long-lasting stable conditions, both laterally and temporally, that are the norm for drift accumulation.

Medium scale (i.e. depositional seismic units)

Internal architecture within a drift is generally complex, as a result of local variation in processes and accumulation rates linked to changes in current activity. In many cases, the history of drift construction is marked by an alternation of periods of sedimentation and periods of erosion or non-deposition. Medium- and small-scale features reflect these changes.

(1) *Seismic units.* Most of the larger drifts will comprise a series of broadly lenticular, upwardly-convex, seismic units.

(2) *Migration direction.* The stacking of units shows migration in downcurrent to oblique direction, coincident with the elongation direction of the drift as a whole. Any lateral migration direction is likely to be influenced by the Coriolis Force (to the right in the Northern Hemisphere, and to the left in the Southern Hemisphere), providing the right morphological context, current direction and latitude.

(3) *Reflector terminations.* Downlapping and sigmoid progradational reflector patterns are typical, whereas a top-lapping pattern is less common.

Small scale (i.e. seismic facies)

In greater detail, the nature of individual seismic facies reflect changes in both depositional processes and in sediment types. They are not unique to contourite drifts and also depend very closely on the methods employed for seismic acquisition and processing. However, once a drift origin has been established, using a combination of seismic and other characteristics, much interesting detail can be gleaned from this small-scale approach.

(1) *Seismic facies.* A wide variety of seismic facies are typical of contourites, most of which are equally present in turbidite and/or hemipelagic systems. These include (i) semi-transparent, reflector-free intervals, (ii) continuous, sub-parallel, moderate- to low-amplitude reflectors, (iii) regular, migrating-wave, moderate- to low-amplitude reflectors, (iv) irregular, wavy to discontinuous, moderate-amplitude reflectors, and (v) an irregular, continuous, single high-amplitude reflector. We tentatively suggest that this order of seismic facies (i to v) reflects increasing strength in the bottom current regime (Fig. 6). Particular seismic facies associations may be more diagnostic of contourite systems, although this area also needs more work.

(2) *Seismic facies cyclicity.* Recent work (e.g. **Stow** *et al.*, **Faugères** *et al.* and **Roveri**) has revealed a common cyclical pattern in some drifts between a more transparent facies (T) and a moderate-amplitude continuous reflector seismic facies (R). Preliminary interpretation suggests seismic facies (R) reflects a greater proportion of silt/sand content contourites, more hiatuses and condensed sedimentation sections, due to increased bottom current intensity. Seismic facies (T), by contrast, is due to low silt/sand content within a more continuous and homogeneous muddy contourite section, reflecting decreased bottom current intensity. The driver for this cyclical model (Fig. 7) is most likely bottom current variation linked to climate change.

Morphological features and bedforms

High-resolution sesimic records, from echosounder, sidescan and deep-tow seismic techniques, coupled with swath bathymetric mapping and seafloor photography, have helped document the principal surface morphological features of contourite systems. 3D seismic reflection profiling, followed by time-slice seismic attribute mapping, has enabled us to interpret the same sorts of features at depth within the sedimentary succession.

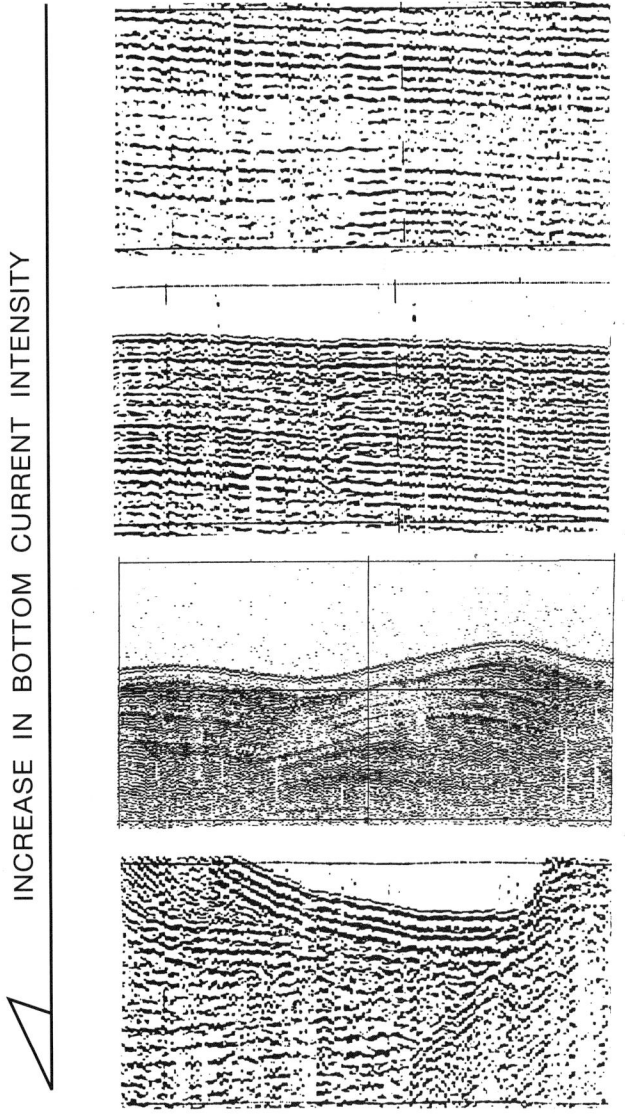

Fig. 6. Principal seismic facies found in contourite drifts and their inferred relationship to bottom-current velocity.

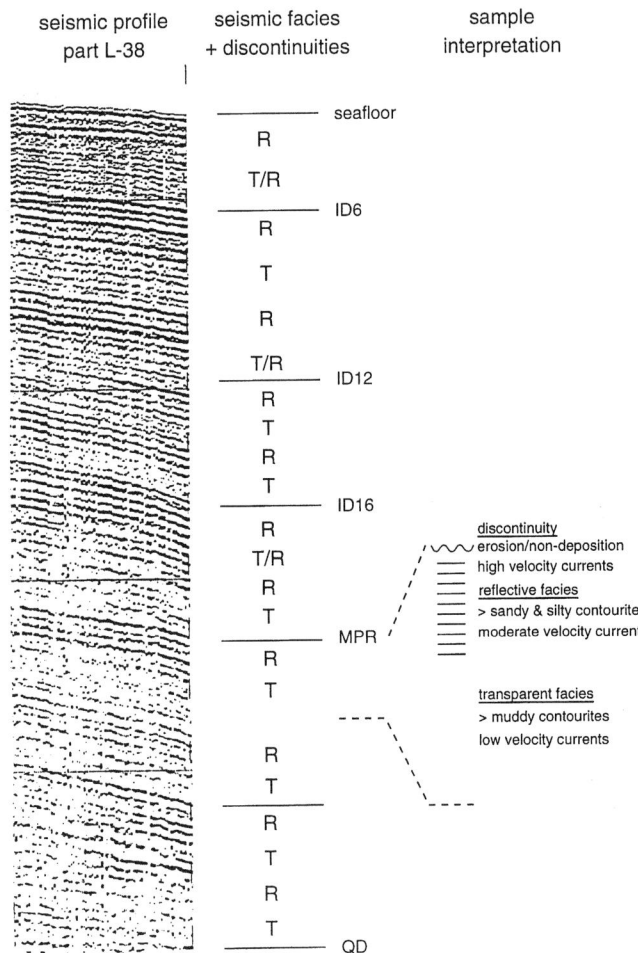

Fig. 7. Model for cyclic variation in seismic facies (based on **Stow** *et al.* and **Llave** *et al.*).

Erosional features that result from strong bottom current activity include scours and moats at various scales, semi-indurated mud horizons, coarse-lag surfaces, gravel and rock pavements, and both large- and small-scale longitudinal furrows. Depositional bedforms observed under moderate to weaker bottom currents include giant sediment waves, large-scale waves and dunes (often barchanoid), and a host of small-scale waves, dunes, ripples and surface lineation features as found beneath any unidirectional current regime. The presumed relationship of these elements to bottom current velocity is illustrated in Figure 8. None of them is exclusive to bottom current systems alone.

Although originally considered diagnostic of contourite systems, regular migrating giant sediment waves are now known to be a very common feature in both contourite and turbidite systems. In addition, a regular wave-like pattern may develop as a result of slope deformation or downslope creep, and in response to the wholesale upward migration or escape of sediment porewaters. The latest attempts to differentiate between these different processes of wave formation are discussed by Wynn & Stow (in press).

Contourite sediment facies

Tremendous advances have been made in the characterization of contourites since they were first described from a modern oceanographic setting nearly 40 years ago, although the majority of that work has remained in a marine setting. Some 50 different legs of the DSDP through ODP progammes have drilled drift and mixed-drift sites, with over 100 sites having recovered contourites. A great many more conventional cores have been taken from these and other drifts around the world during a variety of national and international campaigns. We can therefore be very confident that our present facies models for contourites are well founded, while recognizing that new contourite types may yet be discovered and some of those most recently described still require further elaboration. However, there is less consensus about the nature of bottom current; turbidity current interaction and of the bottom current reworked turbidite facies model. The original 'contourites' described by Hollister and co-workers were most probably of this latter type. The debate surrounding fossil contourites is outlined in a later section.

The wide variety of contourite facies that have been recognized on the basis of variations in grain size and composition are listed and briefly described below and illustrated schematically in Figure 9. Many of the papers in this volume present good photographs of most of these facies, but see in particular papers by **Akhurst** *et al.*, **Knutz** *et al.* and **Stow** *et al.* (Faro).

(a) Siliciclastic contourites (muddy, silty, sandy and gravel-rich variations);
(b) Shale-clast/shale-chip contourites (all compositions possible);
(c) Volcaniclastic contourites (muddy, silty, sandy and gravel-rich variations);
(d) Calcareous bioclastic contourites (calcilutite, calcisiltite, calcarenite and calcirudite variations);

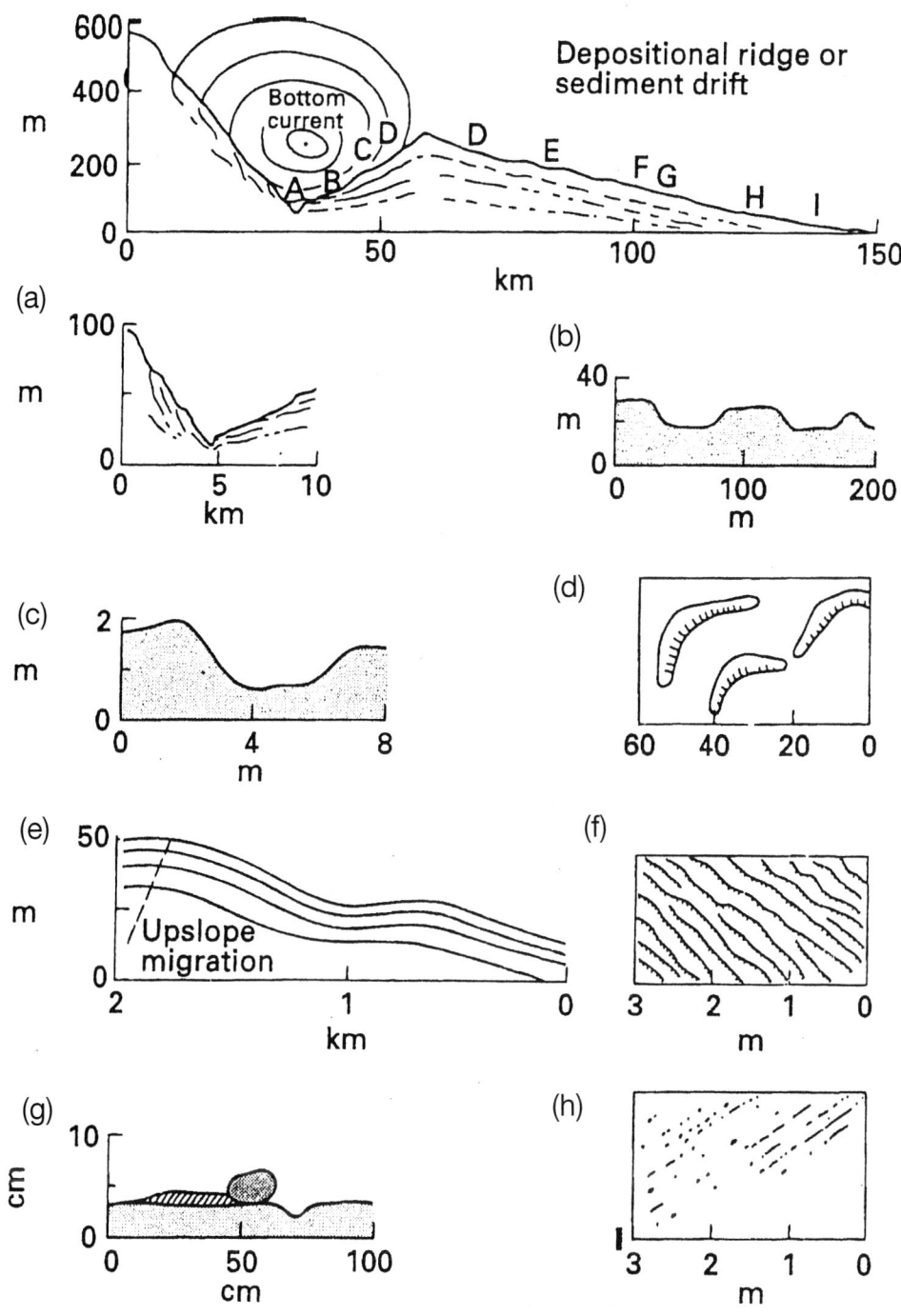

Fig. 8. Erosional features and depositional bedforms typically found on the seafloor beneath bottom-current systems.

(e) Siliceous bioclastic contourites (mainly sand grade recognized);
(f) Manganiferous muddy contourites (+ manganiferous nodules/pavements);
(g) Other contourite related facies ('shallow-water' contourites, reworked turbidites).

Muddy contourites are homogeneous, poorly bedded and highly bioturbated, with rare primary lamination (partly destroyed by bioturbation), and irregular winnowed concentrations of coarser material. They have a silty-clay grain size, poor sorting and a mixed terrigenous (or volcaniclastic), biogenic composition. The components are in part local, including a pelagic contribution and in part far-travelled.

Silty contourites (also referred to as *mottled silty contourites*) commonly show bioturbational mottling to indistinct discontinuous lamination, and are gradationally interbedded with both muddy and sandy contourite facies. Sharp to irregular tops and bases of silty layers are common, together with thin lenses of coarser material. They have a poorly sorted clayey-sandy silt size and a mixed composition.

Sandy contourites occur as both thin irregular layers and as much thicker units within the finer grained facies, and are generally thoroughly bioturbated throughout (e.g. **Stow et al.** (Barra)). In some cases, rare primary horizontal and cross-lamination is preserved (though partially destroyed by bioturbation), together with irregular erosional contacts and coarser concentrations or lags. The mean grain size is normally no greater than fine sand, and sorting is mostly poor due to bioturbational mixing, but more rarely clean and well sorted sands occur. Both positive and negative grading may be present. A mixed terrigenous-biogenic composition is typical, with evidence of abrasion, fragmented bioclasts and iron-oxide staining.

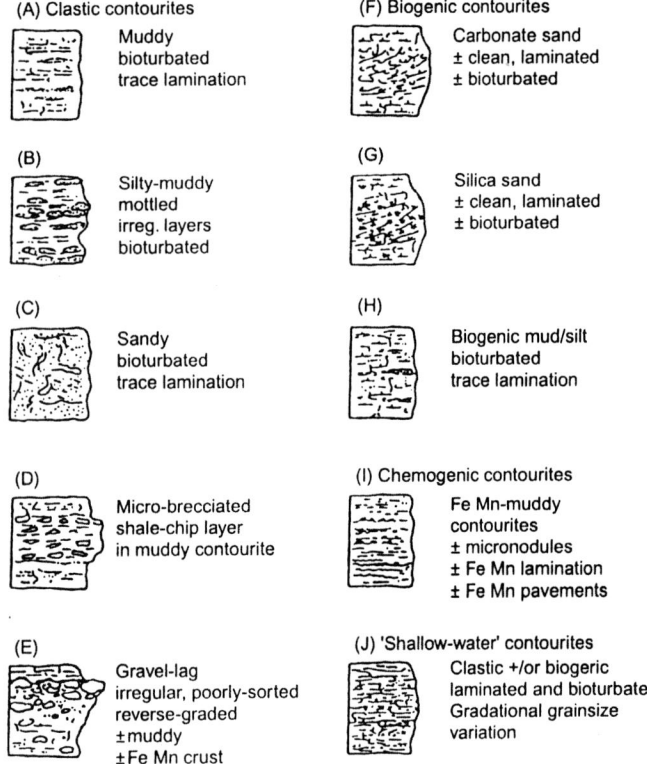

Fig. 9. Contourite facies models for clastic, biogenic, chemogenic and 'shallow-water' contourites (modified from Stow *et al.* 1996).

Gravel rich contourites are common in drifts at high latitudes as a result of input from ice-rafted material (e.g. **Howe *et al.*, Akhurst *et al.* and Laberg *et al.*). Under relatively low-velocity currents, the gravel and coarse sandy material remains as a passive input into the contourite sequence and is not subsequently reworked to any great extent by bottom currents. Gravel lags indicative of more extensive winnowing have been noted from both glacigenic contourites and from shallow straits, narrow moats and passageways, where gravel pavements are locally developed in response to high velocity bottom current activity.

Shale-clast or shale-chip layers in both muddy and sandy contourites have been recognized from relatively few locations to date (e.g. **Faugères *et al.*** (Vema)). They result from substrate erosion under strong bottom currents, where erosion has led to a firmer substrate and, in some cases, burrowing on the omission surface has helped break up the semi-firm muds.

Calcareous and siliceous biogenic contourites occur in regions of dominant pelagic biogenic input, including open ocean sites and beneath areas of upwelling. In most cases bedding is indistinct, but may be enhanced by cyclic variations in composition, and primary sedimentary structures are poorly developed or absent, in part due to thorough bioturbation as in siliciclastic contourites. In rare cases, the primary lamination appears to have been well preserved. The mean grain size is most commonly silty clay, clayey silt or muddy sandy, poorly sorted and with a distinct sand size fraction representing the coarser biogenic particles that have not been too fragmented during transport. The composition is typically pelagic to hemipelagic, including nannofossils and foraminifera as dominant elements in the calcareous contourites and radiolaria or diatoms dominant in the siliceous facies. Many of the biogenic particles are fragmented and stained with either iron oxides or manganese dioxde. There is a variable admixture of terrigenous or volcaniclastic material.

Manganiferous contourites are those in which manganiferous or ferro-manganiferous rich horizons are common. This metal enrichment may occur as very fine dispersed particles, as a coating on individual particles of the background sediment, as fine encrusted horizons or laminae, or as micronodules. It has been observed in both muddy and biogenic contourites from several drifts (eg. **Faugères *et al.*** and **Vema & Columbia**).

Contourite-related facies. It is clearly important to recognise that bottom currents will influence to a greater or lesser extent other deep-water sediments, particularly pelagic, hemipelagic, turbiditic and glacigenic, both during and after deposition. Where the influence is marked and deposition occurs in a drift, then the sediment is termed *contourite*. Where the influence is less severe, such that features of the original deposit type remain dominant, then the sediment is said to have been influenced by bottom currents, as in *bottom current reworked turbidites*. Some silt-laminated facies, as well as the thin, clean, cross-laminated sands originally described by Hollister & Heezen (1972) from NE American margin, are most likely of this type. The features that we suggest best characterize reworked turbidites are summarised in Table 2, together with those of normal (depositional) contourites. These differ from the criteria proposed recently by several workers based purely on ancient turbidite successions (e.g. Stanley 1988; Mutti *et al.* 1992; Shanmugam *et al.* 1993, Shanmugam 2000), but we believe they are more reliable because of the inherent uncertainties of interpretation when dealing with ancient series.

Some of the sediments that have been described recently from mixed drift systems, such as those on the Antarctic Peninsula margin (**Rebesco *et al.***), as well as others from shallow-water, upper-slope to outer-shelf settings (**Roveri *et al.*** and **Sivkov *et al.***), are of a rather different facies. They show clear, but somewhat irregular lamination coupled with bioturbation throughout, and a poor to moderately well sorted, silty grain size. This may represent a hybrid turbidite-contourite facies type and/or a shallow-water contourite facies.

Contourite sequences and current velocity

Muddy, silty and sandy contourites, of siliciclastic, bioclastic, volcaniclastic or mixed composition, commonly occur in composite sequences or partial sequences a few decimetres in thickness (typical range 0.2–3 m). The ideal or complete sequence shows overall negative grading from muddy through silty to sandy contourites and then positive grading back through silty to muddy contourite facies (Fig. 10). Such sequences of grain size and facies variation are now widely recognized from many drifts although, as with the ideal turbidite sequences, partial sequences of different thickness are equally common. Following the turbidite analogy of notation for the *Bouma*, *Stow* and *Lowe* turbidite sequences (see Stow *et al.* 1996), we propose here that a useful advance for the shorthand description of the contourite sequence is to use the notation C1–5 as follows: C5, upper muddy contourite division; C4, upper mottled silty contourite division; C3, middle sandy contourite division; C2, lower mottled silty contourite division; C1, lower muddy contourite division.

Thus a complete sequence *of any composition* is referred to as C1–5. In a vertical succession of repeated sequences, there is a seamless transition from C5 of the underlying sequence to C1 of the overlying sequence. This should be arbitrarily taken at the mid-point of the C5/C1 couplet. Base-only partial sequences are referred to as C1–2 or C1–3, and top-only sequences as C3–5 or C4–5 as appropriate. Rather than introducing new division notation for the rare occurrence of other contourite facies within the sequence, it seems more sensible to highlight these departures from the standard sequence verbally. The base-only sequences that pass up into a gravel-lag and non-depositional surface as

Table 2. *Main characteristics of muddy contourites, sandy contourites and bottom current reworked turbidites (from Stow et al. 1998a)*

	Muddy contourites (terrigenous or biogenic)	Sandy contourites (terrigenous or biogenic)	Reworked turbidites (any composition)
Occurrence	thick uniform sequences of fine-grained sediment in deep-water settings interbedded with turbidites and other resedimented facies on inferred continental margins	thin to medium beds in muddy contourite sequences, rarely thick/v.thick units reworked tops of sandy turbidites in interbedded sequences coarse lag in deep-sea channels and straits	in any normal turbidite setting where strong, permanent bottom currents have been active
Structure	dominantly homogeneous, bedding not sharply defined, but cyclicity common bioturbational mottling generally common to dominant distinct burrows (typical deep-water assemblage) present in many places coarse lag concentrations (especially biogenic) reflect composition of coarse fraction in mud primary silt/mud lamination – rare, but no regular sequence as in turbidites sharp and erosive contacts common in parts	generally bioturbated and burrowed throughout with little primary structure remaining parallel and cross-lamination more rarely preserved (often with bioturbation) no regular structural sequence as in turbidites may show reverse grading near top, with sharp/erosive contacts common	lower divisions of turbidite may be preserved, with the upper divisions either removed completely or modified by reworking bioturbation/burrowing common through reworked top reverse grading and irregular lag concentrations bi-directional cross-lamination, may be clean micro-cross-laminated silts with bioturbation sharp erosive contacts may occur within turbidite sequence
Texture	dominantly silty mud frequently high sand content (0–15%) of biogenic tests in clastic contourites medium to poorly sorted, ungraded, no offshore textural trends may show marked textural difference from interbedded turbidite if transport distances are different	silt to sand-sized, more rarely gravel may be relatively free of mud and well sorted in some cases tendency to low or negative skewness values no offshore trends	removed/non-deposition of fines significant textural differences from underlying turbidite (e.g. cleaner, better sorted, reverse grading + lag, negative skewness)
Fabric	mud fabric – typically more parallel alignment of clays than for turbidites, but not well present in fossil contourites primary silt laminae or coarse lag deposits show grain orientation parallel to the current (along-slope)	indication of grain orientation parallel to the bottom current (along-slope) or more randomised by bioturbation other features (eg structures) also indicate alongslope flow, where preserved	interbedded, reworked turbidite layers may show widely bimodal grain orientations or a more random playmodel fabric
Composition	mixed contourites have combination of biogenic and terrigenous material (may be distinct from interbedded turbidites) terrigenous material dominantly reflects nearby land/shelf source with some along-slope mixing and small amount of far travelled material (no down-slope trends)	mixed biogenic/terrigenous composition typical terrigenous composition dependent on local source biogenic material from pelagic, benthic and resedimented sources, typically fragmented and iron-stained organic-carbon content very low	composition entirely reflects that of turbidite, with part of fine fraction removed long exposure and winnowing may lead to chemogenic precipitation (probably rare) organic-carbon content very low
Sequence	typically arranged in decimetric cycles of grain-size and/or compositional variation with sandy contourites see model (Fig. 2) – partial sequences also common	typically arranged in decimetric cycles of grain-size and/or compositional variation with muddy contourites see model (Fig. 2) – partial sequences also common	presents a typical turbidite sequence (ie top-absent or top reworked) does not occur within standard cyclic contourite sequence

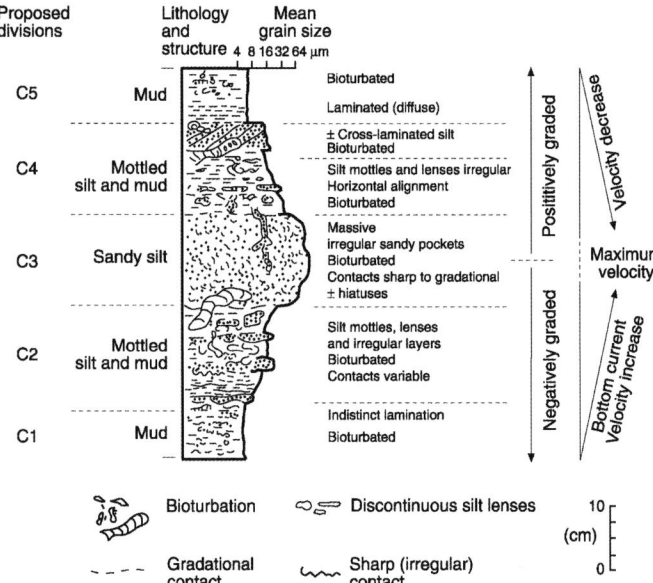

Fig. 10. Composite contourite facies model showing grain size variation through the standard mud-silt-sand contourite sequence, linked to variation in bottom current velocity (modified from Stow *et al.* 1996). The notation C1–C5 for the different facies divisions is introduced in this paper.

found in the Rockall Trough (**Howe *et al.***) should be described as C1–3 *with a gravel top*. Likewise, the top-only contourites of the Rockall Trough and Faroe–Shetland Channel (Stoker *et al.* 1998a; **Howe *et al.*** and **Akhurst *et al.***) can be referred to as C3–5 *with a sharp erosive base*.

The origin of the C1–5 sequence is related *either* to long-term fluctuations in the mean current velocity, *and/or* to variation in sediment supply. Stacked sequences indicate cyclic variation in the forcing variables. Although not enough data exist to be certain of the time scale of these cycles, some evidence points towards 5000–20 000 cycles for certain marginal drifts of terrigenous to mixed composition. In bioclastic successions, the cyclic facies pattern has a longer time-scale (20 000–40 000 years) in the few examples from which we have good dating, and is closely analogous to the Milankovitch cyclicity recognised in many pelagic and hemipelagic successions. It is therefore believed to be driven by the same mechanism of orbital forcing of climate that then effect changes in bottom current velocity.

Differentiating the relative importance of current velocity versus sediment supply is by no means simple. The most thorough approach is to analyse variation in bulk sediment mean grain size, and then to consider the co-variance or not of compositional attributes (e.g. terrigenous/biogenic ratios, benthic/planktonic ratios, percentage of coarse sand/gravel and shale chips, presence/absence of far-travelled components, clay/silt ratio), current indicators (e.g. scour surfaces, lamination, lag horizons, coarse-grained lenses, shale chip concentrations), and bioturbation intensity coupled with ichnofacies types. These data need to be collated and compared for different sites over the same drift in order to distinguish regional from local effects, and to observe down-current trends.

Considerable advances have been made, however, in utilizing simpler proxies for bottom current velocity. The most important of these is mean grain size of the sortable silt (10–63 microns) fraction (Robinson & McCave 1994; McCave *et al.* 1995), which has been most recently adapted to remove the effects of ice-rafted sediment supply in NE Atlantic drifts by Bianchi (2000). A more indirect proxy is to use variation in magnetic susceptibility to mirror the flux of terrigenous (magnetic) components. Further refinement in the reconstruction of paleocurrent variation is to undertake a detailed analysis of mass accumulation rates, as demonstrated by Hall & McCave (2000) using Th-230 excess oxygen systematics coupled with sortable silt grain size on the Iberian margin.

The link between contourite sequences and changes in paleoclimate and paleocirculation is an extremely important one. Where such sequences can be correctly decoded then we can build up a more accurate understanding of the palaeo-ocean and its environment. The occurrence of widespread hiatuses in the deep-ocean sediment record is best related to episodes of particularly intense bottom currents. More locally, such strong currents result in significant sediment winnowing and the accumulation of sand, gravel and shale-clast contourites.

Recognition of contourites and application to the ancient record

Identification of turbidites in both modern and ancient series is generally clear cut. They are single event deposits with well defined characteristics. Identification of contourites, however, is more complex. Bottom currents affect to a greater or lesser extent ambient sedimentation by other processes (pelagic, hemipelagic and turbiditic) so that a blend of characteristics is the common result. The slow and continuous nature of contourite accumulation means that primary features are often blurred or removed by secondary effects, especially bioturbation. Consequently, the recognition of contourites can never be a 'quick-fix' based on simple sediment or seismic appearance, but must always involve careful consideration of a range of characteristics and conditions. The three-stage approach to contourite identification favoured here is slightly refined and summarized from earlier work (Lovell & Stow 1981; Stow *et al.* 1998a), and presented in Table 3. Typical characteristics of muddy and sandy contourites in comparison with bottom current reworked turbidites are given in Table 2.

The application of these criteria, the facies and seismic models derived from modern systems, to ancient series exposed on land has been generally poor. The early erroneous interpretation of many fine grained turbidite successions as contourites has now been well documented (Stow & Lovell 1979; Pickering *et al.* 1989; Stow *et al.* 1998a). Several recent studies have addressed the problem of bottom current reworking of turbidites (Stanley 1988; Mutti *et al.* 1992; Shanmugam *et al.* 1993; Shanmugam 2000; Stow *et al.* 1998a) with some significant advances. But there remains much debate over the detailed sedimentary structures that can be attributed to one process or the other. It may be, in fact, that weak turbidity currents and strong bottom currents (e.g. benthic storm events) have very similar effects on the bottom sediment, so that their distinction on the basis of sediment characteristics alone will not be possible. This whole issue requires further work.

There are a small but growing number of examples of fossil contourites that fit most of our criteria for identification. These include parts of the Cretaceous Talme Yafe Formation in Israel (Bein & Weiler 1976), the Ordovician Jiuxi Drift (Duan *et al.* 1993) and Pingliang Drift (Gao *et al.* 1995) in China, the Paleogene Lefkara Formation in Cyprus (Kahler & Stow 1998), and the Neogene Misaki Formation in south central Japan (Stow *et al.* 1998b). All but the first of these are presented, with additional data and refinement, in this volume (**Stow *et al.*** (Cyprus and Japan); **Luo *et al.*** (China)). In addition, the paper by **Ito** describes Plio-Pleistocene sandy contourites within the turbidite-dominated Kasuza Group of southern Japan. We suggest that these may be a good example of a bottom current influence on fine-grained turbidites.

Implications and further research

There are of course many reasons to study bottom currents and contourites, not least because they represent such an important

Table 3. *Criteria for the recognition of contourites in both modern and ancient systems (from Stow et al. 1998a)*

Stage 1: Small-scale (field, borehole or lab)
- Do the sediments have the range of features shown in Table 2 or as described in the text?
- Where there is a possibility of mixed turbidite/contourite sequences, can a distinction be made between the two facies on the basis of character and/or palaeocurrent evidence?
- Is there sufficient evidence to discount deposition from fine-grained turbidity currents? Particular care must be taken for inferred reworked turbidtes.
- Where there is a possibility of mixed hemipelagite–pelagite/contourite sequences, is there sufficient evidence for the influence of bottom currents during sedimentation?
- Can any cyclicity present be related to variation in bottom current velocity rather than to variations in terrigenous input or biogenic productivity?

Stage 2: Medium-scale (drift, formation or region)
- Do regional trends in facies occurrence, palaeocurrent directions, textures, mineralogical or geochemical tracers exist that would support a bottom current origin?
- Is there any other evidence of bottom current activity such as unconformities, condensed sequences, regional variation in thickness, drift geometry, etc?
- Is it possible to reconstruct the shape and 3D geometry of the whole sedimentary body? and, if so, are the elongation and propagation trends parallel or perpendicular to the inferred margin?
- Are the associated facies, paleontological data and rates of accumulation compatible with a contourite interpretation?

Stage 3: Large-scale (system, ocean or continent)
- Do the conclusions from Stages 1 and 2 above fit with what is known from other independent lines of evidence concerning major oceanographic or palaeoceanographic features and continental reconstructions?
- What kind of bottom current systems exist at present or might have existed in the study area at the time of deposition, taking into account constraints imposed by known palaeoclimatic conditions and inferred basin location and geometry?

component of the still little known deep ocean basins and their margins. Furthermore, they hold an important key to the decoding of paleocirculation records encapsulated in oceanic drifts and corresponding hiatuses, that are closely linked to past climatic change. Thick units of sandy contourites together with bottom-current reworked sandy turbidites are potentially important as hydrocarbon reservoirs where suitably buried in association with source rocks. The nature and effects of bottom currents on margin stability and subsea engineering projects need to be carefully evaluated.

Based on this overview and compendium of examples, together with the work of IGCP 432 over the past few years, we highlight below some of the key directions for future research in the contourite field.

(1) Integration of the physical oceanographic and sedimentological approaches – joint study of benthic storms and their effects, the nature of oceanic gateways, and long-term measurements of bottom currents coupled with their impact on the seafloor.
(2) Particular study of modified contourite drift-turbidite systems, sheeted drifts interbedded in margin settings, and shallow-water/marginal sea drifts.
(3) Further elaboration of the seismic facies of contourite sytsems and their relationship to flow velocity, and of the cyclic model of seismic facies presented here.
(4) Integration of biological oceanographic and sedimentological approaches – joint study of contourite ichnofacies and rates of accumulation, contourite sequences and periodicity, and of the nature and extent of sandy contourites.
(5) Further work on the links between contourite sequences and cycles and bottom current velocity fluctuation, and on potential proxies for characterising these changes.
(6) Collaboration between different specialists in the hunt for and decoding of fossil contourite successions exposed on land.

References

BEIN, A. & WEILER, Y. 1976. The Cretaceous Talme Yafe Formation, a contour current shaped sedimentary prism of calcareous detritus at the continental margin of the Arabian Craton. *Sedimentology*, **23**, 511–532.

BIANCHI, G. G. 2000. Iceland-Scotland overflow fluctuations over the past 82 ka using a novel approach. *31st International Geological Congress, Abstracts Volume,* Rio de Janeiro, Brazil.

DUAN, T. Z., GAO Z. Z., ZENG, Y. F. & STOW, D. A. V. 1993. A fossil carbonate contourite drift on the Lower Ordovician palaeocontinental margin of the middle Yangtze Terrane, Jiuxi, northern Hunan, southern China. *Sedimentary Geology*, **82**, 271–284.

ELLWOOD, B. B. & LEDBETTER, M. T. 1977. Antarctic bottom water fluctuations in the Vema Channel: effects of velocity changes on particle alignment and size. *Earth and Planetary Science Letters*, **35**, 189–198.

FAUGÈRES, J.-C. & STOW, D. A. V. 1993. Bottom-current-controlled sedimentation: a synthesis of the contourite problem. *Sedimentary Geology*, **82**, 287–297.

FAUGÈRES, J.-C., GONTHIER, E. & STOW, D. A. V. 1984. Contourite drift moulded by deep Mediterranean outflow. *Geology*, **12**, 296–300.

FAUGÈRES, J.-C., STOW, D. A. V., IMBERT, P., VIANA, A. R. & WYNN, R. B. 1999. Seismic features diagnostic of contourite drifts. *Marine Geology*, **162**, 1–38.

GAO, Z. Z., ERIKSSON, K. A., HE, Y. B., LUO, S. S. & GUO, J. H. 1998. *Deep-Water Traction Current Deposits*. Science Press, Beijing, New York. 57–105.

GAO, Z. Z., LUO, S. S., HE, Y. B., ZHANG, J. S. & TANG, Z. J. 1995. The Middle Ordovician contourite on the west margin of Ordos. *Acta Sedimentologica Sinica*, **13**(4), 16–26. (in Chinese with English abstract).

GILBERT, I. M., PUDSEY, C. J. & MURRAY, J. W. 1998. A sediment record of cyclic bottom-current variability from the NW Weddell Sea. *Sedimentary Geology*, **115**, 185–214.

GONTHIER, E., FAUGÈRES, J.-C. & STOW, D. A. V. 1984. Contourite facies of the Faroe drift, Gulf of Cadiz. *In*: STOW, D. A. V. & PIPER, D. J. W. (eds) *Fine-Grained Sediments: Deep-Water Processes and Facies*. Geological Society, London, Special Publications, **15**, 275–292.

HEEZEN, B. C., HOLLISTER, C. D. & RUDDIMAN, W. F. 1966. Shaping the continental rise by deep geostrophic contour currents. *Science*, **152**, 502–508.

HOLLISTER, C. D. 1993. The concept of deep-sea contourites. *Sedimentary Geology*, **82**, 5–11.

HOLLISTER, C. D. & HEEZEN, B. C. 1972. Geologic effects of ocean bottom currents. *In*: GORDON, A. L. (ed.) *Studies in Physical Oceanography – A Tribute to George Wust on his 80th Birthday*. Gordon & Breach, New York, **2**, 37–66.

HOWE, J. A., STOKER, M. S. & STOW, D. A. V. 1994. Late Cenozoic sediment drift complex, NE Rockall Trough, North Atlantic. *Palaeoceanography*, **9**, 989–999.

KAHLER, G. & STOW, D. A. V. 1998. Turbidites and contourites of the Palaeogene Lefkara Formation, Southern Cyprus. *Sedimentary Geology*, **115**, 215–231.

LOVELL, J. P. B. & STOW, D. A. V. 1981. Identification of ancient sandy contourites. *Geology*, **9**, 347–349.

MALDONADO, A. & NELSON, C. H. 1999. Marine geology of the Gulf of Cadiz. *Marine Geology, Special Issue*, **155**.

MCCAVE, I. N., HOLLISTER, C. D. & NOWELL, A. R. M. 1988. *Deep Ocean Sediment Transport: HEBBLE collected reprints 1980–1987*. Woods Hole Oceanographic Institute, Woods Hole.

MCCAVE, I. N., MANIGHETTI, B. & ROBINSON, S. G. 1995. Sortable silt and fine sediment size/composition slicing: parameters for paleocurrent speed and paleoceanography. *Paleoceanography*, **10**, 593–610.

MIENERT, J. 1998. European North Atlantic Margin (ENAM): Sediment pathways, processes and flux. *Marine Geology, Special Issue*, **152**.

MUTTI, E. *et al.* 1992. *Turbidite Sandstones*. AGIP/Istituto di Geologia, Universita di Parma, Parma, 236 pp.

NOWELL, A. R. M. & HOLLISTER, C. D. 1985. Deep Ocean Sediment Transport – Preliminary Results of the High Energy Benthic Boundary Layer Experiment. *Marine Geology*, **66**.

PICKERING, K. T., HISCOTT, R. N. & HEIN, F. J. 1989. *Deep-Marine Environments: Clastic Sedimentation and Tectonics*. Unwin Hyman, London.

PUDSEY, C. J., BARKER, P. F. & HAMILTON, N. 1988. Weddell Sea abyssal sediments: a record of Antarctic Bottom Water flow. *Marine Geology*, **81**, 289–314.

REBESCO, M. & STOW, D. A. V. in press. Seismic expression of contourites and related deposits. *Marine Geophysical Researches Special Issue*, in press.

ROBINSON, S. G. & MCCAVE, I. N. 1994. Orbital forcing of bottom-current enhanced sedimentation on Feni Drift, NE Atlantic, during the mid-Pleistocene. *Palaeoceanography*, **9**, 943–972.

SHANMUGAM, G. 2000. Fifty years of the turbidite paradigm (1950's to 1990's): Deep-water processes and facies models – a critical perspective. *Marine & Petroleum Geology*, **17**, 285–342.

SHANMUGAM, G., SPALDING, T. D. & ROFHEART, D. H. 1993. Process sedimentology and reservoir quality of bottom-current reworked sands (sandy contourites): An example from the Gulf of Mexico. *American Association of Petroleum Geologists Bulletin*, **77**, 1241–1259.

SHANMUGAM, G., SPALDING, T. D. & ROFHEART, D. H. 1995. Deep marine bottom current reworked sands (Pliocene and Pleistocene): Ewing Bank 826 Field, Gulf of Mexico. *Proceedings SEPM Core Workshop*, **20**, 25–54.

STANLEY, D. J. 1988. Turbidites reworked by bottom currents; Upper Cretaceous examples from St. Croix, US Virgin Islands. *Marine Science*, **22**, 79 pp.

STOKER, M. S., AKHURST, M., HOWE, J. A. & STOW, D. A. V. 1998a. Sediment drifts and contourites on the continental margin off NW Britain. *Sedimentary Geology*, **115**, 33–52.

STOKER, M. S., EVANS, D. & CRAMP, A. 1998b. *Sedimentation, Mass Wasting and Stability*. Geological Society, London, Special Publication, **129**.

STOW, D. A. V. 1979. Distinguishing between fine-grained turbidites and contourites on the deep-water margin off Nova Scotia. *Sedimentology*, **26**, 371–387.

STOW, D. A. V. 1982. Bottom currents and contourites in the North Atlantic. *Bull. Inst. Geol. Bassin d'Aquitaine*, **31**, 151–166.

STOW, D. A. V. 1994. Deep-sea processes of sediment transport and deposition. *In*: PYE, K. (ed.) *Sediment Transport and Depositional Processes*. Blackwell, Oxford, 257–291.

STOW, D. A. V. 2001. Deep-Sea Sediment Drifts. *In*: STEELE, J. H., TUREKIAN, K. K. & THORPE, S. A. (eds) *Encyclopedia of Ocean Sciences*. Academic Press, London.

STOW, D. A. V. & LOVELL, J. P. B. 1979. Contourites: their recognition in modern and ancient sediments. *Earth Sciences Reviews*, **14**, 251–291.

STOW, D. A. V. & FAUGÈRES, J.-C. 1993. Contourites and Bottom Currents. *Sedimentary Geology, Special Volume* **82**, 1–310.

STOW, D. A. V. & FAUGÈRES, J.-C. 1998. Contourites, Turbidites and Process Interaction. *Sedimentary Geology, Special Issue*, **115**.

STOW, D. A. V. & TABREZ, A. 1998. Hemipelagites: facies, processes and models. *Geological Society Special Publication*, **129**, 317–338.

STOW, D. A. V. & MAYALL, M. 1999. Deep-water Sedimentary Systems: New Models for the 21st Century. *Marine and Petroleum Geology, Special Issue* **17**.

STOW, D. A. V., FAUGÈRES, J.-C. & GONTHIER, E. 1986. Facies distribution and textural variation in Faroe drift contourites: velocity fluctuation and drift growth. *Marine Geology*, **72**, 71–100.

STOW, D. A. V., READING, H. G. & COLLINSON, J. D. 1996. Deep Seas. *In:* READING, H. G. (ed.) *Sedimentary Environments (3rd ed.)*. Blackwell Science, Oxford, 395–454.

STOW, D. A. V., FAUGÈRES, J.-C., VIANA, A. & GONTHIER, E. 1998a. Fossil contourites: a critical review. *Sedimentary Geology*, **115**, 3–32.

STOW, D. A. V., TAIRA, A., OGAWA, Y., SOH, W., TANIGUCHI, H. & PICKERING, K. T. 1998b. Volcaniclastic sediments, process interaction and depositional setting of the Miocene–Pliocene Miura Group, SE Japan. *Sedimentary Geology*, **115**, 351–382.

VIANA, A., FAUGÈRES, J.-C., STOW, D. A. V. & IMBERT, P. 1998a. Bottom-current controlled sand deposits: a review from modern shallow to deep water environments. *Sedimentary Geology*, **115**, 53–80.

VIANA, A. *ET AL*. 1998b. Hydrology, morphology and sedimentology of the Campos continental margin, offshore Brazil. *Sedimentary Geology*, **115**, 133–158.

WYNN, R. B. & STOW, D. A. V. in press. Recognition and interpretation of deep-water sediment waves. *Marine Geology, Special Issue*.

Contourites of the Nova Scotian continental rise and the HEBBLE area

I. N. McCAVE[1], R. C. CHANDLER[2], S. A. SWIFT[3] & B. E. TUCHOLKE[3]

[1]*Department of Earth Sciences, University of Cambridge, Cambridge, UK (e-mail: mccave@esc.cam.ac.uk)*
[2]*Alvin Operations, Woods Hole Oceanographic Institution, Woods Hole, MA 02543, USA*
[3]*Department of Geology and Geophysics, Woods Hole Oceanographic Institution, Woods Hole, MA 02543, USA*

Abstract: The Nova Scotian continental rise is swept by a Deep Western Boundary Current system comprising layers of Labrador Sea Water overlying a core of Norwegian Sea Overflow Water at depths of 3100–3900 m, and below about 4600 m a cold stream of southern-source water. Seismic-reflection data show that the rise contains sediments transported downslope in channels and debris lobes, but there is also evidence of current-controlled deposition and erosion in the post-Eocene sequence. The rise is now mantled by Holocene contourites that have accumulated at a rate of *c.* 6 cm ka^{-1} and are <1 m thick. Bottom photographs show a zonation in current effects and bedform types, with longitudinal ripples and strong currents prevalent at 4800–5000 m, smaller bedforms and progressively weaker currents up to *c.* 4000 m, and mostly tranquil seafloor above 4000 m. Bedform scales and orientations also suggest significant short-term (hours to weeks) variability in current velocity but a mean contour-following flow to the southwest at longer time scales (months to years). These structures are not preserved in the sediment because of pervasive bioturbation and the uppermost layers have negligible preservation potential. The sediments display clear current controlled effects in their grain-size structure involving both percentage of (foraminiferal) sand, and size and percentage in the 10–63 µm range, the 'sortable silt'. There is a sand-rich zone at 4800–4900 m and below 5000 m, and a decreasing silt/clay ratio from 5100 m up to 4000 m. Although much of the sedimentary sequence probably has been emplaced by downslope processes, it has been significantly modified by the Deep Western Boundary Current. Particularly strong and variable currents which rework sediments below *c.* 4800 m probably are engendered by interaction of Gulf Stream eddies with the DWBC. Although strong currents and upstream input from turbidites and debris flows might be thought to favour a coarse-grained deposit, the facies at the HEBBLE (High Energy Benthic Boundary Layer Experiment) site is muddy contourite with ≤ 12% sand.

This was to have been Charley Hollister's paper. He led many cruises to the HEBBLE area between 1979 and 1986, and insights gained from studies of the area are an important part of his scientific legacy. We might have expected to have to help out with production of this paper via some last minute cajoling, but we certainly did not expect the sad task of doing it without him. So this is the best we can do, without his insight, and we dedicate it to Charley's memory.

Geological and oceanographic setting

The continental margin of Nova Scotia consists of strata emplaced after the late Early Jurassic rifting of the North Atlantic (Jansa & Wade 1975; King & MacLean 1976; Wade & MacLean 1990). The sedimentary prism is about 3 km thick beneath the lower continental rise and thickens landward to an average of 10–11 km beneath the continental slope and shelf (Tucholke *et al*. 1982). The continental slope, ranging in depth from about 200 to >2000 m, has an average slope of 3.6°. The continental rise continues more gently to the abyssal plain near 5000 m at slopes of less than 1.5°. Numerous small submarine channels incise the continental slope starting at the shelf-slope break, and many continue across the rise to the abyssal plain (Fig. 1).

The continental shelf off Nova Scotia extends from the Northeast Channel in the west to the Laurentian channel in the east, broadening (125 to 250 km) and shallowing eastward (180 to 100 m). Lateral variations in deep structure beneath the Nova Scotian shelf have affected the morphology and geologic development of the margin throughout its history. Two features in particular are important. The first feature is a change in structure of basement underlying the shelf east and west of 61°30′W. Here, gently seaward-dipping basement beneath the LaHave Platform in the west changes to thinned and extended continental crust in the Sable and Abenaki Sub-basins in the east. Post-rift sediment thickens westward from <1 km on the LaHave Platform to >18 km in the Sable Sub-basin (Wade & MacLean 1990). These differences appear to correlate with present seafloor morphology (Uchupi & Swift 1991). To the east, the continental slope is relatively steep (5–7°), submarine canyons with relief >1000 m incise the slope and upper rise, and the middle and lower rise are constructed of coalesced fans. To the west, the slope dips gently seaward (2°), canyons have been filled so that seafloor relief rarely exceeds 150–200 m, and no distinct fans are recognized.

A second deep feature that influences shallow structure and morphology is the development of a 30–100 km wide belt of closely spaced evaporite diapirs and pillows beneath the upper continental rise between *c.* 2000 m and 3000–4000 m water depth (Fig. 4; Wade & MacLean 1990). Halokinesis of these Upper Triassic–Lower Jurassic deposits began in the Middle Jurassic and continues to the present time. Only a few diapirs penetrate the >8 km of strata to disturb the seafloor, but general uplift produces a shallowing of seafloor dip above the salt province and a steepening just basinward (Swift 1987).

The Nova Scotian Rise below about 2000 m is swept by a deep western boundary current (DWBC) mainly of North Atlantic Deep Water (NADW) but with some southern source high-silica water below about 4600 m (Pickart 1991). The NADW comprises a core of Norwegian Sea Overflow Water (with components from Denmark Strait and Charlie–Gibbs Fracture Zone) between about 3100 and 3900 m distinguished by temperature and salinity (T/S) and chlorofluorocarbon (CFC) characteristics (Pickart 1991; Smethie 1993; Smethie *et al.* 2000). The mean speed of the NADW core from source is *c.* 1.5 cm s^{-1} estimated from CFC F-11/F-12 ratio but, because of recirculation *en route*, this significantly underestimates local flow speeds.

The HEBBLE (High Energy Benthic Boundary Layer Experiment, Nowell *et al.* 1982) site is located on the lower Nova Scotian continental rise between 4810 and 4830 m depth (Fig. 1B). It lies beneath a southwesterly eddy-driven flow of cold Antarctic-source water in a 'cold filament' 100 km wide centred on the 4900 m isobath (Weatherly & Kelley 1982, 1985*a*; Hogg 1983). The site is also from time to time beneath the Gulf Stream and its warm-core rings. The high eddy kinetic energy from this source is

From: STOW, D. A. V., PUDSEY, C. J., HOWE, J. A., FAUGÈRES, J.-C. & VIANA, A. R. (eds)
Deep-Water Contourite Systems: Modern Drifts and Ancient Series, Seismic and Sedimentary Characteristics.
Geological Society, London, Memoirs, **22**, 21–38. 0435-4052/02/$15.00 © The Geological Society of London 2002.

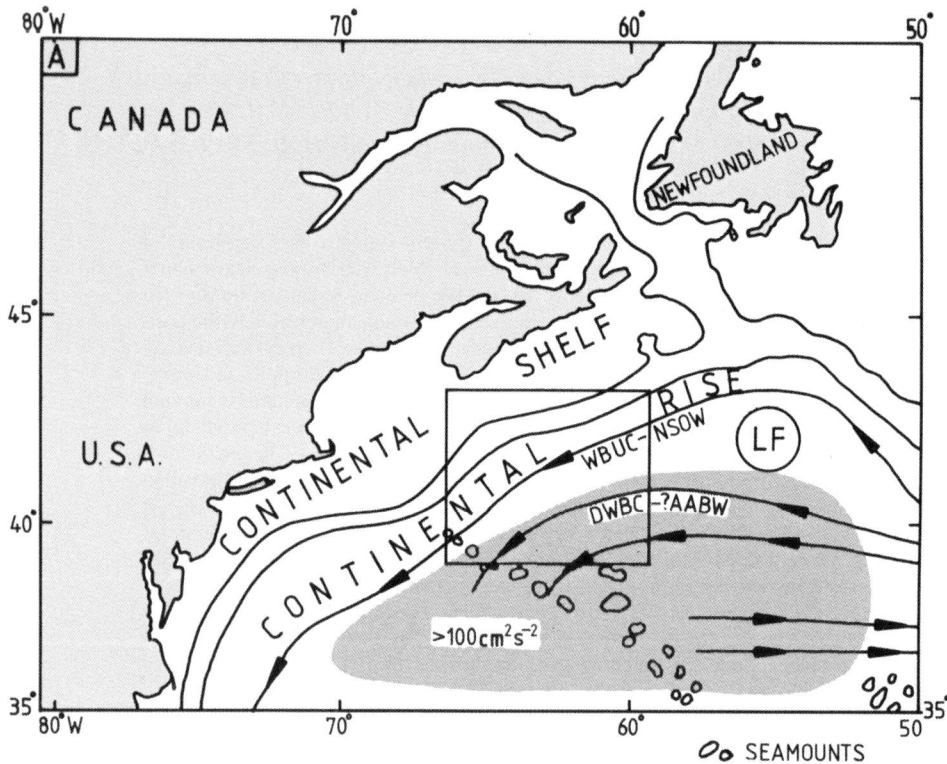

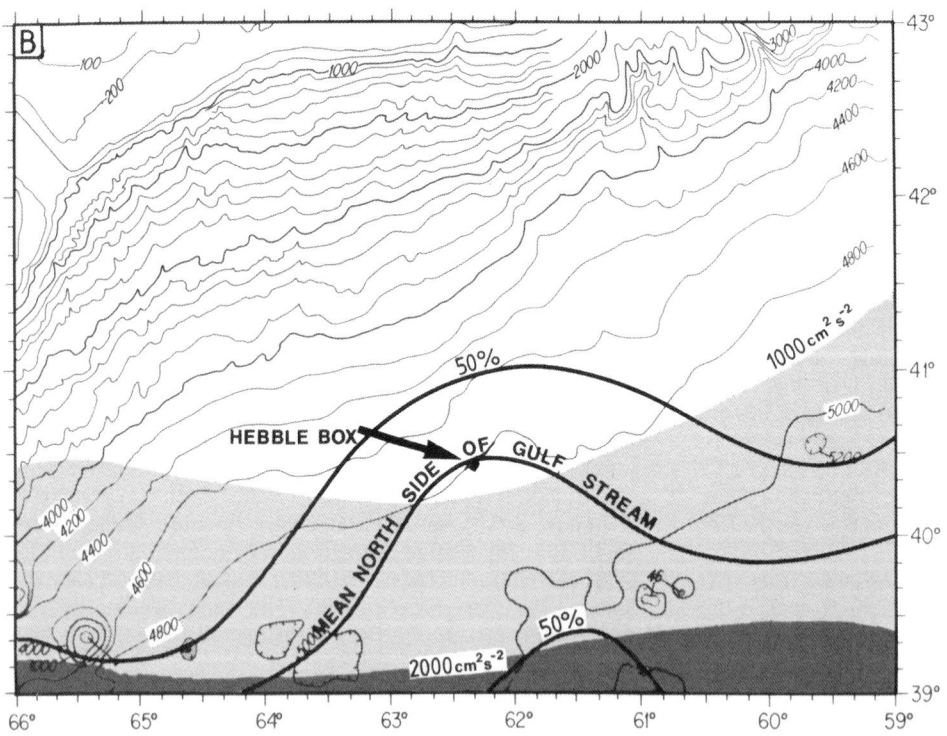

Fig. 1. Location of the study area on the Nova Scotian continental rise. (**A**) The Western Boundary Undercurrent of Norwegian Sea Overflow Water and a deeper western boundary current of Antarctic Bottom Water with the region of highest abyssal eddy kinetic energy shaded (Schmitz 1984). 'LF' is the Laurentian Fan. (**B**) Detail of the square in (A) showing bathymetry of the rise, the mean position of the north side of the Gulf Stream with its limits for 50% of the time (Fisher 1977) and the surface eddy K_E (Richardson 1983). (The HEBBLE 'box' is a 2 × 4 km area marked by transponders for the duration of the experiment in which detailed investigations were conducted.)

felt at the bottom and is expressed there in periodic reversal and enhancement of the southwesterly abyssal flow. Periods of extremely rapid flow approaching 40 cm s^{-1} have been recorded (Richardson *et al.* 1981; Weatherly & Kelley 1982, 1985*b*; Grant *et al.* 1985; Gross & Williams 1991) (Fig. 2). (**Note**: the maximum flow speeds of up to 74 cm s^{-1} recorded by Richardson *et al.* (1981) are probably in error. Dr M. Wimbush (pers. comm.) mentioned this in 1986 and recently (31/3/00) wrote to INMcC '... there was a spurious behaviour in some of our current meters in which the micro-switch, which was activated on each current-meter rotor revolution, would sometimes show evidence of 'contact bounce'. This would cause it to count 'twice' on each revolution. That made me suspect some of our current measurements might after all have been in error by a factor of two'). A value of 37 cm s^{-1} for the maximum would be in line with the longer records given in Figure 2. These 'benthic storms' have major effects on the

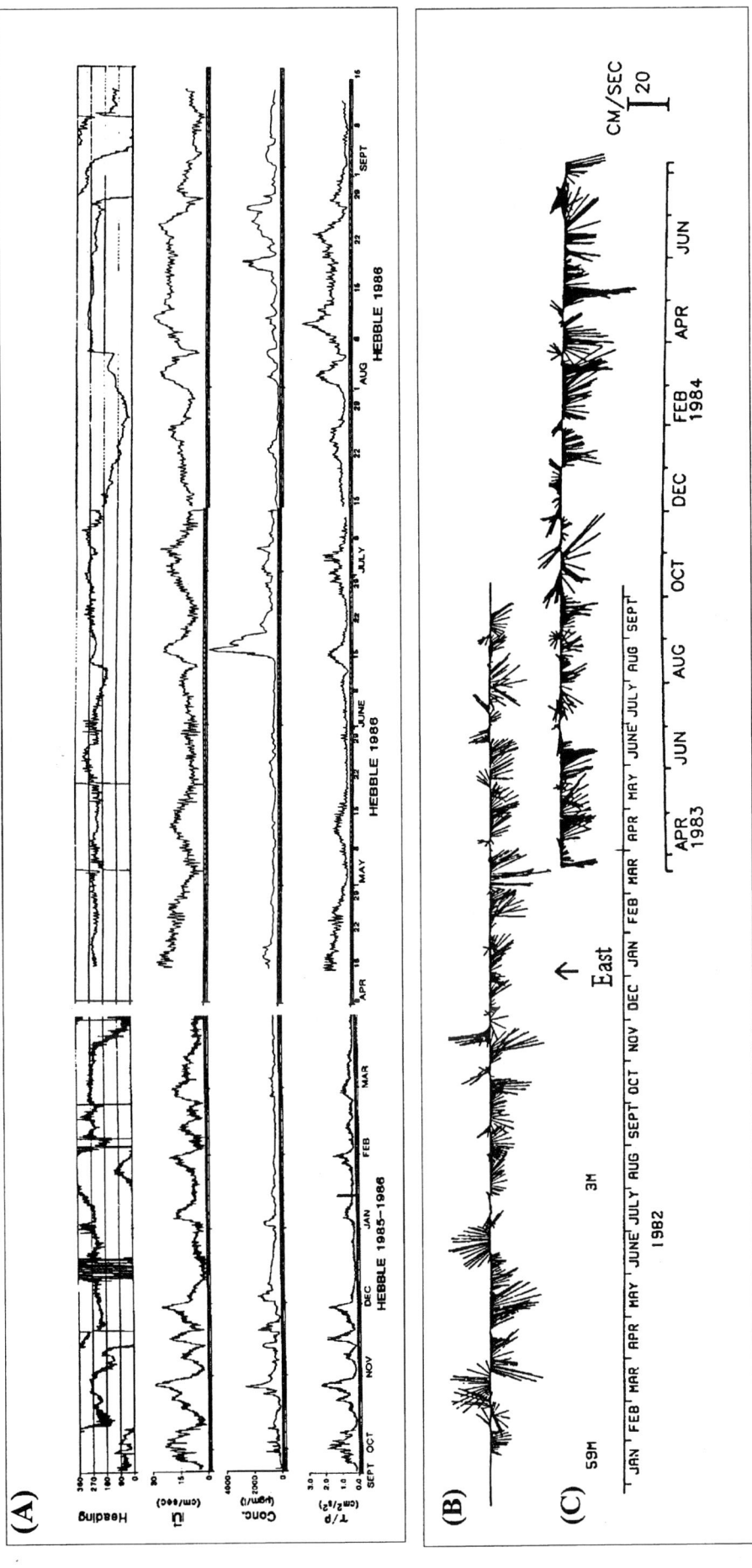

Fig. 2. (A) Time series of BASS tripod instrument data from the HEBBLE site at 40° 26.84′N, 62° 21.85′W, 4812 m depth, spanning September 1985–September 1986. Direction and speed of the 30 min average vector velocity measurements are at 4.9 m.a.b. 'Conc' is suspended sediment concentration at 2.2 m.a.b. Estimates of stress are from turbulent kinetic energy obtained from direct measures of the total velocity vector variance from 30-min averages of 2-Hz measurements at 0.73 m.a.b. (compiled from Gross & Williams 1991). (B) Current meter record from the HEBBLE site, 4810–4840 m. It is a composite of three records: (1) 59 m above bottom at 40° 27′N, 62° 22′W obtained between 23/1/82 and 9/7/82; (2) 3 m above bottom at 40° 27′N, 62° 20′W between 9/7/82 and 15/6/83; and (3) 19 m above bottom at 40° 24′N, 62° 16′W obtained between 8/6/83 and 18/9/99. (Weatherly & Kelley 1985*b*). (C) Current meter record from 97 m.a.b. at 40° 29′N, 62° 03.6′W, 4886 m depth March 1983–August 1984 (Welsh *et al.* 1991).

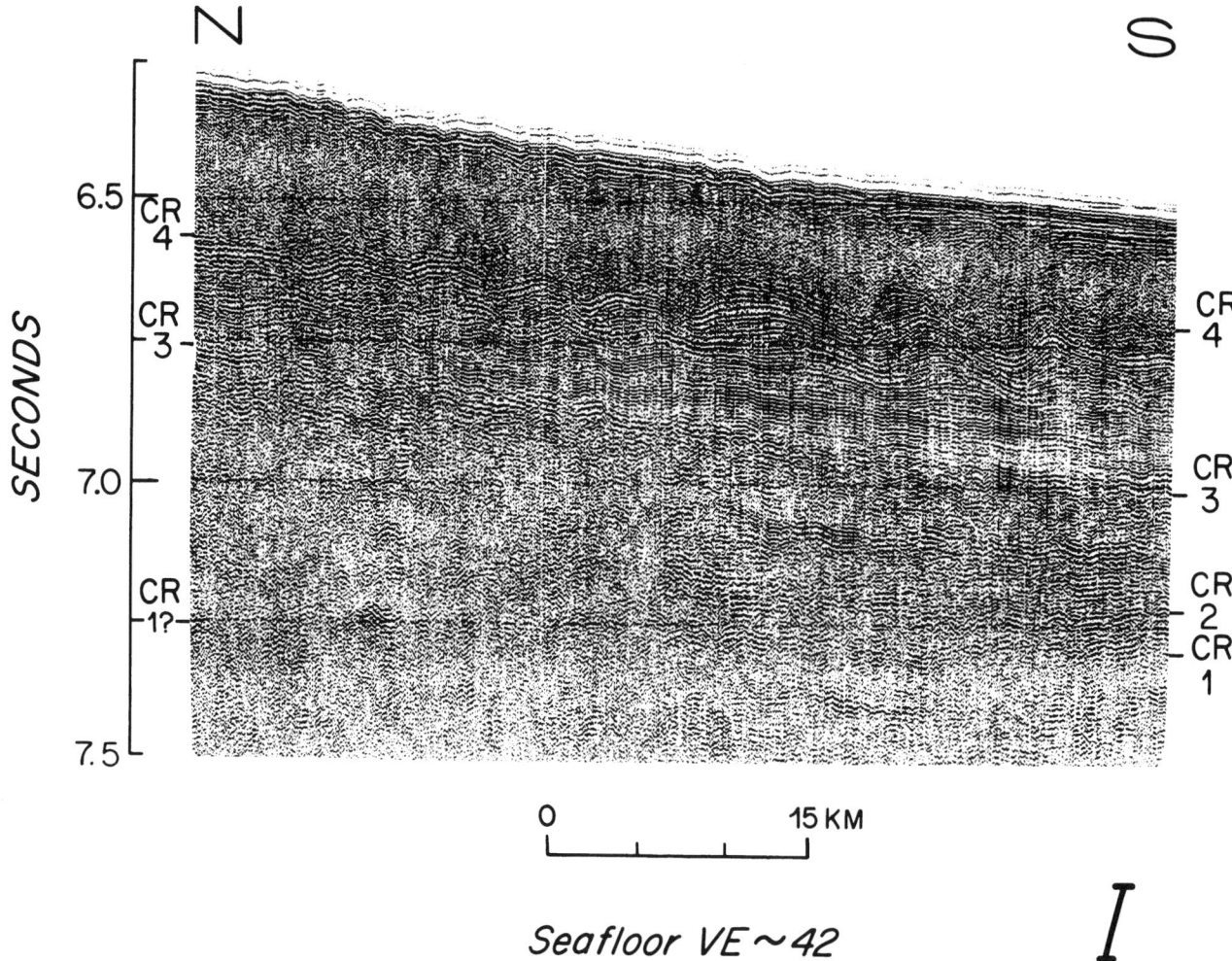

Fig. 3. A single-channel, watergun reflection profile (80–170 Hz) illustrates the seismic stratigraphy on the lower rise seaward of the diapir province. Profile is located about 50 km E of the HEBBLE site (from Swift 1987). Horizons CR1 and CR2 are Miocene(?) unconformities eroded by abyssal currents. The CR2-CR3 interval forms half of the lower rise section above A^u and comprises stacked channel and debris flow deposits. The CR3-CR4 interval is a Pliocene (?) deposit located on the lower rise east of HEBBLE characterized by laterally continuous reflections and sediment waves. Swift (1987) argues that the unit was formed by deposition of fine-grained sediments from bottom currents flowing towards the southwest. The interval above CR4 comprises three laterally distinct and mostly incoherent seismic units.

surface few centimetres of seafloor sediments, which are extremely mobile and subject to very rapid erosion and deposition (Hollister & McCave 1984; McCave et al. 1984; Gross & Williams 1991; Gross & Dade 1991). In a matter of days to weeks, seafloor conditions can change from a state of tranquil deposition and sediment reworking by benthic organisms to a state of extreme storm scour.

Seismic stratigraphy

Reflection sequences in seismic profiles suggest that abyssal currents have episodically eroded unconformities on the continental rise off western Nova Scotia since the late Eocene to early Oligocene, but deposits from such currents appear to be a minor component of the stratigraphic section near the HEBBLE site. In the absence of ocean drill sites in the basin northeast of the New England Seamounts, stratigraphic control is obtained by correlation along long seismic lines (Swift et al. 1986; Ebinger & Tucholke 1988). There is dense industry well and seismic coverage on the Scotian shelf, but the stratigraphy cannot be correlated confidently seaward onto the lower rise because of the intervening 55 to 75 km wide belt of salt diapirism (Swift 1987; Wade & MacLean 1990).

At the HEBBLE site, c. 5 km of sediment overlies Middle Jurassic ocean crust and thins basinward beneath the Sohm Abyssal Plain (Tucholke et al. 1982; Swift et al. 1986; Ebinger & Tucholke 1988; Wade & MacLean 1990). During the late Eocene-Oligocene, abyssal currents deeply eroded the entire continental rise (Tucholke & Mountain 1979) and the resulting surface, horizon A^u, is a prominent seismic marker off Nova Scotia (Swift et al. 1986; Ebinger & Tucholke 1988). The section above A^u forms a third of the total sediment section. Based on reflection characteristics and shape of isopachs, Swift (1987) identified four significant unconformities above A^u but interpreted only one seismic sequence (CR3–CR4) as a probable contourite deposit. Figure 3 shows a seismic dip section that illustrates the stratigraphy above A^u just east of the HEBBLE site. The deeper horizons CR1 and CR2 are Miocene unconformities eroded by abyssal currents that overlie units whose isopachs and internal reflections suggest deposition by down-slope processes. The shallower horizon CR3 is characterized by a change in depositional style rather than erosion. Between CR2 and CR3, reflections are commonly chaotic, whereas reflections above CR3 are continuous and include sediment waves (Fig. 3). Unlike all other units, the CR3-CR4 interval thickens seaward and is limited to the lowermost rise (Fig. 4). Swift (1987) interpreted the interval as a middle Pliocene contourite unit deposited from a southwest-

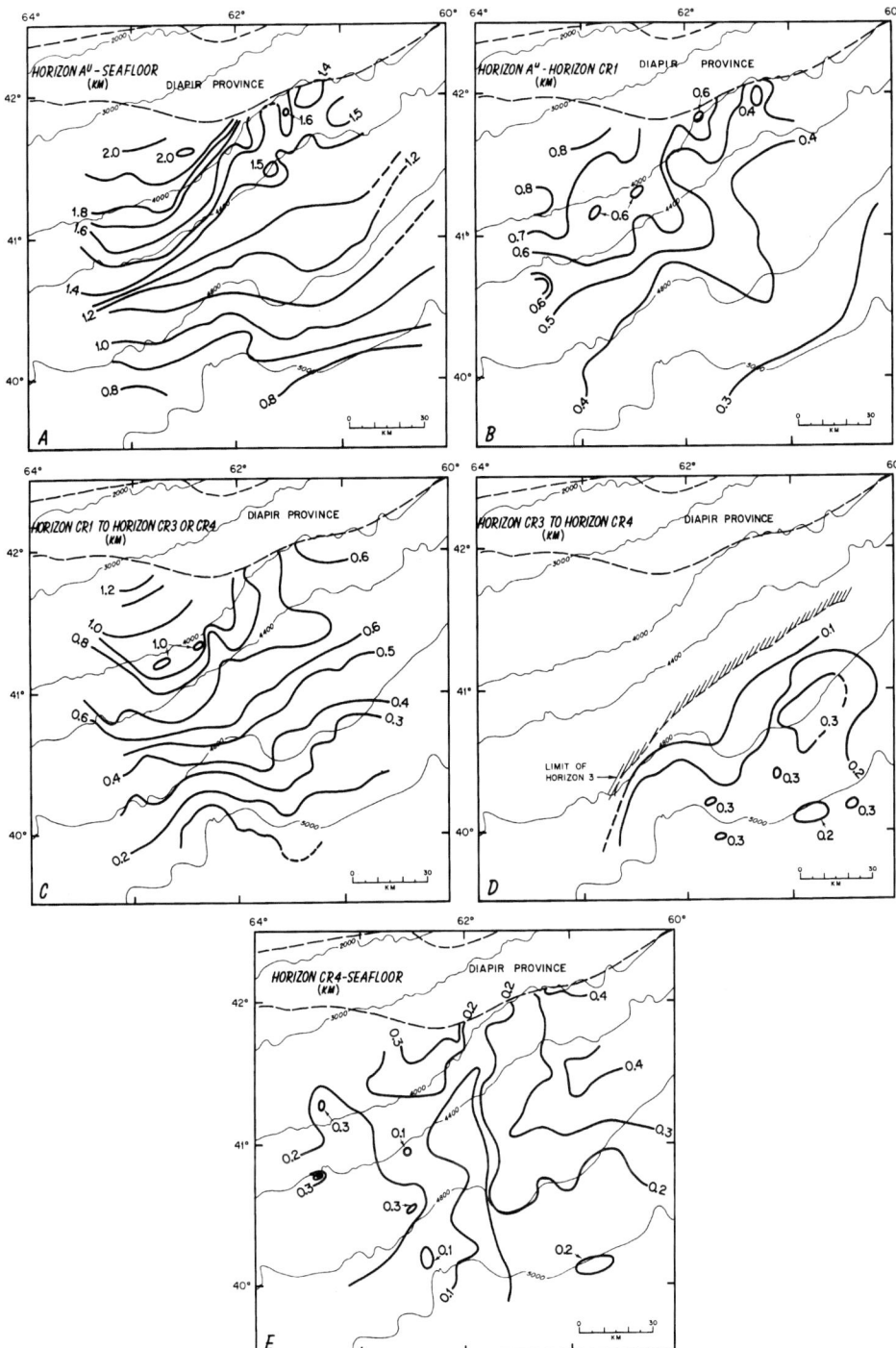

Fig. 4. Isopachs (km) for Neogene seismic units above horizon A^u on the lower rise. The form and seismic character of the lower three units (A, B & C) are consistent with emplacement by down-slope processes. The contourite deposit between CR3 and CR4 (D) is confined to the lowermost rise east of the HEBBLE site. The interval above CR4 is divided into three laterally distinct deposits mainly comprising thick mass-movement deposits.

following bottom current following a pulse in sediment supply from the eastern Canada onto the Laurentian Fan. The seismic sequence above CR4 consists of three internally incoherent subsequences that are in spatially separated depocentres (Fig. 4) and are differentiated vertically by thin bands of continuous reflections. The seismic character, shape, and large size of these sequences led Swift (1987) to suggest that they were olistostromes shed from the upper rise in response to late Pliocene and early Pleistocene seafloor uplift in the diapir province. These masswasting deposits are 100–400 m thick and they are responsible for the present shape of the lower rise.

Shallow stratigraphy and seafloor morphology: 3.5 kHz profiles

High resolution seismic profiles and side-scan sonar images indicate that the seafloor physiography around the HEBBLE site formed during the Pleistocene by deposition from debris flows and turbidity currents in low-relief channels. Based on a Deep Tow survey near the HEBBLE site, Shor & Lonsdale (1981) mapped six seafloor channels with thalweg depths of 3–20 m, widths of 500–1000 m, and spacing of 3–4 km. The channels trend downslope towards 150° and can be traced for >30 km across the lower continental rise. After the channels were formed, the thalwegs were incised by turbidity currents and then draped with a thin (1–2 m thick) layer of mud. Long-range GLORIA side-scan images show that these channels merge, bifurcate and are discontinuous over distances greater than 100 km (Hughes Clarke et al. 1992). This channel pattern extends from the upper slope to the abyssal plain over a wide area off western Nova Scotia. Berry & Piper (1993) estimate that sediment accumulated on the middle rise (3200–4200 m) at 0.85 m ka^{-1} during glacial periods and 0.18 m ka^{-1} during interglacials. They conclude that channel development during the Pleistocene correlated with glacial maxima. As in the sequence mapped by Shor & Lonsdale, they argued that several channels were filled by debris flows during the deglaciation at 15–13.5 ka, subsequently eroded by turbidity currents, and then capped with hemipelagic drape.

Damuth et al. (1981) mapped echocharacter in 3.5 kHz subbottom profiler records and determined that the HEBBLE site is in a region of seafloor channels and turbidity current deposits about 30 km west of a region of sediment waves. Swift (1985) found that this transition between primarily down-slope processes to the west and current reworking to the east is a boundary that extends c. 370 km from the continental slope (c. 500 m water depth) to the abyssal plain (c. 5000 m depth) (Figs 5 & 6). He attributed the boundary to lateral changes in shelf topography and sediment delivered to the shelf edge by late Pleistocene ice sheets, whereas Uchupi & Swift (1991) suggested that the boundary relates to the westward decrease in continental slope gradient and seafloor relief.

Age control on the lower rise is poor, but contourite deposits east of the boundary probably began accumulating during the late Pleistocene and continued to accumulate through the last glacial maximum into the Holocene. Seaward of the diapir province, current-reworked deposits are least 40–60 m thick in the 3.5 kHz records (Fig. 5). In seismic profiles from this region (e.g. Fig. 3), laterally coherent reflections extend from the seafloor down to 0.05–0.10 s or about 80–165 m sub-bottom. Long term accumulation rates are unknown here, but a depth of 165 m probably encompasses the last few glacial cycles. Using Berry & Piper's rate of 0.85 m ka^{-1} for glacial periods on the upper rise, accumulation of contourites started at the latest 94 ka ago and probably rather earlier.

The facies boundary shown in Figure 6 is much older. Seaward of the diapir province, this boundary coincides with the western edge of the easternmost lobe comprising the seismic interval between CR4 and the seafloor (compare Figs 4 & 5). These lobes lack coherent reflections and appear to be either olistostromes (Swift 1987) or turbidite fans deposited during the Pleistocene without significant reworking by currents. The boundary also coincides with the western limit of the CR3–CR4 interval (Fig. 4) that Swift (1987) interpreted as contourites deposited in the Pliocene. Although deeper seismic intervals are not so clearly defined by the boundary, isopachs for all units appear to change trend across the facies boundary (Fig. 4).

The facies boundary can be traced north to the along-strike change in morphology and structure of the continental slope near 62°W (Fig. 6; Swift 1985; Uchupi & Swift 1991). To the west of the boundary, most slope canyons have topographic relief less than 200 m, channels on the rise have few levees, and down-slope transport is poorly focused (Hughes Clark et al. 1992). The region to the east is more typical of the western North Atlantic where canyon relief is 600–1000 m, leveed channels extend from the canyons to the abyssal plain, and the continental rise is organised into coalescing fans. These differences in morphology cause differences in the way sediments are deposited on the continental rise. To the west, we infer that turbidity currents and debris flows are small-scale but that their deposits are widely dispersed across the rise, particularly during glacials. During interglacials (the Holocene), contourite deposits accumulate slowly (c. 4 cm ka^{-1}) on the lower rise, yielding a thickness undetectable in seismic profiles.

To the east, turbidity currents and debris flows originate in the slope canyon systems and are funneled into channels crossing the rise. During times of high sediment flux from the shelf edge, fans develop rapidly on the lower rise. At times of low flux, most sediment in downslope flows is confined to channels crossing the rise; thus, outside these channels abyssal currents can rework and reshape lower rise sediments into contourite deposits such as those detected east of the facies boundary.

Bottom currents interpreted from seafloor photography

The seafloor on the Nova Scotian continental rise in and around the HEBBLE site is one of the most intensely photographed regions of a deep continental margin on the globe. More than 2000 bottom photographs were acquired between 1962 and 1983, with the majority taken during the HEBBLE site surveys and experiments in 1979–1983. The photographs provide both good areal coverage (Figs 7 & 8) and a temporal perspective on abyssal current activity over a 22-year period.

Tucholke et al. (1985) analysed the bottom photographs to estimate the magnitude and time scales of bottom current intensity and variability. Using observed features of microtopography and bedform development, they refined the relative scale of current strength of Hollister & McCave (1984) and defined the variation in current intensity across the continental rise (Fig. 7). Above about 4000 m, the seafloor is heavily tracked and trailed and benthic organisms were commonly photographed. From about 1500–3200 m the seafloor consists of bioturbated and consolidated clay that is relatively free of loose debris. There are very few direct indications of reworking by currents, but we cannot rule out the possibility of occasional rapid flow. It may be that natural cohesion in these fine-grained sediments, together with grain binding by organic by products of biological activity, protects the seafloor against reworking at least during short period increases in current intensity. Between 3200 and 4000 m, evidence for a tranquil seafloor is clearer with photographs showing abundant bioturbation as well as scattered fine organic and mineral debris (Fig. 9A), indicating the absence of geologically significant currents.

Interpreted strength of bottom currents increases dramatically with seafloor depth below about 4000 m (Fig. 7), and below c. 4500 m the seafloor is universally smoothed by currents. There is a correlative decrease in bottom-water clarity with increasing seafloor depth, and at 4900–5100 m the seafloor was not visible in about half the bottom photographs at camera ranges of only 1–3 m. Instrumental measurements and direct sampling of suspended sediment in the nepheloid layer at these depths (Biscaye et al. 1980; McCave 1983; Gardner et al. 1985) show both strong variability in suspended-sediment load and increasing maximum loads with increasing seafloor depth. Considered together, all these features suggest intermittent currents of increasing maximum strength with increasing depth. Direct current measurements (Richardson et al. 1981; Weatherly & Kelley 1985b; Gross & Williams 1991; Welsh et al. 1991), (Fig. 2) substantiate this pattern and provide quantitative constraints. Currents at five measurement locations between 4158 m and 5022 m increase in speed downslope from a mean of 4 cm s^{-1} (14 cm s^{-1} maximum) to a mean of 16 cm s^{-1} (37 cm s^{-1} maximum). Farther seaward, a slight decrease in speed (11 cm s^{-1} mean, 21 cm s^{-1} maximum) was measured over the edge of the Sohm Abyssal Plain (halved speeds given by Richardson et al. 1981, see **Note** above from Wimbush).

Current direction and a qualitative estimate of directional variability at different time scales were also interpreted from bottom photographs by Tucholke et al. 1985) (Figs 7 & 8). This analysis assumed that larger bedforms take more time to form and thus represent longer-term averages of current direction, whereas fine-scale features respond quickly to current changes and thus demonstrate the range of short-term directional variability. The

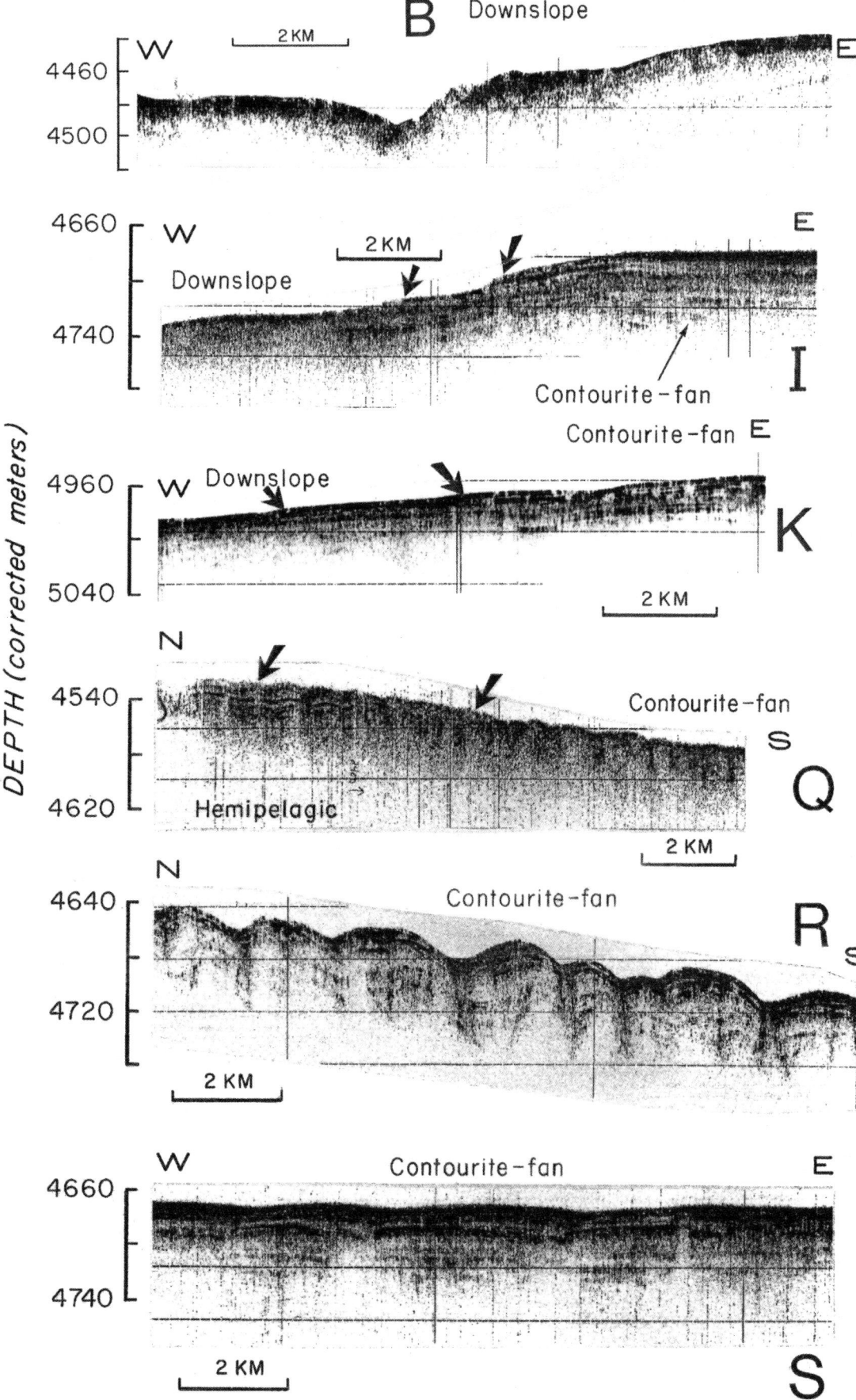

Fig. 5. Echosounding (3.5 kHz) facies on the lower Nova Scotian rise include 'downslope' and 'contourite-fan' associations. (Location of profiles shown on Fig. 6). Profile B crosses a small channel in downslope facies. I and K are profiles across the eastern boundary of the downslope association. Arrows indicate reflector outcrops. Q–S are profiles of contourite-fan association. Apparent mudwave wavelength increases southward from S to R and westward from R to S (profiles from Swift 1985).

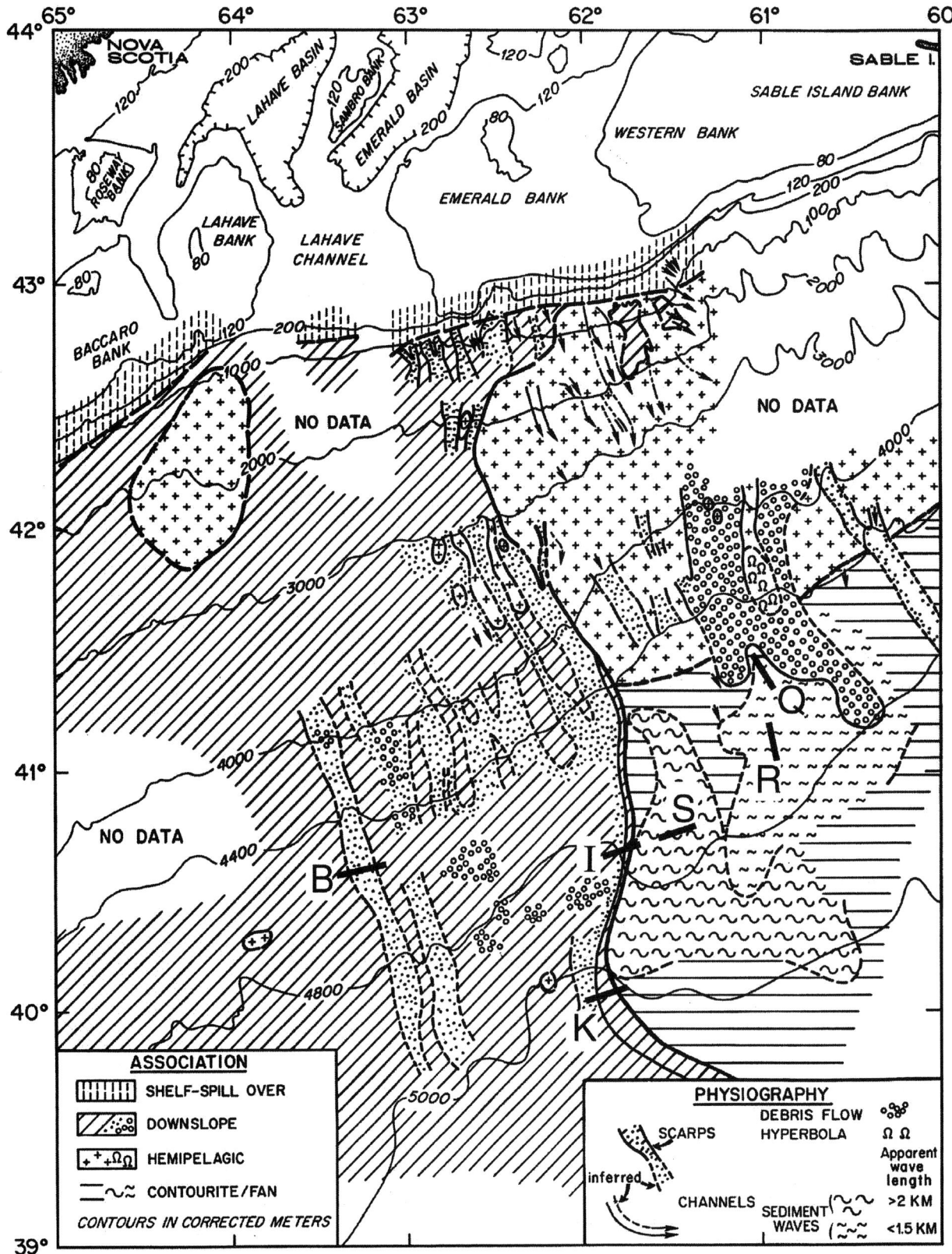

Fig. 6. The distribution of 3.5 kHz echocharacter associations shows a distinct boundary running from the continental slope to the abyssal plain near 62°15′W (from Swift 1985). Heavy line separating associations is broken where boundary is uncertain. Channels (dots and arrows) and debris flows (circles) are downslope deposits that could be identified by echocharacter and physiography. Arrows indicate channels having apparent widths less than c.1 km. Channel boundaries and arrows are dashed where uncertain. 'Horseshoe' pattern indicates hemipelagic deposits with hyperbolic echo character. Wavy patterns mark contourite-fan associations divided into two wavelength classes.

largest bedforms photographed are longitudinal ripples (LRs) (Figs 9D–F & 10) which parallel the direction of flow and have roughly symmetrical, triangular cross-sections (Flood 1981; Tucholke 1982; McCave et al. 1984; Swift et al. 1985). LRs often appear in bottom photographs as apparently consolidated sediment that is tool marked by bedload debris (Fig. 9F), and thus they probably are not evanescent features but persist for longer timescales, perhaps months to years. It has been suggested that

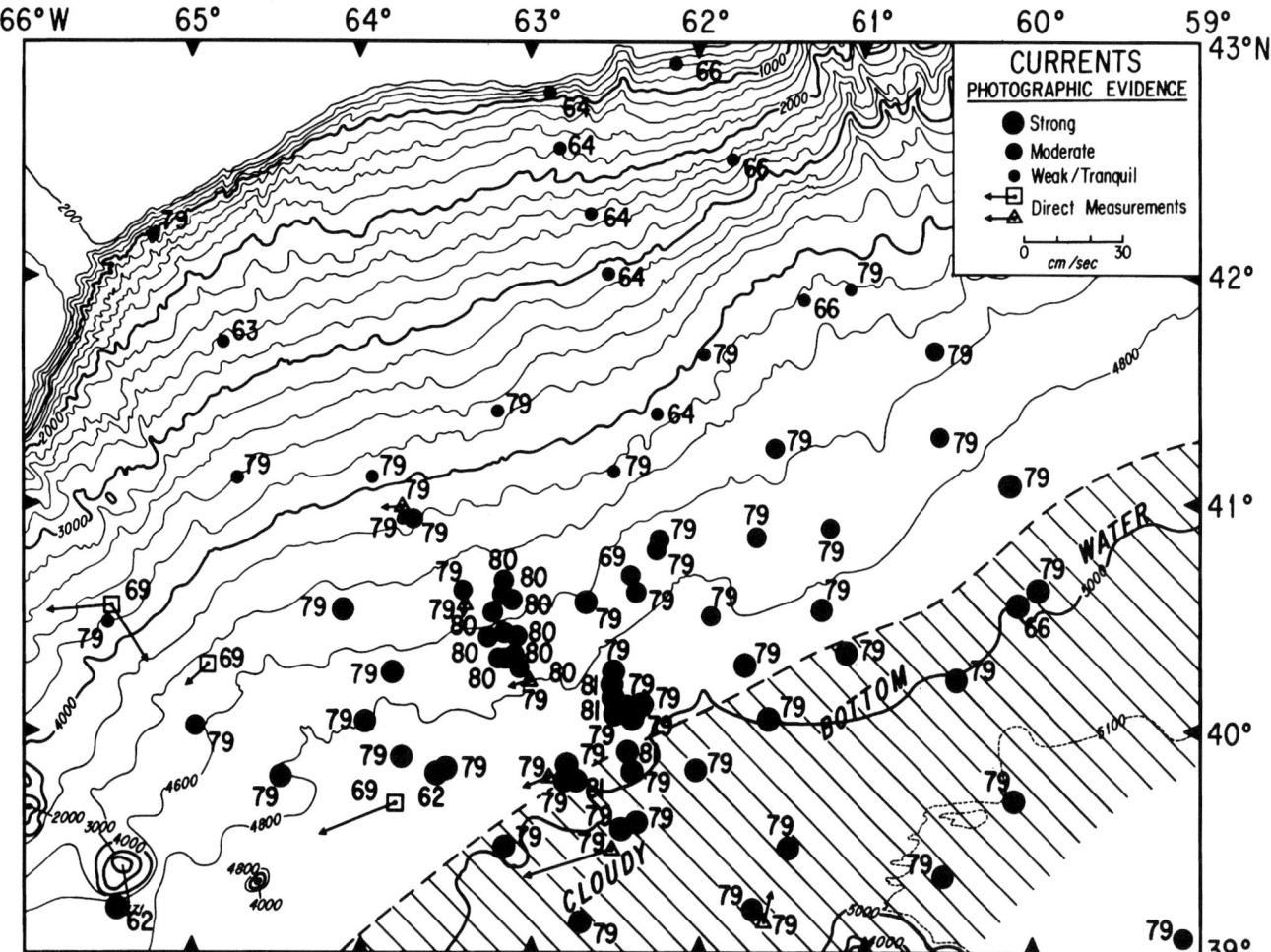

Fig. 7. Current intensity interpreted from bottom photographs on the Nova Scotian continental rise (from Tucholke *et al.* 1985). Numbers beside dots are the last two digits of the year in which the photographs were taken. Direct current measurements are indicated by squares (magnitude and direction of maximum one-minute current velocity (Zimmerman 1971)) and triangles (magnitude and direction of mean current velocity over entire recording period (Richardson *et al.* 1981). In the zone of cloudy bottom water, the seafloor is obscured by suspended particulate matter in about half the bottom photographs.

they are analogous to vector-averaging current meters, initially forming behind obstacles such as biological mounds and developing in the direction of mean flow as variable currents alternately construct or degrade the two sides of the bedform (Tucholke 1982, 1986). An alternative view is that they are initially formed rapidly by deposition from a concentrated suspension following a storm and reflect a helical flow pattern in the bottom boundary layer (McCave *et al.* 1984). LRs appear mostly between 4800 and 5000 m on the lower rise, and they are oriented within *c.* 15° of the local bathymetric contours (Fig. 8a). Where the heads or tails of the bedforms have been photographed, they indicate flow directed to the southwest. In sum, the LRs suggest long-term (months to years), geologically significant, contour-parallel flow that is consistent with southwest-directed circulation of the 'cold filament' along the lower rise.

'Mound-and-tail' features 2–10 cm high comprise most intermediate-scale bedforms photographed (Fig. 9B & C). Because of their up- and down-stream asymmetry, these bedforms are good current-direction indicators. They demonstrate consistent contour-parallel flow to the southwest (Fig. 8B), probably on average time scales shorter than those required for LR formation. Small-scale bedforms, typically <2 cm high and up to a few centimeters long, occur in a wide range of morphologies ranging from fine toolmarks and lineations (Fig. 9C & F), through crags-and-tails (Fig. 9D), cornices on larger bedforms (Fig. 9D & E), and transverse or barchan-like ripples in unconsolidated bedload (Fig. 9F). Current orientations (±180°) and absolute directions determined from these features show a wide range, although the mean trend of directions is along bathymetric contours toward the southwest. The varied bedforms in this grouping probably are formed or are modified on time scales of hours to weeks. Thus, they suggest significant short-term variability in current velocities, superimposed on the longer-term contour-parallel flow indicated by larger bedforms. Despite this rich assemblage of sediment bedforms, some of which provide a record in the uppermost few centimetres of the sea bed, that record is not preserved because the uppermost layers are constantly being reworked and the preserved layer below that is thoroughly bioturbated by echinoids and polychaetes (Baldwin & McCave 1999).

Sediment properties of the continental rise

Surficial sediments (topmost *c.* 1 cm) and subsurface sediments (to 77 cm depth) across the continental rise were analysed by Driscoll *et al.* (1985). Most sediments are muds and sandy muds according to Folk's (1974) classification. Mean sand content is 12.4%, silt is 50.1% and clay (defined as <4 μm, > 8 φ) is 37.5%.

Sediment texture: sands

The pattern of sediment textural variations across the lower slope and rise is one of sand-enriched zones flanked by zones of finer

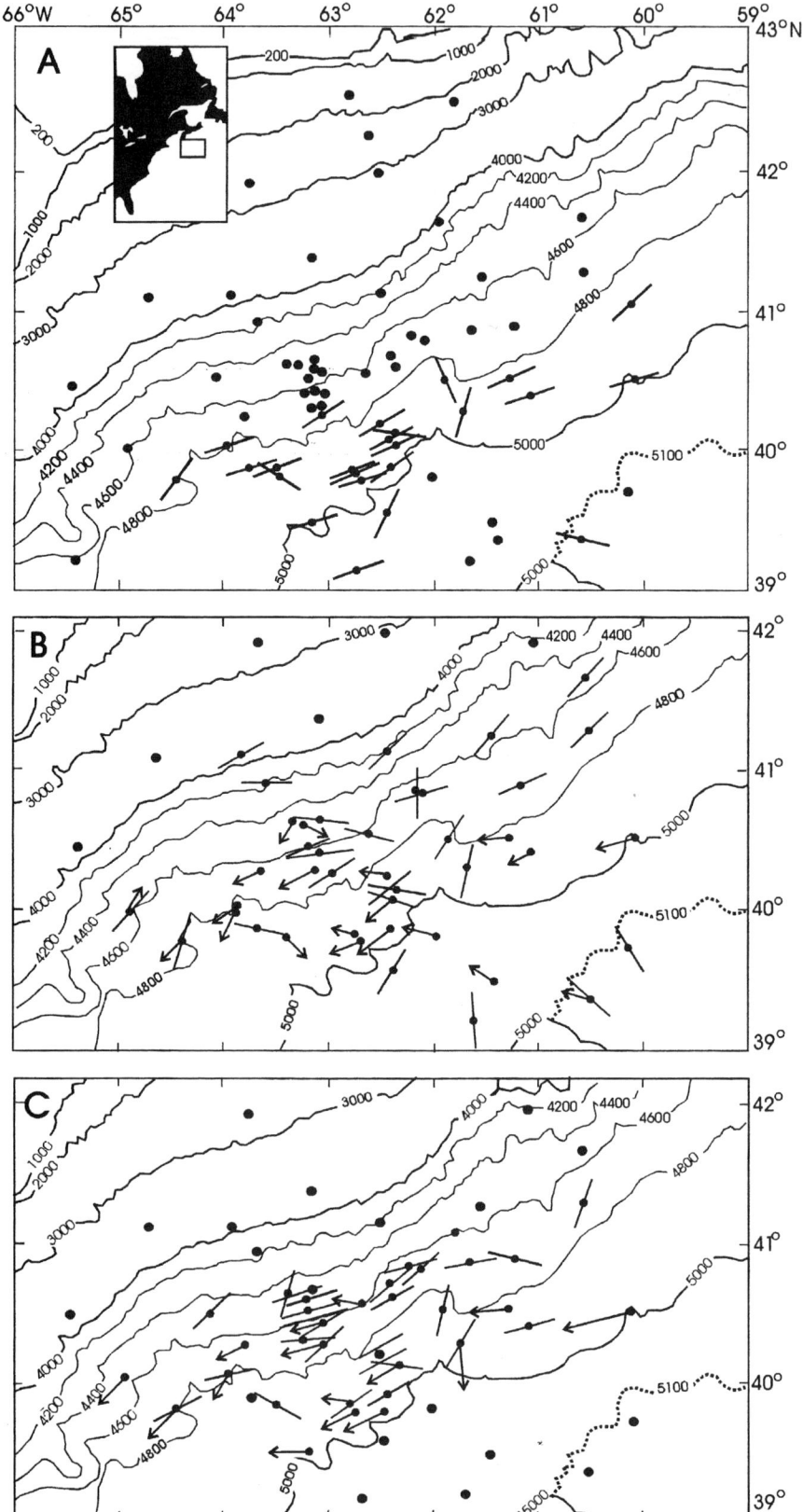

Fig. 8. Locations and orientations of bedforms observed in bottom photographs (from Tucholke *et al.* 1985). (**A**) Longitudinal ripples. Each LR photographed is represented by one line. Dots without lines show where no LRs were observed. (**B**) Intermediate scale bedforms (*c.* 2–10 cm high). Lineation-only measurements are indicated by lines. Dots without lines show where current-produced bedforms of this scale were not observed. (**C**) Bedforms smaller than *c.* 2 cm high. Lines show where only lineation (±180°) and not direction of currents could be determined. Other symbols as in (B).

Fig. 9. Examples of seafloor photographs taken on the Nova Scotian continental rise (from Tucholke *et al.* 1985). Arrows show true north and each photograph covers an area on the order of one square meter. (**A**) Tranquil, biologically reworked seafloor (3950 m). (**B**) Elongated mounds, probably originally of biological origin, on current-smoothed seafloor (4473 m). (**C**) Strong current smoothing of elongated mound-and-tail bedforms with superimposed small crags-and-tails and and tool marks (4829 m). (**D**) Low-amplitude longitudinal ripple (LR) with cornice on southeastern side (4868 m). Smaller crags-and-tails and tool marks are oriented differently than the LR and are interpreted to be formed by short-term currents deviating from the longer-term direction indicated by the LR. (**E**) Longitudinal ripple with small crags-and-tails and tool marks superimposed (4987 m). (**F**) Longitudinal ripple of well consolidated silty clay that has been tool-marked by obliquely transported bedload particles (5081 m). Small barchan ripples of light-coloured fine sand appear on the flank of the LR.

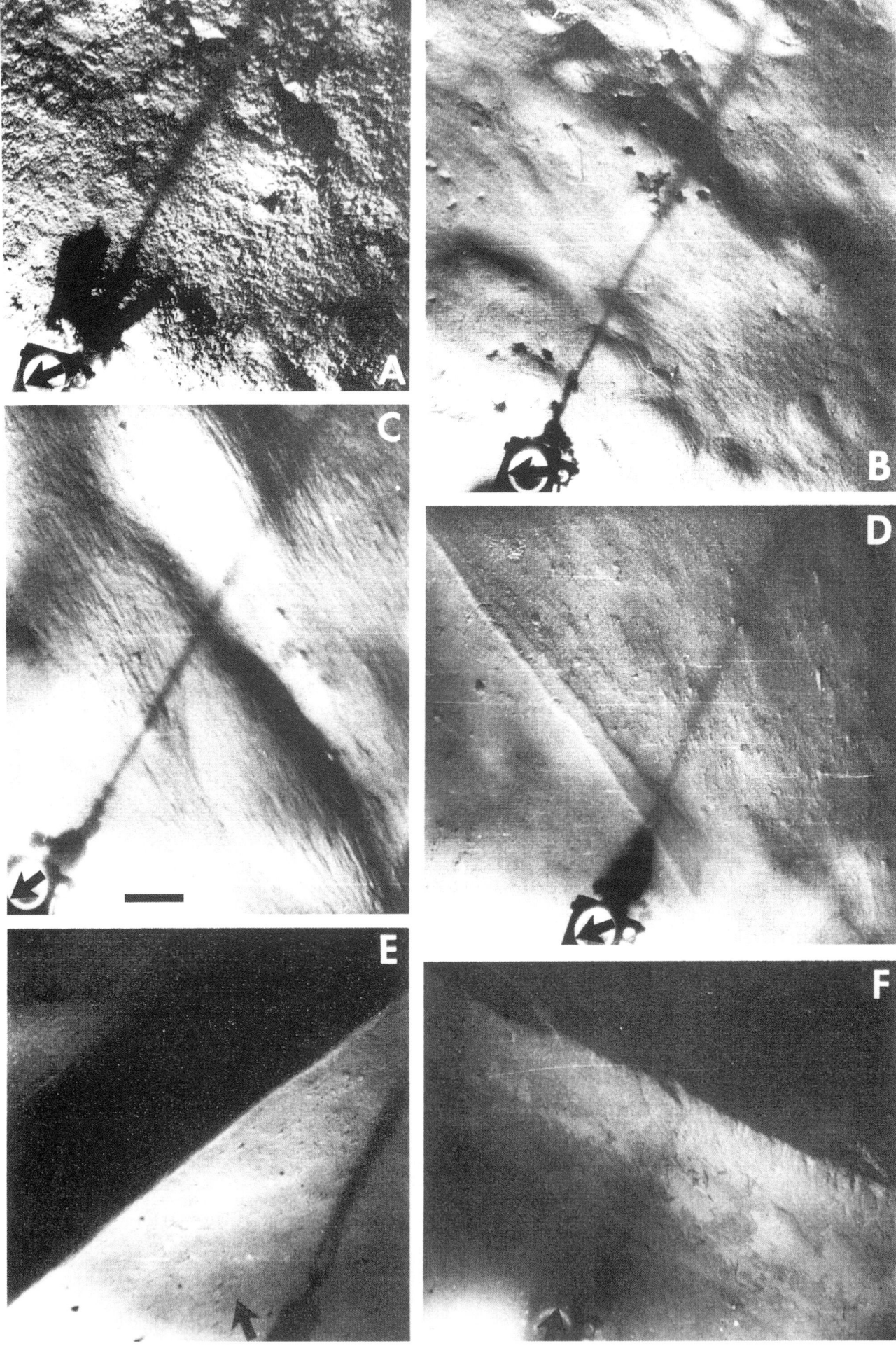

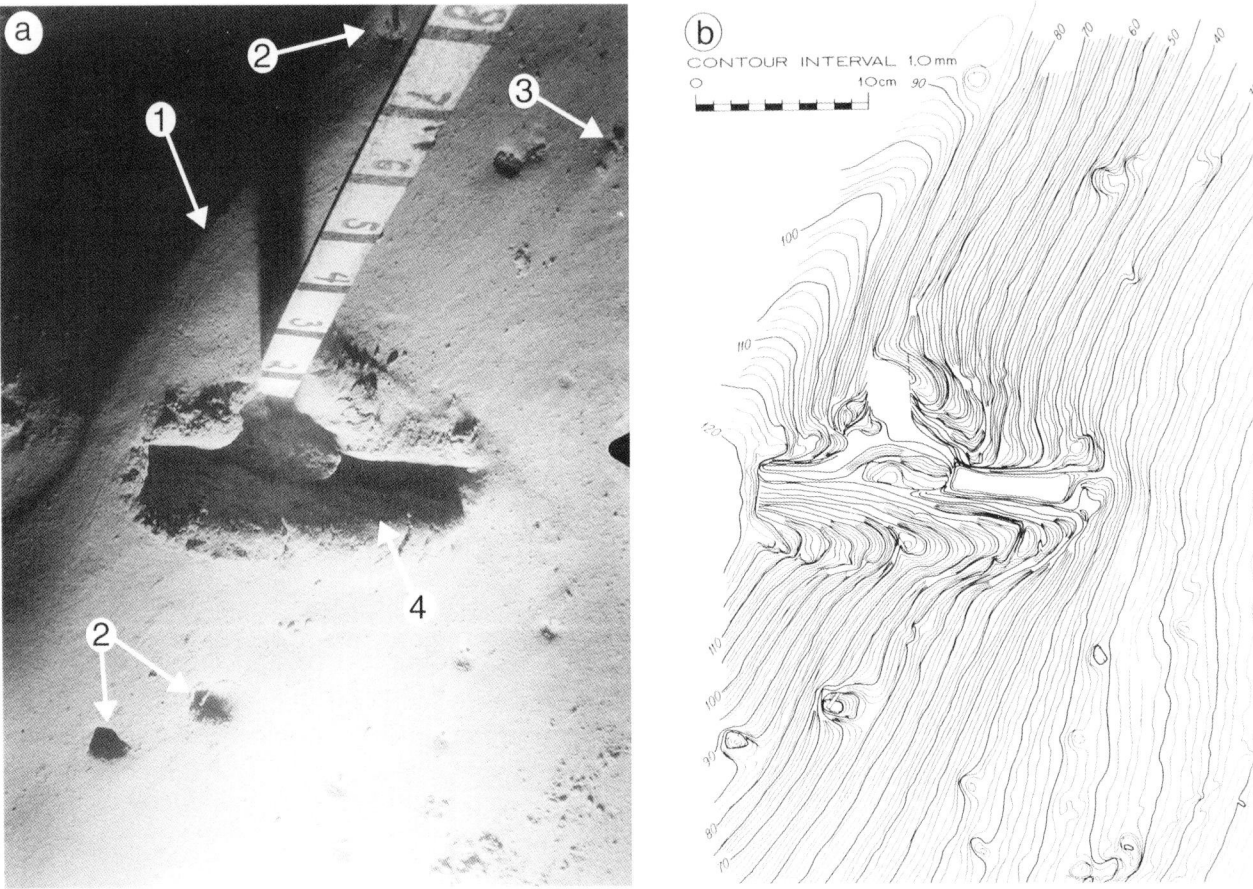

Fig. 10. (**A**) Left image of a stereo pair of a longitudinal ripple at 4986 m (39° 53′N, 62° 25′W). 1, ripple crest; 2, biological pits; 3, stalk; 4, mud deformed by scale-bar mounted below cameras. (**B**) Contour map of (A) relative to lowest point in the stereo image. Interval 1 mm. Total relief is 120 mm. (Hollister found a large stereophotogrammetry instrument used for terrain mapping at a defence-related institution in Washington and persuaded them to process his bottom stereophoto pairs!).

sediments (Fig. 11A). There are two distinct sandy zones. One is well defined and centred between the 4800 and 4900 m isobaths, and the second is situated below 5000 m. The coarsest samples in the shallower sandy zone are located at or slightly below the 4800 m contour. Sand content decreases rapidly upslope and more gradually downslope.

Just west of 63°W in this sandy zone is a complicated area exhibiting a wide range of grain size distributions. The complexity undoubtedly is partly due to the fact that this zone has the highest sampling density on the rise. However, the samples also are located in and around Triffid Channel (dotted line in Fig. 11) which has a relief of 30–60 m and a well defined levee on its west side. Although side-scan sonar observations suggest that the channel is presently inactive (Shor & Lonsdale 1981), there is still a strong effect in the grain-size distribution, with channel fill containing >10% sand. The coarse fraction in the channel is composed primarily of planktonic foraminifera (77–94%), with smaller amounts of benthic foraminifera (1–6%), and terrigenous material (mainly quartz grains, 0–21%). The average coarse fraction composition in all samples on the rise is 66% planktonic foraminifera, 2% benthic foraminifera and 32% terrigenous material. Thus, the channel sediments are enriched in planktonic foraminiferal tests and poor in mineral sand. In the debate over the origin of sandy contourites it is notable that these sediments, despite the high current speeds to which they are intermittently subjected and source of sand upstream in the turbidites of the Laurentian Fan, would not be considered 'sandy' other than by formal definition of >10% of <63 μm material.

Fine fraction

A low-silt/high-clay (S/C <1) band is prominent at depths of approximately 4100 to 4700 m (Fig. 11B). Upslope from this band the silt/clay ratio is generally higher, locally reaching a maximum of four. Downslope from the band the silt/clay ratio also steadily increases, reaching a maximum in excess of three in the deepest sandy zone (5000 m). From 4100 m down to the proximal edge of the Sohm Abyssal Plain there is a clear and consistent trend toward more silt and less clay.

This trend is also clear in frequency-curve analyses (Driscoll et al. 1985, fig. 5A & B). Details of the fine-sediment, size-frequency distributions reveal distinct silt and clay modes. The silt fraction itself contains modes at c. 4 and c. 16 μm (see also McCave 1985). The height and size of the coarser silt peak are closely correlated, and in the HEBBLE area the modal size of the peak is inversely correlated with percent clay ($r = -0.93$, McCave 1985). In deeper water the relative importance of the 16 μm silt peak increases. The silt peak height increased from <15% per ϕ above 4500 m to >25% per ϕ below 5000 m.

Within the high clay band, and east of about 63°W (Triffid Channel), clay content is uniformly high at 50–60%. Silt and sand content is also fairly uniform at 40–50% and <10% respectively. In contrast, west of Triffid Channel the clay percentages are 10–30% lower (Fig. 11C) and the silt contents are correspondingly higher. Thus, as already observed in the sand fraction, Triffid Channel appears to exercise significant control on grain size distribution.

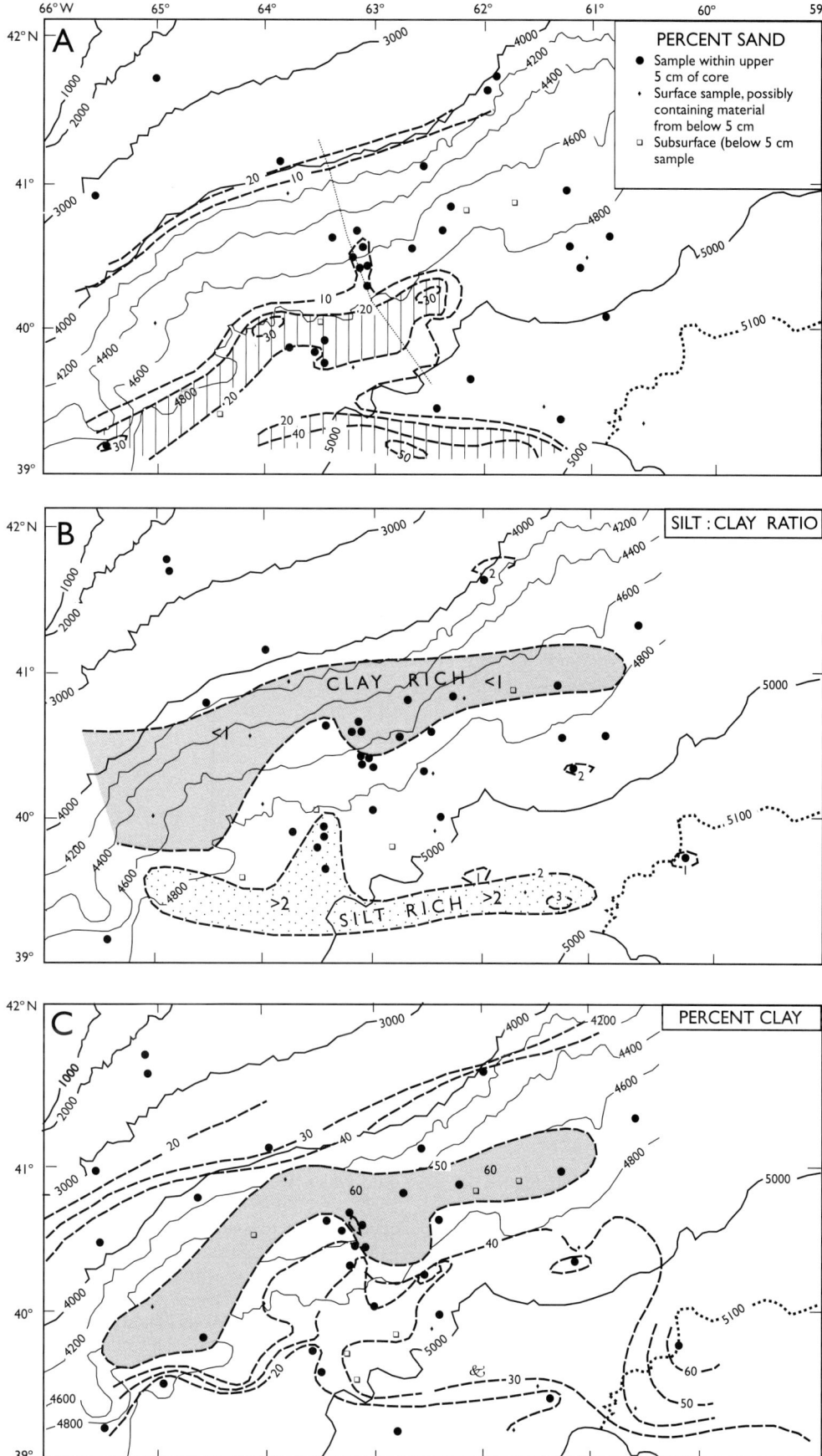

Fig. 11. Sedimentary parameters (from Driscoll *et al.* 1985). (**A**) Percent sand (>63 μm) in surface samples. Dotted line near 63°W shows position of Triffid Channel. Bathymetry modified from Shor (1984). (**B**) Values of silt/clay ratio in surface samples. (**C**) Percent of clay (<4 μm) in surface samples.

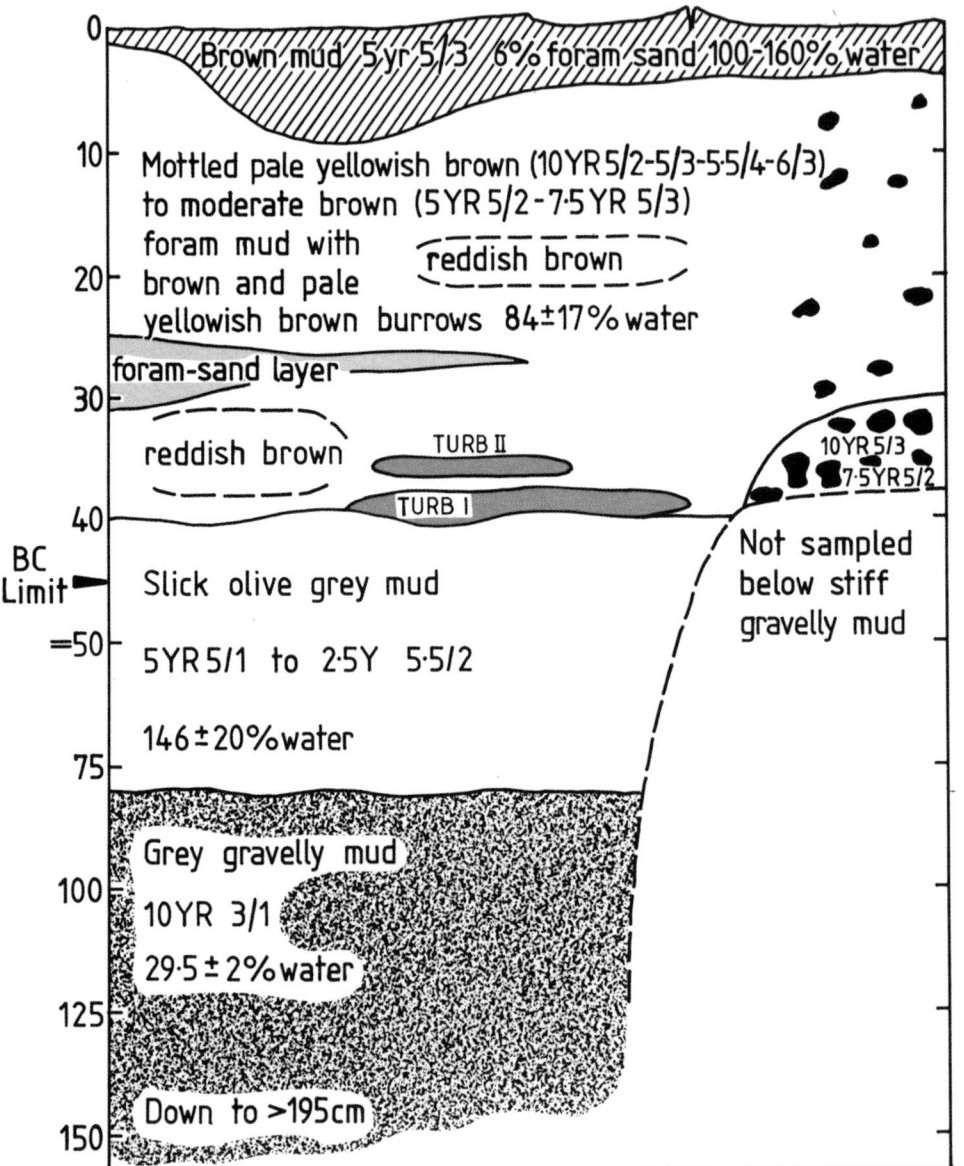

Fig. 12. Summary stratigraphy of the HEBBLE area. Water contents are expressed as salt-corrected percent of dry weight in gravity core LGC-1 from Dr A. Silva. Turbidites and the stiff gravel layer between 30 and 40 cm are shown but the turbidite at 20 cm is omitted. Note scale change below 50 cm.

Sediment properties: HEBBLE area

Lithology

Unit 1. The surficial stratigraphy is summarised in Figure 12. There are four principal units. Unit 1 is the mobile surface HEBBLE mud, a brown clayey silt (mud) of low strength (<0.4 kPa) and water content >47% wet weight, covering the whole area. Its thickness varies randomly throughout the area from 0.5 to 11.3 cm, averaging *c.* 4 cm. The amount of sand, which is 90% foraminiferal, is rather variable with a geometric mean and ±1 standard deviation of $6.1^{+10.3}_{-3.8}$%.

Unit 2. This unit is of sandy mud, the sand being chiefly planktonic foraminifera and *c.*10.0%. The material is stiffer than the surface sediment (shear strength up to 4 kPa, McCave *et al.* 1984) and has <47% water content. The unit is extensively burrowed and the upper part with current lamination and fine-scale burrows has little preservation potential, so the resulting deposit is pervasively bioturbated (Fig. 13), (Baldwin & McCave 1999). The lower part of unit 2 also contains a few turbidites. Knorr 96 box core 10, which had a longitudinal ripple in the HEBBLE Mud (McCave *et al.* 1984), contained a green fine terrigenous sand turbidite at *c.* 20 cm depth (Fig. 13), and a coarse sand and gravel turbidite at the base of the core (*c.* 40 cm depth).

Unit 3. In the base of a few box cores there was a slick olive-grey mud which is also encountered in the gravity core LGC-1 (40° 26.95′N 62° 20.78′W) where it extends down to 74 cm. It contains burrows with foraminiferal fillings but is not bioturbated to the extent of Unit 2. Foraminifera examined by Dr B. Corliss (pers. comm. 1983) show a gradation from warmer to cooler species downcore with the base of the Holocene between 60 and 70 cm and no fauna below 90 cm. The Holocene sedimentation rate thus averages about 6 cm ka^{-1}. This rate agrees with other values from similar water depths on this margin (Ericson *et al.* 1961; Zimmerman 1972) but is a factor of three less than values on the upper continental rise (Berry & Piper 1993).

Grain size of the <63 μm fraction of Unit 1

Form of distribution. Grain size data of the disaggregated mud fraction were obtained by Coulter Counter (1.6–63 μm) and pipette (% <2 μm). The basic form of the distributions measured by Coulter Counter is trimodal. There is a fine mode with a peak

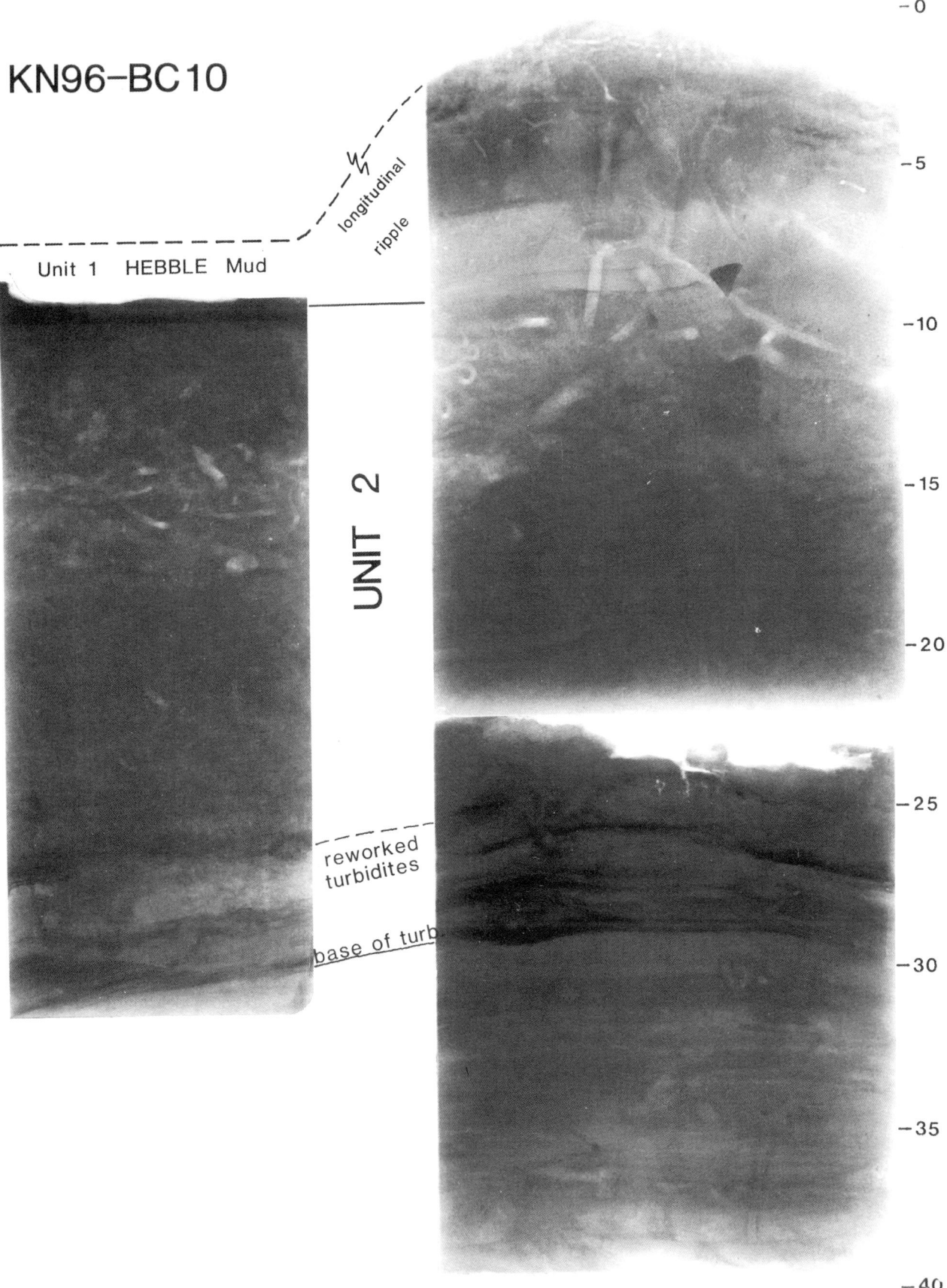

Fig. 13. X-radiographs of KN96-BC10. The main section under the crest of the longitudinal ripple was made in two sections with unavoidable disturbance of the lower part. The section to the left was from the lower flanks of the ripple about 15 cm away; 2 cm of HEBBLE mud has been scraped from the surface before the slab corer was inserted. Darker layers are more dense, generally sandy (foraminifera). Note presence of granules and pebble under the ripple crest that are considered by McCave (1988) to be glacial material that has been biologically pumped upwards, and immobile under modern currents. The bioturbation of these deposits is tiered with fine-scale burrows in the upper part and large scale reworking by echinoids and polychaetes below (Baldwin & McCave 1999). The upper layers bearing current lamination have virtually zero preservation potential and the lower 'historical layer' is what becomes the stratigraphic record at this high-energy site.

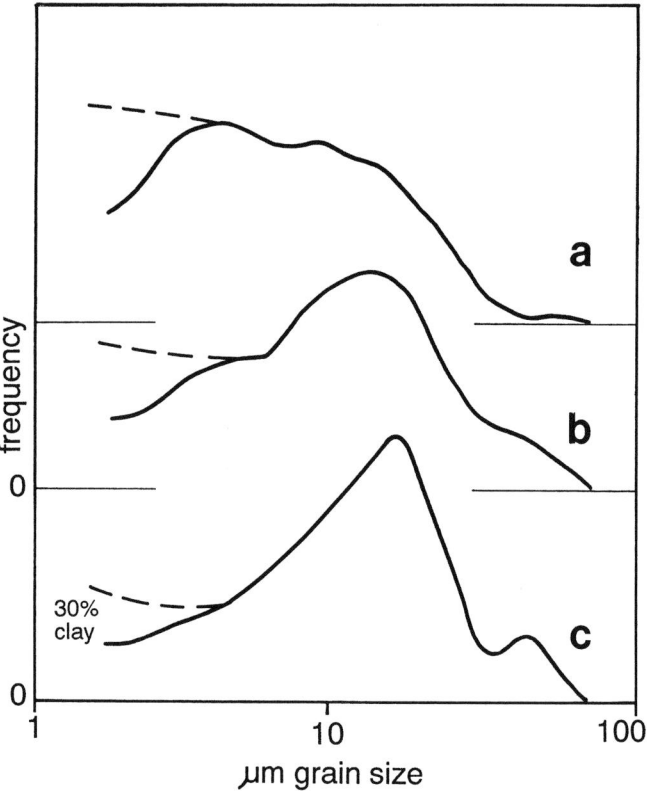

Fig. 14. Size distribution of the surface *c*.1 mm measured by Coulter Counter. Note that size increases to the right. (**A**) from 4500 m depth site, core KN103 BC1 (40° 53.6′N, 63° 43.9′W) showing a pronounced fine mode. (**B**) HEBBLE area in 1983 (core KN103 BC6) after moderate flow showing trimodal structure. (**C**) HEBBLE area in 1982 (core KN96 BC25) after strong flow showing unimodal structure with truncated (by sieving) sand tail (after McCave 1985).

between 3 and 4 µm, a coarse mode usually between 10 and 17 µm, and a truncated ramp at the coarse end (>30 µm) of the distribution (e.g. Fig. 14B, BC6, 0–1 mm). This truncated ramp is the fine tail of the sand distribution. The fine mode is the coarse end of the clay size range and declines between 3 and 1.6 µm because of an under-counting artifact.

The coarse mode size and peak height were found to be well correlated. As the modal size increases it becomes relatively more abundant, and as the coarse mode size increases the percentage of clay in the samples decreases, a remarkably close relationship with $r = 0.93$. A powerful physical control is suggested by these relations.

These data, as with the pipette data for the whole continental rise given above, show fractionation during deposition of the fines in which deposition under stronger flows yields less clay and increases both the percentage and modal size of the silt. This suggestion is borne out by the samples from a shallow site (4500 m depth) where the currents are weaker (Richardson *et al.* 1981). As is evident from Figure 14A, the coarse mode at that site is very fine (9–10 µm) and is less abundant than the fine mode, a situation which never occurs at 4820 m at the HEBBLE site.

Temporal variation. The silt-size spectra in the HEBBLE area were consistent in shape in the July 1982 (Knorr 96) data, being unimodal, often with the truncated sand tail (Fig. 14C). All four of the June 1983 (Knorr 103) samples were clearly polymodal (Fig. 14 B). It suggests that the HEBBLE mud sampled in 1983 was not the same as that sampled during 1982, just a year earlier, and that at least one episode of erosion and redeposition had occurred during the intervening period.

Discussion and conclusions

Contour-following abyssal currents have influenced sedimentation patterns on the Nova Scotian continental rise since the late Eocene to early Oligocene, but their relative importance as inferred from seismic stratigraphy has varied significantly and is not well documented quantitatively. The HEBBLE studies, concentrated on present oceanographic conditions and on mostly surficial sediments in the uppermost few centimeters of the seabed, show that current-controlled sedimentation has been dominant in geologically recent times, mainly the Holocene. The primary evidence is from seafloor photographic and sediment grain size studies coupled with current-meter records.

Seafloor photographs show a strong depth-dependent zonation in current intensity and suggest relatively tranquil conditions above about 4000 m. Curiously, the classic 'Western Boundary Undercurrent' of Norwegian Sea Overflow Water is thought to occur in this shallower depth range (Hogg 1983; Pickart 1991) where it is weakly developed. Below 4000 m currents steadily increase in average intensity and have a mean contour-parallel flow to the southwest, but they are unsteady in direction on short time scales. At depths of 4900–5000 m the seafloor lies beneath the mean axis of the Gulf Stream, and the high eddy kinetic energy of this current undoubtedly contributes to episodically intensifying, impeding, or reversing the mean southwest flow in the 'cold filament' comprising the Deep Western Boundary Current (Welsh *et al.* 1991; Ezer & Weatherly 1991; Peggion & Weatherley 1991).

Abyssal-current zonation suggested in bottom photography on the Nova Scotian rise is directly supported in sediment grain size data. Dominantly contour-parallel patterns in (1) sand percent; (2) silt/clay ratio; (3) silt mode peak-height; (4) clay per cent; and (5) grain size principal components are readily observed (Driscoll *et al.* 1985), with the occurrence of a dominant sandy zone at 4800–5000 m being a principal feature. Erosional events with peak speeds >15–20 cm s^{-1} are required to increase the occurrence of thin winnowed sand laminae and raise the overall sand percentage in this contour-parallel zone. The southeastward increase in abyssal eddy kinetic energy (eddy K_E) toward the axis of the Gulf Stream enhances peak current speeds in this high-sand zone and undoubtedly contributes to its development.

Silt/clay ratio decreases steadily upslope. Because the silt and clay form a uniform cohesive mud whose critical erosion velocity is controlled by cohesion, not grain size, the fine components of the mud are not fractionated by erosion. Under erosion all the fine components are suspended and it is unlikely that the silt is preferentially left behind. Concentration of silt in the fine fraction is thus a product of preferential deposition. For example, if deposition occurs from a flow of 10 cm s^{-1}, it will have a higher silt/clay ratio than if it occurs from a flow of 5 cm s^{-1} (McCave & Swift 1976). Most of the measured flow speeds are depositional (i.e. less than 20 cm s^{-1}), but the average current speeds increase downslope to a maximum at 4900–5000 m. Thus one would expect the observed downslope increase in silt content and silt/clay ratio on the basis that deposition occurred on average from faster flows.

Although the dominant control by bottom currents in modern times and possibly throughout the Holocene is apparent, there are superimposed, less well developed cross-slope grain size trends that might suggest the influence of downslope sediment transport. This is particularly apparent in and around Triffid Channel where increased sand fraction occurs in the channel and higher silt/clay ratios appear southwest of it. However, because the sand fraction in the channel has a much higher ratio of foraminiferal tests to mineral grains than expected from comparison with sediments over the remainder of the rise, it appears that much of the sand is explained by concentrating effects of abyssal currents in a depression rather than by downslope transport. On the other hand, the high silt/clay ratio downstream (southwest) of the channel could be an effect of channel-bank overflow. It is a matter of debate whether downslope processes have been recently active but are subsidiary to current control, or whether the cross-slope grain size patterns are relict and reworked from an earlier (e.g. glacial) time when they may have had a more dominant role. We incline to the latter view.

Based on research supported by the Office of Naval Research. Tucholke acknowledges support from the Henry Bryant Bigelow Chair in Oceanography. Swift received support from ONR contract N00014-98-1-0506. We are grateful for the most constructive reviews of David Piper and John Howe. Cambridge Earth Sciences number 6030, WHOI contribution number 10391.

References

BALDWIN, C. T. & McCAVE, I. N. 1999. Bioturbation in an active deep-sea area: implications for models of trace fossil tiering. *PALAIOS*, **14**, 375–388.

BERRY, J. A. & PIPER, D. J. W. 1993. Seismic stratigraphy of the central Scotian rise: a record of continental margin glaciation. *Geo-Marine Letters*, **13**, 197–206.

BISCAYE, P. E., GARDNER, W. D., ZANEVELD, J. R. V., PAK, H. & TUCHOLKE, B. 1980. Nephels! have we got nephels! *EOS, Transactions, American Geophysical Union*, **61**, 1014.

DAMUTH, J. E., TUCHOLKE, B. E. & SHOR, A. N. 1981. Bathymetry and near-bottom sedimentation processes of the Nova Scotia continental rise. *EOS, Transactions, American Geophysical Union*, **62**, 892.

DRISCOLL, M. L., TUCHOLKE, B. E. & McCAVE, I. N. 1985. Seafloor zonation and sediment texture on the Nova Scotian lower continental rise. *Marine Geology*, **66**, 25–41.

EBINGER, C. J. & TUCHOLKE, B. E. 1988. Marine geology of Sohm Basin, Canadian Atlantic margin. *American Association of Petroleum Geologists Bulletin*, **72**, 1450–1468.

ERICSON, D. B., EWING, M., WOLLIN, G. & HEEZEN, B. C. 1961. Atlantic deep-sea sediment cores. *Geological Society of America Bulletin*, **72**, 193–286.

EZER, T. & WEATHERLY, G. L. 1991. Small-scale spatial structure and long-term variability and near-bottom layers in the HEBBLE area. *Marine Geology*, **99**, 319–328.

FISHER, A. 1977. Historical limits of the northern edge of the Gulf Stream. *Gulfstream*, **3**, 6–7.

FLOOD, R. D. 1981. Longitudinal triangular ripples in the Blake–Bahama Basin. *Marine Geology*, **39**, M13–M20.

FOLK, R. L. 1974. *Petrology of Sediment Rocks*. Hemphill Publishing Co., Austin.

GARDNER, W. D., BISCAYE, P. E., ZANAVELD, J. R. V. & RICHARDSON, M. J. 1985. Calibration and comparison of the LDGO nephelometer and OSU transmissometer on the Nova Scotian continental. *Marine Geology*, **66**, 323–344.

GRANT, W. D., WILLIAMS, A. J. & GROSS, T. F. 1985. A description of the bottom boundary layer at the HEBBLE Site: low-frequency forcing, bottom stress and temperature structure. *Marine Geology*, **66**, 219–241.

GROSS, T. F. & DADE, W. B. 1991. Suspended sediment storm modelling. *Marine Geology*, **99**, 343–360.

GROSS, T. F. & WILLIAMS, A. J. 1991. Characterization of deep-sea storms. *Marine Geology*, **99**, 281–301.

HOGG, N. G. 1983. A note on the deep circulation of the western North Atlantic: its nature and causes. *Deep-Sea Research*, **30**, 945–961.

HOLLISTER, C. D. & McCAVE, I. N. 1984. Sedimentation under deep sea storms. *Nature*, **309**, 220–225.

HUGHES CLARKE, J. E., O'LEARY, D. W. & PIPER, D. J. W. 1992. Western Nova Scotia continental rise: relative importance of mass wasting and deep boundary-current activity. *In*: POAG, C. W. & DE GRACIANSKY, P. C. (eds) *Geologic Evolution of Atlantic Continental Rises*. Van Nostrand Reinhold, New York, 266–281.

JANSA, L. F. & WADE, J. A. 1975. Geology of the continental margin off Nova Scotia and Newfoundland. *In*: VAN DER LINDEN, W. J. M. & WADE, J. A. (eds) *Offshore Geology of Eastern Canada*. Geological Survey of Canada Paper **74–30**(2), 51–105.

KING, L. H. & MACLEAN, B. 1976. *Geology of the Scotian Shelf*. Geological Survey of Canada Paper **74–31**.

McCAVE, I. N. 1983. Particulate size spectra, behaviour and origin of nepheloid layers over the Nova Scotian Continental Rise. *Journal of Geophysical Research*, **88**, 7647–7666.

McCAVE, I. N. 1985. Sedimentology and stratigraphy of box cores from the HEBBLE site on the Nova Scotian Continental Rise. *Marine Geology*, **66**, 59–89.

McCAVE, I. N. 1988. Biological pumping upwards of the coarse fraction of deep-sea sediments. *Journal of Sedimentary Petrology*, **58**, 148–158.

McCAVE, I. N. & SWIFT, S. A. 1976. A physical model for the rate of deposition of fine-grained sediment in the deep sea. *Geological Society of America Bulletin*, **87**, 541–546.

McCAVE, I. N., HOLLISTER, C. D., DEMASTER, D. J., NITTROUER, C. A., SILVA, A. J. & YINGST, J. Y. 1984. Analysis of a longitudinal ripple from the Nova Scotian continental rise. *Marine Geology*, **58**, 275–286.

NOWELL, A. R. M., HOLLISTER, C. D. & JUMARS, P. A. 1982. High Energy Benthic Boundary Layer Experiment: HEBBLE. *EOS, Transactions American Geophysical Union*, **63**(31), 594–595.

PEGGION, G. & WEATHERLY, G. L. 1991. On the interaction of the bottom boundary layer and deep rings. *Marine Geology*, **99**, 329–342.

PICKART, R. S. 1991. Water mass components of the North Atlantic Deep Western Boundary Current. *Deep-Sea Research*, **39**, 1553–1572.

RICHARDSON, M. J., WIMBUSH, M. & MAYER, L. 1981. Exceptionally strong near-bottom flows on the continental rise of Nova Scotia. *Science*, **213**, 887–888.

RICHARDSON, P. L. 1983. Eddy kinetic energy in the North Atlantic from surface drifters. *Journal of Geophysical Research*, **88**, 4355–4367.

SCHMITZ, W. J. 1984. Abyssal eddy kinetic energy in the North Atlantic. *Journal of Marine Research*, **42**, 509–536.

SHOR, A. N. 1984. *Bathymetry*. Ocean Margin Drilling Program, Regional Atlas Series, Atlases 2, 3. Marine Science International, Woods Hole, Mass.

SHOR, A. N. & LONSDALE, P. 1981. HEBBLE site characterization: downslope processes on the Nova Scotia lower continental rise. *EOS, Transactions of the American Geophysical Union*, **62**, 892.

SMETHIE, W. M. 1993. Tracing the thermohaline circulation in the western North Atlantic using chlorofluorocarbons. *Progress in Oceanography*, **31**, 51–99.

SMETHIE, W. M., FINE, R. A., PUTZKA, A. & JONES E. P. 2000. Tracing the flow of North Atlantic Deep Water using chlorofluorocarbons. *Journal of Geophysical Research*, **105**, 14297–14323.

SWIFT, S. A. 1985. Late Pleistocene sedimentation on the continental slope and rise off western Nova Scotia. *Geological Society of America Bulletin*, **96**, 832–841.

SWIFT, S. A. 1987. Late Cretaceous-Cenozoic development of the outer continental margin, south-western Nova Scotia. *Bulletin of the American Association of Petroleum Geologists*, **71**, 678–701.

SWIFT, S. A., HOLLISTER, C. D. & CHANDLER, R. S. 1985. Close-up stereophotographs of abyssal bedforms on the Nova Scotian continental rise. *Marine Geology*, **66**, 303–322.

SWIFT, S. A., C. EBINGER, & TUCHOLKE, B. E. 1986. Seismic stratigraphic correlation across the New England Seamounts, western North Atlantic Ocean. *Geology*, **14**, 346–349.

TUCHOLKE, B. E. 1982. Origin of longitudinal triangular ripples on the Nova Scotian continental rise. *Nature*, **296**, 735–737.

TUCHOLKE, B. E. 1986. Analysis of a longitudinal ripple from the Nova Scotian continental rise: Comment. *Marine Geology*, **72**, 371–373.

TUCHOLKE, B. E. & MOUNTAIN, G. S. 1979. Seismic stratigraphy, lithostratigraphy and paleo sedimentation patterns in the North American Basin. *In*: TALWANI, M., HAY, W. & RYAN, W. B. F. (eds) *Deep Drilling in the Atlantic Ocean: Continental Margins and Paleoenvironment*. Maurice Ewing Series 3, American Geophysical Union, Washington, D. C., 58–86.

TUCHOLKE, B. E., HOLLISTER, C. D., BISCAYE, P. E. & GARDNER, W. D. 1985. Abyssal current character determined from sediment bedforms on the Nova Scotian continental rise. *Marine Geology*, **66**, 43–47.

TUCHOLKE, B. E., HOUTZ, R. E. & LUDWIG, W. J. 1982. Sediment thickness and depth to basement in western North Atlantic basin. *Bulletin of the American Association of Petroleum Geologists*, **66**, 1384–1395.

UCHUPI, E. & SWIFT, S. A. 1991. Plio-Pleistocene slope construction off western Nova Scotia, Canada. *Cuadernos de Geologia Iberica*, Special Issue no. **15**, 15–35.

WADE, J. A. & MACLEAN, B. C. 1990. The geology of the southeastern margin of Canada, Part 2. Aspects of the geology of the Scotian Basin from recent seismic and well data. *In*: KEEN, M. J. & WILLIAMS, G. L. (eds) *Geology of the Continental Margin of Eastern Canada*. Geological Survey of Canada, Geology of Canada, **2**, 167–238.

WEATHERLY, G. L. & KELLEY JR., E. A. 1982. 'Too Cold' bottom layers at the base of the scotian Rise. *Journal of Marine Research*, **40**, 985–1012.

WEATHERLY, G. L. & KELLEY JR. E. A. 1985*a*. Two views of the cold filament. *Journal of Physical Oceanography*, **15**, 68–81.

WEATHERLY, G. L. & KELLEY JR. E. A. 1985*b*. Storms and flow reversals at the HEBBLE site. *Marine Geology*, **66**, 205–218.

WELSH, E. B., HOGG, N. G. & HENDRY, R. M. 1991. The relationship of low-frequency deep variability near the HEBBLE site to Gulf Stream fluctuations. *Marine Geology*, **99**, 303–317.

ZIMMERMAN, H. B. 1972. Sediments of the New England continental rise. *Geological Society of America Bulletin*, **83**, 3709–3724.

ZIMMERMAN, H. B. 1971. Bottom currents on the New England continental rise. *Journal of Geophysical Research*, **76**, 5865–5876.

The Greater Antilles Outer Ridge: development of a distal sedimentary drift by deposition of fine-grained contourites

BRIAN E. TUCHOLKE

Woods Hole Oceanographic Institution, Woods Hole, Massachusetts 02543, USA (e-mail: btucholke@whoi.edu)

Abstract: The Greater Antilles Outer Ridge, located north and northwest of the Puerto Rico Trench, is a deep (>5100 m), distal sediment drift more than 900 km long and up to 1 km thick. It has been isolated from sources of downslope sedimentation throughout its history and is formed of clay- to fine silt-size terrigenous sediments that have been deposited from suspended load carried in the Western Boundary Undercurrent, together with 0–30% pelagic foraminiferal carbonate. Because of the fine, relatively uniform grain size of the sediments, the outer ridge consists of sediments that are seismically transparent in low-frequency reflection profiles. Sediment tracers (chlorite in sediments and suspended particulate matter, reddish clays in cores) indicate that at least a portion of the ridge sediments has been transported more than 2000 km from the eastern margin of North America north of 40°N. The outer ridge began to develop as early as the beginning of Oligocene time when strong, deep thermohaline circulation developed in the North Atlantic and the trough initiating the present Puerto Rico Trench had cut off downslope sedimentation from the Greater Antilles. The fastest growth of the outer ridge probably occurred beginning in the early Miocene, about the same time that large drifts such as the Blake Outer Ridge were initiated along the North American margin. Since that time, the most rapid sedimentation has been along the crest of the northwestern outer ridge where suspended load is deposited in a shear zone between opposing currents on the two ridge flanks.

Many contourite deposits occur on continental margins or in close proximity to sources of clay- to sand-size sediment fed downslope in gravity flows. In these environments it can be difficult to quantify and differentiate the geological effects of downslope processes from the effects of bottom current processes, which generally act along slope. In contrast, sediment drifts that are physiographically isolated from downslope sources do not exhibit these complications. Aside from a pelagic component settling from the overlying water column, they are pure contourite sediments, and they exhibit structures and characteristics developed solely from current-controlled sedimentation.

The Greater Antilles Outer Ridge (GAOR) north of the Puerto Rico Trench in the western North Atlantic (Fig. 1) is a well developed example of a distal and isolated contourite drift. It is a large sedimentary feature that has been totally disconnected from downslope sediment sources since middle to late Eocene time. It lies at depths greater than about 5100 m and thus is mostly below the calcite lysocline. Consequently the outer ridge consists dominantly of fine-grained sediments that have been deposited from suspended load carried in abyssal currents, with only a minor component of pelagic sedimentation. Although the Greater Antilles Outer Ridge was studied in detail in the early 1970s (Tucholke et al. 1973; Tucholke & Ewing 1974; Tucholke 1975), there has been almost no subsequent research in this region. In this synthesis I review the modern geological and oceanographic setting of the outer ridge together with its depositional history since middle Eocene time, based largely on this earlier work.

Geological setting and physiography

The Greater Antilles Outer Ridge overlies middle Cretaceous ocean crust north of the easternmost Bahama Banks and the Puerto Rico Trench. It is bounded to the northwest by the southern Hatteras Abyssal Plain at depths of c. 5520 m and to the northeast by the deep (>5800 m) Nares Abyssal Plain (Fig. 1). Vema Gap, which is a conduit for turbidity currents flowing from the Hatteras to Nares abyssal plain, lies along part of the northern margin of the ridge. To the west, the Silver Abyssal Plain isolates the outer ridge from downslope deposition of sediments derived from the Bahama Banks. A small sedimentary drift, the Caicos Outer Ridge, lies near the base of the Bahama Banks. A slightly elevated sedimentary sill extends east from Caicos Outer Ridge to connect with the northwesternmost GAOR. An elevated sill (Navidad sill) also connects the southeasternmost Bahama Banks to the central part of the GAOR; this elevation is largely due to a basement bulge that has been created as ocean crust is deformed and obliquely subducted in the Puerto Rico Trench. The trench itself bounds the southern side of the GAOR, isolating it from sediments moving in gravity flows from the Greater Antilles.

The GAOR can be divided into two provinces, each with somewhat different geological characteristics and depositional history. East of about 67°W the ridge parallels the Puerto Rico Trench for more than 450 km and is 5300–5400 m deep along its axis. Much of the ridge's physiographic elevation is due to the fact that it overlies the crustal outer-high seaward of the trench. The bathymetry is irregular, partly because high-amplitude basement peaks interrupt the sedimentary sequence and partly because current-controlled deposition is uneven around these obstacles (Figs 2 & 3). Drift deposits that define the ridge thin markedly east of about 62.5°W, but the actual eastward extent of these deposits is uncertain.

The GAOR west of 67°W trends north and northwest for about 450 km and has minimum depths of 5100–5200 m. Here the physiographic expression of the ridge is due entirely to accumulation of drift deposits. There is no underlying bulge in basement topography and basement interruptions of the sedimentary sequence are rare (Figs 4–7).

Finer-scale morphology of the ridge varies between relatively smooth seafloor and sediment waves. The sediment waves in some places are regular (Fig. 7), but more commonly they have variable form, wavelength and amplitude. Small flat ponds occasionally occur in depressions between sediment waves, and they appear to have formed from local mass wasting and turbidity currents. The best developed of these is in relatively complex topography on the northeast flank of the ridge near 22.5°N, 67.5°W (Fig. 1). There the sediment waves have been modified into a dendritic drainage system extending toward the Nares Abyssal Plain, forming a network of 'layered valleys' (Tucholke 1975). This kind of ponding and layering is restricted entirely to the western sector of the GAOR (see Fig. 13).

Outer ridge sediments

Sediments forming the GAOR are homogeneous, terrigenous, brown, gray and reddish clays (<2 μm) and fine silts that are locally enriched in carbonate (Figs 8 & 9). They exhibit little

From: STOW, D. A. V., PUDSEY, C. J., HOWE, J. A., FAUGÈRES, J.-C. & VIANA, A. R. (eds)
Deep-Water Contourite Systems: Modern Drifts and Ancient Series, Seismic and Sedimentary Characteristics.
Geological Society, London, Memoirs, **22**, 39–55. 0435-4052/02/$15.00 © The Geological Society of London 2002.

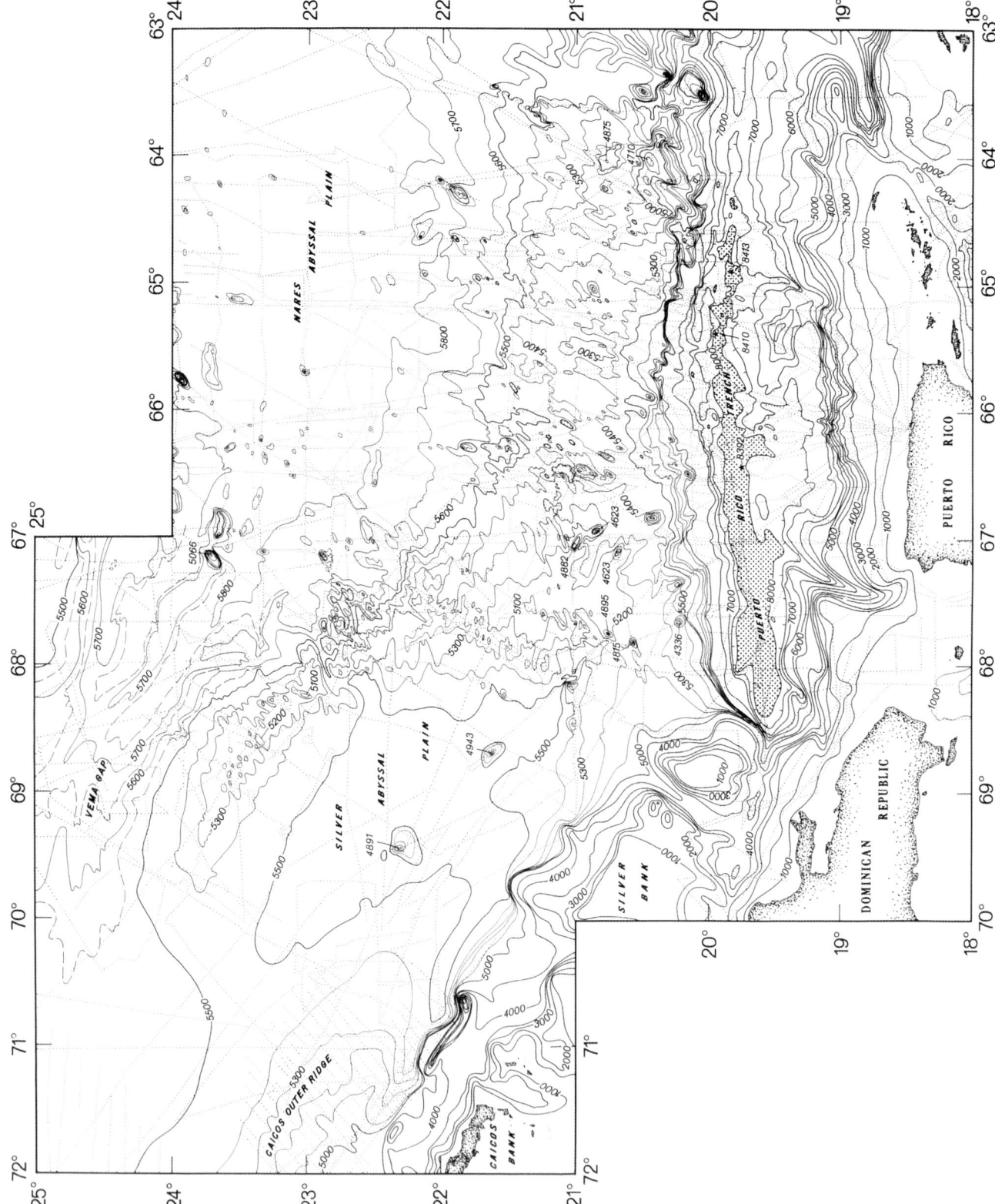

Fig. 1. Bathymetry of the Greater Antilles Outer Ridge and vicinity (from Tucholke et al. 1973). The outer ridge extends east–west north of the Puerto Rico Trench and to the NW between Silver Abyssal Plain and Vema Gap. Contours are in corrected metres.

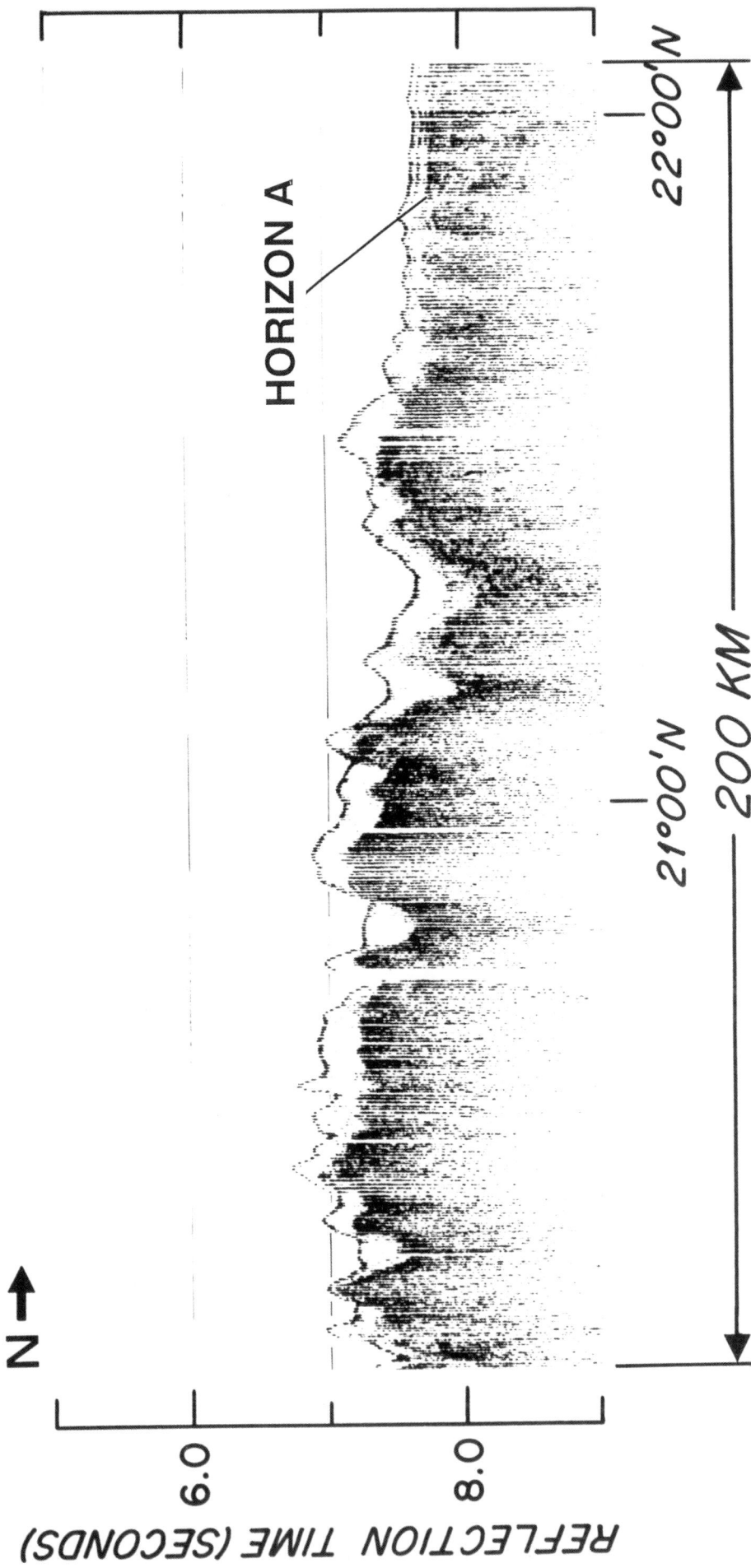

Fig. 2. Seismic reflection profile extending north–south across the Greater Antilles Outer Ridge to the southernmost Nares Abyssal Plain along 62°52'W, between 20°13'N and 22°04'N (R/V Conrad Cruise 8). The top of the middle to upper Eocene Horizon A complex is labeled where it is observed under the edge of Nares Abyssal Plain. At this far-eastern end of the GAOR, seismically transparent outer ridge sediments directly overlie oceanic basement. Figure from Tucholke & Ewing (1974).

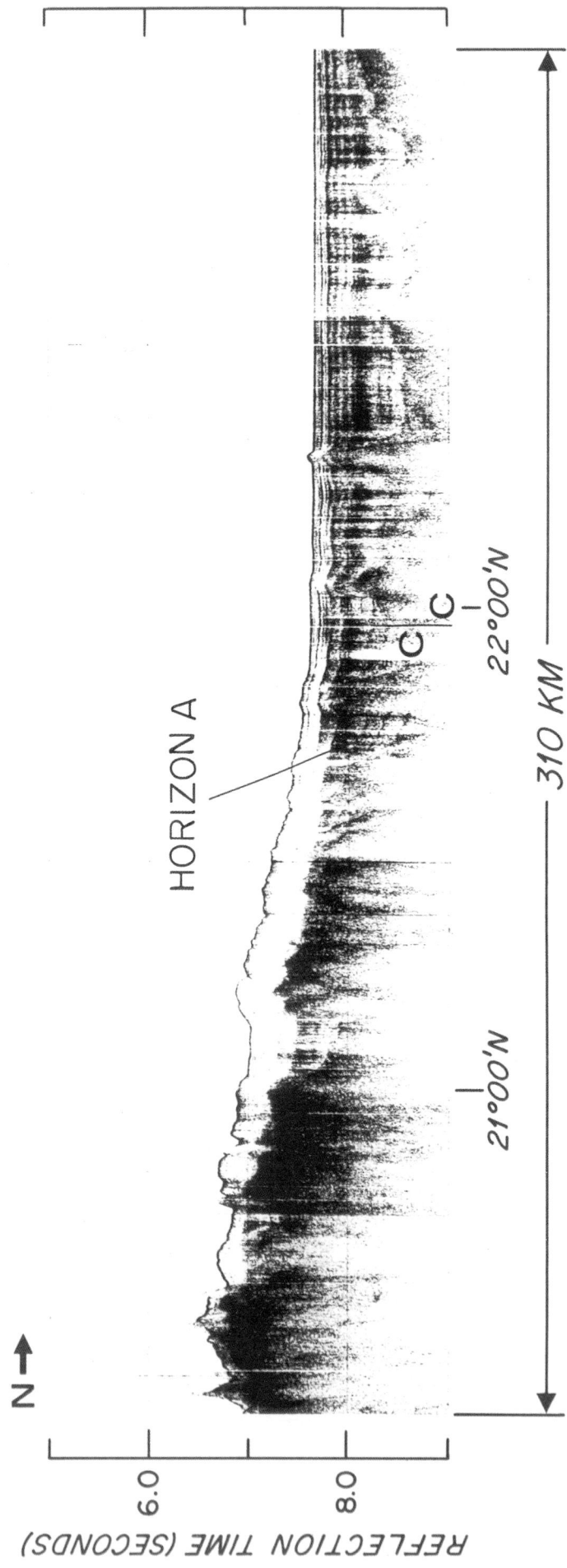

Fig. 3. Seismic reflection profile across the Greater Antilles Outer Ridge and southern Nares Abyssal Plain near 64°W (R/V Conrad Cruise 10). Location in Figure 8. The Horizon A complex extends beneath the outer ridge but is interrupted by numerous basement peaks. Note that seismically transparent sediments have accumulated faster than the abyssal plain sediments and the outer ridge has prograded onto the abyssal plain. Figure from Tucholke & Ewing (1974).

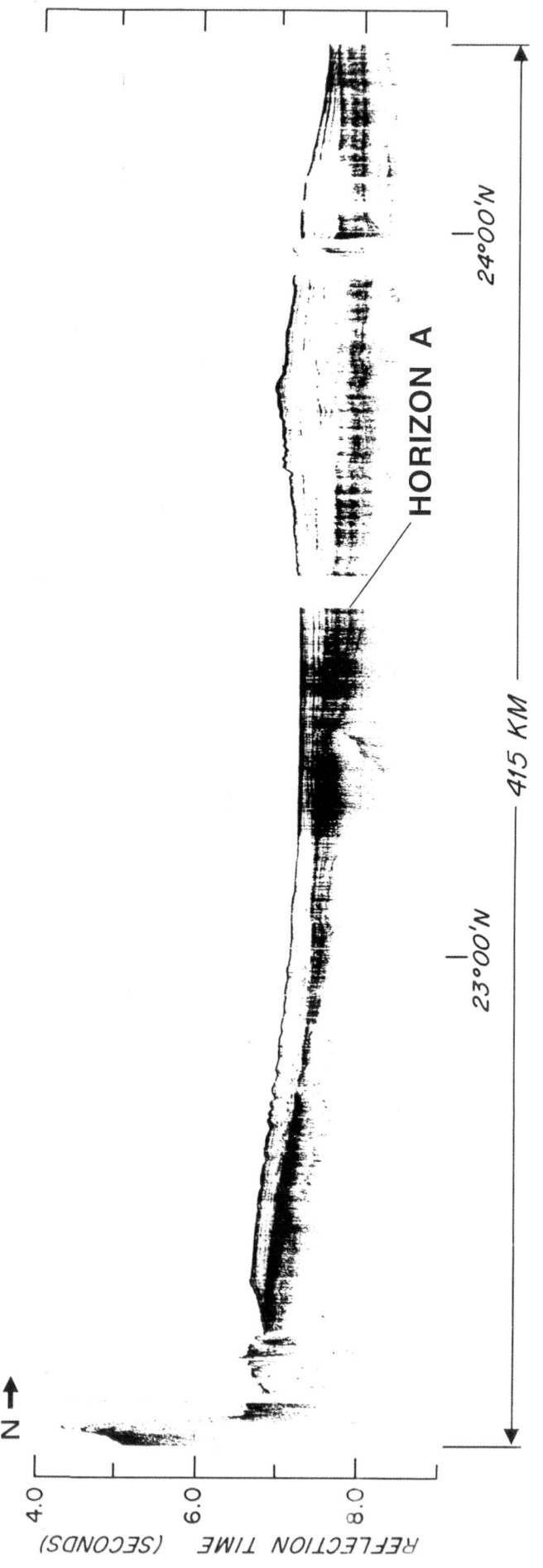

Fig. 4. Seismic reflection profile extending from the base of the Bahama Banks across the Caicos Outer Ridge, Silver Abyssal Plain, and northwestern Greater Antilles Outer Ridge to Vema Gap (R/V Conrad Cruise 10). Location in Figure 8. Seismically transparent sediments forming the core of the GAOR appear above the flat-lying Horizon A complex and the ridge subsequently expanded outward, interfingering with abyssal plain sediments. Note that the Horizon A complex beneath the Caicos Outer Ridge is truncated in an angular unconformity (Horizon A^U) that was eroded by abyssal currents in late Eocene to Oligocene time. Figure from Tucholke & Ewing (1974).

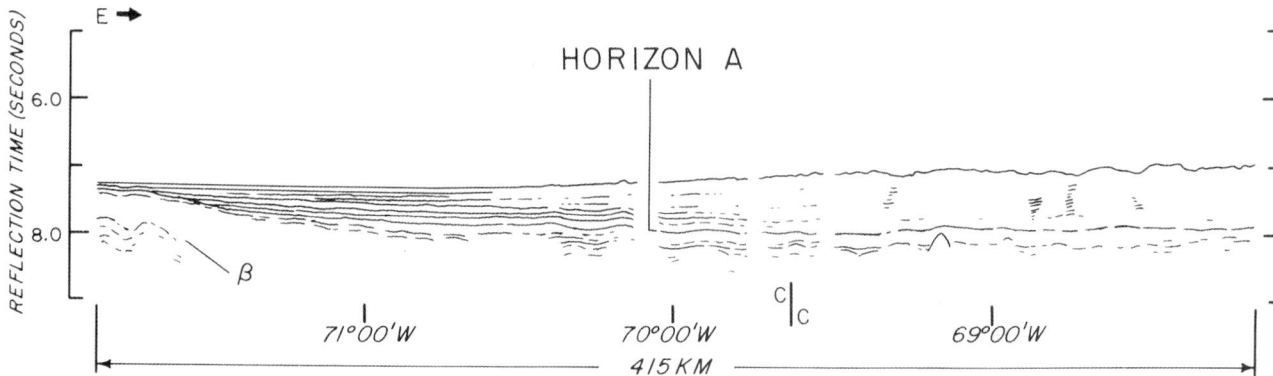

Fig. 5. Tracing of a seismic reflection profile across the southern Hatteras Abyssal Plain and along the axis of the northwestern Greater Antilles Outer Ridge (R/V Vema Cruise 22). Location in Figure 8. Sediments of the Horizon A complex and deeper beds completely cover basement beneath this section of the outer ridge. Seismically transparent outer-ridge sediments have prograded northwestward, interfingering with turbidites of the Hatteras Abyssal Plain. Weak, stacked reflections within the transparent sediments mark the locations of layered valleys developed between sediment waves. Figure from Tucholke & Ewing (1974).

textural stratification except where carbonates occur, although colour contacts between units are often distinct. Layers of sediment coarser than clay size are rare, and where present they are thin (<1 cm) and contain concentrations of ash derived from the Antilles, carbonate, or manganese micronodules. Cored sediments show extensive reworking by burrowing organisms. This is apparent from color mottling and in carbonate-enriched patches worked downward from higher-carbonate zones into underlying clay-rich intervals.

Grain size is very uniform, averaging 80% clay (<2 μm) and 20% silt (2–62 μm) with a variation of 10%. Sand content (>62 μm) rarely exceeds 5% and the sand consists entirely of biogenic and authigenic components (foraminifera, manganese micronodules, fish teeth, siliceous sponge spicules, and rare radiolaria and diatoms, in order of decreasing abundance). None of the silt or sand components in cores are organized into identifiable structures that might be attributed to current-controlled deposition. The fine grain size of the dominant clay component also precludes megascopic identification of current-produced bedforms.

Distinct silt layers occur only in local turbidite ponds between sediment waves, as for example in the layered valleys noted above. The silts form micaceous laminae 1–2 mm thick, interbedded with clays. Individual turbidites in these areas, identified from textural and colour criteria, are usually less than 10 cm thick.

Total carbonate content of the outer-ridge sediments normally ranges from 0–30%, averages 10–15%, and is dominated by foraminifera with usually minor amounts of coccoliths and unidentified detrital carbonate. Carbonate concentrations vary cyclically downcore, and the variations are interpreted to be controlled by a combination of changes in surface-water productivity and seafloor dissolution, with the latter probably related to changes in flux or temperature of bottom water flowing around the outer ridge. The surface-water productivity and bottom-water dissolution effects appear not to be correlated in phase (Tucholke 1975).

The clay-fraction mineralogy is comprised of the clay minerals montmorillonite, illite, chlorite and kaolinite. It is notable that chlorite concentrations in outer ridge sediments are richer than expected (Fig. 10); weathering in sediment source areas at these tropical to sub-tropical latitudes tends to produce kaolinite- and montmorillonite-rich sediments (Biscaye 1965; Hathaway 1972). The chlorite enrichment has been attributed to southward transport of higher-latitude, mechanically weathered sediments by abyssal currents (Tucholke 1975). Taking into account the mineralogy and average concentration of the silt fraction, bulk mineralogy of the terrigenous component in outer ridge sediments is 89% layered silicates, 7% quartz, and 4% plagioclase.

Sediment accumulation rates on the GAOR have been determined both from radiocarbon dating and from foraminiferal zonations (Tucholke 1975). The fastest accumulating sediments are along the crest of the outer ridge west of 67°W, where rates range from about 6 cm ka^{-1} up to more than 30 cm ka^{-1}. Lower rates of 3–4 cm ka^{-1} prevail on the adjacent ridge flanks. Comparably low rates also occur on the eastern ridge crest and north flank. On the south flank of the eastern GAOR and along the north wall of the Puerto Rico Trench, unfossiliferous to sparsely fossiliferous, ash-rich sediments of lower Eocene to upper Miocene age have been piston-cored. The very low sedimentation rates indicated by these data probably result from a combination of sediments being displaced into the trench by gravity flows and isolation of the south flank from deep currents with significant sediment load (Tucholke & Eittreim 1974).

Hydrography and abyssal currents

Data from hydrographic sections show that the bottom water over the GAOR has temperatures colder than 1.7°C (mostly colder than 1.6°C) and salinities less than 34.875 ‰ (Tucholke *et al.* 1973) (all temperatures given here are potential temperatures). Water with these characteristics is Antarctic Bottom Water (AABW) and has an ultimate source in the circum-Antarctic (Wright & Worthington 1970). Isotherms and isohalines both are spaced more closely between about 1.6°C and 1.8°C than in the rest of the deep-water column, and this gradient marks the transition with overlying North Atlantic Deep Water (NADW).

Water masses over the GAOR are part of the deep current system, classically termed the Western Boundary Undercurrent (WBUC), that flows from north to south along the continental margin of eastern North America (Fig. 10). Shallower currents in this system (*c.* 3000 to 4900 m depths) move NADW derived largely from the Norwegian Sea, but deeper currents contain an increasing component of AABW. The AABW is thought to be incorporated from flows entering the western North Atlantic along the west flank of the Mid-Atlantic Ridge; the AABW circulates clockwise around the Bermuda Rise and also reaches northwestward as far as the Newfoundland-Nova Scotia continental rise where it is recirculated southward at the base of the WBUC (McCave & Tucholke 1986).

Dynamic calculations based on hydrographic measurements over the GAOR indicate that the bottom water enters the region in a southeast-directed flow along the base of the Bahama Banks, circulates clockwise around the northwestern part of the outer ridge, and flows east along the north flank of the eastern outer

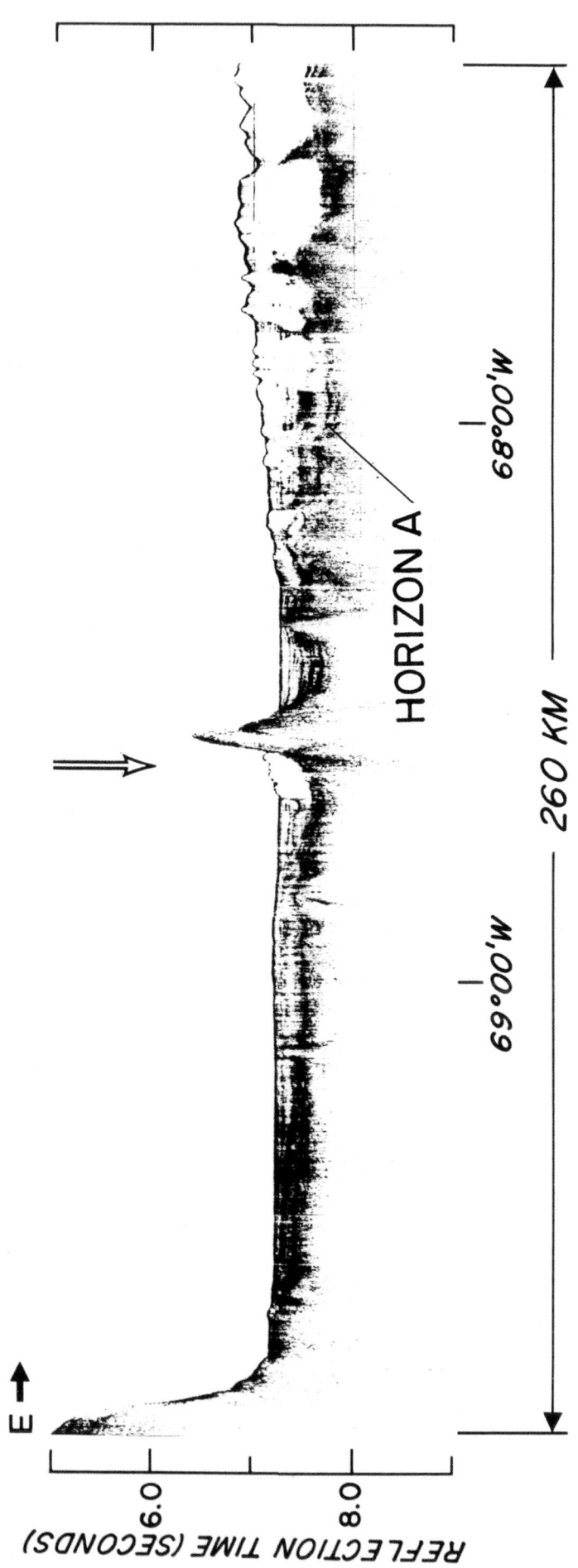

Fig. 6. Seismic reflection profile extending from the base of the Bahama Banks across the southern Silver Abyssal Plain and onto the central Greater Antilles Outer Ridge (R/V Conrad Cruise 10). Location in Figure 8. Irregular sediment waves and local, stacked reflections marking layered valleys appear in the seismically transparent outer ridge sediments. A small current-deposited drift (arrow) flanks the seamount in the southern Silver Abyssal Plain. Figure from Tucholke & Ewing (1974).

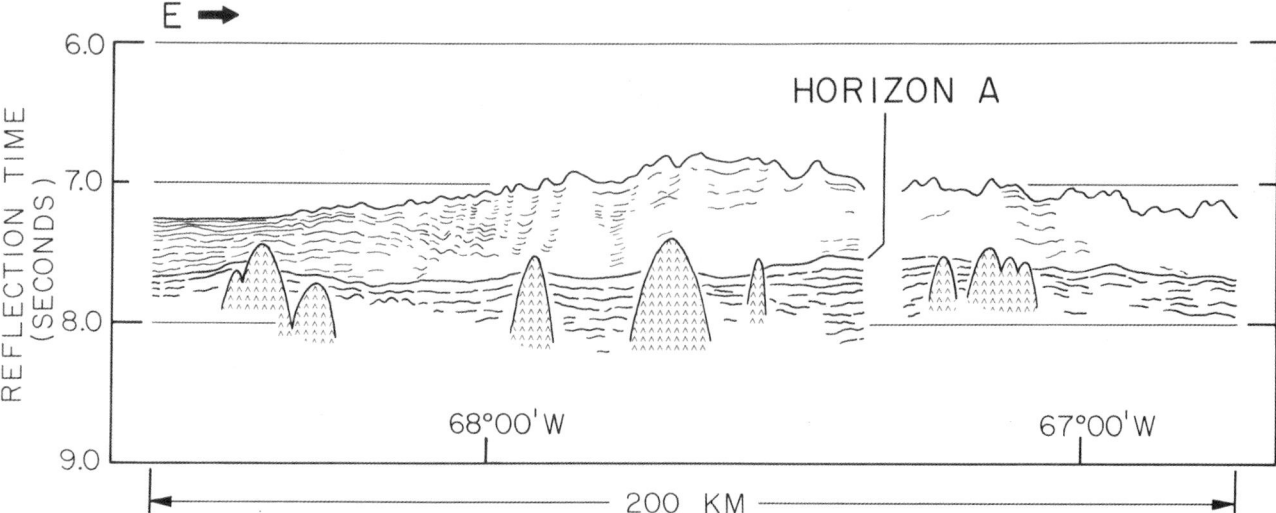

Fig. 7. Tracing of a seismic reflection profile across the central Greater Antilles Outer Ridge (R/V Conrad Cruise 10). Location in Figure 8. Basement peaks locally interrupt the Horizon A complex. Migrating sediment waves and reflections from intervening layered valleys appear in the seismically transparent outer-ridge sediments. Figure from Tucholke & Ewing (1974).

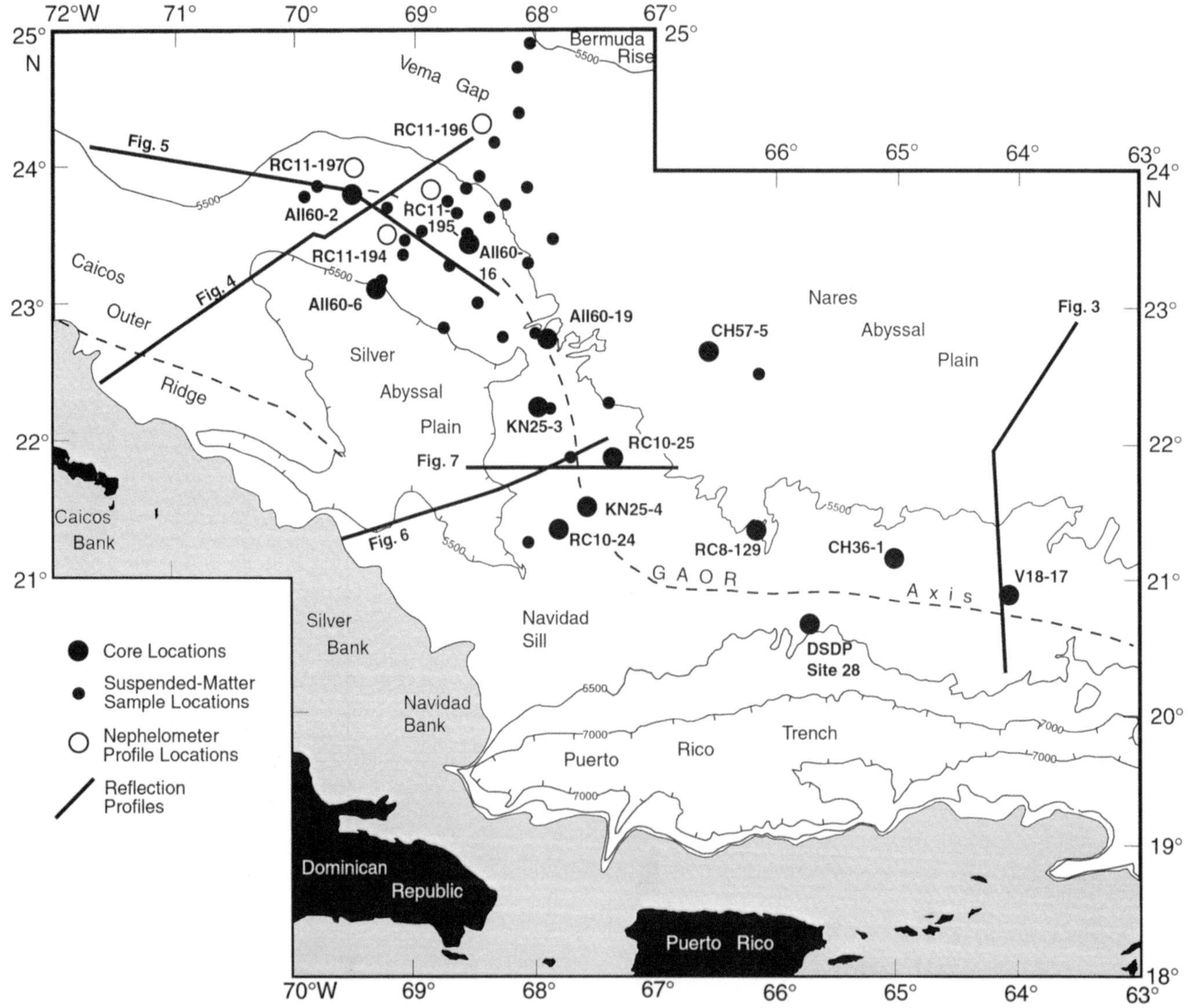

Fig. 8. Simplified bathymetric map showing locations of seismic reflection profiles in Figures 2 to 7, locations of cores illustrated in Figure 9, and locations of suspended particulate matter sampling stations and nephelometer profiles in Figure 12. DSDP Site 28 (Bader *et al.* 1970) recovered two cores of brown clay from the outer ridge at subbottom depths of 59–78 m, and six cores from deeper, stratified sequences below 169 m subbottom. Contours are in corrected metres, with depths less than 5000 m shaded. The axes of the Greater Antilles Outer Ridge and Caicos Outer Ridge are indicated by dashed lines.

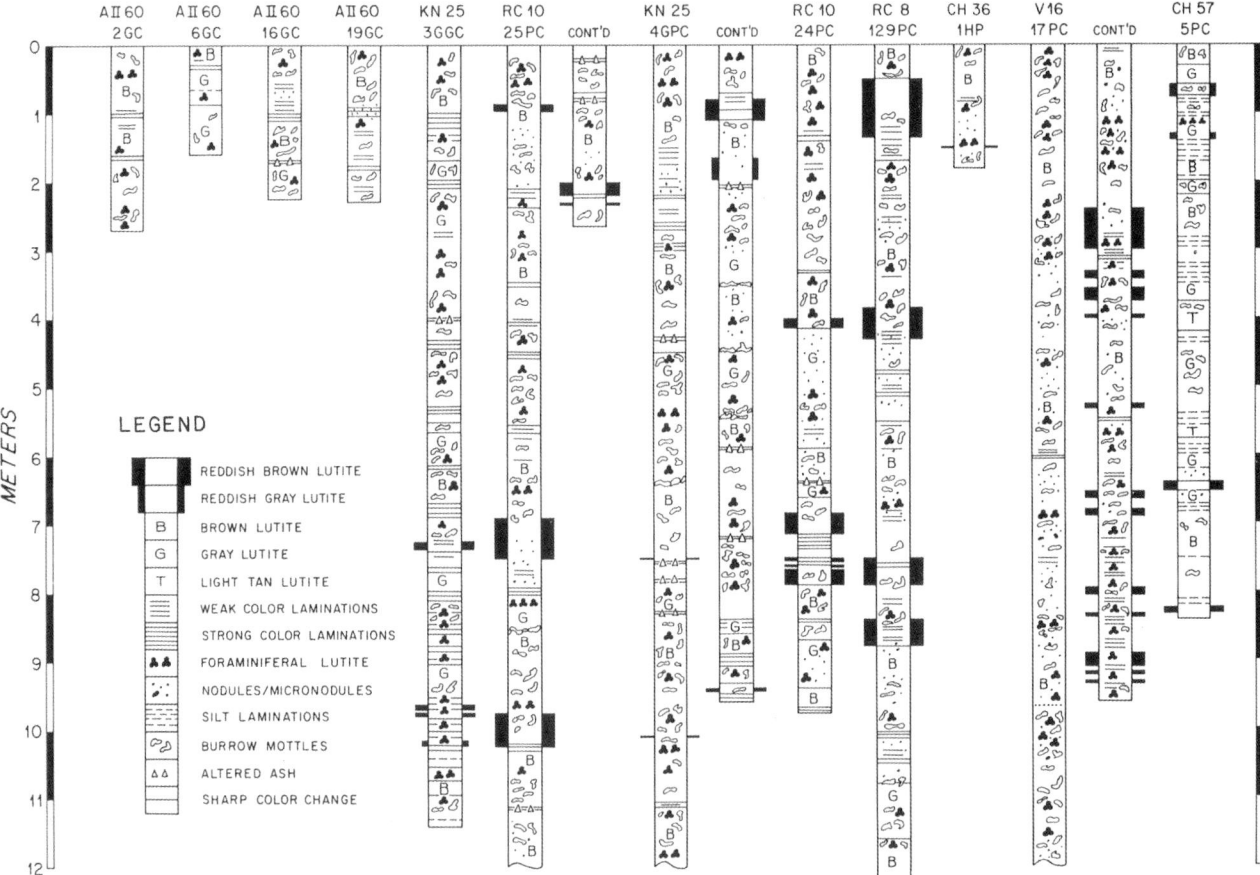

Fig. 9. Lithological summary of cores taken from the Greater Antilles Outer Ridge, arranged from west (left) to east and contrasted against one core containing fine-grained turbidites from the Nares Abyssal Plain (far right). Core locations in Figure 8. The outer-ridge cores are dominantly bioturbated clay (lutite) with varying admixtures of foraminiferal carbonate, usually 0–30%. GC, gravity core; GGC, giant gravity core; PC, piston core; GPC, giant piston core; HP, heat-probe gravity core. Figure from Tucholke (1975).

ridge (Fig. 11). Part of this circulation passes southeastward over Navidad sill and then flows east along the south wall of the Puerto Rico Trench. A northwesterly flow of AABW on the northern side of Vema Gap (southern margin of the Bermuda Rise) opposes southeasterly flow on the north flank of the northwestern outer ridge; it appears to divert some of the outer ridge flow to the north along the eastern edge of the Hatteras Abyssal Plain as part of the clockwise AABW circulation around the Bermuda Rise.

Direct current measurements and interpretations of bottom currents from seafloor photographs support the directional patterns of abyssal circulation around the GAOR derived from hydrographic calculations (Fig. 11). Measured currents on the north flank of the eastern outer ridge show dominantly easterly flow at <2 to 10 cm s^{-1} with short-term reversals (Fig. 11, location 2), while currents farther north on the southern Nares Abyssal Plain show both westerly and easterly currents at about the same speeds (location 1). The westerly flow in the Nares measurements may be part of the AABW circulation that courses around the Bermuda Rise. The boundary between this and the easterly flow on the northern ridge flank may shift episodically, thus accounting for flow reversals in both areas.

Measured currents on the southwest flank of the western outer ridge are strongly unidirectional to the northwest over a measurement period of about six months, while currents on the northeast flank flow in the opposite direction, to the southeast, over the same period (Fig. 11). Current speeds in both locations range from 2–17 cm s^{-1}, with maximum three-day-average speeds of 13–15 cm s^{-1}. Current-direction indicators from bottom photographs are available mostly from camera stations over the western outer ridge. They confirm a contour-parallel flow that circulates clockwise around the western outer ridge and then eastward along the north flank of the eastern outer ridge.

Suspended sediment

The character and distribution of the suspended sediment load in abyssal currents over the GAOR are known from a limited number of light-scattering (nephelometer) profiles and from filtration of bottom-water samples collected in large-volume Niskin bottles (Figs 8 & 12) (Tucholke & Eittreim 1974, Tucholke 1975). The nephelometer profiles show a near-bottom nepheloid layer that is concentrated in the AABW (colder than 1.8°C) but that also extends into overlying NADW up to about 1.9°C (Fig. 12). Concentrations of suspended matter in the nepheloid layer range from <10 µg l^{-1} up to 63 µg l^{-1}. The concentrations do not follow a smooth pattern with depth as suggested by the nephelometer profiles but instead indicate that there may be considerable small-scale patchiness in sediment distribution within the nepheloid layer. Nonetheless, both maximum concentrations and depth-averaged concentrations follow the contour of the light-scattering profiles and suggest a core of enhanced sediment load in AABW at about 1.6°–1.7°C, i.e. at depths of c. 5050–5400 m.

AABW flow that passes over Navidad sill (Fig. 11) is accompanied by a nepheloid layer that appears as a mid-water maximum at about 1.5°–1.9°C in the southern part of the Puerto Rico Trench (Tucholke & Eittreim 1974). Maximum suspended matter concentrations, based on in situ calibration of the nephelometer (Biscaye

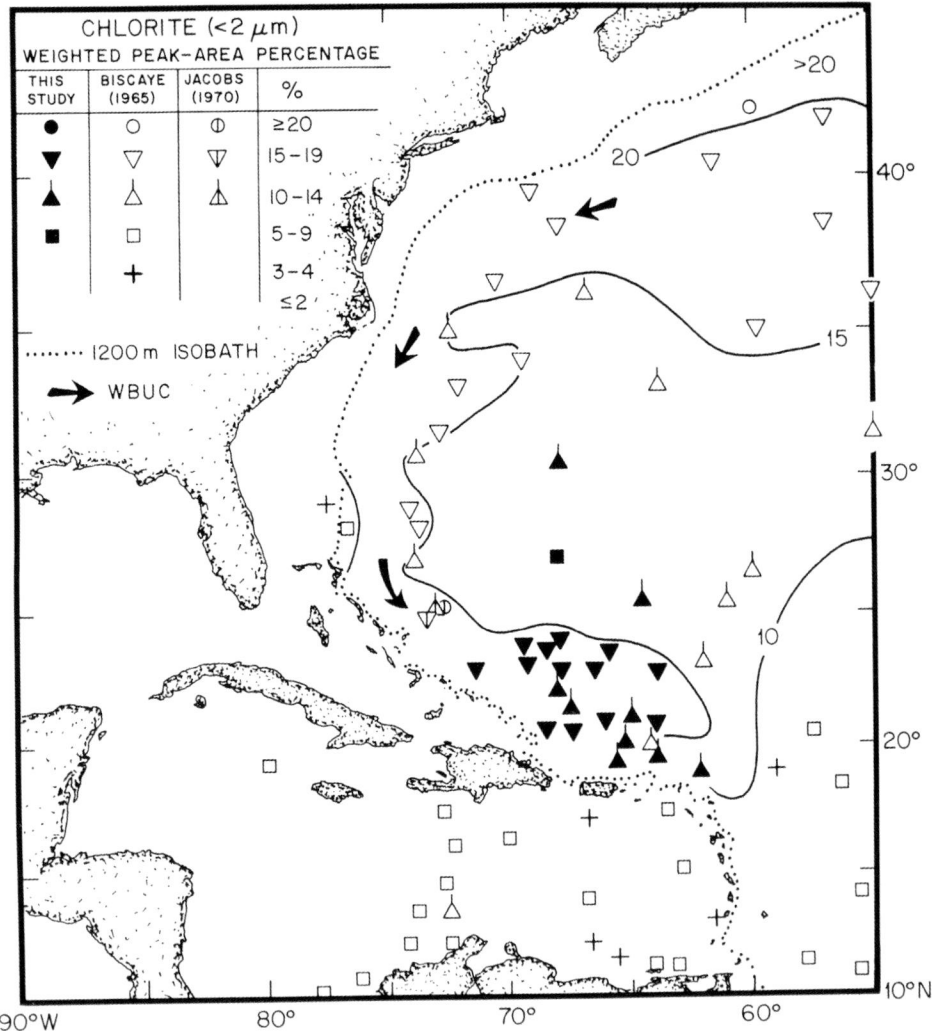

Fig. 10. Chlorite abundance in the clay-size (<2 μm) fraction of surface and near-surface sediments in the western North Atlantic, determined from X-ray diffraction (from Tucholke 1975). Compositions (weighted peak-area percentage) were determined using peak areas and weighting factors of Biscaye (1965). The chlorite-enriched tongue extending southward along the continental margin of North America and to the Greater Antilles Outer Ridge is interpreted to indicate long-distance sediment transport by the Western Boundary Undercurrent (WBUC).

& Eittreim 1974), are up to 30 μg l^{-1} over the western part of the trench and decrease down current to the east. A weak flow of AABW appears to circulate westward into the trench and along its north wall (south wall of the eastern outer ridge). This flow is accompanied by a poorly developed bottom nepheloid layer (Tucholke & Eittreim 1974), and it is not likely to be a significant source of sediment for the outer ridge.

Mean grain size of suspended matter sampled over the GAOR as determined in scanning electron microscope (SEM) samples is in the order of 3–4 μm, slightly coarser than that of the underlying outer-ridge sediments. However, the SEM technique probably underestimates the very fine-grained component (<0.2 μm), so this apparent difference may be artificial (Tucholke 1975).

Composition of the <2 μm fraction is more than 90% clay mineral platelets, with feldspar, quartz, and fragments of calcareous nannoplankton and biogenic silica constituting the remainder. The same components make up the silt-size fraction, but only 30–50% of this fraction is layered silicates. X-ray diffraction of bulk suspended-matter samples shows that chlorite is enriched in the samples (about 2 parts in 10), much as is observed in the underlying sediment. Unlike the seafloor sediment, however, illite appears to be strongly enriched and montmorillonite depleted. It is possible, though not certain, that this may be an artifact caused by incomplete development of the montmorillonite mineral lattice, and thus poor X-ray detection, in the very small-volume suspended matter samples analysed (Tucholke 1974).

Seismic stratigraphy and sediment distribution

As would be expected from the fine, uniform grain size and lack of coarse beds, impedance contrasts in GAOR sediments are small. Consequently, the sediments mostly are seismically transparent in conventional reflection profiles obtained at frequencies in the range of 10's–100's Hz (Figs 2–7). Weak internal reflections locally suggest migration of sediment waves as the outer ridge was constructed (Figs 5 & 7); most of these reflection packets are similar to the layered valleys noted earlier, and they probably exist because gravity flows locally remobilized sediments and redeposited them in sorted or graded beds.

The transparent layer that forms the GAOR overlies a markedly different seismic reflection sequence that consists of flat, reflective beds (Figs 3–7). These beds are part of the 'Horizon A complex' that is widespread in the western North Atlantic and consists of upper lower to lower middle Eocene biosiliceous sediments and cherts (Horizon A^C), as well as middle to possibly upper Eocene turbidites (Horizon A^T) (Tucholke & Mountain 1979). Sediments within and below this complex are thickest beneath the western outer ridge, where they bury most of the basement topography (Figs 4–7). Beneath the outer ridge east of 67°W this sequence thins and is commonly interrupted by basement peaks (Fig. 3), and east of about 63°W it is absent except under the Nares Abyssal Plain (Fig. 2). Cores recovered from the stratified sequence at DSDP Site 28 near 66°W (Fig. 8) contain clays, silts, chalk and chert (Bader *et al.* 1970).

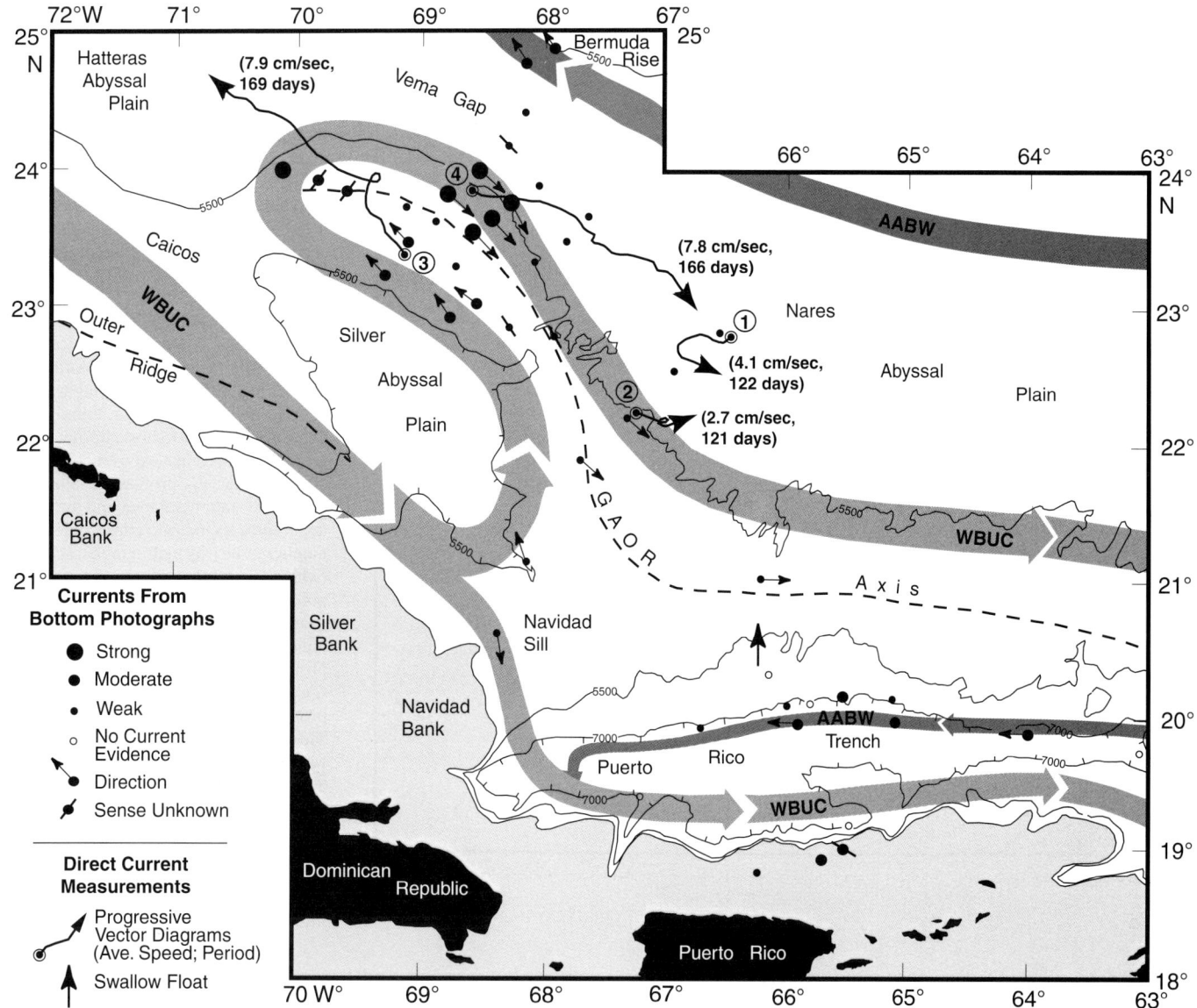

Fig. 11. Summary of abyssal current patterns in the region of the Greater Antilles Outer Ridge, based on all available data including hydrographic calculations (Tucholke *et al.* 1973; Tucholke & Eittreim 1974). Base map as in Figure 8. The shaded arrows show generalized circulation of the Western Boundary Undercurrent and of Antarctic Bottom Water entering from the South Atlantic. Current-metre measurements (circled numbers) are summarized as progressive vector diagrams with the starting point at the measurement location; durations of measurements in days and average current speed for the durations are indicated in parentheses. Measurements were made 15 m above the seafloor at locations 1 and 2, and 100 m above the seafloor at locations 3 and 4.

The transition from seismically laminated pre-Horizon A sediments to seismically transparent post-A sediments differs beneath the eastern and the western GAOR. Under the central and eastern outer ridge the change is abrupt (Figs 3, 6 & 7); over time the transparent layer accumulated more rapidly than sediments in the Nares Abyssal Plain and it has slowly prograded several tens of kilometres northward into the abyssal plain (Fig. 3). Under the northwest end of the outer ridge, deposition of laminated sediments persisted for 100–200 m above the Horizon A complex, interfingering with a small lens of transparent sediment that was probably the nascent core of this limb of the outer ridge (Fig. 4). As the ridge here was constructed, the zone of interfingering expanded outward in all directions from the ridge core, and the ridge has prograded more than 100 km to the northwest (Fig. 5).

Isopachs of sediment thickness above the Horizon A complex (Fig. 13) clearly reflect the physiographic form of the GAOR. Maximum sediment thickness is up to 0.9–1.0 km beneath the axis of the western outer ridge but only about 0.5 km beneath the eastern outer ridge. The adjacent Nares Abyssal Plain, which accumulates sediments from very distal turbidity currents passing through Vema Gap, has an average thickness of only 0.3–0.4 km of post-Eocene sediments. The Silver Abyssal Plain west of the outer ridge has slightly thicker post-Eocene sediments because it is proximal to a sediment source in the southeast Bahama Banks. At the extreme northwest end of the GAOR, isopachs show that the outer-ridge trend of thickened sediments extends into stratified sediments at the southern end of Hatteras Abyssal Plain (Fig. 13). This pattern suggests that sediment accumulation there is not entirely from turbidity currents but is enhanced by an additional component of current-controlled deposition. Because the seafloor in this area is flat, a slight regional depression of underlying basement is required to accommodate the thickening.

The reflective, flat-lying character of sediments within and below the Horizon A complex indicates that they were deposited from downslope gravity flows, and the source area is thought to be the northeastern Antilles arc (Tucholke & Ewing 1974). The abrupt upward change from these reflective sediments to seismically transparent sediments beneath the eastern GAOR is interpreted to coincide with middle to late Eocene initiation of the

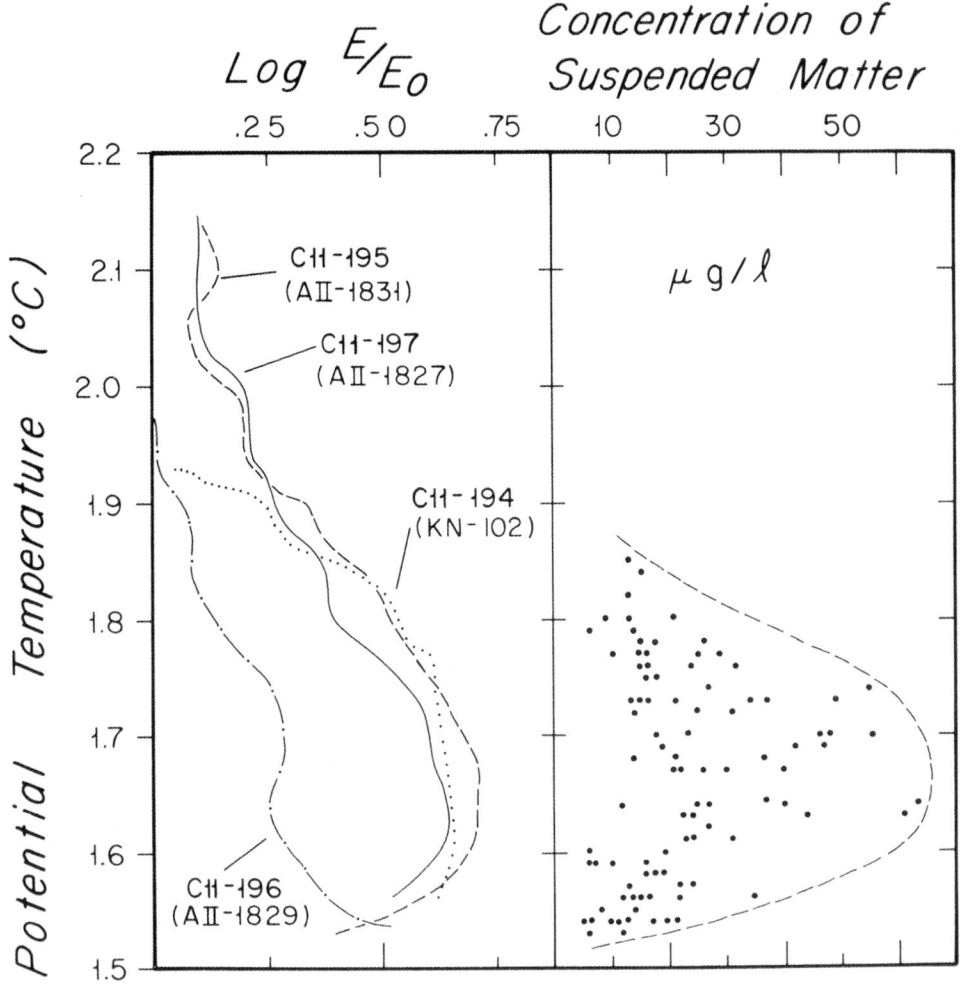

Fig. 12. Summary of light-scattering profiles made by the Lamont-Doherty nephelometer (left) and concentrations of suspended particulate matter measured by filtration of large-volume Niskin bottle samples (right) in the bottom nepheloid layer over the Greater Antilles Outer Ridge (from Tucholke 1975). Sampling locations are given in Figure 8. Data are plotted against potential temperature; temperatures below 1.8°C are Antarctic Bottom Water and the overlying water is North Atlantic Deep Water. Film exposure, E, in the light-scattering profiles is normalized against exposure in the clearest water, E_O; profiles are labeled by cruise and station number, and the labels in parentheses indicate cruise and station number of nearby hydrographic stations (within 22 km) that were used to provide potential temperature data. Note the robust near-bottom nepheloid layer indicated by three of the light-scattering profiles and by the suspended matter samples. The C11-196 light-scattering profile is from the center of Vema Gap. This profile shows a depletion of suspended matter in the bottom nepheloid layer; it falls outside the core of both the WBUC flow around the Greater Antilles Outer Ridge and the westerly AABW flow in the northern part of Vema Gap.

trough that subsequently developed into the present Puerto Rico Trench. Since that time the north slope of the trench and south slope of the eastern outer ridge have been isolated from any significant sediment source, and an average of less than 200 m of sediment has accumulated there over the past c. 40 Ma (Fig. 13).

Development of the Greater Antilles Outer Ridge

Current-controlled deposition that formed the core of the GAOR first appears above the Horizon A complex and thus post-dates the middle to late Eocene. Mountain & Miller (1992) found evidence on the southern Bermuda Rise that a pulse of strong, southern-source deep circulation occurred in late Paleocene time, but if these currents were active in the area of the GAOR their geological record appears to have been overwhelmed by the downslope sedimentation. A major oceanographic shift in the North Atlantic to a regime wherein sedimentation was influenced or controlled by deep circulation began in late Eocene to early Oligocene time (Tucholke & Mountain 1979). The onset of this regime is recorded in a widespread unconformity (Horizon A^U) that was eroded by abyssal currents along the continental margin of eastern North America and the base of the Bahama Banks. The unconformity can be observed truncating beds of the Horizon A complex under Caicos Outer Ridge (Fig. 4). Although there is no direct stratigraphic control on the age of the oldest GAOR sediments, it is reasonable to infer that the core of the outer ridge contains sediments eroded from the continental margin during this event.

Current-controlled deposition that initiated the GAOR probably was focussed beneath the eastern outer ridge by the crustal outer-high north of the Puerto Rico Trench, which diverted part of the WBUC into a flow around its northern margin (Fig. 14a). Interaction of this current with the numerous basement peaks protruding through the Horizon A complex most likely triggered deposition of local drifts, much like the isolated drift flanking the seamount in the southern Silver Abyssal Plain (Fig. 6) and the drifts presently on the far-eastern outer ridge (Fig. 2). As these drifts grew and merged, the increasing physiographic expression of the ridge would have forced the abyssal currents into a more organized system, with a distinct branch of the flow circulating around the drift (Fig. 14b). Along the Antilles margin south of the trench and in the area of the present northwestern outer ridge, downslope sedimentation characterized by seismically chaotic or laminated deposits dominated over current-controlled deposition.

The northwestward growth of the GAOR into the area of the present Hatteras Abyssal Plain is roughly constrained by interpretation of the seismic stratigraphy to have occurred by Miocene time (Fig. 14c) (Tucholke & Ewing 1974). Abyssal current intensity along the North American margin appears to have decreased during the Oligocene so that deposition rather than erosion began to predominate, and by early to middle Miocene time the cores of large sediment drifts such as the Blake Outer Ridge and Chesapeake Drift were forming there (Mountain & Tucholke 1985). Growth of the northwestern GAOR may be coincident with the initiation of these drifts.

The factors that caused this limb of the outer ridge to extend to the northwest remain unclear. One possibility is that diversion of the WBUC around the western end of the existing ridge core led

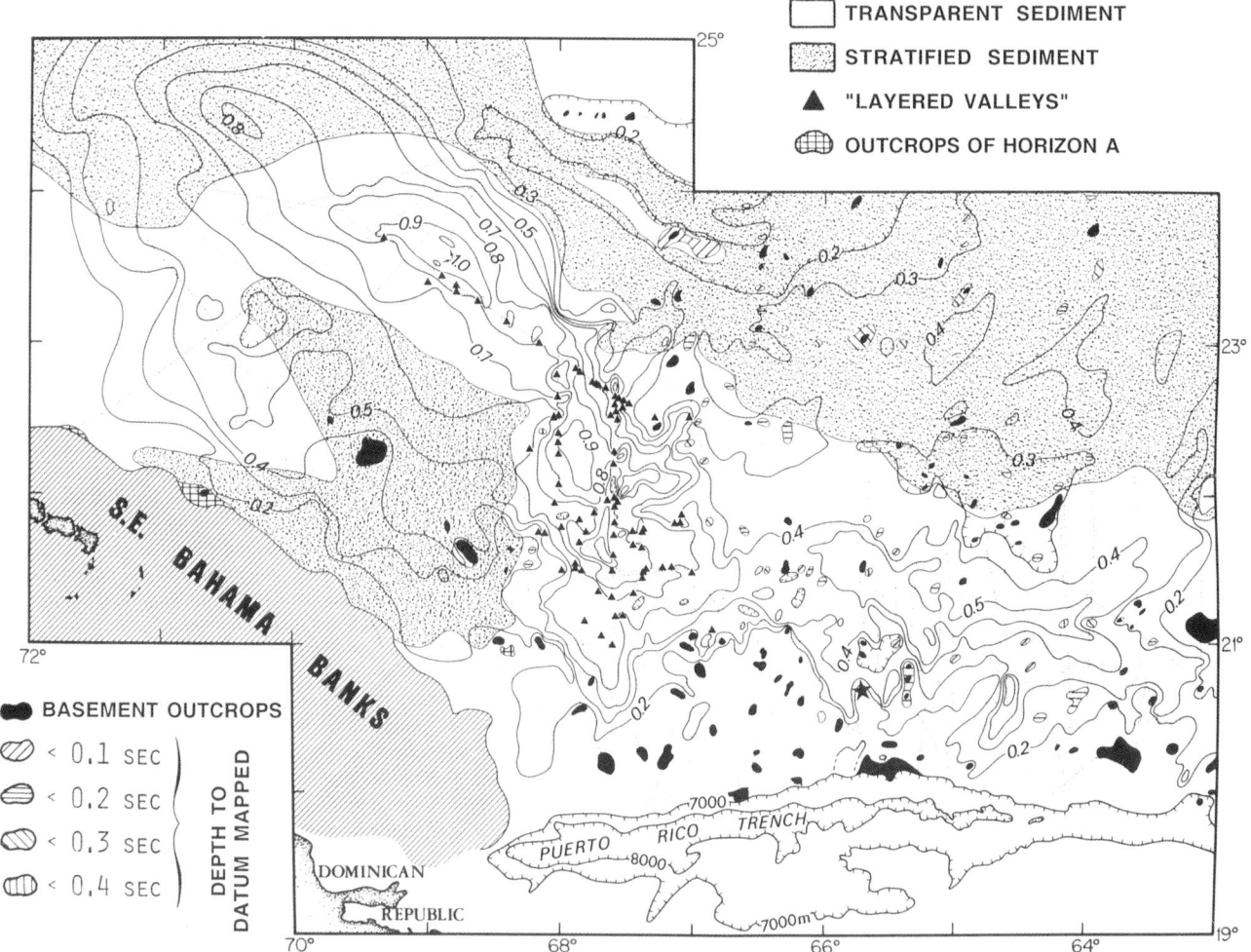

Fig. 13. Mapped sediment thickness above the stratified Horizon A complex (contours in seconds two-way travel time) and above basement peaks that interrupt Horizon A (see code at lower left). Control provided by seismic reflection profiles is shown by light dotted lines. The star shows the location of DSDP Site 28. Shaded areas show where sediments in the mapped interval are seismically stratified; these are mostly restricted to the abyssal plains. The southern limit of mapped thicknesses is the 7000 m contour along the north wall of the Puerto Rico Trench. The Horizon A complex crops out in three small areas at the base of the Bahama Banks and at the southern end of the Silver Abyssal Plain (crosshatch pattern) where abyssal currents eroded the complex and have prevented deposition since early Oligocene time. Note that sediment thickness patterns correlate closely with the physiographic form of the Greater Antilles Outer Ridge. Layered valleys within the transparent layer are restricted entirely to the western outer ridge. Figure from Tucholke & Ewing (1974).

to a zone of current shear between northwest- and southeast-directed currents (Fig. 14b, c), and rapid deposition of suspended load beneath the shear zone formed the northwestern outer ridge (Fig. 15). In an alternate scenario, depositional patterns may have been affected by interaction of the WBUC and AABW entering the area from the southeast. Ice volume increased substantially in the middle Miocene (e.g. Miller *et al.* 1991), and it may have led to increased flux of AABW into the western North Atlantic. Amplified westward flow of AABW through Vema Gap would have opposed the eastward-flowing WBUC, and it could either have diverted the WBUC or created a shear zone against this current that triggered deposition to form the northwestern GAOR.

Discussion and conclusions

Most sedimentary drifts in continental-margin settings have direct or nearly direct downslope sources of sediment; bottom currents have reworked and transported these sediments to varying degrees, depending primarily on grain-size distribution and current intensity. The resulting drifts are complex aggregations of beds with mixed signatures of cross-slope and along-slope sedimentary processes. In contrast, the Greater Antilles Outer Ridge has been deposited entirely from suspended load carried in abyssal currents. Thus its characteristics can provide important insights into purely current-controlled sedimentary processes over long time and spatial scales.

The enrichment of chlorite in GAOR sediments (Fig. 10) provides strong evidence that a significant fraction of the sediment has been transported for great distances, up to 2000 km or more, from the North American continental margin north of Cape Hatteras. Hollister (1967) used the reddish colour of sediments derived from Permian–Carboniferous red beds in the Canadian Maritime Provinces as a tracer of northern-source sediments carried by bottom currents along the US continental rise, and the reddish brown and reddish gray clays on the GAOR (Fig. 9) may also reflect a continuation of this dispersal pattern. In the dominant, non-biogenic portion of GAOR sediment, grain size averages more than 80% clay (<2 μm), with the remainder consisting of very fine silt. These sediments were sampled only to 10–20 sub-bottom in piston cores but their seismic signature is the same as deeper, seismically transparent sediments that form the Oligocene(?)-Miocene core of the outer ridge, and it is likely that

A. LATE EOCENE TO EARLY OLIGOCENE

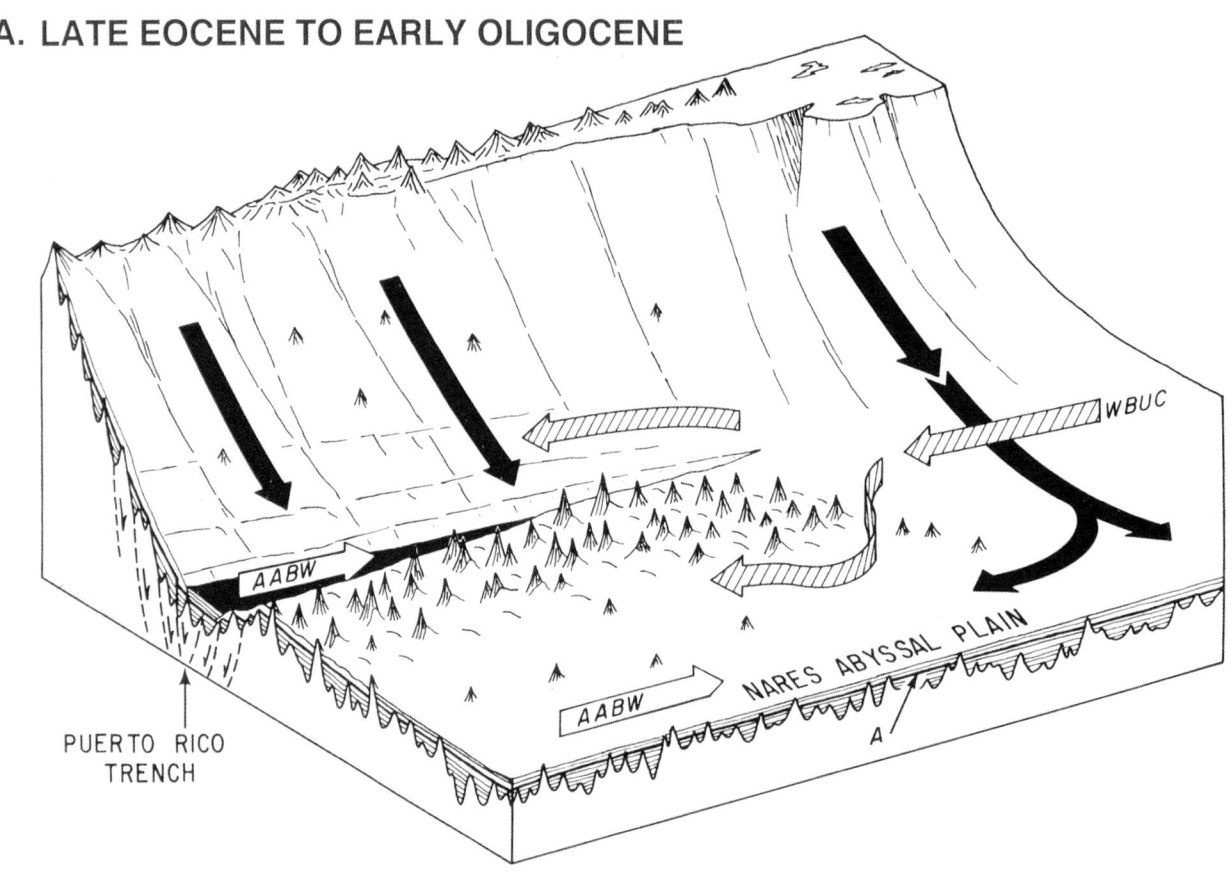

B. LATE OLIGOCENE

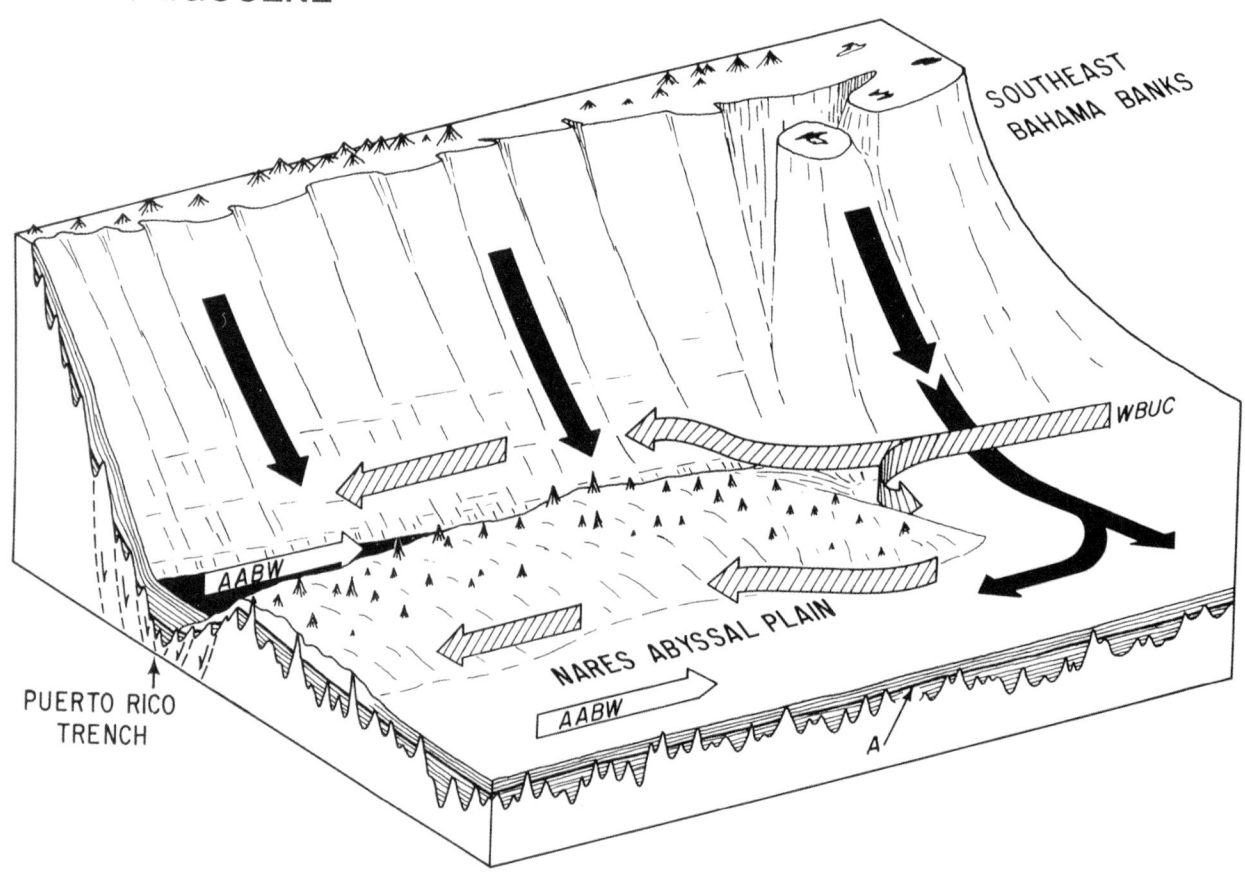

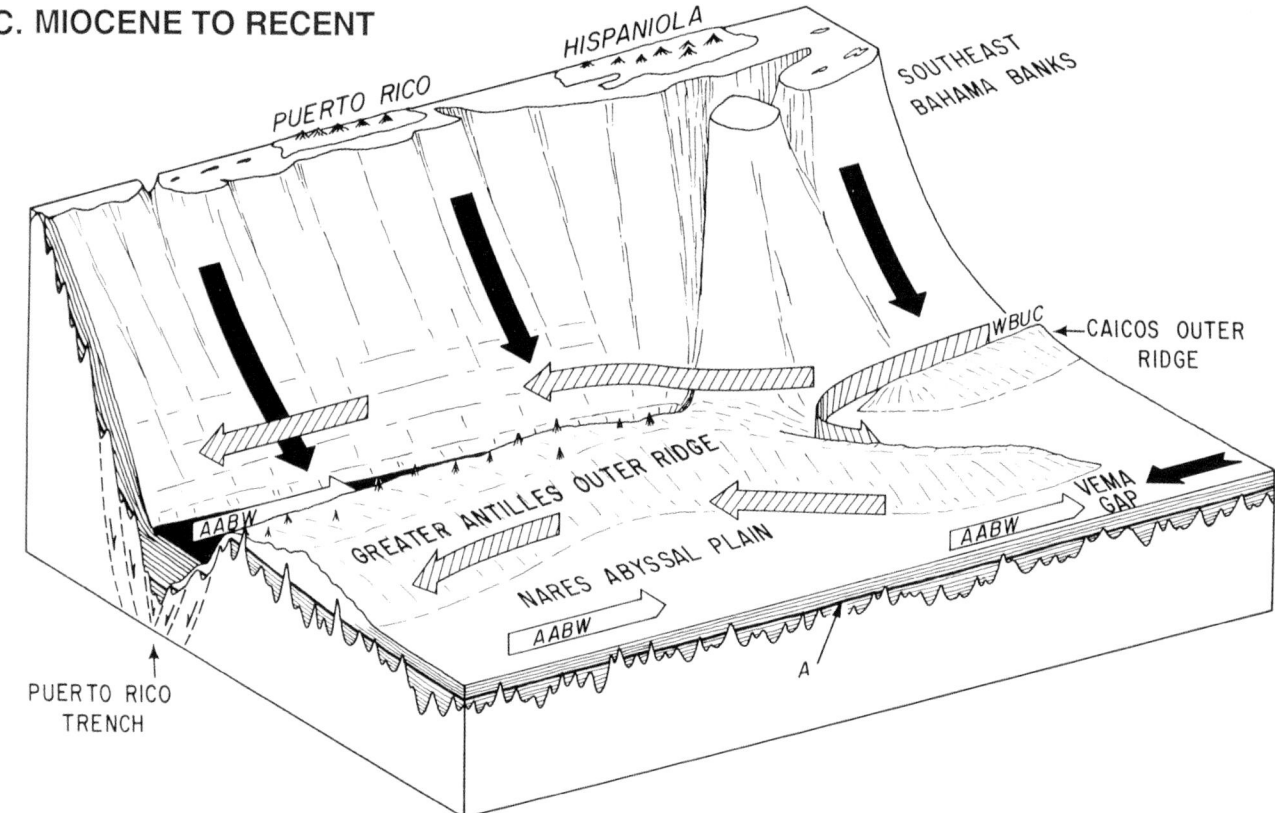

Fig. 14. Schematic summary interpreting the evolution of the Greater Antilles Outer Ridge (view to the southwest, covering the approximate area of Figure 1). (**a**) Late Eocene to early Oligocene: The developing Puerto Rico Trench cut off downslope sedimentation (black arrows) from the Greater Antilles and created a crustal outer-high that diverted part of the newly developed Western Boundary Undercurrent. Current-controlled deposition along the irregular topography of this bulge formed the core of the eastern outer ridge. Downslope gravity flows from the Bahama Banks continued to deposit stratified sediments in the area of the western outer ridge and on the surrounding abyssal plains. (**b**) Late Oligocene: Growth of the GAOR may have diverted WBUC flow enough that a shear zone developed between opposing currents at the western end of the ridge, stimulating ridge growth northwestward into the abyssal plain. (**c**) Miocene to Recent: By this time the western extension of the GAOR was well established, and ridge-crest sediments accumulated rapidly between opposing flows that followed the ridge flanks. The Caicos Outer Ridge also was deposited along the margin of the Bahama Banks, blocking downslope sediment supply to the Hatteras and Nares abyssal plains and diverting it to the Silver Abyssal Plain. Figure adapted from Tucholke & Ewing (1974).

most of the ridge has similar grain-size characteristics. From all these considerations, we infer that on a time scale of at least c. 20 Ma. the Western Boundary Undercurrent has been competent to transport very fine silt and clay-size sediments for thousands of kilometres. Composition and grain-size distribution in modern particulate-matter samples from bottom water over the GAOR are very similar to those in the seafloor sediments, so present-day dynamic conditions of sedimentation appear to be similar to average conditions over the longer-term construction of the outer ridge.

From observations in bottom photographs, current-produced bedforms seem to be neither well nor widely developed in GAOR sediments. Because of their fine grain size the sediments have high cohesion once deposited, and they are not easily eroded by ambient currents that have maximum speeds of only 15–17 cm s^{-1}. The lack of significant bedload transport probably hinders the production of pronounced bedforms. The only distinct lenses of coarse (silt-size) sediment known on the outer ridge occur in 'layered valleys' between sediment waves. These appear to be deposited from small turbidity currents generated by failure of rapidly deposited sediments on the adjacent sediment waves. Thus, unlike the conditions at sedimentary drifts along continental margins, sediments are moved in gravity flows as a result of, rather than as a prelude to, current-controlled sedimentation. The large-scale abyssal sediment waves on the GAOR are widespread but they do not seem often to develop in regular wave trains. However, this observation is not robust because there is relatively sparse echosounding and seismic reflection coverage, and denser data sets (e.g. multibeam bathymetry) could well reveal extensive development of coherent sediment waves. In some places sediment waves clearly do show both regular patterns and a long history of growth, dating well back into the Miocene (Fig. 7).

Abyssal current directions around the GAOR closely parallel bathymetric contours (Fig. 11), within certain limits. Direct current measurements over periods of months reflect the significant influence that even small topographic gradients have in organizing the flow. Currents along the flanks of the GAOR at locations 2–4 (Fig. 11), for example, follow the local contours (with short-term excursions) even though the seafloor slope is only c. 0.5°–1.0°. At location 1 on the Nares Abyssal Plain, however, the seafloor slope is less than 0.05° and there is no consistent direction in measured currents. Current directions determined from bedforms, tool marks, and other sedimentary features in bottom photographs are the most consistently contour-parallel. As already noted, current erosion or molding of the cohesive outer ridge sediments is difficult, so short-term flow variability is unlikely to modify the bed significantly. The bedforms probably represent average flow over periods of at least months to perhaps many years, and they come closer to characterizing flow that has long-term geological effects than do direct current measurements.

In the area of the Greater Antilles Outer Ridge, the primary controls on position and growth of sedimentary drifts have been current interaction with varying seafloor topography (which controls current vectors) and suspended particulate load.

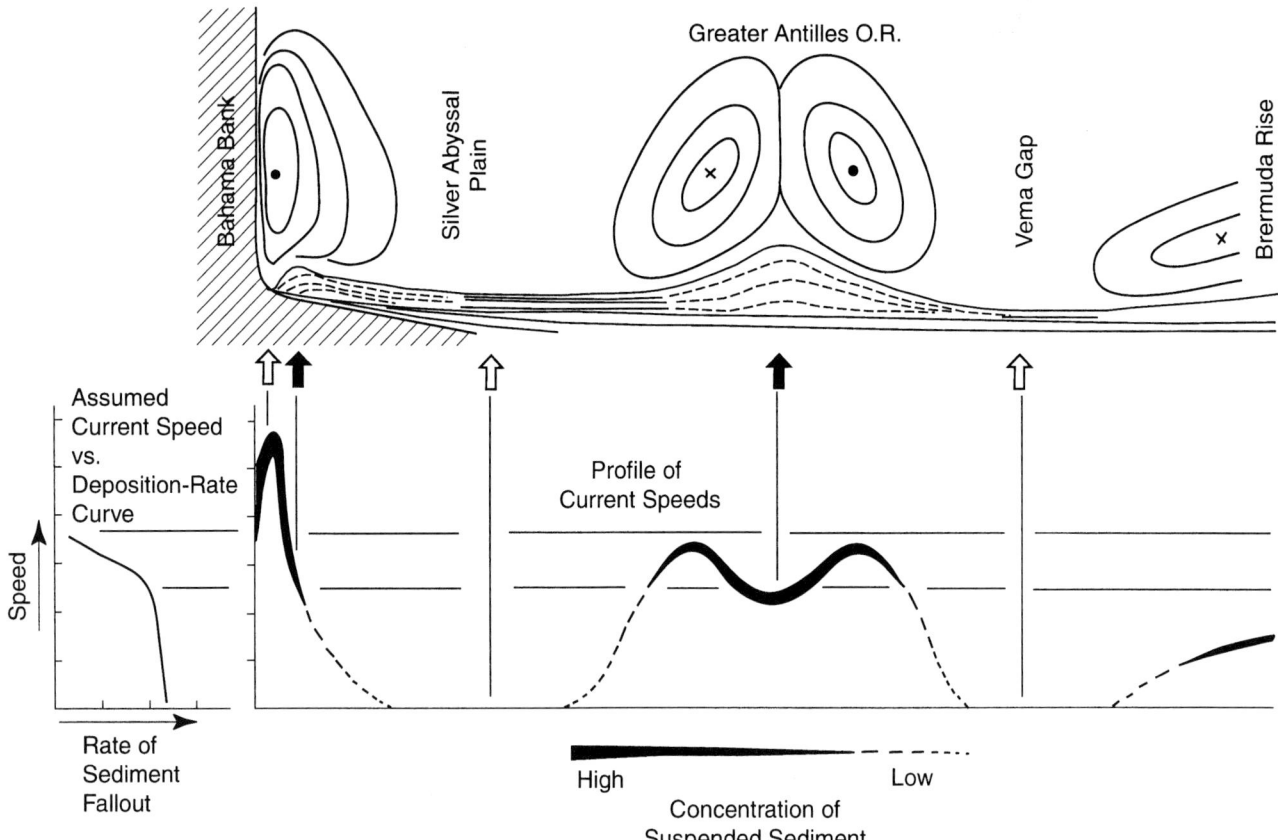

Fig. 15. Simple model for current-controlled deposition along a section extending northeast from the Bahama Banks to the southernmost Bermuda Rise, adapted from Tucholke & Ewing (1974) (compare with the seismic section in Figure 4). The top panel shows idealized flow of the WBUC along the Bahama Banks and around the Greater Antilles Outer Ridge, and flow of AABW in the northern part of Vema Gap (dots represent flows out of the page, X's are flows into the page). At bottom right is schematic current speed with superimposed concentrations of suspended matter. The current-speed versus deposition-rate curve at lower left assumes rapid sediment fallout over a small speed range, but the shape of the curve is not critical. Rapid sediment deposition from currents requires both significant suspended load and low current speeds. Thus, although suspended-matter concentration is high in the WBUC close to the Bahama Banks, high current speeds prevent deposition at the base of the banks (open arrow) while decreasing speeds away from the banks allow deposition on Caicos Outer Ridge (filled arrow). Over the GAOR, a combination of high suspended load and lower current speeds in the shear zone between opposing flows allows rapid deposition on the ridge crest (filled arrow); higher speeds with approximately the same suspended load attenuate deposition on the ridge flanks. Deposition from abyssal currents is minimal over the Silver Abyssal Plain and in Vema Gap where both current speeds and suspended load are low (open arrows).

Interaction of the WBUC with irregular topography formed the initial core of the GAOR on the outer high north of the Puerto Rico Trench probably in the Oligocene (Fig. 14). The deceleration and interactions of threads of currents over this confused topography most likely initiated deposition from a substantial suspended load. As the ridge grew, opposing NW- and SE-directed currents developed on its western end; lower current speeds in the shear zone between these flows allowed rapid deposition of suspended load to form the northwestern limb of the outer ridge as it exists today (Figs 14c & 15). To the southwest, in the WBUC along the Bahama Banks, suspended load may be high but currents are topographically intensified against the steep banks, so an erosional/non-depositional zone exists at the base of the banks inboard of the Caicos Outer Ridge. Shear zones with low current speeds between opposing flows exist over the Silver Abyssal Plain and in Vema Gap, but no coherent drift deposits have accumulated in these locations. This is probably explained by the fact that the shear zones are not stabilized by topography and thus are migratory, and by the absence of significant particulate suspended load away from the cores of the main currents.

My studies of current-controlled deposition in the region of the Greater Antilles Outer Ridge extend back to my PhD thesis research, undertaken under the tutelage of Charles D. Hollister. Charley was a constant source of enthusiasm, as well as personal and scientific inspiration, both in those early days and over the many intervening years of our continuing friendship. I hope that this brief synthesis reflects at least some small insight into the broad research questions of abyssal sedimentation that he inspired and promoted throughout his career. Geological and geophysical studies of the Greater Antilles Outer Ridge were funded by the US National Science Foundation and the Office of Naval Research. This overview was prepared with support from the Henry Bryant Bigelow Chair in Oceanography at Woods Hole Oceanographic Institution. Woods Hole Oceanographic Institution Contribution No. 10392.

References

BADER, R. G., GERARD, R. D., BENSON, W. E. ET AL. 1970. Site 28. *In:* BADER, R. G. ET AL. *Initial Reports of the Deep Sea Drilling Project.* US Government Printing Office, Washington, DC, 125–143.

BISCAYE, P. E. 1965. Mineralogy and sedimentation of Recent deep-sea clay in the Atlantic Ocean and adjacent seas and oceans. *Geological Society of America Bulletin*, **76**, 803–832.

BISCAYE, P. E. & EITTREIM, S. L. 1974. Variations in benthic boundary layer phenomena: Nepheloid layer in the North American Basin. *In:* GIBBS, R. J. (ed.) *Suspended Solids in Water.* Plenum Press, New York, 227–260.

HATHAWAY, J. C. 1972. Regional clay mineral facies in estuaries and continental margin of the United States East Coast. *Geological Society of America Memoir*, **133**, 293–316.

HOLLISTER, C. D. 1967. *Sediment distribution and deep circulation in the western North Atlantic*. PhD thesis, Columbia University.

MCCAVE, I. N. & TUCHOLKE, B. E. 1986. Deep current-controlled sedimentation in the western North Atlantic. *In*: VOGT, P. R. & TUCHOLKE, B. E. (eds) *The Geology of North America, Volume M, The Western North Atlantic Region*. Geological Society of America, Boulder, Co., 451–468.

MILLER, K. G., WRIGHT, J. D. & FAIRBANKS, R. G. 1991. Unlocking the ice house: Oligocene-Miocene oxygen isotopes, eustasy, and margin erosion. *Journal of Geophysical Research*, **96**, 6829–6848.

MOUNTAIN, G. S. & MILLER, K. G. 1992. Seismic and geologic evidence for early Paleogene deepwater circulation in the western North Atlantic. *Paleoceanography*, **7**, 423–439.

MOUNTAIN, G. S. & TUCHOLKE, B. E. 1985. Mesozoic and Cenozoic geology of the US Atlantic continental slope and rise. *In*: POAG, C. W. (ed.) *Geologic Evolution of the United States Atlantic Margin*. Van Nostrand Reinhold, New York, 293–341.

TUCHOLKE, B. E. 1974. Determination of montmorillonite in small samples and implications for suspended-matter studies. *Journal of Sedimentary Petrology*, **44**, 254–258.

TUCHOLKE, B. E. 1975. Sediment distribution and deposition by the western Boundary Undercurrent: The Greater Antilles Outer Ridge. *Journal of Geology*, **83**, 177–207.

TUCHOLKE, B. E. & EITTREIM, S. 1974. The Western Boundary Undercurrent as a turbidity maximum over the Puerto Rico Trench. *Journal of Geophysical Research*, **79**, 4115–4118.

TUCHOLKE, B. E. & EWING, J. I. 1974. Bathymetry and sediment geometry of the Greater Antilles Outer Ridge and vicinity. *Geological Society of America Bulletin*, **85**, 1789–1802.

TUCHOLKE, B. E. & MOUNTAIN, G. S. 1979. Seismic stratigraphy, lithostratigraphy and paleosedimentation patterns in the North American Basin. *In*: TALWANI, M., HAY, W. & RYAN, W. B. F. (eds) *Deep Drilling in the Atlantic Ocean: Continental Margins and Paleoenvironment*. Maurice Ewing Series **3**, American Geophysical Union, Washington, DC, 58–86.

TUCHOLKE, B. E., WRIGHT, W. R. & HOLLISTER, C. D. 1973. Abyssal circulation over the Greater Antilles Outer Ridge. *Deep-Sea Research*, **20**, 973–995.

WRIGHT, W. R. & WORTHINGTON, L. V. 1970. *The water masses of the North Atlantic Ocean: A volumetric census of temperature and salinity*. Serial Atlas of the Marine Environment Folio **19**. American Geographical Society, New York.

The Lofoten Drift, Norwegian Sea

JAN SVERRE LABERG[1], TORE O. VORREN[1] & STIG-MORTEN KNUTSEN[2]

[1]*Department of Geology, University of Tromsø, N-9037 Tromsø, Norway*
[2]*Norsk Hydro a.s. PO Box 31, N-9401 Harstad, Norway*

Abstract: The Lofoten Contourite Drift is located below *c.* 1000 m water depth on the continental slope off northern Norway. It has a maximum thickness of about 360 m and correlation to published seismic stratigraphy implies a Neogene age. The drift probably originated from the deposition of suspended sediments derived from winnowing of the shelf and upper slope. The geometry of the uppermost part of the drift is characterised by maximum thickness at the mound crest, which implies that it is probably presently active. Within gravity core JM97-950/1 (1160 m water depth), the upper 86 cm represents sediments resulting from Holocene winnowing of the shelf and upper slope. The underlying sediments, to 5.5 m core depth, were deposited in a late Weichselian glacimarine environment as revealed by its content of ice-rafted debris and the dating results. Late Weichselian drift growth is consistent with a circulation system transporting surface water northeastward along the continental margin of Norway also during the Weichselian. The downslope sediment input to the study area was probably relatively low during glacial periods due to the evolution of the nearby mainland area, the Lofoten Islands. These islands may have acted as a sediment barrier causing large fluvial and/or glacial drainage systems from central Fennoscandia to be routed south and north of the study area. As a result, alongslope sediment transport has provided the main sediment input to this part of the continental slope.

In this paper, we present high-resolution seismic data (analogue sparker and 3.5 kHz records), a multi-channel seismic line together with a 5.5 m long gravity core from the continental margin off northern Norway (Fig. 1) revealing a contourite drift, the Lofoten Drift (Table 1). Our aim is to discuss the origin, age and growth of the Lofoten Drift. This contribution is a further development of Laberg *et al.* (1999).

The area of study is located in the northeastern Norwegian–Greenland Sea, where the water depth varies from about 300 m at the shelf break to more than 3000 m in the Lofoten Basin (Figs 1 and 2). The continental margin in the study area is characterised by a relatively steep slope gradient (shelf break to *c.* 1100 m water

Table 1. *Principal characteristics of the Lofoten Drift*

Location:	Norwegian Sea
Setting:	lower continental slope (from *c.* 1000 m water depth)
Age:	Miocene – Recent
Drift type:	mounded, elongate drift (*sensu* Stoker *et al.* 1998) mounded, giant elongated, separated drift (*sensu* Faugères *et al.* 1999)
Dimensions:	360 m thick, at least 90 km long (along drift axis)
Seismic facies:	layered, continuous, parallel or slightly divergent reflections of medium amplitude
Sediment facies:	sandy mud, partly laminated and comprising clasts both randomly distributed and concentrated within cm thick intervals

depth) of about 7.5°. Downslope, there is a well defined slope break below which the gradient is gentler, *c.* 3° (Fig. 3).

Oceanographic setting

Present day surface circulation in the eastern Norwegian–Greenland Sea is dominated by the Norwegian Current transporting Atlantic water towards the northeast (Mosby 1968). The Norwegian Current is composed of three distinct streams (two continental margin and one coastal branches) which join into one single current west of the Lofoten Islands (Fig. 1). Here the strongest measured currents are in excess of 1.1 m s^{-1} at 15 m water depth (Poulain *et al.* 1996). At intermediate water depths, the Arctic Intermediate Water intrudes from the northern Icelandic slope (possibly also further north) between the Norwegian Current and the bottom water and prohibits mixing of these two water masses (Blindheim 1990). Current meter measurements confirm the presence of a continental slope current west of the Lofoten Islands (down to 2000 m water depth) with mean values in the range 0.2–0.4 m s^{-1} and directed towards the northeast, approximately parallel to the isobaths (Heathershaw *et al.* 1998). Model results for along-slope velocity off Lofoten show an increase of nearly an order of magnitude from the slope area at 2000 m water depth up to the outer continental shelf (200 m water depth) (Heathershaw *et al.* 1998). Bottom water, below 2000 m in the Lofoten Basin enters the area from the northwest across

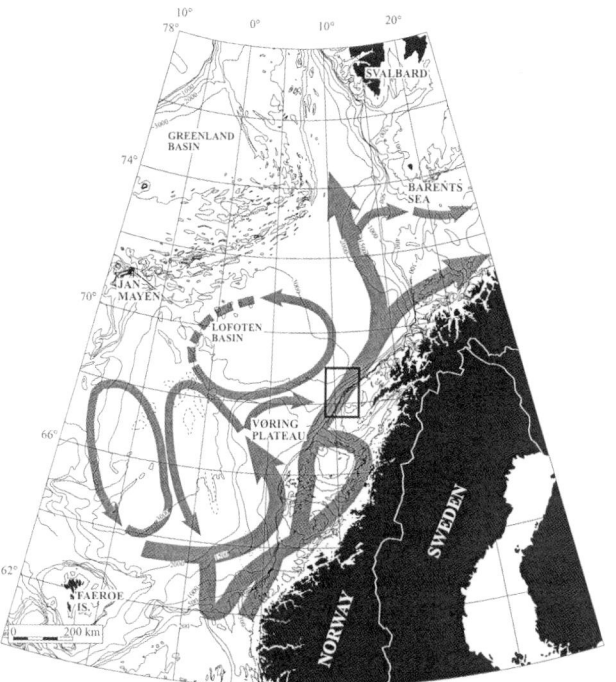

Fig. 1. Bathymetric map of the Norwegian Sea (from Perry *et al.* 1980). The arrows give the direction of the surface water circulation in the Norwegian–Greenland Sea and have been adapted from Poulain *et al.* (1996). The location of Figure 2 is shown by the frame.

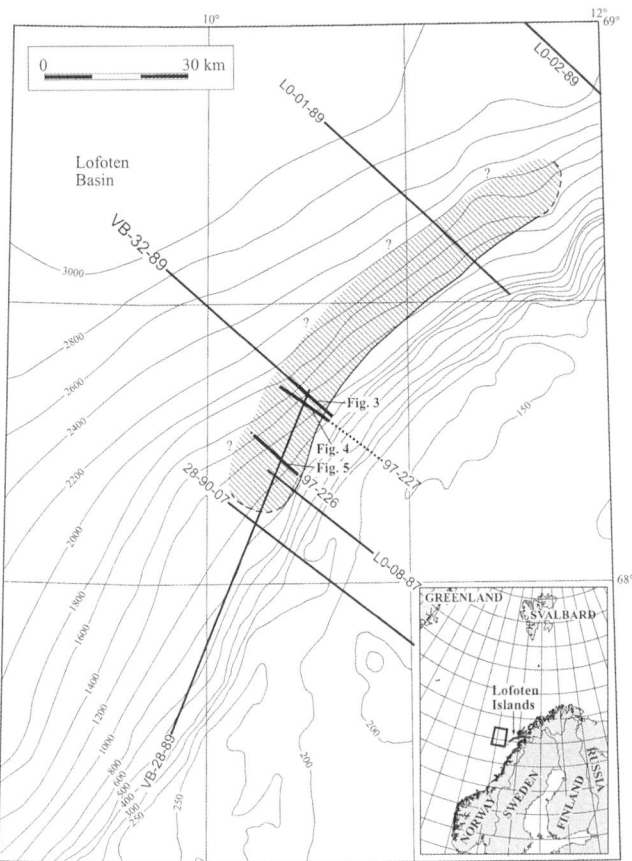

the mid-ocean ridge north of Jan Mayen (Swift & Koltermann 1988).

The Norwegian Current was established in its present form at about 9800 BP (Hald & Aspeli 1997). A comparable circulation pattern probably also existed during the last interglacial (Eemian) interval (Henrich *et al.* 1989). Until recently, the last glacial interval in the Norwegian–Greenland Sea, the Weichselian, was inferred to be characterized by a cold circulation cell (Kellogg 1980). This interpretation has now been challenged by several studies. Hebbeln *et al.* (1995) documented advection of relatively warm water from the North Atlantic Ocean up along the continental margin of Norway and into the Fram Strait in two short-term events, at 27 to 22.5 and 19.5 to 14.5 ka ago. Dokken & Hald (1996) reported evidence of six periods of sea-ice break up during isotope stages 4, 3 and 2, probably caused by inflow of North

Fig. 2. Bathymetric map of the studied part of the continental margin off Norway including the location of the Lofoten Drift (hatched). The seismic data base is indicated; multichannel seismics are given by bold lines and the Sparker/3.5 kHz profiles by dotted lines. The location of Figures 3, 4 and 5 is also shown. The bathymetry is from Perry *et al.* (1980).

Fig. 3. Segment of the downslope-oriented multichannel seismic profile VB-32-89 (10–100 Hz frequency) across the Lofoten Drift. The base drift reflection corresponds to the intra Miocene reflection of Blystad *et al.* (1995). The drift can be divided into four subunits bounded by the reflection unconformities A, B, C and D. The internal unit signature is characterised by layered, continuous, parallel or slightly divergent internal seismic reflections of medium to high amplitude. A palaeo-moat associated with reflection B and the erosional scarp on the lower part of the drift are indicated. See Figure 2 for location.

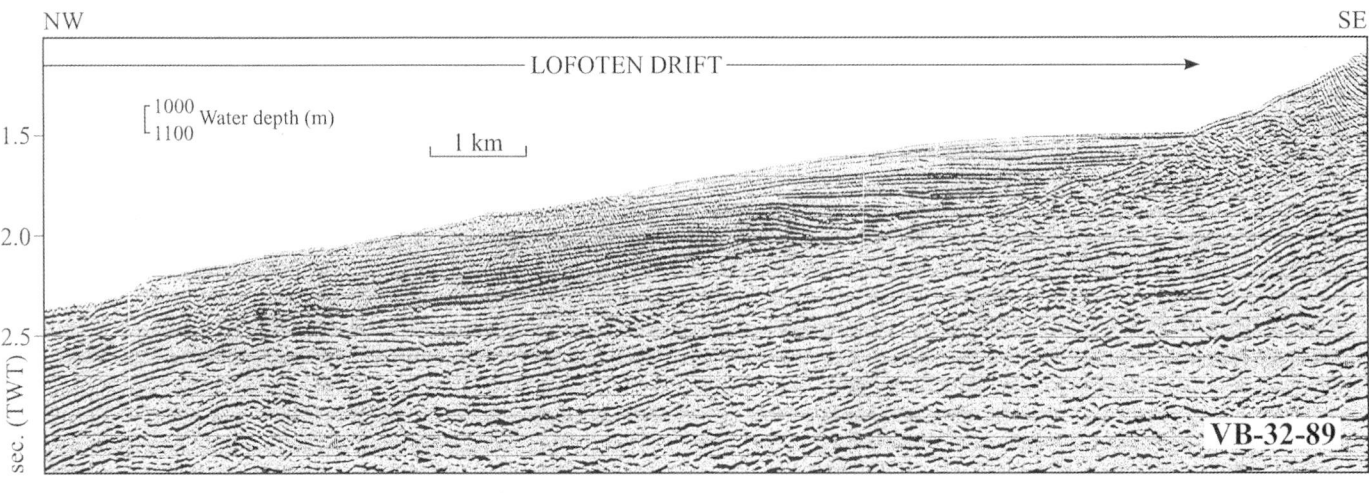

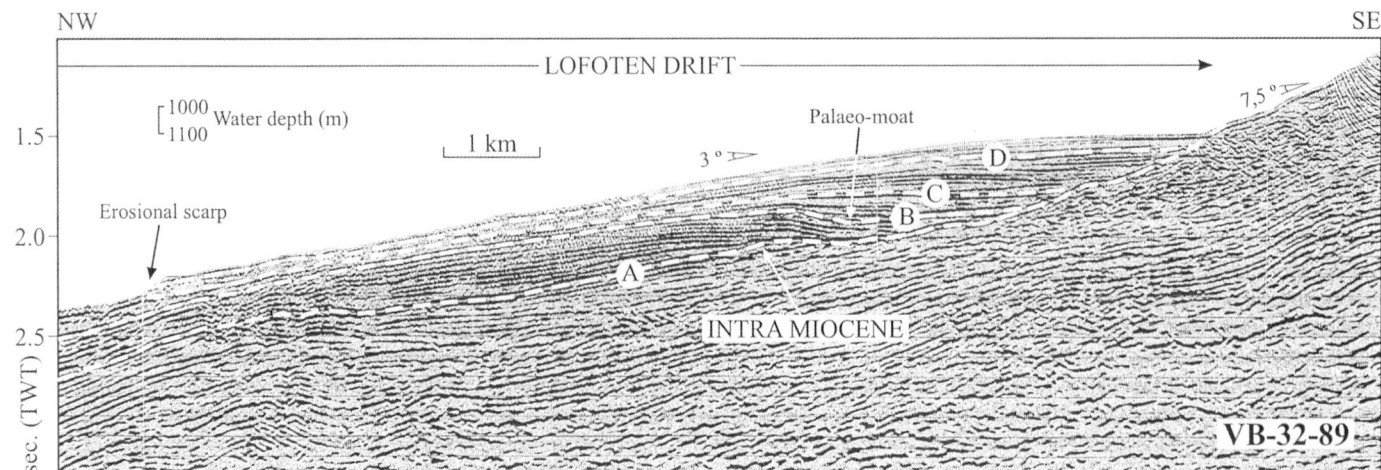

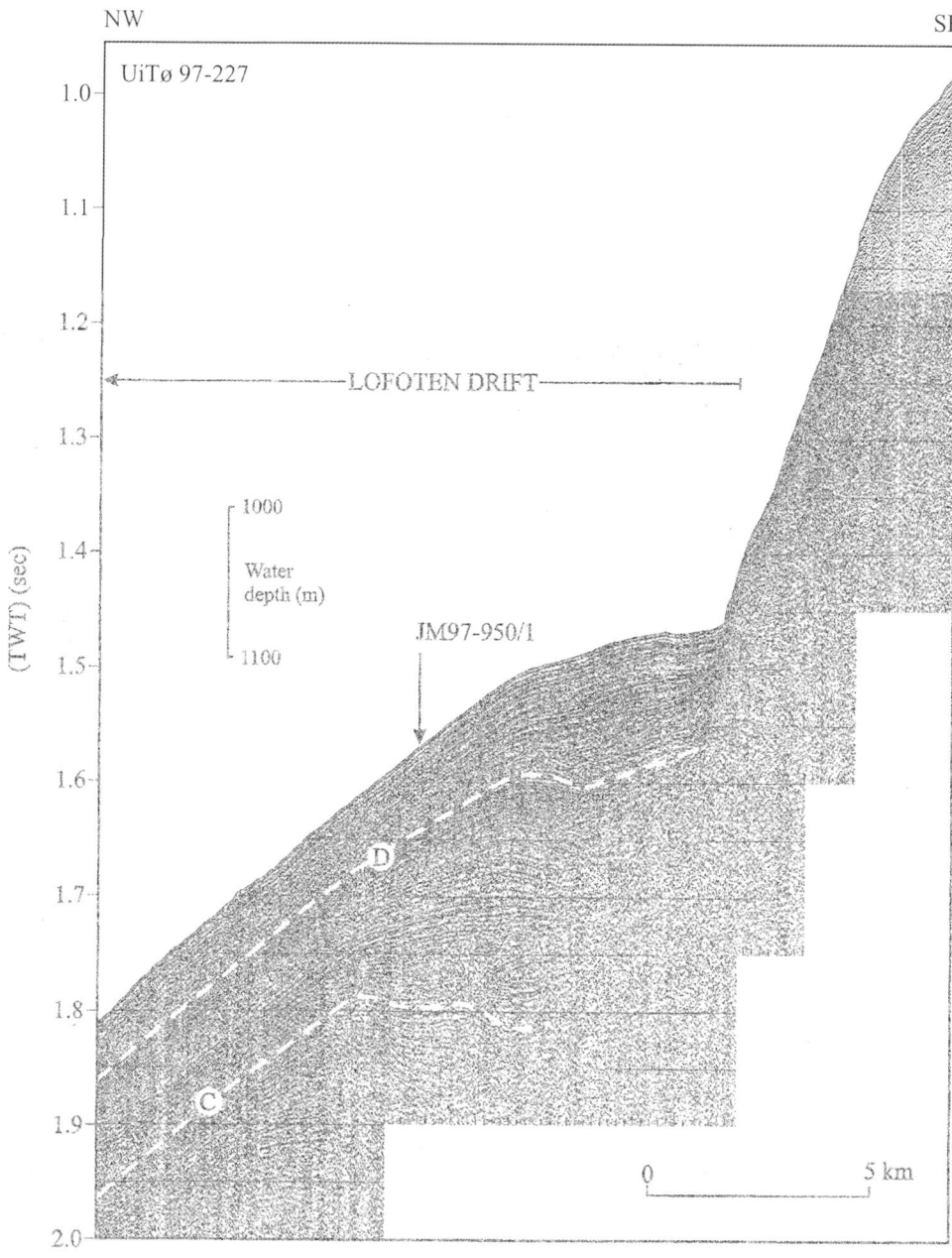

Fig. 4. Part of Sparker profile UiTø 97-227 (100–700 Hz frequency) across the Lofoten Drift illustrating the layered, continuous, parallel or slightly divergent internal seismic signature of the two youngest seismic units. For location, see Figure 2.

Atlantic surface water. Hebbeln *et al.* (1998) found that a permanent meridional current system similar to the present-day condition existed in the Norwegian–Greenland Sea through the last 200 000 years. Strong climatic variations are related to changing intensities of the major north- and southward flowing currents (Hebbeln *et al.* 1998).

Modern particle flux studies, based on results from several sediment traps deployed along the path of the Norwegian Current, reported an average total annual particle flux of 21.3 g m^{-2}. The lithogenic flux comprised 7.6 g m^{-2}. The lithogenic particles settling were dominated by fine clay, and the majority probably originated by resuspension of material from the shelf and continental slope off Norway (Honjo 1990).

Stratigraphic context

The studied part of the continental margin has a thin Cenozoic succession (Vorren *et al.* 1998). Relatively little is known about the age and origin of the sediments on the continental margin west of the Lofoten Islands. The only data available are regional seismic lines. The base-drift reflection was dated by alongslope correlation from the seismic lines published by the Norwegian Petroleum Directorate (Blystad *et al.* 1995). This correlation implies a Miocene age for the initiation of the drift (Fig. 3).

Seismic characteristics

The Lofoten Drift is located on the lower continental slope, from about 1000 m water depth, downslope from the steepest part of the continental slope (Fig. 2). At the upslope drift termination there is a well defined change in sea-floor gradient (Fig. 3), while the downslope drift boundary is more gradual and could not be defined from the present data base. The maximum drift relief is about 50 m above the main sea-floor (Fig. 3), and the drift axis is contour-parallel, oriented northeastward (Fig. 2). Alongslope, the Lofoten drift can be followed for at least 90 km based on seismic

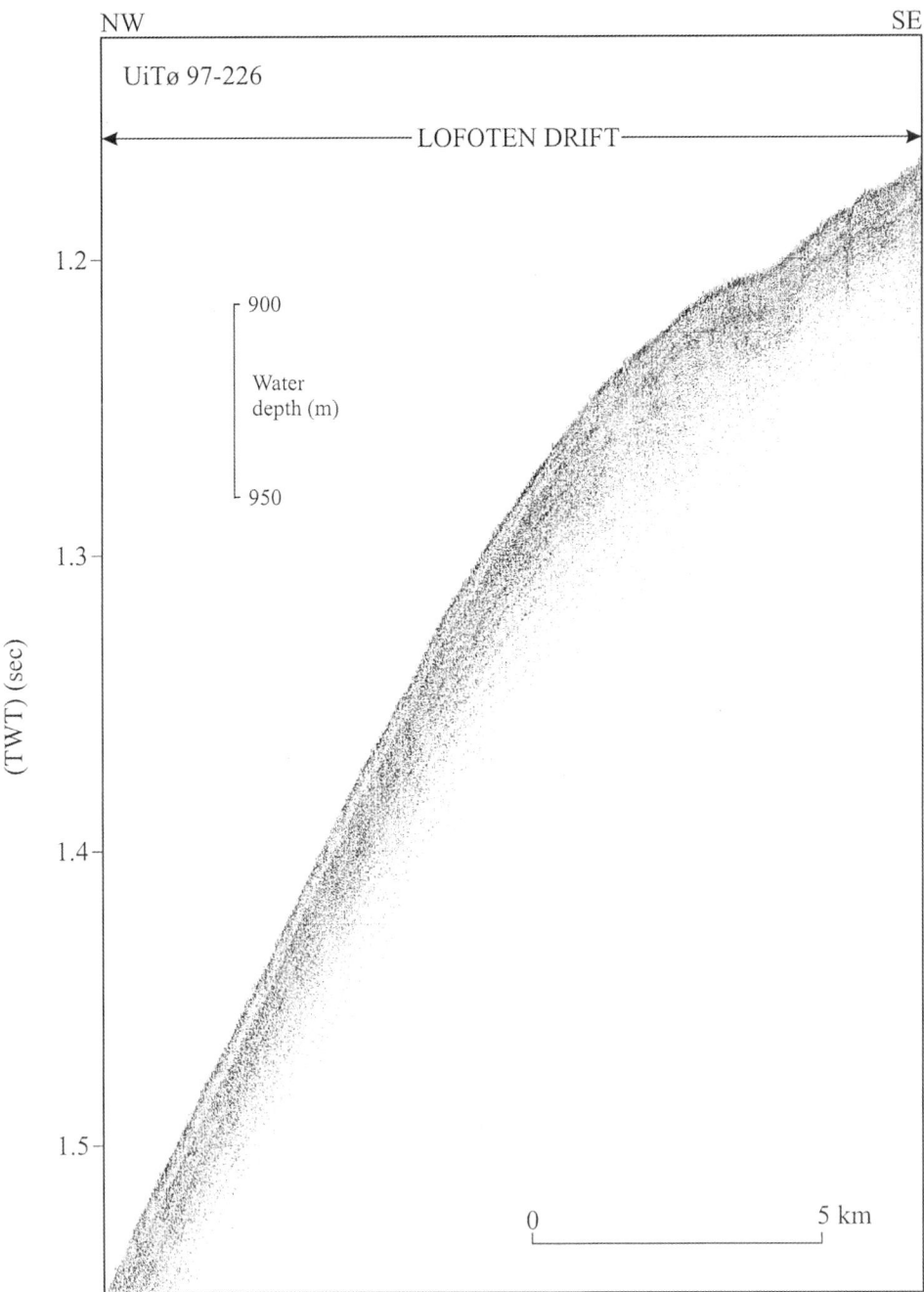

Fig. 5. 3.5 kHz profile UiTø 97-226 across the Lofoten Drift. Within the upper part of the youngest seismic unit, subunits of slightly decreasing thickness away from the mound crest are seen. For location, see Figure 2.

profiles (Fig. 2). On multichannel seismic data the base of the drift deposit is defined by a regional, high-amplitude reflection, and a maximum drift thickness of c. 400 ms (equivalent to 360 m using an average interval velocity of 1800 m s^{-1} for the drift) is inferred.

The internal seismic signature of the Lofoten Drift is characterized by layered, continuous, parallel or slightly divergent internal seismic reflections of medium amplitude, as described from comparable settings (e.g. Faugères & Stow 1993; Stoker *et al.* 1998). This reflection configuration reveals a progressive upslope accretion onto the continental slope. The drift deposit can be divided into four units bounded by high amplitude unconformities (Fig. 3). Analogue, single-channel sparker data display in more detail the layered, continuous, parallel or slightly divergent internal seismic signature of the two youngest seismic units (Fig. 4). Within the upper part of the youngest seismic unit a 3.5 kHz profile displays upper continuous subunits of slightly decreasing thickness away from the mound crest (Fig. 5).

Sediments: core description and facies

A 5.5 m long gravity core, JM97-950/1 was acquired from the upper part of the drift, at about 1160 m water depth (Fig. 4). The core comprises four main units (Fig. 6). Unit I (0–10 cm) is a dark, massive yellow-brown sandy mud. Unit II (10–86 cm) comprises light brownish grey sandy mud. Faint lamination and some bioturbation have been identified within this unit. Unit III (86–109 cm) includes grey and dark grey sandy mud layers, up to 5 cm thick, some containing mud clasts (light grey/light brownish grey) up to a few cm in diameter. Unit IV (109–554 cm) is a grey/light brownish grey sandy mud, partly laminated comprising clasts both randomly distributed and concentrated within cm thick intervals (Figs 6 & 7). The clasts are more frequent in the lower part of the unit (> 350 cm). Monosulphidic knots are relatively frequent in the upper part (< 350 cm) of the core becoming less pronounced downcore.

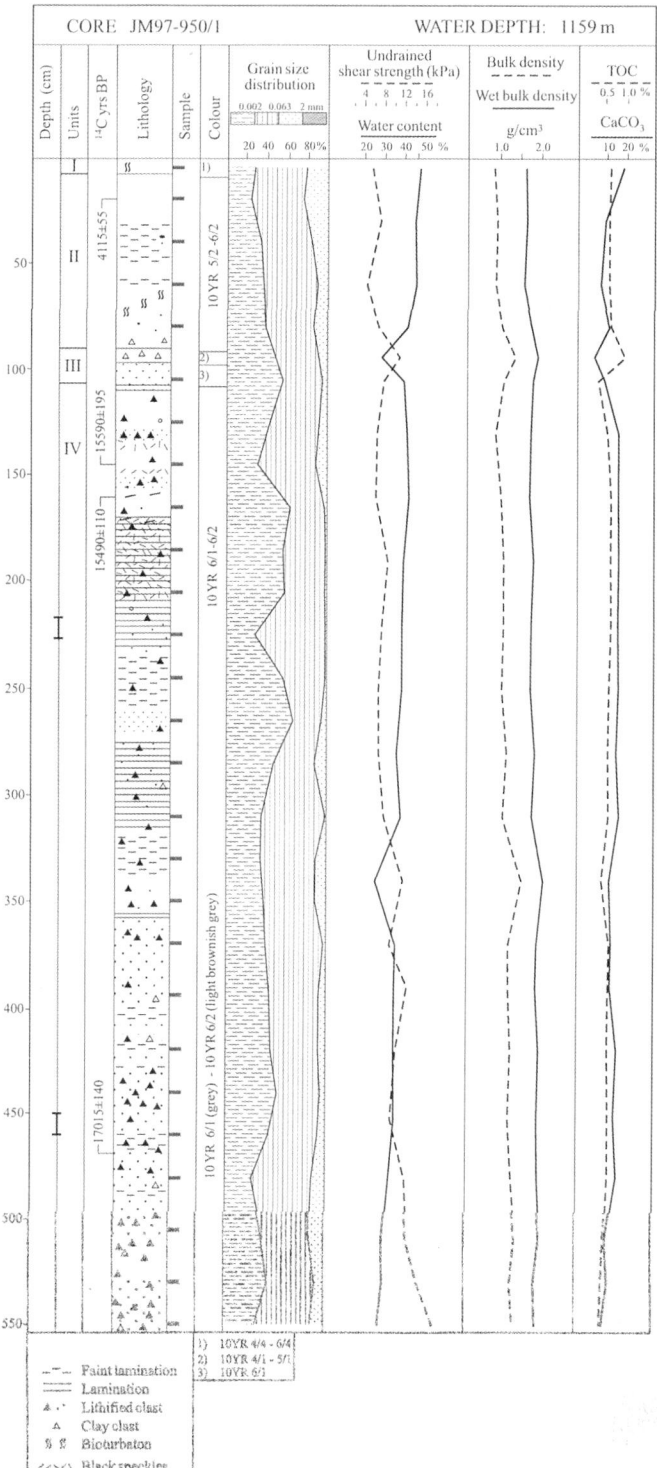

Fig. 6. Stratigraphy and lithology of gravity core JM97-950/1 (located in Figure 4) showing colour, grain size distribution, undrained shear strength, water content (% of wet weight), bulk density, wet bulk density, total organic carbon content (TOC), $CaCO_3$ content and ^{14}C AMS dating results. The location of X-radiographs (Figure 7) is indicated.

The grain size distribution, physical properties and total organic carbon/$CaCO_3$ content are relatively uniform throughout the core (Fig. 6). Grain size analyses show c. 30–50% clay, 30–50% silt and 2–15% sand. The undrained shear strength is low at the top of the core, 5 kPa, increasing gradually to 18 kPa at the base, while the water content decreases from about 47% at the top to 28% in the lowermost part of the core. The $CaCO_3$ content is relatively high within unit I, 18%, decreasing to 9% within unit II, and 6% in unit III. Unit IV is characterized by a relatively high $CaCO_3$ content in the upper part, c. 14%, decreasing to 9% at the base of the core (Fig. 6).

Four accelerator mass spectrometry (AMS) dates were provided by the Laboratories in Trondheim/Uppsala (Lab. Ref. TUa-2644–2647). All dates have been corrected for the marine reservoir age of 440 a (Mangerud & Gulliksen 1975). An age of 4115 ± 55 ^{14}C years BP was reported on *N. Pachyderma (d)* from 21–23 cm depth. From the grey/light brownish grey sandy mud, unit IV in core JM97-950/1, dates on *N. Pachyderma (s)* gave: 15 590 ± 195 ^{14}C years BP at 147–149 cm depth, 15 490 ± 110 ^{14}C years BP at 166–168 cm depth, and 17 015 ± 140 ^{14}C years BP at 461–463 cm depth. There are no signs of sediment reworking at 147 to 168 cm core depth implying that the apparent age reversal at this depth is a result of the level of precision of the dating method.

Discussion

The Lofoten Drift is located from c. 1000 m water depth, immediately downslope from the steepest part of the continental slope. The surface water circulation within the study area, established in its present form in the early Holocene (Hald & Aspeli 1997), has been found to cause winnowing of the slope sediments down to a water depth of 600–800 m (Kenyon 1986). We speculate that suspended sediments derived from the surface water winnowing process, by settling and mixing is included into the underlying Arctic Intermediate Water. Due to the lower flow speed of the Arctic Intermediate Water some of the suspended sediment is deposited. The youngest part of the contourite drift could thus have been deposited as a result of mixing of Atlantic surface water and Arctic Intermediate Water during the Holocene. Little *in situ* winnowing by the weak Arctic Intermediate Water over the drift itself is indicated by the unsorted Holocene sediments, c. 30% clay, c. 50–65% silt and c. 10–15% sand. This interpretation is also supported by the modern particle flux studies by Honjo (1990), who found that the lithogenic particles settling in the lower continental rise of the eastern Lofoten Basin probably originated by resuspension from the shelf and continental slope off Norway.

The sediment physical properties, a gradual increase in undrained shear strength and decrease in water content, show that the core comprises a continuous, normally consolidated sediment record. No periods of erosion or non-deposition have been identified. The ^{14}C AMS dating results together with the absence of ice rafted debris indicates that the upper 86 cm represents sediments derived from the Holocene winnowing process of the shelf and upper slope. Downcore, the dating results show that this part of the drift was deposited during the late Weichselian. The occurrence of sedimentary and lithified clasts within units III and IV reveals an input of ice-rafted material in a glacimarine environment An increase in clast content below 350 cm core depth may be due to more intense iceberg rafting, a lower sedimentation rate or both. Drift growth during the late Weichselian is consistent with a circulation system transporting surface water northeastward along the continental margin of Norway during the Weichselian as suggested by Hebbeln *et al.* (1998).

Contourites in the Norwegian–Greenland Sea have previously been identified from seismic data in the Fram Strait (Eiken & Hinz 1993) and short core samples from the Barents Sea continental margin (Yoon & Chough 1993). Recently a contourite drift, the Nyk Drift, was reported from the Norwegian continental margin slightly south of the Lofoten Drift (Laberg *et al.* in press). The Nyk Drift has seismic characteristics very similar to the Lofoten Drift but contrary to the Lofoten Drift it is found interbedded with Saalian and late Weichselian glacigenic debris flow sediments deposited when the Fennoscandian Ice Sheet was

JM97-950/1

216-228 cm core depth

450-462 cm core depth

Fig. 7. X-radiographs of (**a**) the laminated sandy mud at 216 to 228 cm core depth and (**b**) the sandy mud at 450 to 462 cm core depth which comprises lithified clasts both randomly distributed and concentrated within cm thick intervals.

present at the shelf break (Henriksen & Vorren 1996; Stuevold & Eldholm 1996).

However, within the study area, relatively little sediment may have been deposited by downslope processes on the continental slope during glacial periods. This may be due to the evolution of the nearby mainland area, the Lofoten Islands. During the early Cretaceous, the Lofoten area developed as a horst while the surrounding areas subsided below a several kilometres thick Lower Cretaceous succession (Løseth & Tveten 1996). A late Cenozoic compressional phase resulted in the uplift and erosion of the entire margin (67°30'N to 69°30'N), just east of the present shelf break (Løseth & Tveten 1996). The combination of these events may have caused the Lofoten land area to act as a sediment barrier. The probable consequence was that the large fluvial and subsequent glacial drainage systems from central Fennoscandia were routed south and north of the study area, i.e. into the southern Barents Sea and to the area off mid Norway. As a result, local terrestrial sediment input to the studied part of the continental slope has probably been restricted due to a small source area (Fig. 8).

This is a contribution to the ENAM II Program (European North Atlantic Margins – Quantification and modelling of large scale sedimentary processes and fluxes, 1996–1999) and financial support from EC MAST III project MAS3-CT95-0003 is gratefully acknowledged. The Norwegian Petroleum Directorate kindly provided the multi-channel seismic data base. The high-resolution seismic data and the gravity core were collected by the University of Tromsø and we gratefully acknowledge the contribution of the crew of RV Jan Mayen. We would also like to thank T. Midtun, J. P. Holm and A. Igesund who produced the figures and M. S. Stoker who corrected the English text and made valuable suggestions for improvements of the manuscript. Thanks are also due to G. G. Bianchi and J. A. Howe for their important comments.

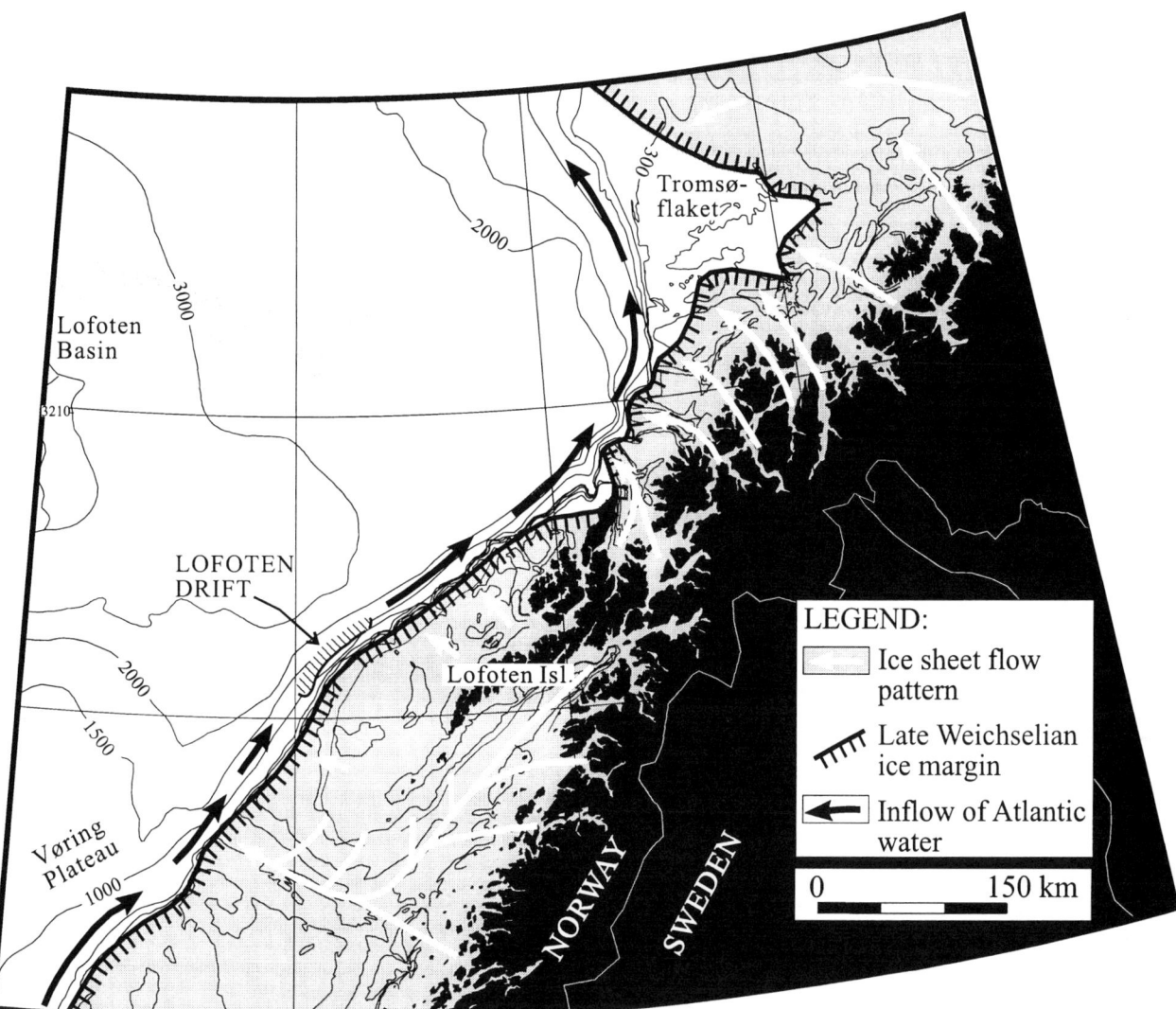

Fig. 8. A model for the palaeoenvironment during the growth of the Lofoten Drift. Due to the evolution of the nearby mainland area, the Lofoten Islands the downslope sediment input to the study area was probably relatively low during glacial periods. These islands may have acted as a sediment barrier causing large fluvial and/or glacial drainage systems from central Fennoscandia to be routed south and north of the study area. As a result, alongslope sediment transport has provided the main sediment input to this part of the continental slope. See text for further discussion.

References

BLINDHEIM, J. 1990. Arctic Intermediate Water in the Norwegian Sea. *Deep-Sea Research*, **37**, 1475–1489.

BLYSTAD, P., BREKKE, H., FÆRSETH, R. B., LARSEN, B. T., SKOGSEID, J. & TØRUDBAKKEN, B. 1995. Structural elements of the Norwegian continental shelf. Part II: The Norwegian Sea region. *Norwegian Petroleum Directorate, Bulletin*, **8**.

DOKKEN, T. & HALD, M. 1996. Rapid climatic shifts during isotope stages 2–4 in the Polar North Atlantic. *Geology*, **24**, 599–602.

EIKEN, O. & HINZ, K. 1993. Contourites in the Fram Strait. *Sedimentary Geology*, **82**, 15–32.

FAUGÈRES, J.-C. & STOW, D. A. V. 1993. Bottom current controlled sedimentation: a synthesis of the contourite problem. *Sedimentary Geology*, **82**, 287–297.

FAUGÈRES, J.-C., STOW, D. A. V., IMBERT, P. & VIANA, A. 1999. Seismic features diagnostic of contourite drifts. *Marine Geology*, **162**, 1–38.

HALD, M. & ASPELI, R. 1997. Rapid climatic shifts of the northern Norwegian Sea during the last deglaciation and the Holocene. *Boreas*, **26**, 15–28.

HEATHERSHAW, A. D., HALL, P. & HUTHNANCE, J. M. 1998. Measurements of the slope current, tidal characteristics and variability west of Vestfjorden, Norway. *Continental Shelf Research*, **18**, 1419–1453.

HEBBELN, D., DOKKEN, T., ANDERSEN, E. S., HALD, M. & ELVERHØI, A. 1995. Moisture supply for northern ice-sheet growth during the Last Glacial Maximum. *Nature*, **370**, 52–55.

HEBBELN, D., HENRICH, R. & BAUMANN, K.-H. 1998. Palaeoceanography of the last interglacial/glacial cycle in the Polar North Atlantic. *Quaternary Science Reviews*, **17**, 125–153.

HENRICH, R., KASSENS, H., VOGELSANG, E. & THIEDE, J. 1989. Sedimentary facies of glacial-interglacial cycles in the Norwegian Sea during the last 350 ka. *Marine Geology*, **86**, 283–319.

HENRIKSEN, S. & VORREN, T. O. 1996. Late Cenozoic sedimentation and uplift history on the mid-Norwegian continental shelf. *Global and Planetary Change*, **12**, 171–199.

HONJO, S. 1990. Particle fluxes and modern sedimentation in the polar oceans. *In*: SMITH, W. O. (ed.) *Polar Oceanography, Part B*. Academic Press, Inc., London, 687–739.

KELLOGG, T. B. 1980. Palaeoclimatology and palaeoceanography of the Norwegian–Greenland Sea: Glacial–interglacial contrasts. *Boreas*, **9**, 115–137.

KENYON, N. H. 1986. Evidence from bedforms for a strong poleward current along the upper continental slope of Northwest Europe. *Marine Geology*, **72**, 187–198.

LABERG, J. S., VORREN, T. O. & KNUTSEN, S.-M. 1999. The Lofoten Contourite Drift off Norway. *Marine Geology*, **159**, 1–6.

LABERG, J. S., DAHLGREN, T., VORREN, T. O., HAFLIDASON, H. & BRYN, P. in press. Seismic analyses of Cenozoic contourite drift development in the Northern Norwegian Sea. *In*: REBESCO, M. & STOW, D. A. V. (eds) *Seismic Expression of Contourites and Related Deposits*. Marine Geophysical Researches, Special Issue, in press.

LØSETH, H. & TVETEN, E. 1996. Post-Caledonian structural evolution of the Lofoten and Vesterålen offshore and onshore areas. *Norsk Geologisk Tidsskrift*, **76**, 215–230.

MANGERUD, J. & GULLIKSEN, S. 1975. Apparent radiocarbon ages of Recent marine shell from Norway, Spitsbergen and Arctic Canada. *Quaternary Research* **5**, 263–273.

MOSBY, H. 1968. Surrounding seas. *In*: SØMME, A. (ed.) *Geography of Norden*. J. W. Cappelens forlag, Oslo, Norway.

PERRY, R. K., FLEMING, H. S., CHERKIS, N. Z., FEDEN, R. H. & VOGT, P. R. 1980. *Bathymetry of the Norwegian-Greenland and Western Barents Sea*. Naval Research Laboratory-Acoustics Division, Environmental Sciences Branch, Washington DC.

POULAIN, P.-M., WARN-VARNAS, A. & NIILER, P. P. 1996. Near-surface circulation of the Nordic Seas as measured by Lagrangian drifters. *Journal of Geophysical Research*, **101**, C8, 18,237–18,258.

STOKER, M. S., AKHURST, M. C., HOWE, J. A. & STOW, D. A. V. 1998. Sediment drifts and contourites on the continental margin off northwest Britain. *Sedimentary Geology*, **115**, 33–51.

STUEVOLD, L. M. & ELDHOLM, O. 1996. Cenozoic uplift of Fennoscandia inferred from a study of the mid-Norwegian margin. *Global and Planetary Change*, **12**, 359–386.

SWIFT, J. H. & KOLTERMANN, K. P. 1988. The Origin of Norwegian Sea Deep Water. *Journal of Geophysical Research*, **93**, 3563–3569.

VORREN, T. O., LABERG, J. S. *ET AL*. 1998. The Norwegian-Greenland Sea continental margins: morphology and late Quaternary sedimentary processes and environment. *In*: ELVERHØI, A. (ed.) *Glacial and Oceanic History of the Polar North Atlantic Margins*. Quaternary Science Review, **17**, 273–302.

YOON, S. H. & CHOUGH, S. K. 1993. Sedimentary characteristics of Late Pleistocene bottom current deposits, Barents slope off northern Norway. *Sedimentary Geology*, **82**, 33–45.

Sediment drifts and contourite sedimentation in the northeastern Rockall Trough and Faroe–Shetland Channel, North Atlantic Ocean

JOHN A. HOWE[1], MARTYN S. STOKER[2], DORRIK A. V. STOW[3] & MAXINE C. AKHURST[2]

[1]*Scottish Association for Marine Science & the University of the Highlands and Islands project, Dunstaffnage Marine Laboratory, Oban, Argyll PA34 4AD, Scotland, UK*
[2]*British Geological Survey, Murchison House, West Mains Road, Edinburgh EH9 3LA, Scotland, UK*
[3]*Southampton Oceanography Centre, School of Ocean and Earth Science, Southampton University, Empress Dock, European Way, Southampton SO14 3ZH, UK*

Abstract: Seismic reflection profiles, shallow cores and seabed photography from the continental margin off NW Britain reveal the variety of bottom current influenced sedimentation in the northern Rockall Trough and Faroe–Shetland Channel. Types of sediment drifts identified include: (1) elongate drifts, both single, and multi-crested; (2) sheeted drift forms, varying from gently domed to flat-lying; and (3) isolated patch drifts, including moat-related drifts. Associated fields of localized sediment waves are developed with the elongate and gently domed, broad sheeted drifts. The contrasting style of sediment drift development reflects the complex interaction between bottom current regime, sediment supply and the bathymetry of the continental margin. The majority of the mounded/gently domed drifts occur in the northern Rockall Trough, with sheetform drifts commonly confined to the Faroe–Shetland Channel, a narrow basin which is an area of net sediment export rather than drift accumulation. Small patch drifts are present in both basins. Muddy, silty muddy and sandy contourites have been recognized from sediment cores sampling the uppermost parts of the drift sequences. Based on their glaciomarine character, the mid- to high-latitude contourites are referred to as glacigenic contourites. Both partial and complete contourite sequences are preserved; the former consist largely of sandy (mid-only) and top-only contourites. Modern sandy contourites have also been identified from seabed photographs on the Hebrides Slope. The contourites are recognized as a rippled mobile sand layer, reworked from a poorly sorted glaciomarine parent deposit.

The northeastern Rockall Trough and Faroe–Shetland Channel are UK deep-water regions currently being explored by the hydrocarbon industry with the potential for offshore development. In this context, after over a decade of surveying using seismic reflection profiles and cores, a wealth of information has become available on the sediment drift geometry and bottom current deposits in the area. A number of papers have been published on the wide variety of bottom current influenced features; from the overall drift complexes of Howe *et al.* (1994), Stoker *et al.* (1998) and Stoker (1998) to the sheeted drifts of Akhurst (1991) and sediment waves of Howe (1996) and Richards *et al.* (1987). The diverse styles of sediment drift observed across this region are a result of the complex interplay between the bathymetry of the margin and the persistent, vigorous bottom current flow.

This paper serves to briefly overview the areas immediately to the north and south of an important deep-ocean gateway in the NE Atlantic, the *Shetland–Rockall gateway*. This narrow but complex conduit allows for the southward passage of bottom water from the Norwegian Sea into the Rockall Trough and Iceland Basin. More detailed papers are included separately in this volume on the Faroe–Shetland Channel (Akhurst *et al.* 2002) and on different parts of the Rockall Trough (Knutz *et al.* 2002; Stow *et al.* 2002).

Geological and oceanographic setting

The Rockall Trough and Faroe–Shetland Channel are two deep-water areas that separate the Hebrides and West Shetland shelves from the Rockall Plateau and Faroes Shelf (Fig. 1). Between the two basins is the prominent northwesterly trending Wyville–Thomson Ridge. The Rockall Trough deepens to the southwest from 1000 m to over 4000 m, although the floor of the trough is generally gently inclined and smooth. The basin of the northern trough contains the Rosemary Bank, Hebrides Terrace and Anton Dohrn seamounts, the bases of which have become scoured by intense bottom current activity. Along the eastern margin, slope fans have become developed, notably adjacent to the Hebrides Terrace seamount, the Barra Fan and southeast of the Wyville–Thomson Ridge, the Sula Sgeir Fan (Stoker 1998). Slope angles along the margins are relatively gentle averaging 1.5–3° on the Hebrides Slope although locally can reach 28° on the Geikie Escarpment and 15° on the Wyville–Thomson Ridge. Compared with the Rockall Trough, the Faroe–Shetland Channel is a narrow, restricted basin, deepening from 1000 m in the southwest to 1700 m in the northeast where it opens out into the Norwegian Basin. To the southwest, the channel turns northwest to form the Faroe Bank Channel, joining into the Iceland Basin. The floor of the channel is gently inclined to the northeast, with a low relief (< 150 m) central ridge (Akhurst 1991). Gentle slope angles of 1–3° are present along the eastern margins of the channel.

Deep-water circulation on the continental margin involves a number of water masses (Fig. 1). Southwest flowing, Norwegian Sea Deep Water (NSDW) fills the Faroe–Shetland Channel to depths of 500 m. A small amount of NSDW overflows the western end of the Wyville–Thomson Ridge and flows south into the Rockall Trough. It becomes mixed with Labrador Sea Water and Antarctic Bottom Water forming North Atlantic Deep Water (NADW) which flows southwards along the western margin of the Rockall Trough (Dickson & Kidd 1986). The eastern flank of the Rockall Trough is bathed by a northward-flowing slope-current down to depths of 1000 m (Booth & Ellett 1983; Huthnance 1986). Some elements of Mediterranean Deep Water have been detected in the slope current in its high silica and salinity values (Hill & Mitchelson-Jacob 1993), and some entrainment of the deeper NADW is possible. Termed Northeast Atlantic Water (NEAW), this slope-current continues north until it is deflected and accelerated to the west by the Wyville–Thomson Ridge.

Current-meter data provide evidence on the velocities of the bottom currents within the area. NSDW flow, at the centre of the Faroe–Shetland Channel in a water depths of 900 m has been recorded at 13 to 22 cm s^{-1}, with maximum flow of 33 cm s^{-1} measured (Akhurst 1991). On the Hebrides Slope in the north,

From: STOW, D. A. V., PUDSEY, C. J., HOWE, J. A., FAUGÈRES, J.-C. & VIANA, A. R. (eds)
Deep-Water Contourite Systems: Modern Drifts and Ancient Series, Seismic and Sedimentary Characteristics.
Geological Society, London, Memoirs, **22**, 65–72. 0435-4052/02/$15.00 © The Geological Society of London 2002.

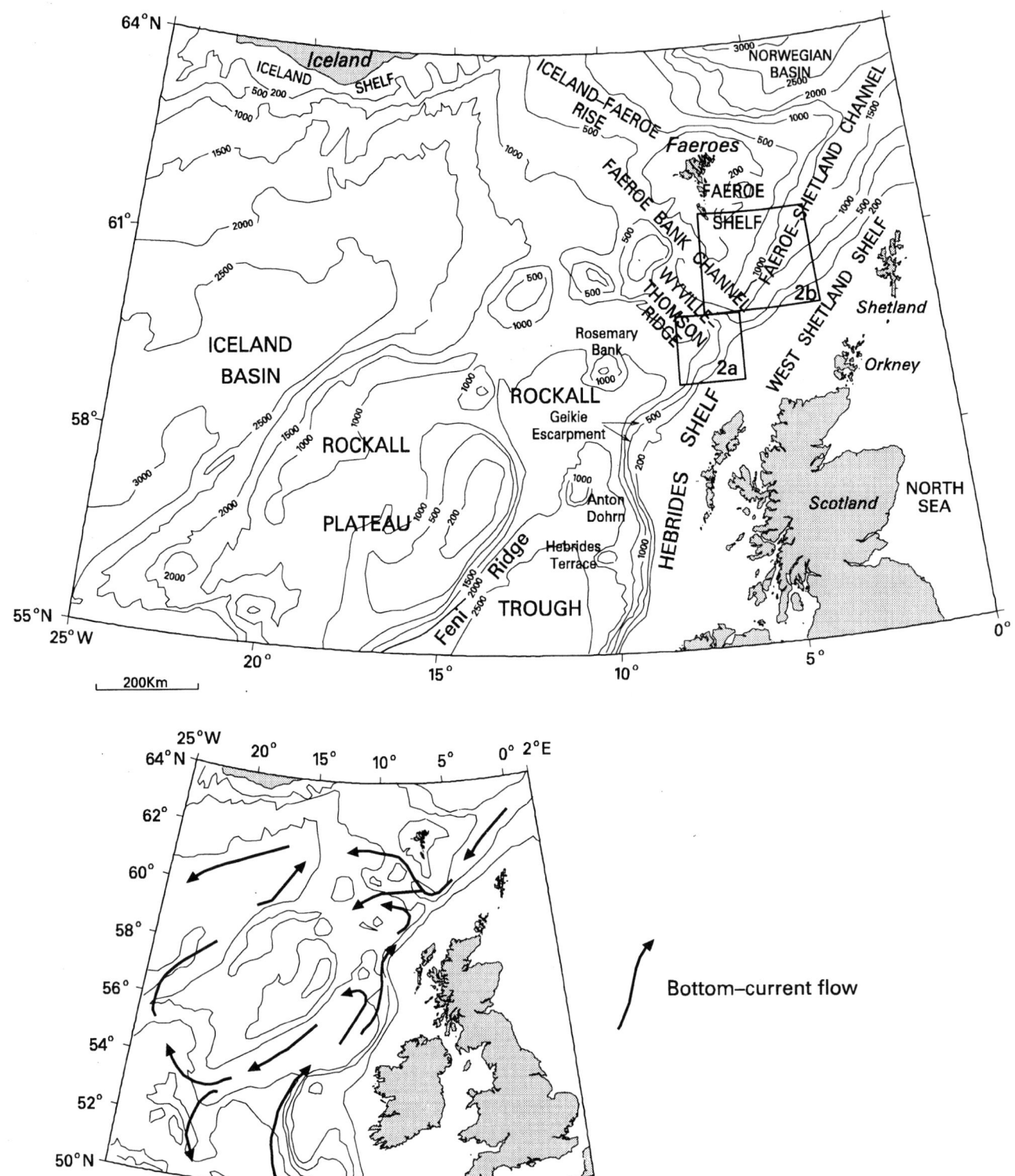

Fig. 1. Bathymetric and oceanographic setting of the northeastern Rockall Trough and Faroe–Shetland Channel.

current velocities of 26 to 48 cm s^{-1} have been detected, in water depths of 403 m and 468 m. Alongslope to the south, flows of 15 to 25 cm s^{-1} have been recorded in water depths 1035 m and 457 m, with a consistent alongslope flow to the northeast (Howe & Humphrey 1995).

Seismic characteristics

A number of different types of sediment drift have been identified in the northern Rockall Trough and Faroe–Shetland Channel, these can be summarized as *elongate* (including *double* and *multi-crested*), *broad-sheeted*, and *isolated* (or 'patch' drift). Also noted are bottom current influenced sediment waves and erosional features such as the moated regions, adjacent to the drifts and channels. The mounded drift forms are confined to the northern Rockall Trough, whilst the sheeted drift and erosional channels are characteristic of the Faroe–Shetland Channel. The seismic character of these features varies considerably across both regions. Location of the seismic reflection profiles is indicated on Figure 2.

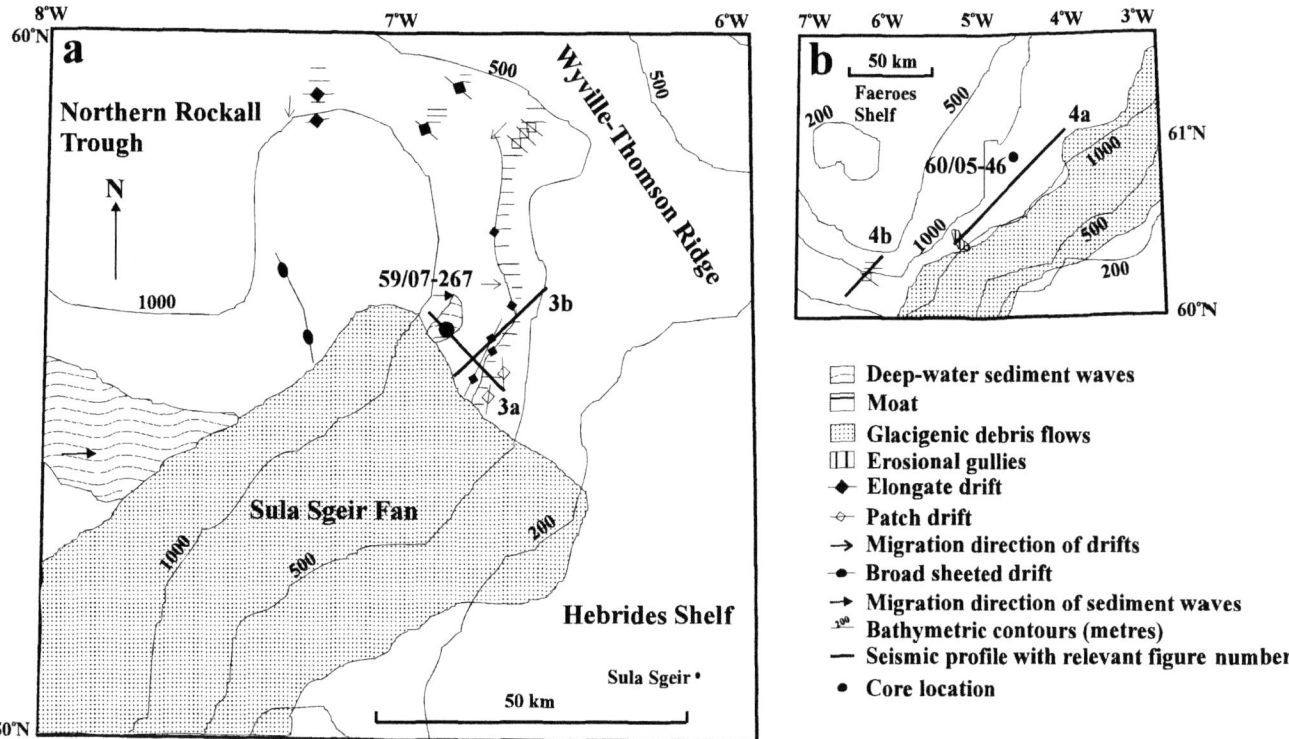

Fig. 2. Location of seismic reflection profiles and cores in the northern Rockall Trough and Faroe–Shetland Channel.

Sediment drifts in the northern Rockall Trough

The elongate drift is located adjacent to the base of the Hebrides Slope and Wyville–Thomson Ridge. The drift is up to 300 m thick, 20 km wide and has a vertical relief of up to 150 m and is asymmetric in cross-section. In seismic reflection profiles, the drift displays parallel, well-layered internal reflectors, with no truncated horizons, although some pinch-out and divergence is present. The drift has a slight upslope migration direction, visible from the reflectors on the drift crest. The high-resolution boomer profile illustrated (Fig. 3a) displays sediment waves on the flank of the elongate drift. The waves are developed in a small (20 km²) region at the base of the Hebrides Slope in response to the along slope currents. Apparent wavelengths are 2 km with a height of 20 m (crest to trough). The seismic units reflect the activity of the waves with the lower, highly reflective sequence the more migratory and hence the most active. The upper unit is more transparent and less migratory, more aggradational (Howe et al. 1994; Howe 1996). The waves become active in the uppermost highly reflective unit, showing preferential erosion and deposition on the apparent southeastern flanks of the waves (Fig. 3a).

The basic elongate drift form of the northern Rockall Trough is also subject to a number of varieties and modifications, reflecting changes in bottom current pathways and activity. The seismic reflection profile illustrated (Fig. 3b) displays the common along slope modification of the drift crest. The single crested drift present to the south has become a multi-crested drift, with three separate but active drifts. The multi-crestal form has developed from a single elongate drift, visible in the now buried seismic sequences of the drift (Stoker et al. 1998; Stoker 1998).

Sheet-form drifts are also present across the region. In the northern Rockall Trough the sheet has slight relief (60 m over tens of kilometres), giving rise to a broad-sheeted drift form of 490 m thick (Howe et al. 1994; Stoker 1998). In this example, a field of sediment waves (550 km²), has developed on the flank of the drift. The waves are now inactive and become buried towards the south by the glacigenic slope apron of the Sula Sgeir Fan (Stoker 1995).

Sediment drifts in the Faroe–Shetland Channel

In the Faroe–Shetland Channel sheeted drift forms are more prevalent, reflecting perhaps the more erosional, non-depositional regime of the area. Sheeted drifts in this area are characterized by parallel, laterally continuous seismic reflectors (Fig. 4a). The drift sequence is thin, when compared to the northern Rockall Trough, with 80–100 m of sediments. In some localized areas the drift sequence has become thinned over palaeotopographic highs. Low angle truncations of the reflectors may also suggest short-lived erosive phase during drift construction (Akhurst 1991). Towards the SW the drift becomes progressively thinner, as the channel shallows and the current increases (Fig. 4a). Erosional gullies are also developed indicating vigorous current activity along the channel. Although the mounded drift form is atypical for the region, small patch drifts do occur as localized features. Adjacent to the junction of the Faroe–Shetland Channel and Faroe Bank, a small 100 m thick and 5 km wide drift has developed (Fig. 4b). An erosional notch has also developed on the slope side of the associated moat, however away from the influence of the slope the sheet-form drift geometry resumes towards the centre of the channel.

Sediment characteristics

A number of short cores have been recovered from both the NE Rockall Trough and the Faroe–Shetland Channel. Analysis of these indicate that the sediment drifts are dominated by contouritic fine-grained muds and sands of a glaciomarine origin (Stoker et al. 1989, 1991, 1993; Akhurst 1991; Howe et al. 1994; Howe 1995, 1996). Three main facies are recognized within these sediments based upon grain size characteristics, sedimentary structures and composition. These are Muddy Sand, Sandy Mud and Mud Facies (Table 1), each of which may include lithic clasts of glaciomarine origin, which are either widely dispersed or in concentrated winnowed horizons (Stoker et al. 1998).

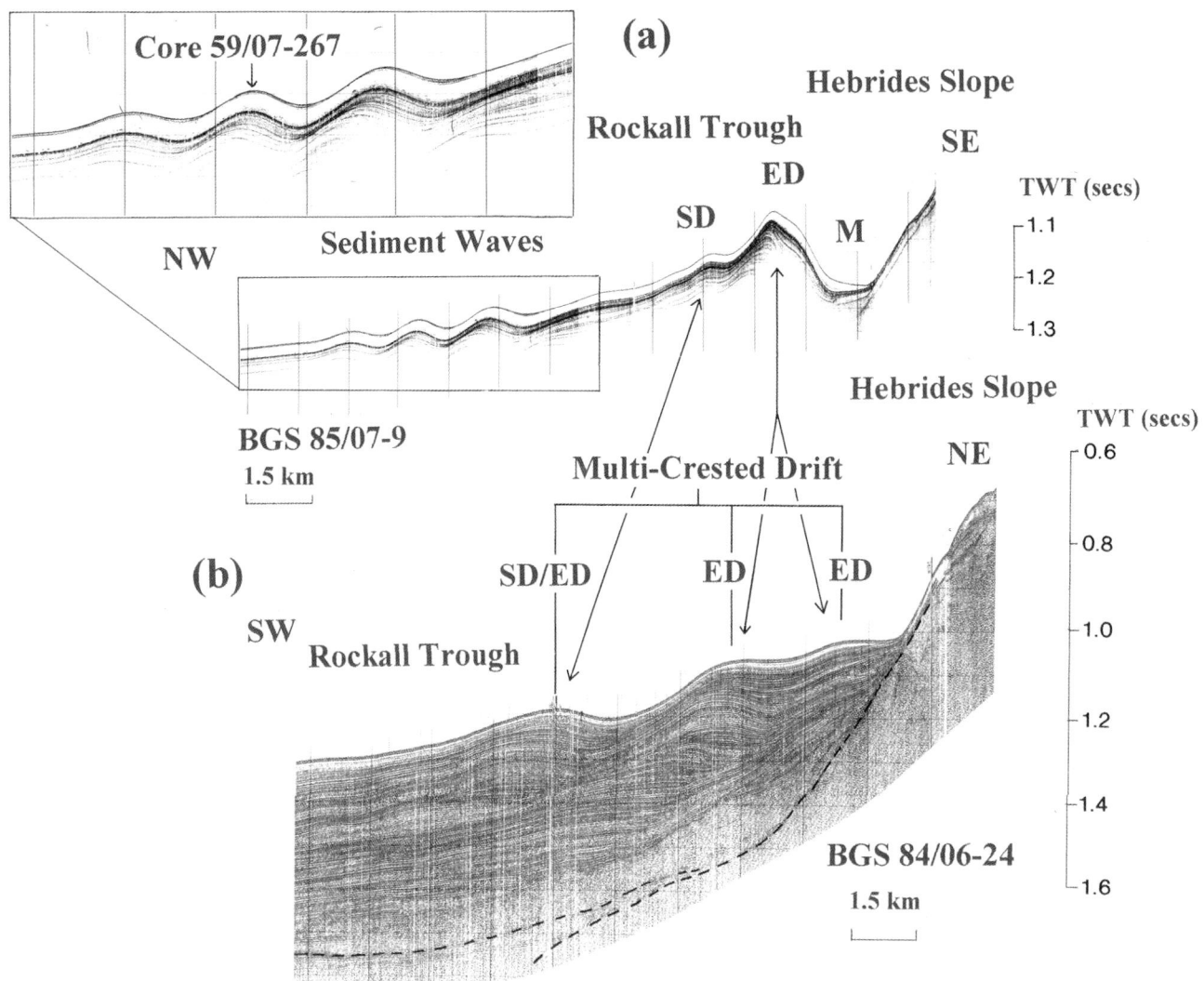

Fig. 3. (**a**) BGS boomer line 85/07-9, showing contouritic sediment waves and elongate sediment drift adjacent to the Hebrides Slope, northeastern Rockall Trough. Location of core 267 indicated. SD, subsidiary drift; ED, elongate drift; M, moat. (**b**) BGS line 84/06-24, showing the development of a multi-crested drift. Abbreviations as for 3(a). Dotted line represents base of Neogene–Quaternary sediments (Adapted from Stoker *et al.* 1998).

Northern Rockall Trough contourites

A muddy sand facies, up to 28 cm thick, which is moderately well sorted, greyish yellow, intensely bioturbated, and with comminuted shell material occurs in the all cores examined. Cross bedding is only rarely preserved. A similar muddy sand layer has also been noted in the lower section of a single core from the moated area adjacent to the elongate drift (Howe *et al.* 1994). This facies is interpreted as a sandy contourite with 30–80% medium- to fine-grained sand and <45% mud (Fig. 5).

Underlying the muddy sand layer in all of the cores is a sandy mud. This is up to 55 cm in thickness, poorly sorted, olive brown in colour, with sporadic gravel clasts. Lamination is visible (both core section and x-radiograph) in some intervals. Small-scale fining- and coarsening-upward units, a few tens of centimetres in thickness, are also common. Bioturbation is intense throughout. This facies is a silty muddy contourite with up to 70% silt and 10–40% sands and some minor gravels.

A poorly sorted, intensely bioturbated, homogeneous mud facies occurs in the lower sections of the cores. The mud is composed of up to 95% silt-clay, <5% sand and variable but low amounts of gravel. The mud lacks any evidence of bottom current

Table 1. *Summary core information, adapted from Stoker* et al. *(1998)*

Sediment Facies	Grain Size	Sedimentary Structures	Thickness	Interpretation
Mud facies	Mud with <1% lithic clasts	Small-scale (<5 mm) homogenizing bioturbation and larger burrow systems	Up to 2.28 m thick	Muddy contourites
Sandy mud facies	Slightly sandy mud and sandy mud with <1% lithic clasts	Small-scale (<5 mm) homogenizing bioturbation and larger burrow systems	Up to 3 m thick	Silty muddy contourites
Muddy sand facies	Muddy sand or slightly gravelly muddy sand with 3% lithic clasts	Small-scale (<5 mm) homogenizing bioturbation and larger burrow systems. Cross bedding rare	Less than 0.5m thick	Sandy contourites

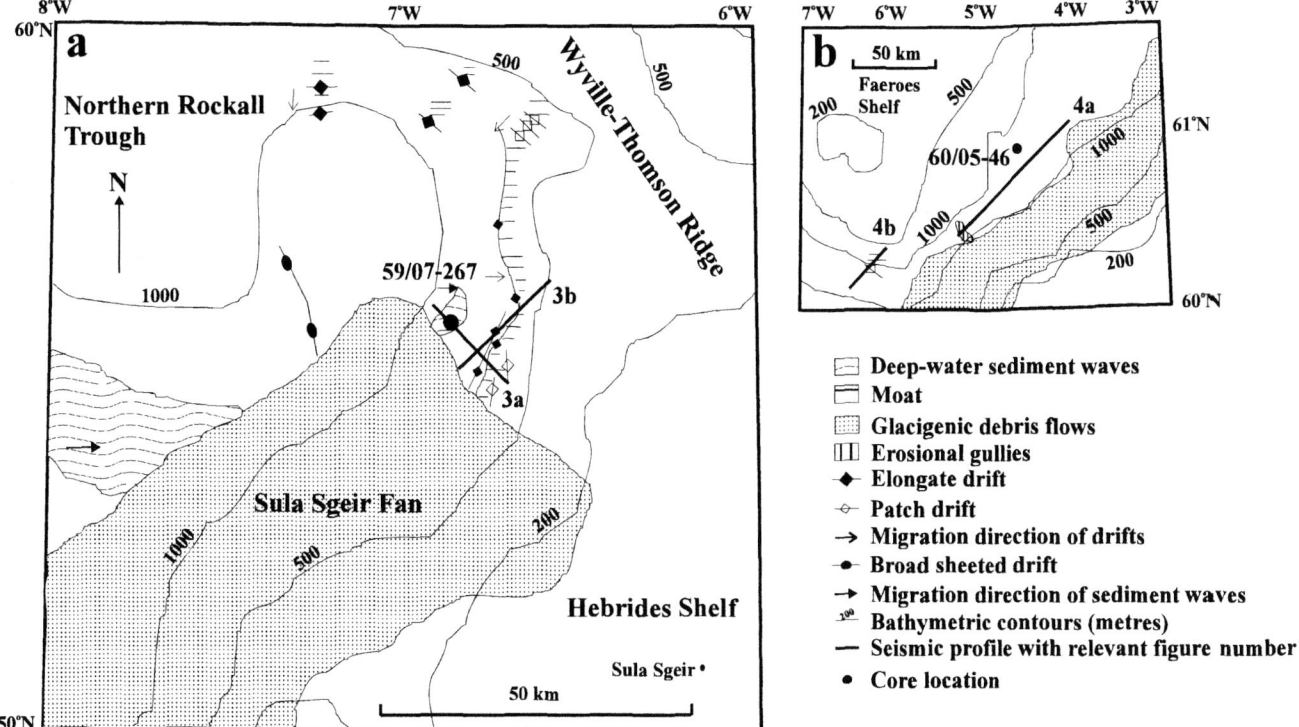

Fig. 2. Location of seismic reflection profiles and cores in the northern Rockall Trough and Faroe–Shetland Channel.

Sediment drifts in the northern Rockall Trough

The elongate drift is located adjacent to the base of the Hebrides Slope and Wyvillle–Thomson Ridge. The drift is up to 300 m thick, 20 km wide and has a vertical relief of up to 150 m and is asymmetric in cross-section. In seismic reflection profiles, the drift displays parallel, well-layered internal reflectors, with no truncated horizons, although some pinch-out and divergence is present. The drift has a slight upslope migration direction, visible from the reflectors on the drift crest. The high-resolution boomer profile illustrated (Fig. 3a) displays sediment waves on the flank of the elongate drift. The waves are developed in a small (20 km²) region at the base of the Hebrides Slope in response to the along slope currents. Apparent wavelengths are 2 km with a height of 20 m (crest to trough). The seismic units reflect the activity of the waves with the lower, highly reflective sequence the more migratory and hence the most active. The upper unit is more transparent and less migratory, more aggradational (Howe et al. 1994; Howe 1996). The waves become active in the uppermost highly reflective unit, showing preferential erosion and deposition on the apparent southeastern flanks of the waves (Fig. 3a).

The basic elongate drift form of the northern Rockall Trough is also subject to a number of varieties and modifications, reflecting changes in bottom current pathways and activity. The seismic reflection profile illustrated (Fig. 3b) displays the common along slope modification of the drift crest. The single crested drift present to the south has become a multi-crested drift, with three separate but active drifts. The multi-crestal form has developed from a single elongate drift, visible in the now buried seismic sequences of the drift (Stoker et al. 1998; Stoker 1998).

Sheet-form drifts are also present across the region. In the northern Rockall Trough the sheet has slight relief (60 m over tens of kilometres), giving rise to a broad-sheeted drift form of 490 m thick (Howe et al. 1994; Stoker 1998). In this example, a field of sediment waves (550 km²), has developed on the flank of the drift. The waves are now inactive and become buried towards the south by the glacigenic slope apron of the Sula Sgeir Fan (Stoker 1995).

Sediment drifts in the Faroe–Shetland Channel

In the Faroe–Shetland Channel sheeted drift forms are more prevalent, reflecting perhaps the more erosional, non-depositional regime of the area. Sheeted drifts in this area are characterized by parallel, laterally continuous seismic reflectors (Fig. 4a). The drift sequence is thin, when compared to the northern Rockall Trough, with 80–100 m of sediments. In some localized areas the drift sequence has become thinned over palaeotopographic highs. Low angle truncations of the reflectors may also suggest short-lived erosive phase during drift construction (Akhurst 1991). Towards the SW the drift becomes progressively thinner, as the channel shallows and the current increases (Fig. 4a). Erosional gullies are also developed indicating vigorous current activity along the channel. Although the mounded drift form is atypical for the region, small patch drifts do occur as localized features. Adjacent to the junction of the Faroe–Shetland Channel and Faroe Bank, a small 100 m thick and 5 km wide drift has developed (Fig. 4b). An erosional notch has also developed on the slope side of the associated moat, however away from the influence of the slope the sheet-form drift geometry resumes towards the centre of the channel.

Sediment characteristics

A number of short cores have been recovered from both the NE Rockall Trough and the Faroe–Shetland Channel. Analysis of these indicate that the sediment drifts are dominated by contouritic fine-grained muds and sands of a glaciomarine origin (Stoker et al. 1989, 1991, 1993; Akhurst 1991; Howe et al. 1994; Howe 1995, 1996). Three main facies are recognized within these sediments based upon grain size characteristics, sedimentary structures and composition. These are Muddy Sand, Sandy Mud and Mud Facies (Table 1), each of which may include lithic clasts of glaciomarine origin, which are either widely dispersed or in concentrated winnowed horizons (Stoker et al. 1998).

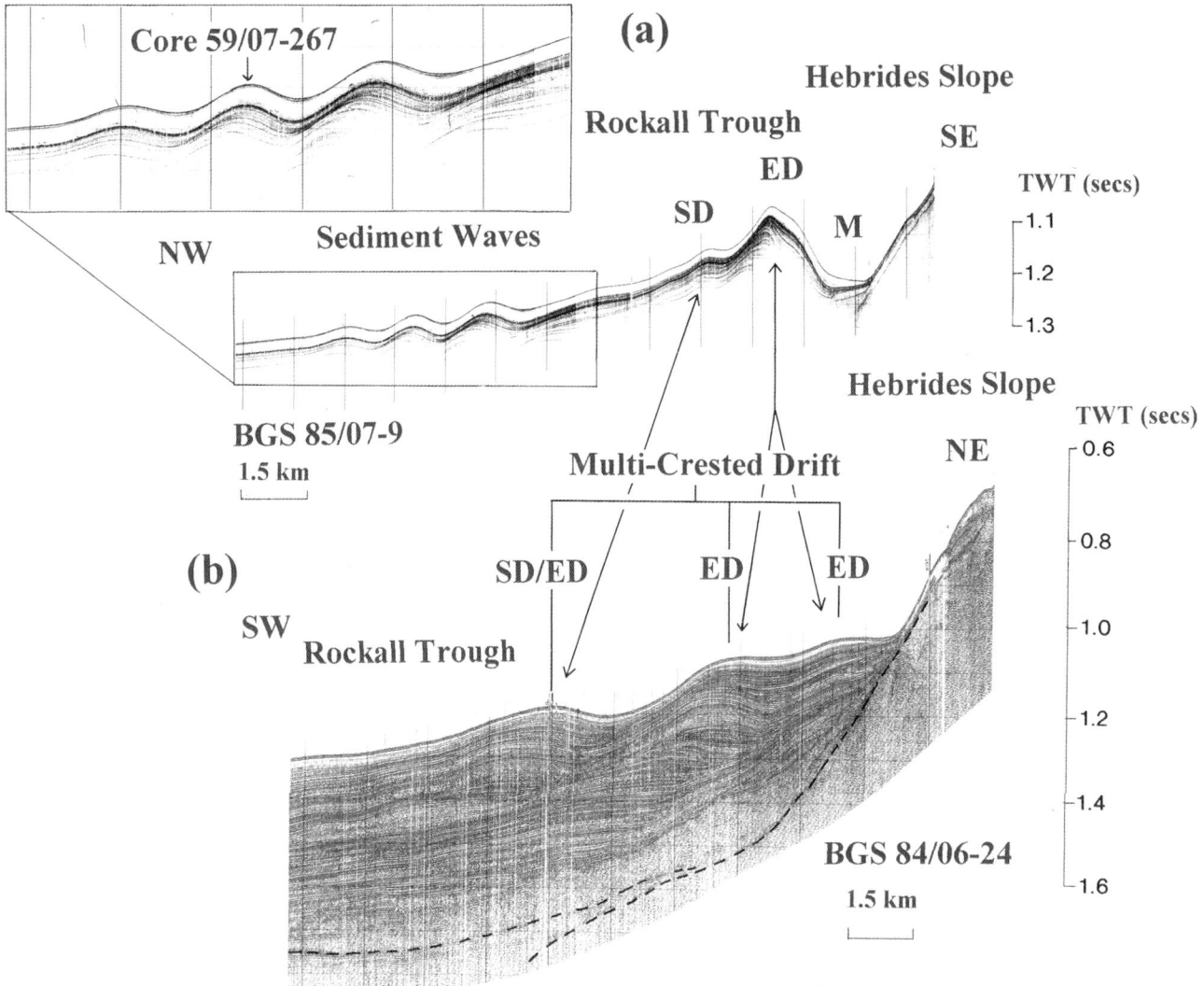

Fig. 3. (**a**) BGS boomer line 85/07-9, showing contouritic sediment waves and elongate sediment drift adjacent to the Hebrides Slope, northeastern Rockall Trough. Location of core 267 indicated. SD, subsidiary drift; ED, elongate drift; M, moat. (**b**) BGS line 84/06-24, showing the development of a multi-crested drift. Abbreviations as for 3(a). Dotted line represents base of Neogene–Quaternary sediments (Adapted from Stoker *et al.* 1998).

Northern Rockall Trough contourites

A muddy sand facies, up to 28 cm thick, which is moderately well sorted, greyish yellow, intensely bioturbated, and with comminuted shell material occurs in the all cores examined. Cross bedding is only rarely preserved. A similar muddy sand layer has also been noted in the lower section of a single core from the moated area adjacent to the elongate drift (Howe *et al.* 1994). This facies is interpreted as a sandy contourite with 30–80% medium- to fine-grained sand and <45% mud (Fig. 5).

Underlying the muddy sand layer in all of the cores is a sandy mud. This is up to 55 cm in thickness, poorly sorted, olive brown in colour, with sporadic gravel clasts. Lamination is visible (both core section and x-radiograph) in some intervals. Small-scale fining- and coarsening-upward units, a few tens of centimetres in thickness, are also common. Bioturbation is intense throughout. This facies is a silty muddy contourite with up to 70% silt and 10–40% sands and some minor gravels.

A poorly sorted, intensely bioturbated, homogeneous mud facies occurs in the lower sections of the cores. The mud is composed of up to 95% silt-clay, <5% sand and variable but low amounts of gravel. The mud lacks any evidence of bottom current

Table 1. *Summary core information, adapted from Stoker* et al. *(1998)*

Sediment Facies	Grain Size	Sedimentary Structures	Thickness	Interpretation
Mud facies	Mud with <1% lithic clasts	Small-scale (<5 mm) homogenizing bioturbation and larger burrow systems	Up to 2.28 m thick	Muddy contourites
Sandy mud facies	Slightly sandy mud and sandy mud with <1% lithic clasts	Small-scale (<5 mm) homogenizing bioturbation and larger burrow systems	Up to 3 m thick	Silty muddy contourites
Muddy sand facies	Muddy sand or slightly gravelly muddy sand with 3% lithic clasts	Small-scale (<5 mm) homogenizing bioturbation and larger burrow systems. Cross bedding rare	Less than 0.5m thick	Sandy contourites

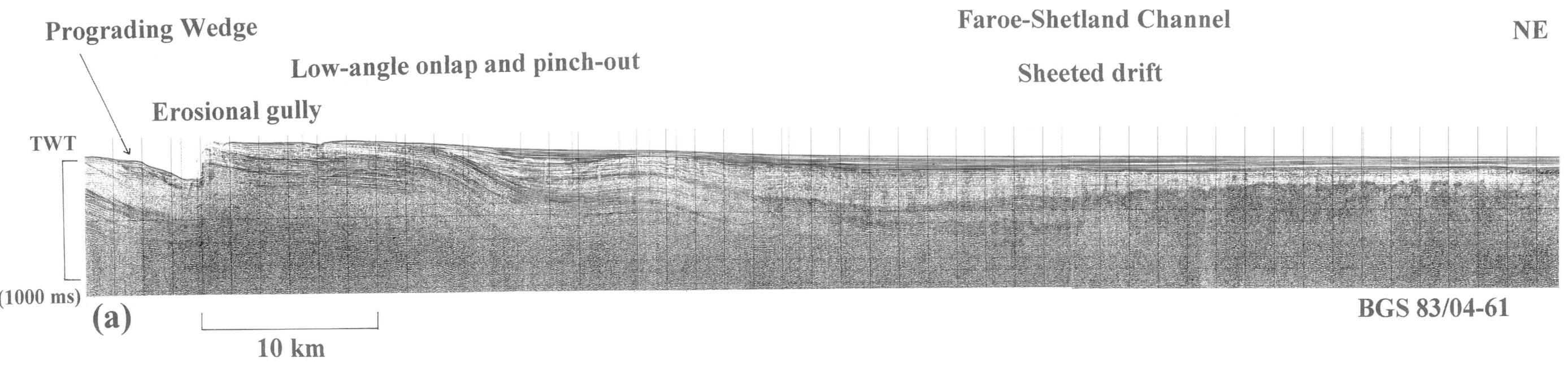
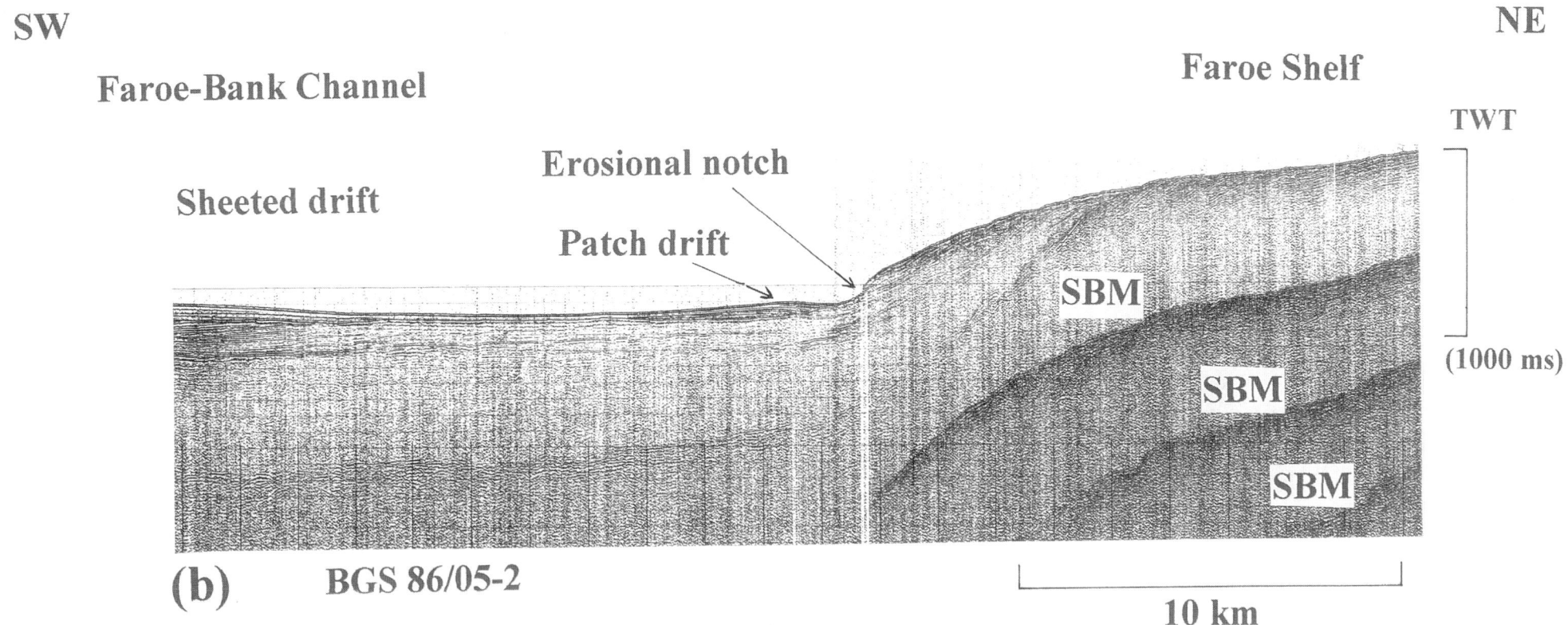

Fig. 4. (a) BGS line 83/04-61, showing the sheeted drift and erosional gulley in the Faeroe–Shetland Channel. A prograding wedge is developed in the erosional. Low angle onlap and pinch-out of the contouritic sediments become more pronouced, as currents increase towards the SW. (b) BGS line 86/05-2 showing sheeted drift and patch drift in the Faeroe–Bank Channel. An erosional notch has been cut by strong current activity at the base of the slope (Adapted from Stoker *et al.* 1998).

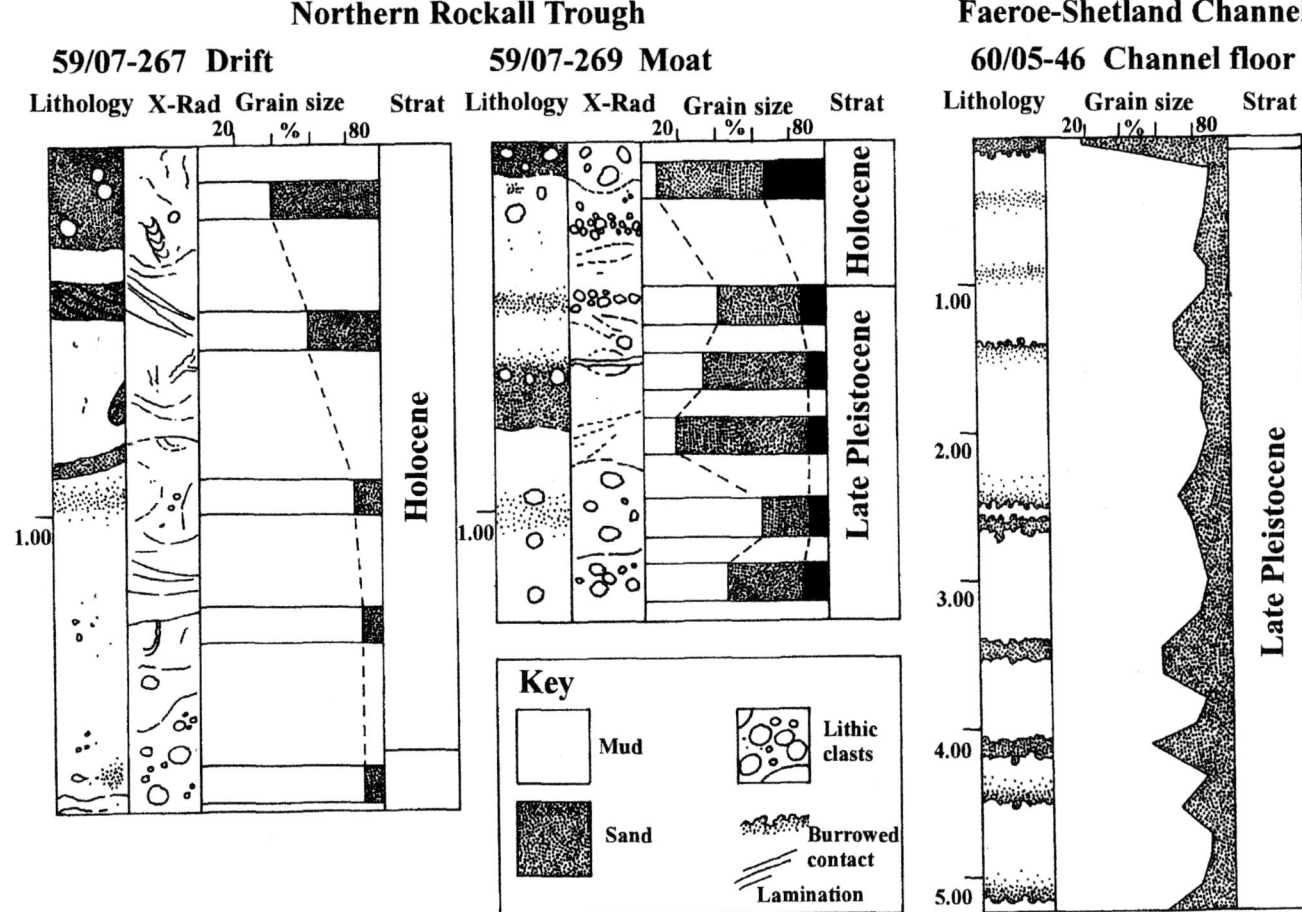

Fig. 5. Core characteristics illustrating the styles of contourite lithology, x-radiographs (Rockall Trough only) of cores 59/07-267, from elongate drift 59/07-269 from a moated area both from Rockall Trough and 60/05-46 from the basin floor (sheeted drift) of the Faroe–Shetland Channel. Stratigraphy shown on is based on biostratigraphic analysis (Adapted from Howe *et al.* 1994; Howe *et al.* 1998; Akhurst 1991).

influence and is best interpreted as being of hemipelagic origin (Howe 1995). However, some subtle grain size variation and indistinct lamination in parts may suggest the influence of weak bottom currents.

Present within the contourite and hemipelagite facies' are thin (<10 cm), poorly sorted, sandy pebbly muds and muddy gravels, composed of 10–40% fine- to coarse-grained sands, 50–90% muds and 10–60% gravels. These sediments are interpreted as ice-distal glaciomarine deposits.

Faroe–Shetland Channel contourites

Generally the cores from the Faroe–Shetland Channel have a poorly sorted glaciomarine character (Akhurst 1991). Lithic dropstone clasts and fine-scale homogenizing bioturbation are ubiquitous. Larger (>5 mm) burrow structures are common throughout, cutting through the earlier bioturbated network (Fig. 5).

A muddy sand facies occurs in beds 15–50 cm thick. Although containing >50% fine-grained sand, the mean grain size of the olive grey sediments is medium- to coarse-grained silt. Lithic clasts are common throughout. Upper contacts are always gradational into the sandy mud facies; lower contacts may be gradational or relatively unbioturbated horizons, with some evidence of erosion. These muddy sands are interpreted to be sandy (and mottled silty) contourites.

Sandy muds are grey to olive grey in colour and up to 3 m thick. Contacts are gradational, and bioturbation is common throughout with diverse burrow types. The sandy muds have a mean grain size of fine- to very fine-grained silt, and typically form part of coarsening- to fining-upward sequences between muddy sand and mud facies. The sandy muds are interpreted as muddy and silty contourites.

Dark olive to olive grey muds are very fine-grained sediments and include distinctive *Zoophycos* burrows throughout. The muds are up to 2.28 m thick, and appear homogeneous. However, cyclical variation in mean size, between fine-grained silt and clay size, suggest a bottom current influence during deposition. In most cases, contacts are gradational, although there is an erosive contact where directly overlain by muddy sand facies. These muds are interpreted as muddy contourites.

The coarsening- to fining-upward sequences of sandy, silty muddy and muddy contourites are common in the Quaternary sediments of the Faroe–Shetland Channel, in addition to the background glaciomarine sedimentation. These sequences can be correlated for more than 50 km along the floor of the Faroe–Shetland Channel (Akhurst 1991) demonstrating the regional scale of bottom current deposition and the accumulation of contourites in a sheet-form geometry. The subtle cyclical variation in mean grain size within the muddy contourites illustrates that the cryptic influence of bottom currents. Where the erosive effects have been locally intensified, coarsening-upward sequences, deposited as bottom current velocity increased, were eroded away as the current velocity reached a maximum. These fining-upward sequences with an erosive base (Akhurst 1991) have been identified in both the NE Rockall Trough and Faroe–Shetland Channel. Where the waning phase only of a bottom current is preserved, this is termed a 'top-only' contourite.

Fig. 6. Seabed photographs from two sites in 863 m of water showing modern reworking of sandy contourites on the Geikie Escarpment of the Hebrides Slope (see Figure 1 for location). (**a**) Rippled contouritic sands with scouring of moats and crag-and-tail development around clasts. (**b**) Large dropstone clast (0.5 m × 0.3 m) with well developed gravel lag lies in a field of rippled contouritic sands. Field of view 1.3 m × 3.0 m to top of photograph (Adapted from Howe & Humphery 1995).

If a record of the maximum current flow is preserved this deposit is termed a 'mid-only' contourite (Akhurst 1991; Stoker et al. 1998). Top-only contourites are noted where the sedimentary sequence is thinned over palaeotopographic highs. Sediment accumulation has become condensed by an increase in winnowing and hence non-deposition currents are locally restricted.

Seabed photographs

Sea-bed photographs have been collected from the northern Rockall Trough. Two photographs are illustrated from 863 m of water on the Hebrides Slope (see Figure 2 for location). The camera used was a full frame, 35 mm survey camera, fitted with a 35 mm focal-length lens. Core, box and grab samples have been recovered from the area the camera sites. Generally the samples are moderately sorted fine-medium-grained sands (approximately 50% of the sample in the size range 0.1–0.4 mm), with a high siliciclastic content (Howe & Humphrey 1995). The sands are modern, active, current-reworked sandy contourites.

Bottom current features from seabed photographs

The camera sites display a variety of features that can be attributed to both current activity and 'normal' hemipelagic sedimentation. Current features include ripples, crag-and-tail structures, scours filled with coarser material, and small current produced lineations. All are consistent with a northeasterly flowing alongslope bottom current (Fig. 6).

Ripples. Ripples are small-scale, asymmetrical, linguoid forms with wavelengths of 7–15 cm and amplitudes of 2–3 cm.

Crag-and-tail structures. A tail of sediment is produced in the lee of an obstacle, usually a lithic clast, and can extend up to 5 cm. Tails reflect the release of sediment from suspension, with the reduced current strengths in the lee of the obstruction.

Current scours. Scouring is usually associated with crag-and-tail structures. In mixed grain size sediment populations, lithic clasts develop an upstream moat and coarser grains remain within this as finer particles are winnowed. In cohesive clays and finer silts the moat develops as a depression extending around the obstacle, the cohesive nature of the sediment preventing winnowing.

Current lineations. Lineations are small, (<1 cm) subtle structures present in all photographs, showing current activity. They appear to reflect small particles projecting from, the sediment surface and producing tiny crag-and-tail and scour crescents.

Discussion

Styles of sediment drift

The continental margin of NW Britain is an area of complex morphology including; shallow-steep slopes, basins, seamounts and ridges. These combine to influence the flow of bottom current systems in the area. This complexity encourages the development of a variety of sediment drift types: mounded elongate drifts with sediment waves, sheeted drifts and patch and moat-related drifts. These drift types conform to those documented by Faugères et al. (1993). However, certain drift forms are more indicative of the bottom current regime under which they developed.

(a) The elongate drifts characteristic of the northern Rockall Trough, developed under a lateral gradient of decreasing current strength away from a core of a strong, high velocity bottom currents. The coexistence of the drift with migratory sediment wave fields and the slow migration of the drift itself indicates a long term stability of the current systems in the region. The multi-crested elongate drifts indicate progressively more complex bottom current pathways, perhaps influenced by developing drift topography itself. The resulting 'drift complex' arises from the interaction of several water masses, together with complex bathymetry.

(b) The sheeted drifts in the Faroe–Shetland Channel formed as a result of a vigorous current system (Norwegian Sea Deep Water). The shallowing and narrowing of the channel towards the SW directs the flow and funnels bottom currents, thus maintaining high velocities across the whole channel floor. The sheeted drift thins to the SW as a result of non-deposition, winnowing and erosion, notably over highs in the channel floor. The surface of the drift is smooth, but with some irregular scouring in parts, and covered with a thin sandy contourite layer. Some mounded drifts and associated moats are found in the Faroe–Bank Channel however, the region appears to have been an area of net sediment export.

(c) Areas of erosion and non-deposition are also a characteristic of the influence of bottom current systems. Localized erosion is visible on seismic records expressed as moats and scours associated with elongate drifts, patch drifts and enhanced around pre-existing topographic features. Widespread regions of erosion and non-deposition also occur where highest bottom current velocities are sustained across a broad area for significant periods of time. Erosive/non-depositional boundaries of this sort can be recognised as subparallel discontinuities within a contourite drift separating units of normal accumulation with a slow lateral migration.

Models of contourite deposition

Facies models for contourite accumulation display muddy, silty and sandy contourites, with increasing grain sizes reflecting an increase in bottom current velocity. These 'classic' contourite lithofacies typically occur in a coarsening-upwards to fining-upwards sequence extending from a few tens of centimetres to a few metres in thickness (Stow et al. 1986). Incomplete contourite sequences are more commonly preserved, a situation similar to turbidite sequences. In the northern Rockall Trough and Faroe–Shetland Channel both partial and complete contourite sequences are identified. In partial sequences winnowing and erosion has downcut and removed the lower divisions of silty and muddy contourites, but the middle sandy and upper fine-grained divisions remain are termed *top-only contourites* (Akhurst 1991; Stoker et al. 1998). Further examples illustrate where an erosive episode has left a sandy contourite resting directly on hemipelagic–glacigenic sediment, which indicates current switch-off rather than gradual deceleration producing a fining-upward contourite sequence. These isolated sandy contourites are termed *mid-only contourites* (Stoker et al. 1998).

Contourite sequences tend to reflect the regional sedimentation pattern. In mid to high latitude areas, an ice-rafted component is typically present. Where these deposits have become reworked by bottom currents, these are referred to as *glacigenic contourites*. Glacigenic contourites are present in both the northern Rockall Trough and especially the Faroe–Shetland Channel. These sediments display many of the features of the parent glaciomarine deposit, including dropstones, extremely poor sorting and a composition of both coarse and fine fractions that is clearly supplied through ice rafting. The overprint of bottom current activity is evident in terms of coarsening-upward to fining-upward couplets, winnowing, sharp erosional contacts, and rare lamination. Some particularly coarse-grained, poorly sorted, pebbly muds do not show clear evidence of deposition by bottom currents, their coarse nature masking their origins. These may well be interpreted as glacigenic hemipelagites with abundant oversize dropstones. Interbedded sequences of this nature reflect, an episodic supply of coarse ice-rafted material to the margin, rather than a simple lag concentration by bottom current winnowing.

Across the Hebrides Slope, the layer (5–50 cm) of sandy contourite deposition is quite extensive. Seabed photographs show regions of finely rippled contouritic sand with localized, exposed patches of bioturbated, hemipelagite or contouritic glaciomarine muds (Howe & Humphery 1995). Unpublished side-scan sonar data show the sands as a mobile layer being actively transported along the slope parallel to the contours by a strong northerly flowing slope current (Graham, pers comm., 1994). The ripples seen in the photographs are good indicators of current conditions at the seabed. Small, asymmetrical, linguoid ripples typically form in fine sands in current speeds of 20–50 cm s^{-1}. They develop periodically, and transverse to current flow. They show that although the current is continuous, its velocity may fluctuate (Howe & Humphrey 1995).

This paper is published with the permission of the Director of the British Geological Survey, NERC.

References

AKHURST, M. C. 1991. *Aspects of late Quaternary sedimentation in the Faroe–Shetland Channel, northwest UK continental margin.* British Geological Survey Technical Report **WB/91/2**.

AKHURST, M. C., STOW, D. A. V. & STOKER, M. S. 2002. Late Quaternary glacigenic contourite, debris flow and turbidite process interaction in the Faroe–Shetland Channel, NW European Continental Margin. *In*: STOW, D. A. V., PUDSEY, C. J., HOWE, J. A., FAUGÈRES, J.-C. & VIANA, A. R. (eds) *Deep-Water Contourite Systems: Modern Drifts and Ancient Series, Seismic and Sedimentary Characteristics.* Geological Society, London, Memoirs, **22**, 73–84.

BOOTH, D. A. & ELLETT, D. J. 1983. The Scottish continental slope current. *Continental Shelf Research*, **2**, 127–146.

DICKSON, R. R. & KIDD, R. B. 1986. Circulation in the Southern Rockall Trough, oceanographic setting of site 610. *Initial Reports of the Deep-Sea Drilling Project*, **114**, 1061–1074.

FAUGÈRES, J.-C., MEZERAIS, M. L. & STOW, D. A. V. 1993. Contourite drift types and their distribution in the North and South Atlantic Ocean basins. *Sedimentary Geology*, **82**, 189–203.

HILL, A. E. & MITCHELSON-JACOB, E. G. 1993. Observations of a poleward flowing saline core on the continental slope, West of Scotland. *Deep-Sea Research*, **40**, 1521–1527.

HOWE, J. A. 1995. Sedimentary processes and variation in slope-current activity during the last glacial-interglacial episode on the Hebrides Slope, Northern Rockall Trough, North Atlantic Ocean. *Sedimentary Geology*, **96**, 201–230.

HOWE, J. A. 1996. Turbidite and contourite sediment waves in the Northern Rockall Trough, North Atlantic Ocean. *Sedimentology*, **43**, 219–234.

HOWE, J. A. & HUMPHERY, J. D. 1995. Photographic evidence for slope-current activity, Hebrides Slope, NE Atlantic Ocean. *Scottish Journal of Geology*, **30**(2), 107–115.

HOWE, J. A., STOKER, M. S. & STOW, D. A. V. 1994. A Late Cenozoic sediment drift complex, northeast Rockall Trough, North Atlantic. *Palaeoceanography*, **9**, 989–999.

HOWE, J. A., HARLAND, R., HINE, N. M. & AUSTIN, W. E. N. 1998. Late Quaternary stratigraphy and palaeoceanographic change in the northern Rockall Trough, North Atlantic Ocean. *In*: STOKER, M. S., EVANS, D. & CRAMP, A. (eds) *Geological Processes on Continental Margins: Sedimentation, Mass Wasting and Stability.* Geological Society, London, Special Publications, **129**, 269–286.

HUTHNANCE, J. M. 1986. Rockall slope current and shelf edge processes. *Proceedings of the Royal Society of Edinburgh*, **88B**, 83–101.

KNUTZ, P. C., JONES, E. J. W., HOWE, J. A., VAN WEERING, T. J. C. & STOW, D. A. V. 2002. Wave-form sheeted contourite drift on the lower Barra fan, NW UK continental margin. *In*: STOW, D. A. V., PUDSEY, C. J., HOWE, J. A., FAUGÈRES, J.-C. & VIANA, A. R. (eds) *Deep-Water Contourite Systems: Modern Drifts and Ancient Series, Seismic and Sedimentary Characteristics.* Geological Society, London, Memoirs, **22**, 85–97.

RICHARDS, P. C., RITCHIE, J. D. & THOMSON, A. R. 1987. Evolution of deep-water climbing dunes in the Rockall Trough – implications for overflow currents across the Wyville-Thomson Ridge in the (?) Late Miocene. *Marine Geology*, **76**, 177–183.

STOKER, M. S. 1995. The influence of glacigenic sedimentation on slope-apron development on the continental margin off NW Britain. *In*: SCRUTTON, R. A., STOKER, M. S., SHIMMIELD, G. B. & TUDHOPE, A. W. (eds) *The Tectonics, Sedimentation and Palaeoceanography of the North Atlantic Region.* Geological Society, London, Special Publications, **90**, 159–177.

STOKER, M. S. 1998. Sediment drift development on the continental margin off NW Britain. *In*: STOKER, M. S., EVANS, D. & CRAMP, A. (eds) *Geological Processes on Continental Margins: Sedimentation, Mass Wasting and Stability.* Geological Society, London Special Publications, **129**, 229–254.

STOKER, M. S., HARLAND, R., MORTON, A. C. & GRAHAM, D. K. 1989. Late Quaternary stratigraphy of the northern Rockall Trough and Faroe–Shetland Channel, northeast Atlantic Ocean. *Journal of Quaternary Science*, **4**, 211–222.

STOKER, M. S., HARLAND, R. & GRAHAM, D. K. 1991. Glacially influenced basin plain sedimentation in the southern Faroe–Shetland Channel, northwest United Kingdom continental margin. *Marine Geology*, **100**, 185–199.

STOKER, M. S., HITCHEN, K. & GRAHAM, C. C. 1993. *United Kingdom Offshore Regional Report: The Geology of the Hebrides and West Shetland Shelves and Adjacent Deep-Water Areas.* HMSO for the British Geological Survey, London.

STOKER, M. S., AKHURST, M. C., HOWE, J. A. & STOW, D. A. V. 1998. Sediment drifts and contourites on the continental margin off northwest Britain. *Sedimentary Geology*, **115**, 33–51.

STOW, D. A. V., FAUGÈRES, J.-C. & GONTHIER, E. 1986. Facies distribution and textural variation in Faro Drift contourites: velocity fluctuation and drift growth. *Marine Geology*, **72**, 71–100.

STOW, D. A. V., ARMISHAW, J. E. & HOLMES, R. 2002. Holocene contourite sand sheet on the Barra Fan slope, NW Hebrides margin. *In*: STOW, D. A. V., PUDSEY, C. J., HOWE, J. A., FAUGÈRES, J.-C. & VIANA, A. R. (eds) *Deep-Water Contourite Systems: Modern Drifts and Ancient Series, Seismic and Sedimentary Characteristics.* Geological Society, London, Memoirs, **22**, 99–119.

Late Quaternary glacigenic contourite, debris flow and turbidite process interaction in the Faroe–Shetland Channel, NW European Continental Margin

MAXINE C. AKHURST[1], DORRIK A. V. STOW[2] & MARTYN S. STOKER[1]

[1]*British Geological Survey, Murchison House, West Mains Road, Edinburgh EH9 3LA, UK*
[2]*SOES–SOC, Southampton University, Southampton SO14 3ZH, UK*

Abstract: The Faroe–Shetland Channel is an important conduit or gateway for the southward flow of cold bottom waters formed in the Norwegian Sea. This Norwegian Sea Overflow Water (NSOW) finds several spillover channels across the Wyville–Thomson Ridge, eventually descending into the northern Rockall Trough and Iceland Basin. The Neogene channel floor succession predominantly displays a broad sheeted drift geometry. Bottom current scours and channels were apparently inherited from an episode of enhanced bottom current activity in late Oligocene/early Miocene. The late Quaternary channel-floor succession is dominated by distal glaciomarine sediments, derived from the shelf and slope during glacial stages and mostly transported by ice-rafting. Glacigenic debris flows and minor turbidity currents were also active across the slope region. Consequently, the principal channel-floor facies are glacigenic contourites that show extensive bioturbation, rare primary structures, mixed composition and marked grain size variation. These features indicate the important influence of cyclical fluctuations in bottom current velocity throughout both stadial and interstadial or interglacial periods. However, the concentration of sandy contourites, erosive surfaces and top-only contourites during interstadials/interglacials and during phases of marked cooling or warming testify to the significance of climate-control on contourite deposition.

The Faroe–Shetland Channel is a NE trending, narrow, funnel-shaped, deep-water trough off NW Scotland, separating the West Shetland Shelf to the southeast from the Faroes Shelf to the northwest. The Channel opens out towards the Norwegian Sea in the NE and abuts against the Wyville–Thomson Ridge in the south where it turns to the NW and becomes the Faroe Bank Channel (Fig. 1).

The Faroe–Shetland Channel is the present-day bathymetric expression of an older sedimentary basin known as the Faroe–Shetland Basin, which has been at the centre of considerable exploration effort by the oil industry over the past 15 years. This followed early discovery of the Devono-Carboniferous Clair Field on the adjacent West Shetland Shelf, and has led more recently to the discovery of several Paleocene reservoirs near the southern end of the basin, including the Schiehallion, Foinhaven, Loyal and Suilven oil fields.

As a consequence of this economic interest, there is a plethora of data from the region, mostly from the pre-Quaternary section and much of which is still confidential. However, there have also been surficial mapping programmes, by the British Geological Survey (BGS) and the Geological Survey of Denmark and Greenland, and environmental surveys in relation to offshore exploration and production issues.

This paper presents a summary of some of these data from the

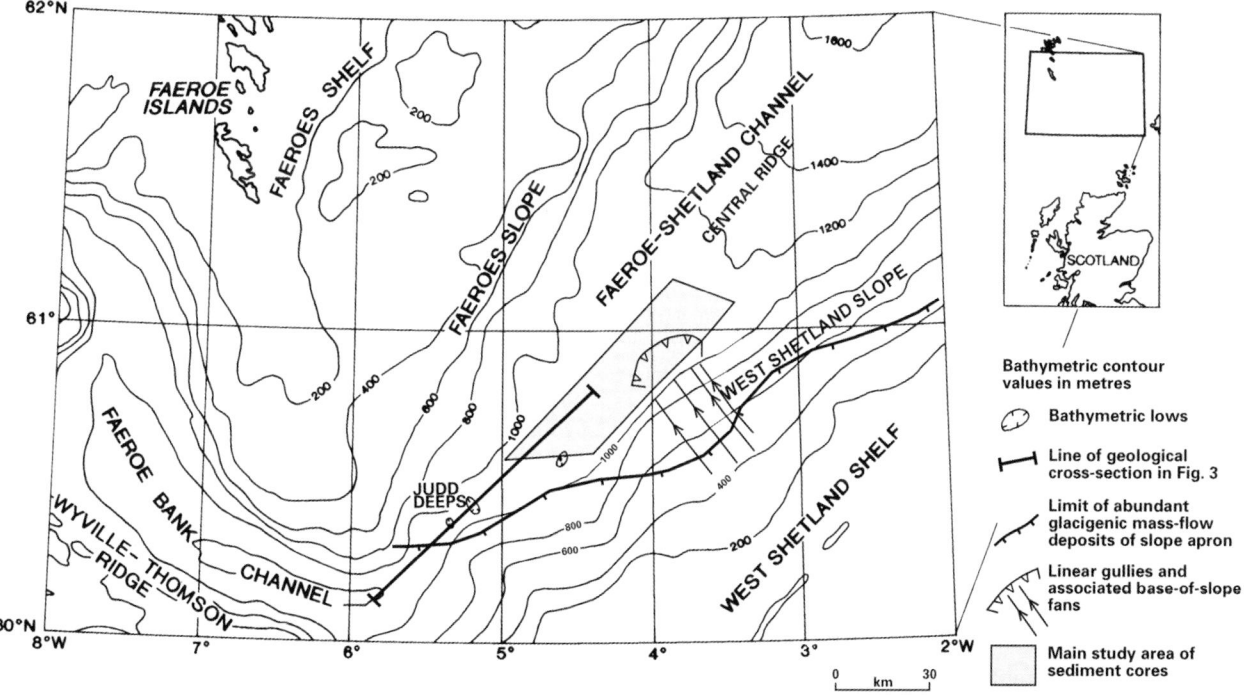

Fig 1. Location map and bathymetry of the SW Faroe–Shetland Channel region, showing the main area of study of the sediment cores. The limit of glacigenic mass-flow slope-apron deposits is taken from Stoker (1999), and the linear gullies and associated fans is adapted from Stoker *et al.* (1993) Masson (2001) and Bulat & Long (in press). Note that debris flows interbedded with the basinal section occur within the channel (e.g. see Fig. 7).

From: Stow, D. A. V., Pudsey, C. J., Howe, J. A., Faugères, J.-C. & Viana, A. R. (eds)
Deep-Water Contourite Systems: Modern Drifts and Ancient Series, Seismic and Sedimentary Characteristics.
Geological Society, London, Memoirs, **22**, 73–84. 0435-4052/02/$15.00 © The Geological Society of London 2002.

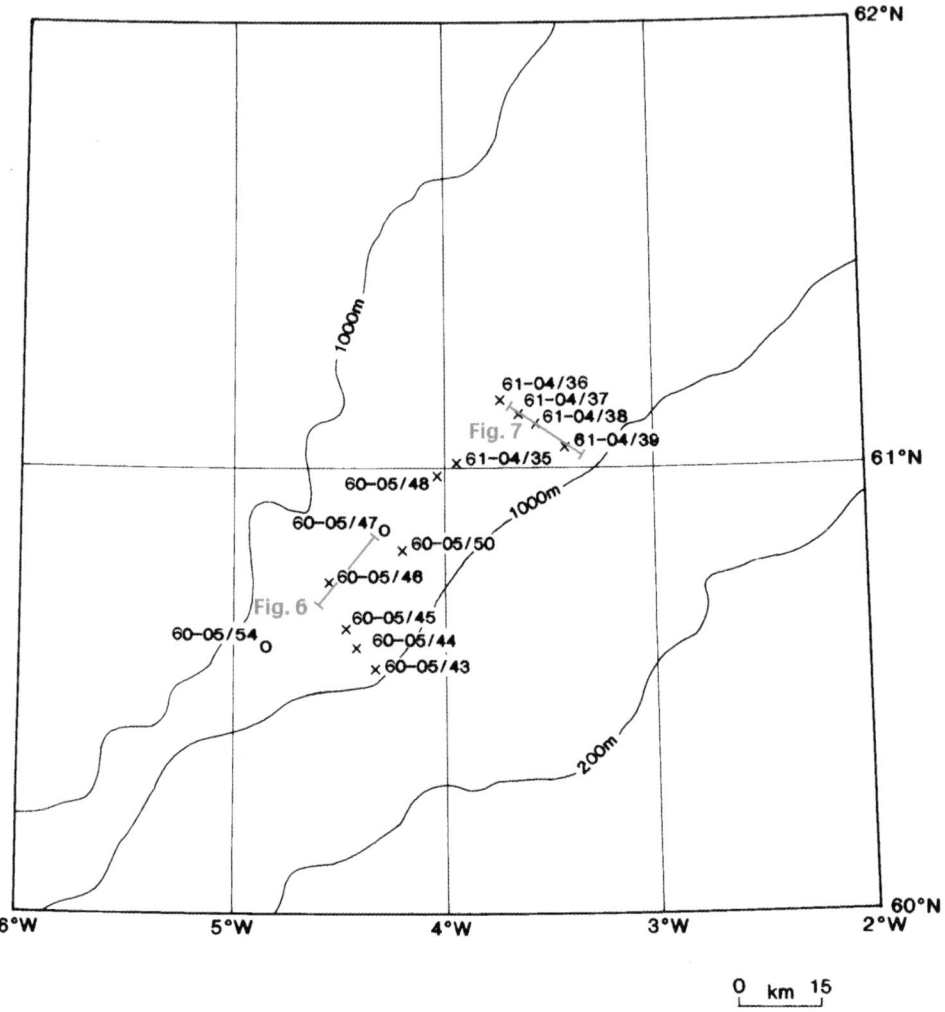

Fig 2. Location of vibrocores, recovered by British Geological Survey, used in this study and seismic profiles shown in Figures 6 and 7.

+ = Vibrocore samples examined
o = Disturbed vibrocore samples

southwest end of the Faroe–Shetland Channel in order to elucidate the nature of late Quaternary and Recent processes and sedimentation in the channel. Regional and oceanographic data are summarised from an extensive literature. Seismic studies, undertaken as part of a regional geological survey by BGS, used 40 cubic inch airguns, 1.5 and 2 kJ sparkers, 5 and 6 kV deep-tow sparkers, and 6 kV deep-tow boomer sources, together with an acoustic velocity of 1.6 km s^{-1} for interpretation of the Quaternary section. Sediment data are from a series of short vibrocores recovered from the Faroe–Shetland Channel (Fig. 2), which were then subjected to careful laboratory analyses. For the sedimentological information, we have drawn extensively on an earlier BGS report by the senior author (Akhurst 1991).

Geological and oceanographic setting

Geological setting and palaeoceanography

The Faroe–Shetland Basin is a narrow fault-bounded basin lying parallel to the northwest UK passive continental margin. It has probably existed as a depocentre since the late Palaeozoic following a Permo-Triassic rifting episode (Doré et al. 1999; Roberts et al. 1999), and has been a particularly important locus for sediment accumulation from the Mesozoic to Recent. The Mesozoic succession is some 7 km thick, of which up to 5 km are Upper Cretaceous deep-water sediments deposited as the basin subsided as a result of crustal extension.

During the Paleocene, basaltic volcanic lavas were extruded along the northwest margin whereas up to 2 km of deep-water sediments accumulated in the basin itself. A further 1.2 km sediment pile now represents the Eocene deposits, although the original thickness has been much reduced by synsedimentary tectonism and an episode of strong bottom current erosion during the latest Oligocene/early Miocene interval (Stoker 1990a; Stoker et al. in press). This intensification of bottom current circulation, which led to channel incision into the trough floor, followed the development of deep-water pathways linking the Arctic and Atlantic oceans via the Norwegian–Greenland Sea, and consequent southward flow of cold Arctic waters (Miller & Tucholke 1983; Zeigler 1988).

It is the plate-tectonic evolution of the Norwegian–Greenland Sea that strongly influenced the development of deep-ocean connections; specifically the opening of the Fram Strait (Northern Gateway), and the subsidence of the Greenland–Scotland Ridge (the Southern Gateway) (Jansen & Raymo 1996; Thiede & Mhyre 1996). It has been suggested that deep-water circulation in the Norwegian–Greenland Sea may have been initiated during late

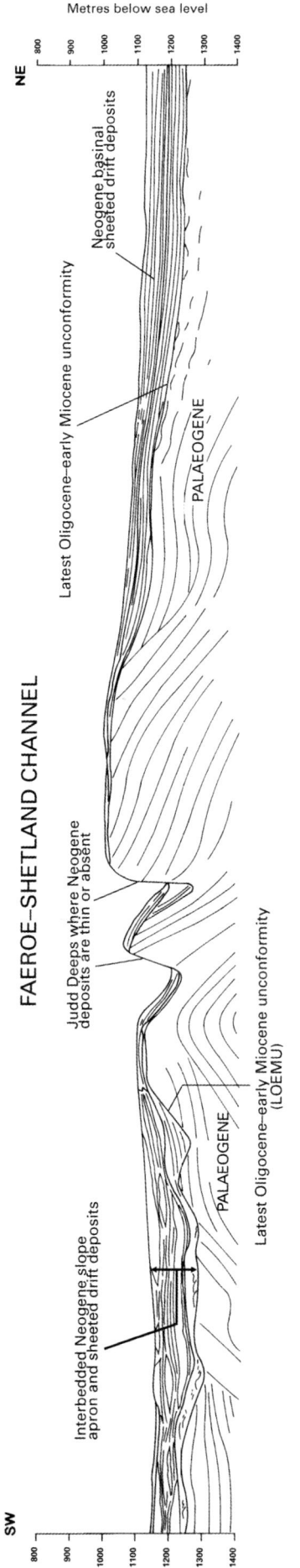

Fig. 3. Interpreted geological cross-section across part of the Faroe–Shetland Channel (modified from Stoker 1990b). For location of section see Figure 1.

Eocene/early Oligocene time (Berggren & Schnitker 1983; Zeigler 1988; Davies *et al.* 2001). However, the modern pattern of deep-water exchange may be a Neogene phenomenon, initiated in the Miocene as the Fram Strait developed a true deep connection, and the Greenland–Scotland Ridge became fully submerged (Eldholm 1990; Jansen & Raymo 1996). The Faroe–Shetland Channel, with the Faroe–Bank Channel, lies at the southeast end of the Greenland–Scotland Ridge, and represents the deepest passageway across the Southern Gateway.

The effects of bottom current activity in the Faroe–Shetland Channel are manifest in the style of deep-water sedimentation associated with the Neogene succession. Overlying the latest Oligocene–early Miocene unconformity (LOEMU of Stoker 1999) is a sequence of Neogene sheeted drift deposits (Stoker *et al.* 1998) that range from about 100 m thick to locally absent over Palaeogene inversion structures on the basin floor at the southwest end of the channel (Fig. 3). This contrasts with the adjacent slope apron on the West Shetland Slope that locally exceeds 300 m thickness. The slope apron prograges into the basin from the West Shetland margin and interdigitates with the thinner basin-floor sheeted drift deposits (Fig. 3). There is clear evidence of non-deposition and local erosion throughout this part of the basinal drift sequence. Farther to the northeast, as the Channel deepens, the Neogene sequence appears to thicken substantially and may be several hundred metres thick (Davies *et al.* 2001).

Oceanographic setting (Fig. 4)

At the present day, cold (–0.5–1°C), southward flowing Norwegian Sea water fills the Faroe–Shetland Channel to a depth of approximately 500 m. This is separated from warm (> 9°C), northward-flowing Atlantic water at the surface by an intermediate layer 100–200 m thick. In fact, surface circulation in the Faroe–Shetland Channel forms an anticlockwise gyre, the core of Norwegian Sea Overflow Water (NSOW) is along the northwest side of the channel and deep eddy currents have been noted within 100 m of the seafloor (Dooley & Meinke 1981; Saunders & Gould 1989; Saunders 1990).

Current meter data available from deep moorings in the Faroe–Shetland Channel, provided by the British Oceanographic Data Centre, typically show mean weekly velocities of 20–40 cm s^{-1} directed towards the southwest or west–southwest. Individual readings can be much greater, and reverse flow interpreted as large eddies is also noted.

Bathymetry

The Faroe–Shetland Channel separates the West Shetland and Faroes margins. The shelf break of the West Shetland Shelf lies at about 200 m water depth, whereas the principal gradient change south of the Faroes Shelf is closer to 300 m (Fig. 1). Slope angles on both flanks are relatively shallow, from 1–3°.

The Faroe–Shetland Channel slopes very gently from a depth of just over 1000 m in the southwest (Fig. 1) to some 1700 m in the northeast, a distance of just less than 450 km. It presents an elongate funnel shape, opening from a width of some 25 km (between the 1000 m isobaths) in the southwest to over 120 km at the northeast end where it merges with the Norwegian Sea. At its southwest end it abuts the Wyville–Thompson Ridge, where it turns sharply towards the northwest and becomes the Faroe–Bank Channel (Fig. 1).

Detailed bathymetric study of the floor reveals that north of about 60° 30′N it has a relatively flat cross-sectional profile, with a low central ridge along part of the axial region as well as trough-parallel bathymetric lows or minor channels in both axial and marginal positions. South of 60° 30′N, the bathymetry is more

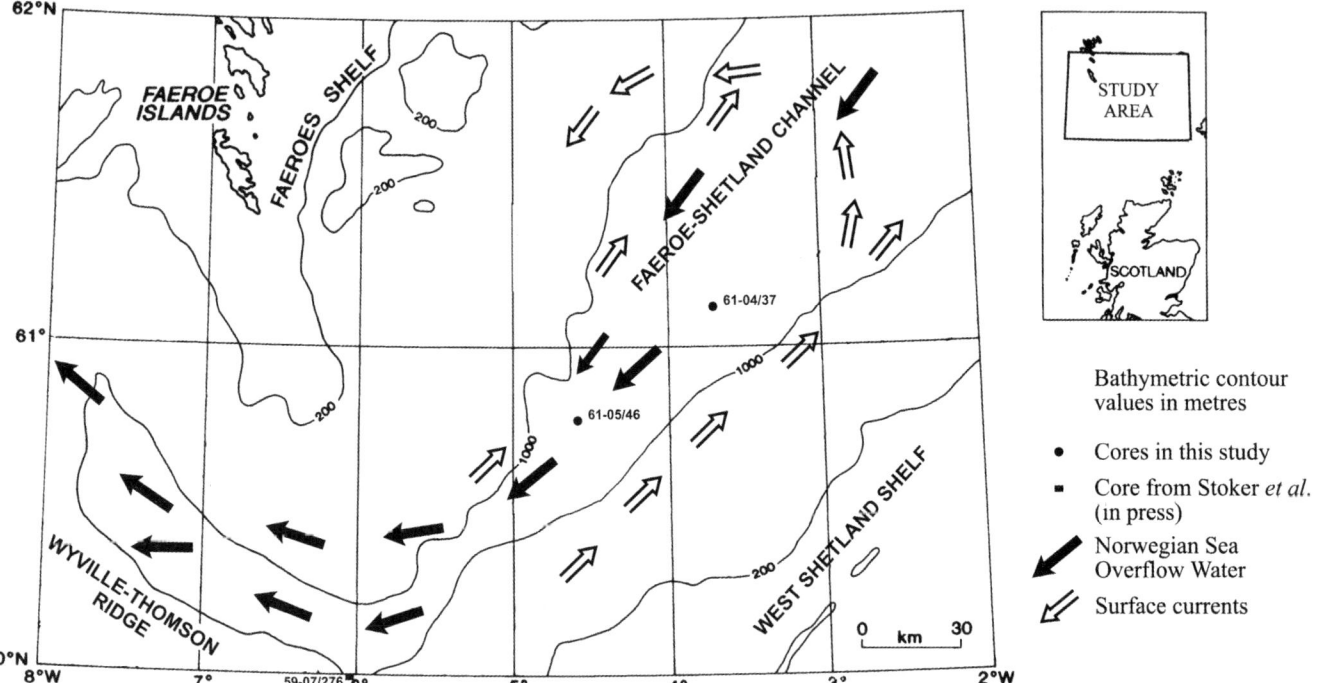

Fig 4. Present-day deep-water and surface circulation in the Faroe–Shetland Channel region.

complex and several enclosed deeps, the Judd Deeps, form a highly sculpted channel floor (Figs 1 & 3). The formation of these deeps may be linked to very strong bottom current erosion in late Oligocene–early Miocene time (Stoker et al. in press).

Stratigraphic context

The Neogene succession on the West Shetland margin has been assigned to the Nordland Group (Stoker 1999). On the shelf and slope, the Nordland group has been divided into Lower, Middle and Upper Nordland units that represent, respectively, Miocene, Pliocene to middle Pleistocene, and middle Pleistocene to Holocene deposits. However, the recognition of these units in the basinal succession remains problematic, particularly at the southwest end of the Channel where the Neogene strata are relatively thin. Thus, for the purpose of this study, especially with respect to their seismic characteristics (see Figs 6 & 7), the basinal strata are here referred to as Neogene (Nordland Group) undivided.

By way of contrast, the sediment cores used in this study have provided a high-resolution subdivision of the uppermost layers of the Neogene succession. Short (< 6 m long) vibrocores recovered from the Faroe–Shetland Channel have been accurately dated using a combination of oxygen isotopes, carbon 14 analysis and microfossil assemblages (Fig. 5; Akhurst 1991). The oldest material recovered for this study is considered to be from the last interglacial (Eemian).

Seismic characteristics of the Neogene basinal succession

In the Faroe–Shetland Channel, the Neogene basinal sequence is mostly characterized by high-frequency, laterally continuous, parallel, sub-horizontal reflectors on both airgun and sparker profiles (Figs 6 & 7), that reflect an overall sheeted drift geometry (Stoker et al. 1998). The base of the sequence locally onlaps the underlying Palaeogene strata where the LOEMU forms an irregular surface (Fig. 6a, b). Such basin-floor onlap is most spectacularly demonstrated in the area of the Judd Deeps (Fig. 3). Within the Judd Deeps, the Neogene reflectors display a downlapping relationship indicative of lateral accretion. Farther to the northeast, internal convergence of reflectors over Palaeogene highs and over the flanks of the low central ridge is indicative of depositional thinning and, more rarely, very low-angle truncation of reflectors suggests localised short-lived erosive episodes (Fig. 7).

On the West Shetland slope, the Neogene slope apron deposits are locally > 300 m thick over parts of the slope south of 60° 30′N. The bulk of the slope apron displays a seismic character indicative of slope progradation and downslope movement (Stoker 1995, 1999; Bulat & Long in press), although eroded remnants of formerly more extensive, Miocene, elongate mounded drifts are buried beneath the prograding wedge on the middle and lower slope (Stoker 1999). The prograding deposits generally comprise transparent and chaotic seismic units bounded by laterally continuous, high amplitude reflectors (Fig. 3). High-resolution profiles show characteristic lensoid units (Fig. 7), which commonly display irregular tops and locally erosive bases. These sediments are interpreted as debris flow deposits (Stoker 1990a, b; Holmes 1990; Stoker et al. 1991; Stoker 1995), the most recent of which were derived from late Pleistocene ice sheets located at the edge of the West Shetland Shelf (Stoker 1995). The debris flows interdigitate downslope with the, seismically well-layered basinal sequence and, locally, extend onto the channel floor (Fig. 3). Where the debris flows and basin-floor sediments are interbedded, the former tends to increase the thickness of the basinal succession (Figs 3 & 7). High-resolution seafloor mapping has revealed areas of downslope-oriented linear gullies associated with base-of-slope debris-flow fans between 60° 30′ and 61° 0′N (Masson 2001; Bulat & Long in press). These linear, highly erosive features are most probably the result of turbidity currents (Fig. 1). Thus, material transported by ice-rafting, debris flow and turbidite processes was delivered onto the channel floor in the area of study.

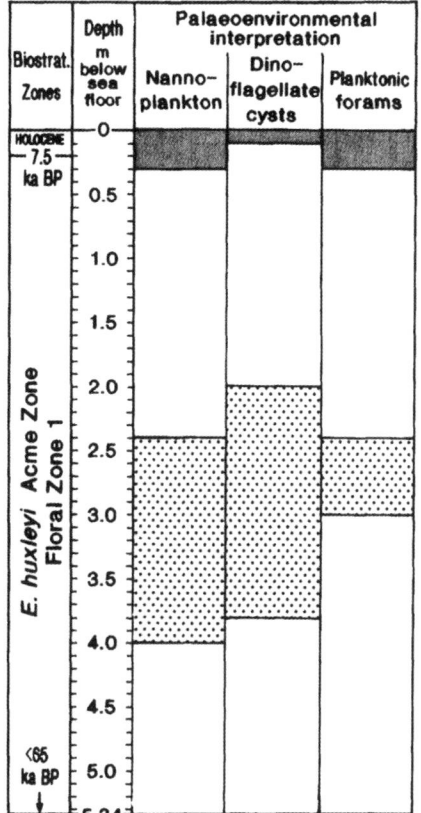
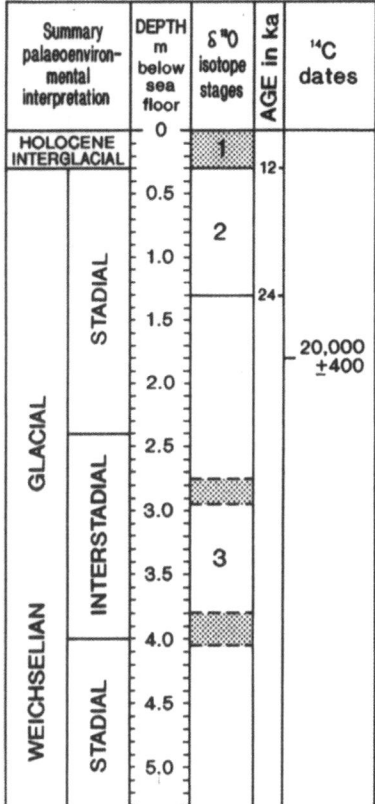
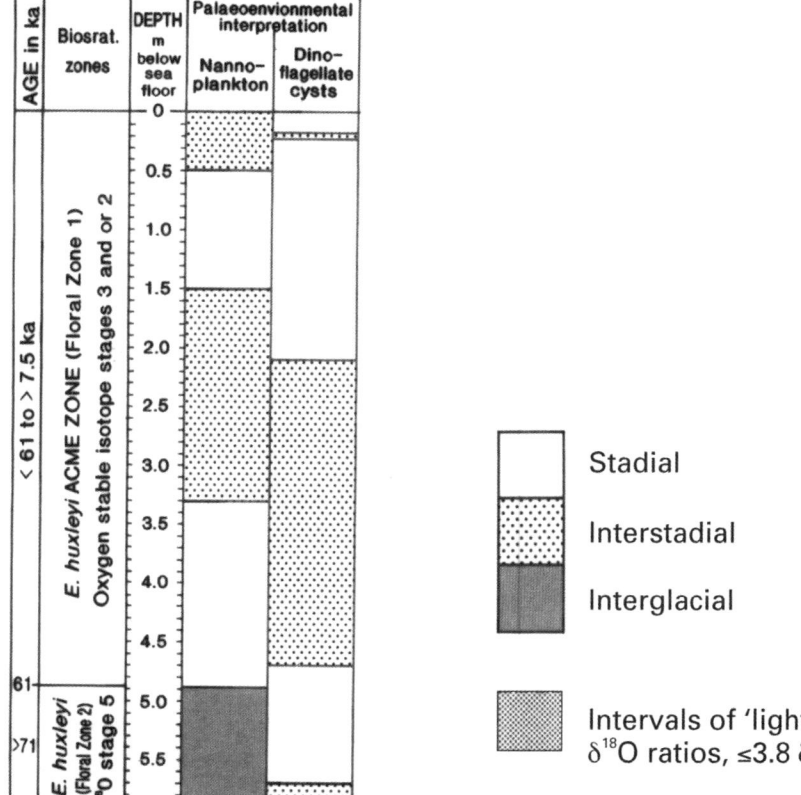

Fig 5. Detailed late Quaternary–Holocene stratigraphy of cores 46 and 37 from the Faroe–Shetland Channel.

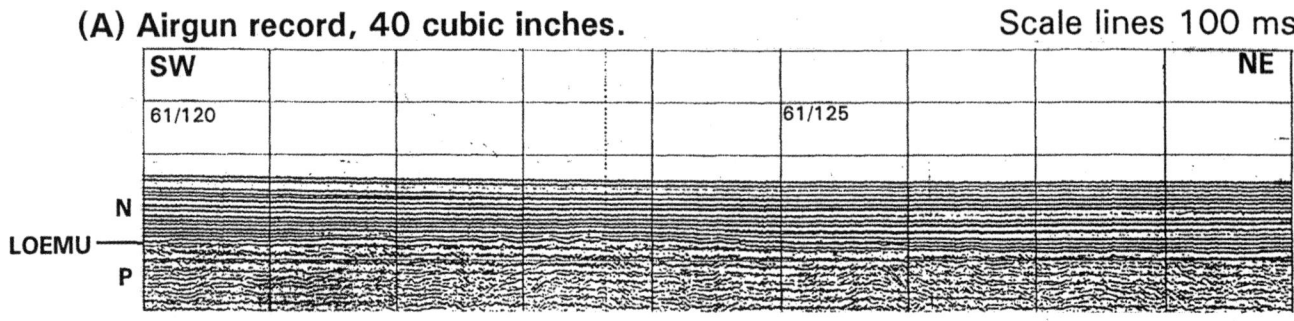

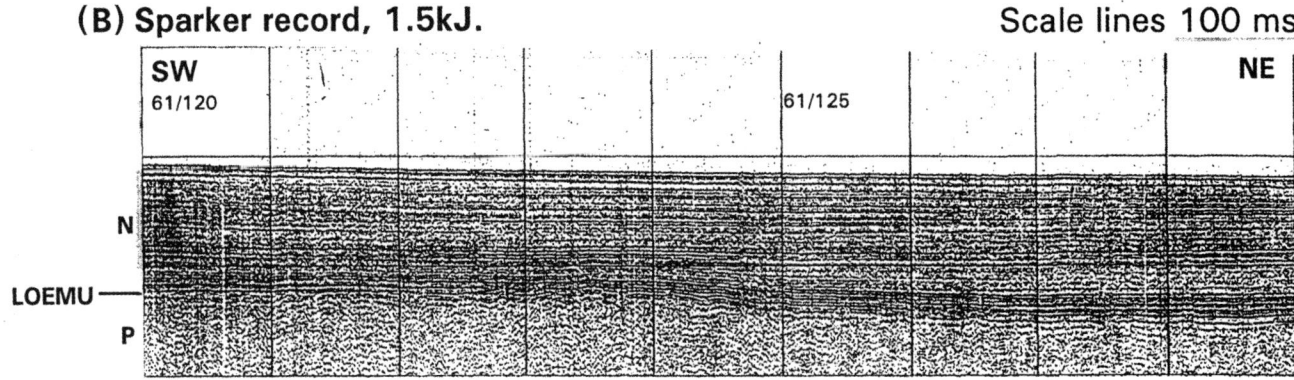

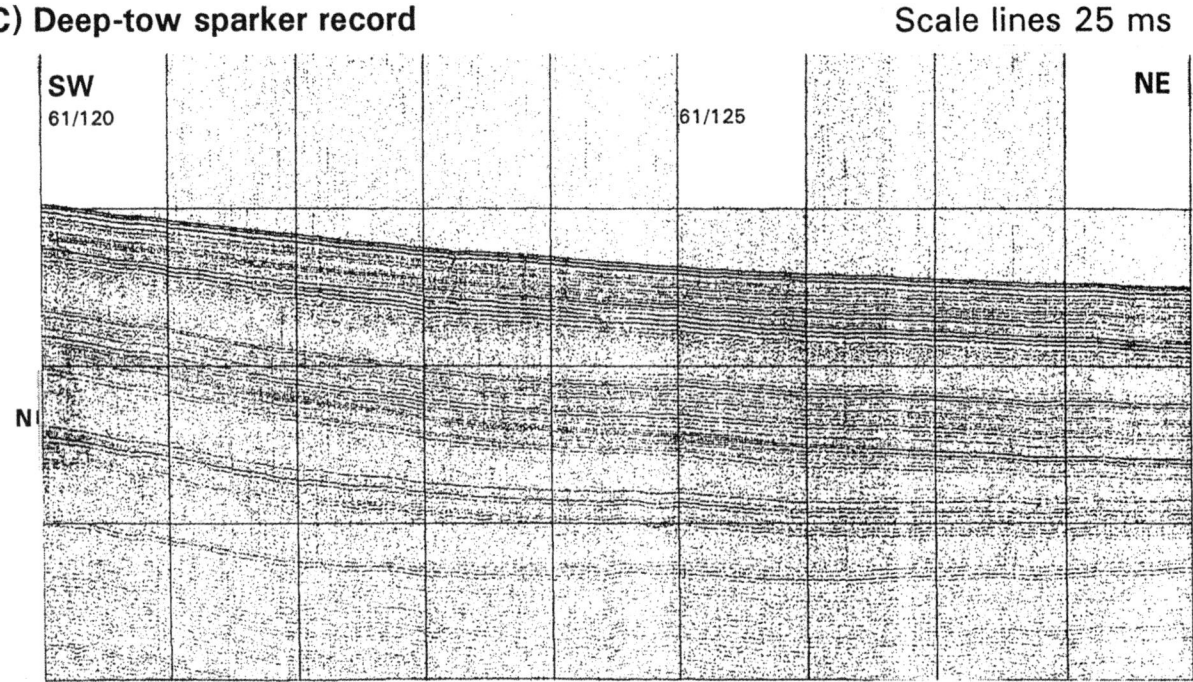

Fig 6. Typical seismic profiles from the floor of the Faroe–Shetland Channel (BGS Survey 83/04, line 61, fixes 120–128), showing characteristic airgun, sparker and deep-tow sparker sources. Note onlap of Neogene strata onto irregular basal unconformity (LOEMU) in profiles A and B. Profile is located in Figure 2.

Sedimentary characteristics of Upper Neogene (Quaternary) basinal deposits

Sediment facies

Five distinct facies have been identified in the Faroe–Shetland Channel based on observations from the eleven vibrocores used for this study. These are listed below with their relative abundance (%) in the sections examined, and illustrated in Figure 8. Gravel clasts and coarse sand grains are common throughout, and can be readily interpreted as glacial Ice Rafted Dropstones (IRD).

Mud	*26%*
Sandy mud – unsorted	*50%*
– sorted	*10%*
Muddy sand and gravelly muddy sand	*8%*
Dark grey mud	*1%*
Pale laminated mud	*5%*

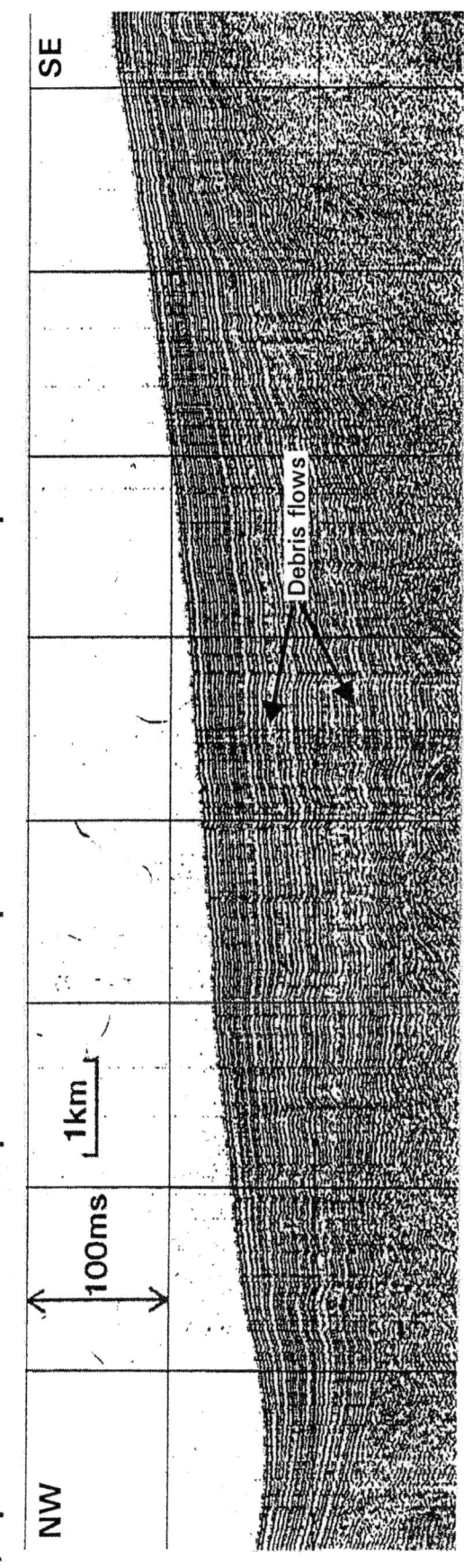
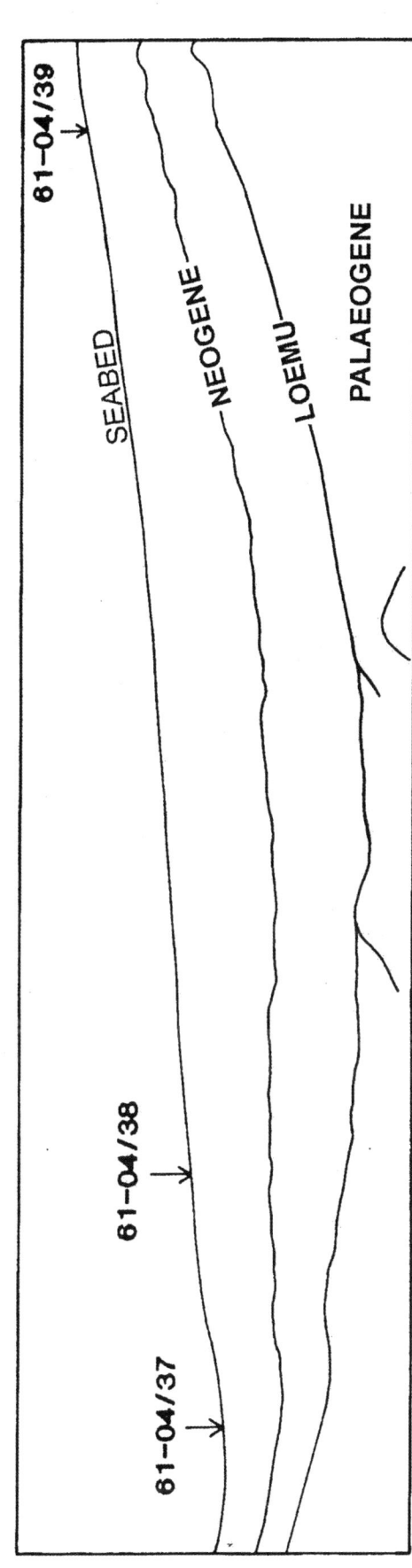

Fig 7. Sparker seismic section (BGS Survey 84/05, line 43) across part of the Faroe–Shetland Channel. Note broad lenticularity of the Faroe–Shetland Channel sequence over the Upper Tertiary section, indicative of a contourite sheet-drift geometry. Note also: 1) the marked thinning of the sedimentary section in the central portion of the Faroe–Shetland Channel, related to non-deposition and erosion under an active bottom current system; and 2) the general thickening of the basinal section where interbedded, lensoid, debris flows are present. LOEMU, latest Oligocene–early Miocene unconformity. Profile is located in Figure 2.

Fig 8. Core photographs of typical sediment facies from Faroe–Shetland Channel cores. See Figure 2 for core locations and abbreviations. (a) Mud Facies, vibrocore 61-04/38VE, 3.88 to 4.12 m below sea floor; (b) Sandy Mud Facies, vibrocore 60-05/43VE, 2.41 to 2.64 m below sea floor; (c) Muddy Sand Facies, vibrocore 60-05/50VE, 4.34 to 4.57 m below sea floor; (d) Dark Mud Facies, vibrocore 60-05/43VE, 4.73 to 4.97 m below sea floor; (e) Pale Mud Facies, vibrocore 60-05/48VE, 0.68 to 0.91 m below sea floor; (f) Pale Mud Facies, vibrocore 60-05/35VE, 1.18 to 1.41 m below sea floor.

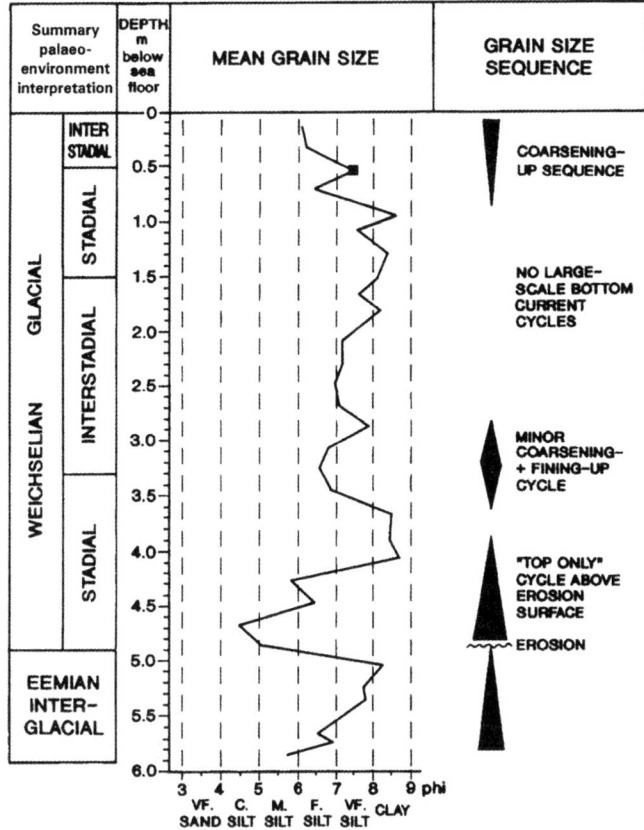

Fig 9. Mean grain-size variation, grain-size sequences and erosive horizons together with best interpretation of vibrocore stratigraphy noted in a) vibrocore 61-05/46VE and b) vibrocore 61-04/37VE. See Figure 2 for core location.

The main characteristics that distinguish each of these facies are their colour, grain size, sedimentary structures, ichnofacies and nature of bedding. They are all dominantly siliciclastic in composition, with a very minor foraminiferal/nannofossil component. The IRD gravel clasts include an assortment of sedimentary, igneous and metamorphic rocks consistent with derivation from the Scottish mainland. Akhurst (1991) provides a detailed overview of the ichnofacies and further work on this topic is in progress (A. Wetzel and others), although we do not consider it further in this paper.

Mud facies: olive grey to dark olive grey when fresh; mainly fine-grained (clay-silt size) with < 1% IRD; mostly structureless and thoroughly bioturbated with distinctive burrow traces common (Fig. 8a); bedding typically gradational (0.10–2.28 m thick), but with sharp contacts in places. *Interpretation* – glaciomarine hemipelagites and muddy contourites.

Sandy mud facies: grey to olive grey when fresh; slightly sandy to sandy muds with variable % IRD; mostly structureless and thoroughly bioturbated with distinctive burrow traces common (Fig. 8b); bedding typically gradational (0.08–3 m thick), but with sharp contacts in places. *Types* – unsorted sandy muds in which the sand fraction shows the full range of grain size; sorted sandy muds in which the sand fraction has a distinct mode in the fine-very fine sand size; sandy muds with micro-shale clast horizons. *Interpretation* – glaciomarine hemipelagites, mottled silt-mud contourites, and micro-shale clast contourites.

Muddy sand facies: dark grey to greyish brown when fresh; poorly sorted very muddy sand size with variable % IRD (in some cases relatively high); mostly structureless and thoroughly bioturbated with distinctive burrow traces common (Fig. 8c); bedding typically gradational (0.10–0.68 m thick) with bioturbated contacts, basal contact generally more distinct than upper, and some erosion evident. *Types* – muddy sand and gravelly muddy sand. *Interpretation* – glaciomarine sandy hemipelagites and sandy contourites.

Dark mud facies: very dark grey brown to dark grey when fresh; mud or slightly sandy mud size with rare IRD and micro-shale clasts (Fig. 8d); structureless to diffuse planar laminated, with little bioturbation; thin distinct beds (3–10 cm thick). *Interpretation* – unclear.

Pale laminated mud facies: pale brown to greyish brown when fresh; mainly fine-grained (silty clay to slightly sandy mud size) with < 1% IRD and soft textured; well laminated (Fig. 8e) and graded laminated units, some contorted lamination, overall normal grading, bioturbation near the tops of beds, sharp at bases, with some distinctive burrow traces (Fig. 8f); isolated thin to medium thick beds (< 0.30 m thick). *Interpretation* – fine-grained turbidites.

Vertical and horizontal distribution

The dominant facies making up some 94% of the section examined are the muds, sandy muds and muddy sands. These are typically arranged in a broadly cyclic manner from mud to sandy mud to sand facies and then back through sandy mud to mud facies (Stoker *et al.* 1998). This pattern is clearly visible on grain size profiles (Fig. 9), where the grain size oscillation cycles are between 0.3 and 1.2 m thick, and is very typical of the standard contourite facies model (Stow *et al.* 1986, 1996).

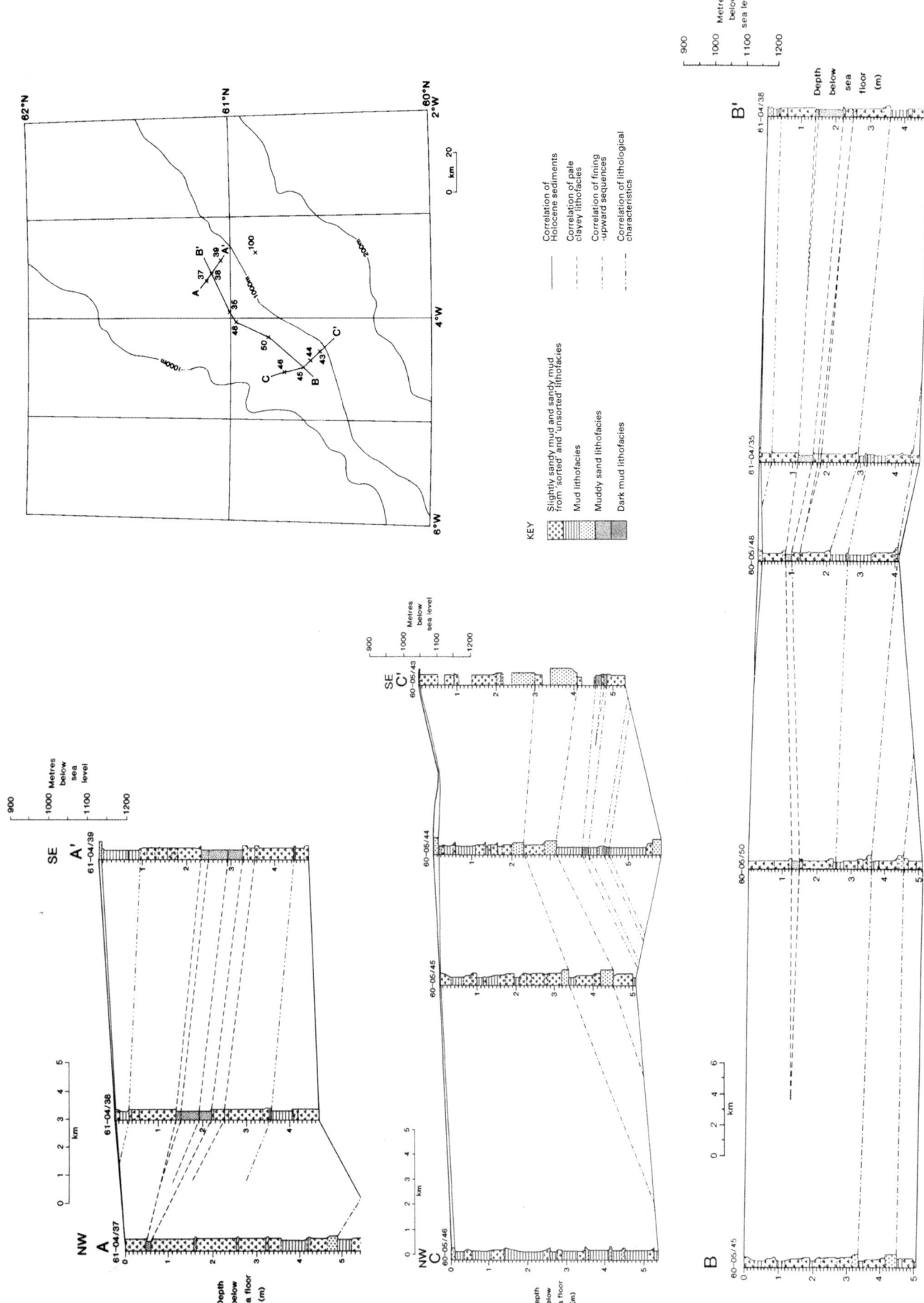

Fig 10. Panel cross-sections showing lateral correlation and vertical distribution of facies from the studied cores.

Some variability in sequences is observed with the fining-up portions of the cycles (i.e. top-only contourites) being common towards the base of the cored section and a coarsening-up cycle (i.e. base-only contourites) occurring at the top of most cores. This latter is capped by the thin Holocene muddy sand unit. Erosive horizons commonly occur coincident with the coarsest grain size and at the base of the fining-up cycles.

Lateral correlation between ten of the study cores has been possible to a limited extent over the study area (Fig. 10). Principal aspects of the lateral distribution of facies are:

- Regional correlation of the surficial Holocene sandy facies as a thin veneer (in places either very thin and non-recovered or absent, e.g. cores 37 and 50).
- The pale-brown laminated turbidite facies can be correlated between cores in the northeastern part of the study area, being thickest in core 39 and decreasing in thickness both across the channel (to cores 38 to 37) and along the channel (to cores 35, 48 and 50), this is consistent with lateral supply from the southeast probably close to core 39, and has been correlated with a debris flow event on the slope (Akhurst 1991).
- Only local correlation of the dark grey mud facies, i.e. between the two cores (43 and 44) in which it occurs.
- Two prominent sandy contourites at the base of the fining-up units (top-only contourites) can be correlated between cores 43, 44 and 45 in the southern transect, and further north to cores 50, 48, 35 and possibly 37 (for the top sand only).
- Correlation of the sandy mud facies and grain size cycles is much less definitive and not attempted here (but see Akhurst 1991, for a fuller discussion).

Sedimentation rates

Dating of selected cores used in this study has been carried out using a combination of biostratigraphic analysis (nannofossils, planktonic foraminifers and dinoflagellate cysts) and limited oxygen isotope data. Average sedimentation rates for the upper 5 m of section vary from about 3 cm ka^{-1} (core 46) to 7 cm ka^{-1} (core 37). Even removing the turbidite beds from the calculation, the highest rates (about 10 cm ka^{-1}) are found in core 37. Taking the Holocene section alone, the mean accumulation rates are from 2–5 cm ka^{-1}.

Discussion

Depositional processes and evidence for contourites

Glaciomarine sedimentation is clearly prevalent through most of the Quaternary section studied in the cores. IRD are everywhere present apart from in the thin Holocene veneer and in localised turbidites. The cores studied from the Faroe–Shetland Channel are mostly, therefore, within the distal glaciomarine environment, where glaciomarine hemipelagic sedimentation dominated and, in this case, was clearly modified by the action of strong bottom currents. Other non-glacial processes are more minor in their contribution.

Although seismic evidence suggests that debris flow processes constructed much of the slope succession, minor turbidity current input is evident locally as indicated by the linear gullies (Fig. 1). Akhurst (1991) has interpreted the thin dark grey clay facies as the distal component of a glacigenic debris flow, partly based on the presence of micro-shale clasts. However, we now suggest that these clasts, as well as those observed as thin horizons within the sandy mud facies are more likely to result from bottom current erosion and local redeposition. Certain other pebbly mud beds with sharp bases might be interpreted as distal debrites, but it is difficult to distinguish these from glacigenic hemipelagites with increased IRD.

The action of bottom currents has been a much more important influence on sedimentation throughout the interval examined for this study. The evidence for this is manifold:

- The known presence of a strong bottom current flow through the Faroe–Shetland Channel throughout the Neogene period at least.
- Seismic patterns that indicate widespread deposition of sheeted drifts, as well as broad areas of erosion.
- A range of sediment facies typical of (glacigenic) contourites including continuous bioturbation with very rare primary sedimentary structures, grain size cyclicity, winnowing of fines and sorting of fine sands, mixed composition and reworking of microfossil assemblages.
- Clear cyclic arrangement of facies and grain size in vertical section yielding both complete and partial contourite sequences, together with clearly erosive boundaries.
- Good lateral correlation of facies cycles, particularly the sandy contourite intervals, over at least 50 km along the Faroe–Shetland Channel.

Drift geometry and sediment export

The SW Faroe–Shetland Channel is characterized principally by the development of sheeted drifts coupled with widespread erosion (see also Stoker et al. 1998). The sheet-like drift geometry formed as a result of vigorous bottom current flow maintained across the whole channel floor as it narrows towards the southwest. The sheeted drift thins towards the southwest and over highs in the channel floor, due to non-deposition, winnowing and erosion. Local scours are evident over an otherwise smooth surface. Elongate mounded drifts, largely buried beneath the slope apron, are only locally preserved along the Faroe–Shetland Channel margins (Stoker 1999). These largely relict (Miocene) features contrast markedly with the flanks of the Channel farther to the northeast, beyond the limit of the study area, where elongate mounded drifts of Plio-Pleistocene age are commonly preserved on the West Shetland slope (Stoker 1999).

The principal role of bottom currents in developing the sheeted drift geometry in the SW Faroe–Shetland Channel basin-floor succession has been one of restricted and/or non-deposition throughout the Plio-Pleistocene interval. Although glaciomarine sediment input across the West Shetland slope would have been relatively large at times during the Plio-Quaternary period, the finer fraction of any glaciomarine hemipelagic and IRD material was effectively maintained in suspension in the southward flowing bottom currents. At times of enhanced bottom current activity, the floor of the Faroe–Shetland Channel was subject to erosion as well as non-deposition. Consequently the Faroe–Shetland Channel has served as an important region of sediment export to the south.

Contourites and climate

In trying to relate fluctuation in bottom current activity to climate change through the late Pleistocene to Holocene period, it is apparent that the relationship is not a simple one. In addition, dating of the cores is still patchy and incomplete so that the relationships outlined below must be considered preliminary.

The most recent warm period (the Holocene) is characterised by a thin sandy contourite veneer as well as by seafloor erosion. This suggests enhanced bottom current activity during the most recent interglacial period, as has been proposed from many other studies in the North Atlantic (Stow et al. 1986; Stoker et al. 1989; Faugères & Stow 1993; Stoker et al. 1998; Armishaw et al. 1998, 2000). Careful analysis of the cores used in this study show a similar increase in bottom current activity during earlier interstadial and interglacial periods (e.g. core 36, Duan et al. 1994), but also a marked increase in top-only contourites and erosive horizons during both rapid warming and cooling phases. This is in line with several earlier studies (e.g. Duplessy *et al.* 1988; Dowling & McCave 1993).

Grain size cyclicity together with other evidence for contourite deposition is also apparent through the clearly glacial periods. We therefore concur with Akhurst (1991), Stoker *et al.* (1991) and Knutz *et al.* (2001) that bottom current activity persisted through the glacial NE Atlantic, but was less intense during glacial stages.

We should like to acknowledge technical and secretarial support from our respective institutions, in particular G. Tulloch, D. Russel and E. Gillespie of the British Geological Survey, and K. Davies of Southampton Oceanography Centre. MCA and MSS wish to acknowledge the Director, British Geological Survey (NERC) for permission to publish this work which was funded by the Department of Energy (now the Department of Trade and Industry); DAVS acknowledges tenure of a Royal Society Industrial Fellowship. Numerous colleagues have given freely of their views and advice during this work, as have two anonymous reviewers on an earlier version of the manuscript.

References

AKHURST, M. C. 1991. *Aspects of late Quaternary sedimentation in the Faroe–Shetland Channel, NW UK continental margin.* BGS Technical Report **WB/91/2**.

ARMISHAW, J. E., HOLMES, R. W. & STOW, D. A. V. 1998. Morphology and sedimentation on the Hebrides Slope and Barra Fan, NW UK continental margin. *In*: STOKER, M. S., EVANS, D. & CRAMP, A. (eds) *Geological Processes on Continental Margins: Sedimentation, Mass-wasting and Stability.* Geological Society, London, Special Publications, **129**, 81–104.

ARMISHAW, J. E., HOLMES, R. & STOW, D. A. V. 2000. The Barra Fan: A bottom-current reworked, glacially-fed submarine fan system. *Marine Petroleum Geology*, **17**, 219–239.

BERGGREN, W. A. & SCHNITKER, D. 1983. Cenozoic marine environments in the North Atlantic and Norwegian-Greenland Sea. *In*: BOTT, M. H. W., SAXOV, S., TALWANI, M. & THIEDE, J. (eds) *Structure and development of the Greenland-Scotland Ridge: new methods and concepts.* Plenum Press, New York, 495–584.

BULAT, J. & LONG, D. in press. Images of the seabed in the Faroe–Shetland Channel from commercial 3D seismic data. *Marine Geophysical Researches*.

DAVIES, R., CARTWRIGHT, J., PIKE, J. & LINE, C. 2001. Early Oligocene initiation of North Atlantic Deep Water formation. *Nature*, **410**, 917–920.

DOOLEY, H. D. & MEINCKE, J. 1981. Circulation and water masses in the Faroese Channels during Overflow '73. *Deutsche Hydrographishe Zeitschrift*, **34**, 4–54.

DORÉ, A. G., LUNDIN, E. R., JENSEN, L. N., BIRKELAND, A., ELIASSEN, P. E. & FICHLER, C. 1999. Principal tectonic events in the evolution of the northwest European Atlantic margin. *In*: FLEET, A. C. & BOLDY, S. A. R. (eds) *Petroleum Geology of Northwest Europe: Proceedings of the 5th Conference.* Geological Society, London. 41–61.

DOWLING, L. M. & MCCAVE, I. N. 1993. Sedimentation on the Feni Drift and late glacial bottom-water production in the Rockall Trough. *Sedimentary Geology*, **82**, 79–87.

DUAN, T., STOW, D. A. V. & MURRAY, J. 1994. Late Quaternary sedimentation influenced by bottom current, gravity flow and glacigenic processes, southern Shetland Channel, NW British continental margin. *In*: *Proceedings of the 14th International Sedimentation Congress*, Recife, Brazil, abstract volume, S2-7.

DUPLESSY, J. C., SHACKLETON, N. J., FAIRBANKS, R. G., LABEYRIE, L., OPPO, D. & KANNEL, N. 1988. Deep-water source variation during the last climatic variation. *Paleoceanography*, **3**, 343–360.

ELDHOLM, O. 1990. Palaeogene North Atlantic Magmatic-Tectonic Events: Environmental Implications, *Memorie della Societa Geological Italiana*, **44**, 13–28.

FAUGÈRES, J.-C. & STOW, D. A. V. 1993. Bottom current-controlled sedimentation: a synthesis of the contourite problem 1993. *Sedimentary Geology* **82**, 287–297

HOLMES, R. 1990. *Foula 60°N–04°W Quaternary geology*. British Geological Survey 1: 250 000 map series.

JANSEN, E. & RAYMO, M. E. 1996. New Frontiers On Past Climates. *In*: JANSEN, E., RAYMO, M. E. & BLUM, P. *et al. Proceedings ODP, Initial Reports*. **162**, College Station, TX (Ocean Drilling Program), 5–20.

KNUTZ, P. C., AUSTIN, W. E. N. & JONES, E. J. W. 2001. Millennial-scale depositional cycles related to British Ice Sheet variability and North Atlantic palaeocirculation since 45 kyr B.P., Barra Fan, UK margin. *Paleoceanography*, **16**, 53–64.

MASSON, D. G. 2001. Sedimentary processes shaping the eastern slope of the Faroe–Shetland Channel. *Continental Shelf Research*, **21**, 825–857.

MILLER, K. G. & TUCHOLKE B. E. 1983. Development of Cenozoic abyssal-circulation south of the Greenland-Scotland Ridge. *In*: BOTT, M. H. P., SAXOV, S., TALWANI, M., & THIEDE, J. (eds) *Structure and development of the Greenland-Scotland Ridge*. Plenum Press, New York. 549–589.

ROBERTS, D. G., THOMPSON, M., MITCHENER, B., HOSSACK, J., CARMICHEAL, S. & BJARNSETH, H.-M. 1999. Palaeozoic to Tertiary Rift and Basin Dynamics: mid-Norway to the Bay of Biscay – a new context for hydrocarbon prospectivity in the deep water frontier. *In*: FLEET, A. C. & BOLDY, S. A. R. (eds). *Petroleum Geology of Northwest Europe: Proceedings of the 5th Conference*. Geological Society, London, 7–40.

SAUNDERS, P. M. 1990. Cold outflow from the Faroe Bank Channel. *Journal of Physical Oceanography*, **20**, 29–43.

SAUNDERS, P. M. & GOULD, W. J. 1989. Current measurements around the Faroe Islands in 1986 and 1987. *Institute of Oceanographic Sciences Deacon Laboratory Report*, **261**.

STOKER, M. S. 1990a. *Judd, 60°N–06°W, Solid geology*. British Geological Survey 1:250 000 map series.

STOKER, M. S. 1990b. *Judd, 60°N–06°W, Quaternary geology*. British Geological Survey 1:250 000 map series.

STOKER, M. S. 1995. The influence of glacigenic sedimentation on slope-apron development on the continental margin off Northwest Britain. *In*: SCRUTTON, R. A., STOKER, M. S., SHIMMIELD, G. B. & TUDHOPE, A. W. (eds) *The Tectonics, Sedimentation and Palaeoceanography of the North Atlantic Region*. Geological Society, London, Special Publication, **90**, 159–177.

STOKER, M. S. 1999. *Stratigraphic nomenclature of the UK North West Margin. 3. Mid- to late Cenozoic stratigraphy*. British Geological Survey, Edinburgh.

STOKER, M., AKHURST, M., HOWE, J. A. & STOW, D. A. V. 1998. Sediment drifts and contourites on the continental margin off NW Britain. *Sedimentary Geology*, **115**, 33–52.

STOKER, M. S., HARLAND, R. & GRAHAM, D. K. 1991. Glacially influenced basin plain sedimentation in the southern Faroe–Shetland Channel, NW UK continental margin. *Marine Geology*, **100**, 185–199.

STOKER, M. S., HARLAND, R., MORTON, A. C. & GRAHAM, D. K. 1989. Late Quaternary stratigraphy of the northern Rockall Trough and Faroe–Shetland Channel, northeast Atlantic Ocean. *Journal of Quaternary Science*, **4**, 211–222.

STOKER, M. S., HITCHIN, K. & GRAHAM, D. K. 1993. *UK Offshore regional report: the geology of the Hebrides and West Shetland shelves and adjacent deep-water areas*. HMSO, BGS, London.

STOKER, M. S., LONG, D. & BULAT, J. in press. A record of mid-Cenozoic strong deep-water erosion in the Faroe–Shetland Channel. *In*: MIENART, J. & WEAVER, P. P. E. (eds) *European margin Sediment Dynamics: Sidescan Sonar and Seismic Images*. Springer Verlag, Berlin.

STOW, D. A. V., FAUGÈRES, J.-C. & GONTHIER, E. 1986. Facies distribution and textural variation in Faroe Drift contourites: velocity fluctuation and drift growth. *Marine Geology*, **72**, 71–100.

STOW, D. A. V., READING, H. G. & COLLINSON, J. 1996. Deep seas. *In*: READING, H. G. (ed.) *Sedimentary Environments and Facies*. Blackwell Scientific Publications, Oxford, 380–442.

THIEDE, J. & MHYRE, A. M. 1996. Introduction to the North Atlantic-Arctic Gateways: Plate Tectonic-Palaeoceanigraphic history and significance. *In*: THIEDE, J., MHYRE, A. M., FIRTH, J. V., JOHSNON, G. L. & RUDDIMAN, W. F. (eds) *Proceedings ODP Scientific Results*, **151**. College Station, TX (Ocean Drilling Program), 3–23.

ZEIGLER, P. A. 1988. *Evolution of the Arctic-North Atlantic and the western Tethys*. American Association of Petroleum Geologists, Tulsa, Memoirs, **43**.

Wave-form sheeted contourite drift on the Barra Fan, NW UK continental margin

PAUL C. KNUTZ[1], E. JOHN W. JONES[2], JOHN A. HOWE[3], TJEERD J. C. VAN WEERING[4] & DORRIK A. V. STOW[5]

[1]*Department of Earth Sciences, Cardiff University, PO Box 914, Cardiff CF1 3YE, UK*
[2]*Department of Geological Sciences, University College London, London WC1E 6BT, UK*
[3]*SAMS, University Highlands & Islands, Oban, Argyll PA34 4AD, Scotland, UK*
[4]*Netherlands Institute for Sea Research, 1790 AB Den Burg, Texel, The Netherlands*
[5]*Southampton Oceanography Centre, Waterfront Campus, Southampton University, Southampton SO14 3ZH, UK*

Abstract: The lithology of a 30 m long piston core (MD95-2006) and high-resolution, seismic profiles from the lower Barra Fan, Rockall Trough, reveal a sheeted drift form with internal sediment waves deposited over the last glacial–interglacial cycle. Deposition of these mainly fine-grained deposits was controlled by a combination of downslope and alongslope transport mechanisms that interacted with the positive topography created by debrite lobes on the lower fan. The core penetrates a small field of sediment waves (wavelength approx. 1 km, height 3–6 m), which onlap a debrite lobe dated to the last glacial maximum. The sedimentary sequence shows: (1) silty-muddy contourites deposited during the mid-Devensian (Marine Isotope Stage 3), (2) glacimarine hemipelagites and sandy turbidites deposited between 26 and 18 C^{14} ka BP, followed by a short phase of erosion and redeposition by bottom currents, and (3) glacimarine hemipelagites and silty-muddy contourites representing the glacial to Holocene transition. On the distal fan edge, a drift sequence with upslope-migrating sediment waves (wavelengths approx. 3 km, height 15–30 m) onlaps the tongue of a previous slide event (pre-Devensian?). These bedforms were probably generated by decelerating, low-density glacigenic turbidity currents, but pirated by contour-following bottom currents on the distal part of the drift.

Contourite drifts in continental slope settings have often been related to a process continuum whereby sediments supplied by downslope transport mechanisms are progressively subjected to winnowing and preferential settling by slow moving, but persistent, geostrophic currents (Heezen *et al.* 1966; Tucholke & Laine 1983; Stanley 1993; Masse *et al.* 1998). But because the transfer from downslope gravity-driven currents to contour-tracing bottom currents may occur gradually over tens to hundreds of kilometres it is difficult to establish reliable facies models, which can lead to identification of downslope-alongslope process interaction. A particular aspect of this problem is how sandy contourites can be distinguished from bottom current reworked fine-grained turbidites (Stow *et al.* 1998). Integration of high-resolution seismics across drift sequences covered by long core stratigraphies is critical to the elucidation of the complex relationship between downslope and alongslope processes in a deep marine setting. Here we present detailed seismic and sediment facies data and attempt an interpretation of the depositional processes that resulted in the wave-form sheeted drifts on the lower Barra fan in the Rockall Trough (Fig. 1).

Core MD95-2006 (depth 2120 m) penetrated 30 m into a small sediment wave field on the northern fringe of the Barra Fan. This region is covered by several high-frequency 6 kV deep-tow boomer profiles collected by the British Geological Survey as part of their offshore reconnaissance mapping programme. In addition, we present a high-resolution sleeve-gun profile across the distal fan edge, obtained through the European ENAM2 research programme (Fig. 2).

Geological and oceanographic setting

The Rockall Trough forms a narrow northeast extension of the East Atlantic basin that trends along the continental margin of the British Isles (Fig.1). It is confined to the west by the Rockall Plateau and to the north by the Wyville–Thomson Ridge, which forms a sill at < 500 m depth between the northern Rockall Trough and the Faroe–Shetland Channel. Water depths range from 4000 m in the southwest part, where the basin deepens into the Porcupine Abyssal Plain, to less than 1000 m in the northeast.

Sedimentation induced by bottom currents, centred on the flanks of the Rockall Trough, is known to have occurred since the late Eocene (Jones *et al.* 1970; Miller & Tucholke 1983; Stow & Holbrook 1984). The mid-upper Cenozoic succession of the eastern margin is characterised by onlapping sheeted and mounded drifts, in many cases showing development of upslope migrating sediment waves (Stoker *et al.* 1998). The sediment drifts along the margin are separated by two slope-apron fan systems, the Sula Sgeir and Barra-Donegal fans, principally constructed by mass-flows supplied by glacimarine processes during glacial periods (Stoker 1995; Vorren & Laberg 1997).

The seismo-stratigraphic structure, comprising four major slide events that form the bulk architecture of the Barra Fan was demonstrated by Holmes *et al.* (1998). Recent alongslope and downslope sedimentation patterns on the upper to middle fan were described by Armishaw *et al.* (1998) using detailed analyses of seafloor morphology. Howe (1996) discussed the origin of sediment waves on the lower Barra Fan and in the NE Rockall Trough based on high-resolution seismic profiles and gravity cores.

The lower Barra Fan presently lies under the influence of a deep poleward current derived from North Atlantic Deep Water (NADW), which is partially mixed with Labrador Sea Water (LSW) from above and Antarctic Bottom water (AABW) from (McCartney 1992). In the southern Rockall Trough this water mass, centred at 2000–3000 m depth, has been recorded with peak flow velocities between 27 and 39 cm s^{-1} (Dickson & Kidd 1986). A component of Norwegian Sea Deep Water (NSDW) intermittently overflowing the Wyville–Thomson Ridge may also contribute to the semi-cyclonic pattern of bottom water circulation in the northern Rockall Trough (Ellett & Roberts 1973; Dickson & Kidd 1986). At depths down to about 1000 m, the Barra Fan is affected by a strong northward-flowing slope current linked to the North Atlantic Current (Kenyon 1986). On the Hebrides Slope, peak flow velocities of this water mass have been measured at 15–25 m s^{-1} (Howe & Humphery 1995).

From: STOW, D. A. V., PUDSEY, C. J., HOWE, J. A., FAUGÈRES, J.-C. & VIANA, A. R. (eds)
Deep-Water Contourite Systems: Modern Drifts and Ancient Series, Seismic and Sedimentary Characteristics.
Geological Society, London, Memoirs, **22**, 85–97. 0435-4052/02/$15.00 © The Geological Society of London 2002.

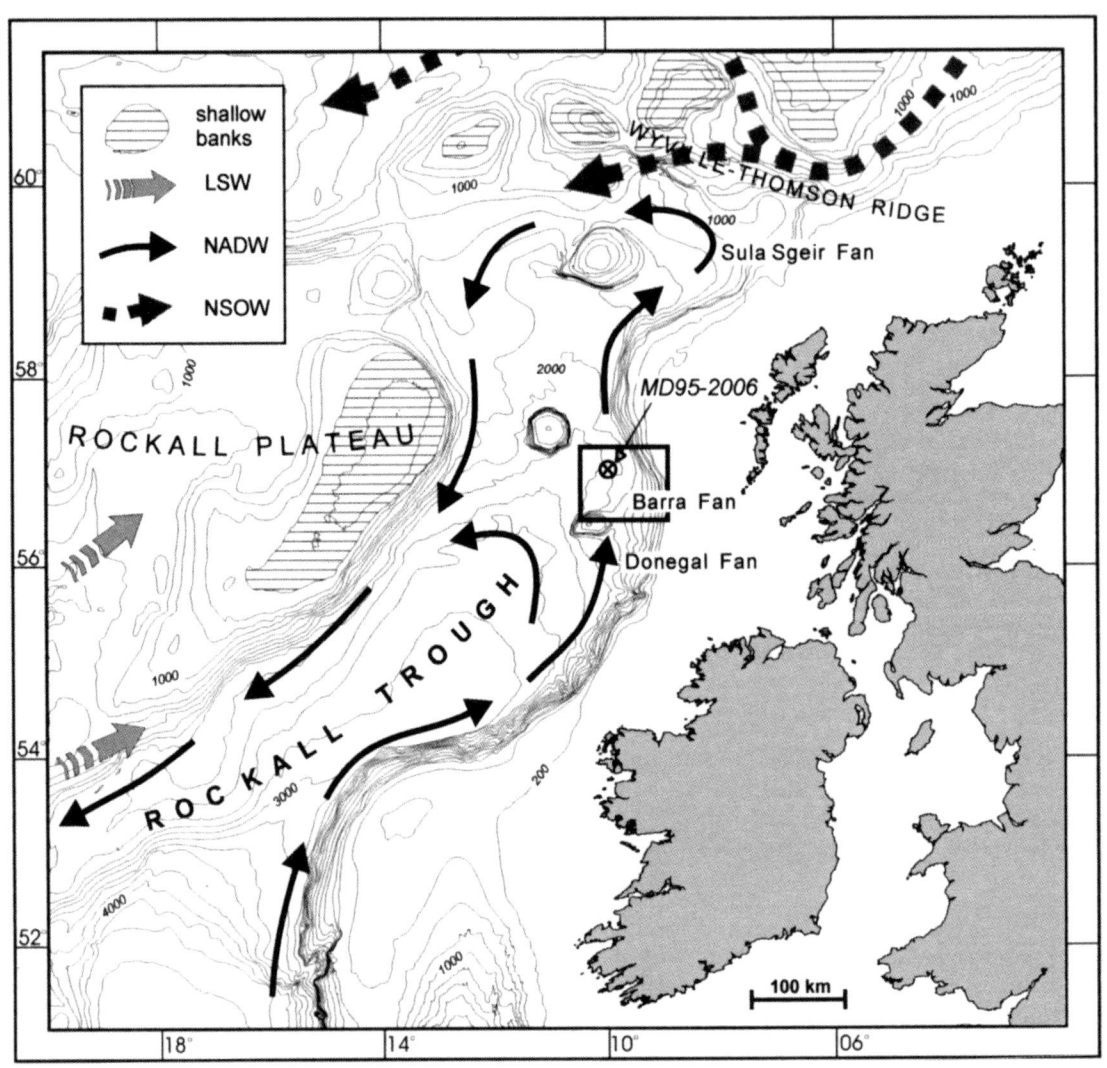

Fig. 1. Deep-water sources and circulation pattern in the Rockall Trough. LSW, Labrador Sea Water; NADW, North Atlantic Deep Water; NSOW, Norwegian Sea Overflow Water. Core site of MD95-2006 is indicated. Black box represents study area shown in Figure 2.

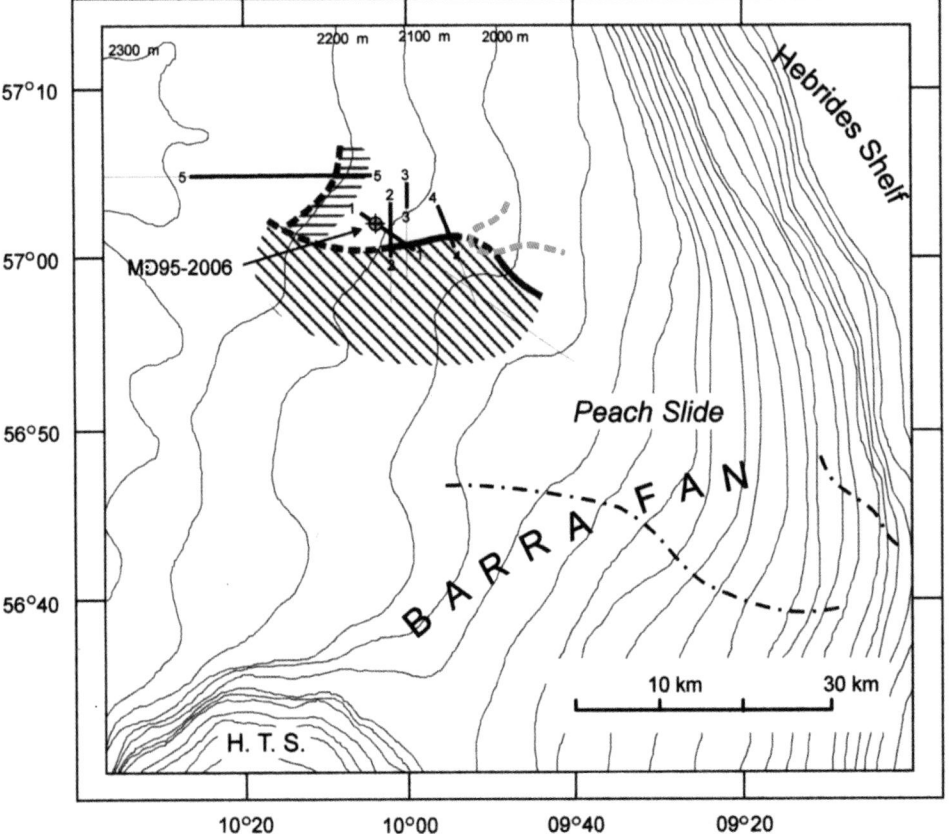

Fig. 2. Position of seismic lines and MD95-2006 in relation to debris flow lobes on the northern Barra Fan. Diagonal lines depict the scarp formed by debris flow 3 (DF 3) while the horizontal lines demarcate the scarp of an older slide event associated with debrite sequence 2 (DF 2) (Holmes *et al.* 1998). The grey broken line shows the possible outline of a younger debrite lobe that impinges on the margin of DF 3. Major scarps bounding the Peach Slide are shown with thin stippled lines. H.T.S. is the Hebrides Terrace Seamount.

Sediment facies and interpretation

A detailed lithological description of MD95-2006 based on sediment texture, structures, colour, qualitative estimates of calcium carbonate content and degree of bioturbation, has allowed us to define seven distinct sediment facies (Fig. 3). Interpretation of depositional environments is based on facies characteristics, magnetic susceptibility patterns and grain size analyses (Fig. 4). The facies recognized are as follows:

Facies A: silty to fine sandy mud, light olive brown, high carbonate content, intense bioturbation, no dropstones.

Facies B: structureless silty mud, olive brown to greyish brown, low-medium carbonate content, moderate to intense bioturbation, no dropstones.

Facies C: heterogeneous silty to sandy mud, dark greyish brown, low carbonate content, abundant Fe–sulphides, weak to absent bioturbation, few dropstones.

Facies D: structureless clayey mud, olive brown, very low carbonate content, abundant Fe–sulphides, weak bioturbation, rare dropstones.

Facies E: silty to fine sandy mud and silt, olive grey, variable carbonate content, moderate to intense bioturbation, diffuse layering in silts, rare dropstones.

Facies F: silty mud, dark gey to olive grey, variable carbonate content, abundant Fe–sulphides, weak to moderate bioturbation, rare dropstones.

Facies G: graded silt laminae and thin sand layers, sharp-based, sands up to coarse granule size, with immature composition and glacially-weathered lithic clasts.

Contourites

The top part of the core comprising facies A–B is interpreted as a late Glacial to Holocene contourite development (Fig. 4). The silty and mottled character of facies A indicates intensive sediment winnowing, presumably reflecting the modern current regime. A similar silty-sandy mud facies has been described from seven gravity cores recovered along the seismic profile 1, a short distance upslope from MD95-2006 (Howe 1996). Facies B is a non-distinct, homogeneous mud with a hemipelagic character, but

Fig. 3. Representative core photographs of the different sediment facies in core MD95-2006. Scale bar is 20 cm long.
(1) Facies B – hemipelagite passing up into muddy contourite (core depth 0.8–1.1 m).
(2) Facies F – silty/muddy contourite-hemipelagite (core depth 11.6–11.9 m)
(3) Facies F – silty/muddy contourite with diffuse lamination (core depth 27.3–27.6 m)
(4) Facies F – silty contourite (mottled, indistinctly laminated) within muddy contourite facies (core depth 22.25–22.45 m)
(5) Facies E – silty contourite layers (mottled, indistinctly laminated) within silty muddy contourite facies (core depth 24.45–24.75 m)
(6) Facies E – silty contourite layers (mottled, indistinctly laminated) within silty muddy contourite facies (core depth 24.75–25.05 m)
(7) Facies G – graded silt-laminated turbidtes (core depth 21.2–21.5 m)
(8) Facies G – graded thin-bedded sandy turbidites (core depth 14.45–14.75 m)

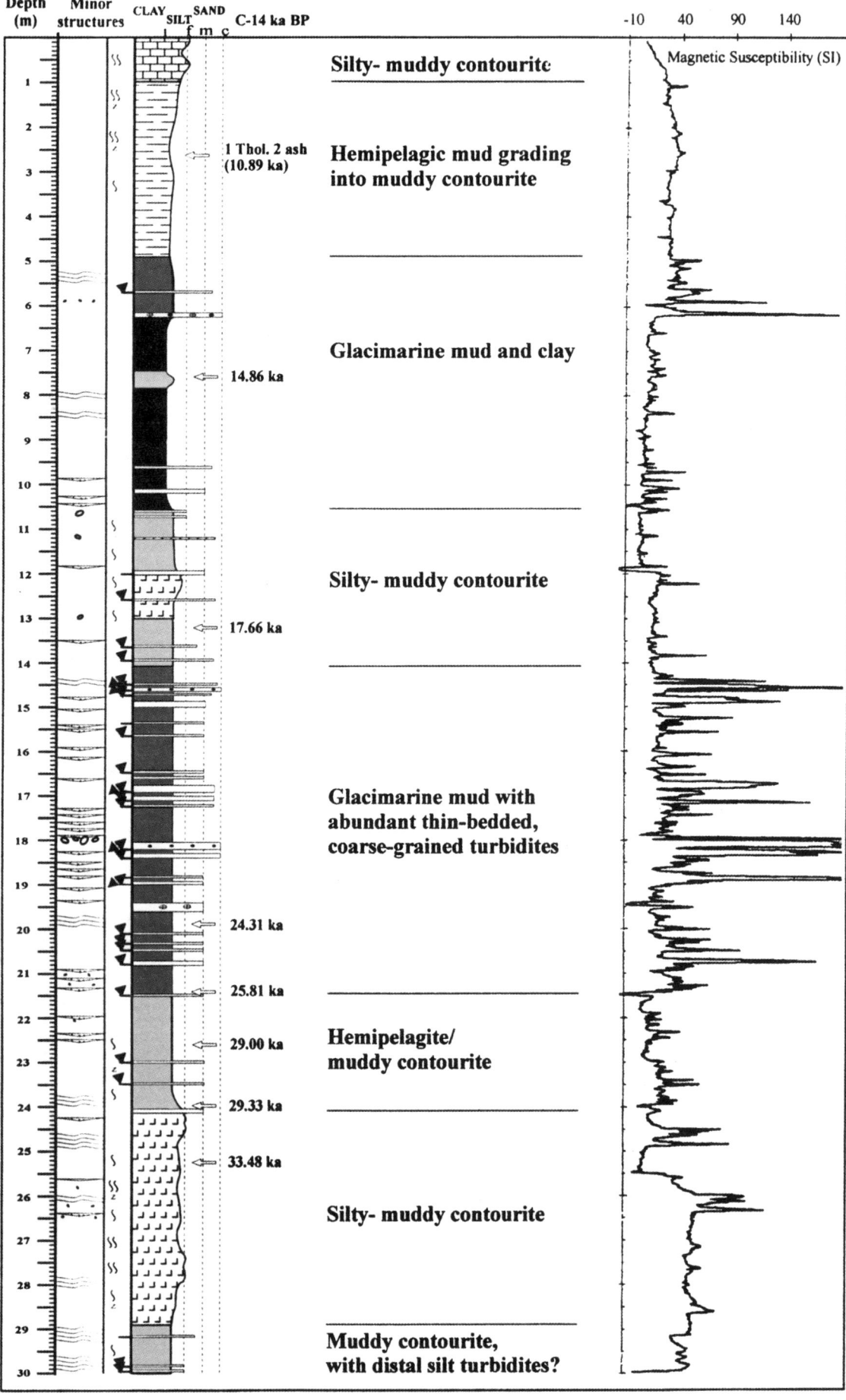

Facies A	Silty to fine sandy mud Light olive brown CaCO₃ content high (>20%) Bioturbation intense Dropstones absent
Facies B	Silty mud, massive Olive brown to greyish brown CaCO₃ content low to medium (3–15%) Bioturbation moderate to intense Dropstones absent
Facies C	Silty to sandy mud, heterogeneous Dark greyish brown CaCO₃ content low (<5%) Fe–sulphides abundant Bioturbation weak to absent Dropstones occasionally present
Facies D	Clayey mud, massive Olive brown CaCO₃ content very low (<3%) Fe–sulphides abundant Bioturbation weak Dropstones rare
Facies E	Silty to fine sandy mud Olive grey CaCO₃ content variable (3–10%) Bioturbation moderate to intense Dropstones rare
Facies F	Silty mud Dark grey to olive grey CaCO₃ content variable (3–7%) Fe–sulphides abundant Bioturbation weak to moderate Dropstones rare
Structures	Graded sand layer with sharp base Coarse grained with pebbles and mud clasts Graded sand layer with sharp base Inverse grading in lower part Medium grained Diffuse muddy sand layer Silt laminae/lenses Floating stones/pebbles Bioturbation: -intense -moderate -zoophycos traces

Fig. 4. Detailed lithology, magnetic susceptibility and interpretation of depositional environments of core MD95-2006.

the upward increasing grain size trend (Fig. 6b) suggests it represents a gradual development from hemipelagic to bottom current influenced sedimentation. Facies E–F are interpreted as glacigenic silty and muddy contourites, following the definitions of Stow & Piper (1984) and Stoker et al. (1998). Facies E displays a variable mottled silty texture characteristic of a silty-muddy contourite, occasionally with thin sandy contourite intervals. The more uniform texture of facies F suggests a hemipelagic depositional environment.

Grain size analyses from core MD95-2006 reveal high frequency fluctuations of the coarse clastic silt component through facies A–B and E–F (Fig. 6b). These silt cycles are apparently unrelated to glacimarine source variations as determined by ice-rafted debris (IRD) content but co-vary with high abundances of planktonic foraminiferans, particularly in the lower part of the core. The silty intervals are inferred to represent periods of sediment winnowing by strong (>15 cm s^{-1}) bottom currents (McCave et al. 1995b). The alternating depositional patterns in MD95-2006 are associated with variations in flow intensity of North Atlantic Deep Water and corresponds to stadial-interstadial cycles in the North Atlantic climate record (Knutz et al. 2001).

The high background level of magnetic susceptibility through facies A–B and facies E (Fig. 4) could be an effect of current winnowing and concentration of high-density, magnetic minerals in the silt fraction (e.g. Ledbetter 1984). However, shifts in background values may also be related to fine-clastic source variations (for instance input from point-sourced glacimarine plumes versus input from the shelf-nepheloid layer). The decrease in magnetic susceptibility towards the core top (facies A) reflects an increase in biogenic carbonate diluting the fine-clastic carriers of the magnetic signal. Decimetre thick layers of well-sorted and fairly homogeneous silt to fine sand at 12.0 and 24.0 m core depth, corresponding to abrupt decreases in magnetic susceptibility, may represent sandy contourite deposits. However, the identification of such bottom current related features from sandy turbidites is difficult without data on regional facies and grain size trends (Stow et al. 1998).

Glacimarine sediments and turbidites

Facies C and D are interpreted as glacimarine muds formed by hemipelagic settling and gravity flows (Fig. 4). A large proportion of clay size material may have been supplied by meltwater overflow plumes. The graded silt laminae and medium to coarse-grained sand layers of facies G are interpreted as glacigenic turbidites based on their clear turbidite structures, immature composition and the presence of glacially weathered lithic clasts. Turbidites are most common in facies C and D but also occur less abundantly in facies F. The pattern of downslope sedimentation through the core is clearly displayed by high-frequency, high-amplitude magnetic susceptibility peaks produced by the high abundance of Ti–magnetite bearing basalts in these layers. The cyclic pattern of magnetic susceptibility peaks within facies C reflects the magnitude of gravity flow events because basal concentrations tend to increase with layer thickness and average grain size. At some levels, ice-rafted debris may contribute to the magnetic susceptibility signal but this source is mostly masked by the presence of sandy turbidites.

Grain size data

The lithology of MD95-2006 is dominated by silty clay with clay (<2 μm) ranging between 30 and 75% (Fig. 6b). A general increase in clay content is displayed from about 26 m up to 6 m core depth followed by a decrease towards the core top. The muddy contourite intervals defined by facies A, B and E are generally more silt rich (> 40 wt%) than the bulk part of the grain

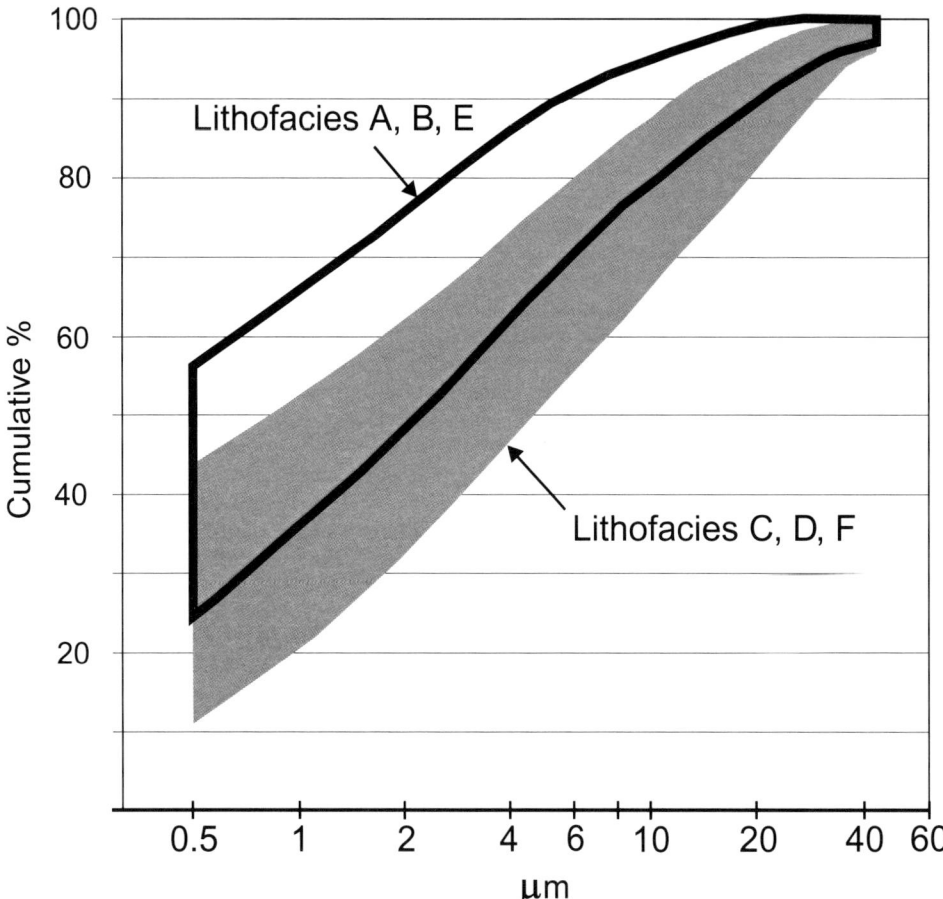

Fig. 5. Cumulative grain size domains from MD95-2006. Grey area, Facies A, B, E; Enclosed area, Facies C, D, F.

size spectrum (Fig. 5). There is, however, a large overlap between grain sizes from silty-muddy contourite and glacimarine hemipelagic intervals which may reflect the fine-scale variation in grain sizes within these facies. Hence, it is not feasible to identify contourite-related facies in MD95-2006 on the basis of cumulative grain size curves alone. A more characteristic feature of the silty-muddy contourites lies in the fluctuating concentrations in coarse silt, particularly evident in the lower part of the core.

Seismic characteristics

Four stratified sedimentary units A1, A2, B and C can be identified in seismic profile 1 across the core site of MD95-2006 (Fig. 6a), and in adjacent seismic sections (Fig. 7a–c). These units can be tied to the lithology of MD95-2006 (Fig. 6b). Debrite sequence 3 (DF 3) of the Peach Slide (Holmes et al. 1998) is recognized by its chaotic reflection pattern and hummocky surface, typically expressed by weak, broadly spaced, hyperbolic reflections. The E–W trending edge of DF 3 can be traced on several, closely-spaced shallow seismic profiles, oriented perpendicular to the margin of the Peach Slide (Fig. 2). Correlation with the dated core sequence suggests that DF 3 was deposited between 18–21 C^{14} ka BP. The presence of a younger debrite tongue that appears to have overrun DF 3 from an easterly direction is indicated on Figure 2. A mega-debrite unit DF 2, associated with a previous slide event, impinges below a large drift sequence about 8 km NW from the MD95-2000 core site (Figs 2 & 8).

Unit A1 forms a semi-transparent drape covering debrites and older stratified units on the lower fan. Close to the DF 3 scarp, this unit shows a more variable seismic signature characterised by wavy stratification and migrating, clinoformal bedforms that internally comprise discontinuous, medium to high-amplitude reflectors, typically truncated at the top (Figs 6 & 7). The wave-forming reflectors converge downslope into two weak, parallel, continuous reflectors, which can be traced across the region (Fig. 8). Unit A1 also contains a distinct mounded sediment drift in the upper part (Fig. 7c).

Sediment waves in unit A1 appear to mimic the morphology of *unit A2*, which form a local seismic unit that extends from the toe of DF 3 and pinches out between units A1 and B. Unit A2 is characterized by sets of upslope prograding to aggrading, medium to high-amplitude reflectors, producing cannibalizing bedforms, similar to those observed in unit A1. The boundary between A1 and A2 is disconformable, with signs of truncation on the downslope flank of the wavy bedforms (Figs 6 & 7).

Unit B contains a series of at least five weak, parallel and continuous reflectors that appear to dip below DF 3 (Figs 6a & 7a). This unit can be traced further downslope on seismic line 5 as a parallel-stratified unit that condenses, or is eroded at the top, above the scarp of DF 2 (Fig. 8). The seismic resolution is not sufficient to trace unit B beyond the margin of DF 2. It is, however, likely that this unit corresponds to the upper part of the wavy drift sequence showing reflector patterns of onlapping channel-fill and upslope migrating sediment waves.

Unit C is present near the depth limit of seismic penetration on the boomer profile but can be discerned by at least two semi-continuous, sub-parallel, reflectors (Fig. 7a–c). Unit C is not detectable on seismic line 5 but is assumed to underlie unit B (to the east, above DF 2). The strong reflector below unit B may form an internal feature of unit C or represent the contact between unit C and the underlying debrite.

Interpretation of seismic units

Unit A1 forms a regional drape that can be traced across the lower fan into the basin of the Rockall Trough and is equivalent to the Gwaelo sequence (Stoker 1995). In MD95-2006 this unit is represented by an upward-coarsening, fine clastic sequence

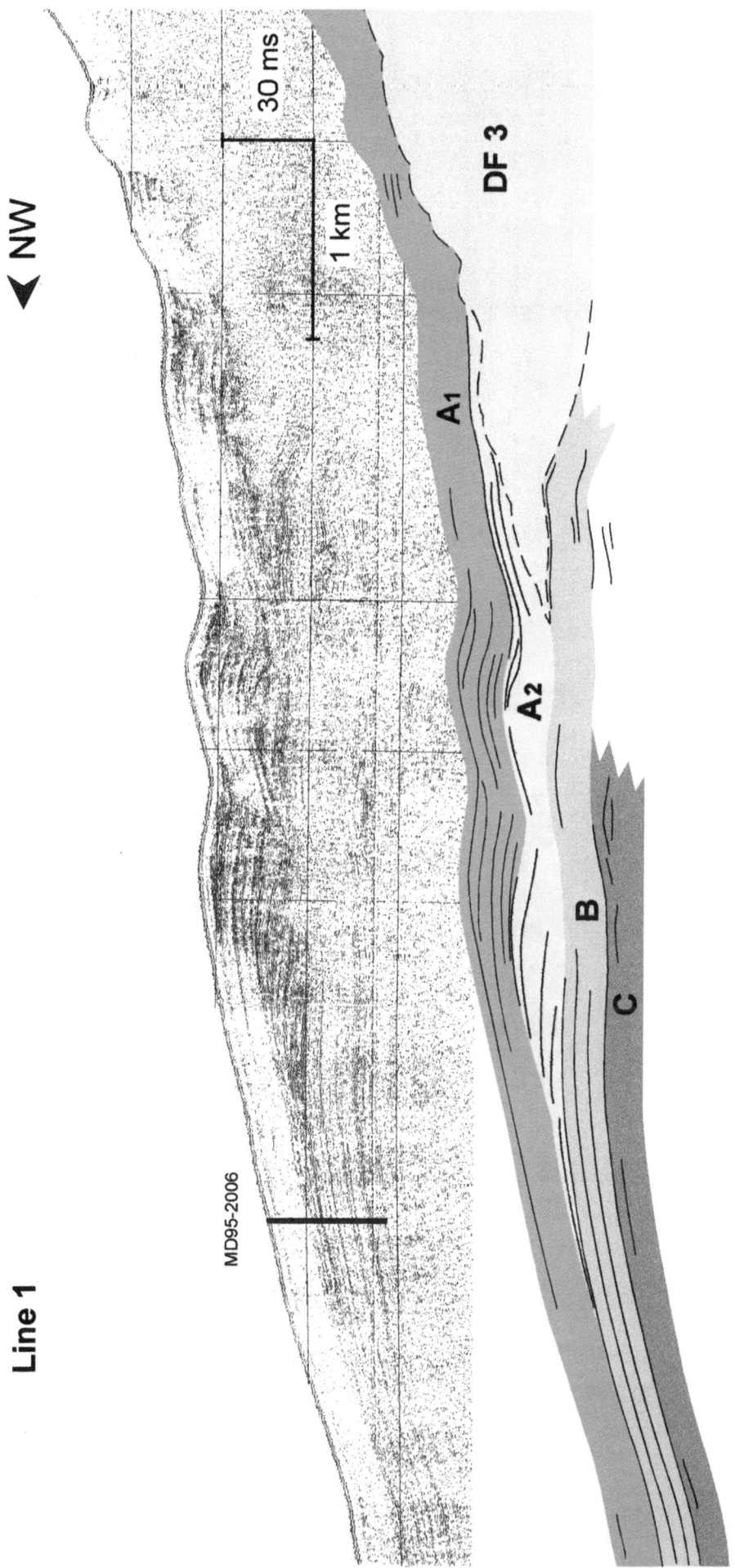

Fig. 6. (a) Deep tow boomer profile 1 showing position of MD95-2006 and interpretation of seismic units. See Figure 2 for location of line. (b) Sedimentological data from MD95-2006 and correlation with seismic line 1. Paleoclimatic events Younger Dryas (Y–D) and Heinrich (H) events 1 to 4 are indicated. 30 ms TWT is equivalent to 23 m penetration based on an acoustic velocity of 1550 m s^{-1}.

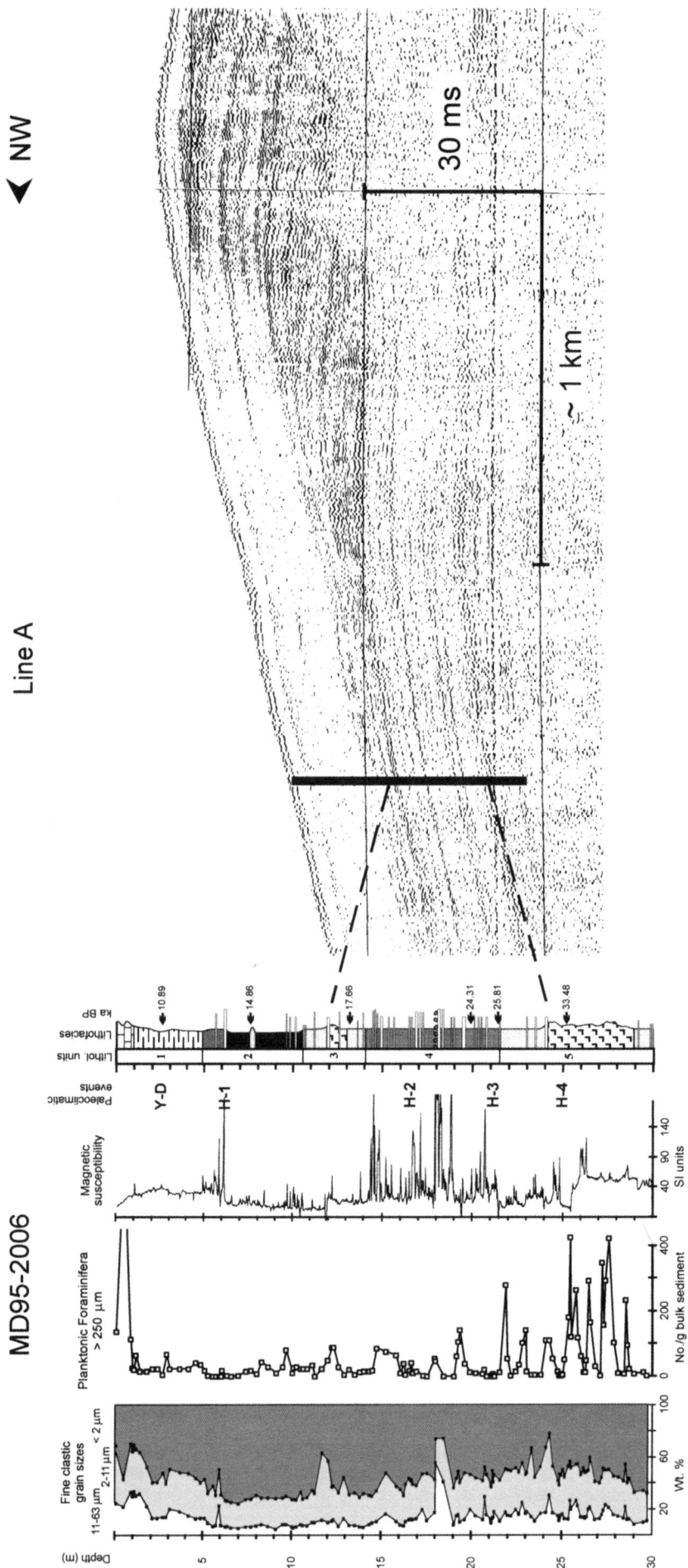

Fig. 6b.

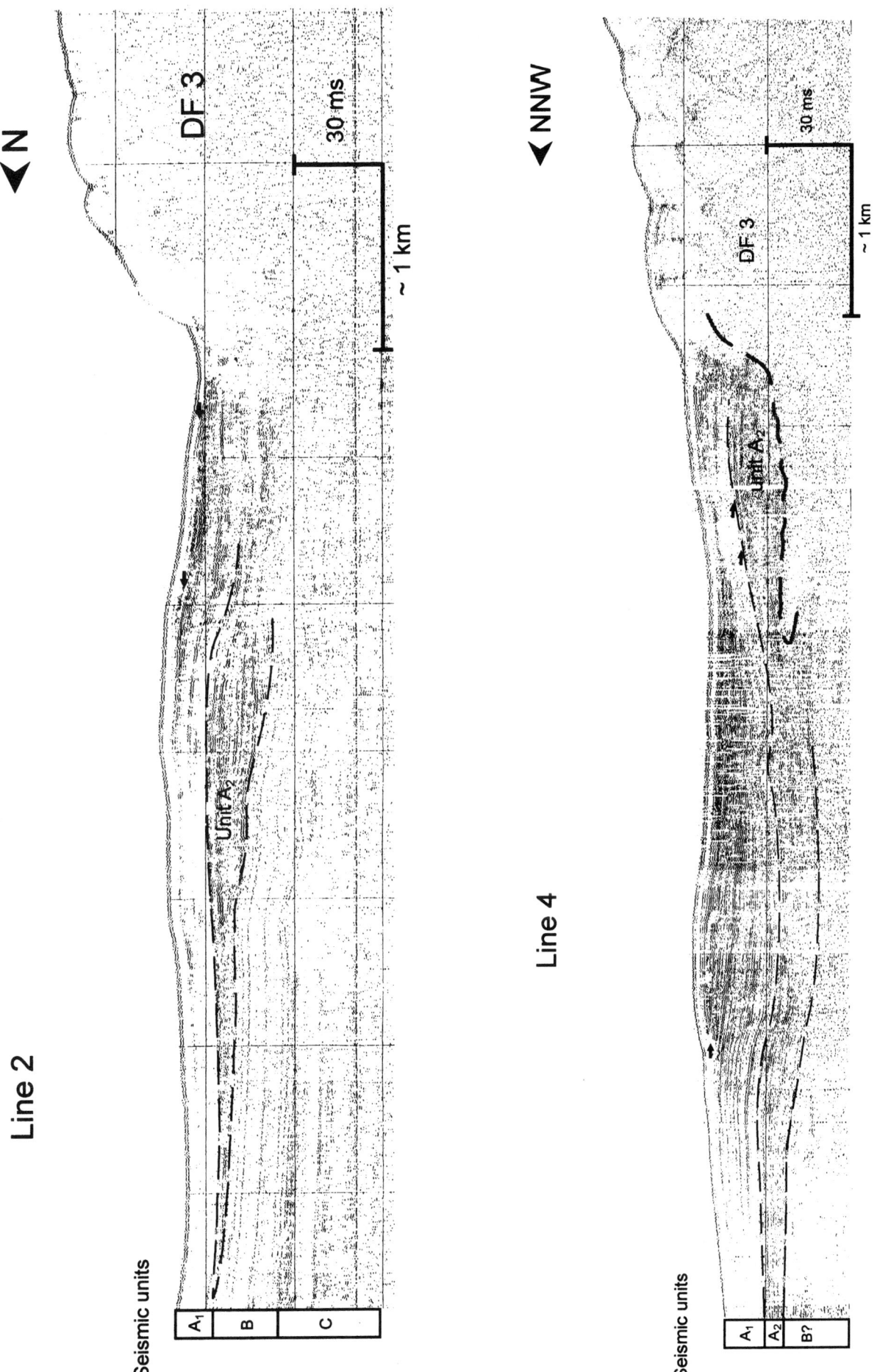

Fig. 7. (**a–c**) Deep tow boomer profiles 2–4 with seismic units shown. Arrows indicate erosional features related to bottom currents. See Figure 2 for locations.

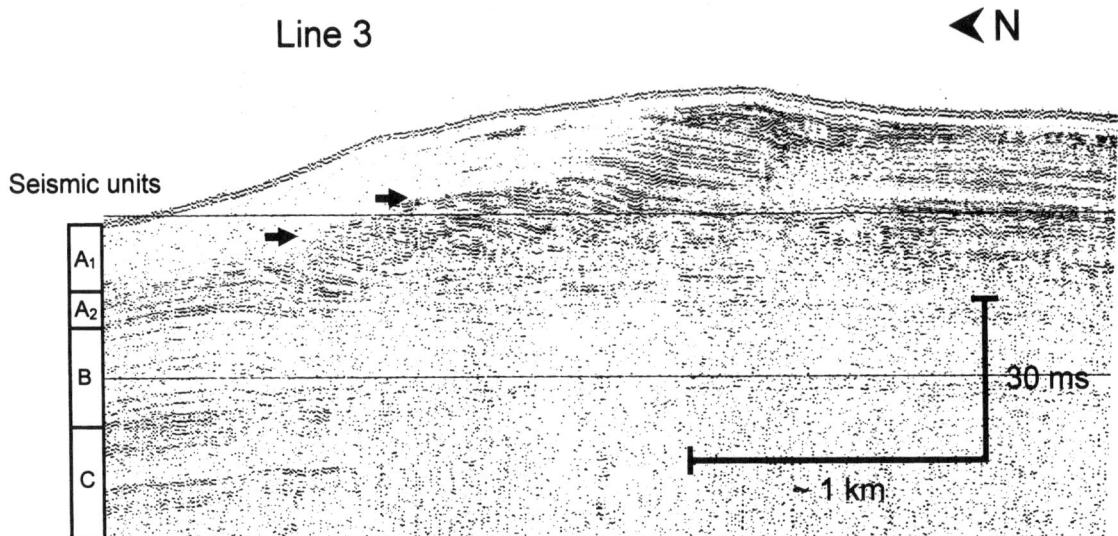

Fig. 7c.

comprising lithological units 1–2 (Fig. 6b), interpreted as the development from a glacimarine hemipelagic to a silty-muddy contourite environment representing the last deglaciation and early Holocene. The two internal reflectors in unit A1 are related to thin-bedded turbidites and ice-rafting events linked to climatic cooling during Heinrich event I and the Younger Dryas (Howe 1996; Knutz et al. 2001). An asymmetric, mounded reflector pattern in the upper part of unit A1 (Fig. 7b) may represent a local silty-sandy contourite drift developed during the Holocene. The geometry of unit A1 and the erosive bedforms at the base of DF 3 (Fig. 7a) suggest the presence of a channel formed by bottom currents routed along the edge of the debris flow.

Unit A2 is interpreted as sandy to silty-muddy contourites on the basis of the discontinuous reflector configuration and migrating, cannibalizing bedforms. The medium-high amplitude wave-forming reflectors in units A1 and A2 have not been penetrated by coring but the downslope pinch-out of unit A2 is represented by a silty-muddy contourite interval in MD95-2006 (Fig. 6b). The sediments of unit A2 may have been deposited by across-fan bottom currents eroding and reworking the exposed surface of DF 3.

Unit B is interpreted as a muddy glacimarine sequence with continuous parallel reflectors associated with thin sandy turbidites of facies D deposited between 26–18 C^{14} ka BP (Fig. 4). These turbidites may represent thin depositional lobes produced by unchannelised, decelerating, gravity flows that reached the lower fan through a series of distributary canyons located across the mid-upper slope region (Armishaw et al. 1998). The release of gravity flows from proglacial environments on the shelf edge could have been triggered by either (1) slope instability caused by loading of ice lobes on the shelf edge (Mulder & Moran 1995) or, (2) periodic meltwater discharge (Hesse et al. 1997) associated with phases of glacial advance and retreat on the Hebridean margin. The fine-grained tail of these gravity flows may have been carried further downslope as low-density turbidity currents that interacted with the pre-existing drift topography and alongslope contour currents to form the large sediment waves present on the distal fan (Fig. 8).

The weak reflectors within unit C (Fig. 7a) correspond to the silty-muddy contourite represented by facies E. This unit may also correspond to the aggrading drift sequence that onlaps the DF 2 sequence (Fig. 8). Seismic facies patterns and topographic features on the lower Barra Fan suggest that two pathways of bottom currents were established subsequent to deposition of DF 3 (18–21 C^{14} ka BP) (Fig. 9): (1) lower NADW tracing the edge of the distal fan and producing erosional scars in the Glacial–Holocene section of the drift sequence associated with DF 2; (2) obliquely downslope oriented bottom currents producing the small-scale wave field along the edge of DF 3. The latter flow component is possibly controlled by NADW deflected across the fan and funnelled through a wide channel, bordered to the south–southwest by the positive topography of DF 3. However, a bottom flow regime across the Barra Fan may also be influenced by deep eddies produced by the strong upper slope current (Howe et al. 1994).

Discussion and conclusions

The Late Pleistocene drift sequence on the lower Barra Fan shows evidence of rapid depositional changes influenced by glacimarine slope processes, hemipelagic sedimentation and deep water circulation in the Rockall Trough. Within this climatically controlled depositional sequence upslope migrating sediment waves (or mud waves) have developed parallel or oblique to topographic steps formed by debrite lobes. Similar large-scale bedforms, described from a variety of continental margin settings, have been related to both downslope sedimentation events and alongslope bottom currents (Jacobi et al. 1975; Damuth 1979; Roberts & Kidd 1979; Normark et al. 1980). In both cases, upcurrent migration of bedforms tends to dominate (McCave et al. 1982; Flood & Shoor 1988). The formation of such sediment waves has been compared to that of fluvial antidunes developed under flow conditions with Froude numbers >1 as a result of internal standing waves (Fox et al. 1968; Normark et al. 1980). An alternative model (Flood 1988) suggests that mud waves are generated by lee-waves caused by upstream obstructions to the flow. While the original lee-wave model could only explain formation of mud waves oriented perpendicular to the flow, an extended version by Blumsack & Weatherly (1989) predicts that fluctuations in the mean current flow can produce maximum bedform growth at oblique angles to the flow direction. This is in accord with observations on mud waves in the Argentine Basin (Weatherly 1993).

The wave fields on the Barra Fan appear to have developed in response to the positive topography generated by debrite units DF 2 and DF 3. Howe (1996) suggested that the sediment waves within seismic unit A1 were deposited from by low-density turbidity currents, a conclusion based on the presence of two

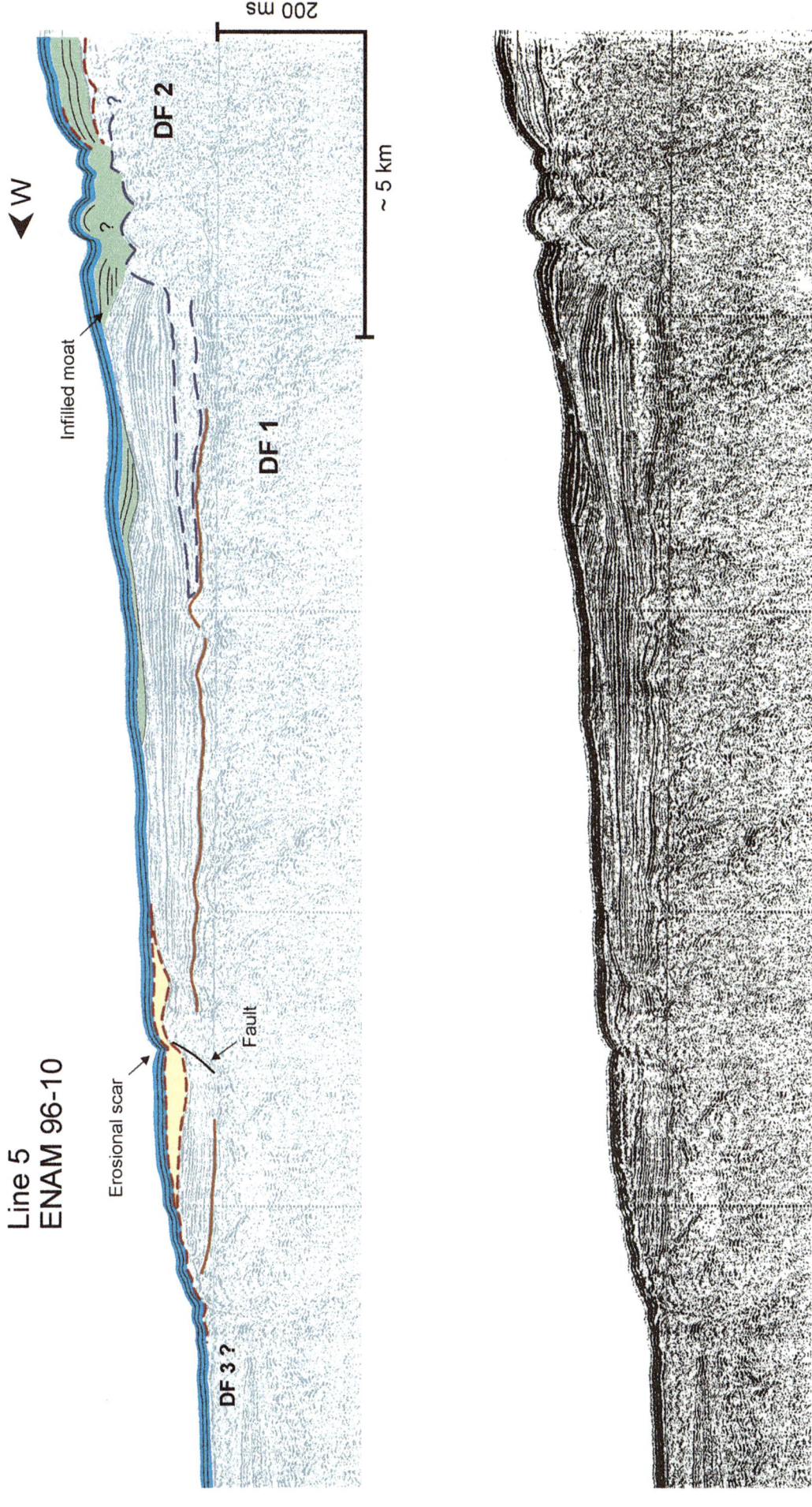

Fig. 8. High-resolution sleeve-gun profile (line 5) with interpretation. Red stippled line indicates possible unconformities. Brown line demarcates base of well-stratified drift sequence. Blue stippled line outlines debris flow sequence 2. Blue draping unit is equivalent to unit A1, while green unit corresponds to unit B. Yellow unit depicts infilling of erosional scars.

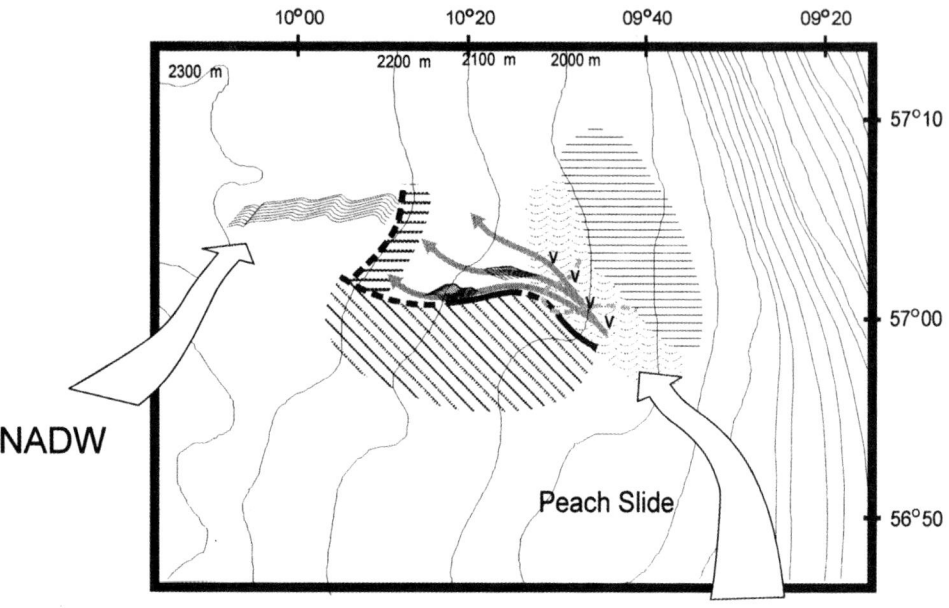

Fig. 9. Seismic patterns and inferred bottom current pathways on the northwestern edge of the Barra Fan. Wavy lines indicate areas where medium-high amplitude, discontinuous reflectors extend to the seafloor. Horizontal lines indicate area where well-stratified seismic facies dominates below the sea floor. V-marks indicate locations of erosional furrows on seismic profiles. Dark horizontal and diagonal lines demarcate the debris flow lobes shown in Figure 2.

graded sand layers observed in shallow cores as well as the orientation of the wave-crests with respect to the fan. Downslope currents flowing oblique to the slope could have produced the observed differential erosion-deposition pattern (deposition on the upcurrent wave crest and erosion on the downslope facing flanks) in accordance with the revised lee-wave model of Blumsack & Weatherly (1989). However, the dominant facies of MD95-2006 and seismic features in the vicinity of the core site suggest that units A1-A2 were deposited in response to complex, semi-permanent bottom currents routed across the mass-flow controlled topography of the lower fan.

Late glacial to Holocene bottom currents appear to have produced substantial erosion and reworking of stratified drifts and underlying debrites. Much of this material may have been redeposited downslope as sandy-silty contourites associated with the medium-high amplitude, discontinuous reflectors of seismic units A1 and A2.

The flow component responsible for the development of the late glacial-Holocene contourite drifts is possibly derived from NADW diverted across the Barra Fan. The topography of DF 3 would have obstructed the flow path of this water mass and led to increasing competence of the bottom currents that were routed along the debris flow scarps. Unit A2 may thus have resulted from reworking and downslope transfer of sediments derived from the exposed surface of DF 2. A diverted flow of NADW across the fan is also inferred by Armishaw et al. (1998, 2000) based on the common occurrence of sandy contourites on the mid-upper slope of the Barra Fan. Alternatively, deep eddies produced by mixing between the upper and lower slope current systems may have influenced bottom current dynamics on the lower fan and contributed to drift morphologies, which are not concordant with the general flow direction (Howe 1996).

The large wave field oriented perpendicular or slightly oblique to DF 2 shows a transition from aggrading waves to upslope-migrating waves. The lower section of the drift may have developed in response to a north-flowing contour current tracing the edge of DF 2, while slower deposition, or erosion, occurred across the surface of DF 2. The prograding bedforms superimposed on the aggrading drifts were probably produced by the same glacigenic gravity flows as recorded in the lithology of MD95-2006. During their passage downslope these may have transformed into turbidity currents with most of the coarse fraction deposited upslope from DF 2. The fading of migrating bedforms away from the fan and the corresponding development of an erosional scar on the distal part of the drift complex suggest that fine-grained turbidity currents became progressively pirated by alongslope bottom currents.

An erosive base of unit A1 suggests that strong bottom currents were active in the Rockall Trough around 16–18 ^{14}C ka BP. This supports previous core studies from the southern Rockall Trough, reporting sedimentological evidence of late glacial deep water flow (Dowling & McCave 1993; McCave et al. 1995a). Dowling & McCave (1993) suggested that this current regime was generated by local production of deep waters in the Rockall Trough, similar to the convective process which occurs today in the Labrador Sea.

Large sediment drift morphologies are conventionally thought to develop over millions of years in response to semi-permanent bottom currents and a steady-state sediment supply (McCave & Tucholke 1986). The high sedimentation rates observed in MD95-2006 and the likelihood that the underlying sequence was deposited during the penultimate glaciation or the early Devensian, suggests that the Barra Fan drift formed over the last 100–200 ka. This rapid drift development was facilitated by high fluxes of glacimarine sediments from the Hebrides Shelf margin, and the presence of topographic traps generated by debrite lobes. Further mapping of seismic and sediment facies is necessary to understand the complex interaction of downslope and alongslope currents, which appear to have shaped the wavy sediment drifts on the lower Barra Fan.

This research was mainly carried out as part of the senior author's PhD programme at Cardiff University, in association with the British Geological Survey. The results are published with permission of the Director of the British Geological Survey. DAVS acknowledges tenure of a Royal Society Industry Fellowship with BP-Amoco.

References

ARMISHAW, J. E., HOLMES, R. W. & STOW, D. A. V. 1998. Morphology and sedimentation on the Hebrides Slope and Barra Fan, NW UK continental margin. *In*: STOKER, M. S., EVANS, D. & CRAMP, A. (eds) *Geological Processes on Continental Margins: Sedimentation, Mass-wasting and Stability*. Geological Society, London, Special Publications, **129**, 81–104.

ARMISHAW, J. E., HOLMES, R. W. & STOW, D. A. V. 2000. The Barra Fan: A bottom current reworked, glacially-fed submarine fan system. *Marine & Petroleum Geology*, **17**, 219–238.

BLUMSACK, S. L. & WEATHERLY, G. L. 1989. Observations of the nearby flow and a model for the growth of mudwaves. *Deep-Sea Research*, **36**, 1327–1339.

DAMUTH, J. E. 1979. Migrating sediment waves created by turbidity currents in the northern South China Basin. *Geology*, **7**, 520–523.

DICKSON, R. R. & KIDD, R. B. 1986. *Deep circulation in the southern Rockall Trough – the oceanographic setting of site 610*. Initial Reports Deep Sea Drilling Project, **94**, US Government Printing Office, Washington DC, 1061–1074.

DOWLING, L. M. & MCCAVE, I. N. 1993. Sedimentation on the Feni Drift and late Glacial bottom water production in the northern Rockall Trough. *Sedimentary Geology*, **82**, 79–87.

ELLETT, D. J. & ROBERTS, D. G. 1973. The overflow of Norwegian Sea Deep Water across the Wyville-Thomson Ridge. *Deep-Sea Research*, **20**, 819–835.

FLOOD, R. D. 1988. A lee-wave model for deep-sea mudwave activity. *Deep-Sea Research*, **35**, 973–983.

FLOOD, R. D. & SHOR, A. N. 1988. Mud waves in the Argentine Basin and their relationship to regional bottom circulation patterns. *Deep-Sea Research*, **35**, 943–971.

FOX, P. J., HEEZEN, B. C. & HARIAN, A. M. 1968. Abyssal antidunes. *Nature*, **220**, 470–472.

HEEZEN, B. C., HOLLISTER, C. D. & RUDDIMAN, W. F. 1966. Shaping of the continental rise by deep geostrophic contour currents. *Science*, **152**, 502–508.

HESSE, R., KLAUCKE, I., RYAN, W. B. F. & PIPER, D. J. W. 1997. Ice-sheet sourced juxtaposed turbidite systems in Labrador Sea. *Geoscience Canada*, **24**(1), 3–12.

HOLMES, R., LONG, D. & DODD, L. R. 1998. Large-scale debrites and submarine landslides on the Barra Fan, west of Britain. *In*: STOKER, M. S., EVANS, D. & CRAMP, A. (eds) *Geological Processes on Continental Margins: Sedimentation, Mass-wasting and Stability*. Geological Society, London, Special Publications **129**, 67–79.

HOWE, J. A. 1996. Turbidite and contourite sediment waves in the northern Rockall trough, north Atlantic Ocean. *Sedimentology*, **43**(2), 219–234.

HOWE, J. A. & HUMPHERY, J. D. 1995. Photographic evidence for slope-current activity, Hebrides Slope, NE Atlantic ocean. *Scottish Journal of Geology*, **30**, 107–115.

HOWE, J. A., STOKER, M. S. & STOW, D. A. V. 1994. Late Cenozoic sediment drift complex, North East Rockall Trough, North Atlantic. *Paleoceanography*, **9**(6), 989–999.

JACOBI, R. D., RABINOWITZ, P. D. & EMBLEY, R. W. 1975. Sediment waves of the Moroccan continental rise. *Marine Geology*, **19**, 61–67.

JONES, E. J. W., EWING, M., EWING, J. J. & EITTREIM, S. L. 1970. Influence of Norwegian Sea overflow water on sedimentation in the northern North Atlantic and Labrador Sea. *Journal of Geophysical Research*, **75**, 1655–1680.

KENYON, N. H. 1986. Evidence from bedforms for a strong poleward current along the upper continental slope of NW Europe. *Marine Geology*, **72**, 187–189.

KNUTZ, P. C., AUSTIN, W. E. N. & JONES, E. J. W. 2001. Millennial scale depositional cycles related to British Ice Sheet variability and North Atlantic palaeocirculation since 45 kyr BP, Barra Fan, UK margin. *Paleoceanography*, **16**, 53–64.

LEDBETTER, M. T. 1984. Current speed in the Vema Channel recorded by particle size of sediment fine fraction. *Marine Geology*, **58**, 137–149.

MASSE, L., FAUGÈRES, J.-C. & HROVATIN, V. 1998. The interplay between turbidity and contour current processes on the Columbia Channel fan drift, Southern Brazil Basin. *Sedimentary Geology*, **115**, 111–132.

MCCARTNEY, M. S. 1992. Recirculation components to the deep boundary current of the northern North Atlantic. *Progress in Oceanography*, **29**(4), 283–383.

MCCAVE, I. N. & TUCHOLKE, B. E. 1986. Deep current-controlled sedimentation in the western North Atlantic. *In*: VOGT, P. R. & TUCHOLKE, B. E. (eds) *The geology of North America, Vol. M, The western North Atlantic region*. Geological Society of American Memoirs, Boulder, Colorado, 451–468.

MCCAVE, I. N., HOLLISTER, C. D., LAINE, E. P., LONSDALE, P. F. & RICHARDSON, M. J. 1982. Erosion and deposition on the eastern margin of the Bermuda Rise in the late Quaternary. *Deep-Sea Research*, **29**, 535–561.

MCCAVE, I. N., MANIGHETTI, B. & BEVERIDGE, N. A. S. 1995a. Circulation in the glacial North Atlantic inferred from grain-size measurements. *Nature*, **374**, 149–152.

MCCAVE, I. N., MANIGHETTI, B. & ROBINSON, S. G. 1995b. Sortable silt and fine sediment size composition slicing – parameters for paleocurrent speed and paleoceanography. *Paleoceanography*, **10**(3), 593–610.

MILLER, K. G. & TUCHOLKE, B. E. 1983. Development of Cenozoic abyssal circulation south of the Greenland-Scotland Ridge. *In*: BOTT, M. H. P., SAXON, S., TALWANI, M. & THIEDE, J. (eds) *Structure and development of the Greenland-Scotland Ridge*. Plenum Press, New York, 549–589.

MULDER, T. & MORAN, K. 1995. Relationship among submarine instabilities, sea-level variations, and the presence of an ice-sheet on the continental shelf – An example from the Verril Canyon area, Scotian Shelf. *Paleoceanography*, **10**(1), 137–154.

NORMARK, W. R., HESS, G. R., STOW, D. A. V. & BOWEN, A. J. 1980. Sediment waves on the Monterey Fan levee: a preliminary physical interpretation. *Marine Geology*, **37**, 1–18.

ROBERTS, D. G. & KIDD, R. B. 1979. Abyssal sediment wave fields on the Feni Ridge, Rockall Trough: long-range sonar studies. *Marine Geology*, **21**, 175–184.

STANLEY, D. J. 1993. Model for turbidite-to-contourite continuum and multiple process transport in deep marine settings: examples in the rock record. *Sedimentary Geology*, **82**, 241–255.

STOKER, M. S. 1995. The influence of glacigenic sedimentation on slope-apron development on the continental margin off Northwest Britain. *In*: SCRUTTON, R. A., STOKER, M. S., SHIMMIELD, G. B. & TUDHOPE, A. W. (eds) *The Tectonics, Sedimentation and Palaeoceanography of the North Atlantic region*. Geological Society, London, Special Publications, **90**, 159–177.

STOKER, M. S., AKHURST, M. C., HOWE, J. A. & STOW, D. A. V. 1998. Sediment drifts and contourites on the continental margin off northwest Britain. *Sedimentary Geology*, **115**(1–4), 33–51.

STOW, D. A. V. & PIPER, D. J. W. 1984. Deep-water fine-grained sediments: facies models. *In*: STOW, D. A. V. & PIPER, D. J. W. (eds) *Fine-Grained Sediments: Deep-Water Processes and Facies*. Geological Society, London, Special Publications, **15**, 611–646.

STOW, D. A. V., FAUGÈRES, J.-C., VIANA, A. & GONTHIER, E. 1998. Fossil contourites: a critical review. *Sedimentary Geology*, **115**, 3–31.

TUCHOLKE, B. E. & LAINE, P. 1983. Neogene and Quaternary development of the lower continental rise off the central U.S. east coast. *In*: WATKINS, J. S. & DRAKE, C. L. (eds) *Studies in Continental Margin Geology*. American Association of Petroloeum Geologists, Memoirs, 295–305.

VORREN, T. O. & LABERG, J. S. 1997. Trough mouth fans – Palaeoclimate and ice-sheet monitors. *Quaternary Science Reviews*, **16**(8), 865–881.

WEATHERLY, G. L. 1993. On deep-current and hydrographic observations from a mudwave region and elsewhere in the Argentine Basin. *Deep-Sea Research*, **40**, 939–961.

Holocene contourite sand sheet on the Barra Fan slope, NW Hebridean margin

DORRIK A. V. STOW[1], JULIE E. ARMISHAW[2] & RICHARD HOLMES[3]

[1]*SOES, Southampton Oceanography Centre, Southampton University, Southampton SO14 3ZH, UK*
(e-mail: davs@soc.soton.ac.uk)
[2]*Robertson Research International, Llanrhos, Llandudno, Gwynedd, UK*
[3]*British Geological Survey, West Mains Road, Edinburgh EH9 3LA, UK*

Abstract: Surficial sediment analyses, bottom photographs and current-meter data, coupled with studies of sidescan and 3.5 kHz echodata clearly show the imprint of a strong, seasonally-affected, slope current on the Hebrides shelf and slope. The present day shelf sediments are in fact relict, coarse-grained material (gravel with boulders) of Pleistocene glacial derivation, which have been modified during the Holocene by winnowing, sea-floor polishing and transport of the finer fraction across the shelf to the northwest. On the outer shelf and upper slope the sharp change from gravel to sand is marked by a physiographic-textural boundary, herein termed the *sandline*, which occurs at a depth of between 170 m and 300 m. Further down the continental margin on the lower slope, the lower limit of the sand-rich facies is marked by the *mudline*. This typically occurs at a depth of around 1200 m, and marks the depth of substantially increased clay-sized material. Above the mudline the long-term (Holocene to present), time-integrated signature of bottom current flow has resulted in the relatively slow accumulation of a mid-slope sandy contourite deposit, the Barra contourite sand sheet. This covers an area of 1000–1500 km^2 with an estimated sand volume of 30 000 m^3. Below the mudline the clay-rich deposits represent a hemipelagic drape with only minor bottom current influence and intense reworking by benthic organisms.

The Barra fan on the NW Hebridean slope apron is an important example of a glacially-fed, trough-mouth fan. Taken together with the contiguous Donegal fan, it forms the largest depocentre off NW Britain covering an area of about 7000 km^2 from the shelf-break to a depth of around 2000 m in the Rockall Trough (Fig. 1). This system has been the focus of detailed study by the authors

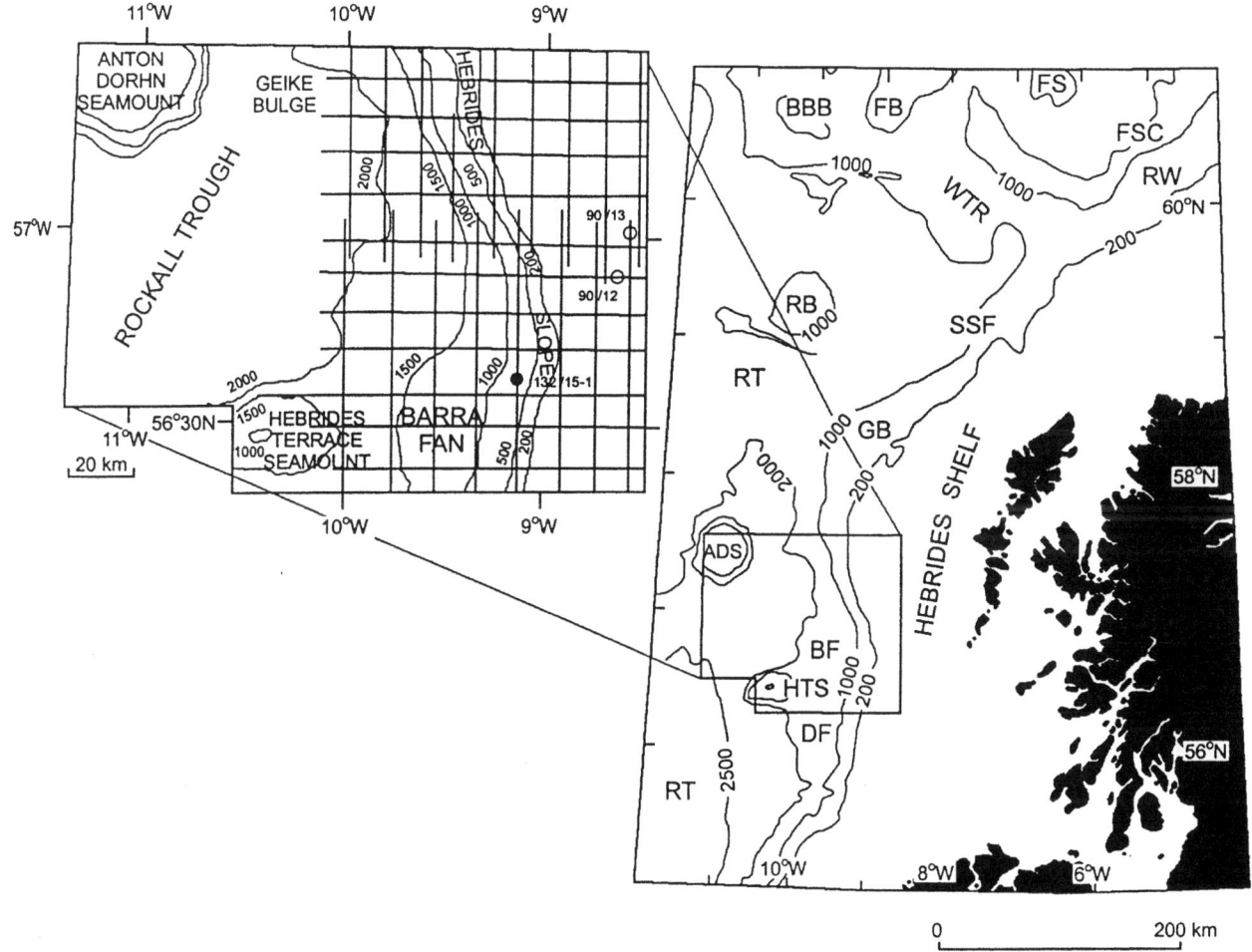

Fig. 1. Location of the study area, bathymetric setting (contours in metres) and track plots of regional 3.5 kHz seismic lines. South of *c.* 57°N, swath bathymetry data are available between 150 and 2000 m water depth for most of the study area. BBB, Bill Bailey's Bank; FB, Faroe Bank; FS, Faroe Shelf; WTR, Wyville–Thomson Ridge; RW, Rona Wedge; FSC, Faroe–Shetland Channel; RT, Rockall Trough; RB, Rosemary Bank; SSF, Sula Sgeir Fan; GB, Geike Bulge; ADS, Anton Dohrn Seamount; BF, Barra Fan; HTS, Hebrides Terrace Seamount; DF, Donegal Fan. (From Armishaw *et al.* 1998).

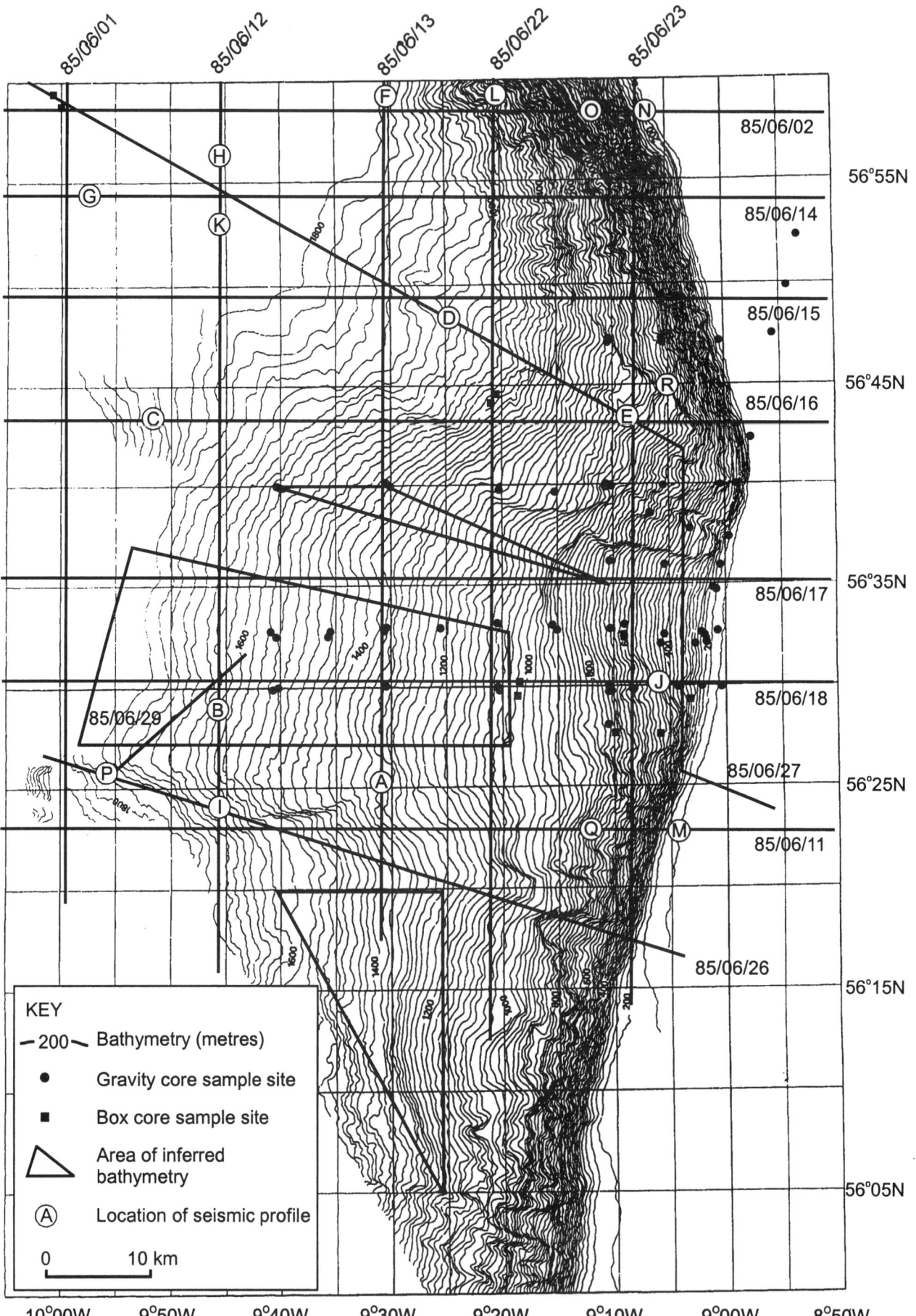

Fig. 2. Digitised bathymetry of the Hebrides Slope based on swath bathymetric mapping. Contours are at 20 m spacing. Black symbols mark the locations of core sites within the study area; dark lines show 3.5 kHz echosounder profiles. Boxed areas mark regions of inferred bathymetry, where survey data was of poor quality.

and other colleagues, that has particularly emphasized the interaction of alongslope, downslope and hemipelagic processes in deposition of the Quaternary–Recent succession (Howe 1995, 1996; Armishaw 1998; Armishaw et al. 1998, 2000; Holmes et al. 1998).

The present study briefly synthesizes some of this earlier work, and then uses detailed petrological analysis in conjunction with visual evidence from seabed photographs, geoacoustical properties of sediments and sedimentary structures to identify the long-term, time-integrated signature of the bottom current flow system on the Hebridean margin. The work also elucidates other active sedimentation processes, including offshelf spillover and downslope-delivery of sediment, and delineates the main sedimentary provinces on the margin that appear to be marked by the presence of a mudline on the mid- to lower-slope, and a sandline on the region between the shelfbreak and upper slope. The in situ observations from seabed photographs enables detailed seasonal changes in current flow and intensity to be recorded. The delineation of facies, in turn is used to define areas of erosion, non-deposition and winnowing, and deposition.

Sampling of cores was conducted over a two year period from 1994 to 1996 from 77 gravity core stations (0.10 to 2.7 m penetration depths) and 10 box core stations, yielding a suite of 102 core samples within the region (Fig. 2). Shipboard and laboratory logging followed standard procedures, including photography and x-radiography of all cores. Subsamples of each box core were collected for measurements of sonic velocity, grain size, carbonate content and total organic carbon content at 2 to 4 cm intervals down the core.

In the laboratory box-core subsamples and split gravity cores were analysed for texture, structure and composition. Textural characteristics of shallow subsurface deposits were determined from gravity core log descriptions and x-radiograph interpretations of half cores taken with a Scanray AC 120L device at settings of 65–75 kV for 2 to 2.5 minutes. Multiple particle size analyses at 2 to 4 cm intervals of silt and clay (<63 μm) fractions were undertaken using a Micrometrics Sedigraph 5000ET at the Department of Geography, University of Edinburgh. For the sand fraction (<4.0φ) a combination of wet- and dry-sieving was used.

Over 700 bottom photographs were obtained using the Proudman Oceanographic Laboratory UMEL deep-sea survey camera. A time series of shots were taken during a two year period, between March 1994 and May 1996 at stations along a southern transect between depths of 140 and 1500 m and stations along a northern transect between depths of 140 and 2000 m (Fig. 2). The seabed area photographed is trapezoidal in shape and about 3 m front to back. The width across the bottom frame is approximately 120 cm and that across the top is 250 cm (Humphery, pers. comm., 1996).

Remote survey data collected included Simrad EM12–120 (12 kHz) multibeam swath bathymetry, as well as 7.5 and 3.5 kHz echosounder seismic profiles.

Geological and oceanographic setting

The area of the continental shelf and slope considered in this study extends from a shoreward depth of about 120 m, to water depths in excess of 2000 m on the lower slope (Figs 1 & 2; Table 1). The shelf and slope system consists of a complex Neogene to Holocene sedimentary prism that extends oceanward between the structural highs of the Geike Bulge in the north and Donegal Platform to the south. The gently dipping shelf has an average gradient of < 0.5° with a somewhat irregular topography of deep inner shelf basins and ridges. The width of the shelf is typically 100 km or more. The slope system is dominated by the Barra Fan complex, prograding westward into the centre of the study region and encroaching on the Hebrides Terrace Seamount.

Within the region the gross hydrography may be distinguished by three main circulatory systems (Booth & Ellett 1983). The outer shelf is characterized by seasonally variable southeasterly weak residual currents (2–4 cm s^{-1}) in spring and summer and stronger northerly flowing currents (4–10 cm s^{-1}) in autumn and winter (Ellett et al. 1986), reflecting the a broadening of the slope current across the shelf during autumn and winter. The circulation over the shelf appears to be chiefly driven by near-diurnal, clockwise-rotating currents, internal tides and wave-induced currents (Huthnance 1986; Gordon & Huthnance 1987).

Table 1. *Principal characteristics of the Barra Contourite Sand Sheet*

Location	Hebridean margin, NE Atlantic
Setting	mid-slope setting, 300–1200 m water depth
Age	Holocene
Drift type	thin contourite sand sheet forming most recent part of mixed facies sheeted drift system
Dimensions	sand sheet 80–100 km × 15–25 km, 0.05–0.4 m thick in study area but more extensive along slope, sheeted drift complex of indeterminate size along slope
Seismic facies	thin sand sheet not fully resolved on seismic records, but gives moderate amplitude subparallel reflectors on 3.5 kHz records, mixed sheeted drift shows interlayered variable seismic facies
Sediment facies	thin sandy contourite facies, over silty and muddy contourites hemipelagites and glaciomarine sediments

A broad region of the mid-slope is dominated by a 36 km wide, fast-moving, north to northeasterly flowing slope current, carrying water of Atlantic origin to the Norwegian Sea (Fig. 3). The principal driving forces of the slope current are meridional pressure gradients, with minor effects of wind-driven processes. Current meter measurements of the current north of Scotland indicate typical speeds of 20 cm s^{-1}, with mean monthly speeds of 13 and 20 cm s^{-1} (Turrell et al. 1992) recorded from along the shelf edge west of the Hebrides.

Although the occurrence of shelf-slope exchange of water is not uncommon the precise mechanisms whereby water is exchanged are highly complex and are not yet fully understood (Ellett, pers. comm., 1995). West of Scotland, little exchange seems to occur during the summer months, as water near the sea floor becomes isolated and so remains cool and dense. However, with the onset of autumnal gales there is an increase in mixing of shelf waters (Edelsten et al. 1976; Booth & Ellett 1983). Cross-slope exchange velocities are generally slow, with an average velocities of 2 cm s^{-1} (Huthnance 1986).

The lower slope is affected by the northerly flow of the North Atlantic Deep Water which forms a continuation of the cyclonic loop that circulates in the southern trough, although it may include elements of North East Atlantic Water from the mid- and upper-slope. Part of this latter water mass derives from the influence of Mediterranean outflow (Harvey 1982). Current activity above the deep water mass is very variable in strength and direction.

Bathymetry

A detailed bathymetric chart of the Barra fan slope region has been constructed from multibeam swath-bathymetry echosounder data (Simrad EM12–120, 12 kHz). High quality data were collected for most of the area allowing contours to be drawn at 10 m intervals (Figs 2 & 3, contours shown at 20 m intervals for greater clarity).

Average slope angles across the region vary from 4–5° north of the Peach Slide to 1–3° on the uninterrupted seaward-dipping plain south of the Slide. Where the slope is highly incised by a network of steep-sided canyons in the north of the study area, slope angles may locally exceed 16°, although this is relatively rare. In the central region the most pronounced topographic feature is the Peach Slide (Holmes 1994; Holmes et al. 1998) which has an area of over 2000 km^2 and slide-transfer volume in the order of 80 km^3. Much of the mid and lower slope in the central and southern parts of the study area appear to be relatively smooth.

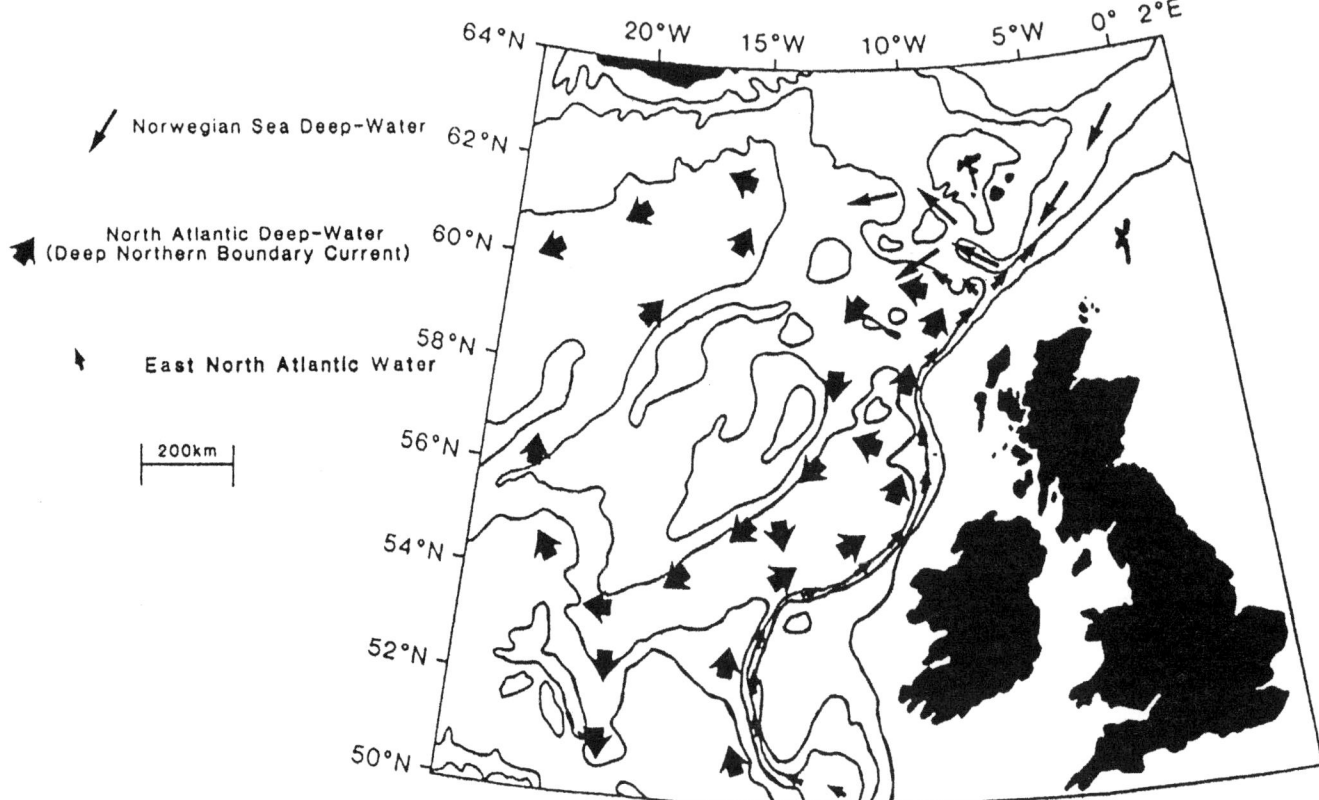

Fig. 3. Present day bottom water circulation pattern in the Rockall Trough and surrounding area, with Norwegian Sea Deep-Water overflowing across the Wyville–Thomson Ridge, North Atlantic Deep-Water flowing as an anticlockwise gyre in the basin, and a slope-current of Eastern North Atlantic Water.

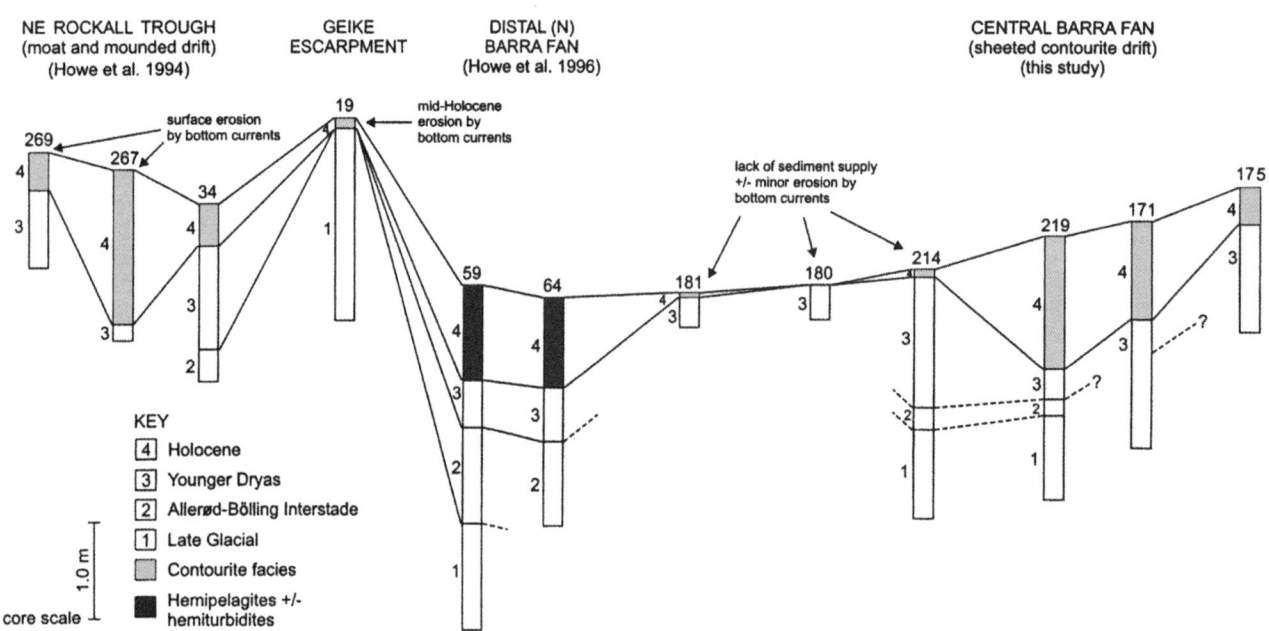

Fig. 4. Stratigraphic correlation panel for the NW Hebridean margin from the NE Rockall Trough to the central Barra fan, based on Howe *et al.* (1994), Stoker *et al.* (1993) and Armishaw *et al.* (2000).

Stratigraphic context

All the slope cores from the study area lie within the topmost part of the Mid-Pleistocene to Holocene seismostratigraphic unit, that generally shows strong basinward progradation during the glacial intervals (Stoker *et al.* 1993, 1994; Stoker in press). The thin Holocene drape and drift section cannot be resolved on seismic records but has been extensively cored. Provisional dating and correlation of the cores has been attempted by combining limited Pb210 and C14 radiometric analyses, the study of microfossil assemblages (foraminifers, nannofossils and dinoflagellate cysts) and the relative abundance of ice-rafted debris (IRD) (Armishaw *et al.* 2000).

On this basis, the sandy layer of variable thickness that occurs at the top of most cores can be assigned a Holocene age (Fig. 4). This is similar to cores from further North on the Rockall Trough

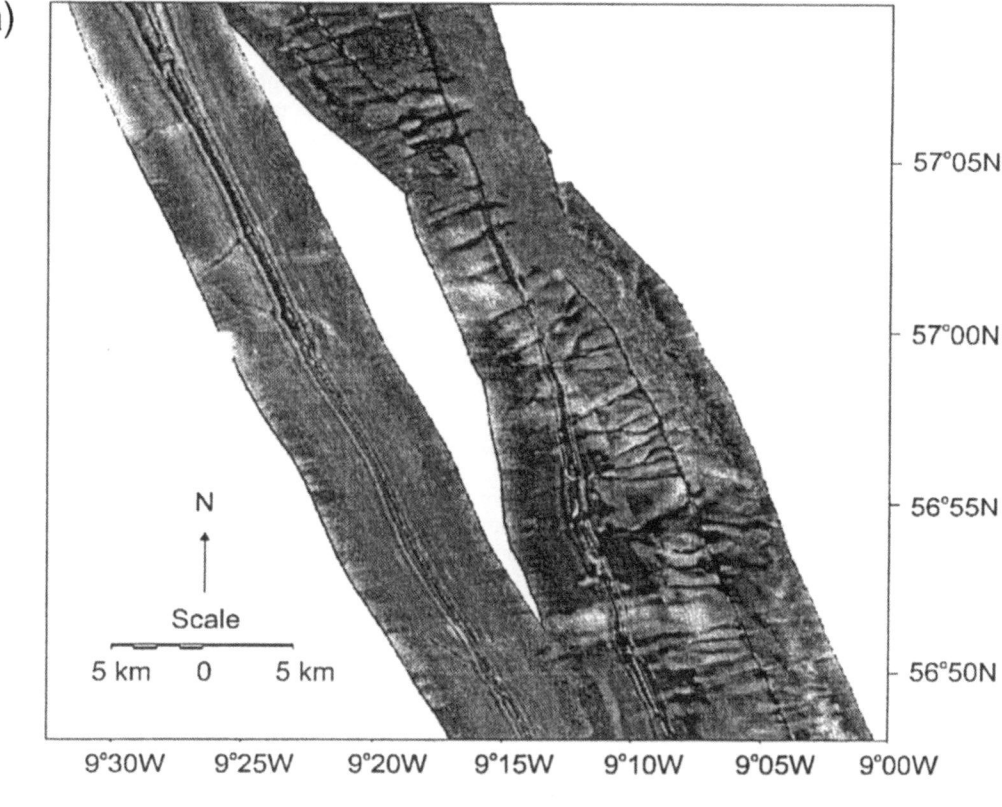

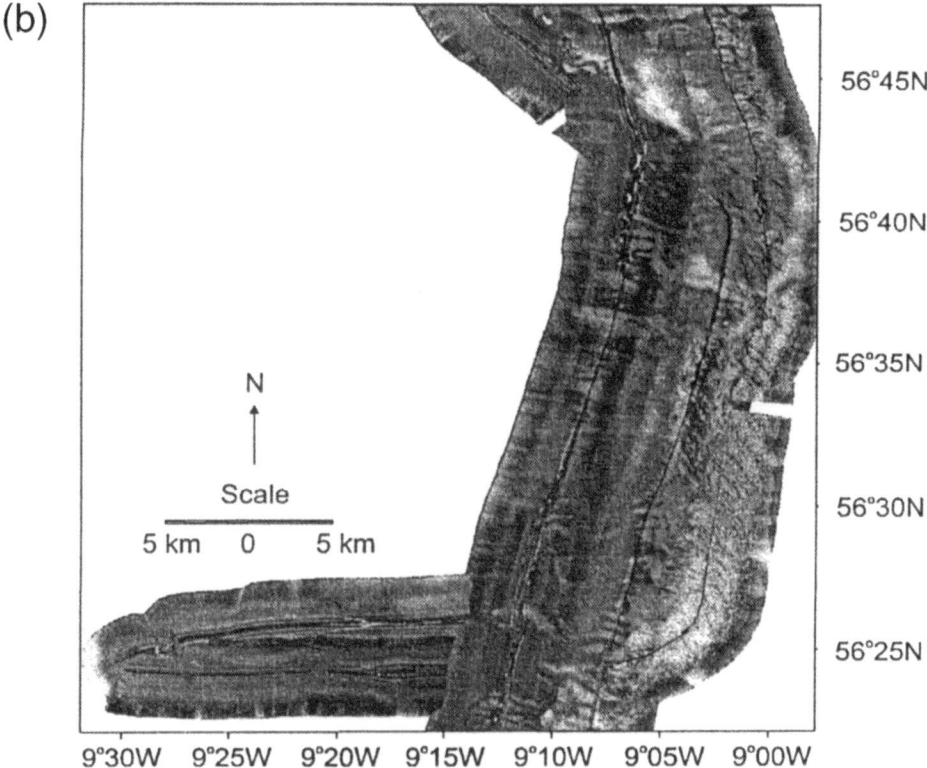

Fig. 5. TOBI side-scan mosaic of the Barra fan. (**a**) Northern region highlighting the dissected nature of the upper slope by erosive canyons. (**b**) Southern region incorporating the Peach Slide and showing the NE–SW trending irregular, variably degraded ice scour marks on the shelf and upper slope. Both regions show a broad mid-slope zone of relatively uniform medium-grey backscatter in some parts apparently masking original underlying elements. This uniform zone is the Barra contourite sand sheet.

slope (Howe *et al.* 1994; Howe 1996), and correlates with a muddy or silty facies typical of the lower slope (Knutz *et al.* 2002) and basin floor. In some cases, the sandy section probably represents a condensed Holocene sequence or part of the early Holocene only, as a result of winnowing and erosion under high bottom current activity. However, we do not yet have adequate stratigraphic control from the study area to confirm this.

Seismic characteristics

Seafloor morphology (Fig. 5)

Combined use of swath bathymetry, high resolution seismic profiles and sidescan sonar images has allowed detailed characterisation of seafloor morphology. Armishaw *et al.* (1998) identified seven principal morphological elements on the outer shelf and slope:

- Ice scour features
- Canyons and channels
- Slides
- Debris-flow masses
- Lobes
- Elongate to wavy bedforms
- Smoothed sediment surfaces

It is only the last two of these that we attribute to bottom current activity. Narrow elongate features characterized by light to dark alternations in TOBI backscatter intensity extend a few kilometres alongslope over parts of the mid and upper slope. Possible sediment waves with across-slope crest orientation are less clearly observed. More distinctive are broad areas of uniform speckled grey, medium intensity backscatter that extend in a somewhat discontinuous, mid-upper slope belt over the much of the study area (approximately 500 km^2). Pale ghosts of other elements are in some places discernible beneath this uniform pattern, whereas just outside the main uniform zone, other elements become dominant with a slightly blurred form. We interpret this uniform backscatter zone as typical of the Barra sand sheet covering a large area of the mid-upper slope. Where it thins markedly the sidescan images of other elements are slightly blurred.

3.5 kHz echocharacter (Figs 6 and 7)

Fourteen distinct seismic facies have been recognised in the study area and their distribution mapped (Armishaw 1998). For the purposes of this paper, we focus only on the four major classes that are broadly based on the earlier echo-character classification schemes of Damuth (1975, 1978, 1980) and Pratson & Laine (1989), and modified according to our own observations.

Class I. Distinct, acoustically stratified seismic facies are characterized by sharply defined, laterally continuous, acoustically layered seafloor reflectors underlying a distinct seabed reflector. This seismic facies class can be further subdivided into four echofacies on the basis of number and clarity of acoustic strata.

Class II. Irregular seismic facies are characterized by a distinct, or indistinct seabed surface reflector and lenticular subbottom reflectors around acoustically unstratified transparent to semi-transparent zones. Two sub-divisions have been recognised and although they are classified separately according to differences in underlying acoustic characteristics the two facies are intergradational and are not always clearly distinguished.

Class III. Wavy echoes are based upon a geometric classification in which the term 'wavy-echo' should not be confused with sedimentary bedforms or processes. The class includes regular to irregular, low to high amplitude overlapping hyperbolae, standing and migrating waves and single mounds or grouped levees.

Class IV. Prolonged seismic facies are characterized by a very prolonged to semi-prolonged seabed reflector underlain by a diffuse blackened zone caused by very high acoustic backscattering with negligible or no subbottom reflectors.

These seismic facies have all been observed by previous investigators in similar continental margin regimes, and the interpretations put forward in those studies are generally applicable to the West Hebridean slope apron system. However, our sedimentary interpretations (below) have been further calibrated by comparison with detailed study of more than 100 cores, extensive bottom photographic coverage, bathymetry and side-scan sonar data.

The distinct seismic facies (Class I) are very widespread throughout the region especially the mid-lower slope and mostly show a drape morphology. All cores recovered from these facies are dominated by hemipelagic sediments. We therefore interpret these seismic facies as mainly fine-grained hemipelagites, with minor interbedded turbidites and/or more coarse-grained glaciomarine input, especially where the reflectors are of higher amplitude and penetration is. Slight changes in apparent thickness and broad lenticularity of subbottom reflectors indicate the influence of bottom currents during deposition, so that sediments in these areas will be partly contouritic.

The irregular seismic facies (Class II) are also very widespread especially on the upper-mid slope, showing transparent to semi-transparent layers and lenses intercalated with acoustically stratified deposits. Cores from the study area include fine hemipelagites, sandy contourites and mixed grade glaciomarine deposits. Elsewhere on the margin cores that penetrate the lenticular transparent zones have recovered debrites (Stoker et al. 1992). We therefore interpret these seismic facies as representing a variety of interbedded facies including hemipelagites, contourites, glaciomarine sediments and debrites.

The wavy seismic facies (Class III) are less widely distributed and more varied in nature than either Class I or II. The regular, migrating waves are best interpreted as either contouritic or turbiditic in origin (see also Howe 1996), whereas the less regular waves are more probably isolated contourite waves or patch drifts. The low relief ridges that parallel cross-slope channels are clearly low-relief levees with one or more back-levee turbidite sediment waves in some cases. Thickness variations of the thin upper seismic unit represents differential sediment accumulation probably caused by the interaction of bottom currents with an irregular slope topography.

The prolonged seismic facies (Class IV) are also varied in nature but of restricted occurrence. They include coarse-grained (sand and gravel) relict glacial deposits on the shelf and uppermost slope that have been subject to intense shelf reworking and spillover processes during the Holocene; slide and slump deposits restricted to steep slopes, showing little acoustic penetration and a semi-prolonged seafloor reflector because of the steep angles involved; and an irregular seafloor (probably coarse-grained) caused by intense bottom current flow and erosion.

The seismic facies map (Fig. 6) shows that the acoustic character of the sea floor throughout the study area is highly variable. This heterogeneity is particularly noteworthy on both the upper and lower continental slope, and indicates a diversity of sea-floor deposits and processes in an area of complex bathymetric relief. The overall trends of the facies are perpendicular to the continental margin; however specific evidence for trends parallel to the continental margin are also displayed. Further evidence for alongslope trends does not directly come out in the seismic facies map but is expressed as thickness variations (drifts and sheets), erosive zones, moats and lag zones and rare bottom current waves and drifts. These features are observed in cores, bottom photographs and side-scan sonar data. The Holocene Barra sand sheet alone is not resolved at this scale of study, but when combined with the late Pleistocene bottom current affected unit can be seen as a thin sheeted drift (Fig. 5).

Sediment facies

Seafloor photography

Photographs of the seafloor illustrate a variety of features attributable to current activity, normal marine hemipelagic sedimentation, and benthic fauna. Observations of the seafloor across a northern and southern slope transect provides clear evidence for: (a) a seaward-fining sequence of surface sediment facies; (b) seasonally affected bottom current flow and erosion between depths of 140 and 1000 m; and (c) deposition at depths below 1500 m evidenced by non-current induced features and a densely populated and variable benthic community.

There is a clear decrease in mean grain-size of the surface sediment cover with depth downslope. Sediments on the outer shelf and upper slope down to 300 m are composed of up to 50% coarse gravelly sand to large cobbles and boulders (2–20 cm diameter, maximum 50 cm) (Fig. 6a). Sand cover with less gravel

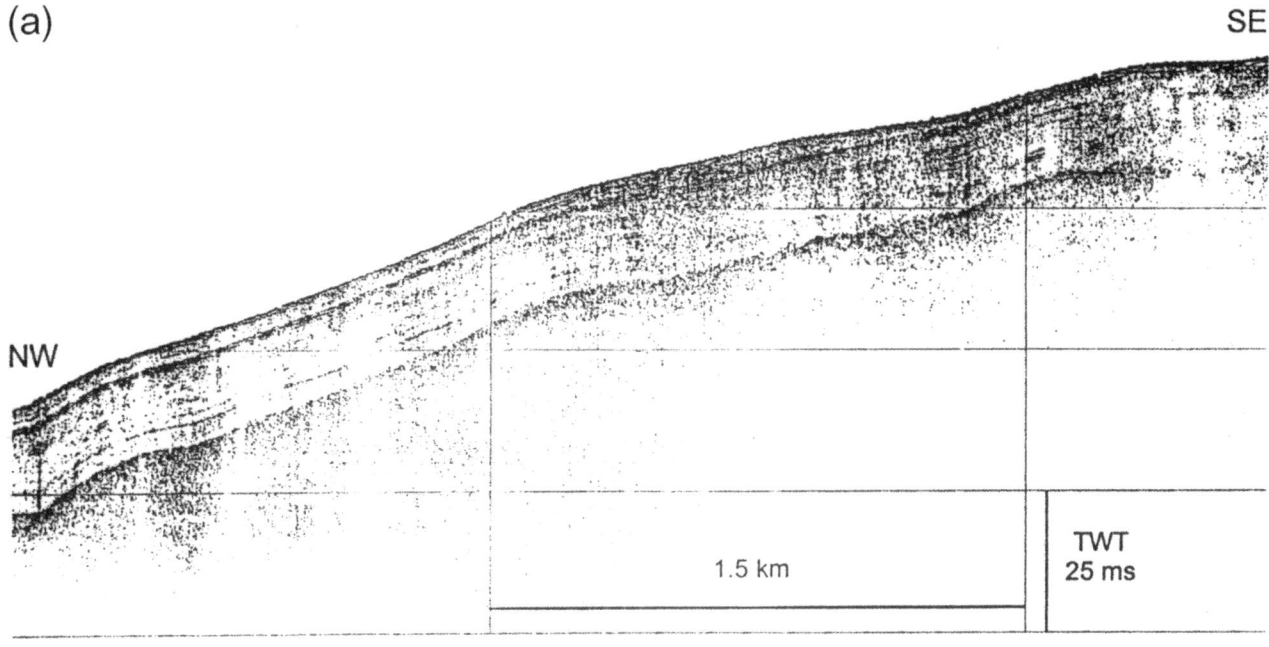

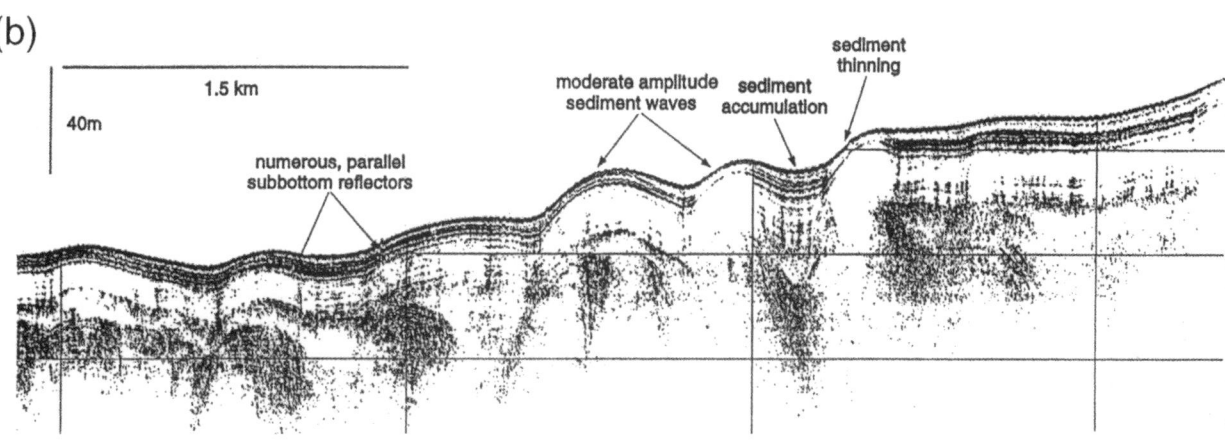

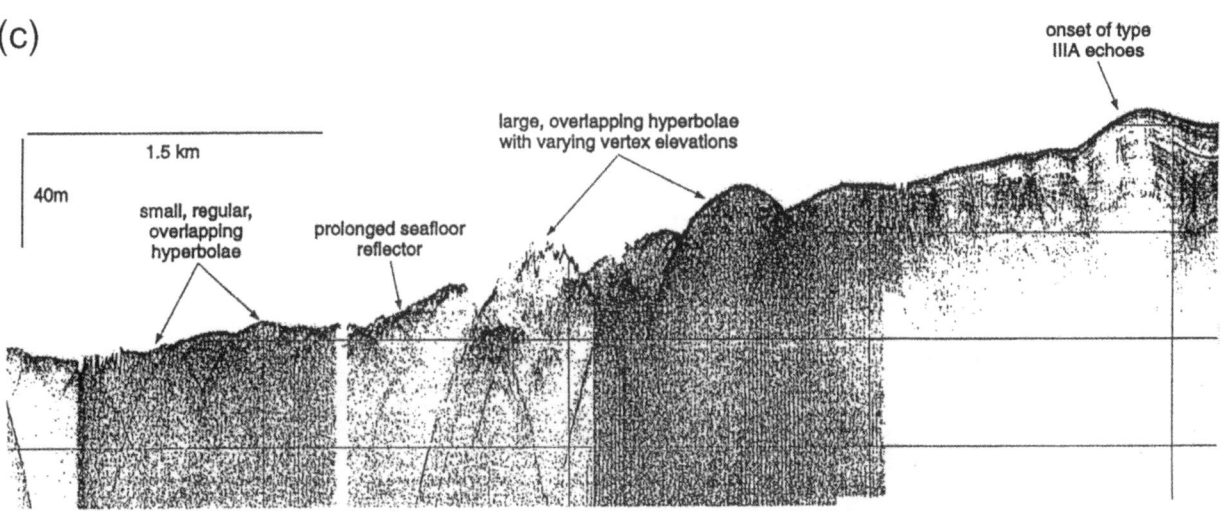

Fig. 6. Selected 3.5 kHz seismic profiles showing features of alongslope sedimentation on the mid and lower slope (profiles A, B, C); downslope sedimentation on the upper to mid slope (profiles D, E); and the effects of seafloor polishing and spillover on the outer shelf to upper slope (profiles F, G, H). Profiles are located on Figure 2, and show examples of seismic facies as follows: A, facies IB; B, facies IIIA; C, facies IVD; D, facies IVC; E, facies IIA and IIB; F, facies IVA; G and H, facies IVB.

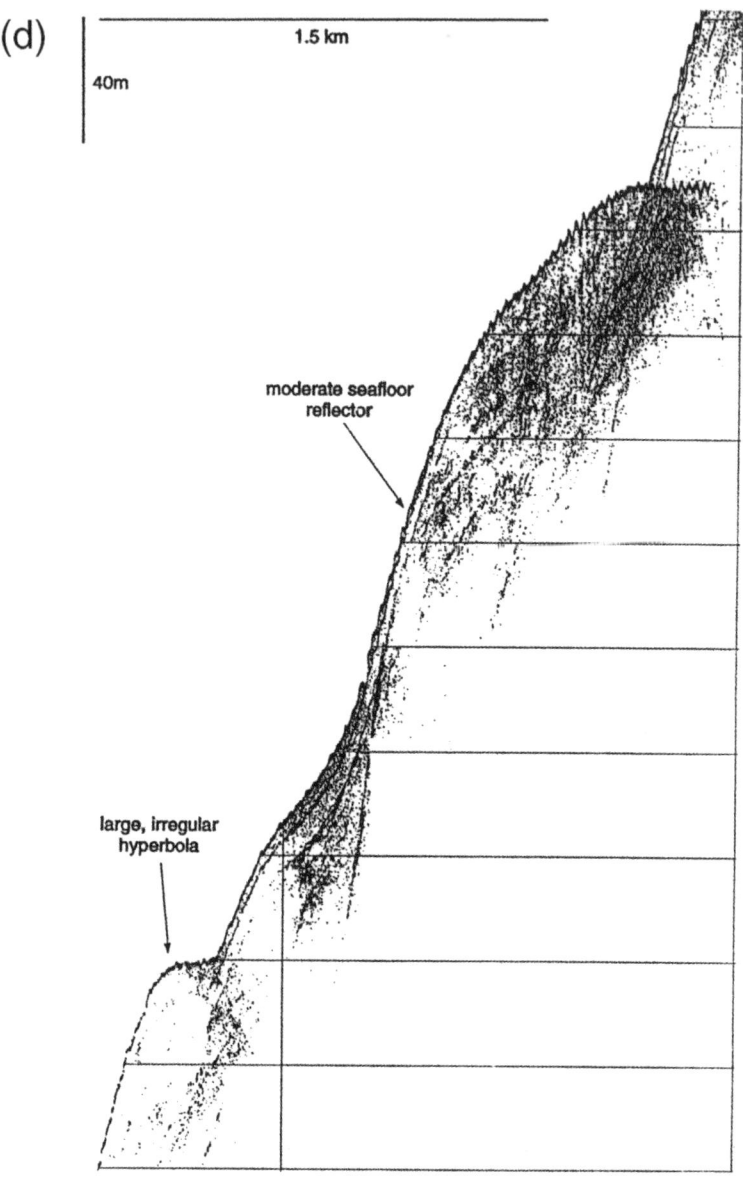

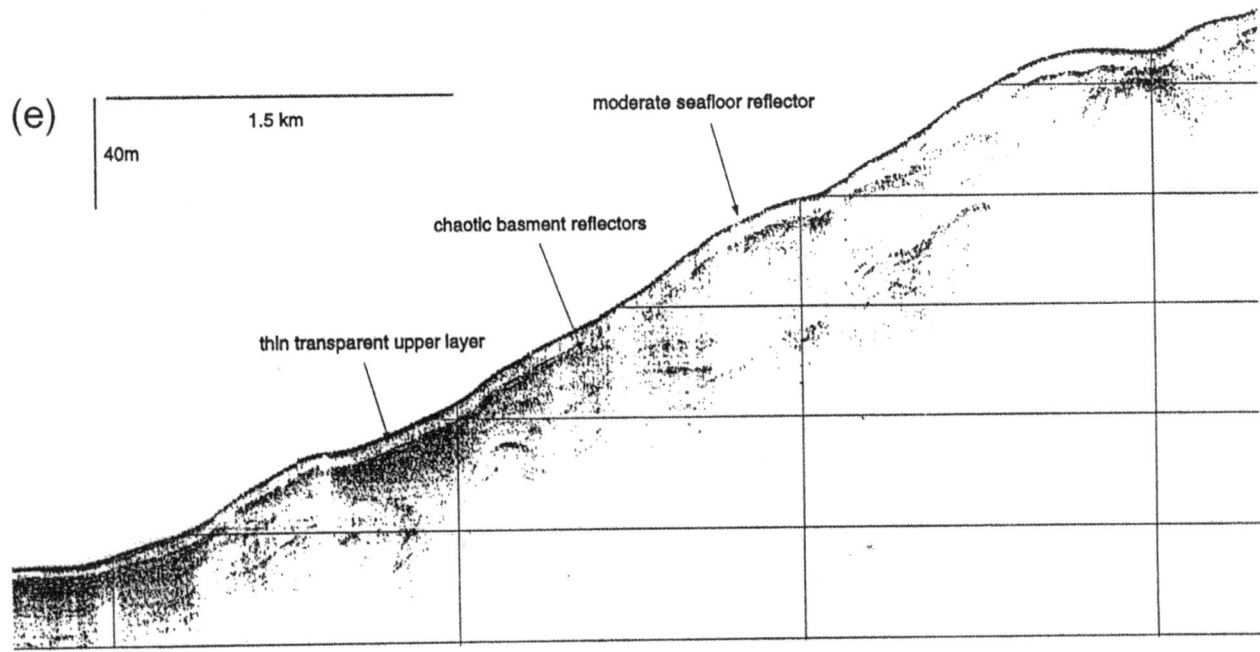

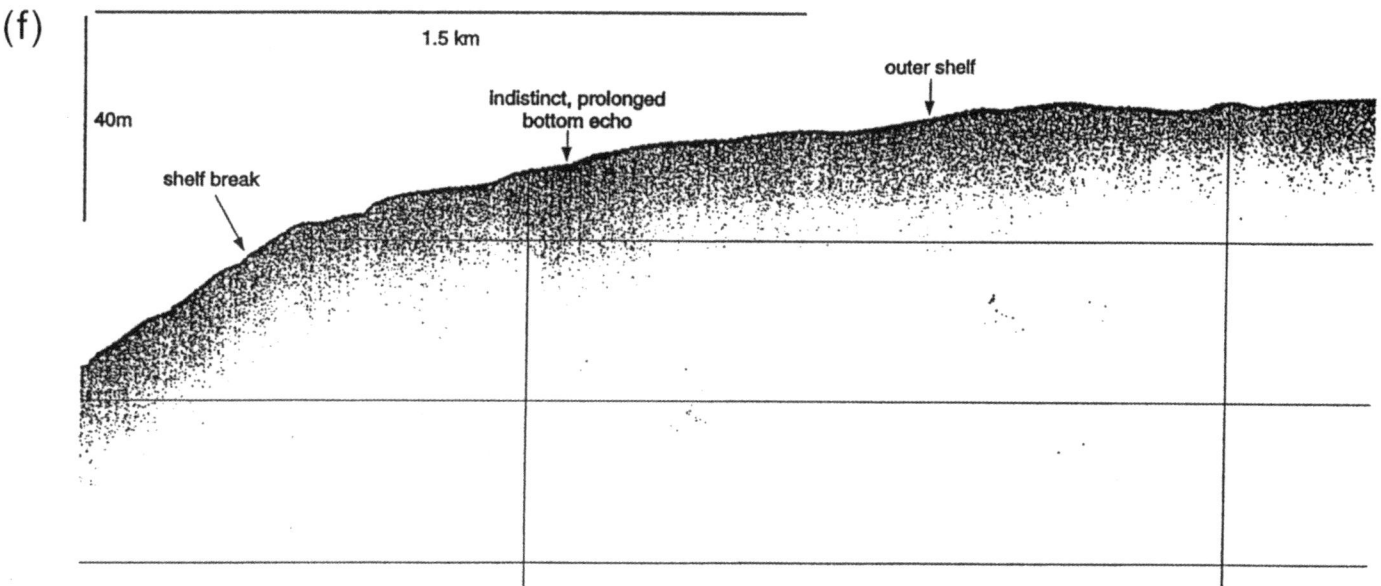

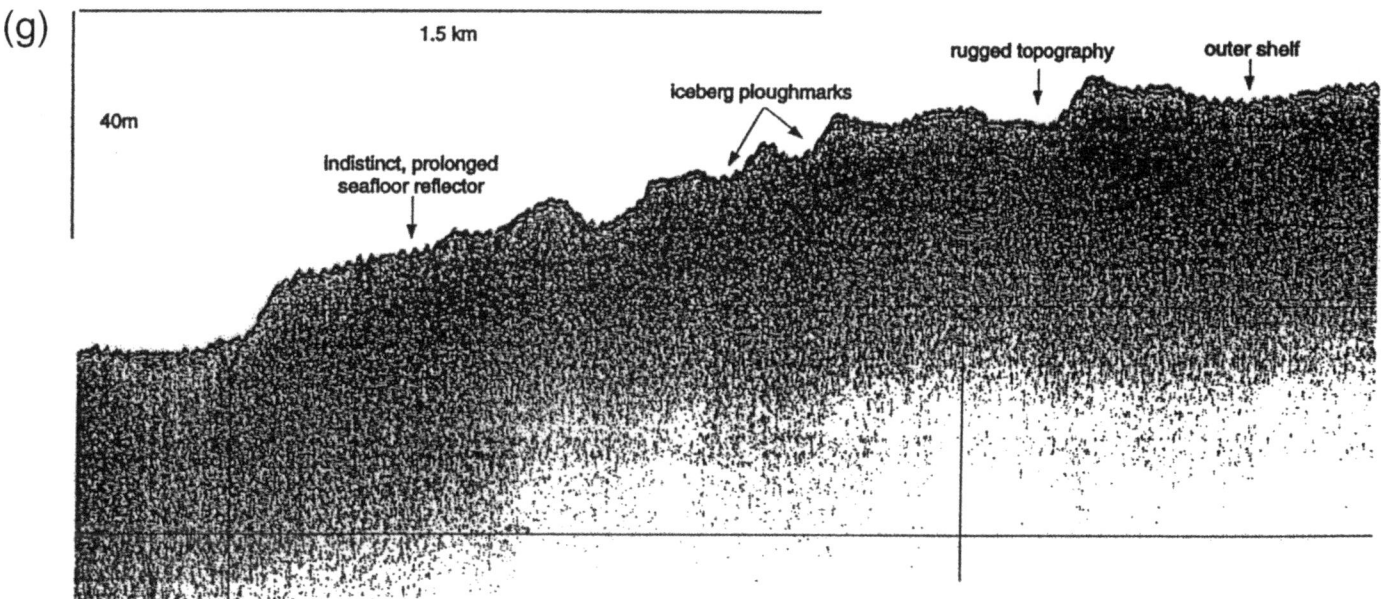

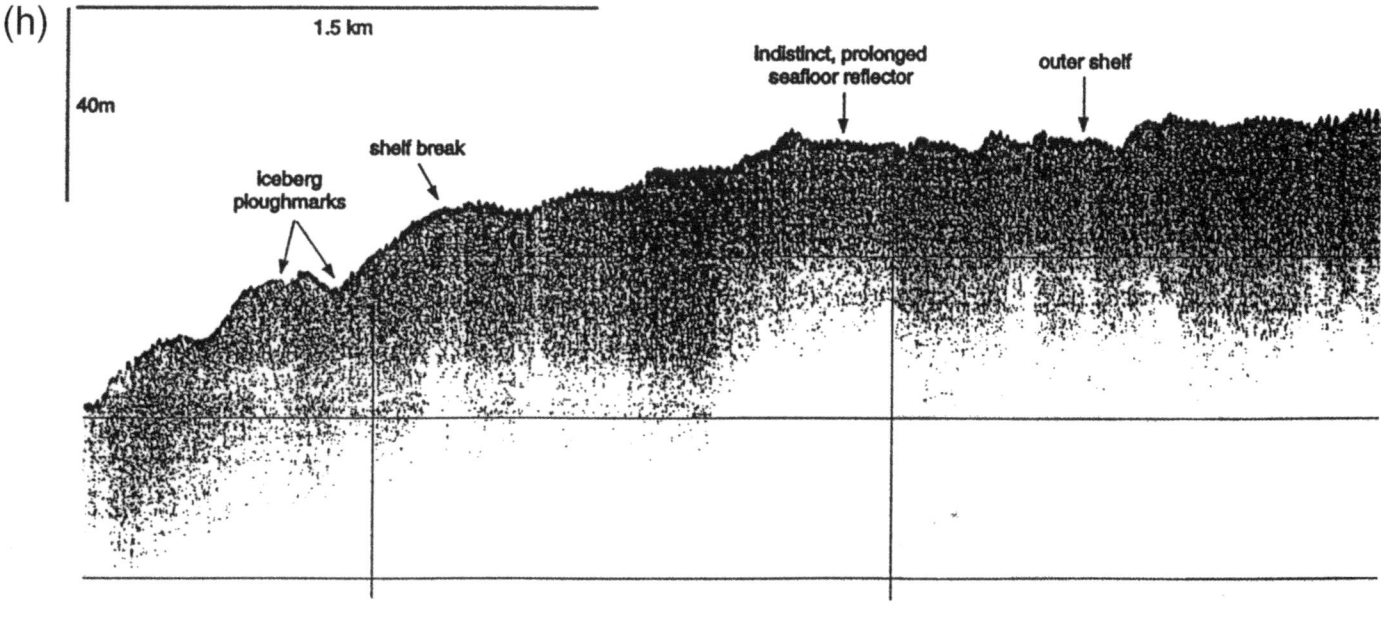

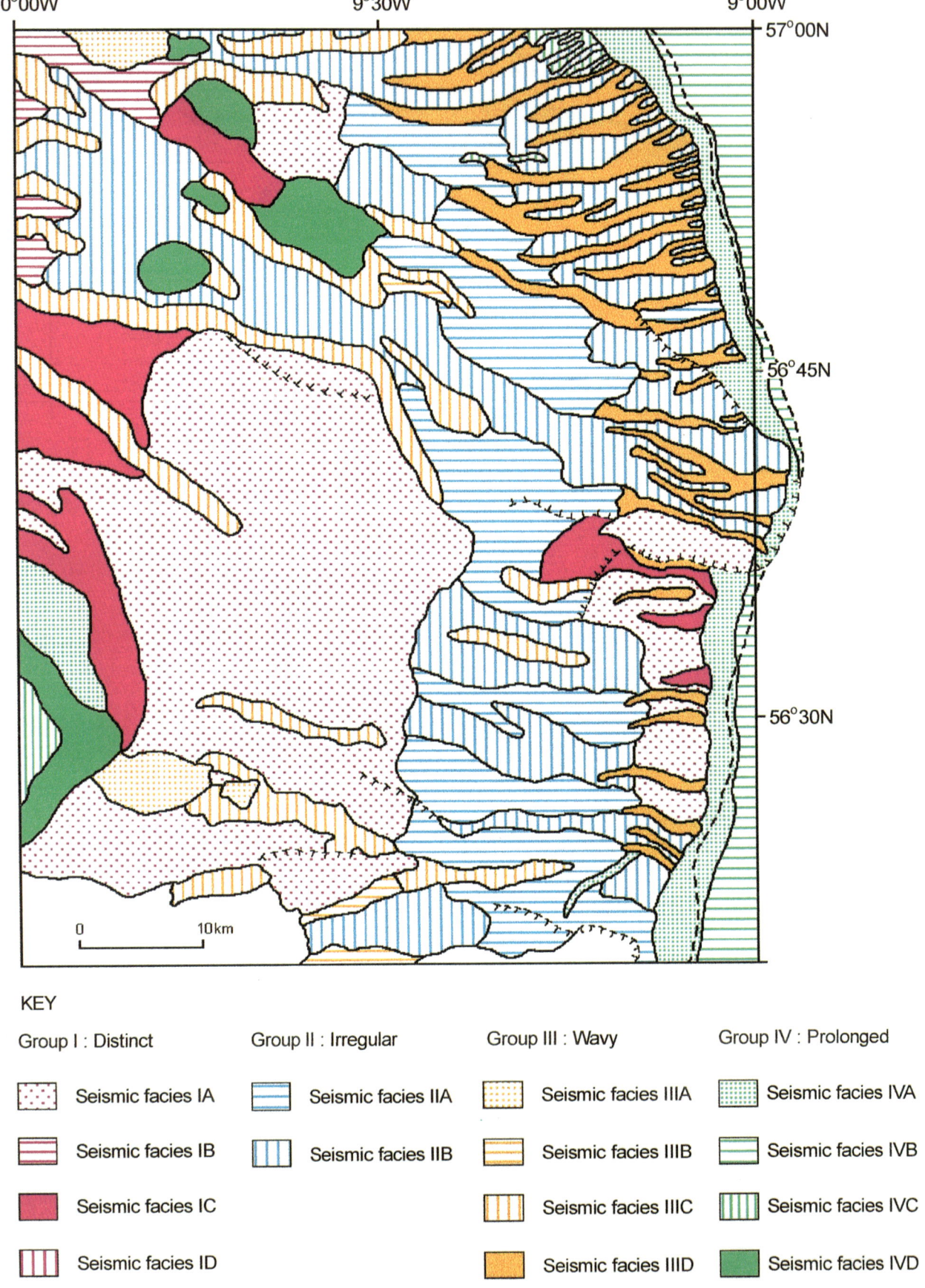

Fig. 7. Map of 3.5 kHz seismic facies over the Barra fan and slope region. Selected examples of the seismic facies are shown in Figure 6. Location and track lines are shown in Figure 2.

persists to a depth of approximately 700 m (Fig. 6b). From 700 m the sand becomes notably siltier to depths of 1000 m (Fig. 6c) with rare gravel patches and boulders (maximum 30 cm diameter). Photographs from the deepest stations shows a muddy sediment cover which is distinctly granular in parts (Fig. 6d).

Current induced features are characteristic of sites between 140 and 1000 m, including sand and silt ripples, sand waves, and obstacle marks. The most common and extensive current features observed are sand and silt ripples. These range from patterns of confused, asymmetric sand ripples on the shelf and upper slope (Fig. 7a) to well-aligned straight crested sandy-silt ripples at 1000 m (Fig. 7b). Ripple crest orientation is variable, depending on the position and depth at which they are found. The majority indicate a north to north–northeast flowing current (Fig. 7a), or more rarely northeast or northwest flow. On the upper slope between water depths 140 to 200 m the presence of a well developed field of sand waves (Fig. 7c) is strong evidence for high velocity currents within the shelfbreak region. On the outer shelf and upper slope less extensive scour crescents and crag-and-tail structures are good indicators for persistent current conditions at the seabed, with velocities of at least $12-15$ cm s^{-1} required for their formation (Lonsdale & Hollister 1979).

Observations from photographs at depths of 1500 m and below (Fig. 7d) show features that can be attributed to processes other than current activity. Photographs show a fine-grained homogeneous sediment with an intensely bioturbated surface and an associated fauna, that can be interpreted as a hemipelagic drape.

Principal sediment facies

On the basis of short cores collected from the study area, Armishaw et al. (2000) identified seven principal facies (Table 2). These were named on the basis of an interpretation of the major processes involved in their deposition and included:

Facies Group A: A, sandy contourite; A1, silty sandy contourite; A3, muddy contourite.

Facies Group B: B, glaciomarine dumpstone; B1, glaciomarine hemipelagite.

Facies Group C: C, hemipelagite; C1, hemipelagite with rare dropstones.

We would add a further facies group comprising two facies to this list to include the upper slope and outer shelf gavels and gravelly sands observed in seafloor photographs as well as from previous studies of surface grab samples (Figs 8 and 9).

Facies Group D: D, relict gravels; D1, relict coarse gravelly sands.

The Barra sand sheet is made up of *Facies A* sandy contourites that occur as a thin surface layer between 0.05 and 0.40 m thick in mid-upper slope cores (Fig. 10). In cores from the Peach Slide, the sandy facies has been displaced to depths as much as 1540 m. The sand is mainly structureless and bioturbated, in some cases with a more or less well-developed reverse grading to the surface. It is a poorly sorted, mainly very fine-grained sand, with a mixed biogenic-terrigenous composition, fragmented bioclastic particles and common iron-staining. Fuller details of facies groups A–C are provided by Armishaw et al. (1998, 2000) and their distribution across the slope illustrated in Figure 11. Further details of the surface facies are given below.

Surface facies – Petrology and texture

Grain size analyses reveal four major textural components (gravel, sand, silt and clay) on the outer shelf and slope (Fig. 12).

The regional sedimentation pattern across the slope shows that the margin is texturally graded from a coarse gravel-rich shelf cover to a predominantly clay-rich sediment on the lower slope (Fig. 13). The proportion of gravel-sized material decreases sharply seaward from the shelf (10% with patches of > 50% gravel) to about 300 m (<1%), below 300 m the proportion of gravel remains relatively constant to the lower slope. The most extensive sediment cover on the middle slope is sand. The greatest proportion of sand-sized material is found between the outer shelf and mid-slope region (70–85%) down to about 1000 m water depths, where the proportion drops to 10–20%. Below 1000 m the proportion of sand remains relatively constant to over 2000 m. Silt is the most variable textural component on the shelf and slope and varies in proportion to as little as 10% on the upper slope to 40–50% at depths of 1500 m. Below 1500 m the proportion decreases to between 10 and 30%. Clay increases rapidly from negligible amounts on the upper slope to over 40% at about 1500 m, unlike that of silt, the proportion of clay then continues to increase gradually to over 60% exceeding 2000 m on the lower slope.

Sediment collected at depths between 140 and 300 m is texturally classified as gravelly sand, that between 300 and 1000 m as sand, that between 1000 and 1500 m as a silt-sand-clay mixture, and deeper than 1500 m as a silty clay. Trends shown in the textural variations are generally similar to the contemporary boundaries mapped on the British Geological Survey 1:250 000 seabed sediment map (James et al. 1990). Textural variations in the present study also reveal some important north–south contrasts (Figs 12 and 13). On the northern Barra Fan the presence of sand is recorded down to a depth of at least 1500 m. In contrast on the southern Fan region the presence of sand is seen to a maximum depth of 1200 m. Detailed particle size analyses of the present data reveal that the proportion of sand-sized material in the surface sediments on the northern Barra Fan is up to 25% at depths of about 1500 m, whereas the proportion of sand-sized material in cores from 1500 m on the southern Barra Fan is on average less than 10%. Similarly the proportion of silt- and clay-sized material is also different from north to south. Samples from the northern Barra Fan at depths of 1500 m contain on average 45% silt and <35% clay-sized material, whereas those from similar depths on the southern Barra Fan contain on average 20% silt and >60% clay. The median grain-size shows a diminution from gravel intermixed with coarse and medium sand (0.5 to 3.0 ϕ) on the outer shelf and upper slope to depths of about 300 m, to fine and very fine sand (2.0 to 4.0 ϕ) and coarse to medium silt grades (4.0 to 6.0 ϕ) to depths of 1000 m, with the predominance of clay-sized (>8.0 ϕ) material below.

Surface facies – Composition and mineralogy

The general seaward fining trend is consistent with a gradual change in composition with depth, as shown by the analysis of the sand and silt fraction (Fig. 14). The proportion of terrigenous material decreases from 60–75% on the shelf and upper slope to 15–30% on the lower slope which is broadly coincident with a decrease in the sand and silt fractions. This gradual change in the proportion of terrigenous material is coupled with an associated increase in the proportion of planktonic material from 25–40% on the upper slope, to 80% on the lower slope. Terrigenous/planktonic ratios range from 0.88 to a maximum of 4.16, with an average of greater than 2.0 over most of the study region. Regional trends consist of increasing terrigenous/planktonic ratios seaward, with a significant increase in the ratio in depths greater than 2000 m. The only areas of approximately equal proportions of sand-sized terrigenous and planktonic grains are on the outer shelf and upper slope, in depths down to about 700 m.

Compositional changes with depth are also well illustrated by the distribution of foraminiferal tests and the proportions

Fig. 8. Seafloor photographs depicting a seaward fining sequence of coarse gravels and sands on the outer shelf and upper slope to fine silt and clay on the lower slope. (**a**) Camera station N200. Seafloor composed of large cobbles and boulders with interstitial coarse and medium grained sand. (**b**) Camera station S300. Seafloor composed of coarse to fine gravel with interstitial sand. (**c**) Camera station N700. Seabed composed of fine sand. Symmetrical sand-ripples suggest oscillating currents in a NNW–SSE trend. (**d**) Camera station S1000 depicting silty sands reworked by weak NW-flowing currents indicated by small-amplitude ripples. (**e**) Camera station N1500. Seafloor is composed of silty clay and clay and is reworked by a mixed benthic population, giving it a granular texture. (**f**) Camera station N2000. Seafloor is composed of silty clay and clay and is reworked by a mixed benthic population, giving it a granular texture. The seabed area photographed is trapezoidal in shape and c. 120 cm across the bottom and c. 250 cm across the top of the frame (Humphery, pers. comm., 1996).

of iron-stained, silt- and sand-sized material. The percentage of foraminiferal tests increases from less than 10% on average on the upper slope at 700 m to 20–30% at about 1000 m depth where the proportion remains constant down to 2100 m. Seaward changes in the proportions of iron-stained and glauconitic material are broadly similar and reveal a gradual decrease from up to 10% on the outer shelf and upper slope to <0.5% at depths below 2000 m. Across the slope depth-related compositional associations are predominantly parallel to the shelfbreak and slope isobaths. Simple graphic plots illustrate that (a) the relative proportion of

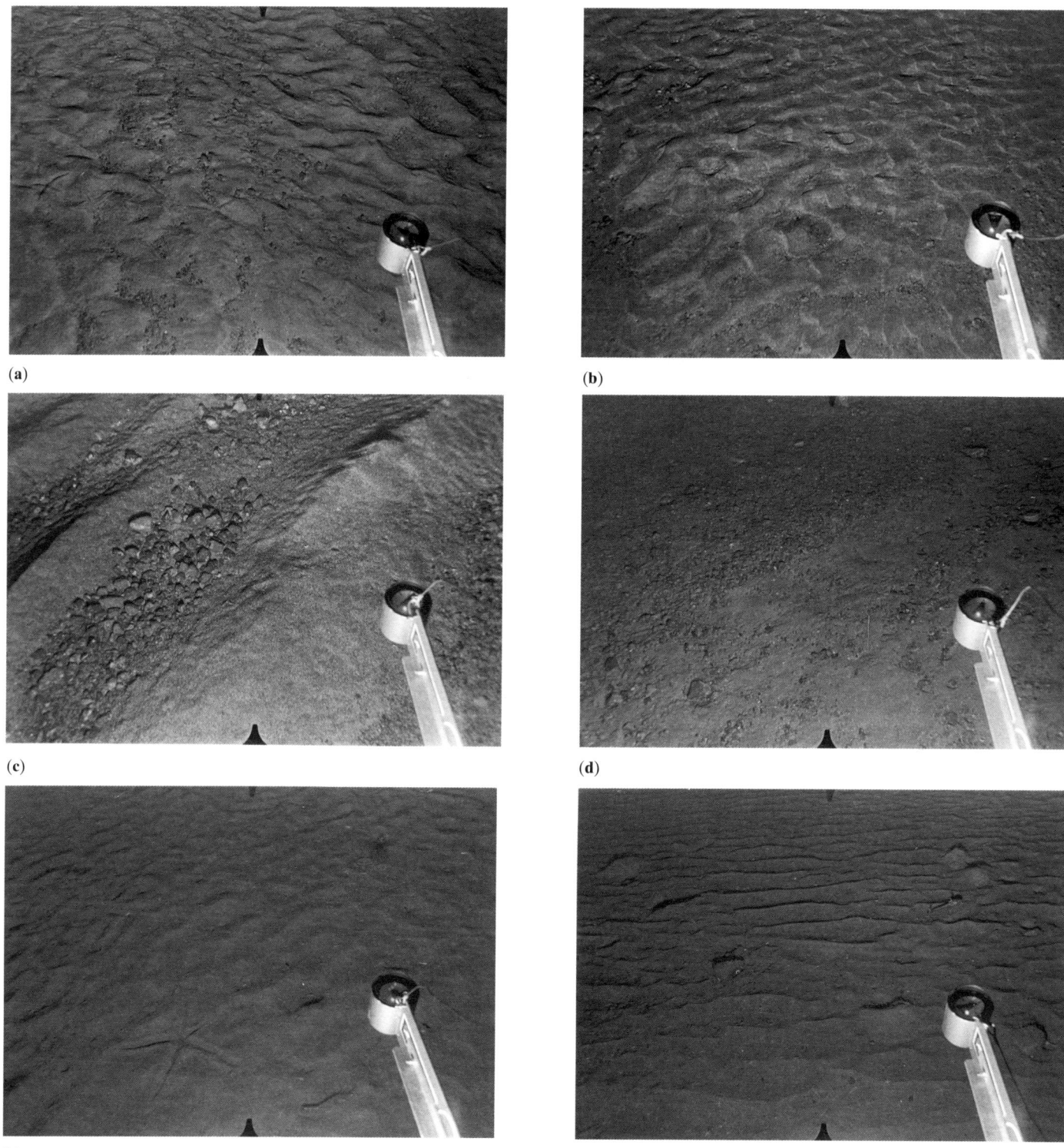

Fig. 9. Bottom photographs indicating bottom-current flow on the shelf and slope. (**a**) Camera station S140 depicting asymmetric and confused sand ripples under the influence of a north to north–northeast flowing current. (**b**) Camera station S200 (August 1995) depicting mainly asymmetric sand ripples under the influence of a northwesterly flowing current, and a weaker set of interference ripples from a more northeasterly flowing current. (**c**) Camera station S200 (March 1995) showing N–S aligned deep, sand waves on the upper slope. (**d**) Camera station S300 depicting a mixed gravel and sand substrate, with evidence of some alignment under the influence of a NNE flowing current. (**e**) Camera station S700 depicting a fine sandy substrate with asymmetric to symmetrical ripples under the influence of a north to north–northwest flowing current. (**f**) Camera station S1000. Field of symmetrical, straight-crested sand ripples, interspersed with small (1–3 cm) and large (up to 20 cm) pebbles and boulders causing the development of crag-and-tail structures. The seabed area photographed is trapezoidal in shape and c. 120 cm across the bottom and c. 250 cm across the top of the frame (Humphery, pers. comm., 1996).

terrigenous fragments and water depth are inversely correlated; (b) the relative proportion of planktonic material and depth correlate positively; and (c) the total proportion of sand-sized particles and terrigenous material correlate positively.

Discussion

Seafloor polishing and sand spillover

From seabed photographs within the study area outer shelf gravels and gravelly sands are seen to be present between depths

Table 2. *Principal sediment facies of the Barra fan study area (modified from Armishaw et al. 2000)*

Sedimentary characteristics	Facies A sandy contourite	Facies A1 silty–sandy contourite	Facies A2 muddy contourite	Facies B glaciomarine dumpstone	Facies B1 glaciomarine hemipelagite	Facies C hemipelagite (much bioturbation)	Facies C1 hemipelagite (+dropstones)	Facies D relict gravel	Facies D1 relict coarse gravelly sand
Colour	olive grey to dark greyish brown	olive grey to dark greyish brown	olive grey to dark olive grey	light olive grey to olive grey	dark grey to dark greyish brown	dark olive grey to dark grey	olive grey to dark olive grey	olive grey to dark grey (variable)	olive grey to dark grey (variable)
Contacts: top	sharp	sharp-gradational	sharp or bioturbated	bioturbated	gradual or bioturbated	grad/sharp/bioturbated	sharp or gradational	Sharp (seafloor)	Sharp (seafloor)
base	sharp	sharp-gradational	gradual or bioturbated	sharp or gradational	gradual or bioturbated	grad/sharp/bioturbated	gradual/bioturbated	sharp-gradational	sharp-gradational
Sedimentary structures	rare negative grading	coarse lenses and pockets, rare negative grading	rare lenses, pockets, laminae and grading	structureless, chaotic with coarse lenses and clasts	structureless, chaotic with coarse lenses and clasts	homogeneous and bioturbated	mostly homogeneous and bioturbated	Structureless (negative grading)	Structureless (negative grading)
Ice-rafted debris	generally absent	rare dropstones	rare dropstones	abundant dropstones < 5 cm diameter	intermittent dropstones < 10 cm diameter	very rare dropstones	rare dropstones	Relict IRD	Relict IRD
Bioturbation type	*Chondrites Mycelia*	*Planolites Mycelia*	*Planolites Mycelia*	*Zoophycos, Planolites Mycelia*	*Mycelia*	*Mycelia*	*Zoophycos*	Unknown	Unknown
Bioturbation intensity	slight to intense	slight to high	slight to moderate	moderate	slight to intense	moderate	moderate to high	Unknown	Unknown
Carbonate content	5–99% biogenic 5–10%	5–90% biogenic 5–10%	5–60% biogenic 10–20% detrital 25–50%	30–60% biogenic 5–30% detrital 5–30%	5–99% biogenic 20–40%	30–60% biogenic 5–30%	30–90% biogenic < 5–10% detrital 30–65%	< 10%	< 10%
Total organic carbon	0.1–0.6%	0.3–0.75%	0.35–0.75%	not known	not known	not known	0.25–0.6%	0 %	0 %
Grain size: mean	4.5–4.7	5.9–7.1	8.1–8.4	6.9–9.5	7.2–8.6	9.6	6.9–7.4	No data	No data
median	2.8–2.9	6.1–9.0		4.9–9.1	8.9–9.5		8.1–9.4		
sorting	2.7–3.0	3.0–3.5	0.9–2.4	2.4–3.0	1.4–2.9	2.9	1.6–4.2		
modes	2 (9)	2–3 (8–9)	6–7 (8–10)	2–7 (9–11)	9–11	8–10	5–6 (2–3)		
skewness	0.8–0.81	−0.1 – −0.9	0.68	−0.6	−0.8 – −3.9	−0.8	−0.5 – −0.9		
(all data in phi units)									
Interpretation	sandy contourite strong bottom currents sheeted drift deposit	silty–sandy contourite moderate bottom currents	muddy contourite weak bottom currents	glacial dumping from rapid meltout of floating ice	glacial fallout from floating ice and hemipelagic dispersion	hemipelagic sedimentation at very low rates	hemipelagic sedimentation and minor glacial input	Glacial to glaciomarine + shelf reworking	Glacial to glaciomarine + shelf reworking

Fig. 10. Selected core photographs from the Barra contourite sand sheet – sandy contourite facies. Top to left.

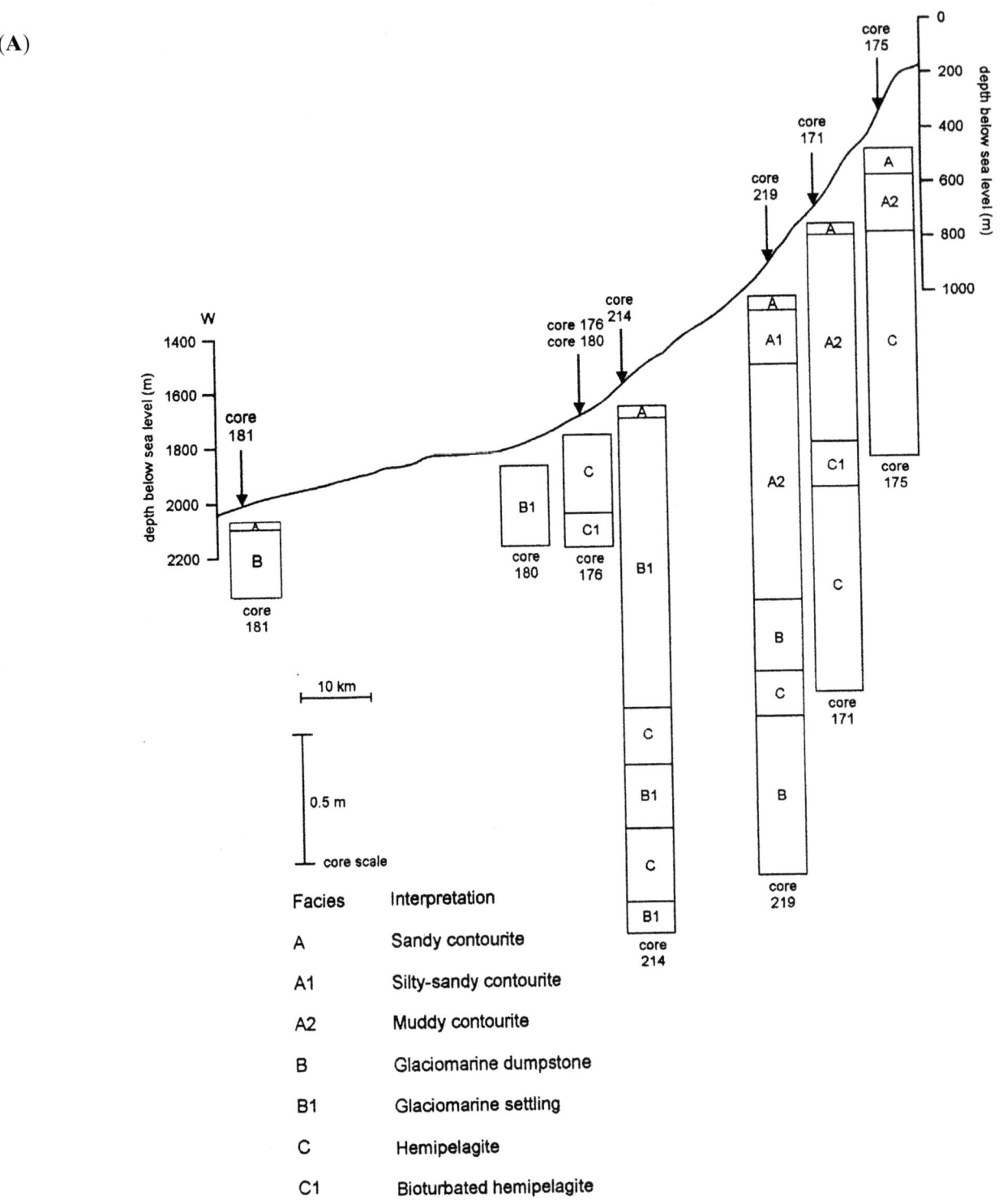

Fig. 11. Slope profiles showing core sections and facies distribution across the central (**A**) and northern (**B**) Barra fan. (From Armishaw et al. 2000).

of about 140 to 300 m. These comprise a mixture of cobbles, boulders, gravel with minor sand and shell detritus; the cobbles vary in size from 5 to 50 cm in diameter in the northern region of the slope and from 2 to 30 cm in diameter to the south of the region. Given the depth of the gravel deposits below sea level, and the limited extent and strength of the shelf circulatory system it is unlikely that the present day hydraulic and morphologic regime is actively delivering coarse-grained sediment to the outer shelf region. Consequently the outer shelf facies represents the modern overprint on a relict Pleistocene shelf environment. The presence of cobbles and boulders is related to glacial, fluvial and coastal processes that prevailed when the coastline extended seaward to a position close to the shelf edge during Pleistocene eustatic low stands. A similar gravel band at the lip of the shelf is also seen on the Norwegian shelf and is a characteristic feature of glaciated margins.

The formation of the outer shelf sand and gravel facies in the present study is very similar to that of the outer shelf/slope sands described by Viana et al. (1998) from several other continental margins including those from the USA Atlantic margin (Stanley et al. 1981; Blake & Doyle 1983) and the Scandinavian margin (Kuijpers et al. 1993; Yoon & Chough 1993) where they are

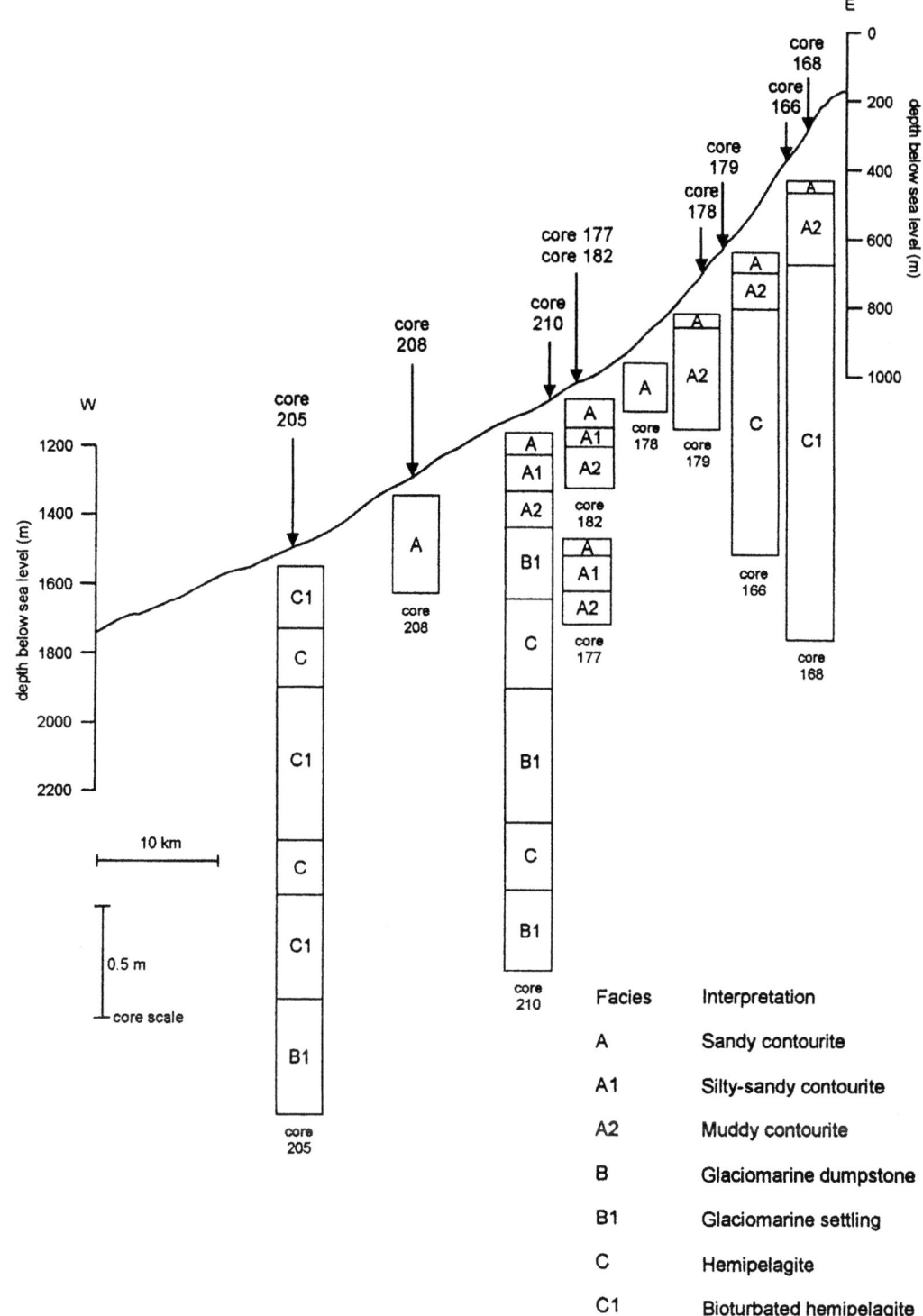

described as bottom current reworked sand accumulations. The combination of processes and relatively shallow water depths involved in their deposition precludes them from being described as contourite or even shallow water contourite deposits. Typical bedforms range from sand and gravel wave fields to smaller-scale, randomly-oriented sand ripples which are believed to result from the passage of a geostrophic outershelf/slope current, combined with the action of large current eddies, storm waves, tides and internal waves. This process combination can be referred to as *seafloor polishing* (Viana *et al.* 1994, 1998).

The removal of sands and some gravel from the shelf and their redistribution to the slope is related to the onshelf penetration of surficial geostrophic currents and associated eddies and is known as shelf *spillover* (Stanley *et al.* 1981). The controlling factors for the formation and preservation of such deposits requires a complex combination of elements. The principal hydrodynamic regime requires the combined effects of a relatively strong geostrophic upper slope current with a marked element of cross-shelf exchange of water, along with the effects of tide-induced bottom currents and the downward propagation of shelf eddies. In the present study the gross circulation is dominated by the strong, persistent north to northeasterly flowing slope current, and

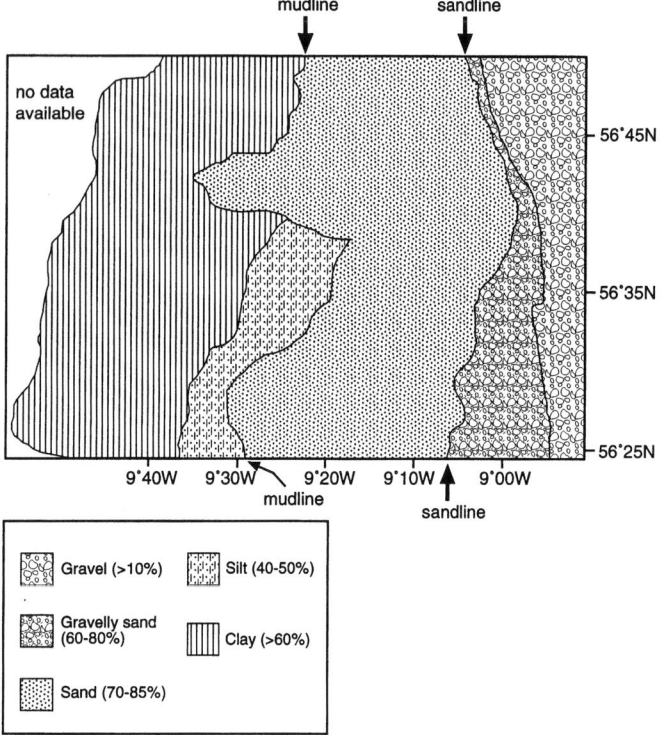

Fig. 12. Map of surficial sediment grain size class from the Barra fan region based on grain size analyses of core top samples. The broad sand belt between the sandline and mudline is referred to as the Barra contourite sand sheet. The marked offset in the position of the mudline at around latitude 56°40' is due to downslope lateral displacement in the Peach Slide.

although the precise mechanisms of cross-shelf exchange are not yet fully understood from the region, there is a marked increase in the mixing on the shelf during autumn and winter due to autumnal gales (Edelsten *et al.* 1976). In light of the lack of modern sediment input from external sources the chief supply of spillover sediment is the relict shelf deposits from Pleistocene sea-level low stands and transgressive episodes (Ferentinos 1976) and the winnowing of glaciomarine sediments on the outer shelf which has led to significant reduction in the original sediment thickness.

Barra contourite sand sheet

Below about 300 m water depth and up to around 1500 m on the northern Barra Fan and 1200 m on the southern Barra Fan, there is a broad area of sandy facies covering the slope surface. This we have termed the *Barra contourite sand sheet*, and interpret it as a mid-depth sandy contourite. Such deposits, as reviewed by Viana *et al.* (1998), are formed under major geostrophic flows in water depths in excess of 300 m and up to about 2000 m. The Barra contourite sand occurs as a thin (3–40 cm) surface sheet over an area of between 1000 and 1500 km^2 within the study region. We estimate its total volume in this area alone at approximately 30 000 m^3.

The presence of symmetric and asymmetric sand-silt ripples on the slope down to depths of 1000 m on the surface of the contourite sheet indicate a persistent, fast-moving alongslope current flow, with flow velocities of over 30 cm s^{-1}. The accumulation of such contourite deposits requires the presence of either relatively strong, semi-permanent geostrophic currents flowing at intermediate depths within the water column, or very strong surficial geostrophic currents that are able to influence the seafloor at these depths (Viana *et al.* 1998). The supply of sand to the slope

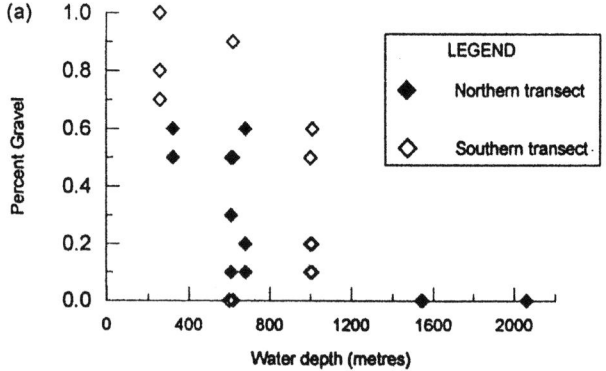

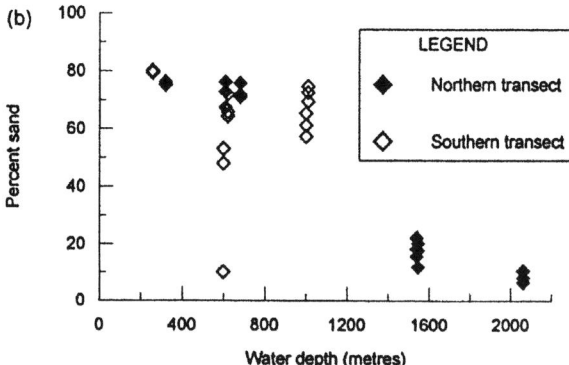

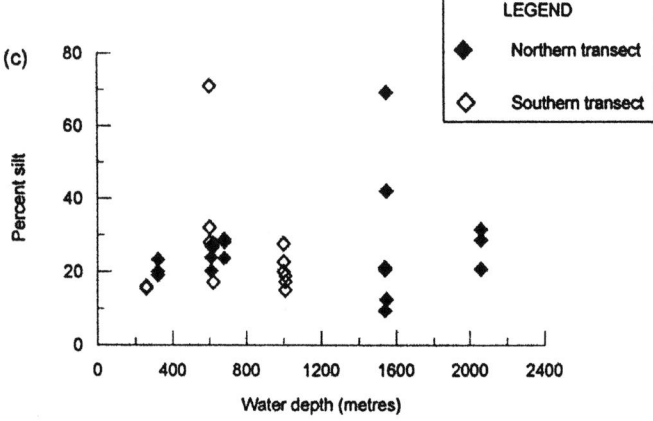

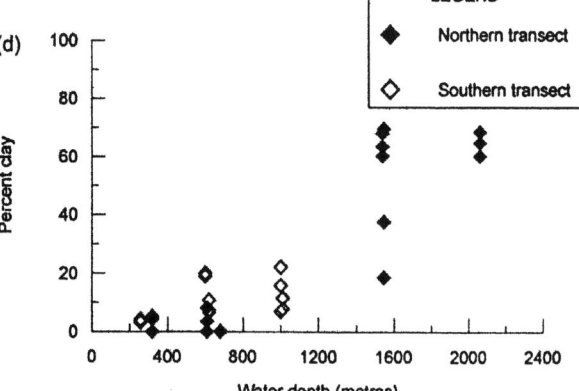

Fig. 13. Graphic plots showing the four major textural classes v. depth. Each point represents at least two analyses from a core top sample.

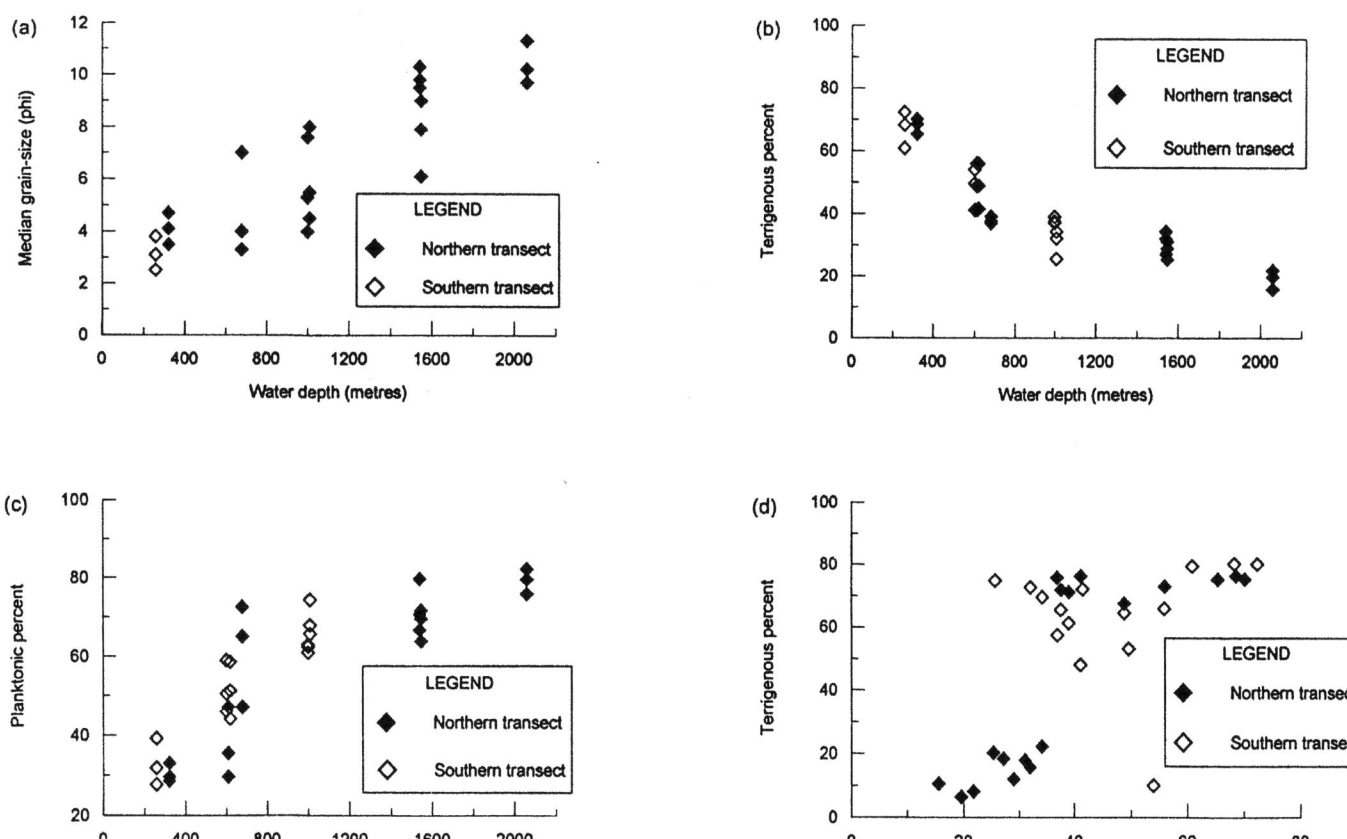

Fig. 14. Graphic plots showing trends of selected petrologic and compositional parameters: (**a**) median grain size v. clay content; (**b**) terrigenous content v. depth; (**c**) planktonic content v. depth; (**d**) terrigenous content v. sand content.

Fig. 15. Model of seafloor polishing and offshelf sand spillover under the influence of a variety of bottom current and downslope gravity processes. (From Stow & Mayall 2000, after Viana *et al.* 1998).

comes from a variety of sources both internal and external including bottom current erosion upstream of the drift, the pirating of offshelf spillover, pelagic biogenic fallout and *in situ* winnowing. The present day hydraulic regime is clearly evidenced by photographs of the seafloor surface. The time series of shots taken over a period of over two years quite clearly shows the seasonal effects on the generally northerly advecting slope current on the Hebrides Slope, and also the fluctuation in both current strength and orientation with depth and location on the slope. Over a two year period photographs indicate a NNW–SSE oscillating flow effective during the winter and spring months and a much reduced flow during the summer. Booth & Ellett (1986) attribute the decrease of a slope current during the summer months to the notable lack of cross-shelf water exchange during the summer and autumn.

Petrologic parameters of the Barra sand sheet also provide an effective means of recognizing the more subtle, long-term (Holocene to recent) influence and interaction of bottom current activity and downslope delivery of sediment on the Hebridean margin that are not otherwise apparent in direct observations of the seafloor. On-going, off-shelf spillover processes on the outer shelf and beyond the shelfbreak, and the influence of bottom current activity, are reflected in the intermediate texture (between shelf gravels and lower slope muds) and mixed terrigenous-biogenic composition of the surface sands. Between depths of approximately 300 and 1200 m the terrigenous sand component is largely attributed to shelf spillover. By contrast, the presence of diverse assemblages of fragmented benthonic and planktonic tests, coupled with an overall upward increase in grain-size through the sand sheet, are characteristic of vertical supply, followed by alongslope transportation and reworking of the surface sediment by vigorous bottom currents.

The lower limit of the transitional facies is delineated by the mudline (Stanley & Wear 1978; Stanley & Freeland 1978; Stanley

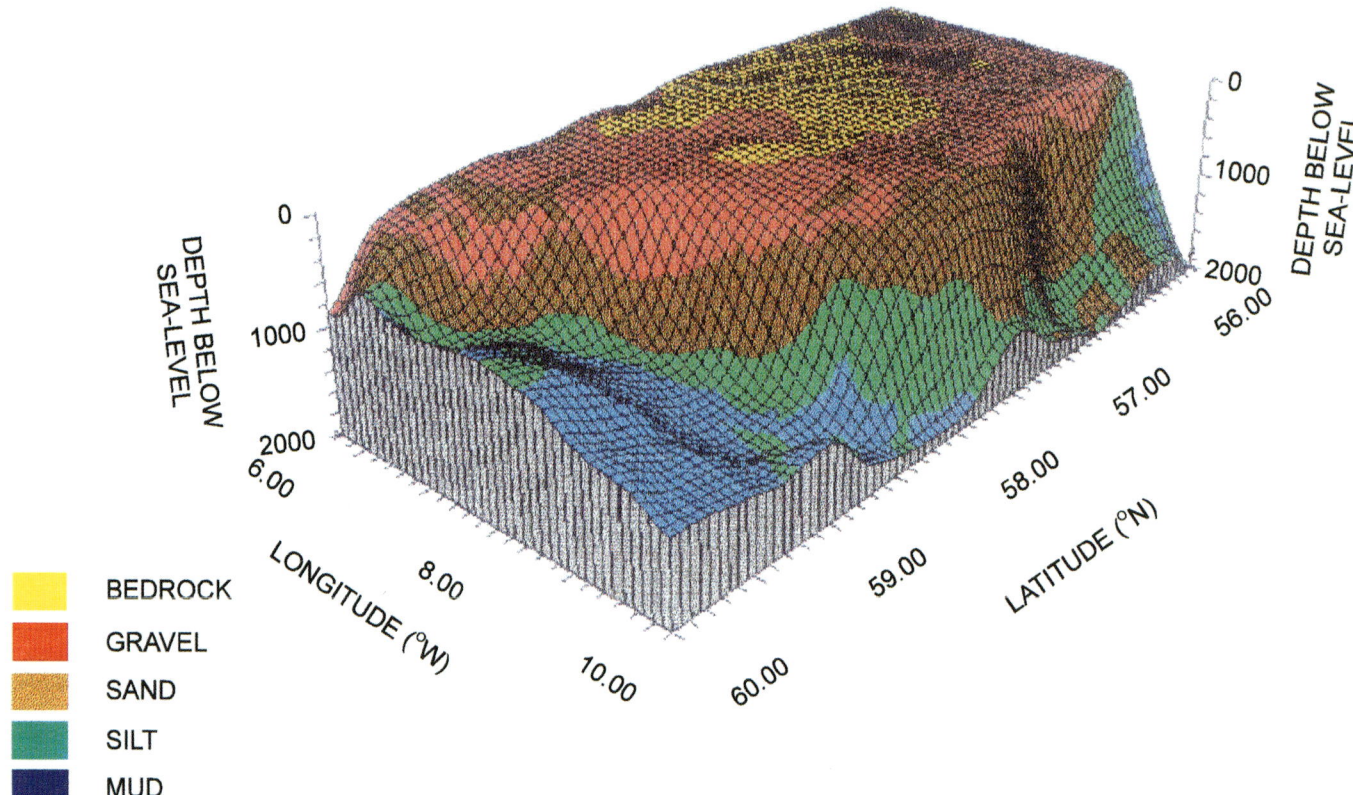

Fig. 16. 3D perspective image of the sea-bed topography derived from processed bathymetric data from the Hebrides Slope between 56°N and 60°N, including data from the study area between 56°N and 57°N. The colour overlay shows surface sediment types (grain size classes) derived from detailed sediment analysis of core top samples. The data show a seaward-fining sequence from gravel on the shelf and upper slope to mud on the lower slope. The broad area of sand lies between an upper slope sandline and lower slope mudline – and is known as the Barra contourite sand sheet.

et al. 1983) (Figs 12 & 13), where the amount of clay no longer increases significantly with depth. At this depth the clay content generally exceeds 60% and the median grain size of samples is generally finer than 8 phi. The presence of non-current-induced features on the seabed are largely attributed to normal marine hemipelagic settling. Observations from photographs clearly indicate a high benthic population which has resulted in an intensely reworked surface and granular texture in some parts. The decrease in the amplitude of current induced bedforms passing down through the transitional facies is attributed to the progressive decrease in current velocity down the lower slope, and the onset of the slope facies marks the decrease in frequency of currents exceeding threshold velocities needed to erode fine sediment. The present limit of the mudline it seems is therefore likely to identify a separation of energy zones; above the mudline marks an erosive or winnowing zone for fine grained sediment whereas sediments below mark a zone of non-erosion/deposition.

Recognizing a Mudline and Sandline

The transition from a gravel-rich outer shelf to a clay-rich lower slope is marked by the presence of two significant facies boundaries that result principally from a combination of hydological and physiographic conditions. The boundary between the outer shelf/upper slope gravels and mid-slope sands is defined here as the *sandline*. The sandline is the first major facies boundary to be encountered as we progress down the slope and represents the level to which a coarse grain size (gravel) spike is present on the slope. Above the sandline the outer shelf/upper slope facies is dominated by coarse sands and gravels up to boulder size; below it the proportion of gravel-sized material is <1% and the major grain-size component is sand (70–85%). The position of the sandline across the region varies between 170 and 300 m. It is rarely coincident with the shelfbreak and tends to be present at a greater depth on the northern Barra Fan, particularly where it is associated with the presence of a canyon-incised steep upper slope.

The second of the two facies boundaries is the *mudline* as previously defined by Stanley & Wear (1978). The mudline marks the lower limit of the Barra sand sheet below which the amount of clay no longer increases significantly with depth. At this depth the clay content generally exceeds 60% and the median grain size of samples is generally finer than 8 phi. Across the region the mudline is present on the northern Barra fan at depths in excess of 1500 m, and on the southern fan to a maximum depth of 1200 m. The mudline here reflects an important boundary between an area of long-term winnowing and non-deposition of fines and their net accumulation in the absence of a strong bottom-current system. By comparison with earlier studies of the mudline on the North American continental margin (Stanley & Freeland 1978; Stanley & Wear 1978; Stanley *et al.* 1983), we note that its depth on the Hebridean margin is generally greater, reflecting the presence of a persistent mid-slope, high-energy, bottom current system along this margin.

Grateful thanks must be expressed to the masters, officers and crew who participated wholeheartedly in all the cruises to the Hebridean slope margin as part of the Natural Environment Research Council Land Ocean Interaction Study: Shelf Edge Study programme. Thanks are also extended to J. Humphery for information and preparation of seafloor photographs, to G. Tulloch as sample curator at the British Geological Survey, Edinburgh, and to B. Marsh and K. Davis for their patient help with the figures. DAVS acknowledges tenure of a Royal Society Industry Fellowship with BP during the writing of this paper. JEA acknowledges the receipt of a NERC-CASE studentship at Southampton University in conjunction with the British Geological Survey in Edinburgh. Many colleagues are to be thanked for their insightful discussions, as well as the two reviewers of an earlier version of the manuscript.

References

ARMISHAW, J. E. 1998. *Bottom current accumulation and sediment facies on the Hebridean slope*. PhD thesis, University of Southampton.

ARMISHAW, J. E., HOLMES, R. W. & STOW, D. A. V. 1998. Morphology and Sedimentation on the Hebridean Slope and Barra Fan NW, UK Continental Margin. *In*: STOKER, M. S., EVANS, D. & CRAMP, A. (eds) *Geological Processes on Continental Margins Sedimentation, Mass-wasting and Stability*. Geological Society, London, Special Publications, **129**, 81–104.

ARMISHAW, J. E., HOLMES, R. W. & STOW, D. A. V. 2000. The Barra fan: a bottom-current reworked, glacially-fed submarine fan system. *Marine & Petrological Geology*, **17**, 219–238.

BLAKE, N. J. & DOYLE, L. J. 1983. Infaunal-sediment relationships at the shelf-slope break. *In*: STANLEY, D. J. & MOORE, G. T. (eds) *The Shelf-Break: Critical Interface on Continental Margins*. Soc. Econ. Paleontologists Mineralogists Spec. Pub. **33**, 381–389.

BOOTH, D. A. & ELLETT, D. J. 1983. The Scottish continental slope current. *Continental Shelf Research*, **2**, 127–146.

DAMUTH, J. E. 1975. Echo-character of the western equatorial Atlantic floor and its relationship to the dispersal and distribution of terrigenous sediments. *Marine Geology*, **18**, 17–45.

DAMUTH, J. E. 1978. Echo-character of the Norwegian-Greenland Sea: Relationship to Quaternary sedimentation. *Marine Geology*, **28**, 1–36.

DAMUTH, J. E. 1980. Use of high-frequency (3.5kHz) echograms in the study of near-bottom sedimentation processes in the deep-sea: a review. *Marine Geology*, **38**, 51–75.

EDELSTEN, D. J., ELLETT, D. J. & EDWARDS, A. 1976. Preliminary results from current measurements at the Scottish continental shelf-edges. *ICES* 1976 (**C: 12**), (mimeo).

ELLETT, D. J., EDWARDS, A. & BOWERS, R. 1986. The hydrography of the Rockall Channel – an overview. *Proceedings of the Royal Society of Edinburgh*, **88B**, 61–81.

FERENTINOS, G. K. 1976. Sediment distribution and transport processes on the outer continental shelf of the Hebridean Sea. *Marine Geology*, **20**, 41–56.

GORDON, R. L. & HUTHNANCE, J. M. 1987. Storm-driven continental shelf waves over the Scottish continental shelf. *Continental Shelf Research*. **7**(9), 1015–1048.

HARVEY, J. G. 1982. Theta-S relationships and water masses in the eastern North Atlantic. *Deep-Sea Research*, **29**, 1021–1033.

HOLMES, R. W. 1994. *Seabed topography and other geotechnical information for the Shelf Edge Study 55°N–60°N NW of Britain*. British Geological Survey, Technical Report **WB/94/15**.

HOLMES, R. W., LONG, D. & DODDS, L. R. 1998. *In*: STOKER, M. S., EVANS, D. & CRAMP, A. (eds) *Geological Processes on Continental Margins Sedimentation, Mass-Wasting, and Stability*. Geological Society, London, Special Publications, **129**, 81–104.

HOWE, J. A. 1995. Sedimentary processes and variation in slope-current activity during the last glacial-interglacial episode on the Hebrides Slope, Nothern Rockall Trough, North Atlantic Ocean. *Sedimentology*, **43**, 219–234.

HOWE, J. A. 1996. Turbidite and contourite sediment waves in the northern Rockall Trough, North Atlantic Ocean. *Sedimentology*, **43**, 219–234.

HOWE, J. A., STOKER, M. S. & STOW, D. A. V. 1994. A Late Cenozoic sediment drift complex, North-East Rockall Trough, North Atlantic. *Palaeoceanography*, **6**, 989–999.

HUTHNANCE, J. M. 1986. The Rockall slope current and shelf-edge processes. *Proceedings of the Royal Society of Edinburgh*, **88B**, 83–101.

JAMES, J. W. C., BOOTH, S. J. & WRIGHT, S. A. 1990. *Peach (56N, 10W): Sea Bed Sediments*. British Geological Survey 1:250,000 Offshore Map Series.

KUIJPERS, A., WERNER, F. & WONG, H. K. 1993. Sandwaves and other large-scale bedforms as indicators of non-tidal surge currents in the Skagerrak off Northern Denmark. *Marine Geology*, **111**(3/4), 209–222.

LONSDALE, P. & HOLLISTER, C. D. 1979. A near bottom traverse of the Rockall Trough: Hydrographic and geological inferences. *Oceanologica Acta*, **2**, 91–105.

PRATSON, L. F. & LAINE, E. P. 1989. The relative importance of gravity-induced versus current-controlled sedimentation during the Quaternary along the mideast U.S. outer continental margin revealed by 3.5 kHz echo character. *Marine Geology*, **89**, 87–126.

STANLEY, D. J. & FREELAND, G. L. 1978. The erosion-deposition boundary in the head of Hudson submarine canyon defined on the basis of submarine observations. *Marine Geology*, **26**, M37–M46.

STANLEY, D. J. & WEAR, M. C. 1978. The "mud-line": and erosional-depositional boundary on the upper continental slope. *Marine Geology*, 28 **M19–M29**.

STANLEY, D. J., SHENG, H., LAMBERT, D. N., RONA, P. A., MCGRAIL, D. W. & JENKYNS, S. J. 1981. Current-influenced depositional provinces, continental margin off Cape Hatteras, identified by petrologic method. *Marine Geology*, **40**, 215–235.

STANLEY, D. J., ADDY, S. K. & BEHRENS, E. W. 1983. The mudline: variability of its position relative to shelfbreak. *In*: STANLEY, D. J. & MOORE, G. T. (eds) *The shelfbreak: Critical Interface on Continental Margins*. Soc. Econ. Paleontologists Mineralogists Special Publication **33**, 381–389.

STOKER, M. S. (in press) Late Neogene development of the UK Atlantic margin. Geological Society, London, Special Publications.

STOKER, M. S., HITCHEN, K. & GRAHAM, C. G. 1993. *United Kingdom offshore regional report: the geology of the Hebrides and West Shetland shelves, and adjacent deep-water areas*. HMSO for the British Geological Survey, London.

STOKER, M. S., LESLIE, A. B., SCOTT. W. D., BRIDEN, N. M., HINE, N. M., HARLAND, R., WILKINSON, I. P., EVANS, D. & ARDUS, D. A. 1994. A record of the late Cenozoic stratigraphy, sedimentation and climate change from the Hebrides Slope, NE Atlantic Ocean. *Journal of the Geological Society, London*, **151**, 235–249.

STOW, D. A. V. & MAYALL, M. 2000. Deep-water sedimentary systems: new models for the 21st century. *Marine & Petroleum Geology*, **17**, 125–135.

TURRELL, W. R., HENDERSON, E. W., SLESSER, G., PAYNE, R. & ADAM, R. D. 1992. Seasonal changes in the circulation of the northern North Sea. *Continental Shelf Research*, **12**(2), 257–286.

VIANA, A. R., FAUGÈRES, J-C. & STOW, D. A. V. 1998. Bottom current controlled sand deposits- A review from modern shallow to deep water environments. *Sedimentary Geology*, **115**, 53–80.

VIANA, A. R., KOWSMANN, R. O. & CADDAH, L. F. G. 1994. *Architecture and oceanographic controls on the sedimentation of the Campos Basin continental slope*. 14th ISC, Abstracts Volume.

YOON, S. H. & CHOUGH, S. K. 1993. Sedimentary characteristics of Late Pleistocene bottom-current deposits, Barents Sea, slope off northern Norway. *Sedimentary Geology*, **82**(1/4), 33–46.

Muddy contourites in the Baltic Sea: an example of a shallow-water contourite system

VADIM SIVKOV[1], VLADIMIR GORBATSKIY[2], ALEXEY KULESHOV[1] & YURY ZHUROV[1]

[1]*Atlantic Branch, Shirshov Institute of Oceanology, Russian Academy of Sciences, Russia*
[2]*Krylov Shipbuilding Research Institute, Saint Petersburg, Russia*

Abstract: Bottom currents in the Baltic Sea have had a pronounced effect on the nature and distribution of sediments throughout the Holocene. Due to the intermittent nature of water exchange between the North and the Baltic Seas, the flow of bottom currents is impermanent. Well-developed nepheloid layers are commonly associated with these bottom currents, and provide evidence that active re-suspension and sediment transport is taking place. Periodically intensified inflow from the North Sea through the Baltic Sea gateway has led to erosion of narrow elongate channels, and associated deposition of small elongate patch drifts and contourite levees. The latter two are best developed on the left flanks of the channels. As a result of strong lateral sediment transport, the overall thickness of muddy contourite layers increases towards the steeper slopes of the sea which are situated on the right-hand side of the inflows. Muddy contourites of the Baltic Sea are predominantly terrigenous in composition, made up of soft, black sulphidic muds that are enriched in organic carbon and manganese. Both enrichments are caused by high biological productivity and periodic stagnation of near-bottom waters. Atypically for most oceanic contourites, these shallow-water organic-rich contourites are finely laminated and generally unbioturbated.

We were invited to contribute this paper to the *Atlas of Contourites* (Stow *et al.* 2002) in order to provide a clear example of a shallow-water contourite system for comparison with the deeper-water oceanic contourites documented in most of the other contributions. There is, of course, much debate as to whether such drift sediments should be termed 'contourites' rather than shallow-water bottom current deposits, as has been advocated recently by Stow *et al.* (1998) and Viana *et al.* (1998) amongst others. Nevertheless, in this paper we have used the term 'contourite' throughout, generally with the shallow-water prefix.

The Baltic Sea water exchange within the North Sea is greatly restricted by the Danish Straits; the narrow and shallow gateway between the two seas. A freshwater surplus in the Baltic causes a surface outflow of brackish water which, in turn, generates a deep current inflow of saline water into the Baltic Sea. The inflowing waters form a series of pools of stratified deep-water in the Baltic Sea, periodically connected by dense, gravity forced bottom currents (Stigebrandt 1995). The importance of near-bottom circulation in determining the nature of the sediment record in the region has become widely recognized. The influence of bottom currents on the transportation and deposition of sediments in the Baltic Sea has been amply documented by Larsen & Kögler (1975), Kögler & Larsen (1977), Sviridov & Sivkov (1992), Sivkov & Sviridov (1994), Emelyanov *et al.* (1995), Sivkov *et al.* (1995), Stryuk *et al.* (1995), Gritsenko & Sivkov (1997), Endler (1998*a*), Sivkov *et al.* (1998), Emelyanov & Gritsenko (1999), Sviridov & Emelyanov (2000) and Emelyanov (2001).

This paper reports on a federal program *Baltic Sea Research* (BALTICA) that has been in progress since 1997 under the direction of the Atlantic Branch of the Shirshov Institute of Oceanology of the Russian Academy of Sciences (ABIORAS). We have examined data collected during the R/V PROFESSOR SHTOKMAN cruises 34 and 44. We have also used some results of the *Gotland Basin Experiment* (GOBEX) initiated in 1993 as a multidisciplinary experiment to investigate the oceanographic, chemical, biological and sedimentological situation in the Eastern Gotland Basin of the Baltic Sea (Emeis 1998). This programme was formulated during a scientific conference at Warnemünde, Germany, sponsored by the European Committee on Ocean and Polar Sciences (ECOPS) on the future of Baltic Sea research.

Acoustic bottom and sub-bottom profiling was carried out during each cruise using an ELAC (type NBS) echo-sounder with a frequency range of 20–30 kHz. Navigation data were collected from a GPS receiver (Raystar-920 Navigator). In the framework of GOBEX activity, a GeoChirp sub-bottom profiling system (GeoAcoustics) was used, with a frequency range of 2–8 kHz (Endler 1998*a*). Based on preliminary analysis of acoustic reflector geometry, the GOBEX expeditions occupied coring locations at which gravity cores were taken. Both classical core descriptions and digital images of the cores were compiled (Meyer 1998). Observations were located across the axis of inflow, where suspended sediment load in near-bottom water is generally high and the effects on sedimentation and erosion is likely to be more marked.

Continuous *in situ* measurement of light scattering and water temperature of the water column was made using an immersible nephelometer developed in ABIORAS. Light scattering by suspended particles, in the 580–700 nm wavelength range, was measured at an angle of 90°. Hydrographic measurements were carried out using both the immersible and towed fish Neil Brown-III (NBIS) CTD probe and the Ocean Seven 316 Probe (IDRONAUT, Italy). Measurements of sound scattering and current velocities were made with an Acoustic Doppler Current Profiler (ADCP, working frequency 300 kHz). This device uses four beams from acoustic antennae providing highly accurate measurements of velocities (up to 1 cm s^{-1}) and closely spaced resolution (up to 1 m) of sound scattering.

Geological and oceanographic setting

Development of the Baltic Sea

The Baltic Sea is a typical internal (intracontinental) sea lying entirely on continental crust, although its coastline only rarely shows a relationship with the pre-Cenozoic substratum. Every Pleistocene ice-sheet in Europe has developed in the northwestern part of the Fenno-Scandinavian crystalline shield. Their generation and development has depended on climatic conditions while their growth has proceeded mainly towards the south and east. Each time this process has occurred it has led to further destruction of weathered cover as well as exposure of the unweathered substratum. During the Pleistocene, therefore, the whole Baltic Basin and its margins were being slowly destroyed (Mojski 1995), and its central part became occupied by a very large lake known as the Baltic Ice Lake (BIL) at around 12 700–12 500 BP. The Baltic Sea eventually developed in place of this lake via a complex series of changes (Kramarska *et al.* 1995; Blazhchishin 1998). The principal events of this evolution were as follows, although dates given are approximations: 12 700–10 200

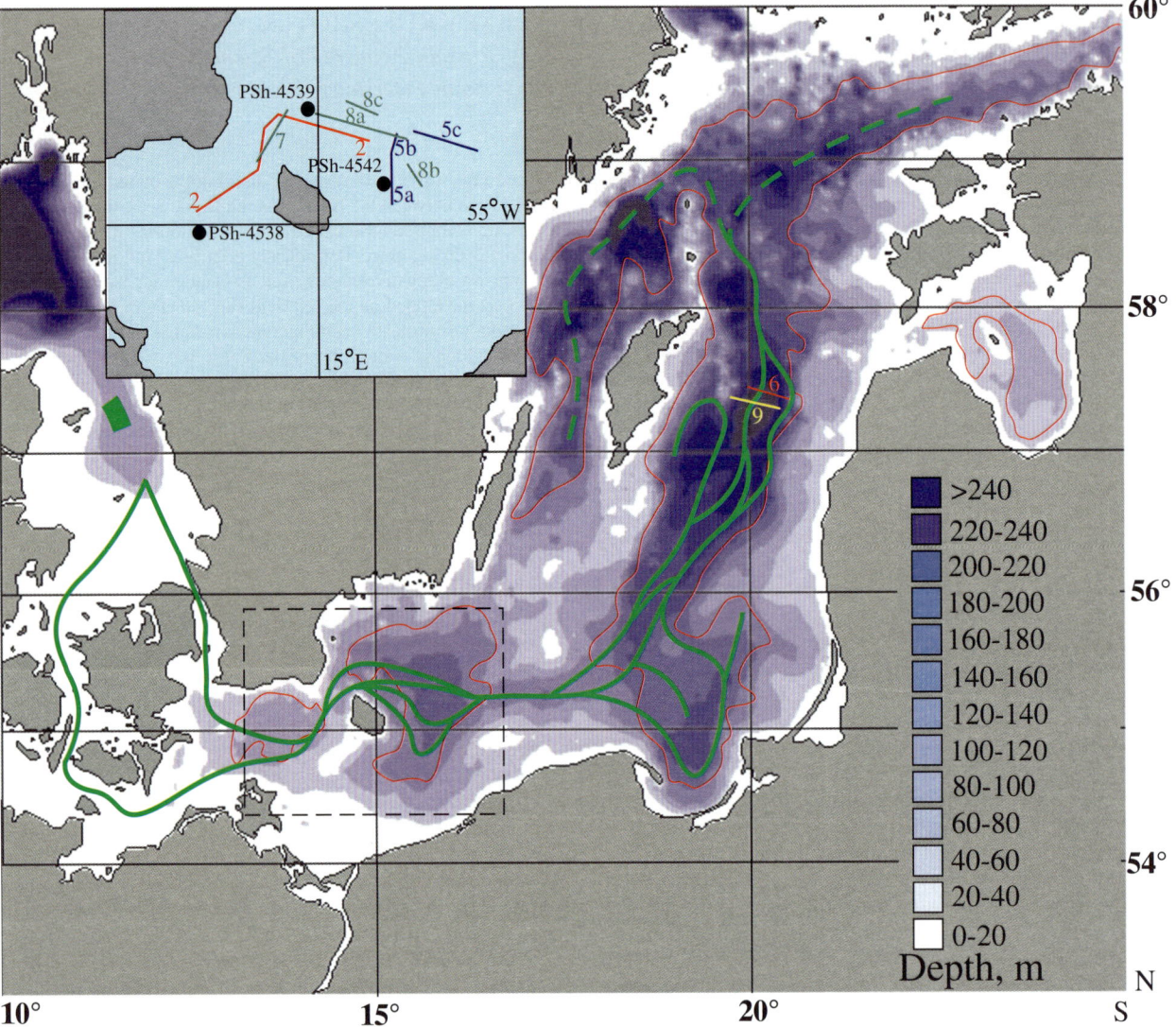

Fig. 1. The regional setting, bathymetry (white-blue scale), pathways of the North Sea bottom current inflows (green lines with arrows), location of the muddy basins (red lines) and present study sections (colour lines), cores and hydrographical stations (black circles).

BP, Baltic Ice Lake; 10 200–9500, Yolida Sea; 9500–8800, Ancylus Lake; 8800–7600, Mastogloia Sea; 7600–4300, Litorina Sea transgression; 4300–present, Baltic Sea.

Oceanographic setting

The Baltic Sea is one of the major estuarine systems on earth with a net long-term freshwater supply amounting to about 15 000 m^{3}s^{-1} and a surface area inside the entrance sills of about 370 000 km^2 (Stigebrandt 1995). Topographically the Baltic Sea has the character of a fjord, with a broad-crested sill located in the mouth (Fig. 1). Within the bathymetric structure of the southern and central Baltic (the Baltic Proper), two areas may be distinguished: a deep-water area and a shallow water area, separated by a gentle slope.

A dense, saline, bottom current from the North Sea flows through the Danish Straits and enters the Baltic Sea. Most of the inflowing water is interleaved in the halocline, which dips from the Danish Straits towards the northeast. The bottom water then flows towards the Central and North Baltic in generally sluggish currents that are intensified along the eastern margins of a series of sub-basins by the action of the Coriolis Force, and also as they pass through narrow passages or gateways between the sub-basins. The salinity distribution of the inflow water changes with the distance from the entrance sills. For topographical reasons, however, there is no continuous bottom current through the gateway (Fig. 2), but flow occurs periodically between sub-basins in which deep water is temporarily stored. This has been likened to a river transporting water between adjacent lakes (Stigebrandt 1995).

Nepheloid layers

Near-bottom boundary layers of homogeneous temperature and salinity are related to bottom frictional effects and, in the western part of the Baltic, associated with bottom currents and advective layers of turbid storm-induced inflow water sandwiched between the bottom and the overlying transparent midwater (Fig. 3). Because of its high water content the uppermost nepheloid layer is transported easily by lateral currents and so forms anomalies in the grain-size distribution of near-bottom suspended matter (Fig. 4). High contrast spots of sound scattering rising from the bottom boundary layer were observed by the ADCP (Fig. 5), and echograms demonstrate intensive disturbance of the sound

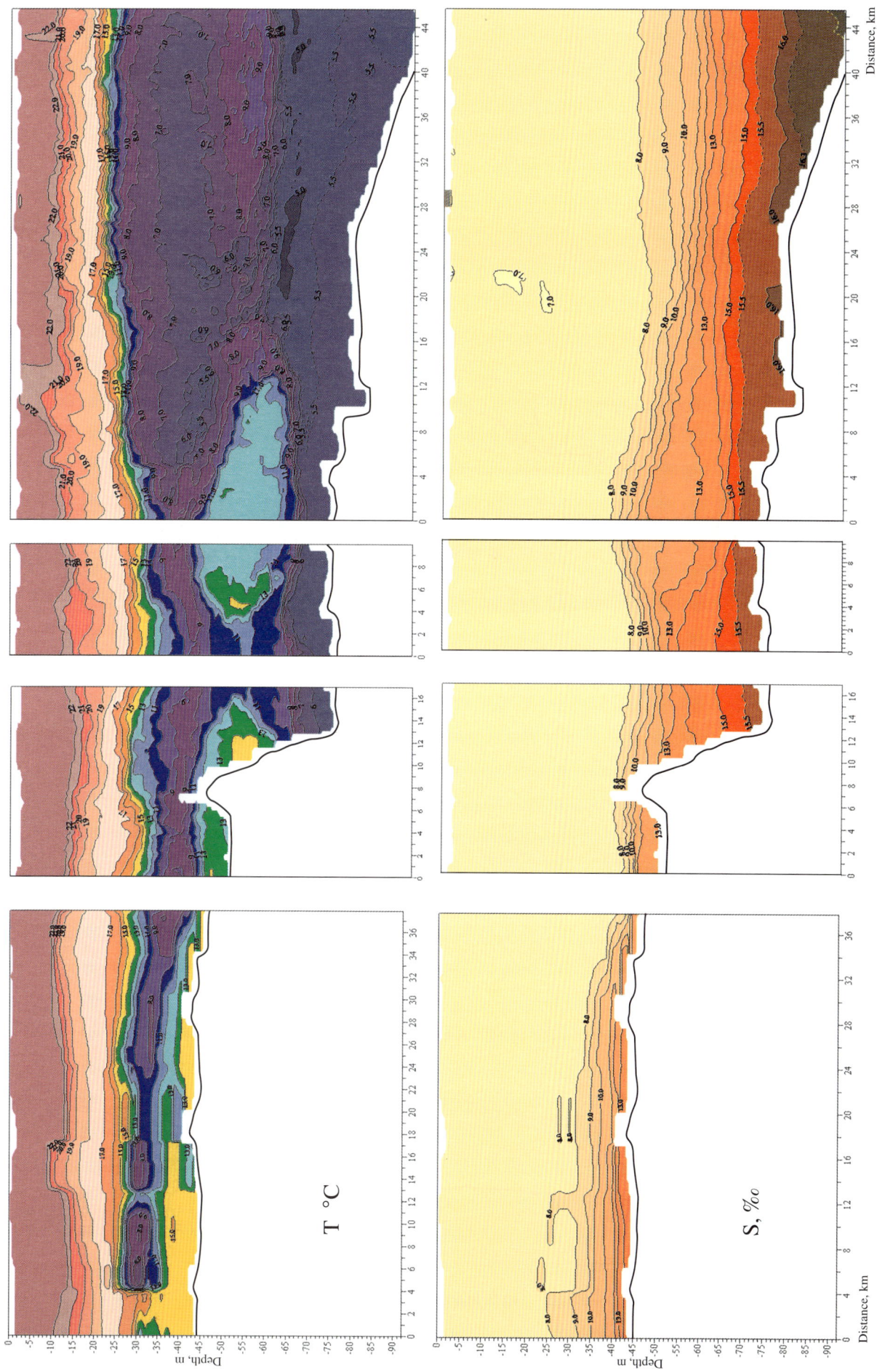

Fig. 2. Intermediate intrusion of the North Sea water (NSW) into the Baltic Sea: upper – temperature, lower – salinity (see fig.1, section 2 made during 34th cruise of R/V 'Professor Shtokman', August 1997).

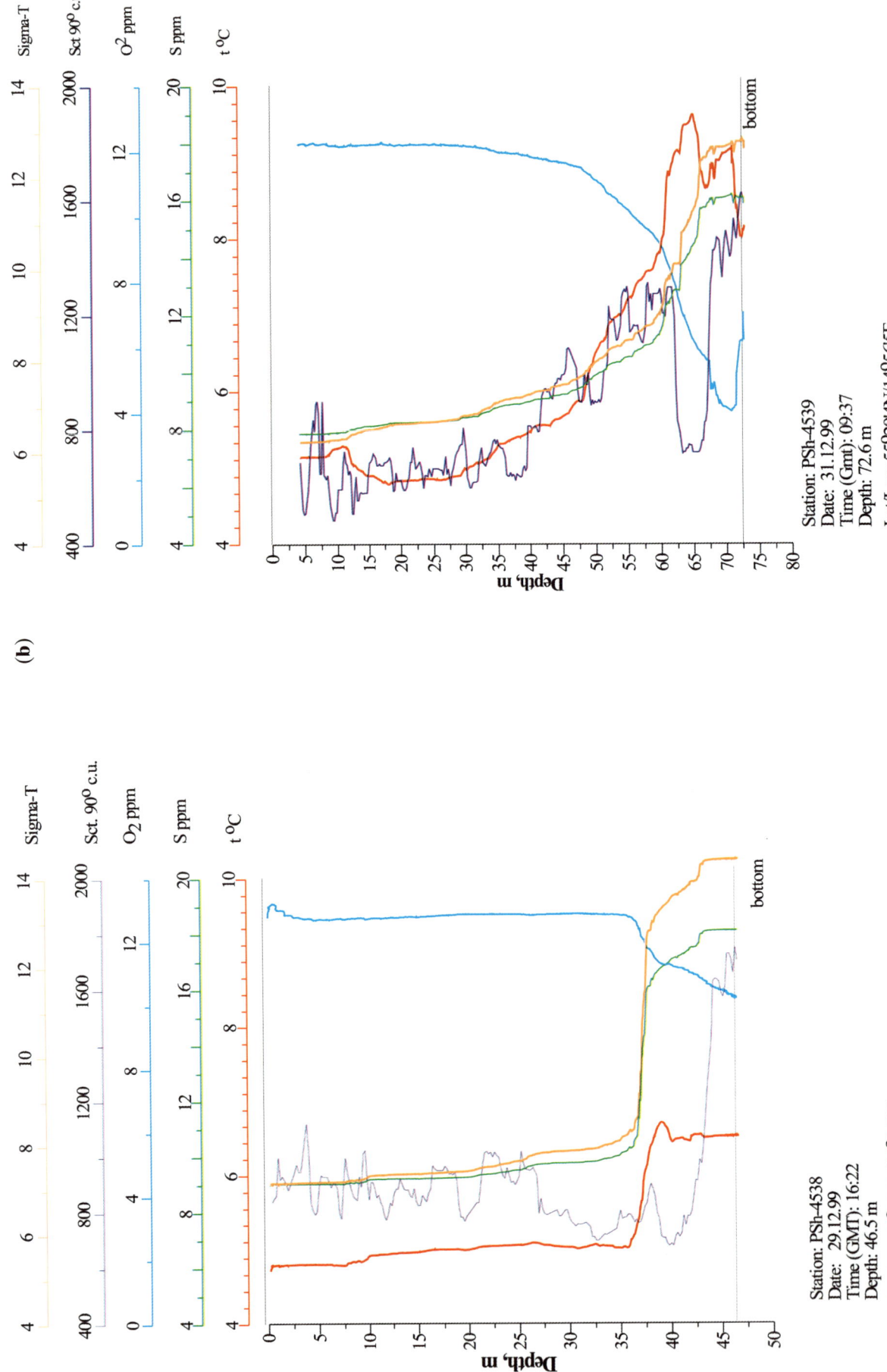

Fig. 3. Hydrographical structure of the water column in the western Baltic during an inflow event: (**a**) station PSh-4538, (**b**) station PSh-4539 (made during 44th cruise of R/V 'Professor Shtokman', December 1999, see fig. 1). Legend: Sigma-T – density, Sct. 90 c.u. – light scattering under 90° (in conventional units), O_2 ppm – oxygen contents, S – salinity, t°C – temperature.

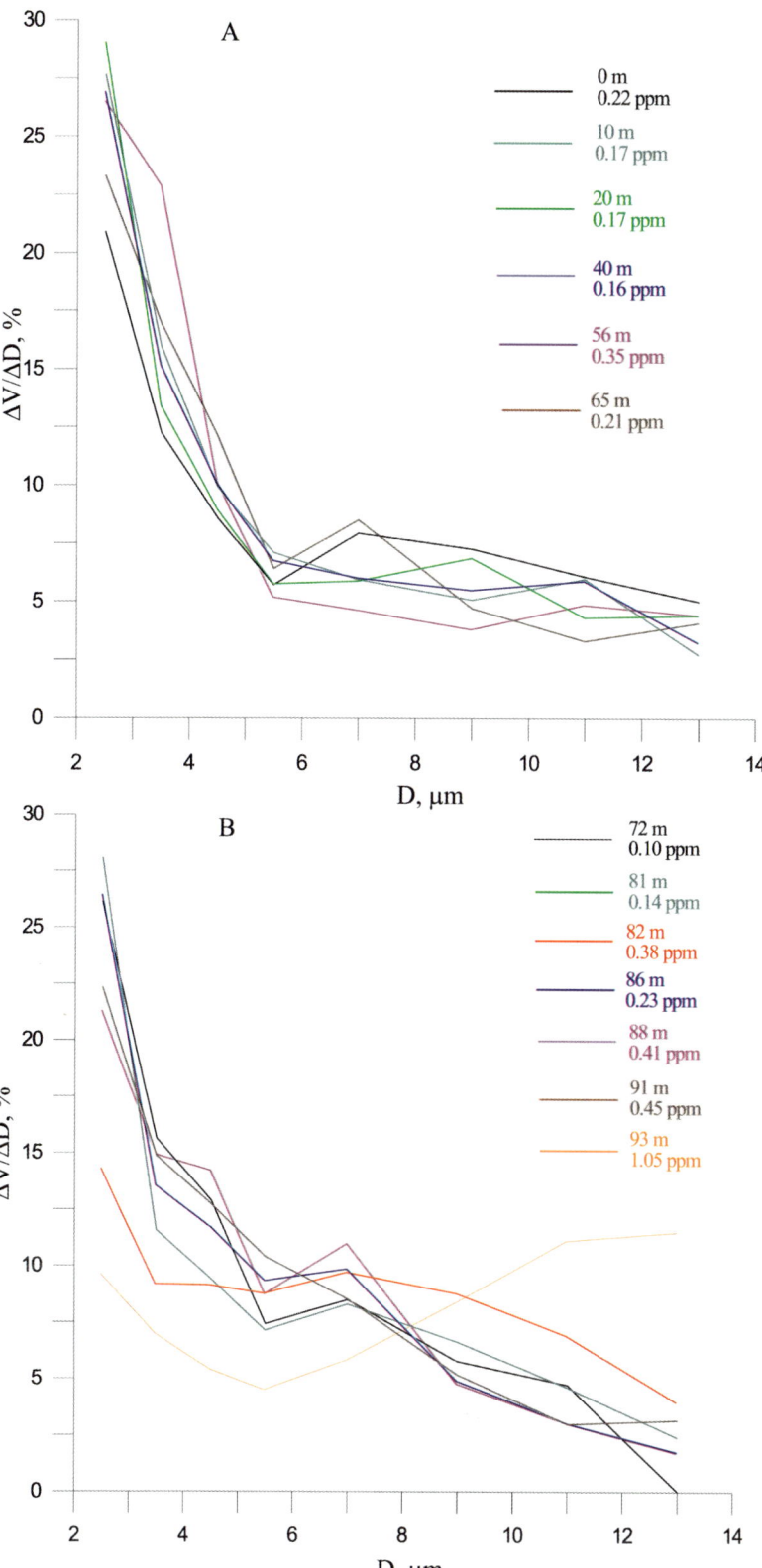

Fig. 4. Particle-size distributions of suspended matter (measured by Coulter Counter) for the station PSh-4542 (depth 94 m) situated in western part of the Baltic Sea (see fig. 1) at the time of an inflow-event of North Sea Bottom Water (winter of 1999–2000): (**a**) upward from halocline, (**b**) downward from halocline. Legend: V – total volume of particles measured in ppm, D – equivalent spherical diameter of particles. Upper numbers near the distributions are levels of water samples in meters; lower numbers are total volumes of particles measured in ppm. A coarsening in the particle-size distribution near the bottom, and probably at the level 82 m, is observed in the samples with the increased particle concentrations originated by inflowing currents.

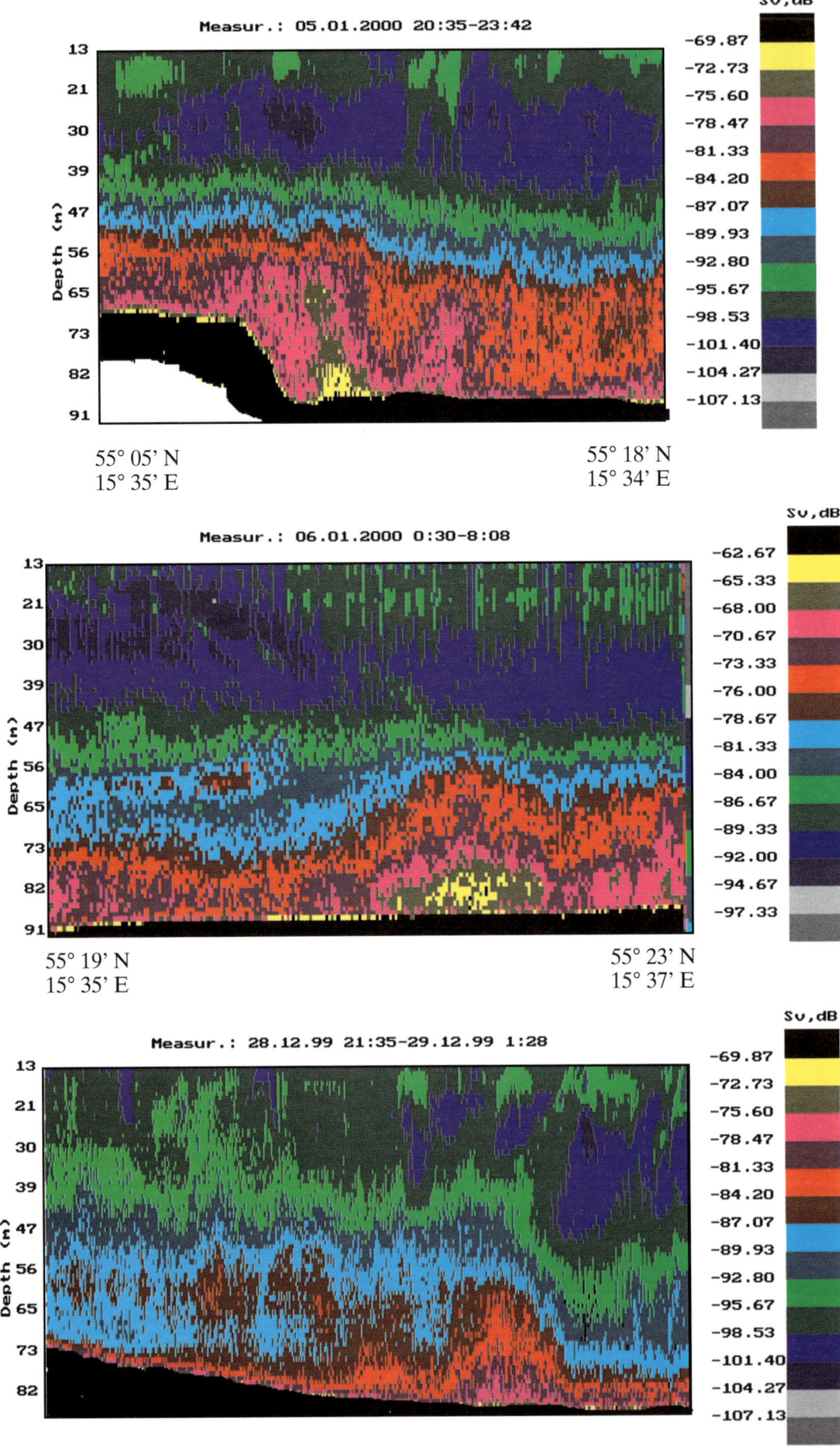

Fig. 5. Spots of sound scattering rising up from the bottom boundary layer observed by the ADCP and associated with suspended matter concentration. Intensity of sound scattering (in dB) is shown by colour scale, black colour corresponds to the bottom. (Location of the ADCP-profiles is shown in fig.1, sections 5a, 5b, 5c).

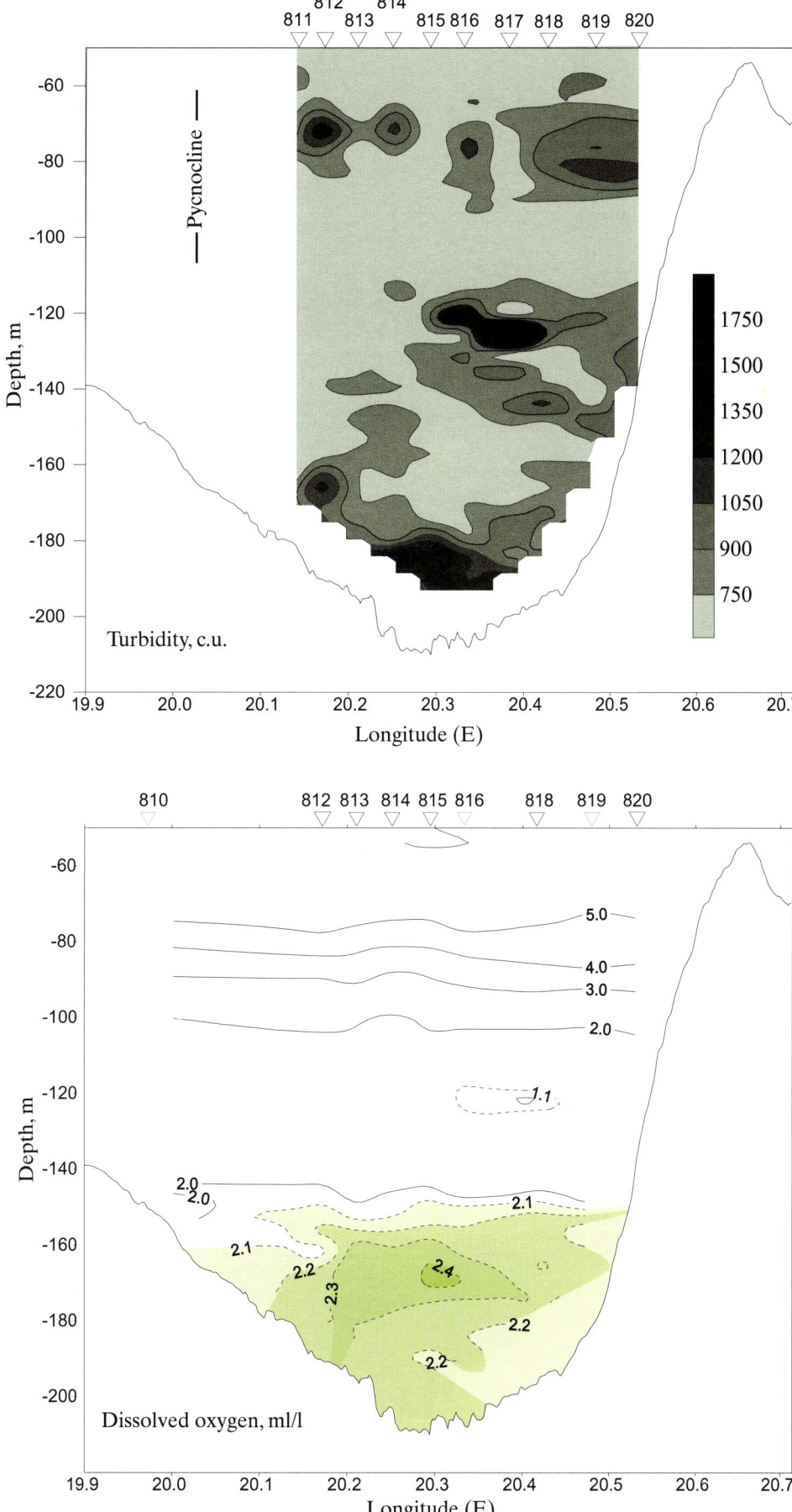

Fig. 6. Hydrographical features in the Gotland Deep (Central Baltic) during major inflow of North Sea water (fig. 1, section 6 made in GOBEX-cruise of R/V 'Alexander von Humboldt', August 1994): a, light scattering (turbidity) under 90° (in conventional units); b, dissolved oxygen; c, salinity; d, density; e, temperature.

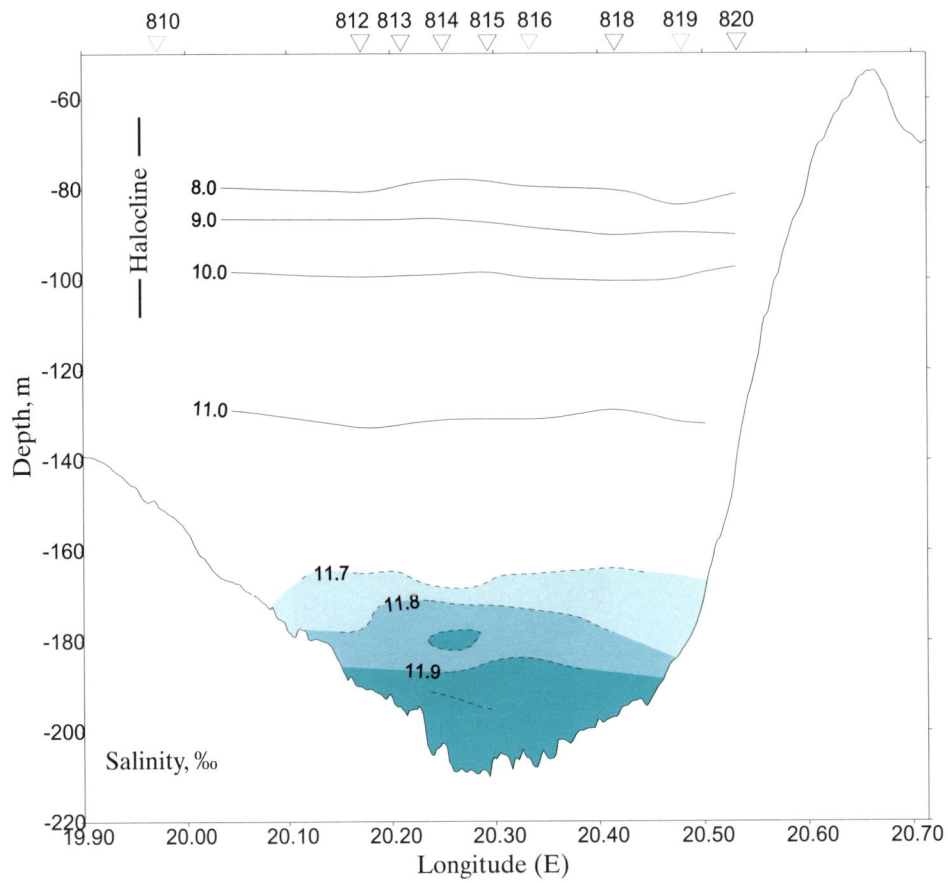
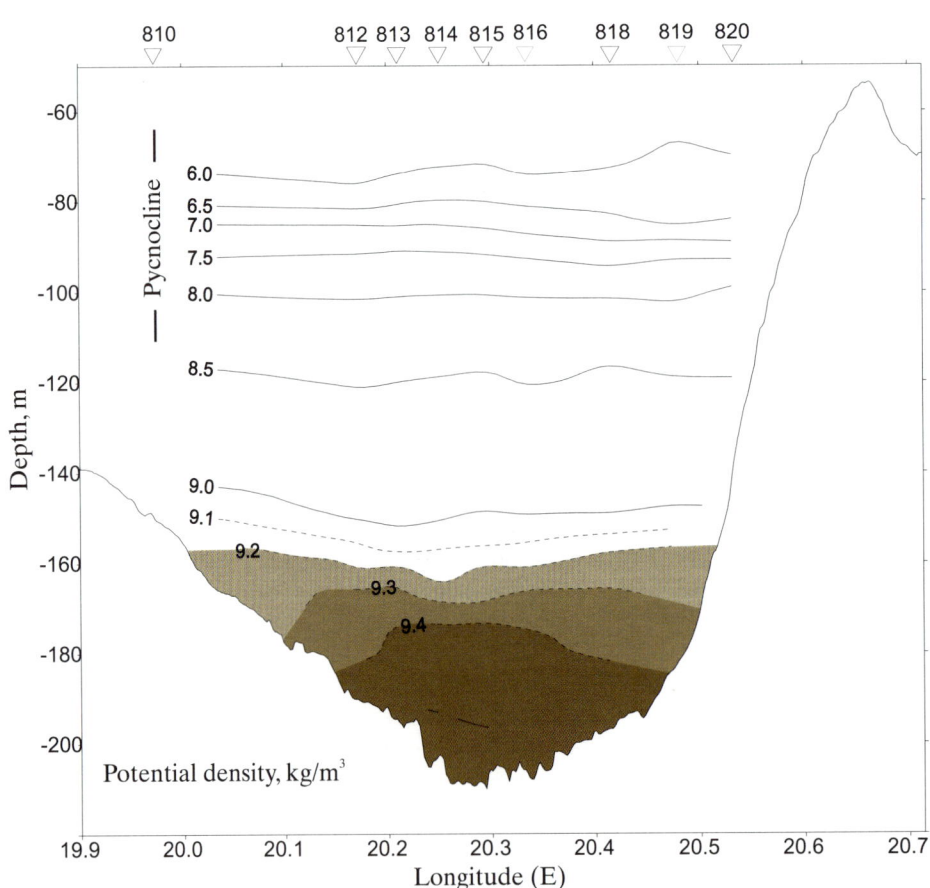

Fig. 6 (c, d).

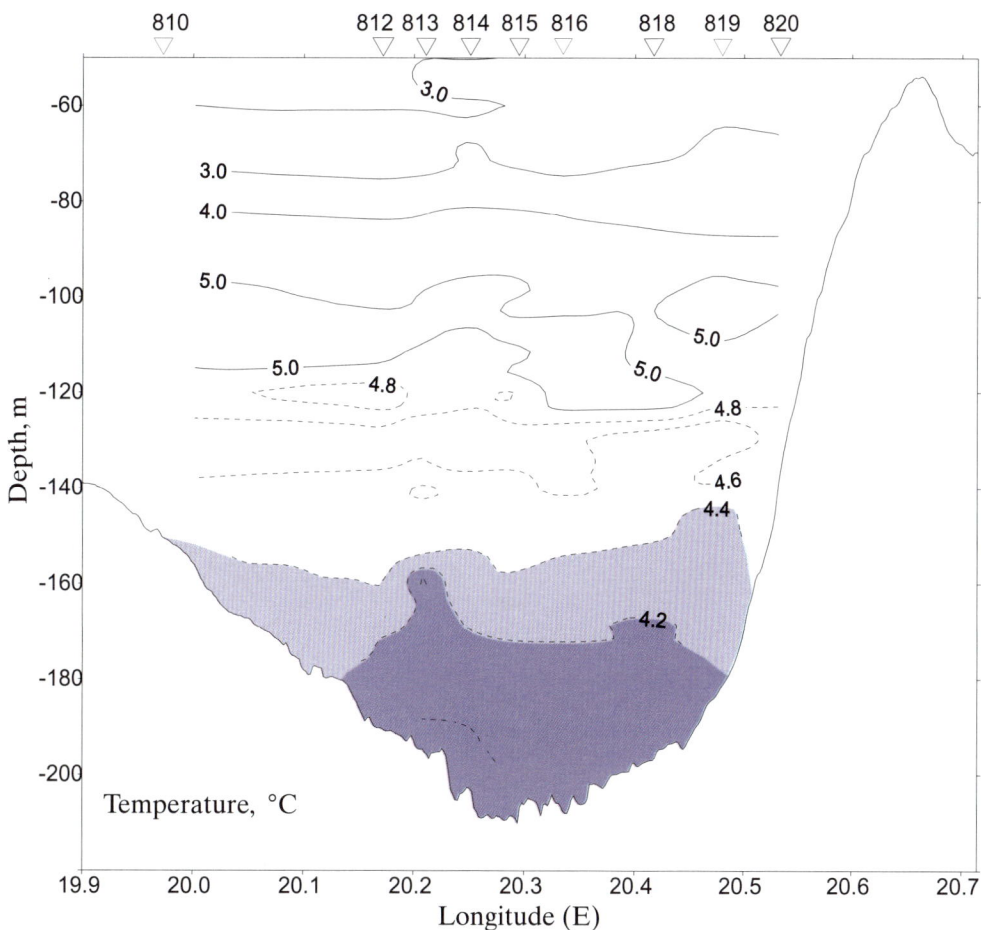

Fig. 6 (e).

scattering field in the vicinity of these spots. The spots are elongated in the direction of the ship's track over 4–5 km. The origin of such spots is presumably related to the local re-suspension of the bottom sediment, especially during peak inflow events when the concentration of suspended matter may be up to 26.9 mg l^{-1} in the near-bottom nepheloid layer (Stryuk *et al.* 1995).

Such well-developed nepheloid layers are commonly associated with bottom currents, and thus provide evidence that re-suspension and sediment transport also takes place in the Central part of the Baltic Sea (Fig. 6). Sporadic events were recorded with peak values between 16 and 25 cm s^{-1} at a level 5 m above the bottom (Hagen & Feistel 1996). Such events are believed to be responsible for the erosion of the bed to form a bottom nepheloid layer. We explain the origin of intermediate nepheloid layers by a lateral injection of particles, that is detached up-stream from the surrounding seafloor relief, and its advection along an isopycnal surface.

Both, local re-suspension (or non-deposition) and advection along isopycnal surfaces probably interact producing complicated nepheloid patterns, whose non-homogeneity makes them difficult to observe. The roughness of the slope relief focuses the energy of currents (for example as topographical eddies) and makes the erosion (non-deposition) of sediments more probable. The shift of water salinity (density) at the upper part of the halocline prevents downward transport of the suspended matter into deeper water. Concentration of suspended particles is elevated in the vicinity of the upper part of the halocline, perhaps due to re-suspension as the result of the focused energy of currents, intrahalocline eddies and internal waves where the halocline intersects the slope. A geochemical source of suspended particles acts in the water column at the boundary between layers of reducing and oxidizing conditions (redox-cline) (Emelyanov 1981), and masks the nepheloid layers induced by the hydrodynamic processes.

Seismic characteristics

Careful analysis of high-resolution, shallow-penetration echosounder and Chirp profiles reveals a clear variation in the thickness of sediment accumulation related to bottom-current pathways. In the first instance, the flow of bottom water through the Baltic Sea gateway is controlled by a system of basement banks and depressions or valleys (Fig. 7). Constriction of the flow between the banks typically leads to reduced sediment accumulation, as a result of sediment transport, non-deposition and erosion. Narrow channels cut within the broader valleys presumably result from local intensification of a stable bottom current system. Thicker sediment accumulation forms as contourite levees or elongate patch drifts adjacent to the channels. These drifts are 1–10 m in height and up to 10 km in width and are best developed on the left flanks of the channels (Fig. 8).

The overall thickness of drift accumulation increases towards the steeper slopes of the margin, situated on the right of the inflowing currents (Endler 1998*b*) (Fig. 9). A maximum thickness of about 12 m is reached just at the base of the eastern slope. The muddy contourites of these shallow Baltic Sea drifts normally have low acoustic attenuation and are very homogeneous on the echogram records. This is perhaps due to the presence of gas, which, above a critical concentration, forms bubbles.

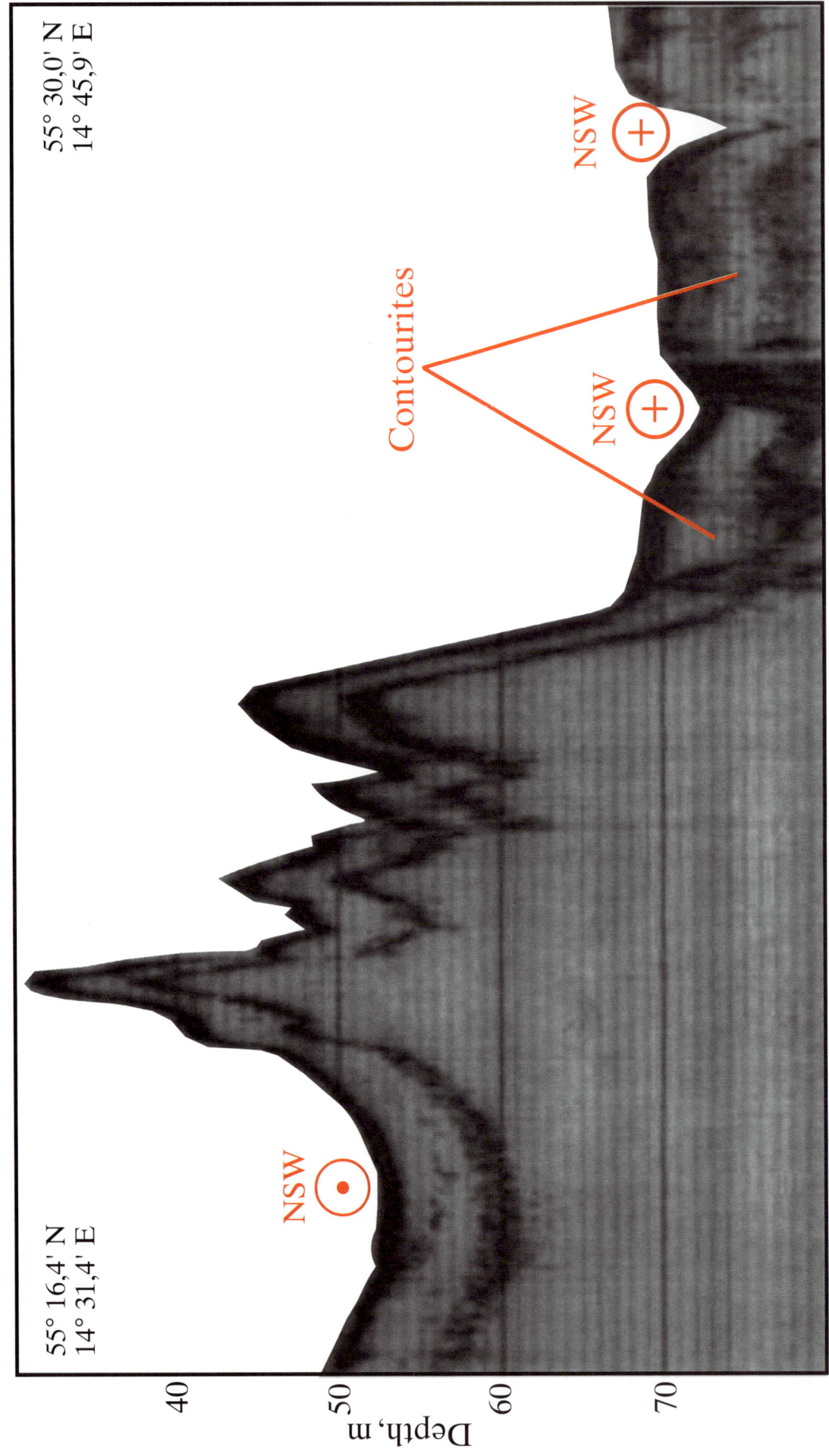

Fig. 7. Flow paths of the bottom water in the gateways of the western Baltic region (fig. 1, section 7). NSW is the North Sea Water.

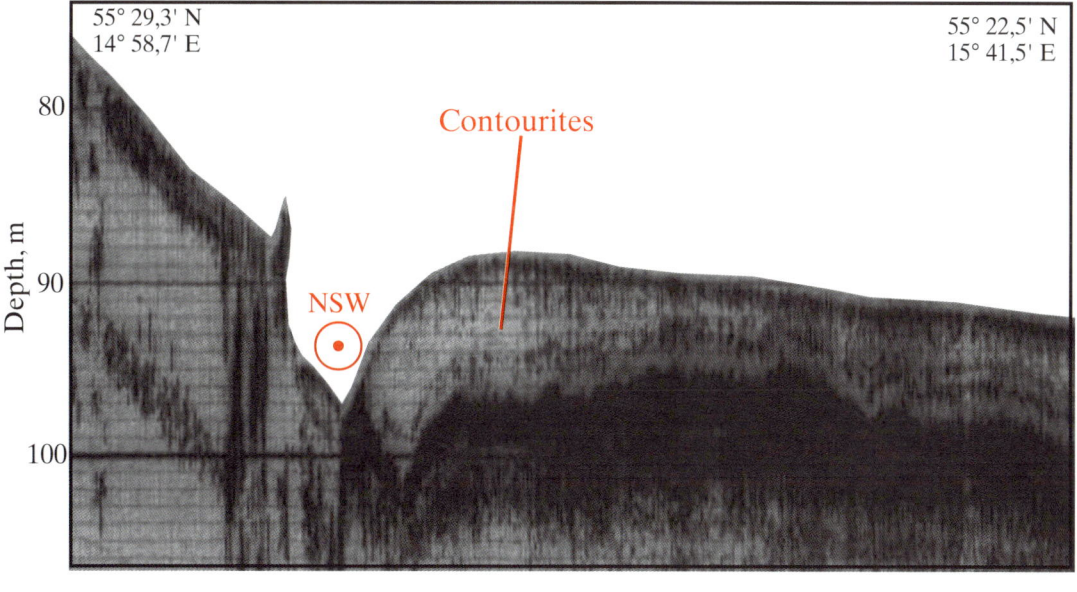
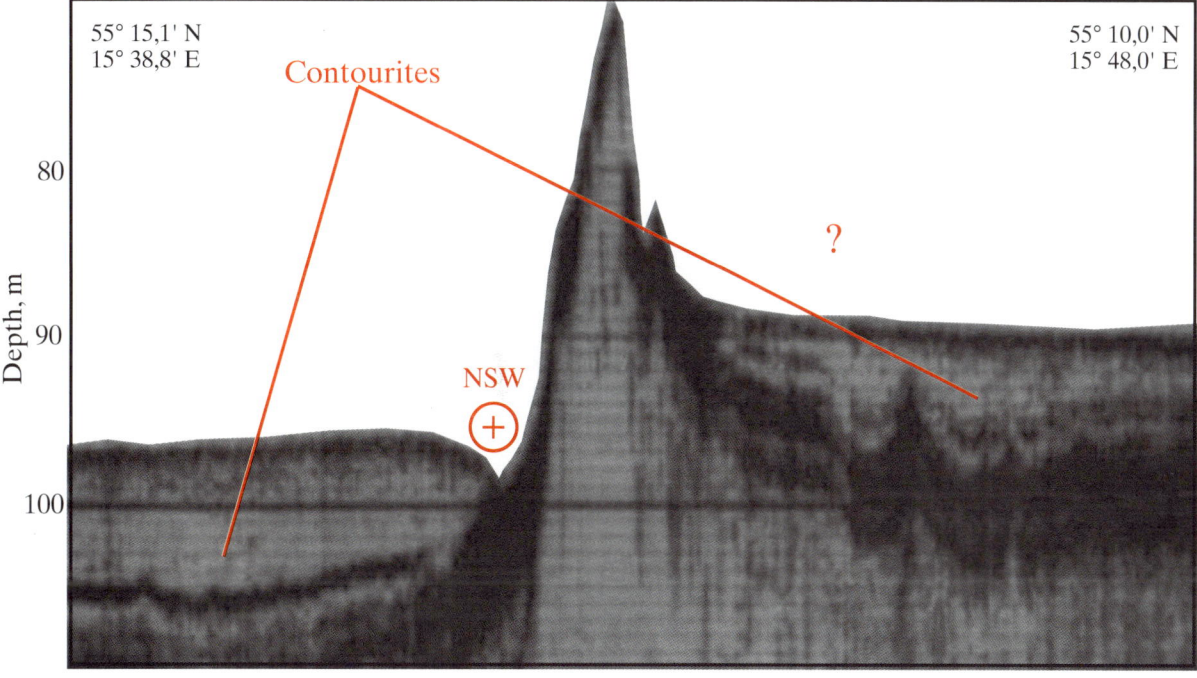
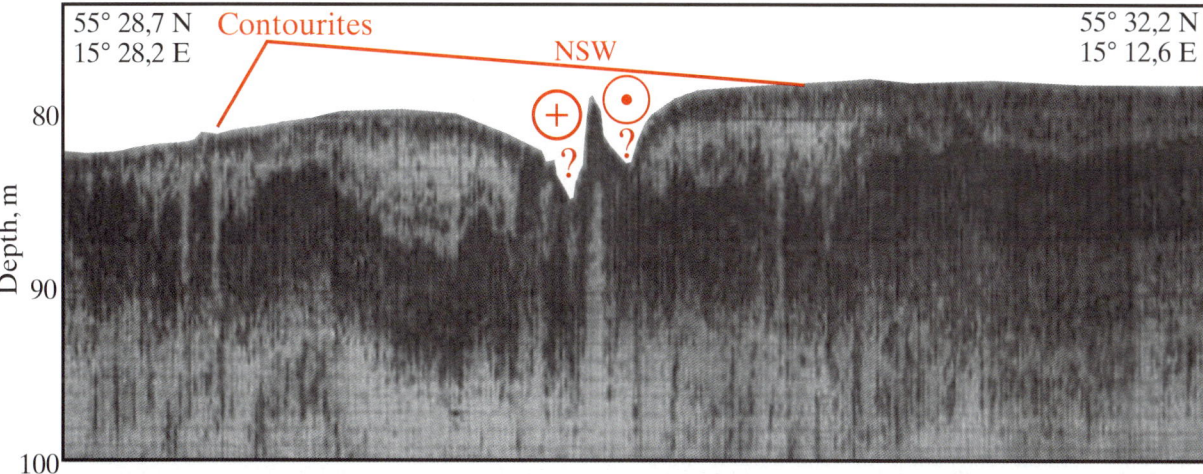

Fig. 8. Seismic profiles showing shallow-water contourite drifts and erosional channels; (acoustic bottom and sub-bottom profiling were performed during 44th cruise of R/V 'Professor Shtokman', December 1999–January 2000, using ELAC (type NBS) echo-sounder; location of the sections 8a, 8b, 8c see fig. 1). NSW is the North Sea Water.

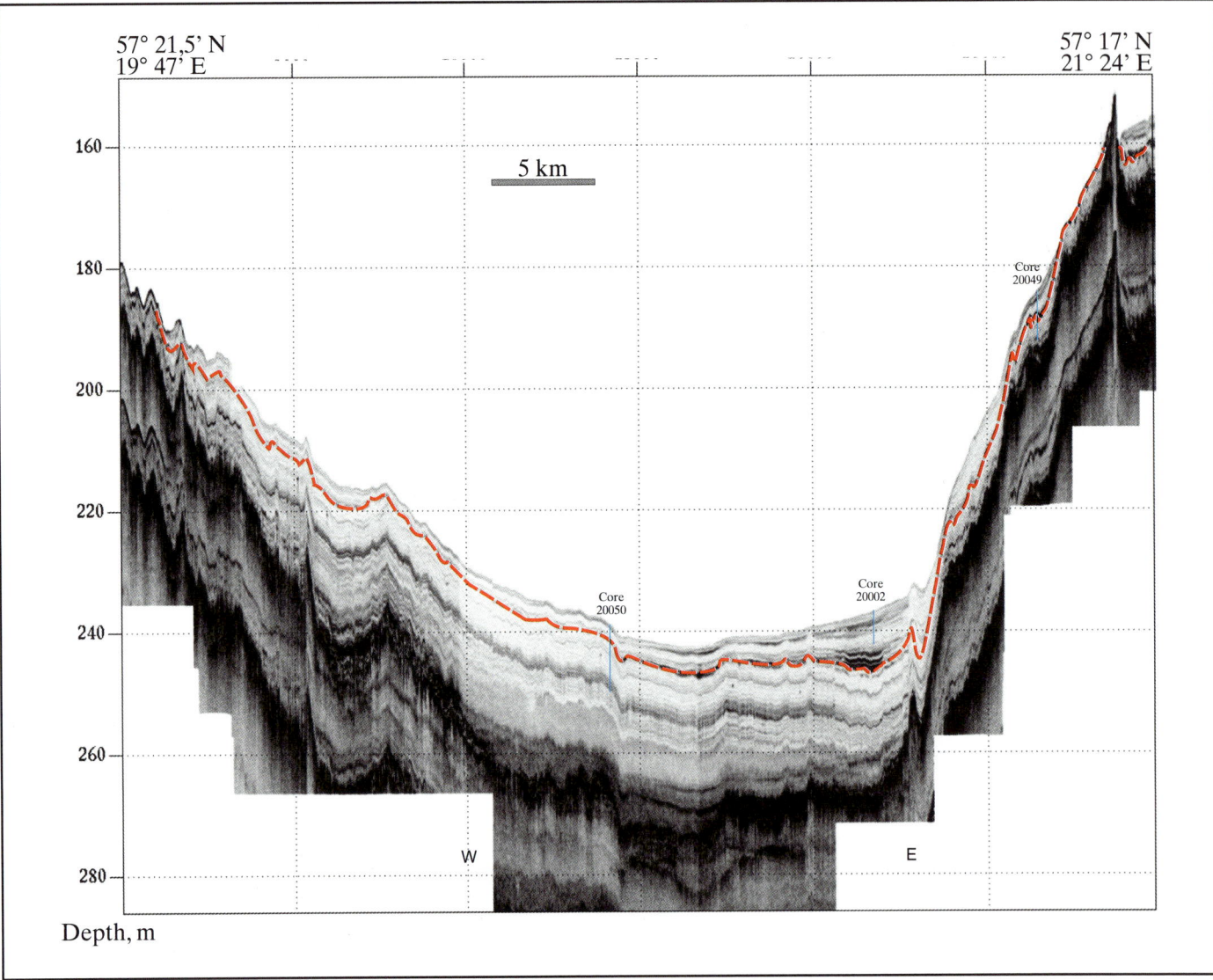

Fig. 9. Acoustic image of layered sediments – transversal section through the Gotland Deep (the Central Baltic) by Endler (1998₁), with indication of location of GOBEX-long sediment cores. Red dotted line is the base of Litorina-to recent marine mud.

Sediment characteristics

Three long GOBEX piston cores recovered from the Baltic Sea gateway region (Fig. 10) show the general characteristics and variation in thickness of sediments from the area. Only sediments from the Litorina Sea and Baltic Sea (i.e. post-7600 years BP, annotated 'L' and 'pL' on Figure 10) are considered to have been influenced by bottom currents. These can therefore be considered as shallow-water contourites. They show little change in composition or other characteristics throughout this period suggesting the environment has not changed very much during the last 7000 years (Fig. 10).

Below a brown oxidized layer, 0.1 to a few centimetres thick, the colour of the mud is mostly olive–grey with irregular patches or bands stained by black iron monosulphide. The black sulphidic mud is very soft, and enriched in both organic carbon and manganese. Both enrichments are caused by high biological productivity and periodic stagnation of near-bottom waters (Kögler & Larsen 1977). These shallow-water muddy contourites are, in fact, a typical flocculated marine sapropelitic silty clay or clayey silt characterized by a high content of organic matter. The content of organic carbon varies from 2 to 5 % in surface samples.

The distinctive fine lamination observed in cores appears to be due to small vertical variation in organic carbon content, but with little other variation in either composition or grain size. There is some indistinct bioturbation evident through the laminated section but, on the whole, burrowing infauna has been sparse. The underlying lithological units, mainly related to deposition in ancestral freshwater lakes, show a much greater degree of bioturbation as well as a paler colour demonstrating low organic-carbon content.

The three core sites illustrated in Figure 10 show marked thickness variation in the development of the shallow-water contourite facies. The core with the thinnest expression (Core 20049) also has a distinct hiatus in the sediment record at the base of the contourite section. This was recovered from a channel location.

Discussion

The Baltic Sea and Baltic Sea gateway can be considered as a relatively small-scale and shallow-water example of bottom-water circulation and sedimentation that is closely analogous to that found in the ocean basins. The gateway region, where bottom

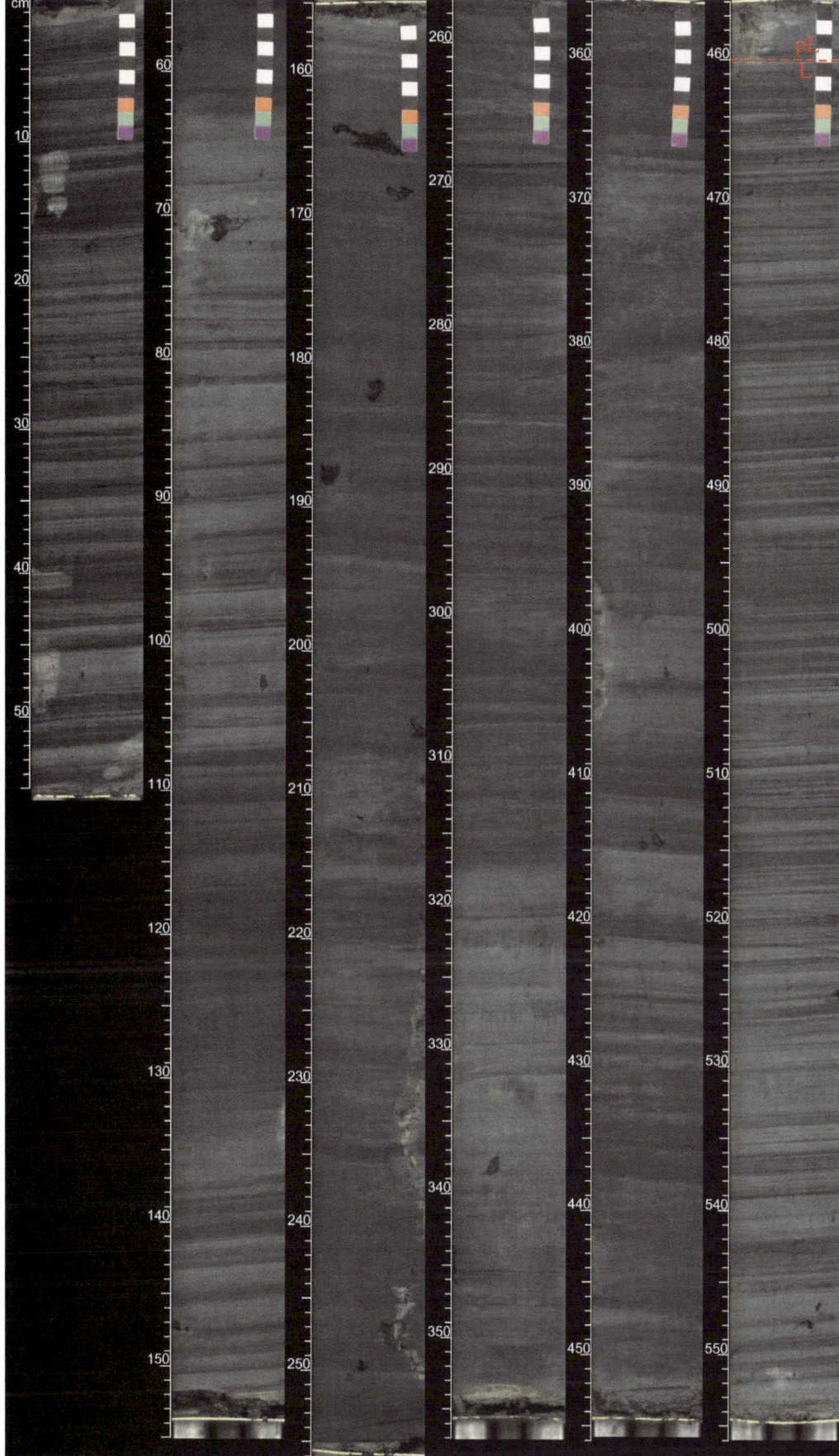

Fig. 10. Typical GOBEX-long sediment cores (location see fig. 9) with Litorina-to recent marine mud, i. e. shallow-water contourites. Lithostratigraphy – by digital images (Meyer, 1998) and logging data, i.e. p-wave-velocity and wet bulk density (Endler 1998b). BIL, Baltic Ice Lake; I, Yolida Sea; A, Ancylus Lake; L, Litorina Sea (to recent).

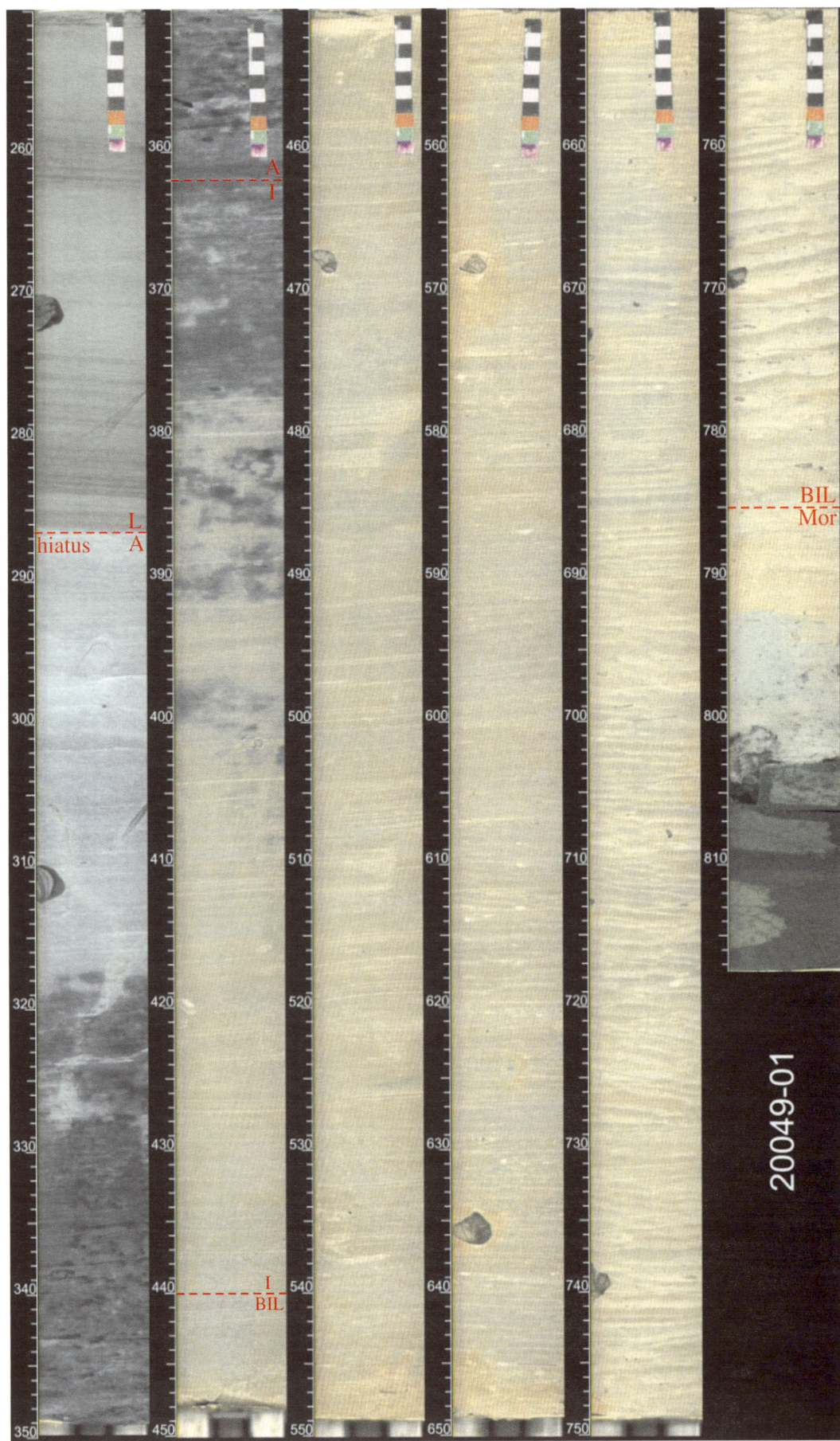

Fig. 10(b).

Fig. 10(c).

currents have been intensified as a result of lateral and/or vertical restriction of the inflow water, are characterized by non-deposition and erosion and hence by a much reduced sediment thickness. Narrow channels are carved by particularly intensive strands of bottom-current flow. These are flanked by contourite levees where accumulation has been dominant and the section is therefore thicker.

The nature of the contourite facies is significantly different from most oceanic contourites in that a diffuse style of lamination is well-developed throughout, and bioturbation is absent to sparse. A similar facies type has been proposed previously as a shallow-water contourite (Stow *et al.* 1996, 1998), and has also been observed in certain high-latitude mixed turbidite–contourite drifts west of the Antarctic peninsula.

Based on previous understanding of Baltic Sea evolution we suggest that these shallow-water muddy contourites mainly originated during periods of active penetration of saline oceanic waters into the Baltic Sea basin. This first occurred during the Litorina Sea Phase between 7600 and 4500 years ago. It has continued, somewhat intermittently, through the Baltic Sea phase to the present day. Further work is in progress that will provide a more systematic and interdisciplinary investigation of the shallow-water contourites in the Baltic Sea.

The authors thank Professor Dorrik Stow (Southampton Oceanography Centre) for his invitation to participate in this issue, and for editorial work on the English. This paper could not have been prepared without the active support of several people, so we would like to thank Dr. Vadim Paka, Dr. Nikolay Golenko, Dr. Mikchail Rudenko, Elena Kuzmina (Atlantic Branch, Shirshov Institute of Oceanology, Russian Academy of Sciences), Professor Kay-Christian Emeis and Dr. Rudolf Endler (Baltic Sea Research Institute, Warnemünde, Germany).

References

BLAZHCHISHIN, A. I. 1998. *Palaeogeography and Evolution of Late Quaternary Sedimentation in the Baltic Sea.* Kaliningrad, 'Yantarny skaz', 160 pp. (in Russian).

EMEIS, K.-C. 1998. Gotland Basin Experiment (GOBEX): Status report on investigations concerning benthic processes, sediment formation and accumulation. *In*: EMEIS, K.-C. & STRUCK, U. (eds) *Gotland Basin Experiment (GOBEX) Status Report on Investigations concerning Benthic Processes, Sediment Formation and Accumulation.* Baltic Sea Research Institute, Warnemünde, Meereswissenschftliche Berichte/Marine Science Reports, **34**, 2–9.

EMELYANOV, E. M. 1981. The manganous carbonate aluminosilicate lithologic-geochemical region of the Gotland and the Landsort basins. *In*:

LISITSYN, A. P. & EMELYANOV, E. M. (eds) *Sedimentation in the Baltic Sea*. Nauka Press, Moscow, 136–180 (in Russian).

EMELYANOV, E. M. 2001. Biogenic components and elements in sediments of the Central Baltic and their distribution. *Marine Geology*, **172**, 23–41.

EMELYANOV, E. M. & GRITSENKO, V. A. 1999. About the role of the near-bottom currents in the formation of the bottom sediments in the Gotland Basin, Baltic Sea. *Oceanologiya*, **39**(5), 776–786 (in Russian).

EMELYANOV, E. M., TRIMONIS, E. S., SLOBODYANIK, V. M. & NIELSEN, O. B. 1995. Sediment thickness and accumulation rates. *Aarhus Geoscience*, **5**, 81–84.

ENDLER, R. 1998a. Acoustic Studies. *In*: EMEIS, K.-C. & STRUCK, U. (eds) *Gotland Basin Experiment (GOBEX) Status Report on Investigations concerning Benthic Processes, Sediment Formation and Accumulation*. Warnemünde, Baltic Sea Research Institute: Meereswissenschftliche Berichte/Marine Science Reports, **34**, 21–34.

ENDLER, R. 1998b. Multy-sensor core logs of GOBEX gravity cores. *In*: EMEIS, K.-C. & STRUCK, U. (eds) *Gotland Basin Experiment (GOBEX) Status Report on Investigations concerning Benthic Processes, Sediment Formation and Accumulation*. Warnemünde, Baltic Sea Research Institute: Meereswissenschftliche Berichte/Marine Science Reports, **34**, 38–54.

GRITSENKO, V. A. & SIVKOV, V. V. 1997. Some results of numerical modeling of suspension-carrying deep currents in the Baltic sea: sedimentological aspects. *In*: CATO, I. & KLINGBERG, F. (eds) *Proceedings of the Fourth Marine Geological Conference: 'The Baltic'*, Uppsala, SGU, Ser. Ca 86, 57–60.

HAGEN, E. & FEISTEL, R. 1996. Lenses of relative saline deep water in the eastern Gotland Basin? *In*: HAGEN, E. (ed.) *GOBEX-Summary Report*. Warnemünde, Baltic Sea Research Institute: Meereswissenschftliche Berichte/Marine Science Reports, **19**, 34–37.

KÖGLER, F.-C. & LARSEN, B. 1977. The West Bornholm basin in the Baltic Sea: geological structure and Quaternary sediments. *Boreas*, **8**, 1–22.

KRAMARSKA, R., USCINOWICZ, S. & ZACHOWICZ, J. 1995. Quaternary. *In*: Geological Atlas of the Southern Baltic, 1:500000 (ed. in chief J. E. Mojski), Sopot-Warszawa, pp. 22–30.

LARSEN, B. & KÖGLER F.-C. 1975. A submarine channel between the deepest parts of the Arkona and the Bornholm basins in the Baltic Sea. *Deutsche Hydrographische Zeitschrift*, **28**(6), 274–276.

MEYER, M. 1998. Digital images of GOBEX cores. *In*: EMEIS, K.-C. & STRUCK, U. (eds) Gotland Basin Experiment (GOBEX) Status Report on Investigations concerning Benthic Processes, Sediment Formation and Accumulation. Warnemünde, Baltic Sea Research Institute: Meereswissenschftliche Berichte/Marine Science Reports, **34**, 36–37.

MOJSKI, J. E. 1995. *Structural conditions of Pleistocene ice-sheet development*. Geological Atlas of the Southern Baltic, 1:500 000, Sopot-Warszawa, 20–22.

SVIRIDOV, N. I. & EMELYANOV, E. M. 2000. Facies-lithological complexes of the Quaternary deposits of the Central- and South-eastern Baltic. *Lithology and Minerals*, **3**, 246–267 (in Russian).

SVIRIDOV, N. I. & SIVKOV, V. V. 1992. Use of seismo-acoustic data to investigate bottom currents of the South-western Baltic Sea. *Oceanologiya* (English Translation), **32**(5), 651–656.

SVIRIDOV, N. I., SIVKOV, V. V., CHRISTIANSEN, C. & LYKKE-ANDERSEN, H. 1995. Near-bottom currents in the South-Western Baltic evaluated from seismo-acoustic data. *Aarhus Geoscience*, **5**, 119–126.

SIVKOV, V., SKYUM, P. & CHRISTIANSEN, C. 1995. The distribution of the suspended matter in relation to the hydrographical conditions. *Aarhus Geoscience*, **5**, 111–117.

SIVKOV, V., EMEIS, K.-C., ENDLER, R., ZHUROV, Y. & KULESHOV, A. 1998. Observations of the nepheloid layers in the Gotland Deep (August 1994). *In*: EMEIS, K.-C. & STRUCK, U. (eds) *Gotland Basin Experiment (GOBEX) Status Report on Investigations concerning Benthic Processes, Sediment Formation and Accumulation*. Warnemünde, Baltic Sea Research Institute: Meereswissenschftliche Berichte/ Marine Science Reports, **34**(34), 84–91.

SIVKOV, V. V. & SVIRIDOV, N. I. 1994. The relation between erosional-accumulative forms of bottom relief and near-bottom currents in the Bornholm deep. *Oceanologiya* (English Translation), **34**(2), 266–270.

STIGEBRANDT, A. 1995. The large-scale vertical circulation of the Baltic Sea. *In*: *Conference Proceedings 'First study Conference on BALTEX'*, (Visby, Sweden, August 28–September 1, 1995, 28–47.

STOW, D. A. V. 1982. Bottom currents and contourites in the North Atlantic. *In*: *Actes Colloque International CNRS Bordeaux, Sept. 1981*. Bulletin Institute Geol. Bassin d'Aquitaine, Bordeaux, **31**, 151–166.

STOW, D. A. V., READING, H. G. & COLLINSON, J. D. 1996. Deep Seas. *In*: READING, H. G. (ed.) *Sedimentary Environments*, 3rd edition. Blackwell Science, Oxford, 395–454.

STOW, D. A. V., FAUGÈRES, J-C., VIANA, A. & GONTHIER, E. 1998. Fossil contourites: a critical review. *Sedimentary Geology*, **115**, 3–32.

STOW, D. A. V., PUDSEY, C. J., HOWE, J., FAUGÈRES, J-C. & VIANA, A. (eds) 2002. *Deep-Water Contourite Systems: Modern Drifts and Ancient Series, Seismic and Sedimentary Characteristics*. Geological Society, London, Memoirs, **22**.

STRYUK, V., SKYUM, P. & CHRISTIANSEN, C. 1995. Sources, supply and distribution of suspended matter. *In*: Geology of the Bornholm Basin. *Aarhus Geoscience*, **5**, 105–110.

VIANA, A., FAUGÈRES, J-C., STOW, D. A. V. & IMBERT, P. 1998. Bottom-current controlled sand deposits: a review from modern shallow to deep water environments. *Sedimentary Geology*, **115**, 53–80.

Faro–Albufeira drift complex, northern Gulf of Cadiz

DORRIK A. V. STOW[1], JEAN-CLAUDE FAUGÈRES[2], ELIANE GONTHIER[2], MICHEL CREMER[2], ESTEFANIA LLAVE[3], F. J. HERNÁNDEZ-MOLINA[4], LUIS SOMOZA[3] & V. DÍAZ-DEL-RÍO[5]

[1] SOES-SOC, University of Southampton, Southampton SO14 3ZH, UK (e-mail: davs@soc.soton.ac.uk)
[2] Departmente de Geologie et Oceanographie, Universite de Bordeaux I, Avenue des Facultes, 33405 Talence, France
[3] Geología Marina, Instituto Tecnológico Geominero de España, Ríos Rosas 23, 28003, Madrid, Spain
[4] Ciencias del Mar, Univ. De Cádiz, 11510, Puerto Real, Cádiz, Portugal
[5] Instituto Español de Oceanografía, C/Puerto Pesquero s/n, 29640, Fuengirola, Málaga, Spain

Abstract: The northern margin of the Gulf of Cadiz is swept by Mediterranean Outflow Water between about 500 and 1000 m water depth. This warm, saline, thermohaline, bottom current attains velocities in excess of 1 m s^{-1} through the narrow and relatively shallow Gibraltar gateway, and then descends and slows as it moves towards the north and west around the Iberian margin. It was established in its present form in the latest Miocene, following tectonic re-opening of the Gibraltar gateway, and has since helped to sculpt the slope region in conjunction with downslope processes and diapiric intrusion. The principal area of contourite deposition, up to 600 m in thickness, is the Faro–Albufeira drift complex in a mid-slope setting some 30 km south of Faro. This comprises an elongate low-mounded drift (Faro–Albufeira) and adjacent broad sheeted drifts (Faro and Bartolomeu Dias Planaltos), flanked and partly dissected by deep, erosional, bottom-current channels and buried channels. The seismic character is one of progradational-aggradational depositional units with laterally extensive sub-parallel reflectors, widespread discontinuities and a large-scale cyclicity in seismic facies. The upper 10 m of cored section comprises muddy, silty and sandy contourites of mixed terrigenous and biogenic composition, that show small-scale cyclicity in grain size and associated sedimentary features. Rates of accumulation varied from < 1 to 14.5 cm ka^{-1} (cores), and 3.5 to 29.5 cm ka^{-1} (seismics). The large and small-scale cyclicity noted can be related to fluctuation in bottom current velocity related to climate and sea-level changes, although the precise correlation between these events remains uncertain.

The Faro–Albufeira drift complex is located in a mid-slope setting on the northern margin of the Gulf of Cadiz. The influence of Mediterranean Outflow Water (MOW) on contourite drift construction along this margin was first recognized by Vanney & Mougenot (1981) and Mougenot & Vanney (1982), who distinguished a series of relatively small elongate drifts south of the Iberian peninsula, including the Faro and Albufeira Drifts. Detailed sedimentological and seismic studies of the Faro Drift were then carried out in the early 1980s (Faugères et al. 1984; Gonthier et al. 1984; Stow et al. 1986, amongst others). The now standard contourite facies model first emerged from this work, by comparing Faro Drift contourites with those of other North Atlantic contourites in particular (Stow 1982; Stow et al. 1986). These early studies also characterised growth of the Faro Drift by progradation and aggradation at the northern end of the Faro–Cadiz Planalto (e.g. Faugères et al. 1985a, b). This platform area, together with the neighbouring Bartolomeu Dias Planalto, we now interpret as a broad sheeted drift, and describe the whole as the *Faro–Albufeira drift complex* (Table 1, Fig. 1).

This contribution brings together a range of different datasets that have been collected over the past 20 years in the Faro–Albufeira region (Figs 2 & 3). There is a dense network of single-channel and multichannel seismic reflection profiles across the region, that can be correlated with oil company boreholes drilled on the adjacent shelf and slope. Analysis of these data is currently in progress, with some abstracts and preliminary papers recently published (e.g. Llave et al. 2000, in press). Extensive sediment data were gathered in 1982 by the French oceanographic vessel, *RV Noroit*, including 300 km of 3.5 kHz seismic profiles, 24 piston/gravity core sites, and five sites for seafloor photography. Most of these data have been published previously and are referred to as appropriate. Considerable research activity continues at present in the Gulf of Cadiz in general, including swath bathymetric and deep-tow sidescan sonar mapping, mainly to the south of our present study area, as well as interpretation of a suite of giant piston cores collected in 1999 by the *Marion Dufresnes* research vessel.

Table 1. *Principal characteristics of the Faro–Albufeira Drift Complex*

Location	northern margin of Gulf of Cadiz, eastern N Atlantic
Setting	mid-slope setting, 500–900 m water depth
Age	latest Miocene/Pliocene to Recent
Drift type	elongate mounded drift and sheet complex closely associated contourite/turbidite channel network
Dimensions	mounded drift 40–80 km × 15–20 km, max 600 m thick, but only low relief (< 50 m) above the adjacent sheeted drift sheeted drift 60–80 km × 40–60 km, max 400–500 m thick
Seismic facies	progradational to aggradational seismic depositional units, laterally extensive, low-high amplitude sub-parallel reflectors, widespread discontinuities – erosional to non-depositional, large-scale cyclicity in seismic facies characteristics
Sediment facies	muddy, silty and sandy contourite facies, mixed terrigenous (dominant) and biogenic composition, small-scale grain-size cyclicity evident through top 10 m cored

Geological and oceanographic setting

Geological framework

The Gulf of Cadiz, located in the eastern sector of the central North Atlantic, forms a deeply concave indentation between the African and European continents (Fig. 1). The Faro–Albufeira drift complex is located on the slope above the transition between the Gloria transform zone, which marks the African–Eurasian plate boundary in the Atlantic realm, and the western front of the Betic-Rif orogenic belt, represented by the Gibraltar Arc. Kinematic studies of the African and Eurasian plates (Dewey et al. 1989; Olivet 1996) show that the area was situated in a N–S convergence setting from the Late Cretaceous to the Tortonian.

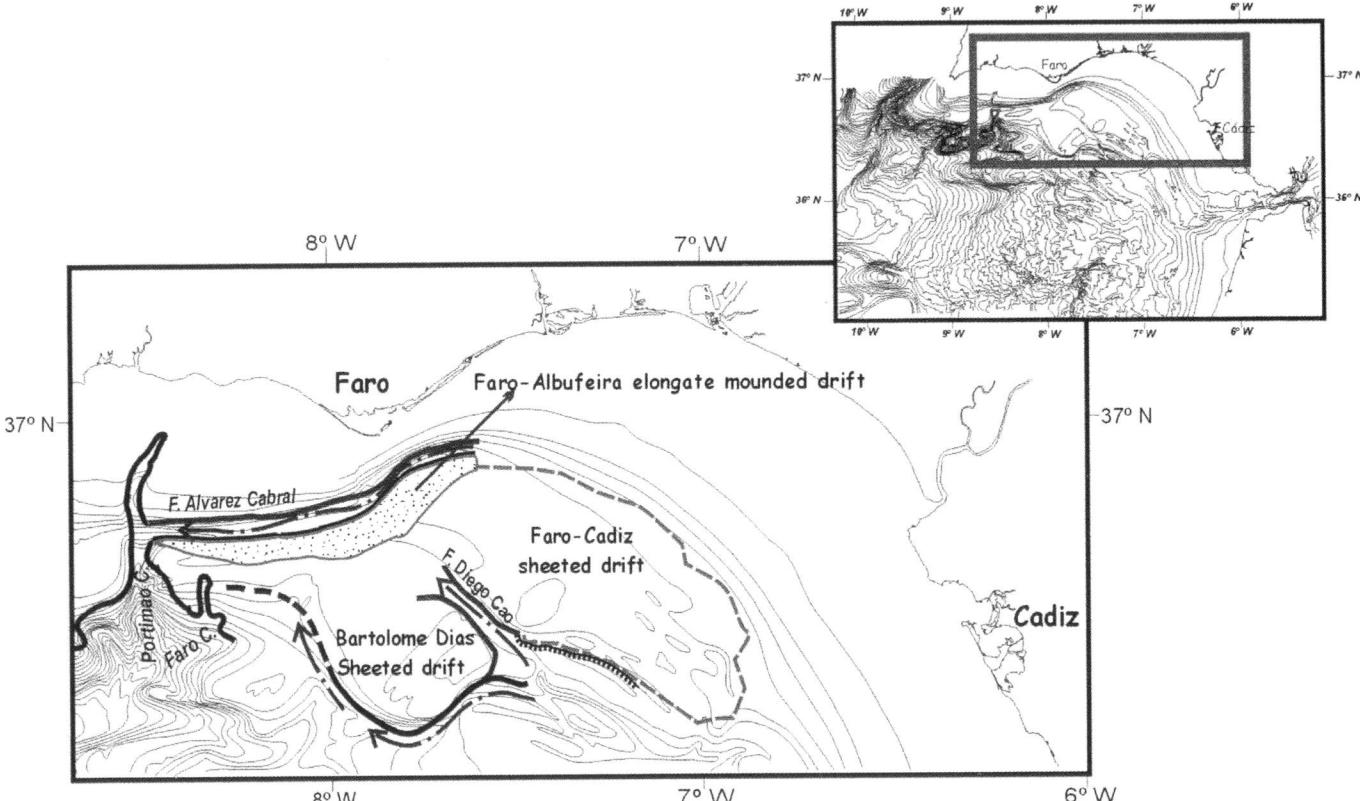

Fig. 1. Faro–Albufeira drift complex location maps. General location on the south Iberian continental margin, Gulf of Cadiz, together with regional bathymetry. Detailed inset shows location and nomenclature for parts of the drift complex (outline as dashed line). Stippled region shows single mounded elongate drift form recognised in this study, formerly named separately as the Faro and Albufeira drifts.

From Late Miocene to the present, the convergence between Africa and Eurasia has become a transpressive NW–SE regime (Argus et al. 1989) (Fig. 4).

Evolution of the Gulf of Cadiz is marked by three successive phases (Somoza et al. 1999) (Fig. 5): (1) construction of a passive margin of Mesozoic age, related to opening of the North Atlantic; (2) development of a compressional regime during the Late Eocene to Early Miocene, related to the closure of the Tethys Alpine sea; and (3) the Miocene foredeep evolution, associated with formation of the Betic–Rif orogen and opening of the Western Mediterranean basin. This stage was characterized by collision of the Betic–Rif accretionary front with the passive margins of the Iberian Peninsula and Africa, which involved the emplacement of a large tectonic olistostrome over the Gulf of Cadiz margin during the Middle Miocene. Since the Late Tortonian, oblique convergence and extensional collapse have progressively given way to more stable conditions during the Upper Pliocene and Quaternary (Maldonado & Nelson 1999; Maldonado et al. 1999).

Drift development is believed to have begun in the latest Miocene to early Pliocene, at first smoothing and covering a topographically irregular surface, as well as being influenced by early tectonic activity in the region. Diapirism is still active throughout the Gulf of Cadiz, with both salt and mud diapirs showing surface expression and breakthrough.

Oceanographic setting

The circulation pattern in the Gulf of Cadiz is presently characterised by an exchange of water masses through the Strait of Gibraltar, here referred to as the Gibraltar gateway (Fig. 6). This exchange involves flow into the Atlantic Ocean of Mediterranean Outflow Water (MOW) near the bottom, and an influx to the Mediterranean of Atlantic water (AI) at the surface. MOW is warm (>13°C) and saline (>36.4‰) with a relatively low oxygen content (4.1–4.6 ml l^{-1}), whereas AI is a turbulent, less saline cool-water mass (Ambar et al. 1976, amongst others). The Gibraltar gateway has controlled the dynamics of water mass exchange over time, modulating that exchange between the Gulf of Cadiz and Alboran Sea and amplifying the Quaternary high-frequency sea-level changes noted in the Alboran Sea basin.

The MOW itself comprises two different water masses that shallow and mix as they pass through the Gibraltar gateway (Fig. 6): Mediterranean Intermediate Water (MIW) and Mediterranean Deep Water (MDW). MIW is generated in the Western Mediterranean basin, flows through the Alboran Sea, bifurcating around the Alboran Islands, where it attains a velocity of about 5–10 cm s^{-1} around the base of the Spanish continental margin. MDW is generally more sluggish in the Mediterranean, with a velocity of about 2 cm s^{-1} around the base of the African slope, before rising towards the sill of Gibraltar gateway (Lacombe & Tchernia 1972; Millot 1987; Parrilla & Kinder 1987; Perkins et al. 1990). MOW is constrained as it passes through the Gibraltar gateway at a water depth of as little as 200 m, and hence reaches a peak velocity measured at approximately 300 cm s^{-1}, decreasing to 180 cm s^{-1} immediately west of the gateway.

MOW then descends, spreads out and is deflected northwards by the Coriolis effect, forming a strong bottom current moving at a water depth of 300–1500 m along the Iberian slope (Fig. 6). There is an overall decrease in velocity, concomitant with a northward and westward broadening and deepening of the water mass, from around 30–40 cm s^{-1} in the vicinity of Faro Drift to 10–20 cm s^{-1} at the west end of the Gulf. In places it divides into several branches due to the influence of bottom topography, and is funneled along deep submarine canyons and valleys (Madelain 1970; Caralp 1988; Ochoa & Bray 1990; Nelson et al. 1993, 1999; Beringer & Price 1999). The interaction of MOW has led to

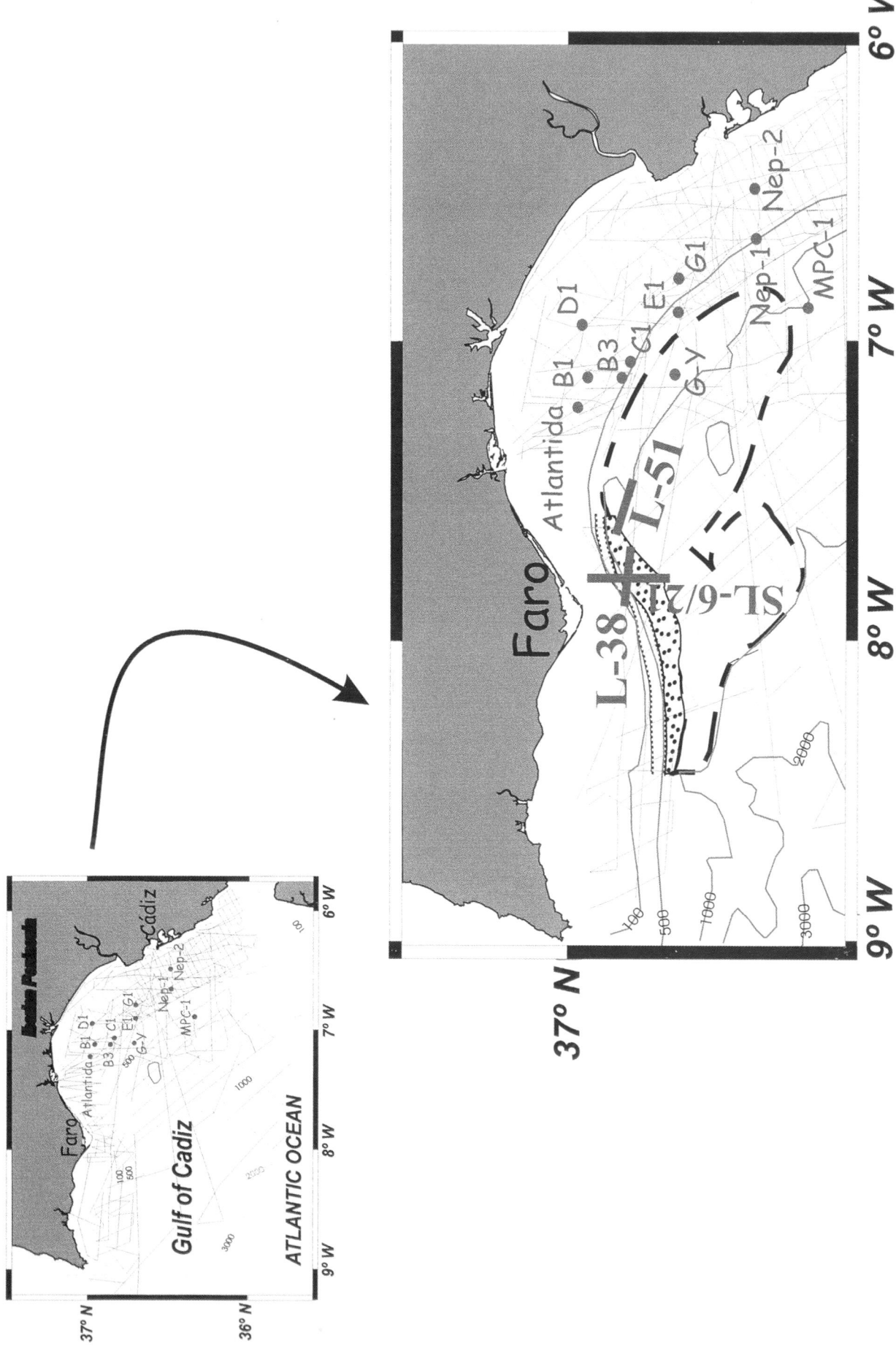

Fig. 2. Seismic and borehole database for the Gulf of Cadiz and study areas. The highlighted portions of L-38 and L-51 are illustrated in Figure 11.

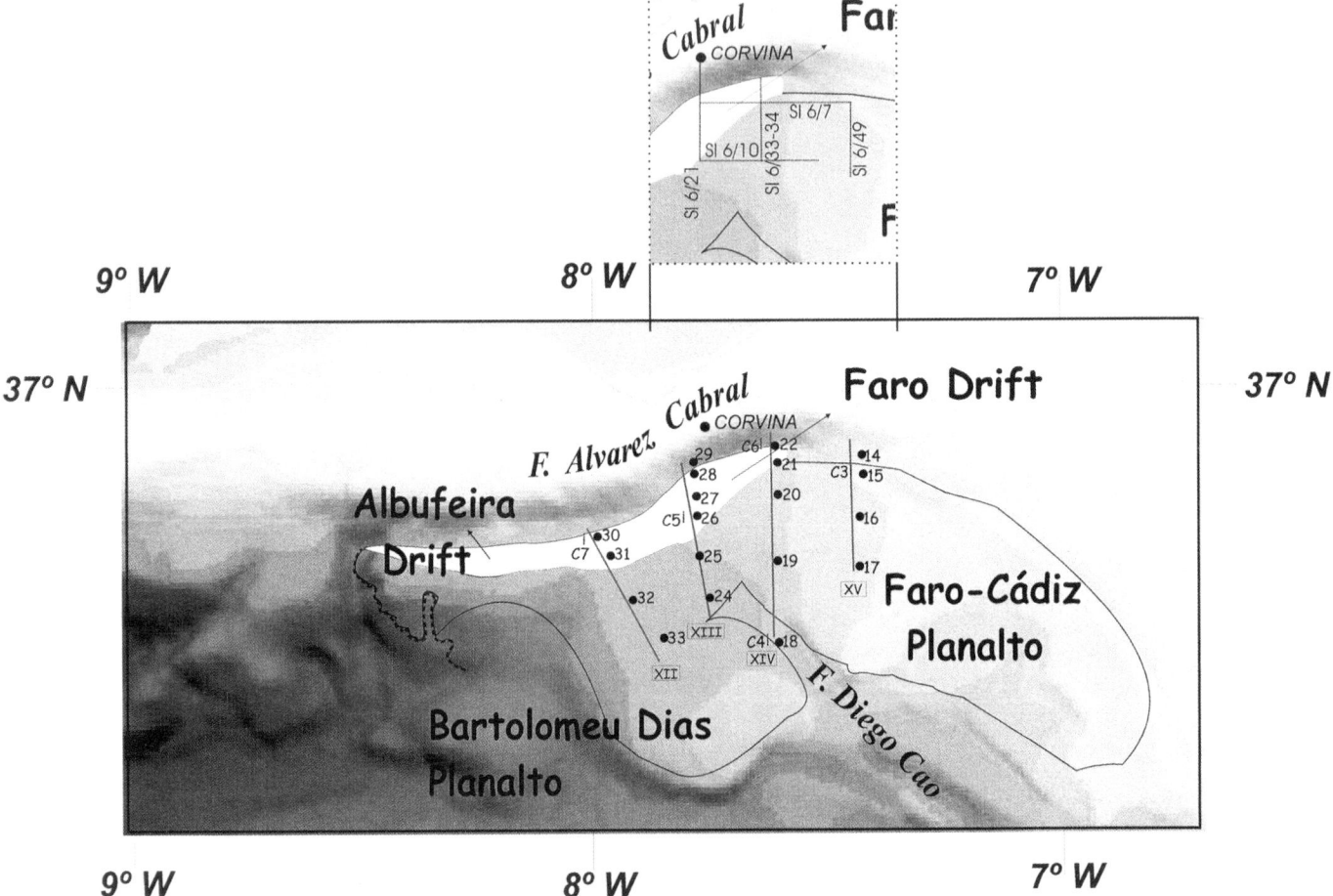

Fig. 3. Database used in Faugères et al. (1985a, b) and Stow et al. (1986). (**A**) seismic profiles and location of the Corvina borehole, used for stratigraphic calibration. (**B**) cores (28), camera stations (C4) and 3.5 kHz profiles (XIV).

construction of a series of contourite drifts (Gonthier et al. 1984; Nelson et al. 1993, 1999). Atlantic Surface Water flows towards the SE above MOW.

Palaeoceanography

Reconstruction of the evolving palaeocirculation pattern during the period of drift growth is still a matter of considerable discussion. Careful decoding of seismic and borehole data through the drifts may help resolve this problem in the future, but our current knowledge is largely theoretical. We can assume that the modern circulation pattern started to develop after the re-opening of the Gibraltar gateway at the end of the Messinian salinity crisis in the Mediterranean (Nelson et al. 1993). During the Lower Pliocene, an estuarine-type water-mass exchange though the Gibraltar gateway (inflow of Atlantic intermediate water and outflow of Mediterranean surface water), may have developed as a response much more humid conditions in the Mediterranean region than at present (Thunell et al. 1991). However, marked global cooling from around 2.4 Ma triggered a shift to more arid conditions in the Mediterranean region, resulting in an anti-estuarine water-mass exchange between the Mediterranean and the Atlantic similar to the present-day situation (Loubere 1987; Thunell et al. 1990, 1991). Since 2.4 Ma, the water-mass exchange is believed to have undergone significant variations in relation to climatic and sea-level changes, though the precise nature and timing of these changes remains unknown (Huang & Stanley 1972; Diester-Haas 1973; Grousset et al. 1988; Vergnaud-Grazzini et al. 1986; Caralp 1988, 1992; Nelson et al. 1993).

Bathymetry

The Gulf of Cadiz has a variable width shelf (15–40 km) with a shallow gradient (0.32–0.2°), and a shelf-break located at a water depth of 140–200 m (Figs 1 & 7). Beyond this, the physiography is quite complex, with some portions of the slope gradient being relatively steep (from 2–8°), but overall showing a low gradient outward bulge with a slope of around 1.5° in the upper part to 0.5° in the lower parts. This general profile is broken by broad platform areas, covered by sheeted and mounded drifts, downslope and alongslope directed channels, and local diapiric relief.

The Faro–Albufeira mounded drift extends in a generally E–W direction along the northern margin of the Gulf, and increases in crestal depth from about 500 m in the east to 700 m in the west. Actual relief above the adjacent sheeted drift platform region to the south is everywhere less than 50 m, though the depth to the northern channel varies up to 220 m. The Faro Canyon was originally believed to cut across the Faro Drift at its western extremity, thereby separating it from the Albufeira Drift. However, more detailed bathymetry now shows, at most, a slight bathymetric dip in the longitudinal profile of the combined drift system, such that the Faro Canyon appears to extend only from its southern flank. The true western extremity is marked by the well-developed Portimao Canyon. To the north, it is bounded by a slightly sinuous valley, the Alvarez Cabral channel, that deepens from zero relief at its upstream end to a maximum mid-drift relief of around 220 m. To the south and east (or upstream end) the drift merges with the Bartolomeu–Dias and Faro–Cadiz platform area that is deeply incised by the Diego Cao channel, and flanked to the south by the even deeper Guadilquivir channel.

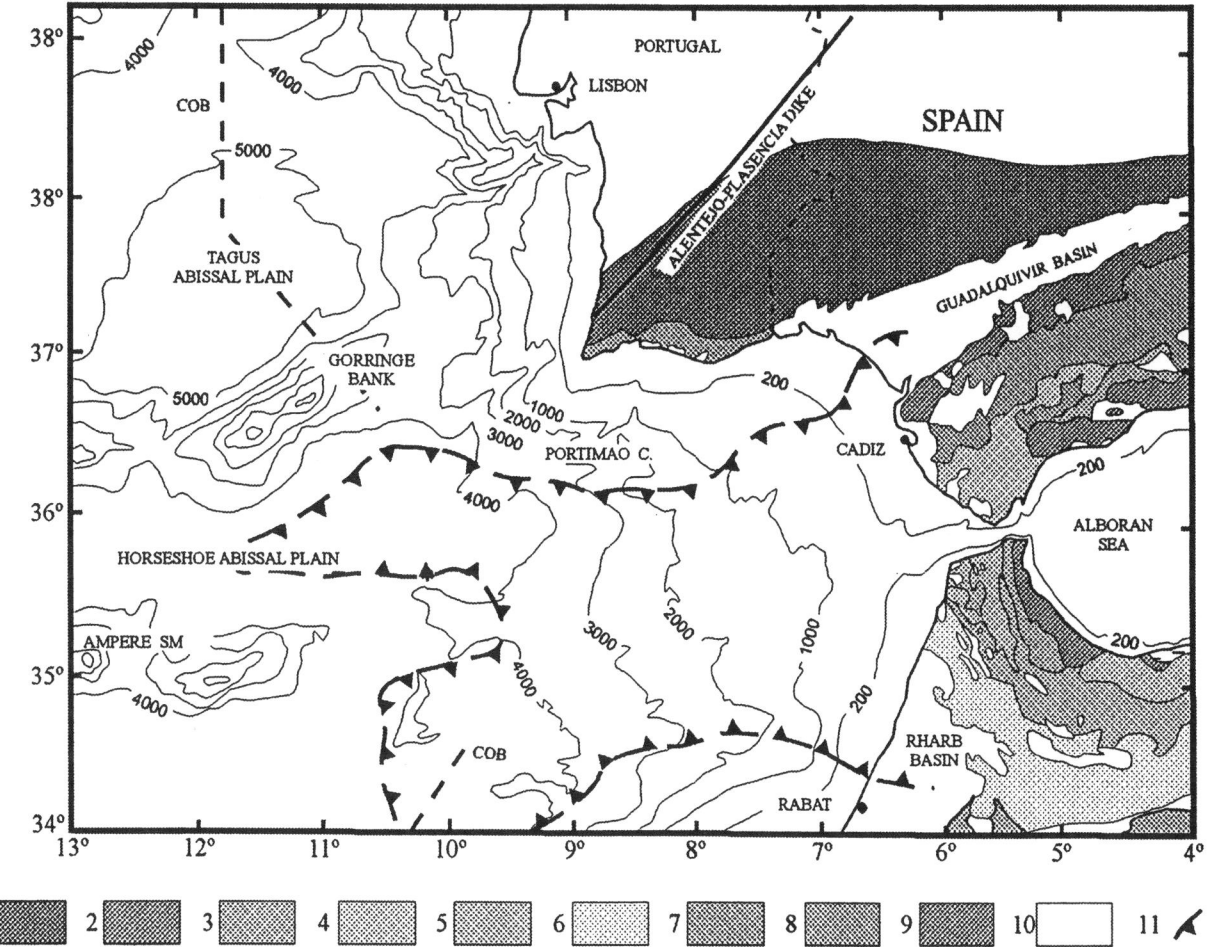

Fig. 4. Geological setting and simplified bathymetry of the Gulf of Cadiz and surrounding region. 1, Hercynian Massif; 2, Betic-Rif Internal Zones; 3, Complex dorsal; 4, Fysch units; 5, Meso- and Intra-Rifian units; 6, Pre-Rifian units; 7, Betic External Zone; 8, Mesozoic palemargins of Iberia and Africa; 9, Guadalquivir olistostrome; 10, Neogene sedimentary basins; 11, Olistostrome front. COB, continent-ocean boundary; bathymetry in metres. (After Maldonado et al. 1999).

Stratigraphic context

We can ascertain the principal stratigraphic events within the Gulf of Cadiz by careful correlation of seismic profiles with a number of oil company boreholes now drilled into shelf sediments north of our study area (Faugères et al. 1985a; Maldonado et al. 1999; Llave et al. 2000) (Figs 8–10). One example of this correlation is shown in Figure 10, in which seismic profile SL 6-21 has been correlated with the Corvina borehole (from Faugères et al. 1985a).

The top of the more highly tilted basement rocks (Jurassic–Cretaceous) occurs close to the Cretaceous–Tertiary boundary and is marked by a generally weak reflector (our *reflector K*), only notable because it marks a clear angular unconformity. Overlying Palaeogene sediments have also been affected by tectonic movements and are marked at their top by a second unconformity, which shows less discordance but a stronger and more distinct reflector (our *reflector O*). This is the top of the acoustic basement noted by Faugères et al. (1985a) and corresponds to an important late Oligocene unconformity, pre-Aquitanian and post-Stampian in age. A third important basin-wide discontinuity marks the onset of drift development in the region (our *reflector M*) and can be correlated with further tectonic adjustments in the Gulf which resulted in re-opening of the Gibraltar gateway. This is dated to within the latest Miocene Messinian stage.

The Faro–Albufeira drift complex, therefore, began to grow in the south during the late Messinian, with major drift accumulation and progradation to the north and west during the Plio-Quaternary. Sequence stratigraphic analysis through the drift complex allows recognition of six third order sequences above *reflector M*, and further subdivision of each of the two Quaternary sequences into at least four fourth order sequences. These are described below under *Seismic Characteristics* and shown in Figure 8.

Dating and correlation of the 19 cores recovered so far from the Faro–Albufeira drift complex is documented by Stow et al. (1986) amongst others, based on studies of pelagic foraminifers and oxygen isotopes (Fig. 9). The dating is not everywhere as precise as might be expected, probably as a result of extensive bottom current reworking, but the onset of the third cold phase (W3) of the Wurmian glacial stage (isotopic stage 2) is clearly present in most cores. This can be dated to 28 000 Ka. Several of the cores show significant hiatuses, and in one core from the northern valley there is a thin Holocene veneer over harder, more compacted sediment that could not be dated.

Seismic characteristics

The present seismic database for the area (Fig. 2) includes medium resolution (Sparker 3, 4 and 7.5 kJ, and Airgun), high resolution (Geopulse 300 and Uniboom), and very high resolution (3.5 kHz) seismic profiles. The maximum depth of penetration to *reflector O* beneath the drift complex is 1.5 s TWT (equivalent to about 1.1–1.4 km of section), whereas maximum thickness of the drift itself is about 0.8 s TWT to *reflector M* (equivalent to around 600–650 m).

The pre-drift unit (Unit I of Faugères et al. 1985a; Fig. 10) is

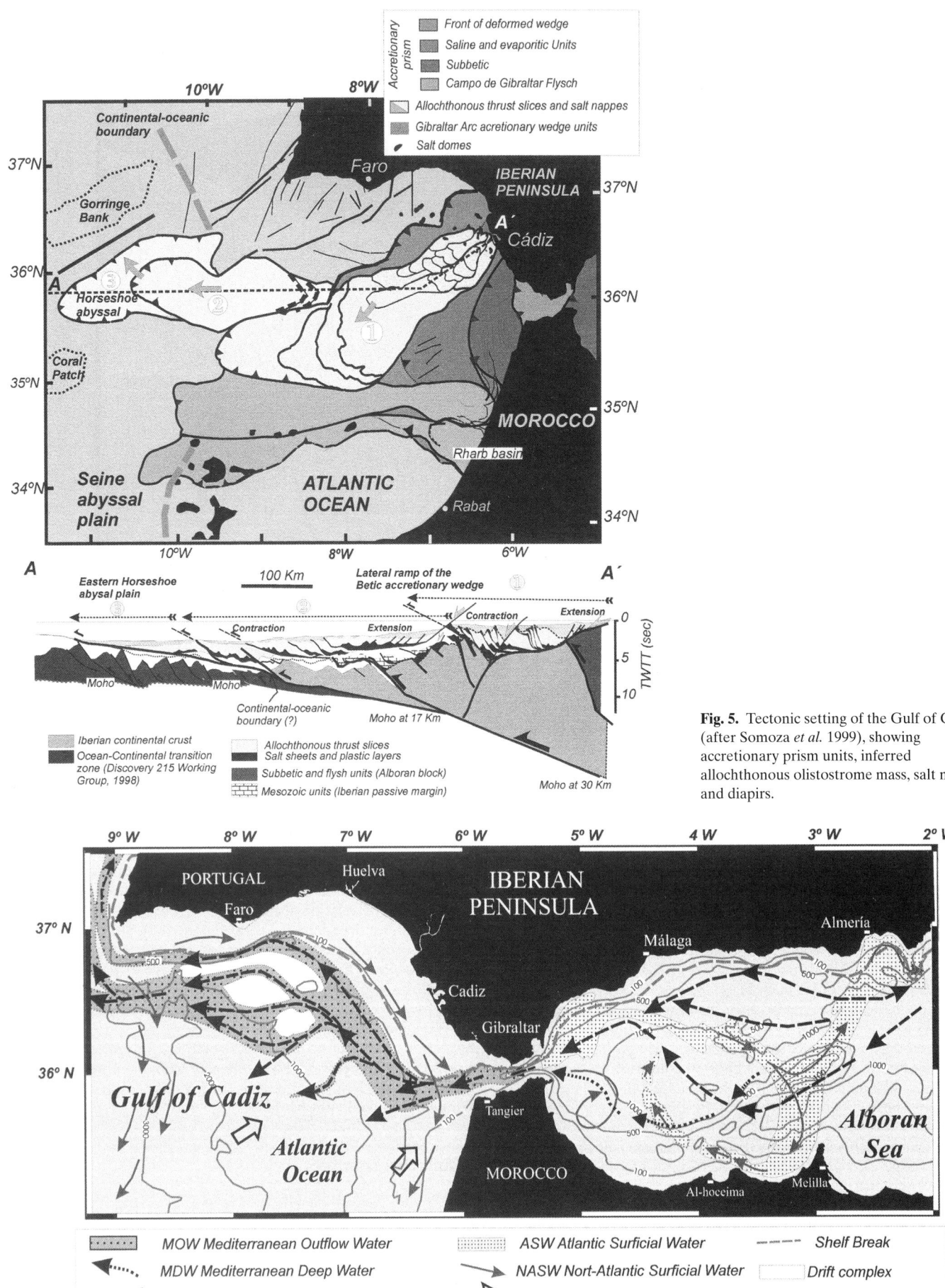

Fig. 5. Tectonic setting of the Gulf of Cadiz (after Somoza *et al.* 1999), showing accretionary prism units, inferred allochthonous olistostrome mass, salt nappes and diapirs.

Fig. 6. Oceanographic setting showing deep-water and surface-water circulation in the Alboran Sea, through the Gibraltar gateway and in the Gulf of Cadiz.

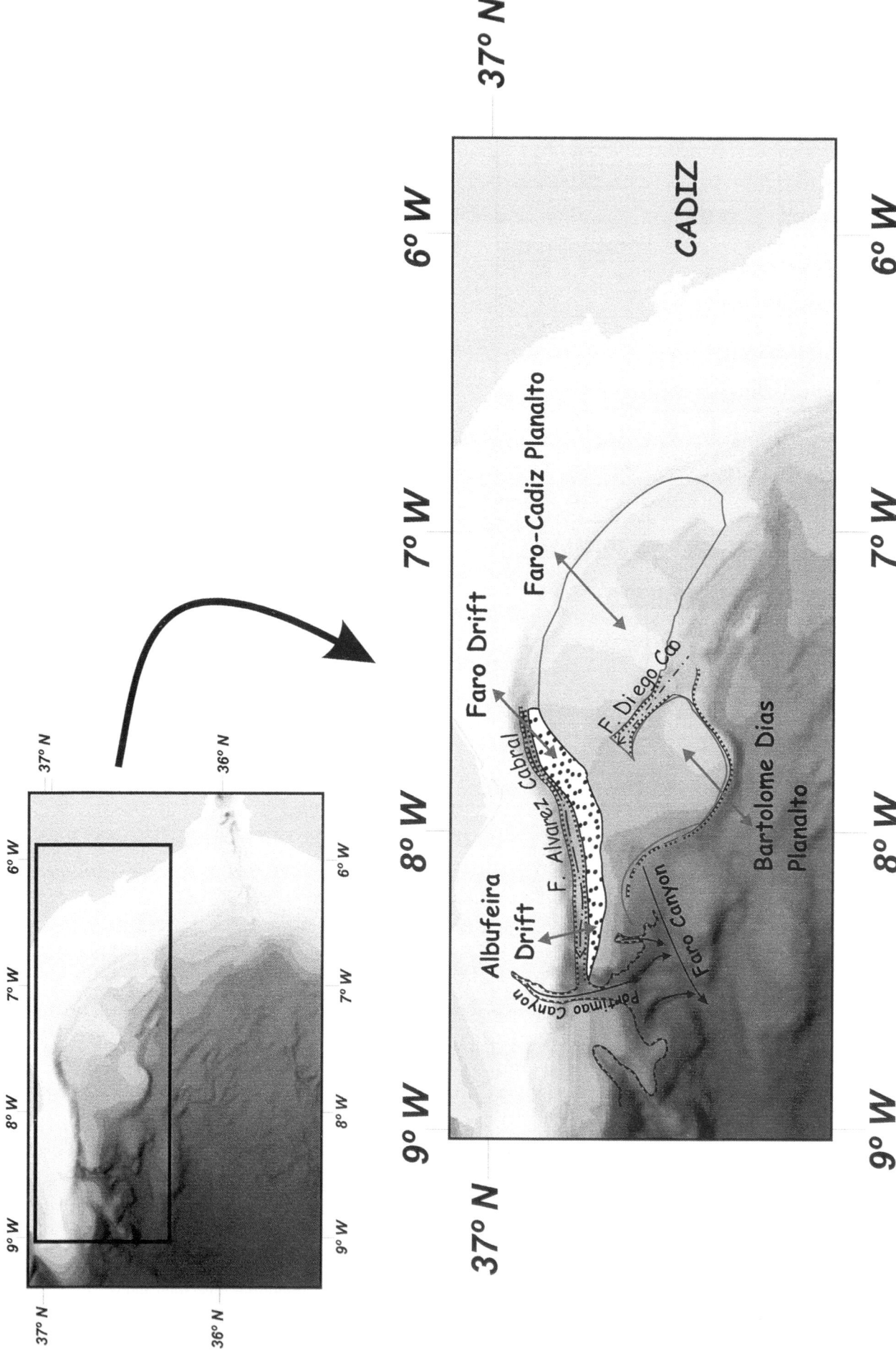

Fig. 7. Regional bathymetry for the Gulf of Cadiz and study area.

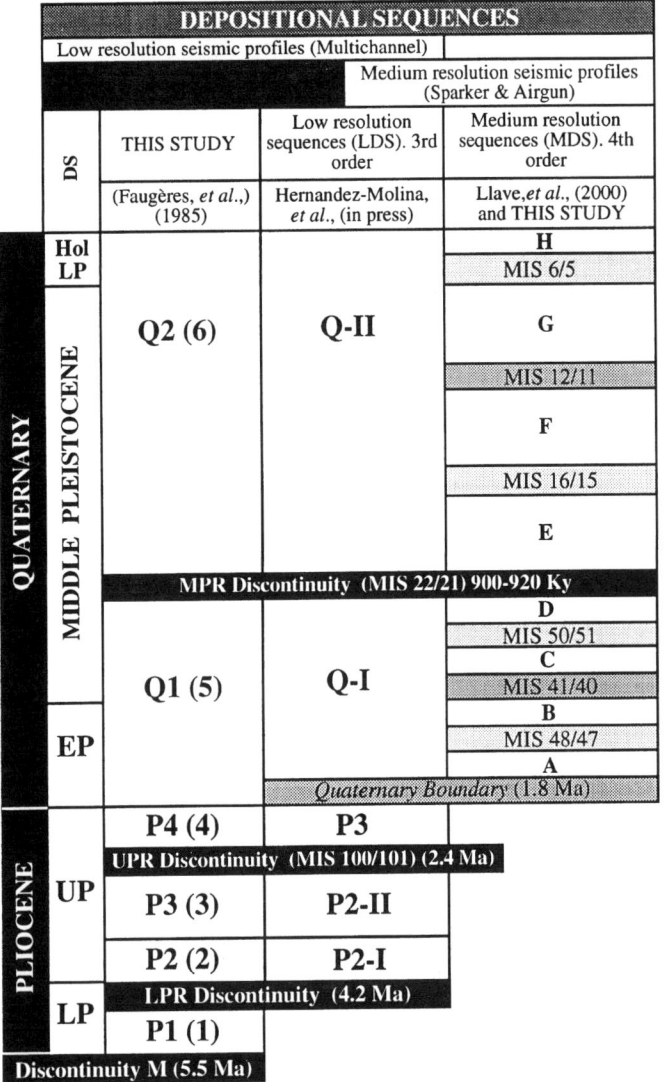

Fig. 8. Drift stratigraphy for the Plio–Quaternary section based on seismic analysis of low and medium resolution multichannel, sparker and airgun profiles. Nomenclature for this study (P1–4, Q1–2) compared with that of Faugères *et al.* (1999) and Llave *et al.* (2000). Principal and minor discontinuity surfaces shown together with approximate age and isotopic stage (MIS).

between 0.3 and 0.6 s thick, and shows generally low-amplitude, irregular and discontinuous reflectors. There is some onlap fill of small depressions and an overall smoothing of the underlying irregular topography. *Reflector O* at the top of this unit marks a regular, distinct, discontinuity surface; it is a high-amplitude reflector with a notable basinward dip.

The drift complex (Unit II of Faugères *et al.* 1985a; (Figs. 10 and 11) is generally around 0.5 s to a maximum of 0.8 s TWT in thickness (400–650 m). It shows more continuous reflectors, from low to high amplitude, aggradational to sigmoid progradational geometry, with onlap to downlap reflector terminations towards the north. As the drift has migrated against and along the northern slope there has been a progressive increase in relief of the Fosse Alvarez Cabral channel, coupled with its upslope migration and some erosion of the continental slope. In more detail, six distinct third order seismic sequences can be identified within the drift complex: P1–4 of inferred Pliocene age, and Q1–2 of Quaternary age. From base to top, these are described briefly below.

P1 Sequence: moderate amplitude, continuous reflectors, oblique progradation then pinching-out and becoming less distinct to the north.

P2 Sequence: moderate amplitude, continuous reflectors; more or less constant thickness, transgressive over P1 with apparent erosion by the northern channel; slight undulations reflect last phase of tectonic instability in the area.

P3 Sequence: moderate amplitude, continuous reflectors, showing oblique progradation to the north, but overall regressive with respect to P2; first sign of low relief drift-channel geometry.

P4 Sequence: moderate to high amplitude reflectors sandwich a zone of low amplitude; thickness relatively constant, but progradation more marked with accentuation of the mounded drift relief, a steeper northern slope and distinct channel form.

Q1 Sequence: alternation of zones of moderate-high and low amplitude reflectors, showing marked progradation and thickening from south to north; downlap reflector terminations into the northern channel, which shows evidence of non-deposition and erosion alternating with periods of accumulation.

Q2 Sequence: very similar to Q1 showing strong sigmoid progradational geometry coupled with aggradation, and an internal cyclicity of zones of high to low amplitude reflectors.

The cyclic alternation of low to high reflectivity couplets, noted in *P4, Q1* and *Q2*, can be considered as fourth order sequences (A to H in Figs 8 & 11). *Q1* and *Q2* are separated by a particularly high amplitude reflector, that is locally erosive, and that can be correlated with the Mid-Pleistocene Revolution at 900–920 ka (see later discussion). This is referred to as *reflector MPR*.

Diapirism has been active in the Gulf of Cadiz since at least the Miocene, and is still active today as witnessed by locally elevated seafloor and exposure of both mud and salt diapirs. These occur as small isolated more or less circular features, more irregular large diapiric zones and in linear trends. In all cases they have had a marked effect on bottom current flow and hence on sediment distribution observed in seismic records. The crestal regions above diapiric intrusions are subject to generally reduced sedimentation and erosion, locally with thicker drift accumulation and erosive moats. Where breakthrough and exposure of the soft diapiric material has occurred, then a crestal depression or linear channel may be sculpted.

The channels associated with the drift complex also show a range of seismic characteristics. The northern channel shows strong lateral migration coupled with basal aggradation. Channel sub-bottom reflectors are continuous and of moderately high amplitude, showing little difference in seismic facies from those of the adjacent drift, apart from a marked thinning of depositional sequences indicative of non-deposition and erosion. There is strong incision into the northern (slope) flank, as well as much evidence of slumping. The southern flank shows a prograding deposition and less slumping. The southern (Diego Cao) channel is more markedly erosive, though diminishing in relief as it debauches onto the Batolomeu Dias platform. It appears to have developed along a linear trend of diapirs. The Portimao canyon in the west is a typically incised, downslope directed, turbidity current channel system. The Faro canyon may have been similar in origin, but has subsequently been buried by drift infill. Smaller filled channels of unclear origin are observed in seismic records towards the western end of the Faro drift (Fig. 11).

3.5 kHz seismic profiles have been studied from parts of the Faro Drift (Faugères *et al.* 1985a, b; Fig. 3). These show a maximum penetration up to about 40 m and three distinct echo-facies: (a) a strong, thick bottom reflector and no clear sub-bottom reflectors; (b) a moderately strong bottom reflector over less distinct irregular sub-bottom reflectors; and (c) a weaker bottom reflector with multiple, sub-parallel sub-bottom reflectors. Broadly, these are distributed according to inferred bottom current intensity – type (a) in the channels, type (b) on the flanks

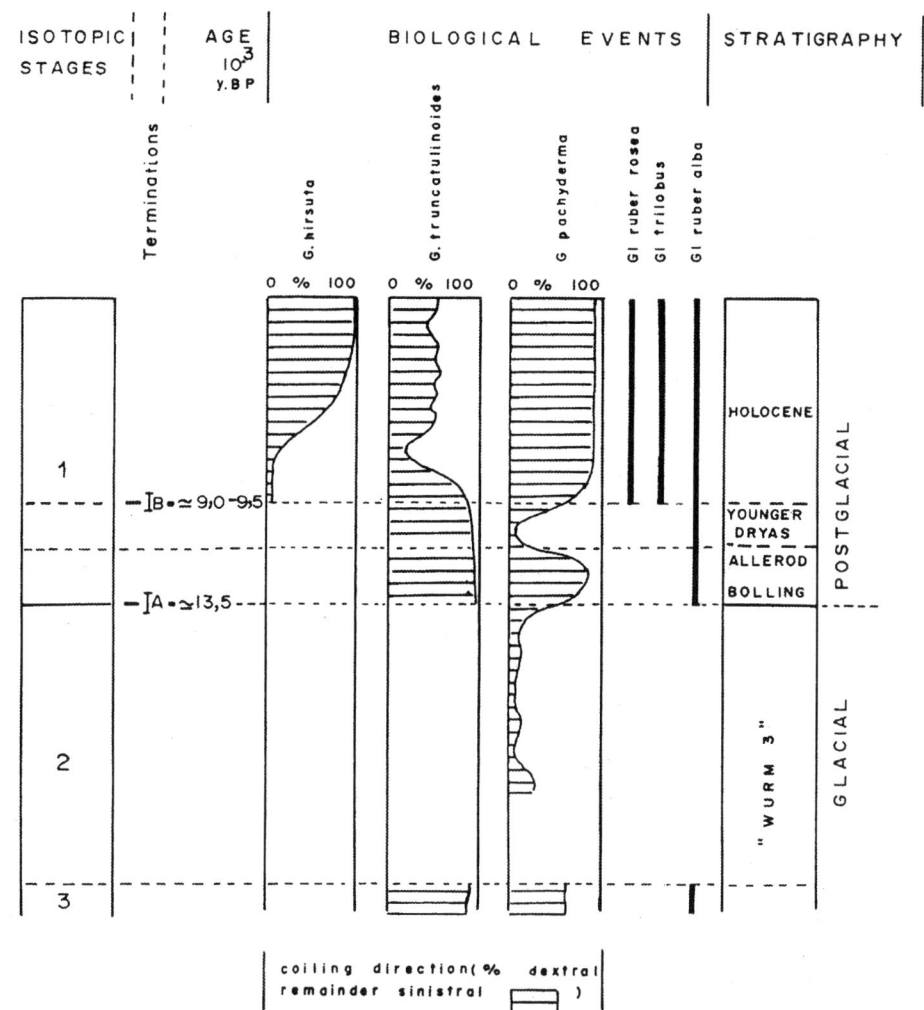

Fig. 9. Drift stratigraphy for the late Quaternary cored section based on biostratigraphic and oxygen isotope studies. (After Stow *et al.* 1986).

and type (c) over the main part of the drift complex. Where sub-bottom reflectors are apparent, these reveal the distribution of the most recent (late Quaternary–Holocene) sedimentation over this part of the drift complex, as well as its overall morphology.

Profile XV at the proximal end of the drift shows minimal development of the Alvarez Cabral channel at the foot of the continental slope and a low crestal portion of the drift that slopes gently down to the more extensive Faro platform. Diapiric highs occur in the central portion and at the southern edge of the platform, adjacent to a deeply incised Diego Cao channel, and in each case show marked reduction in sediment thickness over the diapir itself.

Profiles XIV and XIII, lying respectively 15 and 30 km to the west, reveal a well developed northern channel, a broad low-mounded drift relief and progressive decrease in depth and erosion of the southern Diego Cao channel. By *Profile XII*, the southern channel has completely disappeared and drift relief is relatively low. All three profiles show that the main site of deposition has been over the crest and northern flank of the drift, progradation has occurred towards the north, and minor erosion or non-deposition has taken place over the southern flank.

Sediment characteristics

Sea-floor photographs

A series of bottom photographs serve to show the principal variation in seafloor characteristics across the drift and in the adjacent valleys (Fig. 12). The northern valley shows a hard heterogeneous substrate with distinct current lineation of gravel material at its downstream (western) end. At its upstream end, the substrate is more muddy and bioturbated, with scattered shell fragments and other coarse debris, as well as local colonies of crinoids and crabs. Drift top sites are generally muddy and bioturbated, either with no evidence for current activity or showing weak current lineation and mud-draped ghosts of ripples. The southern valley shows much evidence of recent current activity, with a variety of asymmetric ripple and megaripple forms on a bigenic sandy substrate. In every case the orientation of the bedforms indicates bottom current flow consistent with that due to the MOW.

It is important to note that in none of the cores taken at sites showing photographic evidence of current activity were any traces of current structures preserved. Continuous and active bioturbation had served to completely destroy such primary sedimentary structures.

Sediment facies

The sediments of the Faro–Albufeira drift complex are nearly all interpreted as contourites. The principal facies include: very fine-grained muddy contourites, mottled silty-muddy contourites, and silt to fine-sand contourites (Fig. 13). These are arranged in somewhat irregular coarsening-up to fining-up cyclic sequences, mostly from 0.25 to 1.5 m in thickness. The chief distinguishing features of these contourites are:

- Rarely preserved primary sedimentary strutures, including poorly developed lamination and sharp, erosive contacts;

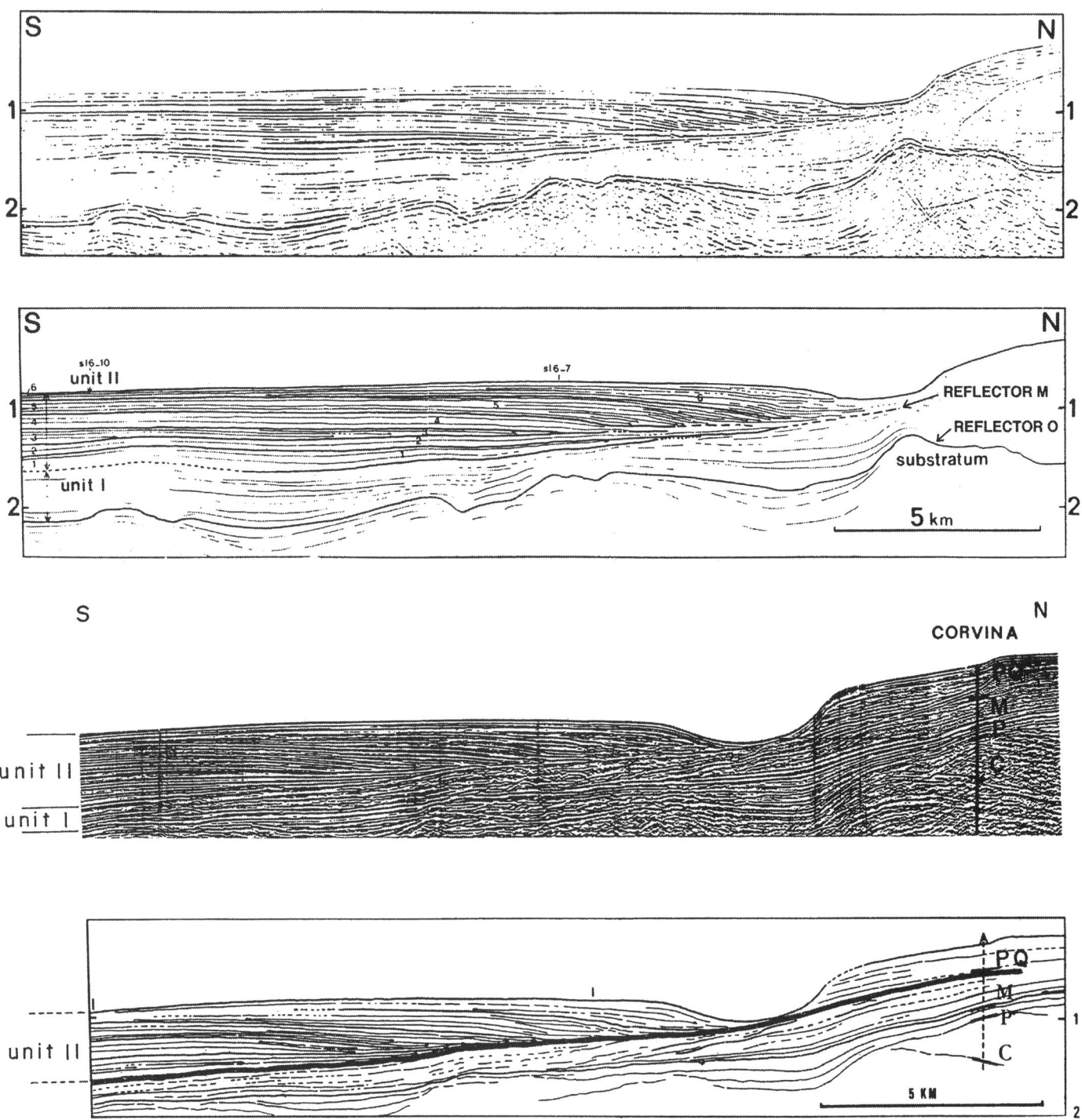

Fig. 10. Seismic reflection (multichannel) profiles (N-S orientation) from proximal and medial portions of the Faro–Albufeira drift (from Faugères *et al.* 1985*a*, Stow *et al.* 1986). (**A**) part of Line 6–34 with interpretation; (**B**) part of line 6–21, with interpretation and borehole correlation. See Figure 3 for profile location.

- Intense and pervasive bioturbation that has been continuous with deposition, and that shows ichnofacies variation associated with grain size changes in facies;
- Generally fine grain size (mean mostly < 63 µm) and poor to moderate sorting, with sandy silts showing slightly better sorting and grain-size distributions indicative of current influence;
- Mixed biogenic-terrigenous composition in variable proportions, with 20–50% biogenic carbonates (planktonic and benthonic foraminifers, bivalves, ostracods) and 50–80% compositionally immature siliciclastic material; fragmentation of biogenics and iron-coating of grains are common.

Turbidites were recovered from slope core 29 and from the mouth of the northern valley, where they constitute 20% and 40% respectively of the succession. On the drift itself, a single thin turbidite is present in each of cores 31 and 32, representing only 2% of the section. All turbidites are very distinctive in terms of coarser grain size at their base, normal grading, primary structures, and a different more terrigenous composition.

The distribution of these principal facies, together with grain size variation, is shown in Figure 14 for all four drift transects currently available. The vertical succession of facies in each of the cores is characterized by a repetition of cyclic contourite

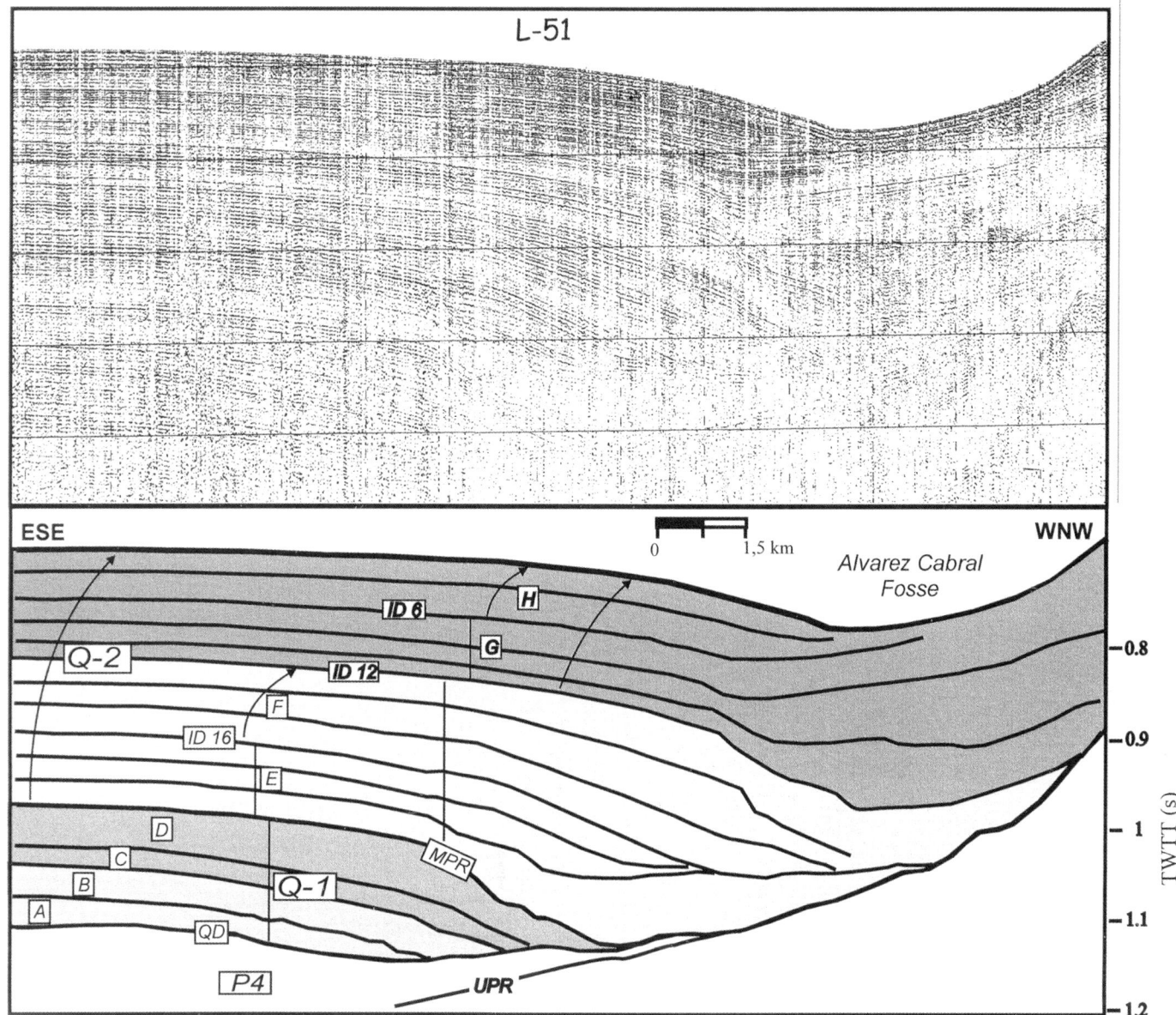

Fig. 11. Seismic reflection (sparker) profiles (approximate E-W orientation) from proximal and medial portions of the Faro–Albufeira drift (from Llave *et al.* 2000). (**A**) Part of Line L-51 with interpretation; (**B**) Part of Line L-38 with interpretation. See Figure 2 for profile location.

sequences, which appear somewhat irregularly distributed due to variation in thickness and completeness of the sequences. However, three cycles in the upper 3 m of section are apparent in many of the cores, and the sandy silt peaks of these cycles are referred to as *Peaks I, II and III*. The topmost *Peak III* occurs at the present sediment surface, so that particular 'cycle' comprises only the lower coarsening-up sequence. These peaks are correlateable over many of the cores and can be dated approximately, by biostratigraphic and isotopic means, as 14 000–15 000, 9 000–10 000 and 0–2 000 a^{-1} BP. In most cores, there is a further less well developed grain size peak at around 28 000 a^{-1} BP, whereas in the valley cores to the south there are several earlier undated peaks that appear to correspond only with muddy contourites on axis of the drift.

Sedimentation rates and hiatuses

The presence of hiatuses is noted in about 40% of the cores recovered from the drift complex, including both drift margin/channel sites as well as drift top sites. Where it has been possible to date these cores, the average sedimentation rates for the past 28 000 years are extremely low, i.e. generally < 1.5 cm ka^{-1} and everywhere < 4 cm ka^{-1}. For the other sites, there is a wide range of rates from about 3 to 15 cm ka^{-1} (Table 2).

Based on seismic records and the seismic depositional units identified (Fig. 8), we can infer accumulation rates for the past 5.5 Ma for both the mounded drift and sheeted drift sections (Table 3). Without applying any correction for increased sediment compaction with depth, mean accumulation rates for the mounded drift are seen to have increased markedly through time, from 35 m Ma^{-1} in the late Pliocene to an average of 103.5 during the early Quaternary, and to 295 m Ma^{-1} in the latest Quaternary. This might be expected from the rapid progradation that has occurred, with none or very little accumulation evident prior to 2.4 Ma.

Approximately the reverse seems to be true for the sheeted platform drift. Here the rates during the whole of the Pliocene ranged from 82.5 to 188.6 m Ma^{-1} (an average of nearly 120 m Ma^{-1}), decreasing to 82.5 and 55 m Ma^{-1} through the early and late Quaternary respectively.

The most significant discontinuities marked by widespread

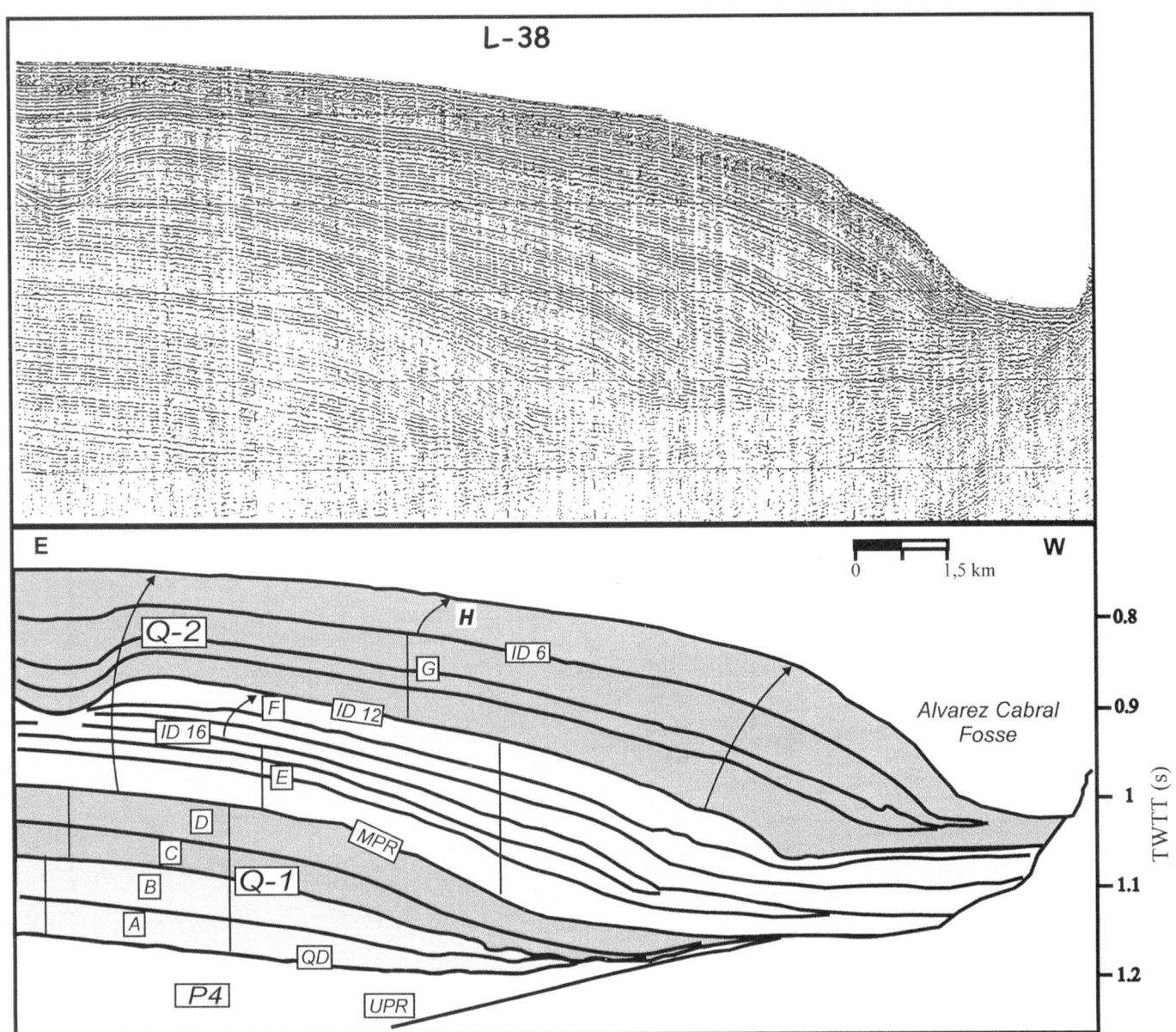

Fig. 11(B).

Table 2. *Sedimentation rates for the last 28 000 a in cores from the Faro–Albufeira drift complex. The presence of hiatuses is also noted. For core locations see Figure 3*

Cores	Sedimentation rates m Ma^{-1} (cm ka^{-1})	Hiatus present
KC8214	120 (12)	
KC8215	< 10 (< 1)	yes
KC8216	40 (4)	
KC8217	>130 (> 13)	
KC8218	< 10 (< 1)	yes
KC8219	35 (3.5)	
KC8220	60 (6)	
KC8221	120 (12)	
KC8222	< 10 (< 1)	yes
KC8224	40 (4)	yes
KC8225	< 15 (< 1.5)	yes
KC8226	110 (11)	
KC8227	145 (14.5)	
KC8230	?	yes
KC8231	35 (3.5)	yes
KC8232	30 (3)	
KC8233	60 (6)	

Table 3. *Accumulation rates for the last 5.5 Ma based on seismic records from the Faro–Albufeira mounded drift and sheeted drift systems. A seismic velocity of 1500 m s^{-1} has been taken to convert seismic Two-Way-Travel-Time (TWT) to thickness, and no adjustment has been made for compaction*

Mounded drift Seismic units	Age at base Ma	Thickness s (TWT)/m	Accumulation rate m Ma^{-1} (cm ka^{-1})
Q2 H-G	0.42	0.165/124	295 (29.5)
F-E	0.90	0.176/132	275 (27.5)
Q1 D-C	1.30	0.066/49.5	124 (12.4)
B-A	1.70	0.044/33	83 (8.3)
P4	2.40	0.033/25	35 (3.5)
P1–3	5.50	Thin to absent	Zero to very low
Sheeted drift Seismic units			
Q2 H-E	0.90	0.066/49.5	55 (5.5)
Q1 D-A	1.70	0.088/66	82.5 (8.25)
P4	2.40	0.176/132	188.6 (18.86)
P3	3.00	0.110/82.5	137.5 (13.75)
P2	4.20	0.088/66	82.5 (8.25)
P1	5.50	0.220/165	127 (12.7)

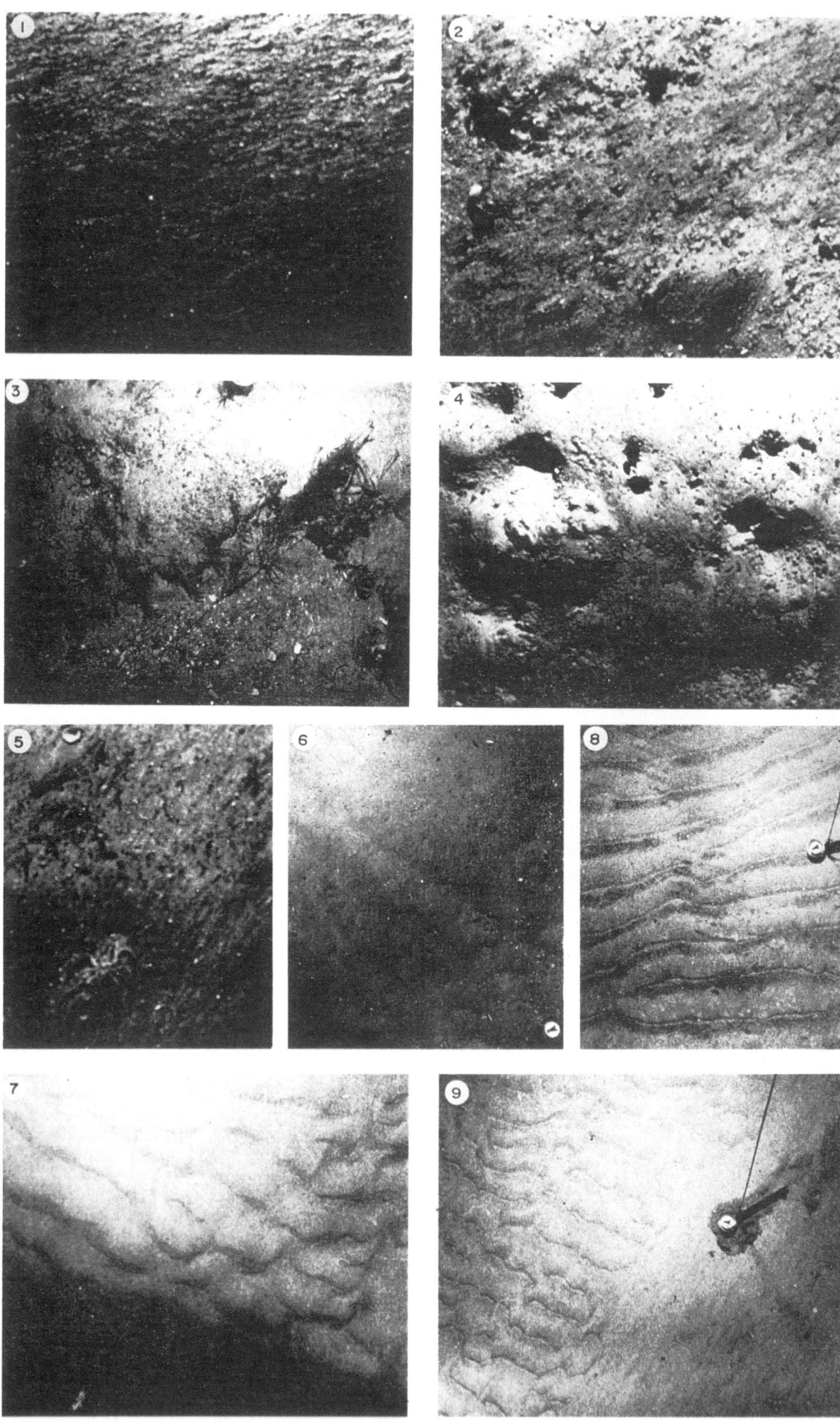

Fig. 12. Bottom photographs from Camera Stations C3–C7 (see Fig. 3 for location). All photos after Faugères et al. (1985b). 1, Northern valley (C7). Heterogeneous seafloor with current lineation shown by alignment of gravel material. Width of view approx. 50 cm. 2, Northern valley (C6). Heterogeneous seafloor with current-aligned shell fragments and other coarse debris over mud. Bioturbational mounds also show current elongation. Width of view approx. 50 cm. 3, Northern valley (C6). Heterogeneous and irregular seafloor with patchy distribution of shell fragments and other coarse debris over mud. Small colony of crinoids and crabs. Width of view approx. 50 cm. 4, Drift top (C5). Muddy seafloor with bioturbational mounds, probably formed from excavating crabs; no evidence of current activity. Width of view approx. 30 cm. 5, Drift top (C3). Muddy seafloor with scattered shell fragments, small bioturbational mounds and weak current lineation, flow direction ENE–WSW. Width of view approx. 40 cm. 6, Drift top (C5). Thin veneer of bioturbated mud with rare scattered shell fragments overlying ghosts of former (?sand) ripples. Flow direction SSE–NNW. Width of view approx. 90 cm. 7, Southern valley (C4). Sandy seafloor showing part of megaripple bedform overrun by a system of smaller-scale linguoid to sinuous-crested ripples. Width of view approx. 100 cm. 8, Southern valley (C4). Sandy seafloor with system of straight to curved-crested asymmetric ripples, flow E–W. Width of view approx. 110 cm. 9, Southern valley (C4). Sandy seafloor with current lineation apparently overrunning a system of linguoid ripples. Both systems show flow direction from SE–NW. Width of view approx. 150 cm.

erosion and non-deposition are inferred at approximately 0.9 Ma, 2.4 Ma and 4.2 Ma. Less important discontinuities are most likely represented by less widespread hiatuses inferred between the other third and fourth order depositional sequences (Fig. 8). Condensed sequences are everywhere apparent overlying diapiric intrusive zones, with much expanded deposition in the intervening regions. Erosion is associated with both diapiric crests and channels.

Discussion

The Faro–Albufeira drift complex is a clear example of an elongate low mounded drift coupled with a dissected sheeted drift, that can be closely linked to a strong bottom current system developed from the MOW. Careful analysis of diverse datasets from the area has permitted, we believe, a robust interpretation of drift growth since the late Miocene, as well as of facies development and distribution over the past 28 000 years. We briefly review these aspects below, and highlight further questions that would best be answered by a programme of shallow drilling in the Gulf of Cadiz closely allied with detailed sidescan mapping and oceanographic measurements.

Drift origin and development

Several factors have led to drift development in the Gulf of Cadiz: (1) the onset of MOW following re-opening of the Gibraltar gateway at the end of the Miocene; (2) the existing morphology in the northern Gulf, with a broad platform abutting a relatively steep slope, serving both to focus bottom currents and trap their sediment load; (3) the Coriolis Force that is responsible for diverting MOW northwards and confining the water mass against the continental slope; (4) an abundant sediment supply from erosion within the Gibraltar gateway, continental runoff and downslope resedimentation from the Iberian peninsula, and primary biogenic productivity yielding mainly microfossil remains; (5) a slow decrease in bottom current velocity away from the Gibraltar gateway as a result of flow spreading and deepening, that has allowed for the deposition of sediment load.

The initial period of drift growth (seismic sequences *P1–3*) was as a broad sheeted drift that spread out across the Faro slope platform and prograded towards the north and northwest as it aggraded in thickness. As the drift relief built up (seismic sequence *P4*) so a broad depression or channel form was created against the slope in the north. This served to refocus the bottom flow leading to erosion or non-deposition in the channel axis and on the slope margin, and differential sedimentation over the evolving drift. Continued progradation coupled with marked aggradation of the mounded Faro–Albufeira drift characterized the Quaternary period (seismic sequences Q1–2). This was associated with minor progradation towards the west and elongation of the mounded form. The northern channel migrated upslope, continued to deepen and to constrain the bottom flow, hence augmenting current velocity locally. Erosion of the northern slope margin and deposition on the southern drift margin occurred. A more detailed analysis of drift evolution in relation to bottom current flow is presented by Faugères *et al.* (1985*a*).

Whereas we can visualize in this way the progressive development of the northern channel (Alvarez Cabral) in response to drift progradation, it is uncertain to what extent and at what periods it may also have been excavated by turbidity currents. Certainly turbidites are present in its more recent fill. We are less clear about the nature and origin of the southern channel (Diego Cao) which deeply dissects the sheeted drifts and then shallows and debauches onto the Bartolomeu Dias platform. Certainly this presently channels a very active stream of MOW and associated sediment load, as evidenced by channel floor bedforms and a

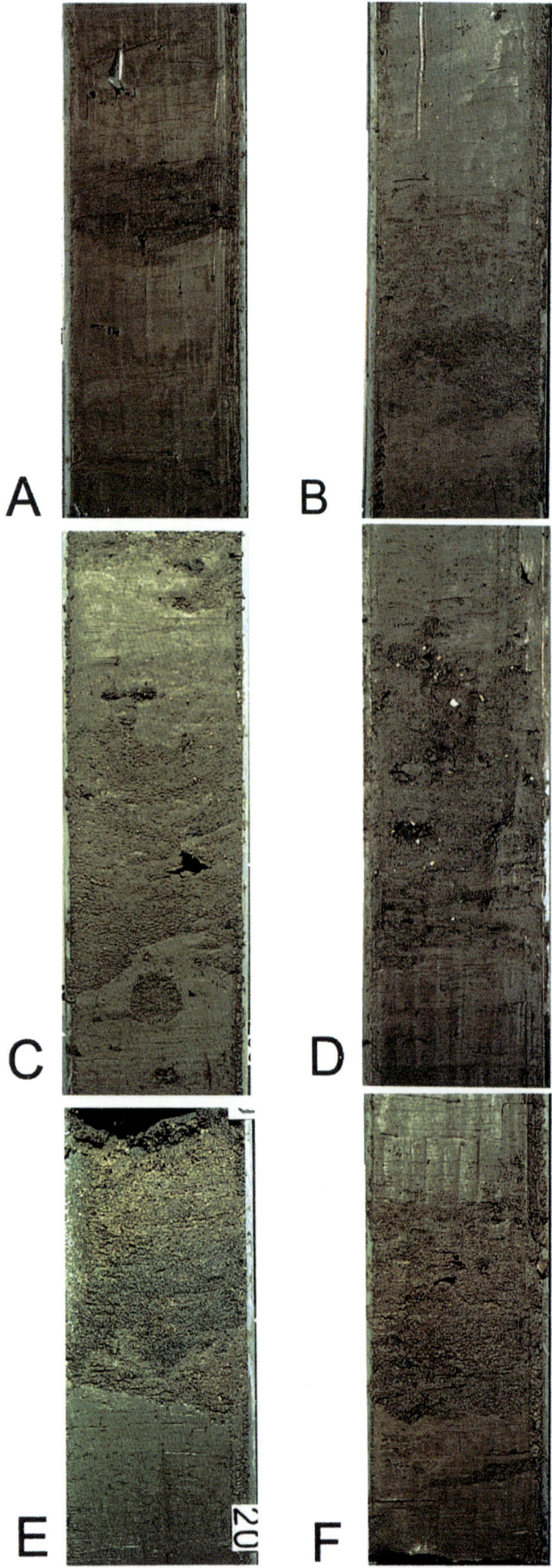

Fig. 13. Photographs of contourite facies from Faro–Albufeira drift cores. These illustrate a range of facies from muddy contourites (**A**), through various silt-mottled intervals (**B, C, D, F**) within a silty muddy contourite facies, to sandy contourite (**E**). Core width 10 cm; core number as shown to right of centimetre scale. See Figure 3 for location.

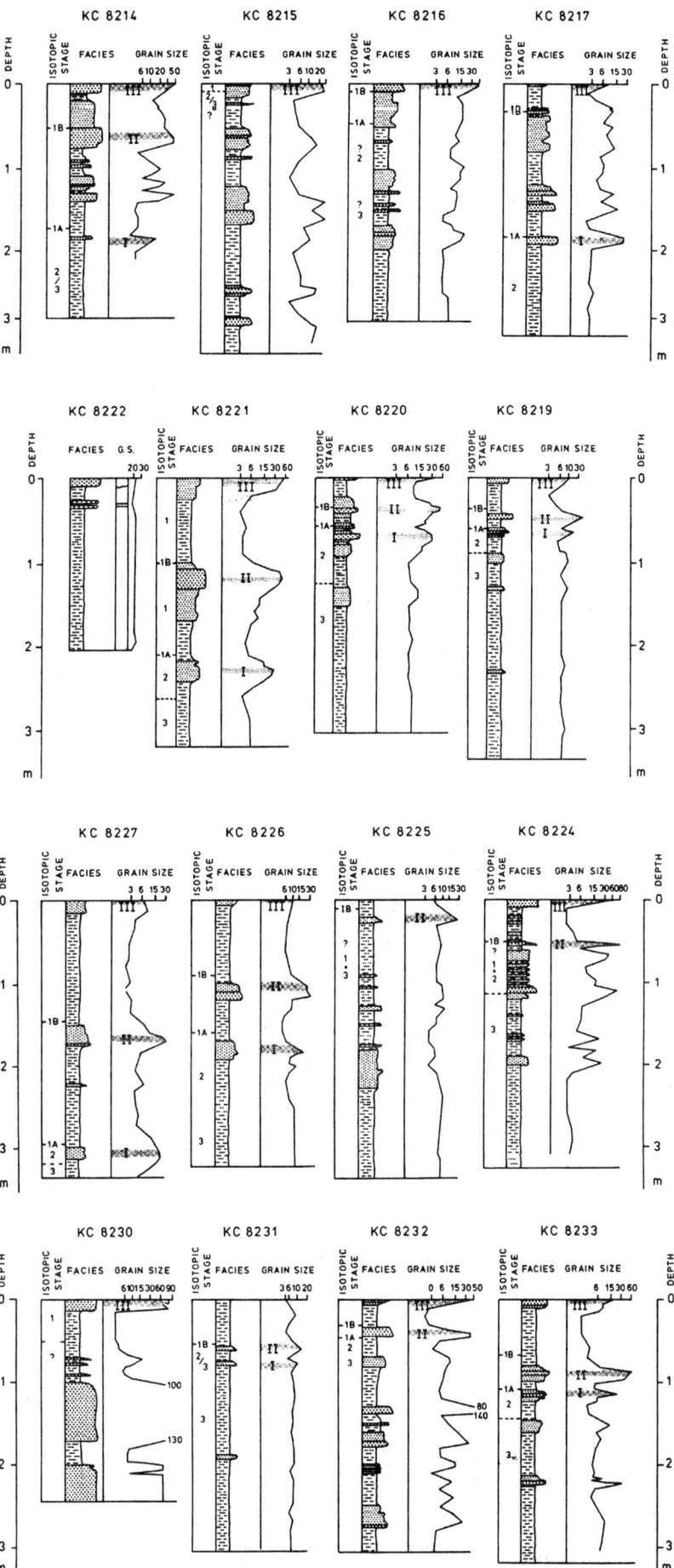

Fig. 14. Facies distribution across four drift transects (after Stow *et al.* 1986). Summary data on lithology, grain size and isotope stratigraphy shown for top 3.5 m only of each core. In most cases the remaining cored section is dominantly of muddy contourite facies.

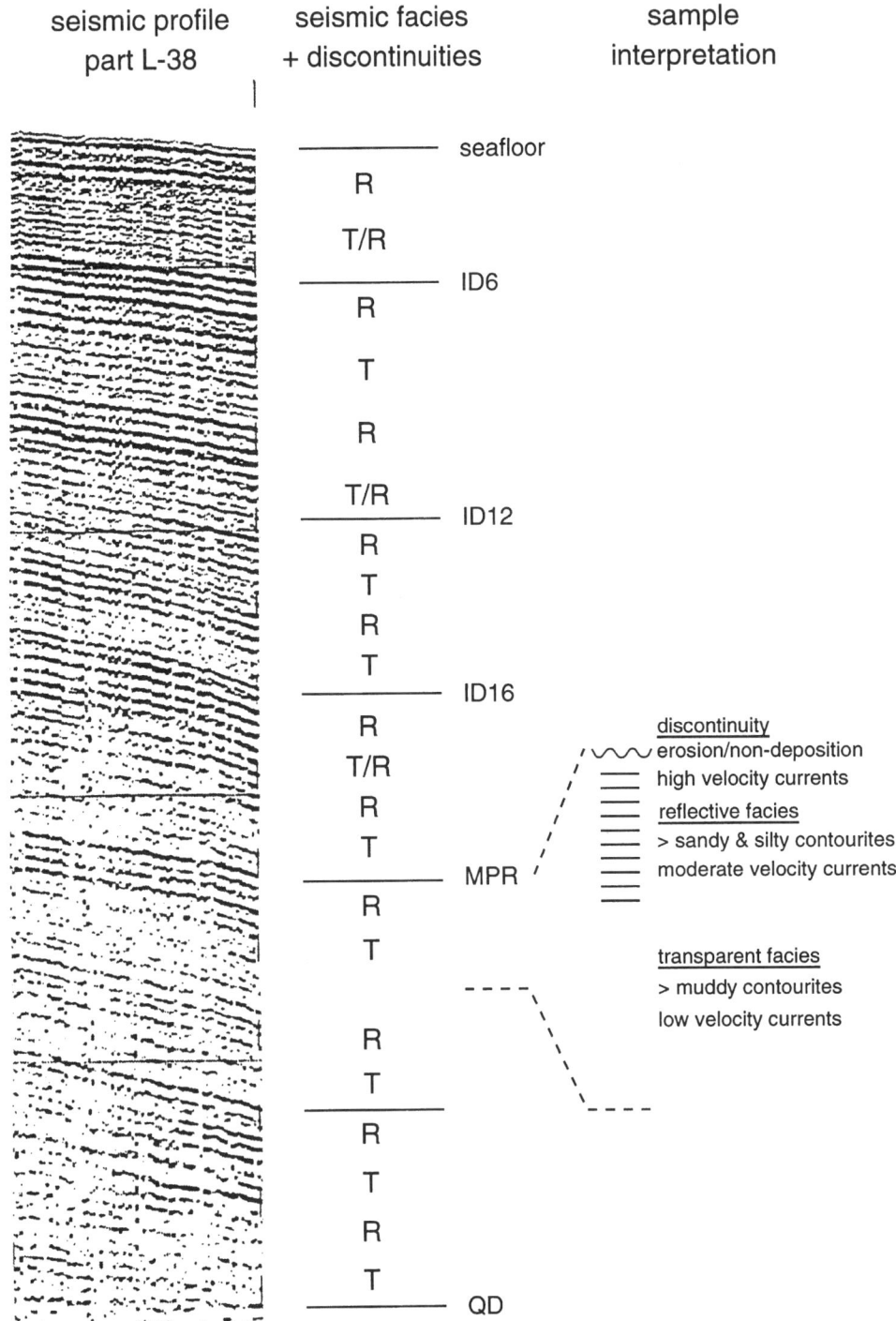

Fig. 15. Towards a model for interpreting seismic facies cyclicity in contourite successions.

condensed seismic sequence. It appears likely that it was excavated into the soft core of a linear diapiric trend by bottom current activity.

The Portimao Canyon and associated tributary complex that marks the western extremity of the drift complex is a typical and apparently active downslope turbidity current channel system, although oceanographic data show that it also serves to channelize a portion of the MOW and direct this downslope. The Faro Canyon was originally of turbidity current origin, but has now been cut off from its slope source and partially infilled by the westward prograding Faro drift. Origin of the two smaller, completely infilled channels that cut across the nose of the drift is still unclear.

Seismic facies cyclicity and sea-level change

The moderately high resolution sparker profiles through the latest Pliocene and Quaternary drift section show a very distinct alternation of a more transparent facies (T) with higher amplitude continuous reflector seismic facies (R). These are especially well defined through the mounded portion of the drift complex, where they can be recognized within the fourth order seismic sequences A–H.

In the absence of direct borehole evidence, we provisionally interpret seismic facies R as due to higher silt/sand content contourites, occurring in a series of discrete beds or zones within a more muddy contourite facies, and also with more hiatuses and

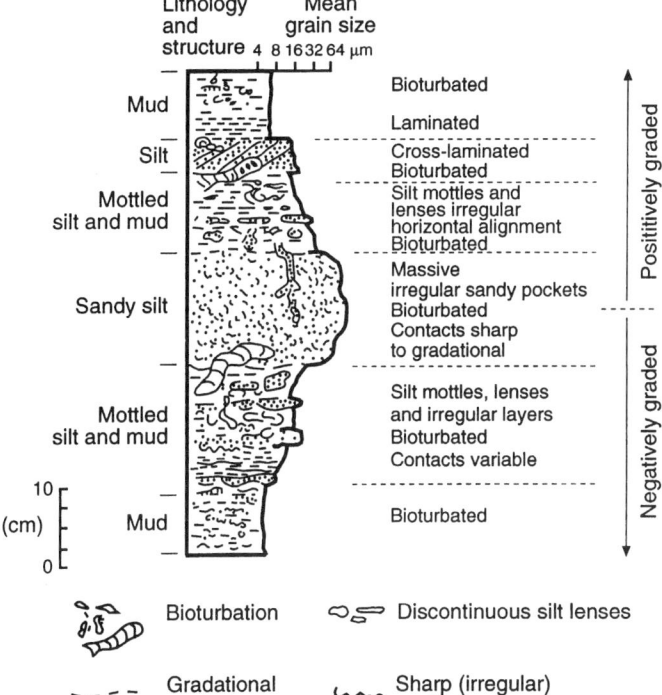

Fig 16. Standard sediment facies model for contourites, originally derived following work on Faro drift cores.

condensed sedimentation sections. This all reflects increased bottom current intensity. The more widespread discontinuities noted (e.g. UPR, MPR) are interpreted as the result of still higher bottom current velocities causing regional deposition and erosion. Seismic facies T, by contrast, is due to low silt/sand content within a more continuous and homogeneous muddy contourite section, reflecting decreased bottom current intensity.

There are about 11 such TR cycles through the Quaternary section. This slight uncertainty in number derives from the fact that the bubble pulse makes it difficult to resolve of the uppermost seismic unit H, whereas decreasing resolution at depth makes the TR cycles in units A–D less distinct. However, we interpret the observed seismic cyclicity as the result of a high amplitude climatic/eustatic cyclicity that has directly affected bottom current intensity over the drift complex. The 11 cycles would therefore correspond, at least approximately, to the principal, climatically-induced, isotopic stages identified through the Quaternary (e.g. Berger *et al.* 1994, Shackleton *et al.* 1990). We illustrate this relationship in Figure 15.

Precise dating of our TR cycles by correlation with shelf boreholes is not sufficiently accurate to allow their positive correlation with particular highstand or lowstand system tracts. Our best estimates show a temporal cyclicity of close to 100 000 years for the late Quaternary–Holocene (Q2), that might suggest eccentricity cycles, and closer to 200 000 years for the early Quaternary (Q1). There is also presently conflicting opinion on the effect that sea-level change would have on bottom current flow through the Gibraltar gateway, and insufficient evidence at this stage to resolve the issue.

Contourite facies cyclicity and bottom-current velocity

The cyclic variation in lithology and texture of contourites from the drift cores have been interpreted as largely due to variation in bottom current intensity, rather than changes in sediment supply, primary productivity, dissolution or diagenesis (Gonthier *et al.* 1984; Stow *et al.* 1986). The main arguments supporting this conclusion are:

- Current induced sedimentary structures are more evident in sandy rather than muddy contourites;
- Sediment hiatuses and/or low rates of sedimentation in adjacent valleys are coincident with the sandy peaks;
- There is a concomitant increase in both biogenic and terrigenous sand fractions in the sandy peaks, and of both benthonic and planktonic foraminifers in the biogenic fraction;
- There is a significant presence of reworked and fragmented benthonic foraminifers and shelly debris, as well as pelagic material;
- There is a relative compositional homogeneity at all sites over the drift, with exotic mineral assemblages only present in the clearly recognizable turbidites;
- The longitudinal down-drift trends in grain size as well as regional differences, such as the coarser grain sizes in the southern channel, cannot be readily explained by localized turbidity current input.

We do, however, recognize certain features that favour at least some influence of increased primary productivity contributing to the sand fraction of peak II, and increased terrigenous supply contributing to peak I.

This type of facies and textural cyclicity is now recognised as a characteristic feature of contourite successions worldwide (Fig. 16). The temporal scale for the Faro drift cycles is rather irregular, based on the relatively short lengths of cored section, and typically ranges between 5 000 and 15 000 years. We do not see any clear correlation of these with known climatic/oceanographic changes documented elsewhere, although Stow *et al.* (1986) do note the possible significance of pulses of temperature increase over the past 15 000 years (Sarnthein *et al.* 1982), including the hypsithermal temperature maximum beginning at 8700 years BP. Longer borehole sections through different parts of the Faro–Albufeira drift complex would undoubtedly refine the picture to date and help resolve some outstanding questions.

References

AMBAR, R., HOWE, M. R. & ABDULLAH, M. I. 1976. A physical and chemical description of the Mediterranean Outflow in the Gulf of Cadiz. *Deutsch Hydrog. Z.*, **29**, 58–68.

ARGUS, D. F., GORDON, R. G., MARTÍNEZ-RUIZ, F., BARAZA, J. & GALIMONT, A. 1999. Pliocene-Pleistocene sedimentary facies at site 976: depositional history in the northwestern Alboran Sea. *In*: ZAHN, R., COMAS, M. C. & KLAUS, A. (eds) *Proceedings of the Ocean Drilling Program, Scientific Results*. Ocean Drilling Program, College Station, TX, **161**, 57–68.

BERGER, W. F., YASUDA, M. K., RICKERT, T., WEFER, G. & TAKAYAMANT, T. 1994. Quaternary time scale for Ontong Java Plateau: Milankovitch template for Ocean Drilling Program Site 806. *Geology*, **22**, 463–467.

BERINGER, M. O. N. & PRICE, F. 1999. A review of the physical oceanography of the Mediterranean outflow. *Marine Geology*, **155**, 63–82.

CARALP, M. H. 1988. Late Glacial to Recent Deep-Sea Benthic Foraminifera from the Northeastern Atlantic (Cadiz Gulf) and Western Mediterranean (Alboran Sea). Palaeoceanographic Results. *Marine Micropalaeontology*, **13**, 265–289.

CARALP, M. H. 1992. Paléohydrologie des bassins profunds nord-marocain (East et Ouest Gibraltar) au Quaternary terminal: apport des foraminifères benthiques. *Bull. Soc. Géol. France*, **163**(2), 169–178.

DEWEY, J. F., HELMAN, M. L., TURCO, E., HUTTON, D. H. W. & KNOTT, S. D. 1989. Kinematics of the western Mediterranean. *In*: COWARD, M. P., DIETRICH, D. & PARK, R. G. (eds) *Alpine Tectonics*. Geological Society, London, Special Publications, **45**, 265–283.

DIESTER-HAAS, L. 1973. No current reversal at 10.000 B.P. in the Strait of Gibraltar. *Marine Geology*, **15**, M1–M9.

FAUGÈRES J.-C., GONTHIER E. & STOW D. A. V. 1984. Contourite drift moulded by deep Mediterranean outflow. *Geology*, **12**, 296–300.

FAUGÈRES, J.-C., CREMER, M. & MONTEIRO, H. 1985a. Essai de reconstitution des processus d'edification de la ride sedimentaire de Faro (marge sud-Portugaise). *Bull. Inst. Géol. Bassin d'Aquitaine, Bordeaux*, **37**, 229–258.

FAUGÈRES, J.-C., FRAPPA, M., GONTHIER, E., RESSEGUIER, A. & STOW, D. A. V. 1985b. Modele et facies de type contourite a la suface d'une ride sedimentaire edifiee par des courants issus de la veine d'eau mediterraneenne. *Bull. Soc. Geol. France*, **8**(1), 35–47.

FAUGÈRES, J.-C., STOW, D. A. V., IMBERT, P. & VIANA, A. 1999. Seismic features diagnostic of contourite drifts. *Marine Geology*, **162**, 1–38.

GONTHIER, E. G., FAUGÈRES, J.-C. & STOW, D. A. V. 1984. Contourite facies of the Faro Drift, Gulf of Cadiz. *In:* STOW, D. A. V. & PIPER, D. J. W. (eds) *Fine-Grained Sediments: Deep-Water Processes and Facies*. Geological Society, London, Special Publication, **15**, 275–292.

GRAFENSTEINM, R. VON., ZAHN, R., TIEDEMANN, R. & MURAT, A. 1999. Planktonic δ18 Records at sites 976 and 977 Alboran Sea: stratigraphy, forcing and paleoceanographic implications. *In:* ZAHN, R., COMAS, M. C. & KLAUS, A. (eds) *Proceedings of the Ocean Drilling Program, Scientific Results*. Ocean Drilling Program, College Station, TX, **161**, 469–479.

GROUSSET, F. E., JORON, J. L., BISCAYE, P. E., LATOUCHE, C., TREVIL, M., MAILLET, N., FAUGÈRES, J.-C. & GONTHIER, E. 1988. Mediterranean Outflow through the Strait of Gibraltar since 18,000 Years B. P.: Mineralogical and geochemical Arguments. *Geo-Marine Letters*, **8**, 25–34.

HAQ, B. U., HARDENBOL, J. & VAIL, P. R. 1987. Chronology of fluctuating sea levels since the Triassic. *Science*, **235**, 1156–1167.

HUANG, T. C. & STANLEY, D. J. 1972. Western Alboran Sea: Sediment dispersal, Ponding and reversal of Currents. *In:* STANLEY, D. J. (ed.) *The Mediterranean Sea: a natural sedimentation Laboratory*. Dowen, Hutchinson and Ross, Stroudsurg, 521–559.

LACOMBE, H. & TCHERNIA, P. 1972. Caracteres hydrologgiques et circulation des eaux en Mediterraée. *In:* STANLEY, D. J. (ed.) *The Mediterranean Sea: A Natural Sedimentation Laboratory*. Dowen, Hutchinson and Ross, Stroudsurg, 25–36.

LOUBERE, P. 1987. Changes in mid-depth North Atlantic and Mediterranean circulation during the Late Pliocene: Isotope and sedimentologic evidence. *Marine Geology*, **77**, 15–38.

LLAVE, E., HERNÁNDEZ-MOLINA, F. J., SOMOZA, L., DÍAZ-DEL-RÍO, V. & GARCÍA, A. C. 2000. Análisis de estratigrafia sísmica y secuencial para el estudio de ciclicidad en el Faro drift durante el cuaternario. *V Congreso geológico de España*, Geotemas, **1**(4), 183–186.

MADELAIN, F. 1970. Influence de la topographie du fond sur l'écoulement méditerranéen entre le détroit de Gibraltar et le Cap Saint-Vincent. *Cah. Océanogr.*, **22**, 43–61.

MALDONADO, A. & NELSON, C. H. 1999. Interaction of tectonic and depositional processes that control the evolution of the Iberian Gulf of Cadiz margin. *Marine Geology*, **155**, 217–242.

MALDONADO, A., SOMOZA, L. & PALLARES, L. 1999. The Betic orogen and the Iberian-African boundary in the Gulf of Cadiz: geological evolution (central North Atlantic). *Marine Geology*, **155**, 9–43.

MOUGENOT, D. & VANNEY, J. 1982. Les rides de contourites Plio-Quaternaires de la pente continentale sud-portugaise. *Bull. Inst. Geol. Bassin d'Aquitaine*, **31**, 131–139.

MILLOT, C. 1987. Circulation in the Western Mediterranean Sea. *Oceanologica Acta*, **10**(2), 143–149.

NELSON, C. H., BARAZA, J. & MALDONADO, A. 1993. Mediterranean undercurrent sandy contourites, Gulf of Cadiz. Spain. *Sedimentary Geology*, **82**, 103–131.

NELSON, C. H., BARAZA, J., MALDONADO, A., RODEROT, J., ESCUTIA, C. & BARBER, J. H. 1999. Influence of Atlantic inflow and Mediterranean outflow currents on late Quaternary sedimentary facies of the Gulf of Cadiz continental margin. *Marine Geology*, **155**, 99–130.

OCHOA, J. & BRAY, N. A. 1991. Water mass exchange in the Gulf of Cadiz. *Deep-Sea Research*, **38**, Supplement I. S465–S503.

OLIVET, J. L. 1996. La Cinématique de la Plaque ibérique. *Bulletin Centres Recherches Exolor. Prof. Elf Aquitaine*, **20**(1), 131–195.

PARRILLA, G. & KINDER, T. H. 1987. Oceanografía física del mar de Alborán. *Bol. Int. Esp. Oceanogr.*, **4**, 133–165.

PERKINS, H., KINDERS, T., VIOLETTE, P. 1990. The Atlantic Inflow in the Western Alboran Sea. *Journal of Physical Oceanography*, **20**.

SARNTHEIN, M. ET AL. 1982. Atmospheric and oceanic circulation patterns off NW Africa during the past 25 million years. *In:* VON RAD U. ET AL. (ed.), *Geology of the NW African Continental Margin*. Springer, Berlin, 545–602.

SHACKLETON, N. J., BERGER, A., PELTIER, W. R. 1990. An Alternative astronomical calibration on the Lower Pleistocene time scales based on ODP site 677. *Transactions of the Royal Society Edinburgh, Earth Sciences*, **81**, 251–261.

SOMOZA, L., MAESTRO, A. & LOWRIE. A. 1999. Allochthonous Blocks as Hydrocarbon Traps in the Gulf of Cadiz. *Offshore Technology Conference, Houston, Texas, 3–6 May:* 571–577.

STOW, D. A. V. 1982. Bottom currents and contourites in the North Atlantic. *Bull. Inst. Geol. Bassin d'Aquitaine*, **31**, 151–166.

STOW, D. A. V., FAUGÈRES, J.-C. & GONTHIER, E. 1986. Facies Distribution and textural variation in Faro Drift contourites: velocity fluctuation and drift growth. *Marine Geology*, **72**, 71–100.

THUNELL, R., WILLIAMS, D., TAPPA, E., RIO, D. & RAFFI, I. 1990. Pliocene-Pleistocene stable isotope record for ODP site 653, Thyrrhenian Sea: implications for the paleoenvironmental history of Mediterranean. *Proceedings of the Ocean Drilling Program, Scientific Results*. Ocean Drilling Program, College Station, TX, **107**, 387–399.

THUNELL, R., RIO, D., SPROVIERI, R. & VERGNAUD-GRAZZINI, C. 1991. An overview of the post-Messinian paleoenvironmental history of the Mediterranean. *Palaeoceanography*, **6**(1), 143–164.

VANNEY, J. R. & MOUGENOT, D. 1981. La plate-forme continentale du Portugal et les provinces adjacentes: analyse geomorphologique. *Memórias dos Serviços Geológicos de Portugal*, **28**.

VERGNAUD-GRAZZINI, C., CARALP, M., FAUGÈRES, J.-C., GONTHIER, E., GROUSSET, F., PUJOL, C. & SALIÈGE, J. F. 1989. Mediterranean outflow throug the Strait of Gibraltar since 18.000 y.B. P. *Oceanologica Acta*, **12**(4), 305–324.

WUST, G. 1961. On the vertical circulation of the Mediterranean Sea. *Jour. Geophy. Research*, **66**, 321–327.

ZAZO, C. 1999. Interglacial sea levels. *Quaternary International*, **55**, 101–113.

The Ceuta Drift, Alboran Sea, southwestern Mediterranean

ERCILLA, G., BARAZA, J., ALONSO, B., ESTRADA, F., CASAS, D. & FARRÁN, M.

CSIC, Instituto de Ciencias del Mar, Paseo Maritimo de la Barceloneta 37–49, 08003 Barcelona, Spain

Abstract: The Ceuta Drift is an elongated-terrace feature (up to 100 km long, 28 km wide, 400 m relief, and 700 ms thick) located in the southwestern Alboran Sea, close to the Gibraltar Strait. It extends between 200 and 700 m water depth, parallel to the Moroccan slope. The drift stratigraphy is defined by the vertical stacking of at least five seismic units bounded by discontinuities: onlap and downlap surfaces at the bottom, and erosive surfaces at the top. Sedimentologically, the most recent deposits are defined by the vertical succesion, from bottom to top, of contouritic sandy muds, muds, and silty clays. The Ceuta Drift began to develop during the early Quaternary, when the pre-existing sea-floor morphology favoured the formation of an offshoot current system from the Mediterranean water masses. The action of this current together with sea-level changes have controlled the growth pattern of this drift.

Geological and oceanographic setting

The Ceuta Drift is located in the southwestern Alboran Sea (SW Mediterranean), a 150 km wide, 300 km long, and up to 2000 m deep Neogene age extensional basin located between the Spanish and African forelands (Fig. 1). The present structure, seafloor morphology and structural boundaries of the Alboran Sea are due to Miocene tectonism (Comas *et al.* 1992), which resulted in a complex geological configuration of the area, defined by marginal hinterlands, and several subbasins separated by morphologic highs (Fig. 1). The present-day highs are mainly of a volcanic nature but during the Middle Miocene a field of mud diapir ridges and mud volcanoes, now buried, developed in the western Alboran Sea (Pérez-Belzuz *et al.* 1997). Tectonics appear to be the main factor controlling the morphology, distribution and evolution of these diapirs, which now extend from the Moroccan margin to the northeast of the Ceuta Drift (Pérez-Belzuz *et al.* 1997).

The Alboran Sea is a very interesting area from an oceanographic point of view because it is where Mediterranean and Atlantic waters first meet after crossing the Gibraltar Strait (Fig. 2). The Atlantic water, lighter, colder and less saline, enters the Alboran Sea at the surface describing two anticyclonic gyres, not always present, one in the western sector and one in the eastern sector (Millot 1987; Heburn & La Violette 1990). This water mass is about 30 km wide and develops a current with a maximum velocity of 1.3 m s^{-1} (La Violette 1987). The dense and highly saline Mediterranean water is much less studied, and the scant researches indicate that it flows westwards towards the Gibraltar Strait as a narrow (20 km wide) current. This sweeps the base of the Moroccan slope as a bottom current (The Mediterranean Undercurrent), reaching speeds of about 5 to 12 cm s^{-1} (Gascard & Richez 1985). The constricted nature of the Gibraltar Strait leads to an acceleration of this current so that measured velocities at the eastern end of the strait, along the southern flank, range between 100 and 300 cm s^{-1} (Heezen & Johnson 1969; Bryden & Stommel 1982; The Donde Va Group 1984).

Morphology

The Ceuta Drift develops in the southwestern sector of the Moroccan slope, between 200 and 700 m water depth (Fig. 1). It is limited to the south and west by the Moroccan shelf, to the east by the Xanuen Bank, to the north by the Western Alboran Basin, and to the northwest by the Ceuta Canyon. The head of this canyon incises the Moroccan shelf deposits and extends seawards towards the NE cutting through the drift deposits. It then displays an open curved pathway towards the NW reaching the seafloor of the Gibraltar Strait (Fig. 1).

Table 1. *Principal characteristics*

Location	Southwestern Alboran Sea (Mediterranean)
Setting	Southwestern Moroccan slope (200 to 700 m water depth)
Age	Quaternary
Drift type	Elongated mound
Dimensions	100 km long, 28 km wide, 400 m high, 700 ms thick
Seismic facies and attributes	Five seismic units bounded by unconformities, and mainly composed of stratified facies. In oblique section the stratified reflectors prograde seaward, downlapping toward the basin and onlapping toward the coast in the lower boundary, and are cut by erosive surfaces in the upper boundary. In longitudinal section, the reflectors converge towards the drift margins. The stratified facies are locally interrupted by cut and fill features related to ancient paleovalley courses.
Sediment facies and attributes	Fining-upward sequences of contourite sandy muds, muds and silty clays, with a terrigenous sand fraction composition; these sediments are occasionally interrupted by millimetric levels of bioclasts. The structureless aspect is interrupted by discrete levels of faint parallel- and cross- lamination.

A detailed bathymetry of the Ceuta Drift does not exist except for its northwestern end, where a multibeam bathymetry is available (Fig. 3). Nevertheless, the general bathymetric chart, clearly shows this drift as a prominent feature on the Moroccan slope, where it forms an elongated terrace up to 28 km wide, 100 km long and with a relief about 400 m except in the northern end where it forms a mound up to 20 km wide and 240 m high. In cross-section, the terrace displays a convex upward profile characterized by relatively low gradients (0.5°) on the top, which contrast with the higher gradients (2.5° to 5°) of the surrounding slopes (Fig. 4). The drift axis has a NW–SE trend and can be traced parallel to the Moroccan margin. The detailed multibeam bathymetry of the northwestern end of the drift shows an asymmetrical convex-upward cross-section with a smoother eastern slope than the western side, and with a rounded northern slope (Fig. 3). This multibeam bathymetry also shows that the pathway of the Ceuta

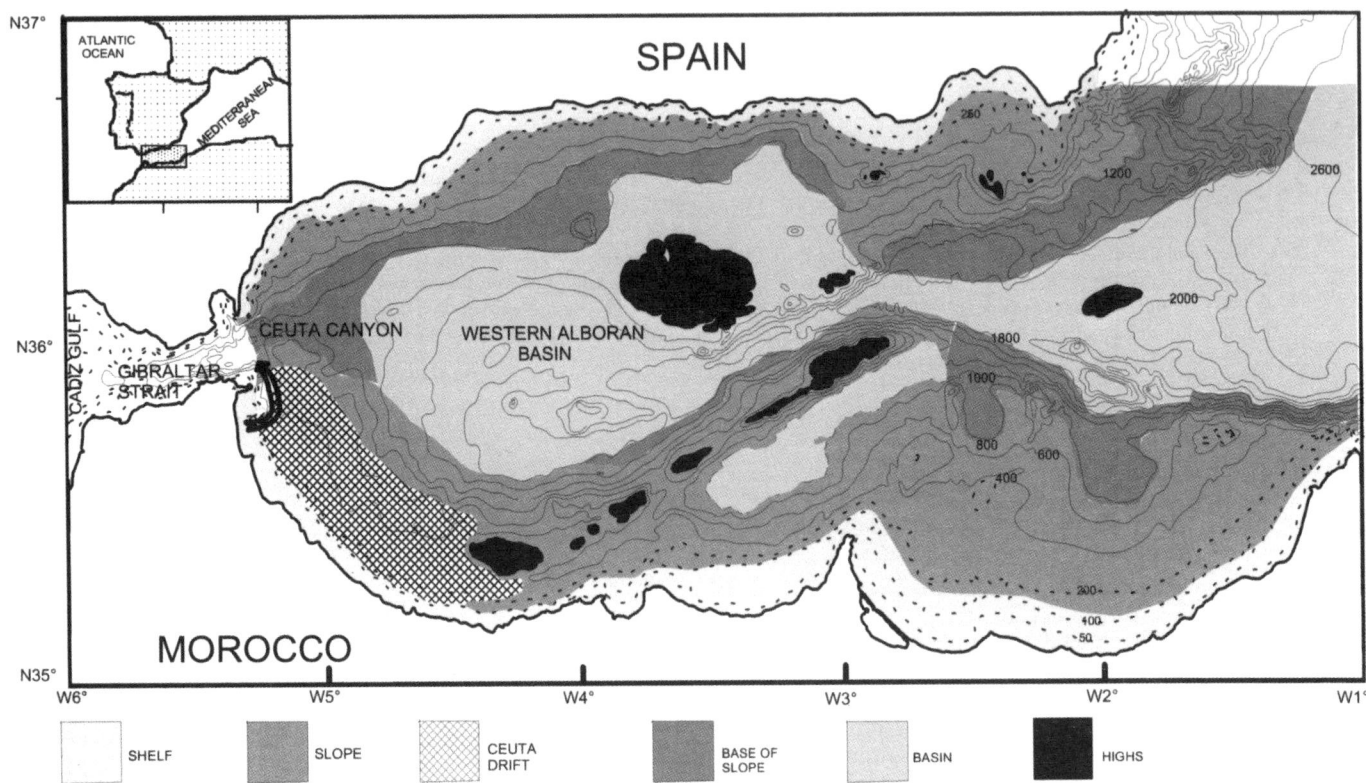

Fig. 1. Bathymetric map (contours in metres) of the Alboran Sea showing the location of the Ceuta Drift. This map also displays the physiographic provinces defined in the Alboran Sea (shelf, slope, base-of-slope and basins), the most relevant morphological highs, and the submarine canyon (Ceuta Canyon) that incises the slope and the base-of-slope. Note how the Ceuta Drift forms a prominent elongated terrace on the Moroccan slope which interrupts the lateral continuity of this province.

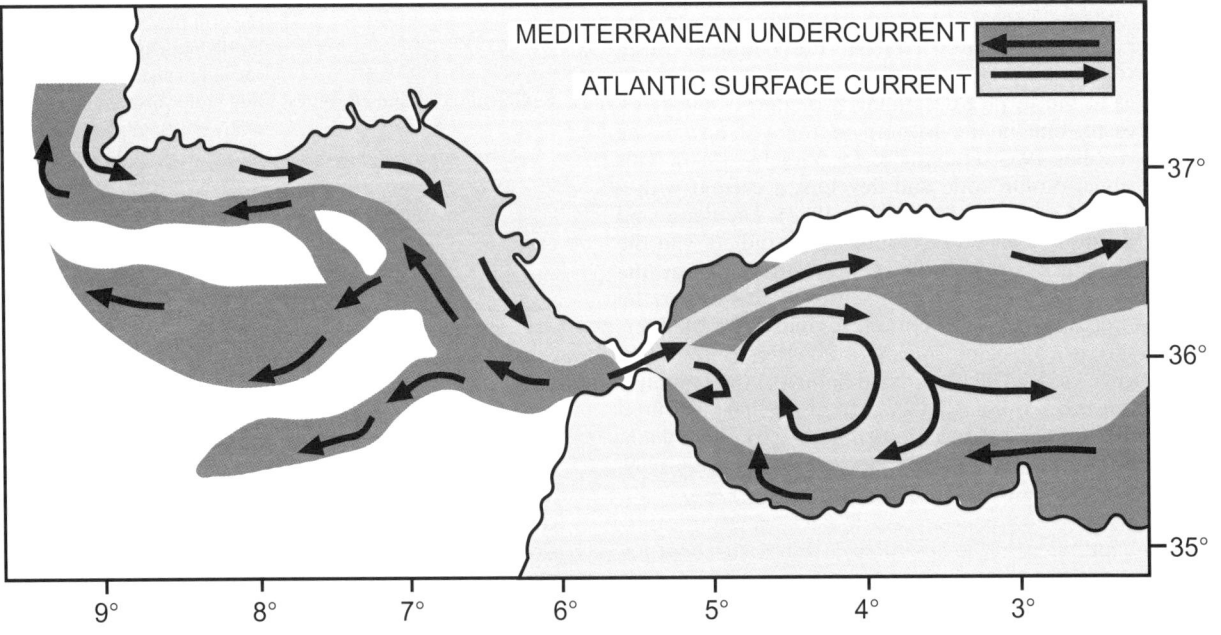

Fig. 2. Schematic diagram showing the prevailing current patterns in the Alboran Sea and the Gulf of Cadiz. The light and cold Atlantic mass of water, (indicated by a light gray pattern) circulates on the surface parallel to the coast of the Gulf of Cadiz and enters the Alboran Sea. After crossing the Gibraltar Strait, part of the Atlantic Water describes one anticyclonic gyre on the western sector of the Alboran Sea, and the remaining continues eastward until a second anticyclonic gyre occurs in the eastern Alboran sector. The warmer and more dense saline Mediterranean waters (indicated by a dark gray pattern) flow westward along the southern sector of the Alboran Sea and sweeps the base of the Moroccan slope as a deep current.

Canyon is slightly sinuous. The seaward slope of the Ceuta Drift connects with a basin floor surface of very low gradients (0.04° to 0.13°) whose flat-lying profile changes northward (toward the Gibraltar Strait) and southward to a slightly concave-up profile (Fig. 3).

Stratigraphic framework

The stratigraphic framework of the southwestern Moroccan margin and the adjacent Western Alboran Basin has received little attention in the literature compared to that of the Spanish

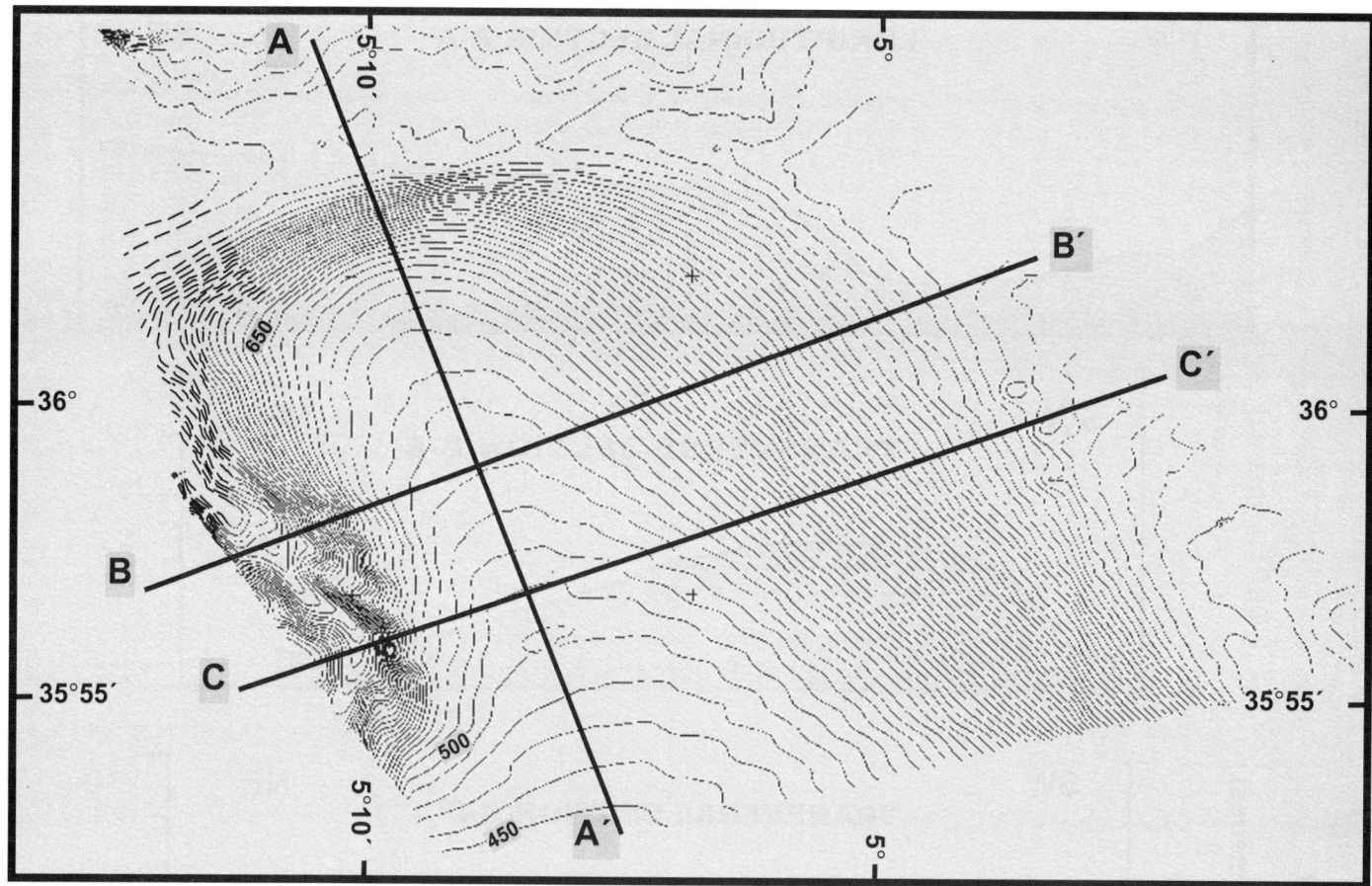

Fig. 3. Detailed bathymetry of the northwestern area of the Ceuta Drift obtained with the multibeam system Simrad EM-12 (contour interval 10 m). The northern end of the drift displays a rounded margin. The eastern drift flank is smoother than its western flank. This is because the Ceuta Canyon has eroded the drift in its northwestern region. Note how the canyon displays a sinuous pathway until it reaches the floor of the Gibraltar Strait. The lines A-A', B-B' and C-C' refer to the sections displayed in Figure 4.

margin. Seismic stratigraphic studies (Tesson *et al.* 1987) have indicated two sequences underlain by a regional unconformity above the acoustic basement: sequence B (Pliocene) and sequence A (Quaternary). These Plio-Quaternary sequences are bounded by a strong, laterally continuous reflector that corresponds to the seismic reflector M (i.e. the Messinian surface) identified in the northwestern Mediterranean (Ryan *et al.* 1970). More detailed stratigraphical studies, based on high resolution seismics carried out in the Western Alboran Sea (Pérez-Belzuz *et al.* 1995; Pérez-Belzuz 1999) reveal up to four sequences within the Plio–Quaternary sediments. These sequences are: 1, lower Pliocene; 2, upper Pliocene; 3, lower Quaternary; and 4, upper Quaternary.

Throughout the Pliocene and Quaternary, the southwestern Alboran Sea has been dominated by terrigenous sedimentation (El Moumni 1994). These deposits are characterized by transparent and discontinous stratified seismic facies in the Pliocene sequence and stratified facies of high amplitude and high lateral continuity in the Quaternary sequence. In both cases, there are prograding and aggrading growth patterns affected by active faulting. The development of these deposits has been controlled by an interplay of glacio-eustatic sea-level changes, physiographic configuration and the oceanographic pattern, whereas the preserved stratal pattern and geometry have been influenced by tectonics (Maldonado *et al.* 1992; Ercilla *et al.* 1994*a*; Pérez-Belzuz *et al.* 1995).

Seismic characteristics

The data set comprises six single-channel high-resolution seismic profiles obtained with sleeve guns (120 cubic inches) (Fig. 5). These profiles show that the base of the Ceuta Drift is marked by a pronounced strong seismic reflector, named P2 reflector, which represents the boundary between the Pliocene and the Quaternary deposits (Campillo *et al.* 1992). The seismic stratigraphy of this drift is made up by the vertical stacking of several seismic units bounded by discontinuities (Figs 6, 7, 8, 9 & 10). Some of these units terminate sharply against the slope boundaries of the drift, interrupting their lateral continuity (Figs 6, 9 & 10). The correlation of these discontinuities and units with those discontinuities and sequences defined by Pérez-Belzuz (1999) in the adjacent areas, reveals that the lower Quaternary sequence (sequence 3) comprises two seismic units (here named 3a & 3b), and the upper Quaternary sequence (sequence 4) comprises three seismic units (here named 4a, 4b, & 4c). The lower and upper seismic sequences are bounded by the Q1 reflector (Campillo *et al.* 1992). At the northern end of the drift, close to the Gibraltar Strait, a greater number of seismic units and bounding discontinuities have been identified (Fig. 7).

The seismic units of the Ceuta Drift are composed of stratified facies of high lateral continuity and amplitude, except in the western area, where a chaotic facies is identified. In this area, the deposits of seismic units 3a, 3b and 4a display numerous erosive and negative surfaces, characterized by a high acoustic amplitude and filled by chaotic and stratified facies with an onlap-fill configuration (Figs 6, 8 & 9). Close to the Ceuta Canyon, the erosive surfaces show well-defined U and V cross-sections of 230 ms thick and 3 km wide, and they probably represent ancient courses of the Ceuta Canyon (Figs 6 & 8). In the rest of the southwestern area these erosive surfaces are not so prominent, but they are shallower, more amalgamated, and have less well-defined features (hundreds of metres wide and about 25 ms of relief) (Figs 9 & 10).

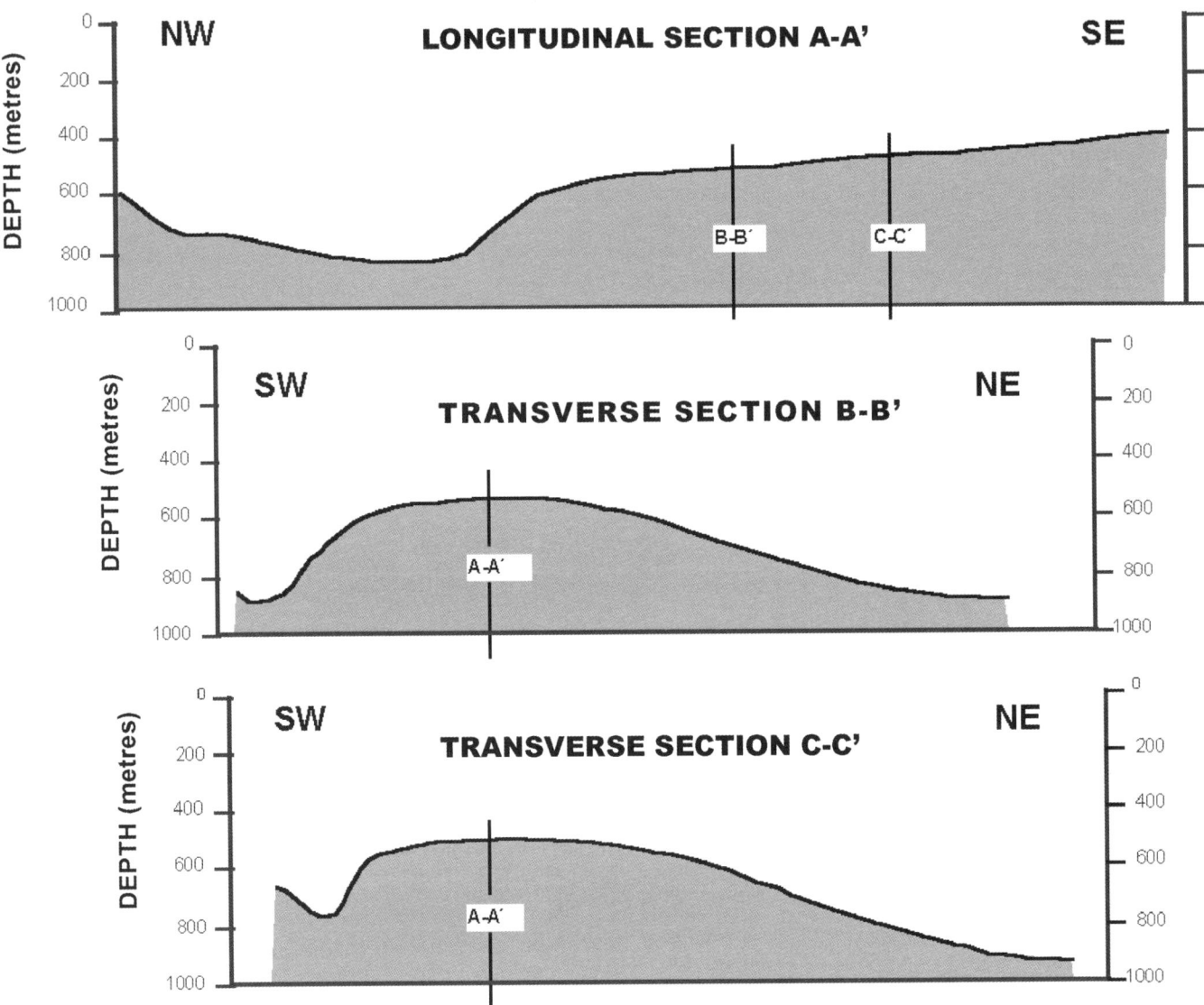

Fig. 4. Longitudinal and transverse sections on the northwestern area of the Ceuta Drift showing the asymmetrical convex-upward sections that characterize the drift.

The isopach map of the Ceuta Drift reveals a maximum thickness of 700 ms (two-way time) with a unique depocenter, and an elongated sediment distribution that remains true to the present-day morphology of the drift (Fig. 11).

In both longitudinal and oblique sections (with respect to the direction of the drift axis), the seismic units that form the Ceuta Drift display a lenticular geometry that thins toward the margins. Internally, the stratified facies shows an aggrading pattern with converging reflectors that downlap onto the lower boundaries. The upper boundaries are represented by erosive surfaces on the slope margin of the drift, whereas on the central-flatter area the upper boundary is a concordant surface (Figs 6 & 7). In transverse sections also, the seismic units display a lenticular geometry with a thickness that thins both seaward and landward. Internally, the deposits have a prograding pattern with reflectors that onlap landward and downlap seaward onto the lower boundary, and are eroded by an irregular upper boundary (Figs 8, 9 & 10).

The flat or concave upward surfaces bounding the distal sector of the drift are generally defined by an irregular basinfloor surface of high amplitude that erodes stratified deposits, except in the nortwestern area (close to the Gibraltar Strait) where it erodes chaotic facies (Figs 6, 7, 8, 9 & 10).

Sediment characteristics

Five gravity cores up to 3 m long recovered form the Ceuta Drift revealed the Late Pleistocene and Holocene sedimentology (Fig. 5). Three textural types are identified: sandy muds, muds and silty clays (Figs 12 & 13). The sandy muds consist of 25% sand, 47% silt and 28% clay, have a mean grain size from 6 to 6.6 phi, and are characterized by very poor sorting. They are identified in those cores located in the westernmost sector of the Ceuta Drift. The colours of the sandy muds are predominantly greyish olive green (5GY3/2) and greyish olive (10Y4/2). These sediments are visually distinctive from the muds and silty clays, and show a structureless aspect with moderate bioturbation.

The muds are poorly sorted consisting of 54% silt, 42% clay and < 4% sand, and have a mean grain size from 6.1 to 8.3 phi. The colour is olive grey (5Y4/1). The silty clays consist of 58% clay, 38% silt and < 4% sand, and have a mean grain size > 8 phi. The colours are olive grey (5Y4/1) and moderate olive brown (5Y4/4). Their sorting is poor and the colour similar to the muds. X-radiographs of the muds and silty clays reveal a structureless aspect occasionally interrupted by faint parallel- and cross-lamination and a mycellium-type bioturbation (Fig. 13A). The radiographs also reveal the presence of millimetric coarser-grained levels composed of shelly debris (mainly molluscs).

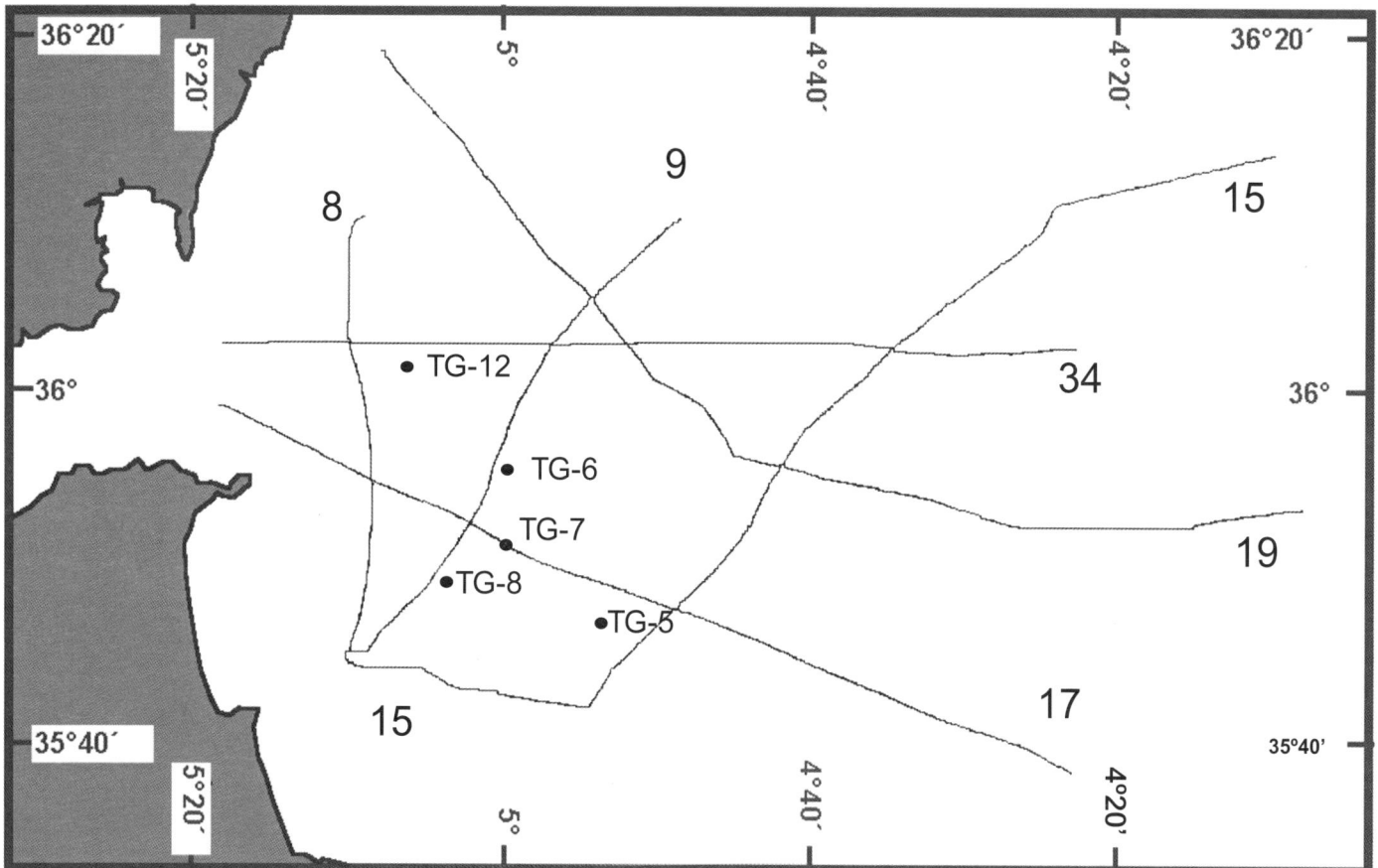

Fig. 5. Location of the seismic lines and sediment cores (*TG*) which have been studied. The seismic lines were obtained with an airgun monochannel system, and the sediment cores were recovered using a gravity corer. Seismic profiles numbered 8, 9, 15,17 and 34, all cross the drift along different courses, whereas profile 19 is located outside the boundaries of the drift and is only used for stratigraphic control.

All three sediment types have a sand fraction with terrigenous composition (90%) mainly composed of light minerals, although in some cases the sandy muds have a mixture of 40–60% terrigenous and biogenic material (Fig. 13B). In all cases, the biogenic components of the sand fraction are made up of mainly broken planktonic and bethonic foraminifers, molluscs and serpulids. The calcium carbonate content ranges between 16 and 29 %.

The thickness and vertical distribution of the three sediment types indicate that the muds are predominant (between 60 and 90% of the core length) and are interrupted by levels of sandy muds (from 39 to 50 cm thick) and thinner silty clays (10 cm thick). Nevertheless, the tops of these successions are always characterized by the presence of levels (up to 40 cm thick) of silty clays, whereas the presence of sandy muds occurs mainly in the lower parts of the recovered succesions (Fig. 12). The lower and upper boundaries of the sandy mud intervals are mostly sharp whereas those of the silty clay intervals have gradational contacts. These sediment succesions are interrupted by irregularly-shaped or planar coarse-grained intervals with abundant biogenic debris, mainly broken mollusc shells, that are up to several decimeters thick. These coarse grained intervals occur more frequently on the south and west sectors of the Ceuta Drift.

Discussion

Although there is relatively little known about the circulation of the deep Mediterranean water masses, not only in the Alboran Sea but in the whole Western Mediterranean, the Ceuta Drift probably represents one of the best examples of deposition under the influence of these bottom currents. In fact, it provides important evidence for the influence of the Mediterranean deep current on the deposition and distribution of sediments in the southwestern area, and for the long time scale over which this current has been active.

With respect to its morphology and the classification of drifts made by Faugères *et al.* (1999), the Ceuta Drift can be considered an elongated mounded drift. Seismic stratigraphy reveals that this drift began to develop during the lower Quaternary. At that time, the morphology of the Alboran Sea was characterized by the presence of an important field of mud diapirs in the Western Alboran Basin (Fig. 14). Some of these diapirs extended to the southwestern Moroccan margin in the form of parallel ridges (Perez-Belzuz *et al.* 1997). We suggest that the presence of these diapirs, which acted as topographic barriers, together with the presence of the Gibraltar Strait and the semi-confined morphology of the Alboran Sea, all favoured the formation of a thread current system that became accelerated due to the corridor effect between the diapir ridges and the southwestern Moroccan slope, and by the funnel effect of the Gibraltar Strait. The principal controlling factor in the genesis of the Ceuta Drift appears to be the pre-existing sea-floor morphology, which has conditioned the pathway of the Mediterranean deep water.

The presence of a seismic pattern that shows little change during the Quaternary, attests to the long-term stability of the Mediterranean deep current. Likewise, the stratal pattern and some types of discontinuities suggest that the amount of sediment that feeds the drift was conditioned by factors such as sea-level changes. The deposits making-up the Ceuta Drift seem to be related to linear supply from the adjacent margin, as suggested by

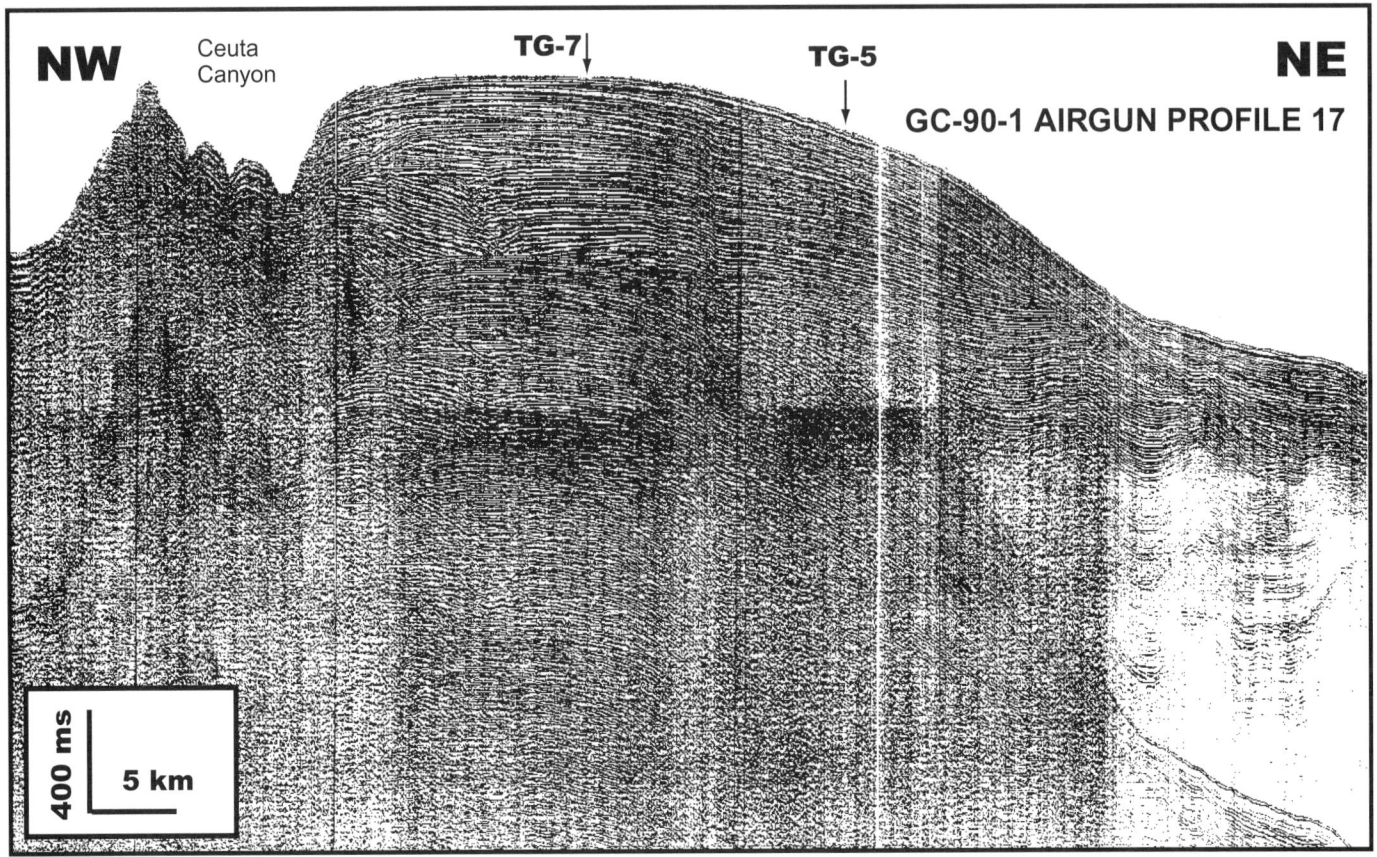

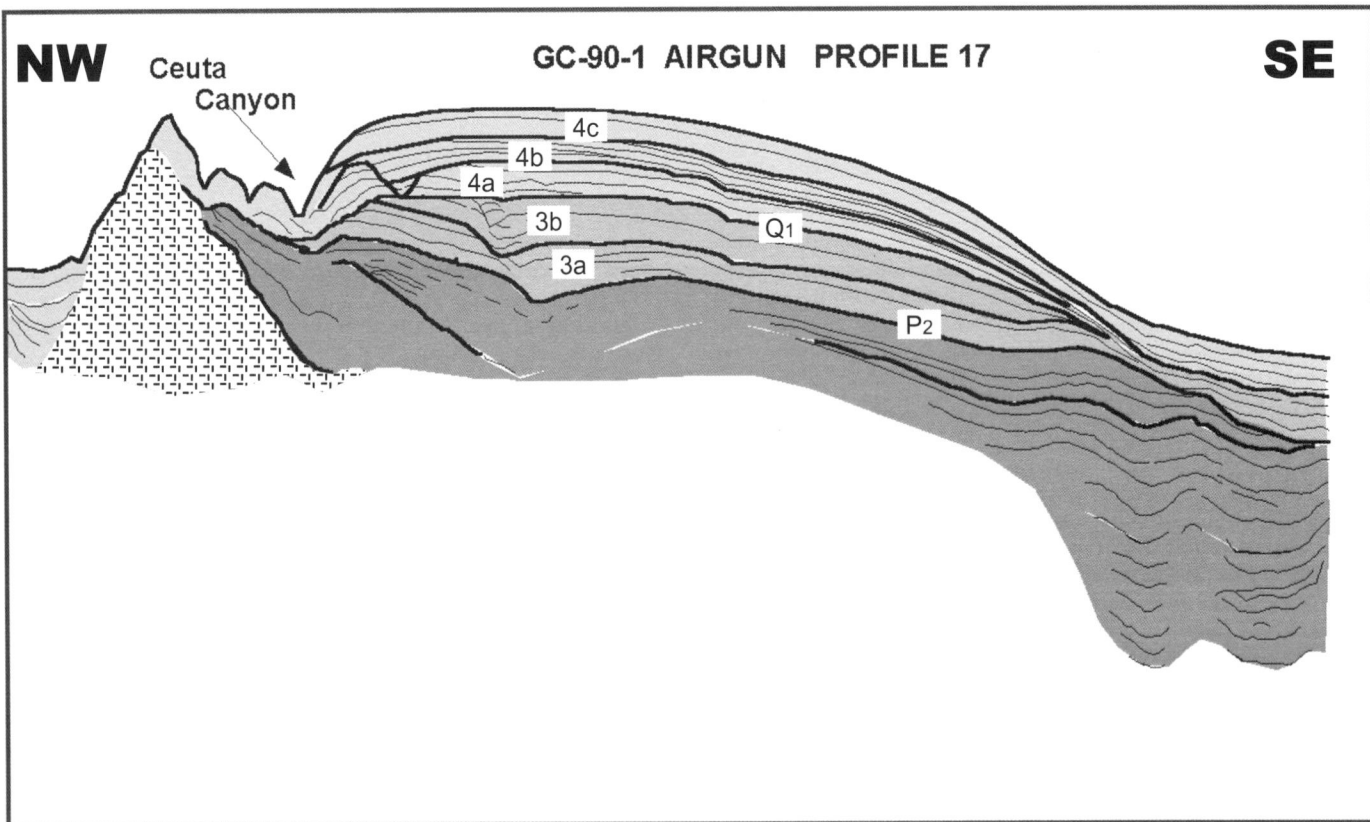

Fig. 6. Airgun seismic profile and line drawing longitudinal to the Ceuta Drift across the wider, central section. The drift has a flat-topped morphology. It is about 45 km wide and 300 m high, strongly asymmetrical with a smooth east flank and a steep western flank emphasized by the presence of the erosional Ceuta Canyon. Internally, the drift is composed of up to five seismic units (3a to 4c) with stratified facies, separated by discontinuity surfaces easily identified on the eastern and western margins of the drift, that correlate to paraconformities on the central part. Erosional and cut-and-fill structures resembling a paleocanyon course are identified on seismic units 3a to 4a, whereas seismic units 4b and 4c are actually eroded by the present Ceuta Canyon. The western limit of all seismic units is a sharp erosional scarp, whereas they progressively thin towards the eastern limit of the drift, pinching-out and disappearing, except the more recent seimic unit 4c. Profile location on Figure 5. P2, boundary between Pliocene and Quaternary deposits; Q1, boundary between lower and upper Quaternary seismic units.

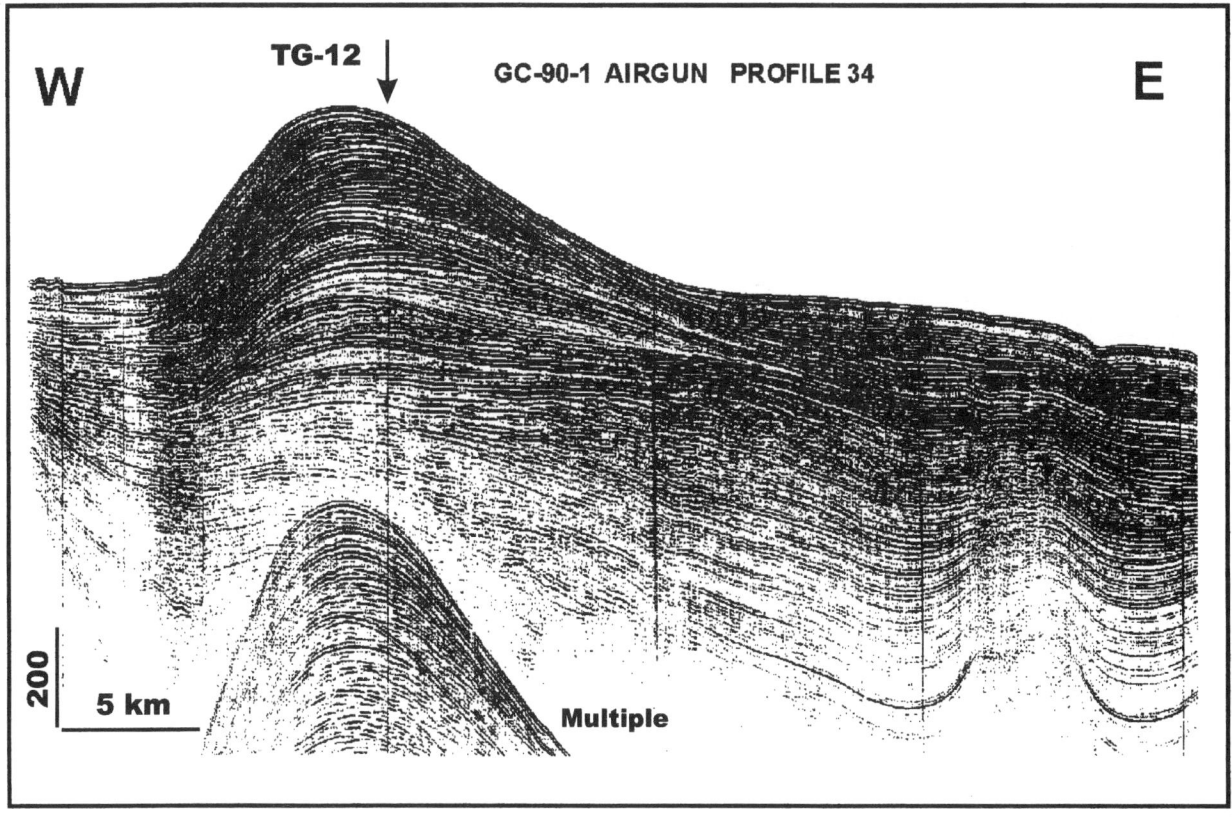

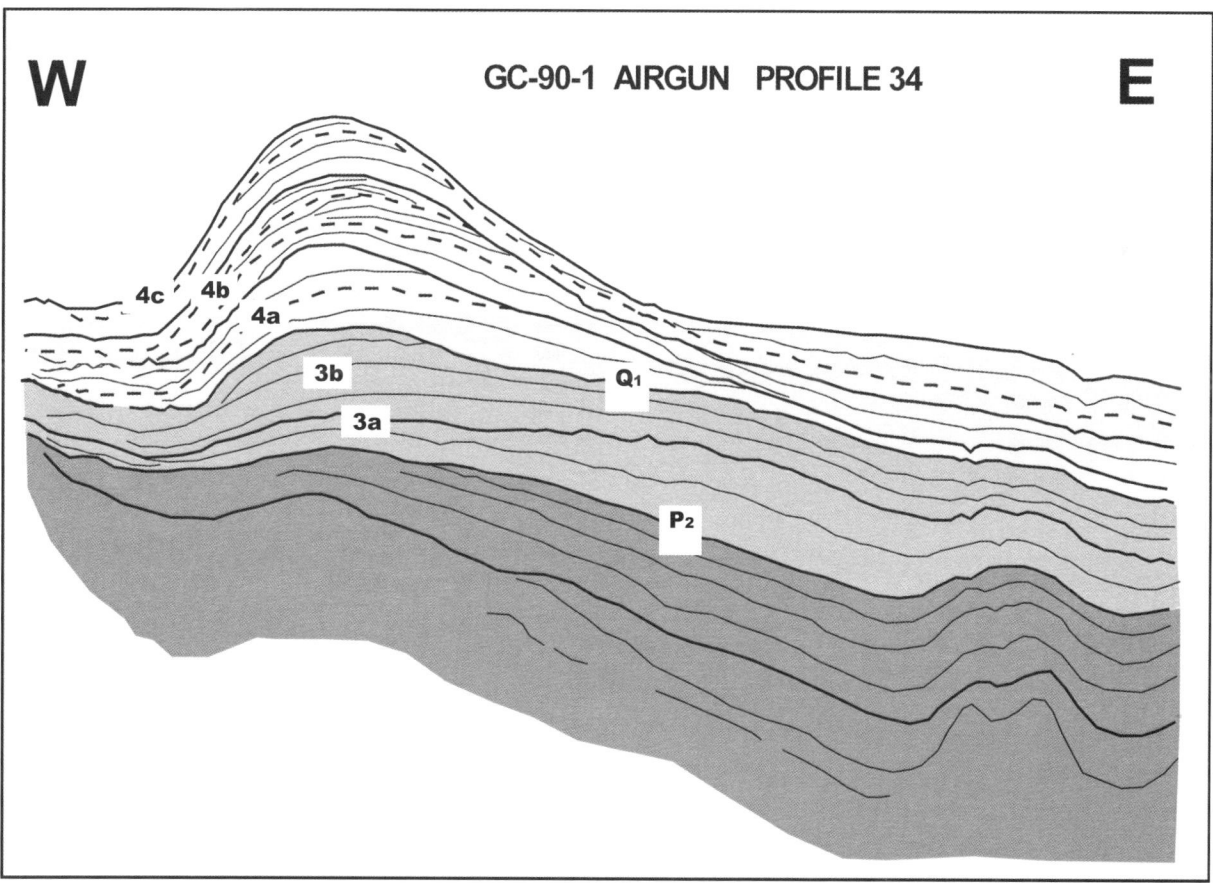

Fig. 7. Airgun seismic profile and line drawing transverse to the Ceuta Drift across its narrower, northern tip. There, the drift has a mounded topography, is about 18 km wide and 240 m high, asymmetrical with a slightly steeper western side. The drift is internally composed of five main seismic units with stratified facies, separated by discontinuity surfaces and named by correlation with those defined on the adjacent northern Alboran slope by Perez-Belzuz *et al.* (1997). The lower seismic units 3a and 3b develop above the P2 reflector, which represents the base of the Quaternary deposits. The rest of seismic units 4a, 4b and 4c are upper Quaternary in age, and they are composed of seismic subunits also bounded by discontinuity surfaces (discontinuous lines) which are only defined in this sector of the drift. Due to this, the seismic units 4a, 4b and 4c are treated as single units because they are correlated throughout the Ceuta Drift. Profile location on Figure 5. P2, boundary between Pliocene and Quaternary deposits; Q1, boundary between lower and upper Quaternary seismic units.

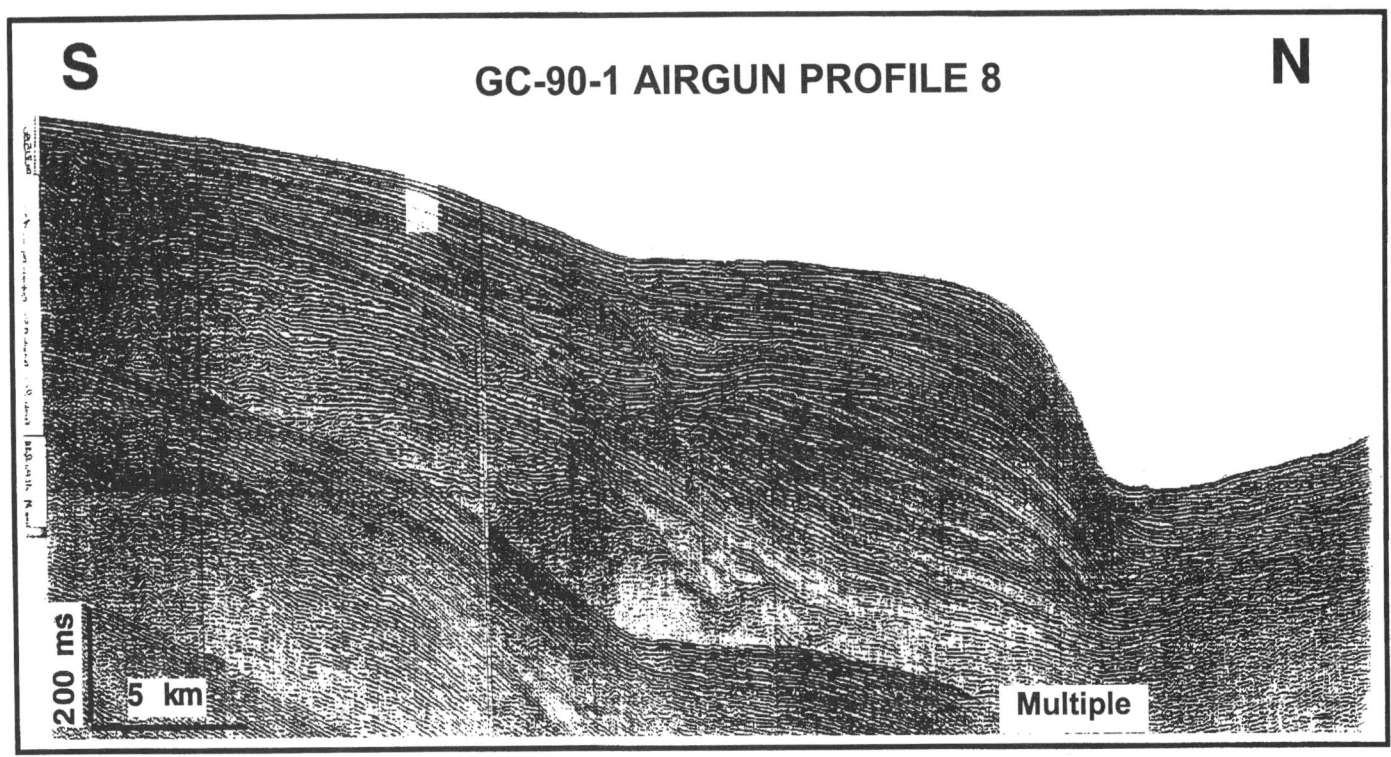

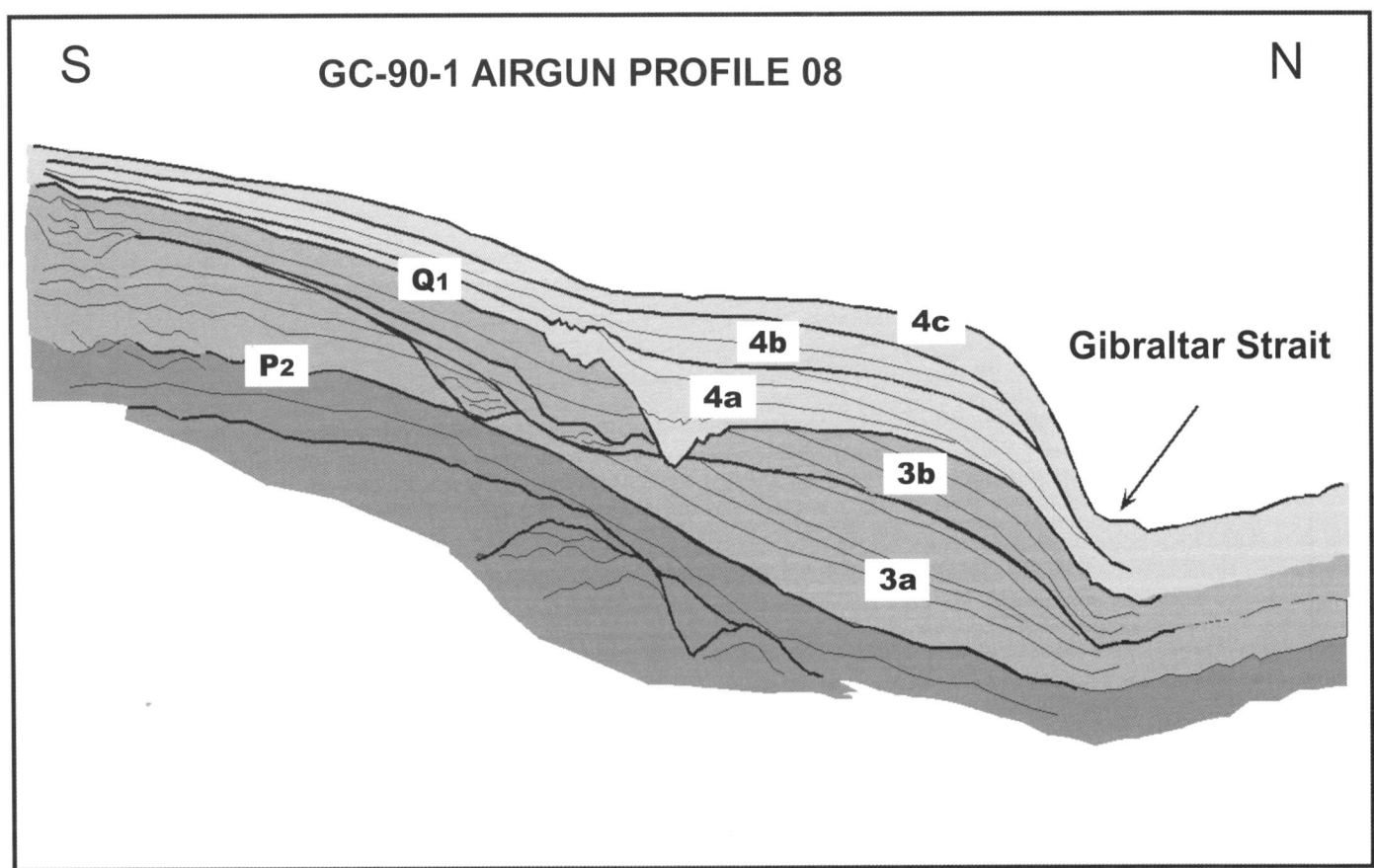

Fig. 8. Airgun seismic profile and line drawing oblique to the Ceuta Drift that crosses its northern sector, close to the Gibraltar Strait. The drift appears as a positive-relief feature, smoothly dipping towards a steep N flank. Seismically, the Ceuta Drift deposits are defined by five seismic units (3a, 3b, 4a, 4b, 4c) internally composed of stratified facies whose reflectors downlap and onlap the lower boundary, and are truncated by the upper boundary. Seismic units 3a, 3b and 4a show palaeovalley features filled with chaotic and stratified facies with an onlap-fill configuration. These palaeovalleys would represent ancient courses of the Ceuta Canyon. The surface bounding the northern sector of the Ceuta Drift displays a slightly concave-upward profile, and is internally composed of chaotic facies with reflections of high amplitude. P2, boundary between Pliocene and Quaternary deposits; Q1, boundary between lower and upper Quaternary seismic units.

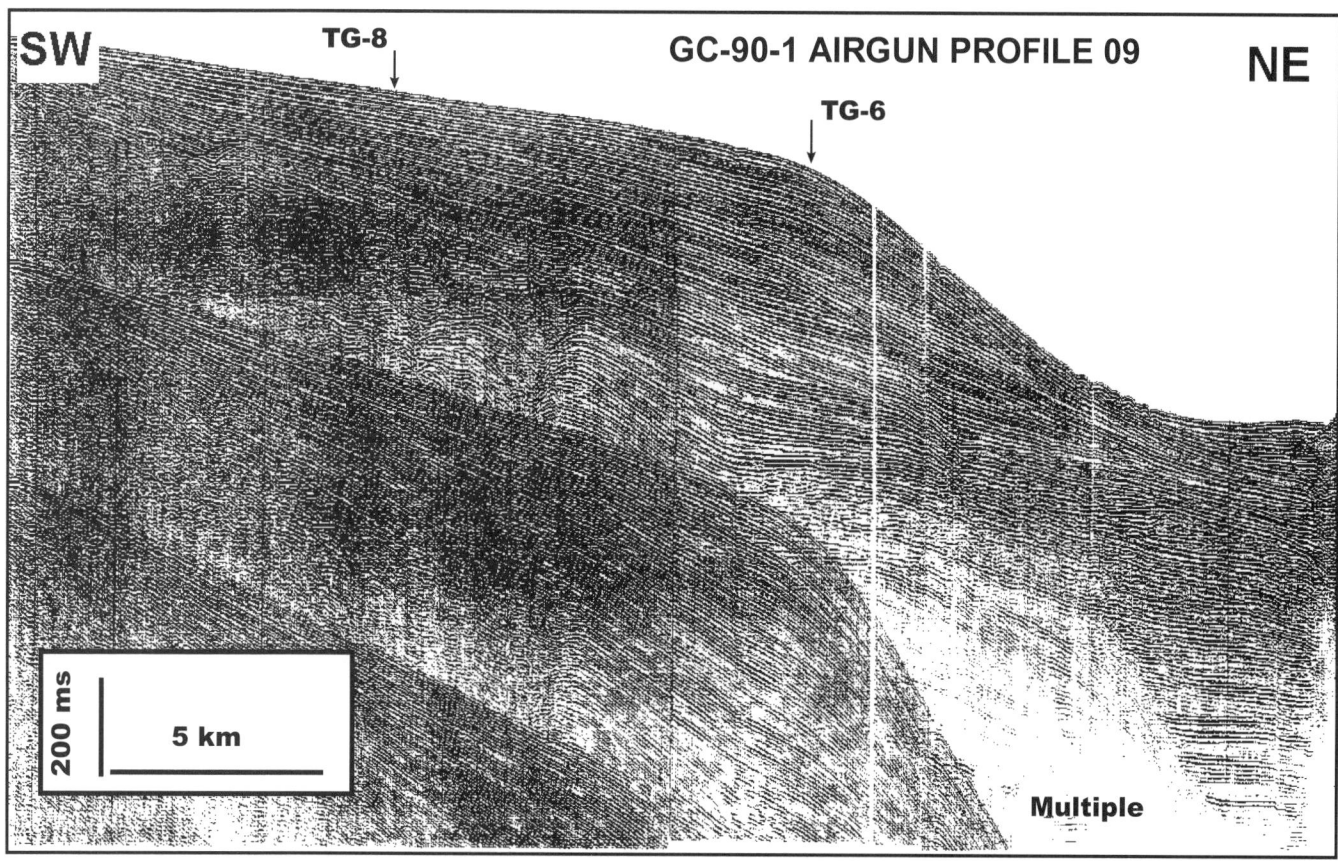

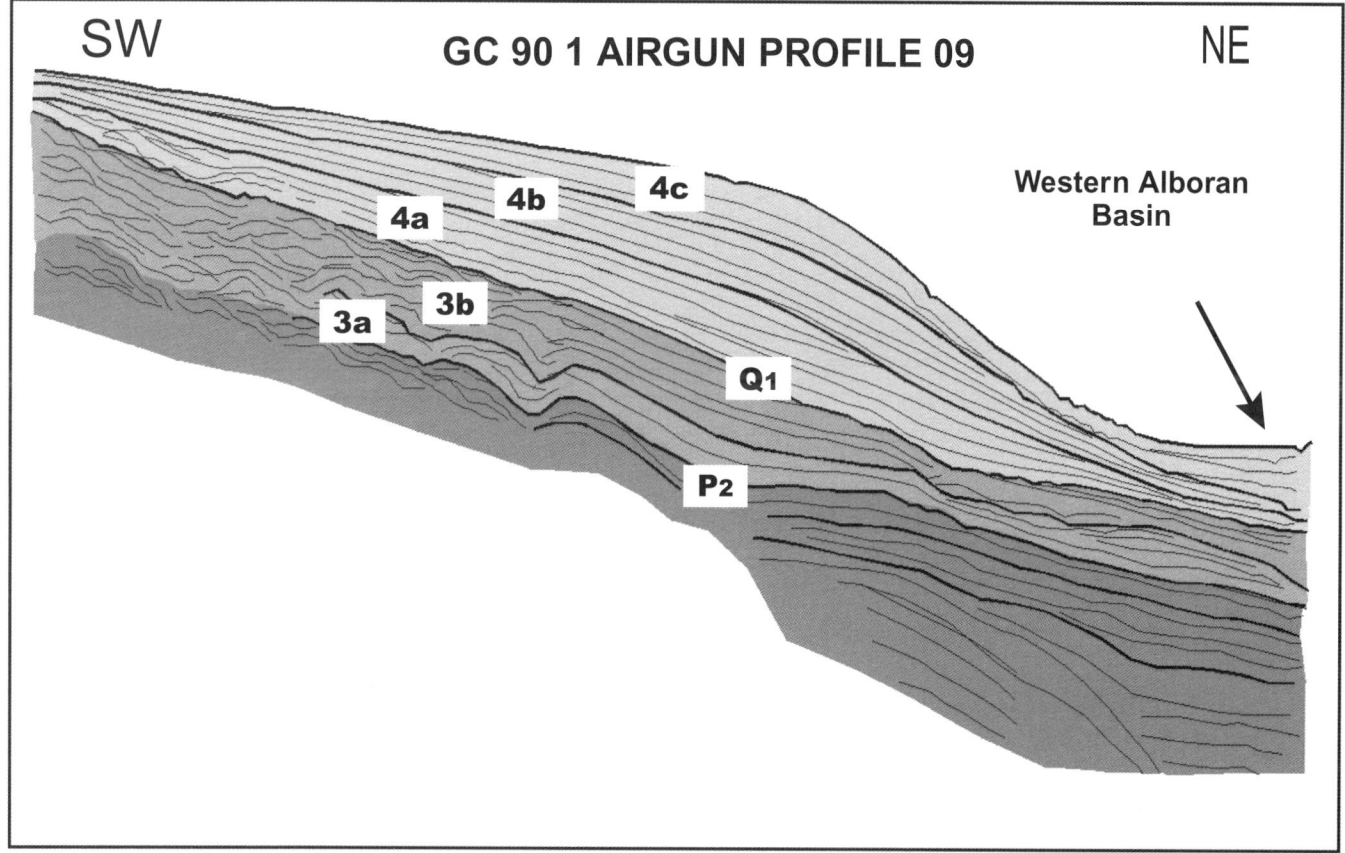

Fig. 9. Airgun seismic profile and line drawing oblique to the Ceuta Drift across its central sector. The drift appears as a positive-relief feature, smoothly dipping towards a steep NE flank. Internally, it is composed of up to five seismic units which consist of stratified and chaotic facies on the lower portion of the drift (seismic units 3a and 3b), parallel stratified facies that change to chaotic facies only on the innermost part of the drift (seismic unit 4b), or just parallel stratified seismic facies (seismic units 4b and 4c). The seismic units are separated by erosional unconformities on the top, and by downlap and onlap surfaces on the bottom. Profile location on Figure 5. P2, boundary between Pliocene and Quaternary deposits; Q1, boundary between lower and upper Quaternary seismic units.

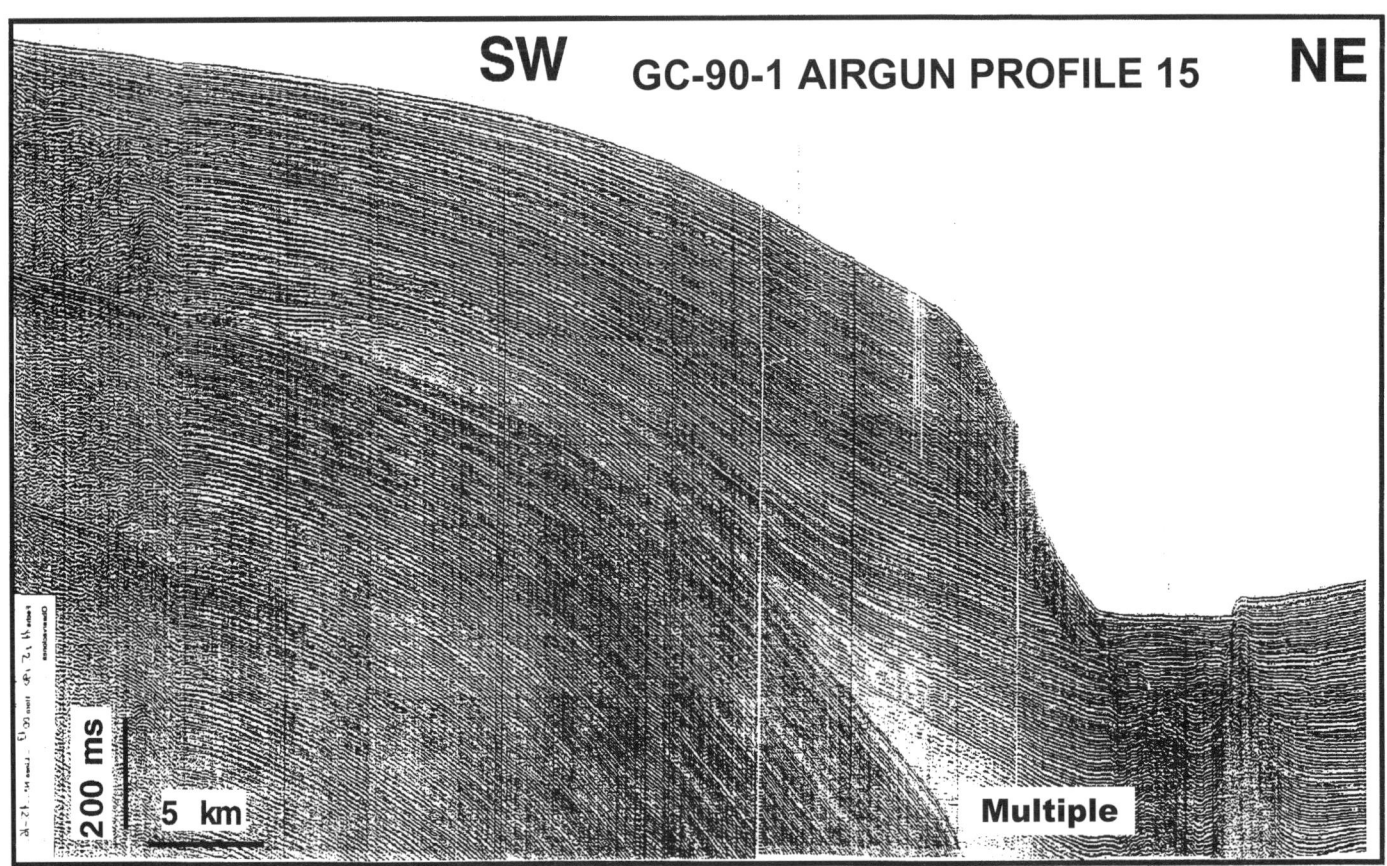

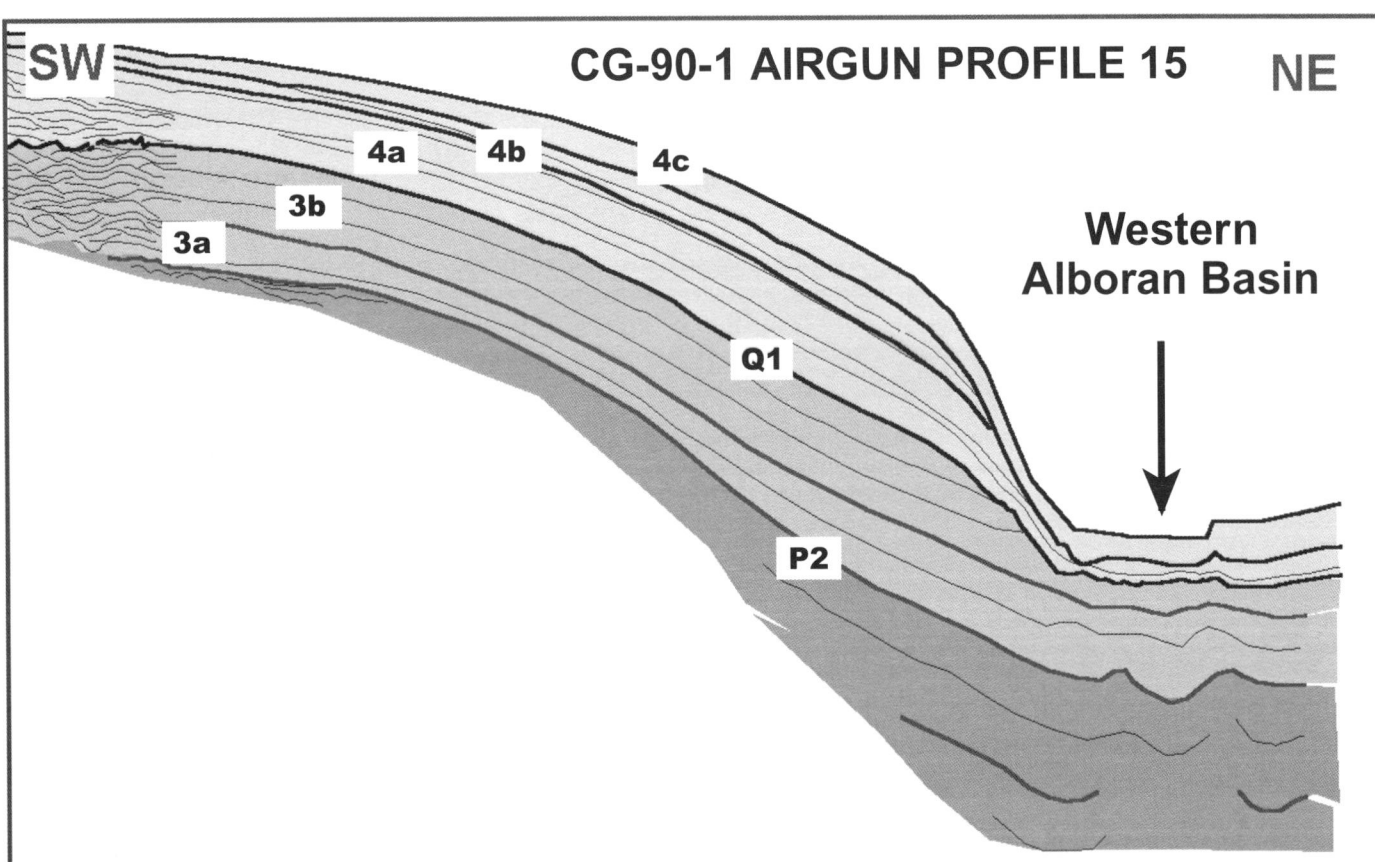

Fig. 10. Airgun seismic profile and line drawing oblique to the Ceuta Drift across its southern sector. The drift appears as a positive-relief feature, smoothly dipping towards a very steep NE flank. Internally, it is composed of up to five seismic units with parallel facies that change to chaotic facies towards the inner part of the drift. The seismic units are separated by erosional unconformities and the seaward lateral continuity of all but the seismic unit 4c are erosionally truncated on the outer margin of the drift. Profile location on Figure 5. P2, boundary between Pliocene and Quaternary deposits; Q1, boundary between lower and upper Quaternary seismic units.

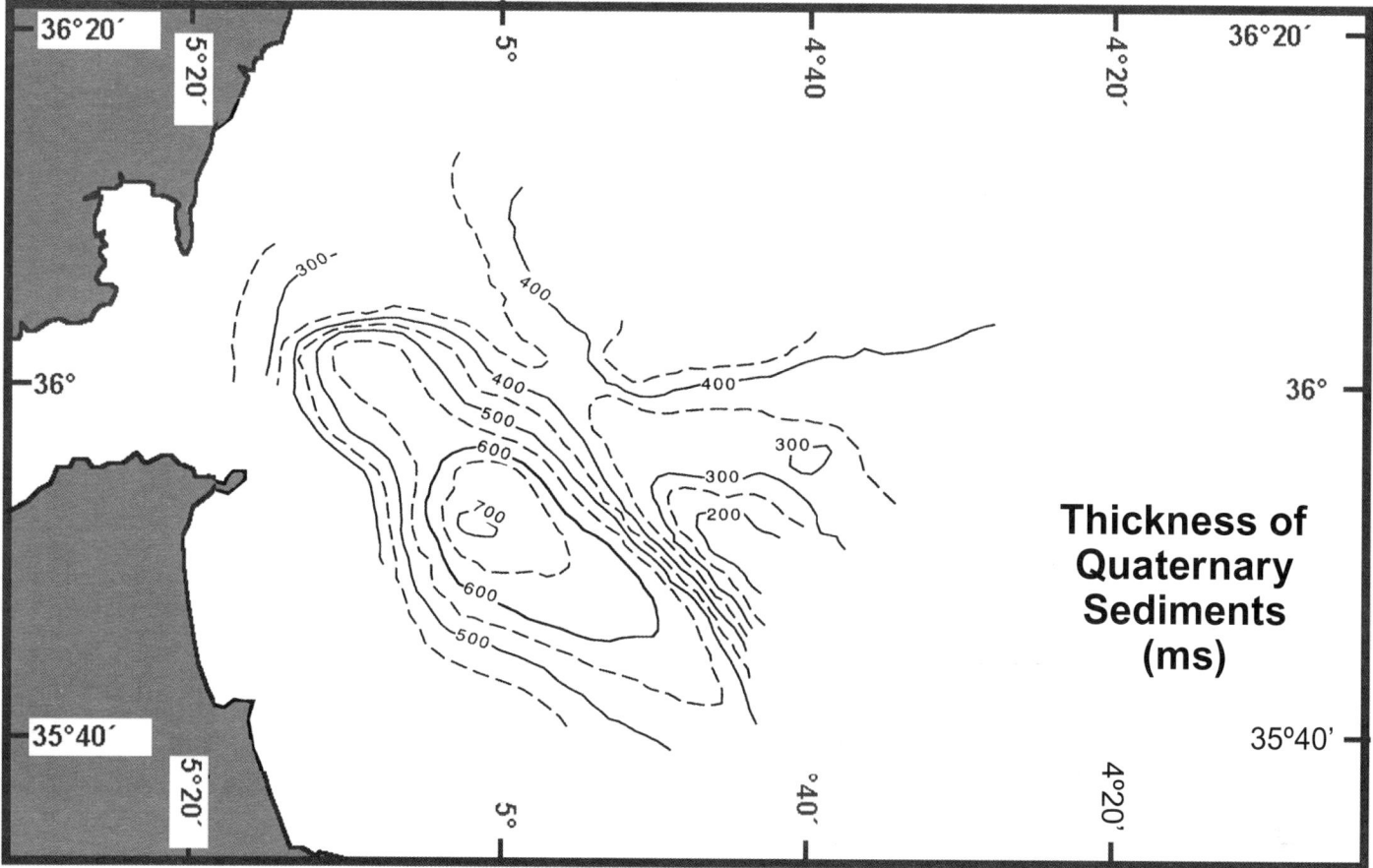

Fig. 11. Isopach map of the Ceuta Drift and adjacent areas. This map represents the thickness of the Quaternary deposits, i.e. those sediments deposited overlying the seismic reflector P2 (boundary between Pliocene and Quaternary deposits).

the drift thickness that nearly parallels the coastline (compare Figs 8 & 9). This distribution is similar to those margins making-up other parts of the Western Mediterranean (Farran & Maldonado 1990; Ercilla et al. 1994b; Ercilla & Alonso 1996; Chiocci et al. 1997). The style of growth of the drift in cross-section (progradational) and the type of boundaries of the seismic units (erosive and downlap), all suggest that the main phase of drift development was controlled by the seaward displacement of the coastline. Sedimentation during the Quaternary on the Western Mediterranean margins, in general, has been controlled by sea-level changes, these margins being mainly inundated by sediments during lowstand stages (Farran & Maldonado 1990; Ercilla et al. 1994a; Ercilla & Alonso 1996; Chiocci et al. 1997). Moreover, the seismic analysis has revealed that the seismic sequences defined in the Ceuta Drift (at least sequence 4) are correlatable with those lowstand sequences making-up the northern Alboran margins and surrounding basins (Ercilla et al. 1994; Pérez-Belzuz et al. 1997).

We suggest that the Ceuta Drift results from the action of the Mediterranean deep current redistributing and depositing sediment supplied mainly during the seaward migration of the coastline. This development occurs in spite of the presence of a major downslope feature represented by the Ceuta Canyon at the western end of the Ceuta Drift. This canyon, that cuts through the drift deposits, has acted as a conduit for downslope sediment transport to the Western Alboran Basin (Huang et al. 1972) during the Quaternary. Ancient palaeocourses of this canyon have been filled by the Ceuta Drift deposits, indicating the importance of alongslope processes during this time. Likewise, the redistribution and deposition by bottom currents is reflected by the onlap terminations in the cross-sections of the Ceuta Drift, and by its geometry (lenticular-shape), stratal pattern (converging seaward and landward) and type of boundaries (downlap and erosive) in its longitudinal section (Figs 8 & 9).

At the NW end of Ceuta Drift the Mediterranean deep current reaches higher velocities (between 100 and 300 cm s^{-1}; Heezen & Johnson 1969; Bryden & Stommel 1982; The Donde Va Group 1984) and its influence on deposition is more striking, producing chaotic facies and a larger number of discontinuous (erosive) surfaces (Figs 7 & 8). The greater erosive power of the Mediterranean deep current in this sector of the drift is also reflected by the basinfloor surface that bounds the drift, which displays a concave, irregular profile, with a high acoustic amplitude, eroding into the sub-bottom surficial chaotic deposits (Figs 3, 7 & 8).

The recognition of contourites in core sections is problematic, particularly regarding their distinction from hemipelagites and muddy turbidites in fine-grained successions (Stow & Lovell 1979; Howe 1995; Stoker et al. 1998). There have been several attempts to establish definitive criteria for their recognition (Stow & Lovell 1979; Stow & Holbrook 1984; Stow et al. 1998). Based on these sedimentological criteria (texture, sorting, sand fraction composition and structures), the sequences of muds with intercalations of sandy muds and silty clays recovered from the surface of the Ceuta Drift clearly represent contourite deposits. The texture of these sediments indicates poor sorting and the presence of local concentrations of coarse, biogenic tests within an ungraded and homogeneous sequence, together with some faint lamination suggests at least weak current control (Fig. 13A). The mainly terrigenous composition of the sand fraction also reflects a dominant land source followed by alongslope mixing (Fig. 13B). The

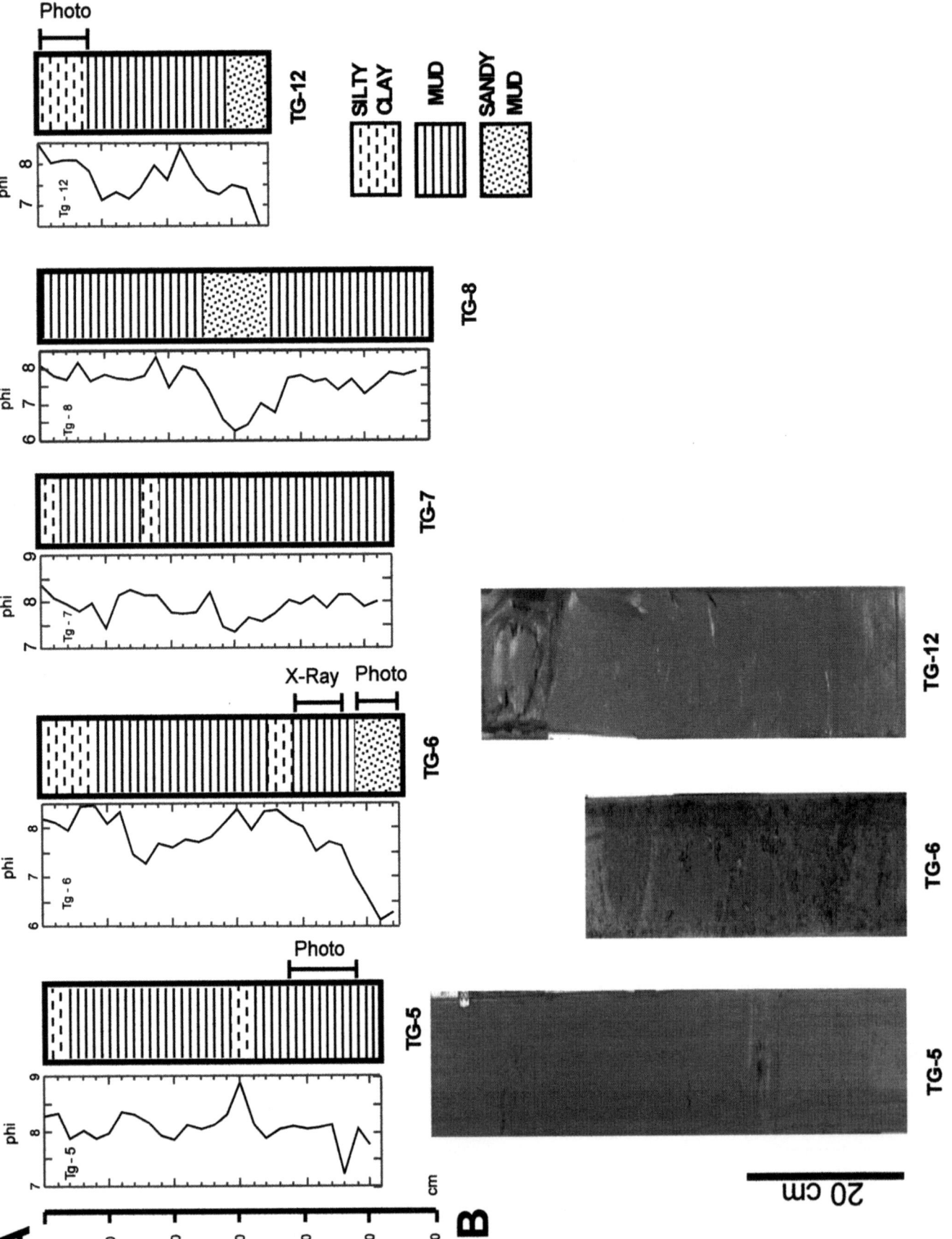

Fig. 12. (A) Lithostratigraphy of the sediment cores with the vertical distribution of the mean grain-size and (B) photographs of core sections from three textural types identified: muds (TG-5), sandy muds (TG-6) and silty clays (TG-12). The muds represent the predominant sediment whereas the sandy muds occur mainly on the bottom and the silty clays on the top. Core location on Figure 5.

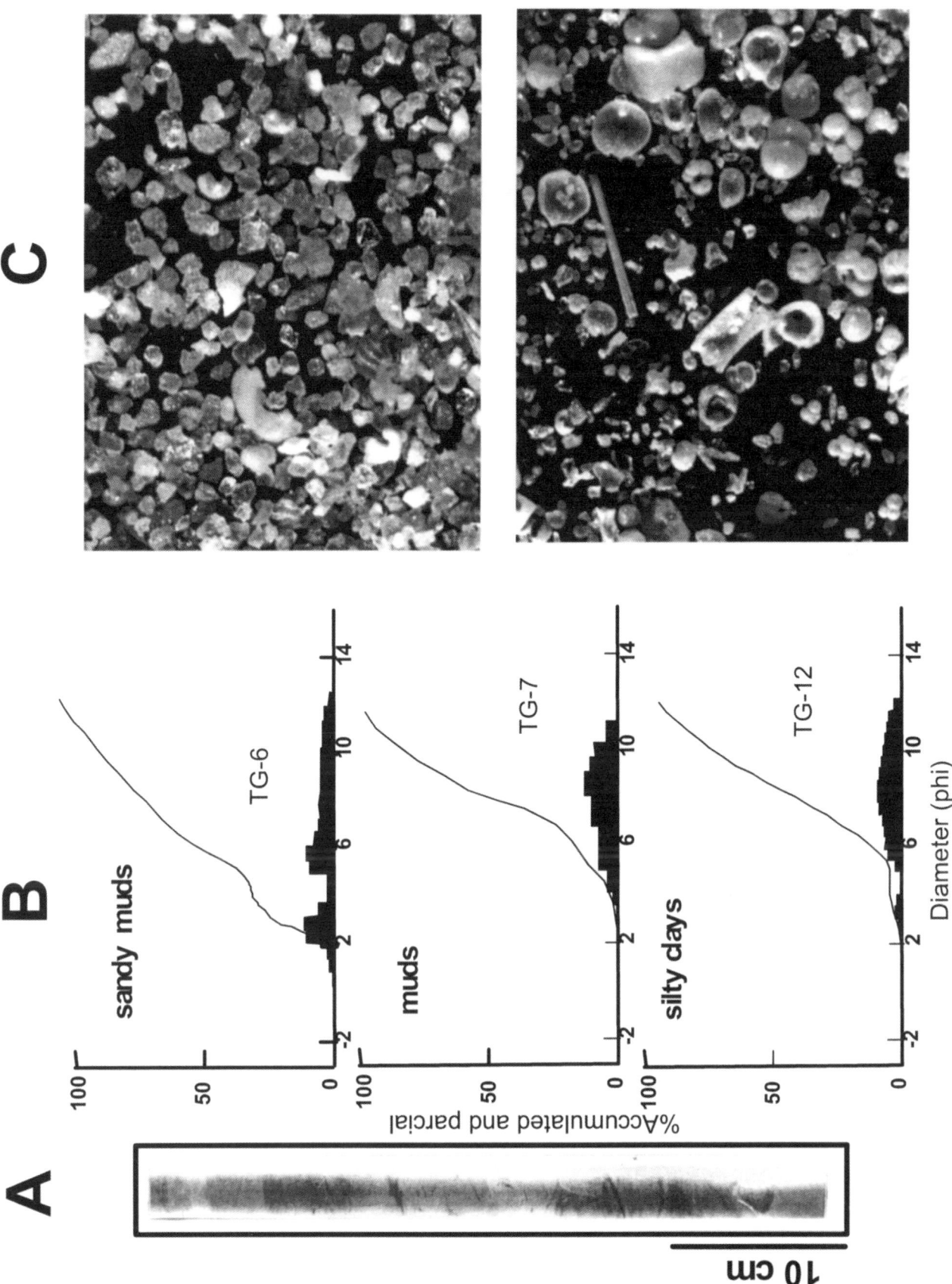

Fig. 13. X-radiograph, grain size distributions and sand fraction composition. (**A**) Positive x-radiograph of a 40 cm-long section, recovered from the northern end of the Ceuta Drift. This sediment is characterized by parallel and cross lamination and a mycellium-type bioturbation. (**B**) Examples of grain-size distribution curves typical of the three textural types identified: sandy muds, muds and silty clays. (**C**) Photographs of the sand fraction showing its main components. The sand fraction of the contourite sediments is mainly terrigenous (upper photo) although occasionaly the sandy muds display a mixed composition (lower boundary).

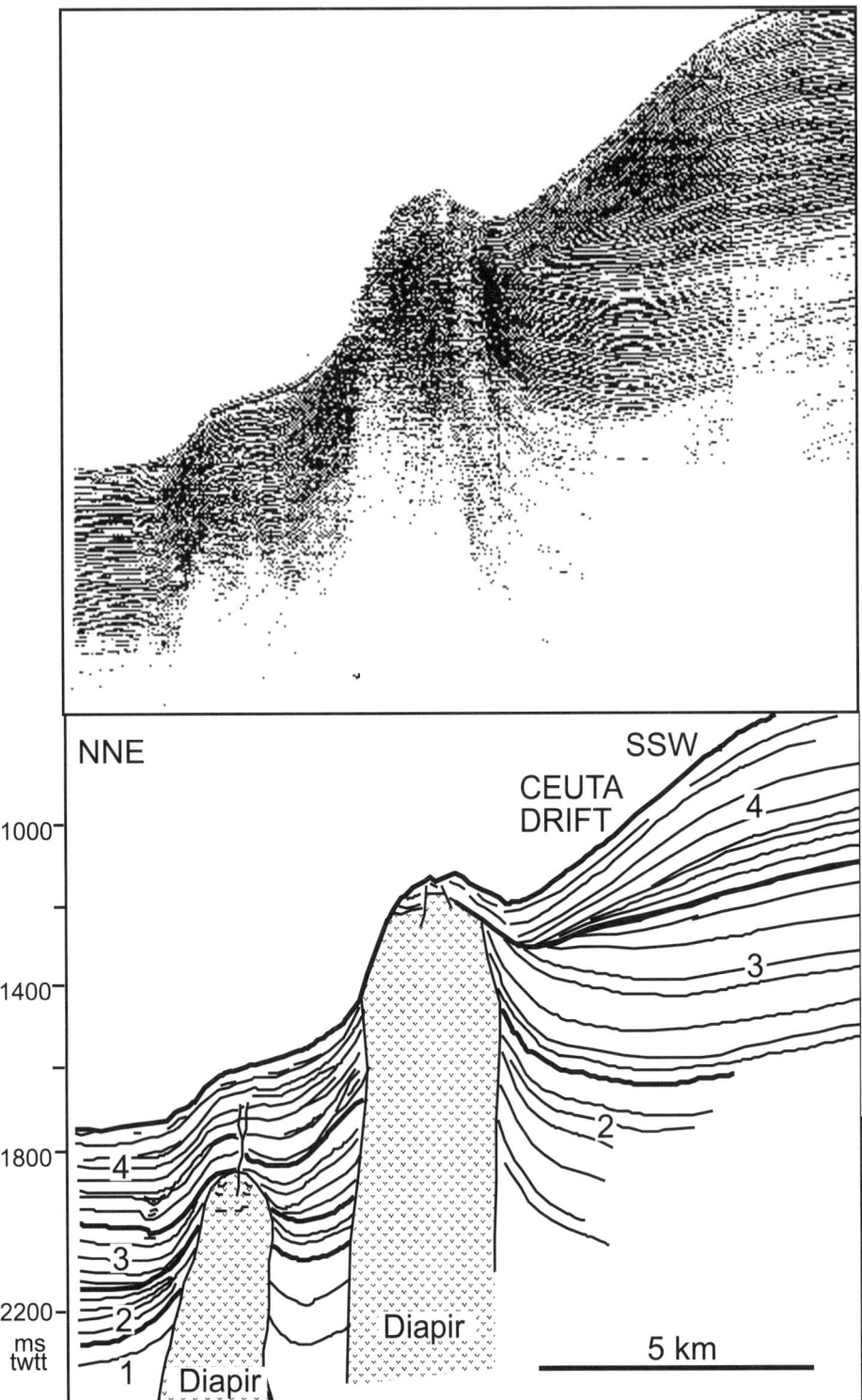

Fig. 14. High-resolution seismic profile with line drawing where the lateral relationships between the Pliocene and Quaternary Ceuta Drift deposits can be observed. Note the subvertical contact in the southern part of the highest diapir which outcrops above the seafloor, and the unconformity between the Pliocene (seismic unit 3) and Quaternary (seismic unit 4) deposits. The stratified facies of the Ceuta Drift comprise reflectors that converge toward the crest of the diapir, with internal downlapping terminations. (Modified from Pérez-Belzuz et al. 1997).

presence of mainly fine-grained sediments also suggests a low energy depositional environment.

The common occurence of sandy mud mainly in the lower part of the cores and of the silty clays at the top may be related to sea-level change (Fig. 12). Consequently, the sandy muds would be deposited during the last lowstand stage, when the shoreline was closer to the drift and coarser sediments were able to reach the slope areas of the margin and be deposited under the action of the Mediterraneran deep current. Deposits with similar coarse textures have also been identified along the northern Alboran margin, where they have been interpreted as lowstand deposits (Ercilla et al. 1994). The predominance of silty clays at the tops of cores would be related to the present highstand stage, when most of the sediment is being trapped on the shelf and only finer sediments reach the distal areas of the margin.

We dedicate this contribution to the memory of J. Baraza who unfortunately died during the writing of this manuscript. We would like to thank J. A. Howe and an anonymous reviewer for corrections and beneficial comments on the manuscript. Funding for this reseach was provided by the Spanish Interministerial Commission for Science and Technology (CICYT). This research was carried out in the framework of the Projects, Márgenes Continentales y Cuencas Profundas: Registro Sedimentario de la Variabilidad Paleoambiental y Palaeoclimática (Ref. AMB-95-0196), Morfoestructura del Mar de Alborán (Ref. AMB92-1315-E), and Marsibal (Ref. REN2000-0336).

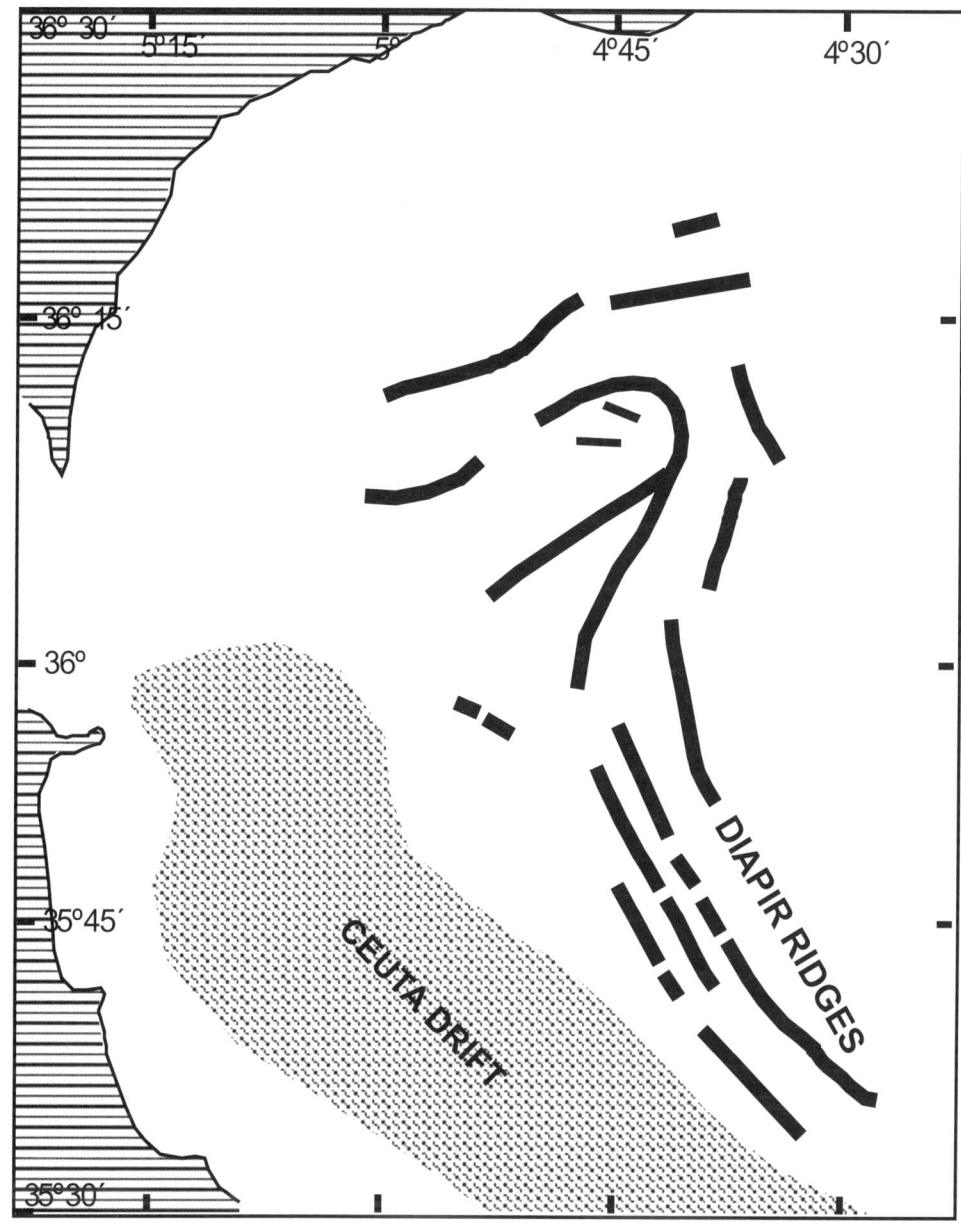

Fig. 15. Map showing the distribution of the crests of the diapirs in the Western Alboran Sea, and the location of the Ceuta Drift in the southwestern sector. The southernmost diapir crests are parallel to the main direction of the Ceuta Drift, and to the Moroccan slope. The presence of the diapirs have favoured the drift development between them and the Moroccan slope. (Modified from Pérez-Belzuz et al. 1997).

References

BRYDEN, L. B. & STOMMEL, H. M. 1982. Origin of the Mediterranean outflow. *Journal of Marine Research*, **40**, 55–71.

CAMPILLO, A. C., MALDONADO, A. & MAUFFRET, A. 1992. Stratigraphic and tectonic evolution of the western Alboran Sea: Late Miocene to Recent. *Geo-Marine Letters*, **12**, 165–172.

CHIOCCI, F. L., ERCILLA, G. & TORRES, J. 1997. Middle-Late Pleistocene stratal architecture of Western Mediterranean margins as the result of the stacking of lowstand deposits. *Sedimentary Geology*, **112**, 195–217.

COMAS, M. C., GARCIA-DUEÑAS, V. & JURADO, M. J. 1992. Neogene extensional tectonic evolution of the Alboran Basin from MCS data. *In*: MALDONADO, A. (ed.) *The Alboran Sea. Geo-Marine Letters*, **12**, 157–164.

EL MOUMNI, B. 1994. *Contribution a l'etude des paleoenvironnements sédimentaires au quaternaire terminal en Mer d'Alborán. Cas de la marge méridional (marge marocaine)*. PhD thesis, Méknes University, Méknes.

ERCILLA, G. & ALONSO, B. 1996. Siliciclastic sequence stratigraphy of passive and tectonically active western mediterranean margins during the Quaternary: The role of global versus local controlling factors. *Journal of Geological Society*, **117**, 125–137.

ERCILLA, G., ALONSO, B. & BARAZA, J. 1994a. Post-Calabrian sequence stratigraphy of the northwestern Alboran Sea. *Marine Geology*, **120**, 249–265.

ERCILLA, G., FARRÁN, M., ALONSO, B. & DIAZ, J. I. 1994b. Pleistocene progradtional growth pattern of the northern Catalonia continental shelf (northwestern Mediterranean). *Geo-Marine Letters*, **14**, 264–271.

FARRÁN, M. & MALDONADO, A. 1990. The Ebro continental shelf: Quaternary seismic stratigraphy and growth patterns. *Marine Geology*, **95**, 333–352.

FAUGÈRES, J.-C., STOW, D. A. V., IMBERT, P. & VIANA, A. R. 1999. Seismic features diagnostic of contourite deposits. *Marine Geology*, **162**, 1–38.

GASCARD, J. C. & RICHEZ, C. 1985. Water masses and circulation in the western Alboran Sea and the Strait of Gibraltar. *Progress in Oceanography*, **15**, 157–216.

HEBURN, G. W. & LA VIOLETTE, P. E. 1990. Variation in the structure of the anticyclonic gyres found in the Alboran Sea. *Journal of Geophysical Research*, **9**, 1599–1613.

HEEZEN, B. C. & JOHNSON, C. D. 1969. Mediterranean undercurrent and microphysiography west of Gibraltar. *Bulletin de l'Institut Océanographique Monaco*, **69**, 1–95.

HUANG, T. C., STANLEY, D. J. & STUCKENRATH, R. 1972. Western Alboran Sea: Sediment dispersal. Ponding and reversal currents. *In*: STANLEY, D. J. (ed.) *The Mediterranean Sea: A natural sedimentation laboratory*. Dowden, Hutchinson & Ross, Stroudsburg, Pennsylvania, 521–559.

LA VIOLETTE, P. E. 1987. Portion of the western Mediterranean circulation experiment complete. *Eos Transactions, AGU*, **68**(9), 123–124.

MALDONADO, A., CAMPILLO, A. C., MAUFFRET, A., ALONSO, B., WOODSIDE, J. & CAMPOS, J. 1992. Alboran Sea late Cenozoic tectonic and stratigraphic evolution. *Geo-Marine Letters*, **12**, 179–186.

MILLOT, C. 1987. Circulation in the western Mediterranean Sea. *Oceanologica Acta*, **10**, 143–149.

PÉREZ-BELZUZ, F., ALONSO, B. & ERCILLA, G. 1995. Modelizaciones en 3-D de los depocentros sedimentarios del Plio-Cuaternario en el margen septentrional y cuenca del Mar de Alborán: Significado Geológico. *In*: *XIII Congreso Español de Sedimentologia*. 26 de Junio al 2 de Julio. Teruel. España. Comunicaciones, 95.

PÉREZ-BELZUZ, F., ALONSO, B. & ERCILLA, G. 1997. History of mud diapirism and trigger mechanism in the western Alboran Sea. *Tectonophysics*, **282**, 399–422.

PÉREZ-BELZUZ, F. 1999. *Evolución Geológica del Mar de Alborán durante el Plio-Cuaternario: Sedimentación y Tectonismo*. PhD thesis, Universidad de Barcelona, Spain.

RYAN, W. B. F., STANLEY, D. J., HERSEY, J. B., FAHLQUIST, D. A. & ALLAN, T. D. 1970. The tectonics and geology of the Mediterranean Sea. *In*: MAXWELL, A. F. (ed.) *The Sea. Ideas an Observation on Progress in the Study of the Seas*. Wiley-Intersc. New York, **4**, 387–492.

STOW, D. A. V. & LOVELL, J. P. B. 1979. Contourites: their recognition in modern and ancient sediments. *Earth Science Reviews*, **14**, 251–291.

STOW, D. A. V. & HOLBROOK, J. A. 1984. North Atlantic contourites: an overview. *In*: STOW, D. A. V. & PIPER, D. J. W. (eds) *Fine-Grained Sediments: Deep-Water Processes and Facies*. Geological Society, London, Special Publication, **15**, 245–256.

STOW, D. A. V., FAUGÈRES, J-C., VIANA, A. & GONTHIER, E. 1998. Fossil contourites: a critical review. *Sedimentary Geology*, **115**, 3–31.

STOKER, M. S., AKHURST, M. C., HOWE, J. A. & STOW, D. A. V. 1998. Sediment drift and contourites on the continental margin off northwest Britain. *Sedimentary Geology*, **115**, 33–51.

TESSON, M., GENSOUS, B. & LABRAIMI, M. 1987. Seismic analysis of the southern margin of the Alboran Sea. *Journal of African Earth Sciences*, **6**, 813–821.

THE DONDE VA GROUP, 1984. Donde va? an oceanographic experiment in the Alboran Sea. *Eos*, **65**(36), 682–683.

# The Sicilian gateway: anatomy of the deep-water connection between East and West Mediterranean basins

MICHAEL S. REEDER[1,3], GUY ROTHWELL[2] & DORRIK A. V. STOW[1]

[1]*School of Ocean and Earth Sciences, University of Southampton, Southampton Oceanography Centre, European Way, Empress Dock, Southampton SO14 3ZH, UK*
[2]*Challenger Division, Southampton Oceanography Centre, Southampton SO14 3ZH, UK*
[3]*Present address: Gaffney, Cline & Associates, Bentley Hall, Blacknest, Alton, Hampshire GU34 4PU, UK (e-mail: mreeder@gaffney-cline.com)*

Abstract: The Sicilian gateway is a narrow, deep, interconnected series of basins, sill valleys and passageways that cuts across the broad, shallow Sicilian–Tunisian Platform in the Central Mediterranean. This deep connection allows dense Levantine Intermediate Water (LIW) formed in the Eastern Mediterranean to flow in a westerly direction through the gateway and exit into the Tyrrhenian and Balearic basins of the Western Mediterranean. LIW is replaced by a strong surface flow of Modified Atlantic Water (MAW). A complex and still active tectonic regime has been an important control on the development of physiography and on the style and distribution of sediments across the Platform.

Within the deep gateway basins, turbidites, debrites and megabeds are intercalated with a background of predominantly muddy and calcareous, hemipelagic and contourite sediments. Evidence for the influence of bottom currents on sedimentation is seen in the construction of small mounded drifts and irregular patch drifts, in zones of scouring and non-deposition, in local photographic evidence of a current-smoothed or rippled seafloor, and as a subtle combination of features present in the background sediments. These include: extensively reworked microfossil assemblages, rare diffuse lamination, coarse lenses of mixed composition within a pervasively bioturbated sediment, and relatively high rates of accumulation.

Between longitudes 10° and 16° E, the Mediterranean Sea shallows to a broad, silled platform known as the Strait of Sicily or the Sicilian–Tunisian Platform (Fig. 1). This broad shallow-water platform covers an area of approximately 250 000 km² between the landmasses of Sicily bordering to the north and Tunisia to the south, with a minimum separation distance of just 70 km (Marsala, Sicily to Cap Bon, Tunisia) and a maximum separation of over 440 km between SE Sicily and the Gulf of Sirte on the North African margin. It is cut through by an interconnected series of narrow elongate basins and deep sills that together form the Sicilian gateway. This gateway allows for the exchange of water masses between the east and west Mediterranean basins and has therefore played an important role in the sedimentation and oceanography of the Central Mediterranean region.

This paper presents an overview of the geological and oceanographic setting of the Sicilian gateway, and examines both seismic and sedimentary evidence for the role of bottom currents in the deposition and erosion of sediments across the region. In fact, the bottom current signature is not everywhere very evident and in many cases appears to be masked by interbedded turbidite/debrite and pelagic/hemipelagic facies. A summary of the principal characteristics of the Sicilian gateway is given in Table 1.

We have drawn together work on the sedimentary system carried out during the 1970's and early 1980's, as well as more recent oceanographic data. More detailed examination has been made of four giant piston cores and high-resolution 3.5 kHz seismic profiles collected during the 1995 *Marion Dufresne* expedition (*MD81*). This latter work is reported in more detail in a PhD thesis by the senior author (Reeder 2000) and by Reeder *et al.* (in press).

Geological and oceanographic setting

Geological framework

The Central Mediterranean region has a complex tectonic framework (Fig. 2) that is not yet fully understood and subject to several contrasting interpretations (e.g. Illies 1981; Winnock 1981; Finetti 1984; Cello *et al.* 1985; Jongsma *et al.* 1985; Boccaletti *et al.* 1987; Cello 1987; Catalono *et al.* 1995). The most significant result

Table 1. *Principal characteristics of the Sicilian gateway*

Table 1	**Principal characteristics of the Sicilian gateway**
Location	Central Mediterranean, separating the east and west Mediterranean basins
Setting	Series of fault-bound deep interconnected basins and valleys cutting across the shallow Sicilian–Tunisian Platform; basins > 1000 m, sill valleys around 400 m water depth
Age	Gateway assumed approximately present form in early Pliocene and has allowed water mass exchange since that time
Drift types	Small elongate mounded drifts and patch drifts in parts; more commonly contourites intercalated with downslope facies (sheet drift or mixed drift systems)
Dimensions	Gateway: 600 km long, 65 km wide (maximum) Basins: 50–100 km long, 15–25 km wide Sill valleys: 25–50 km long, 10–25 km wide Drifts: small mounds 5 × 20 km maximum, basinwide sheets
Seismic facies	Moderate-high amplitude sub-parallel reflectors, close to more widely spaced; downslope seismic facies intercalated with basinal seismic facies
Sediment facies	Interbedded turbidite, debrite, megabed and hemipelagite/contourite facies; subtle indications of contourite influence on background sediment.

From: STOW, D. A. V., PUDSEY, C. J., HOWE, J. A., FAUGÈRES, J.-C. & VIANA, A. R. (eds)
Deep-Water Contourite Systems: Modern Drifts and Ancient Series, Seismic and Sedimentary Characteristics.
Geological Society, London, Memoirs, **22**, 171–189. 0435-4052/02/$15.00 © The Geological Society of London 2002.

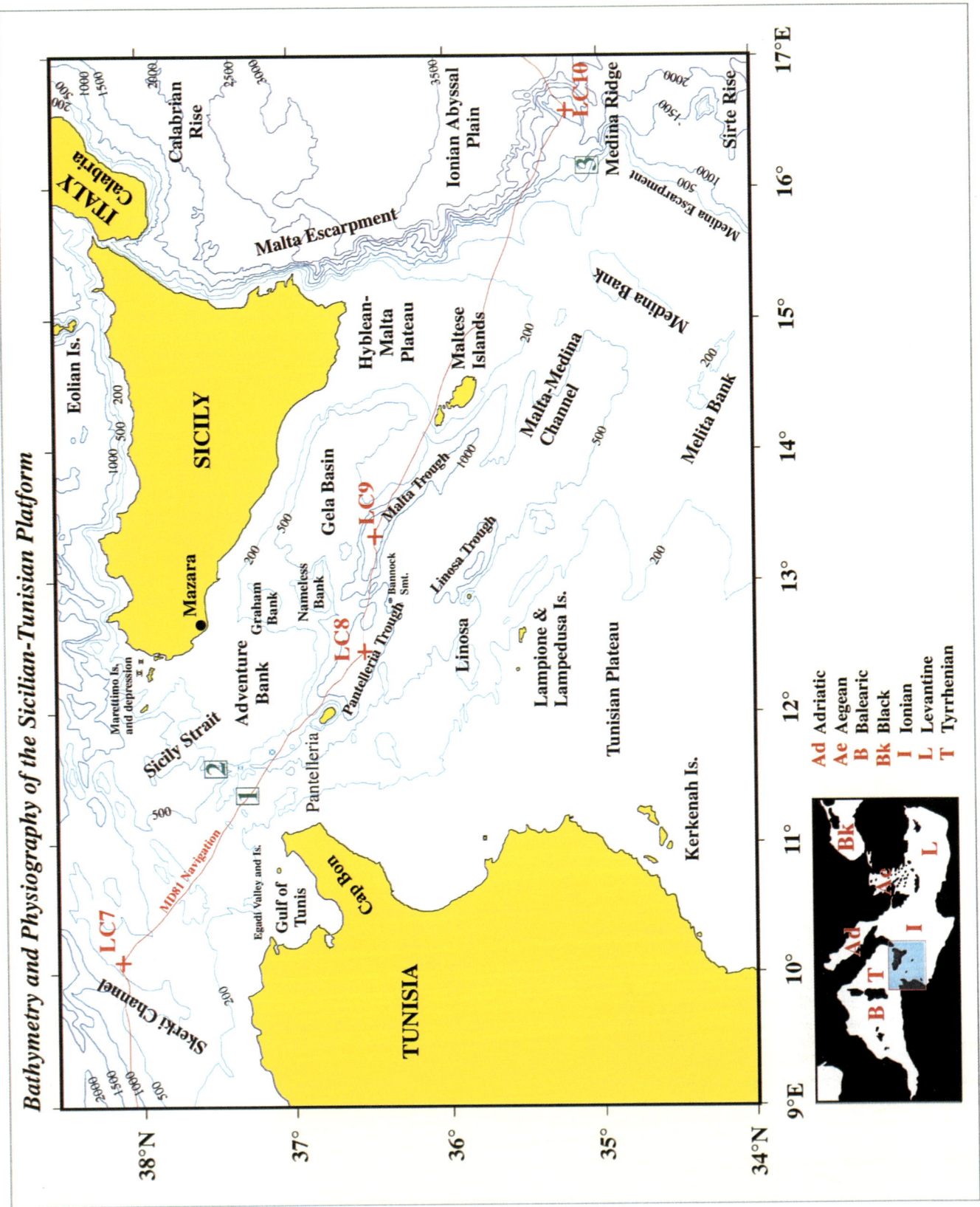

Fig. 1. General location, physiography and bathymetry of the Sicilian–Tunisian Platform. The map also shows the course of the Marion Dufresne 81 expedition (red line), giant piston core sites (red crosses), and the location of narrow valleys across the Sicily (1, 2) and Malta (3) sills.

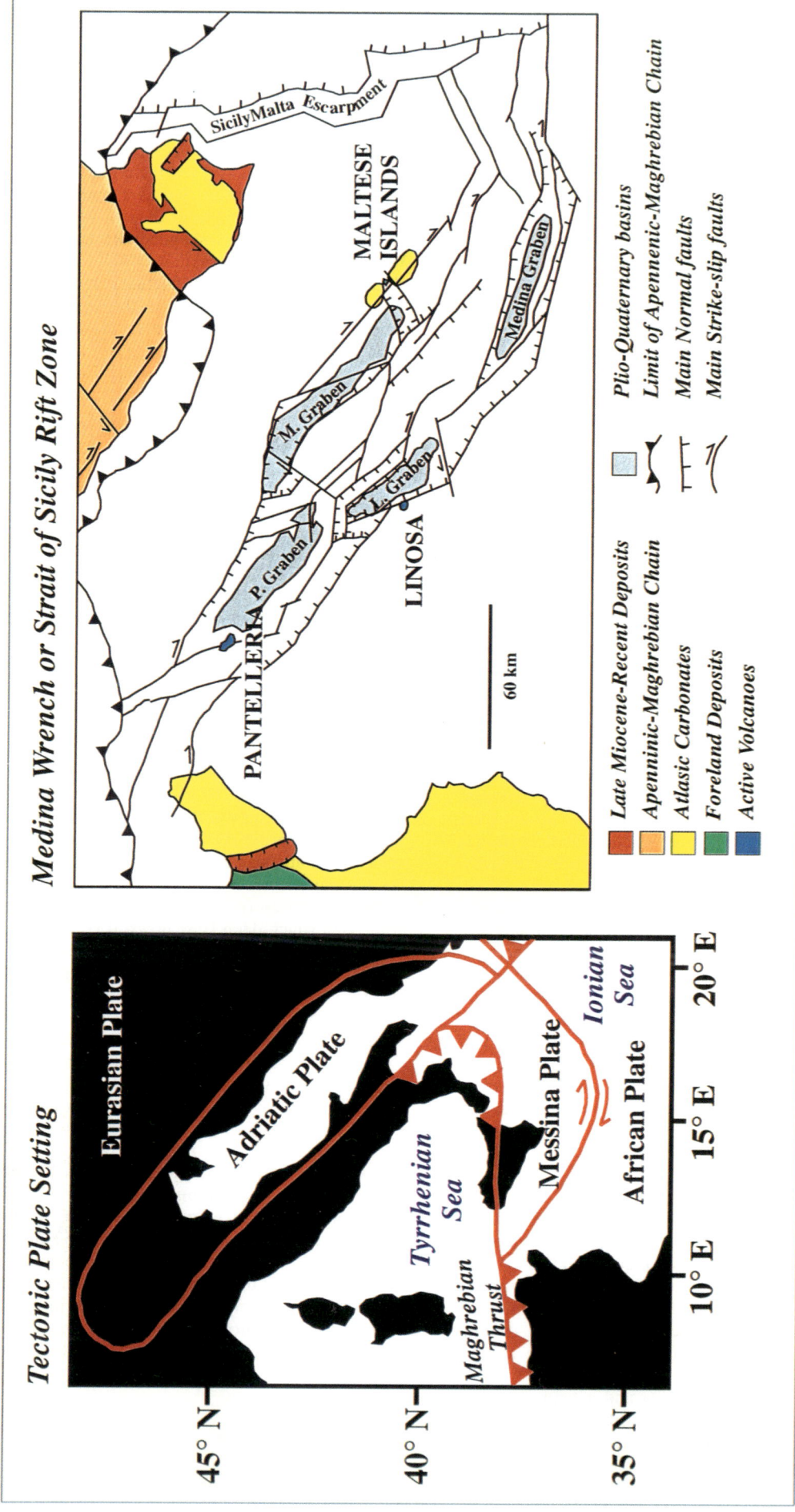

Fig. 2. Geological setting of the Sicilian gateway, showing the complex interaction of the Eurasian and African tectonic plates leading to creation of the Messina and Adriatic microplates (after Jongsma *et al.* 1985). The Messina microplate has been rotated creating a region of extensive dextral strike-slip faulting across the Sicilian–Tunisian platform known as the Medina Wrench Zone (after Cello 1987). This zone with its deep troughs and sill valleys is known as the Sicilian gateway.

of these complex tectonic movements, as far as the study area is concerned, was the creation of deep, transtensional grabens and associated transpressional horsts within the Sicilian–Tunisian Platform, and the two bounding sills/escarpments of the eastern and western platform margins. The most widely accepted tectonic history of the region is summarized below.

During early Tertiary time, the northern edge of the African Foreland collided with the southern edge of the Eurasian plate. This resulted in partial subduction of the African Foreland beneath the Eurasian plate and southwards thrusting of the Maghrebian Arc across central Sicily during the Oligocene–Miocene period. As collision continued, the Ionian Abyssal Plain to the east was subducted beneath the Calabrian Foreland forming the steep Malta and Medina Escarpments with a total vertical displacement of between 2 and 3 km to the east. Major NW–SE trending faults were created during this time across the Sicilian–Tunisian Platform together with a conjugate series of secondary faults oriented approximately WSW–ENE.

Resistance to subduction by the continental crust of the Hyblean–Malta Plateau led to small NW–SE faulted blocks being rotated anti-clockwise by 15°, giving a wide variety of fault fabrics and structural styles across the Sicilian–Tunisian Platform. This rotation created a shear zone, named the Medina Wrench Zone (also known as the Strait of Sicily Rift Zone), showing an overall dextral movement. The zone extends at least 800 km from the NW extremity of the Sicily Strait to the eastern end of the Medina Ridge, and is characterized by a series of deep en-echelon troughs. The Messina micro-plate has been formed from African continental crust and has the subduction zone of the Calabrian Arc to the North and the dextral strike-slip Medina Wrench Zone to the south (Cello 1987; Jongsma et al. 1987).

There are five main centres of volcanism in the region that are believed to be related to the extension and thinning of continental crust in the Medina Wrench Zone and to subduction of the African plate beneath the Eurasian plate (Di Paolo 1973; Grandjacquet & Mascle 1978; Boccaletti et al. 1984; Jongsma et al. 1985; Calanchi et al. 1989; Argnani 1993; Colantoni et al. 1993, amongst others). The five centres are: (1) the major volcano of Mount Etna on the eastern margin of Sicily, (2) the island of Pantelleria, (3) the island of Linosa, (4) the Eolian islands to the north of Sicily, and (5) localized submarine activity.

Seismic activity, though relatively less than in the eastern Mediterranean at present, occurs in relation to wrench and subduction tectonics and as a result of volcanic activity. On the basis of high-resolution seismic records, there has been much active faulting of recent and/or subrecent origin throughout the platform area. Vertical displacement along faults is commonly in excess of mean sedimentation rates, i.e. greater than 20 cm ka^{-1} (Maldonado & Stanley 1976). Locally, there is intense folding and distortion of near-surface reflectors, indicative of compression; elsewhere, the faulting is apparently normal and extensional.

Oceanographic setting

The present-day circulation over the Sicilian–Tunisian Platform is thermohaline in nature, essentially driven by the formation and sinking of dense water in the Eastern Mediterranean (Fig. 3). This deep saline water mass, known as Levantine Sea Intermediate Water (LIW), flows from east to west through the Sicily gateway and is replaced by less dense surface water (Modified Atlantic Water, MAW) passing over the platform and into the Eastern Mediterranean Sea.

LIW has an average salinity of 38.7‰ and a temperature of approximately 14°C (Stanley et al. 1975; Moretti et al. 1993) and sinks in the Levantine Basin forming an anticlockwise gyre. This deep current flows in a westerly direction towards the Sicilian–Tunisian platform, through which its direction is controlled by the gateway bathymetry. At the Sicily Channel the current is split in two by a small central ridge between two narrow pathways, one shallowing to 365 m on the Tunisian side and the other to 430 m deep on the Sicilian side (Astraldi et al. 1996). The annual mean high velocity on exit through the eastern passage is greater than 35 cm s^{-1}, whereas a value of only 10 cm s^{-1} has been measured in the western passage. Bottom water velocities recorded over the main platform region are 15 cm s^{-1} (winter) and 5 cm s^{-1} (summer) (Marani et al. 1993; Astraldi et al. 1996). On exiting the Sicily gateway, the main part of the LIW flows northwards into the Tyrrhenian Sea as a result of bottom topography and the Coriolis Effect (Astraldi et al. 1996).

The surface water mass, generally known either as Atlantic Water (AW) or Modified Atlantic Water (MAW), has a salinity of 37.4‰ and temperature that fluctuates seasonally between 13° and 23°C. It flows generally eastwards through the gateway region at a velocity between 10 and 25 cm s^{-1}, typically, and >30 cm s^{-1} at times, although large eddies are also observed and one distinct strand of the current flows nearer the Tunisian coast. MAW has a base at 100–200 m, with mixing between LIW and AW occurring up to a water depth as shallow as 60 m (Stanley et al. 1975; Manzella et al. 1990; Moretti et al. 1993; Astraldi et al. 1996).

Bathymetry

The Sicilian–Tunisian platform has an irregular and varied bathymetry (Figs 1 & 4). At its western extremity it slopes down towards the Balearic Abyssal Plain and the islands of Sardinia and Corsica of the Western Mediterranean. Its eastern margin is denoted by the Malta Sill and the steep, approximately N–S-trending Medina and Sicily–Malta Escarpments which fall sharply to the Ionian Abyssal Plain and Sirte Rise of the Eastern Mediterranean Basin. Between these, approximately 47% of the platform area has a water depth shallower than 200 m, comprising broad flat shelves and banks along the Tunisian and Sicilian margins. A further 50% of the area comprises an irregular, typically gently sloping zone between 200 m and 600 m water depth, including the Gela foredeep basin south of Sicily. Several islands are emergent in the central part of the Sicilian–Tunisian Platform, including the small carbonate islands of Malta, Lampedusa and Lampione, and the volcanic islands of Pantelleria and Linosa.

The Sicily gateway itself is the 3% of the region deeper than 600 m through which deep LIW flows. It comprises three deep fault-bounded en-echelon troughs, the Pantelleria, Malta and Linosa Troughs, the broad Malta–Medina Channel and the narrow gaps or passageways across the Sicily and Malta sills. Together, these trend in a NW–SE orientation and incise the central part of the Sicilian–Tunisian Platform.

The Pantelleria Trough is the most westerly of the three grabens and is approximately 80 km in length and 30 km in width, with a relatively flat floor in excess of 1300 m water depth. The island of Pantelleria is to the NW of the trough, and is bounded by high, irregular slopes of 5–30° on the NE flank and 6–11° on the southwestern flank. The southeastern end is closed by the Bannock Seamount. *The Malta Trough* lies to the NW of the carbonate horst supporting the islands of Malta, Gozo and Comino and forms a narrow (18 km wide) elongate (150 km long) that reaches a depth of over 1700 m (1721 m at core site LC9). *The Linosa Trough* is located to the north of the volcanic island of Linosa. It is 75 km long, 15 km wide and over 1600 m deep. It opens to the southeast into the broader, shallower *Malta–Medina Channel*. The depths of the three sills (as numbered on Fig. 1) are 365 m, 430 m and 370 m respectively.

Stratigraphic context

During latest Eocene time, there existed a 300 km wide slab of Neotethyan oceanic crust between southern Europe (Corsica,

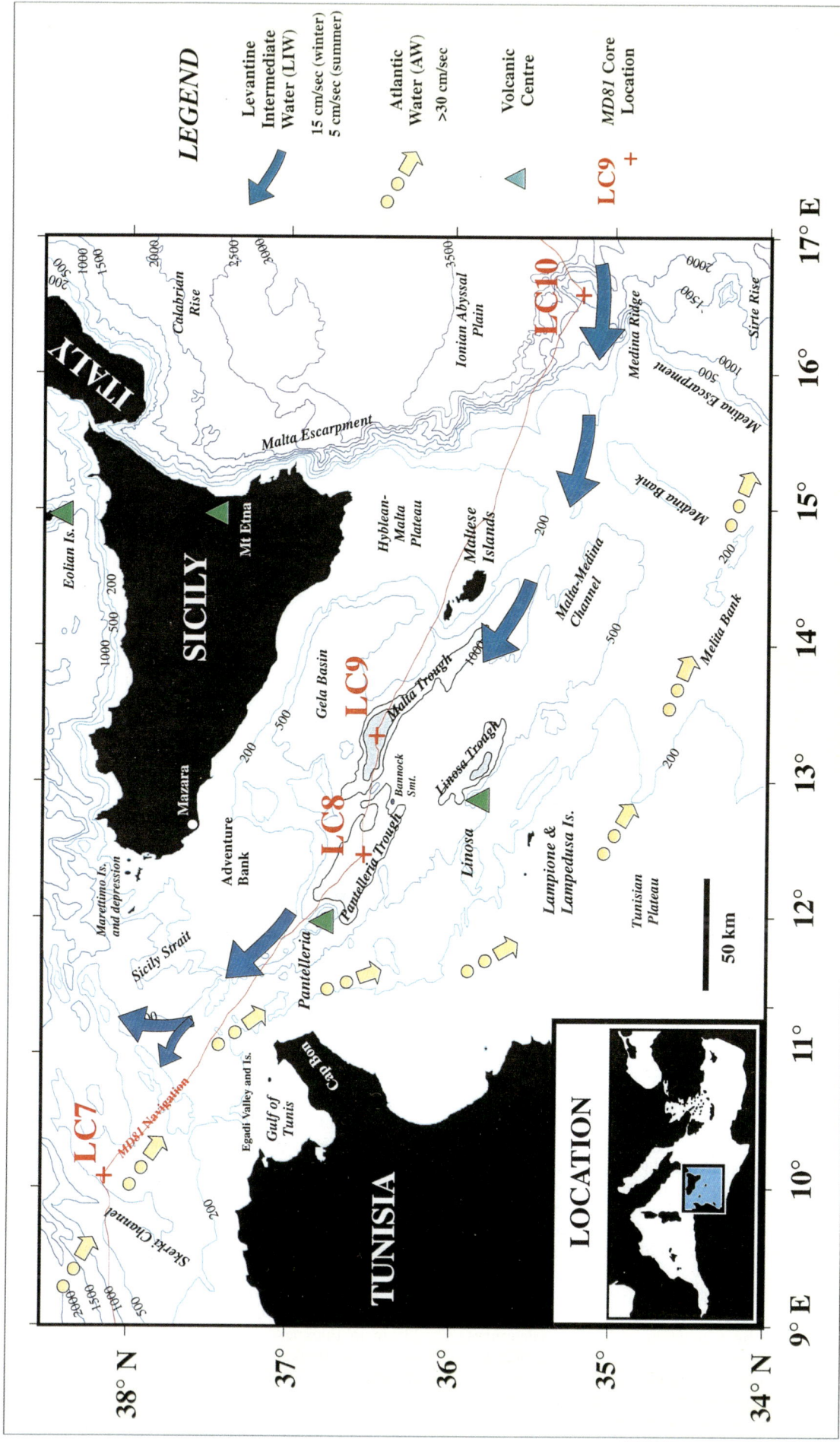

Fig. 3. Oceanographic setting and water mass exchange through the Sicilian gateway. Deep westward flowing LIW is contained within the deep troughs and sill valleys; shallow MAW follows the North African coastline (modified from Manzella et al. 1990; Moretti et al. 1993; Astraldi et al. 1996).

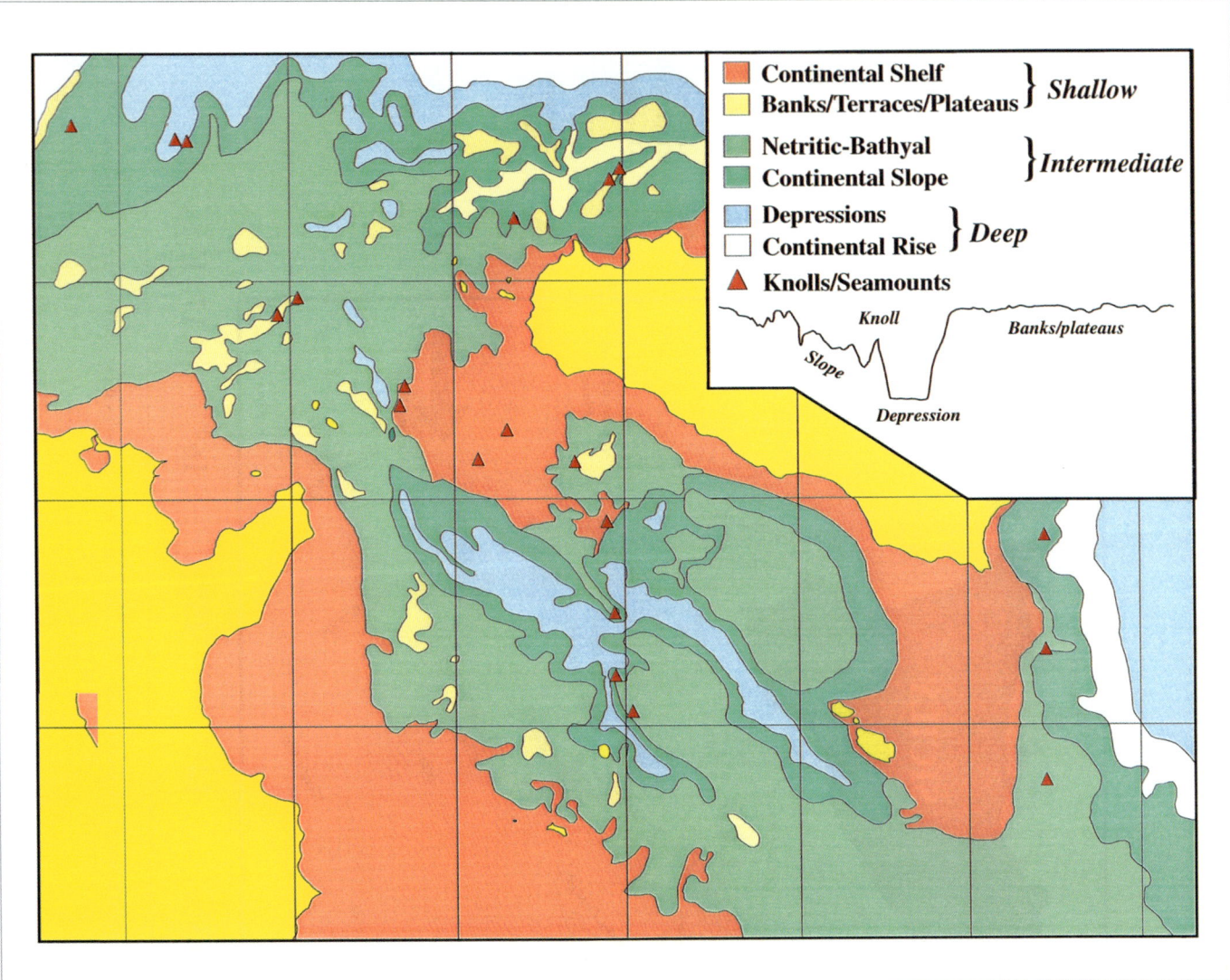

Fig. 4. Physiographic regions of the Sicilian–Tunisian Platform determined from bathymetry and topographic features (modified from Maldonado & Stanley 1976, 1977).

Sardinia and the Ballearic islands) and northern Africa (Tunisia and including most of Sicily). Complex closure together with re-positioning of plate boundaries took place through the Oligocene and most of the Miocene (Dewey *et al.* 1989), so that the present day tectonic configuration did not exist before the latest Miocene to early Pliocene.

Although there is an absence of well control through the sedimentary successions of the gateway region, the top-Messinian horizon is readily identified on Sparker profiles, so that the thickness of the Plio-Quaternary section can be estimated (using a sonic velocity of 1800 m s^{-1} through these relatively unconsolidated sediments). This varies markedly in both the shallow and intermediate depth areas, from 0–225 m (0–0.25 s TWT) on the platform, and from 0–720 m (0–0.8 s TWT) in the intermediate region. The deeper basins have a more uniform, thicker Plio–Quaternary section, in some cases up to 1260 m thick (1.4 s TWT).

The cores available for this study reach a maximum length of just over 31 m, and therefore provide a high-resolution record of part of the Pleistocene–Holocene succession only. This part of the sequence can be well dated by a combination of micropalaeontological analysis, oxygen–isotope stratigraphy, tephra chronology and, for core LC10 only, sapropel chronology from the eastern Mediterranean basin (Fig. 5, see Reeder 2000). The sill region successions (LC7 and LC10) are the most slowly accumulated, extending back to about 700 ka and 550 ka respectively. The trough sequences (LC8 and LC9), by contrast, were deposited much more rapidly, each within the last 60 ka. Radiocarbon dating of the shorter cores (maximum around 8 m long) recovered during earlier cruises yielded basal ages < 50 ka.

Seismic characteristics

This study had access to a limited number of medium (Sparker) and high-resolution (3.5 kHz) seismic profiles across the Sicilian–Tunisian platform area, sufficient, however, to illustrate the complexity and variability of sedimentation in the region (Figs 6–8). For the purposes of this study, tectonic basement is generally visible either as a more or less deformed (folded, faulted, thrust-faulted) pre-Messinian seismic unit or, locally, as a distinctive and irregular post-Messinian volcanic complex.

Overlying this basement, the shallow platform regions (continental shelves, banks) show a reduced Plio–Quaternary section, generally less than 0.1 s but locally up to 0.25 s (TWT) or about 225 m in thickness. The intermediate regions vary from a thin regular to thicker irregular Plio–Quaternary cover, averaging around 0.4 s (TWT, 360 m). The Gela Foredeep, for example, shows a heavily faulted clastic sediment fill of a foreland basin, at least 0.5 s thick (TWT, 450 m) in parts, but with pinchouts over fault highs as well as thickening on the downthrown side.

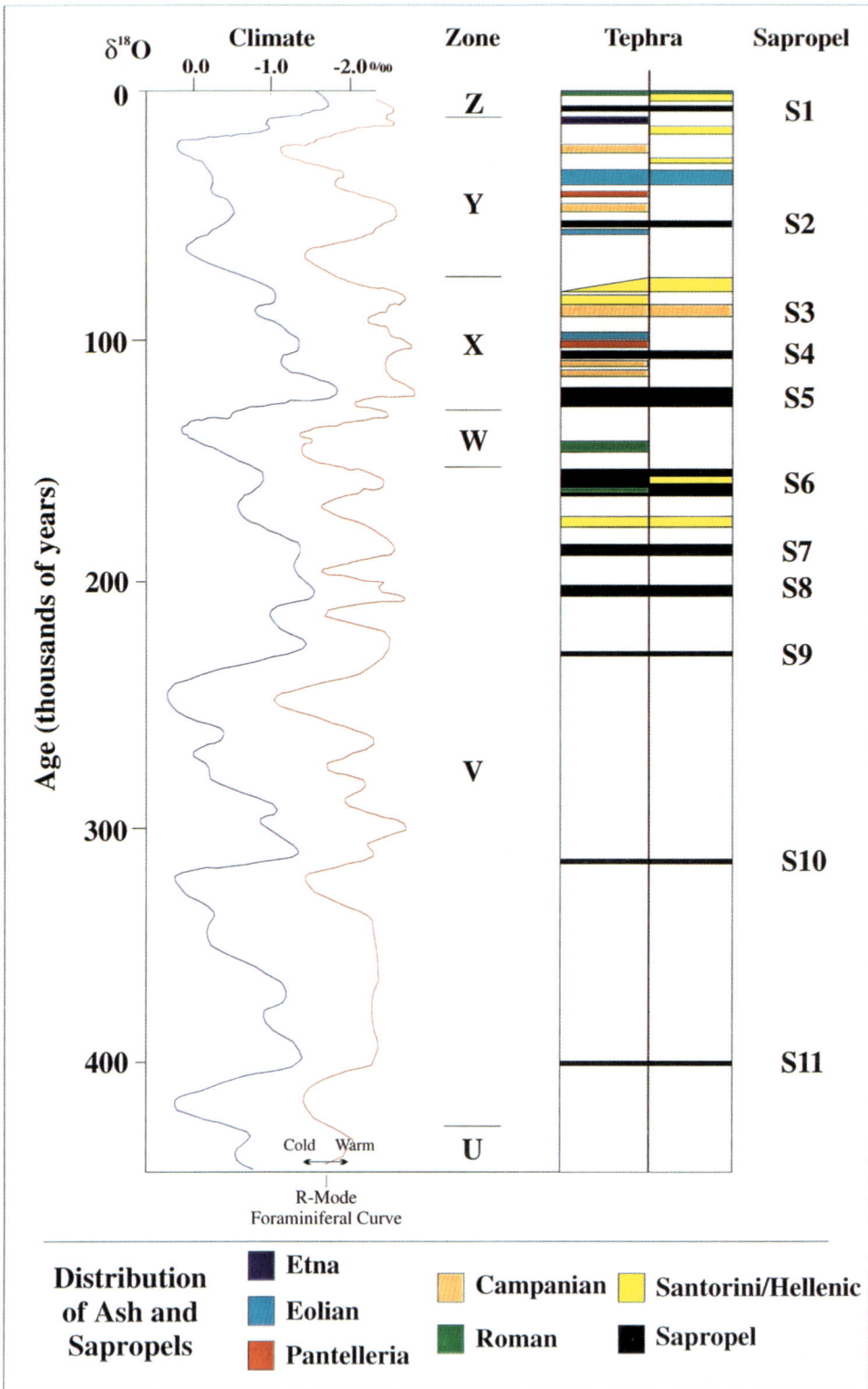

Fig. 5. Stratigraphic framework for the Pleistocene–Holocene section recovered in giant piston cores from the Sicilian gateway region, showing climatic zones based on oxygen isotopes, tephra and sapropel layers (modified after Keller *et al.* 1978).

The deep narrow basins and channels as well as the Malta escarpment show steep, faulted margins with irregular sediment cover and extensive slide-slump seismic facies evident in parts. The basin fills comprise regular parallel to sub-parallel, moderate amplitude reflectors, over 1.4 s thick (TWT, 1260 m) in places. The dominant thin, closely-spaced reflectors on 3.5 kHz profiles are interspersed with thick parallel-sided acoustically transparent layers, especially in the Pantelleria Trough and the deeper part of the Malta Trough. From direct correlation with cores, these are shown to be debrite-turbidite megabeds (Reeder *et al.* 2000).

Within the narrow Sicily Channel that forms the northern part of the Sicily gateway, there is a small elongate mounded drift flanked on either side by erosive/non-depositional moats. This forms the upper 0.25 s (TWT) or at least 200 m of section. Smaller patch drifts and localised scour, including the exposure of bedrock on the seafloor, are also evident (Marani *et al.* 1993; Bowles *et al.* 1993).

Sediment characteristics

Physiographic regions and facies

Early work by D. J. Stanley and co-authors (Stanley *et al.* 1975; Maldonado & Stanley 1976, 1977), based on the analysis of

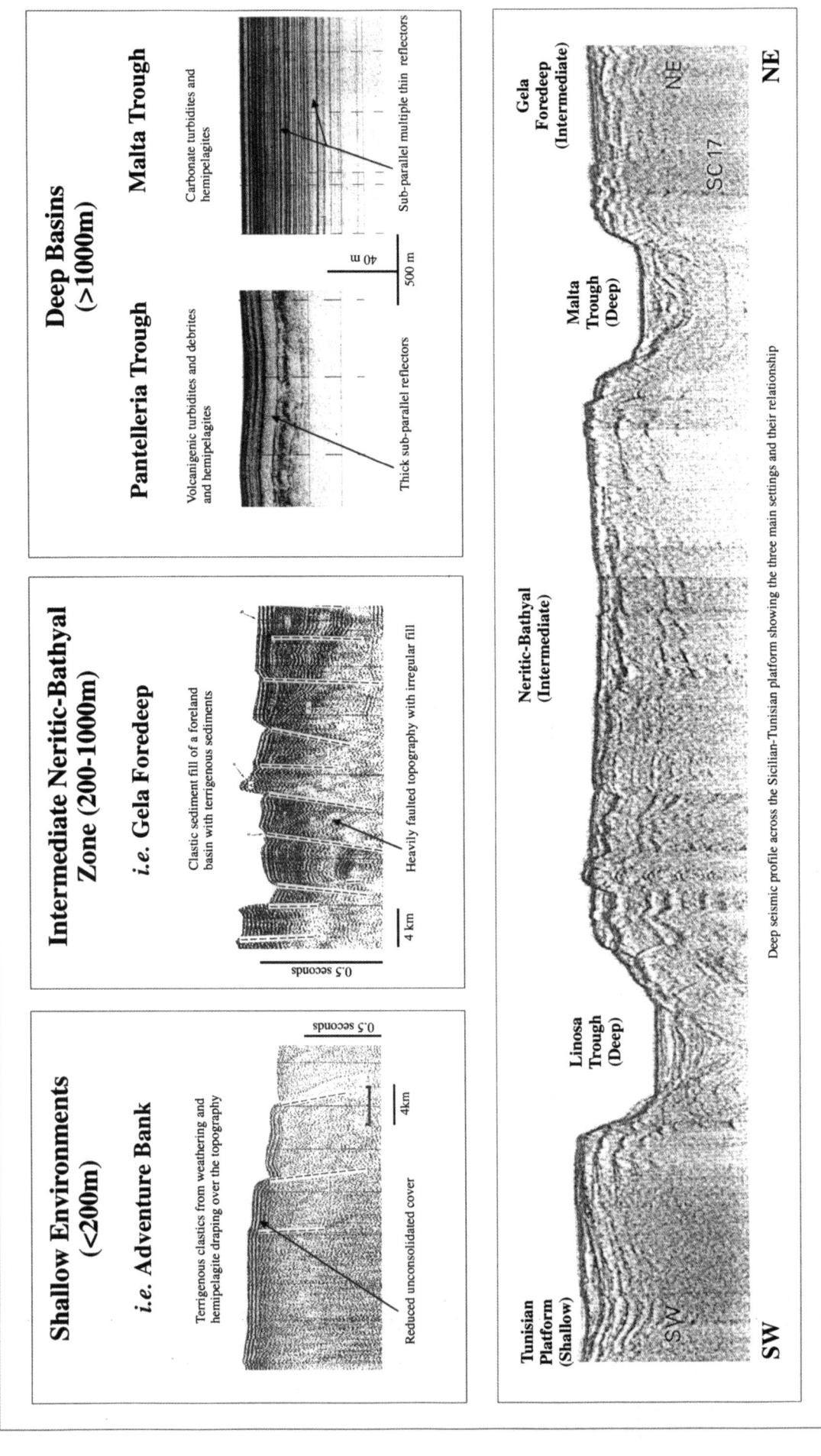

Fig. 6. Seismic images of the three main physiographic regions of the Sicilian–Tunisian Platform. (a) The shallow environment characterized on the Adventure Bank (Sicilian margin) by terrigenous sediments and on the Tunisian margin by platform carbonates (section after Maldonado & Stanley 1976). The physiographic region forms approximately 47% of the Sicilian–Tunisian Platform area. (b) The heavily-faulted horst and graben topography of the neritic-bathyal intermediate environment, forming approximately half of the regional physiographic area (after Maldonado & Stanley 1976). (c) Sub-parallel reflectors characterizing the sub-surface of flat, deep troughs (3% of area), the fill of which is dominated by sediments from the local islands and pelagic settlings (from the MD81 cruise). (d) The relationship between the three physiographic regions (after Calanchi et al. 1989).

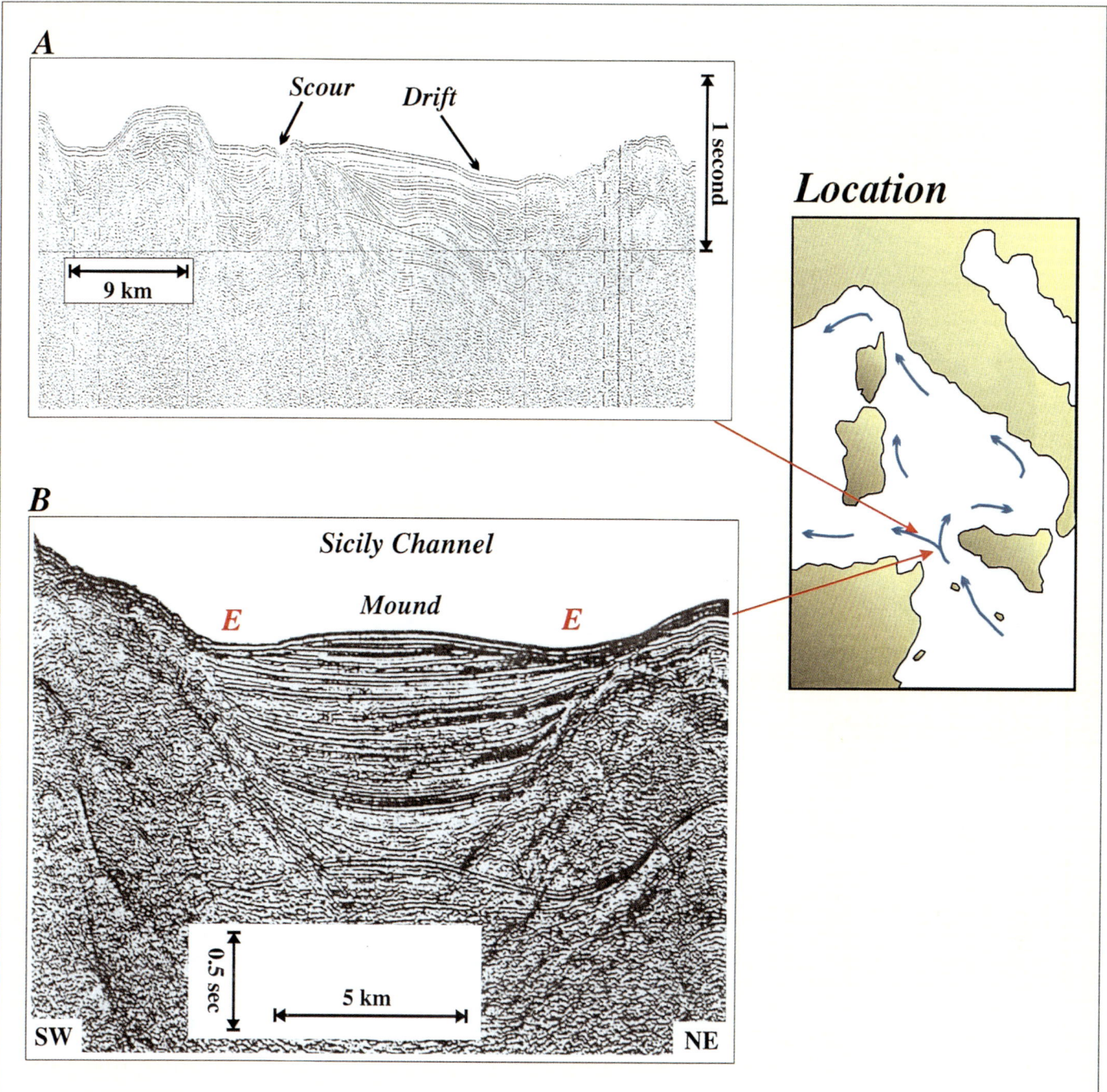

Fig. 7. Seismic evidence for erosion and deposition by bottom currents in the Sicilian gateway region, after Bowles *et al.* (1993) (A) and Marani *et al.* (1993) (B).

seafloor photographs, 3.5 kHz and Sparker seismic profiles and some 32 piston cores, provided the first good overview of sedimentation across the whole platform region. The principal physiographic regions they identify are the shallow banks (with local topographic highs), the intermediate depth platform (including neritic-bathyal, canyon and slope), and the deep basins (Fig. 4, Section 3), each with their own distinctive sediment facies and faunas. Subsequent work by Colantoni *et al.* (1993), and Reeder (2000, Reeder *et al.* in press) focused on sedimentation in the deeper basins and in the eastern and western sill slope regions.

The principal facies found across the whole region are listed below, and their distribution illustrated in Figure 9.

(1) Rock and gravel – including exposed bedrock on the seafloor and coarse gravel lag deposits, mainly siliciclastic, mixed with some bioclastic; typically found along rocky coastlines, on volcanic banks and outer shelf highs.
(2) Coarse calcareous sands – containing high proportions of bioclastic detritus, found mainly on the shallow platforms; typically concentrated by current winnowing.
(3) Sands and silts – generally bioclastic or volcanigenic, distributed on both the shallow platforms and in the deep basins, resulting from airfall tephra and turbidity currents.
(4) Muds – comprising the most abundant of the platform facies, more or less calcareous biogenic in nature, deposited by hemipelagic, contourite and turbidite processes.

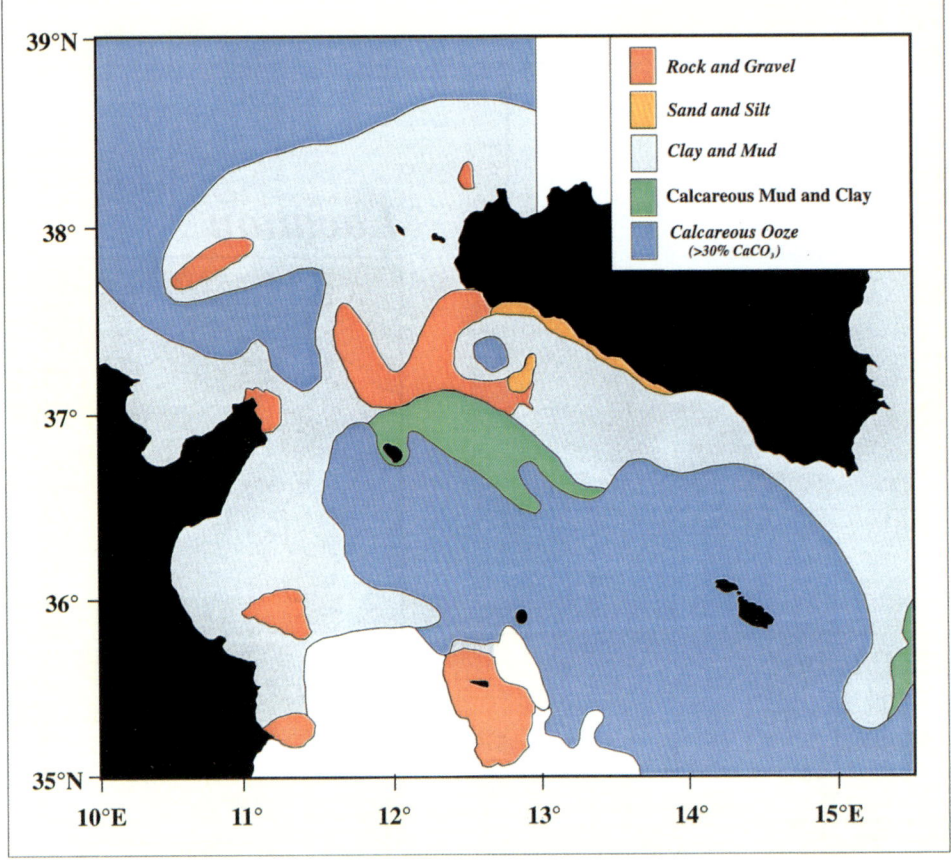

Fig. 8. Map showing surficial sediment distribution across the Sicilian–Tunisian Platform (after Stanley & Maldonado 1976).

(5) Sapropels and organic oozes – dark-coloured, fine-grained, organic-carbon rich sediments that accumulated under periodic anoxic conditions in the eastern Mediterranean Basin; present only in the easternmost sill region (LC10), and as a dark hemipelagic mud in the Linosa Trough.

(6) Volcanic ash – occurring as graded layers of mud, sand and gravel size, present across the whole platform, but particularly preserved as primary air-fall tephra layers and as secondary, mixed bioclastic-volcaniclastic turbidites on the deep basin floors.

Facies associations and depositional processes

Distinctive facies associations and depositional processes characterise each of the principal environments.

The shallow platform area is the most varied, typified locally by exposed bedrock, gravel substrate and volcanic ashfall deposits, and more generally by a complex mosaic of coarse calcareous sands, mixed composition sands and silts, and bioturbated muds. Sediments are supplied from coastal erosion, river drainage, direct volcanic fall, benthic macrofauna and planktonic microfauna. They are distributed by the normal range of shelf processes, including waves, tides and shelf currents, as well as hemipelagic accumulation in more protected regions. Winnowing by strong, shallow water, bottom currents is responsible for pockets of coarse bioclastic and mixed sand facies.

The intermediate zone is dominated by homogeneous, bioturbated, calcareous mud-rich facies, resulting mainly from slow hemipelagic accumulation. Locally, there are different facies, including turbidite sands and silts, mostly in slope channels, sapropel layers on the eastern slope towards the Ionian Basin and Sirte Rise, and thin airfall volcaniclastic horizons interbedded throughout the region. On steeper slopes the sediments are subject to slide and slump remobilization. Cores LC7 and LC10 from the western and eastern slopes respectively are illustrated in Figures 10 and 11.

The deep basins and sill channels are characterized by regularly interbedded sand, silt and mud facies and, more rarely, gravels (Figs 12 and 13). These are deposited by the normal range of deep-water processes including debris flows, turbidity currents, bottom currents and hemipelagic fall. Slump deposits are locally present around basin margins, whereas megabeds made up of debrite-turbidite couplets are known to extend across the whole basin floor. Turbidites in the Pantelleria Trough are mostly derived from a volcanic source and mixed with variable amounts of bioclastic debris. Those recovered from the Malta Trough are dominantly bioclastic in composition.

Sedimentary evidence for the influence of bottom currents on both the mud and some of the sandy horizons is generally very subtle, based on the following aspects.

- Whereas much of the western slope core (LC7) can be interpreted as being of hemipelagic origin, bottom current influence in parts is indicated by pervasive bioturbation coupled with rare diffuse lamination, mixed composition, broken bioclastic debris, and a distinctive oscillation in mean grain size.
- The local concentration of bioclastic sandy horizons with sharp tops and an absence of turbidite mud caps (in the Strait Narrows channel and in core LC7), suggests the introduction of material by turbidity currents and its subsequent winnowing by bottom currents.
- The presence of partially reworked microfossil assemblages in slope cores LC7 and LC10, as well as highly reworked assemblages in basin cores (LC8 and LC9), suggests bottom current reworking.
- The calcareous muddy intervals between turbidite layers in

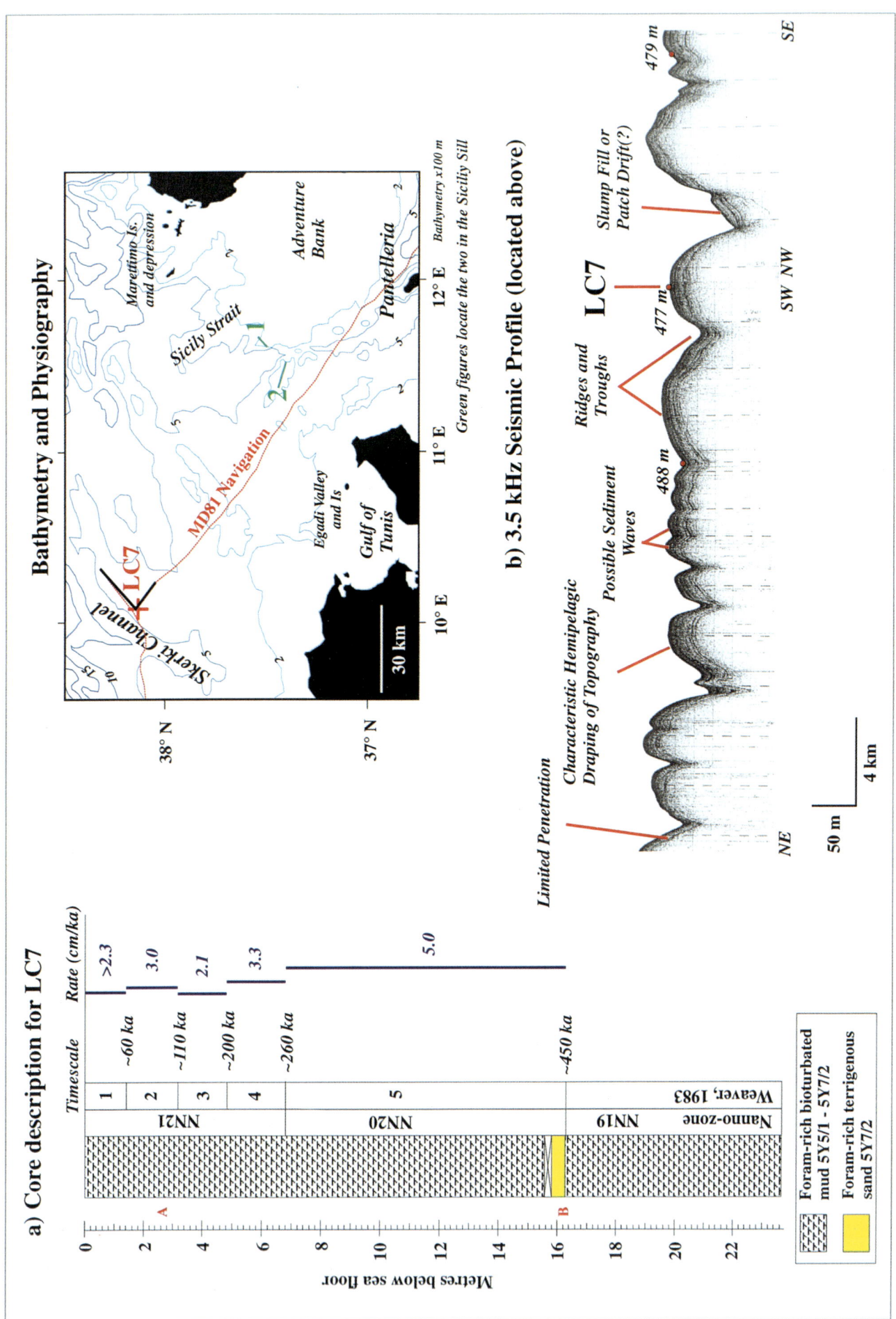

Fig. 9. Sediment and 3.5 kHz seismic characteristics of the western Sicilian–Tunisian Platform, on the slope west of the Sicily sill. (a) LC7 core log – A and B show the location of core photographs in Figure 13. (b) 3.5 kHz seismic profile passing through the core site, shown by dark line on the location map above.

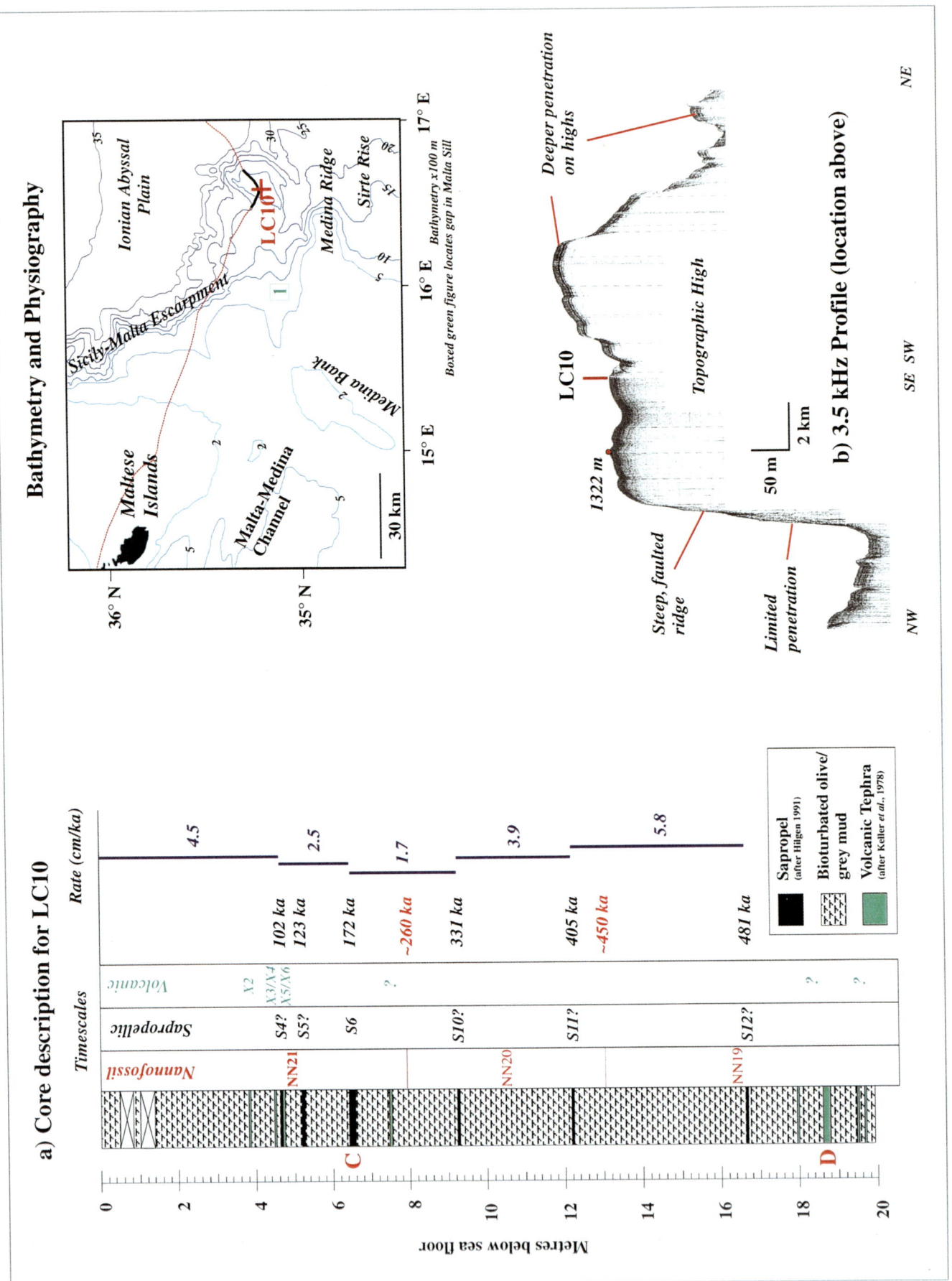

Fig. 10. Sediment and 3.5 kHz seismic characteristics of the eastern Sicilian–Tunisian Platform, on the slope east of the Malta sill. (**a**) LC10 core log – C and D show the location of core photographs in Figure 13. (**b**) 3.5 kHz seismic profile passing through the core site, shown by dark line on the location map above.

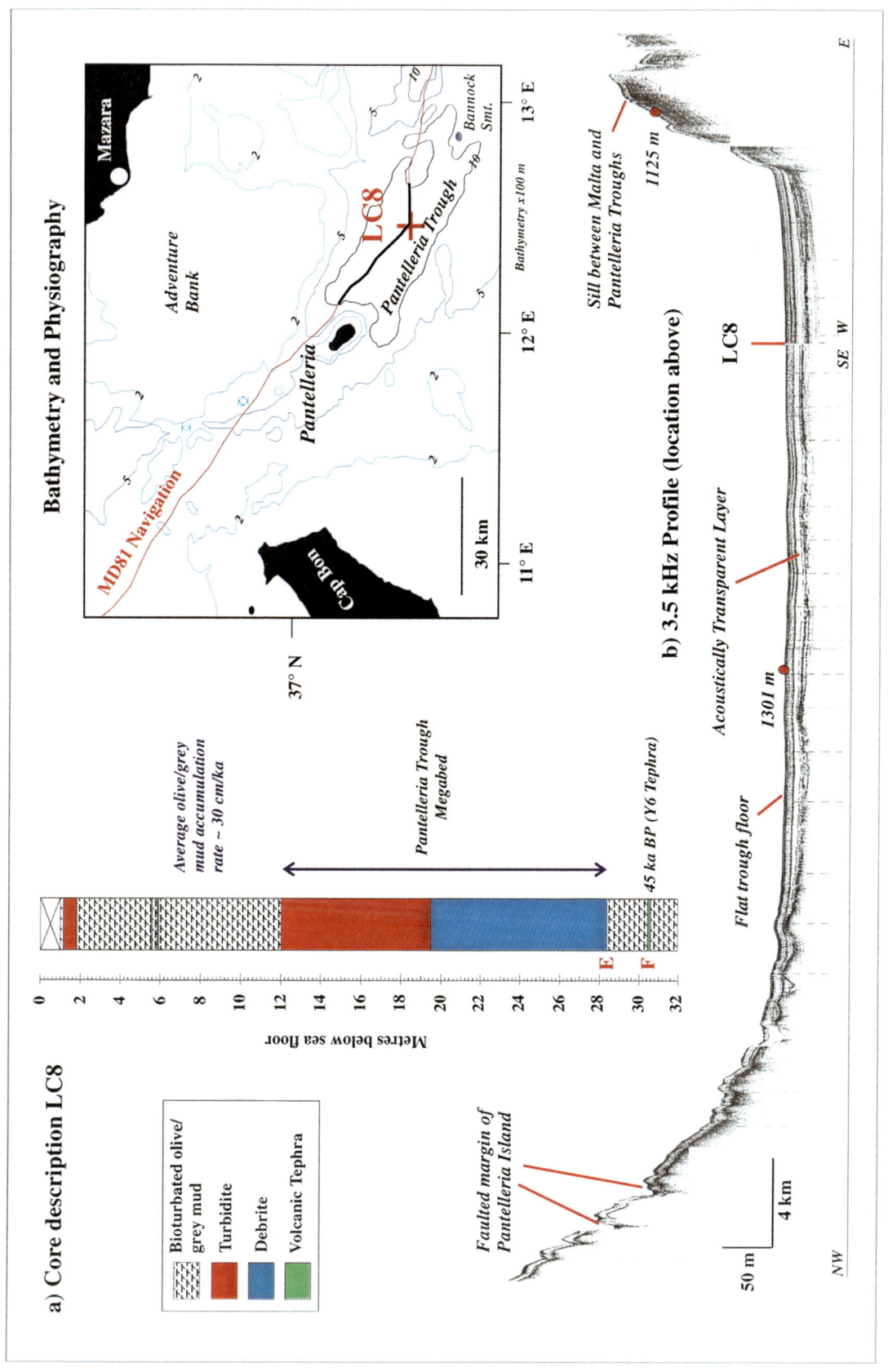

Fig. 11. Sediment and 3.5 kHz seismic characteristics of the Pantelleria Trough in the central Sicilian–Tunisian Platform. (**a**) LC8 core log – E and F show the location of core photographs in Figure 13. (**b**) 3.5 kHz seismic profile passing through the core site, shown by dark line on the location map above.

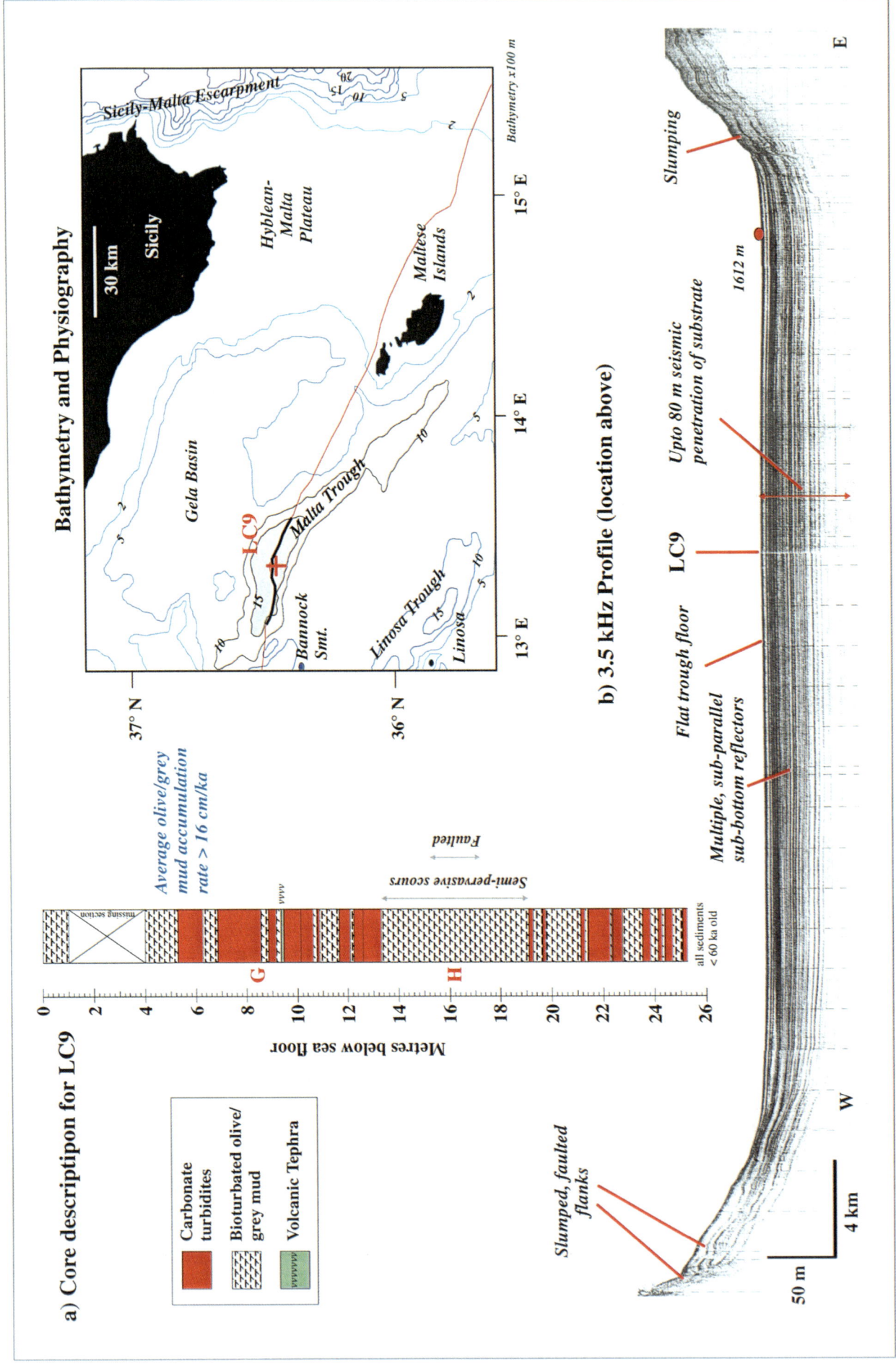

Fig. 12. Sediment and 3.5 kHz seismic characteristics of the Malta Trough in the central Sicilian–Tunisian Platform. (**a**) LC9 core log – G and H show the location of core photographs in Figure 13. (**b**) 3.5 kHz seismic profile passing through the core site, shown by dark line on the location map above.

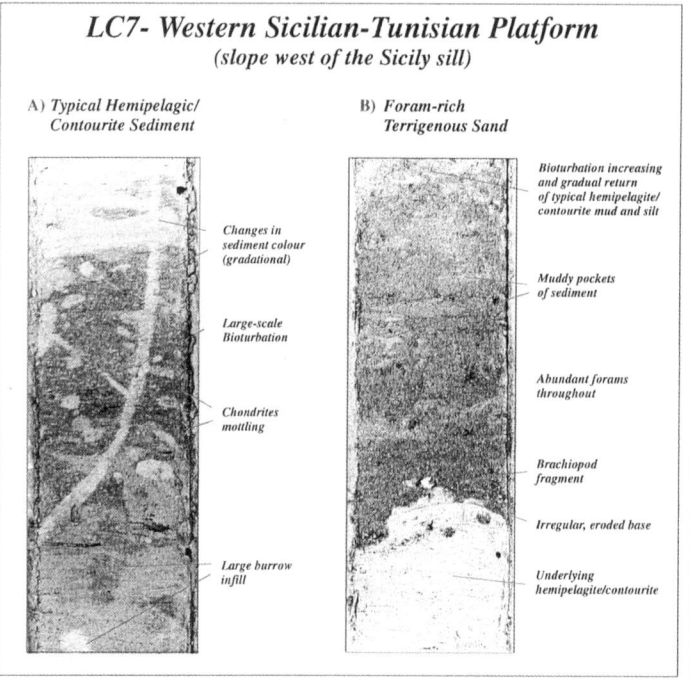
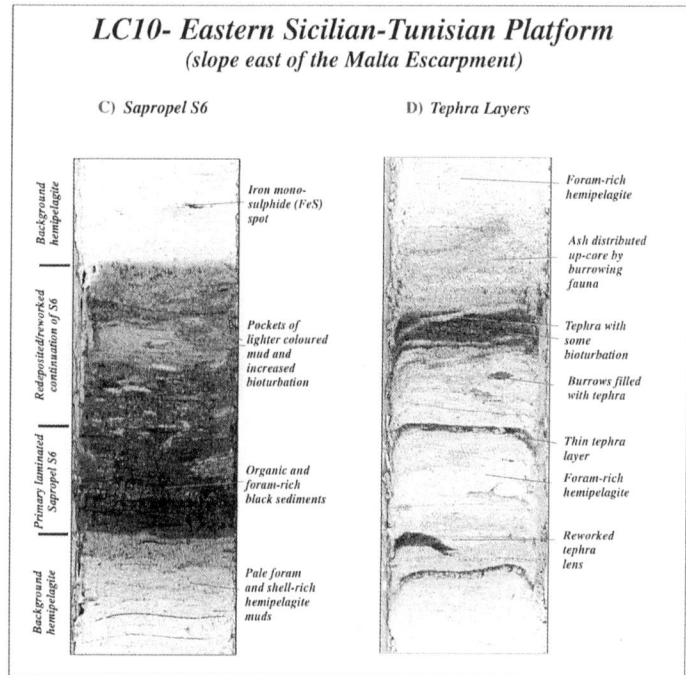
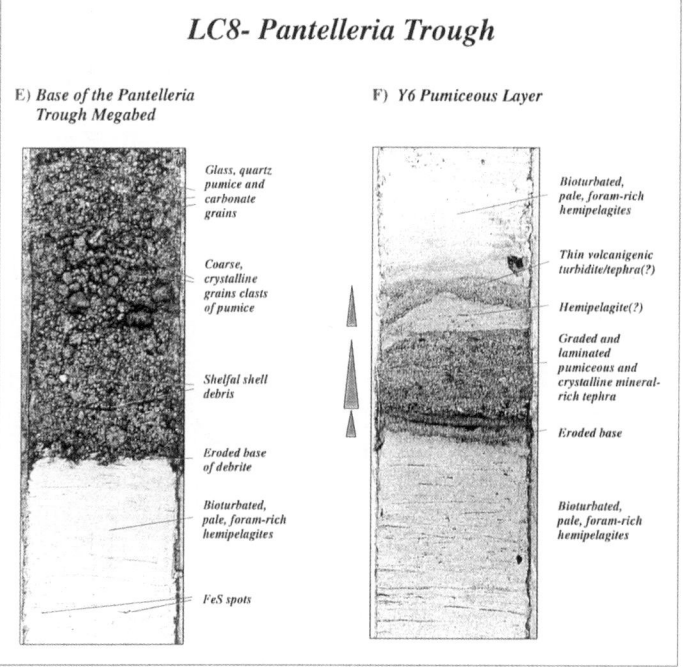
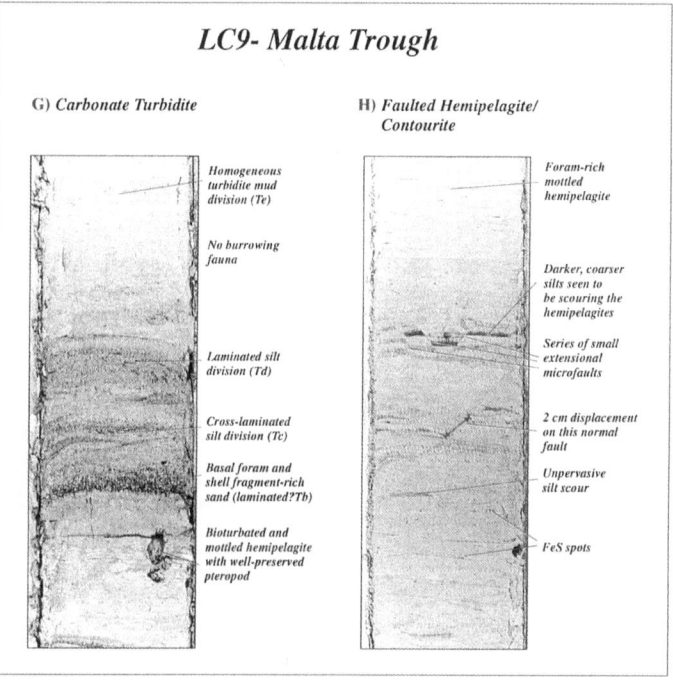

Fig. 13. Detailed core photographs from cores LC7 (**a**, **b**), LC10 (**c**, **d**), LC8 (**e**, **f**) and LC9 (**g**, **h**). Located on Figures 9–12.

the basin cores, especially core LC9 (i.e. background sediments), display variable amounts of bioturbation, scoured discontinuous lamination and lenses of foraminiferal and fragmented shell-rich silts and sands, all suggestive of bottom current influence.

- Reeder *et al.* (in press) further argue that the relatively high rates of background sedimentation found in basin cores (> 16–30 cm ka^{-1}) suggest that the regional hemipelagic sedimentation has been augmented by bottom currents. They invoke a process of bottom-current flow lofting to help trap the contourite sediments in the basins.
- Whereas most of the seafloor photographs reported by Maldonado & Stanley (1976) show little evidence of current activity in the deep basins, they do record probable bottom current smoothing of an otherwise bioturbated surface within

the Malta Trough. The shallow platform regions, by contrast, show considerable evidence for shallow bottom current winnowing and concentration of coarse bioclastic debris.

Sedimentation rates and hiatuses

Based on seismic records, the average rates of sedimentation for the Plio-Quaternary section vary from < 4 cm ka^{-1} on the shallow platform, around 6–8 cm ka^{-1} in the intermediate zone and from 16–25 cm ka^{-1} in the deeper basins. However, the shelf region, in particular, would have been subject to episodic emergence and erosion during glacial sealevel lowstands, so that any sedimentary sequence will be full of hiatuses and condensed sections.

Data from cored sections (Maldonado & Stanley 1976; Reeder

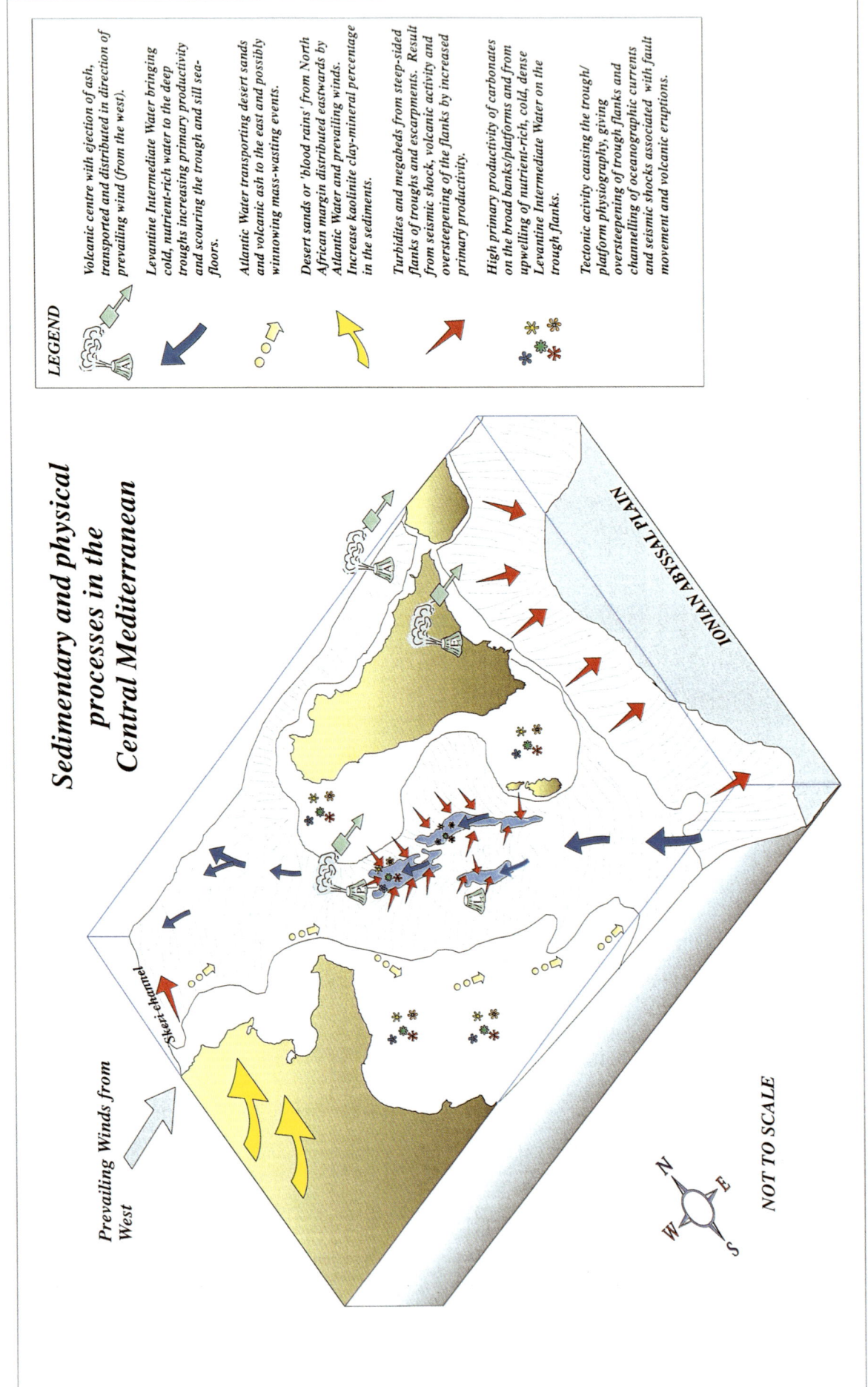

Fig. 14. Schematic representation of the sedimentary and physical processes operating in the Sicilian gateway, Central Mediterranean region.

SYNTHESIS OF SEDIMENTARY AND GEOLOGICAL PROCESS ACROSS THE SICILIAN-TUNISIAN PLATFORM DURING THE LATE QUATERNARY

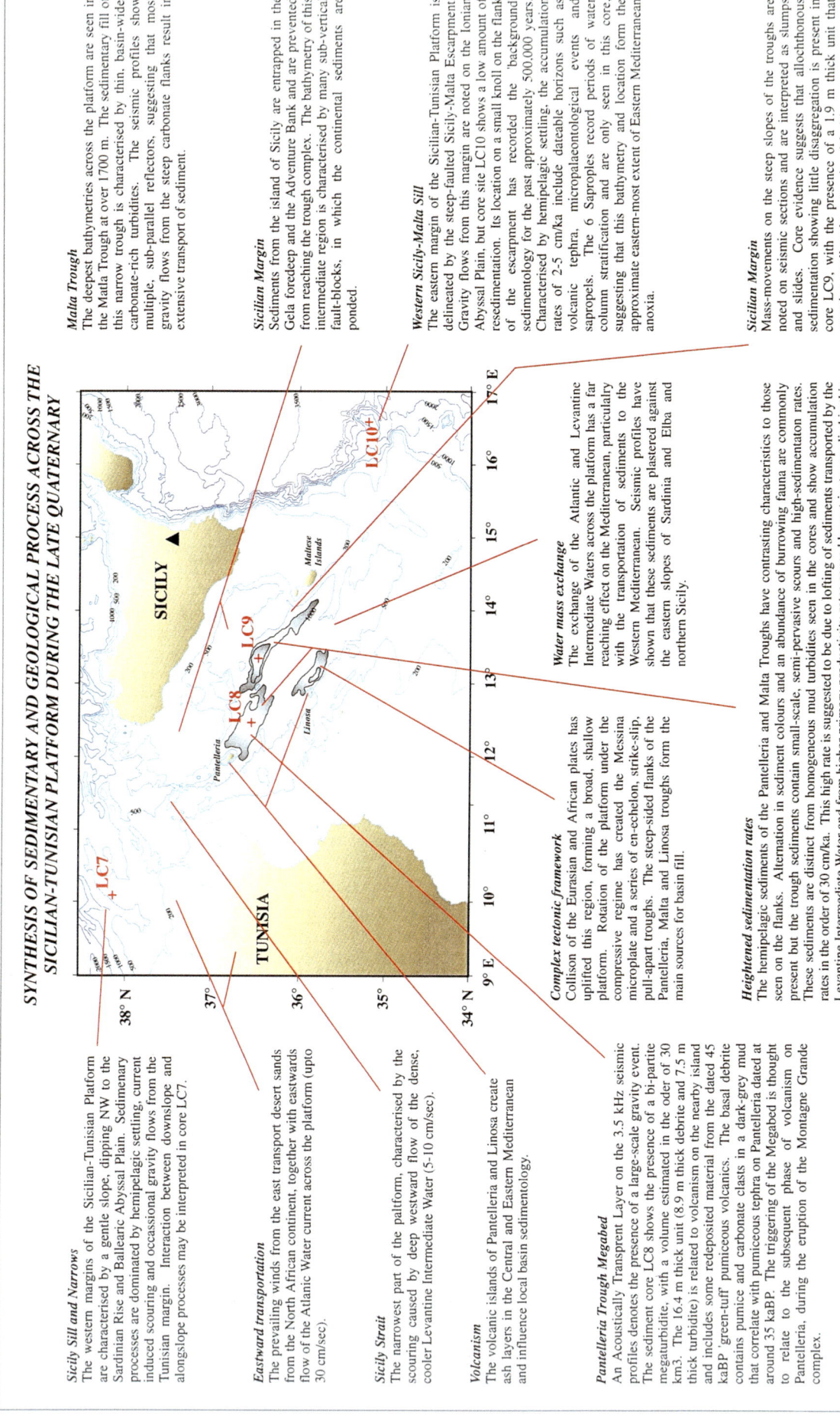

Sicily Sill and Narrows
The western margins of the Sicilian-Tunisian Platform are characterised by a gentle slope, dipping NW to the Sardinian Rise and Ballearic Abyssal Plain. Sedimenary processes are dominated by hemipelagic settling, current induced scouring and occasional gravity flows from the Tunisian margin. Interaction between downslope and alongslope processes may be interpreted in core LC7.

Eastward transportation
The prevailing winds from the east transport desert sands from the North African continent, together with eastwards flow of the Atlanic Water current across the platform (upto 30 cm/sec).

Sicily Strait
The narrowest part of the paltform, characterised by the scouring caused by deep westward flow of the dense, cooler Levantine Intermediate Water (5-10 cm/sec).

Volcanism
The volcanic islands of Pantelleria and Linosa create ash layers in the Central and Eastern Mediterranean and influence local basin sedimentology.

Pantelleria Trough Megabed
An Acoustically Transpent Layer on the 3.5 kHz seismic profiles denotes the presence of a large-scale gravity event. The sediment core LC8 shows the presence of a bi-partite megaturbidite, with a volume estimated in the oder of 30 km3. The 16.4 m thick unit (8.9 m thick debrite and 7.5 m thick turbidite) is related to volcanism on the nearby island and includes some redeposited material from the dated 45 kaBP 'green-tuff' pumiceous volcanics. The basal debrite contains pumice and carbonate clasts in a dark-grey mud that correlate with pumiceous tephra on Pantelleria dated at around 35 kaBP. The triggering of the Megabed is thought to relate to the subsequent phase of volcanism on Pantelleria, during the eruption of the Montagne Grande complex.

Complex tectonic framework
Collision of the Eurasian and African plates has uplifted this region, forming a broad, shallow platform. Rotation of the platform under the compressive regime has created the Messina microplate and a series of en-echelon, strike-slip, pull-apart troughs. The steep-sided flanks of the Pantelleria, Malta and Linosa troughs form the main sources for basin fill.

Heightened sedimentation rates
The hemipelagic sediments of the Pantelleria and Malta Troughs have contrasting characteristics to those seen on the flanks. Alternation in sediment colours and an abundance of burrowing fauna are commonly present but the trough sediments contain small-scale, semi-pervasive scours and high-sedimentaton rates. These sediments are distinct from homogeneous mud turbidites seen in the cores and show accumulation rates in the order of 30 cm/ka. This high rate is suggested to be due to lofting of sediments transported by the Levantine Intermediate Water and from higher primary productivity rates caused by nutrients supplied in this cold, denser water mass.

Water mass exchange
The exchange of the Atlantic and Levantine Intermediate Waters across the platform has a far reaching effect on the Mediterranean, particularly with the transportation of sediments to the Western Mediterranean. Seismic profiles have shown that these sediments are plastered against the eastern slopes of Sardinia and Elba and northern Sicily.

Malta Trough
The deepest bathymetries across the platform are seen in the Matla Trough at over 1700 m. The sedimentary fill of this narrow trough is characterised by thin, basin-wide, carbonate-rich turbidites. The seismic profiles show multiple, sub-parallel reflectors, suggesting that most gravity flows from the steep carbonate flanks result in extensive transport of sediment.

Sicilian Margin
Sediments from the island of Sicily are entrapped in the Gela foredeep and the Adventure Bank and are prevented from reaching the trough complex. The bathymetry of this intermediate region is characterised by many sub-vertical fault-blocks, in which the continental sediments are ponded.

Western Sicily-Malta Sill
The eastern margin of the Sicilian-Tunisian Platform is delineated by the steep-faulted Sicily-Malta Escarpment. Gravity flows from this margin are noted on the Ionian Abyssal Plain, but core site LC10 shows a low amount of resedimentation. Its location on a small knoll on the flank of the escarpment has recorded the 'background' sedimentology for the past approximately 500,000 years. Characterised by hemipelagic settling, the accumulation rates of 2-5 cm/ka include dateable horizons such as volcanic tephra, micropalaeontological events and sapropels. The 6 Sapropels record periods of water column stratification and are only seen in this core, suggesting that this bathymetry and location form the approximate eastern-most extent of Eastern Mediterranean anoxia.

Sicilian Margin
Mass-movements on the steep slopes of the troughs are noted on seismic sections and are interpreted as slumps and slides. Core evidence suggests that allochthonous sedimentation showing little disaggregation is present in core LC9, with the presence of a 1.9 m thick unit that contains small-scale, extensional faulting.

Fig. 15. General synthesis of the geological, sedimentary and physical processes operating in the Sicilian gateway, Central Mediterranean region.

et al. 2000), present a somewhat different picture through part of the Pleistocene and Holocene interval. The shallow platform cores show some of the highest rates (52 cm ka^{-1}) between about 17 and 25 ka, and then a truncation at the top of the core in some cases. However, part of this apparent hiatus may be due to sediment loss during the coring process. The intermediate depth zone also shows relatively high rates (16–40 cm ka^{-1}, average 25 cm ka^{-1}) across the main part of the platform, but much lower rates for the mainly hemipelagic sections in slope cores LC7 and LC10 (2–6 cm ka^{-1}). The deeper basin cores are quite variable, mostly averaging 20–25 cm ka^{-1}, but with values of 50, 64 and 128 cm ka^{-1} through intervals with many and thick turbidites. As mentioned above, the rates of background sedimentation calculated for cores LC8 and LC9 are > 16 cm ka^{-1} and 30 cm ka^{-1} respectively.

Discussion and conclusions

The different oceanic gateways around the world that currently serve both to compartmentalise and to connect various parts of the ocean basins and marginal seas, each present their own unique geological, oceanographic and sedimentary record (Stow & Morri 2000). Equally, however, they display many common attributes, chief amongst which appear to be their physiographic and geological complexity and the resulting variety of depositional processes and facies that occur.

The Sicilian Platform currently acts as a partial topographic barrier between the eastern and western Mediterranean basins, having taken on approximately its present form following the Messinian salinity crisis at the end of the Miocene. Most of the area is either shallow water (< 200 m) platform or of intermediate depth (200–600 m approximately). Both these physiographic provinces display highly irregular topography as a result of much neotectonic activity. They are cut through by a series of deep (> 1000 m), interconnected, fault-bound troughs that are linked with valleys incised into the shallow Malta and Sicily sills at either end of the platform.

We refer to this narrow passageway as the *Sicilian Gateway* (Figs 14 and 15) as it controls the exchange of water masses between the eastern and western basins. Levantine Sea Intermediate Water, formed by evaporation and sinking in the Levantine Sea, is channelled through the deep troughs, and is replaced by a surface flow of Modified Atlantic Water. Flow velocity of the bottom current is enhanced by the restricted topography, especially through the sill valleys, while interaction with local topographic barriers leads to flow disturbance and increased mixing with the overlying water mass. Upward mixing of the bottom nepheloid layer thereby occurs in a process known as *bottom current flow lofting* (Reeder *et al.* in press).

Sedimentation across the shallow platform is influenced by several different material sources and by a complex range of shallow-water processes that have resulted in a mosaic of coarse to fine-grained facies of biogenic, siliciclastic and volcaniclastic derivation. Strong bottom currents related to Modified Atlantic Water inflow, as well as to tides and waves, are responsible for the patchy distribution of relatively coarse-grained, mixed composition sands and gravels. Deposits of the intermediate depth zone are dominated by bioturbated hemipelagic muds. The slope sites to the east and west, respectively, of the Malta and Sicily sills show mainly slow (2–5 cm ka^{-1}) hemipelagic sedimentation, being influenced by bottom currents in the west and by basin stagnation and associated sapropel formation in the east. Airborne volcanic tephra as well as Saharan dust are dispersed towards the east by the prevailing winds.

Sedimentation in the Sicilian Gateway troughs and sill valleys is characterized by downslope gravity flows, resulting from frequent volcanic and seismic activity on their flanks that displace platform carbonates and/or volcanic debris, and by high rates of hemipelgic/contourite accumulation (16–30 cm ka^{-1}). These rates are augmented by the lofting process described above.

The interpretation of contourite deposition is more subtle and equivocal than for turbidites (Stow 1994; Stow *et al.* 1996). The principal lines of evidence for gateway contourites are recognized at three scales of study.

(1) At the large scale, oceanographic data clearly demonstrate bottom current flow from east to west through the Sicilian gateway. (2) At the medium scale, seismic data show mounded drift development in the narrow valley across the Sicily sill, as well as other areas with bottom current scoured gullies, mounds and bedforms. Contourite accumulation in the main gateway troughs is intercalated (as thin sheet or mixed drifts) with the downslope facies. (3) At the small scale, core analysis reveals extensive reworking of microfossil assemblages, minor scours, diffuse and lenticular lamination, and cyclic grain size profiles within the background bioturbated calcareous mud facies. To what extent these are hemipelagic versus contourite muds is uncertain. Bottom current winnowing of coarse bioclastic sands introduced by turbidity currents is also evident.

This complex sedimentation pattern has been influenced by the interaction of volcanic, seismic, topographic, sea-level and climatic controls (Fig. 15). During its Pleistocene history the Sicilian Gateway has periodically acted as a barrier to water exchange, principally during times of lowered sea-level. This has led to erosion and reworking of exposed platform regions and to increased downslope input into the deep basins, including some of the megabeds observed. There was also progressive starvation of oxygen from the bottom of the water column in the eastern Mediterranean basin and hence the accumulation of a series of sapropel layers. The Malta sill represents approximately the western cut-off point for the development of bottom water anoxia and sapropel deposition. The rest of the gateway remained sufficiently well ventilated, via a more open Sicily sill, to preclude stagnation but, presumably also, to restrict contourite accumulation to highstand system tracts when through-flow resumed.

MSR acknowledges tenure of an NERC Research Studentship award while undertaking the research for this paper. DAVS acknowledges tenure of a Royal Society Industrial Fellowship with BP-Amoco. Both express thanks for general support to their respective institutions, and to the reviewers of an earlier version of the manuscript.

References

ARGNANI, A. 1993. Neogene tectonics of the Strait of Sicily. *UNESCO reports in Marine Science*, **58**, 55–60.

ASTRALDI, M., GASPARINI, G. P., SPARNOCCHIA, S., MORETTI, M. & SANSONE, E. 1996. The characteristics of the water masses and water transport in the Sicily Strait at long time scales. *Bulletin de l'Institute Océanographique*, **17**, 95–115.

BOCCALETTI, M., CELLO, G. & TORTORICI, L. 1987. Transtensional tectonics in the Sicily Channel. *Journal of Structural Geology*, **9**, 869–876.

BOCCALETTI, M., NICOLICH, R. & TORTORICI, L. 1984. The Calabrian Arc and the Ionian Sea in the dynamic evolution of the Central Mediterranean, *Marine Geology*, **55**, 219–245.

BOWLES, F. A., LAMBERT, D. N. & RICHARDSON, M. D. 1993. Sediment patterns within the trough separating the Tunisian and Sicilian platforms. *UNESCO reports in Marine Science*, **58**, 129–134.

CALANCHI, N., COLANTONI, P., ROSSI, P. L., SAITTA, M. & SERRI, G. 1989. The Strait of Sicily continental rift systems: Physiography and petrochemistry of the submarine volcanic centres, *Marine Geology*, **87**, 55–83.

CATALANO, R., DI STEFANO, P., SULLI, A. & VITALE, F. P. 1996. Paleogeography and structure of the central Mediterranean: Sicily and its offshore, *Tectonophysics*, **260**, 291–323.

CELLO, G. 1987. Structure and deformation processes in the Strait of Sicily "rift zone". *Tectonophysics*, **141**, 237–247.

CELLO, G., CRISCI, G., MARABINI, S. & TORTORICI, L. 1985. Transtentive tectonics in the Strait of Sicily: structural and volcanological evidence from the island of Pantelleria. *Tectonics*, **1**, 311–322.

COLANTONI, P., TRAMONTANA, M. & ALBERINI, C. 1993. Some notes on recent turbiditic sedimentation in the Pantelleria Basin (Sicily Channel), *UNESCO reports in Marine Science*, **58**, 147–152.

DEWEY, J. F., HELMAN, M. L., TURCO, E., HUTTON, D. H. W. & KNOTT, S. D. 1989. Kinematics of the western Mediterranean. *Geol. Soc. London Spec. Publ.* **45**, 265–283.

DI PAOLO, G. M. 1973. The island of Linosa (Sicily Channel), *Bulletin Volcanologique*, **37**, 149–174.

FINETTI, I. 1984. Geophysical study of the Strait of Sicily Channel rift zone, *Bollettino di Geofisica Teorica Ed. Applicata*, **15**, 263–341.

GRANDJACQUET, C. & MASCLE, G. 1978. The structure of the Ionian Sea, Sicily, and Calabria-Lucana. *In*: NAIRN, A. E. M., KANES, W. H. & STEHLI, F. G. (eds) *The Ocean Basins and Margins*, **4B**, Plenum Press, New York, 257–329.

ILLIES, J. H. 1981. Graben formation – the Maltese islands – a case history, *Tectonophysics*, **73**, 151–168.

JONGSMA, D., VAN HINTE, J. E. & WOODSIDE, J. M. 1985. Geologic structure and neotectonics of the North African Continental Margin south of Sicily, *Marine and Petroleum Geology*, **2**, 156–179.

JONGSMA, D., WOODSIDE, J. M., KING, G. C. P. & VAN HINTE 1987. The Medina Wrench: A key to the kinematics of the central and eastern Mediterranean over the past 5 Ma, *Earth and Planetary Science Letters*, **82**, 87–106.

KELLER, J., RYAN, W. B. F., NINKOVICH, D. & ALTHERR, R. 1978. Explosive volcanic activity in the Mediterranean over the past 200 000 a as recorded in deep-sea sediments, *Geological Society of America Bulletin*, **89**, 591–604.

MALDONADO, A. & STANLEY, D. J. 1976. Late Quaternary sedimentation and stratigraphy in the Strait of Sicily, *Smithsonian Contributions to the Earth Sciences*, **16**.

MALDONADO, A. & STANLEY, D. J. 1977. Lithofacies as a function of depth in the Strait of Sicily, *Geology*, **5**, 111–117.

MANZELLA, G. M. R. 1994. The seasonal variability of the water masses and transport through the Strait of Sicily, *Coastal and Estuarine Studies*, **46**, 33–45.

MANZELLA, G. M. R., HOPKINS, T. S., MINNETT, P. J. & NACINI, E. 1990. Atlantic Water in the Strait of Sicily, *Journal of Geophysical Research*, **95**, C2, 1569–1575.

MARANI, M., ARGNANI, A., ROVERI, M. & TRINCARDI 1993. Sediment drifts and erosional surfaces in the central Mediterranean: seismic evidence of bottom-current activity, *Sedimentary Geology*, **82**, 207–220.

MORETTI, M., SANSONE, E., SPEZIE, G. & DE MAIO, A. 1993. Results of investigations in the Sicily Channel (1986–1990), *Deep-Sea Research II*, **40**, 1181–1192.

REEDER, M. S. 2000. *Megaturbidites and the late Quaternary sedimentology of the Eastern and Central Mediterranean Sea*. PhD thesis, University of Southampton.

REEDER, M. S., STOW, D. A. V. & ROTHWELL, R. G. in press. Sedimentation in the Sicilian Gateway, central Mediterranean Sea. *Marine Geology*.

STANLEY, D. J., MALDONADO, A. & STUCKENRATH, R. 1975. Strait of Sicily depositional rates and patterns, and possible reversal of currents in the late Quaternary. *Paleogeography, Paleoclimatology, Paleoecology*, **18**, 279–291.

STOW, D. A. V. 1994. Deep sea processes of sediment transport and deposition. *In*: Pye, K. (ed.) *Sediment transport and depositional processes*. Blackwell Scientific Publications, Oxford, 257–291.

STOW, D. A. V. & MORRI, C. 2000. Anatomy of deep oceanic gateways: architectural elements, processes and facies. *In: 31st International geological Congress*, Brazil, Abstract Volume.

STOW, D. A. V., READING, H. G. & COLLINSON, J. D. 1996. Deep Seas. *In*: READING, H. G. (ed.) *Sedimentary environments: processes, facies and stratigraphy*. Blackwell Science Ltd, Oxford, 395–453.

WINNOCK, E. 1981. Structure du block pelagien. *In*: WEZEL, F. C. (ed.) *Sedimentary basins of Mediterranean margins*. Institute Geology, University of Urbino, Italy, 445–464.

Sediment drifts of the Corsica Channel, northern Tyrrhenian Sea

MARCO ROVERI

Istituto di Geologia Marina – CNR, Via Gobetti 101, 40129, Bologna, Italy (e-mail: roveri@igm.bo.cnr.it)
Present address: Dipartimento di Scienze della Terra, Università di Parma, Viole delle Scienze 157A, 43100 Parma, Italy (e-mail: roveri@unipr.it)

Abstract: A sediment drift complex, resulting from the activity of the northward flowing Levantine Intermediate Water, occurs in intermediate water depth (300–600 m) on the eastern flank of the Corsica Basin. Because of the effect of topographic constriction, bottom currents are here accelerated, reaching velocities that are sufficient to erode, transport and redistribute fine-grained sediment. The sediment drift complex shows a variety of depositional and erosional features that appear very similar to oceanic examples. The development of such features appears to be mainly controlled by an interaction between the bottom current regime and slope topography. Seismic geometries and core data show that such features have grown since middle Pliocene time under a long-term stable bottom-current regime. Short-term variability of current efficiency, a concept including the combined effects of current speed, sediment availability and local topography, as a result of climate and sea-level changes, is recorded by the cyclical superposition of small-scale depositional units.

Table 1. *Principal characteristics*

Location	Tyrrhenian Sea, Mediterranean
Setting	Levantine Intermediate Current; narrow basin with depositional areas coincident with main axis of flow
Age	Middle Pliocene to Recent
Drift type	Elongate, mounded drift (separated) to sheeted drift (plastered)
Dimensions	length 30 km, width < 10 km
Seismic characteristics	Elongated sediment mounds and associated deep moats, laterally continuous reflectors converging toward the moat axis and in the area downslope of mound crests
Sediment characteristics	Bioturbated silty mud with variable biogenic content

The Mediterranean Sea is usually considered as a basin having a low-energy, bottom current regime with respect to oceanic basins that are swept by more powerful geostrophic currents. Accordingly, sedimentary features clearly related to bottom current activity are not commonly described from the continental margins of the Mediterranean basin. Moreover, due to the young age and intense tectonic and volcanic activity of the Mediterranean area, slope and basin settings are more frequently dominated by gravity flows, thus masking the products of bottom current processes. However, due to the recent development of high-resolution seismic tools, several examples of bottom current related sedimentary bodies and erosional features have been recently recognized in the Mediterranean, that appear very similar to oceanic drifts (Carter & McCave 1994; Howe *et al.* 1994; Howe 1996; McCave & Tucholke 1986; Stoker *et al.* 1998; Faugères *et al.* 1999). These features occur in small sub-basins or passageways where the interaction with local topography is likely to induce bottom current acceleration (Marani *et al.* 1993; Nelson *et al.* 1993; Roveri *et al.* 1994). The best examples are from intermediate water depths (<1000 m), suggesting that the main bottom current affecting Mediterranean continental margins is derived from Levantine Intermediate Water (LIW). Current-meter data are scarce but few available velocity values are in the order of 10 to 20 cm s^{-1} within Corsica and Sicily channels. The Corsica Basin provides particularly good examples of sediment drifts that developed in a stratigraphic interval ranging from the late-Pliocene to Recent. Depositional and erosional features of the eastern slope of Corsica Basin have been previously related to sediment instability on a steep slope (slump folds, sediment creep, etc. see Stanley 1980). However, the general physiographic context and the large to small-scale seismic geometries suggest, instead, a series of actively growing bedforms related to long-lived, bottom current activity. External shapes and internal geometries are very similar to sediment drifts of the Atlantic Ocean (Faugères *et al.* 1999) and oceanographic data, although scarce in the study area, allow such an interpretation (Artale & Gasparini, 1990; Astraldi *et al.* 1990; Marani *et al.* 1993).

This paper illustrates the seismic geometries of the Corsica Basin sediment drifts and discusses their spatial variability related to topographic changes and the factors controlling their evolution through the time. Limited short-core data are available for seismic calibration of, at least, the surficial sediment.

Geologic and oceanographic setting

The Corsica Basin is a narrow and shallow, N–S trending basin separating the Corsica shelf (to the west) from the Elba Ridge (to the east) and connecting the deep Tyrrhenian and Ligurian basins to the south and north, respectively (Fig. 1). The northern termination of the Corsica Basin is represented by the Capraia sill (430 m depth). The basin width at the 200 m isobath gradually decreases toward the north, from 30 km near Pianosa Island to 10 km across the Capraia sill. The deepest, axial part of the basin is relatively smooth and flat and dips gently (0.3°) to the south.

The Corsica Basin is one of the oldest Tyrrhenian basins, with a sedimentary fill made up of more than 4000 m of Oligocene to Recent deposits (Gabin 1972; Viaris de Lesegno 1978; Bacini Sedimentari 1979; Zitellini *et al.* 1986). The basin first formed in a compressional setting and was then reactivated in an extensional regime during the late Miocene to early Pliocene opening of the Tyrrhenian Sea (Selli & Fabbri 1981; Bartole *et al.* 1991). Since the mid-Pliocene, the whole northern Tyrrhenian area underwent a phase of subsidence and overall deepening which persisted during the Quaternary. According to Pascucci *et al.* (1999), post-rift subsidence in the Corsica Basin started in the late Messinian–early Pliocene time.

The Corsica Basin has an asymmetrical shape related to both its structural setting and sedimentary fill. The basin axis is closer to the steeper (3°–4°) eastern slope, which is controlled by buried high-angle extensional faults. The western slope is gentler (1°–1.5°) and seismic profiles show that the basin axis shifted towards the east during Plio–Quaternary times due to the higher sediment input and faster progradation of Corsica margin; here,

From: Stow, D. A. V., Pudsey, C. J., Howe, J. A., Faugères, J.-C. & Viana, A. R. (eds)
Deep-Water Contourite Systems: Modern Drifts and Ancient Series, Seismic and Sedimentary Characteristics.
Geological Society, London, Memoirs, **22**, 191–208. 0435-4052/02/$15.00 © The Geological Society of London 2002.

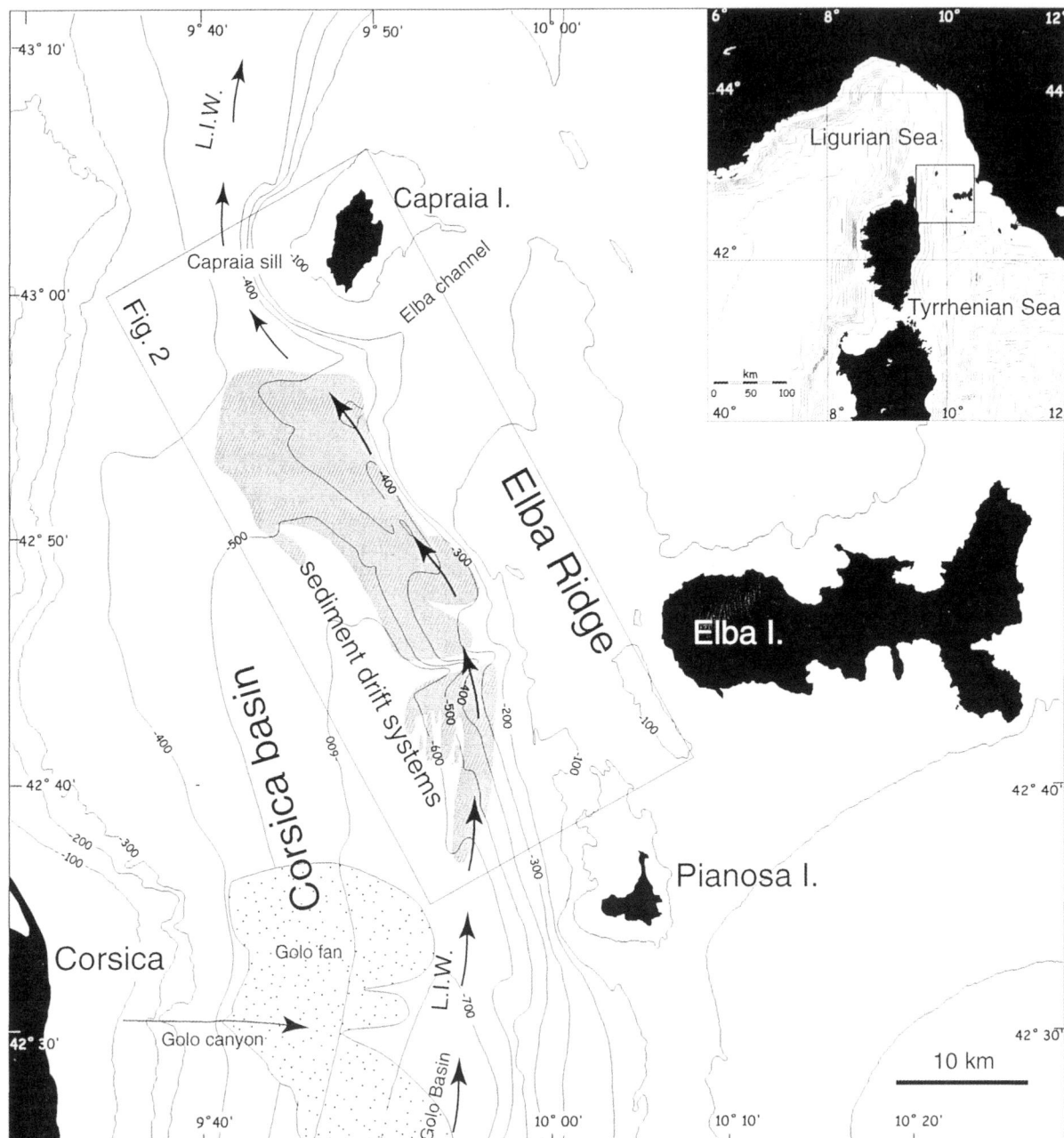

Fig. 1. Location map of the study area. The dark grey area indicates the sediment drift complex developed on the eastern flank of the Corsica Channel; the pale grey area shows the extent of the Golo fan, the major turbidite system in the area.

several deltaic lobes and small turbiditic systems occur (e.g. Golo Fan, see Figure 1; Stanley *et al.* 1980; Bellaiche *et al.* 1994).

The eastern slope is characterized by a relatively low sediment input. Due to the complex morphological setting of the Tuscan margin (Pascucci *et al.*, 1999), sediments from the Apennines are trapped within the shelf or carried toward the Tyrrhenian and Ligurian basins through structurally controlled NW–SE trending depressions located eastward of the Corsica basin and acting as primary conduits for sediment transfer. As a consequence, the eastern slope of Corsica Channel is almost starved, not only during sea-level highstand, but even during sea-level fall and lowstand. Evidence for Plio–Quaternary deltaic and turbidite systems are indeed lacking along the entire eastern margin. The small canyon (or gully) cutting the eastern slope (see Fig. 2) is not connected to a subaerial drainage system, nor have turbidite deposits been found at its basinward end. Terrigenous input is thought to be mainly delivered by bottom currents reworking basinal and slope deposits and/or redistributing fine-grained, turbulent fractions of gravity flows entering the basin from the western margin.

The large-scale pattern of Mediterranean water circulation has been well-known since the studies of Wust (1961), and subsequent work has mainly been concerned with refining the general model (see Millot 1987). The Mediterranean Sea is characterized by three main water masses: the surficial Atlantic Water (AW), entering from the Strait of Gibraltar and flowing eastward; the Levantine Intermediate Water (LIW), forming in the eastern Mediterranean, moving westward at intermediate depths (200–800 m) and outflowing from Gibraltar; and the Western Mediterranean Deep Water (WMDW), confined to deeper basins.

The warm (>13°C) and saline (>38‰) Levantine Intermediate Water is known to flow northward along the Corsica Channel at depths between 200 and 600 m. Current-meter data show that the mean annual velocity of the LIW is in the order of 8 cm s^{-1}; during winter, mean flow velocity increases up to 17 cm s^{-1} with peaks of 35 cm s^{-1} (Artale & Gasparini 1990). Due to basin topography and

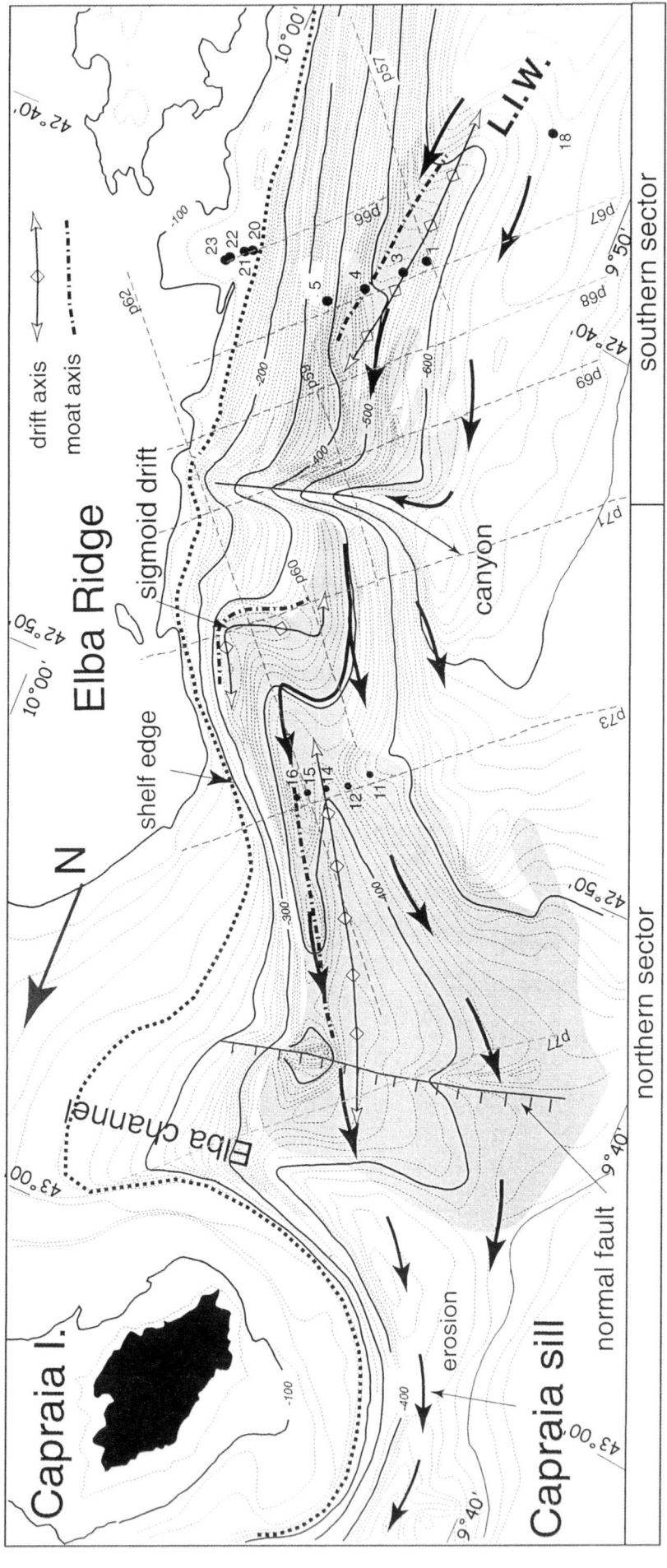

Fig. 2. Enlarged map of the eastern flank of the Corsica Channel with detailed bathymetry (contours at 10 m) and main morphological elements.

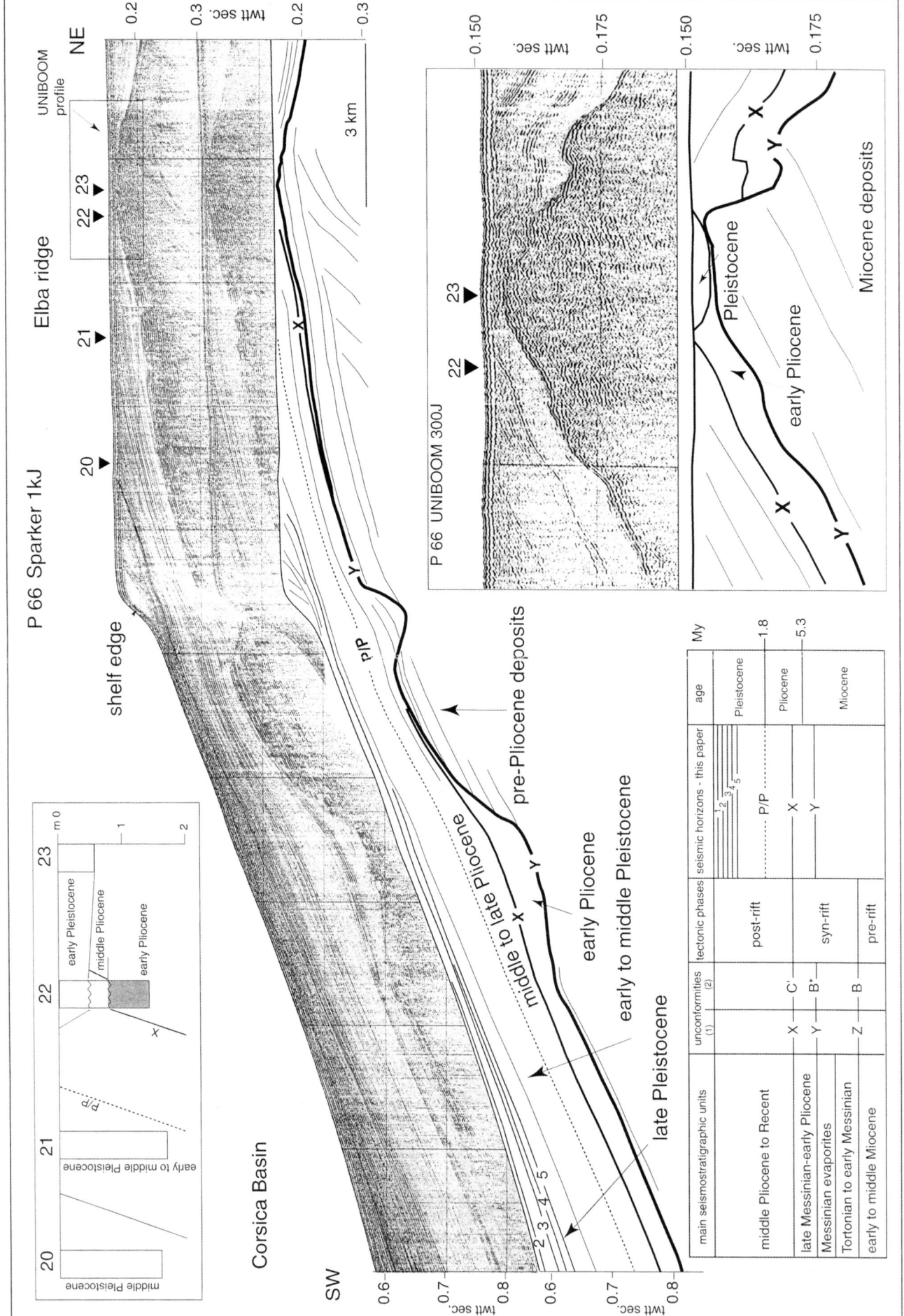

Fig. 3. General stratigraphic framework of the eastern margin of Corsica basin, based on a core transect along seismic profile P66.

Coriolis effect, the flow is forced against the eastern flank of the basin and here a significant influence on sediment distribution and slope shaping is expected. In the area between Pianosa and Capraia Islands, where basin depth and width considerably decrease, a large and long-lived drift complex has developed along the slope. This complex is made up of several depositional and erosional features primarily related to the interaction between the bottom current regime and sea-floor topography, and to current changes through time.

Bathymetry

A detailed bathymetry (shown in Fig. 2, with a contour interval of 10 m.) of the eastern margin of the Corsica Basin has been reconstructed by interpolating GPS positioned single-beam, high-frequency echosounder data acquired between 1991 and 1996. This area can be subdivided in two sectors with different morphological characteristics, separated by a short canyon that extends at right angles across the slope from the shelf edge.

In the southern sector (Pianosa–Elba), the basin flank trends NNW–SSE slightly oblique (10°) to the inferred axis of LIW flow. The base of slope is in 600 to 700 meters water depth and the slope shows a smooth morphology with dips <4°. An elongated sediment drift flanked by several subsidiary drifts occurs in water depths ranging between 400 and 550 m. The axis of the sediment drifts are parallel to the mean LIW flow; this implies that they develop at a low angle with the regional contour, climbing slightly upslope toward the north.

Because of the northward shallowing of the Corsica basin, in the northern sector (Elba–Capraia) the base of slope is shallower (400–500 m) than in the southern area. A further characteristic of this sector is that the eastern basin flank turns to a NW–SE direction, making a high angle (40°) with the mean axis of the LIW. Slope angles are slightly steeper (4°–5°) than in the southern sector, and the upper slope is here characterized by enhanced erosion. Along the lower slope (300–500 m water depth), a large elongated drift is developed, associated with a deep moat; its axis can be traced northwestward for more than 10 km parallel to the basin margin; the drift disappears in the Elba channel depression. The slope topography is at places complicated by the occurrence of small reliefs, related to the presence of an extensional fault, along which minor sediment drifts have developed. In the southern, shallower part of the northern sector, strong erosion of upper slope deposits characterizes the landward side of the moat, especially in the complex area where slope trend changes. Here a small drift with a sigmoidal shape (parallel to the local contours) has developed in relatively shallow water (300 m depth; see Fig. 2).

Stratigraphic context

Three main regional unconformities are commonly recognized in the marine sequences of the Tyrrhenian basins; they correspond to easily detectable key reflectors that are usually referred to as 'Z' (marking the base of syn-rift deposits – late Miocene), 'Y' (the erosional surface at the top of Messinian evaporites) and 'X' (base of post-rift deposits – middle Pliocene) (Selli & Fabbri 1971; Fabbri et al. 1981; Viaris de Lesegno 1978; Zitellini et al. 1986).

Reflectors 'Y' and 'X' have been recognized and traced throughout the study area. A core transect in the shelf area of the Elba Ridge along seismic profile P66 (Fig. 3) has helped to calibrate the seismic stratigraphic units. The 'X' reflector was penetrated in core 22; bio- and magnetostratigraphic analysis confirmed its middle-Pliocene age (Roveri, Capotondi and Vigliotti, unpublished data; Fig. 3). Evidence of the growth of sediment drifts is found immediately above the 'X' reflector; accordingly, a middle-Pliocene age for the onset of contour current activity in the Corsica basin is here tentatively suggested.

Five well recognizable reflectors (numbered 1 to 5, starting from the top) above 'X' have also been traced throughout the study area. These mark cyclical amplitude and geometrical changes within the drift bodies and can be traced in to deeper areas of the basin allowing correlation with turbidite systems. Bio- and magnetostratigraphic data from the core transect (Fig. 3) suggest a late Pleistocene age for reflector 5, while the younger reflector 1 marks the end of last glaciation (20 000–18 000 years BP). According to these data, sediment drift deposition started in the middle to late Pliocene, but faster growth was attained during the upper half of the Pleistocene.

Whereas the early history of drift development can only be roughly reconstructed, due to the lack of long cores or borehole data, the recent evolution of such systems can be investigated in much more detail through the integration of conventional gravity and piston cores and high-resolution seismic data. Core transects across sediment drifts in the southern and northern sectors show in detail their evolution during the late Quaternary sea-level fluctuations, especially during the last post-glacial sea-level rise. The recognition in a reference basinal core (18) of characteristic biostratigraphic units (ecozones of Capotondi et al. 1999), their calibration with AMS ^{14}C datings and their matching with the magnetic-susceptibilty pattern, allows very detailed correlation (Fig. 4). These correlations define a high-resolution stratigraphic framework made of sedimentary units whose lateral thickness changes mirror the external shape and the large-scale seismic geometries, thus reinforcing their interpretation as actively growing bottom current related features.

Sediments

Drifts and mud waves of the Corsica Channel are essentially muddy sedimentary bodies; this is evidenced by the available shallow cores and is consistent with the great penetration attained by high-frequency acoustic devices. Cores have been collected using a gravity corer and variable barrel length (2–6 m). The most common lithology recovered from the drifts is a foraminifera-rich, strongly bioturbated, silty mud (Fig. 5); primary sedimentary structures related to traction processes have not been observed. Thin coarser grained beds (fine sand) only occur in the stratigraphical interval corresponding to the Younger Dryas cold event; in this case the coarse fraction is mainly represented by biogenic debris (forams and molluscs).

Seismic characteristics

In the study area more than 1000 km of GPS positioned seismic and acoustic profiles and 50 cores have been collected during cruises ET91 (N/O BANNOCK), ET93, ET95 and ET96 (N/O URANIA), carried out by CNR Istituto di Geologia Marina of Bologna. Seismic data consist of a regular grid (< 2 km) of digitally recorded single channel 1 kJ Sparker; 3.5 kHz sonar profiles, 300 J UNIBOOM and 2–7 kHz CHIRP SONAR profiles. However, no long- or short-range side-scan sonar imagery or seabed photographs are yet available from the study area.

The southern sector of the eastern flank of the Corsica basin (Fig. 2) is characterized by regular, slightly convex-upward slope clinoforms, dipping at typical angles of 3.8°; clinoforms can be traced up to the shelf area without significant dip changes. The middle and lower slope are characterized by an elongated, northward trending, slightly asymmetrical mounded drift that shows an external sigmoidal shape (Fig. 6); in cross section the drift has a width of >4 km and a thickness of >150 m; the relief above the seabed is very small (a few metres) at its southern end but increases substantially toward the shallower northern end (>10 m). Internally the drift shows a central portion with diverging

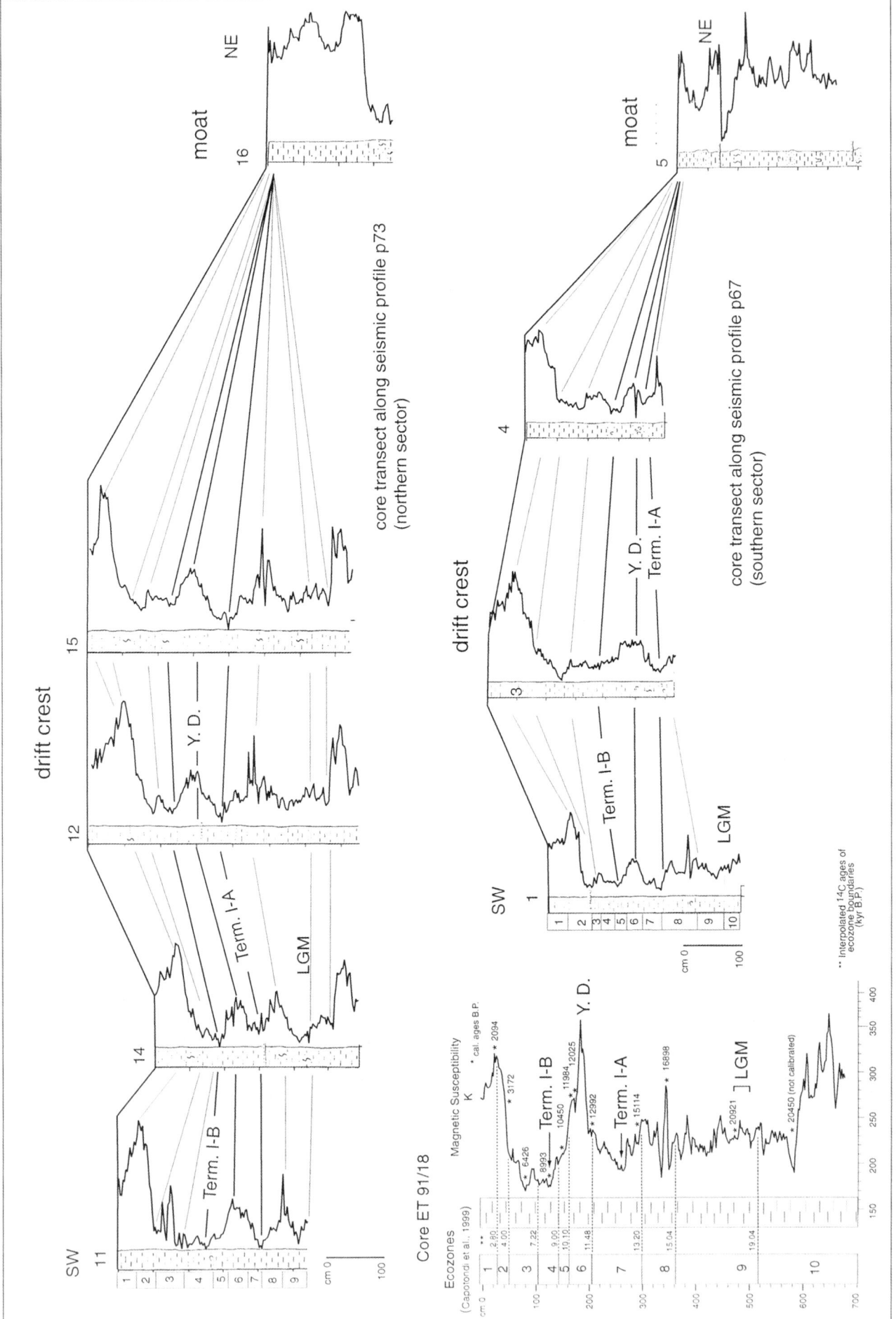

Fig. 4. Core data showing the late-Quaternary stratigraphic framework of the study area. Core transects illustrate good correspondance between the external shape of the drift and the thickness distribution of the most recent deposits.

Fig. 5. Photographs of cores 14 and 3. Core bottom is on the lower left in each core. Core locations in Figure 2.

reflectors and consequent stratigraphic expansion; downslope of the crest reflectors rapidly converge and a stratigraphically reduced section develops. Reflectors converge and onlap upslope into a moat with minor convex-up bodies (subsidiary drifts).

The smooth upper slope above the moat is marked by a significant reduction in thickness of the stratigraphic units compared to the same units in the drift. This suggests erosion and/or non deposition over this part of the slope. In the base of slope area, reflectors are divergent downslope into small sandy turbidite systems fed from the Corsican margin and reaching the basin axis (Fig. 7). Minor convex-upward bodies, here interpreted as subsidiary drifts, occur both on the landward and basinward sides of the main

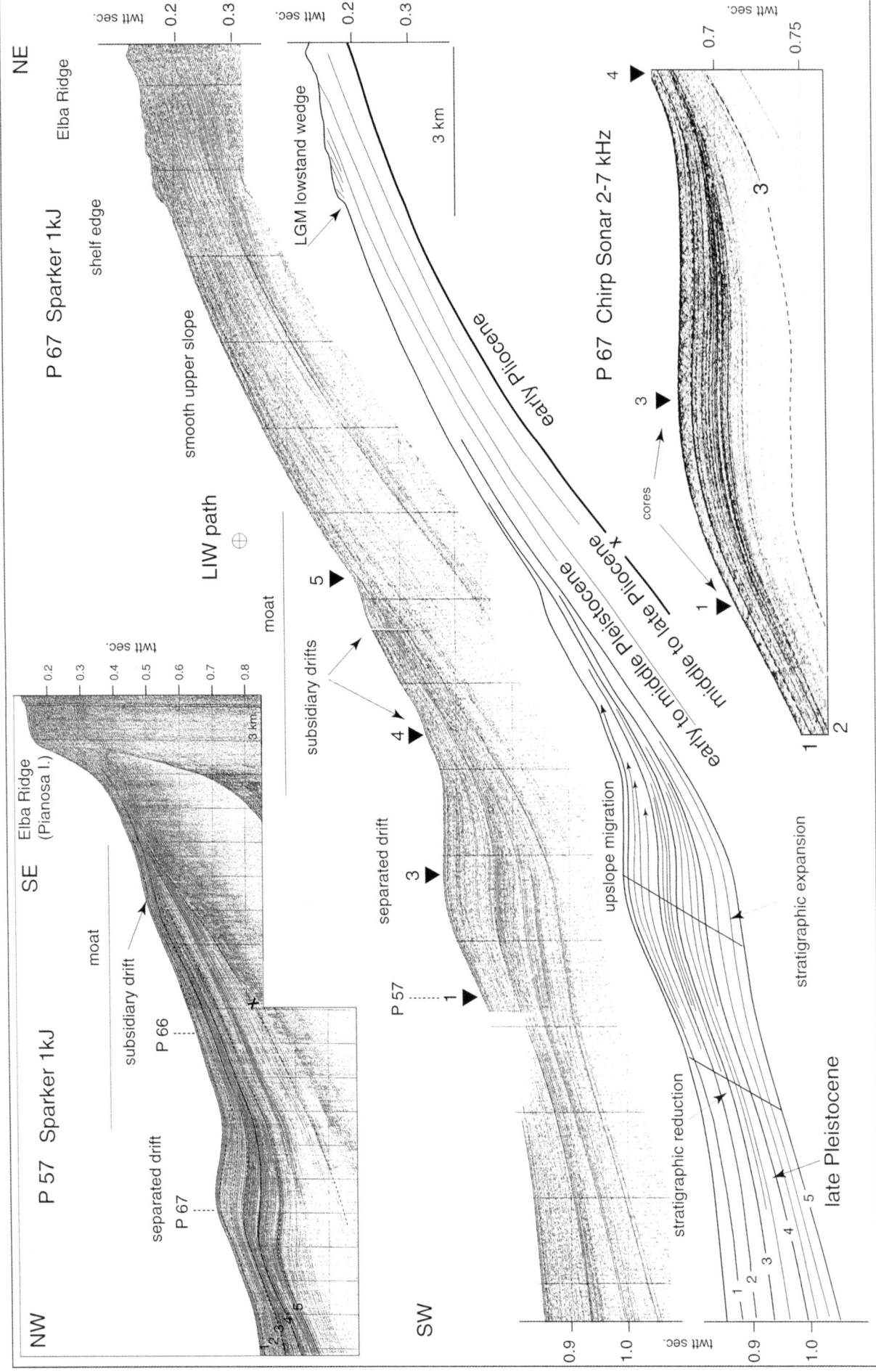

Fig. 6. Elongate drifts of the southern sector. Very high-resolution Chirp-Sonar profile shows the sigmoidal shape of small-scale seismic units within the stratigraphically expanded crest zone of the drift.

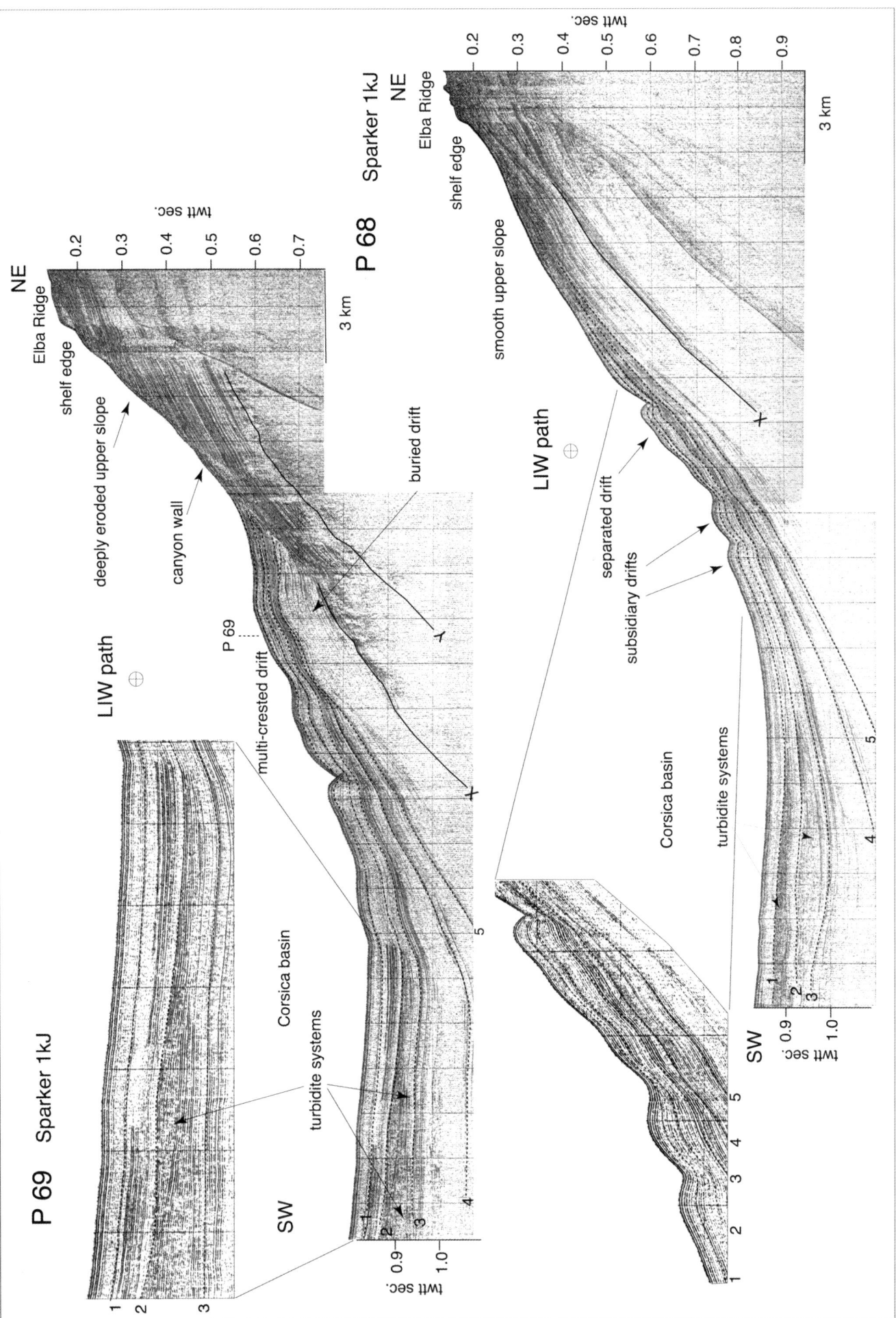

Fig. 7. These profiles illustrate the progressive change in external shape and internal geometries of the drift shown in Figure 6; the northernmost profile (P69) shows a buried simple drift that is replaced upward by the progressive growth of a multi-crested drift which represents the northward evolution of the elongate drift of Figure 6.

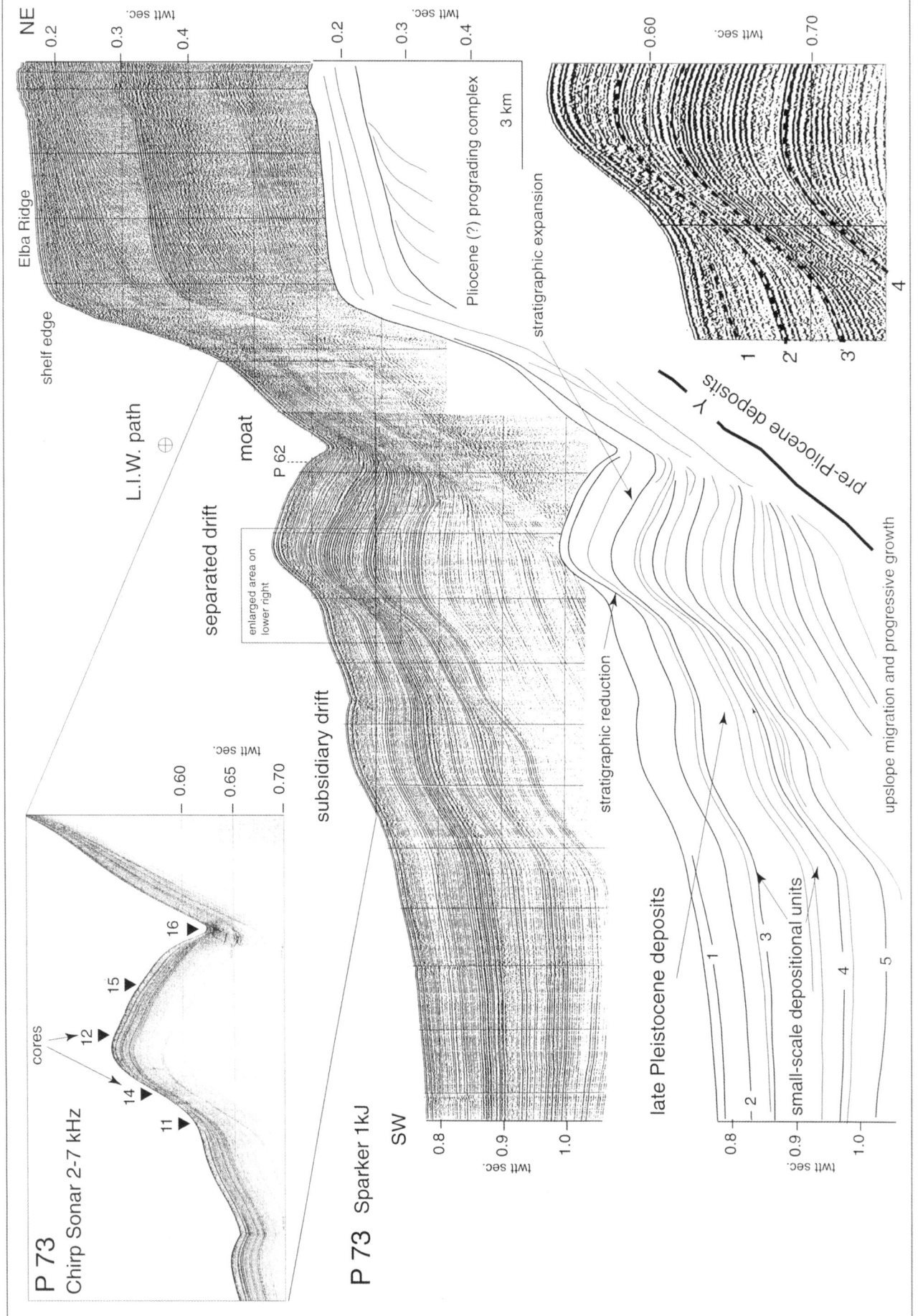

Fig. 8. The elongate asymmetrical drift of the northern sector with a subsidiary drift on its downslope side and a deep moat on its landward side. Note the repetitive pattern of small-scale depositional units that accompanies the growth of the drift.

elongated drift (Fig. 7). These features are parallel to the main drift, having a reduced height and width, and a relief that increases to the north where the elongated drift seems to develop a multi-crested shape (Fig. 7).

Seismic data show that this sedimentary body has grown through time with a clear upslope migration. No major erosional surfaces are recognizable within it, suggesting a relatively stable long-term bottom current regime. Nevertheless, a cyclic pattern given by stacked seismic units with sharp bases and gradual upward increase of reflector amplitude, can be recognized. The abrupt upward transition from high to low-amplitude reflectors defines surfaces of high lateral continuity that are ideal for regional correlation purposes (see surfaces 1 to 5 in seismic profiles). Within each of these small-scale units, reflectors onlap the basal surfaces in an upslope direction. In places these regional surfaces appear to truncate the underlying reflectors although it is not clear whether the truncation is only apparent. The thickness reduction observed in the area behind the crest of the drift may be due to some erosion and/or thinning of the strata.

In the northern sector, a large mounded, elongate drift (>10 km long, 5 km wide and more than 300 m thick; Fig. 8) has developed parallel to the bathymetric contours. The deposit is a strongly asymmetric mound with steeper basinward and gentler landward sides. Its relief above adjacent sea floor reaches 50 m and a narrow moat (up to 70 m deep) parallels the drift on its landward side. These characters allow to classify this deposit as a separated drift (*sensu* Faugères *et al.* 1999). This drift shows the same internal architecture as that found in the southern area. Small-scale aggradational units are bounded by non-depositional or slightly erosional surfaces (see reflectors 1 to 5) that can be correlated between the two sites. Small-scale units are characterized by low to moderately high-amplitude reflectors converging and onlapping against the main surfaces in the moat area. On the downslope side, reflectors converge and steepen forming a stratigraphic reduction zone just behind the drift crest; reflectors then diverge again toward the crest of a subsidiary drift growing onto the back of the main drift (see enlarged box on Fig. 8). Maximum stratigraphic expansion is attained in the area between the crest and the moat; in the latter, as well as in the upper slope area, a dramatic stratigraphic reduction occurs, through both non-deposition and (moderate) erosion. Seismic profiles show an overall increase through time of the main-drift height above the sea floor and a consistent upslope migration; a similar trend characterizes also the subsidiary drift and points to long-term stability of the bottom current regime.

A small sigmoidal drift occurs where the regional slope changes trend; this drift is associated with a moat and with major erosion of the upper-slope deposits on its landward side (Figs 9 and 10). This small, separated drift is a very complex upslope-climbing feature (Fig. 11), and is the youngest among similar SE-migrating features which characterize the slope immediately north of the canyon. Seismic geometries suggest repeated (at least three) phases of growth and upslope migration of the drift and moat, separated by episodes of rapid basinward shift of the entire system (Figs 10 and 11). Following each seaward shift, the abandoned moat is apparently filled by downslope gravity processes (see onlap geometries in Figs 10 and 11) and the slope profile is gradually smoothed. Therefore, while individual drift bodies migrate upslope and have a sigmoidal shape in plan view, according to the local path of contour currents, the slope as a whole prograde in a W–NW direction (i.e. parallel to the large-scale LIW path).

Discussion

The Corsica Basin offers a good example of interaction between bottom currents and topography in controlling the drift morphology. Seismic geometries and stratigraphic data also allow an attempt to discuss the role of sea-level and climate change on drift evolution.

Drift morphology and topography

In the Corsica Basin, changes in drift size and morphology seem to be related both to the water depth and to the interaction between bottom currents and local topographic setting. Topographic restriction both in width and height force the northward flow of LIW to accelerate while crossing the Corsica Basin, reaching velocities that are compatible with erosion and transport of fine-grained sediment. Oceanographic data show that the LIW flows northward against the eastern side of the basin at depths between 200 and 600 m (Astraldi *et al.* 1990). Sediment drifts and erosional features develop in exactly this depth range.

The southern sector is deeper (700 m) and the regional slope strikes at low angle (10°) with the inferred mean LIW flow. In this area, an elongate drift and minor subsidiary drifts have developed with their axes parallel to the flow. The main drift is relatively flat-mounded and migrates with an upslope component.

The small canyon separating the northern and southern sectors shows some interesting features; it has an upper straight part, almost normal to the regional slope, while in its lower reach the canyon axis clearly deviates to the north; in cross-section (see Fig. 12) the canyon appears asymmetrical, with a steeper, erosional wall to the north and a gentler, depositional one to the south. These characteristics seem to reflect the action of northward flowing bottom currents. In addition, the laterally accreting fill without apparent interfingering with turbiditic aggradational deposits, may indicate that the canyon is, in part, influenced by bottom current activity. A chaotic unit observed within its fill is probably related to an occasional failure generated from its walls and/or its headward region.

The area where the regional slope changes trend is extremely complex; immediately north of the canyon, local topography forces the bottom current (or at least its upper part) to turn to the right and then to the left, following the sigmoidal shape of the bathymetric contours. A small, sigmoid-shaped separated drift with a deep erosional moat develops at shallow depths (300 m; Figs 2, 9, 10 and 11). This drift shows a clear upslope migration (i.e. toward SE–E–NE). The slope area between the drift and the canyon shows other similar buried features, suggesting a complex history of growth and upslope migration up to a critical depth (very close to the upper boundary of the LIW), followed by the abandonment of the system and the development of a new drift in a deeper and basinward position; this phase is accompanied by the fill of the previous moat through downslope gravity processes that can be inferred from the presence of clear onlap terminations against the walls of the buried moat (Figs 10 and 11). This mechanism, repeated through time has led to the lateral stacking of at least three distinct drift systems. Unfortunately, the three phases of drift growth and abandonment can not be correlated to the other sectors, due to the very local and diachronous character of their bounding erosional surfaces, although seismic refelctors 1 to 5 can be traced to the most recent drift, suggesting a Pliocene to middle Pleistocene age for the buried ones.

The northern sector is shallower (<500 m) and the slope changes its trend to a NW direction thus making an angle of 40° with the inferred mean flow of LIW. With the progressive northward narrowing of the basin, there is a natural intensification of current velocity. The overall shape of the slope is convex up with a steeper basal part, most likely due to erosional scouring by contour currents impinging onto the basin floor (see Fig. 9). This area is characterized by a large separated drift with main axis parallel to the shelf-edge trend and flanked by a deep moat on its upslope side. The drift has developed on a pre-existing morphology (tilted blocks of late-Miocene extensional phase) and shows an upslope migration.

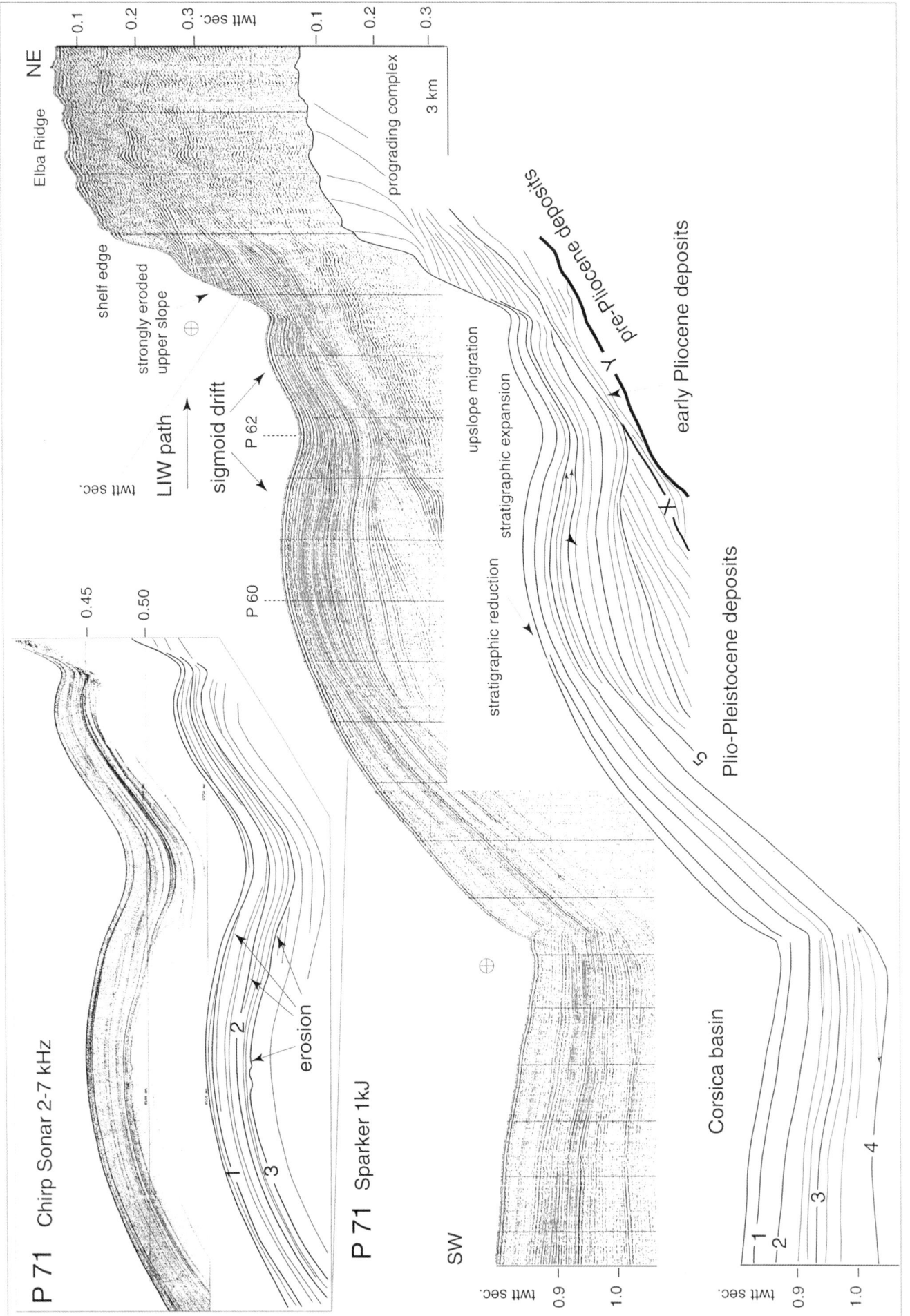

Fig. 9. Seismic profile cutting obliquely the sigmoidal drift developed in the complex area where the basin margin changes orientation.

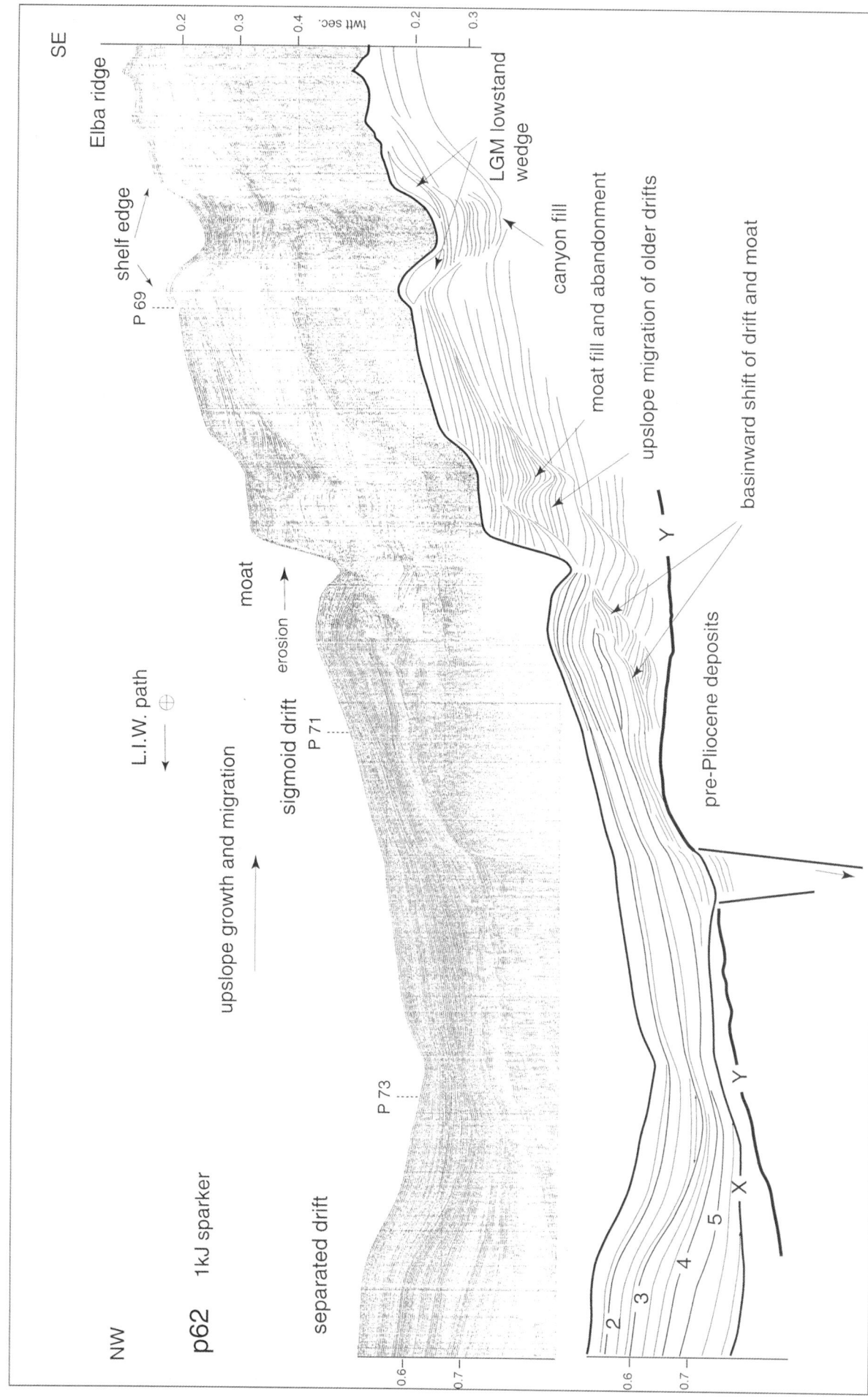

Fig. 10. Geometry of the sigmoidal drift viewed along a direction almost normal to migration. The present-day drift is associated with an erosional moat; similar drift and moat features characterize the slope area to the SE, suggesting basinward shift of the system through the time.

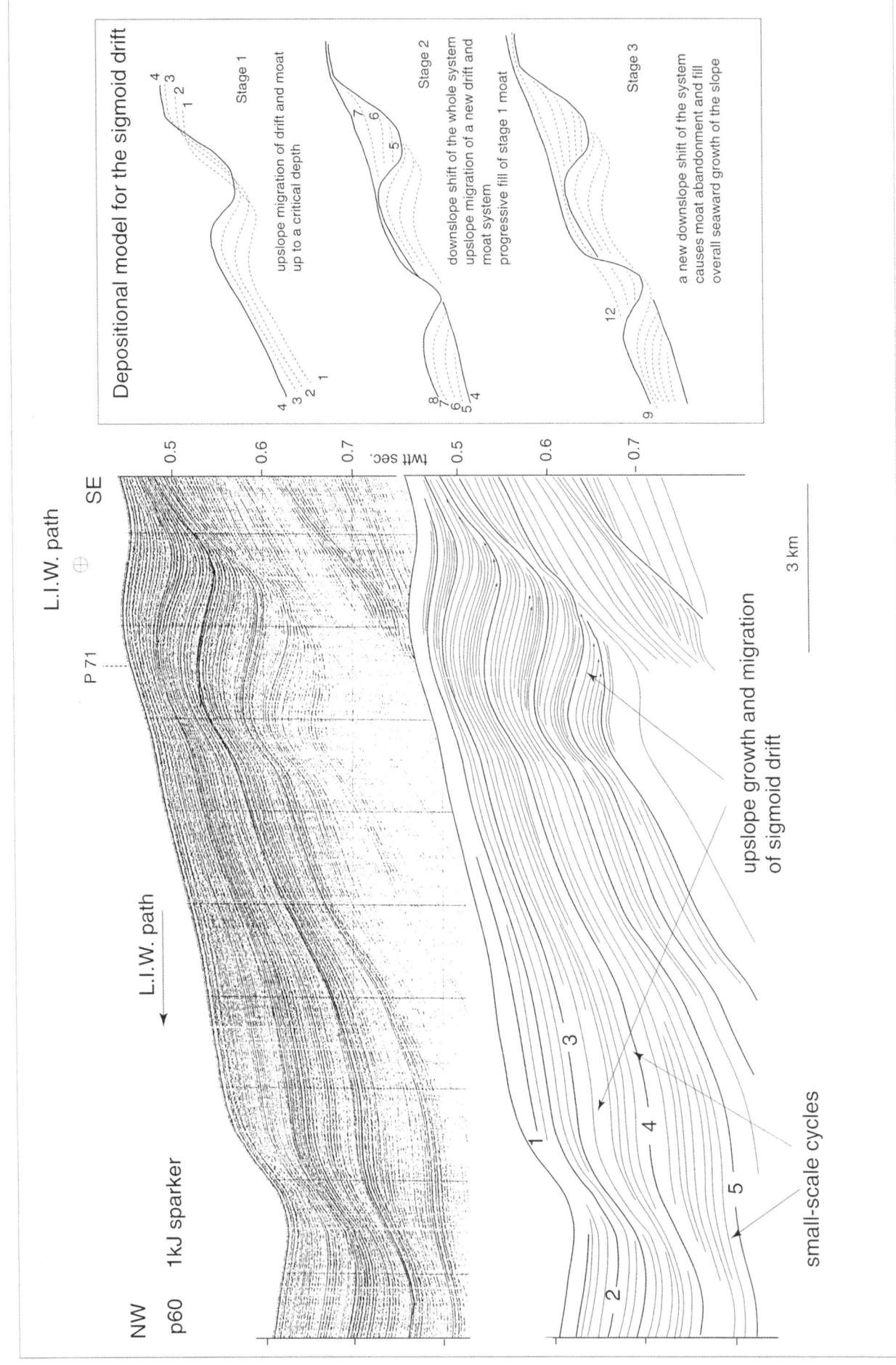

Fig. 11. Details of the small-scale geometrical relationships within the drift reported in Figures 8 and 9 as seen in more basinward sections. Note the complex onlap-downlap terminations of reflectors within small-scale units bounded by higher-amplitude and more continuous reflectors.

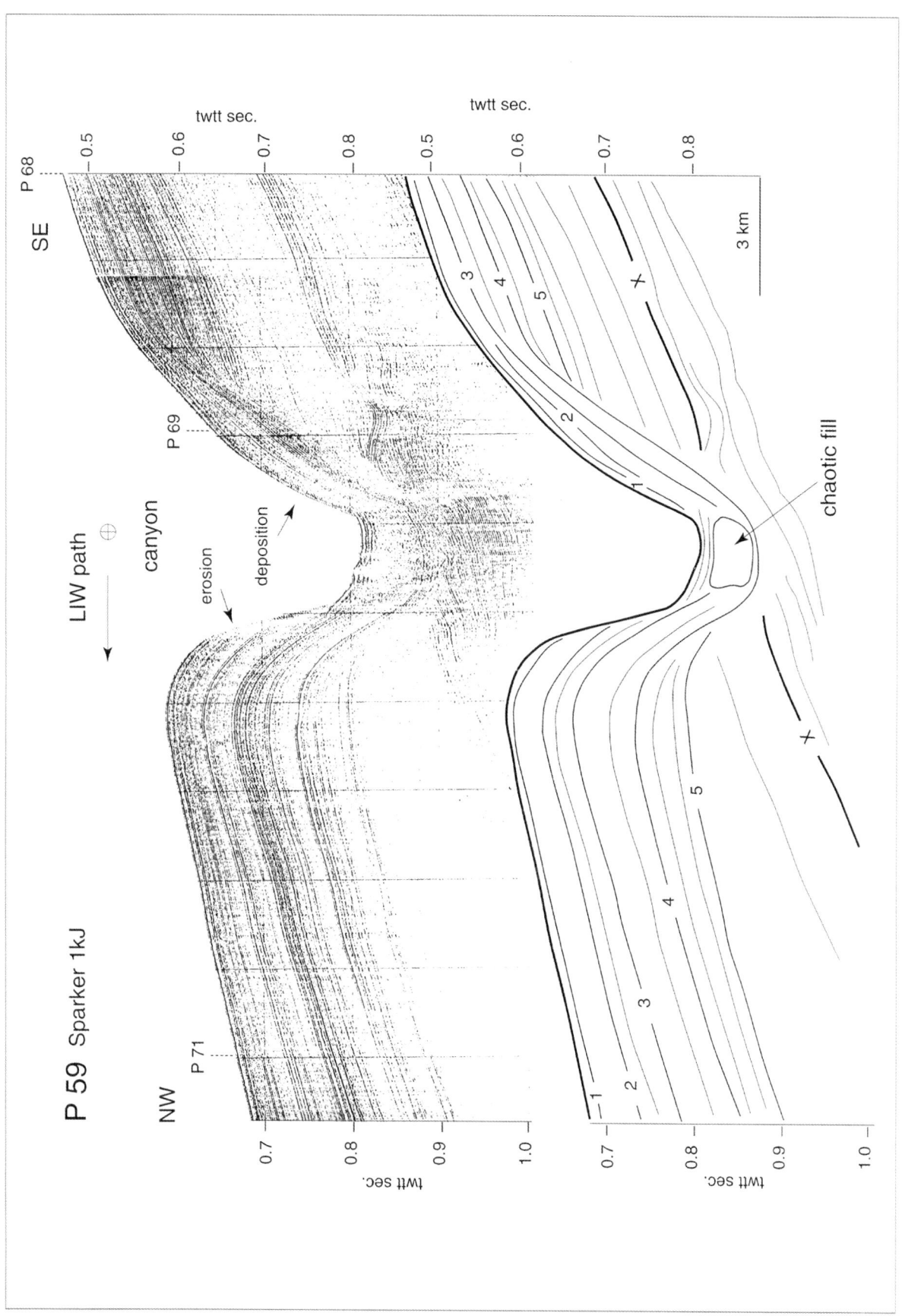

Fig. 12. Seismic profile crossing the canyon that separates the northern and southern basin sectors. Note the erosional character of northern wall and the growth of a small sedimentary body in the southern one.

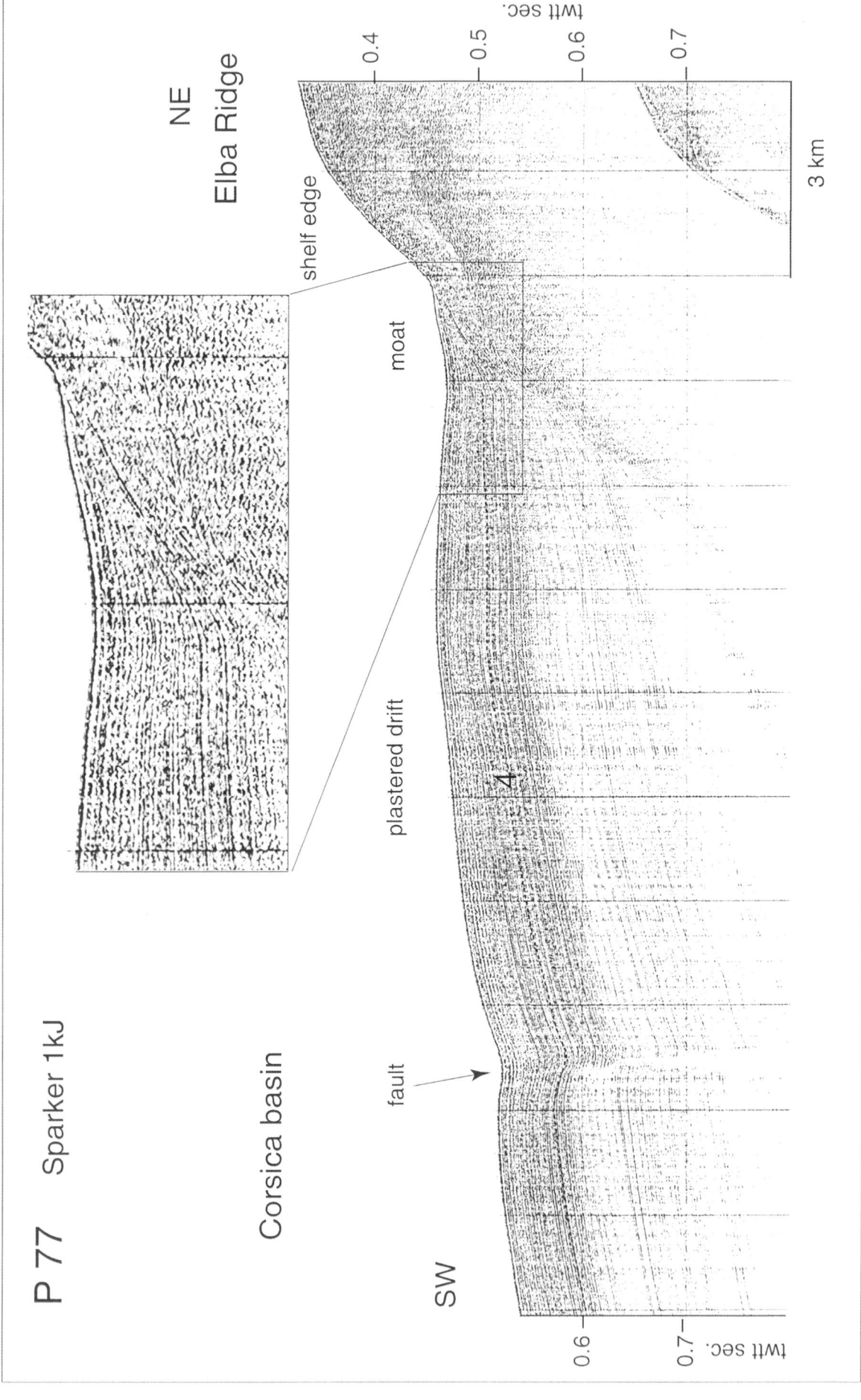

Fig. 13. The expression of the elongate drift of Figure 8 to the north, where it changes to a broad-sheeted shape, probably due to a higher-energy bottom-current regime brought about by the northward narrowing and shallowing of the basin.

The large mounded drift looses relief gradually and passes into a broad-sheeted drift (compare with plastered-type of Faugères *et al.* 1999) near the Capraia sill (Fig. 13). Here the cross-section of the Channel dramatically decreases and no depositional features have been observed; the sill is most likely swept by powerful currents that cause erosion and non-deposition. No data are available for the area immediately to the north, where the Corsica Channel deepens and opens to the Ligurian Sea.

Drift history and paleoceanographic changes

Current-related sedimentary features of the Corsica Basin have actively grown since the middle–late Pliocene, after the opening and progressive deepening of the Tyrrhenian Sea. Their continuous and relatively homegeneous growth suggest that the large-scale circulation pattern of the Mediterranean has not changed subtantially since that time. As a consequence, sediment drifts of the Corsica Basin have a great intrinsic potential for detailed reconstruction of the palaeoceanographic history of the Mediterranean. The time interval encompassed by these depositional features is characterized by important global events, like the onset of northern hemisphere glaciation (middle Pliocene), and more local ones, such as the cyclic deposition of Mediterranean sapropels, the organic-rich layers related to a complex interplay between astronomically driven climate changes and the biotic response (see Cramp & O'Sullivan 1999 for a review of this subject).

The basic stratigraphic architecture of the Corsica drifts, common to oceanic examples (Faugères *et al.* 1999), is characterized by the cyclic superposition of small-scale convex-upward aggradational units (tens of metres thick) bounded by non-depositional or slightly erosional surfaces (surfaces 1 to 5). The intervals of gradual growth and the episodes of erosion or non-deposition can be traced over the whole area, suggesting that drift evolution was controlled by large-scale hydrological events, most likely related to major climatic and/or sea level changes. The discontinuities can be traced into the deepest part of the Corsica Basin where they mark the abandonment of turbidite systems (Fig. 7). This seems to indicate that the evolution of sediment drifts and turbidite systems is controlled by the same factors.

The development of turbidite systems, especially those occurring on passive margins, is essentially controlled by high-frequency (4th and/or 5th order) climatic and eustatic changes. The main growth occurs during falling sea-level and low-stand phases, hence during increasingly cooling climate. If the observed repetitive pattern of Corsica basin turbidite systems growth and abandonment can be related to the Quaternary 100 000 year glacio-eustatic fluctuations, then the apparently coeval growth of sediment drifts implies that either (1) the competence of bottom currents increased during falling sea-level, thereby enhancing sediment erosion, transport and deposition, or (2) more sediment became available for bottom-current transport by the increased frequency of gravity flows along the western slope.

As for point (1), absolute criteria for recognizing current-speed changes within a given contourite system are very difficult to establish (see discussion in Faugères *et al.* 1999). Muddy contourite systems often develop in low to moderate energy current regimes; increases in current speed lead to non-deposition and/or erosion, although non-deposition could also be related to sediment starvation. Similarly, drift growth can be promoted by current deceleration, but also by increased sediment availability. Current speed may increase during sea-level fall due to the reduction of the basinal cross-section. Also, sea-level falls are related to climate cooling and this independently affects the circulation pattern of water masses within the basin.

Like other oceanic basins (Faugères *et al.* 1999), the circulation pattern of the Mediterranean during glacial times is still matter of debate. On a seasonal basis, it is known that LIW speed increases during the winter due to the stronger temperature gradient between northwestern and southeastern Mediterranean; although, this short term observation can hardly be extrapolated to longer periods.

Concerning sediment input changes (point 2), it is worth noting that turbidite systems of the Corsica Basin are fed only by western (i.e. Corsican) sources (Bellaiche *et al.* 1994; Stanley *et al.* 1980), while the eastern ones are dominated by bottom currents. An increased sediment input during sea-level fall and low-stand leads to the growth of turbidite systems in the axial part of the basin; moreover, due to the reduced width of the basin, low density turbidity currents or nepheloid layers as well as the fine-grained tails of large volume high-density gravity flows can be easily captured and redistributed by bottom currents flowing northward against the eastern side of the basin.

The two points are not mutually exclusive, and their combination could well explain the observed large-scale stratigraphic architecture. For all these reasons, the concept of bottom current 'efficiency' is probably more useful in describing contourite drift evolution. Current 'efficiency' is here defined as the capability of a bottom current to construct sediment drifts. The 'efficiency' concept does not necessarily imply high or low current velocities; it rather results from a combination of sediment availability, current speed and local topography and, as a consequence, it strongly depends on the local physiographic setting. Using this concept, the sedimentary history of the Corsica Basin can be described in terms of alternating phases of high and low-efficiency of the systems.

The way these systems actually grow can be tentatively assessed by looking at the evolution of sediment drifts through a short-period time-window (i.e. at the millennial scale climatic fluctuations within the last deglacial phase). Data from conventional cores (Fig. 4) show that sediment drifts have grown more rapidly before and after the two rapid melting events (Terminations 1A and 1B) and in particular since the attainment of present-day sea-level high-stand (5500 years BP). The time interval between the two terminations comprising the cold Younger Dryas event is recorded by a unit of uniform thickness made up of slightly coarser-grained sediment. In fewer cores this interval is instead characterized by erosion. It seems that general circulation has been increasingly more 'efficient' after sea-level rise, thus promoting in the late-Holocene a differential growth that fits very well with the external shape of the drifts, resulting from longer-term evolution.

Calibration of core and seismic data suggests that reflector 1 (i.e. the most recent discontinuity surface) falls in a time interval corresponding to the LGM or to the early deglacial time. These results from the youngest deposits do not contradict the hypothesis, based upon basin-wide correlation of small-scale depositional units, that drift growth mainly occurs during sea-level falls and low-stands. Our cores do not reach the last interglacial (125 ka), hence at the moment we have no data to assess drift growth during the whole glacial phase. Moreover, the correlation between drifts and turbidite systems is still preliminary. Higher resolution seismic data are needed to identify evolutionary sub-stages within the basic depositional units and to obtain a much finer tuning between the development of turbidite and drift systems.

With the available data, we can state that the same important hydrological change hindering the 'efficiency' of the drift system also lead to the abandonment of the turbidite system; this may suggest that a sudden decrease in sediment availability and a general reorganization of water circulation in the Mediterranean took place in the early deglacial time after a longer phase of drift growth during the glacial interval initiated some time after the last interglacial.

This is contribution no. 1876 of Istituto di Geologia Marina. The author wish to thank the Captain and the crew of the R/V Bannock and URANIA for helpful assistance given during cruises in the Corsica Channel. The manuscript greatly benefited from the thorough comments provided by J.-C. Faugères, B. Alonso and F. Trincardi.

References

ARTALE, M. & GASPARINI, G. P. 1990. Simultaneous temperature and velocity measurements of the internal wave field in the Corsican Channel (Eastern Ligurian Sea). *Journal of Geophysical Research*, **95**(C2), 1635–1645.

ASTRALDI, M., GASPARINI, G. P., MANZELLA, G. M. R. & HOPKINS, T. S. 1990. Temporal variability of currents in the Eastern Ligurian Sea. *Journal of Geophysical Research*, **95**(C2), 1515–1522.

BACINI SEDIMENTARI. 1979. *Primi dati geologici sul Bacino della Corsica (Mar Tirreno)*. Atti Convegno Scientifico Nazionale, P. F. Oceanografia e Fondi Marini, 713–727.

BARTOLE, R., TORELLI, L., MATTEI, G., PEIS, D. & BRANCOLINI, G. 1991. *Assetto stratigrafico-strutturale del Tirreno settentrionale: stato dell'arte*. Studi Geologici Camerti, special volume **1991/1**, 115–140.

BELLAICHE, G., DROZ, L., GAULLIER, V. & PAUTOT, G. 1994. Small submarine fans on the eastern margin of Corsica: Sedimentary significance and tectonic implications. *Marine Geology*, **117**, 177–185.

CAPOTONDI, L., BORSETTI, A. M. & MORIGI, C. 1999. Foraminiferal ecozones, a high resolution proxy for the late Quaternary biochronology in the central Mediterranean Sea. *Marine Geology*, **153**, 253–274.

CARTER, L. & MCCAVE, I. N. 1994. Development of sediment drifts approaching an active plate margin under the SW Pacific Deep Western Boundary Undercurrent. *Palaeoceanography*, **9**, 1061–1085.

CRAMP, A., & O'SULLIVAN, G. 1999. Neogene sapropels in the Mediterranean: a review. *Marine Geology*, **153**, 11–28.

FABBRI, A., GALLIGNANI, P. & ZITELLINI, N. 1981. Geologic evolution of the peri-Tyrrhenian sedimentary basins. *In*: WEZEL, I. C. (ed.) *Sedimentary Basins of Mediterranean Margins*.

FAUGÈRES, J.-C., STOW, D. A.V., IMBERT, P. & VIANA, A. 1999. Seismic features diagnostic of contouritic drifts. *Marine Geology*, **162**, 1–38.

GABIN, R. 1972. Resultats d'une etude de sismique reflexion dans le Canal de Corse, et de sondeur de vase dans le Bassin Toscan. *Marine Geology*, **13**, 267–286.

HOWE, J. A. 1996. Turbidite and contourite sediment waves in the Northern Rockall Trough, North Atlantic Ocean. *Sedimentology*, **43**, 219–234.

HOWE, J. A., STOKER, M. S. & STOW, D. A.V. 1994. Late Cenozoic sediment drift complex, northeast Rockall Trough, North Atlantic. *Palaeoceanography*, **9**, 989–999.

MARANI, M., ARGNANI, A., ROVERI, M. & TRINCARDI, F. 1993. Sediment drifts and erosional surfaces in the central Mediterranean: seismic evidence of bottom-current activity. *Sedimentary Geology*, **82**, 207–220.

MCCAVE, I. N. & TUCHOLKE, B. E. 1986. Deep current-controlled sedimentation in the western North Atlantic. *In*: VOGT, P. R. & TUCHOLKE, B. E. (eds) *The Geology of North America: The Western North Atlantic region*. Geological Society of America, Boulder, CO, 451–468.

MILLOT, C. 1987. Circulation in the Western Mediterranean Sea. *Oceanologica Acta*, **10**, 143–149.

NELSON, C. H., BARAZA, J. & MALDONADO, A. 1993. Mediterranean undercurrent 'contourites' in the eastern Gulf of Cadiz. *Sedimentary Geology*, **82**, 103–132.

PASCUCCI, V., MERLINI, S. & MARTINI, I. P. 1999. Seismic stratigraphy of the Miocene–Pleistocene sedimentary basins of the Northern Tyrrhenian Sea and Western Tuscany (Italy). *Basin Research*, **11**, 337–356.

ROVERI, M., BOSCHETTI, A. & PENITENTI, D. 1994. Sedimentary features of a bottom current-dominated slope: the eastern margin of the Corsican Trough (northern Tyrrhenian Sea, Italy). *In*: *15th IAS Regional Meeting Ischia*, Abstract Volume, 356–357.

SELLI, R. & FABBRI, A. 1971. *Tyrrhenian: A Pliocene Deep-Sea*. Accademia Nazionale dei Lincei, Rendiconti, Classe Scienze Fisiche, Matematiche e Naturali, **50**(5), 104–166.

STANLEY, D. J., REHAULT, J. P. & STUCKENRATH, R. 1980. Turbid-layer bypassing model: the Corsican Trough, northwestern Mediterranean. *Marine Geology*, **37**, 19–40.

STOKER, M. S., AKHURST, M. C., HOWE, J. A. & STOW, D. A. V. 1998. Sediment drifts and contourites on the continental margin off northwest Britain. *Sedimentary Geology*, **115**, 33–51.

VIARIS DE LESEGNO, L. 1978. *Etude structurale de la Mer Tyrrhenienne Septentrionale*. These de iiieme cycle, University De Paris.

ZITELLINI, N., TRINCARDI, F., MARANI, M. & FABBRI, A. 1986. Neogene tectonics of the Northern Tyrrhenian sea. *Giornale di Geologia*, **48**, 25–40.

WUST, G., 1961. On the vertical circulation of the Mediterranean Sea. *Journal of Geophysical Research*, **66**, 3261–3271.

The Vema contourite fan in the South Brazilian basin

JEAN-CLAUDE FAUGÈRES[1], S. ZARAGOSI[1], M. L. MÉZERAIS[2] & L. MASSÉ[1]

[1]*Department of Geology and Oceanography (DGO – UMR CNRS 'EPOC', 5805), University of Bordeaux I, Avenue des Facultés, 33405 Talence, France*
[2]*International Offshore Technical Assistance, 5 rue des Chevaliers de Malte, Port Ariane, 34970 Lattes, France*

Abstract: The Vema *contourite fan* is a Neogene mud-rich accumulation (200–400 m thick), fed by Antarctic Bottom Water bottom currents and located downstream of the Rio Grande Rise. It forms one single, mounded, fan-shaped body deposited between two major channels through which the main part of the deep AABW circulation is funneled. A suite of cores and seismic lines have been collected over the whole area. The sediments deposited below the shear zone between the two current branches consist almost exclusively of muddy contourites, either homogeneous in structure or micro-brecciated. Manganiferous deposits occur in the vicinity of the channels and on the channel floors. As a result of the morphological and hydrological background, the contourite drift has prograded mostly downstream. It is composed of several depositional units bounded by widespread discontinuities showing erosional patterns. This geometry results from an alternation of episodes of strong and unstable erosive currents, and periods of relatively weak and stable depositional currents.

The Vema channel-related drift, in the southwestern part of the South Brazilian basin, is a fan-shaped sedimentary body built downstream of a deep channel by active bottom currents (Fig. 1, Table 1). It has been defined as a *contourite fan* by Mézerais *et al.* (1993). This paper summarises previous work on the Vema contourite fan (Mézerais 1991; Mézerais *et al.* 1993; Faugères *et al.* 1998), based on data acquired during the BYBLOS cruise (Faugères 1988), including 3.5 kHz and water-gun seismic lines and six Kullenberg cores.

Geological and oceanographic setting

The Brazil basin is limited to the south by an E–W basement ridge, the Rio Grande rise, to the NW by the Brazilian continental margin, to the north by the Vitoria Trindade ridge and to the east by the Mid-Atlantic Ridge.

The basin is linked to the Argentine Basin by two major submarine valleys, the Vema channel and the Sao Paulo abyssal gap which incise the Rio Grande rise (Le Pichon *et al.* 1971;

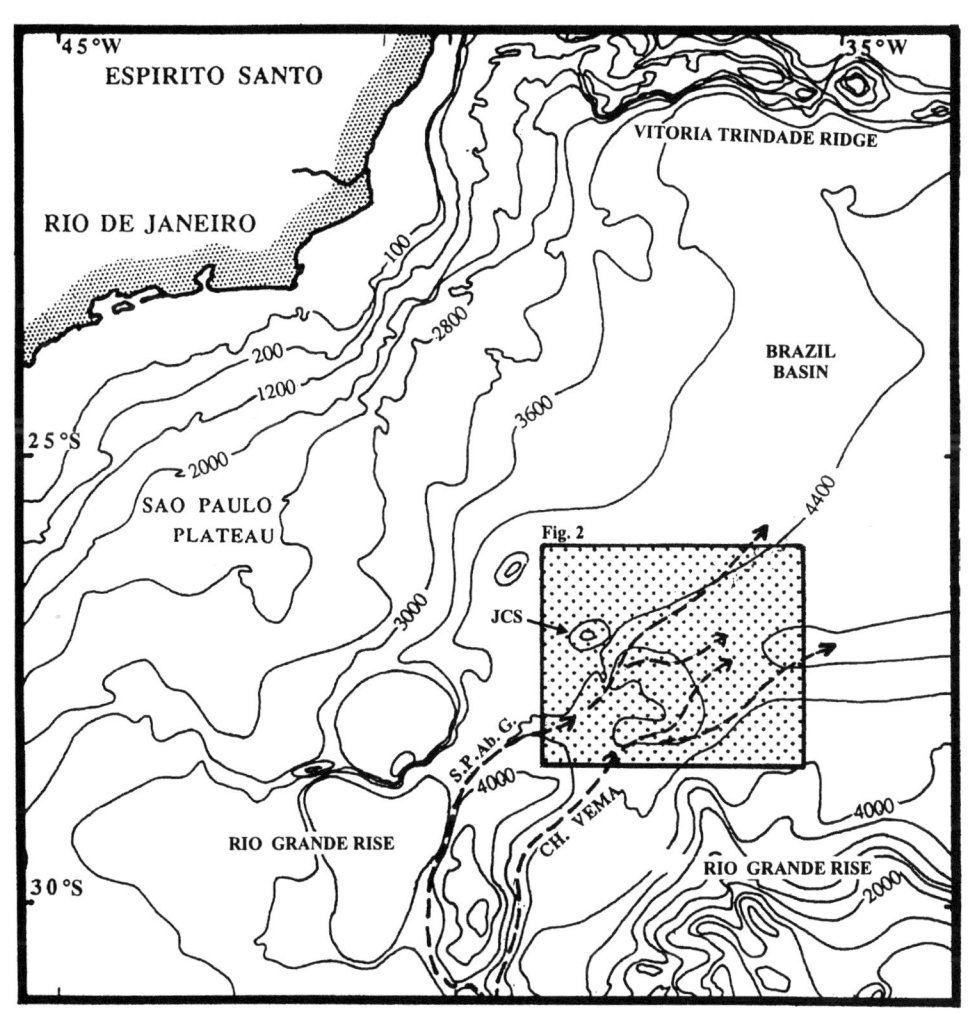

Fig. 1. Location of the area studied in the South Brazilian basin (CH. Vema, Vema channel; SP Ab. G, Sao Paulo Abyssal Gap (or channel); JC S, Jean Charcot Seamount; the dashed arrows show the AABW pathways.

Table 1. *Principal characteristics*

Location:	South Brazilian Basin
Setting:	northern end of the Vema channel at 4400–4600 m water depth
Age:	late Oligocene to Recent
Drift type:	contourite fan
Dimensions:	length 150 km; width 120 km; thickness 200–400 m
Seismic facies:	sheet-like, lobate or channel-levee seismic geometry, for the surficial deposits; lenticular units as remains of larger ancient lobes associated with erosional discontinuities of regional extension, in the subsurface.
Sediment facies:	silty-clay muddy contourites either homogeneous or micro-brecciated, and manganiferous deposits

Johnson *et al.* 1977; Gamboa & Rabinovitz 1981, 1984; Gamboa *et al.* 1983; Mello 1988). The area is swept by strong bottom currents originating from the Antarctic Bottom Water (AABW) and flowing from the Argentine basin towards the north (Wright 1970; Reid *et al.* 1977; Hogg *et al.* 1982; Johnson 1983; Pierre *et al.* 1991; Speer & Zenk 1993). These currents are funnelled through the Vema channel and the Sao Paulo abyssal gap. Current measurements from these channels indicate velocities of up to 30 cm s^{-1} which decrease northward and downstream of the channels, along the South Brazilian rise (Hogg *et al.* 1982). In such a hydrological context, the detailed study of deposit patterns sheds some light on a better understanding of the contourite problem (Faugères & Stow 1993a).

Bathymetry

The contourite deposits are developed downstream of the outlets of these two channels towards the northeast, in an area bounded laterally, to the south and to the northwest, by volcanic ridges (Mézerais 1991; Mézerais *et al.* 1993). The accumulation is roughly fan-shaped (Fig. 2), with an apex located between the channel outlets, and can be identified on the bathymetric map of Cherkis (1983). It lies in 4400–4750 m of water depth and has an approximate length of 150 km and a width of 60 to 120 km increasing downstream. This accumulation is Miocene to Recent (15 Ma) in age and covers an area of about 15 000 km^2 with a maximum thickness of 400 m.

The contourite fan is limited on both sides by channel-like features that are the downstream extensions of the Sao Paulo abyssal gap and the Vema channel, along which the currents are constricted. These two channels converge into a crescent-shaped depression that bounds the contourite fan to the north (Figs 2 and 7).

Stratigraphy

The stratigraphical interpretation of the whole contourite fan body proposed here is supported by a transverse (NW–SE) profile across the accumulation (BC, Fig. 2) and one profile along the accumulation (G1G, Fig. 2). Our interpretation relies upon the seismic character of the different units deposited during the Neogene. The seismic lines were calibrated from DSDP Leg 72 data, Site 515 (Barker *et al.* 1983a,b; Gamboa *et al.* 1983; Johnson & Rasmussen 1984; Mézerais 1991). The upper Quaternary sediment stratigraphy is on nannofossil biostratigraphy, excess Th^{230} activity analyses, carbonate curves and comparison with well-dated cores previously collected in the same area (Johnson *et al.* 1977; Williams & Ledbetter 1979; Ledbetter & Ellwood 1982; Massé 1993).

Seismic characteristics

Seismic reflection profiles

Seismic profile analysis allows us to distinguish four major sediment units in the Neogene deposits of the fan drift (Fig. 3). In

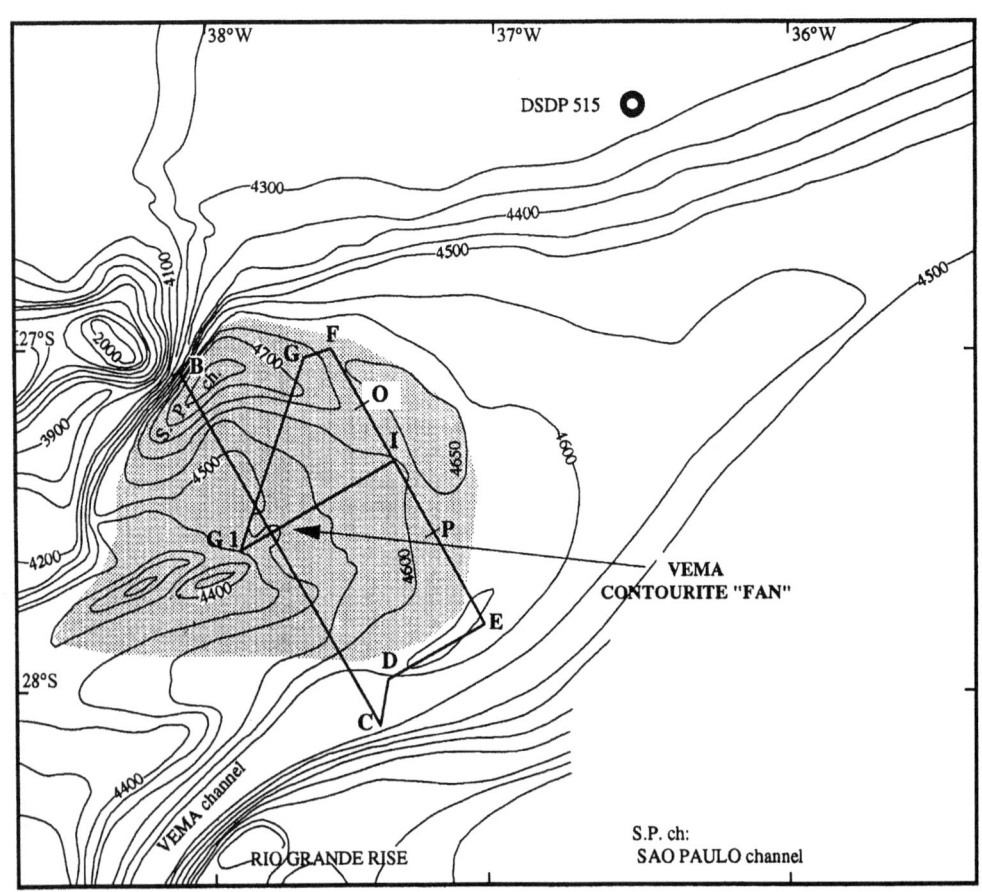

Fig. 2. Bathymetric map of the Vema contourite fan and location of the seismic lines (BC, Fig. 3 and GG1, Fig. 4).

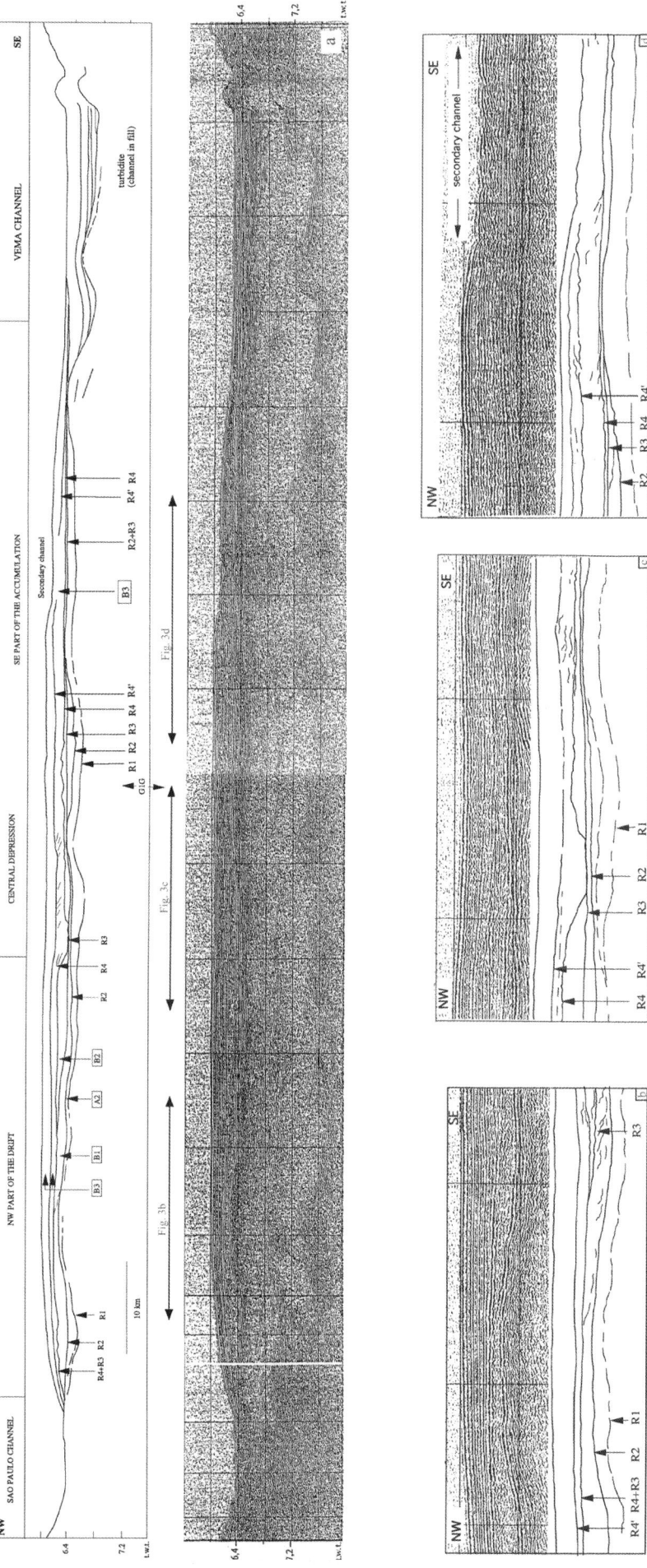

Fig. 3. Strike BC seismic profile, located upstream of the main accumulation (see Fig. 2 for the location). R1, R2, R3, R4 and R4' refer to the main discontinuities, and A2, B1, B2 and B3 to the main sediment units. (**a**) strike BC profile and interpretation, (**b**), (**c**) and (**d**) details of strike BC seismic profile located on (**a**). (**b**) the NW part of the accumulation showing the B3 units (fairly continuous reflections especially above R4') overlying the B1 unit (chaotic facies) and the R3/R4 erosional surface in between; (**c**) the central depression showing the erosional patterns of R4', R4, R3 and R2 and the facies geometry of unit B1 (chaotic) and B2 (gently chaotic at the base and progressively more continous towards the top); (**d**) the SE part of the accumulation and the secondary channel with high amplitude chaotic reflections; note the erosive patterns of R4' and R3.

these units, the observed spectrum of the seismic reflection-types can be defined as a continuum between:

(a) drapes, continuous and parallel, underlaying the areas with high sedimentation rates and very weak currents (unit B3);
(b) transparent seismofacies interpreted as low energy homogeneous deposits (units B1 and B2);
(c) parallel-continuous reflectors affected by small hyperbolae indicating a 'roughness zone' due to a more active circulation producing megaripples and dunes (B3 upper subunit, in the central depression);
(d) chaotic reflections in major and secondary channels (B3 lower subunit) or corresponding to discontinuous deposits on surfaces swept by active currents (B1 and B2 in the SE area);
(e) irregular, high amplitude reflectors that underline the erosional unconformities.

The contourite accumulation (Fig. 3) is separated from the substratum (basal units A1 and A2) by an erosional surface, R2. The accumulation developed above R2, and consists of the four depositional units, BI, B2 and B3 composed of two subunits bounded by strong continuous reflectors, R3, R4 and R4'. R3 and R4 are erosive discontinuities whereas R4' is a rather conformable discontinuity with more gentle erosion.

The ages of the units and major bounding unconformities were calibrated from DSDP site 515. The erosive R2, R3 and R4 are assumed to reflect episodes of intense and widespread current activity at the end of the Oligocene, from the end of the early Miocene to middle Miocene, and end of the late Miocene, respectively. Sedimentological investigations (Faugères et al. 1998) suggest that velocities were much greater during these events than observed at present. R4' would be upper Pliocene in age and could correspond to a weaker erosional episode than the previous ones.

The lower unit B1 (late Oligocene–early Miocene) covers a small area compared to the present-day contourite fan (Fig. 5). It shows seismic reflections downlapping mainly towards the north or the northeast (Fig. 3), with nearly transparent facies and chaotic reflections downstream suggesting active bottom currents at that time. Unit B1 was essentially developed in the northwestern part of the present day accumulation (Figs 3, 4 and 5), in an area sheltered from the influence of the most active currents, in the 'shadow' of the segment of the rise in between the two channel outlets.

Unit B2 (middle to late Miocene) is elongated parallel to the axes of the channels (Fig. 5) and is the eroded remnant of a bigger sediment accumulation. The unit prograextends mainly towards the north–northeast (downlaps onto the reflector R3, Fig. 4), and slightly towards the southeast (Fig. 3). A vertical gradation from gently chaotic at the base of the unit to more continous reflections toward the top, suggests a decrease of the bottom circulation during the unit deposition.

The B3 upper unit, Pliocene to Quaternary in age, gives the fan its present day shape. This unit overlaps the B1 and B2 units laterally and distally and it covers a much wider area than these units (Figs 3, 4 and 5). The currents are probably less active than during the deposition of the previous units. B3 is composed of two sub-units. The lower sub-unit, in the northwest, shows predominantly a high amplitude, high continuity character and prograes towards the Sao Paulo channel (Fig. 3). On the contrary, the southeastern zone is mostly chaotic (locally oblique reflections) with a southeastward migrating trend toward the valley that extends from the Vema channel; the currents are restricted in this valley and a minor channel that runs across the accumulation (Figs 3b, c & 5). The upper sub-unit of B3 determines the present-day morphology of the fan (Fig. 3). It is of uniform thickness (100 to 110 m), prograes and downlaps laterally towards the Vema valley, and distally towards the northeast and the east beyond the underlying units, and rests on the basal surface R2 (Figs 3 and 4).

This upper sub-unit consists of continuous and parallel reflections, locally wavy or with small hyperbolae and suggests a significant change in the energy of deposition, still lower than during the previous episodes.

Echofacies mapping

The 3.5kHz echofacies analyses have been made according to the methodology of Damuth (1975) and Damuth & Hayes (1977) and the interpretation is supported by core lithology (Massé 1993). The contourite fan has been subdivided into several regions, each of which shows distinctive echofacies (Figs 6, 7 and 8).

(a) The *bottom of the Vema channel* is characterized by a prolonged echo, without sub-bottom reflectors (IA, Fig. 6a). This echo is related to surficial ferro-manganese crusts and beds (0 to 1.9 m below sea-floor) which bear evidence of high-velocity bottom currents lasting for millions of years (Massé 1993).
(b) The *flanks of the channels* are the site of local sediment slumps. These slumps are underlined by large, irregular, overlapping hyperbolae (IIIA, not illustrated).
(c) The *main body of the accumulation* (eastern and middle parts) shows a distinct echo-facies with parallel, continuous sub-bottom reflectors (IB, Fig. 6b) without any evidence of high bottom current activity. This facies, observed as deep as 105 m below the sea-floor, is composed of predominantly silty-clayey muds. It should correspond to a rather low flow regime. Locally on the accumulation or laterally on the adjacent rise, it passes to a wavy hyperbolic echo, still with sub-bottom reflectors (IIIB, Fig. 6c), that could reflect slightly stronger flow.
(d) The *distal part of the accumulation* and the *crescent-shaped depression* are marked by echoes with less continuous sub-bottom reflectors or locally a slightly chaotic echo and regular overlapping hyperbolae with vertices more or less tangential to the sea floor (IIID, Fig. 6d). These features are interpreted to be the acoustic response of small-scale erosional bedforms which would indicate currents slightly more active than in the main depositional area, confirmed by the occurrence of abundant mud clasts in the muddy deposits.

All these data suggest that most of the surficial deposits are muddy sediments deposited under a regime of low-energy bottom currents. Only in the major channels are currents fast enough for erosional processes or non-deposition, and for maintaining steep erosive flanks where mass movement occurs.

Sediments: core description

Sediment facies

Six kullenberg cores have been collected: one in the Vema valley floor, four on the fan drift and one on the northern rise close to the drift (Figs 6 and 7). All of them are located in an area swept by the AABW current. No evidence of turbidites has been observed in these late Quaternary deposits. The sediments on the drift accumulation are almost exclusively very fine-grained silty muds (Fig. 9). The median grain size ranges from 3 to 9 μm, with a fraction coarser than 15 μm composed of quartz grains (30%), Ca-Na feldspars (30%) and micas (15%) together with diatoms and corroded fragments of foraminifers. The carbonate contents are very low as deposition takes place below the CCD; they never exceed 25% of the bulk sediment, and are mostly less than 5%. Taking into account the colour and depositional and diagenetic features, three different types of muds can be distinguished. In the Vema valley, manganiferous deposits occur and form a decimetric

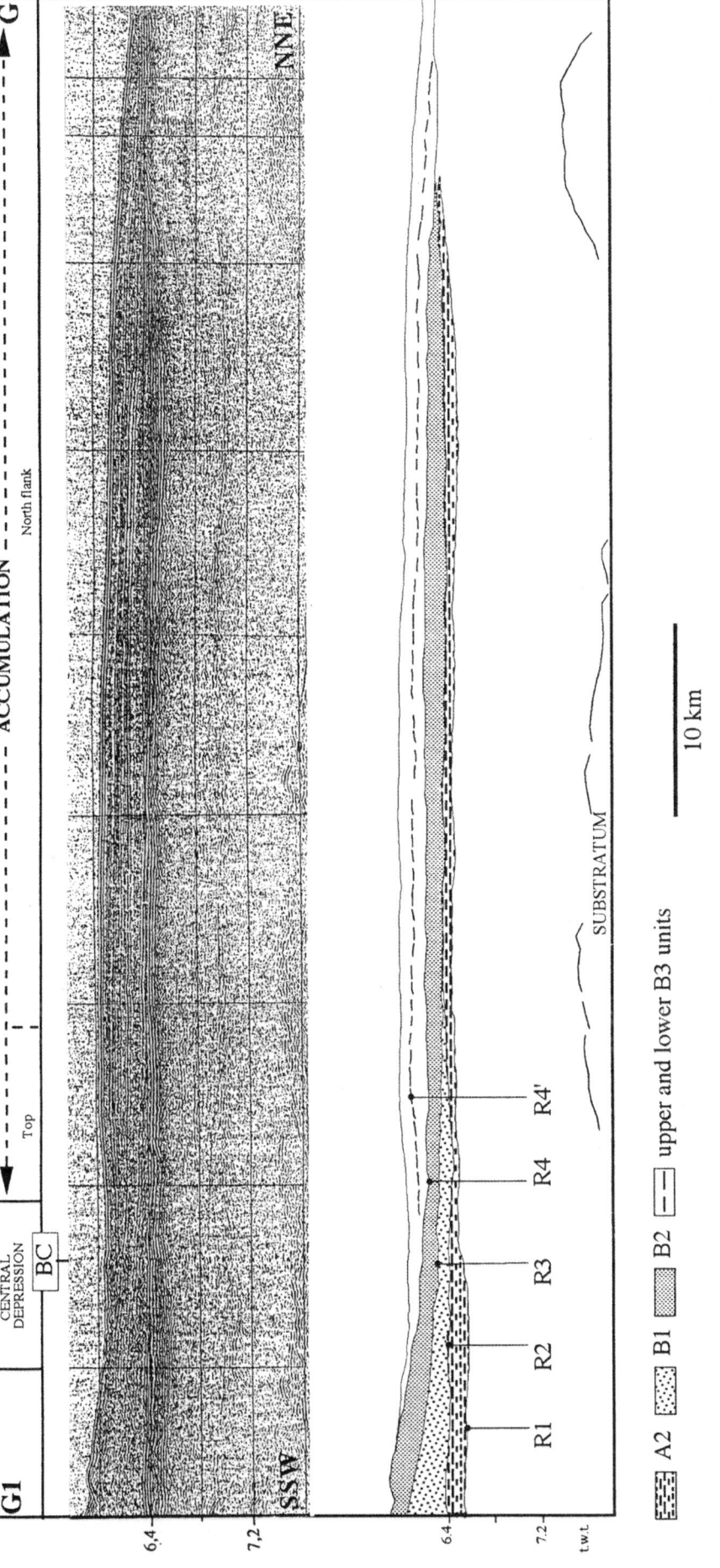

Fig. 4. Dip G1G seismic profile (see Fig. 2 for the location). A2, seismic unit overlain by the contourite fan units (B1, B2 and B3).

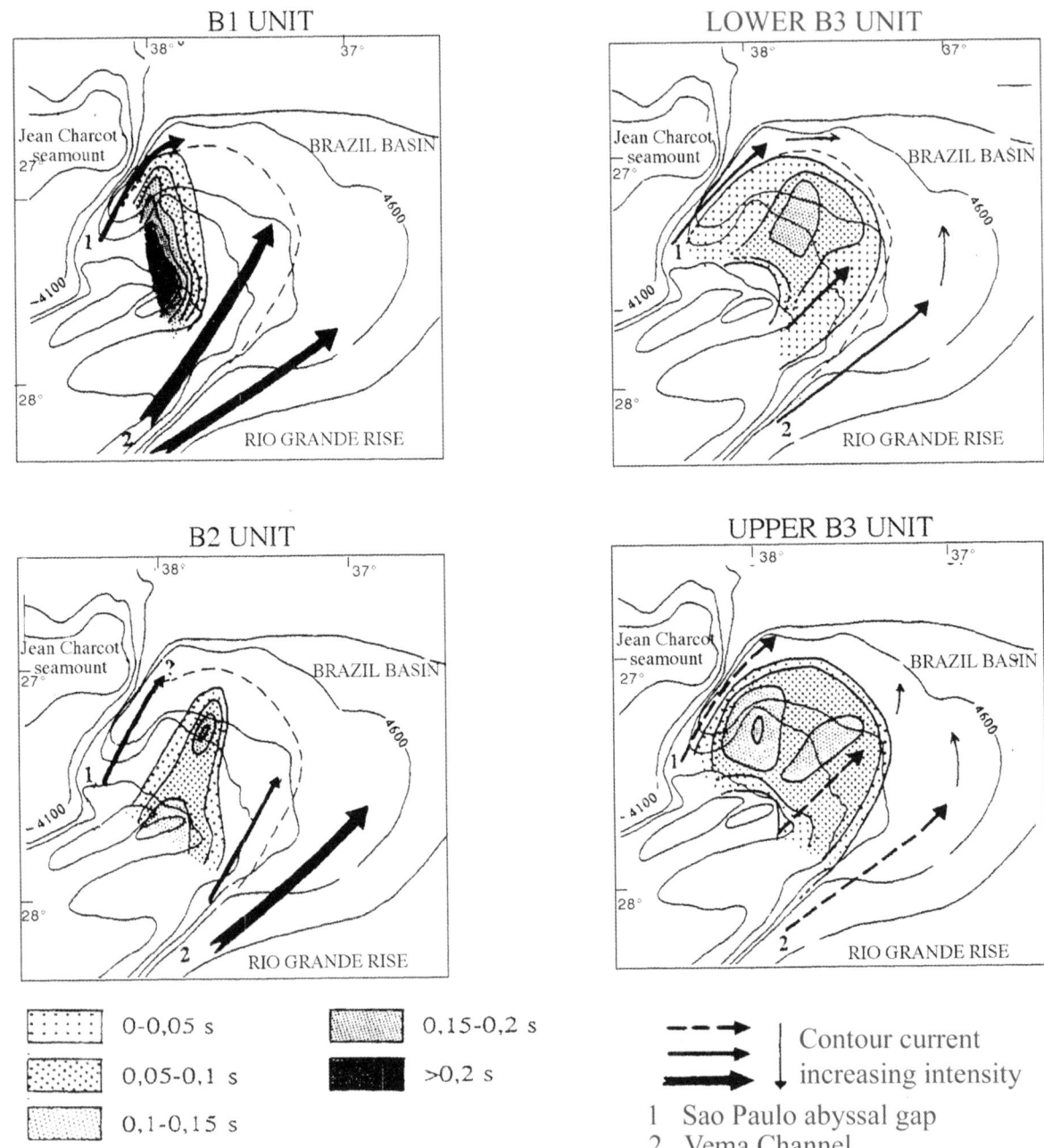

Fig. 5. Isopach map and bottom current intensity for the different sediment units.

crust on the sea-floor (Figs 9 and 10d, e). Millimetric to centimetric layers of manganiferous deposits may also be observed interbedded in the muddy sediments of the fan drift itself (Figs 9 and 10a, c).

The five sediment facies distinguished include:

(a) **Brownish-yellow muds** (Fig. 10a, b) bearing bioturbation marks, and frequently displaying centimetric beds with millimetric to centimetric dark-grey manganese laminations, that appear as very diffuse mottling, slightly darker than the adjacent muds;

(b) **Grey-green homogeneous muds**, slightly bioturbated (Fig. 10a, b);

(c) **Grey-green microbrecciated muds**, very similar to the above muds but showing interbedded millimetric to decimetric layers of muddy clasts composed of the same sediment as the grey-green homogeneous muds (Figs 9 and 10b);

(d) **Manganiferous indurated crust**, typically up to 20 cm thick, and more rarely up to 80 cm, representing millions of years of sedimentation, showing a crudely laminated pattern, with evidence of numerous erosional surfaces and few thin layers or lenses of brownish muds trapped in between (Fig. 10d, e);

(e) **Manganiferous semi-indurated thin layers** (mm to cm) with sharp boundaries that are interbedded within the thick brownish-yellow muds (Fig. 10c).

The brownish-yellow and grey-green muds are interpreted as muddy contourites deposited under low velocity current conditions: the major currents are confined within the two channels and only a sluggish circulation occurs on the fan drift. The facies colour changes and the occurrence of dark manganiferous mottling are interpreted as the result of early diagenetic processes controlled by local fluctuations in the sedimentation rate. The occurrence of microbrecciated mud clasts suggests deposition in a higher energy environment with more active currents. Deposition of manganiferous crusts or layers is related respectively to

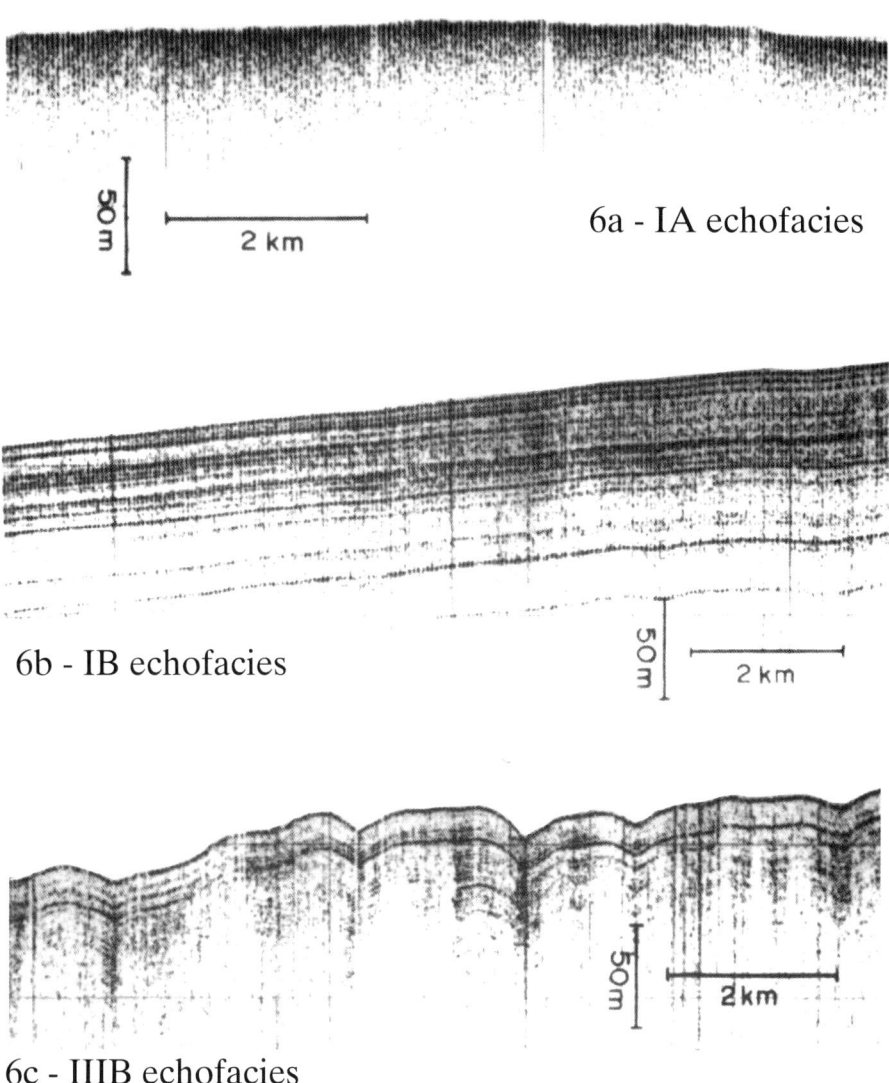

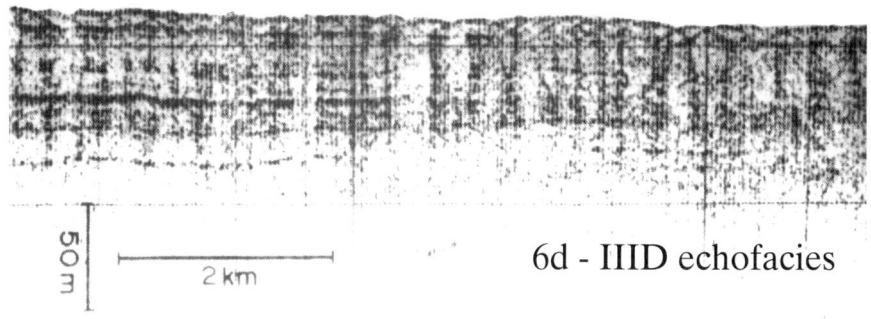

Fig. 6. The main type of 3.5kHz echofacies: (**a**) IA echofacies, prolonged echo, without sub-bottom reflectors; (**b**) IB echofacies, distinct echo-facies with parallel, continuous sub-bottom reflectors; (**c**) IIIB echofacies, wavy echo with sub-bottom reflectors related to sediment waves of various morphology; (**d**) IIID echofacies, regular overlapping hyperbolae with vertices tangential to the sea floor associated with more or less discontinuous sub-bottom reflectors.

environments of very intense circulation like the Vema valley or to short episodes of increased bottom current circulation that overflows the drift accumulation.

Vertical succession: palaeocurrent variation and sedimentation rate

Cores collected on the fan drift or the adjacent rise show a large variety of vertical facies distribution. As a rule, they show alternating brownish-yellow muds and grey-green muds (Figs 9–12). Transitions between the different mud layers can be gradual, sharp or marked by an erosional surface (Fig. 10). Erosional surfaces may also be observed within the brownish-yellow muds where they are often associated with a thin indurated manganiferous layer (Fig. 10c). Such features are evidence of high amplitude and short-term current events. Slight variations in the mud grain-size (measured as percentage of particles greater than 10 μm, Massé *et al.* 1994) have also been interpreted as indicating long-term and low amplitude current fluctuations.

Detailed analyses of the facies pattern (core KS8803, on the adjacent rise, Fig. 11) indicates a major change in the

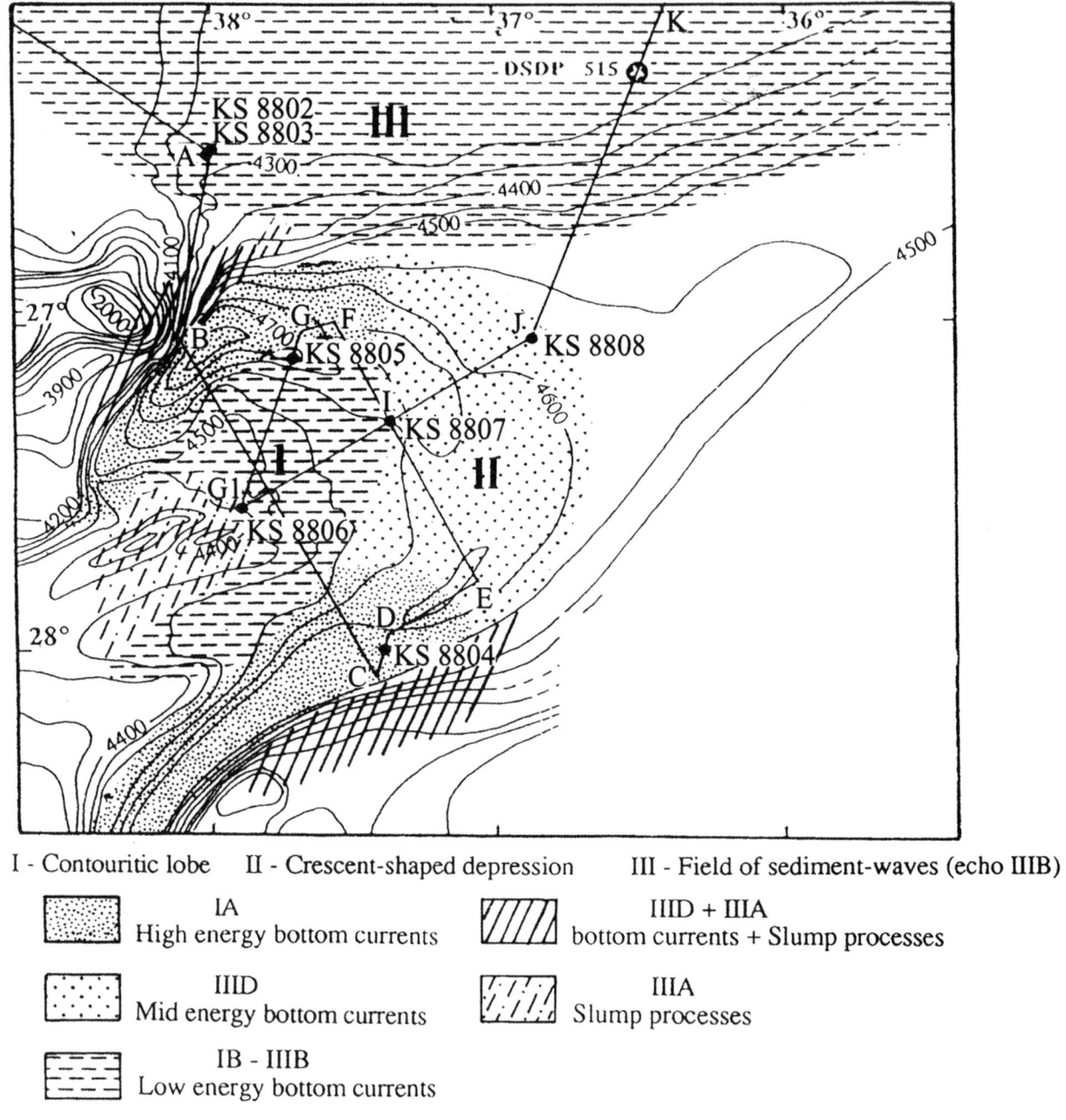

Fig. 7. 3.5kHz echofacies distribution on the Vema contourite fan, and core location.

environmental conditions around 350 ka BP (oxygen isotope stage 9). Before this time (from about 600 ka to 350 ka BP), the sedimentation was characterized by an instability in current activity, with strong flow events recorded as erosional surfaces and occuring approximately every 50 ka. The following period (350 ka BP to present) is marked by globally weaker current activity and long-term fluctuations of lower amplitude and longer duration (100 ka cycle), only recorded by grain size fluctuations. The highest velocities occur preferentially during periods of climatic cooling (Massé et al. 1994).

Most of the cores collected on the drift do not show a sediment record of the environmental conditions as clearly as core KS8803. This is mainly due to deeper erosion during strong current events.

The mean sedimentation rate on the rise is between 2.5 and 2.8 cm ka^{-1} since 350 ka BP, with a sharp decrease after 60 ka BP to 1 cm ka^{-1} (Fig. 12). On the drift (core KS8806, Fig. 11), the mean rate is initially higher at 5.9 cm ka^{-1}, but then shows a similar decrease. Other cores from the drift show recent brownish-yellow muds overlying grey-green microbrecciated muds. The occurrence of a marked erosional surface at the boundary between the two facies, close to the core top (Figs 10b and 13), with probably an important sedimentary gap, prevents any reliable dating of the deposits.

Discussion

The Vema contourite fan is a channel-related drift (Mézerais et al. 1993; Faugères et al. 1993b, 1999). It was deposited downstream of a channel outlet and the sediment deposition is controlled by the bottom current activity. Here, the major currents are presently funnelled through two valleys where transport processes are dominant. These valleys bound an area with low current activity allowing deposition.

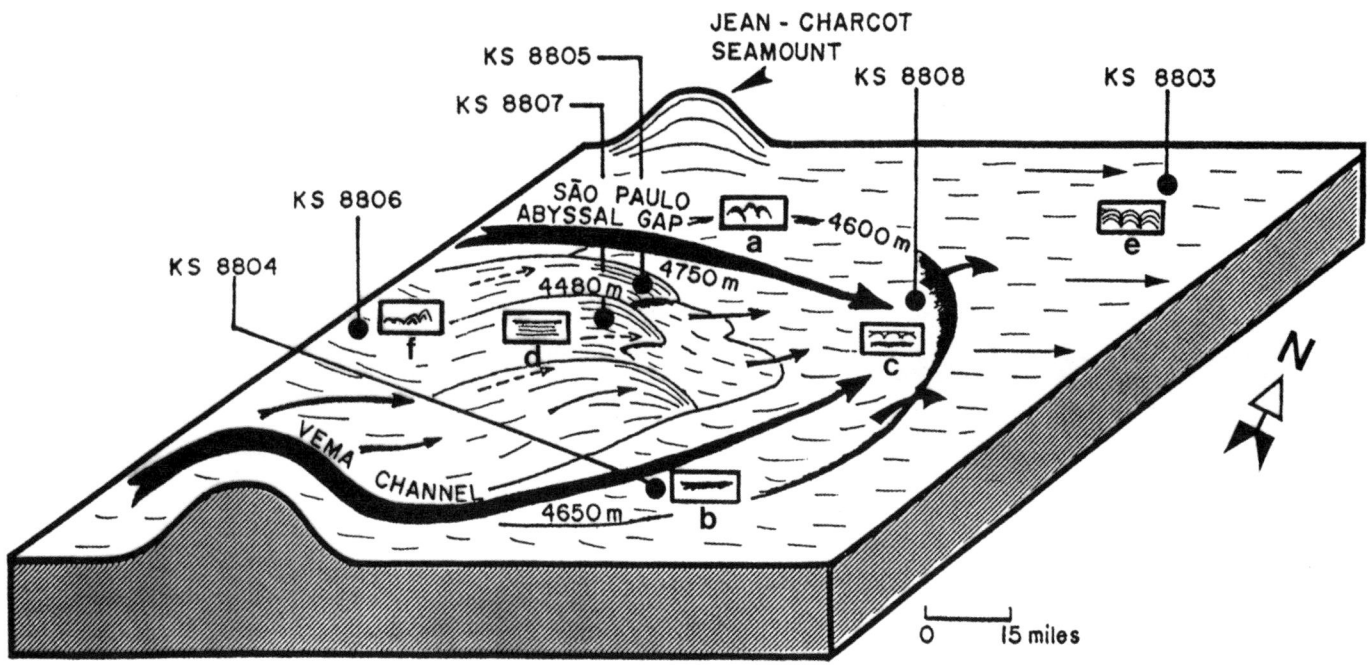

Fig. 8. Scheme showing the drift morphology and the relationship between the current pathways and the echofacies distribution. (**a**) III-A+III-D echofacies; (**b**) I-A echofacies; (**c**) III-D echofacies; (**d**) I-B echofacies; (**e**) III-B echofacies; (**f**) III-A echofacies.

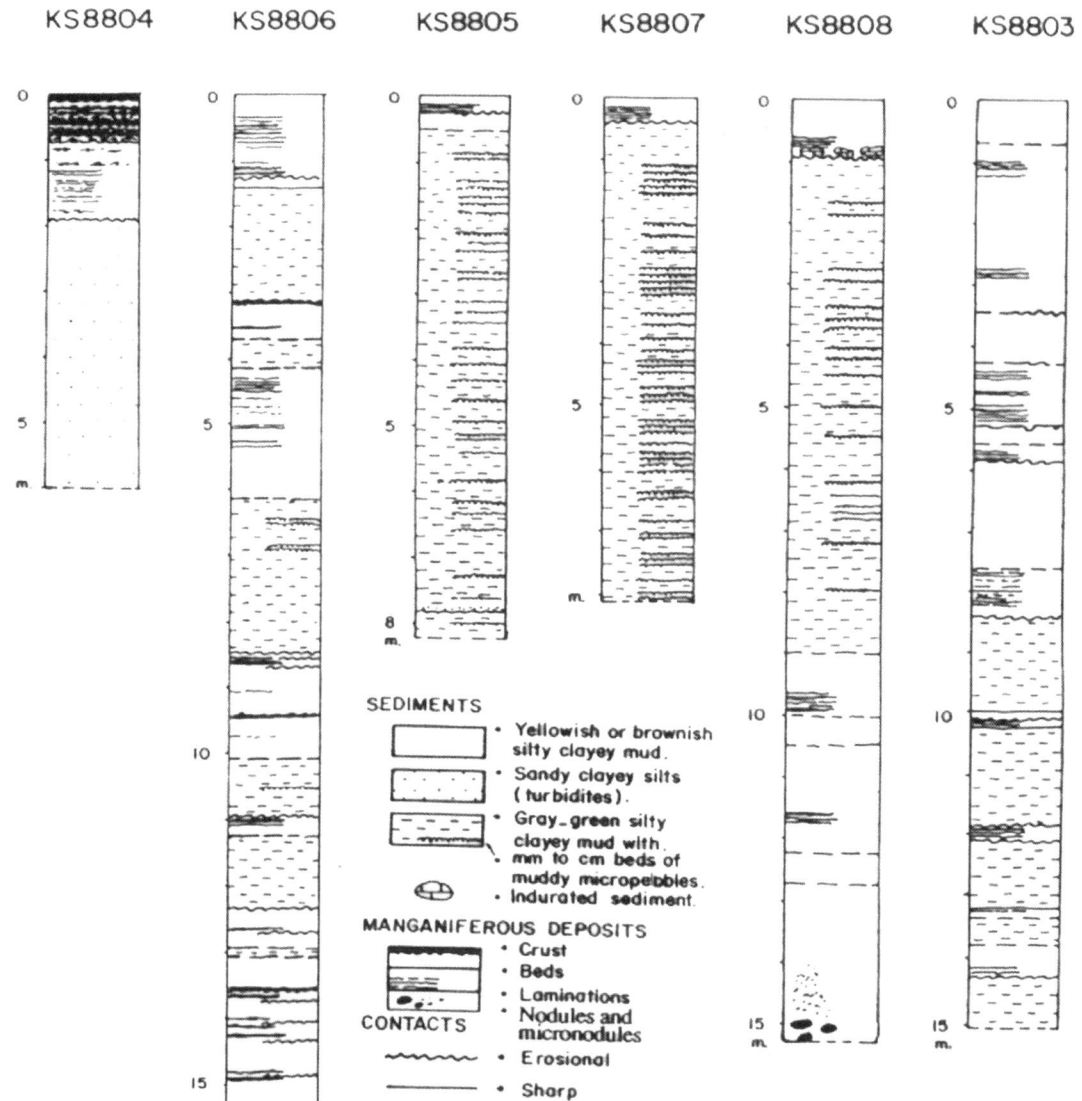

Fig. 9. Core lithology (from Mézerais *et al.* 1993); core location in Figures 7 & 8.

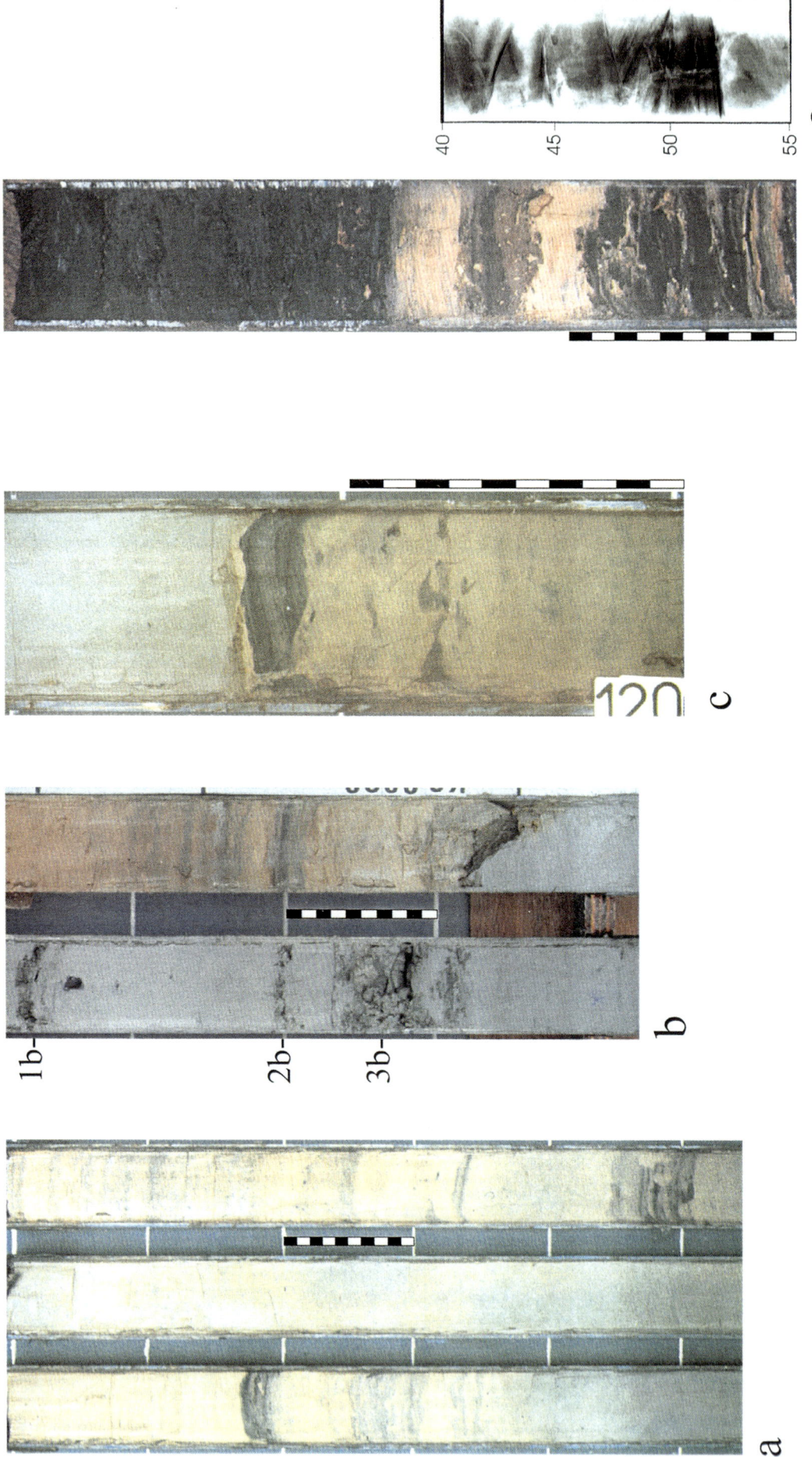

Fig. 10. Muddy contourite facies and manganiferous sediments (the black and white scale represents 10cm). (**a**) muddy contourite (KS8803); brownish-yellow muds associated with indurated manganiferous layers (black cm layers) alternate with greenish muds (KS8808). (**b**) sharp contact (erosive hydrologic event) between brownish-yellow muds and greenish muds showing interbedded layers (1b, 2b & 3b) of microbrecciated muds (KS8808). (**c**) Erosive contact between yellowish and greenish contouritic muds underlined by a centimetric manganiferous crust (KS8806). (**d**) & (**e**) decimetric manganiferous crust on the Vema channel sea-floor (KS8804, from 0 to 15 cm bsf); black manganiferous deposits with remains of yellowish muds in between (**d**) and X-ray radiograph showing details of the erosional surfaces inside the crust (**e**).

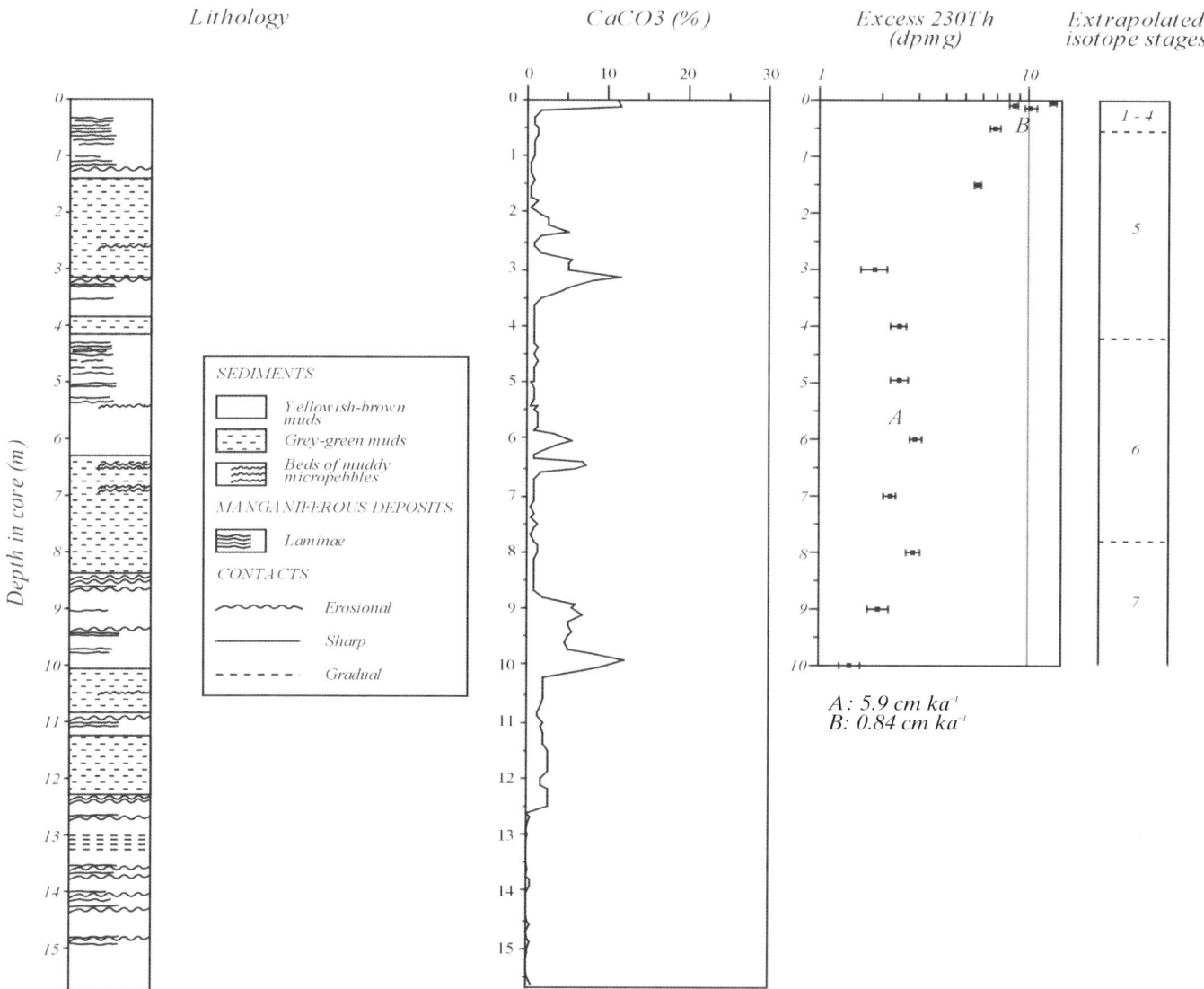

Fig. 11. Lithology and stratigraphic framework of core KS8806. Isotope stages are extrapolated from carbonate curve and excess ^{230}Th data.

The Quaternary deposits consist mainly of silty-clayey muds that are readily classified as muddy contourites. However, we also recognize particular facies that do not fall in the classical contourite models (Stow 1982; Faugères et al. 1984; Gonthier et al. 1984; Stow et al. 1996). The muddy contourites are very homogeneous but show frequent erosional surfaces and color variations (yellowish, brownish, blackish to green) due to diagenetic processes. They include interbedded microbrecciated muddy contourites that reflect erosive activity synchronous with current deposition. Crusts or thin black layers, composed of manganese and iron oxides, and associated with erosional surfaces, is another type of contourite sediment deposited by high velocity currents. These are called manganiferous contourites.

At a longer time scale, from the base of the Miocene to the present day, the seismofacies and architecture of the contourite fan indicate a continuing predominance of current control on sedimentation. Depositional processes take place below the shear zone between the two currents issuing from the Vema channel and Sao Paulo Gap, and the depositional area is located between the sectors swept by these currents: the lower the current activity, the larger the depositional area. The deposits occur as narrow contourite sheets at times when there has been a high volume of bottom water in both channels, and a wide extent of this water laterally and downstream (as for unit B1). When the flow volume is lower they occur in the form of wide levees bounded laterally by two narrow channels (as for B3 sub-units). During such periods, due to the low current activity, the drift relief aggrades and expands laterally between the two main flows which tend to migrate towards the northwest (Sao Paulo channel) and the southeast (Vema channel). This suggests that the Coriolis force does not significantly affect deposition, otherwise the current would have been constrained to the left in Sao Paulo channel and migration would have shown a unique trend. This depositional geometry may be subsequently strongly altered during the erosional events.

The Neogene evolution shows an overall decrease in the energy of the currents, interrupted by erosional episodes. Consequently, the accumulation results from alternating episodes of sedimentation and widespread erosion. The units built during the constructional episodes were probably much bigger than the preserved parts. The present day relief of the contourite fan results from the deposition of the last B3 unit.

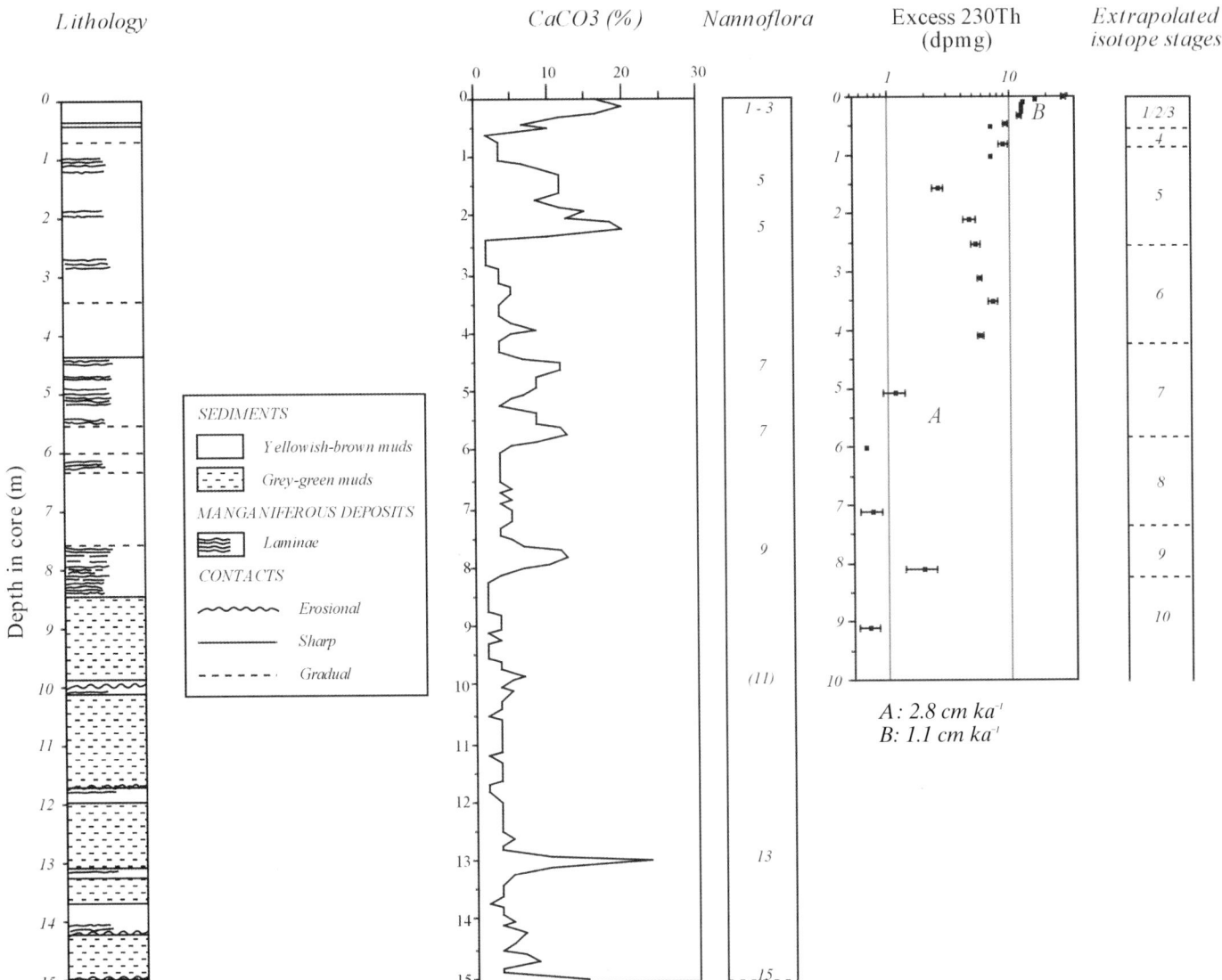

Fig. 12. Lithology and stratigraphic framework of core KS8803. Isotope stages are extrapolated from carbonate curve and excess ^{230}Th data (from Massé *et al.* 1996).

This type of accumulation is characterized by erosional surfaces and discontinuities of regional extent. It lies on an erosional surface and then builds vertically and laterally by the deposition of two types of sedimentary bodies: sheet-like lobate features and sedimentary levees. The ancient lobes appear today as thin, lenticular units of limited extent, as a result of drastic erosion during strong current episodes. Their lateral migration appears to be random. The recent levees are much more widespread as they have not yet undergone any major erosion. Their progradation is well marked by downlap geometries. Such seismic patterns may help to discriminate fan drifts from turbiditic accumulations (Faugères *et al.* 1998, 1999). However, due to the variety of current behaviours, the contourite fan displays a wide range of seismofacies fairly similar to those encountered in turbiditic accumulations.

The seismic geometry and distribution of the deposits are thus mainly controlled by the high variability of bottom current intensity. However, as the depositional area is limited laterally by volcanic ridges and seamounts, the accumulation tends to develop vertically and downcurrent. Its morphology and evolution are then controlled by the tectonic background which limits the lateral migration of the deposits. Variations of sea level and climate may also play a role by controlling the available sediment supply and the rate of deposition.

Some sandy contouritic accumulations can form significant hydrocarbon reservoirs (Carminati & Scarton 1991; Shanmugan *et al.* 1993). A contourite fan located downstream of a channel mouth, like the Vema fan, shows a fan-like morphology and sheet-like, lobate or channel-levee seismic geometry, and then may be mistaken for a sand-rich turbiditic levee. However as the material is fine-grained, silty to clayey muds, it would provide no hydrocarbon reservoir at all and probably not even a good seal. Some of the seismic patterns underlined above may help in the recognition of such contouritic deposits.

The authors are grateful to C. Pudsey and J. Howe for their constructive and critical comments and reviews. This research was mainly supported by funds provided by the UMR number 5805-CNRS (France). This paper is University Bordeaux I, DGO-UMR CNRS 5805, constribution number 1378.

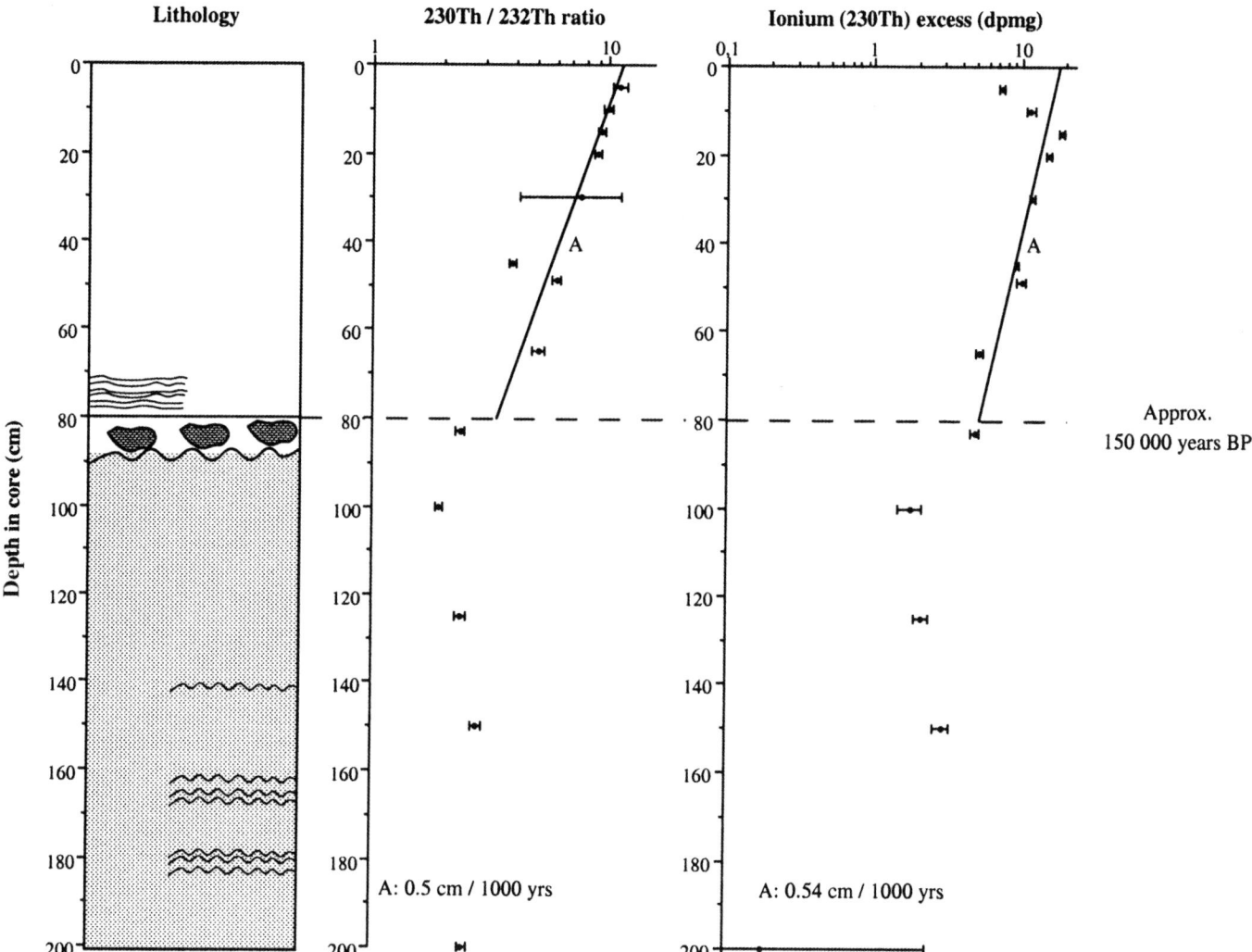

Fig. 13. Lithology and stratigraphic framework of core KS8808. Isotope stages are extrapolated from carbonate curve and excess ^{230}Th data (from Massé, 1993).

References

BARKER, P. F., BUFFLER, R. T. & GAMBOA, L. P. 1983a. *A seismic reflection study of the Rio Grande rise. In*: Initial Reports of the DSDP, **72**, Washington, US Government Printing Office, 499–517.

BARKER, P. F., CARLSON, R. L. ET AL. 1983b. *Brazil basin. Site 515. In*: Initial Reports of the DSDP, **72**, Washington, U.S Government Printing Office.

CARMINATTI M. & SCARTON J. C. 1991. Sequence stratigraphy of the Oligocene turbidite complex of the Campos basin, offshore Brazil. *In*: WEIMAR, P. & LINK, M. H. (eds) *Seismic Facies and Sedimentary Processes of Submarine Fans and Turbidite Systems*. Springer-Verlag, Heidelberg, 241–246.

CHERKIS, N. Z. 1983. Mapa batimétrico da margem continental brasileira. *Projeto REMAC*.

DAMUTH, J. E. 1975. Echo character of the western equatorial Atlantic floor and its relationship to the dispersal and distribution of terrigenous sediments. *Marine Geology*, **18**, 17–45.

DAMUTH, J. E. & HAYES, D. E. 1977. Echo character of the east Brazilian continental margin and its relationship to sedimentary processes. *Marine Geology*, **24**, 73–95.

FAUGÈRES, J.-C. 1988. Mission océanographique Byblos. Bassin du Brésil. *Géochronique*. Société Géologique du France, Bureau de Recherches Géologique et Miniciès, **27**, 8.

FAUGÈRES J.-C. & STOW, D. A. V. 1993. Bottom current controlled sedimentation: a synthesis of the contourite problem. *Sedimentary Geology*, **82**, 287–297.

FAUGÈRES, J.-C., GONTHIER, E. & STOW, D. A. V. 1984. Contourite drift molded by deep Mediterranean outflow. *Geology*, **12**, 296–300.

FAUGÈRES, J.-C., MÉZERAIS, M. L. & STOW, D. A. V. 1993. Contourite drift types and their distribution in the North and South Atlantic Ocean basins. *Sedimentary Geology*, **82**, 189–203.

FAUGÈRES, J.-C., IMBERT, P., MÉZERAIS, M. L. & CREMER, M. 1998. Seismic patterns of a 'contourite fan' – muddy drift related to the Vema Channel (South Brazilian basin) – and a sandy distal deep-sea fan downstream of the Cap Ferret system (Bay of Biscay): a comparison. *Sedimentary Geology*, **115**, 81–110.

FAUGÈRES, J.-C., STOW, D. A. V., IMBERT, P. & VIANA, A. 1999. Seismic features diagnostic of contourite drifts. *Marine Geology*, **162**, 1–38.

GAMBOA, L. A. P. & RABINOVITZ, P. D. 1981. The Rio Grande Rise fracture zone in the western South Atlantic and its tectonic implications. *Earth and Planetary Science Letters*, **52**, 410–418.

GAMBOA, L. A. P. & RABINOVITZ, P. D. 1984. The evolution of the Rio Grande Rise in the southwest Atlantic Ocean. *Marine Geology*, **58**, 35–58.

GAMBOA, L. A. P., BUFFLER, R. T & BARKER, P. F. 1983. *Seismic stratigraphy and geologic history of the Rio Grande Gap and Southern Brazil* basin. *In*: Init. Reports of the DSDP, **72**, Washington, US Government Printing Office, 481–498.

GONTHIER, E., FAUGÈRES, J.-C. & STOW, D. A. V. 1984. Contourite facies of the Faro drift, Gulf of Cadiz. *In*: STOW, D. A. V. & PIPER, D. (eds) *Fine grained sediments: deep water processes and facies*. Geological Society, London, Special Publications **15**, 275–292.

HOGG, N. G., BISCAYE, P., GARDNAR, W. & SCHMITZ, W. J. 1982. On the transport and modification of Antarctic Bottom Water in the Vema channel. *Journal of Marine Research*, **40** (suppl.), 231–263.

JOHNSON, D. A. *et al*. 1983. Palaeocirculations in the southwestern Atlantic. *In*: BARKER, P. F., CARLSON, R. L., JOHNSON, D. A. *ET AL*. Initial Reports DSDP, **72**, Washington, US Government Printing Office, 977–994.

JOHNSON, D. A., LEDBETTER, M. T. & BURCKLE L. H. 1977. Vema channel palaeoceanography: Pleistocene dissolution cycles and episodic bottom water flow. *Marine Geology*, **23**, 1–33.

JOHNSON, D. A. & RASMUSSEN, K. A. 1984. Late Cenozoic turbidite and contourite deposition in the Southern Brazil basin. *Marine Geology*, **58**, 225–262.

LEDBETTER, M. T. & ELLEWOOD, B. B. 1979. Variations in particle alignment and size in sediments of the Vema channel record Antarctic Bottom Water velocity changes during the last 400 000 years. *In*: CRADDOCK, C. (ed.) *Antarctic Geoscience*. University of Wisconsin Press, Madison, 1033–1038.

LE PICHON, X., EWING, J. & TRUCHAN, M. 1971. Sediment transport and distribution in the Argentine Basin. Part 2: Antarctic Bottom Current passage in the Brazil Basin. *In*: AHRENS, L. H., PRESS, F., RUNCORN, S. K. & UREY, H. C. (eds) *Physics and chemistry of the Earth*, **8**, Pergamon Press, New York, 29–84.

MASSÉ, L. 1993. Sédimentation océanique profonde au Quaternaire. Flux sédimentés et paléocirculations dans l'Atlantique Sud-Ouest: Bassin Sud-Brésilien et prisme d'accrétion Sud-Barbade. *Thèse Universitè Bordeaux I*.

MASSÉ, L., FAUGÈRES, J.-C., BERNAT, M., PUJOS, A. & MÉZERAIS, M. L. 1994. A 600 000 year record of Antarctic Bottom Water activity inferred from sediment textures and structures in a sediment core from the Southern Brazil Basin. *Palaeoceanography*, **9**, 1017–1026.

MASSÉ, L., FAUGÈRES, J.-C., PUJOS, C., PUJOS, A., LABEYRIE, L. D., BERNAT, M. 1996. Sediment flux distribution in the Southern Brazil basin during the later Quaternary: the role of deep sea currents. *Sedimentology*, **43**, 115–132.

MELLO, G. A. 1988. Processos sedimentares recentes na bacia do Brazil: setor sudeste-sul. *These de Mestrado, University Fluminense Niteroï* (Brazil).

MÉZERAIS, M. L. 1991. Accumulations sédimentaires profondes turbiditique (deep-sea fan du Cap Ferret) et contouritique (bassin sud-brésilien): géométrie, faciès, édification. *Thèse Universitè Bordeaux I*.

MÉZERAIS, M. L., FAUGÈRES, J.-C., FIGUEIREDO, A. & MASSÉ, L. 1993. Contour current accumulation off Vema Channel mouth, Southern Brazil basin. *Sedimentary Geology*, **82**, 1–4: 173–188.

PIERRE, C., VERGNAUD-GRAZZINI, C. & FAUGÈRES, J.-C. 1991. Oxygen and carbon stable isotope tracers of the water masses in the central Brazil basin. *Deep Sea Research*, **38**, 597–606.

REID, J. L., NOWLIN, W. D. & PATZERT, W. C. 1977. On the characteristics and circulation of the southwestern Atlantic Ocean. *Journal of Physical Oceanography*, **7**, 62–91.

SHANMUGAN, G., SPALDING, T. D. & ROFHEART, D. H. 1993. Process, sedimentology and reservoir quality of deep-marine bottom-current reworked sands (sandy contourites): an example from the Gulf of Mexico. *AAPG Bulletin*, **77/7**, 1241–1259.

SPEER, K. G. & ZENK, W. 1993. The flow of Antarctic Bottom water into the Brazil Basin. *Journal of Physical Oceanography*, **12**, 2667–2682.

STOW, D. A. V. 1982. Bottom currents and contourites in the North Atlantic. *Bulletin Institut Géologique du Bassin Aquitaine*, Bordeaux, **31**, 151–156.

STOW D. A. V., READING, H. G. & COLLINSON, J. D. 1996. Deep clastic seas. *In*: READING, H. G. (ed.) *Sedimentary Environments: Processes, Facies and Stratigraphy*. Blackwell Science, 395–453.

WILLIAMS, D. F. & LEDBETTER, M. T. 1979. Chronology of late Brunhes biostratigraphy and late Pliocene disconformities in the Vema channel (SW Atlantic). *Marine Micropalaeontology*, **4**, 125–136.

WRIGHT, W. R. 1970. Northward transport of Antarctic Bottom Water in the western Atlantic Ocean. *Deep-Sea Research*, **17**, 367–371.

The Columbia Channel–levee system: a fan drift in the southern Brazil Basin

J.-C. FAUGÈRES[1], A. FRANCA LIMA[2], L. MASSÉ[1] & S. ZARAGOSI[1]

[1]*Department of Geology and Oceanography (DGO – UMR CNRS 'EPOC', 5805), University of Bordeaux I, Avenue des Facultés, 33405 Talence, France*
[2]*Instituto Oceanografico-USP, 191 Praçavdo Oceanografico, 05508-900 Sao Paulo, SP, Brésil*

Abstract: The Columbia Channel is a turbiditic channel elongated W–E on the rise of the south Brazilian basin (4200 to 5000 m water depth). The whole area is swept by the northward flowing Antarctic Bottom Water. As a consequence, depositional processes have built a fan drift system. This system displays a levee along the northern flank of the channel while no levee occurs on its southern flank due to the Coriolis effect. The levee (400 km in length and 100 to 200 km in width) is bounded to the north by the Vitoria–Trindade Seamounts. It shows, first, a W–E trend parallel to the channel axis and predominantly turbiditic pattern, and then a S–N trend parallel to the rise contours with a predominant contouritic pattern. Its thickness is up to 1000 m. The distribution of sedimentary processes and associated deposits were investigated on the basis of water gun seismic and 3.5 kHz echosounding profiles, and core lithology. On the lower S–N part of the levee, the deposits consist of muddy contourites. On the shallowest part, turbidites that originate from the upper continental margin in the channel and on the southern part of the levee close to the channel, and from the Vitoria–Trindade Seamounts on the northern part of the levee, are interbedded with contouritic muds, and top-truncated silty turbidites. Areas subjected to turbidity current processes show chaotic to well-stratified, high amplitude reflections, in the subsurface, and more or less prolonged echofacies with or without sub-bottom reflectors, at the seabed. Areas subjected to contour currents show, in the subsurface, transparent seismofacies with some discontinuous low amplitude wavy reflections, and, in the surficial deposits, predominant wavy echofacies with sub-bottom relectors, frequently associated with tangential hyperbolae.

The Columbia fan-drift system in the northern part of the southern Brazil Basin is an example of a mixed drift body formed as a consequence of the interaction of turbidity and contour current depositional processes (Fig. 1, Table 1). This is similar to the Hikurangi fan-drift system off New Zealand (Carter & McCave 1994; Lewis 1994; McCave & Carter 1997). In this paper we summarize and develop earlier work on the Columbia fan-drift (Massé et al. 1998; Faugères et al. 1999), based on data acquired during the BYBLOS cruise (Faugères 1988) including 3.5 kHz and water-gun seismic lines, and six Kullenberg cores. The cores were collected along two 3.5 kHz profiles crossing the system. The shallower western profile runs S–N, normal to the axis of the channel. Cores KS 8820 and KS 8821 are located on the northern levee, core KS 8822 in the axis of the channel, and cores KS 8823 and KS 8824 on the southern edge of the channel. The deeper eastern transect is oriented SW–NE, with core KS 8826 collected on top of the levee.

Geological and oceanographic setting

The Southern Brazil Basin first developed during the Cretaceous and is marked by active transform faults and volcanic lineaments directed E–W (Schobbenhaus et al. 1984). The Columbia Channel is more or less parallel to two of these major lineaments, the Rio de Janeiro lineament in the south and the Vitoria Trindade lineament in the north, and its course seems to be under a strong structural control. The channel is a major feature that displays an overall WNW–ESE orientation and can be traced to 5000 m water depth across the abyssal plain. The shallow-water sediments passing through the channel accumulate on its northern edge, where a smooth levee occurs. In contrast, a gently northward sloping accumulation is observed on the southern flank of the channel.

In the Southern Brazil Basin, Antarctic Bottom Water (AABW) is found at depths greater than 4000 m, and flows northwards (Reid et al. 1977; Reid 1996). It is responsible for the construction of contouritic accumulations on the lower rise and the abyssal plain (Massé 1993; Mézerais et al. 1993; Massé et al. 1994; Faugères et al. 2002). In the vicinity of the Columbia Channel,

Table 1. *Principal characteristics*

Location:	the continental rise and abyssal plain of the south Brazilian basin
Setting:	major channel-levee system along a volcanic seamount chain (4200–4700 m water depth)
Age:	end Oligocene ? to Recent
Drift type:	fan-drift system with a turbiditic levee merging into a contouritic levee downslope
Dimensions:	length 400 km; width 100 to 200 km; thickness about 1000 m
Seismic facies:	transparent reflections and a very thick undulating echofacies with numerous subbottom reflectors as contourite signature
Sediment facies:	turbidites, muddy contourites, top-truncated turbidites

AABW is deflected eastwards along the northern levee and the Vitoria–Trindade Chain (Fig. 15), developing eastward trending rise swells, and enters the northern part of the Brazil Basin through the deep passages crossing the Chain (Mello 1988; Castro 1992).

Two major sources of sediments may provide material to the study area (Massé 1993; Massé et al. 1996): (1) the Brazilian continental slope and rise north of Rio de Janeiro (downslope supply with high kaolinite contents) and (2) the deep Argentine and South Brazil Basin swept by the northward moving AABW current (alongslope supply with very low kaolinite and predominant chlorite and smectite contents). In addition, some sediments may come from the surrounding volcanic seamounts.

Bathymetry

The Columbia Channel (Figs 1 to 7) is a deep valley elongated downslope from 4200 to 5200 m, with a downslope decreasing channel depth (from about 400 to 250 m), a width of about 20 km at the top of the valley flanks, and 4 to 5 km on the very flat valley floor. It is firstly directed WNW–ESE and then moves slightly

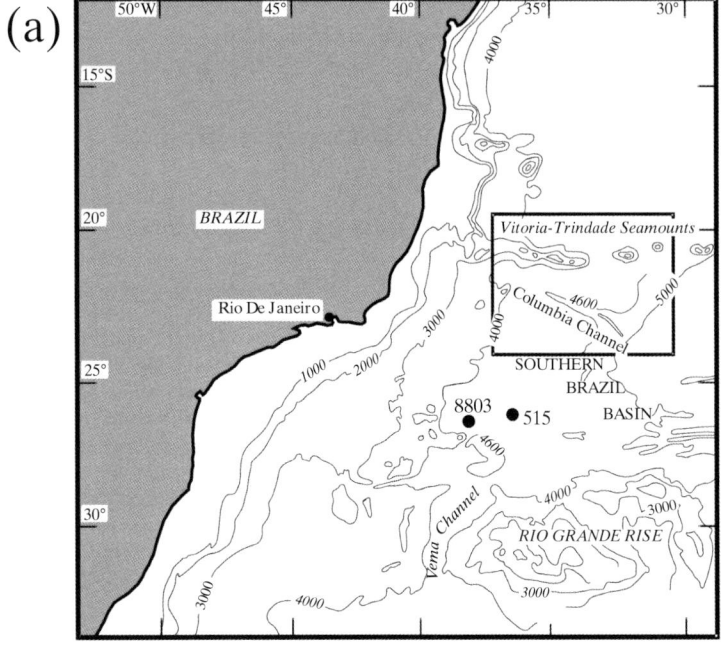

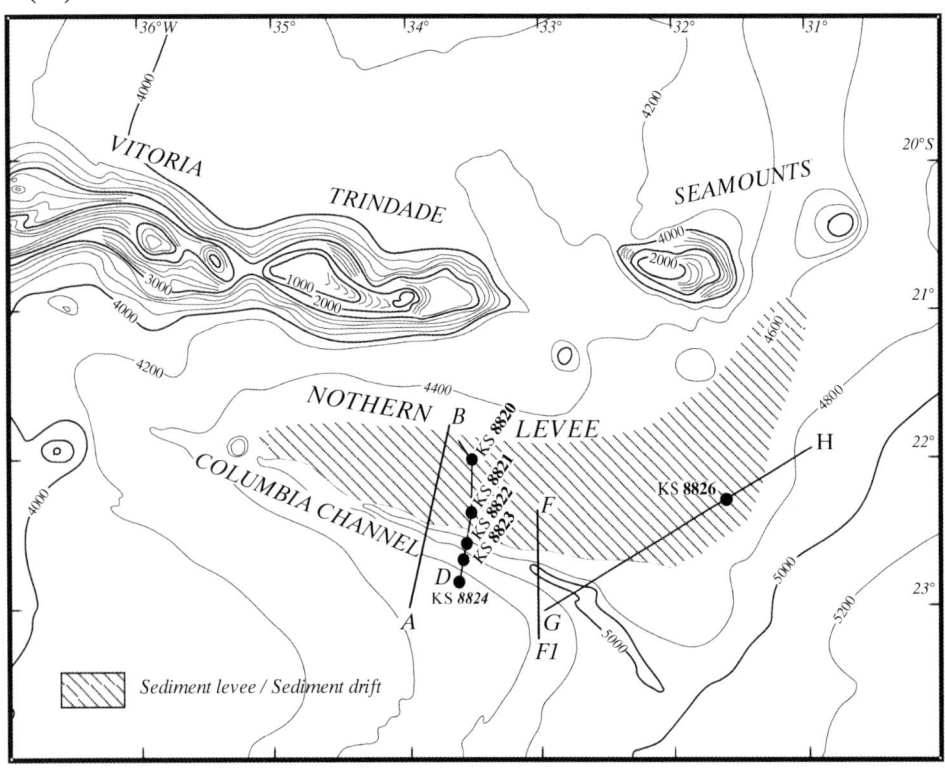

Fig. 1. (a) Location map of the study area, with DSDP site 515 and Byblos core KS8803; (b) bathymetric map of the Columbia Channel–Levee System (depths are in metres), and location of the cores and seismic and 3.5 kHz echosounding profiles (AB, FF1, GH).

towards a NW–SE trend. The irregular flanks show evidence of active current erosion: reflection truncations, flat terraces, and erosive scars linked to sliding processes.

The northern levee developed along the channel is about 400 km long and 100 to 200 km wide and is bounded to the north by the Vitoria–Trindade volcanic seamounts. It has a gentle relief with a fairly flat and regular surface without any secondary channels. At greater depths, the levee shifts towards a SSW–NNE orientation, parallel to the rise contour, and has a more mounded shape with wavy bedforms. There is no levee south of the channel where the rise sea-floor slopes down gently and regularly northwards and eastwards.

Stratigraphy

Due to a lack of carbonate in the sediments deposited below the CCD, and the occurrence of frequent hiatuses and sediment reworking linked to the turbidity currents, in the cores, there is no good control of the core and seismic stratigraphy. A fairly reliable stratigraphy has been established only for core KS 8826. It is supported by nannofossil biostratigraphy, excess 230^{Th} activity analyses, carbonate curves and comparison with cores collected in the Vema contourite fan area (Massé 1993; Faugères et al. 2002). In addition, there is no deep-sea drilling site in the vicinity of the study area to support seismic interpretation.

Seismic characteristics: reflection profile

Seismic reflection profiles

Two water-gun profiles that cross the system (Figs 2 and 3) are presented here. The proposed stratigraphical interpretation is hypothetical as it is only supported by data from DSDP site 515 (Gamboa & Rabinowitz, 1981, 1984; Gamboa *et al.* 1983; Fig. 1a) located far away to the south. Profile AB oriented NNE–SSW (Fig. 2) crosses the shallow part of the system. It shows a striking contrast in seismic pattern between the area south of the channel and the northern levee.

South of the channel, two seismic units can be distinguished above the acoustic substratum:

(a) A lower South Unit 1 (SU1, 600 m in the south to 400 m close to the channel) is subdivided into 2 subunits. SU1a, at the base, is characterized by predominant high amplitude chaotic reflections that merge upwards into high amplitude reflectors showing a progressively better stratification. The SU1b-SU1a boundary is marked by a locally erosional discontinuity (R1) and by a reflection change. SU1b displays well-stratified high amplitude fairly continuous reflections. However the reflectors become locally irregular, with onlap or truncated geometry often associated with thin lenses of chaotic reflections (erosive shallow mini-channels). SU1b is truncated at the top by an erosional, flat and horizontal surface (R2).
(b) An upper South Unit 2 (SU2) increases in thickness southward (400 to 800m) and displays very transparent reflections with small-scale wavy reflections, the amplitude of which may increase towards the South or at the top.

Below the Columbia Channel, the acoustic substratum is not well marked. Basal transparent to chaotic reflections are overlain by irregular reflectors. A shallow channel (ch.2, Fig. 2) appears at approximately the same horizon as the R1 discontinuity. It then migrates northwards still as a shallow feature, and deepens due to sediment accumulation on both flanks and/or erosion and transport on the valley bottom. Both flanks show high amplitude chaotic reflections.

North of the channel, 2 major units can be distinguished:

(a) A basal North Unit (NU1, 350–400 m) covers the acoustic substratum. It is characterized by predominant high amplitude chaotic reflections that merge upwards into high amplitude reflectors progressively showing better stratification. The most prominent reflectors present a gentle southward dip and suggest a N–S progradation of the deposits. This unit is bounded at the top by a discontinuity underlined by a rapid change in the seismic reflections, and local erosive patterns. In the southern part of this unit, the reflections display channel-like geometries (ch. 1, Fig. 2). At the north end of the profile where the unit thickens, correlations between other available seismic profiles allow the interpretation of this discontinuity as equivalent to R1. This implies that NU1 can be correlated with SU1a, and hence the equivalent of SU1b either should have been (partly or totally) eroded before the deposition of the overlying NU2 or corresponds to the lower part of the overlying NU2 Unit.
(b) An Upper North Unit 2 (NU2, 600–800 m) shows, at the base, uniform transparent reflections (SU1b?) that progressively merge upwards into high amplitude chaotic reflections in the south of the unit, and, in the uppermost northern part of the unit, high amplitude irregular and more or less continuous reflectors that, in the north, gently dip southward. Wavy geometries occur at the top, near the channel.

SU2 is characterized by transparent to wavy reflections and interpreted as a contouritic sheet drift. It was most likely deposited since the upper Oligocene on the R2 surface (correlated with the lower-Oligocene AABW erosive event, Gamboa & Rabinowitz 1981, 1984; Gamboa *et al.* 1983; Mézerais 1991; Mézerais *et al.* 1993; Faugères *et al.* 2002). SU1, at the base, could correspond to pelagic hemipelagic sedimentation (SU1a, pre-Eocene deposits?), merging progressively upwards into coarser-grained deposits controlled by turbidity or contour currents (SU1b, lower to middle Eocene ?).

The flat upper North Unit (NU2) compared to the upper south unit (SU2) displays different seismic patterns with predominant chaotic and high amplitude reflections. It is interpreted as a turbiditic sheet. The turbiditic supply seems to originate from either the Columbia Channel or from a more northern source (Vitoria Trindade ?) with turbidity currents responsible for the southward dip of the high amplitude continuous reflectors in the northern upper part of the levee. However, in the lower part of the levee and near the top close to the channel axis, the occurrence of transparent reflections associated with subtle wavy bedding, very similar to the reflections in the SU2 unit, suggest that the contouritic deposits may play a significant role in the building of the levee body. Taking into account the flat surface, it would be more sensible to call this 'northern levee' a 'turbiditic sheet' and even probably a 'mixed turbiditic-contouritic sheet'. The age of the basal R1 discontinuity remains speculative: (?) lower Eocene. Both flanks of the channel display very high amplitude chaotic reflections and a sharp lateral contact with the contouritic deposits to the south and the turbiditic deposits to the north. They are interpreted as coarser-grained turbiditic sediments associated with downslope mass movement.

Profile GH (Fig. 3) crosses the deepest part of the system in a NE–SW direction. It displays a basal SU1' comparable to SU1 unit and bounded at the top by the R2 erosive surface. SU1' runs with similar seismic pattern northeastward, and is called NU1'. However NU1' is bounded at the top by an erosive surface interpreted as equivalent to R1. Above R2, south of the channel, an upper SU2' unit shows seismic reflections identical to those of the SU2 upslope, implying no major change in the contouritic sheet drift down the rise, except for a decrease in thickness. Above R1, north of the channel, a thick mounded unit (NU2', up to 800 m) also presents seismic patterns similar to those of the contouritic SU2 and SU2' (transparent to wavy reflections). It is interpreted as a contouritic levee. This means that the predominantly turbiditic sheet (NU2) observed upslope has merged progressively into a contouritic levee (NU2'). The Columbia Channel first occurrence seems to be synchronous to R1 discontinuity or slightly before. A clear northward migration of the channel axis occurs during the first stages of the channel history, which is opposite to what should be produced by predominant turbidity currents, i.e. the building of a turbiditic levee on the left (north) flank of the channel. This would imply the dominant role of contour currents in building the north basal levee. On the channel flanks, high amplitude continuous reflections associated with more or less discontinuous to chaotic reflections more likely correspond to turbiditic deposits. These deposits extend further away from the channel, and the lateral transition with the contouritic deposits are more gradational than on the previous profile.

Echofacies mapping

The echofacies and the lithology of the deposits from the shallow part of the system are presented on Fig. 4. Various echo types have been distinguished (Figs 4 and 5a) and interpreted according to Damuth (1975) and Damuth & Hayes (1977).

Large irregular hyperbolae with varying vertex elevations (echo-type IIIA) are found in the steeper portions of the channel walls due to a rugged morphology. The axis of the channel is

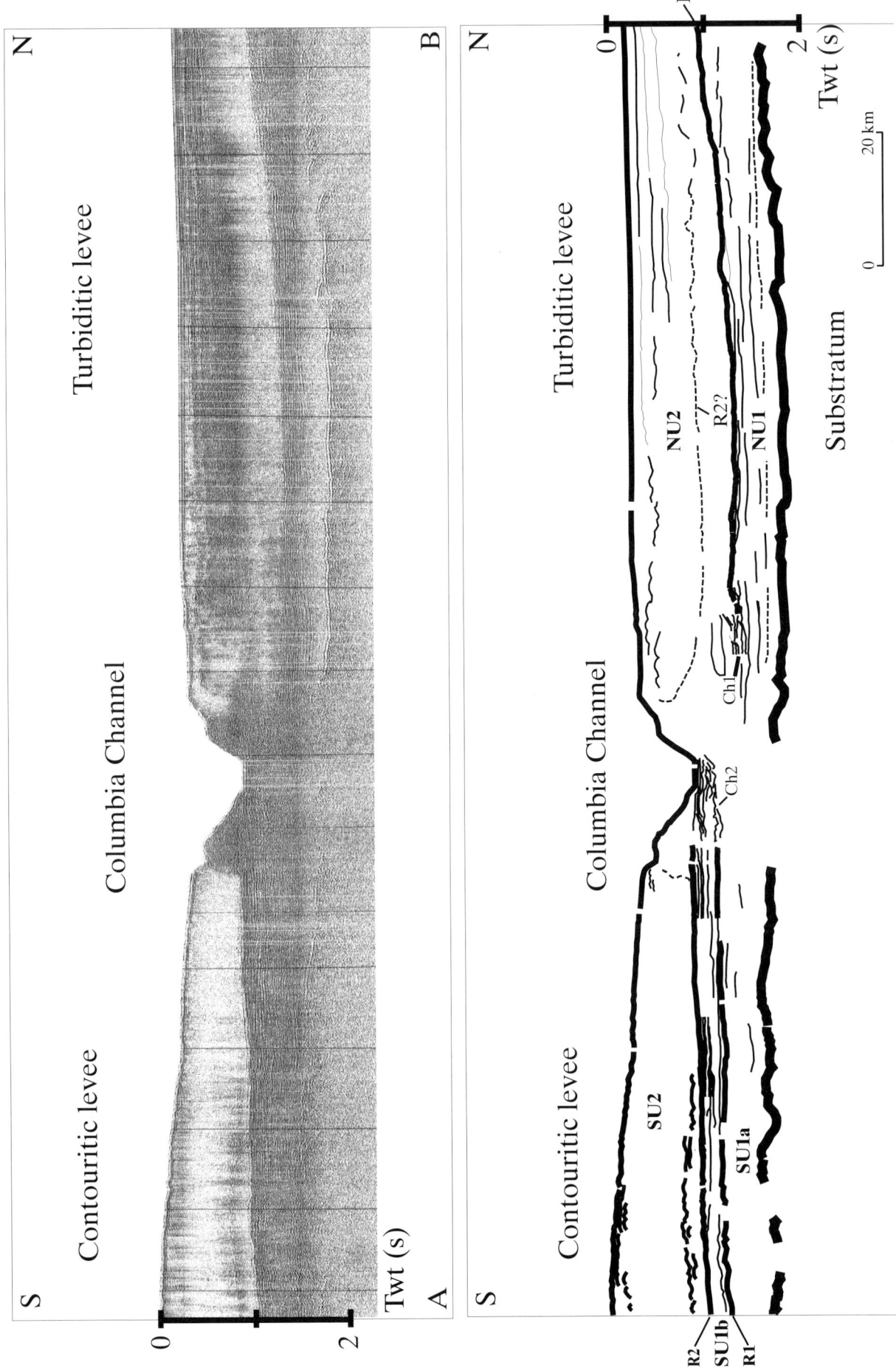

Fig. 2. Seismic line A–B, crossing the shallower part of the channel-levee system.

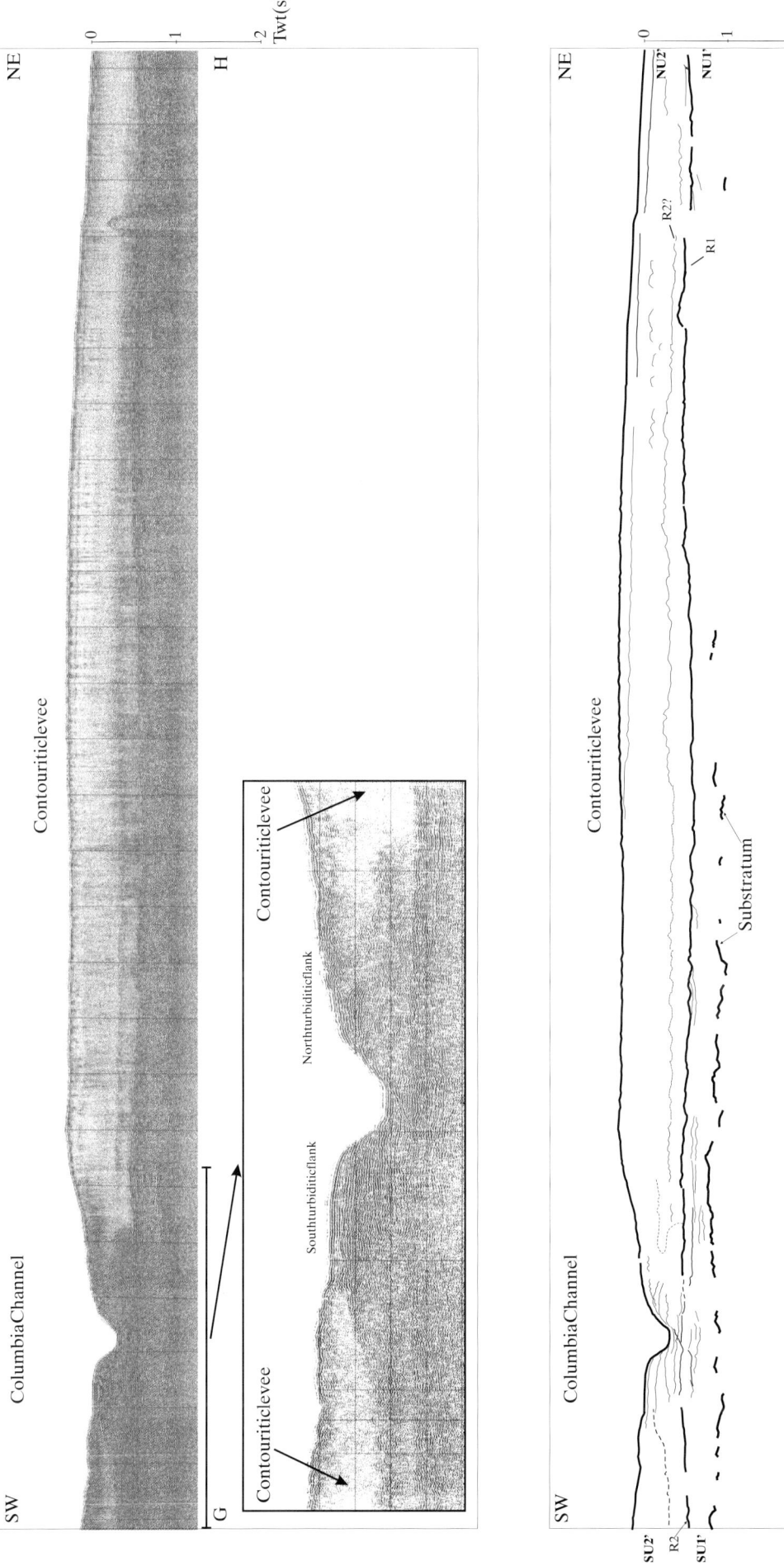

Fig. 3. Seismic line G–H, crossing the deepest part of the channel-levee system.

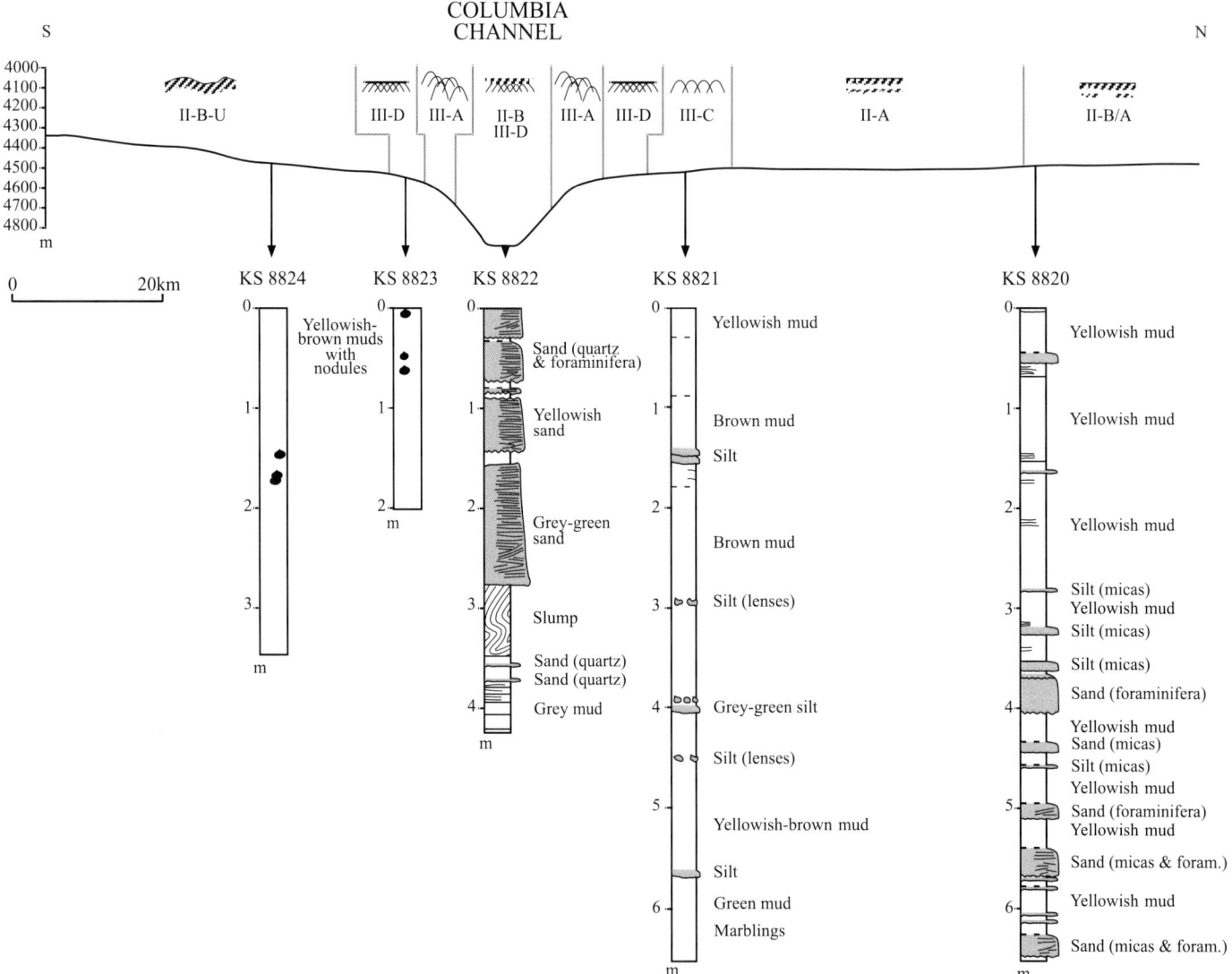

Fig. 4. Echofacies distribution and core lithology along the western transect (DB profile, see Fig. 1a for location).

characterized by the association of a very prolonged bottom echo with no sub-bottom reflectors (echo-type IIB) and hyperbolae tangential to the sea-floor (echo-type IIB-IIID). They indicate abundant sandy deposits (IIB) in an area with active turbidity currents as confirmed by core KS 8822. The upper parts of both flanks display a similar sharp surface echo with hyperbolae tangential to the sea-floor (IIID echo-type). Such echoes on the northern flank may be associated with a decrease in turbidity current energy and sand abundance. On the southern flank, it corresponds to very homogeneous yellowish-brown muds associated with manganese nodules, with neither coarse-grained silt or sand material nor turbiditic deposits (core KS 8823).

A small area further north of the IIID echotype is characterized by regular overlapping hyperbolae with varying vertex elevations above the sea-floor (echo-type IIIC). This may be caused by regularly spaced erosional depositional bed forms and could reflect a combination of overflowing turbidity currents and contour currents, as suggested by core KS 8821 showing muddy deposits with thin and widely spaced silt/sand layers (see Discussion section).

To the north, the major part of the levee is characterized by indistinct prolonged or semi-prolonged echoes. North to south, there is a southward succession of echofacies: firstly IIB/A echo-type, between indistinct prolonged echoes and semi-prolonged echoes with zones of discontinuous parallel sub-bottom reflectors, related to the occurrence of abundant silt/sand layers (KS 8820); then a typical IIA echo-type characterized by semi-prolonged bottom echoes with discontinuous, parallel sub-bottom reflectors that indicate finer-grained deposits; and lastly the IIIC echotype. This succession seems to indicate a decreasing trend in the abundance and/or thickness of sandy layers, and suggests that the Vitoria-Trindade chain is a significant sediment source for the levee.

The distal part of the southern flank of the channel shows an undulating prolonged bottom echo with no sub-bottom reflectors (echo-type IIB-U, Fig. 5a, b). This echo is associated with homogeneous manganiferous yellowish-brown muds (KS 8824), very similar to that of core KS 8823. Consequently, this IIB-U echo could result from regular small erosional to depositional contour-current-generated bedforms (metric to decimetric; Ewing et al. 1973; Embley 1975), whereas the wavy pattern indicates larger sediment waves (average wavelength of 1 km).

To conclude, contour current processes seem to be dominant south of the channel. Turbidity currents are dominant in the channel and on the northern levee where deposits originate from supply transported by the channel or delivered from the Vitoria Trindade chain. Evidence of process interaction is present on the levee.

In the deepest part of the system (Figs 1, 5c–g and 6), the morphology of the channel-levee is asymmetrical with a channel

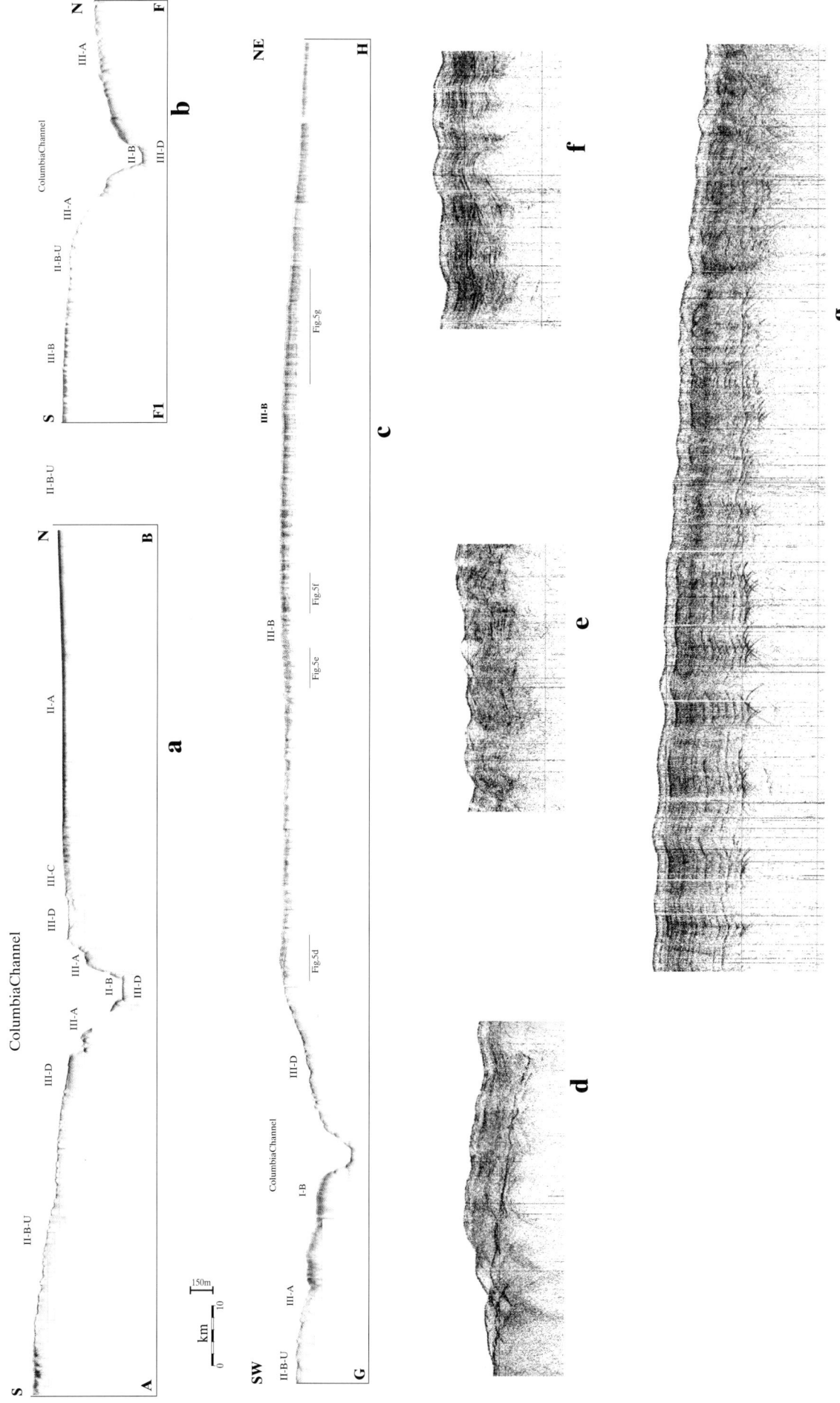

Fig. 5. 3.5kHz profiles (see Fig. 1 for location) and echofacies. (a) Profile (AB) crossing the shallower part of the channel-levee system; (b) Profile (FF1) crossing the mid part of the channel-levee system; (b) Profile (GH) crossing the deeper part of the channel-levee system; (d, e, f & g) Detail of contouritic echofacies, showing an erosive discontinuity underlined by a III-D echofacies and overlained by a wavy echofacies with more or less discontinuous subbottom reflectors (d), wavy surficial echo associated with chaotic subbottom reflectors probably due to sliding disturbance (e), onlaping contact between two generations of sediment waves (f), and typical contouritic III-B echofacies: a very thick undulated echo-type with stationary surficial sediment waves and comformable sub-bottom reflectors, some of them associated with tangential hyperbolae (g).

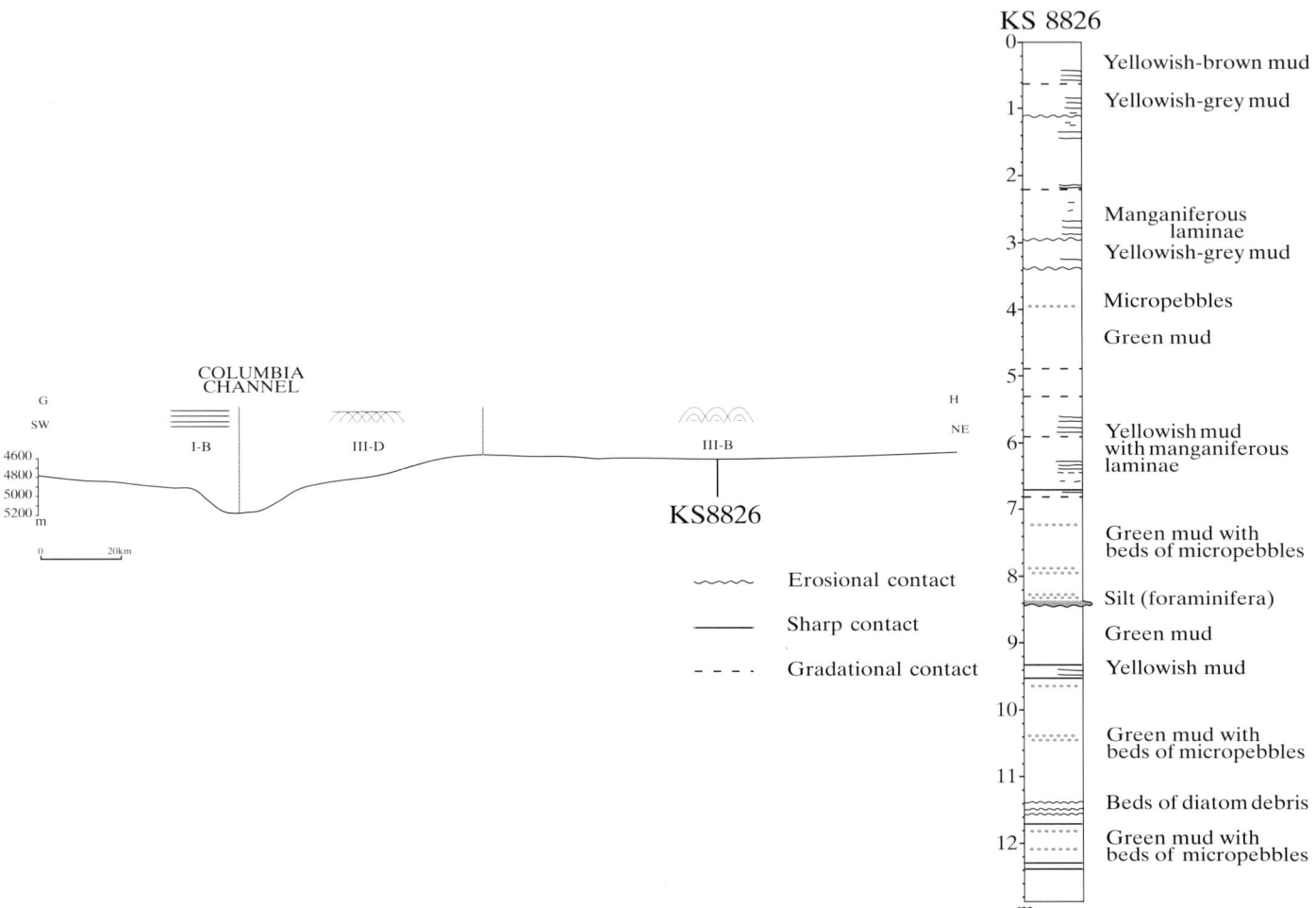

Fig. 6. Echofacies distribution and core lithology (KS 8826) along the eastern transect (GH profile, see Fig. 1a for location).

bottom depth of 5200 m, a prominent northern levee culminating at a depth of 4600 m, and a southern flank with a lower sediment relief (4800 m). The profile shows the predominance of hyperbolic and wavy echoes.

The southern flank of the channel displays a IB echo-type with a sharp bottom echo and continuous, sharp, parallel sub-bottom reflectors, probably associated with mud deposition from very sluggish bottom currents. The II-B indistinct, prolonged echo-type only occurs on the valley bottom (Figs 4, 5 and 6). The northern flank of the channel displays a IIID echo-type similar to that observed along the shallower western profile. It is associated with erosional/depositional bedforms generated by more active geostrophic and/or turbiditic bottom currents. Finally, the major part of the northern levee displays a very thick IIIB undulating echo-type characterized by deep regular single or slightly overlapping hyperbolae with comfortable sub-bottom reflectors, some of them associated with tangential hyperbolae (Fig. 5c, g). This may indicate current-generated stationary sediment waves that are typical of the contouritic levee deposits. As shown by core KS 8826 (Fig. 11), these deposits consist of abundant muds with extremely rare and thin silty layers. Very similar lithology and echofacies were described in the southernmost part of the Brazil Basin, where only muddy contourites are deposited along the path of the AABW currents (Faugères *et al.* 2002). Locally the III-B is strongly disturbed, probably by sediment sliding (Fig. 5e) and episodes of more active currents responsible for erosive surfaces are underlined by high amplitude reflectors associated with tangential hyperbolae (Fig. 5d)

Sediments: core description and facies

The facies observed in the five cores collected on the upper part of the system fall into two categories (Figs 4, 8, 9 and 10): (1) silty-clayey muds, and (2) interbedded sandy turbidites.

Cores KS 8824 and KS 8823 (Fig. 4) on the southern flank of the channel, are characterized by very homogeneous yellowish-brown muds with ferro-manganiferous nodules. The carbonate content is very low (less than 5%). The median grain size is 5 μm (fine silt and clay, Fig. 7), and the sand content very low (< 3.5%). There are no silt-sand turbiditic layers. It is likely that turbidites are deflected towards the north as a result of the Coriolis force, and geostrophic bottom currents are the dominant process in this area. Consequently, these muds are interpreted as contouritic muds rather than 'red' pelagic clay. This is confirmed by the wavy IIB-U echofacies nearby and by the strong lithological and grain size similarities between these deposits and contouritic muds described further south (Fig. 7), at the northern exit of the Vema Channel (Massé 1993; Massé *et al.* 1994, Faugères *et al.* 2002).

Core KS 8822, in the axis of the Columbia Channel, displays very thick (up to 120 cm) classical sandy turbidites (Fig. 4). The material shows a medium carbonate content (20 to 40%), a median grain size between 60 and 200 μm and a sand content always exceeding 40%. The sand fraction is composed of quartz and micas (30 and 10% respectively), planktonic foraminifera (50%) and minor quantities of various bioclasts. Interbedded muds are similar to those described south of the channel (Fig. 7) and are then interpreted as muddy contourites. However, the mud

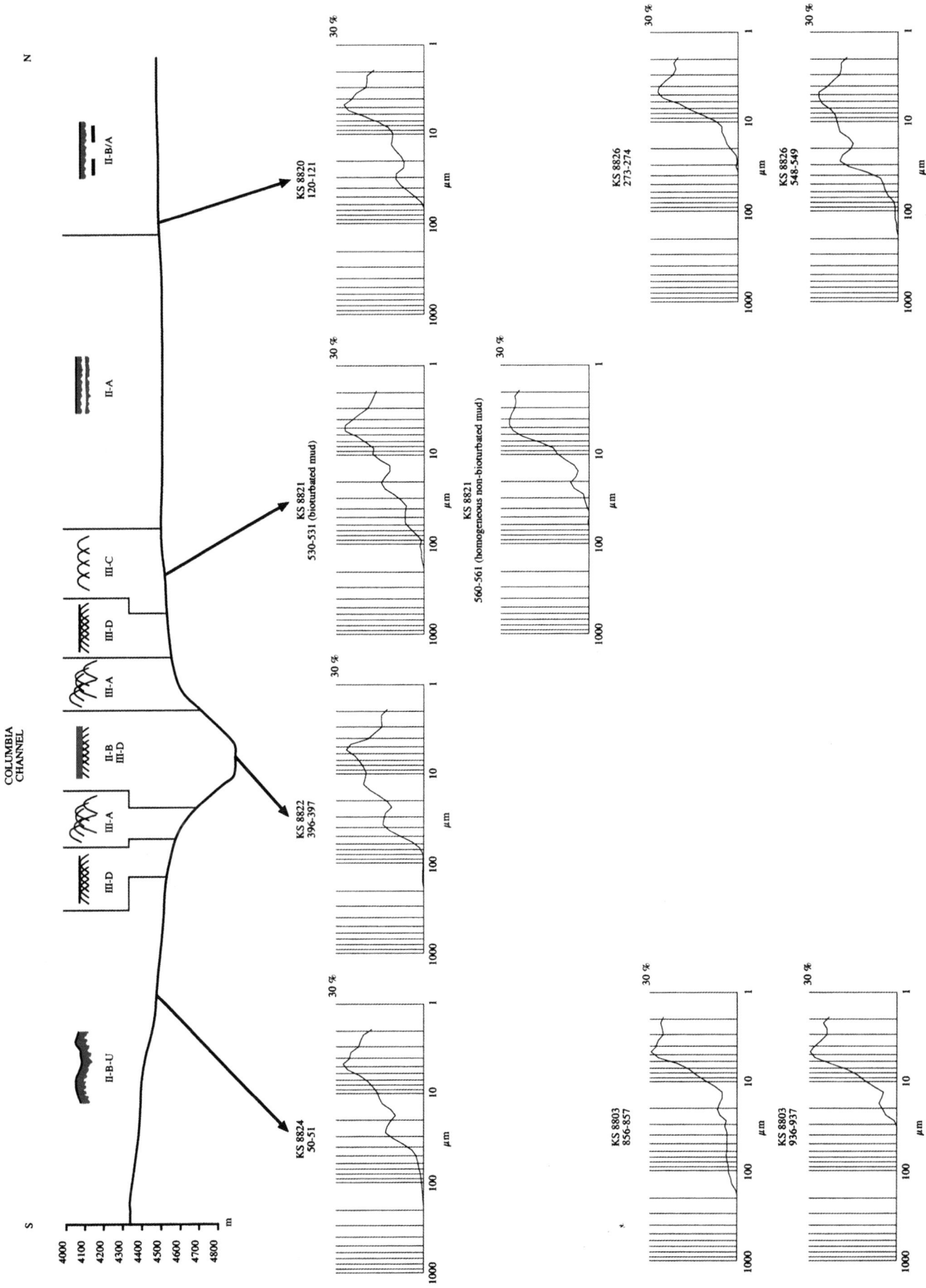

Fig. 7. Grain size frequency curves for the muds. Two samples from core KS 8803 (see Fig. 1a for location) representing the whole range of grain size variability in contouritic muds are included for comparison.

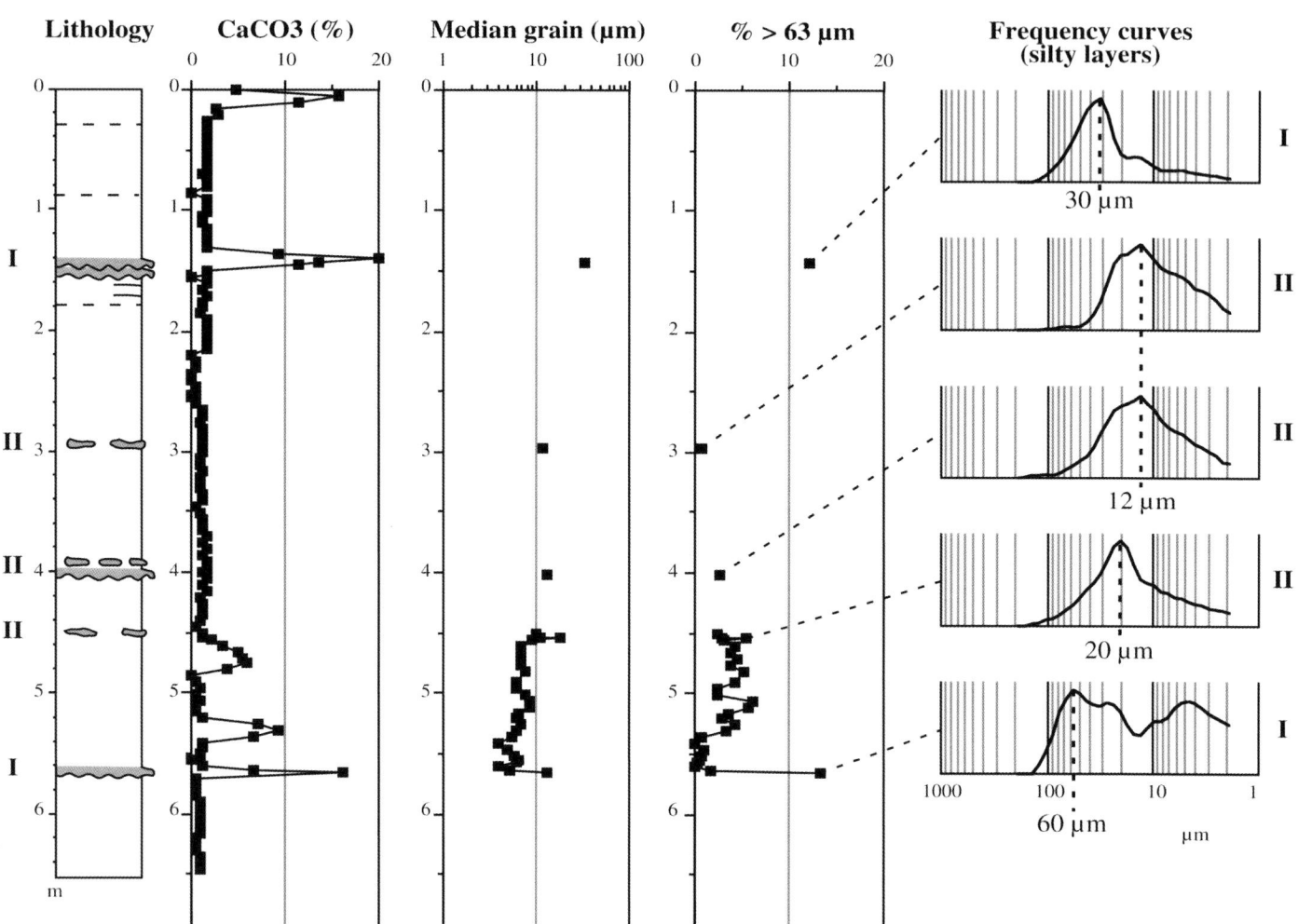

Fig. 8. Lithology of core KS 8821 (northern levee). I and II refer to type I (silty muddy turbidites) and type II (Top truncated silty muddy turbidites) silty layers respectively.

modal peak in KS 8822 is slightly coarser and suggests that a part of the mud material may be derived from turbiditic supply.

Core KS 8821 (Fig. 8), on the northern levee close to the axis of the channel, is characterized by abundant muds similar to those described on the southern flank of the channel (Fig. 7), which could thus be considered as muddy contourites, with a part of the material derived from turbiditic flows. This seems supported by the mineralogical composition (K/I ratio; Massé et al. 1998). In addition to these muds, the core displays thin and widely spaced silty layers (Figs 8 and 9) that fall into two types interpreted as type I: silty-muddy turbidites and type II: top-truncated silty-muddy turbidites resulting from contour current reworking.

Type I, silty-muddy turbidites (145 and 565 cm in Figs 8 & 9), show a basal thin layer (3 cm), with an erosive contact at the base, a significant carbonate content (10 to 20%), a median grain up to 30 μm (coarse silts), and a sand content exceeding 10%. This is overlain, with a gradual transition, by a cm- to dm-thick layer of very homogeneous, non-bioturbated mud, with some coarser layers but no evidence of grading. At the top, these muds are progressively overlain by bioturbated muds with fairly similar texture (Fig. 7). The non-bioturbated mud suggests rapid deposition and is interpreted as the upper division of a silty-muddy turbidite. In contrast, bioturbated muds are interpreted as muddy contourites with a much lower deposition rate.

Type II, top-truncated silty-muddy turbidites (155, 295, 395, 403 and 455 cm in Figs 8 and 9), have a cm-thick lenticular basal bed with an erosive contact, zero carbonate content, a median grain-size never exceeding 20 μm, a modal peak corresponding to fine silts (10 to 20 μm), and a sand content lower than 5%. In contrast to type I silty muddy turbidites, they are not overlain by homogeneous muds. Type II silts at 455 cm are topped by a very thin alternation (1 cm thick) of millimetric silty and muddy layers and characterized by a sharp contact at the top. This is overlain by bioturbated muddy contourites. Some burrows truncate the silt layer and the erosional contact at the base. These type II sequences are interpreted as turbiditic deposits truncated by the action of contour currents (see Discussion section).

Core KS 8820, on the northern levee, well away from the axis of the channel, is characterized by abundant graded sandy turbidites (Fig. 10). However, these turbidites are thinner (up to 50 cm) and more frequent than in the axis of the channel (core KS 8822), and the material is characterized by more variable carbonate contents (5 to 85%) and median grain-sizes (20 to 150 μm). The sand content ranges from 5 to 85%, with modal peaks at 30 to 150 μm, (coarse silt/fine to medium sands). The sand composition is different from that described in the channel with a much lower quartz content (never exceeding 10%), more abundant foraminifera (up to 90%), and up to 50% of fine mica. The

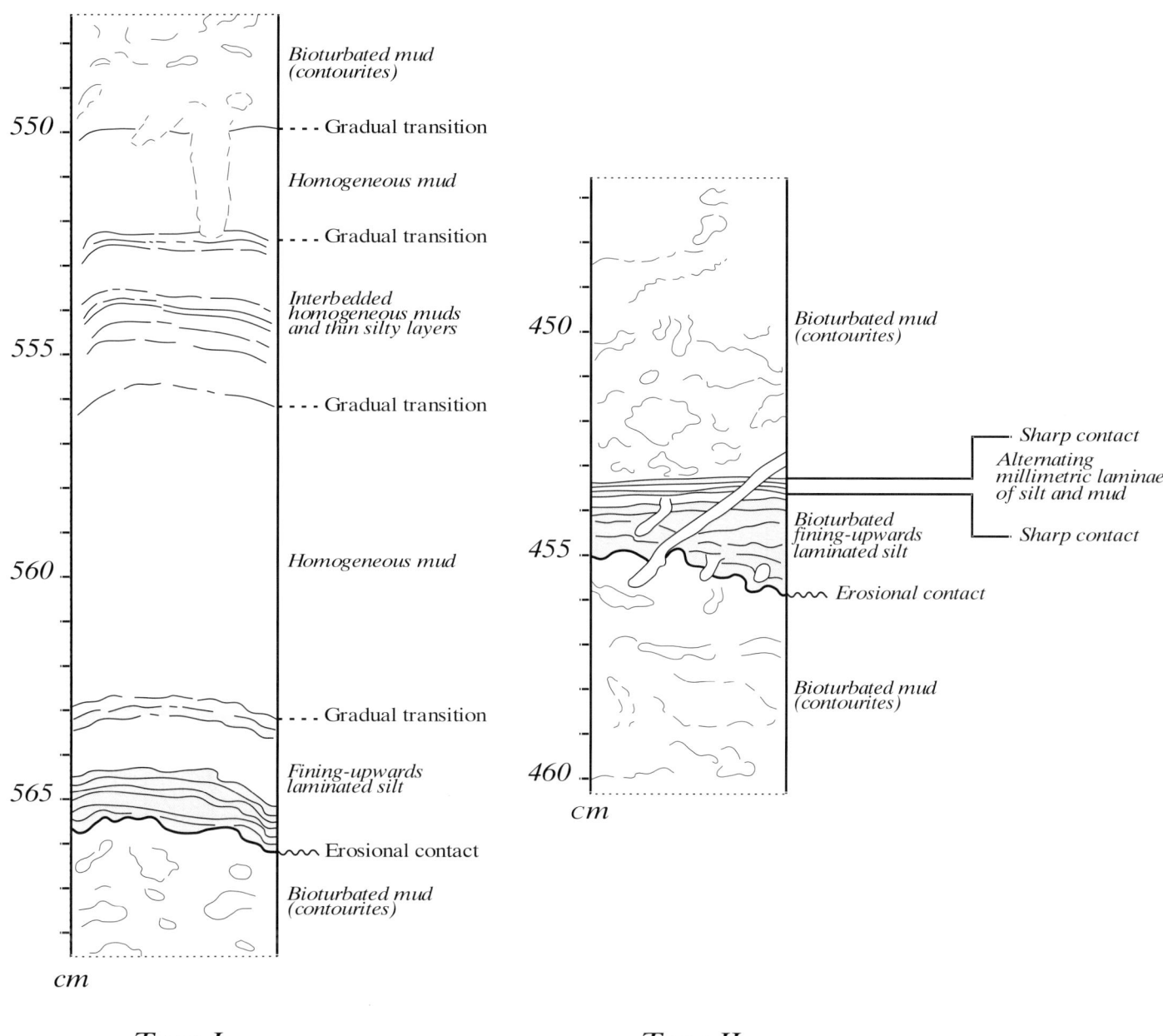

Fig. 9. Interpretative sketches of X-radiographs showing the detailed lithologic characters of type I and type II silt layers (core KS 8821, northern levee).

interbedded muds (Fig. 7) are very similar to those described on the southern flank of the channel, and are assumed to be muddy contourites, although a part of the material may be derived from the turbiditic flows.

The facies observed in core KS 8826 from the top of the deepest part of the levee (Figs 6, 11 and 12) are dominated by silty-clayey muddy contourites. These muds display a grain size mostly ranging between 3 and 7 μm (fine silt and clay-sized material), with a > 10 μm fraction between 10% and 30%, very low carbonate contents (mostly less than 5%, sometimes up to 25 %) and occasional erosional surfaces (Fig. 12). Two major facies can be defined that alternate throughout the core: (1) yellowish-brown muds, bearing bioturbation marks, and frequently displaying manganese enrichments that appear as millimetre-to-centimetre-scale dark grey laminae, ranging from very diffuse mottling, slightly darker than the adjacent muds to well-defined horizons with sharp boundaries; and (2) grey-green homogeneous muds, occasionally displaying millimetric beds of muddy micropebbles (microbrecciated muds) due to short episodes of enhanced bottom currents (Massé 1993). These muds are very similar to the muddy contourites described in the previous cores, and also further south at the northern exit of the Vema Channel where similar facies alternations have been observed (Mézerais 1991; Mézerais et al. 1993; Massé 1993; Massé et al. 1994; Faugères et al. 2002). All these characteristics give evidence of AABW contour current control during deposition. Clay mineralogical data (K/I ratio, Massé et al. 1998), suggest that a part of the muddy material in KS 8826 may be derived from the Columbia Channel turbiditic flows. A unique silty-clay calcareous turbidite occurs at 840 cm depth in the core (Fig. 12). It shows a 10 cm thick sequence with a basal erosive foram-rich bed overlain by white muds (50% carbonate) that suggest a Vitoria Trindade chain biogenic sediment source.

KS 8820 (4476 m)

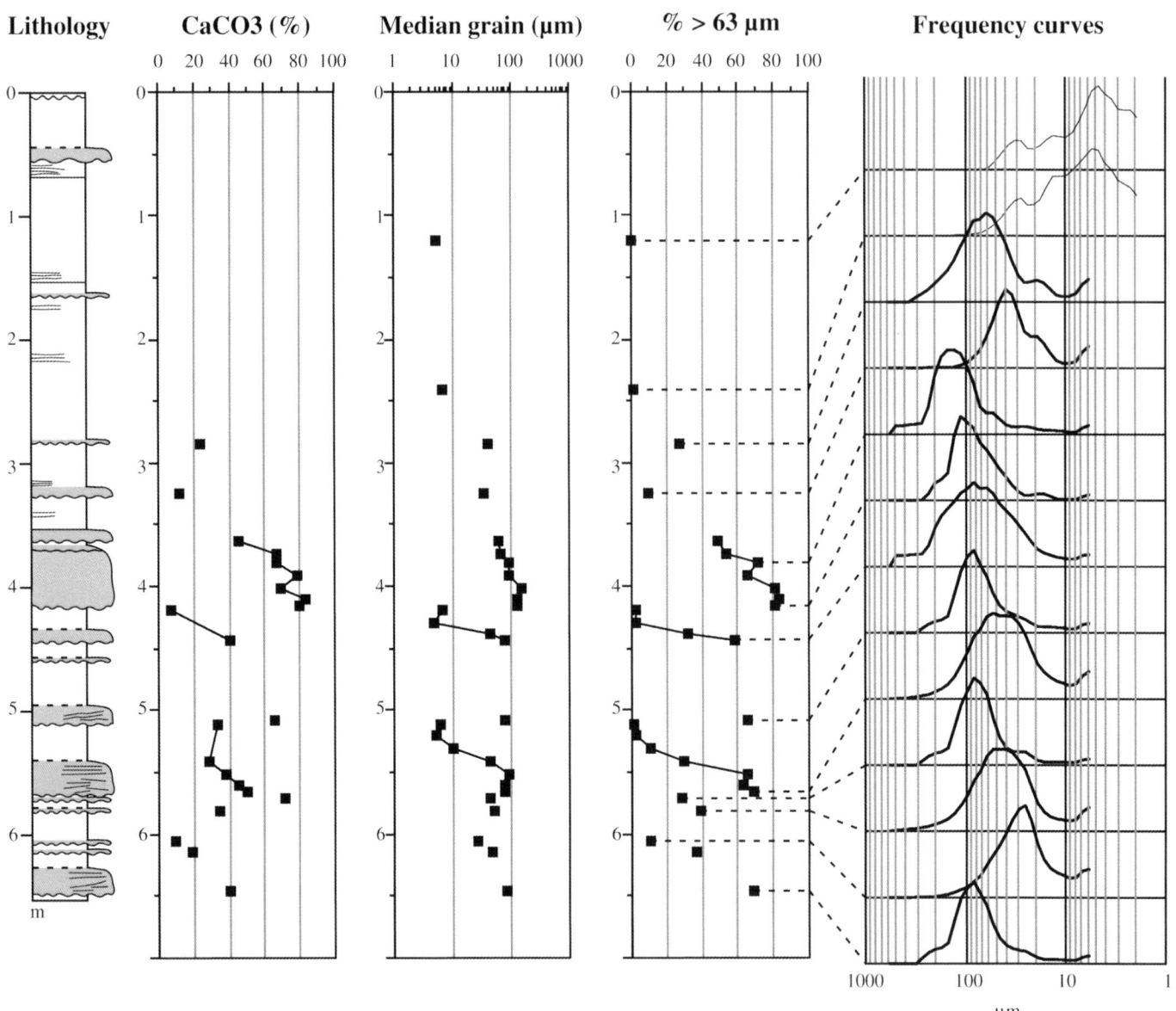

Fig. 10. Lithology of core KS 8820 (northern levee).

Discussion

The lithologic and seismic data demonstrate the contrast existing in the Columbia Channel system between the shallowest area, west of 33°W where the northern levee is parallel to the axis of the channel and characterized by abundant turbiditic deposits, and the deepest area, east of 33°E where the levee extends parallel to the path of bottom currents, and is characterized by dominant muddy contourites.

Contouritic processes and associated deposits

It has been shown in the previous sections that the silty-clay muds could be deposited by contour currents. Such a depositional process for the muds is supported by the clay contents in cores of the shallow (KS 8821) and deepest (KS 8826) part of the levee. Chlorite transported by AABW from high southerly latitudes, is considered as a good tracer of this contour current (Biscaye 1965; Massé 1993; Massé *et al.* 1996; Petschick *et al.* 1996). The C/I ratio in the muds of these two cores is very close to that of the contouritic muds deposited at the northern exit of the Vema Channel. Kaolinite is a good tracer of supply derived from the upper part of the margin and transported donwnslope by gravity processes. Values of the K/I ratio indicate far more lower kaolinite contents than on the middle continental rise, suggesting that only a part of the muddy material is derived from turbiditic flows. Similar contouritic muds with very low carbonate contents (<15%), a modal peak centred around 5 μm (fine silt and clay) are also dominant on the southern flank of the channel (cores KS 8823 and KS 8824, Fig. 7).

Despite the fact that turbiditic deposits are dominant on the shallower part of the northern levee, the whole area is located at depths greater than 4000 m and is swept by the AABW currents. Consequently, the muds deposited between the silty-sandy turbiditic sequences should experience some contour current control

KS 8826 (4567 m)

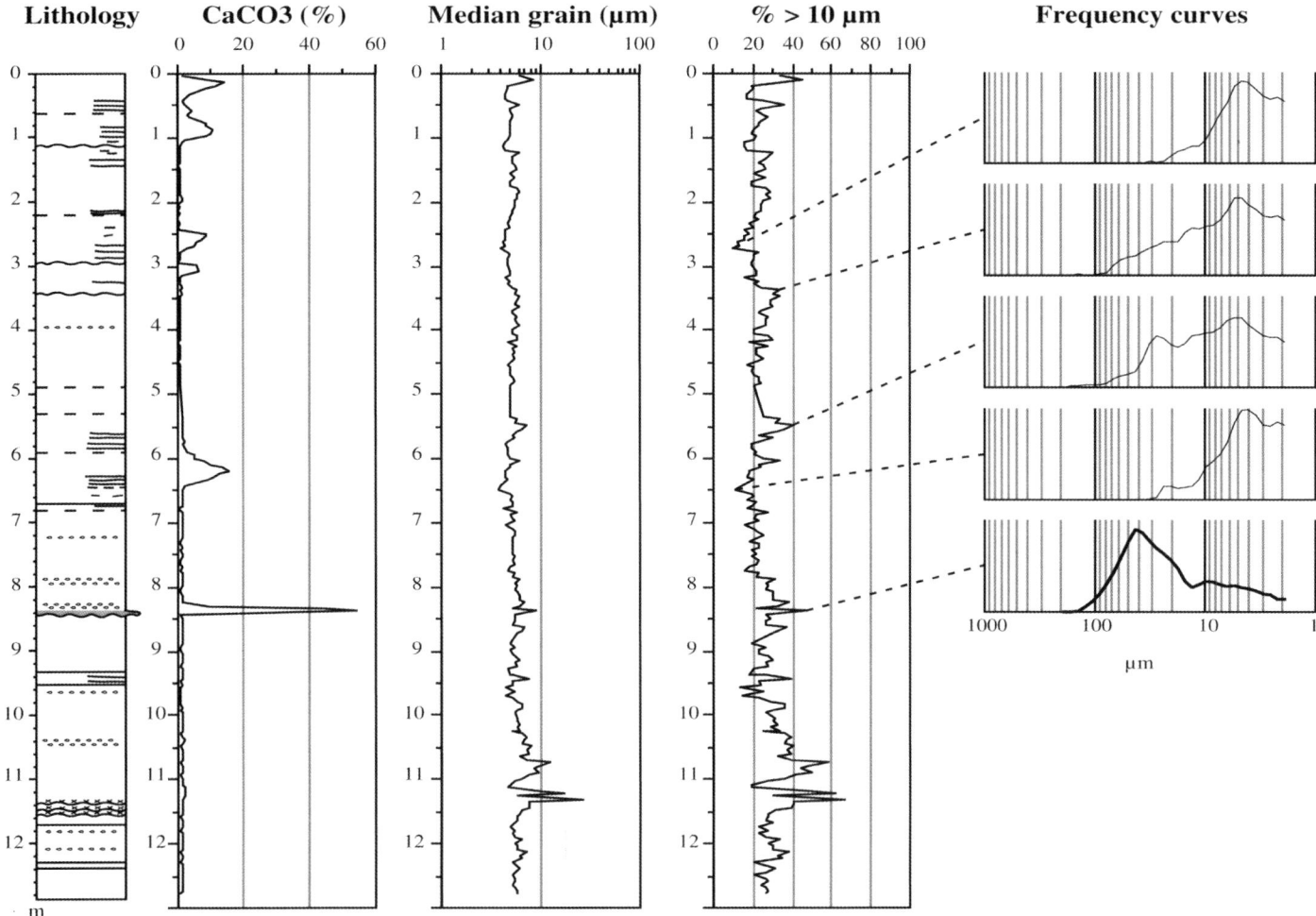

Fig. 11. Lithology of core KS 8826 (northern contouritic levee). See Figure 6 for the significance of lithologic symbols.

and may be interpreted as muddy contourites rather than hemipelagites or turbiditic muds. Distinguishing between them is not easy. However, as in core KS 8821 (Fig. 9), there is a clear lithological difference between the very homogeneous muds gradually overlying some silty beds and interpreted as muddy turbidites, and the bioturbated muds sometimes showing sharp contacts with the underlying silty beds and interpreted as contour-current deposits.

These Quaternary muddy contourites are fairly homogeneous but they show colour variations (yellowish, brownish, blackish to green) due to variation in the rate of terrigenous supply and to diagenetic processes. South of the channel, they are predominantly yellowish and are associated with manganiferous nodules. On the deep contouritic northern levee, they show an alternation of yellowish and greenish layers with microbrecciated muddy contourites, associated with erosional surfaces and thin cm-crusts or thin layers composed of manganese and iron oxides (Figs 11 and 12). The occurrence of some muddy layers enriched in diatoms (Fig. 14a) originating from high southern latitudes (Massé 1993), is evidence of sediment deposition under the AABW control. Most of these contouritic deposits, already observed further south in the Brazilian Basin (Faugères et al. 2002), display peculiar facies and do not fall in the classical contourite models (Stow & Lovell 1979; Stow 1982; Faugères et al. 1984; Gonthier et al. 1984; Stow et al. 1996).

Turbiditic processes and associated deposits

Typical silty-sandy turbiditic deposits (Bouma 1962) are common in cores along the western transect where their composition reveals a clear distinction between (1) quartz-rich turbidites, and (2) quartz-poor, mica- and foram-rich turbidites.

Quartz-rich sandy turbidites are found in the axis of the channel (KS 8822). This material originates from the upper continental rise to the west, where similar material is described (Massé 1993; Massé et al. 1996). Probably due to the great depth of the channel, the turbiditic sand loads transported in the channel are not really involved in overflows on the northern levee as displayed by the rapid disappearance of prolonged bottom echoes from the channel axis towards the north, and the absence of thick sandy turbidites in KS 8821. The finer silty-muddy turbidites observed in this core (type I silty layers) are probably derived from the Columbia Channel as the core is not very far from its axis, fairly abundant quartz grains are observed in the fraction >10 µm, and the K/I ratio close to one is similar to that on the middle continental rise.

Quartz-poor, mica- and foram-rich turbidites are abundant on the northernmost part of the levee (KS 8820). The distribution of echofacies on the northern levee (Fig. 4) with a southward decreasing trend in sand content (from KS 8820 to KS 8821) strongly suggests a distinct sediment source from that observed in

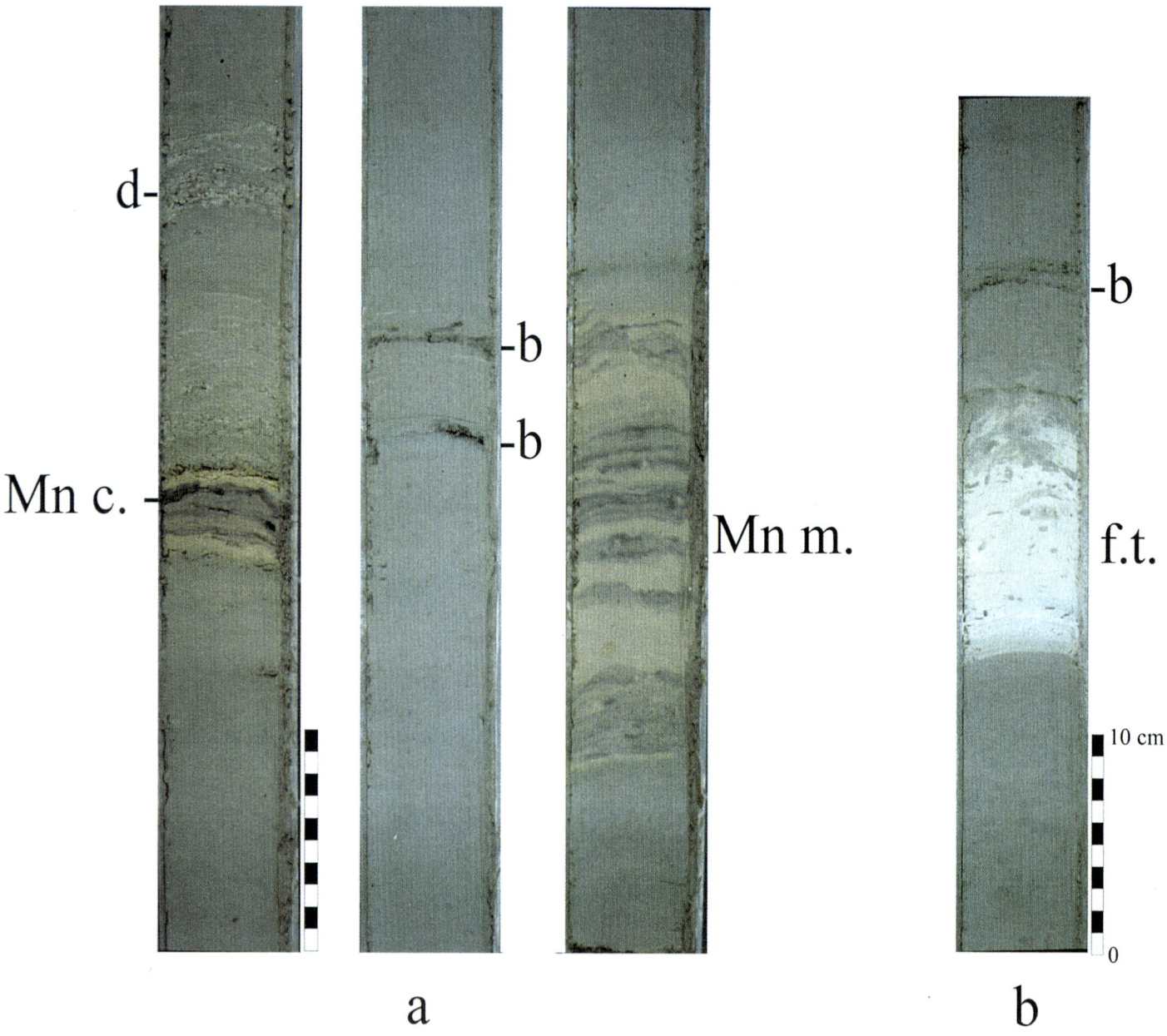

Fig. 12. Photographs of KS 8826 facies (northern contouritic levee): (**a**) yellowish manganiferous mud (Mn c, manganiferous crust; Mn m, manganiferous marbling) alternating with greenish mud showing layers of microbrecciated layers (**b**) and diatom-rich mud (**d**); (**b**) greenish muds with a white foram-rich turbidite (f.t.).

the axis of the channel, for a large part of the levee. It probably comes from the walls of the Vitoria–Trindade Seamounts directly to the north. This chain results from an accumulation of terrigenous material between the isolated seamounts (Fainstein & Summerhayes 1975), now covered by a carpet of pelagic sediments. The foram-rich turbidites certainly come from this area. The origin of the mica-rich material is more speculative: either terrigenous deposits derived from the Vitoria-Trindade chain or from the Columbia Channel, as mica particles may be transported for long distances.

The interaction between turbiditic and contouritic processes

Silty-muddy, top-truncated turbidites (type II silty layers, KS 8821) are the only sediment facies that displays direct evidence of the both turbidity and contour current activity at the same place and time. They show a basal erosive thin layer or lens of laminated silt and silty mud that could be of turbiditic origin, as they have lithological characters similar to the silty muddy turbidites. However as there is a sharp contact at the top (Fig. 9), overlain by bioturbated muds, they are interpreted as turbidite deposits truncated at the top by contour currents. The currents may have been sufficiently active to remove or to prevent the rapid deposition of the floating foram tests and muddy top of the sequence. The result is a top-cut sequence, a finer grain size and removal of the carbonate (from over 10% in the original silty muddy turbidites to 0%). The currents could also be responsible for the formation of the laminated structures. All these features strongly suggest the activity of contour currents during or shortly after the turbidite deposition. This type of deposit is typically what some authors refer to as bottom current reworked turbidites (Stanley 1993; Shanmugam *et al.* 1993*a*, *b*; Viana & Faugères 1998).

One can argue that the turbiditic flow itself could be responsible for winnowing of finer particles. This implies that type II silts are deposited from denser and faster currents than type I. However, this is inconsistent with the finer grain-size of type II silts with

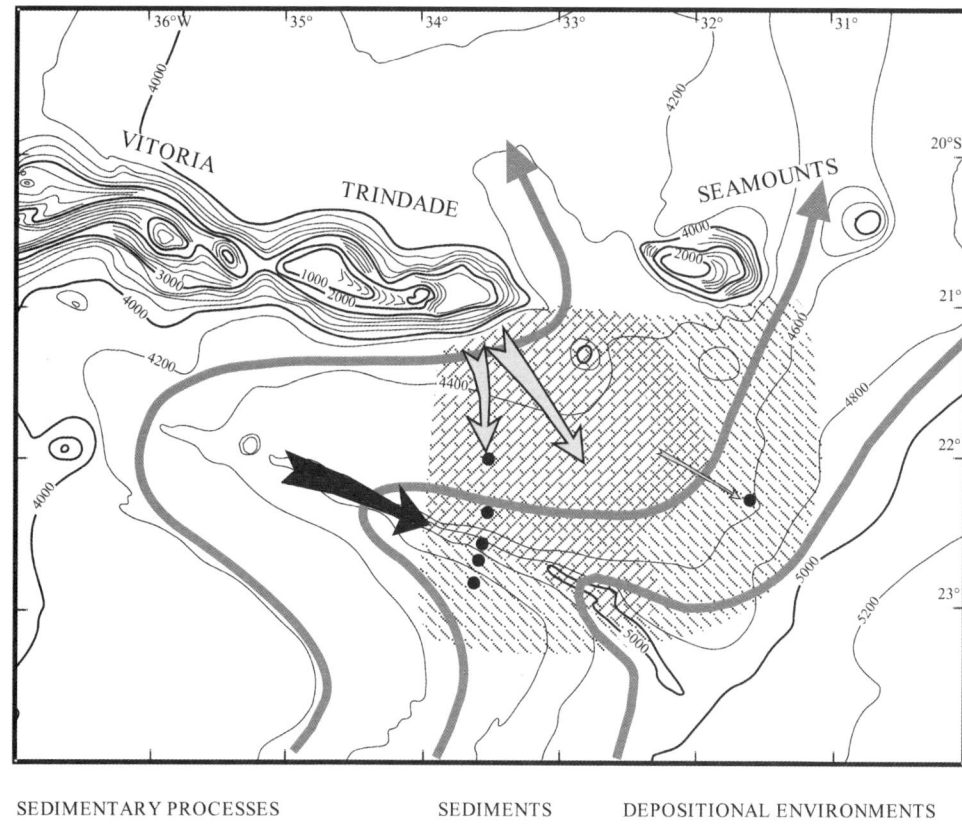

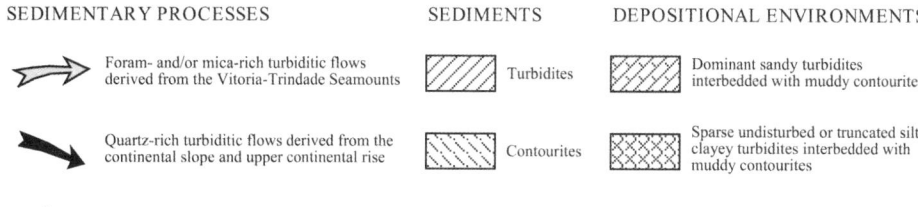

Fig. 13. Synthetic map showing the contour current pathways, and the distribution of sedimentary processes, associated deposits and depositional environments in the study area.

respect to type I silts. If mud deposition occurs above type I silts, then it is likely to occur for similar or slightly finer type II silts. Consequently, we have to assume that the reworking of type II silts is done by contour currents linked to the AABW. The velocity of AABW currents in the SW Atlantic has varied through time (e.g. Jones & Johnson 1984; Ledbetter 1986; Massé et al. 1994). Type II silts would be the result of distal silty-muddy turbidite reworking during periods of enhanced AABW circulation, or even benthic storm activity (Hollister & McCave 1984; Peggion & Weatherly 1991) resulting from eddies generated when the southward moving Brazil Current crosses the Vitoria–Trindade Chain (Schmid et al. 1995), whereas type I silts would be deposited during periods of sluggish AABW circulation.

The distribution of sedimentary processes and associated deposits is summarized on Figure 13.

The authors are grateful to C. Pudsey and J. Howe for their constructive and critical comments and reviews. This research was mainly supported by funds provided by the UMR number 5805-CNRS (France). This paper is University Bordeaux I, DGO-UMR CNRS 5805, contribution number 1377.

References

BISCAYE, P. E. 1965. Mineralogy and sedimentation of recent deep-sea clay in the Atlantic ocean and adjacent seas and oceans. *Geological Society Am. Bull.*, **76**, 803–832.

BOUMA, A. H. 1962. *Sedimentology of some flysch deposits*. Elsevier, Amsterdam.

CARTER, L. & MCCAVE, I. N. 1994. Development of sediment drifts approaching an active plate margin under the SW Pacific Deep Western Boundary Current. *Palaeoceanography*, **9**, 1061–1085.

CASTRO, D. D. 1992. *Morfologia da margem continental Sudeste-Sul Brasileira e estratigrafia seismica do sopé continental*. MSc Thesis, University Federal do Rio de Janeiro, Rio de Janeiro.

DAMUTH, J. E. 1975. Echo character of the western equatorial Atlantic floor and its relationship to the dispersal and distribution of terrigenous sediments. *Marine Geology*, **18**, 17–45.

DAMUTH, J. E. & HAYES, D. E. 1977. Echo character of the east brazilian continental margin and its relationship to sedimentary processes. *Marine Geology*, **24**, 73–95.

EMBLEY, R. W., 1975. *Studies of deep-sea sedimentation processes using high-frequency seismic data*. Thesis, Columbia University, Palisades, N. Y.

EWING, M., EITTREIM, S., EWING, J. & LE PICHON, X. 1971. Sediment transport and distribution in the Argentine Basin, 3. Nepheloid layer and processes of sedimentation. *Phys. Chem. Earth*, **8**, 49–78.

FAINSTEIN, R. & SUMMERHAYES, C. P. 1982. Structure and origin of marginal banks off eastern Brazil. *Marine Geology*, **46**, 199–215.

FAUGÈRES, J.-C. 1988. La campagne Byblos. Rôle des courants profonds dans la sédimentation de la marge est-brésilienne. *Géochronique*, Soc. Géol. France, **27**, 8.

FAUGÈRES, J.-C., GONTHIER, E. & STOW, D. A. V. 1984. Contourite drift molded by deep Mediterranean outflow. *Geology*, **12**, 296–300.

FAUGÈRES, J.-C., STOW, D. A. V., IMBERT, P. & VIANA, A. 1999. Seismic features diagnostic of contourite drifts. *Marine Geology*, **162**, 1–38.

FAUGÈRES, J.-C., ZARAGOSI, S., MÉZERAIS, M. L. & MASSÉ, L. 2002. The Vema contourite fan in the south Brazilian Basin. *In*: STOW, D. A. V., PUDSEY, C. J., HOWE, J. A., FAUGÈRES, J.-C. & VIANA, A. R. (eds)

Deep-Water Contourite Systems: Modern Drifts and Ancient Series, Seismic and Sedimentary Characteristics. Geological Society, London, Memoirs, **22**, 223–238.

GAMBOA, L. A. P. & RABINOVITZ, P. D. 1981. The Rio Grande Rise fracture zone in the western south Atlantic and its tectonic implications. *Earth and Planetary Sciences Letters*, **52**, 410–418.

GAMBOA, L. A. P. & RABINOVITZ, P. D. 1984. The evolution of the Rio Grande Rise in the southwest Atlantic Ocean. *Marine Geology*, **58**, 35–58.

GAMBOA, L. A. P., BUFFLER, R. T. & BARKER, P. F. 1983. Seismic stratigraphy and geologic history of the Rio Grande Gap and southern Brazil Basin. *In Initial Reports of the DSDP*, **72**, Washington US Government Printing Office 481–498.

GONTHIER, E., FAUGÈRES, J.-C. & STOW, D. A. V. 1984. Contourite facies of the Faro Drift, Gulf of Cadiz. *In*: STOW, D. A. V. & PIPER, D. J. W. (eds) *Fine-Grained Sediments: Deep Water Processes and Facies*. Geological Society, London, Special Publications, **15**, 245–256.

HOLLISTER, C. D. & MCCAVE, I. N. 1984. Sedimentation under deep-sea storms. *Nature*, **309**, 220–225.

JONES, G. A. & JOHNSON, D. A. 1984. Displaced Antarctic diatoms in Vema Channel sediments: Late Pleistocene/Holocene fluctuations in AABW flow. *Marine Geology*, **58**, 187–212.

LEDBETTER, M. T. 1986. A late Pleistocene time-series of bottom-current speed in the Vema Channel. *Palaeogeography, Palaeoclimatology & Palaeoecology*, **53**, 97–105.

LEWIS, K. B. 1994. The 1500-km long Hikurangi Channel: trench-axis channel that escapes its trench, crosses a plateau, and feeds a fan drift. *Geo-Marine Letters*, **14**, 19–28.

MASSÉ, L. 1993. Sédimentation océanique profonde au Quaternaire. *Flux sédimentés et paléocirculations dans l'Atlantique Sud-Ouest: Bassin Sud-Brésilien et prisme d'accrétion Sud-Barbade*. Thesis, Universitie Bordeaux I, France.

MASSÉ, L., FAUGÈRES, J.-C., BERNAT, M., PUJOS, A. & MÉZERAIS, M. L. 1994. A 600 000 year record of Antarctic Bottom Water activity inferred from sediment textures and structures in a sediment core from the Southern Brazil Basin. *Paleoceanography*, **9**, 1017–1026.

MASSÉ, L., FAUGÈRES, J.-C., PUJOL, C., PUJOS, A., LABEYRIE, L. D. & BERNAT, M. 1996. Sediment flux distribution in the Southern Brazil Basin during the late Quaternary: The role of deep-sea currents. *Sedimentology*, **43**, 115–132.

MASSÉ L., HROVATIN, V. & FAUGÈRES, J.-C. 1998. The interplay between turbiditic and contouritic processes in the Southern Brazil Basin. *Sedimentary Geology*, **115**, 111–132.

MCCAVE, I. N. & CARTER, L. 1997. Recent sedimentation beneath the deep Western Boundary Current off northern New Zealand. *Deep-Sea Research*, **44**, 1203–1237.

MELLO, G. A. 1988. *Processos sedimentares recentes na Bacia do Brasil: setor Sudeste-Sul*. MSc Thesis, Universitie Federal do Rio de Janeiro, Rio de Janeiro.

MÉZERAIS, M. L. 1991. *Accumulations sédimentaires profondes, par courants de turbidité (Eventail du Cap-Ferret, Golfe de Gascogne) et par courants de fond (débouché du Chenal Vema, Bassin sud Brésilien). – Géométrie, faciès et processus d'édification*. Thesis, Univ. Bordeaux I, France.

MÉZERAIS, M. L., FAUGÈRES, J.-C., FIGUEIREDO JR., A. G. & MASSÉ, L. 1993. Contour current accumulation off the Vema Channel mouth, southern Brazil Basin: pattern of a 'contourite fan'. *In*: STOW, D. A. V. & FAUGÈRES, J.-C. (eds) *Contourites and Bottom Currents*. Sedimentary Geology, **82**, 173–187.

PEGGION, G. & WEATHERLY, G. L. 1991. On the interaction of the bottom boundary layer and deep rings. *Marine Geology*, **99**, 329–342.

PETSCHICK, R., KUHN, G. & GINGELE, F. 1996. Clay mineral distribution in surface sediments in the South Atlantic: sources, transport and relation to oceanography. *Marine Geology*, **130**, 203–229.

REID J. L. 1996. On the circulation in of the South Atlantic ocean. *In*: WEFER, G. *et al.* (eds) *The South Atlantic: Present and Past Circulation*. Springer, New York, 13–44.

REID, J. L., NOWLIN, W. D. & PATZERT, W. C. 1977. On the characteristics and circulation of the southwestern Atlantic Ocean. *Journal of Physical Oceanography*, **7**, 62–91.

SCHMID, C., SCHÄFER, H., PODESTA, G. & ZENK, W. 1996. The Vitoria eddy and its relation to the Brazil Current. *Journal of Physical Oceanography*, **25**, 2532–2546.

SCHOBBENHAUS C., DE ALMEIDA CAMPOS D., DERZE G. R. & ASMUS H. E. 1984. Geologia do Brasil. Texto explicativo do mapa geologico do Brasil e da area oceanica adjacente incluido depositos minerais. Escala 1: 2 500 000. *Publicaçao da Divisao de Geologia e Mineralogia, Departamento Nacional da Produçao Mineral, Ministerio das minas e energia do Brasil*. Ediçao comemorativa do "Cinquentenario".

SHANMUGAM, G., SPALDING, T. D. & ROFHEART, D. H. 1993a. Process sedimentology and reservoir quality of deep-marine bottom-current reworked sands (sandy contourites): an example from the Gulf of Mexico. *AAPG Bulletin*, **77**, 1241–1259.

SHANMUGAM, G., SPALDING, T. D. & ROFHEART, D. H. 1993b. Traction structures in deep-marine, bottom-current reworked sands in the Pliocene and Pleistocene, Gulf of Mexico. *Geology*, **21**, 929–932.

STANLEY, D. J. 1993. Model for turbidite-to-contourite continuum and multiple process transport in deep marine settings: examples in the rock record. *In*: STOW, D. A. V. & FAUGÈRES, J.-C. (eds) *Contourites and Bottom Currents*. Sedimentary Geology, **82**, 241–255.

STOW, D. A. V. 1982. Bottom-current and contourites in the North Atlantic. *Bull. Inst. Geol. Bass. Aquitaine*, Bordeaux, **31**, 151–156.

STOW, D. A. V. & LOVELL, J. P. B. 1979. Contourites: their recognition in modern and ancient sediments. *Earth Science Revues*, **14**, 251–291.

STOW, D. A. V., READING, H. G. & COLLINSON, J. D. 1996. Deep seas. *In*: READING, H. G. (ed.) *Sedimentary Environments: Processes, Facies and Stratigraphy*. Blackwell Science, Oxford, 395–453.

VIANA, A. R., FAUGÈRES, J.-C., STOW, D. A. V. 1998. Bottom current-controlled sand deposits. A review of modern shallow – to deep water environments. *Sedimentary Geology*, **115**, 53–80.

Contour currents, sediment drifts and abyssal erosion on the northeastern continental margin off Brazil

P. O. GOMES[1] & A. R. VIANA[2]

[1] *Amerada Hess Limited, 33 Grosvenor Place, London SW1X 7HY, UK (e-mail: Paulo.Otavio@Hess.com)*
[2] *Petróleo Brasileiro S.A., PETROBRAS E&P, Campos Basin branch, Av. Elias Agostinho, 665 Macae, RJ 27.913-350, Brazil*

Abstract: Seismic reflection data collected as part of the Brazilian Continental Shelf Survey has allowed the recognition of large sediment drifts related to bottom currents along the continental rise off northeastern Brazil. Antarctic Bottom Water (AABW), which flows northward across a physiographic constriction – the hotspot-related Bahia Seamounts – has controlled the deposition of these contourite drifts. Following this pathway since the Middle Oligocene, AABW has also led to the formation of regional unconformities and the excavation of the 800 km long and up to 470 m deep Pernambuco Seachannel. This feature, which represents the 'trunk channel' of a complex submarine drainage system, flows into the Pernambuco Abyssal Plain, building a large fan-like deposit – the Pernambuco Countourite Fan. As observed elsewhere along continental margins, the geometry of the sediment drifts alternates from mounded to sheeted, channel-related or confined, depending on the physiographic setting and on the velocity of the flow. The so-called 'modified drift-turbidite systems', formed by an interplay between downslope sediment gravity flows and alongslope bottom-current-controlled deposition, are well developed near the continental slope region, particularly in the Sergipe Basin, where the São Francisco River has built its deep-sea fan.

The important role of thermohaline oceanic circulation in shaping and reworking the sedimentary apron of continental margins and oceanic basins has been realized in the last years, with the recognition of ancient and recent sediment drifts formed by bottom (contour) currents all over the world. Publications have defined a number of aspects related to bottom-current-controlled sedimentation, including the definition of contourites, depositional processes, facies associations, contourite drift types and their seismic expression. Also considered are the interactions between turbidity and contour current processes, and the correlation of these processes with global eustatic and climatic changes (Locker & Laine 1993; Stow & Faugères 1993, 1998; Faugères *et al.* 1999).

On the Brazilian continental margin, the influence of thermohaline bottom currents on the transportation and deposition of sediments has been investigated in the Campos Basin (Mutti *et al.* 1980; Carminatti & Scarton 1991; Souza Cruz 1995; Viana 1998; Viana *et al.* 1998; Viana & Faugères 1998). Gamboa *et al.* (1983b) recognized seismic features related to deep-sea erosion and progradation, generated by cold deep water flow in the southern Brazil Basin. In the same region, located at the downstream exit of Vema Channel, Mézerais *et al.* (1993) characterized a contourite fan.

On the northeastern continental margin, however, except for some brief discussions by Gorini *et al.* (1982) and Gorini & Carvalho (1984), little work has been conducted on the effects of the deep oceanic circulation on the sedimentary record. A study by Cherkis *et al.* (1992) in the Bahia Seamounts region described an important deep-sea channel (Pernambuco Seachannel) as a pathway for cold bottom waters (AABW) into the eastern Brazil Basin. Gomes & Gamboa (1999) pointed out some morphological, stratigraphic and depositional aspects of that channel.

The multichannel seismic reflection data used in this study were collected in 1992, as part of the 'Brazilian Continental Shelf Survey Plan' (LEPLAC Project), a regional geophysical survey conducted by the Brazilian Navy and PETROBRAS. The acquisition was performed by RV 'Almirante Câmara', with a shot interval of 50 m, using high-pressure air guns as source. Data were recorded with a sampling interval of 4 ms and a record length of 12 seconds. A 2500 m long streamer was used, with 96 channels and a group interval of 25 m. The seismic processing sequence comprised demultiplexing, source-receiver deconvolution, predictive deconvolution, correlation-type velocity analysis, external mute, stacking with dip-move-out correction, finite-difference migration, noise attenuation, frequency filtering and trace equalization.

Geological and oceanographic setting

This study concentrates on the deep oceanic region adjacent to the Jacuípe, Sergipe and Alagoas Brazilian coastal basins, between the 9°30'S and 13°30'S latitudes (Fig. 1). All the seismic features discussed are located on the continental rise, in water depths from 3000 m to 5000 m. This is a typical passive margin developed since the Late Aptian after rifting and continental break-up. The ocean-continent transition processes at the region were studied by Mohriak *et al.* (1995), who presented an abrupt crustal thinning model, controlled by high-angle synthetic faults that converge to a intracrustal horizon, with intense magmatic activity and emplacement of seaward-dipping reflector wedges (Figs 4 & 5). This crustal transition style, along with sediment supply and denudation processes, has strong influences on the shelf-slope morphology. In fact, the study area is characterized by a narrow shelf and a steep slope, which is strongly affected by canyon carving and mass wasting processes (Cainelli 1992). The most prominent erosional features in this setting are the Japaratuba Canyon and the São Francisco Canyon (Fig. 2), which captures the continental drainage of the São Francisco River. This river has built a large deep-sea fan on the lower slope and upper continental rise – the São Francisco Deep-Sea Fan – since the Late Neogene (Cainelli 1992).

Other important features of the continental rise are the E–W fracture zones (Bode Verde and Sergipe fracture zones) and NW–SE volcanic chains (Bahia Seamounts), which merge together to form a natural physiographic barrier on the ocean floor (Fig. 1). The Bahia Seamounts are formed by a group of volcanic ridges and/or isolated elevations, which constitute conspicuous NW–SE chains at the continental rise off Bahia coastal state (Fleming *et al.* 1982a; Cherkis *et al.* 1992). These seamounts are believed to have been emplaced during the South American plate drift over a pair of hotspots, which were concurrently active from the Late Cretaceous to Early Eocene (Bryan & Cherkis 1995). Recent studies based on seismic data support this hotspot model (Gomes 2000).

The most important E–W structure in the study area is the

From: STOW, D. A. V., PUDSEY, C. J., HOWE, J. A., FAUGÈRES, J.-C. & VIANA, A. R. (eds)
Deep-Water Contourite Systems: Modern Drifts and Ancient Series, Seismic and Sedimentary Characteristics.
Geological Society, London, Memoirs, 22, 239–248. 0435-4052/02/$15.00 © The Geological Society of London 2002.

Fig. 1. Bathymetric chart of the northeastern continental margin off Brazil, with the location of LEPLAC geophysical profiles. The study area is singled-out by a dashed rectangle. The yellow lines (A, B and C) are displayed at Figures, 3, 4 and 5. The arrows point out to the Pernambuco Seachannel (modified from Cherkis *et al.* 1989).

Bode Verde Fracture Zone, a double large-offset-fracture zone (Fleming *et al.* 1982b; Gorini *et al.* 1984) that probably resulted from a change in spreading direction during the Late Cretaceous (Cherkis *et al.* 1992). Another E–W structure, here called Sergipe Fracture Zone (Fig. 5), merges with the the northern part of the Bahia Seamounts. According to the concepts of Mello & Dias (1996), this feature can be considered as a short-lived, small-offset-fracture zone.

Another outstanding feature of the continental rise setting is the 800 km long and up to 470 m deep Pernambuco Seachannel (Gorini & Carvalho 1984; Cherkis *et al.* 1989), a sinuous deep-sea channel which follows an approximately N–S course across the Bahia Seamounts region (Cherkis *et al.* 1992), between the 4600 m and 5000 m bathymetric contours (Figs 1 & 2).

The Antarctic Bottom Water (AABW), following the bathymetric contours in a northward flow, is the cold deep-water mass that sweeps the ocean floor in the study area. This bottom current, which is deviated to the left due to the Coriolis effect in the southern hemisphere, is believed to have been active at least since the Oligocene, when the temperature gradient between poles and equatorial regions, plus tectonic reorganization, created the right conditions for development of global thermohaline deep oceanic circulation (Johnson 1985).

In its northward flow along the continental rise off NE Brazil, the AABW encounters a physiographic barrier formed by the Bahia Seamonts and the Bode Verde Fracture Zone. Part of the flow is accelerated through an abyssal gap, leading to the excavation of a confined deep-sea channel (the Pernambuco Seachannel); another flow, also accelerated by physiographic constriction, is deflected toward the slope, under the influence of the Coriolis force, causing lateral channel migration and deep-sea progradation.

Stratigraphic context

In an attempt to analyse the stratigraphic evolution of the region under the sequence stratigraphy paradigm, Gomes (2000) mapped five main regional unconformities over the seismic dataset shown in Figure 1. These deep seismic surfaces, with a local or widespread erosional character (see Discussion), represent the boundaries of stratigraphic sequences with a minimum duration of 9 Ma, which were correlated to the supercycles of Haq *et al.* (1988). The following second order sequences, with their probable correlative global cycles, were individualized: the Upper Cretaceous, Paleocene (Tejas A1+A2), Eocene (TA3+TA4), Oligo–Miocene (Tejas B1), Mid-Neogene (TB2) and Upper-Neogene/Quaternary (TB3) sequences (Fig. 3).

Oceanic basement underlies the Upper Cretaceous sequence, which is overlain by the Upper Cretaceous (UK) unconformity; above this unit, the Paleocene sequence was deposited and partially eroded during the development of the Lower Eocene (LEoc) unconformity; the following Eocene sequence is topped by a strong erosional surface, represented by the Mid-Oligocene (MOlig) unconformity; this unit was covered by the Mid-Neogene sequence which also has a well-defined erosional top – the Lower Miocene (LMioc) unconformity; the sediment package was overlain by the Upper-Neogene/Quaternary sequence (Fig. 3), which extends to the seafloor.

The three older unconformities have biostratigraphic and well-log control, and represent continental rise extensions of the surfaces mapped by Cainelli (1992) in the shelf-slope system of the Sergipe Basin. The two younger unconformities, which could not be tied to shelf events due to slope gravitational processes, were correlated to DSDP wells 23 and 24, located on the Pernambuco Abyssal Plain (Fig. 1).

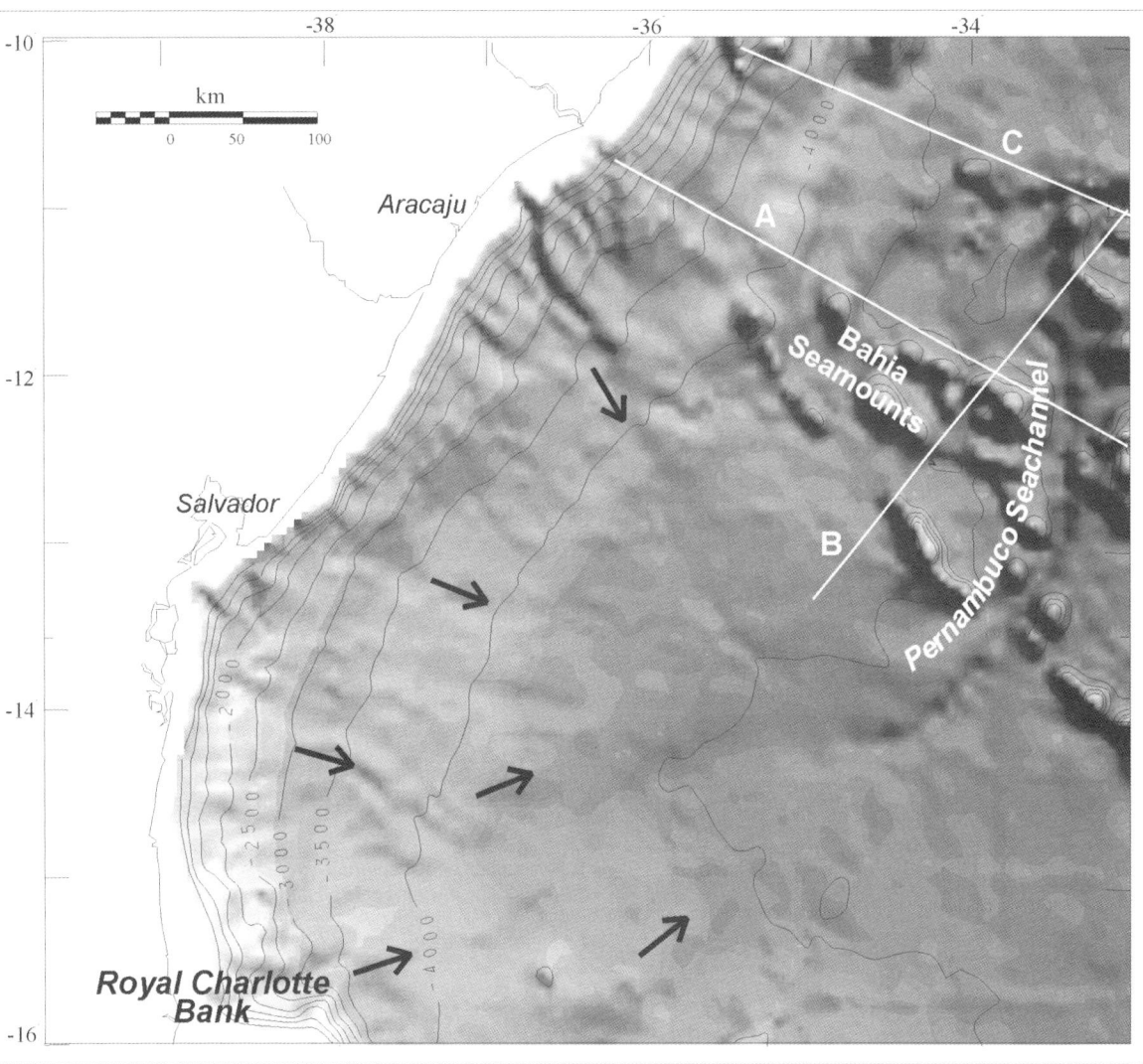

Fig. 2. Detailed bathymetric map of the continental margin around Pernambuco Seachannel. The black arrows indicate the confluence of slope canyons and channels to a main (trunk) channel, forming a submarine drainage system which is controlled by oceanic circulation. The map does not capture the total extension of the seismic profiles.

Seismic characteristics

Abyssal erosion

Seismic data show impressive erosional events related to bottom currents along the continental rise off NE Brazil. The most prominent erosional feature is the Pernambuco Seachannel, which conducts the northward-flowing Antarctic Bottom Water across the Bahia Seamounts region (see Geological and Oceanographic Setting). Rather than an isolated deep-sea channel, the Pernambuco Seachannel (PSC) represents a complex submarine drainage system, with an interconnected network of tributary channels that are linked up both with slope canyons and with other AABW pathways (Fig. 2). The system debouches in the Pernambuco Abyssal Plain, where a large deep-sea fan is built – the Pernambuco Contourite Fan.

Seismic lines A, B and C cross the PSC at three different points, from south to north (Fig. 1). At lines A and B (Figs 4 & 5) the channel is confined between seamounts; line C (Fig. 6) shows one of the maximum incisions (about 470 m deep) carved by the bottom current. An interesting feature is the meandering thalweg of the PSC, which migrates from the western flank (line A) to the center (line B), and from there to eastern flank of the channel (line C). A similar morphology was observed for the giant Northwest Atlantic Mid-Ocean Channel, in the Labrador Sea (Cough & Hesse 1976).

The oldest base of PSC, better observed on line C (Fig. 6), coincides with the inter-regional Mid-Oligocene unconformity. This horizon can be seismically correlated to the unconformity 'A' of Gamboa *et al.* (1983a), and thus to the paleobase of Vema Channel, in the southern Brasil Basin, which footprints the onset of AABW flow in the southwestern Atlantic, c. 30 Ma (Johnson 1985). This Mid-Oligocene unconformity, which can be correlated to the well-known 'Chattian base' eustatic event on the global cycle chart, records the initial erosional activity of the AABW both in the southern and in the northeastern parts of the Brasil Basin.

It is important to emphasize that the erosional–depositional action of the AABW is not only restricted to channel settings, where the flow is accelerated throughout physiographic constrictions. This bottom current affects a broad region of the continental margin, and has been controlling the formation of regional unconformities and the deposition of contourite drifts over the last 30 Ma (see discussion). This concept is well-illustrated on seismic profile A (Fig. 4), which shows extensive unconformities generated by widespread erosion in the deep-sea.

No.	SEQUENCE Age	Supercycle (Haq et al., 1988)	SEISMIC CHARACTERISTICS
6	Upper-Neogene / Quaternary	TB3	These two sequences have a predominant parallel seismic pattern, with moderate amplitude reflections, which can reflect hemipelagic sedimentation, or, in most part of the region, a contour-current-controlled deposition. Large mounded or sheeted contourite drifts can be recognized, as well as large-scale sediment waves ($\lambda = 3$ km) and prograding clinoforms related to bottom current lateral migration. Deep-sea channels excavated near physiographic constrictions truncate these sequences in several places. A regional erosional unconformity bounds these units.
5	Mid-Neogene	TB2	
4	Oligo-Miocene	TB1	This sequence also varies a lot along the study area, showing different thicknesses, seismic characters and types of deposits. Mounded and sheeted sediment drifts are common, reflecting the input of an alongslope bottom-current-controlled deposition. Prograding clinoforms related to lateral channel migration can also be observed, as well as hemipelagic drapes and turbiditic fills (Fig. 6).
3	Eocene	TA3+TA4 (?)	This depositional unit has not an unique pattern along the study area, so that different thicknesses, seismic characters and types of deposits can be found. However, the most common is a type of chaotic/hummocky seismic pattern, formed by low amplitude, discontinuous and contorted reflections, probably related to hydrofracturing and fluid escape processes (Fig. 6). The truncations on the upper boundary of this sequence indicates widespread erosion of the seafloor.
2	Paleocene	TA1+TA2 (?)	Characterized by parallel to sub-parallel seismic patterns, this sequence has higher amplitudes than the underlying sequence, reflecting an increase on the contribution of turbiditic sedimentation (Fig. 6). Chaotic seismic patterns related to slumping and debris-flow processes are common on the slope region. Although erosional at some sites, the upper boudary of this unity is most of the times a correlative conformity.
1	Upper Cretaceous		The reflections within this sequence are characterized by a parallel pattern, with low to moderate amplitudes. In distal areas, these low amplitudes are related to a transparent seismic character, which probably reflects the predominance of hemipelagic deposition (Fig. 6). Towards the slope, turbidites and other sediment gravity flow-related deposits are expected to be dominant. The unconformity that overlies this sequence shows strong evidence of abyssal erosion in restricted points.
Oceanic Basement (Layer 2)			

Fig. 3. Summary chart of ages and seismic characteristics for sequences of the northeastern continental margin off Brazil, showing possible equivalents on the global cycle chart of Haq et al. (1988). The two younger unconformities were only mapped in the continental rise.

Sediment drifts

Contourite drift types in the study area include elongate-mounded, sheeted, channel-related and confined geometries, depending on the physiographic setting and on the velocity of the flow (Faugères et al. 1993, 1999). Together with the so-called 'modified drift-turbidite systems', also important in the region, they form a continuous spectrum of deposits, where one type can evolve into another through time and space[1].

Elongate-mounded drifts, with an average width of 10 km, are

---

[1] Unfortunately, due to a complete absence of core data, there is no direct information about the sedimentology of these sediment drifts. Two wells were drilled in 1969 by the Deep Sea Drilling Project, in a site located hundreds of kilometres away from the region of main contouritic deposition. Furthermore, by the time of DSDP-23/24 core description, there was a strong turbidite paradigm in force, and the early concepts of contourites (Hollister & Heezen 1972) were just about to be introduced.

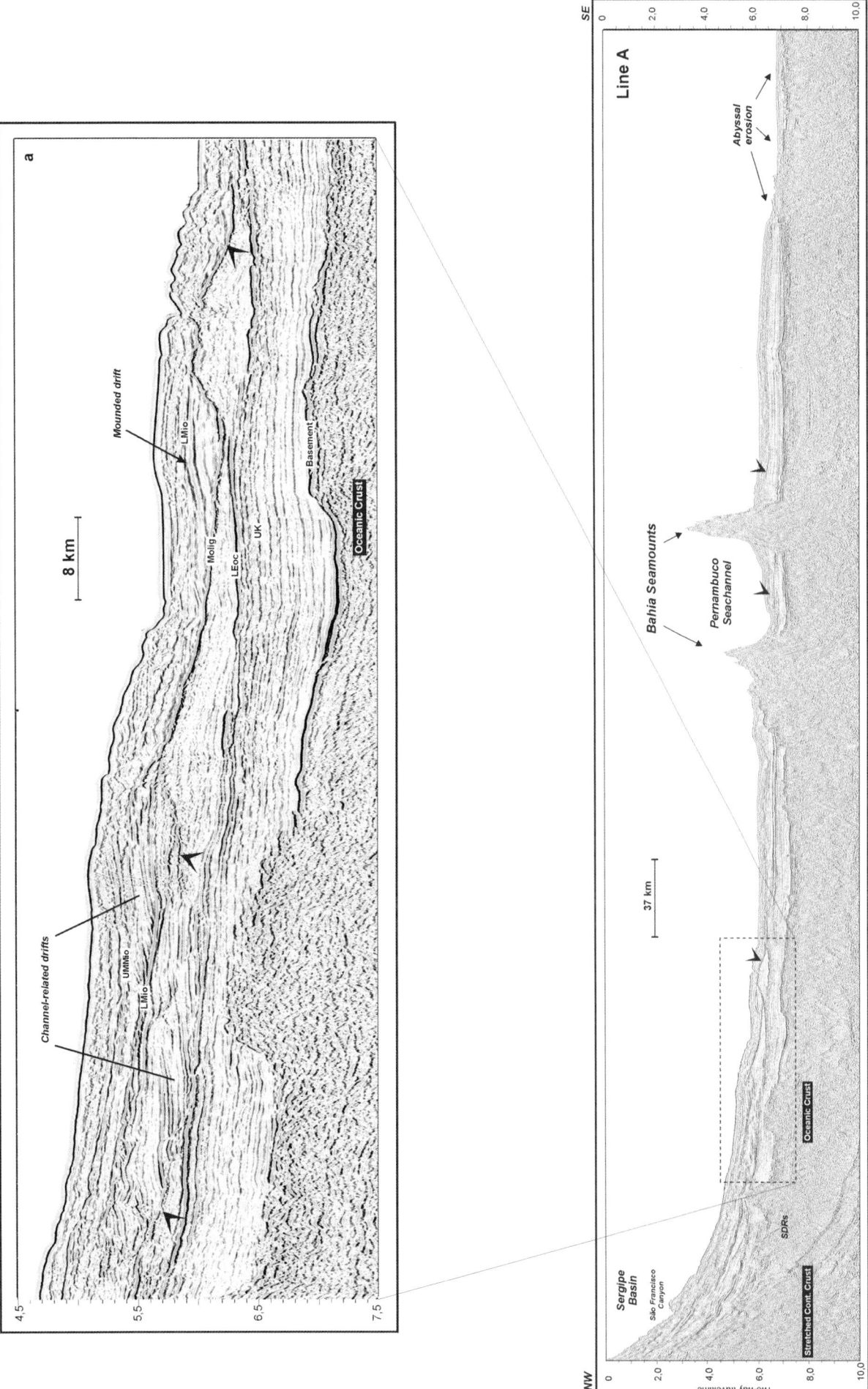

Fig. 4. Dip seismic line A, which crosses the continental margin from the shelf break of the Sergipe basin to the abyssal plain border (see Fig. 1 for location). The arrows point out the Mid-Oligocene unconformity, showing that the erosional activity of the bottom current (AABW), rather than restricted to a confined deep-sea channel setting, affects a broader region of the continental margin. The selected zoom displays remarkable seismic features related to lateral (landward) channel migration of the contour current, over both the Mid-Oligocene and the Lower Miocene unconformities. The geometry of these channel-related drifts resemble that of fluvial point bars. The slope region, where the São Francisco Canyon is carved, is an ideal setting for an interplay between turbidity and contour current processes. One can notice the presence of volcanic plugs and seaward-dipping reflectors wedges (SDRs) at the ocean-continent boundary region (UK, Upper Cretaceous unconf.; LEoc, Lower Eocene unconf.; MOlig, Mid-Oligocene unconf.; LMio, Lower Miocene unconf.; UMMio, Upper Middle Miocene unconf.).

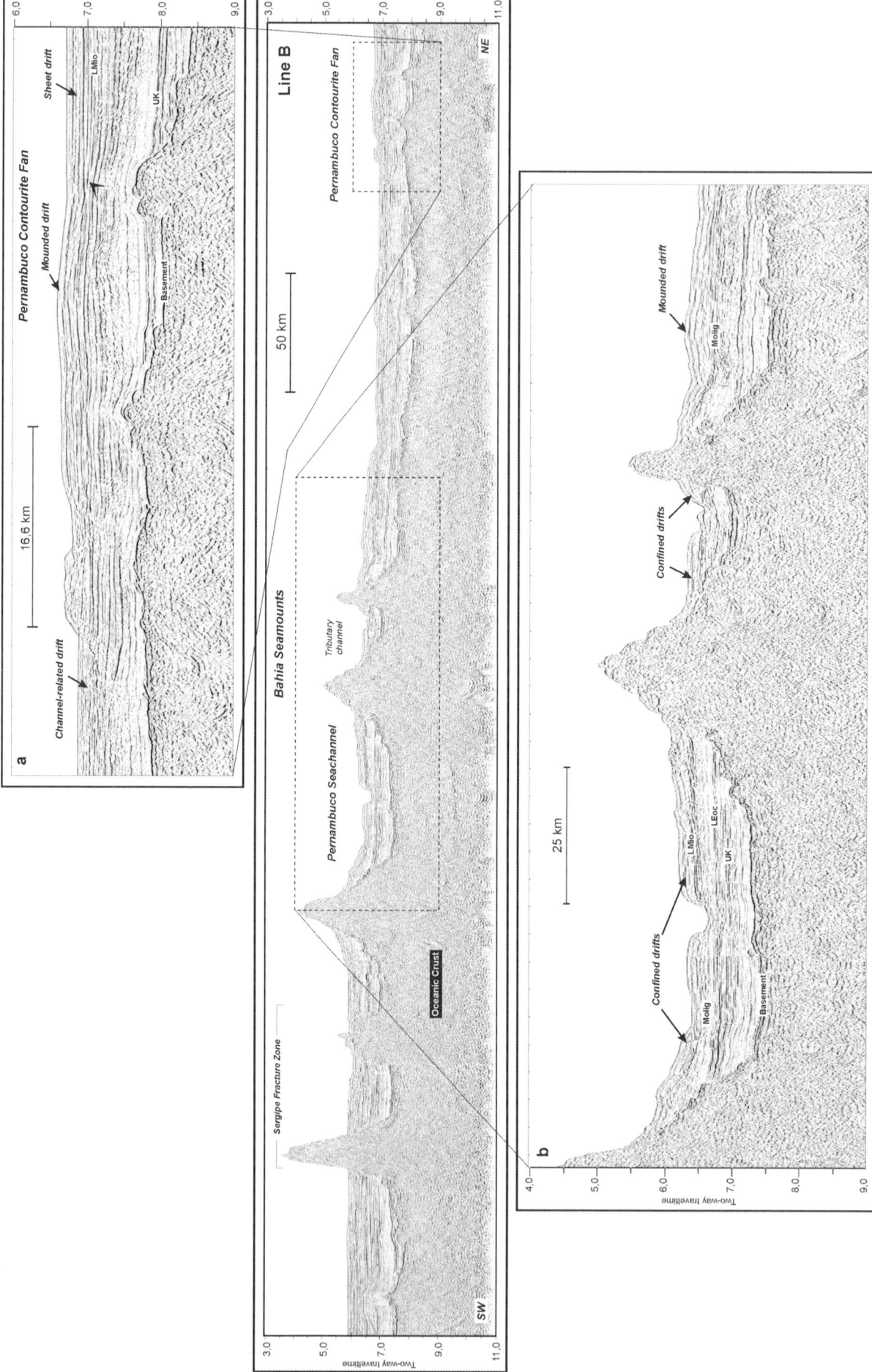

Fig. 5. Strike seismic line B, which traverses the lower continental rise off NE Brazil, showing the Pernambuco Seachannel and its tributary, the Bahia Seamounts, and the Sergipe Fracture Zone (see Fig. 1 for location). Mounded, sheeted, channel-related (a) and confined (b) contourite drifts can be observed. The northwestern edge of the profile crosses the Pernambuco Contourite Fan, where the predominant depositional pattern is aggradational. The arrow at the upper zoom (a) points out to a progradational unit, maybe related to an eventual increase on the velocity of the flow. One can notice that the main channel morphology is reproduced, as a small-scale equivalent, in the tributary channel. The estimated depth of the confined main channel at this crossing is about 450 m (UK, Upper Cretaceous unconf.; LEoc, Lower Eocene unconf.; MOlig, Mid-Oligocene unconf.; LMio, Lower Miocene unconf.; UMMio, Upper Middle Miocene unconf.).

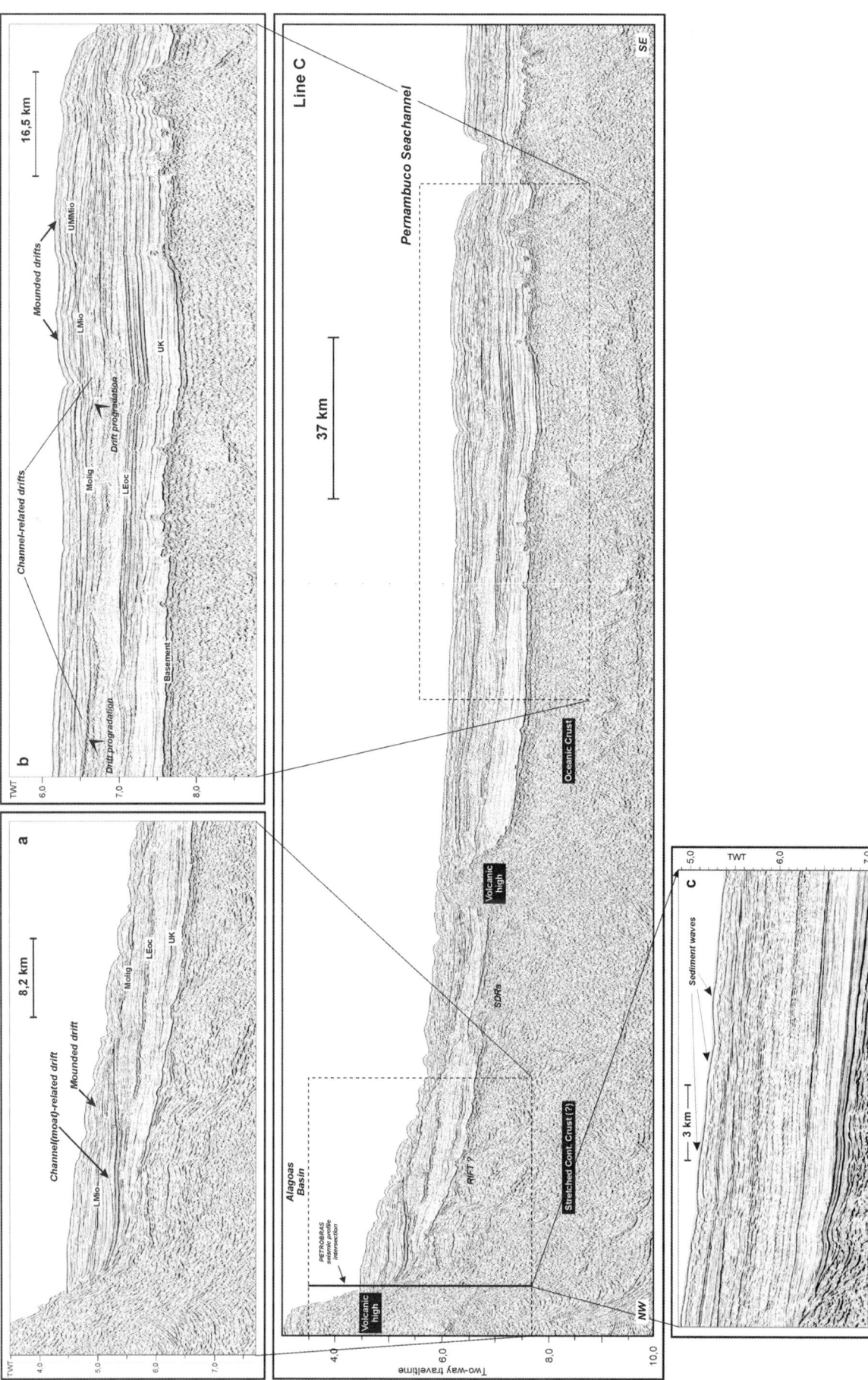

Fig. 6. Dip seismic line C, which traverses the continental margin off Alagoas basin (see Fig. 1 for location). Channel-related drifts and mounded drifts can be observed in both upper zooms. At the case **b**, arrows point out to seismic features related to drift westward progradation over the Mid-Oligocene unconformity. The lower profile (case **c**), which crosses line C, shows typical sediment waves, with an average wavelength of 3 km. The Pernambuco Seachannel, with a 470 m-deep incision at this situation, has a higher left margin, which resembles an asymmetrical levee of a turbidity channel. This similarity, however, is apparent, since the higher margin is not formed by true overbanking deposition, but mainly by drift aggradation/progradation during lateral channel migration, under the influence of Coriolis force. One can observe, at case **b**, a different seismic character for each sequence: an almost transparent aspect at the Upper Cretaceous unit; plane-parallel reflectors at the Palaeocene sequence; chaotic fluid-escape-related character at the Eocene unit; low-frequency and moderate amplitude reflectors at the post-Eocene sequences, with a predominant progradational pattern at the lower unit, which changes to a more aggradational style toward the top units. At the ocean-continent boundary region, volcanic plugs and seaward-dipping reflectors wedges (SDRs) borders a distal rift(?) sequence (UK, Upper Cretaceous unconf.; LEoc, Lower Eocene unconf.; MOlig, Mid-Oligocene unconf.; LMio, Lower Miocene unconf.; UMMio, Upper Middle Miocene unconf.).

well-developed on the left (west) margin of the Pernambuco Seachannel (Figs 5a,b and 6a,b). The best examples can be observed in the uppermost sedimentary unit (Upper-Neogene/Quaternary sequence), particularly on line C (Fig. 6), showing a low-frequency-moderate-amplitude seismic character.

Contourite sheets form extensive low-relief accumulations in regions swept by low-velocity bottom currents. In the study area, typical sheeted drifts can be found near the mouth of PSC, outside the physiographic constriction of the Bahia Seamounts (Figs 5a,b). As the flow velocity increases and decreases from time to time, there is an alternation between the deposition of mound and sheet-like contourite drifts. Both types can be covered by large-scale sediment waves (Fig. 6c).

Channel-related drifts have a remarkable seismic expression on the continental margin off NE Brazil. They are of three types: those deposited on the floor and flanks of channels (Fig. 5a) or moats (Fig. 6a); the westward prograding clinoforms that records the lateral migration of the bottom current under the influence of the Coriolis force (Figs 4a & 6b); the large fan-like deposit formed at the downcurrent exit of the Pernambuco Seachannel, here called *Pernambuco Contourite Fan*, of which only the fringe has been captured on our seismic data (Fig. 5a).

Typical examples of confined drifts can be found in small semi-isolated basins located between seamounts (Figs 4 & 5). Apart from their confinement, the seismic character of these deposits resembles those of elongate-mounded drifts, as pointed out by Faugères *et al.* (1999).

Modified drift-turbidite systems (Faugères *et al.* 1999) are expected to play an important role in the sedimentary evolution of the study area, particularly in the slope region of the Sergipe basin, where the São Francisco Deep-Sea Fan (Cainelli 1992) was deposited (see Geological and Oceanographic Setting). Line A (Fig. 4a), which crosses the edge of this submarine fan, is a suitable site for an interplay between downslope sediment gravity flows and alongslope bottom-current-controlled deposition. Similar process interaction was studied along the US Atlantic margin (Locker & Laine 1992) and on the southeastern Brazilian continental margin (Souza Cruz 1995; Massé *et al.* 1998; Viana & Faugères 1998).

Discussion

The relationship between eustasy and the formation of erosional unconformities on the continental rise is still controversial (Poag 1987; Aubry 1991; Faugères *et al.* 1999). Seismic evidence along Atlantic margins suggest that abyssal erosion is driven by the intensification of deep-water circulation during glaciation periods[2] (Tucholke & Embley 1984; Mountain & Tucholke 1985; Souza Cruz 1995; Gomes 2000). Tectonic events, however, causing regional basement uplifts, can also magnify the erosional capacity of bottom currents. Regardless of the driving mechanism, one must always consider the importance of the physiographic setting.

It can be argued that global oceanic circulation has only played a lead role in abyssal erosion process since the Eocene/Oligocene boundary, when the temperature divergence between surface and bottom waters (Kennet & Shackleton 1976) led to the formation of a cold northward-flowing AABW, around 32 Ma (Johnson 1983). Even though morphological conditions for intermediate to deep water connections between the southern ocean and the North Atlantic were in existence since the Late Cretaceous (Van Andel *et al.* 1977), there is no evidence of a vigorous oceanic circulation in pre-Oligocene times, when the water mass stratification was controlled by salinity (Thierstein 1979). So, one can differentiate a strong post-Eocene thermohaline circulation, controlled by temperature gradient, from a sluggish pre-Oligocene halothermal circulation, driven by salinity changes.

Hence, abyssal erosional surfaces of broad regional extent are less likely to be found prior to the Oligocene. However, as pointed out previously, all the depositional supersequences mapped on the continental margin off NE Brazil are bounded by unconformities (and their correlative conformities), which extend from the shelf to the lower rise and can be correlated to sea-level falls on the global cycle chart.

How can one explain, therefore, the presence of abyssal unconformities, which show evidence of erosional activity at several points of the study area (e.g. UK unconformity, Fig. 6)? A possible clue to solving this question is the different seismic character of the unconformities. The local erosional character of the older ones, contrasting with the regional widespread erosional aspect of the post-Eocene unconformities, may indicate that tectonic events have been partly responsible for pre-Oligocene unconformities, magnifying the erosional activity of the 'sluggish' bottom currents at some sites. Some tectono-eustatic influences, always considered in relative sea-level changes, can be ascribed to changes in plate motion, variation in sea-floor spreading rates and lithospheric flexure, all of these leading to intraplate stresses (Thorne & Watts 1984; Cloething 1986, among many others).

On the other hand, there is no need to invoke possible tectonic influences for the post-Eocene unconformities, since they appear more likely to have a glacio-eustatic control. As discussed earlier, the Oligocene sequence upper boundary is probably associated with the initiation of Antarctic Bottom Water flow in the Brazil Basin. From that event on, all evidence indicates the presence of intermittent deep-sea erosional activity, as a response to strong thermohaline oceanic circulation. Thus, AABW is considered responsible for both the formation of regional unconformities and the deposition of contourite drifts in the last 30 Ma, with a climatic-controlled alternation of erosional and depositional events.

Another relevant point in the present discussion is the differentiation between downslope sediment gravity flows and alongslope bottom current controlled deposition. With virtually no core data or seabed photographs to provide sedimentological information, seismic reflection data remain as the only available means for describing the deposits of the study area. Whereas along the Pernambuco Seachannel and its margins the seismic character is typical of contouritic deposition, in the tributary drainage network region towards the slope (Fig. 2), there is some uncertainty, so that the deposits can be characterized either as contourites or turbidites, or even as process interaction-related deposits (Fig. 4a).

Hence, considering the overall drainage system, including slope canyons, the network of tributary channels and the main deep-sea channel, along with the Pernambuco Contourite Fan and the São Francisco Deep-Sea Fan, this region can be regarded as a truly mixed turbidity current/contour current construction. As a matter of fact, in deep marine settings there is a turbidite-to-contourite continuum and multiple process transport, as pointed out by Stanley (1993). However, such turbidite-contourite interaction still needs better understanding, and represents an important challenge for hydrocarbon exploration in the Sergipe Basin, since it has strong implications for reservoir quality and distribution.

We are very grateful to our PETROBRAS colleagues B. S. Gomes and J. M. de Souza for their support throughout this study. Discussions and collaborative work with L. A. P. Gamboa, S. F. Santos, and R. S. F. D'Ávila were particularly helpful. We also thank D. A. V. Stow and J. Howe for critically reviewing the original manuscript. A special gratitude goes to J. C. Della Fávera, supervisor of the first author's studies at U.E.R.J.

References

AUBRY, M-P. 1991. Sequence Stratigraphy: Eustacy or Tectonic Imprint? *Journal of Geophysical Research*, **96** (B4), 6641–6679.

---

[2] Considering an eustatic curve, these deep-sea erosional events would be placed somewhere between (glacio-eustatic) sea-level falls and early sea-level rises.

BRYAN, P. C. & CHERKIS, N. Z. 1995. The Bahia Seamounts: Test of a hotspot model and a preliminary South American Late Cretaceous to Tertiary apparent polar wander path. *Tectonophysics*, **241**, 317–340.

CAINELLI, C. 1992. *Sequence stratigraphy, canyons, and gravity mass-flow deposits in the Piaçabuçu Formation, Sergipe-Alagoas basin, Brazil.* PhD Thesis The University of Texas at Austin, Austin, Texas.

CARMINATTI, M. & SCARTON, J. C. 1991. Sequence stratigraphy of the Oligocene turbidite complex of the campos Basin, offshore Brazil. *In*: WEIMER, P. & LINK, M. H. (eds) *Seismic Facies and Sedimentary Processes of Submarine Fans and Turbidite Systems*. Springer-Verlag, Berlin, 241–246.

CHERKIS, N. Z., CHAYES, D. A. & COSTA, L. C. 1992. The bathymetry and distribution of the Bahia Seamounts, Brazil Basin. *Marine Geology*, **103**, 335–347.

CHERKIS, N. Z., FLEMING, H. S. & BROZENA, J. M. 1989. *Bathymetry of the South Atlantic Ocean: 3° S to 40° S*. Geological Society America, Boulder, Map Chart Series **MC-069**.

CLOETHING, S. 1986. Intraplate stresses: A new tectonic mechanism for flutuations of relative sea level. *Geology*, Boulder, **14**, 617–620.

COUGH, S. & HESSE, R. 1976. Submarine meandering thalweg and turbidity currents flowing for 4,000 km in the Northwest Atlantic Mid-Ocean Channel, Labrador Sea. *Geology*, **4**, 529–533.

FAUGÈRES, J.-C., MÉZERAIS, M.-L. & STOW, D. A. V. 1993. Contourite drift types and their distribution in the North and South Atlantic Ocean basins. *In*: STOW, D. A. V. & FAUGÈRES, J.-C. (eds) *Contourites and Bottom Currents*. Sedimentary Geology, **82**, 189–203.

FAUGÈRES, J.-C., STOW, D. A. V., IMBERT, P. & VIANA, A. R. 1999. Seismic features diagnostic of contourite drifts. *Marine Geology*, Amsterdam, **162**(1), 1–38.

FLEMING, H. S., GORINI, M. A., PERRY, R. K., CARVALHO, J. C., GRIEP, G. H. & LEITE, O. R. 1982a. Zona de Fratura Bode Verde – uma zona de fratura dupla traçada do meio do oceano à Margem Continental Brasileira. *In*: *XXXII Congresso Brasileiro de Geologia*, Salvador, BA. Anais. . ., Salvador, SBG, Res. Com. 2.

FLEMING, H. S., PERRY, R. K., GORINI, M. A., GRIEP, G. H. & CARVALHO, J. C. 1982b. Montes Submarinos da Bahia – um novo lineamento na margem continental nordeste do Brasil. *In*: *XXXII Congresso Brasileiro de Geologia*, Salvador, BA. Anais. . ., Salvador, SBG, Res. Com. 2.

GAMBOA, L. A. P., BUFFLER, R. T. & BARKER, P. F. 1983a. Seismic stratigraphy and geologic history of the Rio Grande gap and southern Brazil Basin. *In*: BARKER, P. F. & JOHNSON, D. A. (eds) *Initial Reports. DSDP*, Washington, US Government Printing Office, **72**, 481–497.

GAMBOA, L. A. P., GANEY, P. & BUFFLER, R. T. 1983b. Erosion and progradation in the deep sea: Examples from the western South Atlantic. *In*: BALLY, A. W. (ed.) *Seismic expression of structural styles – a picture and work atlas*. AAPG, 3, 1.2.2/19–1.1.2/25.

GOMES, P. O. 2000. Distensão Crustal, Implantação de Crosta Oceânica e Aspectos Evolutivos das Zonas de Fratura e da Sedimentação no Segmento Nordeste da Margem Continental Brasileira. MSc Thesis, *Univ. do Estado do Rio de Janeiro*, Rio de Janeiro.

GOMES, P. O. & GAMBOA L. A. P. 1999. Correntes de contorno e erosão submarina no segmento nordeste da Margem Continental Brasileira. *In*: *VI Cong. Int. Soc. Bras. Geof.*, Rio de Janeiro, RJ. Abstracts (CD-rom), SBGf.

GORINI, M. A. & CARVALHO, J. C. 1984. Geologia da margem continental inferior brasileira e do fundo oceânico adjacente. *In*: *DNPM (Brasil)*. Geologia do Brasil, 473–489.

GORINI, M. A., FLEMING, H. S., PERRY, R. K., GRIEP, G. H. & CARVALHO, J. C. 1982. A influência estrutural na sedimentação da margem continental ao largo dos estados da Bahia, Sergipe e Alagoas. *In*: *XXXII Congresso Brasileiro de Geologia*, Salvador, BA. Anais. . ., Salvador, SBG, Res. Com. 2.

GORINI, M. A., FLEMING, H. S., CARVALHO, J. C., BROZENA, J., GRIEP, G. H., CHERKIS, N. S. & MELLO, S. L. M. 1984. Características morfotectónicas da Zona de Fratura dupla Bode Verde e o seu traçado em direção aos Montes Submarinos da Bahia. *In*: *XXXIII Congresso Internacional de Societie Brasileira de Geofísica*, Rio de Janeiro, RJ. Anais. . ., SBG, 4, 1.615–1.628.

HAQ, B. U., HARDENBOL, J. & VAIL, P. R. 1988. Mesozoic and Cenozoic Chronostratrigraphy and Eustatic Cycles. *In*: WILGUS, C. K. *et al.* (eds) *Sea level changes – an integrated approach*. SEPM 42, Tulsa, 71–108.

HOLLISTER, C. D. & HEEZEN, B. C. 1972. Geological effects of ocean bottom currents: western North Atlantic. *In*: GORDON, A. L. (ed.) *Studies in Physical Oceanography*, **2**. Gordon and Breach, New York, 37–66.

JOHNSON, D. A. 1983. Palaeocirculation of the southwestern Atlantic. *In*: BARKER, P. F. & JOHNSON, D. A. (eds) *Initial Reports DSDP*, Washington, US Government Printing Office, **72**, 977–994.

JOHNSON, D. A. 1985. Abyssal teleconnections II. Initiation of AABW flow in the southwestern Atlantic. *In*: HSÜ, K. J. & WEISSERT, H. J. (eds) *South Atlantic Palaeoceanography*. Cambridge University Press, 243–281.

KENNETT, J. P. & SHACKLETON, N. J. 1976. Oxygene isotope evidence for the development of the psychrosphere 38 Ma ago. *Nature*, **260**, 513–515.

LOCKER, S. D. & LAINE, E. P. 1992. Palaeogene–Neogene depositional history of the middle US Atlantic continental rise: mixed turbidite and contourite depositional systems. *Marine Geology*, **103**, 137–164.

MELLO, S. L. M. & DIAS, M. S. 1996. Magnetoestratigrafia da crosta oceânica entre as zonas de fratura de Ascensão e Bode Verde. *Revista Brasileira de Geofísica*, Rio de Janeiro, **14**(3), 237–252.

MÉZERAIS, M.-L., FAUGÈRES, J.-C., FIGUEIREDO, A. G. JR. & MASSÉ, L. 1993. Contour current accumulation off Vema Channel mouth, southern Brazil Basin: pattern of a "contourite fan". *In*: STOW, D. A. V. & FAUGÈRES, J.-C. (eds) *Contourites and Bottom Currents*. Sedimentary Geology, **82**, 173–187.

MASSÉ, L., FAUGÈRES, J.-C. & HROVATIN, V. 1998. The interplay between turbidity and contour current processes on the Columbia Channel fan drift, Southern Brazil Basin. *In*: STOW, D. A. V. & FAUGÈRES, J.-C. (eds) *Contourites, Turbidites and Process Interaction*. Sedimentary Geology, **115**, 111–132.

MOHRIAK, W. U., RABELO, J. H. L., MATOS, R. D. & BARROS, M. C. 1995. Deep seismic reflection profiling of sedimentary basins offshore Brazil: geological objectives and preliminary results in the Sergipe Basin. *Journal of Geodynamics*, **20**(4), 515–539.

MOUNTAIN, G. S. & TUCHOLKE, B. E. 1985. Mesozoic and Cenozoic Geology of the U. S. Atlantic Continental Slope and Rise. *In*: POAG, C. W. (ed.) *Geologic Evolution of the United States Atlantic Margin*. Van Nostrand Reinhold Co., New York, 293–341.

MUTTI, E., BARROS, M. C., POSSATO, S. & RUMENOS, L. R. G. 1980. Deep sea fan turbidite sediments winnowed by bottom currents in the Eocene of the Campos Basin, Brazilian offshore. *In*: *1st European Meeting of the International Sedimentologists*, **1**, 114 (abstracts).

POAG, C. W. 1987. The New Jersey transect: Stratigraphic framework and depositional history of a sediment-rich passive margin. *In*: POAG, C. W., WATTS, A. B., *et al.* (eds). *Initial Reports. DSDP*, Washington, US Government Printing Office, **95**, 763–817.

SOUZA CRUZ, C. E. 1995. Estratigrafia e sedimentação de águas profundas do Neógeno da Bacia de Campos, Estado do Rio de Janeiro, Brasil. PhD. Thesis, *Univ. Federal do Rio Grande do Sul*, Porto Alegre.

STANLEY, D. J. 1993. Model for turbidite-to-contourite continuum and multiple process transport in deep marine settings: examples in the rock record. *In*: STOW, D. A. V. & FAUGÈRES, J.-C. (eds) *Contourites and Bottom Currents*. Sedimentary Geology, **82**, 241–255.

STOW, D. A. V. & FAUGÈRES, J.-C. 1993. Contourites and Bottom Currents. *Sedimentary Geology*, Special volume **82**.

STOW, D. A. V. & FAUGÈRES, J.-C. 1998. Contourites, Turbidites and Process Interaction. *Sedimentary Geology*, Special volume **113**.

THIERSTEIN, H. R. 1979. Palaeoceanographic implications of organic carbon and carbonate distribution in Mesozoic deepsea sediments. *In*: TALWANI, M., HAY, W. W. & RYAN, W. W. F. (eds) *Deep Drilling Results in the Atlantic Ocean: Continental Margins and Palaeoenvironment*. AGU, Maurice Ewing Series, **3**, Washington, 249–274.

THORNE, J. & WATTS, A. B. 1984. Seismic reflectors and unconformities at passive continental margins. *Nature*, **311**, 365–368.

TUCHOLKE, B. E. & EMBLEY, R. W. 1984. Cenozoic regional erosion of the abyssal sea floor off South Africa. *In*: SCHLEE, J. S. (ed.) *Interregional unconformities and hydrocarbon accumulation*. American Association of Petroleum Geologists, Memoirs, **36**, 145–164.

VAN ANDEL, TJ. H., THIEDE, J., SCLATER, J. G. & HAY, W. W. 1977. Depositional history of the South Atlantic Ocean during the last 125 million years. *Journal of Geology*, **85**, 651–698.

VIANA, A. R. 1998. Le rôle et l'enregistrement des courants océaniques dan les dépôts de marges continentales: La marge du bassin sud-est Brésilen. PhD Thesis. *Université Bordeaux I*, Bordeaux.

VIANA, A. R. & FAUGÈRES, J.-C. 1998. Upper slope sand deposits: the example of Campos basin, a latest Pleistocene–Holocene record of the interaction between alongslope and downslope currents. *In*: STOKER, M. S., EVANS, D. & CRAMP, A. (eds) *Geological Processes on Continental Margin Sedimentation, Mass-Wasting and Stability*. Geological Society London, Special Publications, **129**, 287–316.

VIANA, A. R., FAUGÈRES, J.-C., KOWSMANN, R. O., LIMA, J. A. M., CADDAH, F. G. & RIZZO, J. G. 1998. Hydrology, morphology and sedimentology of the Campos continental margin, offshore Brazil. *In*: STOW, D. A. V. & FAUGÈRES, J.-C. (eds) *Contourites, Turbidites and Process Interaction*. Sedimentary Geology, **115**, 133–157.

Evidence of bottom current influence on the Neogene to Quaternary sedimentation along the northern Campos Slope, SW Atlantic Margin

ADRIANO R. VIANA, CÍZIA M. HERCOS, WALDEMAR DE ALMEIDA JR., JOSÉ LUÍS C. MAGALHÃES & SINARA B. DE ANDRADE

Petróleo Brasileiro S.A., PETROBRAS E&P, Campos Basin branch, Av. Elias Agostinho, 665 Macae, RJ 27.913-350, Brazil (e-mail: aviana@petrobras.com.br)

Abstract: Geophysical and sedimentological data indicate that bottom currents have played a fundamental role in deposition along the Campos continental slope, especially during the Neogene and Quaternary. Sediment drifts are clearly observed in 2D and 3D seismic data. The external geometry and internal reflection pattern of these drifts suggest the predominant action of northward-flowing currents (Southern Ocean Current) along the middle and lower slope (650–1200 m). Upper Quaternary sediments within that zone are composed of highly bioturbated, silty to sandy mud, with rare lamination and no other primary structures. On the upper slope, below the southward flowing Brazil Current, sub-bottom profiles and side-scan sonar records indicate the development of several bedform styles. On the uppermost slope, between 200–300 m, longitudinal lineations and transverse bedforms (2D and 3D dunes) are observed. Sediments are siliciclastic to mixed silty to muddy sand, with rare primary traction structures preserved. Downslope, from 300 to 650 m, linear crested bedforms, a few metres high (2–7 m) are developed. In this zone, an alongslope similarity in the depositional style over more than 50 km suggests bottom current control on sedimentation. The deposits are composed of 1 m of silty muds overlying a decimetric layer of silty sand that grades downslope to a highly oxidized, bioturbated, fine-grained interval. The variability of sediment accumulation rate (2 to 30 cm ka^{-1}) is related to high frequency temporal and local modifications.

Campos Basin is situated along the Brazilian south-eastern continental margin, between latitudes 21°S and 24°S and longitudes 38.5°W and 42°W. It is an important hydrocarbon-producing basin, which has witnessed a very active exploration effort over the past two decades, and one that has moved drilling into progressively deeper water. There is, therefore, a wealth of data that exists, not only from the subsurface, but also on present-day sedimentation and oceanographic conditions.

In this paper we present some geological and geophysical aspects of the contourite drifts developed in the northern portion of Campos Basin, from the upper Miocene to Recent. Data consists of side-scan sonar images, 3.5 kHz profiles, multichannel seismic profiles, and piston cores (Fig. 2). The continental shelf is dominated by carbonate and siliciclastic sands which are exported to the slope by offshelf spillover (Viana & Faugères 1998) and then pirated and redistributed along the upper slope by the western boundary Brazil Current, developing the accumulation of upper slope sands, which are presented separately in this volume (Viana *et al.* 2002). Downslope the sands grade to hemipelagic mud. Over the lower slope and the São Paulo Plateau a pelagic marl covers slope-derived debris flow and slump deposits, and rare late Pleistocene turbidites. Holocene sands are also observed, related to highstand sediment-gravity flows through some of the major submarine canyons (São Tomé canyon, Viana 1998, and Almirante Camara canyon, Machado *et al.* 1998).

Geological and oceanographic setting

Campos Basin is located along a divergent passive margin, which results from the opening of the South Atlantic initiated during the late Jurassic. The basin fill comprises six megasequences: (1) A continental pre-rift megasequence (late Jurassic to early Neocomian); (2) A syn-rift megasequence (early Neocomian to early Aptian); (3) A transitional evaporitic megasequence (middle Aptian to early Albian); (4) A shallow carbonate platform megasequence (early to middle Albian); (5) A marine transgressive megasequence (late Albian to early Tertiary); (6) A marine regressive megasequence (early Tertiary to Recent). Regional unconformities bound these megasequences, apart from the gradational transition from the evaporitic to the shallow water carbonates (Guardado *et al.* 1990). A recent review of the evolution of Brazilian marginal basins is presented by Bruhn (1998). Salt tectonics is the most important structural control on the development of topographic highs and lows which guided the trends of sediment by-pass and accumulation. Sediment starvation due to a high subsidence rate and low sediment influx characterises the marine transgressive megasequence, where several oil-bearing turbidite deposits are found (Guardado *et al.* 1990; Bruhn 1998). The marine regressive megasequence in Campos Basin attains a maximum thickness of sediments in excess of 3600 m, deposited during the Miocene. This includes the most prolific oil-bearing deposits of the Brazilian continental margin.

The Campos margin is characterized by a convex coastline and a protruding shelfbreak following the outline of Cape São Tomé. There is a wide shelf platform (around 100 km average width), with shelfbreak depths ranging from 80 m in the north to 210 m in the south. The shelfbreak is commonly marked by a conspicuous escarpment with an erosional terrace (3 to 15 km wide) at its foot between 200 and 350 m water depth. This has been related to seafloor abrasion by slope boundary currents (Viana & Faugères 1998). The continental slope is carved by numerous submarine canyons and mass movement scars connecting the slope to the São Paulo Plateau, a relatively low gradient area marked by an irregular topography. Such topography is the ultimate result of the interaction between the halokinesis of the Aptian–Albian evaporitic layer and the shallow water-derived sediment load. In this paper we will focus on the zone to the north of the Itapemirim canyon, where the Almirante Camara, the Grussaí and the Tabajara canyons are the most prominent features (Fig.1).

Several changes in global tectonics and climate have played an important role in oceanic circulation along the East Brazilian margin with a consequent impact on the sedimentation. The most important of these are: (1) during the late Albian and Cenomanian, the establishment of a subtropical high-pressure cell over the narrow and still restricted South Atlantic (Parrish & Curtis 1982); (2) in the late Turonian, the establishment of a full connection between the South Atlantic, the North Atlantic and Indian Oceans, thus leading to true oceanic conditions (Dias-Brito 1987); (3) in the middle Eocene, the onset of the regressive megasequence; (4) in the Eo-Oligocene, the intensification of AABW deep-water thermohaline circulation through the Brazil Basin

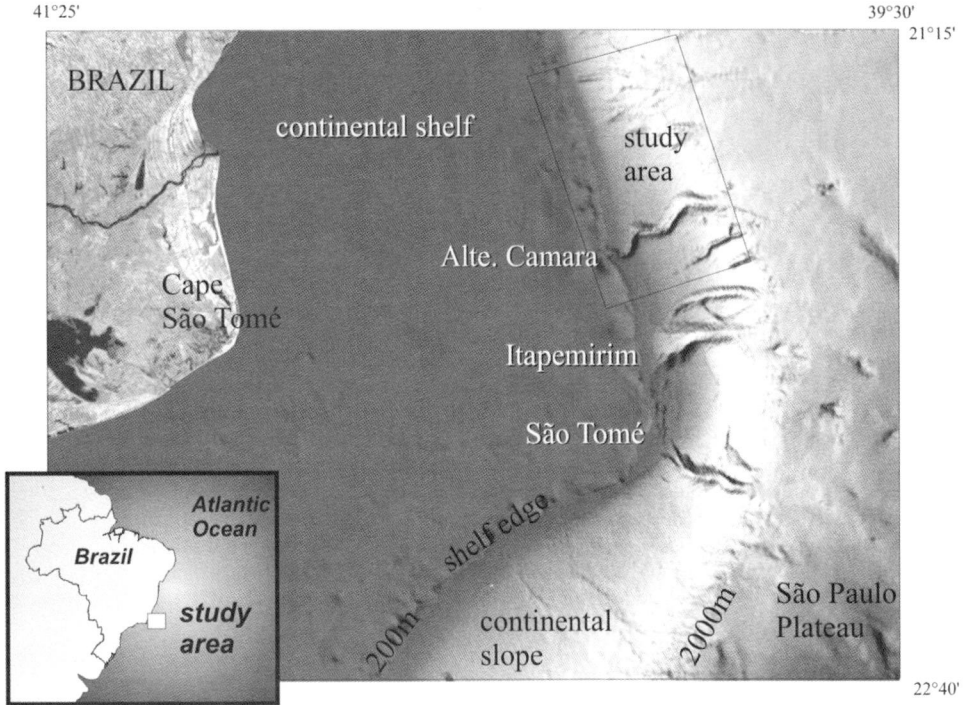

Fig. 1. General physiographical map of the study area.

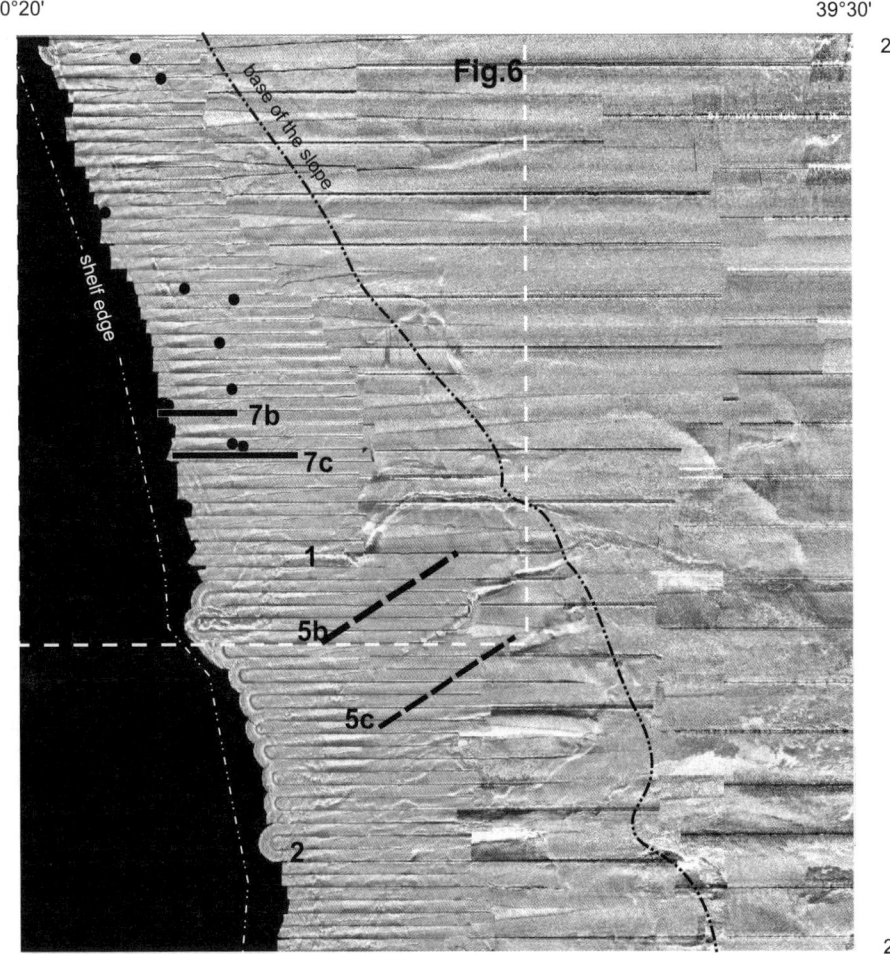

Fig. 2. Side-scan sonar mosaic of the study area. Dashed lines indicate the shelfbreak and the base of the slope positions. 1 and 2 are the submarine canyons showed in Figure 3, respectively 1, Almirante Camara Canyon, and 2, Itapemirim Canyon. Solid circles indicate the location of the studied piston cores. Bold dashed lines indicate the location of the seismic profiles 5b and 5c, and single bold lines 7b and 7c indicate the 3.5 kHz profiles presented in this article. Dashed inset indicates the area of Figure 6.

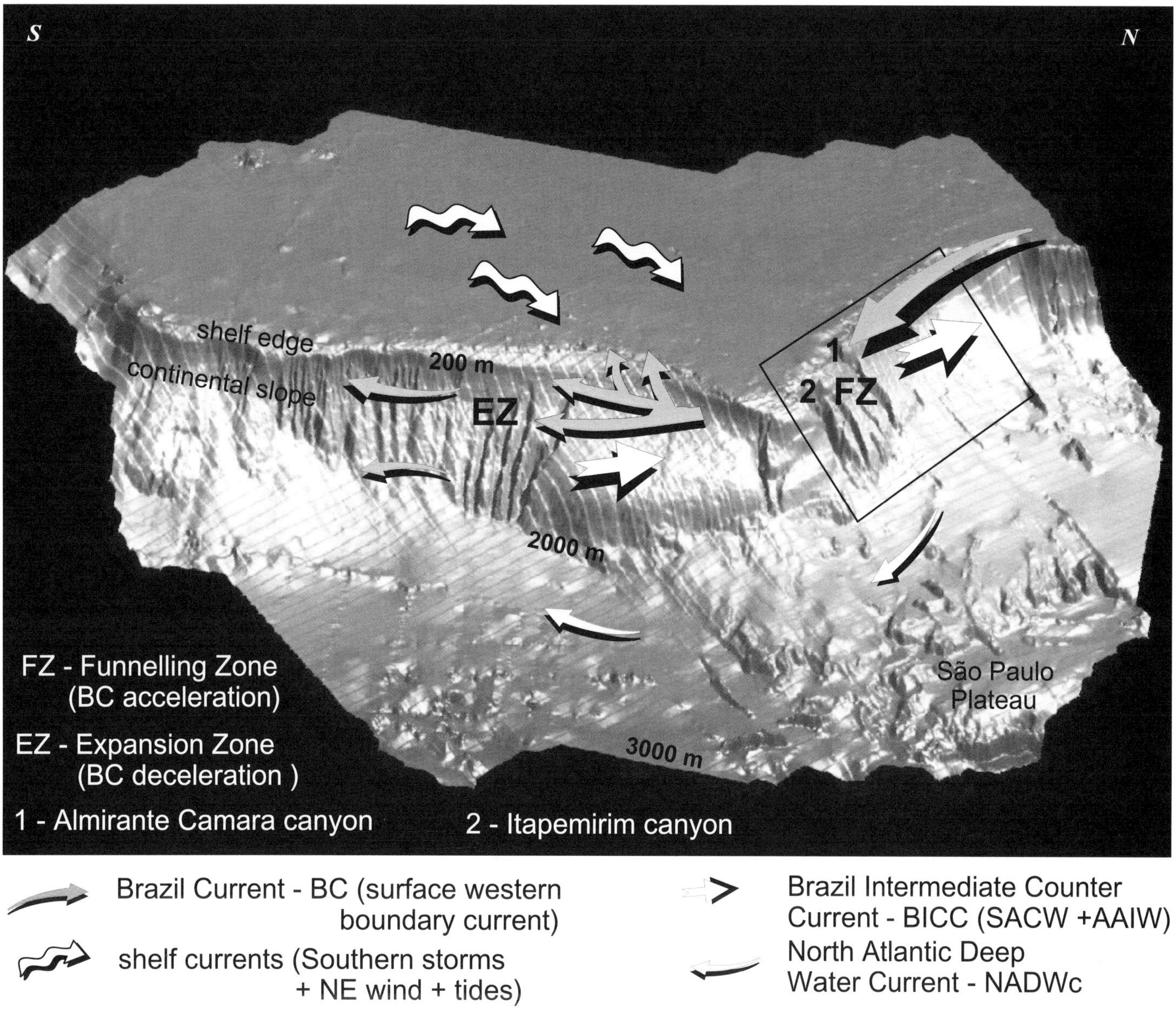

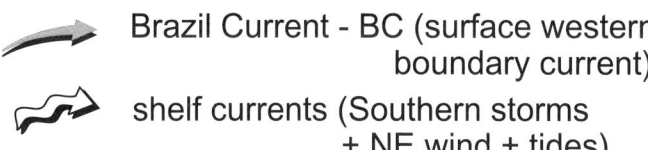

Fig. 3. Shaded 3D bathymetric view of Campos Basin slope. Arrows represent the projection of the main currents. Inset indicates the study area presented in Figure 4.

(Johnson 1985); (5) in the late Oligocene, the opening of the Drake Passage, which established the Antarctic Circumpolar Current and the general circulation pattern of the South Atlantic as it is today (Miller & Fairbanks 1985; Miller *et al.* 1991); (6) in the middle Miocene (19–16.5 Ma BP), an intense erosive activity of bottom waters in the Campos Basin (Souza Cruz, 1995); (7) in the middle/upper Miocene, glaciation of eastern Antarctica can be linked with further strong regional erosion along the Campos margin (Viana *et al.* 1990).

The present-day general circulation in the study area (Figs 2, 3 and 4) is represented by the surface Brazil Current, a southward flowing western boundary current, which attains a velocity of over 1.2 m s^{-1} near to the bottom, less than 30 km downstream from the study area. Below this, the Brazil Intermediate Counter-Current occupies water depths of 400–1000 m and carries to the north the South Atlantic Central Water and the Antarctic Intermediate Water, attaining peak velocities in excess of 30 cm s^{-1} over the middle slope. The lower slope and São Paulo Plateau are covered by North Atlantic Deep Water, carried southward by a sluggish current. In the most distal portion of the basin, in water depths greater than 3500 m, lies Antarctic Bottom Water.

Seismic and stratigraphic record

The construction of the continental slope on the northern Campos Basin continental margin from the Neogene to the Holocene has been characterized by the activity of bottom currents. The main points that support this contention are:

(1) Slope drifts and bottom current related deposits have developed in Campos Basin throughout the Neogene. Earlier activity of bottom currents reworking slope sediments, at least since Eocene time, is reported by Mutti *et al.* (1980), Souza Cruz (1995) and Machado *et al.* (2000) among others.
(2) The onset of particularly strong bottom current activity along the slope occurred in the middle/late Miocene (Serravalian/Tortonian, *c.* 11 Ma), following a major sea-level fall marked by a regional unconformity (Grey Marker, Viana *et al.* 1990).
(3) From the late Miocene to Holocene a southerly derived bottom current regime prevailed on the middle/lower slope, as evidenced by upslope migrating fields of sediment waves and the development of elongated, separated drifts.

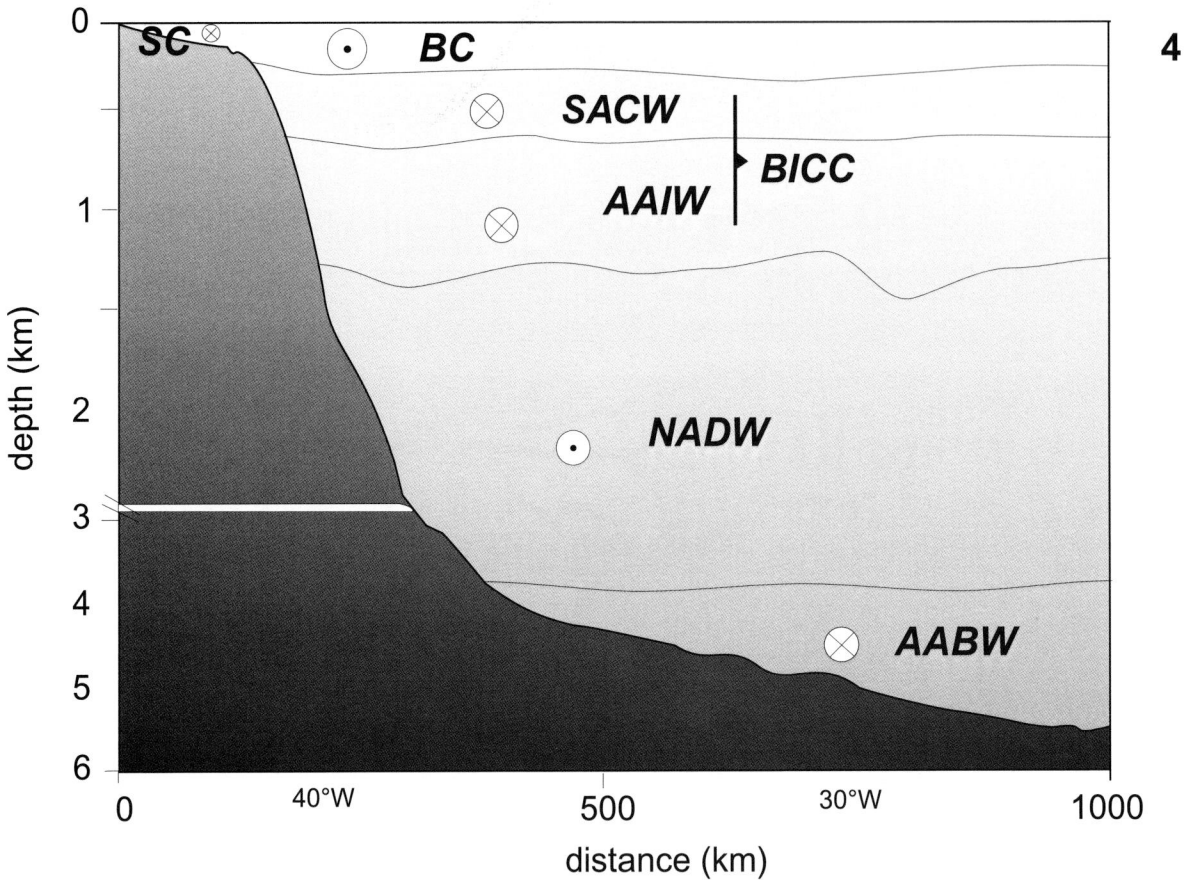

Fig. 4. Schematic dip section with water masses and current distribution. SC, shelf currents; BC, Brazil Current; BICC, Brazil Intermediate Counter-Current; SACW, South Atlantic Central Water; AAIW, Antarctic Intermediate Water; NADW, North Atlantic Deep Water; AABW, Antarctic Bottom Water.

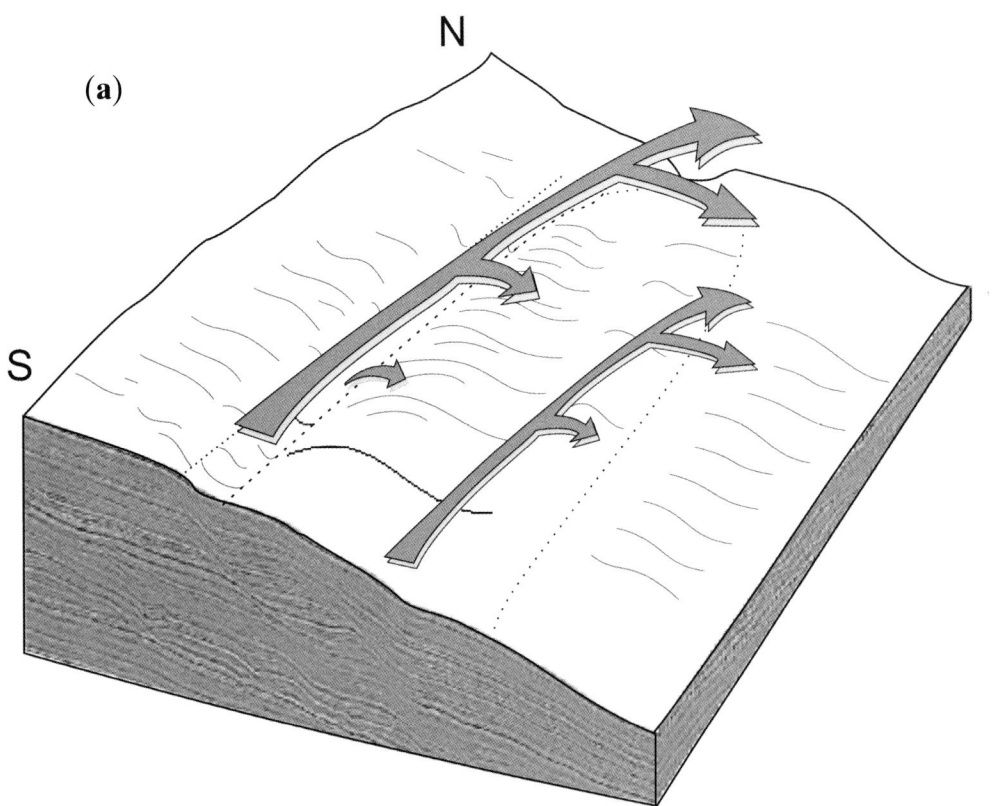

Fig. 5. (**a**) Block diagram illustrating the action of southerly-derived contour currents in sculpting the slope and developing upslope migrating sediment waves and alongslope drifts. The Coriolis force constrains the current towards the left, drifting sediments towards the right (downslope); the basal erosional surface migrates upslope, indicating the general migration trend of the system. (**b**) Multichannel dip seismic profile illustrating well developed upslope migrating sediment waves related to the passage of a southern-derived contour current. (**c**) Multichannel dip seismic profile presenting the result of the action of both northward and southward flowing contour currents; northward currents are responsible for the development of sediment waves and a separated drift and southward currents develop incisions resembling furrows. Location of seismic profiles presented in Figure 2.

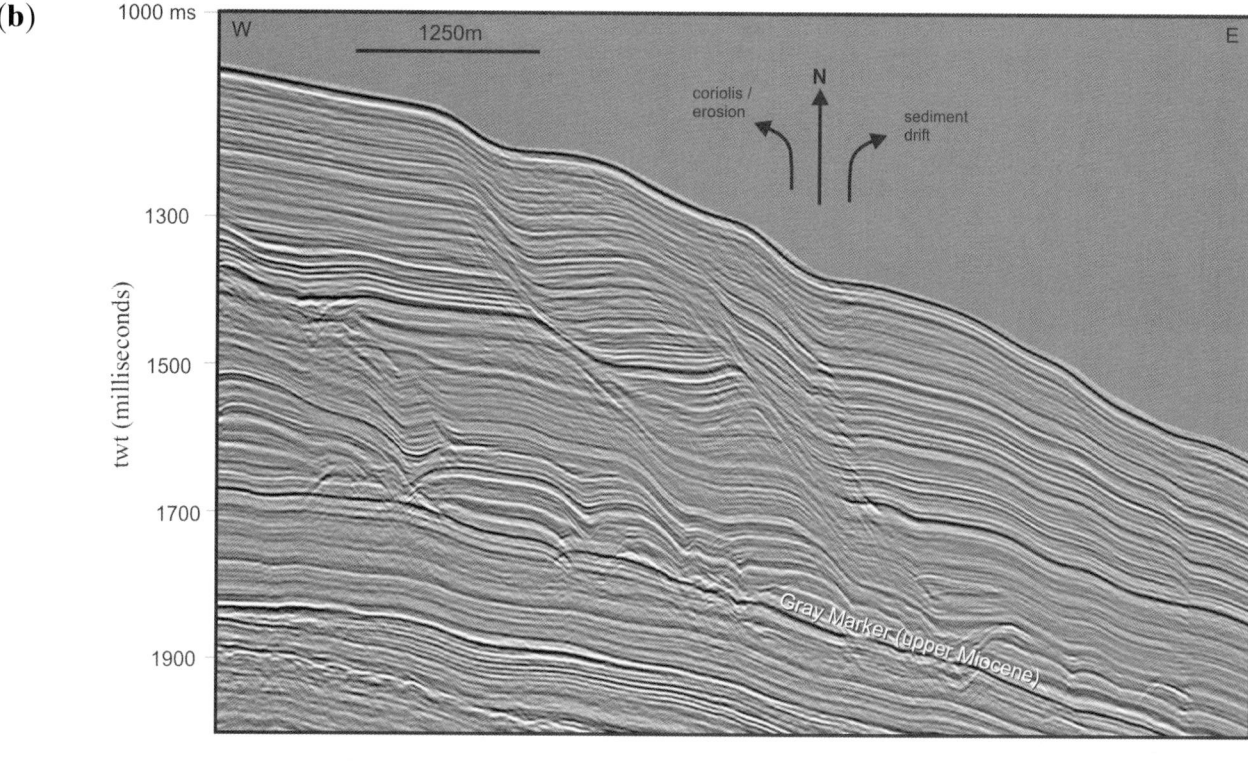

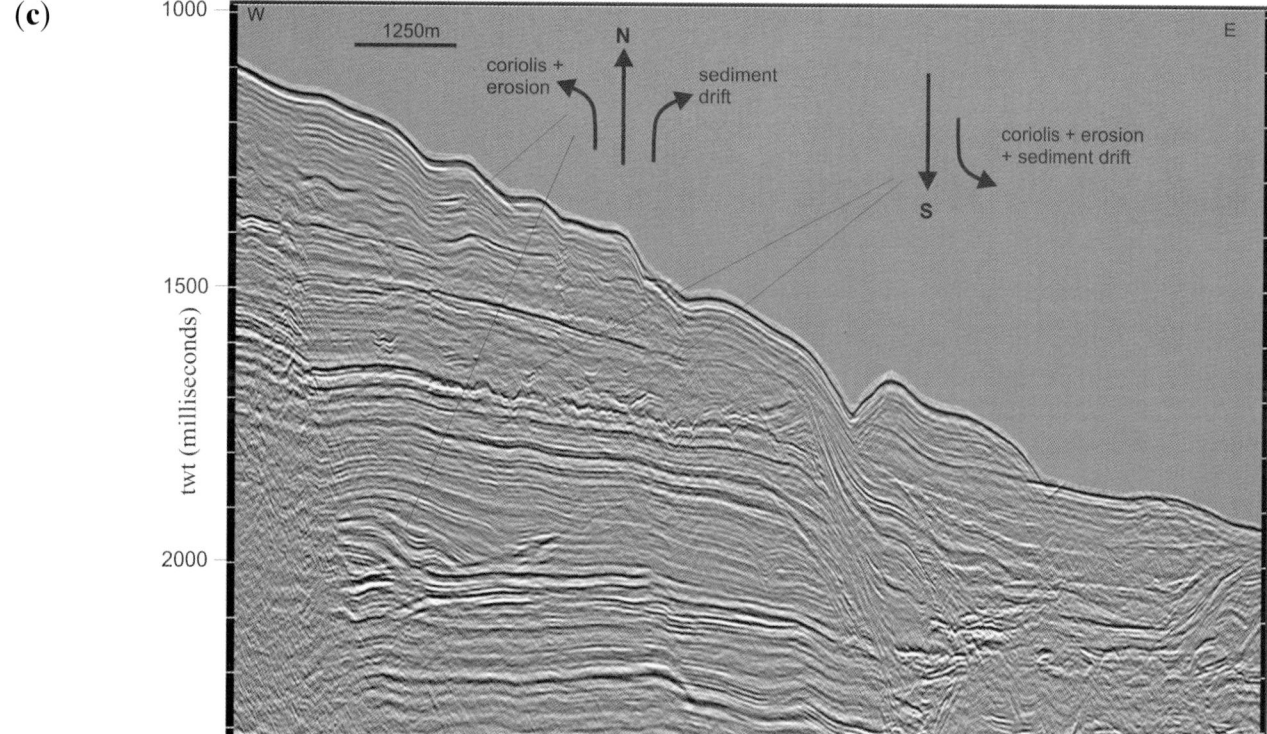

Figures 5a, b and c represent the proposed scheme for explaining some of the seismic features observed, where a dominant northward current in the middle/lower slope region has sculpted the upslope migrating sediment waves. Erosional features present in Figure 5c are mainly related to a southward flowing current. The non-confinement of the flow due to the Coriolis force and the slope gradient favour a downslope drift of the current axis, influenced by an important gravitational component. This is believed to have resulted in the carving of erosional features oblique to the slope.

Sedimentary facies

Analysis of piston cores retrieved to the north of Almirante Camara canyon provides evidence of an alongslope trend of correlatable sedimentary facies (Fig. 6). Based on these cores three zones can be identified whose sedimentation styles seem to respond to a depositional mechanism most probably associated with a marked slope boundary current.

Zone 1 (150–200 m). Cores t1 and t2 are located near the top of the slope, on a terrace cut into the foot of the shelfbreak

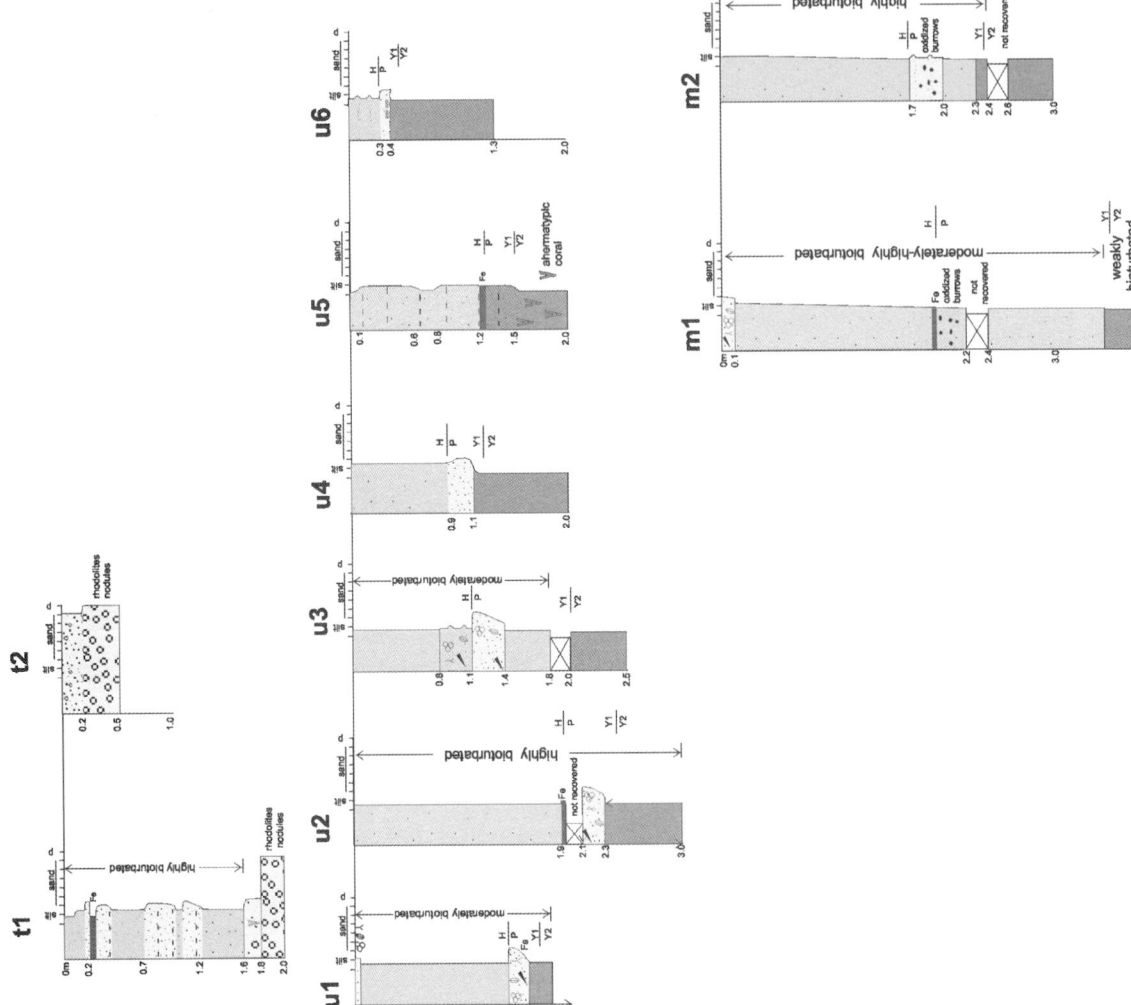
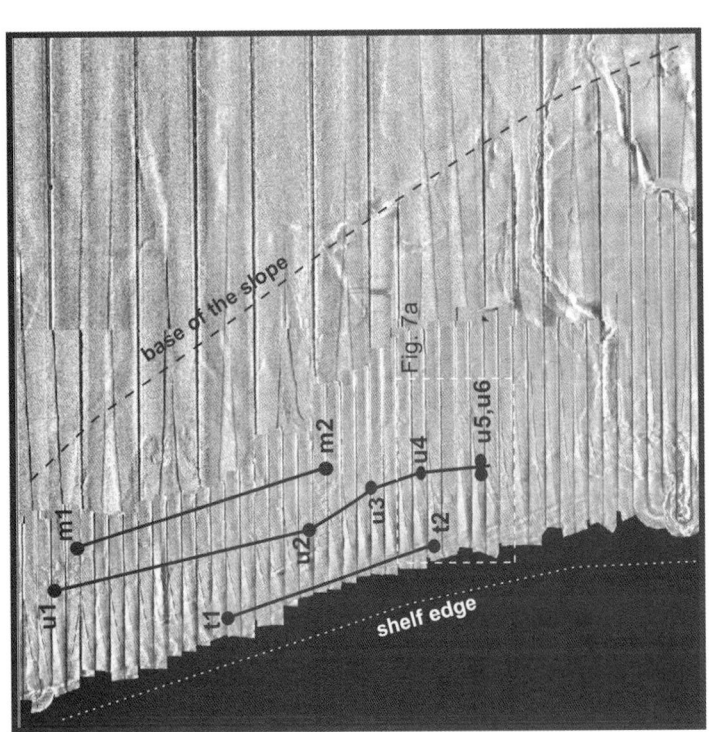

Fig. 6. Side-scan sonar mosaic showing the distribution of the studied cores. Three morphological zones are recognized parallel to the isobaths: uppermost slope terrace (t), upper slope (u) and middle slope (m). Dashed inset indicates the area of Figure 7a.

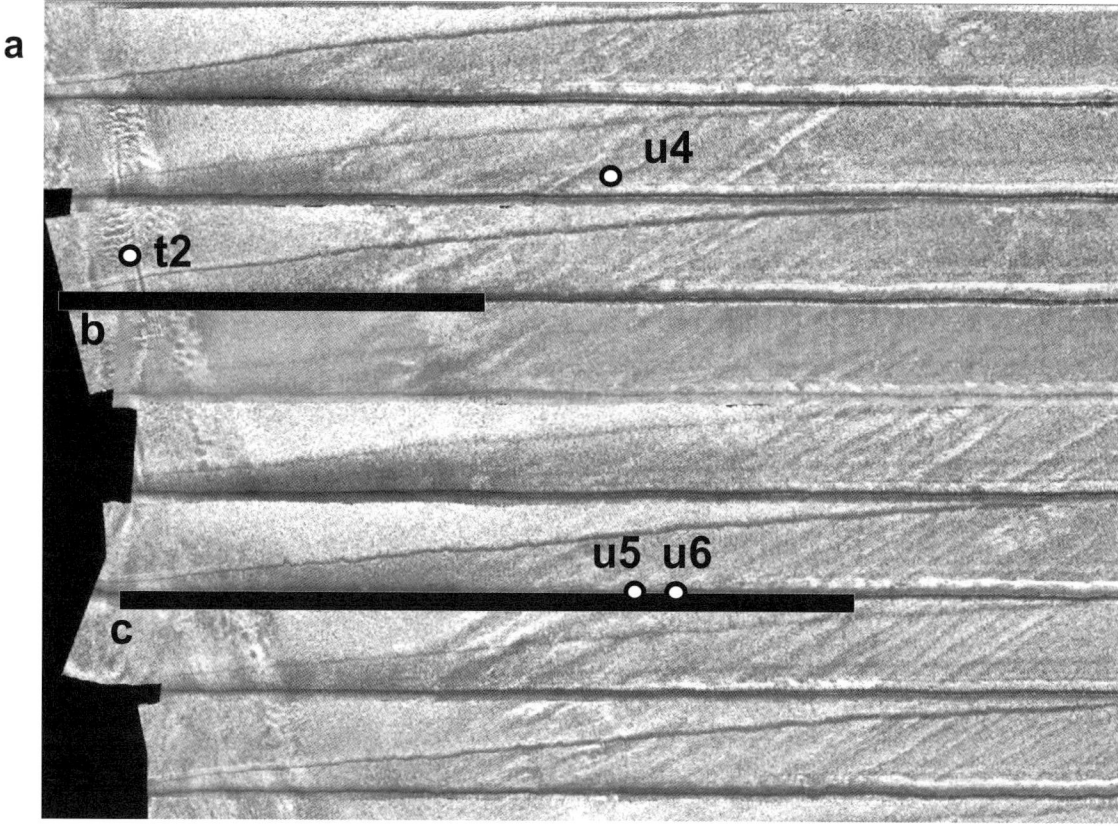

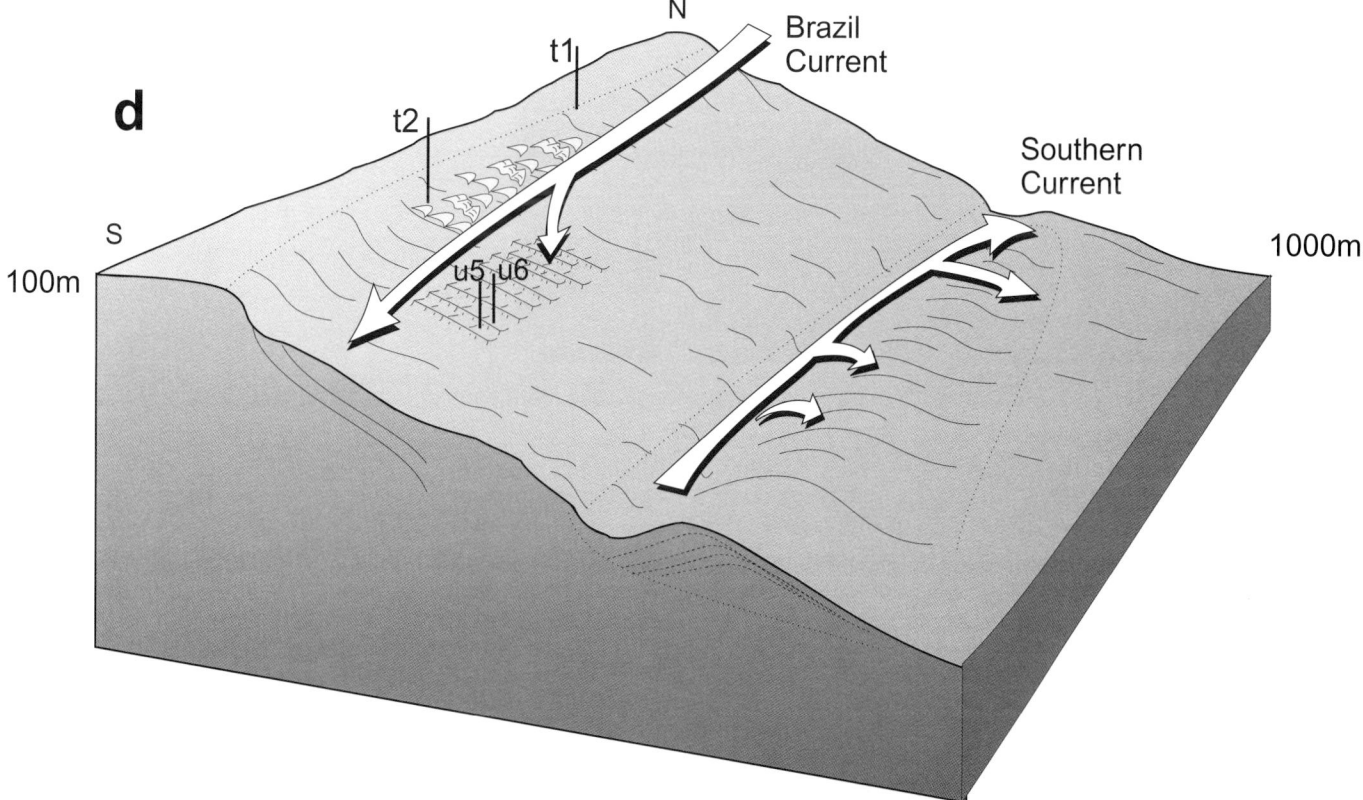

Fig. 7. Detailed side-scan sonar mosaic (**a**) showing the upper slope bedforms related to the Brazil Current action and the location of the 3.5kHz records (**b**, **c**) and piston cores. t2 is located in the region of the 3D dunes near the top of the slope where a narrow terrace is developed. u4, u5, and u6 were retrieved over the southward migrating 2D dunes field. (**d**) Block diagram represents a general picture of the bedform development as a result of different bottom currents in the study area.

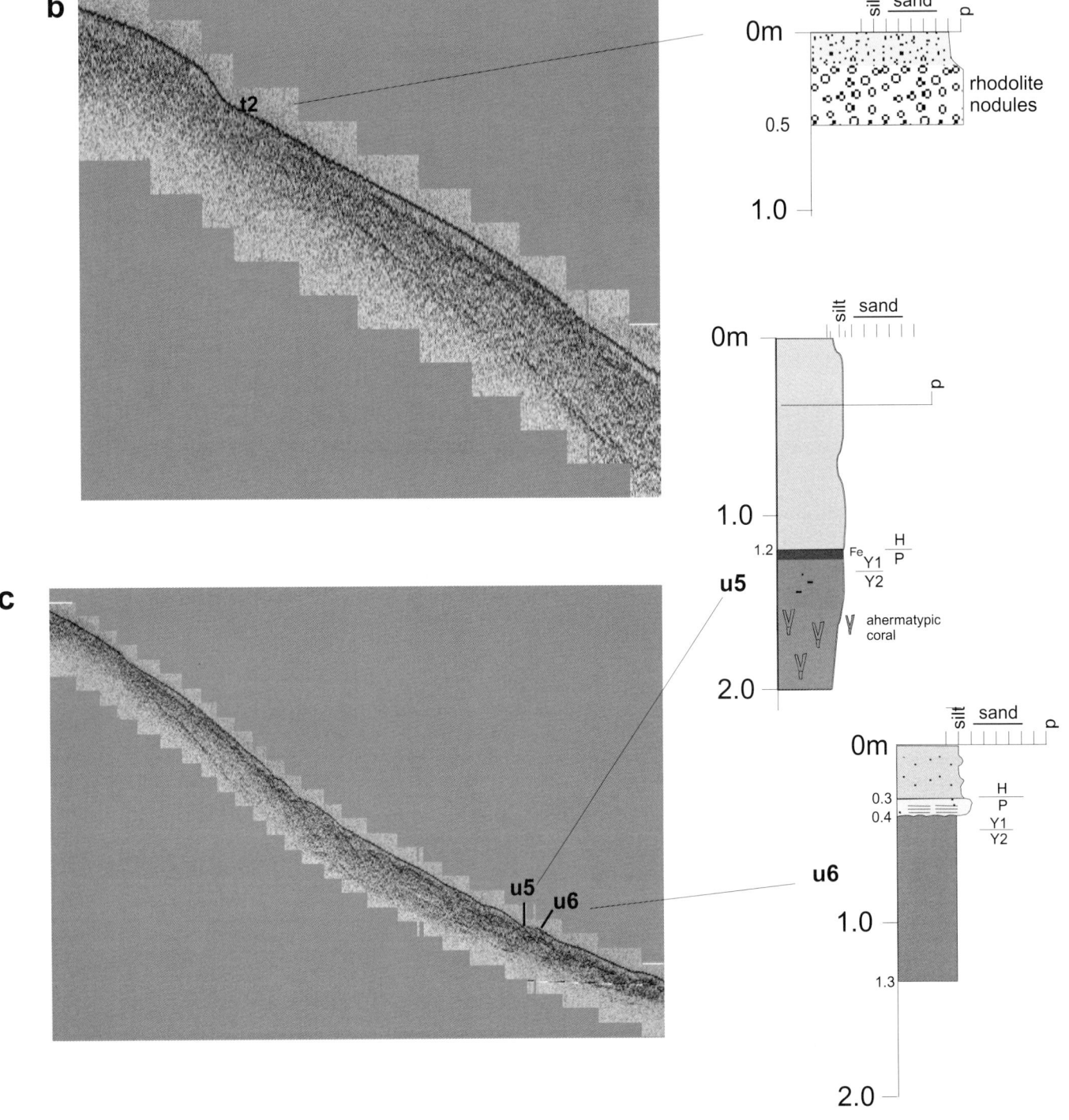

Fig. 7(b, c).

escarpment. Both present a basal layer of rhodolite nodules (gravel to pebble-sized), occasionally associated with mud and bioclastic sand. These are overlain by very fine-grained (t1) to medium/coarse-grained (t2) bioclastic sand. The upper bed of t2 comprises closely interbedded sandy mud and muddy sand layers with poorly defined contacts and no primary structure preserved due to intense bioturbation. A marly layer caps the sandy bed. Correlation with cores retrieved 50 km to the south and discussed in Viana & Faugères (1998) suggest that the top of the rhodolite layer represents the Pleistocene/Holocene passage, with different Holocene accumulation rates for t1 and t2.

Zone 2 (450–650 m). Cores u2, u3, u4, and u5 were retrieved from a little deeper in the upper slope region. They present a basal section composed of hemipelagic greenish gray silty mud. The greenish mud is overlain by a 10 to 20 cm thick moderately to highly bioturbated sand layer, composed of very fine-grained mixed-composition sand (quartz, mica, glauconite and bioclasts) with a variable mud content. The contacts between mud and sand layers are typically masked by bioturbation but, where preserved, they are sharp or erosive. The uppermost layer present in the cores of this zone is marked by a metre thick olive gray hemipelagic mud developed during the Holocene (Vicalvi 1999). Planktonic foraminifera biostratigraphic analysis performed by Vicalvi (1999) indicates that the passage from the Y2 to the Y1 zone, marked by the disappearance of the *Puleniatina obliquiloculata* (40/42 Ka BP) occurs always near the top of the greenish mud. The sand layer occurs at the very end of the Pleistocene

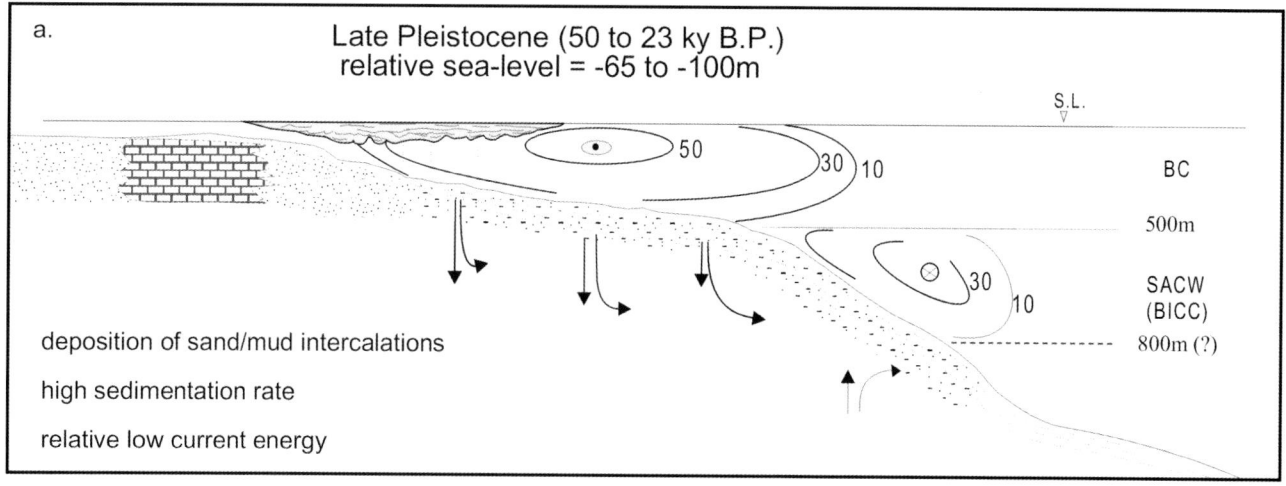

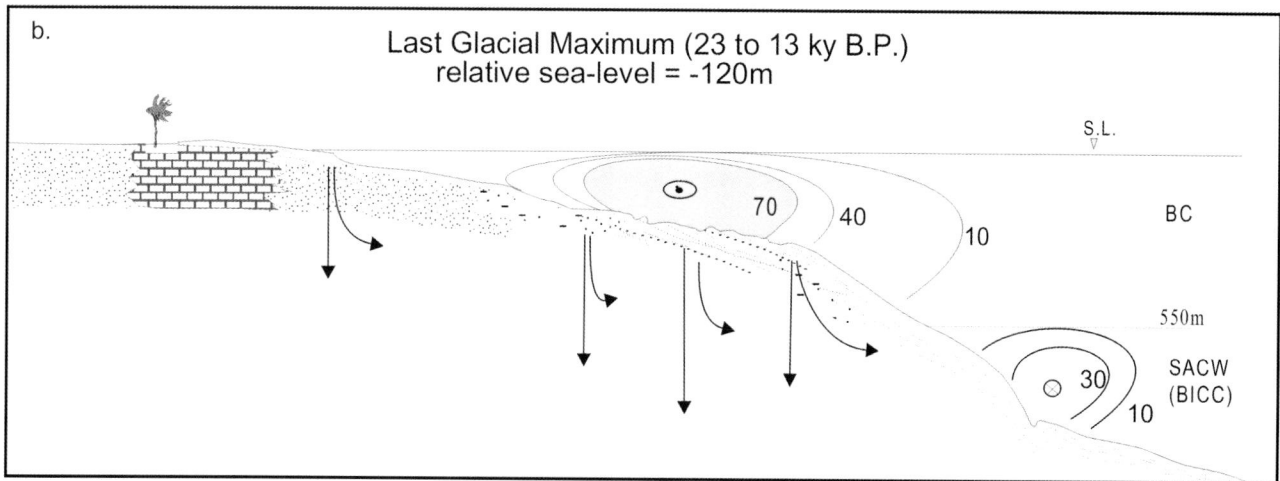

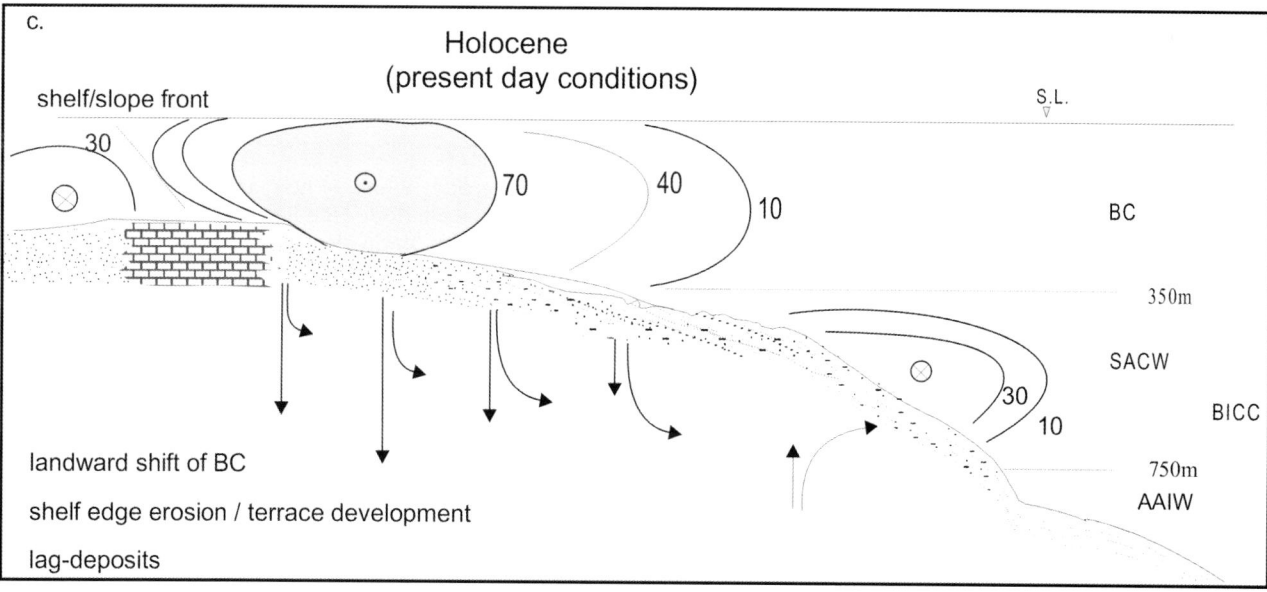

Fig. 8. Schematic reconstruction of latest Pleistocene to Recent Campos surface and intermediate slope circulation, and bottom current impact on sedimentation. Current speeds are hypothetical. Arrows indicate the presumed direction and intensity of sediment transport (from Viana & Faugères 1998).

corresponding to sediments deposited during the Last Glacial Maximum.

Zone 3 (750–850 m). In the cores m1 and m2, retrieved respectively in water depths of 717 and 825 m, the same depositional pattern as in the upper slope cores (Zone 2) is observed, although the sediments are generally finer grained. At the same depth, where the sand layer is observed in the upper slope cores, there is a thin interval of light-coloured sandy mud, extremely bioturbated and with oxidized burrows. This marks the boundary of the Pleistocene and Holocene in the middle slope. The Y1/Y2 boundary (40/42 Ka BP) is also observed near the top of a weakly

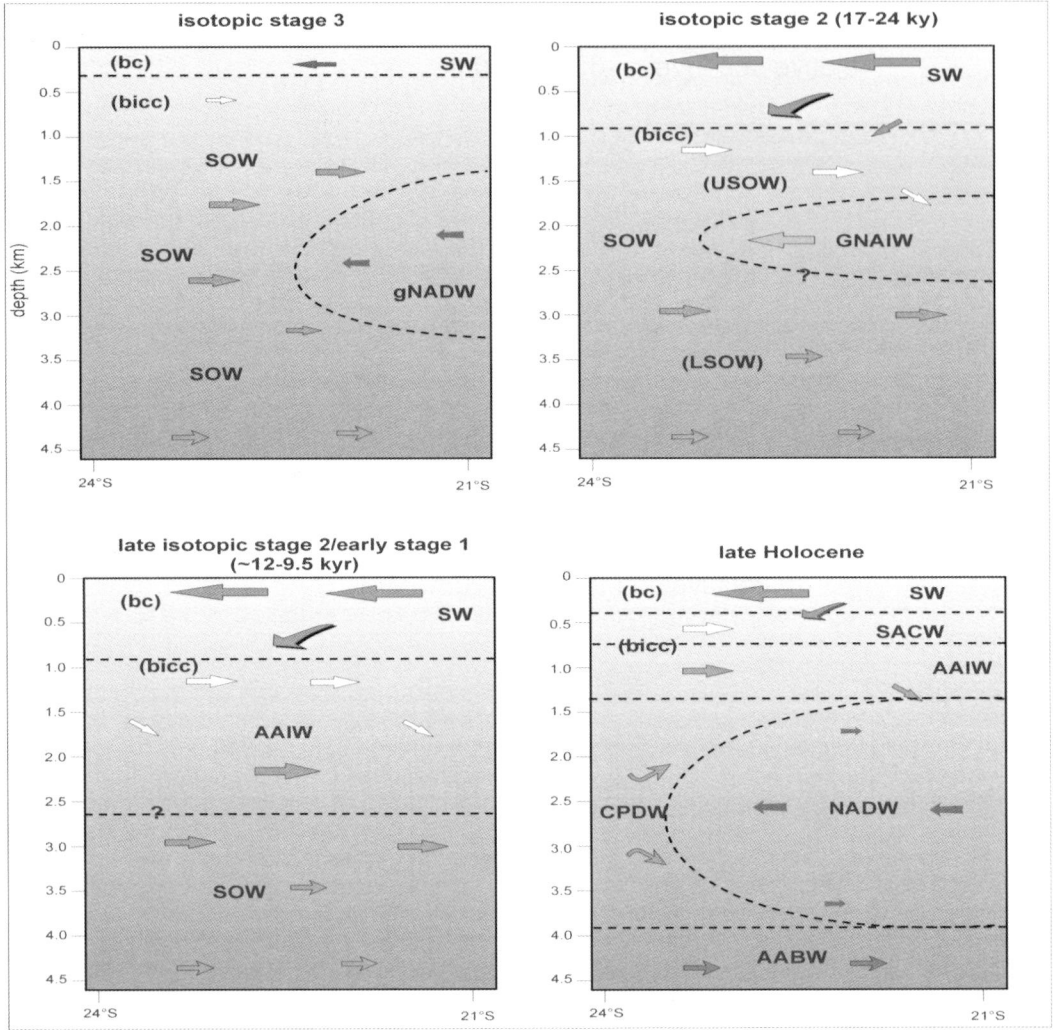

Fig. 9. Diagram representing the main water masses present in Campos Basin during the latest Quaternary as discussed in the text. Arrows indicate relative intensity and direction of flow.

to moderately bioturbated greenish gray mud layer. A general coarsening upward pattern (mud to coarse silt/very fine-grained sand) is observed in the upper olive grey mud layer.

The different surficial facies in zones 1, 2 and 3 can be related to surface circulation of the northerly derived western boundary Brazil Current (BC) that impinges on the sea-floor on the upper slope. This has led to development of different bedforms in response to both oscillation in the position and intensity of the flow, and the nature of the surficial sediment (Fig. 7a). 3D dunes are observed on the present uppermost slope, associated with coarse- to fine-grained sand (t1) (Fig. 7b), and linear-crested bedforms, up to 6 m high, associated with silt to very fine-grained sand (u4, u5) (Fig. 7c) are noted on the upper/middle slope. Truncation of reflectors against the seafloor on the uppermost slope, observed in the 3.5 kHz profiles, is attributed to the erosive action of the Brazil Current and marked in the sonar image as lineations in the sector of the 3D bedforms, represented in the block diagram of Figure 7d.

Discussion

The Neogene to Quaternary sedimentation of the Campos Basin slope is ultimately controlled by the interaction between slope physiography, bottom circulation and sediment supply. Seismic and sedimentary facies, coupled with seafloor images indicate the existence of permanent activity in the bottom currents in the study area. Recently Viana (1998) and Viana & Faugères (1998) proposed a scenario for the last 50 000 years of palaeocirculation on the Campos Basin slope (Figs 8 and 9). That reconstruction was based on the analysis of several piston cores, taking into account sediment texture and composition, sedimentary internal structures, isotopic and biostratigraphic analysis and ^{14}C dating.

The main aspects of the palaeocirculation scheme, summarized in Figures 8 and 9, are:

(1) During the glacial stage 3, between 24 ka and 50 ka BP, corresponding to the last sea level fall, circulation intensity was weak. The palaeo-Brazil Current was only capable of transporting southward fine river-derived suspended sediments from the Paraiba do Sul river, the mouth of which was close to the shelf edge. The middle slope was swept by the upper branch of a northward flowing Southern Ocean Water (SOW) and the lower slope, by a southward flowing glacial North Atlantic Deep Water (gNADW). Periods of alternation between the SOW and the gNADW on the lower slope are suggested (Viana 1998). Coarse-grained sediment deposition in deep-waters is mainly related to gravity flows. Fine-grained contourites/hemipelagites are the dominant deposits as the result of the interaction between surface and intermediate currents and river sediment discharge over the slope. High sedimentation rates are observed related to that period of turbid, sediment-rich slope water. The southward transport of suspended material induced a higher sedimentation rate to the south of São Tomé canyon (78 cm Ka^{-1}) even though the main supply sources were located tens of kilometres to the

north where the sedimentation rate is 34cm Ka^{-1}. Evidence for a rather weak superficial circulation during this period includes: the presence of fine-grained sediments on the uppermost slope, the reduced quantity of sandy sediments which occur in a continuum with muddy sediments, and the absence of widespread erosional scours. During that time, the main river, Paraiba do Sul, seems to have been connected directly to the head of Almirante Camara canyon (Kowsmann & Costa 1979). In such a scenario, most of the coarse-grained river-derived sediment should be transported directly to the deep sea through the canyon, as confirmed by cores and side-scan sonar images (Machado et al. 1998).

(2) During the Last Glacial Maximum/maximum lowstand, between 24 ka and 13 ka, the BC core shifted to its deepest position, between the 350 m and 450 m isobaths, and attained its greatest speeds. Such an increase in the current intensity and depth of action was responsible for the seafloor erosion and the shaping of upper slope terraces and large linear-crested bedforms between 350 and 750 m water depth. Up to 30 m of sediment column were removed (Viana et al. 1998). Over the middle slope an active/intense upper branch of the Southern Ocean Water (SOW) was present, reworking bottom sediments and constructing elongated separated drifts and upslope migrating sediment waves. The lower slope was temporarily occupied by the southernmost (yet reported) position of the southward extension of the Glacial North Atlantic Intermediate Water (GNAIW), a relatively nutrient-depleted, slow-flowing water. The major part of the coarse-grained sediment was transported southward as bed-load by the BC, feeding into canyon heads and developing upper slope sandy contourite deposits (upslope sands model of Viana & Faugères 1998; Viana et al. 2002).

(3) Near the Pleistocene/Holocene boundary, the GNAIW was replaced by a strong intermediate depth circulation (upper SOW/Antarctic Intermediate Water AAIW); a coarse-grained centimetre-thick layer is locally observed (cores u_n) grading downslope to a highly oxidized fine-grained, burrowed layer. This implies a strong increase in the energy of the environment at that time, as a result of a drastic increase in the surface and intermediate circulation.

(4) At the beginning of Holocene the re-appearance of NADW due to major ice melting in the northern hemisphere is inferred on the lower slope, below which a Holocene pelagic marl (ooze) was deposited. The southward intrusion of NADW confined the AAIW to the middle slope, where a condensed section was developed. During the Holocene rise of sea-level, the strong BC core shifted landward. This shift constrained the BC core against the foot of the shelf edge escarpment; muddy sand locally accumulated as a consequence of weakening in the longitudinal transport. 3D bedforms are locally developed at the uppermost slope as a response to local variation in sediment texture and bottom current intensity. A metre thick sediment drape is deposited as the result of highstand current activity over the upper/middle slope.

Thus, the BC acceleration began during the Last Glacial Maximum (18 ka) continuing up to the middle Holocene. The identification of a period of strong bottom current activity indicates that vigorous longitudinal transport processes have taken place in Campos Basin, constructing sedimentary bodies, eroding the seafloor and reworking previously deposited gravity sediments. Distinction of the water masses present at each episode is fundamental to the comprehension of the sediment styles observed on the Campos slope and in the forecasting of depositional geometry and facies distribution of the resultant deposits. However, in spite of the evidence of the impact of bottom currents on sedimentation, some of the figures here presented such as slope sediment waves can alternatively be interpreted as creep-derived features or even as the result of progressive upslope shallow-seated faulting. Numerical and physical simulations can reproduce both processes – bottom current and gravity flows (Syvitsky, pers. comm.) and distinction is still dependent upon the approach and culture of each interpreter as well as on the volume and quantity of available data.

References

BRUHN, C. H. L. 1998. *Petroleum Geology of rift and passive margin turbidite systems: Brazilian and worldwide examples, Part 2: Deep-water reservoirs from the eastern Brazilian rift and passive margin basins.* AAPG International Conference & Exhibition Course 6, Rio de Janeiro.

DIAS-BRITO, D. 1987. A Bacia de Campos no Mesocretáceo: uma contribuição à paleoceanografia do Atlântico Sul primitivo. *Revista Brasileira de Geociências*, **17**, 162–167.

GUARDADO, L. R., GAMBOA, L. A. P. & LUCCHESI, C. F. 1990. Petroleum Geology of the Campos Basin, Brazil: a model for a producing Atlantic-type basin. *In*: EDWARDS, J. D. & SANTOGROSSI, P. A. (eds) *Divergent/passive margin basins.* AAPG Memoir, **48**, 3–79.

JOHNSON, D. A. 1985. Abyssal teleconnections II. Initiation of Antarctic bottom water flow in the southwestern Altantic. *In*: HSÜ, K. J. & WEISSERT, H. J. (eds) *South Atlantic Paleoceanography.* Cambridge University Press, Cambridge, 243–281.

KOWSMANN, R. O. & COSTA, M. P. A. 1979. *Sedimentação Quaternária da Margem Continental Brasileira e das Areas Adjacentes.* Projeto REMAC, Rio de Janeiro, PETROBRAS, **8**.

MACHADO, L. C. R., KOWSMANN, R. O., ALMEIDA JR., W., MURAKAMI, C. Y., SCHREINER, S., MILLER, D. J. & PIAUILINO, P. O. V. 1998. *Modern turbidite system in the Campos basin: key to reservoir heterogeneities.* AAPG International Conference & Exhibition, Rio de Janeiro.

MACHADO, L. C. R., VIANA, A. R. & WINTER, W. R. 2000. Muddy and Sandy Contourites inside Turbidite Reservoir Beds in the Campos Basin, Brazil. *In*: *Annals of the 31st International Geological Congress*, IUGS, Rio de Janeiro.

MILLER, K. G. & FAIRBANKS, R. G. 1985. Oligocene to Miocene carbon isotope cycles and abyssal circulation changes. *In*: SUNDQUIST, E. J. & BROECKER, W. S. (eds) *The Carbon Cylce and Atmospheric CO$_2$: Natural Variations Archean to Present.* Geophysical Monograph, American Geophysical Union, **32**, 469–486.

MILLER, K. G., WRIGHT, J. D. & FAIRBANKS, R. G. 1991. Unlocking the Ice House: Oligocene–Miocene oxygen isotopes, eustary, and margin erosion. *Journal of Geophysical Research*, **96**, 6829–6848.

MUTTI, E., BARROS, N., POSSATO, S. & RUMENOS, L. 1980. Deep-sea fan turbidite sediments winnowed by bottom currents in the Eocene of the Campos Basin – Brazilian offshore. Abstracts, Bochum, 1st European meeting IAS, **114**.

PARRISH, J. T. & CURTIS, R. L. 1982. Atmospheric circulation, upwelling, and organic-rich rocks in the Mesozoic and Cenozoic eras. *Palaeogeography, Palaeoclimatology, Palaeoecology*, **40**, 31–66.

SOUZA CRUZ, C. E. 1995. Estratigrafia e sedimentação de águas profundas do Neogeno da Bacia de Campos. Ph.D. Thesis, Porto Alegre (IG/UFRGS).

VIANA, A. R. 1998. *Le rôle des courants océaniques dans les dépôts de marges continentales: le Bassin Sud-Est Brésilien.* PhD Thesis, Université Bordeaux I, France.

VIANA, A. R. & FAUGÈRES, J.-C. 1998. Upper slope sand deposits: the example of Campos Basin, a latest Pleistocene/Holocene record of the interaction between along-slope and downslope currents. *In*: STOKER, M. S., EVANS, D. & CRAMP, A. (eds) *Geological Processes on Continental Margins: Sedimentation, Mass-Wasting and Stability.* Geological Society, London, Special Publications **29**, 287–316.

VIANA, A. R., KOWSMANN, R. O. & CASTRO, D. D. 1990. *A discordância do Mioceno médio/superior, Um marco regional na Bacia de Campos.* 36th Congresso Brasileiro de Geologia, Natal, Brazil, **1**, 313–323.

VIANA, A. R., FAUGÈRES, J.-C., KOWSMANN, R. O., LIMA, J. A. M., CADDAH, L. F. G. & RIZZO, J. G. 1998. Hydrology, morphology and sedimentology of the Campos Continental Margin, Offshore Brazil. *In*: STOW, D. A. V. & FAUGÈRES, J.-C. (eds) *Contourites, turbidites and process interaction.* Sedimentary Geology, Special Issue, **115**(1/4), 133–158.

VIANA, A. R., ALMEIDA JR., W. & ALMEIDA, C. W. 2002. Late Quaternary upper slope sands – the shallow water sandy contourites of Campos Basin, SW Atlantic margin. *In*: STOW, D. A. V., PUDSEY, C. J., HOWE, J. A., FAUGÈRES, J.-C. & VIANA, A. R. (eds) *Deep-Water Contourite Systems: Modern Drifts and Ancient Series, Seismic and Sedimentary Characteristics.* Geological Society, London, Memoirs, **22**, 261–270.

VICALVI, M. A. 1999. Zoneamento bioestratigráfico e paleoclimático do Quaternário superior do talude da Bacia de Campos e Platô de São Paulo adjacente, com base em foraminíferos planctônicos. PhD Thesis, IG/UFRJ, Rio de Janeiro.

Upper slope sands: late Quaternary shallow-water sandy contourites of Campos Basin, SW Atlantic Margin

ADRIANO R. VIANA, WALDEMAR DE ALMEIDA JR. & CLEIDE WILHELM DE ALMEIDA

Petróleo Brasileiro S.A., PETROBRAS E&P,Campos Basin branch, Av. Elias Agostinho, 665 Macaé, RJ 27.913-350, Brazil
(e-mail: aviana@petrobras.com.br)

Abstract: Upper slope sand deposits comprise a widespread but thin elongate accumulation of coarse to very fine-grained sand resulting from the action of slope boundary currents upon shelf-derived sediments. Sediment distribution on the Campos Basin upper slope responds to the action of the southward-flowing western boundary Brazil Current (BC). Linear, multi-source sediment supply to the slope is provided by shelf overspill due to the action of different forcing mechanisms: tides, storm fronts, and BC current onshelf penetration as gyres and meanders. On the slope, the sediment is pirated and redistributed by the BC. Coarse-grained sediments (pebbles to very coarse sand) are found below the zone of maximum acceleration of the BC. Downstream, fining is observed as a consequence of the morphologically controlled BC deceleration. The resultant accumulation is an elongate (*c.* 70 km long) and thin (< 50 m) wedge-shaped deposit. This depositional model is based on hydrographic, physiographic and sedimentologic characteristics of the modern Campos Basin margin, SE Brazil and characterizes a shallow water contouritic deposit.

The Campos Basin is located on the southeastern Brazilian continental margin, between 21°S and 24.5°S. A marked similarity between the modern coastline and the shelf break is observed, characterized by the seaward projection of the margin at the Cape São Tomé region (Fig. 1). The continental shelf widens and deepens to the south of this cape, ranging from 50 km wide in the north, with a NW–SE trend, to 100 km wide in the south, with a NE–SW trend. The continental slope extends 45 km from the shelfbreak (~120 m) down to the 2000 m isobath. Large submarine canyons are developed perpendicular to the shelfbreak: the Almirante Camara and Itapemirim to the north, and São Tomé to the south (Fig. 1).The shelfbreak is marked by a conspicuous scarp between the 120 and 220 m isobaths, in places with a slope angle of up to 10°, and a series of gullies that locally incise the shelf edge. Mass-movement scars and buried canyon heads smooth the scarp in the southernmost part of the study area.

The northern upper slope is separated from the southern slope by the Sao Tomé submarine canyon (STC) and presents a 10 km wide, flat erosional terrace, the Albacora Terrace, that extends from the base of the shelf break scarp to the 450 m-isobath. In its inner portion the terrace is marked by bottom current related erosional ridges elongated parallel or slightly oblique to the isobaths, and trending south towards the São Tomé canyon head. These ridges are tens of metres high, a few kilometres long and hundreds of metres wide. Furrows are also evident parallel to the isobaths in this region, passing in to a region of sinuous-crested sand waves to the south (Fig. 2). The southern upper slope is marked by a narrow terrace developed at the foot of the shelf edge scarp, at about the 200 m isobath. The upper slope shows a generally concave profile from 250 m to 550 m water depth, on the upper/middle slope boundary. To the far south of the study area, the headwalls of slide scars cut into the shelf edge.

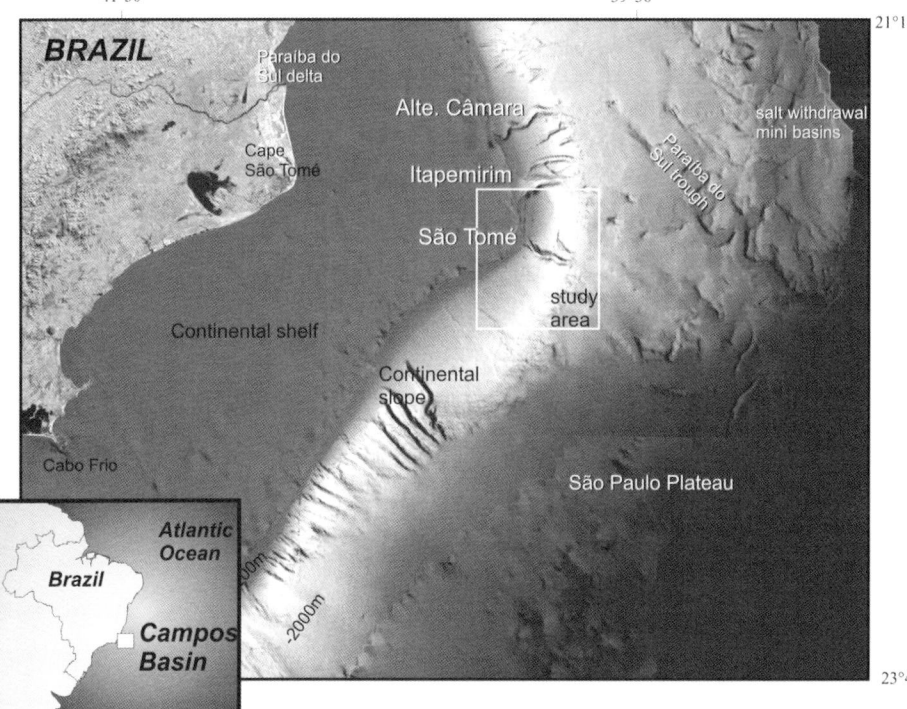

Fig. 1. Shaded bathymetric map presenting the main physiographic features mentioned in the text. White rectangle outlines the study area. Camara, Itapemirim and São Tomé are submarine canyons cited in the text. Note the similarity in shape between the coastline and the shelf edge. Deepest areas to the right reach up to 3000 m deep. Lighter areas correspond to continental slope (depths ranging from 400 m to 2000 m. Detailed bathymetry is presented in Figure 5.

From: STOW, D. A. V., PUDSEY, C. J., HOWE, J. A., FAUGÈRES, J.-C. & VIANA, A. R. (eds)
Deep-Water Contourite Systems: Modern Drifts and Ancient Series, Seismic and Sedimentary Characteristics.
Geological Society, London, Memoirs, **22**, 261–270. 0435-4052/02/$15.00 © The Geological Society of London 2002.

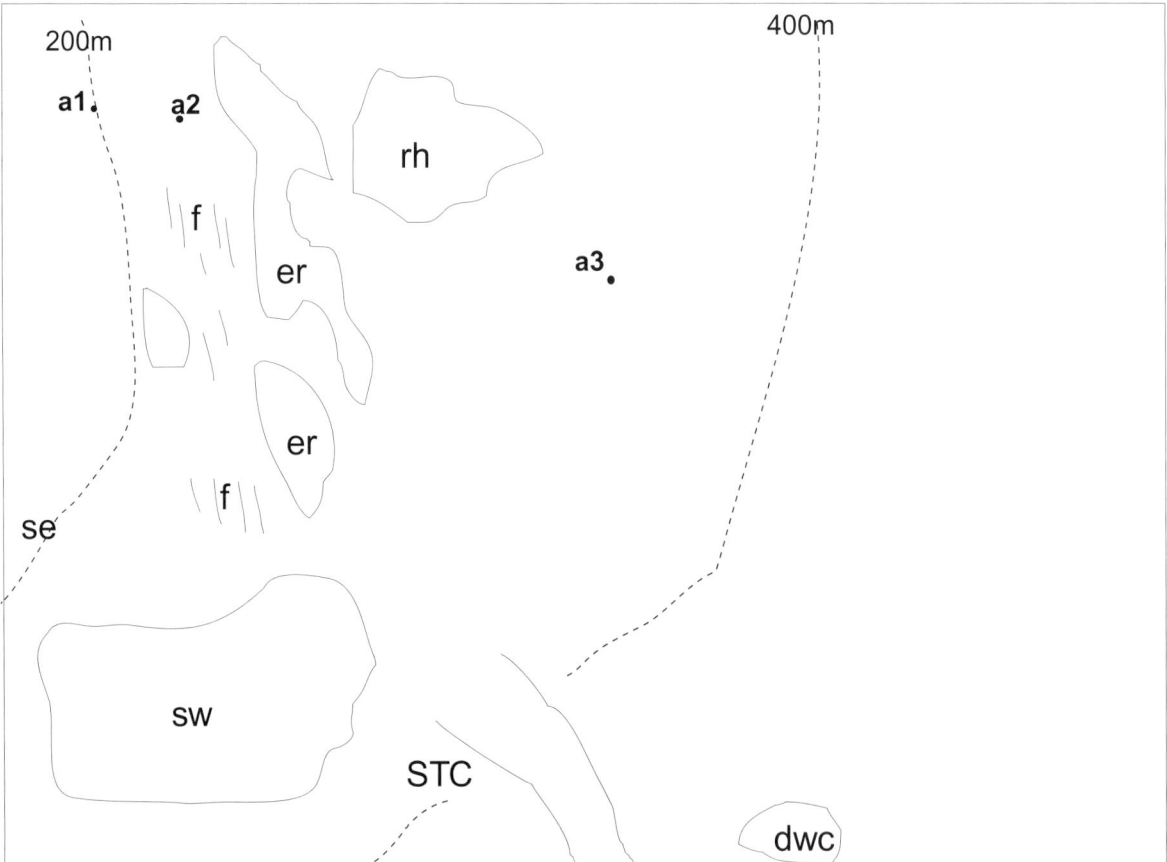

Fig. 2. Mosaic of side scan sonar images illustrating the funnelling zone of the southward flowing Brazil Current (BC). The impact of bottom currents and their southward intensity decrease are expressed as furrows (f) to the north where bottom currents are topographically confined between the shelf edge escarpment (se) and an upper slope erosional ridge (er), passing to 3D sandwaves (sw) to the south, in the expansion zone. Sand migration trend is towards the São Tomé canyon head (STC). a1, a2 and a3 are cores from the funnelling zone presented in Figure 7. dwc, deep water carbonates; rh, rhodolites.

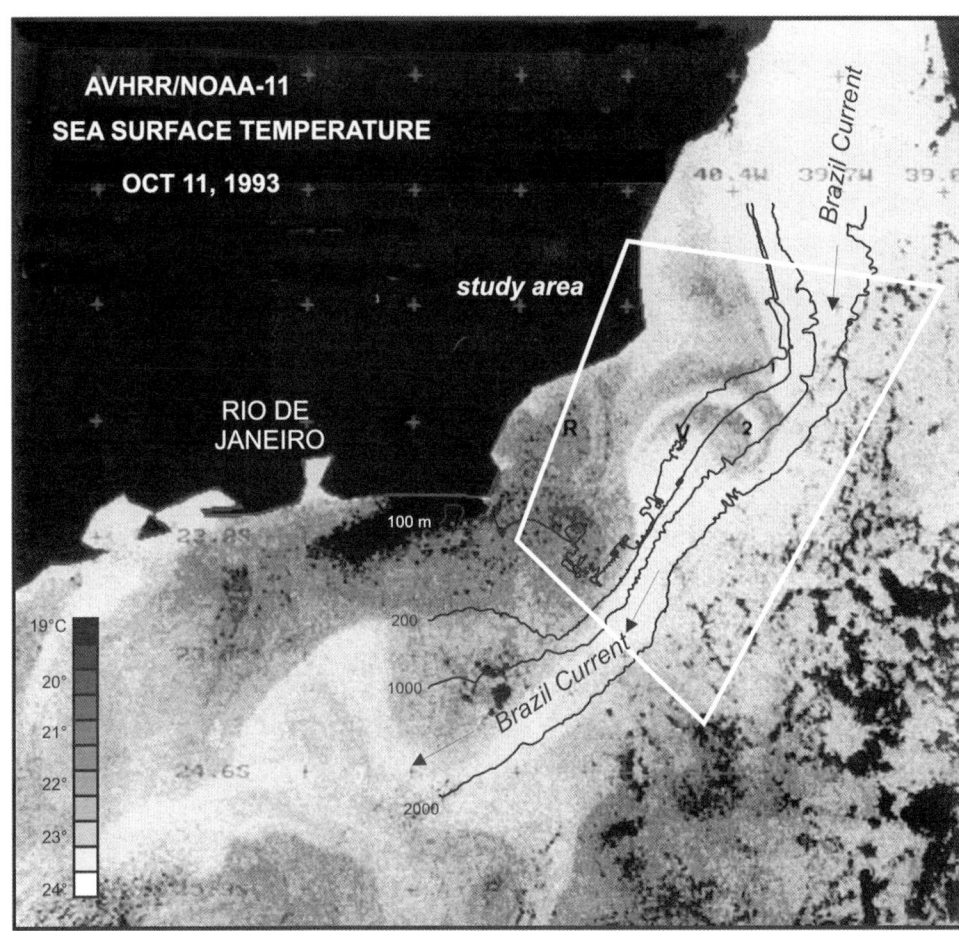

Fig. 3. Sea surface Temperature satellite image illustrating the general surface circulation of the SE Brazilian margin. V – Zone of Brazil Current (BC) eddy generation; 1, Exit of the BC funnelling zone (maximum lateral current narrowing and speed increase); 2, Re-entering of a BC filament in the BC expansion zone (current deceleration and flow perturbation – meanders and eddies).

Oceanographic setting

The hydrographic characteristics of the south-eastern Brazilian margin have been studied by many authors and are listed in Viana (1998) and Viana & Faugères (1998). The surface circulation (Fig. 3) can be separated into shelf currents and the Brazil Current (BC). Shelf currents are the result of combined meteorological and tidal forcing upon the shelf waters. The observations made by Campos & Miller (1995) suggest a northeastward propagation of shelf waters between the shelf break and the coast. The BC is a warm (> 20°C) saline (> 36‰) southward flowing western boundary current driven by the atmospheric circulation of the South Atlantic Gyre. The inner edge of the BC roughly coincides with the shelf break. Below the BC (400–1500 m), the Brazil Intermediate Counter-Current (BICC; Lima 1998) drives to the north both the South Atlantic Central Water (SACW) and the Antarctic Intermediate Water (AAIW). Over the lower slope, the sluggish southward flowing North Atlantic Deep Water is sometimes disturbed by benthic storms, that result in an increased flow velocity and, in some cases, a reversal in the flow direction.

The passage of the cyclonic gyres over the shelf imposes a 'sea-floor polishing' effect (Viana *et al.* 1995) provoking re-suspension and transport of sediments (Fig. 4). Eddies, storm- and tide-driven currents induce the shelf sand to spill over onto the slope, mainly as low concentration gravity flows and secondarily as high concentration fluid and plastic gravity flows. Once on the slope, sediments are subject to the BC and transported along the isobaths (Fig. 5).

Sediment characteristics

Superficial sediments on the outer shelf are composed of siliciclastic sands, carbonate debris and mixed composition material (Fig. 6). Siliciclastic sands extend throughout the outer shelf area developing large fields of sand waves, dunes and megaripples, tens of kilometres long and kilometres to tens of kilometres wide, being developed from the outer shelf to the shelf break. Between the 85 m and 130 m isobaths, sand waves have rectilinear crests (2D geometry in the sense of Harms *et al.* 1982) and show a northeastern transport direction towards the shelf break. This direction corresponds to the resultant trend of the bottom shelf currents. On the upper slope terrace, sets of curvilinear-crested sand waves (3D geometry) are observed migrating towards the southwest, controlled by the Brazil Current passage.

The vertical succession of the latest Pleistocene–Recent upper slope sediments (Fig. 7) comprise from base to top: Facies 1 – a more than 50 m thick interval of closely interbedded sand and mud (S/M ratio < 0.2), resembling the muddy-silty contourites described from the Faro Drift (Gonthier *et al.* 1984), and deposited during the late Pleistocene: 85 – *c.* 25 ka BP (biozone Y2/lower Y1; Ericson and Wollin 1968); Facies 2 – a few metres- to several decimetres-thick interval of pebble to fine-grained sand, erosive into the muddy-silty contourite interval, and deposited during the Last Glacial Maximum (*c.* 18 ka BP) up to the onset of sea-level rise (Latest Pleistocene/early Holocene *c.* 12 ka BP); Facies 3 – at the very top of the sequence, extending downslope to the 700 m isobath, a fining upward few metres to several decimetres thick interval of coarse- to very fine-grained sand, deposited during the Holocene. Maximum thickness of the coarse-grained deposits (Facies 2 and 3) is about 30 m (Fig. 8).

Discussion

Upper slope sand deposits were developed on the modern Campos Basin margin due to particular characteristics of margin morphology and superficial hydrodynamics. The main conditions

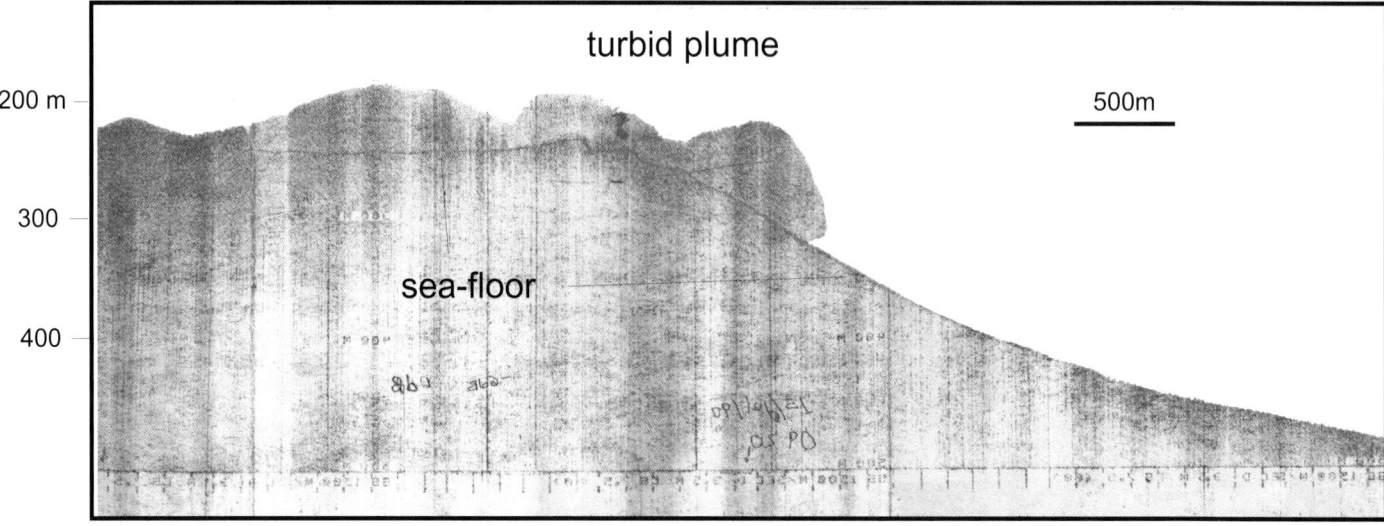

Fig. 4. 12 kHz echo-sounder record showing a turbid plume being detached from the shelf edge in the zone of the Brazil Current eddy activity (point 2 in the satellite image).

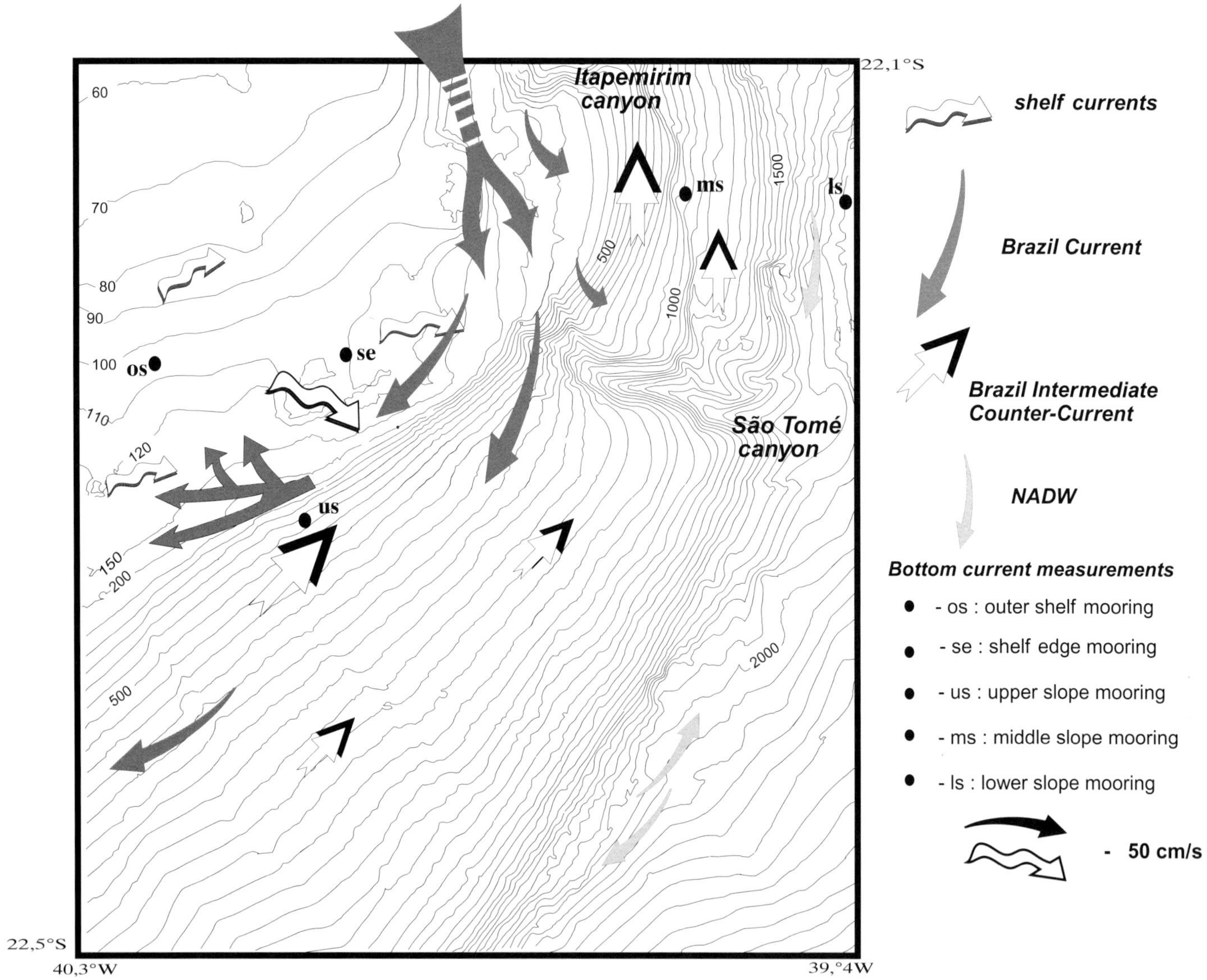

Fig. 5. Sea-floor projection of the main bottom currents: shelf currents are meteorologically controlled (southerly storm fronts and NE trade winds); the Brazil Current is a locally vigourous western boundary surface current with meanders and eddies; the Brazil Intermediate Counter-Current and the North Atlantic Deep Water Current are geostrophic currents locally controlled by topographically-induced long-term oscillations.

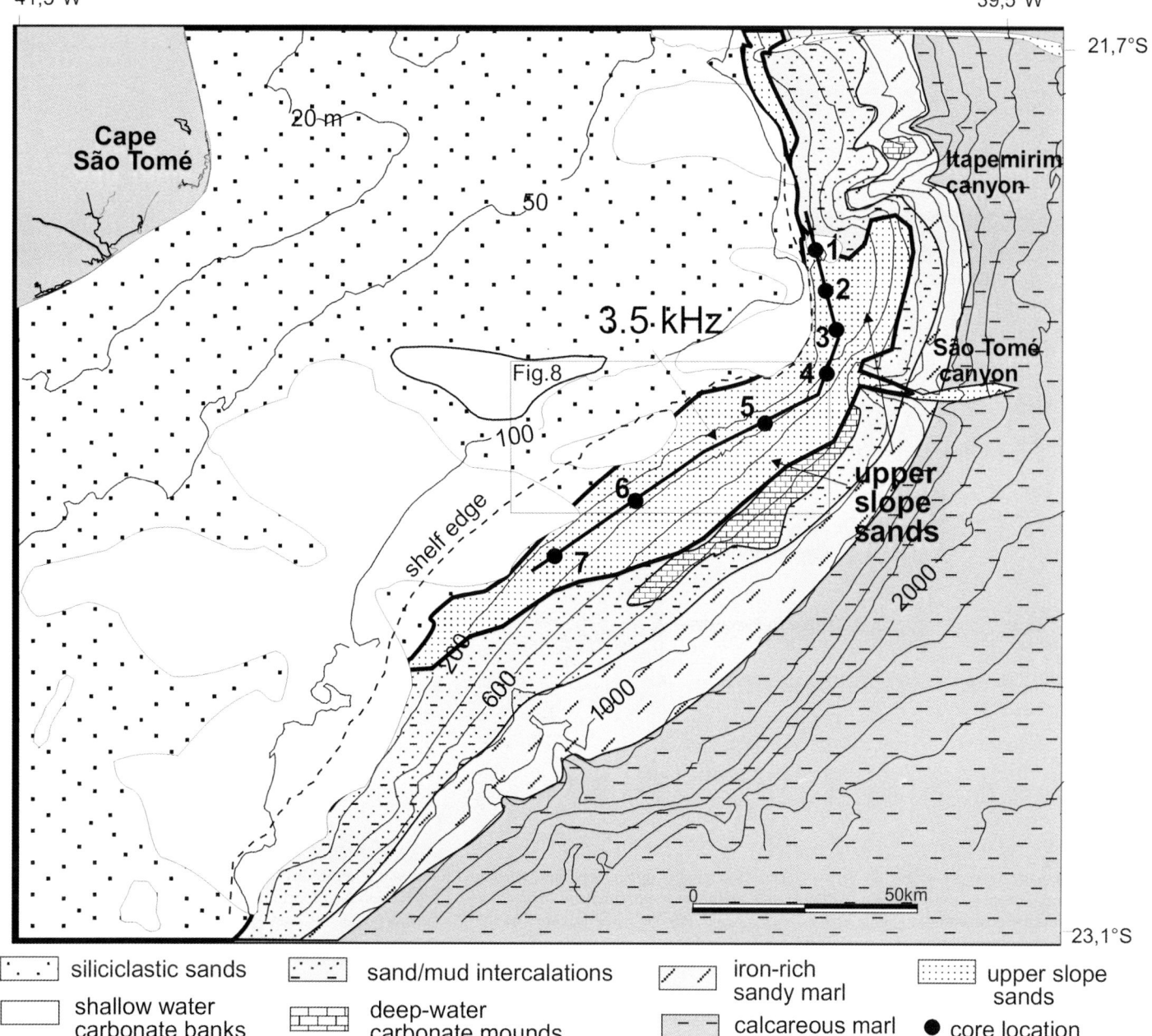

Fig. 6. Surface sediment distribution map highlighting the upper slope sand distribution, and location of the cores and 3.5 kHz record presented in Figure 9.

for development of upper slope sand accumulations are: (1) a convex outer shelf/slope morphology; (2) sandy sediment available at the shelf edge; (3) net offshelf transport of shelf sands induced by shelf edge bottom currents; and (4) the presence of a relatively strong slope boundary current. The interaction between the shelf/slope morphology and the hydrological factors is very important in leading to the deposition of upper slope sands. Two key physiographic sectors are involved: (1) The seaward projection of the shelf edge and the region immediately to the north. The morphology of this zone induces a funnelling effect on the southward circulation of the superficial BC. In this 'funnelling zone' the BC is accelerated to very high speeds and sweeps the shelf edge and the uppermost slope. (2) South of the seaward projection of the shelf, the BC expands and decelerates in response to the change in shelf/slope morphology, which shows a landward inflexion. In this 'expansion zone', the BC meanders and generates eddies whose activity interferes with the shelf currents.

In the 'funnelling zone', the modern sedimentation reflects very high energy environments (Figs 9 and 10a). As a consequence, the shelf edge is swept clean of sediment and the BC introduces coarse sediments onto the upper slope. Gravelly-to-coarse-grained sand deposits, and erosional features are common. Downslope, the sediments become fine-grained (very-fine sand at the transition between upper and middle slope, around 600 m water depth).

In the expansion zone (Figs 9 and 10b), outer shelf environments are dominated by shelf currents which induce the ENE migration of sand waves towards the shelf edge. Oceanic eddies constitute an important mechanism in the sand transport across the outer shelf. Downwelling of shelf waters seems to increase the downslope transfer of shelf sands, which are prolonged in the form of gravity-driven sand fluxes. Sand fluxes behave as successive and continuous grain flows, and, more locally, as high-density turbidity currents. Sandy sediments flow downslope, mainly through canyons and gullies, and also as direct spillover. In contrast to the northern area, where erosive features predominate, the uppermost

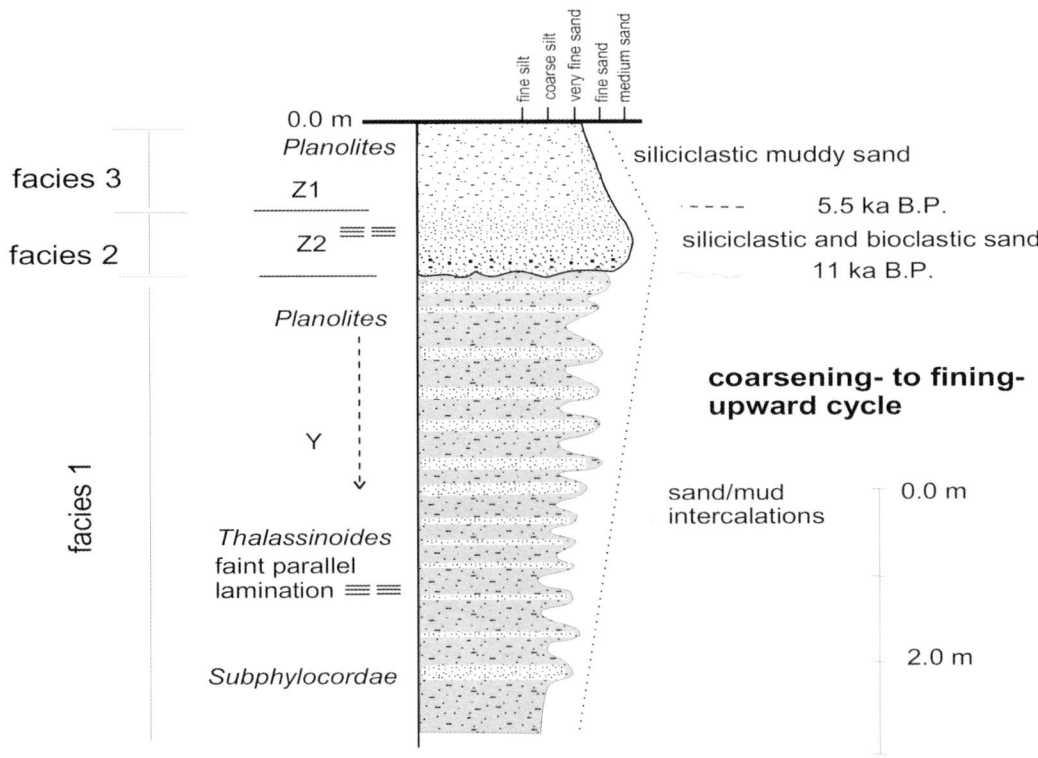

Fig. 7. Schematic log representing the vertical facies succession of upper slope sand accumulation at water depths around 400 m. X, Y, and Z are foraminiferal biozones from Ericson & Wollin (1966). Sediment is extremely bioturbated with ichnofossils also reported. The general characteristics of these deposits are coarsening- to fining-upward cycles, separated one from the other by sharp erosional contacts. Coarse-grained sediments were mainly deposited during the Holocene.

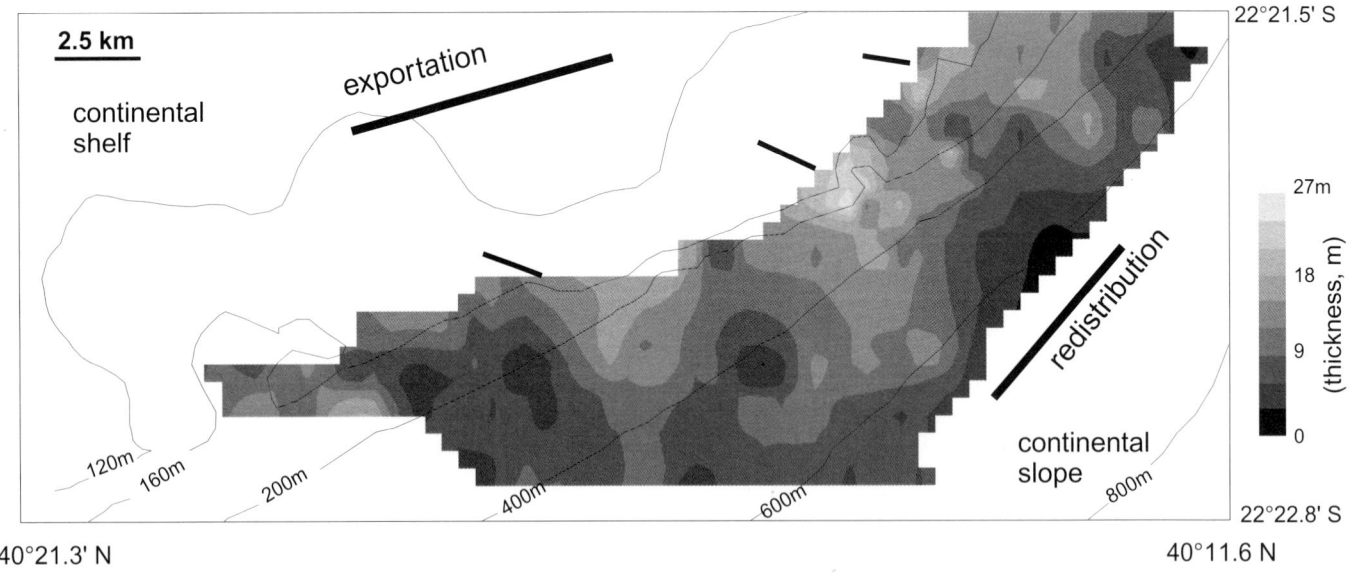

Fig. 8. Isopach map of upper slope sands in the expansion zone, elaborated from 3.5 kHz data. Thicker deposits are found spread along the foot of the shelf edge escarpment. Arrows indicate main trends of sand transport.

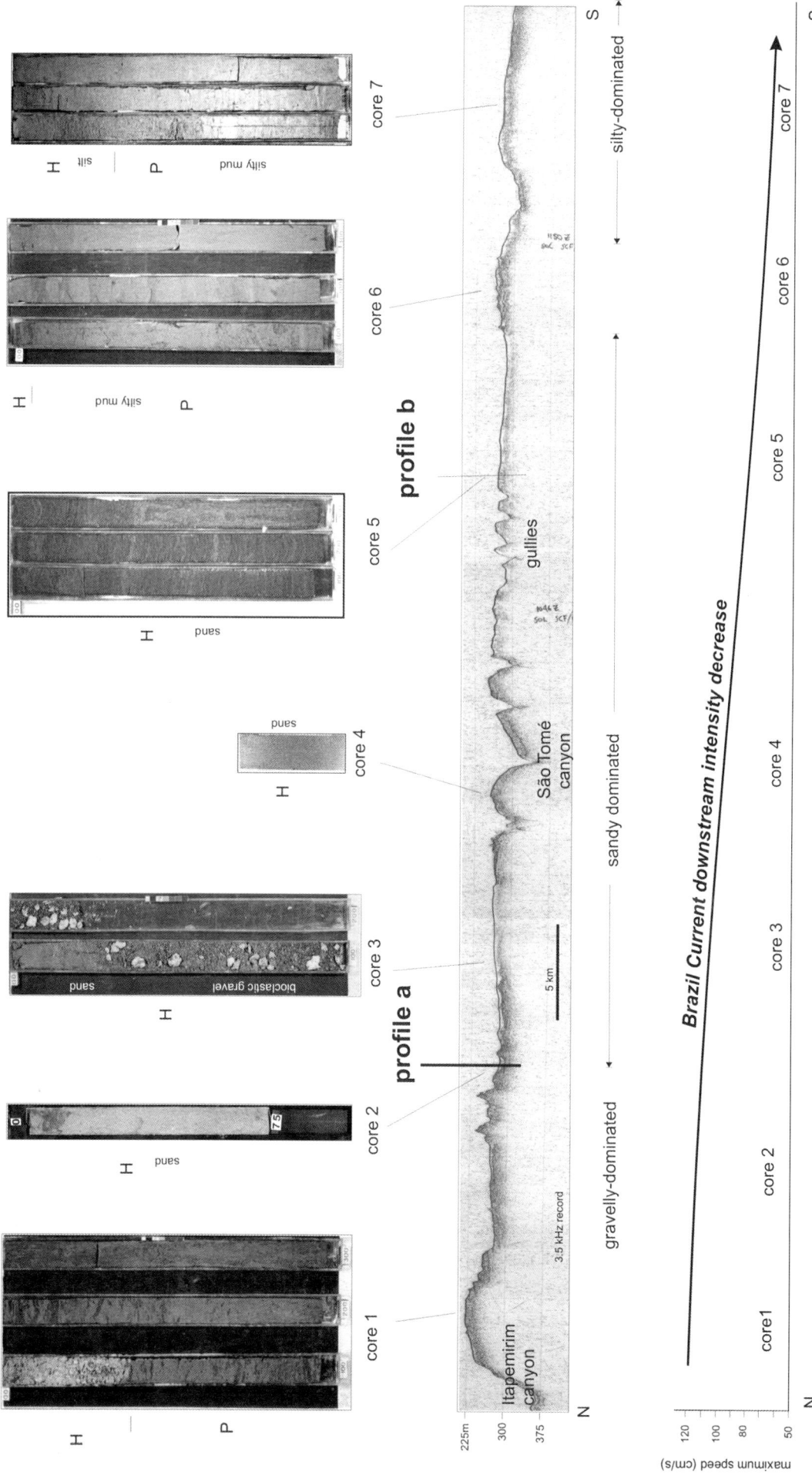

Fig. 9. Integration of core, shallow seismic (3.5kHz) and hydrographic data in an alongslope trend (~350 m water depth) corroborates the influence of the Brazil Current (BC) on sediment distribution over the slope. The shallow transparent zone on 3.5 kHz profile (see location on Fig. 6) represents areas of sand accumulation. A decrease in grain size is observed (cores 1 to 7) related to a decrease in intensity of the BC. In the current funnelling/acceleration zone, coarse-grained sediments are deposited (cores 1 to 3) and the Pleistocene (P)/Holocene (H) boundary is often marked by a sharp/erosional contact (core 1). The São Tomé canyon head is filled by BC-transported sands (core 4, see also side scan sonar image, Fig. 2). Coarse to medium sand was retrieved in the entrance of the expansion zone (core 5), showing planar and cross lamination. Sea-floor erosion (core 6) is related to sediment-starved topographic highs. Fine-grained sediments retrieved in core 7 represent the distal portion of the upper slope sand deposit, related to the absence of shelf sediment input and to a decrease in bottom current activity.

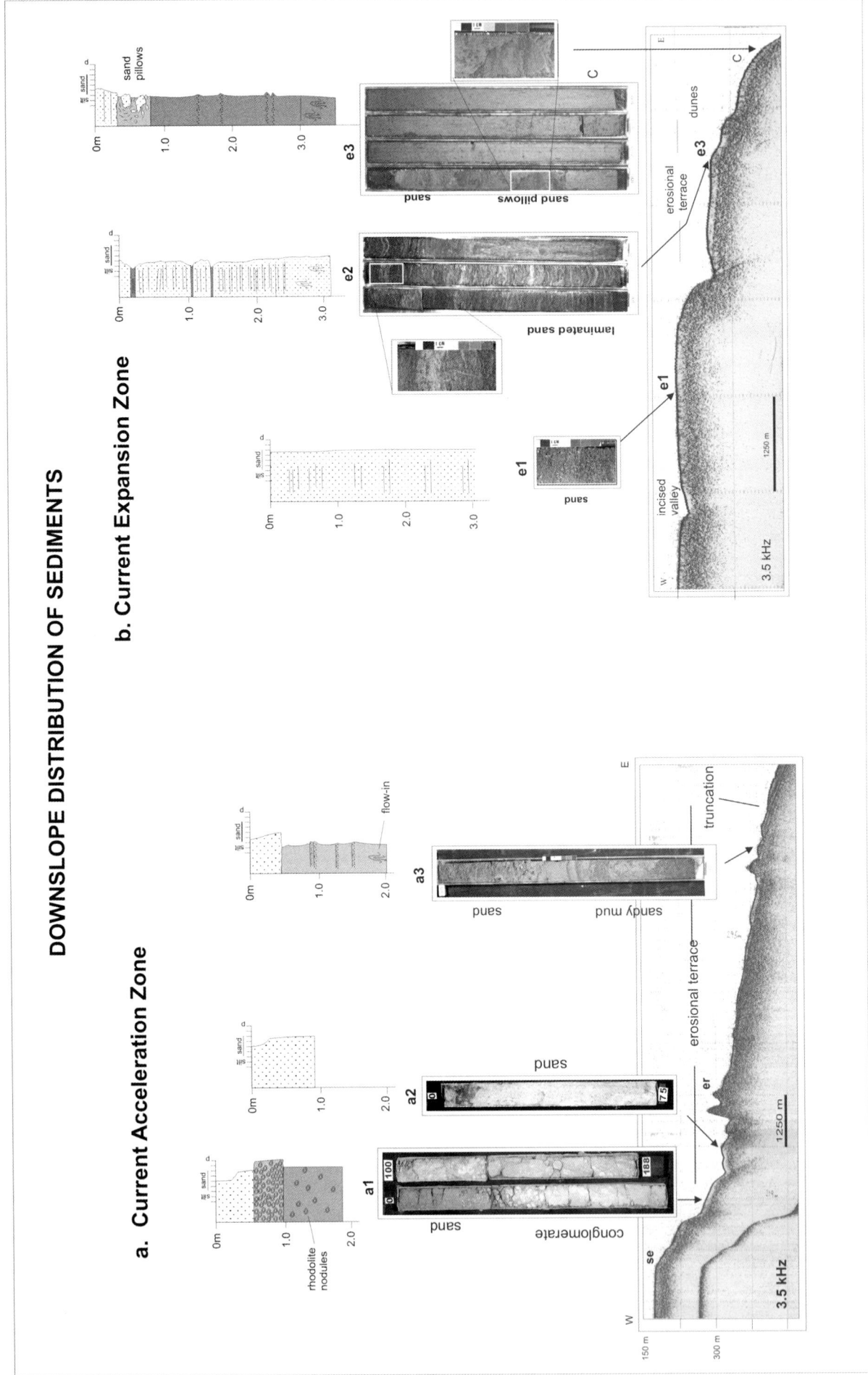

Fig. 10. Downslope 3.5 kHz profiles in the current funnelling zone (**a**) and in the current expansion zone (**b**). The downslope fining is well characterized in both cross sections. In section (**a**), very coarse-grained sediments are related to the export of the adjacent shelf edge carbonates (se). The channelled area between the shelf edge and the erosional ridges (er). The erosional terrace results from the activity of the BC. Truncation is observed in the distal portion of this section. Location of cores a1, a2 and a3 are presented in Fig. 2. Section (**b**) shows incised valleys at the outermost shelf and a narrow erosional terrace at the uppermost slope where large bedforms are developed. Core e2 presents planar and cross lamination indicating the high energy of the environment. Core e3 presents sandy contourites intermingled with gravity flow deposits, characterised by the medium-grained sand in a finer matrix (silty sand).

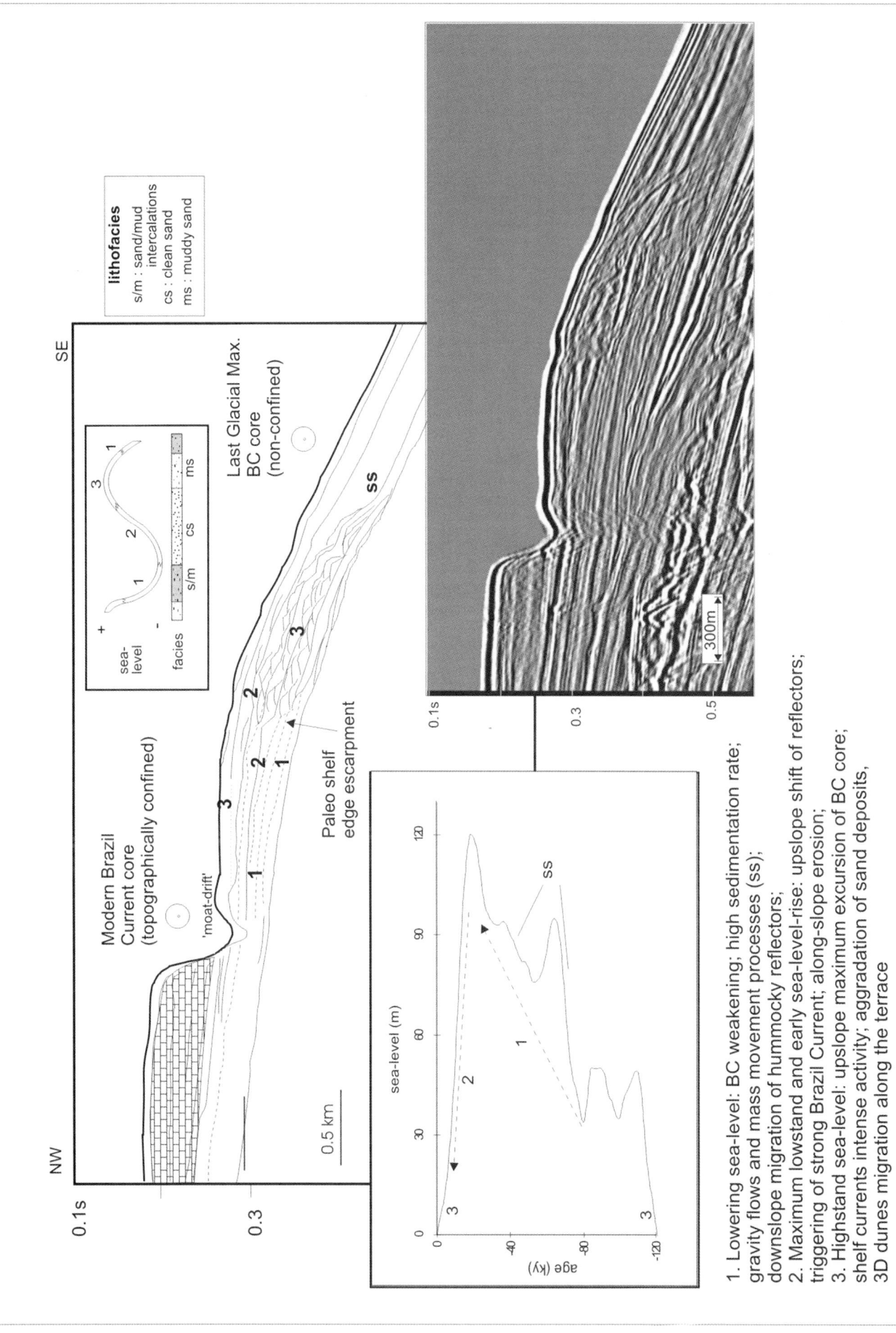

Fig. 11. Seismic-stratigraphic scheme of upper slope sand deposition. Clean sands were deposited during the modern highstand. The relative deceleration of the BC during the modern highstand resulted in the deposition of muddy sands. Sand/mud intercalations correspond to sea-level lowering deposits. Major periods of sand accumulation are found from the early sea-level rise to the present highstand.

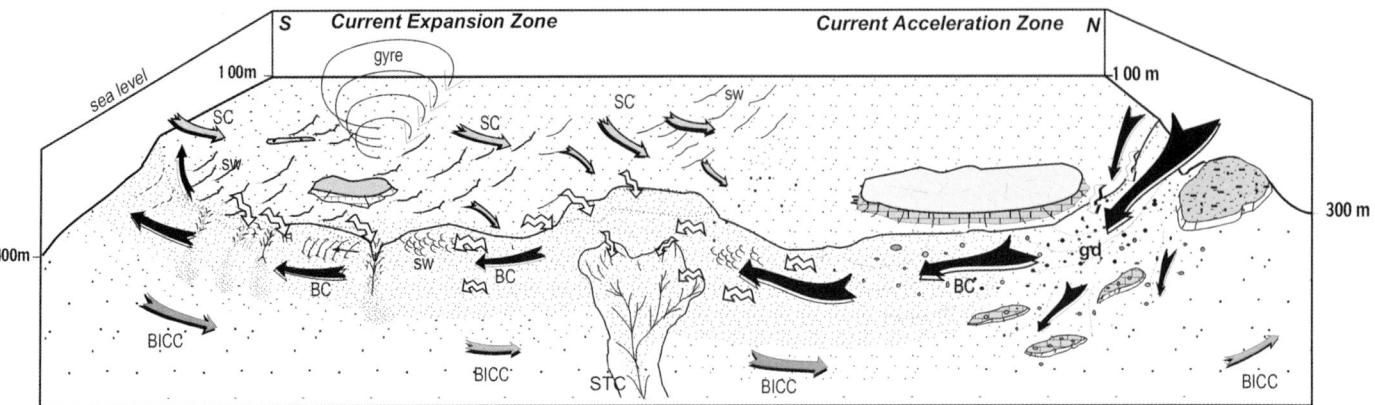

Fig. 12. Schematic representation of the main sedimentary processes related to the passage of bottom currents over the shelf edge and upper slope. BC, Brazil Current; BICC, Brazil Intermediate Counter-Current; SC, Shelf Currents; STC, São Tomé Canyon; sw, sand waves; gd, gravel deposit. Wavy arrows indicate bed-load transfer. From Viana & Faugères (1998).

slope is characterized by bedforms migrating along the trend of the southward flowing BC. They develop curvilinear-crested (3D) sand waves in the head of the gullies and rectilinear-crested (2D) sand waves in the intergully areas. Downslope, the sand is redistributed as sand sheets by the BICC.

The vertical transition to underlying fine-grained sediments is erosive in the funnelling zone, whereas in the expansion zone, it is sharp or gradual. The general vertical facies succession observed in both zones defines a coarsening-fining upward sequence. The general geometry of the upper slope sand deposits is a thin wedge-shaped accumulation, tens of metres thick in the proximal zone, thinning downslope. The wedge is five to ten km wide downslope and extends tens of kilometres alongslope.

The internal seismic pattern of the coarse-grained sandy layers is transparent to discontinuous and hummocky. The fine-grained sand to sand/mud facies shows a hummocky clinoform seismic pattern grading to wavy and then parallel reflections in the steepest southern zone, and a parallel pattern in the flattened northern zone (Fig. 11).

The model here proposed is summarised in Figure 12. It suggests that important sand accumulations occur at the upper slope during highstands of sea-level (Fig. 11). The highstand activity is due to the presence of shelf currents that sweep off the shelf large volumes of sand, and to strong superficial slope-boundary currents, that rework the sand along the slope. These currents were inhibited during sea-level fall, and the shelf currents no longer existed at the maximum lowstand, when the shelf areas were subaerially exposed.

Coarse-grained, well-sorted sand deposits covering relatively large areas make the upper slope sand deposits an attractive play to be investigated for potential hydrocarbon reservoirs. The main difficulty in the application of this model to hydrocarbon reservoir analogues seems to be the vertical and lateral development of sealing facies. The Campos Basin example suggests that the hydrodynamic changes which accompany the climatic/sea-level fluctuations may be severe enough to induce the needed facies alternation. A more systematic study of modern environments must be coupled with research on ancient rock outcrops in order to evaluate the geological record of different expressions of the controlling mechanisms of our model.

The analysis of facies distribution, depositional geometry and physiographic and hydrographic features indicates that the development of upper slope sand bodies results mainly from the alongslope sediment reworking and redistribution by slope boundary currents. Factors important in generating such sedimentation are: (i) relatively strong offshelf-trending bottom currents, (ii) a linear multiple source supply, with direct shelf sediment spillover, (iii) short downslope gravity transport, and (iv) vigorous slope boundary currents. The presence of coarser facies and large erosional features on the northern slope indicates the higher energy of that area corroborating the hydrographic observations. The grain size fining trend along the upper slope sand deposits coincides with Brazil Current downstream intensity decrease demonstrating the direct relationship between the deposit and the Brazil current activity. The fundamental elements of such a model are: a sand-rich shelf; an erosional terrace at the top of the slope; offshelf sand spillover by gravity flows and dune migration; sand rich deposits on the upper slope, and; lateral filling of canyon heads.

References

CAMPOS, E. J. & MILLER, J. L. 1995. Hydrography of the South Brazil Bight as observed during Project COROAS. *EOS Transactions*, American Geophysical Union Fall Meeting, **76**, F324.

ERICSON, D. B. & WOLLIN, G. 1968. Pleistocene climates and chronology in deep-sea sediments. *Science*, **162**, 1227–1234.

GONTHIER, E., FAUGÈRES, J.-C. & STOW, D. A. V. 1984. Contourite facies of the Faro Drift, Gulf of Cadiz. *In*: STOW, D. A. V. & PIPER, D. J. W. (eds) *Fine-Grained Sediments: Deep-Water Processes and Facies*. Geological Society, London, Special Publications, **15**, 245–256.

HARMS, J. C., SOUTHARD, J. B. & WALKER, R. G. 1982. *Structures and sequences in clastic rocks*. SEPM, Calgary.

LIMA, J. A. M. 1998. *Oceanic circulation on the Brazilian shelf break and continental slope at 22°S*. PhD thesis, University of New South Wales, Australia.

VIANA, A. R. 1998. *Le rôle et l'enregistrement des courants océaniques dans les dépôts de marges continentales: la marge du bassin Sud-Est brésilien*. Tese de doutorado, Université Bordeaux 1.

VIANA, A. R. & FAUGÈRES, J.-C. 1998. Upper slope sand deposits: the example of Campos Basin, a latest Pleistocene/Holocene record of the interaction between along-slope and downslope currents. *In*: STOKER, M. S., EVANS, D. & CRAMP, A. (eds) *Geological Processes on Continental Margins: Sedimentation, Mass-Wasting and Stability*. Geological Society, London, Special Publications, **29**, 287–316.

VIANA, A. R., FAUGÈRES, J.-C. & LIMA, J. A. M. 1995. *The role of outer shelf currents in feeding deep-water systems*. 16th IAS Regional Meeting of Sedimentology, Aix-les-Bains, Book of Abstracts, Publication ASF, Paris, **22**, 151.

VIANA, A. R., FAUGÈRES, J.-C. & STOW, D. A. V. 1998. Bottom current controlled sand deposits – a review from shallow to deep water environments. *In*: STOW, D. A. V. & FAUGÈRES, J.-C. (eds) *Contourites, Turbidites and Process Interaction*. Sedimentary Geology, Special Issue, **115**, 53–80.

Contourites on the Agulhas Plateau, SW Indian Ocean: indications for the evolution of currents since Palaeogene times

GABRIELE UENZELMANN-NEBEN

Alfred-Wegener-Institut für Polar- und Meeresforschung, P.O. Box 120161, 27515 Bremerhaven, Germany
(e-mail: uenzel@awi-bremerhaven.de)

Abstract: The area south of South Africa is one of the most important gateway regions for the interchange of watermasses from the Atlantic, Indian and Southern oceans. This results in a very complex flow pattern which up to now has been known only in general terms. For this study, a set of seismic reflection lines from the southern Agulhas Plateau has been analysed. These show strong indications for the effect of bottom currents on sedimentation in the form of sediment drifts, channels, erosional unconformities and sediment waves. These observations have been used to infer the development of palaeocirculation over the southern Agulhas Plateau since Paleogene times. Three different currents have been identified. The oldest observed flows across the Agulhas Plateau from the southwestern tip to the northeast, and probably dates from the early Eocene times. It is believed to comprise an Antarctic Bottomwater component derived from its source in the south. The eastern Agulhas Plateau is characterized by a south-flowing current, which has been active since the Lower Oligocene and appears to have remained stationary to within a distance of 10 km. The Agulhas Retroflection is considered as its most likely source. The third current observed flows along the western flank of the plateau, dates from the Middle Miocene and probably results from an Antarctic Bottomwater component re-circulated via the Cape Basin.

This study presents the results of a small-scale seismic investigation over the Agulhas Plateau. These data are presented and interpreted in the light of the previously known geological and oceanographic setting, and with reference to the limited amount of core data that exists. It is fully recognized that interpretation of seismic data in terms of bottom current or other effects must remain speculative until further varied datasets are available. However, as an important key to southern ocean bottom circulation and watermass exchange, a first attempt is presented here.

Geological and oceanographic setting

The Agulhas Plateau is located about 500 km SE of the Cape of Good Hope in the SW Indian Ocean (Fig. 1). This morphological structure rises up to 2500 m above the surrounding seafloor. The exact tectonic evolution of the Agulhas Plateau is still under debate. Strong indications for a continental origin were found (LaBreque & Hayes 1979; Allen & Tucholke 1981; Martin & Hartnady 1986; Ben-Avraham et al. 1995) with the inference that the plateau formed one structural unit together with the Falkland Plateau and the Maud Rise prior to the opening of the South Atlantic. However, Kristoffersen & LaBreque (1991) suggest a common evolution for the North East Georgia Rise and the Agulhas Plateau and thus an oceanic origin of the plateau. A third theory proposes continental fragments overprinted by excessive volcanism and thus adds the Agulhas Plateau to the world-wide suite of Large Igneous Provinces of predominantly oceanic origin (Uenzelmann-Neben et al. 1999).

Reconstruction of the plateau's sedimentary development starting in the Maastrichtian is complex. As sediment cores have shown, the input of terrigenous material from Africa is small (Tucholke & Embley 1984). The sedimentary succession is characterized by a number of hiatuses through the Paleocene, indicating low sedimentation rates as well as strong bottom-current erosion (Barrett 1977; Tucholke & Carpenter 1977; Dingle & Camden-Smith 1979; Tucholke & Embley 1984; Siesser et al. 1988). An erosional zone or moat appears to encircle the plateau and has been related to the persistent effect of Antarctic Bottom Water (AABW), which has been active in this area since the glaciation of West Antarctica increased markedly in the Mid–Miocene (Camden-Smith et al. 1981; Tucholke & Embley 1984).

But the AABW is not the only current active south of Africa.

The southeast African continental margin represents a critical gateway within the oceanic circulation system where Indian–Pacific Ocean and Atlantic Ocean water masses meet and mix. Here, the surface Agulhas Current, Antarctic Intermediate Water (AAIW), North Atlantic Deepwater (NADW) and AABW are all present in the water column. Consequently, the fluctuating strength of interocean circulation should be sensitively recorded by erosional–depositional processes on the subjacent seafloor over an approximately 4 km depth range (Ben-Avraham et al. 1994).

The Agulhas Current is the western boundary current of the Indian Ocean. The current is highly baroclinic, extends deeper than 2000 m, has surface speeds that often exceed 2 m s^{-1} and a mean transport of 10^8 cm^3 s^{-1} (Winter & Martin 1990; Lutjeharms 1996; De Ruijter et al. 1999). South of Africa the Agulhas Current turns abruptly eastward in a tight retroflection loop and becomes known as the Agulhas Return Current (Fig. 2b). This current flows eastward to the Agulhas Plateau where it forms a major northward loop around the plateau (Camden Smith et al. 1981; Lutjeharms 1996; De Ruijter et al. 1999). The Agulhas Current is very sensitive to bottom topography and its location at the retroflection is very unstable. The Antarctic Intermediate Water (AAIW) follows the same path near South Africa as the Agulhas Current and also shows a retroflection (Fig. 2a; Lutjeharms 1996). It then flows eastward across the Agulhas Plateau. A south–southwestward flow of AABW can be observed along the western flank of the Agulhas Plateau at water depths of 3900–4900 m (Fig. 2a), whereas across the southern flank of the plateau the AABW shows a strong northeasterly flow (Tucholke & Embley 1984). This flow pattern results in an erosional zone at the western rim of the plateau and thinned sediments on the eastern flank.

All three geostrophic currents active in the area of the Agulhas Plateau are sensitive to bottom topography and thus fall within the group of contour currents (Hollister 1993; Faugères & Stow 1993; Redding 1996). Contour currents may winnow finer grained sediments to leave a discontinuous coarse-grained sand/gravel lag or transport sand and mud to construct elongate mounds or sediment drifts (Keary 1993). Each of the three currents has left its imprint in the sedimentary record in terms of contourite deposition, drift construction and zones of erosion or non-deposition. This paper investigates the seismic expression of this sediment record observed on the southern Agulhas Plateau, in an attempt to gather evidence on the Cenozoic evolution of the currents and their past flow patterns.

From: STOW, D. A. V., PUDSEY, C. J., HOWE, J. A., FAUGÈRES, J.-C. & VIANA, A. R. (eds)
Deep-Water Contourite Systems: Modern Drifts and Ancient Series, Seismic and Sedimentary Characteristics.
Geological Society, London, Memoirs, **22**, 271–288. 0435-4052/02/$15.00 © The Geological Society of London 2002.

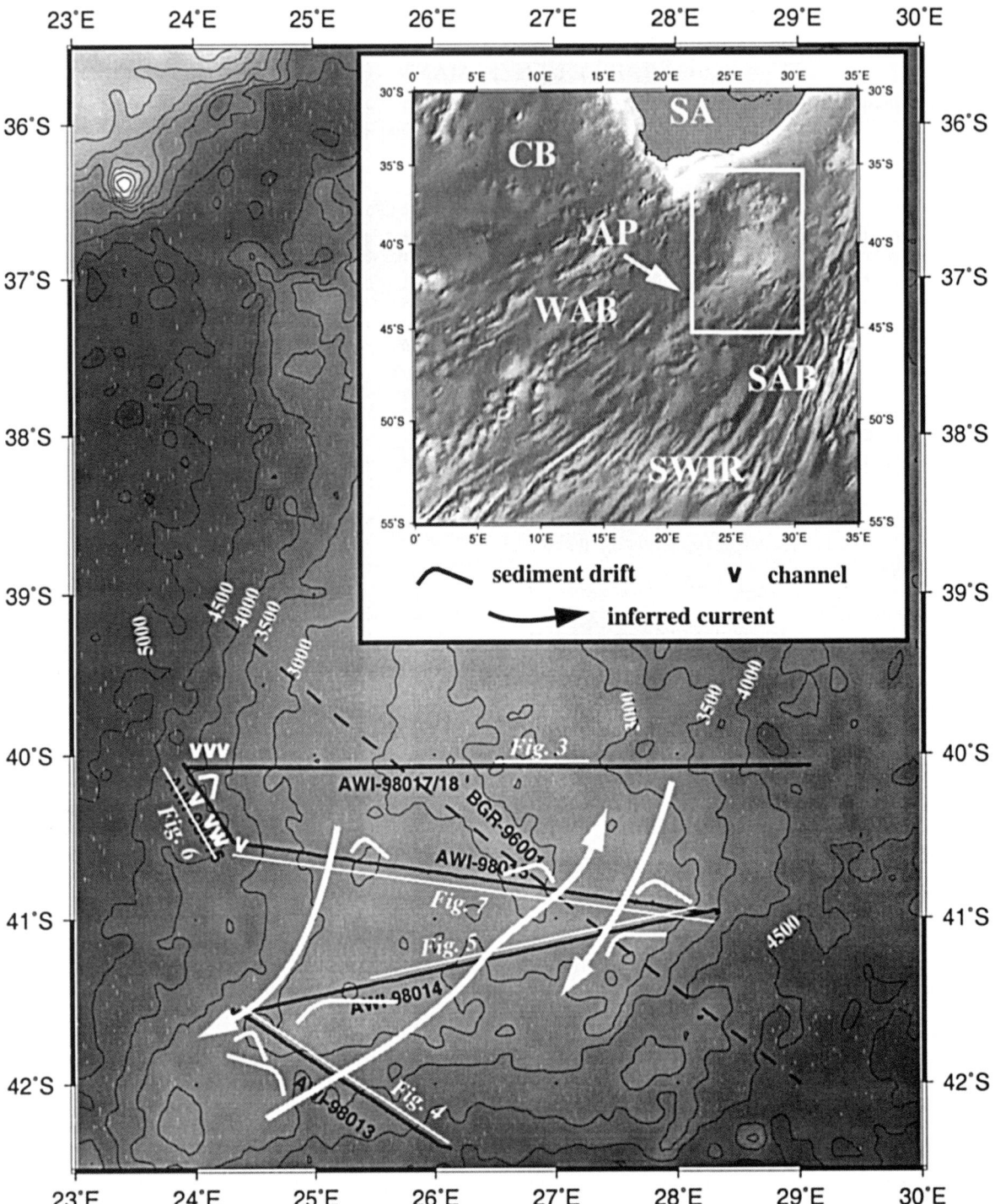

Fig. 1. Bathymetric map of the Agulhas Plateau showing the location of seismic reflection lines and observed sedimentary features (sediment drifts, channels and inferred currents). The insert map shows the location of the Agulhas Plateau relative to South Africa. Bathymetry is satellite derived from Smith & Sandwell (1977). AP, Agulhas Plateau; CB, Cape Basin; SA, South Africa; SAB, South Agulhas Basin; SWIR, Southwest Indian Ridge; WAB, West Agulhas Basin.

Seismostratigraphic framework

The data set presented comprises six high resolution multichannel seismic reflection lines (1550 km total length) gathered in 1997/1998 on the southern Agulhas Plateau (Fig. 1). Two GI-guns™ were used as seismic sources and a 2700 m long streamer (96 channels) for recording the data.

The sedimentary record observed in these lines shows a number of prominent reflections. The stratigraphic framework discussed here was derived via a correlation of our data with the stratigraphy as defined by Tucholke & Carpenter (1977) and Tucholke & Embley (1984). They identified four distinct horizons in their seismic lines which they related to regional hiatuses.

The deepest reflection observed is very smooth with a strong amplitude and good continuity (Maastrichtian in Fig. 3). This reflector is correlated with the top of the acoustic basement of Tucholke et al. (1981). They were able to core the reflector where it crops out at the western flank and high up on the plateau and identified it as a surface of unconformity. The minimum age of the reflector was found to be Maastrichtian, and at least 25 Ma of the

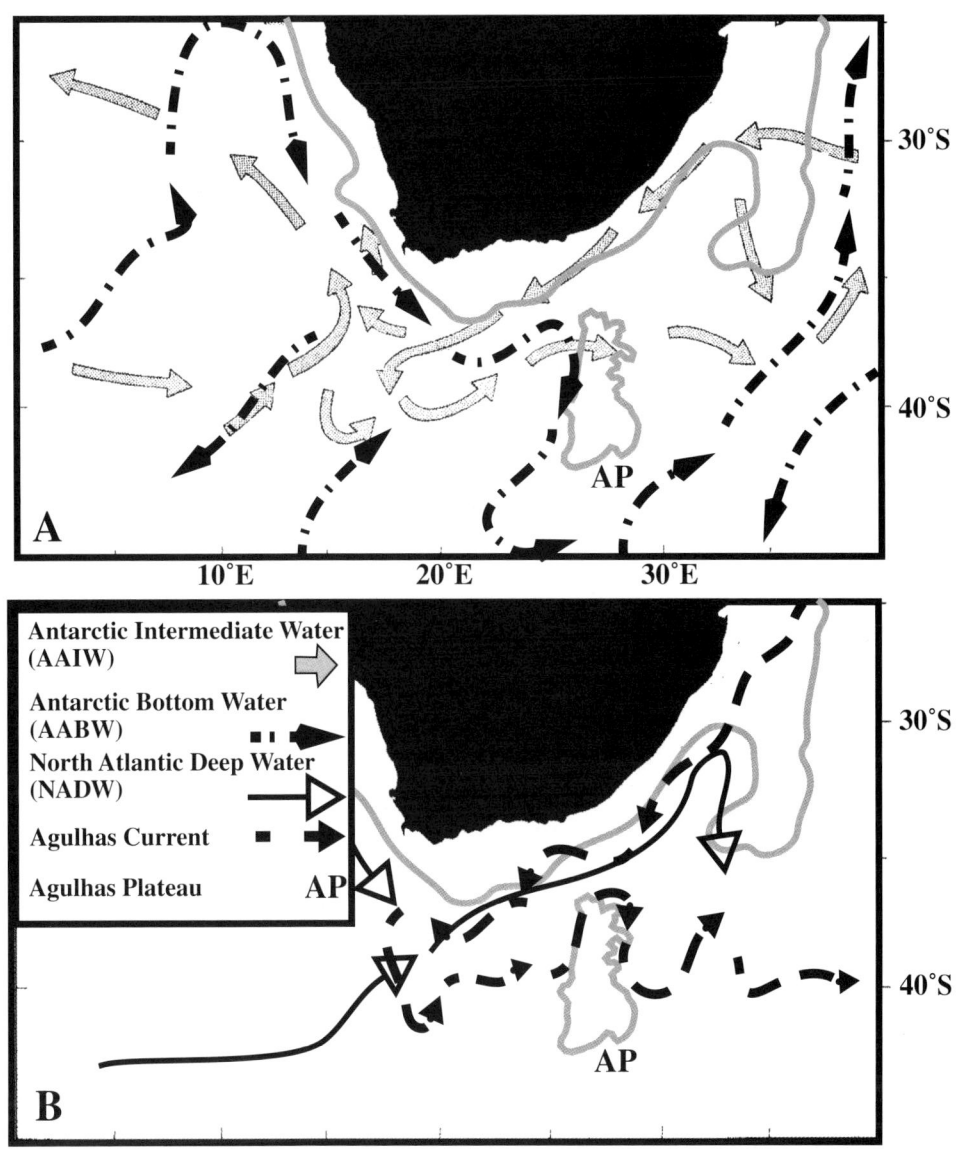

Fig. 2. Present circulation pattern of Antarctic Bottomwater (AABW), Antarctic Intermediate Water (AAIW) and the Agulhas Current (AC) in the South Atlantic and South Indic (modified from Dietrich *et al.* 1975; Faugères *et al.* 1993; Lutjeharms 1996; Niemi *et al.* 2000). A) the deeper watermasses (AABW and AAIW), B) NADW and the surface current (AC). The light gray line shows the 3000 m depth isoline of the Agulhas Plateau (AP).

older sedimentary record was missing (Tucholke *et al.* 1981). Our data reveal this reflector to be an erosional surface. The underlying reflectors, which were identified as originating from lavaflows and basement (Uenzelmann-Neben *et al.* 1999), terminate against the Late Cretaceous reflector thus documenting strong erosion (Figs 3–7).

The oldest intra-Cenozoic hiatus of regional significance occurs at the Paleocene/Eocene boundary. It correlates with a sealevel highstand and low sedimentation rates (Tucholke & Embley 1984). The corresponding reflector shows a strong amplitude and a mostly conformable relationship to both beds above and below (LE in Fig. 3). The second distinct reflector corresponds to an Early/Middle Oligocene hiatus (LO in Fig. 3). This hiatus is interpreted to result from intensified abyssal currents, e.g. the production and spreading of the Antarctic Circumpolar Current (ACC, Tucholke & Embley 1984). The Lower Oligocene reflector is of medium to strong amplitude and in parts represents an unconformity (e.g. Fig. 4, CDPs 3200–4000).

The Middle Miocene reflector (MM in Fig. 3) is characterized by weaker amplitudes and frequent wedge-outs at the seafloor (e.g. Fig. 4, CDP 6200 & Fig. 5, CDP 6400). It has also been interpreted as due to erosion and non-deposition by AABW (Tucholke & Embley 1984). The most important regional hiatus is of upper Miocene/lower Pliocene age and can be attributed to erosion and redeposition of sediments by circumpolar deep water within the ACC (Tucholke & Embley 1984). This hiatus is represented by a strong reflector (LP in Fig. 3) which is often found very close to and thus indistinguishable from the seafloor. The veneer of Agulhas Plateau Plio–Pleistocene sediments on top of the Lower Pliocene reflector is up to several tens of metres thick in places. This is mostly the case in areas of sediment drifts (Figs 4, 5 & 7).

Seismic characteristics of contourites

The shape of contourite drifts is controlled by bottom current intensity and the length of time over which the bottom current processes have operated, by interaction with bottom topography, and by the effect of Coriolis Force (Faugères *et al.* 1999). Contourite sheets show a very broad low-mounded geometry. Their internal seismofacies is characterized by low amplitude discontinuous reflectors or a more or less transparent appearance. Elongate mounded drifts are distinctly elongate and mounded in shape. They are tens to hundreds of kilometres long and several hundred metres thick, and quite distinct from the typically paired levees of turbidity current origin. Sediment waves are large-scale, regular bedforms with wavelengths between one and ten kilometres and amplitudes ranging from ten to a hundred metres. They

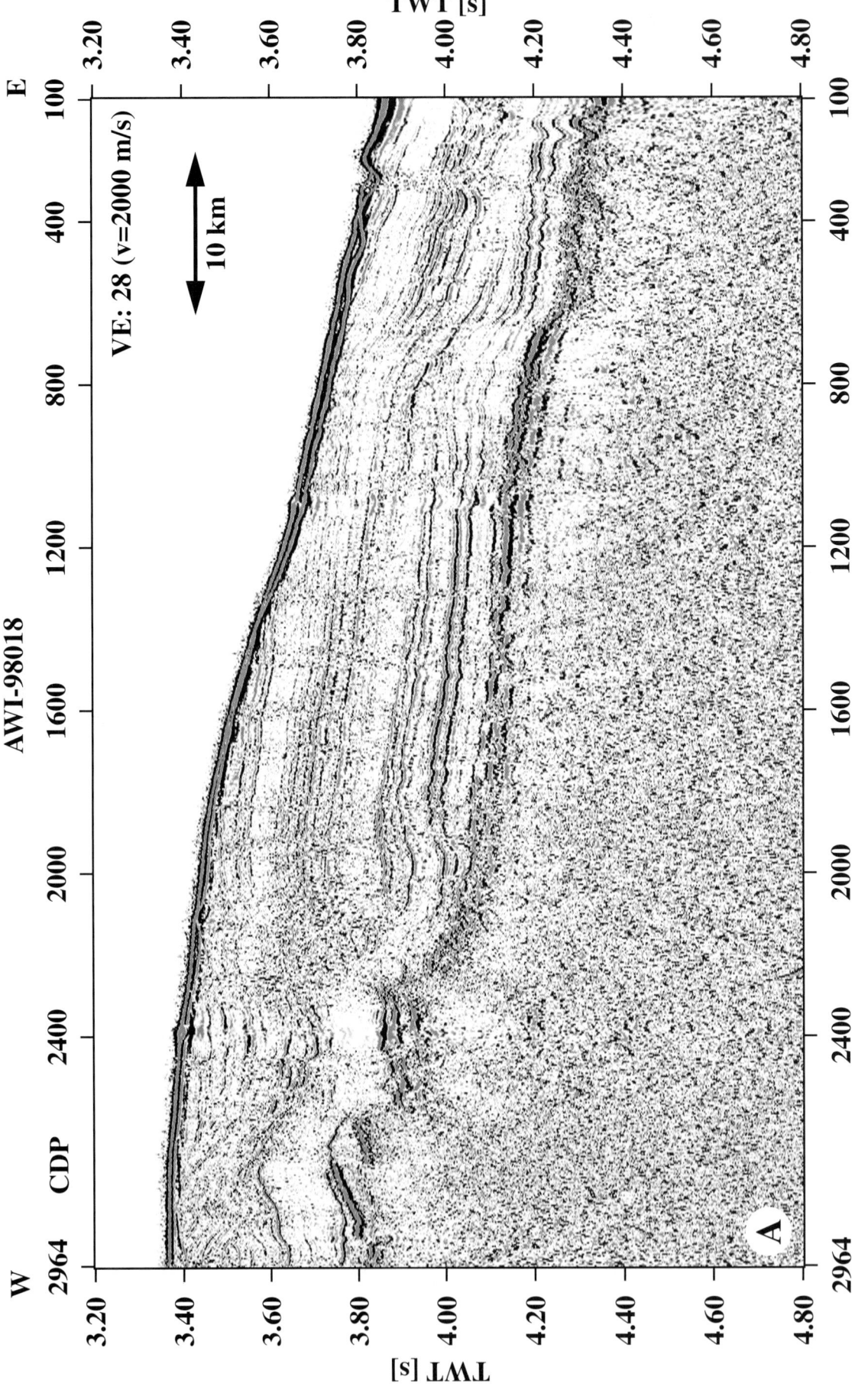

Fig. 3. Line AWI-98018. A) seismic data. B) interpretation. Via a correlation with the stratigraphy as defined by Tucholke & Carpenter (1977) and Tucholke & Embley (1984) a number of reflections could be identified. EC, extrusion centre; LE, Lower Eocene; LO, Lower Oligocene; LP, Lower Pliocene; MM, Middle Miocene.

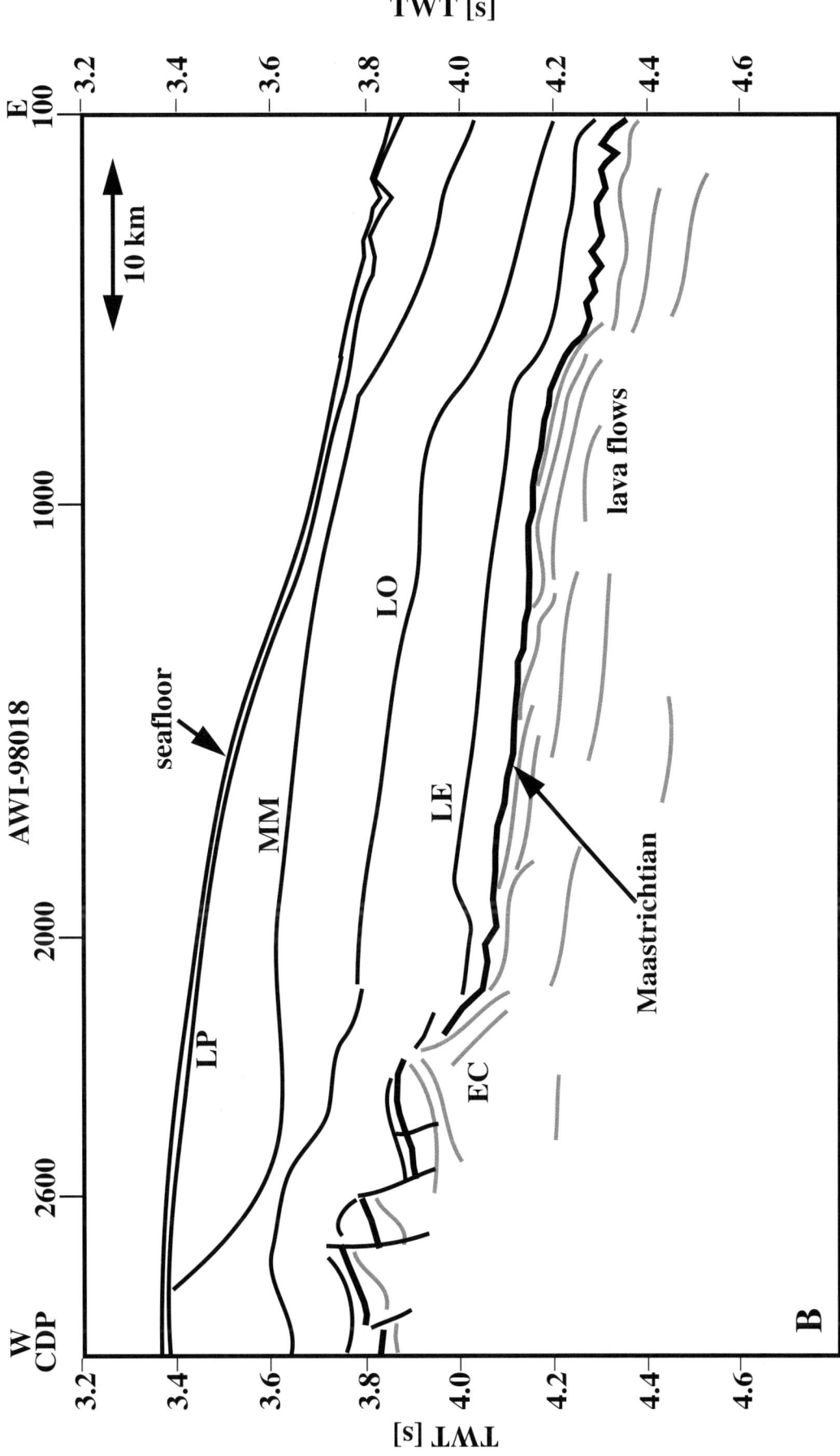

Fig. 3 (B).

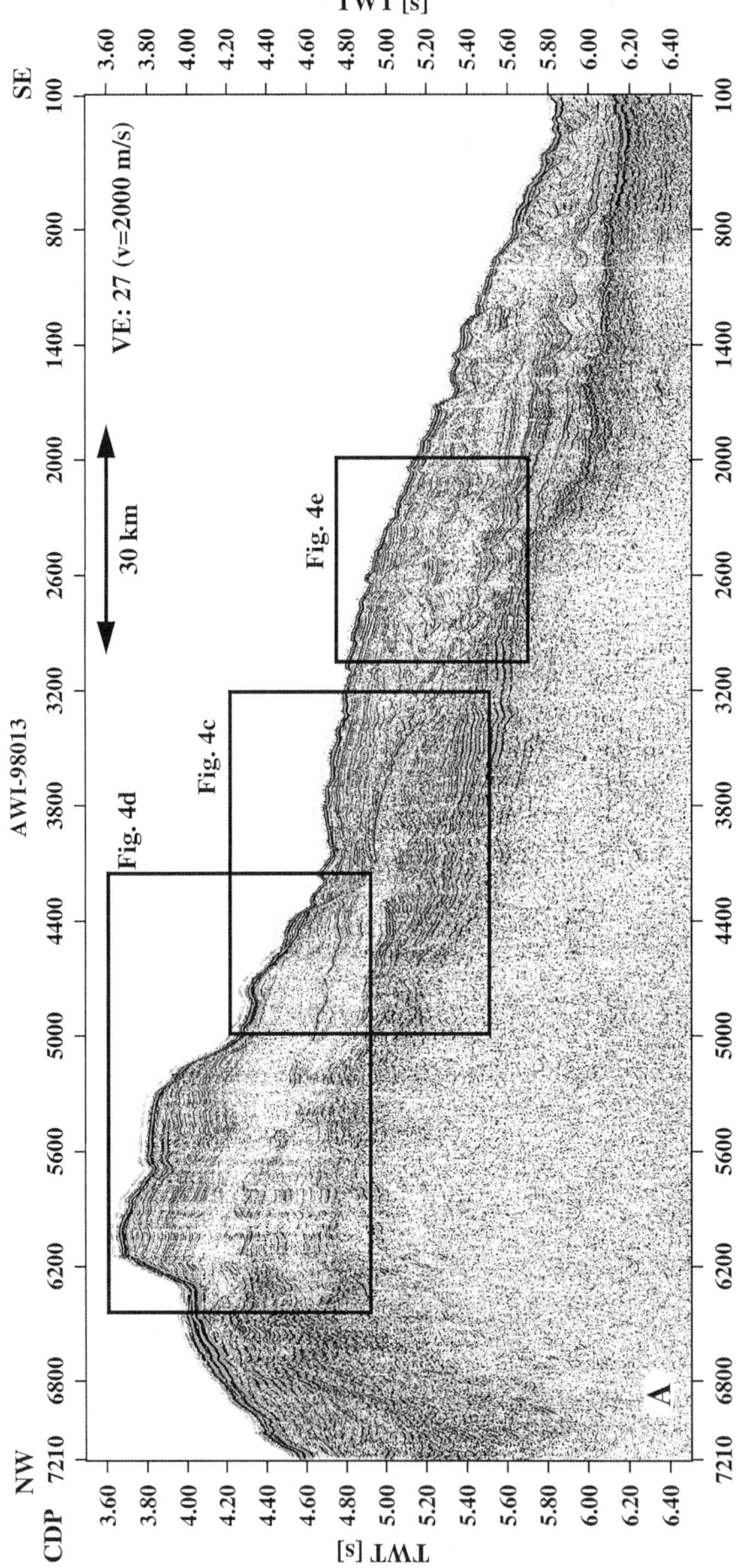

Fig. 4. Line AWI-98013. (A) seismic data, (B) interpretation, (C) detail of the eastern, older part of the sediment drift, (D) detail of the western, younger part of the sediment drift, (E) detail of the sediment wave field. Note the sediment drift on the western part of the profile has migrated westward with time. A south-setting as well as a north-setting current have been inferred to have shaped this drift. The sediment waves show a westward movement as well even if not for the same extent as the drift. Thin lines in (E) denote small faults, dashed lines the axes of some sediment waves. ⊗ indicates a current flowing into the figure whereas ⊙ indicates a current flowing out of the figure plane. EC, extrusion centre; LE, Lower Eocene; LO, Lower Oligocene; LP, Lower Pliocene; MM, Middle Miocene.

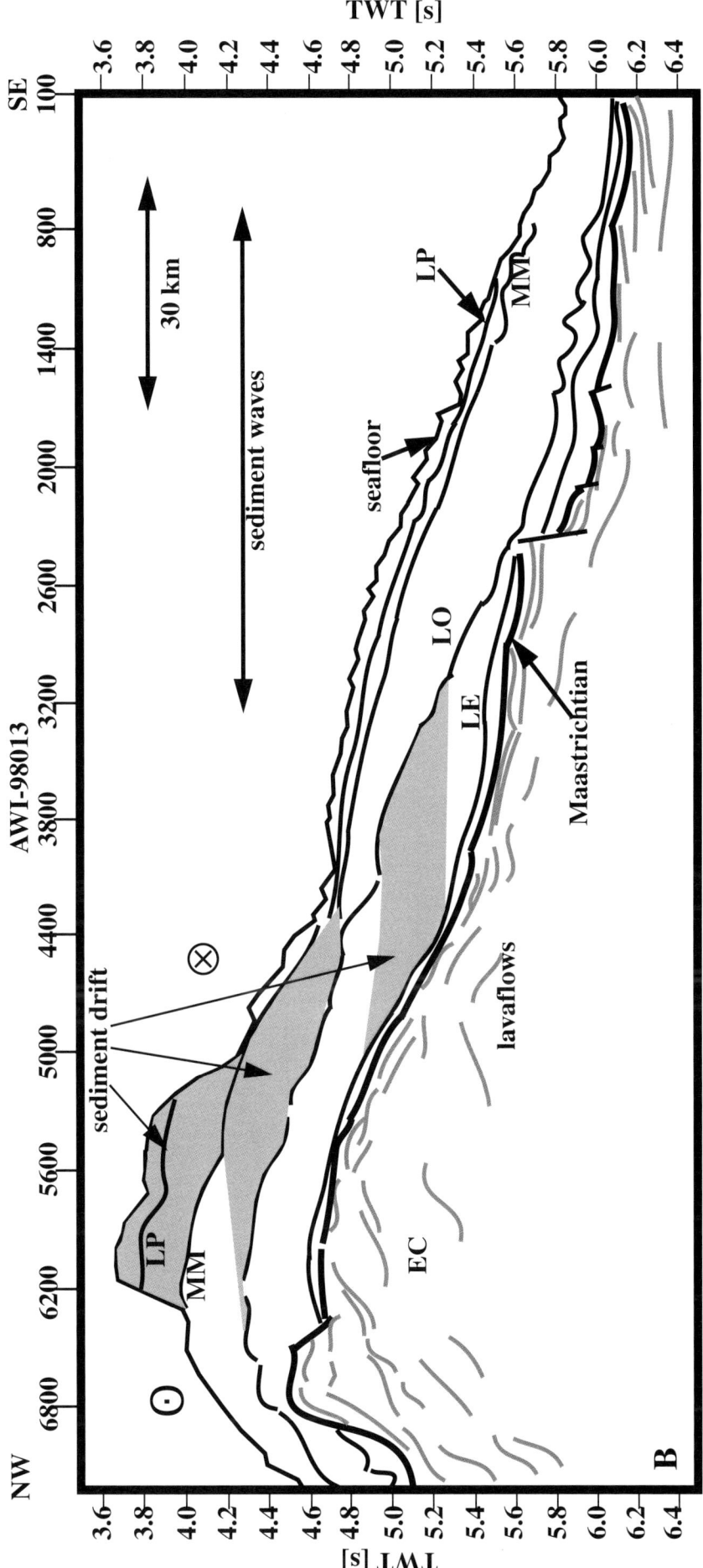

Fig. 4(B).

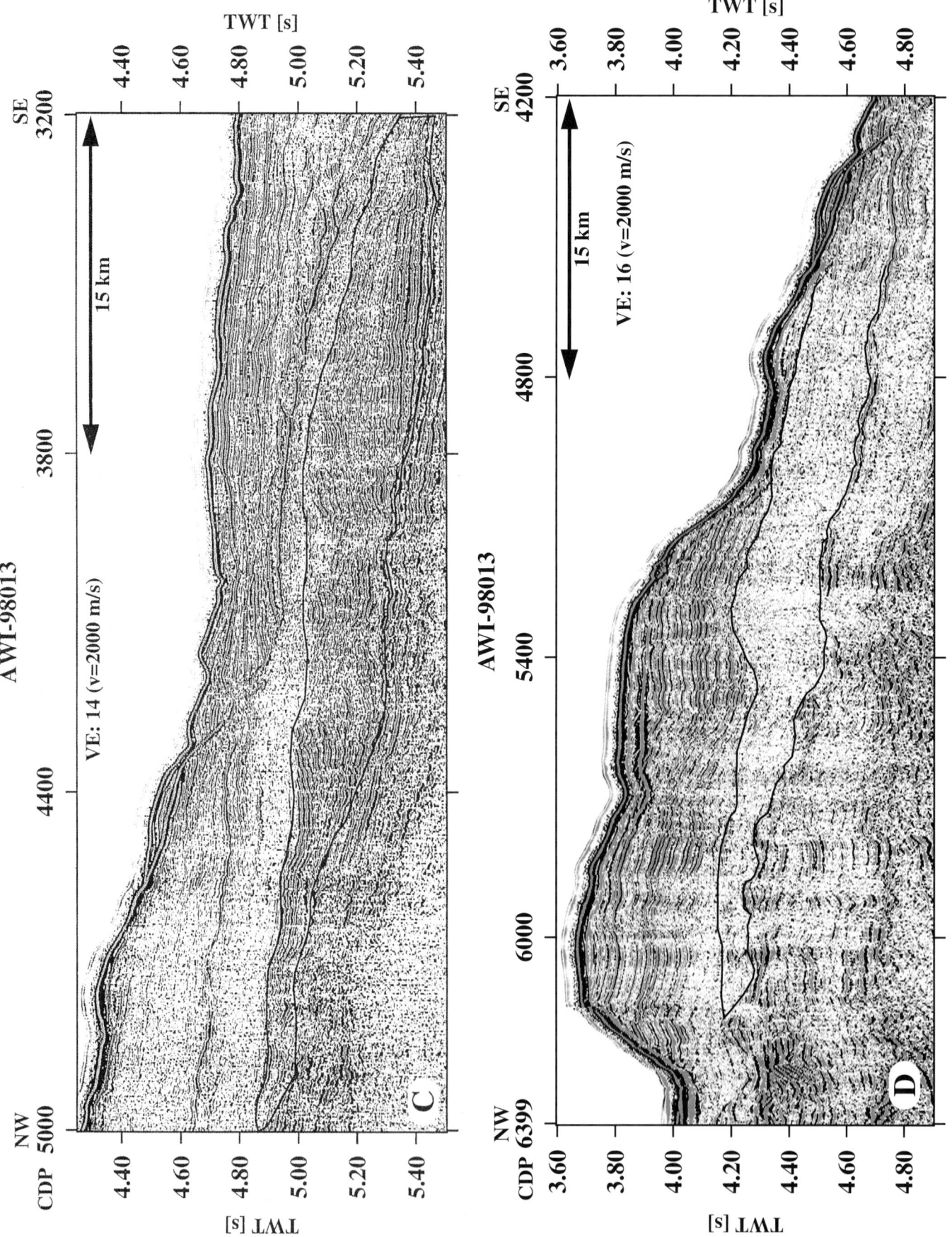

Fig. 4(C, D).

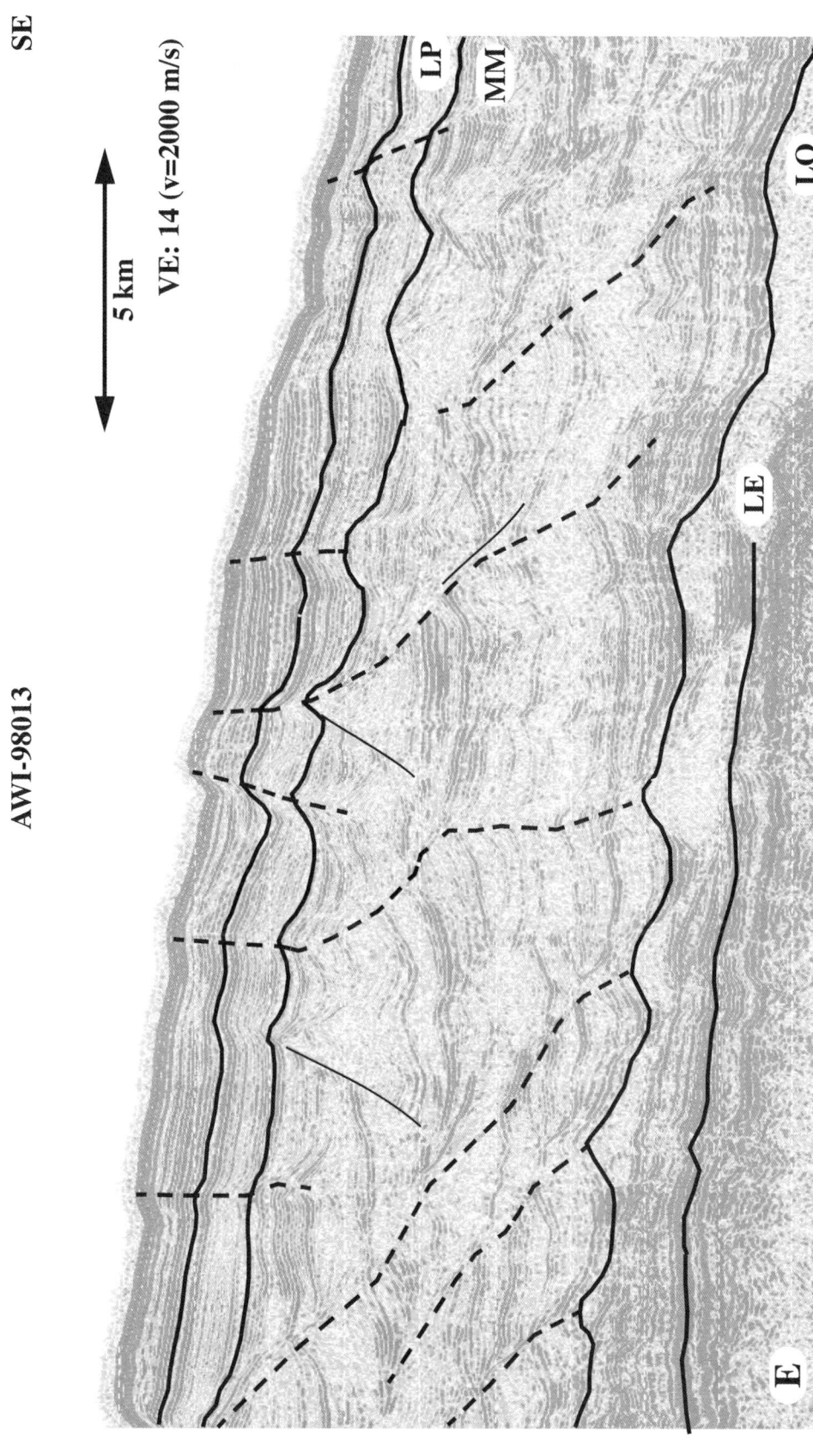

Fig. 4(E).

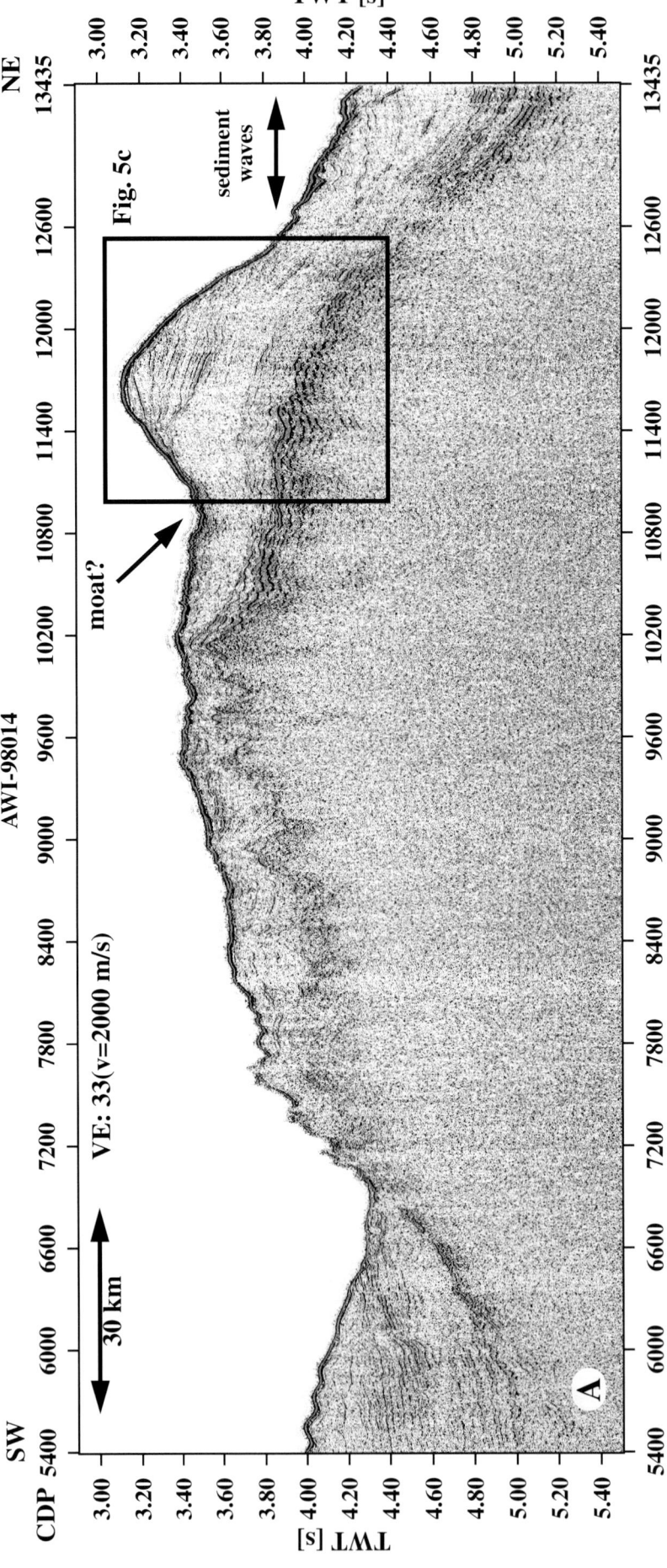

Fig. 5. Line AWI-98014. (A) seismic data, (B) interpretation, (C) detail of sediment drift. Note the depression on the western part of the line where a north setting current eroded the material. A prominent sediment drift can be observed on the eastern part of the line. This drift shows several re-locations. Still, the current which built-up the drift appears to have been fairly stationery since Lower Oligocene times. ⊗ indicates a current flowing into the figure whereas ⊙ indicates a current flowing out of the figure plane. EC, extrusion centre; LE, Lower Eocene; LO, Lower Oligocene; LP, Lower Pliocene; MM, Middle Miocene.

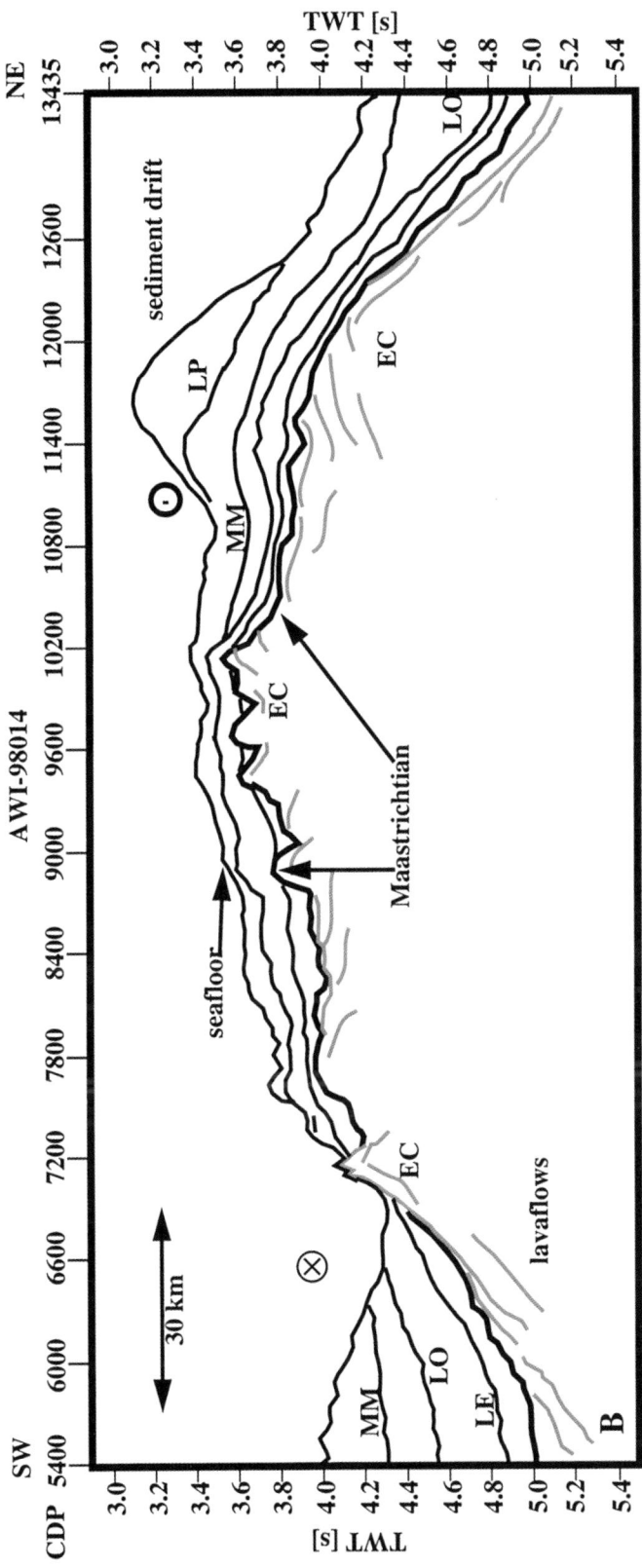

Fig. 5(B).

form in fine-grained sediments and occur over large areas of the deep seafloor as the result of both turbidity current and bottom current processes.

The history of contourite drift growth is marked by alternations of periods of sedimentation and erosion/non-deposition corresponding to instabilities and/or changes in the current regime. This leads to the superposition of depositional units of lenticular geometry, with smooth to irregular bounding surfaces, whose limits correspond to major discontinuities. The principal seismic facies observed are: (a) transparent layers interbedded with zones of seismic reflectors, (b) smooth, parallel reflectors of moderate to low amplitude interbedded with transparent zones, (c) short, discontinuous to chaotic reflectors, (d) sigmoid progradational reflectors, (e) gently wavy reflectors (irregular wavelength >10 km) and regular, migrating sediment waves (wavelength, 1–10 km), and (f) horizontal and low-inclination reflectors truncated at the seafloor or by an internal erosional surface (Faugères et al. 1999).

Seismic characteristics: Agulhas Plateau

Our seismic data from the southern Agulhas Plateau show evidence for the strong current system effective in this area (Tucholke & Embley 1984; Lutjeharms 1996). Recognition of dunes and drifts, channels, erosional unconformities and wedge-out of the reflections at the seafloor document the powerful effect of both the Agulhas Current and the AABW.

A number of sediment drifts can be observed on the southern Agulhas Plateau (Figs 1, 4, 5 and 7). Some of these drifts are characterized by strong parallel reflectors which are truncated at the seafloor or at an internal erosional surface (Figs 4c, CDPs 3200–5000; 4d, CDPs 5000–6300; 7a, CDPs 800–1900 and 4200–5400 and 7c CDPs 5000–5800). Another drift interbedded between two reflective ones appears more transparent (Fig. 4d, CDPs 4600–6200). A stack of lenticular units can be observed (Figs 5c, 7a; CDPs 4200–5400 and 7c, CDPs 9400–10100). These features are all characteristic of contourite drifts (Faugères et al. 1999).

The sediment drifts observed on the Agulhas Plateau are up to 650 m thick and between 5 and 85 km wide. The bases of the drifts are mostly found in seismic units of the Lower Oligocene–Middle Miocene or Middle Miocene–Lower Pliocene. This observation indicates active contour currents on the plateau since Lower Oligocene time, which coincides with the beginning of production and spreading of the Antarctic Circumpolar Current (ACC, Tucholke & Embley 1984).

The oldest drift identified can be found on the western part of line AWI-98013 (Fig. 4, CDPS 2900–6400). The drift appears initially to have been built-up from the east, as the slope is steeper on the eastern side for the units Lower Eocene–Lower Oligocene and Lower Oligocene–Middle Miocene (Figs 4c, CDPs 2900–5000 & 4d, CDPs 4400–6400, respectively). This indicates a north-flowing current. The crest of the drift has migrated westward by about 38 km, suggesting migration of the current axis. From Middle Miocene time, the western slope of the drift appears to have become much steeper, suggesting the influence of a second, south-flowing current (Fig. 4d, CDPs 5000–6300), which we interpret as the beginning of a strong AABW flow.

Another prominent sediment drift can be observed on the eastern part of line AWI-98014 (Fig. 5), with relatively steep slopes on both the western (15°) and eastern (25°) side of the drift. This suggests the present-day influence of a south-flowing contour current. The base of the drift is formed by the Lower Oligocene reflection (Fig. 5, CDPs 11000–12500), and three subsequent re-locations of the drift can be distinguished (Fig. 5c). From Lower Oligocene to Lower Pliocene the drift moved about 750 m to the west. Subsequently, indications for a larger move to the east have been found (10 km, post-Lower Pliocene, Fig. 5c). Since then, the drift migrated again to the west (6 km). Thus, it appears that the

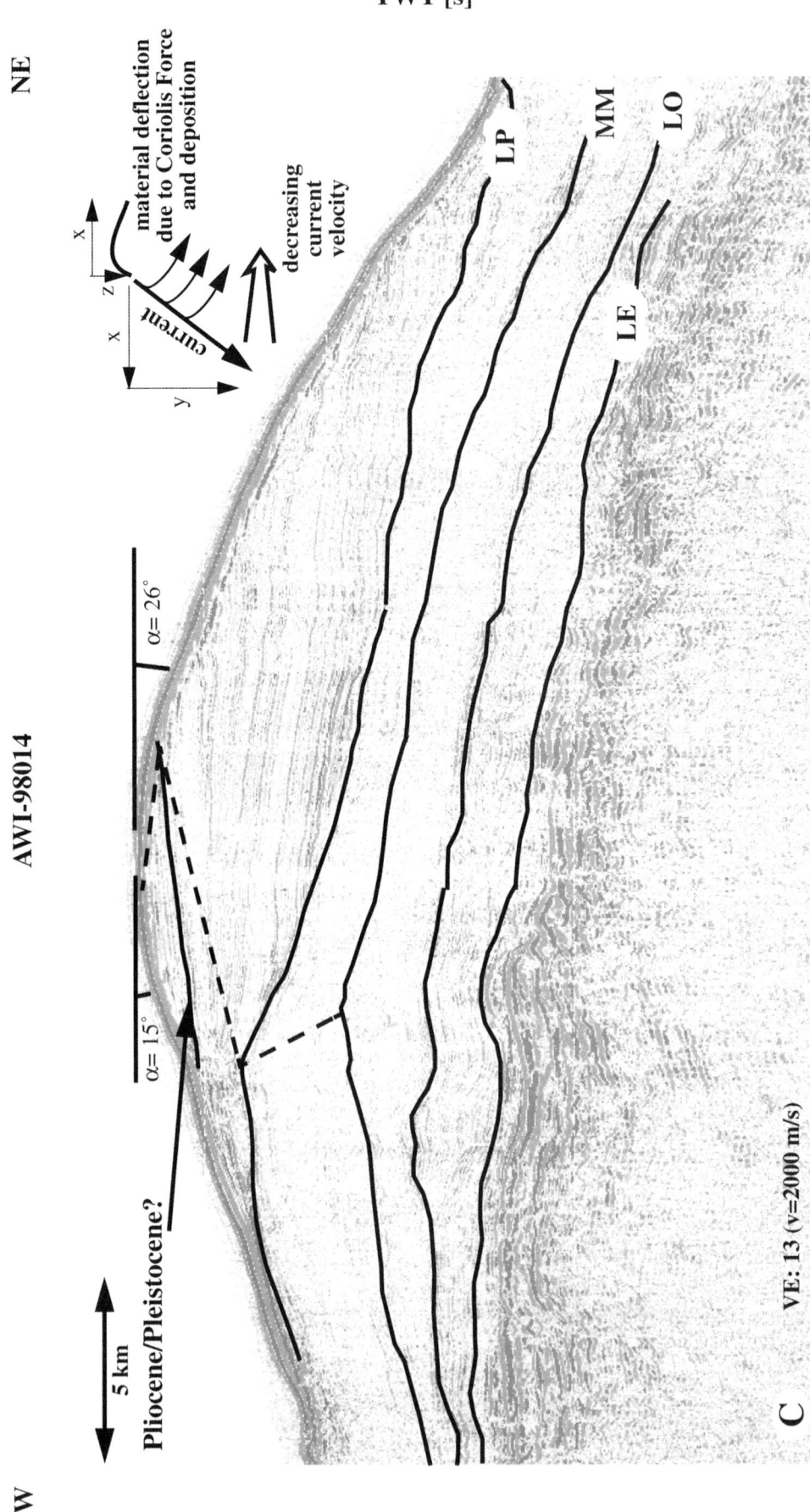

Fig. 5(C).

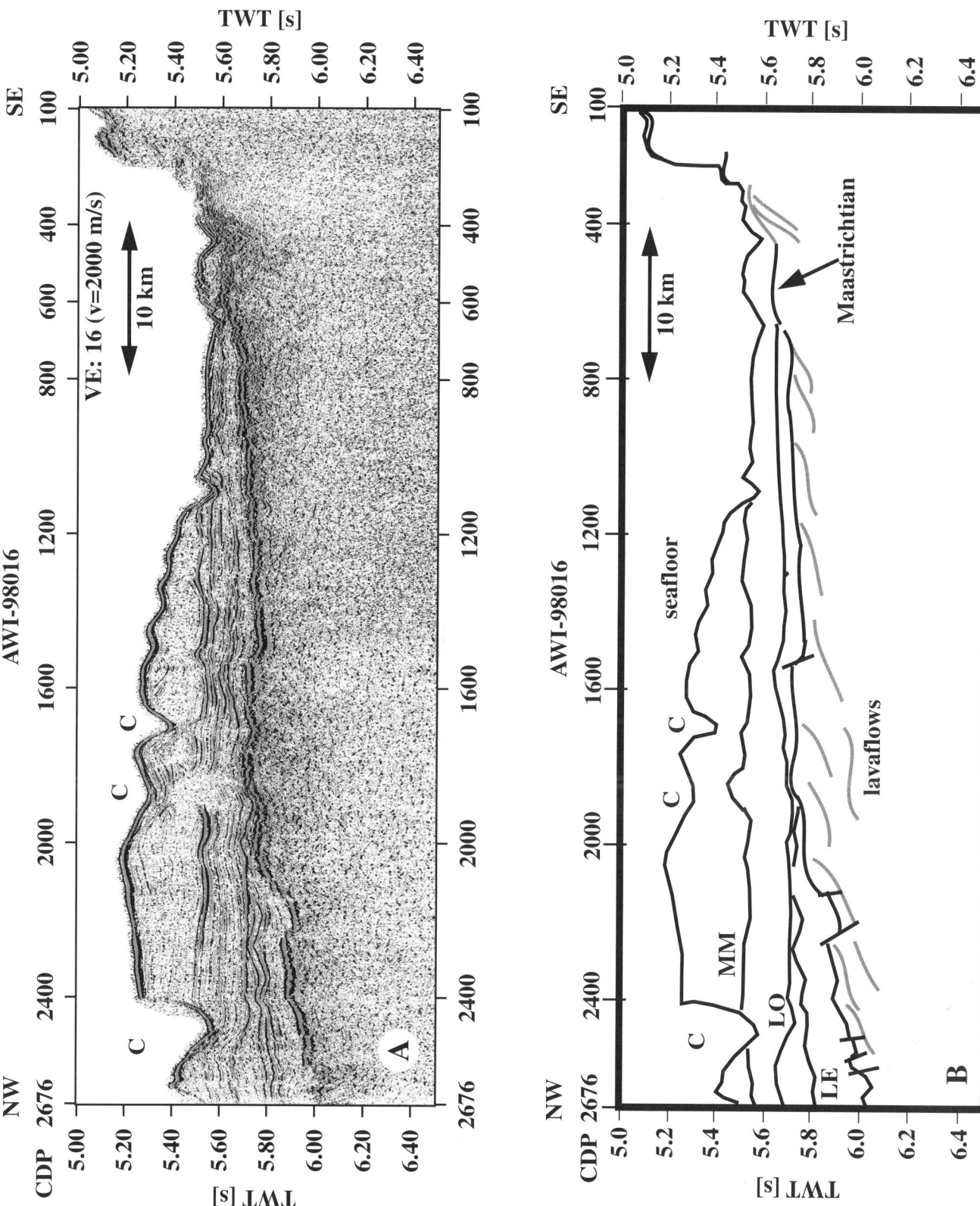

Fig. 6. Line AWI-98016. (A) seismic data, (B) interpretation. Note the channels which cut down into sediments as old as Lower Oligocene–Middle Miocene. C, channel; LE, Lower Eocene; LO, Lower Oligocene; LP, Lower Pliocene; MM, Middle Miocene.

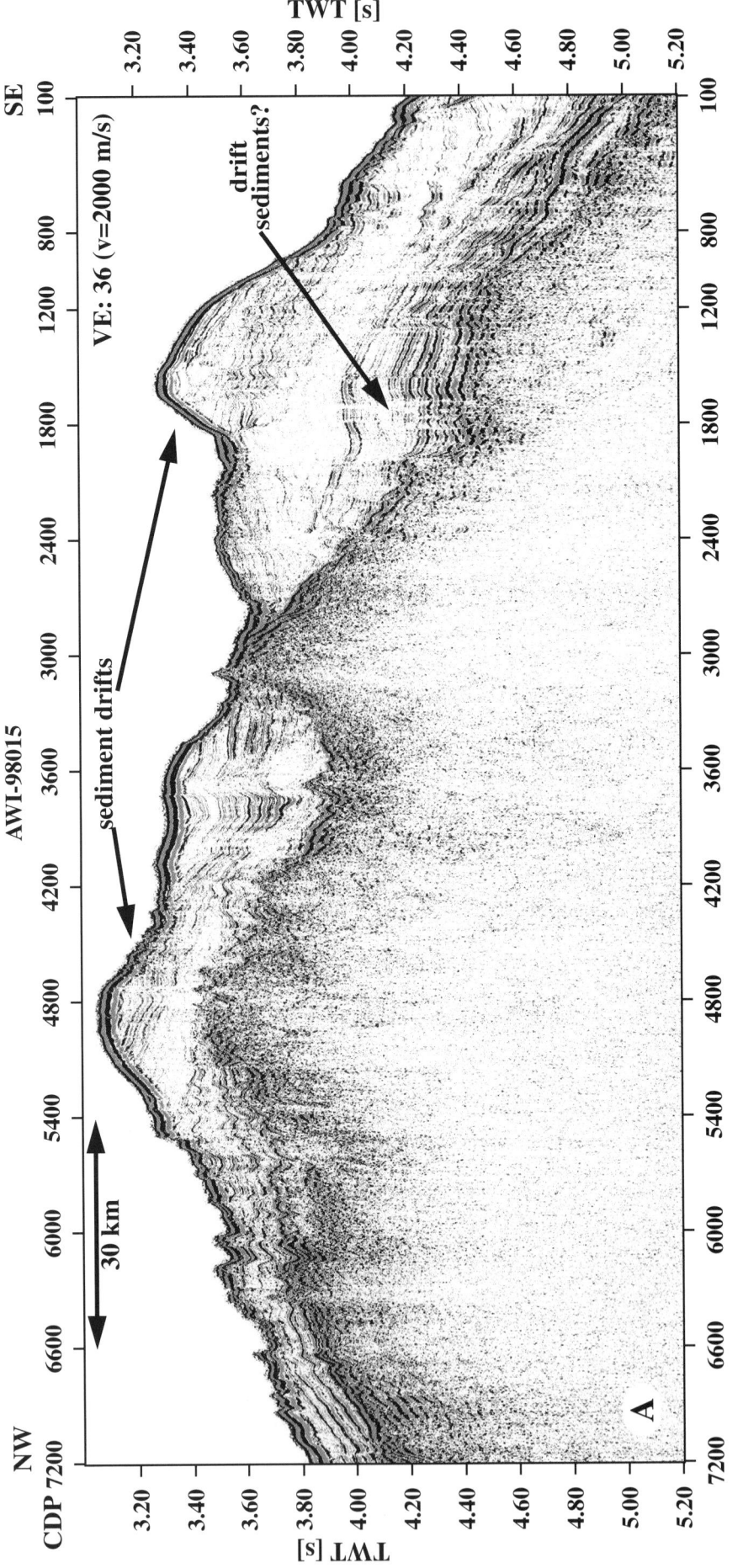

Fig. 7. Line AWI-98015. (A) seismic data, (B) interpretation of eastern part, (C) seismic data, and (D) interpretation of western part of the profile. Several sediment drifts can be identified shaped by both south- and north-setting currents. A distinct depression can be observed as well. ⊗ indicates a current flowing into the figure whereas ⊙ indicates a current flowing out of the figure plane. EC, extrusion centre; LE, Lower Eocene; LO, Lower Oligocene; LP, Lower Pliocene; MM, Middle Miocene.

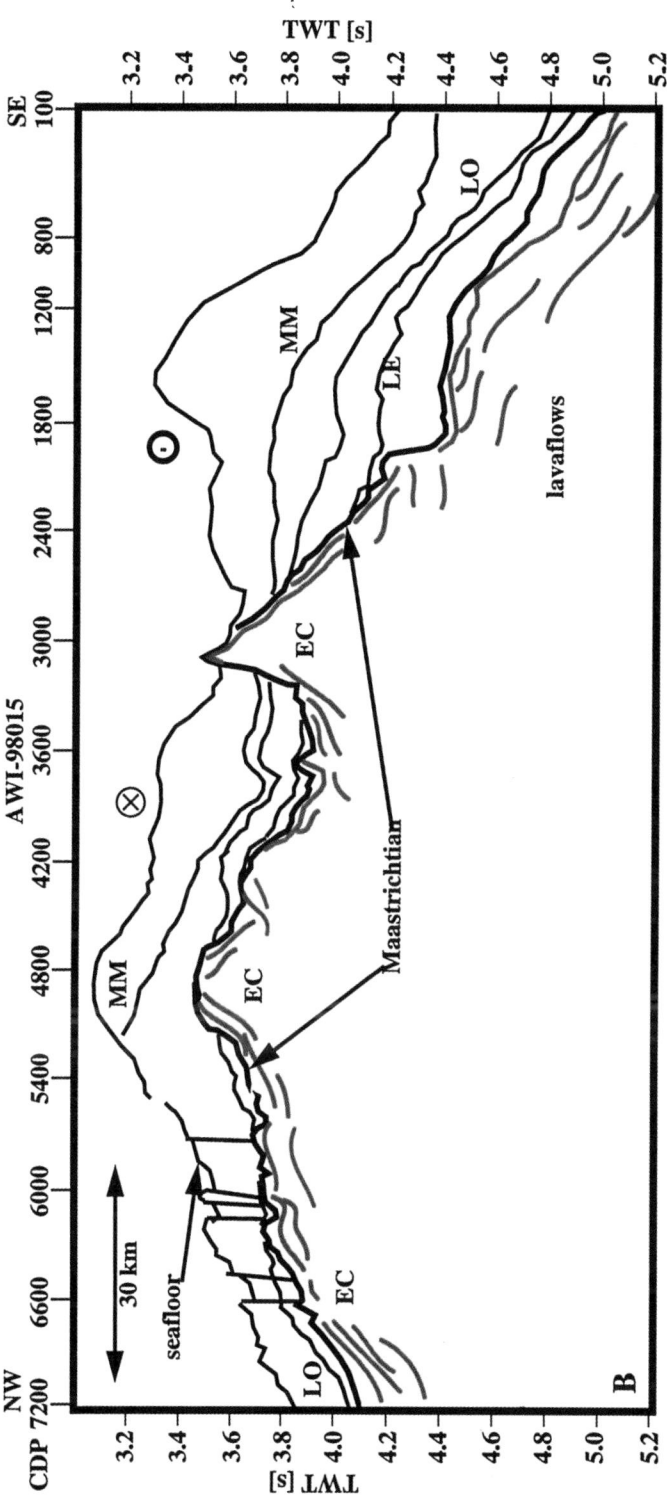

Fig. 7(B).

contour current shaping this drift has been active and maintained its position to within about 10 km since Lower Oligocene times.

Erosion has strongly affected the sedimentary sequences over much of the Agulhas Plateau, as is evident from the age of the outcropping sediments. Only a thin veneer of Quaternary sediments covers parts of the plateau (Tucholke & Carpenter 1977; Barrett 1977; Tucholke & Embley 1984), and in large areas the outcropping sedimentary unit is of Middle Miocene–Lower Pliocene or Lower Oligocene–Middle Miocene age or even older (Figs 3–7). Piston cores taken over the Agulhas Plateau show a well-defined unconformity between Pliocene–Quaternary and older sediments (Tucholke & Carpenter 1977).

A number of channels cutting the sedimentary sequences down into Lower Oligocene–Middle Miocene can be found on the western flank of the Agulhas Plateau (Fig. 6). Their occurrence supports the interpretation of strongly erosive currents. Further evidence for erosion can be found in the large depressions, in some cases with erosion down to basement, as seen on lines AWI-98014 (Fig. 5, CDPs 6000–7500) and AWI-98015 (Fig. 7a, CDPs 2400–3300).

East of the prominent sediment drift on line AWI-98013, a field of sediment waves can be observed (Fig. 4, CDPs 200–3300). The oldest occurrence of these waves is within seismic unit Lower Eocene–Lower Oligocene (east of CDP 3150) where the sediment waves are only poorly defined. The sediment waves appear much clearer and more regular within unit Lower Oligocene–Middle Miocene, showing amplitudes of about 50 m and wavelengths of about 3 km. In part, the sediment waves appear slightly irregular and are affected by small-scale faults, which may be the result of fluid escape or differential compaction, in turn leading to the irregular appearance of the sediment waves. A westward shift of the waves up to 9 km is evident (Fig. 4c). Thus, the sediment waves follow the migration of the crest of the sediment drift although not to the same extent. In Middle Miocene–Lower Pliocene, the sediment waves show a shift back to the east (Fig. 4c). Another set of sediment waves can be found east of the upstanding sediment drift on line AWI-98014 within unit Middle Miocene–Lower Pliocene (Fig. 5, CDPs 12700–13200). There, the amplitude of the sediment waves is about 25 m and they show wavelengths of up to 3.8 km, with no obvious migration.

Discussion

Based on the information gathered from the distribution and orientation (as far as this can be resolved by our seismic lines) of the sediment drifts, the locations of channels and the age of the outcropping sediment according to Tucholke & Carpenter (1977) and Tucholke & Embley (1984), the following model for the current pattern is proposed (Fig. 1). The western Agulhas Plateau has been affected by a south-flowing current since the Mid-Miocene. More speculatively, this current is partly deflected into several distinct streams down the flank of the plateau into the West Agulhas Basin, thereby cutting the observed channels. An alternative explanation might be local slumping and turbidity-current erosion. Across the southern Agulhas Plateau a NE-flowing current is inferred. Tucholke & Embley (1984) found indications for a similar current from bottom photographs. This current might result from a return loop of the flow on the western flank, which then turns northwards to build the sediment drift observed on line AWI-98013, erode the depression on line AWI-98014 and deposit another sediment drift observed on line AWI-98015.

At the present day, AABW is reported to cross the western Agulhas Plateau from north to south before entering the South Agulhas Basin and turning to the east (Fig. 2, Faugères et al. 1993). The inferred south-flowing current depicted from our data is probably, therefore, made up of AABW, and has been active since Middle Miocene times. In leaving the Agulhas Plateau and

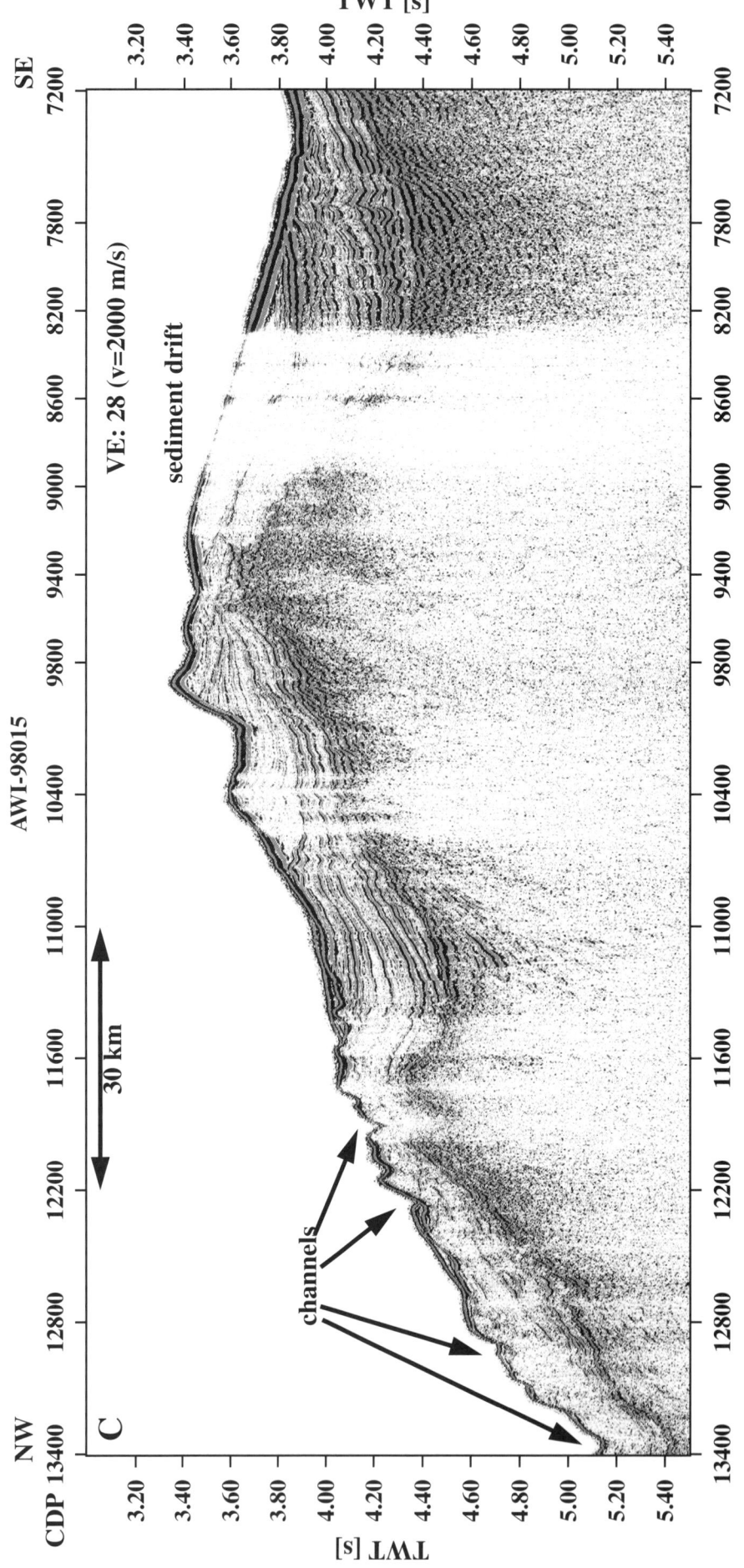

Fig. 7(C).

encountering greater water depths, a branch of the current might have been split and deflected round the southwestern tip of the plateau to flow northeastwards. Tucholke & Embley (1984) also reported a strong northeasterly flow across the southern Agulhas Plateau which they attributed to an AABW component derived directly from an Antarctic source. Since the observed sedimentary drifts indicate this flow to be the oldest on the southern Agulhas Plateau, it is believed to have existed since the early Eocene. It has also migrated some 38 km westwards.

Two sediment drifts on lines AWI-98014 and AWI-98015 indicate a second south-flowing current along the eastern flank of the southern Agulhas Plateau, which has influenced the sedimentary distribution since Lower Oligocene times. There exist two possible explanations for this eastern current. Firstly, the general flow of the AAIW has been described as being west to east across the northern Agulhas Plateau (Fig. 2, Lutjeharms 1996). A branch of this flow might have been deflected to the south. Secondly, the Agulhas Retroflection is known to flow southwards along the eastern flank of the Agulhas Plateau (Fig. 2). An extension of the retroflection farther to the south would lead to the observed sediment drifts and is within limits (Winter & Martin 1990, O. Boebel, pers. comm).

In conclusion, the Agulhas region south of South Africa is one of the most important gateways that allow interchange between water masses of the South Atlantic, Indian and Southern Oceans. This involves a very complex flow pattern which up to now has been known only in general terms. This study of seismic reflection profiles over the Agulhas Plateau has led to some refinement of the general model, in particular with regard to paleocirculation. However, further oceanographic work coupled with sediment coring will be necessary to resolve more definitely which currents have been active.

I am grateful for the support of the captain and crew of M. V. Petr Kottsov who helped us enormously during the expedition. I am further grateful to Prof. K. Hinz, Bundesanstalt für Geowissenschaften und Rohstoffe, for permission to use seismic line BGR-96001. I thank R. Dingle, J-C. Faugères and one anonymous reviewer for their careful and helpful comments. The work was funded by the Bundesministerium für Bildung, Forschung und Technologie under contract No. 03G0532A. The author is responsible for the contents of this paper.

References

ALLEN, R. B. & TUCHOLKE, B. E. 1981. Petrography and implications of continental rocks from the Agulhas Plateau, southwest Indian Ocean. *Geology*, **9**, 463–468.

BARRETT, D. M. 1977. The Agulhas Plateau off southern Africa: A geophysical study. *Geological Society of America Bulletin*, **88**, 749–763.

BEN-AVRAHAM, Z., NIEMI, T. M. & HARTNADY, C. J. H. 1994. Mid-Tertiary changes in deep ocean circulation patterns in the Natal Valley and Transkei Basin, Southwest Indian Ocean. *Earth and Planetary Science Letters*, **121**, 639–646.

BEN-AVRAHAM, Z., HARTNADY, C. J. H. & LE ROEX, A. P. 1995. Neotectonic activity on continental fragments in the southwest Indian Ocean: Agulhas Plateau and Mozambique Ridge. *Journal of Geophysical Research*, **100**, 6199–6211.

CAMDEN-SMITH, F., PERRINS, L.-A., DINGLE, R. V. & BRUNDRIT, G. B. 1981. A preliminary report on long-term bottom-current measurements and sediment transport/erosion in the Agulhas Passage, southwest Indian Ocean. *Marine Geology*, **29**, 81–88.

DE RUIJTER, W. P. M., BIASTOCH, A., DRIJFHOUT, S. S., LUTJEHARMS, J. R. E., MATANO, R. P., PICHEVIN, T., VAN LEEUWEN, P. J. & WEIJER, W. 1999. Indian-Atlantic interocean exchange: Dynamics, estimation and impact. *Journal of Geophysical Research*, **C104**, 208885–20910.

DIETRICH, G., KALLE, K., KRAUSS, W. & SIEDLER, G. 1975. *Allgemeine Meereskunde*. Gebr. Bornträger.

DINGLE, R. V. & CAMDEN-SMITH, F. 1979. Acoustic stratigraphy and current-generated bedforms in deep ocean basins off southeastern Africa. *Marine Geology*, **33**, 239–260.

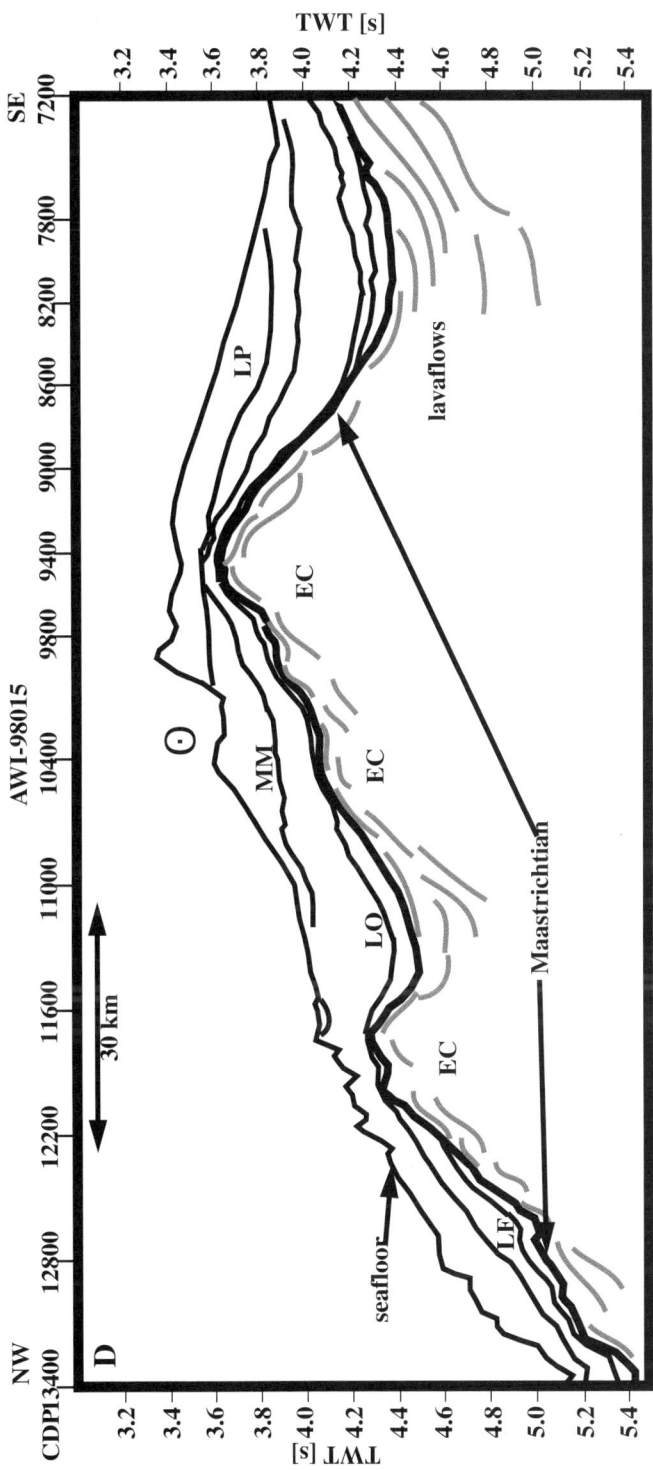

Fig. 7(D).

FAUGÈRES, J.-C. & STOW, D. A. V. 1993. Bottom-controlled sedimentation: A synthesis of the contourite problem. *Sedimentary Geology*, **82**, 287–297.

FAUGÈRES, J.-C., MÉZERAIS, M. L. & STOW, D. A. V. 1993. Contourite drift types and their distribution in the North and South Atlantic ocean basins. *Sedimentary Geology*, **82**, 189–203.

FAUGÈRES, J.-C., STOW, D. A. V., IMBERT, P. & VIANA, A. 1999. Seismic features diagnostic of contourite drifts. *Marine Geology*, **162**, 1–38.

HOLLISTER, C. D. 1993. The concept of deep-sea contourites. *Sedimentary Geology*, **82**, 5–11.

KEARY, P. 1993. *The Encyclopedic Dictionary of the Solid Earth Sciences*. Blackwell Scientific Publications, Oxford.

KRISTOFFERSEN, Y. & LABRECQUE, J. L. 1991. On the tectonic history and origin of the Northeast Georgia Rise. *In:* CIESIELSKI, P. F., Y. KRISTOFFERSEN *et al.* (eds) *Proceedings of the Ocean Drilling Program, Scientific Results*, **114**. College Station, TX, Ocean Drilling Programme, 23–38.

LABRECQUE, J. L. & HAYES, D. E. 1979. Seafloor spreading history of the Agulhas Basin. *Earth and Planetary Science Letters*, **45**, 411–428.

LUTJEHARMS, J. R. E. 1996. The exchange of water between the South Indian and South Atlantic Oceans. *In:* WEFER, G., BERGER, W. H., SIEDLER, G. & WEBG, D. J. (eds) *The South Atlantic: Present and Past Circulation*. Springer Verlag, Berlin, 125–162.

MARTIN, A. K. & HARTNADY, C. J. H. 1986. Plate tectonic development of the southwest Indian Ocean: A revised reconstruction of East Antarctica and Africa. *Journal of Geophysical Research*, **91**, 4767–4786.

NIEMI, T., BEN-AVRAHAM, Z., HARTNADY, C. J. H. & REZNIKOV, M. 2000. Post-Eocene seismic stratigraphy of the deep ocean basin adjacent to the southeast African continental margin: A record of geostrophic bottom current systems. *Marine Geology*, **162**, 237–258.

READING, H. G. 1996. *Sedimentary environments: Processes, facies and stratigraphy*. Blackwell Scientific Publications, Oxford.

SIESSER, W., ROGERS, J. & WINTER, A. 1988. Late Neogene erosion of the Agulhas Moat and the Oligocene position of Subantarctic surface water. *Marine Geology*, **80**, 119–129.

SMITH, W. H. F. & SANDWELL, D. T. 1977. Global seafloor topography from satellite altimetry and ship depth sounding. *Science*, **277**, 1956–1962.

TUCHOLKE, B. E. & CARPENTER, G. B. 1977. Sedimentary distribution and Cenozoic sedimentation patterns on the Agulhas Plateau. *Geological Society of America Bulletins*, **88**, 1337–1346.

TUCHOLKE, B. & EMBLEY, R. E. 1984. Cenozoic regional erosion of the abyssal sea floor off South Africa. *In*: SCHLEE, J. S. (ed.) *Interregional unconformities and hydrocarbon accumulation*. AAPG, Tulsa, Memoir, **36**, 145–164.

TUCHOLKE, B. E., HOUTZ, R. E. & BARRETT, D. M. 1981. Continental crust beneath the Agulhas Plateau, southwest Indian Ocean. *Journal of Geophysical Research*, **86**, 3791–38061.

UENZELMANN-NEBEN, G., GOHL, K., EHRHARDT, A. & SEARGENT, M. J. 1999. Agulhas Plateau, SW Indian Ocean: New evidence for excessive volcanism. *Geophysical Research Letters*, **26**, 1941–1944.

WINTER, A. & MARTIN, K. 1990. Late Quaternary history of the Agulhas Current. *Palaeoceanography*, **5**, 479–486.

The Weddell Sea: contourites and hemipelagites at the northern margin of the Weddell Gyre

CAROL J. PUDSEY

British Antarctic Survey, Madingley Road, Cambridge, UK (e-mail: cjp@bas.ac.uk)

Abstract: Fine-grained contourites and hemipelagites occur in the northern Weddell Sea, deposited from the Weddell Gyre. Where bottom currents are intensified over the slope, sandy and silty contourites have been deposited above 2000 m water depth. Mudwaves are generally uncommon, but are present in Powell Basin. The area has potential for high-resolution palaeoclimate studies (particularly the history of Antarctic Bottom Water production) if difficulties in dating can be overcome.

Geological and oceanographic setting

The Weddell Sea is a large marginal basin some 1100 km from N to S, lying east of the Antarctic Peninsula (Fig. 1). Its southern boundary includes the large Filchner and Ronne Ice Shelves and the coasts of Coats Land and Dronning Maud Land in East Antarctica. The northern boundary is the South Scotia Ridge, a chain of continental fragments and palaeo-island arcs (Barker *et al.* 1991). Oceanic basement in the Weddell Sea ranges from late Jurassic to early Miocene in age (Livermore & Woollett 1993; Livermore & Hunter 1996).

The Antarctic Peninsula is predominantly a Mesozoic magmatic

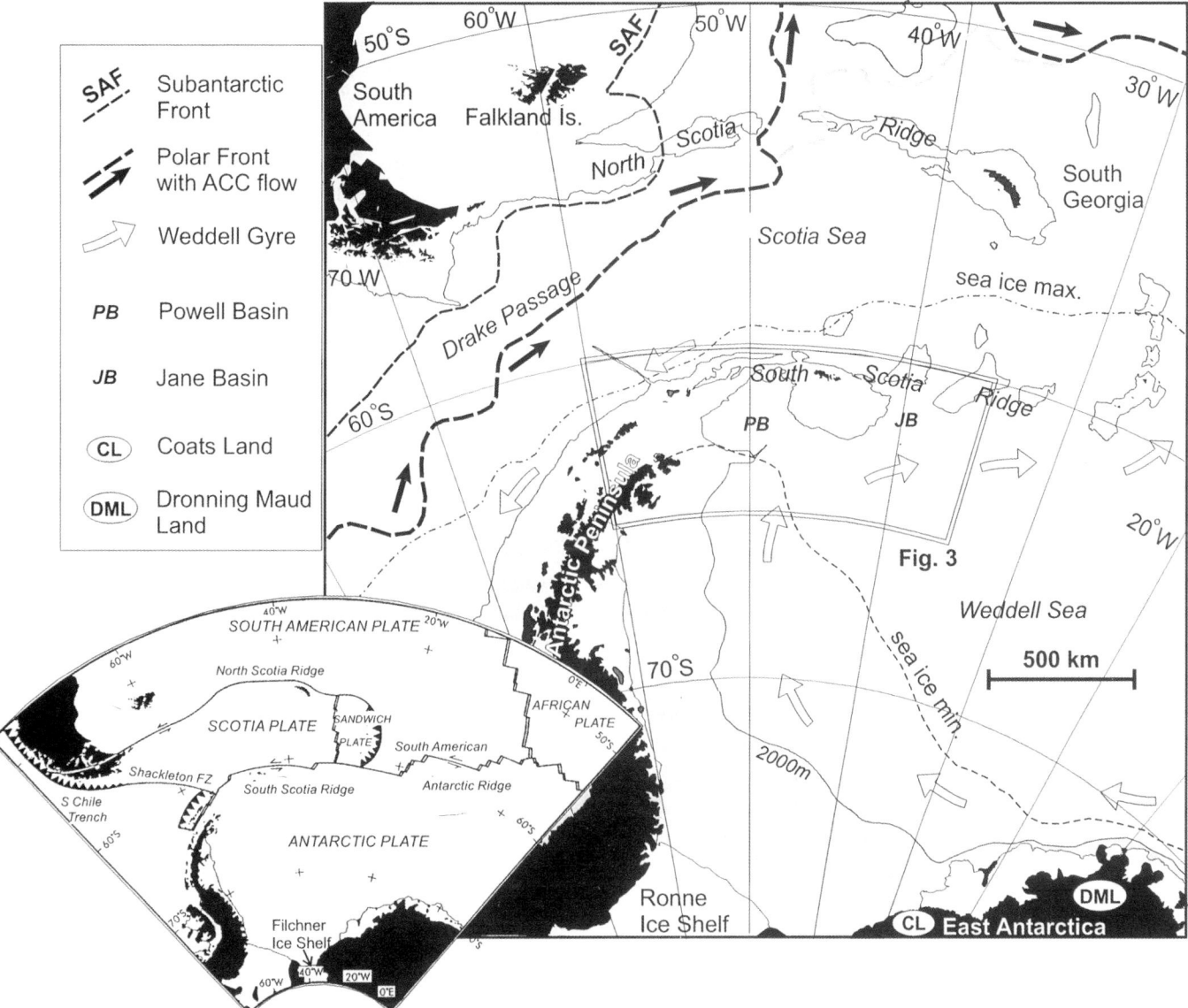

Fig. 1. Location map of the Weddell Sea, showing the deep circulation and the locations of surface-water features. Areas shallower than 2000 m shaded. Inset map from Tectonic Map (1985).

From: STOW, D. A. V., PUDSEY, C. J., HOWE, J. A., FAUGÈRES, J.-C. & VIANA, A. R. (eds)
Deep-Water Contourite Systems: Modern Drifts and Ancient Series, Seismic and Sedimentary Characteristics.
Geological Society, London, Memoirs, **22**, 289–303. 0435-4052/02/$15.00 © The Geological Society of London 2002.

Table 1. *Principal characteristics*

Location	Northern Weddell Sea, including Powell Basin and Jane Basin
Setting	Northern margin of the cyclonic Weddell Gyre: important area of bottom water formation
Age	Oligocene to Recent
Drift type	Plastered and separated drifts, transitional to hemipelagic basin fill
Dimensions	70–100 km wide, several hundred km downcurrent
Seismic characteristics	Laterally continuous parallel to undulating reflectors over most of basin; pinch-out at steep basin margins; plastered drift with buried sediment waves on NW slope, active field of sediment waves in Powell Basin
Sediment characteristics	Mainly fine-grained terrigenous, very poorly sorted, bioturbated; glacial-interglacial cyclicity in texture and biogenic content; sandy-silty contourites on slope

arc which formed along the active margin of Gondwana by subduction of Pacific ocean floor (see papers in Storey *et al.* 1996). The East Antarctic craton includes Archaean to Proterozoic igneous and metamorphic basement, overlain by Palaeozoic to Mesozoic cover (Tingey 1991). The coast is dominated by large fringing ice shelves, except for the islands in the north and part of Coats Land (Fig. 1). During Quaternary glacial stages, grounded ice extended to the continental shelf edge (review in Bentley & Anderson 1998), which resulted in copious sediment supply to the slope and rise (Grobe & Mackensen 1992).

At the western and southern basin margins there are thick (3–5 km) prograded wedges of sediment derived from the glaciated Antarctic continent (Barker & Lonsdale 1991; Oszko 1997). In much of the northern Weddell Sea, sediment thickness is only about 1 km (e.g. Shipboard Scientific Party 1988*a*) though it attains 2–3 km in Powell Basin (King *et al.* 1997). This atlas chapter describes contourite deposition along the northern margin of the Weddell Sea north of 65°S, including Powell Basin and Jane Basin (Fig. 1).

The Weddell Gyre (Fig. 1) is an important component of global thermohaline circulation as it is the source for much of the world ocean's cold bottom water (see Orsi *et al.* 1993; Fahrbach *et al.* 1995; Stow *et al.* 2002). In the northern Weddell Sea, the deepest water mass is Weddell Sea Bottom Water (WSBW, $\theta \leq -0.7°C$; Carmack 1974). Weddell Sea Deep Water ($\theta \leq -0°C$) and the overlying Circumpolar Deep Water ($\theta \geq -0°C$) occupy most of the water column, with Antarctic Surface Water in the uppermost 200 m. Deep current flow has been measured at several sites on the northern side of the Weddell Gyre, from the upper slope to the abyssal plain (Figs 2 and 3).

Bottom water flow at BAS moorings 1, 2 and 3 was to the east or northeast at mean speeds from 2.8 to 11.7 cm s^{-1} and with mean velocity vectors of 1.2 to 11.3 cm s^{-1} (Barber & Crane 1995). Westward flow at mooring 8 implies deep recirculation within Jane Basin. The much larger set of AWI mooring data discussed by Fahrbach *et al.* (1994) showed the gyre circulation to be dominated by the boundary currents at the continental slopes, with flow nearly parallel to the bathymetric contours. Over the interior of the basin, mean currents are variable in direction and generally slower than 1 cm s^{-1} (e.g. moorings 217 and 218, Fig. 3). Moorings 206, 207 and 216 recorded a northerly outflow of water, into Powell Basin and round the tip of the Antarctic Peninsula (Figs 1–3).

The entire Weddell Sea is covered by pack ice for at least part of the year, and the western part remains ice-covered even during the late summer minimum extent (Fig. 1; Gloersen *et al.* 1992). This results in low biological productivity. In a comprehensive study of diatoms in surface sediments, Zielinski & Gersonde (1997) found low diatom numbers in the SE Weddell Sea, complete dissolution in the centre of the basin, but high abundances along the northern margin. Biogenic carbonate is rare in Weddell Sea sediments. Calcareous foraminifera occur on the shelf and slope (Anderson 1975; Mackensen *et al.* 1990; Gilbert *et al.* 1998) but are absent from the deeper parts of the basin.

Occurrence of contourites – site survey data

Figure 3 is a new map compiled from all available seismic and 3.5 kHz acoustic profiles, showing the distribution of contouritic and hemipelagic sediments north of 65°S. The profiles allow distinction between (i) fine-grained contourites and hemipelagites, which have a mounded or draped topography locally with sediment waves, and show deep acoustic penetration; (ii) sandy contourites, which form a plastered drift on the slope at 52°W and show shallow or no acoustic penetration; (iii) turbidites, which are flat and ponded, show shallow to deep acoustic penetration and include small channels.

The northern Weddell Sea, Jane Basin and its westward continuation to the eastern side of Powell Basin contain predominantly muddy contourites or hemipelagites (see later for examples). Acoustic penetration is commonly 40–60 m. At the margins of this area, against the South Orkney Microcontinent, Jane Bank and the ridge south of Powell Basin, the thinning of reflectors towards steep slopes suggests localized non-deposition and current scour, i.e. transitional to separated drift morphology (Pudsey *et al.* 1988, fig. 5B). The transition southeastwards into turbidites at about 64°S occurs approximately at the northernmost surface expression of WNW-trending fracture-zone ridges (Tectonic Map 1985); these ridges are large enough to show up on the 4000 m contour at the eastern edge of Figure 3. Turbidites also fill the central part of Powell Basin, and their feeder channels have been mapped by Howe *et al.* (1998). The area of sediment waves in northwestern Powell Basin was noted by Lawver *et al.* (1994), King *et al.* (1997) and Coren *et al.* (1997), and described by Howe *et al.* (1998). The area of sandy contourites on the slope at 52°W was mapped by Gilbert *et al.* (1998).

Stratigraphic context

Overall sedimentation rates in the northern Weddell Sea, estimated from total sediment thickness and the age of the oceanic basement, range from 22 m Ma^{-1} to 70 m Ma^{-1} (Tectonic Map 1985; King & Barker 1988; King *et al.* 1997). Sediments recovered at ODP Site 697 in Jane Basin (Fig. 3) were dated by a combination of magnetostratigraphy and siliceous biostratigraphy. The combined age-depth plot of Gersonde *et al.* (1990) showed a sedimentation rate of 100 m Ma^{-1} for the interval 250–320 m, decreasing to 28 m Ma^{-1} for the upper 120 m. The timescale for the upper 200 m was revised by Ramsay & Baldauf (1999) to give a sedimentation rate of 44.6 m Ma^{-1}.

It is very difficult to date northern Weddell Sea Quaternary sediments with any precision, though the cores commonly show cyclicity in diatom content and grain size which has been ascribed to glacial-interglacial cycles (see below). The lack of biogenic carbonate precludes the derivation of $\delta^{18}O$ curves, and the likely presence of reworked organic material complicates the interpretation of radiocarbon dates. Magnetostratigraphy, trace element and isotope geochemistry have been used with some success. Within normally magnetised sediments of the Brunhes chron, the Blake Event (117 ka, Tucholka *et al.* 1987) was identified at ODP Site 697 (Shipboard Scientific Party 1988*b*) and in BAS cores 15 and 44 (Fig. 3; O'Brien 1989). The Laschamp and Blake Events were found in AWI cores from Jane Basin by Grünig (1991). Shimmield *et al.* (1994) correlated peaks in biogenic barium in

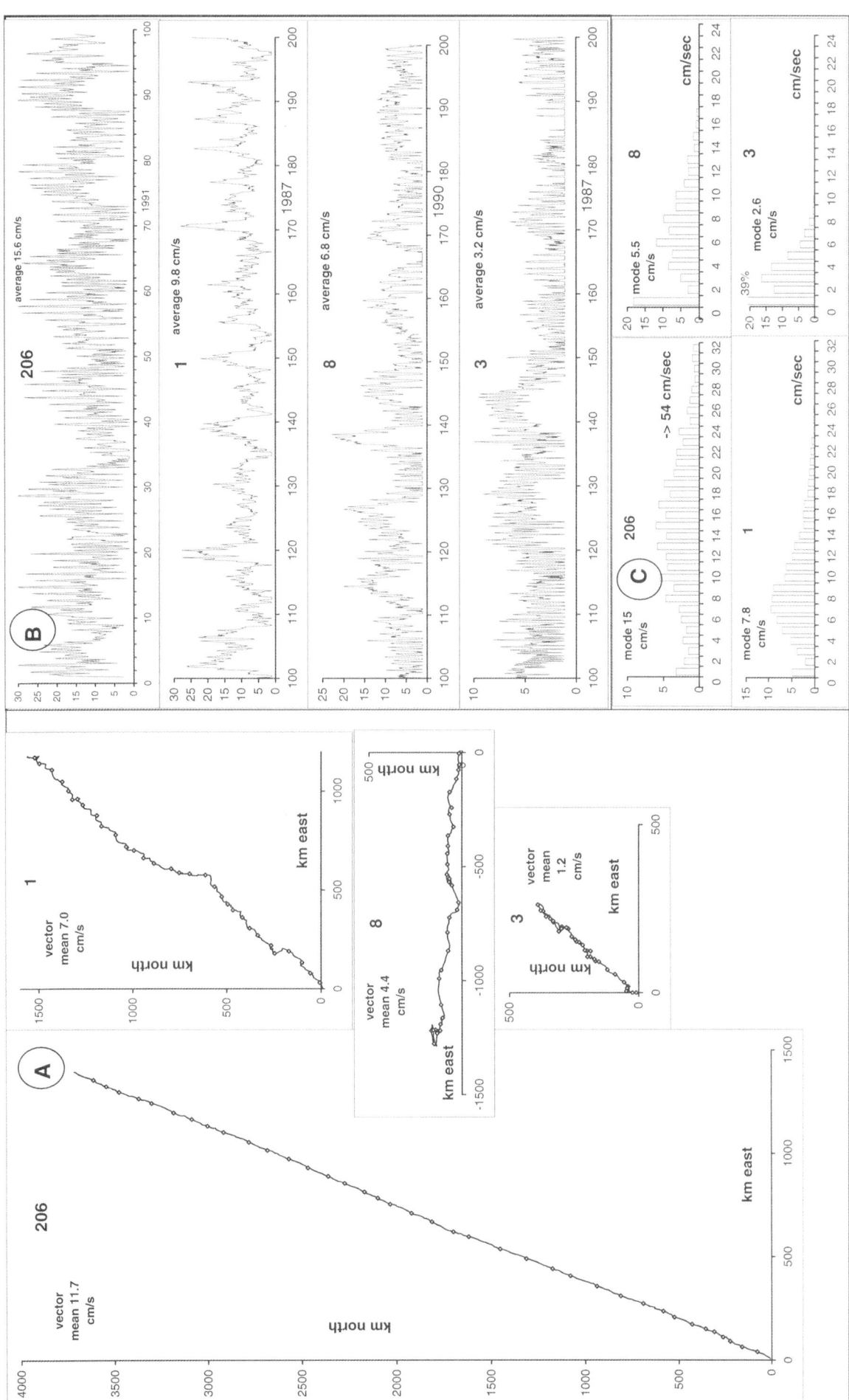

Fig. 2. Current meter data for British Antarctic Survey (BAS) moorings 1, 3 and 8 and Alfred-Wegener Institute (AWI) mooring 206. All meters were positioned 50 m above the seabed. (**A**) progressive vector diagrams from approx. 1 year of data; open symbols at 10 day intervals; vector mean currents also shown. (**B**) speed values for representative 100 day periods, with arithmetic mean speed for the whole record. (**C**) histograms of current speed values; these show the most commonly attained speeds, the proportion of very high speeds, and for how much of the time the meter rotor was stalled (flow slower than 1.5 cm s^{-1}). Flow was strongest and steadiest at mooring 206, on the upper slope. Unsteady flow at moorings 1 and 8 recorded the passage of eddies and benthic storms (speeds over 15 cm s^{-1} maintained for two days or more). Tidal cycles were seen at all sites, with the greatest tidal amplitudes at the shallowest site 206. At mooring 1 the current speed rarely fell below 6 cm s^{-1} and sometimes exceeded 20 cm s^{-1}. At mooring 3 speed was below 1.5 cm s^{-1} for 39% of the year and most of the flow was tidal; speeds over 5 cm s^{-1} were only attained for 4–6 h at a time during a few days per lunar month. Mooring 8 was also dominated by tidal frequencies and the rotor was stalled for 15% of the time.

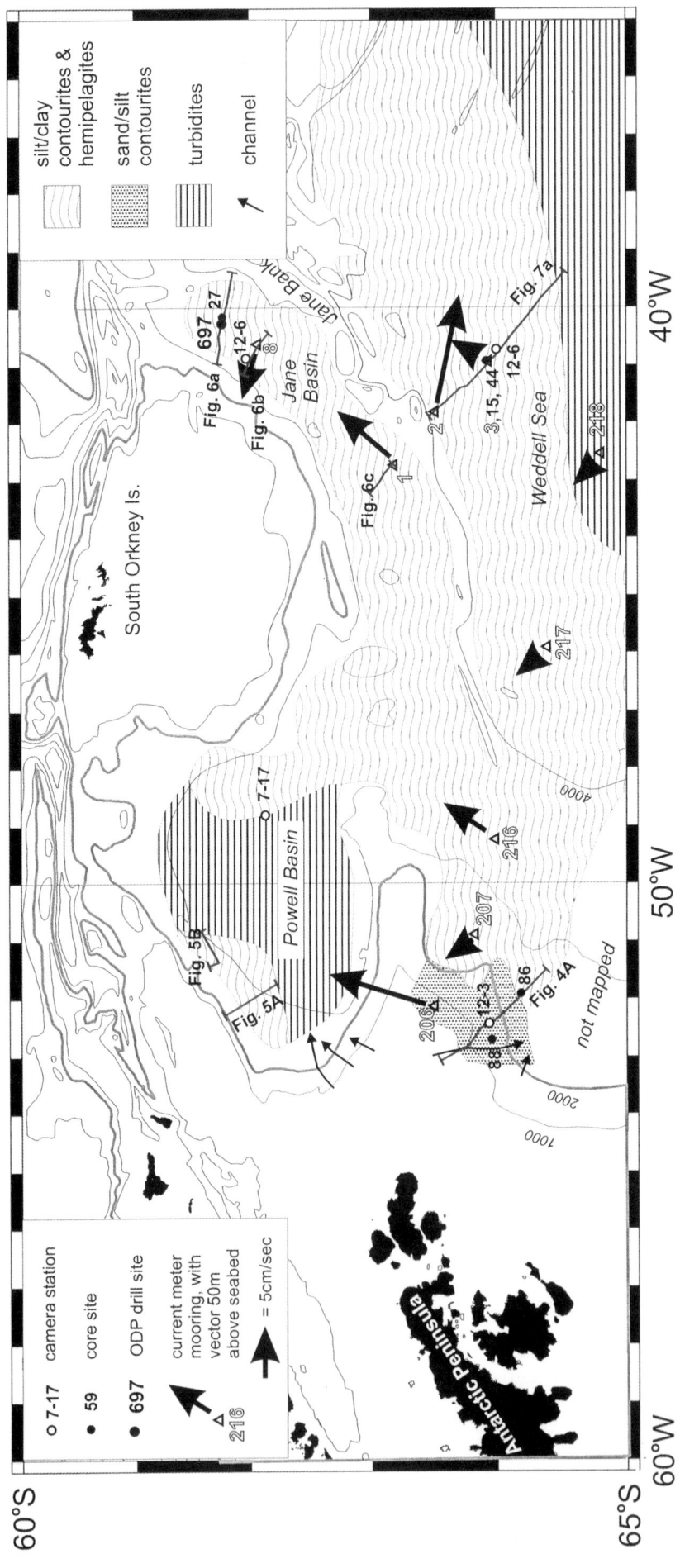

Fig. 3. The distribution of contourites and turbidites, north of 65°S. The bathymetric base map is Tectonic Map (1985) with modification in the SW where more recent data are available. Thick grey contour at 2000 m. Large arrows show current directions at AWI and BAS moorings. Locations of profiles in Figures 4–7 also shown.

BAS cores 27 and 41 (Fig. 3) with marine isotope stages 1 and 5e. Grünig (1991), Brehme (1992) and Frank et al. (1995) measured biogenic barium and $^{230}Th_{excess}$ in cores from about 44°W in Jane Basin, and identified stages 1, 5e and 7. Taken together these results suggest fairly consistent late Quaternary sedimentation rates of 3–4 cm ka^{-1}, down to about 300 ka.

Seismic, 3.5 kHz and swath bathymetric data

We illustrate examples from four areas: the sandy contourite drift on the slope at about 52°W (Fig. 4), the sediment waves in Powell Basin (Fig. 5), and the lower-energy sedimentary environments of Jane Basin and the northern Weddell Sea at 40–42°W (Figs 6 and 7).

Sediment characteristics

Bottom photographs

Hollister & Elder (1969) inferred moderate to strong bottom currents along the western margin of the Weddell Sea, including Powell Basin, from photographs showing oriented lineations and deflected sessile organisms. They also noted murky bottom water, suggesting resuspension of sediment by currents. Bottom photos taken in 1963–1964 on cruises of the USNS *Eltanin* (Goodell 1964, 1965) are still available from the Antarctic Research Facility, Florida State University (Fig. 8).

Core descriptions and facies

Detailed core decriptions, with grain-size data, have been published for the sandy contourite area on the NW Weddell Sea slope (Gilbert et al. 1998), for Jane Basin (Grünig 1991; Brehme 1992; Pudsey 1992) and for the northern Weddell Sea (Pudsey et al. 1988). There is a widely-recognised cyclicity between a fine-grained, laminated or bioturbated terrigenous facies interpreted as glacial, and a coarser-grained, bioturbated, mixed biogenic-terrigenous facies interpreted as interglacial. In the interglacial facies, biogenic silica (diatoms and radiolarians) is more abundant farther north, while biogenic carbonate (mainly planktonic foraminifera) is more abundant at shallower water depths. Deep acoustic penetration corresponds to predominantly fine-grained sediments, while sands were recovered from the slope in the NW Weddell Sea (Figs 3 and 4). As representative examples we illustrate cores 88 and 56 from the NW Weddell Sea slope (Fig. 9), core 27 from Jane Basin (Fig. 10) and core 15 from the northern Weddell Sea (Fig. 11).

Surface sediment heavy mineral composition was presented for most of the Weddell Sea by Diekmann & Kuhn (1999). In the northwestern Weddell Sea and the area south of the South Orkney Microcontinent, a 'western Weddell Sea mixed assemblage' was found, with principal components garnet, pyroxene and hornblende and minor epidote, derived from both the Antarctic Peninsula and East Antarctica. While these authors emphasised ice-rafting processes rather than current transport, they did suggest the influence of quartz-rich suspensions within the Weddell Gyre could be seen as far north as the slope in the NW Weddell Sea. In cores from the NW Weddell Sea slope, all samples contain high proportions of both quartz and lithic grains with 5–10% feldspar (Gilbert et al. 1998). Heavy minerals (c. 1%) are dominated by zircon and tourmaline, suggesting a predominantly igneous provenance. Lithic grains are mainly sedimentary or volcanic in origin, as are the larger lithic clasts identified during core logging. This suite of lithotypes is consistent with an Antarctic Peninsula provenance.

Analytical results

In the NW Weddell Sea, there are small but consistent variations between the sandy units in each of the slope cores. The core tops ('sand 1') contain 5–20% gravel, show some inverse grading and the sand mode is slightly finer than 3ϕ with a large coarse tail (Fig. 12A, 0.1 m). The next sand unit downcore ('sand 2') contains less gravel, is normally graded (except in core 88) and the sand mode is slightly coarser than 3ϕ, although the clay content is higher than in sand 1 (Fig. 12A, 0.5 m). The third sandy unit is again gravel-bearing, has a sand mode coarser than 3ϕ, and has a variable clay content (Fig. 12A, 1.1 m). The muddy units also show some systematic down-core variation; in core 88 the muddy unit from 0.21–0.45 m is unsorted (Fig. 12A, 0.4 m) but from 0.53–0.92 m there is a weak mode in the fine sand range (Fig. 12A, 0.8 m).

Cores from other areas are all fine-grained (generally 60–70% clay) and very poorly sorted; indeed it is impossible to derive the standard grain size parameters of sorting and skewness because the fine end of the grain size distribution is well beyond the usual Sedigraph measurement limit of 13 phi (0.5 microns). Nevertheless there are clear downcore variations in the proportion of silt (Fig. 13). In Jane Basin the interglacial units contain diatoms and radiolarians, and much more ice-rafted debris than the glacial units. Interglacial units are also siltier than glacials, and the presence of a broad mode near the silt-clay boundary suggests weak current sorting, rather than the unsorted size distribution expected from ice-rafting alone.

Discussion

Are the sediments contourites?

The site survey data allow distinction of areas which accumulate predominantly contourites or hemipelagic sediments, from areas of turbidites. The northwestern slope where strong bottom currents have been measured, and Powell Basin where sediment waves are presently active, are unequivocal contourite areas. Jane Basin and the northern Weddell Sea are transitional to hemipelagic areas. The Weddell Gyre as a whole is a very low-energy sedimentary environment, except close to the margins (see Michels et al. 2002).

History of bottom water flow

The first (deepest) appearance of mudwaves in the Powell Basin seismic records may mark the opening of a deep-water pathway into the basin. The transtensional southern edge of the basin is thought to have opened during seafloor spreading in the Late Oligocene (Howe et al. 1998). The inception of deep AABW (Weddell Gyre) flow, and hence the expansion of the Powell Basin mudwave field, may have occurred during the Early Miocene.

ODP Site 697 in Jane Basin recovered sediments of early Pliocene to Quaternary age. Grain-size analysis by Pudsey (1990) revealed long-term changes downcore. In the oldest sediments (4.7 to 3.2 Ma) ice-rafted debris is common with sand forming 4–10%, but there is no evidence for sorting of the fine fraction. From 3.3 to 2.9 Ma an increased proportion of medium to coarse silt implies bottom-current sorting. Cyclic grain-size variations during the past 0.5 Ma reflect glacial-interglacial cycles.

The cyclic variation in grain size downcore led several authors to infer systematic variation in bottom water flow between glacials and interglacials in the late Quaternary. Relative to the present interglacial, flow was considerably slower at the Last Glacial Maximum (Pudsey et al. 1988; Grunig 1991; Brehme 1992; Pudsey 1992). In contrast, flow of the Antarctic Circumpolar Current was stronger at the LGM than today (Pudsey & Howe 2002).

Gilbert et al. (1998) presented no age information for the cores

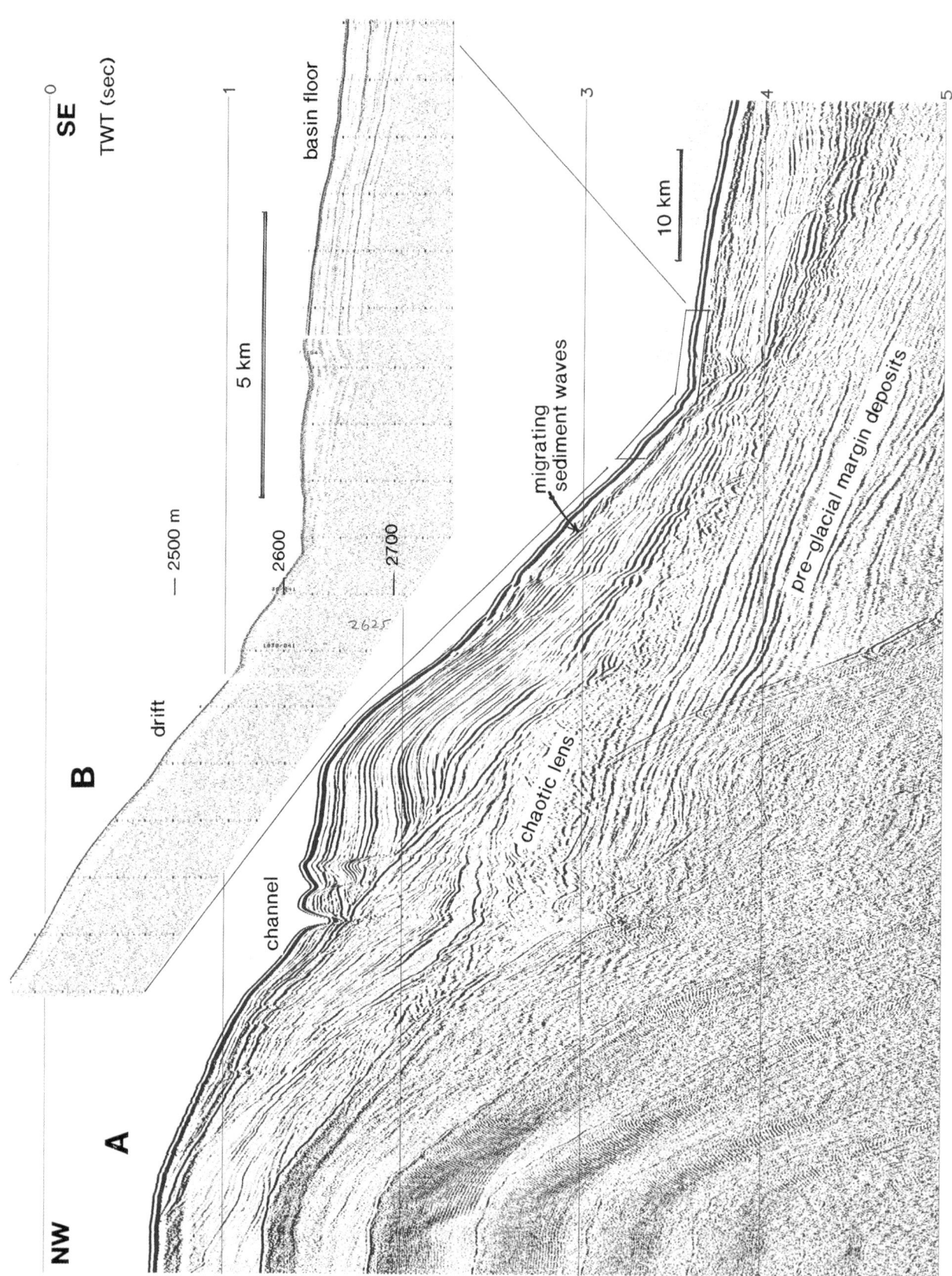

Fig. 4. (**A**) BAS MCS Line 14, location on Figure 3. A plastered drift extends from a prominent N–S trending channel (Gilbert *et al.* 1998) down to a break in slope at 3.6 s TWT. Migrating sediment waves are present within the lower part of the drift. Below the drift, a lens of chaotic seismic character is interpreted as a slumped mass. Beneath this lens, thick SE-dipping parallel-bedded sediments are probably preglacial passive margin deposits. Basement lies at 8.6 s and is thought to be of at least Early Cretaceous age (Barker & Lonsdale 1991). (**B**) 3.5 kHz profile at the base of the drift (box on Fig. 4A). The drift is acoustically opaque and at its base is a sharp facies change to sediments showing 30–40 m acoustic penetration and distinct, parallel to wavy sub-bottom reflectors (echo types IID and III of Gilbert *et al.* 1998), interpreted as muddy contourites.

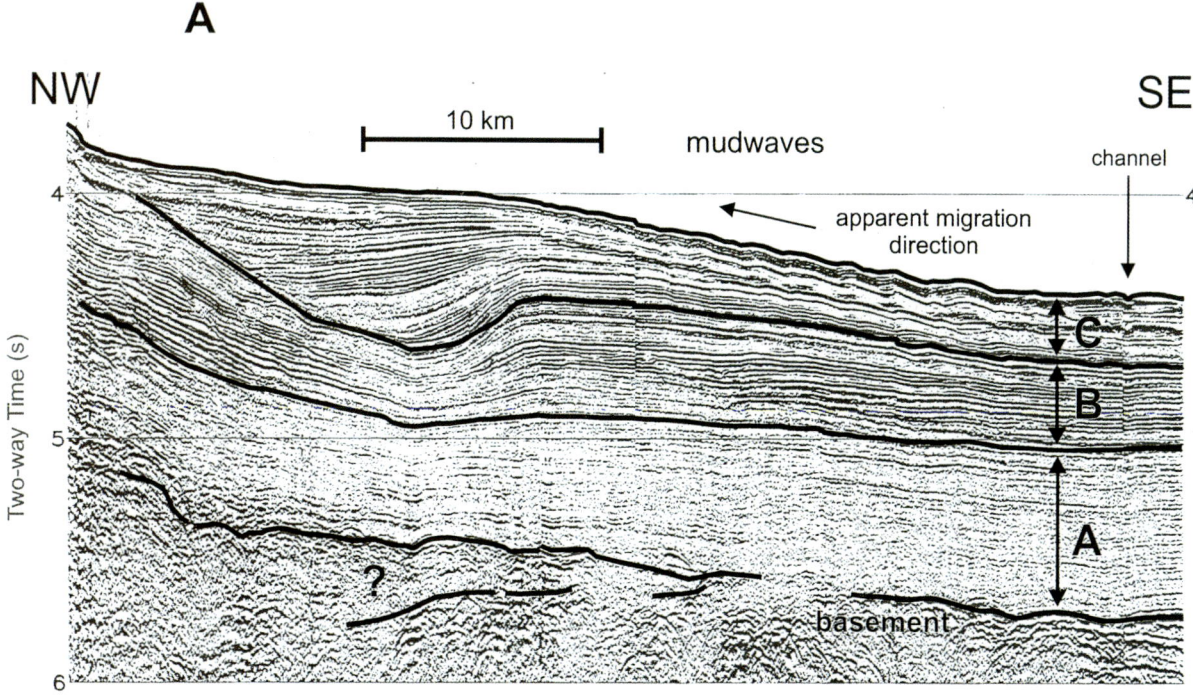

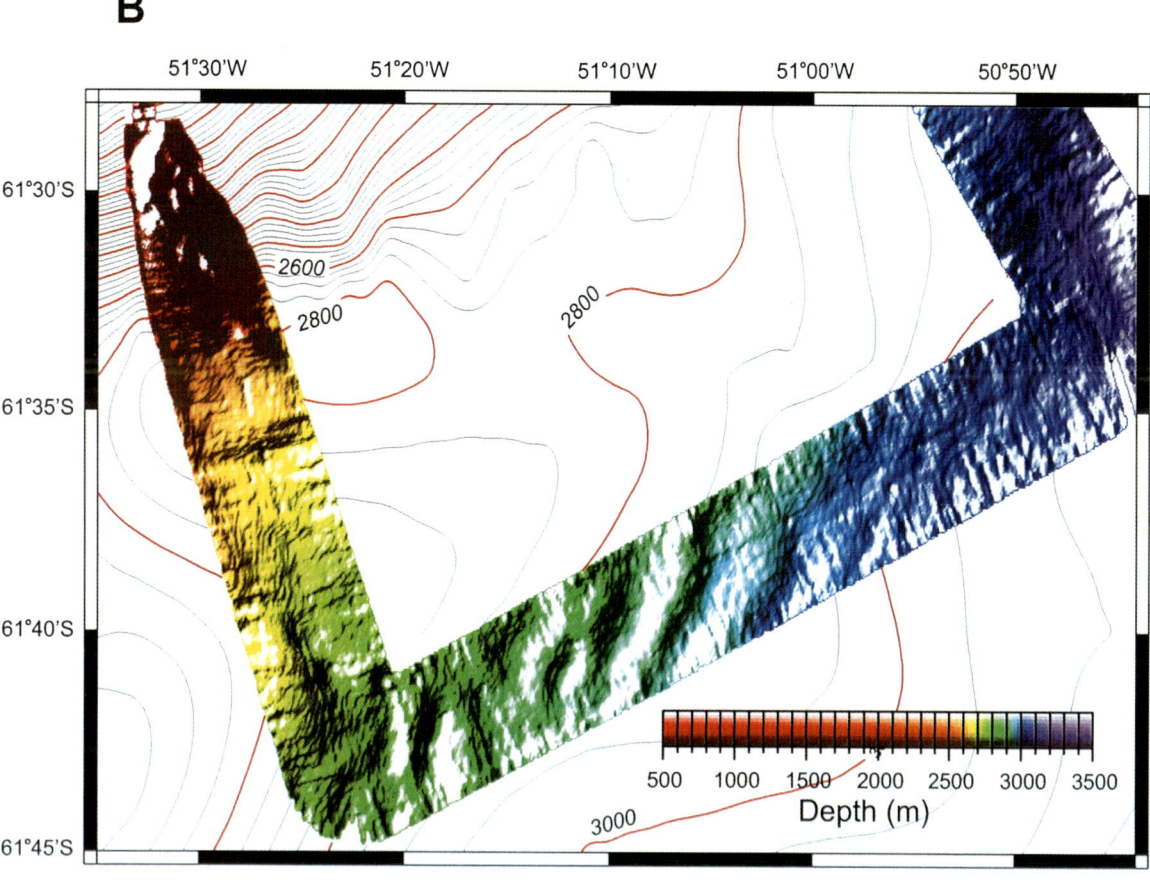

Fig. 5. (**A**) Hesperides Line M05, near-trace record, location on Figure 3. Strong basement reflections at 1–1.5 s below the seafloor represent oceanic crust of mid-Oligocene age (King *et al.* 1997). The upper two seismic units B and C show strong, laterally continuous reflectors. Mudwaves in the southeastern part of unit C have an upslope migration direction. Unit C onlaps the margin sediments of unit B, suggesting alongslope supply by contour currents. From Howe *et al.* (1998). A similar profile was illustrated by King *et al.* (1997). (**B**) Swath bathymetry in northern Powell Basin, location on Figure 3. EM12 sonar data (illumination from direction 300°) show mudwaves from 51°W to 51°20'W. The wave crests trend approximately 030° and are spaced 2–4 km apart, with amplitudes of 20–50 m. Parallel lines at right angles to the ship track are an artefact of sonar beam nulls. Reprinted from *Marine Geology*, v. 149, Howe *et al.*, mudwave activity and current-controlled sedimentation in Powell Basin, northern Weddell, Antarctica, 229–241, © 1998 with permission from Elsevier Science.

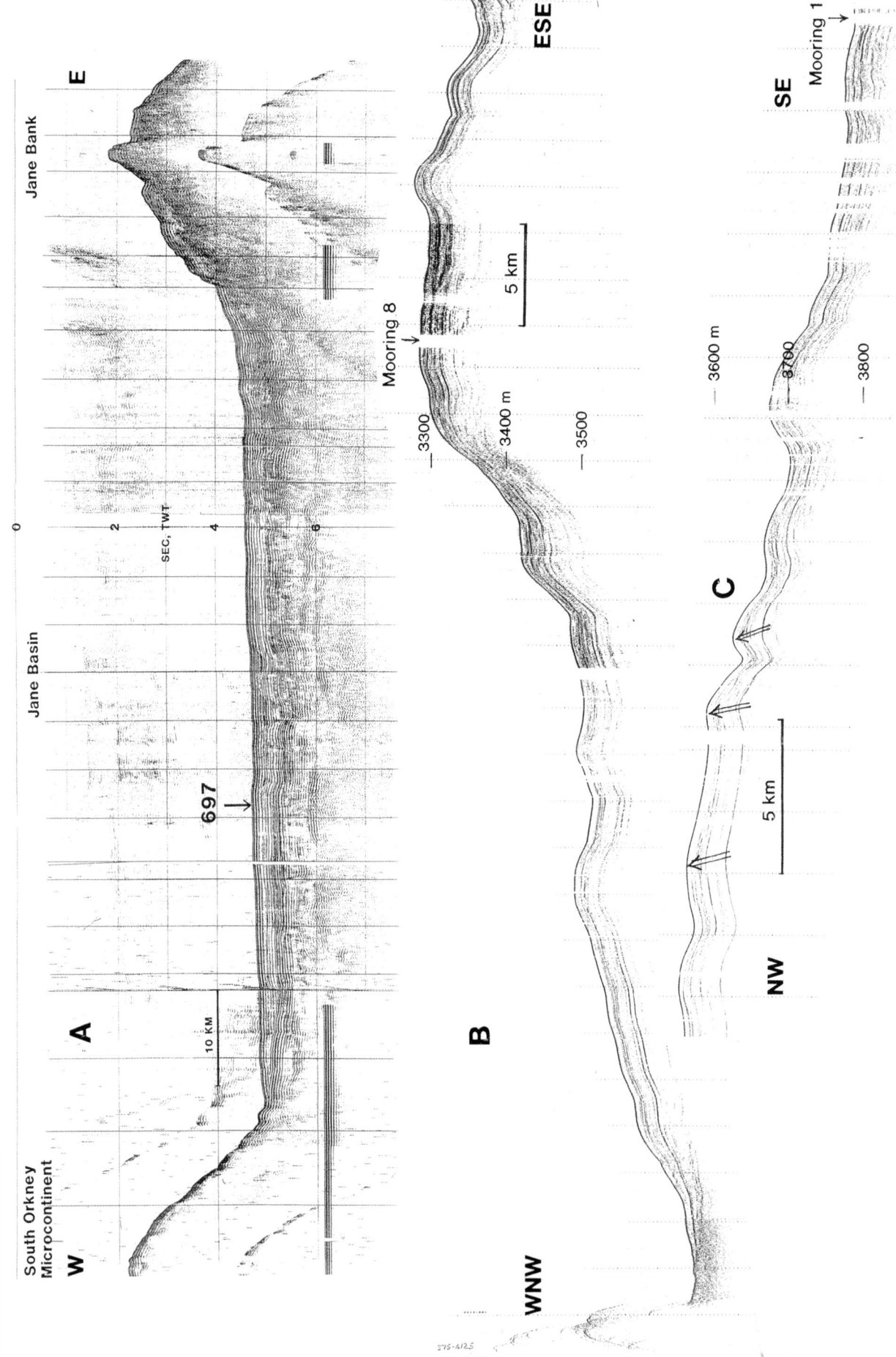

Fig. 6. (**A**) Single-channel seismic profile, northern Jane Basin (Bransfield 890 Line E), location on Figure 3. Basement (Oligocene to early Miocene, King & Barker 1988) is visible in places just above 6 s TWT. The lower half of the sedimentary section has generally discontinuous reflectors and shows evidence for derivation from the South Orkney Microcontinent and the Miocene volcanic arc of Jane Bank (Shipboard Scientific Party 1988b). The overlying sediments pinch out at the base of the South Orkney Microcontinent slope and appear to be hemipelagic. From Pudsey (1990). (**B**) 3.5 kHz profile, WNW of Jane Bank

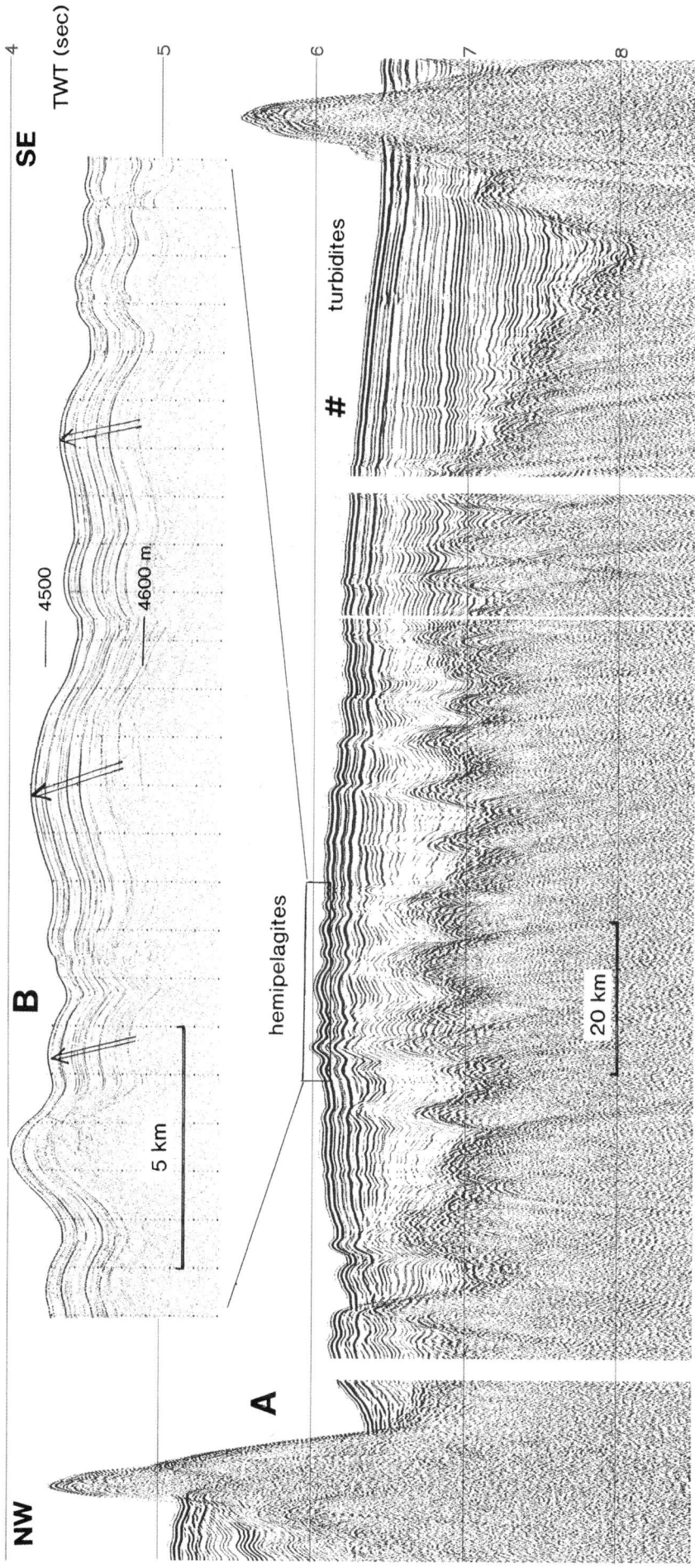

Fig. 7. (**A**) BAS multichannel seismic profile, northern Weddell basin (D154 Line 15), location on Figure 3. Rough oceanic basement ranges in age from early Miocene in the NW to mid-Oligocene in the SE, (from identification of magnetic anomalies; Tectonic Map 1985). The lower part of the sedimentary sequence fills hollows in the basement and is interpreted as turbiditic; this geometry extends up to the seabed to the SE of the # marker. Pudsey et al. (1988) described turbidites in cores from within 20 km of the seamounts, supporting this interpretation. Farther NW, the overlying sediments show a draped geometry and are interpreted as hemipelagic. They pinch out at the base of the ridge at the NW end of the profile. (**B**) 3.5 kHz profile, box on Figure 7A. Parallel, mainly continuous sub-bottom reflectors show a draped geometry with upslope migration of mudwave crests (double arrows).

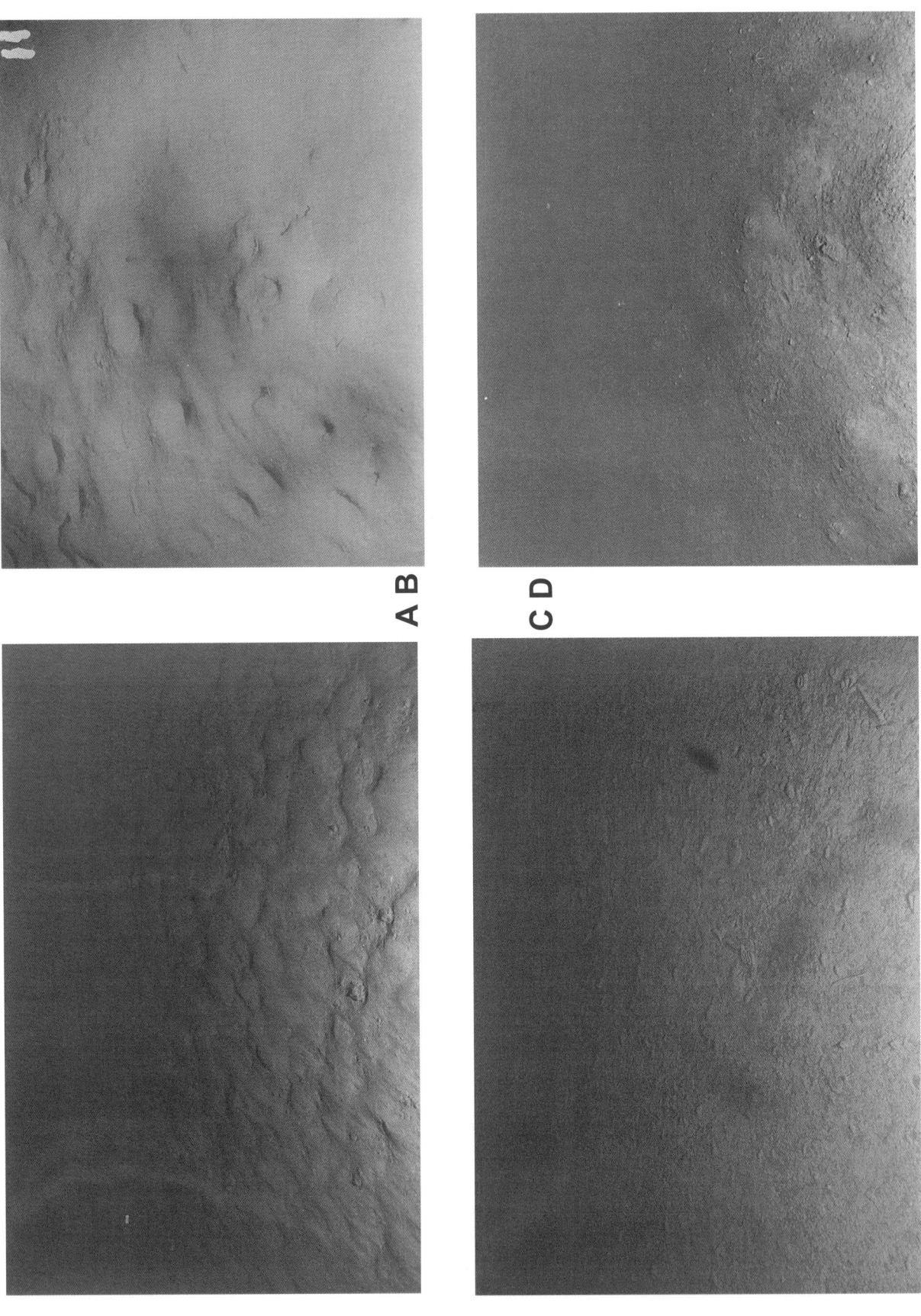

Fig. 8. Seabed photographs; location on Figure 3. Captions and sediment textural data from Goodell (1964, 1965). (**A**) Eltanin 12-3, NW Weddell Sea slope, near core 88. The soft bottom has scattered pebbles and erratics, and relatively strong bottom currents are present as shown by scour marks around pebbles, elongate cut-and-fill structures parallel to the current direction, and rhomboidal ripple marks. The fauna is sparse and largely limited to rare ophiuroids. Grain size distribution of surface sediment: 68.4% sand and gravel, 19.9% silt, 11.7% clay. (**B**) Eltanin 7-17, eastern Powell Basin. Burrowed bottom well covered with echinoid trails and one or two brittle stars. Abundant conical pits. 51.1% sand, 34.6% silt, 14.3% clay. (**C**) Eltanin 12-6, northern Weddell Sea, near core 44. Although fauna is sparse, the bottom is fairly heavily covered with excrement of at least two types. 0.8% sand, 44.2% silt, 55% clay. Muddy, bioturbated, no evidence for currents. (**D**) Eltanin 12-7, Jane Basin. Similar to Fig. 8C; abundant burrows, trails and piles of excrement. Rare ophiuroids. No sediment textural data available.

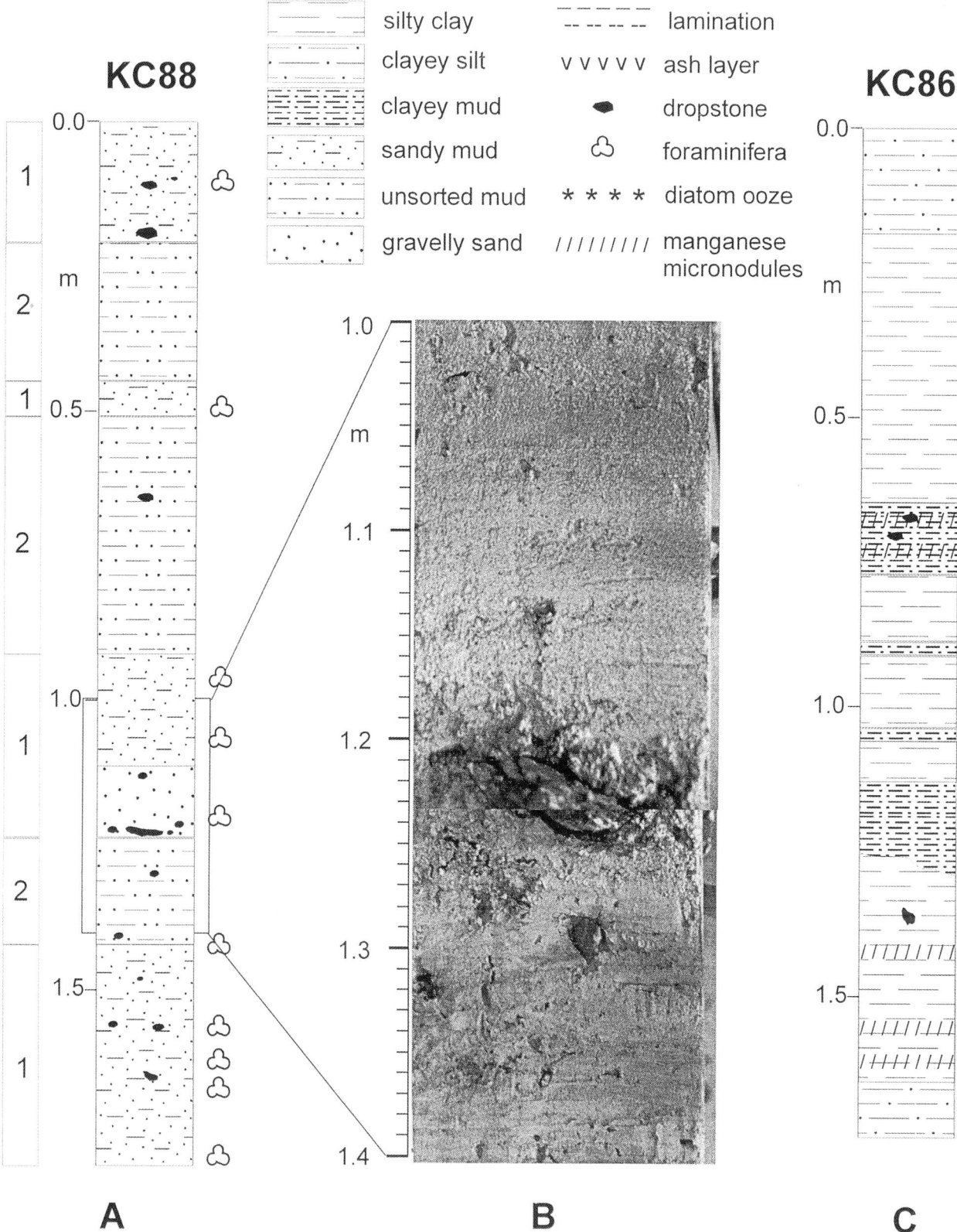

Fig. 9. (**A**) Graphic log of core 88, redrawn from Gilbert *et al.* (1998). Sediments from the slope in this area show a compositional and textural cyclicity between two facies (left column). Facies 1, which occurs at the seabed and in one to three more units down each core, comprises poorly to very poorly sorted fine sands and silty sands with up to 20–30% gravel. Some of these units show weak normal grading or faint parallel lamination. Basal contacts are generally sharp, and some show evidence for erosion. Biogenic silica (diatoms, radiolarians and sponge spicules) constitutes up to 10% of the sediment, and both planktonic and benthonic foraminifera are common, particularly in the subsurface sandy units. Facies 2 comprises extremely poorly sorted muddy sands and sandy muds with 0–2% gravel, about 30% clay and negligible biogenic silica or carbonate. These muddy units gradationally overlie facies 1 sands and are ungraded. Burrow mottling occurs in places; otherwise sedimentary structures are absent. (**B**) Interval 1.0–14 m of core 88, showing sand overlying gravel lag. Large angular basalt dropstone at 1.20–1.24 m. (**C**) Graphic log of core 86, redrawn from Gilbert *et al.* (1998). Sediments from the lower slope and rise in this area contain no biogenic carbonate and only trace amounts of biogenic silica (mainly diatoms). They show a weak cyclic variation in grain size but have not been divided into facies. Surface sediments are relatively coarse, generally containing 30–40% silt, 50–60% clay and a few percent sand. The finer-grained intervals generally contain 20–30% silt and 70–80% clay with little or no sand. The high clay contents of these deposits prohibited derivation of sorting parameters, so differences in degree of sorting between the coarse and fine intervals could not be established.

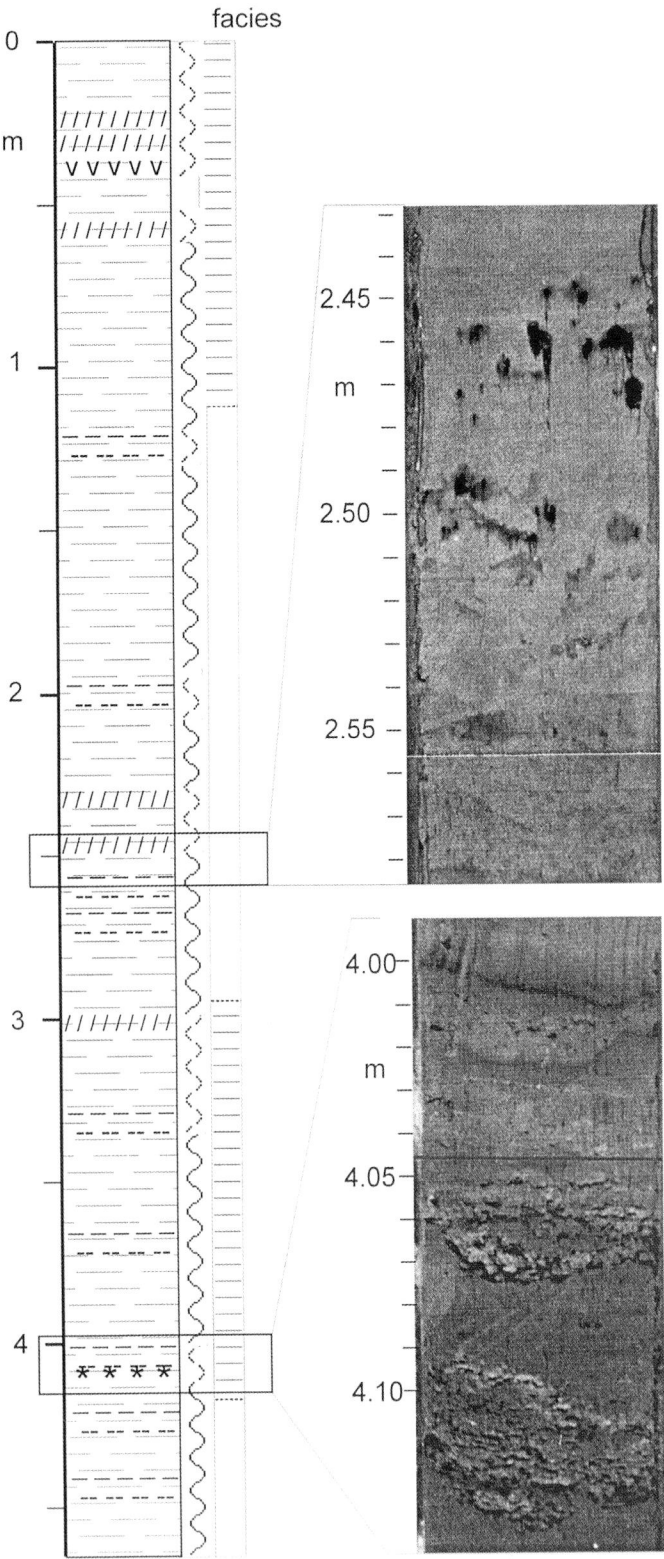

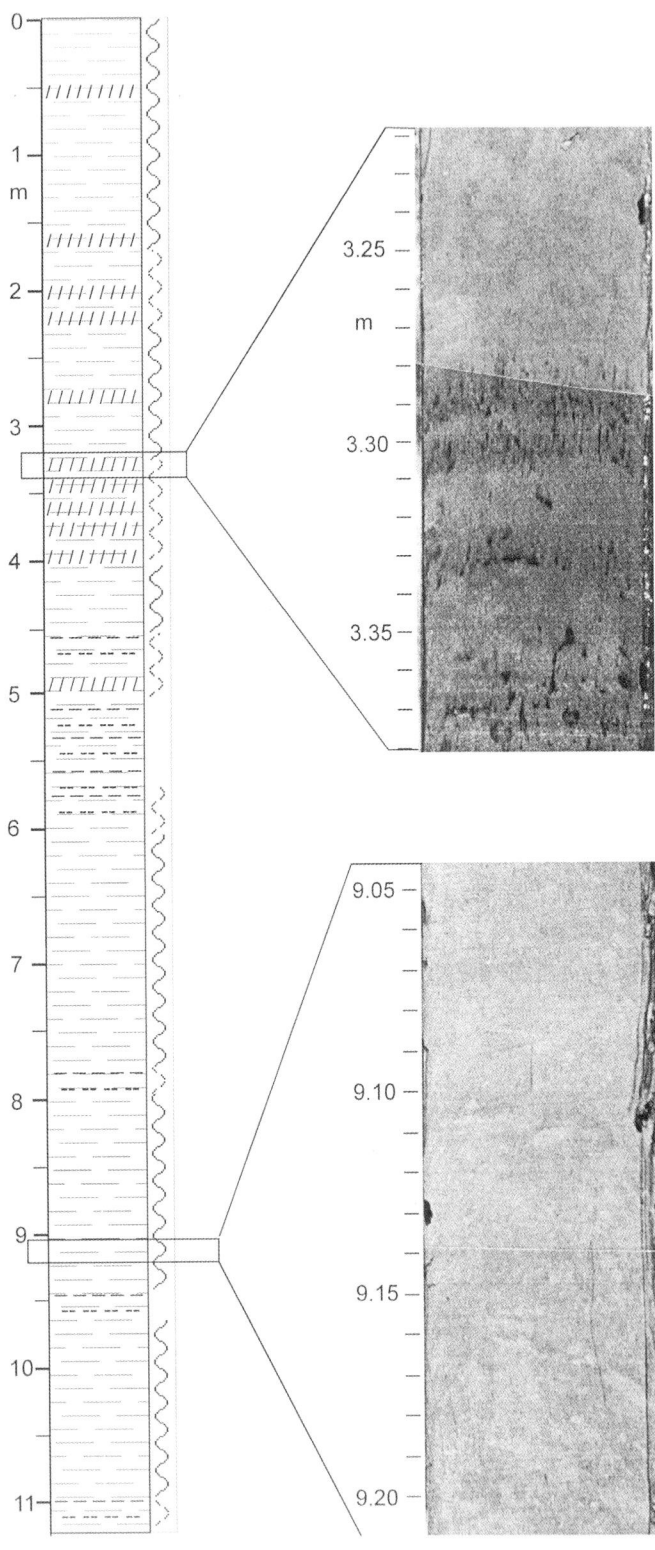

Fig. 10. Graphic log of core 27, redrawn from Pudsey (1992). Location on Figure 3, key in Figure 9. Sediments in Jane Basin can be divided into two facies, diatom-bearing silty clay and clay, shown as dashed ornament and white in facies column. Burrow mottling is present throughout. Diatoms are poorly to moderately preserved, except for a group of ooze laminae at 4.05–4.13 m depth which contain an unusual pristine diatom assemblage (Jordan et al. 1991). Upper photo: Silty clay with burrow mottling, interpreted as glacial. Lower photo: Diatom ooze layers, interglacial stage 5.

Fig. 11. Graphic log of core 15, redrawn from Pudsey et al. (1988). Location on Figure 3. The sediments consist mainly of greyish brown to dark greyish brown burrow-mottled clay and silty clay, barren apart from a trace of diatoms at the surface. There is no visible cyclicity in this core, or in others from the area of undulating seabed (but note the cyclicity in texture, Fig. 13C). Bioturbated dark laminae of manganese micronodules are common in the upper part, and thin silt laminae occur sparsely in the lower part, being much more common in cores from farther south (in the turbidite province, Fig. 3; Pudsey et al. 1988). Upper photo: Dark laminae of manganese micronodules disseminated by burrowing. Lower photo: Intensely burrowed clay.

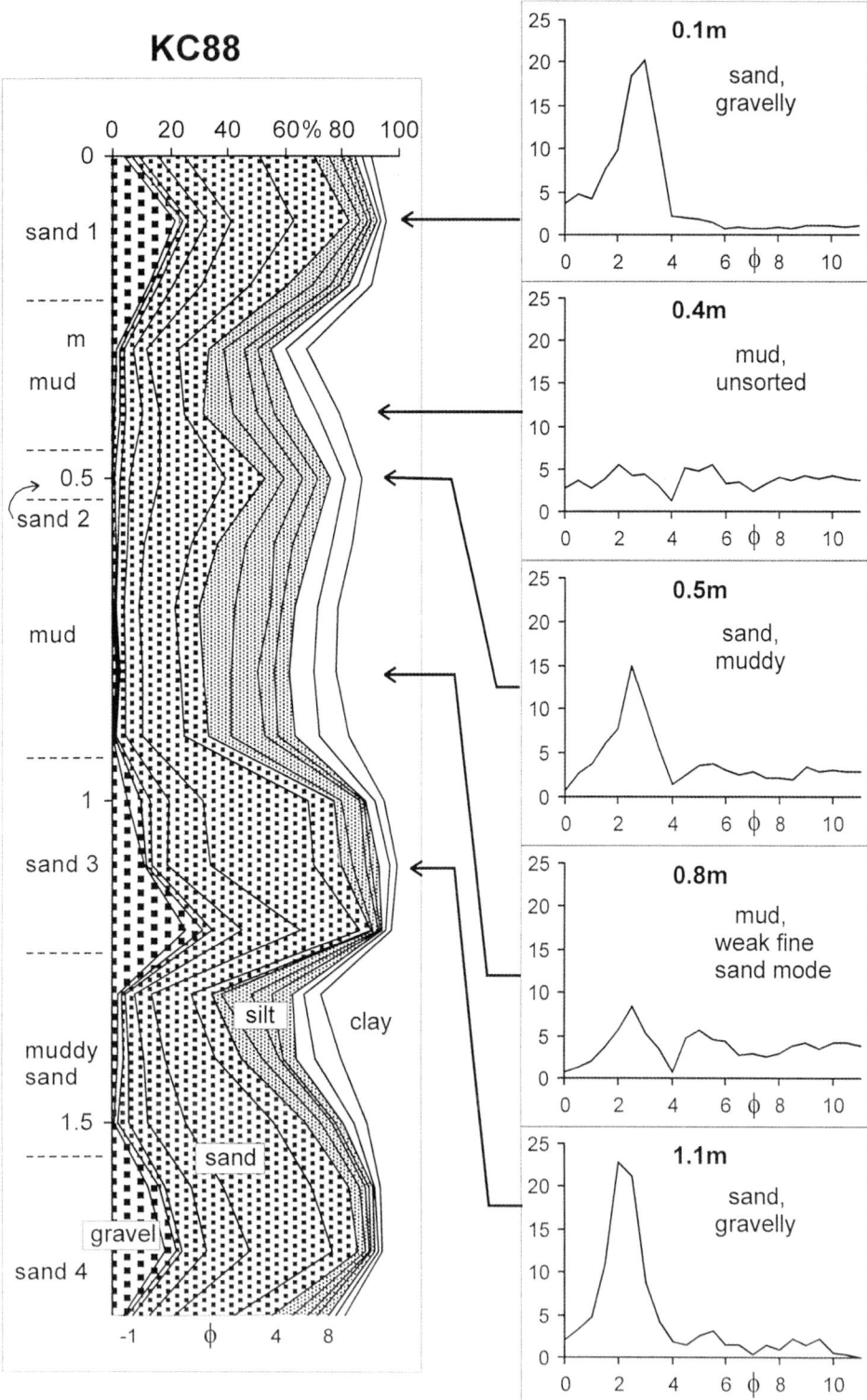

Fig. 12. (**A**) Downcore grain size data for core 88. From Gilbert *et al.* (1998). Histograms show settling tube and Sedigraph data. From the top down the following units are distinguished: sand 1; gravelly, mode finer than 3ϕ, agglutinated foraminiferal fauna, mud; unsorted, barren, sand 2; non-gravelly, muddy, mode coarser than 3ϕ, *Epistominella exigua – Globocassidulina crassa* fauna, mud; with weak fine sand mode, barren, sand 3; gravelly, mode coarser than 3ϕ, *Ehrenbergina glabra – G. crassa – Trifarina angulosa* fauna with abundant planktonic foraminifera, sandy mud; unsorted, *E. exigua – G. crassa – T. angulosa* fauna sand 4; gravelly, *E. exigua* fauna. These benthic foraminiferal assemblages record water chemistry changes which have affected the level of the CCD. Common *Trifarina angulosa* is thought to indicate strong bottom currents (Mackensen *et al.* 1990), a relationship supported by the better sorting in the *T. angulosa*-bearing sand 3.

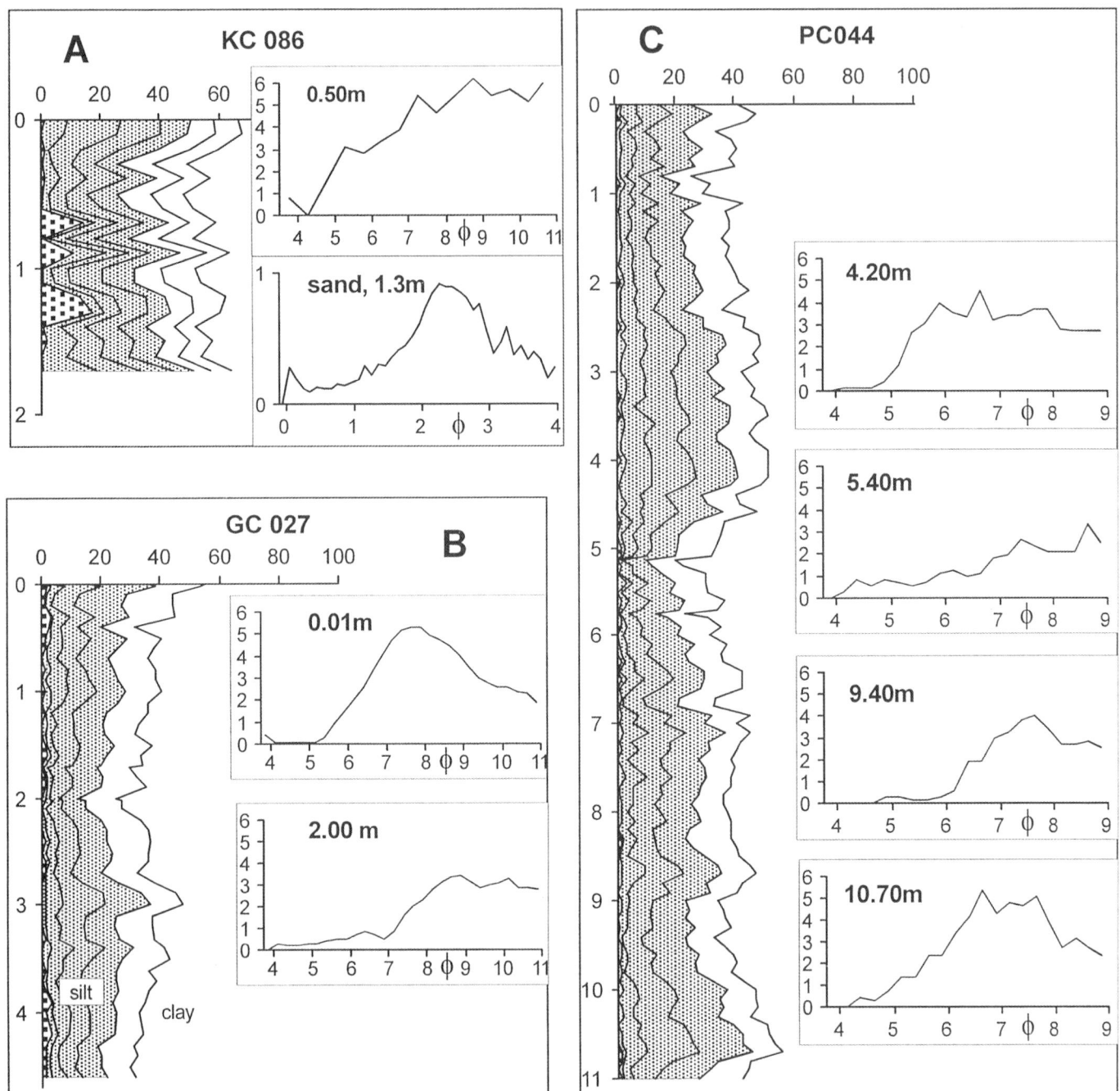

Fig. 13. (**A**) Downcore grain size data for core 86. From Gilbert *et al.* (1998). Frequency curves show settling tube (sand, 0–4 ϕ) and Sedigraph (mud, 4–11 ϕ) data. (**B**) Downcore grain-size data for core 27. From Pudsey (1992). Frequency curves show Sedigraph data. Interglacial silty clays have up to 2–3% sand and a broad mode near the silt-clay boundary (0.01 m inset). Glacial clays have only a trace of sand and no mode (2.00 m inset). (**C**) Downcore grain size data for core 44, very near core 15. From Pudsey (1992). Frequency curves show Sedigraph data. The sediments range from silty clay (up to 40% silt) to clay (less than 20% silt). Sorting is extremely poor, though some samples have a weak mode at 7–8 ϕ. The cyclic variations in grain size are not related to any visible features in the cores, except for the samples at 5.1 m and 5.8 m which are dark bands consisting of almost pure clay.

from the NW Weddell Sea slope, apart from the presence of Quaternary diatoms and foraminifera. The plastered drift is a relatively young feature constructed on a thick sequence (5 s TWT; Fig. 4) of Mesozoic–Cenozoic continental margin sediments. It occupies approximately the upper 1 s (TWT) of the sediment column. The drift may have been initiated either at the onset of bottom water flow in the Early Miocene, or at the start of high sediment supply to the western Weddell Sea in the latest Miocene (Kennett & Barker 1990).

P. Morris assisted with data management and plotted Figures 4A and 7A. A. Maldonado kindly provided Figure 5. E. Fahrbach and G. Rohardt supplied the data for AWI mooring 206. K. Roessig at the Antarctic Research Facility, Florida State University, supplied the photographs in Figure 8. The manuscript was reviewed by C.-D. Hillenbrand and D. Stow.

References

ANDERSON, J. B. 1975. Factors controlling $CaCO_3$ dissolution in the Weddell Sea from foraminiferal distribution patterns. *Marine Geology*, **19**, 315–332.

BARBER, M. & CRANE, D. 1995. Current flow in the north-west Weddell Sea. *Antarctic Science*, **7**, 39–50.

BARKER, P. F., DALZIEL, I. W. D. & STOREY, B. C. 1991. Tectonic development of the Scotia Arc region. *In*: TINGEY, R. J. (ed.) *Geology of Antarctica*. Clarendon Press, Oxford, 215–248.

BARKER, P. F. & LONSDALE, M. J. 1991. A multichannel seismic profile across the Weddell Sea margin of the Antarctic Peninsula: regional tectonic implications. *In*: THOMSON, M. R. A., CRAME, J. A., THOMPSON, J., ET AL. (eds) *Geological Evolution of Antarctica*. Cambridge University Press, Cambridge, 237–241.

BENTLEY, M. J. & ANDERSON, J. B. 1998. Glacial and marine geological evidence for the ice sheet configuration in the Weddell Sea – Antarctic Peninsula region during the Last Glacial Maximum. *Antarctic Science*, **10**, 309–325.

BREHME, I. 1992. Sediment facies and bottom water current on the continental slope in the northwestern Weddell Sea. *Berichte zur Polarforschung*, **110**, 1–127.

CARMACK, E. C. 1974. A quantitative characterisation of water masses in the Weddell Sea during summer. *Deep-Sea Research*, **21**, 431–442.

DIEKMANN, B. & KUHN, G. 1999. Provenance and dispersal of glacial-marine surface sediments in the Weddell Sea and adjoining areas, Antarctica: ice-rafting versus current transport. *Marine Geology*, **158**, 209–231.

COREN, F., CECCONE, G., LODOLO, E., ZANOLLA, C., ZITELLINI, N., BONAZZI, C. & CENTONZE, J. 1997. Morphology, seismic structure and tectonic development of the Powell Basin, Antarctica. *Journal of the Geological Society*, **154**, 849–862.

FAHRBACH, E., ROHARDT, G., SCHRODER, M. & STRASS, V. 1994. Transport and structure of the Weddell Gyre. *Annales Geophysicae*, **12**, 840–855.

FAHRBACH, E., ROHARDT, G., SCHEELE, N., SCHRODER, M., STRASS, V. & WISOTZKI, A. 1995. Formation and discharge of deep and bottom water in the northwestern Weddell Sea. *Journal of Marine Research*, **53**, 515–538.

FRANK, M., EISENHAUER, A., BONN, W. J., WALTER, P., GROBE, H., KUBIK, P. W., DITTRICH-HANNEN, B. & MANGINI, A. 1995. Sediment redistribution versus palaeoproductivity change: Weddell Sea margin sediment stratigraphy and biogenic particle flux of the last 250,000 years deduced from $^{230}Th_{ex}$, ^{10}Be and biogenic barium profiles. *Earth and Planetary Science Letters*, **136**, 559–573.

GERSONDE, R., ABELMANN, A., BURCKLE, L. H., NAMILTON, N., LAZARUS, D., MCCARTNEY, K., O'BRIEN, P., SPIESS, V. & WISE, S. W. 1990. Biostratigraphic synthesis of Neogene siliceous microfossils from the Antarctic Ocean, ODP Leg 113 (Weddell Sea). *Proceedings of the Ocean Drilling Program, Scientific Results*, **113**, 915–936.

GILBERT, I. M., PUDSEY, C. J. & MURRAY, J. W. 1998. A sediment record of cyclic bottom current variability from the northwest Weddell Sea. *Sedimentary Geology*, **115**, 185–214.

GLOERSEN, P., CAMPBELL, W. J., CAVALIERI, D. J., COMISO, J. C., PARKINSON, C. L., & ZWALLY, H. J. 1992. Arctic and Antarctic sea ice, 1978–1987: Satellite passive microwave observations and analysis. *NASA Special Publication*, **SP-511**, 1–290.

GOODELL, H. G. 1964. *Marine Geology of the Drake Passage, Scotia Sea and South Sandwich Trench (USNS Eltanin cruises 1–8)*. Florida State University.

GOODELL, H. G. 1965. *Marine Geology: USNS Eltanin cruises 9–15*. Florida State University.

GROBE, H. & MACKENSEN, A. 1992. Late Quaternary climatic cycles as recorded in sediments from the Antarctic continental margin. *AGU, Antarctic Research Series*, **56**, 349–376.

GRUNIG, S. 1991. Quaternary sedimentation processes on the continental margin of the South Orkney Plateau, NW Weddell Sea (Antarctica). *Berichte zur Polarforschung*, **75**.

HOLLISTER, C. D. & ELDER, R. B. 1969. Contour currents in the Weddell Sea. *Deep-Sea Research*, **16**, 99–101.

HOWE, J. A., LIVERMORE, R. A. & MALDONADO, A. 1998. Mudwave activity and current-controlled sedimentation in Powell Basin, northern Weddell Sea, Antarctica. *Marine Geology*, **149**, 229–241.

JORDAN, R. W., PRIDDLE, J., PUDSEY, C. J., BARKER, P. F. & WHITEHOUSE, M. J. 1991. Unusual diatom layers in Upper Pleistocene sediments from the northern Weddell Sea. *Deep-Sea Research*, **38**, 829–843.

KENNETT, J. P. & BARKER, P. F. 1990. Latest Cretaceous to Cenozoic climate and oceanographic developments in the Weddell Sea, Antarctica: an ocean drilling perspective. *Proceedings of the Ocean Drilling Program, Scientific Results*, **113**, 937–960.

KING, E. C. & BARKER, P. F. 1988. The margins of the South Orkney Microcontinent. *Journal of the Geological Society*, **145**, 317–331.

KING, E.C., LEITCHENKOV, G., GALINDO-ZALDIVAR, J., MALDONADO, A. & LODOLO, E. 1997. Crustal structure and sedimentation in Powell Basin. *AGU, Antarctic Research Series*, **71**, 75–93.

LAWVER, L. A., WILLIAMS, T. & SLOAN, B. 1994. Seismic stratigraphy and heat flow of Powell Basin. *Terra Antartica*, **1**, 309–310.

LIVERMORE, R. A. & HUNTER, R. J. 1996. Mesozoic seafloor spreading in the southern Weddell Sea. *In*: STOREY, B. C., KING, E. C. & LIVERMORE, R. A. (eds) *Weddell Sea Tectonics and Gondwana break-up*. Geological Society, London, Special Publications, **108**, 227–241.

LIVERMORE, R. A. & WOOLLETT, R. W. 1993. Seafloor spreading in the Weddell Sea and southwest Atlantic. *Earth and Planetary Science Letters*, **117**, 475–495.

MACKENSEN, A., GROBE, H., KUHN, G. & FUTTERER, D. K. 1990. Benthic foraminiferal assemblages from the eastern Weddell Sea between 68 and 73°S: distribution, ecology and fossilization potential. *Marine Micropalaeontology*, **16**, 241–283.

MICHELS, K. H., KUHN, G., HILLENBRAND, C.-D., DIEKMANN, B., FÜTTERER, D. K., GROBE, H. & UENZELMANN-NEBEN, G. 2002. The southern Weddell Sea: combined contourite–turbidite sedimentation at the southeastern margin of the Weddell Gyre. *In*: STOW, D. A. V., PUDSEY, C. J., HOWE, J. A., FAUGÈRES, J.-C. & VIANA, A. R. (eds) *Deep-Water Contourite Systems: Modern Drifts and Ancient Series, Seismic and Sedimentary Characteristics*. Geological Society, London, Memoirs, **22**, 305–323.

O'BRIEN, P. D. 1989. *The magnetostratigraphy of marine sediments from Jane Basin, southeast of the South Orkney Microcontinent, Antarctica*. Unpublished PhD thesis, University of Southampton.

ORSI A. H., NOWLIN, W. D. JR. & WHITWORTH, T. III 1993. On the circulation and stratification of the Weddell Gyre. *Deep-Sea Research*, **40**, 169–203.

OSZKO, L. 1997. Tectonic structures and glaciomarine sedimentation in the south-eastern Weddell Sea from seismic reflection data. *Berichte zur Polarforschung*, **222**, 1–153.

PUDSEY, C. J. 1990. Grain size and diatom content of hemipelagic sediments at Site 697, ODP Leg 113: a record of Pliocene–Pleistocene climate. *Proceedings of the Ocean Drilling Program, Scientific Results*, **113**, 111–120.

PUDSEY, C. J. 1992. Late Quaternary changes in Antarctic Bottom Water velocity inferred from sediment grain size in the northern Weddell Sea. *Marine Geology*, **107**, 9–33.

PUDSEY, C. J., BARKER, P. F. & HAMILTON, N. 1988. Weddell Sea abyssal sediments: a record of Antarctic Bottom Water flow. *Marine Geology*, **81**, 289–314.

PUDSEY, C. J. & HOWE, J. 2002. Mixed biosiliceous–terrigenous sedimentation under the Antarctic Circumpolar Current, Scotia Sea. *In*: STOW, D. A. V., PUDSEY, C. J., HOWE, J. A., FAUGÈRES, J.-C. & VIANA, A. R. (eds) *Deep-Water Contourite Systems: Modern Drifts and Ancient Series, Seismic and Sedimentary Characteristics*. Geological Society, London, Memoirs, **22**, 325–336.

RAMSAY, A. T. S. & BALDAUF, J. G. (eds) 1999. *A Reassessment of the Southern Ocean biochronology*. Geological Society, London, Memoirs, **18**.

SHIMMIELD, G. B., DERRICK, S., MACKENSEN, A., GROBE, H. & PUDSEY, C. J. 1994. The history of biogenic silica, organic carbon and barium accumulation in the Weddell Sea and Antarctic Ocean over the last 150,000 years. *In*: ZAHN, R. *et al.* (eds) *Carbon Cycling in the Glacial Ocean: Constraints on the Ocean's Role in Global Change*. Proceedings, NATO ARW, **I 17**, 555–574.

SHIPBOARD SCIENTIFIC PARTY 1988a. Site 694. *Proceedings of the Ocean Drilling Program, Initial Reports*, **113**, 449–525.

SHIPBOARD SCIENTIFIC PARTY 1988b. Site 697. *Proceedings of the Ocean Drilling Program, Initial Reports*, **113**, 705–774.

STOREY, B. C., VAUGHAN, A. P. & MILLAR, I. L. 1996. Geodynamic evolution of the Antarctic Peninsula during Mesozoic times and its bearing on Weddell Sea history. *In*: STOREY, B. C., KING, E. C. & LIVERMORE, R. A. (eds) *Weddell Sea Tectonics and Gondwana break-up*. Geological Society, London, Special Publications, **108**, 87–103.

STOW, D. A. V., FAUGÈRES, J.-C., HOWE, J. A., PUDSEY, C. J. & VIANA, A. R. 2002. Bottom currents, contourites and deep-sea sediment drifts: current state-of-the-art. *In*: STOW, D. A. V., PUDSEY, C. J., HOWE, J. A., FAUGÈRES, J.-C. & VIANA, A. R. (eds) *Deep-Water Contourite Systems: Modern Drifts and Ancient Series, Seismic and Sedimentary Characteristics*. Geological Society, London, Memoirs, **22**, 7–20.

TECTONIC MAP OF THE SCOTIA ARC. 1985. Sheet BAS (Misc.) **3**.

TINGEY, R. J. 1991. The regional geology of Archaean and Proterozoic rocks in Antarctica. *In*: TINGEY, R. J. (ed.) *The Geology of Antarctica*. Clarendon Press, Oxford, 1–73.

TUCHOLKA, P., FONTUGNE, M., GUICHARD, F. & PATERNE, M. 1987. The Blake magnetic polarity event in cores from the Mediterranean Sea. *Earth and Planetary Science Letters*, **86**, 320–326.

ZIELINSKI, U. & GERSONDE, R. 1997. Diatom distribution in Southern Ocean surface sediments (Atlantic sector): implications for palaeoenvironmental reconstructions. *Palaeogeography, Palaeoclimatology, Palaeoecology*, **129**, 213–250.

The southern Weddell Sea: combined contourite–turbidite sedimentation at the southeastern margin of the Weddell Gyre

K. H. MICHELS*, G. KUHN, C.-D. HILLENBRAND, B. DIEKMANN, D. K. FÜTTERER, H. GROBE & G. UENZELMANN-NEBEN

Alfred Wegener Institute for Polar and Marine Research, Columbusstr., D-27568 Bremerhaven, Germany (e-mail: kmichels@awi-bremerhaven.de)

Abstract: Sedimentary processes in the southeastern Weddell Sea are influenced by glacial-interglacial ice-shelf dynamics and the cyclonic circulation of the Weddell Gyre, which affects all water masses down to the sea floor. Significantly increased sedimentation rates occur during glacial stages, when ice sheets advance to the shelf edge and trigger gravitational sediment transport to the deep sea. Downslope transport on the Crary Fan and off Dronning Maud and Coats Land is channelized into three huge channel systems, which originate on the eastern, the central and the western Crary Fan. They gradually turn from a northerly direction eastward until they follow a course parallel to the continental slope. All channels show strongly asymmetric cross sections with well-developed levees on their northwestern sides, forming wedge-shaped sediment bodies. They level off very gently. Levees on the southeastern sides are small, if present at all. This characteristic morphology likely results from the process of combined turbidite–contourite deposition. Strong thermohaline currents of the Weddell Gyre entrain particles from turbidity-current suspensions, which flow down the channels, and carry them westward out of the channel where they settle on a surface gently dipping away from the channel. These sediments are intercalated with overbank deposits of high-energy and high-volume turbidity currents, which preferentially flood the left of the channels (looking downchannel) as a result of Coriolis force. In the distal setting of the easternmost channel-levee complex, where thermohaline currents are directed northeastward as a result of a recirculation of water masses from the Enderby Basin, the setting and the internal structures of a wedge-shaped sediment body indicate a contourite drift rather than a channel levee. Dating of the sediments reveals that the levees in their present form started to develop with a late Miocene cooling event, which caused an expansion of the East Antarctic Ice Sheet and an invigoration of thermohaline current activity.

Geological and oceanographic setting

The Weddell Sea is a large marginal sea of the Southern Ocean, bounded in the south by the large Filchner and Ronne Ice Shelves, in the west by the Antarctic Peninsula, and in the north by the South Scotia Ridge (Fig. 1). To the northeast it opens to the South Atlantic, whereas in the SE Coats Land and Dronning Maud Land form its boundary. These coastal areas are part of the East Antarctic shield, which is built up of crystalline Precambrian basement overlain by undeformed sedimentary rocks of the Devonian to Triassic Beacon Supergroup and by mid-Jurassic tholeiitic intrusions and flood basalts (British Antarctic Survey 1985; Tingey 1991). The coast of Coats Land and Dronning Maud Land is characterized by large fringing ice shelves.

The general oceanographic circulation in the Weddell Sea is dominated by the cyclonic Weddell Gyre, which affects all water masses down to the seafloor (Carmack & Foster 1975a, b; Deacon 1979; Gordon et al. 1981). For a detailed review of the Weddell Sea oceanography see Fahrbach et al. (1998). The following section summarizes the most important features.

The Weddell Sea is known as an important area for bottom-water formation; about 70% of the Antarctic Bottom Water formation is influenced by processes in the Weddell Sea (Carmack & Foster 1977). The uppermost water mass in the southeastern Weddell Sea is the Winter Water (WW), a residual layer (100–200 m thick) formed in winter during sea-ice formation. By mixing with Warm Deep Water (WDW) in 200 to 1500 m water depth, it contributes to the formation of a thin layer of Modified Warm Deep Water (MWDW), located between WDW and WW. The Antarctic Bottom Water (AABW) occurs below 1500 m water depth down to the sea floor and forms the deepest water mass in the eastern and southeastern Weddell Sea. The Antarctic Coastal Current mainly comprises Eastern Shelf Water (ESW), and follows the contours of the coast on its way through the Weddell Sea, until it reaches the Crary Trough where a major branch turns south. ESW and Western Shelf Water (WSW) from the shelf areas west of the Crary Trough circulate under the Filchner and Ronne Ice Shelves, and it is mainly WSW which contributes to the formation of Ice Shelf Water (ISW) by mixing processes. The ISW flows along the western slope of the Crary Trough in water depth of 300 to 800 m and leaves it to the north across the sill, where it contributes to the formation of Weddell Sea Bottom Water (WSBW), by mixing with WDW (Foldvik et al. 1985). The WSBW forms a water mass underlying the AABW in the western and northern Weddell Sea. The downslope flow of ISW and/or WSBW in the southern Weddell Sea influences sediment transport processes, especially in the channels and gullies on the continental slope. Sea ice generally covers more than 80% of the Weddell Sea in the austral winter (Sea Ice Climatic Atlas 1985). Early in the austral spring season a polynya develops along the shelf off Dronning Maud Land and expands southwestward, reaching a width of 200 km (Zwally et al. 1985; Gloersen et al. 1992). Other areas in the southwestern Weddell Sea generally are still covered

Table 1. *Principal characteristics*

Location	Southern Weddell Sea
Setting	Southern margin of the cyclonic Weddell Gyre: important area of bottom water formation
Age	Upper Miocene to Recent
Drift type	Combined contourite-turbidite sedimentation in levees along the western side of turbiditic channels
Dimensions	Several tens of kilometres wide, several hundreds of kilometres long
Seismic characteristics	Low-amplitude, continuous, thin-layered reflectors in wedge-shaped sediment bodies
Sediment characteristics	Fine-grained terrigenous, moderately to well sorted, generally not bioturbated, development mainly during glacials

From: STOW, D. A. V., PUDSEY, C. J., HOWE, J. A., FAUGÈRES, J.-C. & VIANA, A. R. (eds)
Deep-Water Contourite Systems: Modern Drifts and Ancient Series, Seismic and Sedimentary Characteristics.

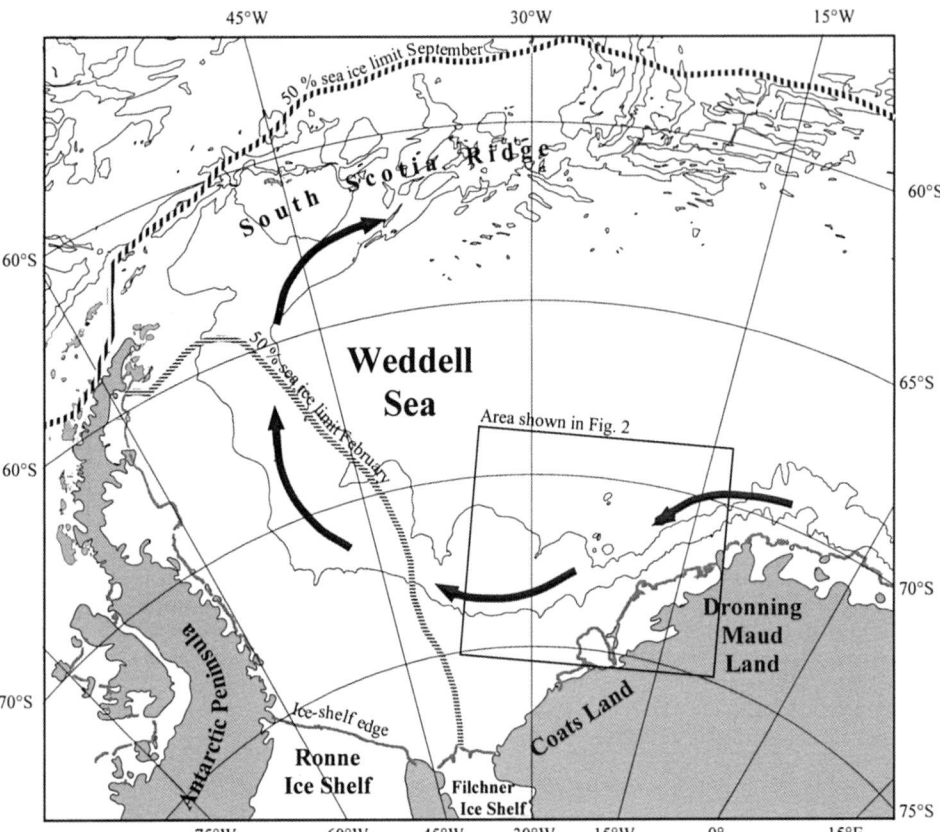

Fig. 1. Map of the Weddell Sea with main geographic features, ocean surface circulation (arrows), and maximum and minimum sea-ice coverage indicated by mean 50% February and September sea-ice limits (data from Zwally *et al.* 1985; extracted from Ocean Data View, Schlitzer 1999). Water depth indicated by 2000 m and 4000 m isobaths. Box marks area shown in Figure 2.

with sea ice to more than 50% during the minimum sea-ice-coverage in mid-February.

Current measurements

Long and short-term current-meter records have been obtained from the southeastern Weddell Sea during the last two decades. The current meters deployed off Cape Norvegia were part of a transect crossing the Weddell Sea to the northern tip of the Antarctic Peninsula. The results for the Cape Norvegia region indicate very strong southwestward currents on the shelf and along the continental slope (Fahrbach *et al.* 1994). For presentation here we have chosen three records of mean daily current speed from moorings off Coats and Dronning Maud Land (Fig. 3, see Fig. 2 for location of moorings; data available by courtesy of G. Rohardt, AWI). An almost year-long current speed record from the shelf off Vestkapp is shown in Figure 3a. Strong southwesterly currents with speeds of up to 24 cm s^{-1} dominate the record. An apparent temporal asymmetry characterizes the speeds and directions. In the first half of 1987 the currents were much stronger and showed less deviation from the southwesterly direction than in the second half of 1987, where current speeds decreased significantly and northerly burst can be found. This is consistent with the observation of Foster & Middleton (1979) and Fahrbach *et al.* (1992) that currents show annual variations with stronger currents in the austral summer.

Figure 3b shows a current-intensity record from the continental slope off Vestkapp (Fahrbach & Rohardt 1988). The data show a fluctuation of alternating southwesterly and northwesterly directions in a time period band of 15 days (more clearly visible in a plot of the six hourly mean current speed; see Fahrbach *et al.* 1992, their fig. 11). Foster & Middleton (1979) discuss basin modes or eddies as possible mechanisms for the fluctuations, whereas Fahrbach *et al.* (1992) assume that wind forcing plays an important role in the generation of these fluctuations.

The data of mooring AWI 213, which was located in a distal channel of the Crary Fan system, show that current direction and intensity can be variable in the basin (Fig. 3c). Two preferential directions of the flow, northeastward and southwestward, can be attributed to the channel alignment. Weber *et al.* (1994) ascribed the dominating northeasterly flow to downstreaming ISW from the Filchner shelf via the Crary Fan channel system, but measurements of water mass properties revealed that the current mainly comprises recirculating water masses originating in the Enderby Basin (Hoppema *et al.* 1998). Hence, the dominating northeasterly current direction probably reflects the local current pattern affected by the topography (Fahrbach *et al.* 1998).

Bathymetry

The shelf in this area is relatively narrow with water depths of 300 to 400 m, dipping gently toward the coast as a result of glacial erosion and glacio-isostatic loading. A distinct shelf break in about 600 m water depth separates the shelf from the steep upper continental slope with inclinations of up to 16°. Along Dronning Maud Land, the midslope includes a terrace dipping seaward at 1.5° between 1500 to 3000 m water depth. The lower slope is formed by the Explora Escarpment (Hinz & Krause 1982; Henriet & Miller 1990), a steep clifflike slope (up to 30°) (Fütterer *et al.* 1990), abruptly rising from the Weddell Abyssal Plain in *c.* 4400 m water depth. Several canyons incise the escarpment, but only the Wegener Canyon at ~14°W is important for regional sediment transport (Fig. 2).

Farther south, along Coats Land, the midslope and lower slope are less steep than the upper slope, until they merge into the Crary Fan, a large deep sea fan seaward of Crary Trough in front of the Filchner Ice Shelf. Here the shelf broadens to almost 400 km. Crary Trough is a shelf feature formed by ice erosion. At the Filchner ice-shelf edge it reaches 1200 m water depth. The transition from the trough to the fan is formed by a sill with water depths of ~600 m. Trough and sill are thought to be the result of both isostatic downwarping beneath the ice masses and glacial

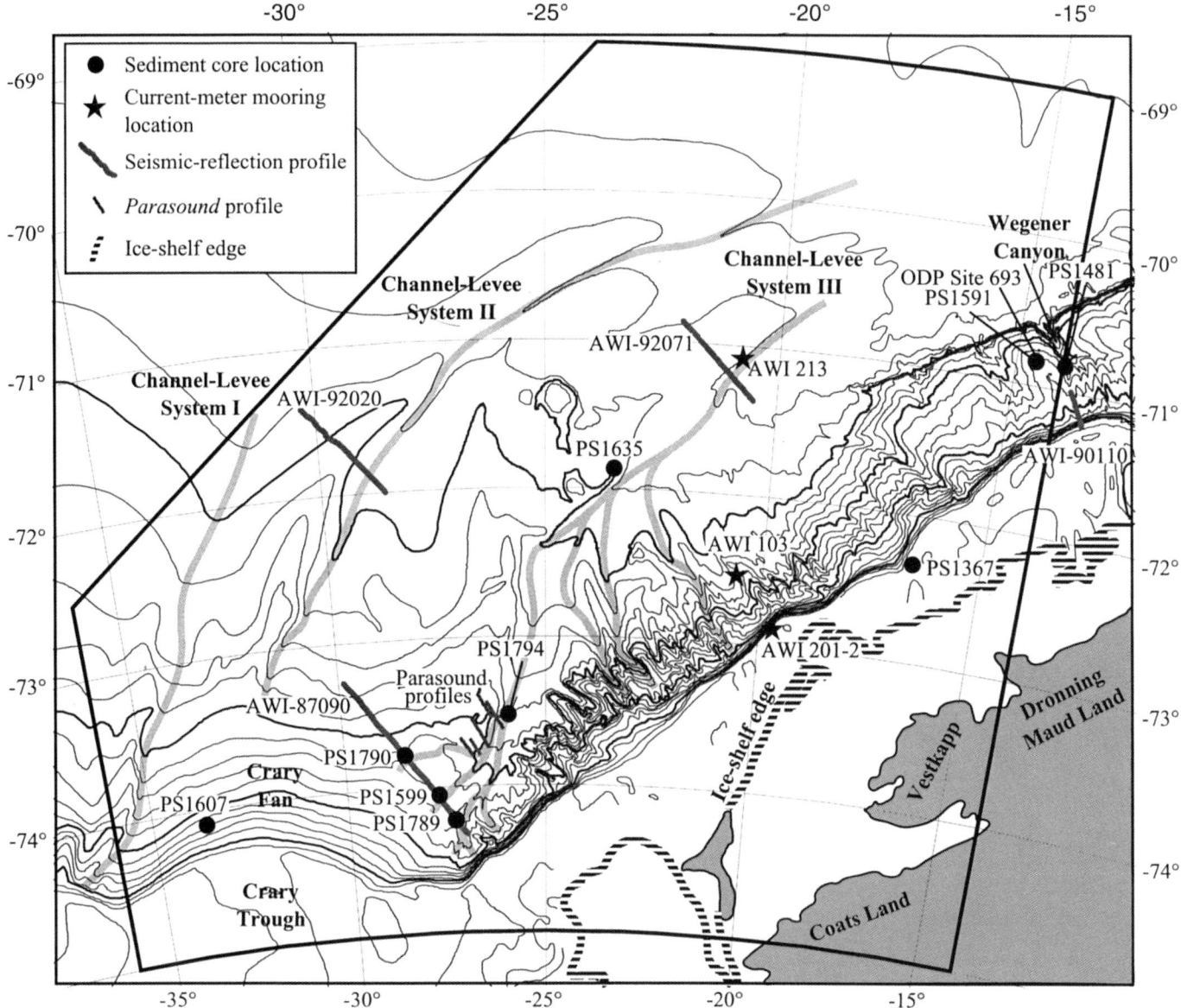

Fig. 2. Map of the southeastern Weddell Sea with main geographic features, courses of Channel-Levee Systems I to III (light grey lines), positions of sediment cores (dots), current-meter moorings (stars), and seismic profiles (dark grey lines) (Schenke *et al.* 1998). Bathymetry is indicated by isobaths in 200 m steps, with thick lines every 1000 m. Box marks area shown in Figures 5a, b & c.

erosion during times when the ice sheet reached further north (Anderson *et al.* 1983; Elverhøi 1981; Elverhøi & Maisey 1983; Fütterer & Melles 1990; Kuvaas & Kristoffersen 1991). The slope of the Crary Fan reaches 4° in the upper part and decreases towards the basin.

The continental slope off Coats Land is intersected by numerous gullies, channels, and small canyons. On the eastern side of Crary Fan, adjacent to the southwestern parts of Coats Land, a 70 km wide NE-inclined terrace dissected by several channels is developed in 2000 m to 3000 m water depth (Weber *et al.* 1994). The channels are flanked by associated ridges on their northwestern sides and merge to a major channel draining to the northeast below 3000 m water depth. Two other channel systems drain the central and western part of the Crary Fan to the north and northeast (Fig. 2). Multichannel seismic investigations showed that the channels become younger from west to east, and that they tend to migrate eastward (Kuvaas & Kristoffersen 1991). The easternmost channel shows a V-shaped cross section in its proximal part, changing to a broad, flat cross section in the distal part. For description of features related to these channels they numbered the channel-levee systems from west to east (Channel-levee System I to III, Fig. 2).

Stratigraphic context

Sedimentation rates in the southeastern Weddell Sea can be estimated for the Neogene from the results of Ocean Drilling Program (ODP) Site 693, which is located on a continental midslope bench off Dronning Maud coast, 10 km from the margin of the Wegener Canyon, in a water depth of 2359 m (Fig. 2). Late Miocene sedimentation rates were c. 24 m Ma^{-1} and increased to 60 m Ma^{-1} in the early Pliocene; Quaternary sedimentation rates were 16 m Ma^{-1} (Gersonde *et al.* 1990). The Pleistocene sedimentation rate for the last ~1 Ma is 10 m Ma^{-1} (Grobe *et al.* 1990b).

Due to a lack of a continuous carbonate content in cores from the southeastern Weddell Sea, sediments cannot be dated using ^{18}O stratigraphy. The likely presence of sediment reworking complicates the use of ^{14}C dating of organic carbon to obtain a stratigraphy for sediment cores. To overcome these difficulties a lithostratigraphy was developed by stacking of sedimentological parameters of 11 cores from the continental slope off Dronning Maud Land (Grobe & Mackensen 1992). The stacks can be correlated with the continuous δ^{18}O records of planktic and benthic foraminifera in core PS1506 (Latitude 68.728°S, Longitude 5.823°W) in the eastern Weddell Sea for the last 300 ka

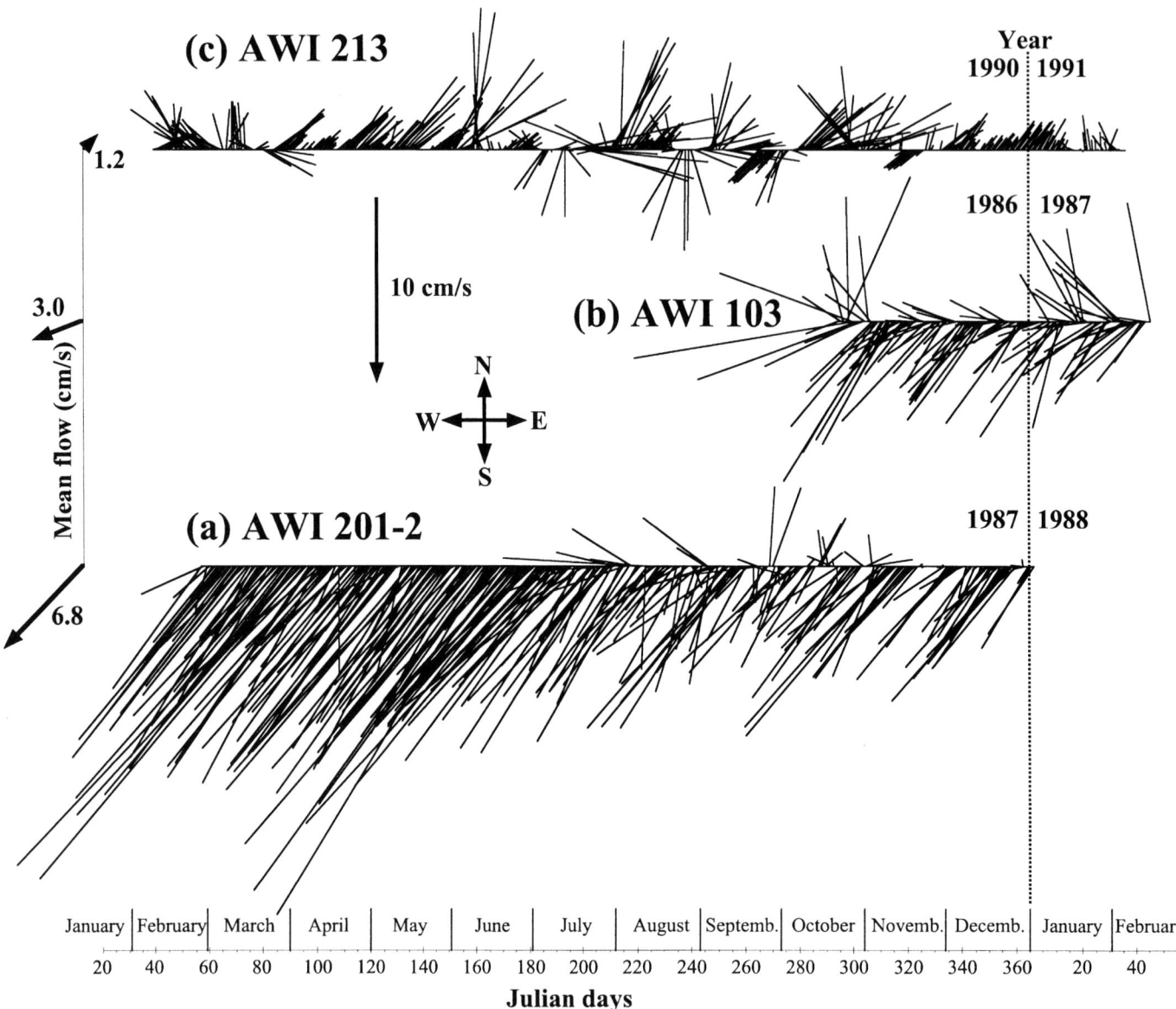

Fig. 3. Stick plot diagrams of the time series of daily mean current speeds from instruments deployed in moorings off Coats and Dronning Maud Land for up to one year (for location of moorings see Fig. 2). (**a**) record from mooring AWI 201–2 on the continental shelf off Vestkapp in 461 m water depth from February 27, 1987 to January 3, 1988 (G. Rohardt, unpublished data). The current meter was located in 380 m of water depth. (**b**) record from mooring AWI 103, deployed in 3415 m water depth at the continental slope off Vestkapp from October 23, 1986 to February 17, 1987 (modified from Fahrbach *et al.* 1992). The current meter was located 4 m above the seafloor. (**c**) record from mooring AWI 213, deployed in 4440 m water depth in a distal channel originating on the eastern Crary Fan, from February 8, 1990 to February 6, 1991 (modified from Rohardt *et al.* 1992). The current meter was located 10 m above the sea floor. Data extracted from the Ocean Circulation Database that can be accessed at www.awi-bremerhaven.de/OZE/ocdb/database.html.

(Mackensen *et al.* 1994) and allow a detailed interpretation of the environmental changes during this period. One of the conditions for the use of this lithostratigraphy, however, is the availability of high-resolution data for the carbonate content, grain size distribution, clay mineralogy, and siliceous microfossil content. The results of this lithostratigraphic approach indicate decreasing mean sedimentation rates with increasing water depth on a profile across the continental slope in the eastern Weddell Sea for hemipelagic sediments during a climatic cycle, with mean values of 5.2 cm ka^{-1} for the upper and 1.3 cm ka^{-1} for the lower slope (Grobe *et al.* 1990*a*). The rates can increase up to 25 cm ka^{-1} close to the shelf in the beginning of each interglacial, whereas lowest values around 0.6 cm ka^{-1} are found during glacials in the basin.

On the eastern Crary Fan, an area which is strongly influenced by contour currents of the Weddell Gyre, glacial sedimentation rates reach values up to 376 cm ka^{-1} (core PS1789 in Fig. 2, water depth is 2411 m). The rates decrease with increasing water depth and distance from the shelf edge to 125 cm ka^{-1} in core PS1599 and 60 cm ka^{-1} in core PS1790 (Fig. 2; Weber *et al.* 1994). On the western Crary Fan, where sediment-acoustic data show debris-flow deposits and slumps, sedimentation rates are in the range of 8 cm ka^{-1} for near-surface sediments in a water depth of 2934 m (cores PS1606); the sedimentation rate decreases to 2 cm ka^{-1} in shallower water depth (1612 m, core PS1607; Melles 1991).

Seismic characteristics: reflection profiles

Five profiles, which show different features of the shape and internal architecture of contourite-influenced sediment bodies, have been chosen for presentation here (Fig. 4a, c–e). These profiles are complemented by a succession of 5 *Parasound* profiles across Channel-Levee Complex III on the eastern Crary Fan (Fig. 4b, for location of seismic and *Parasound* profiles see Fig. 2). A seismic stratigraphy has been developed by Miller *et al.* (1990) on the basis of the results from ODP Site 693 (Leg 113, Barker *et al.*

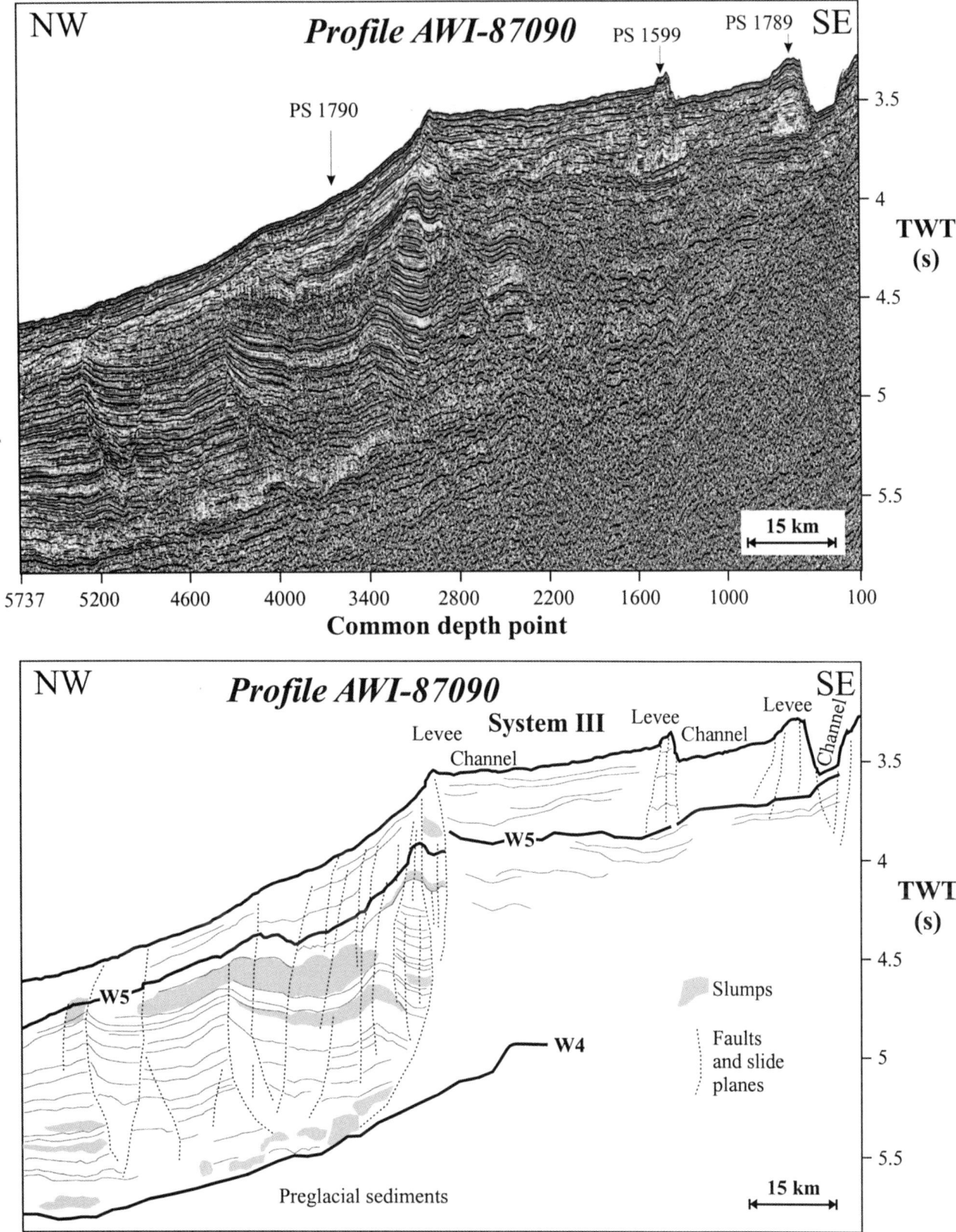

Fig. 4. (**a**) Multichannel seismic profile AWI-87090 across the eastern Crary Fan (modified from Oszkó 1997; for location of the profile see Fig. 2), with core locations indicated above profile. Sediments above unconformity W4 represent the Cenozoic glaciomarine sequence in the Weddell Sea (Miller *et al.* 1990). The sediment sequence in the northwestern part of the profile shows many faults, growth faults and listric shear planes, associated with slumped blocks and slides (gray shaded). In the southeastern part of the profile three channel-levee systems can be seen, two of them with well-developed levees, which show indication of overloading as a result of very high sedimentation rates. Indeed, sedimentation rates of 125 cm ka^{-1} and more than 200 cm ka^{-1} have been measured for last glacial sediments in cores PS1599 and PS1789, respectively (Weber *et al.* 1994). The sequence above unconformity W5 represents upper Miocene to Recent sediments deposited under an increased influence of contour currents (Miller *et al.* 1990).

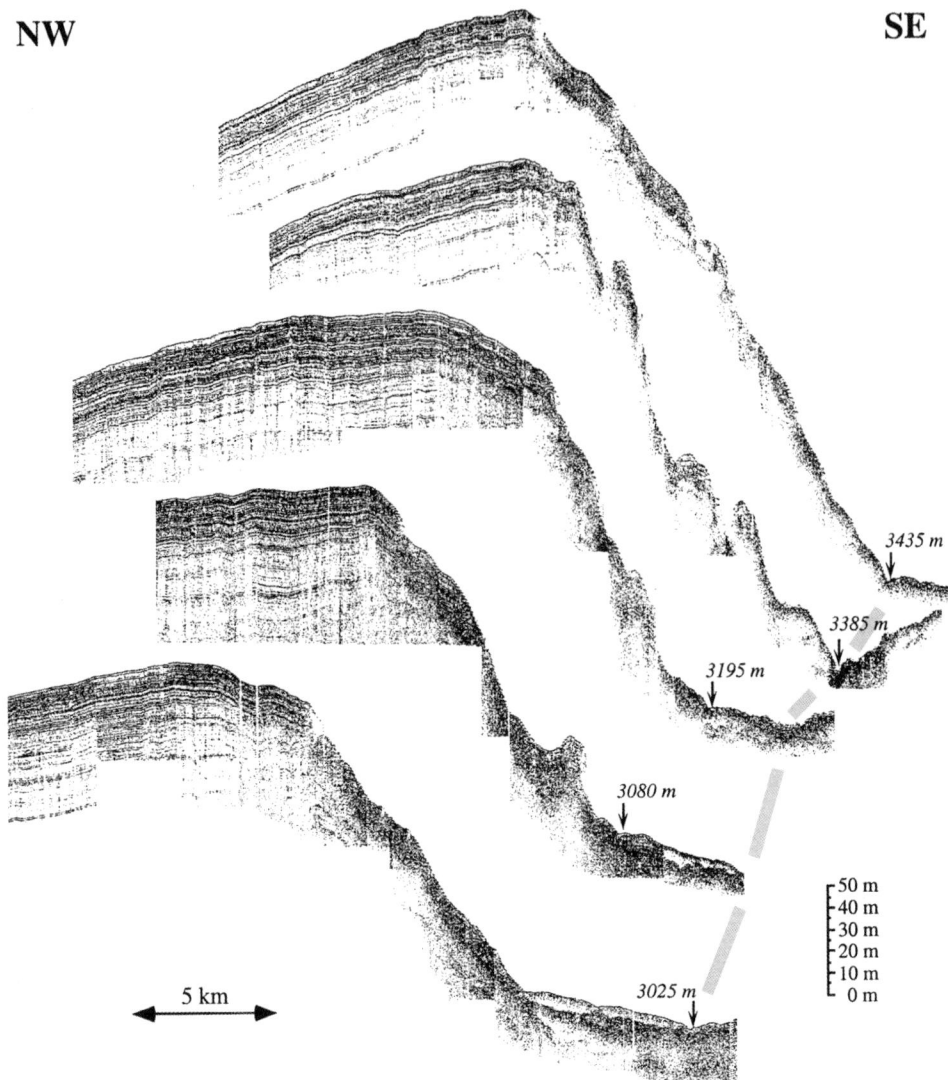

Fig. 4. (**b**) Succession of five *Parasound* profiles from the levee crest into the adjacent channel in water depths of *c.* 3000 to 3450 m of Channel-Levee System III (modified from Kuhn & Weber 1993; for location of the profiles see Fig. 2). Deep penetration and parallel reflectors characterize the levee deposits, thus indicating high sedimentation rates. Channel deposits show a rough topography, prolonged reflectors and some slides, which is typical for coarse sediments and erosive conditions.

1988, 1990). The description of the profiles shown in Figs 4a and c–e is based on this stratigraphy, which has been refined and applied to numerous other seismic profiles in the southeastern Weddell Sea by Kuvaas & Kristoffersen (1991), Moons *et al.* (1992), Oszkó (1997), and Bart *et al.* (1999).

Profile AWI-87090 crosses the eastern Crary Fan from northwest to southeast (Oszkó 1997). In the southeast the profile shows two well-developed channel-levee systems, where the levees form narrow ridges which overtop the adjacent seaward slope (Fig. 4a). A sequence with a thickness of *c.* 0.5 s two-way travel time (TWT) under these ridges is separated from the ambient sediment by faults, forming a wedge-shaped sediment ridge. Sediment overloading as a result of very high sedimentation rates is thought to cause subsidence of these ridges, and suggests very high sedimentation rates on the ridges. Sedimentation rates of 125 to 250 cm ka^{-1} and > 200 cm ka^{-1} have been determined for the period of the last glacial maximum at core locations PS1599 and PS1789 on the two ridges, respectively (Weber *et al.* 1994). The northwestern part of the profile is characterized by the presence of numerous large, synsedimentary, listric shear planes and faults, some of them developed as growth faults, associated with a number of slump blocks and slides. The slope inclination increases significantly seaward of the location where the first deep fault occurs. The faults end at the prominent reflector W4 in more than 1 s TWT sediment depth. W4 represents the base of the Cenozoic glaciomarine sequence in the Weddell Sea (Miller *et al.* 1990). The susceptibility of the sediment to failure can be seen as an indicator for high sedimentation rates in combination with a high pore water pressure.

A succession of five *Parasound* profiles (from Kuhn & Weber 1993) in the downward course of Channel-Levee System III is depicted in Figure 4b. The sediments on the ridge allow deep acoustic penetration and the reflection pattern consists of numerous parallel to subparallel reflectors. The crest of the levee is not bounded by faults, as observed in the proximal region of the Channel III in Profile AWI-87090, so that the reflectors of the levee are undisturbed. The channel bottom and slope show a rough topography with some slide deposits. Prolonged reflectors indicate coarse sediment and an erosive regime. Profile AWI-92020 (Fig. 4c; Oszkó 1997) crosses the Channel-Levee System II on the central Crary Fan in a water depth of *c.* 4000 m. Above a coarse layer of channel deposits two asymmetric levees (the levee of System II and the old eastern levee) developed along Channel II. Another channel (labeled old channel) developed southeast of

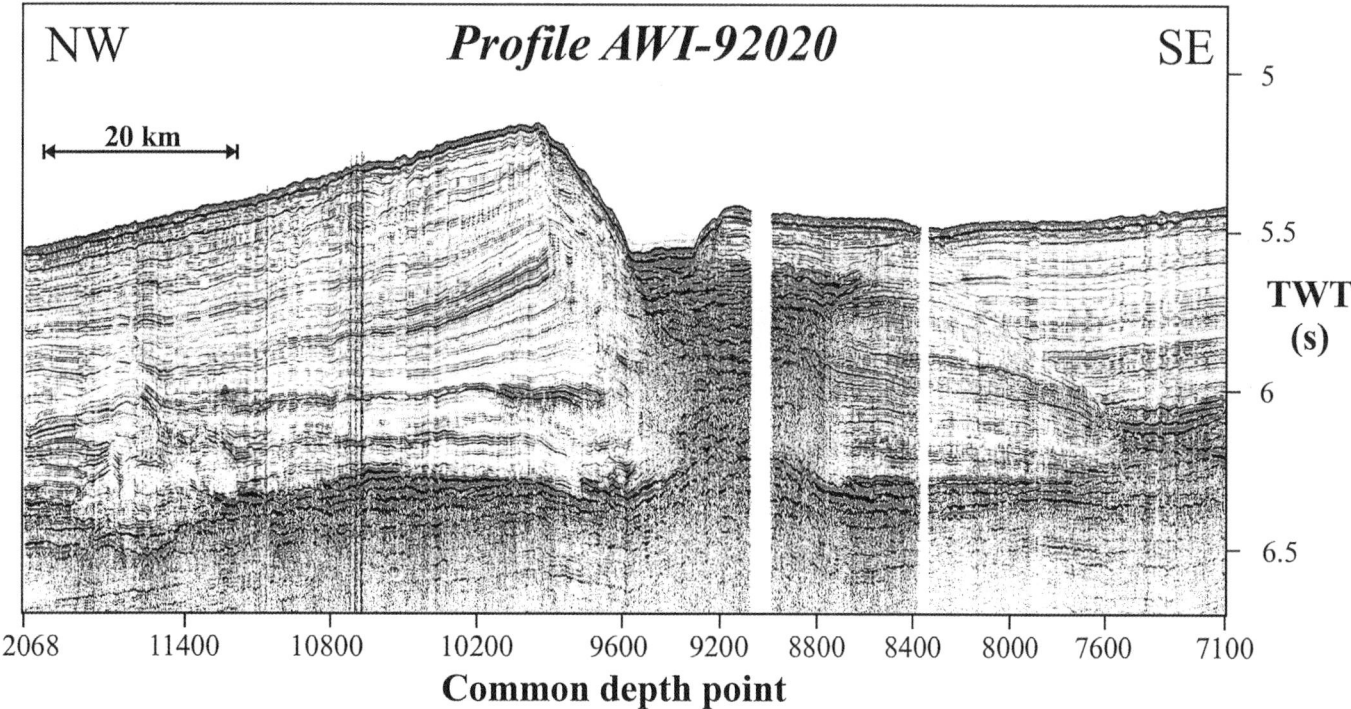

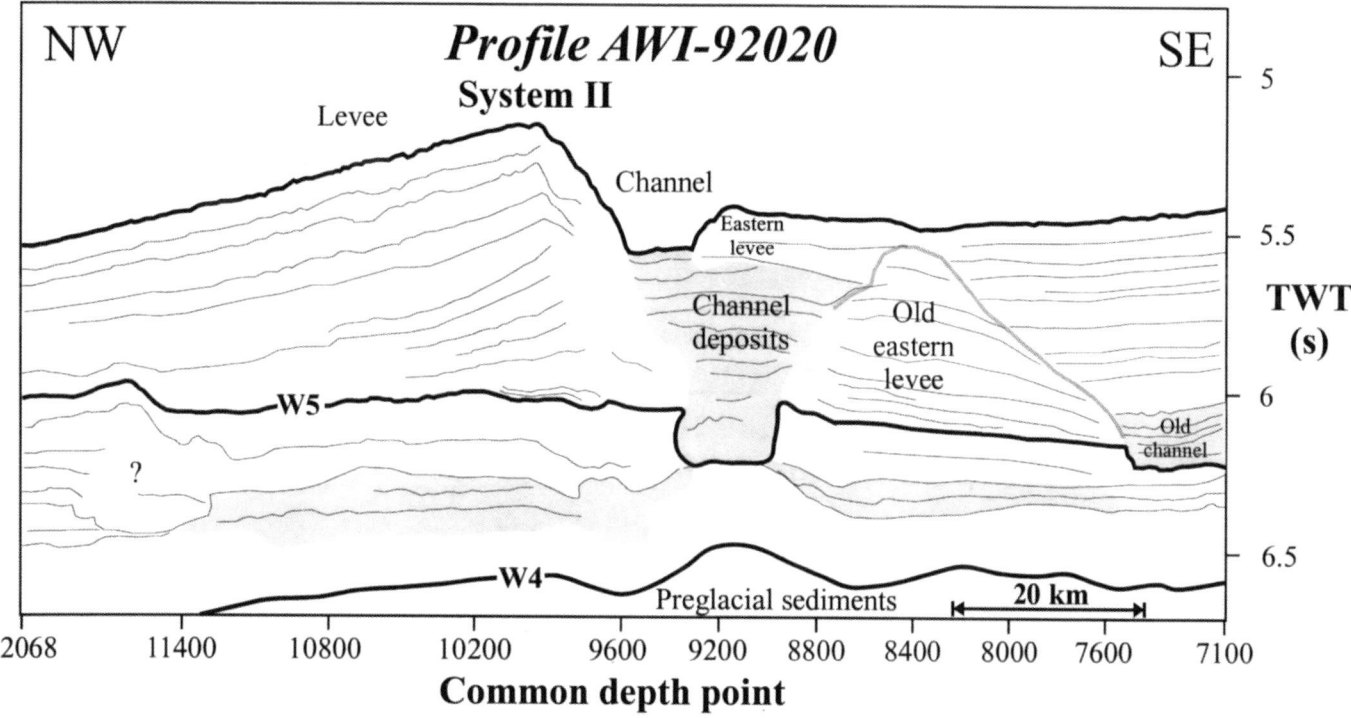

Fig. 4. (**c**) Multichannel seismic profile AWI-92020 across Channel-Levee System II in a water depth of *c*. 4000 m (modified from Oszkó 1997; for location of the profile see Fig. 2). A very well-developed northwestern levee can be seen, showing low-amplitude, highly continuous, thin layered reflectors. The southeastern side of the channel shows only a very small levee, merging into parallel-bedded sediment southeast of it. The subsurface structures indicate that a larger southeastern levee (labeled old eastern levee) was present earlier, but has been buried. Note that position of unconformity W5 is tentative. However, this reflector marks the bases of wedge-shaped channel levees, and thus correlates well with the idea of increased current influence for younger sediments. Coarse channel sediments are shaded grey.

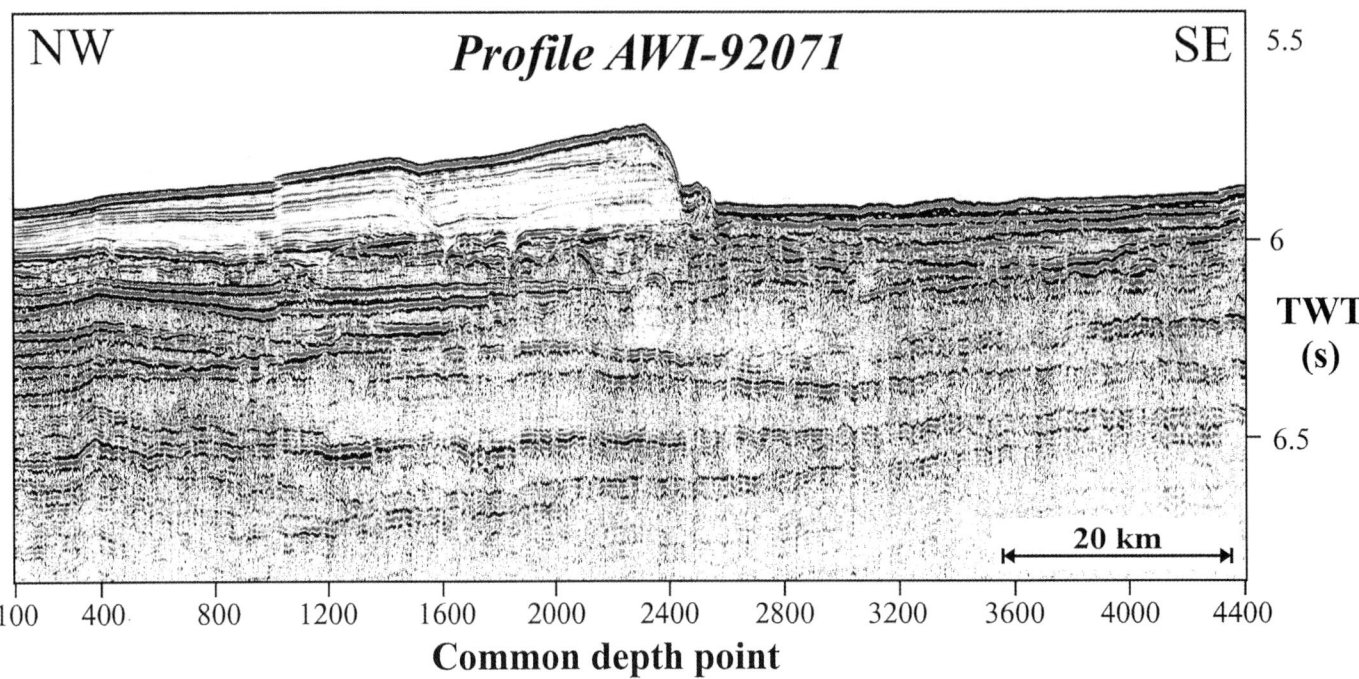

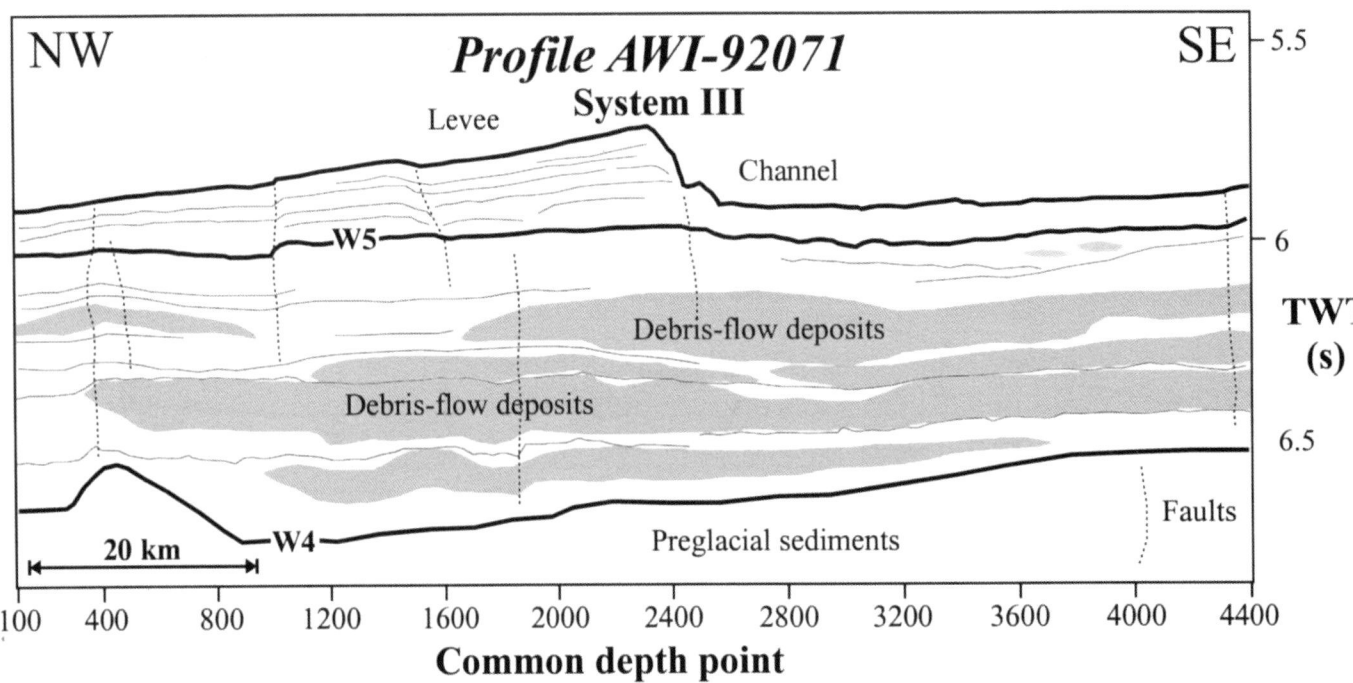

Fig. 4. (**d**) Multichannel seismic profile AWI-92071 across Channel-Levee System III in a water depth of *c*. 4400 m (modified from Oszkó 1997; for location of the profile see Fig. 2). The levee formed above unconformity W5 as a well-developed wedge-shaped sediment unit, showing low-amplitude, highly continuous, thin layered reflectors. The associated channel is broad and flat with very high-amplitude continuous reflectors. Beneath this a sediment sequence charaterized by high-amplitude, moderately continuous reflectors, mixed with acoustically transparent sections, which are thought to be debris-flow deposits (grey shaded), probably represents the early Cenozoic glaciomarine sedimentation.

the eastern levee, but both, old channel and old levee, were subsequently buried from the southeast, maybe by levee sediments of System III. The main channel decreased significantly in size and developed a smaller eastern levee. Bart *et al.* (1999) interpreted the coarse channel deposits between *c*. 6.3 and 6.5 TWT as a chaotic seismic facies resulting from large-volume mass wasting processes of sediments from the Crary Fan and the Dronning Maud Land slopes during the early Pliocene.

Profile AWI-92071 (Fig. 4d; Oszkó 1997) is located across Channel III in a water depth of *c*. 4400 m. A levee-like feature developed as a narrow wedge of low-amplitude, continuous, thin-layered reflectors on top of a *c*. 0.6 s TWT thick sediment sequence characterized by sharp discontinuities and many debris-flow deposits and bounded by unconformities W4 and W5. The seafloor southeast of the wedge-shaped sediment body is very flat and shows highly reflective coarse sediments. There is no indication of an eastern channel restriction. Taking into account a northeasterly current direction (measured nearby in mooring AWI 213), the levee of System III may have developed as a drift body shaped by contourite currents. The transition between a more turbidite-influenced channel-bound levee system to a contourite-current shaped drift takes place between the positions of

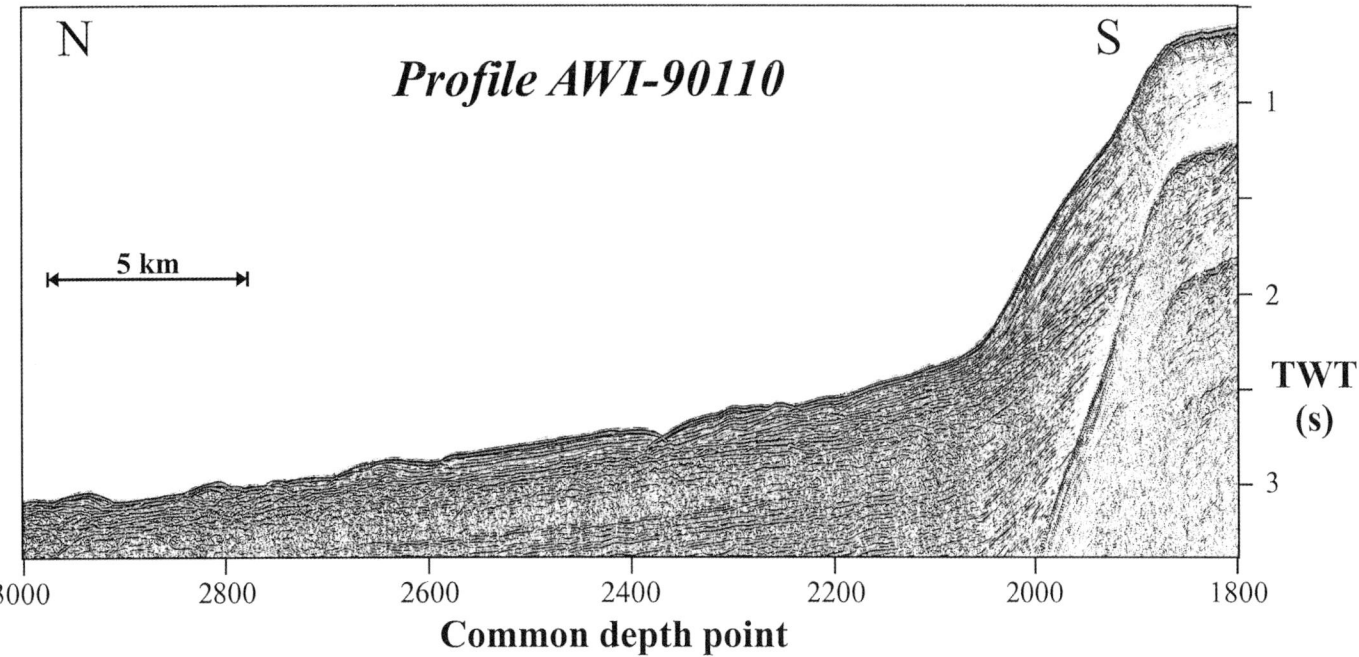

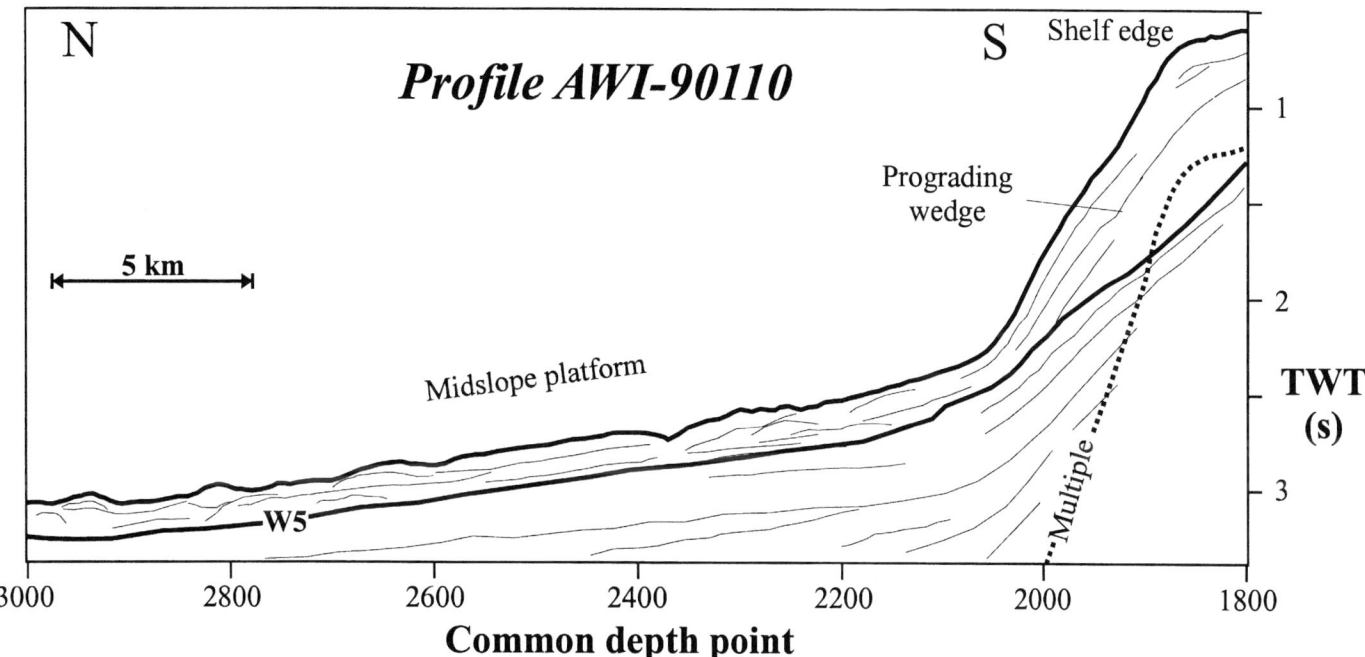

Fig. 4. (**e**) Multichannel seismic profile AWI-90110 across the upper midslope terrace, the upper continental slope and the shelf edge landward of the Wegener Canyon (modified from Oszkó 1997; for location of the profile see Fig. 2). The uppermost unit above unconformity W5 shows a rough undulating surface and high-amplitude, discontinuous reflectors, indicating relative small lens-shaped sediment bodies in the area of the midslope terrace. These merge landward into lower-amplitude, more continuous reflectors, representing the foreset beds of a prograding wedge.

the Parasound profiles and seismic profile AWI-92071, i.e. between 3000 and 4400 m of water depth.

Profile AWI-90110 (Fig. 4e; Oszkó 1997) crosses the upper midslope terrace and the upper continental slope and shelf edge off Dronning Maud Land at $c.$ 14° W. The upper part of the profile shows prograding foreset beds at the shelf edge. The midslope terrace shows a sediment sequence above unconformity W5 which is characterized by an irregular, undulating sediment surface and small lens-shaped sediment bodies, indicating strong current influence and residual sediments. These bodies merge into the foreset beds of the upper continental slope without any apparent unconformity. Very low sedimentation rates of less than 1 cm ka^{-1} for the last 4 Ma in core PS1481, lying in 2452 m water depth in extension of Profile AWI-90110, confirm the formation of residual sediments (Grobe & Mackensen 1992).

Seismic characteristics: 3.5 kHz/Parasound mapping

In the southeastern Weddell Sea, in an area between 14° and 36°W and 69.5° and 75°S, we mapped the penetration depths (Fig. 5b) and echosounder facies types (Fig. 5c) using 3.5 kHz and

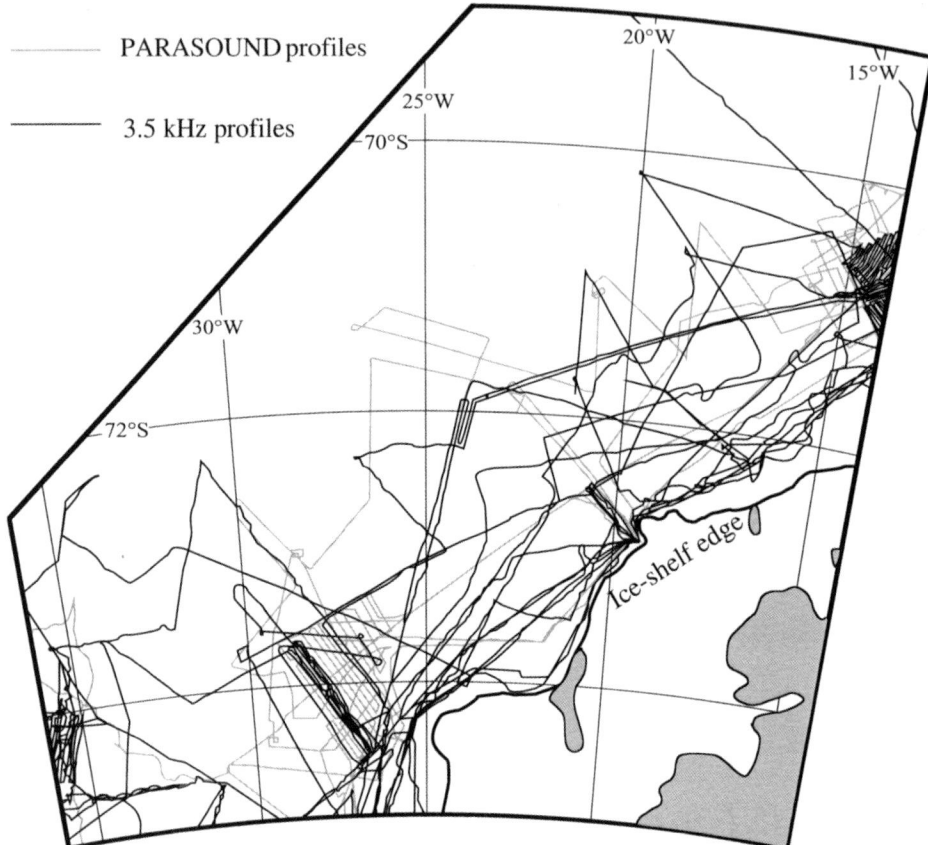

Fig. 5. (a) 3.5 kHz and *Parasound* profile grid in the southeastern Weddell Sea, on which the acoustic penetration and echo-type character maps (Figs 5b & c) are based (from Kuhn *et al.* 1995).

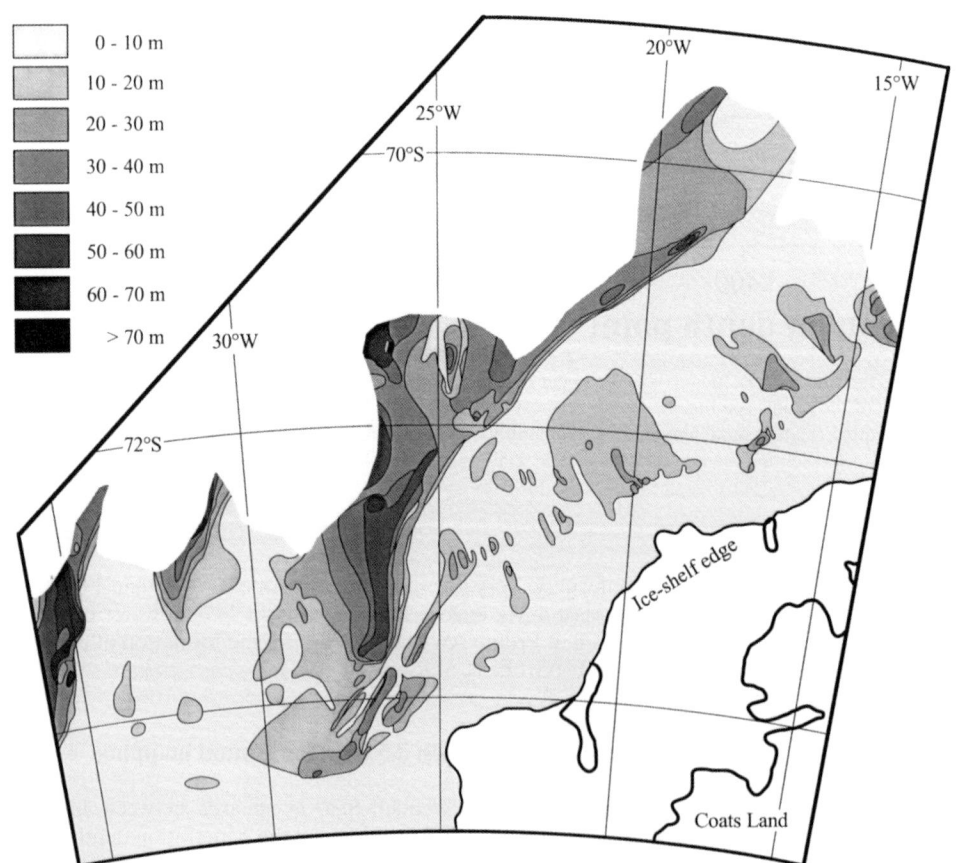

Fig. 5. (b) Acoustic penetration of a 3.5 kHz signal in the southeastern Weddell Sea, based on a p-wave velocity of 1500 m s^{-1} (from Kuhn *et al.* 1995). High penetration depths indicate high sedimentation rates and fine-grained sediments.

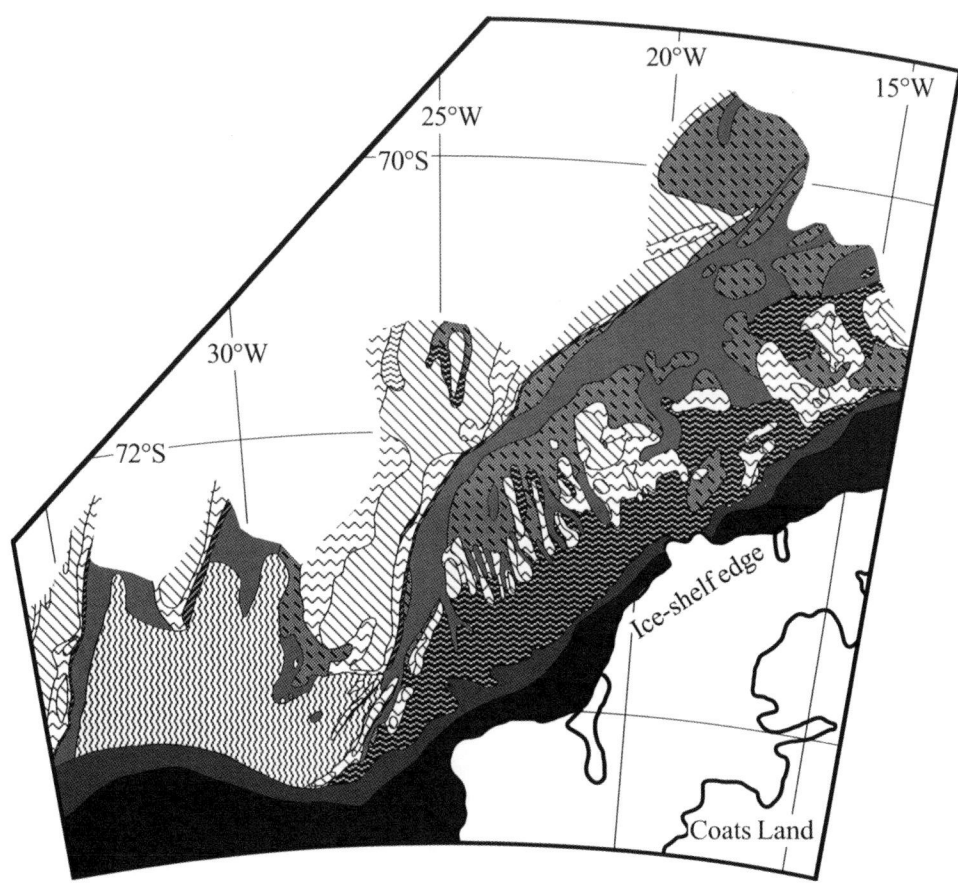

Fig. 5. (c) Classification of the echo-type character of surface sediments in the southeastern Weddell Sea. A key to the patterns, a description and an example of their associated echo types, and an interpretation is given in Figure 5d (compiled from Kuhn *et al.* 1995; Kuhn & Weber 1993; Melles & Kuhn 1993).

Parasound systems. The maps are a compilation of results from Kuhn & Weber (1993) and Melles & Kuhn (1993), extended by data from the Coats and Dronning Maud Land continental shelf, slope and adjacent deep-sea areas. Ten different facies patterns could be distinguished and related to certain sedimentary processes and structures (Fig. 5d). The maps are based on a dense grid of 3.5 kHz and *Parasound* profiles (Fig. 5a). Penetration depth generally is a function of the physical properties of a sediment. For the southeastern Weddell Sea, where terrigenous particles dominate sediment composition, the penetration depth can mainly be attributed to a combination of grain size and water content of the sediment. The grain size is controlled by the glacial-marine transport processes, which can be inferred from the echosounder profiles, whereas the water content is mainly affected by sedimentation rate. The distribution of facies types and penetration depths shows a very clear relation to the channel systems on the Crary Fan and along the continental slope off Coats and Dronning Maud Land. The channels themselves appear as broad paths characterized by prolonged bottom reflectors and shallow penetration depths. On their left side (looking downstream) they are bound by a small band of channel slope facies followed by multi-layered or wavy multi-layered facies and deep acoustic penetration, which slowly decreases with distance from the channels. To the right of the channels penetration depth is generally low. The proximal Crary Fan is characterized by large areas of wedging sub-bottom reflectors and low penetration depths, indicating prevalence of debris-flow deposits.

Sediments: seabed photographs

Seabed photographs have been taken in the Weddell Sea during the cruises of USCGC Glacier and USNS Eltanin in the 1960s. Hollister & Elder (1969) inferred the direction and strength of bottom currents from sediment lineations and the deflection of organisms by currents on oriented photographs at three sites on the shelf next to Crary Trough. All three photographs showed abundant large benthic animals and strong current evidence. Current direction was to the northwest.

Sediments: core description and facies

The surface sediments on the continental shelf in the southeastern Weddell Sea are dominated by poorly-sorted, coarse residual deposits; finer particles generally are kept in suspension in the turbulent water conditions and are carried away by the Antarctic Coastal Current (Elverhøi & Roaldset 1983). Biogenic particles are an integral part of the glaciomarine deposits. The upper continental slope down to a water depth of *c.* 1600 m shows also residual sediments with a sand content of more than 40% (Grobe & Mackensen 1992). Silt and clay is winnowed away by contour currents and transported southwest along the continental slope. With increasing water depth, on the mid and lower slope, the mud content increases until it dominates the sediment. Its grain-size distribution pattern is the result of marine sorting processes acting on a glacially derived, poorly sorted debris (Fütterer & Melles 1990; Melles 1991). Debris-flow deposits are described from the mid and lower slope (Anderson *et al.* 1979; Kuhn & Weber 1993; Oszkó 1997). Clay mineral and heavy mineral assemblages in surface sediments from the Weddell Sea generally reflect the influences of the oceanographic and climatic regimes, the sediment sources and transportation processes (Ehrmann *et al.* 1992; Petschick *et al.* 1996; Diekmann & Kuhn 1999; Diekmann *et al.* 1999).

Highest sedimentation rates are postulated for the transitions

Laminated/stratified facies of the marginal basin and the lower continental slope

Pattern	Description of echo type	Example	Interpretation
	("multi-Layers") several sharp continuous parallel subbottom reflectors, high penetration (30 - 80 m)	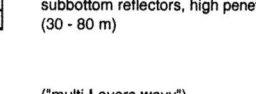	Undisturbed sequence of laminated fine-grained sediments deposited on channel levees mainly by contourite currents, high sedimentation rate
	("multi-Layers wavy") several sharp continuous to discontinuous subbottom reflectors, wavy, migrating layers, high penetration (>30 m)	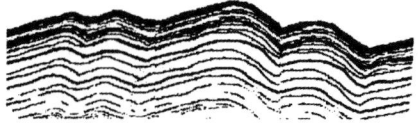	Migrating sediment waves consisting of fine-grained sediments, deposited at high sedimentation rates
	("semi-Prolonged Layers") semiprolonged reflectors, some continuous to discontinuous subbottom reflectors, moderate penetration (10 - 30 m)		Medium-grained sediments, moderate sedimentation rates, in some places erosion
	("multi-Layers discontinuous") several sharp discontinuous subbottom reflectors, hummocky pattern, some diffraction hyperbolae, moderate to high penetration (20 - 40 m)		Fine-grained sediments, deposits disturbed by synsedimentary tectonics or by subbottom topography, moderate to high sedimentation rate

Channel facies of the marginal basin and the lower continental slope

Pattern	Description of echo type	Example	Interpretation
	("Prolonged bottom") strong and prolonged bottom reflector, low penetration (<10 m), no subbottom reflectors, low diffraction hyperbolae (ampl. <40 m)		Coarse grained sediments in channels and along the thalweg of turbidity and density currents, low sedimentation rate
	("Diffraction hyperbolae, slope") steep eastward dipping slope with prolonged bottom echo and diffraction hyperbolae, low penetration (<10 m)	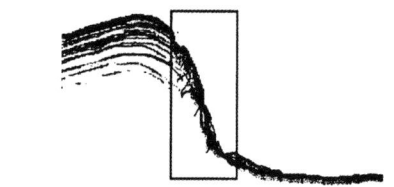	Steep erosional slopes on the western side of channels on the lower slope and in the basin

Shelf and slope facies

Pattern	Description of echo type	Example	Interpretation
	("Diffraction hyperbolae") large irregular partly overlapping hyperbolae with varying amplitudes, some with subbottom reflectors		Rough topography (gullies, channels and ridges) in mid-slope areas of the continental slope
	("Prolonged bottom, shelf slope") steep shelf slope with prolonged bottom echo and few diffraction hyperbolae, low penetration (<10 m)		Coarse upper-slope sediments, fine sediment particles winnowed and removed by currents
	("Prolonged bottom, shelf") strong prolonged bottom echo on the shelf, with many small diffraction hyperbolae, partly rough bottom topography, low penetration (<10 m)		Overconsolidated coarse shelf sediments, ersosion by shelf ice and icebergs

Fan facies

Pattern	Description of echo type	Example	Interpretation
	("Wedging subbottoms") semi-prolonged reflectors, pillow-type layers, transparent pattern, wedging subbottoms, low penetration (0 - 20 m)		Deposits of debris flows and slumps, high sediment input, coarse- to fine-grained unsorted sediments, Crary Fan

| Pattern | Description of echo type | Example | Interpretation |

Fig. 5. (**d**) Description of echo types used in Figure 5c, examples for the echo types from 3.5 kHz lines, and an interpretation regarding sedimentary processes (modified from Kuhn & Weber 1993 and Kuhn *et al.* 1995).

from glacial to interglacial conditions as a result of intensified ice rafting (Fütterer et al. 1988; Grobe & Mackensen 1992), whereas Kuvaas & Kristoffersen (1991) and Weber et al. (1994) propose highest sedimentation rates for the glacial periods when grounded ice extended to the shelf edge. The main topographical units along the southeastern Weddell Sea continental slope and adjacent areas have been sampled by cores. Detailed studies of the sedimentary environment and of sediment provenance have been carried out for the continental shelf (Anderson et al. 1980; Elverhøi & Roaldset 1983), slope, and basin along Dronning Maud Land (Grobe & Mackensen 1992), for Coats Land (Diekmann & Kuhn 1997), the eastern Crary Fan (Weber 1992; Weber et al. 1994), and the middle and western Crary Fan (Melles 1991; Melles & Kuhn 1993; Melles et al. 1995).

In compilation, four major lithofacies types can be distinguished: contourite-turbidite facies, hemipelagic facies, debris-flow facies, and shelf facies (Grobe & Mackensen 1992; Weber et al. 1994; Diekmann & Kuhn 1997). For a more detailed representation of the lithofacies types in different cores from the southeastern Weddell Sea, we split the contourite–turbidite facies into four subtypes: a fine-laminated facies, a coarse-laminated/stratified facies, and two transitional facies types to document the transition of the fine-laminated facies to the hemipelagic facies and the transition of the coarse-laminated/stratified facies to the hemipelagic facies.

Contourite–turbidite facies

The contourite–turbidite facies shows distinct layers of parallel-bedded, in the case of coarser layers sometimes cross-bedded and/or graded, clayey, muddy, and sandy sediment. Lamination or stratification is in the sub-millimetre to centimetre range and bioturbation generally is absent. In fine-bedded laminae the clay content can be as high as 65%. In contrast, cross-bedded layers almost entirely consist of coarse silt and sand. The facies was mainly deposited during glacials on the middle and lower slope and in the marginal basin. For the lithological column of the core figures (Fig. 6a–g) the contourite–turbidite facies has been split into a fine-laminated facies and a coarse-laminated/stratified facies, to allow a more detailed lithological classification. The fine-laminated facies is well developed in cores PS1789, PS1599 and PS1790 on the upper Crary Fan. An example from core PS1790 is shown in Figure 6f. The coarser-laminated or stratified facies is also present in many parts of these cores. An example from a core taken in greater water depth (PS1635) can be seen in Figure 6d.

The contourite-turbidite facies originates in deposition from combined contour- and turbidity-current activity. The comparison of the coarse-laminated/stratified facies from a levee (e.g. in core PS1635, Fig. 6d) with the coarse-laminated/stratified facies in a channel (core PS1794, Fig. 6g) shows that they are very different, with the channel facies having a much higher sand content. This suggests that the levee sediments consist of the fines of turbidity current suspensions.

The configuration of pronounced sediment levees on the northwestern side of the channels indicate that overspill sedimentation from turbidity or density currents, triggered by the Coriolis force, plays an important role in the supply of sediment. In addition, the west- or southwestward directed vigorous contour currents entrain suspension from the turbidity currents and redeposit it downstream. A similar situation prevails on the western side of the Antarctic Peninsula (Rebesco et al. 1996, 1997), although there turbidity currents flow approximately perpendicular to contour currents, rather than in the opposite direction as here. The transitional facies from the fine laminated and the coarse-laminated/stratified facies to the hemipelagic facies are probably of combined contourite–turbidite origin, but environmental conditions allowed for benthic activity, so that bioturbation occurred and the lamination is blurred.

Hemipelagic facies

The hemipelagic facies consists of bioturbated mud with a dominance of the fine fraction (clay contents c. 30 to 50%), occasionally with increased proportions of ice-rafted debris (IRD). Part of the terrigenous sand fraction may contain biogenic opal. Generally strong to very strong bioturbation has blurred the primary sedimentary structures, although a faint stratification may be observed. The facies occurs during interglacials or during moderate interglacial to glacial conditions on the continental slope and in the marginal basin and often forms the surface sediments. An example of this facies in an X-radiograph section from core PS1789 can be seen in Figure 6e.

Debris-flow facies

The debris-flow or slump facies consists of a structureless mixture of coarse- and fine-grained sediments, which show no bioturbation. The facies mainly occurs in glacial intervals in the area of the central Crary Fan. Core 1607 from the western Crary Fan shows this facies under a 2 m thick surface layer of hemipelagic facies (Fig. 6c).

Shelf facies

Holocene shelf sediments comprise a high content of coarse-grained IRD, and a depletion of fine fraction, which is winnowed by the vigorous Antarctic Coastal Current. These residual sediments are underlain by overconsolidated diamictons, representing subglacial deposits from periods when the ice margin was located near the shelf edge. A good example of Holocene shelf facies is encountered in core PS1367 (Fig. 6a) from a water depth of 303 m on the shelf off Dronning Maud Land. This facies provides the source of the huge sediment masses that are removed from the shelf by ice sheets during glacial advances. They are released at the shelf edge where they initiate extensive turbidity-current activity in the channels of the continental slope.

A general predominance of the hemipelagic facies can be observed in many locations on the continental slope during interglacial stages, whereas the laminated facies prevails during glacials. However, conditions are variable along the continental slope in the southeastern Weddell Sea, and knowledge is insufficient so far about processes that are important especially for the formation of the glacial facies types, e.g. possible open water conditions during glacials.

Sediment grain size: summary analytical results

Granulometric differences offer the best criteria for the distinction of the lithofacies types. Silt- and clay-sized particles generally make up 80 to 90% of the contourite-turbidite facies, but significant differences exist for the silt-sized composition of the fine-laminated and the coarse-laminated/stratified type. Fine-laminated sediments show a broad range of silt grain sizes with a maximum in the range of 6 to 7 ϕ, whereas coarse-laminated/stratified sediments generally have a peaked silt-size composition with a sharp maximum around 5 ϕ, which sometimes reaches into the sand-sized range (e.g. cores PS1599 and PS1635, Figs 6b, d). The IRD content can occasionally be high, especially for the fine-laminated facies, but generally is low or absent.

Hemipelagic sediments show a uniformly distributed silt size fraction on the continental slope. In core PS1635 from the basin hemipelagic sediments show a maximum in the 7 to 8 ϕ range, accompanied by an increased IRD content. This points to a depositional mechanism which is not dominated by currents.

The shelf facies shows high contents of gravel and sand,

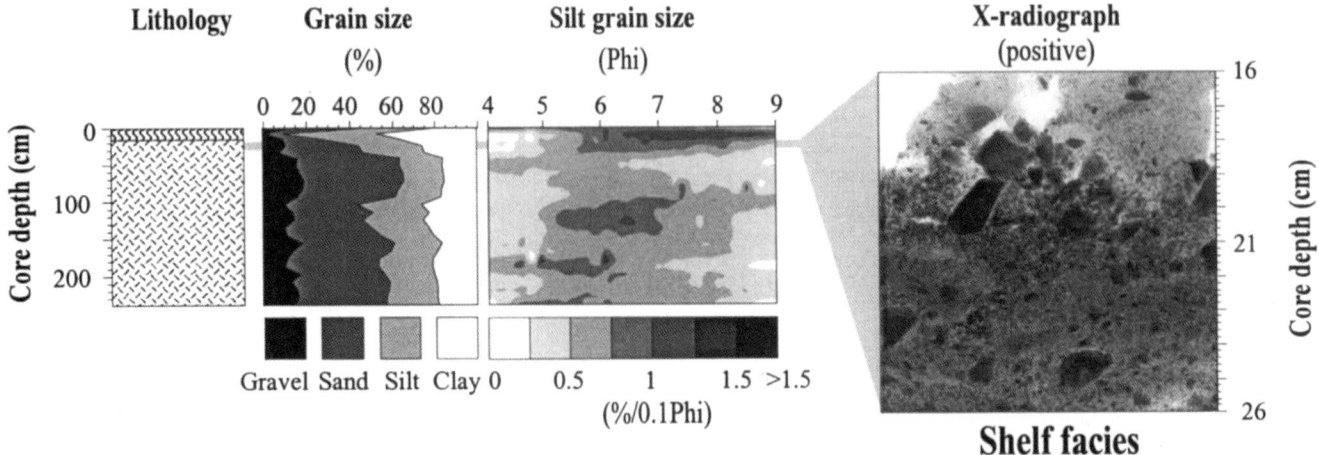

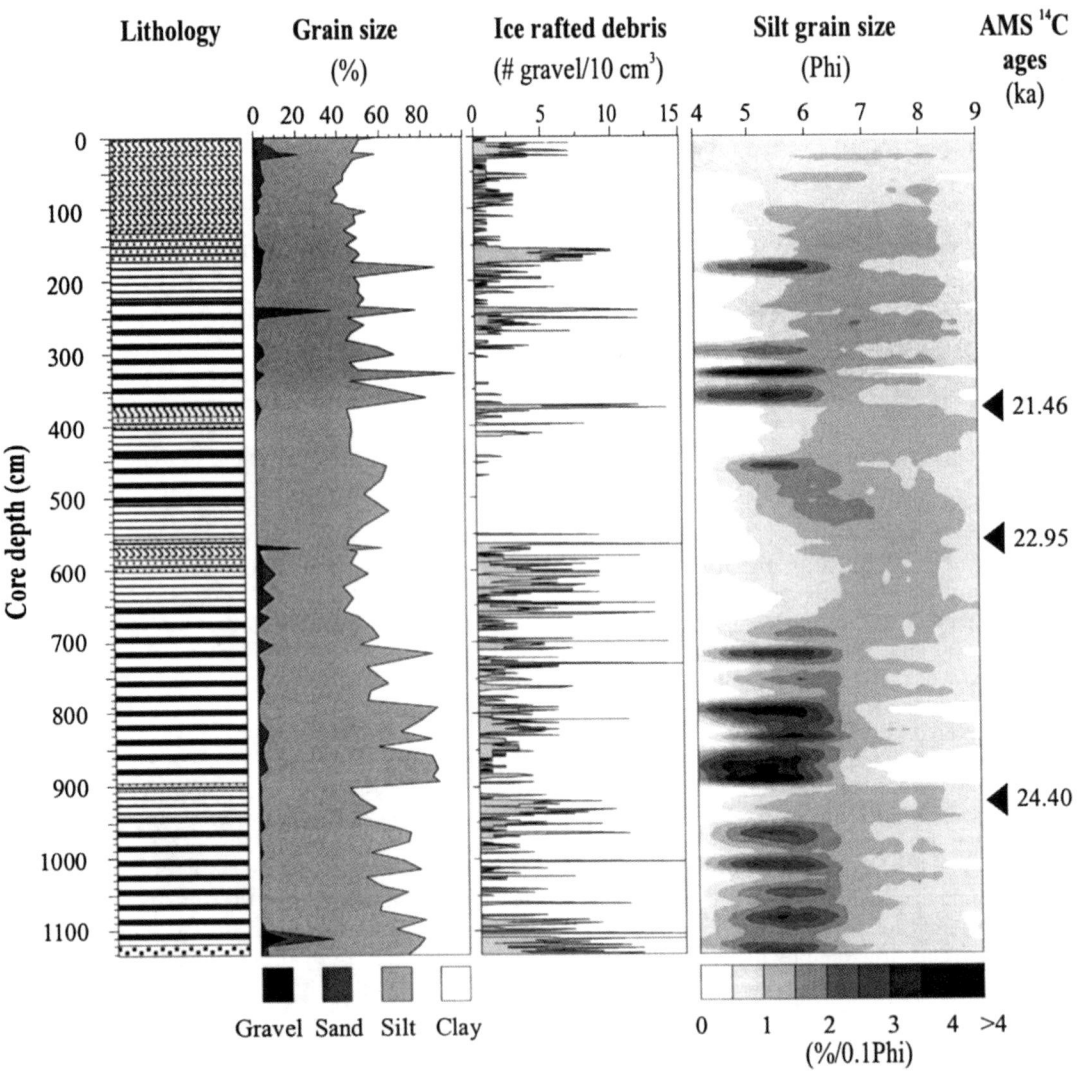

Fig. 6. Graphic logs of sediment cores PS1367 (**a**), PS1599 (**b**), PS1607 (**c**), PS1635 (**d**), PS1789 (**e**), PS1790 (**f**), and PS1794 (**g**). From left to right, the columns show: (i) lithology, classified into 7 different facies types (four major types; the contourite-turbidite facies is split up in four subtypes, (ii) the cumulative grain-size distribution of gravel, sand, silt, and clay versus sediment depth, (iii) the number of gravel-sized IRD particles per 10 cm³ versus core depth (no data available for PS1367), (iv) a contour diagram of the grain size distribution within the silt fraction versus sediment depth (no data available for PS1607 and PS1794). Ages of AMS ^{14}C dated sediment samples are shown for PS1599, PS1789, and PS1790. Age in PS1607 is based on δ^{18}O curve. Age of 780 ka in PS1635 corresponds to a magnetic reversal. X-radiographs of core sections typical of the different facies types are also shown. Note the different depth scale for PS1794 (data from Melles 1991; Weber 1992; Diekmann & Kuhn 1997; Grobe unpublished data).

Sedimentological data and age models for all cores can be downloaded from the information system 'PANGAEA' under www.pangaea.de/PangaVista

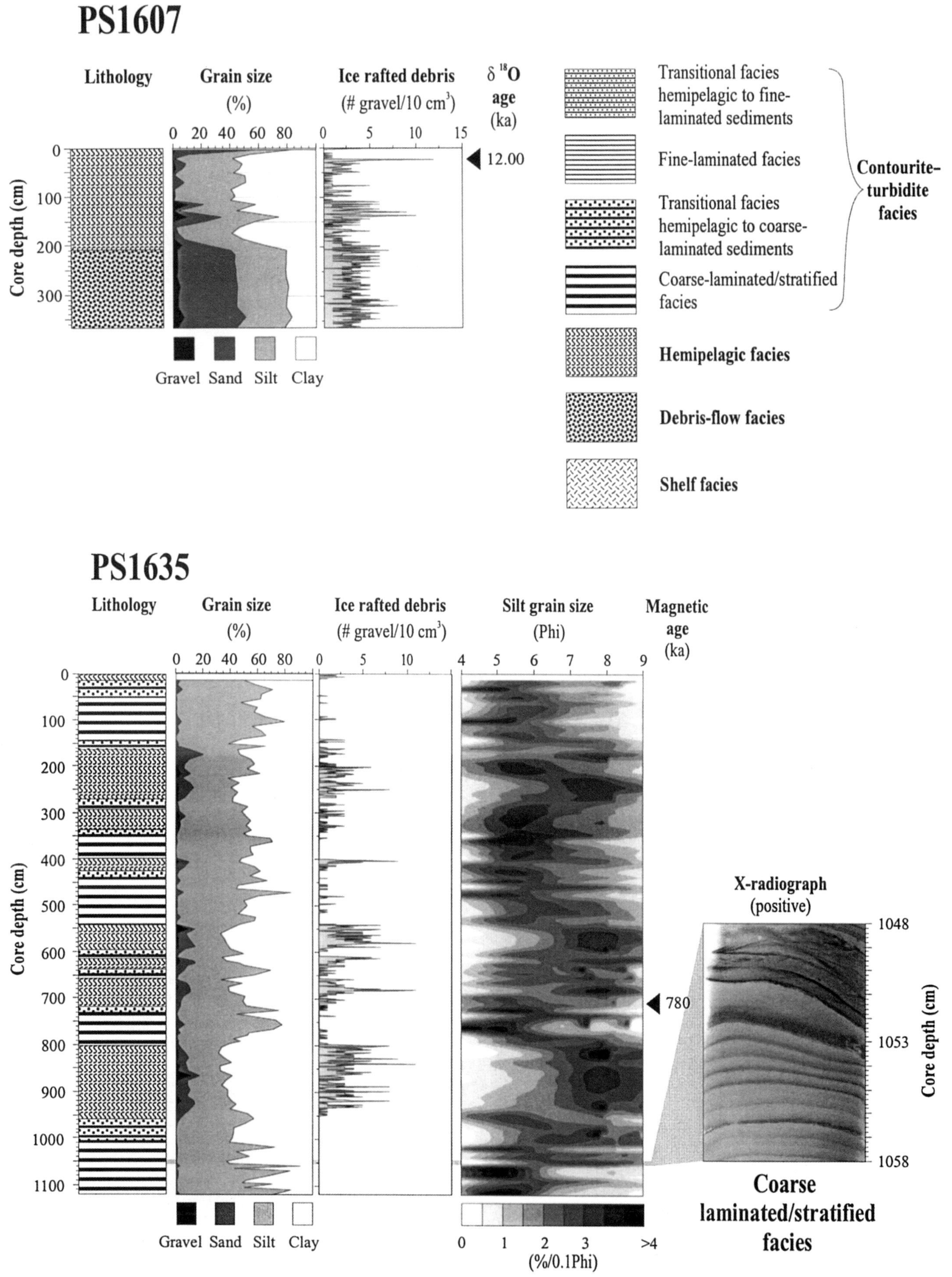

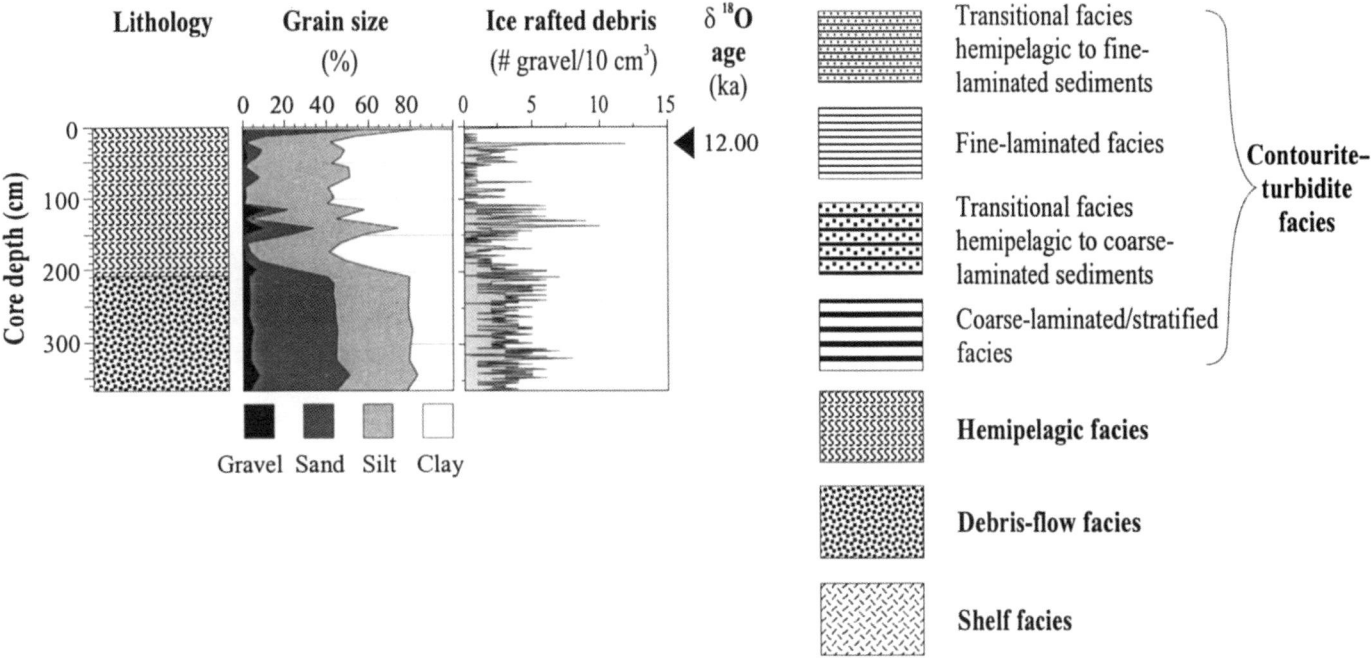

Fig. 6. PS1607 (**c**), PS1635 (**d**).

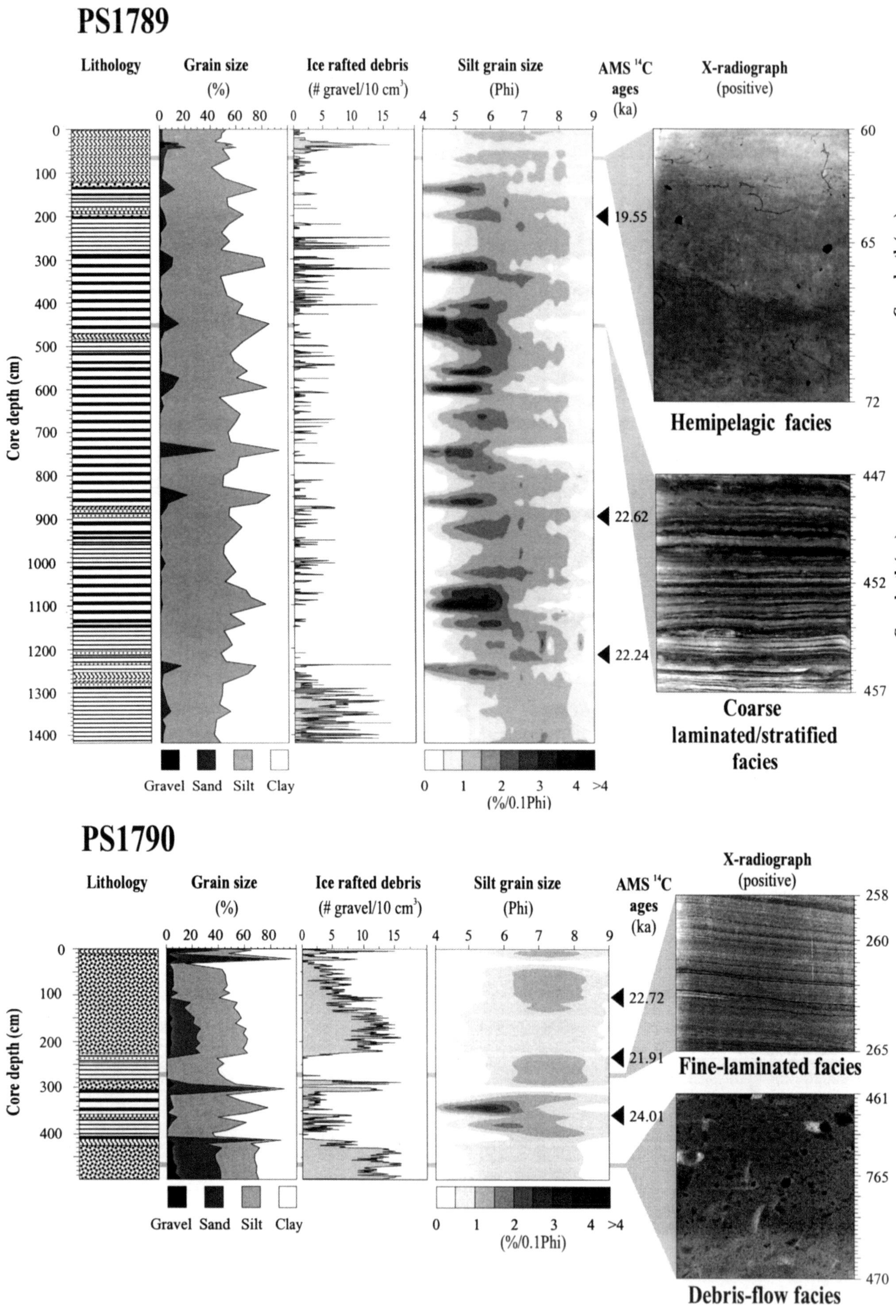

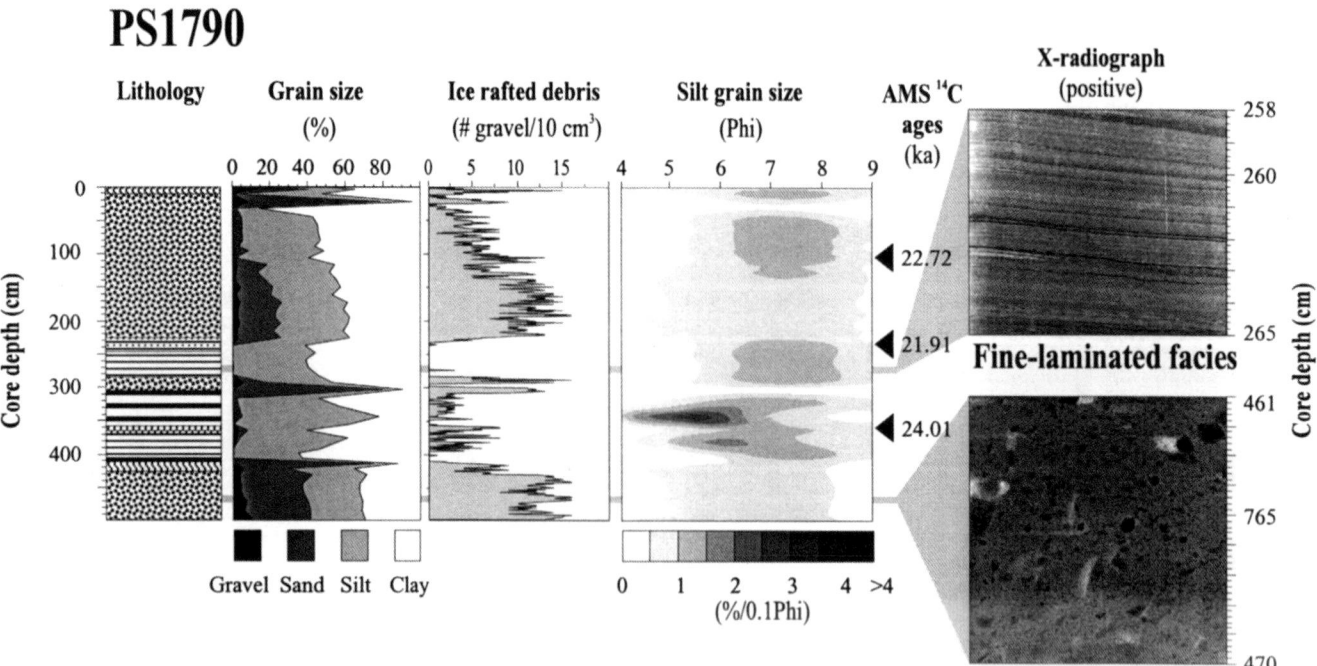

Fig. 6. PS1789 (**e**), PS1790 (**f**).

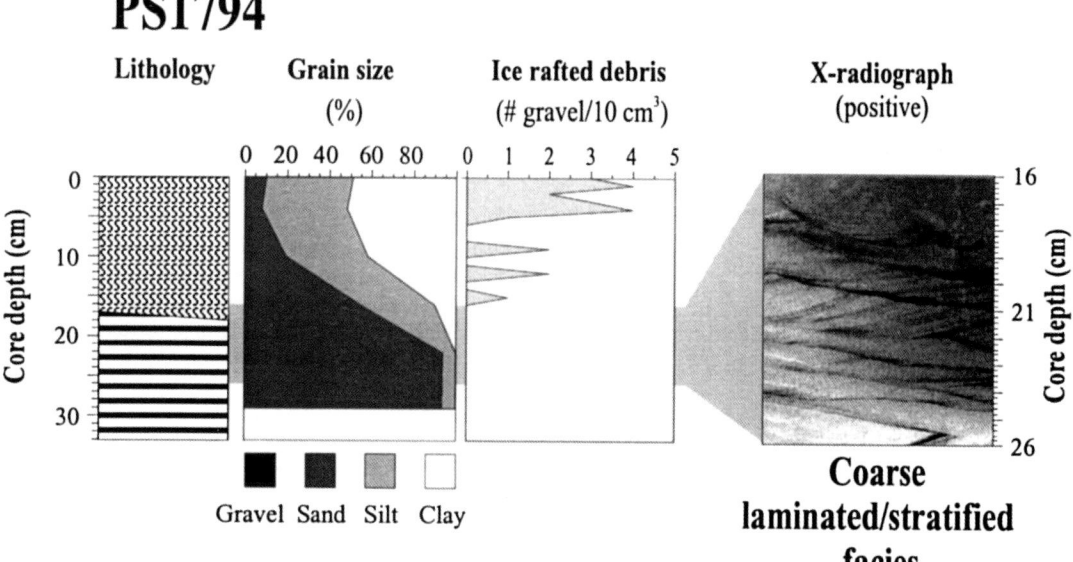

Fig. 6. PS1794 (g).

whereas the distribution of grain sizes in the silt fraction does not show a significant pattern (see core PS1367, Fig. 6a). The granulometric composition of the debris-flow facies is very similar to that of the shelf facies, because debris flows in the Crary Fan area mainly consist of redeposited shelf sediments (see core PS1607 and PS1790, Figs. 6c & f).

Discussion

Are there contourites?

The depositional environment of the laminated and/or clearly-stratified sediments along the continental margin of the southeastern Weddell Sea undoubtedly reflects strong current activity. However, it is difficult to decipher whether the currents were of contouritic or turbiditic origin on the basis of grain size and sedimentary structures, because unequivocal criteria for both sediment types are missing.

Even in the seismic profiles it is very difficult to assign the features of the channel-levee systems to either turbiditic or contouritic influence, though Faugères et al. (1999) listed many features diagnostic of contourite drifts and turbidites. The difficulties arise from the fact that the geostrophic and turbidite currents are directed in almost opposite directions (the turbidite channels run to the northeast, thermohaline currents of the Weddell Gyre are directed to the southwest), so that criteria that can be used for distinction of either type (e.g. downstream coarsening or fining of levee sediments or the progradation of the levee from a downslope direction to a course parallel to the slope) add up or overlay each other. However, without doubt the sediments in the southeastern Weddell Sea form combined or intercalated turbidite and contourite sequences.

Holocene sediments are basically characterized by hemipelagic facies, though the modern continental-slope environment in the southeastern Weddell Sea is known to be moderately to highly energetic as a result of the cyclonic currents of the Weddell Gyre. A possible reason for the lack of lamination and weak stratification of hemipelagic sediments may be bioturbation, which did not occur or was less intense during glacial conditions because of perennial sea-ice cover and significantly higher sedimentation rates. This would not rule out a contouritic and/or turbiditic origin, or at least a contouritic and/or turbiditic influence on the deposition of Holocene hemipelagic sediments.

Hydrodynamic interpretation

The formation of the combined turbidite-contourite levees, as inferred from site survey and sediment data, involves the following processes: Sediment suspensions are generated on the shelf or the upper continental slope by turbidity currents (or the formation of ISW) and move gravitationally down the continental slope, where they are rapidly channelised in a system of channels and gullies. Both turbidity currents and ISW plumes have the ability to entrain sediment and erode along the channel thalweg, depending on the density contrast with surrounding water masses. ISW can reach current velocities exceeding 100 cm s^{-1} (Foldvik & Gammelsrød 1988). The suspensions in the channels are subject to the vigorous current of the Weddell Gyre and to Coriolis forcing, which together cause a part of the suspension to spill over the northwestern side of the channels. By subsequently losing its sediment load with distance from the channel, the overspill sedimentation builds up a levee along the left flank of the channel. The combined action of contour current and Coriolis force almost completely impedes the formation of a levee on the right, eastern side of the channels. In a succession of several channel-levee systems, the overspill sedimentation of one channel fills up the space before the weakly developed levee along the right side of the next channel, and by this suppresses the emergence of an eastern levee. The grain size of levee deposits and the thickness of the laminae or depositional units, as well as the distance a suspension plume travels down the channel, depend on the magnitude of the suspension event and the availability and supply of sediment at the upper slope, which in turn controls the sedimentation rates. The fact that the broad wedge-shaped form of the levees started to develop with the late Miocene invigoration of thermohaline currents in the Antarctic argues for a strong current influence on levee formation.

The sedimentation rates were significantly different during late Quaternary glacial and interglacial periods. The last glacial showed rates up to 30 times higher than Holocene sedimentation rates (Weber et al. 1994), because the sea level was significantly lower and the grounding line of the Antarctic ice sheet in the southernmost Weddell Sea was located at the shelf edge so that glacial debris was directly discharged to the slope (Grobe & Mackensen 1992; Bentley & Anderson 1998; Anderson & Andrews 1999). With the late-glacial retreat of the ice sheet and the Holocene sea-level rise, additional depositional space became available on the shelf where sediment was stored before reaching the slope. During glacial periods, two processes caused a significant increase of sedimentation rates: (1) the reworking of interglacial deposits on the shelf during advances of the ice sheets across the shelf; (2) the release of IRD by calving glaciers when the ice sheet reduced its size during climate warming.

The enormous release of sediment probably triggered mass wasting, slumps, slides, debris flows and turbidity currents. Gravitational processes were more important than the processes involved with the formation of ISW. Further support for this assumption is given by the investigation of water mass properties in the southeastern Weddell Sea, which suggest that northwestward-directed currents measured in certain moorings indicate a recirculation of water masses originating in the Enderby Basin (Hoppema et al. 1998, Fahrbach et al. 1998), rather than ISW flowing downslope in channels.

C. J. Pudsey and J. A. Howe are gratefully acknowledged for their helpful and constructive reviews.

References

ANDERSON, J. B. & ANDREWS, J. T. 1999. Radiocarbon constraints on ice sheet advance and retreat in the Weddell Sea, Antarctica. *Geology*, **27**, 179–182.

ANDERSON, J. B., BRAKE, C., DOMACK, E., MYERS, N. & WRIGHT, R. 1983. Development of a polar glacial-marine sedimentation model from Antarctic Quaternary deposits and glaciological information. *In*: MOLNIA, B. F. (ed.) *Glacial-Marine Sedimentation*. Plenum Press, New York, 233–264.

ANDERSON, J. B., KURTZ, D. D., DOMACK, E. W. & BALSHAW, K. M. 1980. Glacial and glacial-marine sediments of the Antarctic continental shelf. *Journal of Geology*, **88**, 399–414.

ANDERSON, J. B., KURTZ, D. D. & WEAVER, F. M. 1979. Sedimentation on the Antarctic continental slope. *In*: *Geology of Continental Slopes*. Special Publication of the Society of Economic Paleontologists and Mineralogists, **27**, 265–283.

BART, P. J., DE BATIST, M. & JOKAT, W. 1999. Interglacial collapse of Crary Trough-Mouth Fan, Weddell Sea, Antarctica: Implications for Antarctic glacial history. *Journal of Sedimentary Research*, **69**(6), 1276–1289.

BARKER, P. F., KENNETT, J. P. ET AL. 1988. *Proceedings of the ODP, Initial Reports, Leg 113*. Ocean Drilling Program, College Station, TX.

BARKER, P. F., KENNETT, J. P. ET AL. 1990. *Proceedings of the ODP, Scientific Results, Leg 113*. Ocean Drilling Program, College Station, TX.

BENTLEY, M. J. & ANDERSON, J. B. 1998. Glacial and marine geological evidence for the ice sheet configuration in the Weddell Sea-Antarctic Peninsula region during the Last Glacial Maximum. *Antarctic Science*, **10**, 309–325.

BRITISH ANTARCTIC SURVEY 1985. *Tectonic map of the Scotia Arc, 1:3,000,000*. BAS (Misc) 3, Cambridge.

CARMACK, E. C. & FOSTER, T. D. 1975a. On the flow of water out of the Weddell Sea. *Deep-Sea Research*, **22**, 711–724.

CARMACK, E. C. & FOSTER, T. D. 1975b. Circulation and distribution of oceanographic properties near the Filchner Ice Shelf. *Deep-Sea Research*, **22**, 77–90.

CARMACK, E. C. & FOSTER, T. D. 1977. Water masses and circulation in the Weddell Sea. *In*: DUNBAR, M. J. (ed.) *Proceedings of the Polar Oceans Conference*, Montreal, 151–165.

DEACON, G. E. R. 1979. The Weddell Gyre. *Deep-Sea Research*, **26A**, 981–995.

DIEKMANN, B. & KUHN, G. 1997. Terrigene Partikeltransporte als Abbild spätquartärer Tiefen- und Bodenwasserzirkulation im Südatlantik und angrenzendem Südpolarmeer. *Zeitschrift der deutschen Geologischen Gesellschaft*, **148**, 405–429.

DIEKMANN, B. & KUHN, G. 1999. Provenance and dispersal of glacial-marine surface sediments in the Weddell Sea and adjoining areas, Antarctica: ice-rafting versus current transport. *Marine Geology*, **158**, 209–231.

DIEKMANN, B., KUHN, G., MACKENSEN, A., PETSCHICK, R., FÜTTERER, D. K., GERSONDE, R., RÜHLEMANN, C. & NIEBLER, H.-S. 1999. Kaolinite and chlorite as tracers of modern and Late Quaternary deep-water circulation in the South Atlantic and the adjoining southern ocean. *In*: *Use of proxies in palaeoceanography: Examples from the South Atlantic*. Springer, Berlin, 285–313.

EHRMANN, W. U., MELLES, M., KUHN, G. & GROBE, H. 1992. Significance of clay mineral assemblages in the Antarctic Ocean. *Marine Geology*, **107**, 249–273.

ELVERHØI, A. 1981. Evidence for a late Wisconsin glaciation of the Weddell Sea. *Nature*, **293**, 641–642.

ELVERHØI, A. & MAISEY, G. 1983. Glacial erosion and morphology of the eastern and southeastern Weddell Sea shelf. *In*: OLIVIER, R. L., JAMES, P. R. & JAGO, J. B. (eds) *Antarctic Earth Science, Proceedings of the Fourth International Symposium Antarctic Earth Science, Adelaide, Australia 1982*. Australian Academy of Science, Canberra 1983, 483–487.

ELVERHØI, A. & ROALDSET, E. 1983. Glaciomarine sediments and suspended particulate matter, Weddell Sea shelf, Antarctica. *Polar Research*, **1**, 1–21.

FAHRBACH, E. & ROHARDT, G. 1988. Moored instrument data. *In*: FAHRBACH, E. (ed.) *Meteorological and Oceanographic Data of the Winter Winter-Weddell-Sea Project 1986 (ANT V/3)*. Berichte zur Polarforschung, **46**.

FAHRBACH, E., ROHARDT, G. & KRAUSE, G. 1992. The Antarctic Coastal Current in the southeastern Weddell Sea. *Polar Biology*, **12**, 171–182.

FAHRBACH, E., ROHARDT, G., SCHRÖDER, M. & STRASS, V. 1994. Transport and structure of the Weddell Gyre. *Annales Geophysicae*, **12**, 840–855.

FAHRBACH, E., SCHRÖDER, M. L. & KLEPIKOV, A. 1998. Circulation and water masses in the Weddell Sea. *In*: LEPPÄRANTA, M. (ed.) *Physics of Ice-Covered Seas*. Lecture notes from a summer school in Savonlinna, Finland, Helsinki University, 569–603.

FAUGÈRES, J.-C., STOW, D. A. V., IMBERT, P. & VIANA, A. 1999. Seismic features diagnostic of contourite drifts. *Marine Geology*, **162**, 1–38.

FOLDVIK, A., GAMMELSRØD, T. & TØRRESEN, T. 1985. Circulation and water masses on the southern Weddell Sea Shelf. *In*: *Oceanology of the Antarctic Continental Shelf*. Antarctic Research Series, **43**, 5–20.

FOLDVIK, A. & GAMMELSRØD, T. 1988. Notes on southern ocean hydrography, sea ice and bottom water formation. *Palaeoceanography, Palaeoclimatology, Palaeoecology*, **67**, 3–17.

FOSTER, T. D. & MIDDLETON, J. H. 1979. Variability in the bottom water of the Weddell Sea. *Deep-Sea Research*, **26**, 743–762.

FÜTTERER, D. K. & MELLES, M. 1990. Sediment patterns in the southern Weddell Sea: Filchner Shelf and Filchner Depression. *In*: BLEIL, U. & THIEDE, J. (eds) *Geological History of the Polar Oceans: Arctic vs. Antarctic*. NATO/ASI Series C, Kluwer Academic Press, Dordrecht, Netherlands, 381–401.

FÜTTERER, D. K., GROBE, H. & GRÜNIG, 1988. Quaternary sediment patterns in the Weddell Sea: relations and environmental conditions. *Palaeoceanography*, **3**(5), 551–561.

FÜTTERER, D. K., KUHN, G. & SCHENKE, H. W. 1990. Wegener Canyon bathymetry and results from rock dredging near ODP sites 691–693, eastern Weddell Sea, Antarctica. *In*: BARKER, P. F. & KENNETT, J. P. ET AL. (eds) *Proceedings of the Ocean Drilling Program, Scientific Results*. Ocean Drilling Program, College Station, TX, **113**, 39–48.

GERSONDE, R., ABELMANN, A., ET AL. 1990. Biostratigraphic synthesis of Neogene siliceous microfossils from the Antarctic Ocean, ODP Leg 113 (Weddell Sea). *In*: BARKER, P. F. & KENNETT, J. P. ET AL. (eds) *Proceedings of the Ocean Drilling Program, Scientific Results*. Ocean Drilling Program, College Station, TX, **113**, 915–936.

GLOERSEN, P., CAMPBELL, W. J., CAVALIERI, D. J., COMISO, J. C., PARKINSON, C. L. & ZWALLY, H. J. 1992. Arctic and Antarctic sea ice, 1978–1987: Satellite passive microwave observations and analysis. NASA **SP-511**.

GORDON, A. L. 1982. Weddell Deep Water variability. *Journal of Marine Research*, Supplement, **40**, 199–217.

GORDON, A. L., MARTINSON, D. G. & TAYLOR, H. W. 1981. The wind-driven circulation in the Weddell-Enderby Basin. *Deep-Sea Research*, **28A**, 151–163.

GROBE, H., MACKENSEN, A., HUBBERTEN, H.-W., SPIESS, V. & FÜTTERER, D. K. 1990a. Stable isotope record and late Quaternary sedimentation rates at the Antarctic continental margin. *In*: BLEIL, U. & THIEDE, J. (eds) *Geological History of the Polar Oceans: Arctic vs. Antarctic*. NATO/ASI Series C, Kluwer Academic Press, Dordrecht, Netherlands, 539–560.

GROBE, H., FÜTTERER, D. K. & SPIEß, V. 1990b. Oligocene to Quaternary sedimentation processes on the Antarctic continental margin, ODP Leg 113, Site 693. *In*: BARKER, P. F. & KENNETT, J. P. *ET AL.* (eds) *Proceedings of the Ocean Drilling Program, Scientific Results*. Ocean Drilling Program, College Station, TX, **113**, 121–131.

GROBE, H. & MACKENSEN, A. 1992. Late Quaternary climatic cycles as recorded in sediments from the Antarctic continental margin. *In*: KENNETT, J. P. & WARNKE, D. A. (eds) *The Antarctic Palaeoenvironment: A Perspective on Global Change*. Antarctic Research Series, **56**, 349–376.

HENRIET, J. P. & MILLER, H. 1990. Some speculation regarding the nature of the Explora-Andenes Escarpment. *In*: BLEIL, U. & THIEDE, J. (eds) *Geological History of the Polar Oceans: Arctic vs. Antarctic*. NATO/ASI Series C, Kluwer Academic Press, Dordrecht, Netherlands, 163–172.

HINZ, K. & KRAUSE, W. 1982. The continental margin of Queen Maud Land/Antarctica: Seismic sequences, structural elements, and geological development. *Geologisches Jahrbuch*, **E23**, 17–41.

HOLLISTER, C. D. & ELDER, R. B. 1969. Contour currents in the Weddell Sea. *Deep-Sea Research*, **16**, 99–101.

HOPPEMA, M., FAHRBACH, E., RICHTER, K.-U., DE BAAR, H. J. W. & KATTNER, G. 1998. Enrichment of silicate and CO_2 and circulation of the bottom water in the Weddell Sea. *Deep-Sea Research*, **I 45**, 1797–1817.

KUHN, G. & WEBER, M. 1993. Acoustical characterization of sediments by Parasound and 3.5-kHz systems: Related sedimentary processes on the southeastern Weddell Sea continental slope, Antarctica. *Marine Geology*, **113**, 201–217.

KUHN, G., SCHENKE, H.-W. & FÜTTERER, D. 1995. The Weddell Fan: A complex channel/ridge-system in the southern Weddell Sea, Antarctica. *In: The role of palaeoceanographic linkages in the global system*. 5th International Conference on Palaeoceanography, Halifax, Nova Scotia, Canada, 163–164.

KUVAAS, B. & KRISTOFFERSEN, Y. 1991. The Crary Fan: a trough-mouth fan on the Weddell Sea continental margin, Antarctica. *Polar Research*, **7**, 43–57.

MACKENSEN, A., GROBE, H., HUBBERTEN, H. & KUHN. G. 1994. Benthic foraminiferal assemblages and the δ13C-signal in the Atlantik sector of the Southern Ocean: Glacial-to-interglacial contrasts. *In*: ZAHN, R. (ed.) *Carbon Cycling in the Glacial Ocean: Constraints on the Ocean's Role in Global Change*. NATO ASI Series, I 17, Springer, Berlin, 105–144.

MELLES, M. 1991. Late Quaternary palaeoglaciology and palaeoceanography at the continental margin of the southern Weddell Sea, Antarctica. *Berichte zur Polarforschung*, **81**.

MELLES, M. & KUHN, G. 1993. Sub-bottom profiling and sedimentological studies in the southern Weddell Sea, Antarctica: evidence for large-scale erosional/depositional processes. *Deep-Sea Research*, **40**(4), 739–760.

MELLES, M., KUHN, G., FÜTTERER, D. K. & MEISCHNER, D. 1995. Processes of modern sedimentation in the southern Weddell Sea, Antarctica – Evidence from surface sediments. *Polarforschung*, **64**, 45–74.

MILLER, H., HENRIET, J. P., KAUL, N. & MOONS, A. 1990. A fine-scale seismc stratigraphy of the eastern margin of the Weddell Sea. *In*: BLEIL, U. & THIEDE, J. (eds) *Geological History of the Polar Oceans; Arctic vs. Antarctic*. NATO/ASI Series C, Kluwer Academic Press, Dordrecht, Netherlands, 131–161.

MOONS, A., DE BATIST, M., HENRIET, J. P. & MILLER, H. 1992. Sequence stratigraphy of the Crary Fan, southeastern Weddell Sea. *In*: YOSHIDA, Y., KAMINUMA, K. & SHIRAISHI, K. (eds) *Recent Progress in Antarctic Earth Science*. Tokyo, 613–618.

OSZKÓ, L. 1997. Tectonic structures and glaciomarine sedimentation in the south-eastern Weddell Sea from seismic reflection data. *Berichte zur Polarforschung*, **222**.

PETSCHICK, R., KUHN, G. & GINGELE, F. 1996. Clay mineral distribution in surface sediments of the South Atlantic: sources, transport, and relation to oceanography. *Marine Geology*, **130**, 203–229.

REBESCO, M., LARTER, R. D., CAMERLENGHI, A. & BARKER, P. F. 1996. Giant sediment dirfts on the continental rise west of the Antarctic Peninsula. *Geo-Marine Letters*, **16**, 65–75.

REBESCO, M., LARTER, R. D., BARKER, P. F., CAMERLENGHI, A. & VANNESTE, L. E. 1997. The history of sedimentation on the continental rise west of the Antarctic Peninsula. *In: Geology and Seismic Statigraphy of the Antarctic Margin, Part 2*. Antarctic Research Series, **71**, 29–49.

ROHARDT, G., FAHRBACH, E., KRAUSE, G. & STRASS, V. H. 1992. Moored current meter and water level recorder measurements in the Weddell Sea 1986–1990. *Berichte Fachbereich Physik*, **28**, Alfred Wegener Institute, Bremerhaven.

SCHENKE, H. W., DIJKSTRA, S., NIEDERJASPER, F., SCHÖNE, T., HINZE, H. & HOPPMANN, B. 1998. *The New Bathymetric Charts of the Weddell Sea: AWI BCWS*. American Geophysical Union, Antarctic Research Series **75**, 371–380 + map.

SCHLITZER, R. 1999. *Ocean Data View*. World Wide Web Address: http://www.awi-bremerhaven.de/GEO/ODV.

SEA ICE CLIMATIC ATLAS 1985. Vol. 1, Antarctic. Prepared by Naval Oceanography Comand Detachment, Asheville, NSTL, MS 39527-5000.

TINGEY, R. J. 1991. *Schematic geological map of Antarctica, Scale 1:10,000,000 (commentary and map)*. Department of Primary Industries and Energy, Bureau of Mineral Resources, Geology and Geophysics, Bulletin 238. Australian Government Publishing Service, Canberra.

WEBER, M. 1992. Late Quaternary sedimentation at the continental margin of the southeastern Weddell Sea, Antarctica. *Berichte zur Polarforschung*, **109**.

WEBER, M. E., BONANI, G. & FÜTTERER, K. D. 1994. Sedimentation processes within channel-ridge systems, southeastern Weddell Sea, Antarctica. *Palaeoceanography*, **9**, 1027–1048.

ZWALLY, H. J., COMISO, J. C., PARKINSON, C. L., CAMPBELL, W. J., CARSEY, F. D. & GLOERSEN, P. 1985. *Antarctic Sea Ice, 1973–1976: Satellite Passive-Microwave Observations*. NASA Special Publication, **459**.

Mixed biosiliceous–terrigenous sedimentation under the Antarctic Circumpolar Current, Scotia Sea

CAROL J. PUDSEY & JOHN A. HOWE

British Antarctic Survey, Madingley Road, Cambridge, UK (e-mail: cjp@bas.ac.uk)

Abstract: Sediment supply to the Scotia Sea is controlled by the east-flowing Antarctic Circumpolar Current (ACC) with some Weddell Gyre influence in the south. Near-bottom flow is unsteady with frequent changes in flow direction and episodic benthic storms. Near the North Scotia Ridge, mounds of sediment up to 1 km thick have accumulated on lower Miocene ocean floor. The basins farther south contain up to 2 km of sediment which is flat-lying or draped rather than mounded. Sediment cores exhibit a biogenic–terrigenous cyclicity related to glacial-interglacial cycles. Grain-size data suggest that ACC flow was stronger during glacials than interglacials.

Geological and oceanographic setting

The Scotia Sea is a small ocean basin in the SW Atlantic Ocean. It extends from Drake Passage, at 65°W between South America and the Antarctic Peninsula, eastwards to the South Sandwich island arc at 27°W (Fig. 1). It consists of a series of back-arc basins formed during the last 30 Ma on the boundary between the South American Plate and the Antarctic Plate (Fig. 1 inset; Barker *et al.* 1991). Its northern and southern boundaries are the North Scotia Ridge at 54°S and the South Scotia Ridge at 60°S. Both these ridges are discontinuous, reaching sea level only at South Georgia and the South Orkney Islands (Fig. 2); deep-water gaps in the ridges attain depths of 3000–3500 m. Detailed bathymetry of the area is given in Tectonic Map (1985).

Much of the Scotia Sea floor lies at depths of 3000–4500 m and is isolated from major continental-margin sources of sediment (Fig. 1). Only the southern tip of South America and the west coast of the Antarctic Peninsula have near-continuous downslope pathways to the Scotia Sea. Sediment supply, both terrigenous and biogenic, is controlled by oceanic circulation: mainly the east-flowing Antarctic Circumpolar Current (ACC) with some Weddell Gyre influence in the south. Bottom potential temperature ranges from +0.2°C near the North Scotia Ridge to –0.6°C near the South Orkney Islands; the water mass at the seabed is therefore Circumpolar Deep Water in the north and Weddell Sea Deep Water in the south (Locarnini *et al.* 1993).

Eastward flow within the ACC is concentrated at the Subantarctic Front and Polar Front (Fig. 1; Nowlin & Clifford 1982). There are few direct measurements of deep water flow except in Drake Passage, where Bryden & Pillsbury (1977) and Whitworth *et al.* (1982) reported unsteady flow with speeds up to 10 cm s^{-1} at a reference level of 2700 m, in water depths of 3500–3900 m. Flow speed decreases with depth, and near the seabed it can be strongly modified by local topography. At BAS moorings 8, 9, 11, 6 and 5 vector-averaged speeds over one year are 3–5 cm s^{-1} and mean speeds 7–12 cm s^{-1}, at 10 m above the seabed (Figs 1 & 2). These moorings are all in areas of Quaternary sediment deposition.

The seasonal occurrence of sea ice is a significant control on biogenic sedimentation in the Scotia Sea. The minimum (early March) and maximum (August–September) ice edges shown on Figure 1 represent the 10 year average from 1973 to 1982 (Sea Ice Climatic Atlas 1985).

Seismic characteristics

Reflection profiles

Drake Passage formed a deep-water pathway by 22 Ma (Barker & Burrell 1977). The western and central Scotia Sea formed by seafloor spreading from about 29 Ma to 6 Ma, but a wide area of seafloor from Drake Passage to 50°W has remained only thinly sedimented during the Neogene, because vigorous ACC flow has prevented continuous deposition. Seismic reflection profiles show dramatic variations in sediment thickness over distances of tens of km, with deposition clearly controlled by the interaction of bottom currents with basement topography (Fig. 3A; Barker & Burrell 1977). Near the North Scotia Ridge, mounds of sediment up to 1 km thick have accumulated on lower Miocene ocean floor (Fig. 3B; Howe & Pudsey 1999), but between the mounds acoustic basement reaches the seabed. Similar but somewhat broader sediment mounds occur in the central Scotia Sea. The basins in the south contain up to 2 km of sediment which is flat-lying or draped rather than mounded, and contains coherent internal reflectors (Fig. 3C).

3.5 kHz profiles

The echo character map of the central Scotia Sea (Fig. 4) shows depositional and non-depositional areas, and the former were subdivided using depth of penetration and geometry of reflectors (Pudsey & Howe 1998). Current-controlled echo types are illustrated in Figure 5.

Almost all deposition in the Scotia Sea is current-controlled: echo types III, IV and V each form about 30% of the total depositional area. In type III 'lumpy' areas, deposition takes place on small mounds defined by the interaction of moderate currents with topography (Fig. 5D). Variations in sedimentation rate are clear at the edges of mounds. In type IV, moderate currents deposit silty/sandy (or biogenic equivalent) sediments over wide

Table 1. *Principal characteristics*

Location	Scotia Sea, SW Atlantic Ocean
Setting	Antarctic Circumpolar Current; depositional areas away from the axis of strongest flow
Age	Miocene to Recent
Drift type	Plastered and detached drifts
Dimensions	50–100 km
Seismic characteristics	Low mounds or plains of sediment, laterally continuous sub-bottom reflectors which converge towards margins of mounds. Sediment waves sparse and localized
Sediment characteristics	Alternating biogenic-terrigenous, silty, bioturbated; more sand in the north, more clay in the south; glacial–interglacial cyclicity in texture and biogenic content

From: STOW, D. A. V., PUDSEY, C. J., HOWE, J. A., FAUGÈRES, J.-C. & VIANA, A. R. (eds)
Deep-Water Contourite Systems: Modern Drifts and Ancient Series, Seismic and Sedimentary Characteristics.
Geological Society, London, Memoirs, **22**, 325–336. 0435-4052/02/$15.00 © The Geological Society of London 2002.

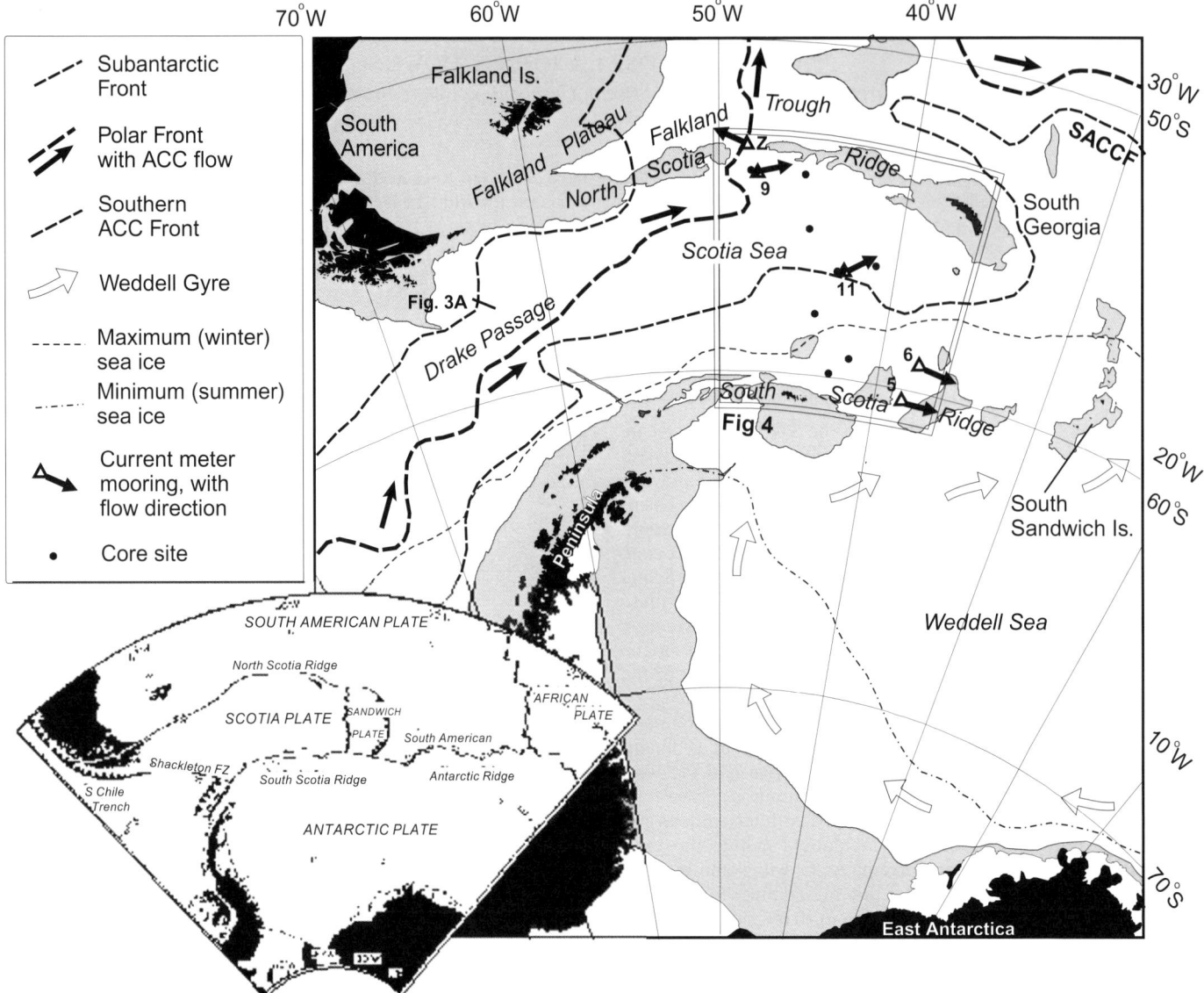

Fig. 1. Location map of the Scotia Sea, showing deep circulation. Areas shallower than 2000 m shaded. Inset map from Tectonic Map (1985). Sea ice limits from Sea Ice Climatic Atlas (1985), position of oceanographic fronts from Orsi *et al.* (1995). The Subantarctic Front separates Subtropical Surface Water from Subantarctic Surface Water. The Polar Front separates Subantarctic Surface Water from Antarctic Surface Water. The Southern ACC Front was defined by Orsi *et al.* (1995) as the southern boundary of Upper Circumpolar Deep Water.

areas (Figs 4 and 5C), with fine material being winnowed out, leading to relatively low sedimentation rates. In type V, moderate to weak currents permit rapid deposition of fine-grained sediments. Type V sediments accumulate in areas protected from strong bottom currents by topography or by great distance from the ACC axis (Fig. 4).

All the seabed types inferred to be influenced by mass flow and downslope processes were grouped in type VI by Pudsey & Howe (1998); type VI represents some 10% of the depositional area. Slumps, debris flows and turbidites occur along the southern margin of the North Scotia Ridge and in the deep basin between Pirie Bank and Bruce Bank. Slumps also occur locally on the margins of other elevated areas (e.g. the east side of Bruce Bank). An area of turbidite deposition extends from South Georgia to a deep basin at 57°S containing several hundred metres of ponded sediment. The limited extent of mass flow deposits suggests that much of the sediment shed southwards from the North Scotia Ridge has been entrained in and redeposited from bottom currents.

Most of the areas of echo type V are low-relief mounds, or smooth areas on gentle regional slopes (north of Pirie Bank and northeast of Bruce Bank). The mound just south of the gap in the North Scotia Ridge (cores 64 and 66, Fig. 5A; Howe & Pudsey 1999) has the highest relief (500 m from crest to western margin). It has developed on the west side of the path of strongest northward ACC flow. Thinner sediments flank the mound, and 'lumpy' sediments (accumulations only a few km across) partly cover the south side of the North Scotia Ridge. More typical are the mounds where cores 77 and 81 are located (Fig. 5B), with relief of 150–200 m from crest to margin.

Stratigraphic context

Although biogenic carbonate occurs only in certain intervals of these cores, precluding measurement of a continuous $\delta^{18}O$ curve, diatoms and radiolarians afford good siliceous biostratigraphy. Chemical (trace element and isotope) methods have also been used for dating. Glacial and interglacial stages down to stage 6 can be identified (Fig. 6).

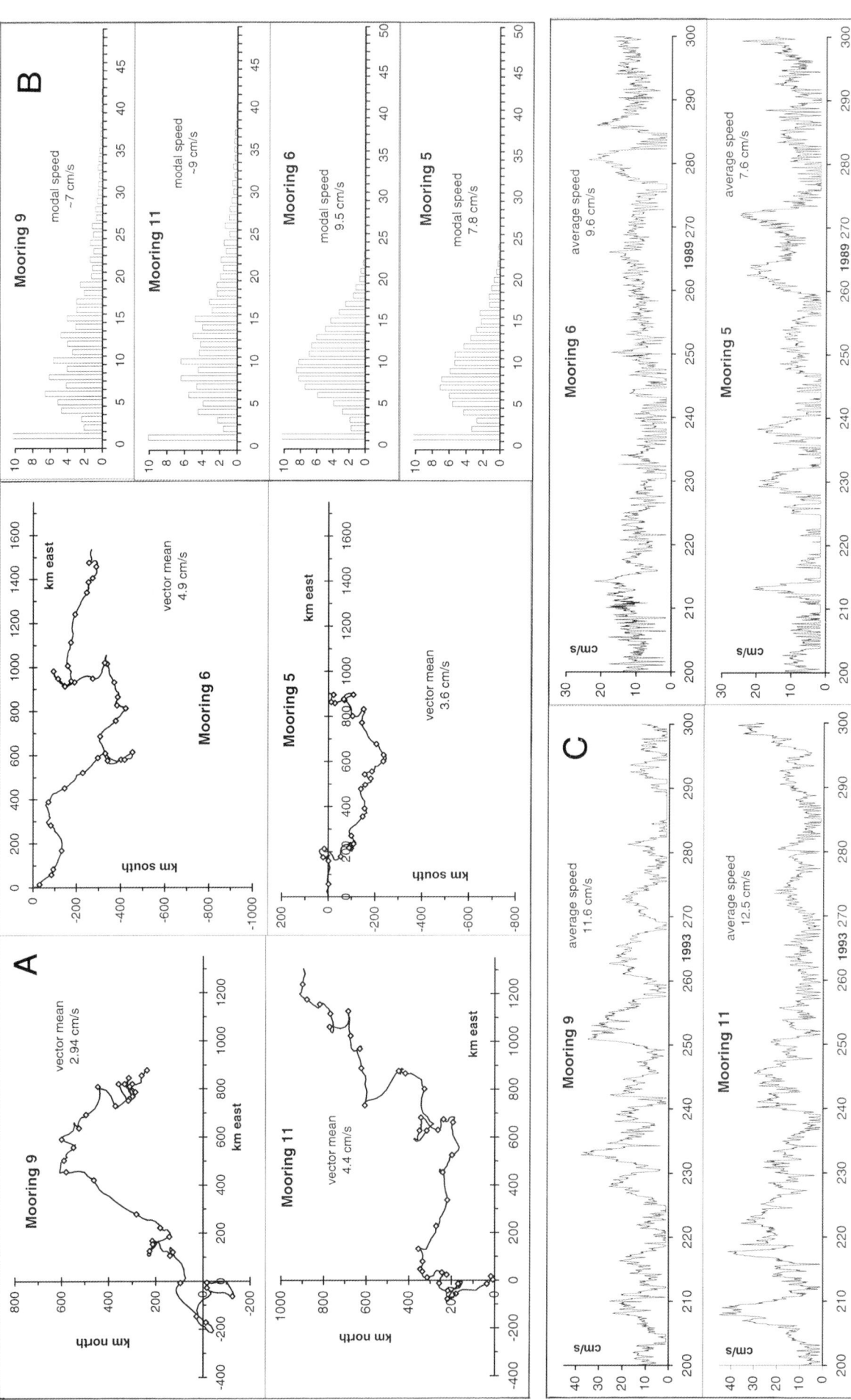

Fig. 2. Current meter data from 10 m above the seabed: mooring locations on Figures 1 and 4. (**A**) Progressive vector diagrams; diamond symbols at ten-day intervals. (**B**) Histograms of current speed values; these show the most commonly attained speeds, the proportion of very high speeds, and for how much of the time the meter rotor was stalled (flow slower than 1.5 cm s^{-1}). (**C**) Speed values for representative 100 day intervals; horizontal axes are in Julian days for years 1989 or 1993. Benthic storms (flow faster than 15 cm s^{-1} for two days or more) are interspersed with periods of very slow flow, when fine sediment particles can settle from suspension. Great variability in current direction, particularly at sites 9 and 11, results in the vector-average current being much slower than the modal or arithmetic mean speeds. Tidal cycles were recorded at all sites.

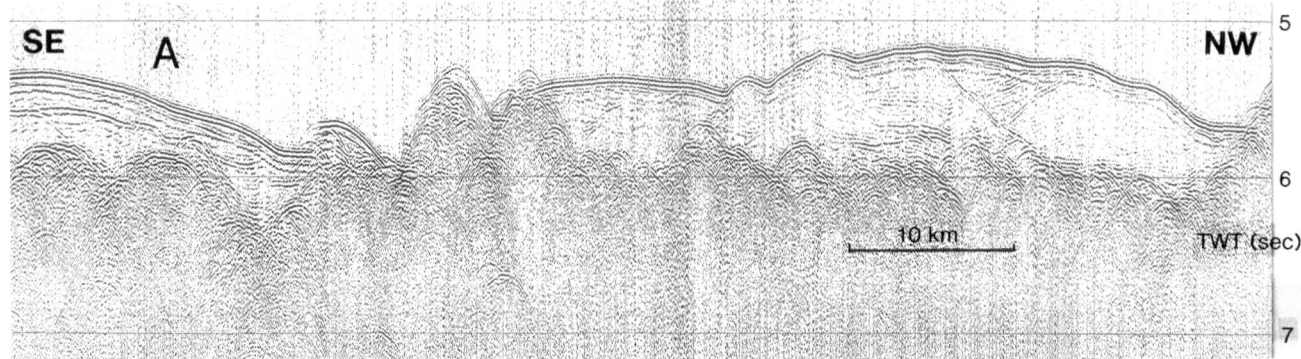
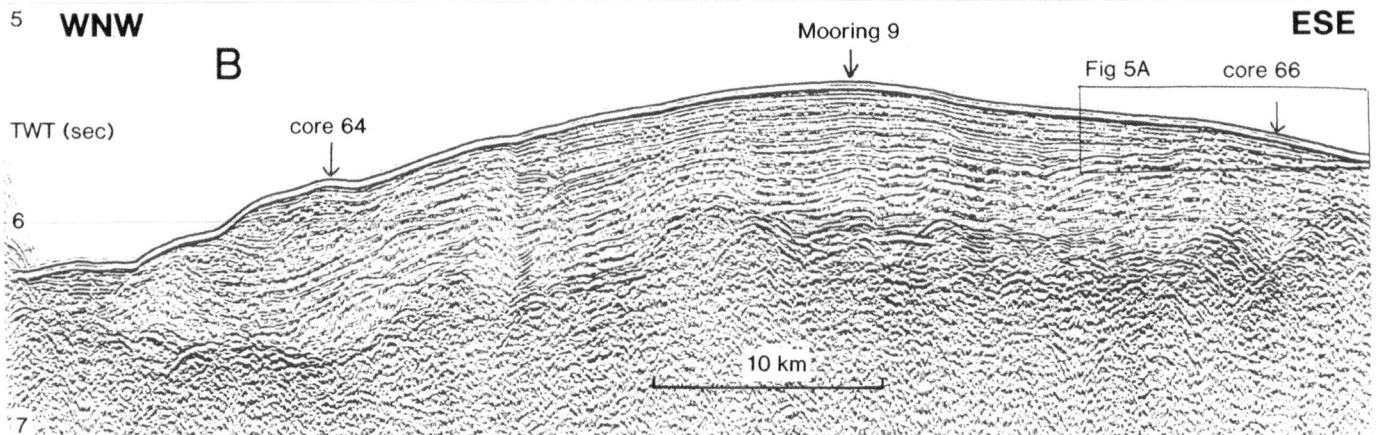
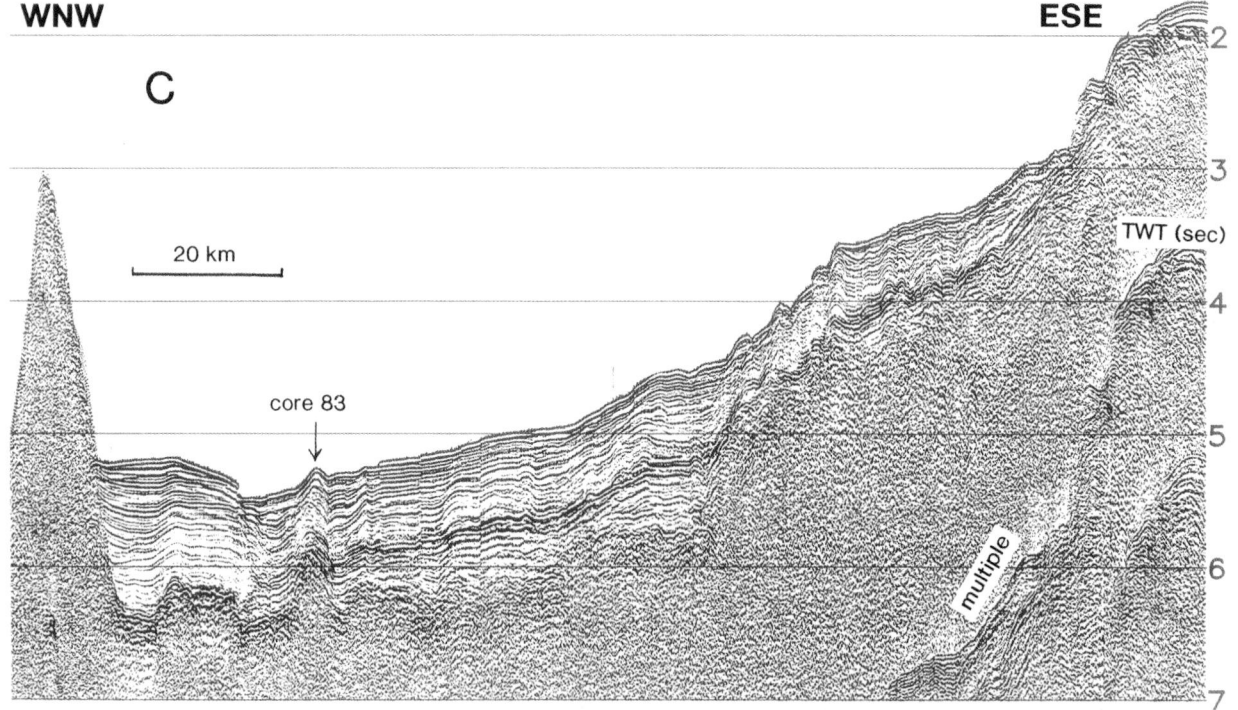

Fig. 3. Seismic reflection profiles, locations on Figures 1 and 4. (**A**) Single-channel profile in northern Drake Passage (Barker & Burrell 1977). Sediment occurs in localised mounds with many internal erosion surfaces showing evidence for strong current control. Ocean crust is 20–23 Ma old. (**B**) Single-channel profile, northern Scotia Sea (Howe & Pudsey 1999). This shows a sediment drift 70 km wide and 750–800 m thick (assuming sound velocity of 2 km s^{-1}) on ocean crust 23 Ma old. Sub-bottom reflectors are parallel and continuous in the centre of the mound, but show evidence for erosion near moats on either side. Drift is confined by basement ridges on either side. (**C**) Unprocessed near-trace plot of multi-channel profile, southern Scotia Sea. Parallel, continuous reflectors at left show low-energy contourites or hemipelagic sediments draped over basement topography. Sediments at right contain internal erosion surfaces, i.e. they show more evidence for current control.

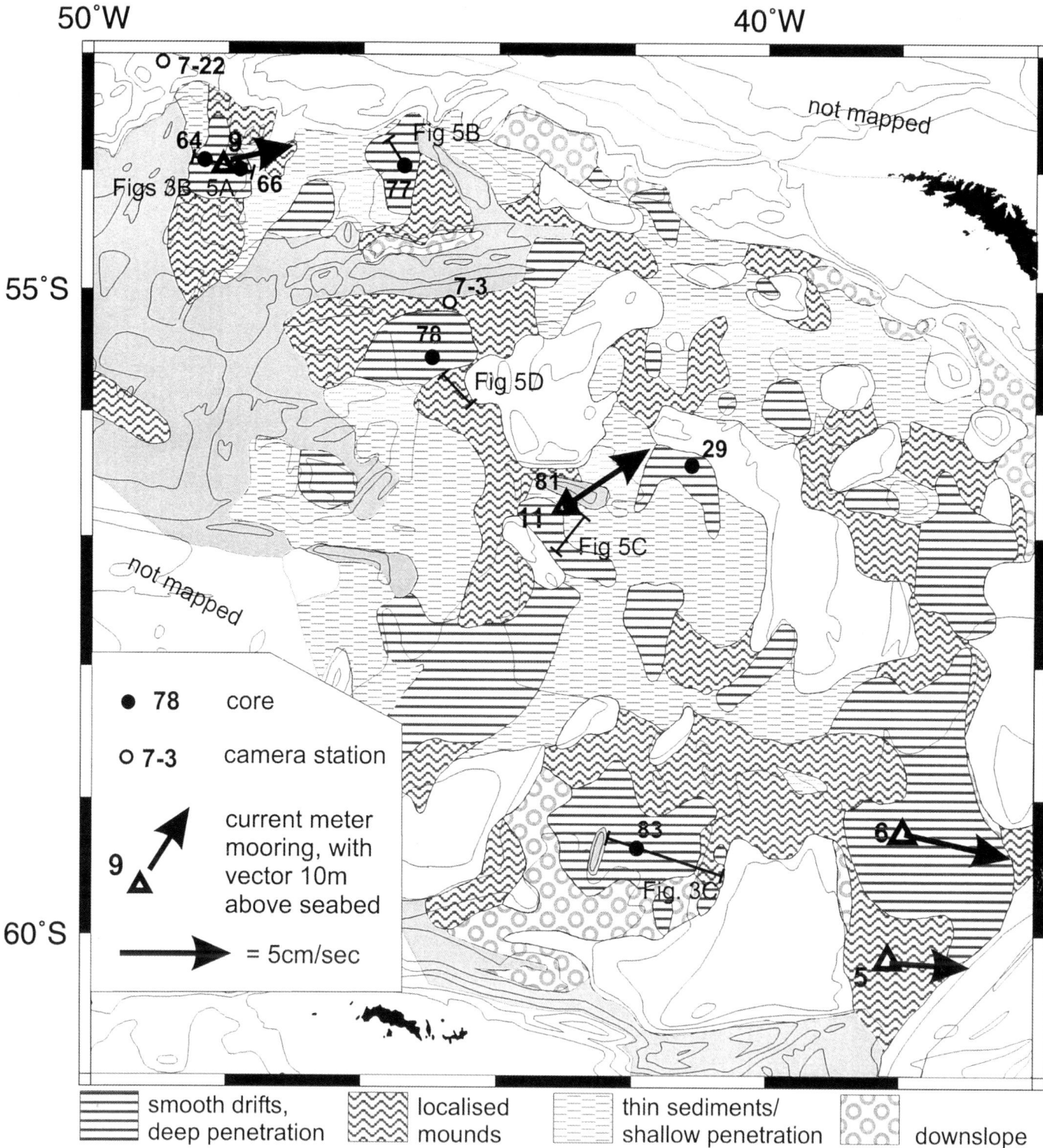

Fig. 4. 3.5 kHz echo character map, from Pudsey & Howe (1998). Non-depositional areas are shown in light green (very rough, hyperbolic diffractions, hundreds of metres of relief; interpreted mainly as basaltic ocean crust, but including seamounts and ridges of uncertain origin) or light brown (smooth seabed with a strong seabed reflector and no sub-bottom penetration, mainly on bathymetric highs, interpreted as fragments of continental crust or now-inactive island arcs of likely Oligocene to Mesozoic age; Barker *et al.* 1991). Depositional areas are divided into four types (types III to VI of Pudsey & Howe 1998): see Figure 5 for examples of the three types interpreted as contourites. All the areas inferred to be influenced mainly by downslope processes have been grouped together in type VI. They include sedimented slopes where the layers have been disrupted by slumping, debris flow deposits, channels and levees, and ponded sediments showing deep sub-bottom penetration and parallel but discontinuous reflectors, interpreted as fine-grained turbidites. Reprinted from Marine Geology, v. 148, Pudsey & Howe, Quaternary history of the Antarctic Circumpolar Current: evidence from the Scotia Sea, 229–241, © 1998, with permission from Elsevier Science.

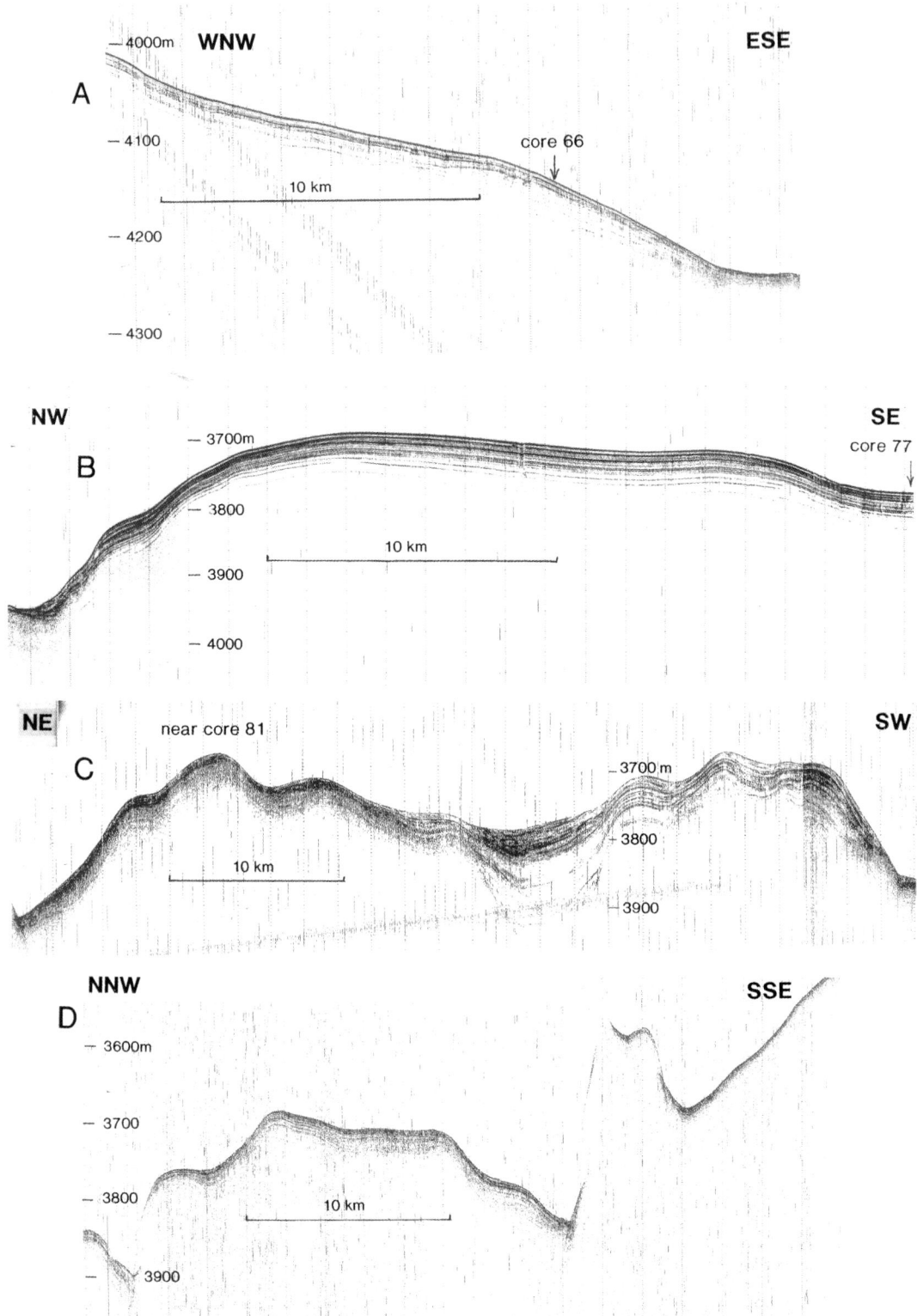

Fig. 5. 3.5 kHz profiles in current-controlled depositional areas; locations on Figure 4. Echo type IV: Undulating or smooth seabed with a distinct, locally prolonged seabed reflector and 10–30 m sub-bottom penetration. Sub-bottom reflectors are continuous, but commonly converge towards the margins of these areas and over elevations within them. The limited sub-bottom penetration suggests relatively coarse grain size, probably silts and sands. The geometry of reflectors suggests current control, with the marginal convergence resulting from non-deposition. Echo type V: Smooth seabed with a distinct (or, rarely, indistinct) seabed reflector, and deep sub-bottom penetration (30–70 m or more). Sub-bottom reflectors are parallel and continuous, draped rather than ponded, and in some areas deeper reflectors may be stronger than the seabed. Deep penetration suggests fine grain size, probably mud or fine-grained biogenic ooze. Geometry of reflectors suggests weak current control, and these sediments could be muddy contourites or hemipelagites. (**A**) Northern Scotia Sea, ESE end of seismic profile in Fig. 3B. Echo type V, with IV at edge of mound near the moat. (**B**) Northern Scotia Sea, NW of core 77. Echo type V. (**C**) Central Scotia Sea, near mooring 11 and core 81. Small basin between two ridges, showing echo types IV and V. (**D**) Echo type III: 'Lumpy' seabed with mounded accumulations of sediment generally 2–10 km across, separated by ridges of basement. Sub-bottom penetration is 20–60 m with distinct, parallel and continuous reflectors which converge towards the margins of mounds. In these areas sediment has been deposited upon irregular basement topography. Deep 3.5 kHz penetration suggests fine grain size, and the mounded form suggests current control. Insufficient sediment has accumulated to bury the basement completely. Central Scotia Sea, near core 78.

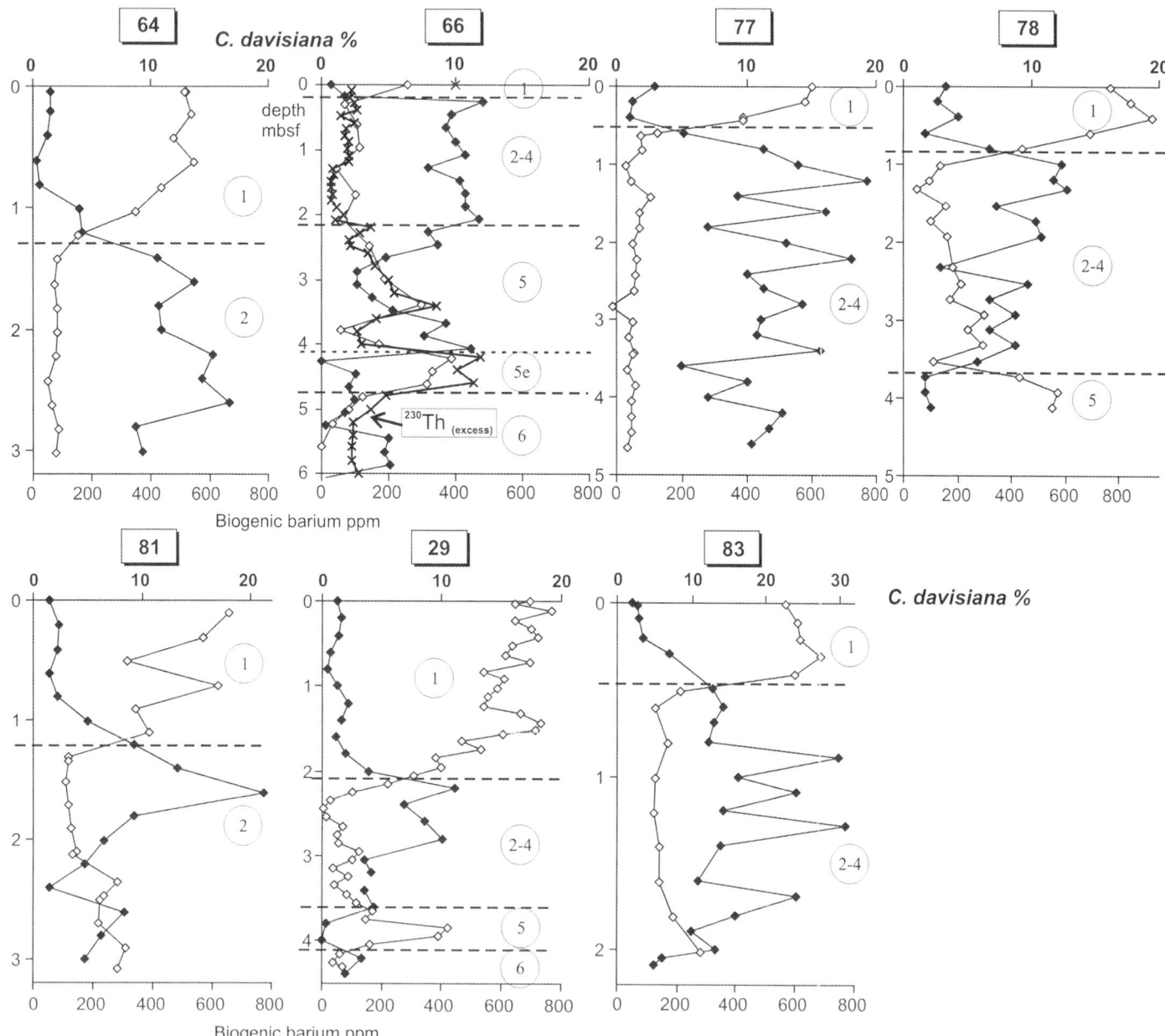

Fig. 6. Biostratigraphy and chemostratigraphy for the core transect. Upper scale (bold) and black symbols show relative abundance of *Cycladophora davisiana* as a percentage of the total radiolarian fauna, based on counts of at least 300 individuals per slide. Lower scale and open symbols show biogenic barium, parts per million (ppm). Lithogenic barium was subtracted from the total using a Ba/Al ratio of 0.0060 (cores 64, 66, 77, 78) or 0.0069 (cores 81, 29, 83; Pudsey & Howe 1998). Encircled figures show the inferred isotope stages (stage boundaries as dashed lines). For core 66, crosses show excess thorium, which has a peak in stage 5e.

Biostratigraphy

The first downcore relative abundance peak of the radiolarian *Cycladophora davisiana* is correlated with Marine Isotope Stage 2, the Last Glacial Maximum at 18 ka (Hays *et al.* 1976; Morley & Hays 1979). In the Scotia Sea, relative abundance of *C. davisiana* is low at the top of each core (generally 1–2%) and increases downwards to a peak of 10–25% of the total radiolarian fauna (Fig. 6). Similarly, downcore variations in the abundance of the diatom *Eucampia antarctica* were used by Jordan & Pudsey (1992) to delineate glacial and interglacial stages in Scotia Sea cores.

Barium stratigraphy and thorium dating

Biogenic barium can be a more stable tracer of palaeoproductivity than organic carbon, biogenic carbonate or opal, particularly in environments where the bottom water is corrosive. Shimmield *et al.* (1994) and Bonn *et al.* (1998) demonstrated that, in the Weddell Sea and Scotia Sea, barium was consistently high in interglacial isotope stages 1, 5e, 7 and 9, and low during glacial stages 2–4, 6, 8 and 10. The cores described here show high biogenic Ba at core tops (generally over 500 ppm) and a downcore decrease to low and uniform values (<200 ppm; Fig. 6). In cores 66, 78 and 29 another interval of high Ba values represents stage 5 (Shimmield *et al.* 1994).

Variations in thorium concentration in the sediment may be interpreted in terms of changes in sedimentation rate (Frank *et al.* 1996). Twenty samples from core 66 were measured using the analytical procedures of Frank *et al.* (1994). The results are consistent with the barium data, showing stage 5e at 4.1–4.7 m (Fig. 6).

The inferred stage boundaries in Figure 6 suggest wide variations in sedimentation rate, particularly for stage 1 (compare cores 66 and 29).

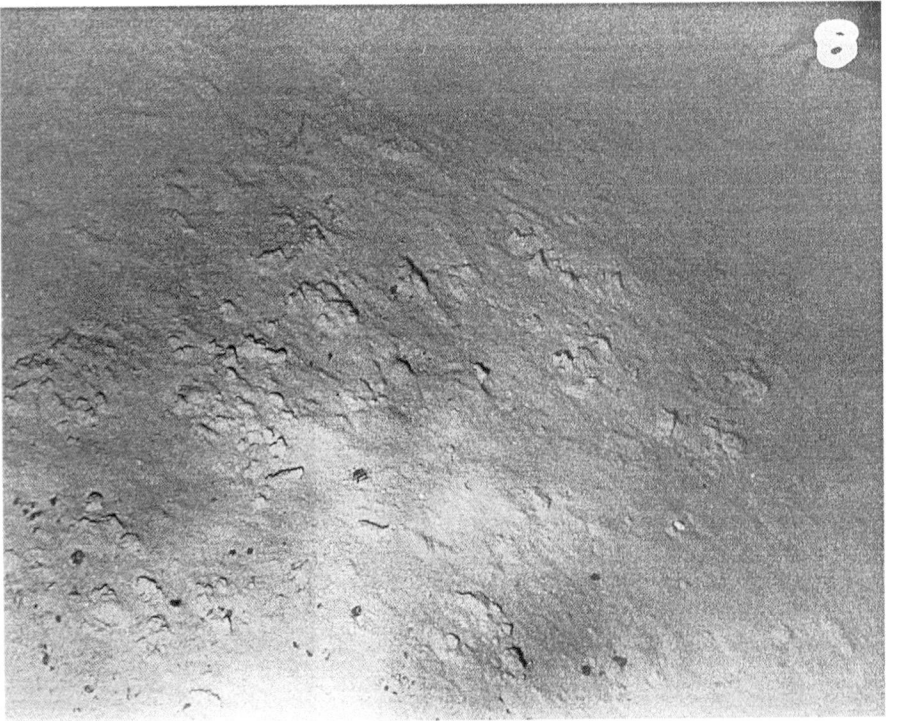

Fig. 7. Bottom photographs, locations on Figure 4. (**A**) Eltanin 7–22, 3146 m depth. This is in an area mapped as non-depositional; note that the 3.5 kHz acoustic profiler cannot resolve a thin sandy to gravelly layer overlying bedrock. The sea bottom is strongly rippled with asymmetrical current ripples. Other pictures at this station show a manganese nodule pavement, a few ophiuroids and echinoids. Sediment texture is 86.8% sand, 11.7% silt and 1.5% clay (Goodell 1964). (**B**) Eltanin 7–3, 3676 m depth, echo type III. Foraminiferal sandy silt; bioturbated sediment surface, no current-generated structures and a sparse fauna. A few small manganese-coated erratic pebbles are present. Sediment texture is 37.8% sand and gravel, 44.1% silt and 18.1% clay (Goodell 1964).

Sediment characteristics

Bottom photographs

Bottom photographs examined by Heezen & Hollister (1971) showed evidence for moderate to strong currents at some stations in the Scotia Sea, but few directions were obtainable. Photographs taken in 1963–1964 on cruises of USNS *Eltanin* (Goodell 1964, 1965) suggest high-energy and low-energy environments at different stations in the Scotia Sea, with widespread bioturbation (Fig. 7).

Core descriptions and facies

All cores exhibit a biogenic–terrigenous cyclicity, with biogenic sediment at the core top passing down into a more terrigenous unit and then into another biogenic unit (Fig. 8).

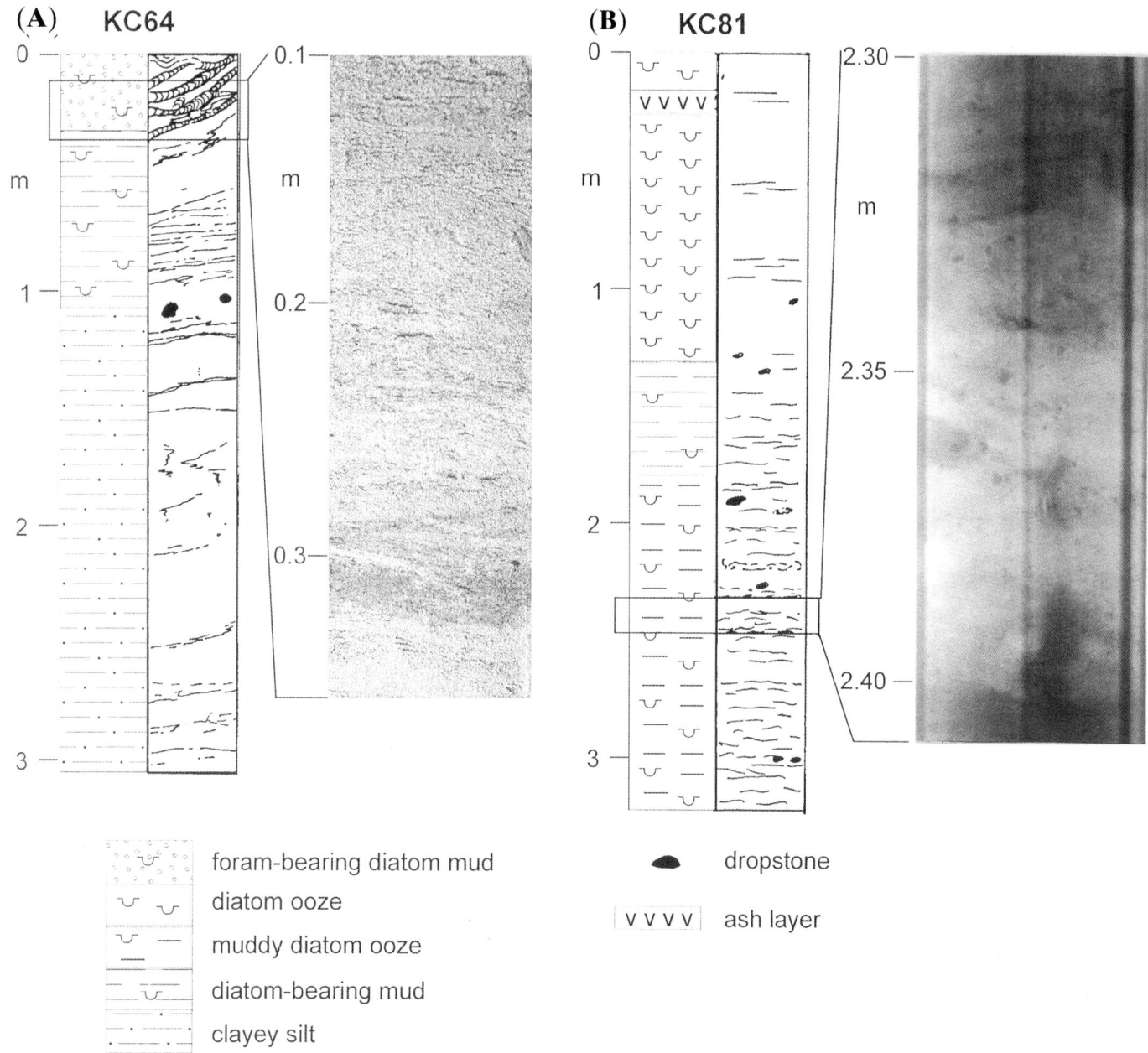

Fig. 8. (**A**) Core 64: lithology (left column) and sedimentary structures (right column), from Howe & Pudsey (1999). Photograph shows lamination and bioturbation at core top. *Zoophycos* burrows are seen in X-radiographs of this and nearby cores. (**B**) Core 81: lithology (left column) and sedimentary structures (right column), with X-radiograph of faint lamination and bioturbation. The vertical lines in the X-radiograph are ridges in the plastic core liner.

The upper biogenic unit is 0.5–2 m thick in most cores. In the northern Scotia Sea (north of 56°S) it consists of dark greyish brown foraminifer-bearing diatom silty mud (0.15–0.7 m thick; Fig. 8) overlying olive grey to greenish grey diatom mud or muddy diatom ooze. The foram-bearing mud is faintly burrow-mottled, with $CaCO_3$ forming 5–17%. Preservation of foraminifera is poor, but a few calcareous nannofossils are also present. The diatom mud is structureless or faintly laminated. Farther south the upper biogenic unit comprises olive to olive grey diatom ooze and mud-bearing diatom ooze. The core-top diatom assemblage is dominated by *Fragilariopsis kerguelensis*, the characteristic open-ocean diatom; preservation is moderate to good. Large centric diatoms and long fragments of *Thalassiothrix* valves are conspicuous in the sand fractions of diatom ooze samples. Faint lamination occurs in this unit in cores 81 and 83. Ice-rafted debris is generally sparse (O'Cofaigh *et al.* 2001).

The terrigenous unit (0.4–2 m thick) consists of diatom-bearing mud with no biogenic carbonate. Colours include dark grey and dark greenish grey, with slightly greener or greyer shades forming faint streaky lamination. Diatom preservation is good in the north, poor in the south, with some reworking of Pliocene and upper Miocene forms in the southern cores. The diatom assemblages are very different from those in the Holocene. *F. kerguelensis* remains the dominant species in the three northern cores, but from core 78 southwards the ice-related diatom *Eucampia antarctica* predominates (Jordan & Pudsey 1992).

The lower biogenic unit is at least 1–1.5 m thick (base not recovered). It comprises grey to dark grey muddy diatom ooze and diatom mud, gradationally interbedded on a scale of 0.1–0.4 m. Core 29 includes muddy diatom ooze layers up to 0.3 m thick with sharp tops and bases. In cores 78, 81 and 83 the lower biogenic unit is conspicuously laminated in various shades of dark greenish grey (Fig. 8B).

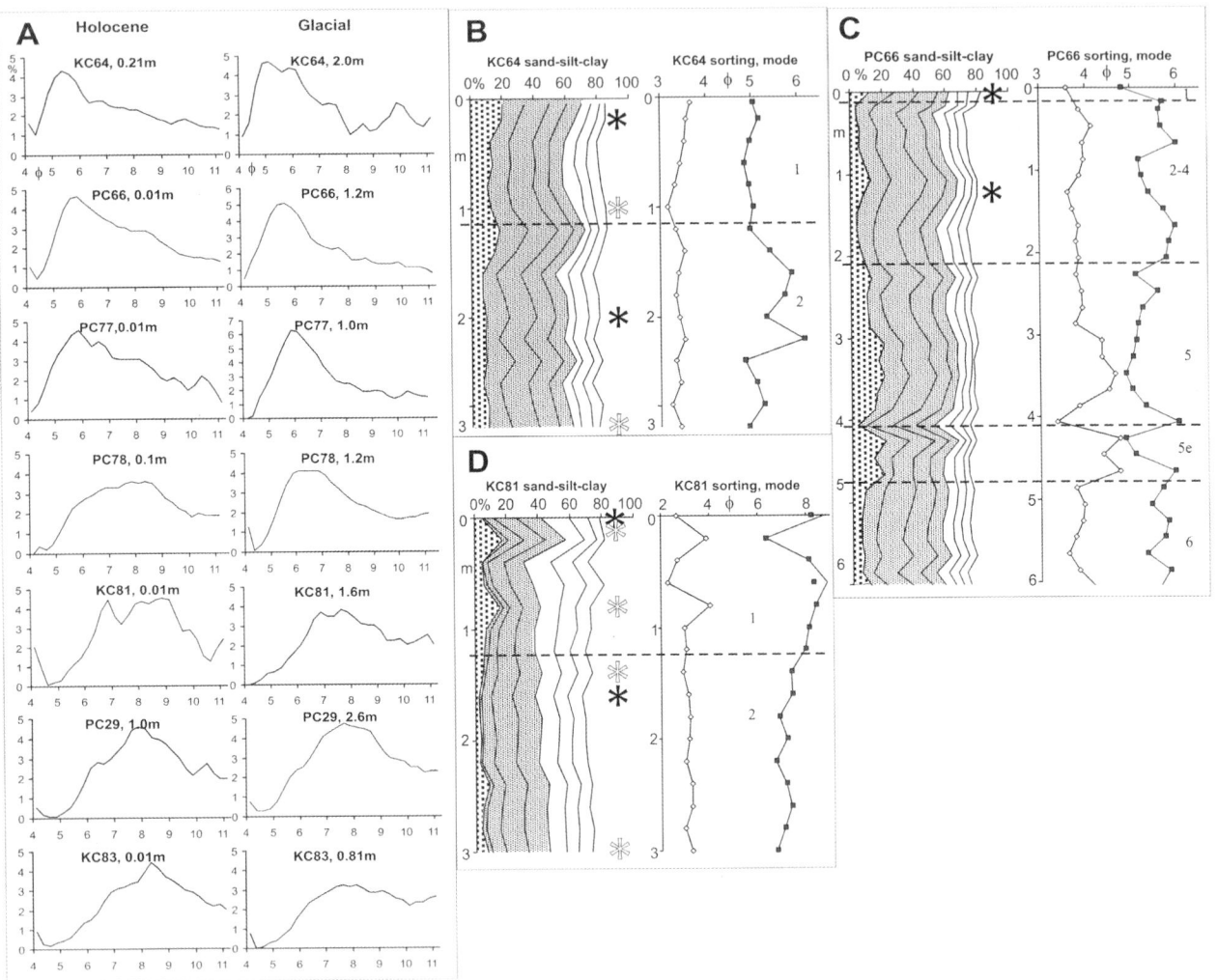

Fig. 9. Grain-size data for the fine fraction, measured by Sedigraph (Pudsey & Howe 1998). MIS boundaries from Figure 6. (**A**) representative size-frequency curves for Holocene and glacial samples in 7 cores, arranged from NW (core 64) to SE (core 83). (**B**) sand-silt-clay plot (with weight % within each 1ϕ interval from 4–11ϕ) for core 64. Sorting (dimensionless) shown as open symbols, mode (phi) shown as filled squares. Black stars mark fine-fraction samples, open stars mark sand samples (Fig. 10). (**C**) Same data for core 66. Interglacial stages 1 and 5 have more sand, poorer sorting and a coarser silt mode than glacial stages 2–4 and 6. (**D**) same data for core 81. In this core interglacial stage 1 has more sand but a finer silt mode than glacial stage 2. The sandy sample at 0.2 m is from a tephra layer.

Summary analytical results

Core-top (Holocene) texture shows an overall fining from NW to SE, i.e. a decrease in sand and silt content and an increase in clay (Pudsey & Howe 1998). South of 56°S a significant proportion of the sand fraction consists of large diatoms which have a low settling velocity, and therefore behave hydrodynamically like silt grains rather than sand. Samples from glacial intervals at all sites contain less sand than in the Holocene, and the sand is almost all terrigenous with a minor amount of radiolarians and large diatoms.

The pattern of fine-fraction deposition is quite complex. Figure 9A shows silt and clay size distributions for representative Holocene and glacial levels in each core. Because of the amount of clay finer than the measurement limit of 11ϕ (commonly 10–30%), calculation of sorting and skewness parameters required estimation of ϕ_{16}, i.e. extrapolation of the fine end of the cumulative frequency curves down to the 16% level. Using the standard Folk parameters, all samples are very poorly sorted ($\sigma_G = 2.05–3.9\phi$) and positively to very positively skewed ($Sk_G = 0.1–0.45$). Downcore data for cores 64, 66 and 81 are also shown in Figure 9.

Sand size distributions from two representative cores are illustrated in Fig. 10. In core tops the sand is very fine and well-sorted with 50–70% in the size range 3.5–4ϕ. In core 64, glacial sediment contains 7% sand which is terrigenous and well sorted with a mode at 3.4ϕ (Fig. 10). In core 81, glacial sediment contains 4–6% sand which is extremely poorly sorted, with common ice-rafted debris (O'Cofaigh et al. 2001).

Discussion

Are the sediments contourites?

The contourite origin inferred from acoustic profiler data is supported by the presence of bioturbation and faint horizontal lamination throughout the cores, and by the absence of sedimentary structures characteristic of turbidites.

In the northern Scotia Sea, where the cored sediments are predominantly terrigenous, Holocene samples contain a mode in the medium-coarse silt range and relatively little clay. This is consistent with our measured current speeds (compare Hollister & McCave 1984; McCave et al. 1995).

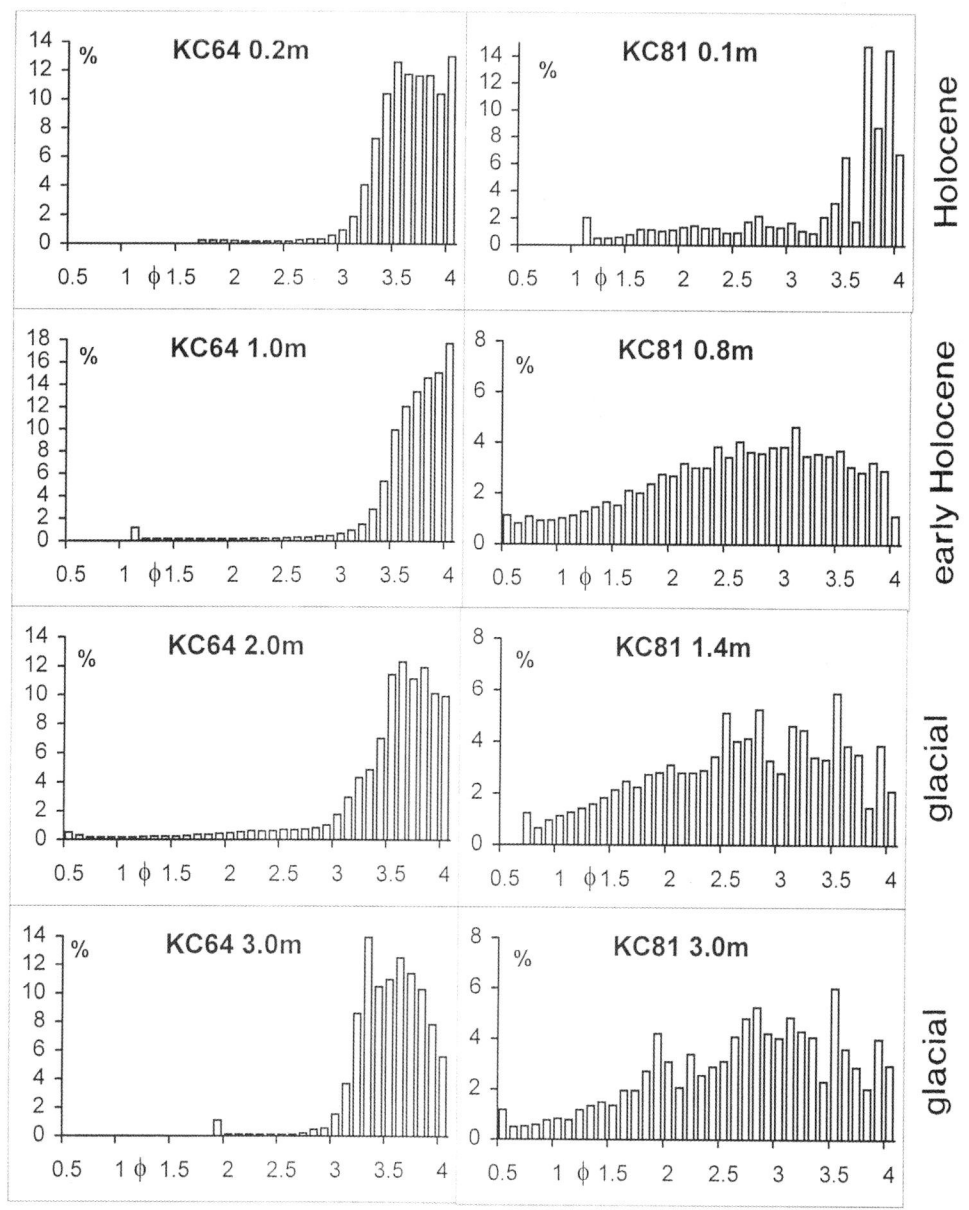

Fig. 10. Size frequency histograms for the sand fraction of cores 64 and 81, measured by settling tube. Samples from core 64 are all very fine, well-sorted sand. In core 81 only the core-top sample is well sorted, the other samples consisting mainly of very poorly sorted sand, characteristic of ice-rafted debris.

Effect of biogenic grains

In the central Scotia Sea (cores 29, 81) the Holocene sand consists predominantly of large diatoms; polymodality of the Holocene fine fraction results from the abundance of biogenic grains. The apparent lack of current sorting is surprising in view of the mean speed of 12.5 cm s^{-1} measured at mooring 11. The hydrodynamic behaviour of diatom oozes is not well known, but it is likely that their erosion, suspension, aggregation and deposition characteristics are different from those of fine terrigenous sediments (compare McCave 1985). Once deposited, diatom ooze containing abundant *Thalassiothrix* spp. may be particularly hard to resuspend as the long frustules mat together.

Unsteadiness of bottom-current flow

It is noteworthy that mudwaves are very rare in the central Scotia Sea. They occur in a small area of the Falkland Trough at 44–45°W, just north of Shag Rocks (Cunningham *et al.* 2002), where they indicate westward bottom current flow along the trough axis. In the Scotia Sea mudwaves do occur within Type III 'lumpy' sediments, but are so scarce that they have not been mapped separately. Pudsey & Howe (1998) speculated that the absence of mudwaves may be related to the unsteady nature of ACC flow. Our current meter observations reveal the unsteadiness of near-bottom flow, with high eddy kinetic energy (K_E) values of 58 and 64 cm^2 s^{-2} at moorings 5 and 6 respectively (Barber & Crane 1995) and 90–100 at moorings 9 and 11. These may be compared with K_E values of only 17–26 at mudwave sites in the central Argentine Basin and 76–127 at other mooring sites near the NW and S margins of that basin (Dickson 1990; Weatherly 1993). It is likely that the high and low ends of the speed spectrum (benthic storms and periods of very slow flow) have at least as much influence on deposition as does the mean speed, but there are as yet insufficient observational data from sediment drifts to be able to quantify palaeo-speed ranges in any detail.

Glacial/interglacial ACC flow

In Holocene interglacial sediments there is a northward increase in the silt/clay ratio. There is also a less distinct northward increase in coarseness of the silt mode (Fig. 9), obscured by the biogenic signal. The proportion of sand-size material does not show a simple south-north trend, again probably because of the influence of biogenic grains (including the slow-settling large diatom fraction). The Holocene sand distributions are all similar with 50–70% of the sand in the 3.5–4φ range.

In LGM sediments, which are predominantly terrigenous, the textural signal is easier to interpret. Although around the Scotia Sea glacially lowered sea level would have exposed additional shelf areas as possible sediment sources, we agree with McCave *et al.* (1995) that at high latitudes, systematic glacial/interglacial changes in grain size of source sediment are unlikely. There is a northward increase in the proportion of sand, and a trend from poorly-sorted sand in the south, interpreted as mainly ice-rafted, to well-sorted sand in the north, interpreted as current-sorted (Fig. 10). The accompanying northward increase in the silt:clay ratio and coarsening and increase in height of the silt mode (Fig. 9) are very clear. These data suggest a gradient in bottom current strength from slow in the south to fast in the north. (Note that the path of strongest deep ACC flow must have been near its present position in the Scotia Sea at LGM, because of the barrier of the North Scotia Ridge; the gap at 48°W remains the first deep gap in the ridge downstream of Drake Passage (Fig. 1)).

This current gradient was from values lower than present-day speeds in the south (as low as 1 cm s^{-1} at the northern margin of the Weddell Gyre; Pudsey 1992), through a region of little glacial–interglacial change in the central Scotia Sea, to values probably faster than today in the north. A LGM mean speed between 13 and 15 cm s^{-1} in the area of mooring 9 was tentatively suggested by Pudsey & Howe (1998). Flow cannot have been much faster than this, or fine-fraction deposition would not have occurred (Hollister & McCave 1984). Flow must have been considerably slower than that recorded at site Z (Fig. 1) which is non-depositional (vector average 17.1 cm s^{-1}, mean speed 21.9 cm s^{-1}; Zenk 1981).

P. Morris assisted with data management and plotted Figures 3A and 3C. A. Cunningham processed the seismic data for Figure 3B. M. Frank supplied the Th data for core 66. K. Roessig at the Antarctic Research Facility, Florida State University, kindly supplied the photographs in Figure 7. Ian Gilbert measured the sand size distributions in core 64. The paper was reviewed by B. Diekmann.

References

BARBER, M. & CRANE, D. 1995. Current flow in the north-west Weddell Sea. *Antarctic Science*, **7**, 39–50.

BARKER, P. F. & BURRELL, J. 1977. The opening of Drake Passage. *Marine Geology*, **25**, 15–34.

BARKER, P. F., DALZIEL, I. W. D. & STOREY, B. C. 1991. Tectonic development of the Scotia Arc region. *In*: R. J. TINGEY (ed.) *Geology of Antarctica*. Clarendon Press, Oxford, 215–248.

BONN, W. J., GINGELE, F. X., GROBE, H., MACKENSEN, A. & FUTTERER, D. K. 1995. Palaeoproductivity at the Antarctic continental margin: opal and barium records for the last 400 ka. *Palaeogeography, Palaeoclimatology, Palaeoecology*, **139**, 195–211.

BRYDEN, H. L. & PILLSBURY, R. D. 1977. Variability of deep flow in the Drake Passage from year-long current measurements. *Journal of Physical Oceanography*, **7**, 803–810.

CUNNINGHAM, A. P., HOWE, J. A. & BARKER, P. F. 2002. Contourite sedimentation in the Falkland Trough, western South Atlantic. *In*: STOW, D. A. V., PUDSEY, C. J., HOWE, J. A., FAUGÈRES, J.-C. & VIANA, A. R. (eds) *Deep-Water Contourite Systems: Modern Drifts and Ancient Series, Seismic and Sedimentary Characteristics*. Geological Society, London Memoirs, **22**, 337–352.

DICKSON, R. R. 1990. *Flow statistics from long-term current-meter moorings; the global dataset in January 1989*. WOCE Report **30**.

FRANK, M., ECKHARDT, J-D., EISENHAUER, A., KUBIK, P. W., DITTRICH-HANNEN, B. & MANGINI, A. 1994. Beryllium 10, thorium 230 and protactinium 231 in Galapagos microplate sediments: implications for hydrothermal activity and paleoproductivity changes during the last 100,000 years. *Paleoceanography*, **9**, 559–578.

FRANK, M., GERSONDE, R., RUTGERS VAN DER LOEFF, M., KUHN, G. & MANGINI, A. 1996. Late Quaternary sediment dating and quantification of lateral sediment redistribution applying ^{230}Th$_{ex}$: a study from the eastern Atlantic sector of the Southern Ocean. *Geologische Rundschau*, **85**, 554–566.

GOODELL, H. G. 1964. *Marine Geology of the Drake Passage, Scotia Sea and South Sandwich Trench (USNS Eltanin cruises 1–8)*. Florida State University.

GOODELL, H. G. 1965. *Marine Geology: USNS Eltanin cruises 9–15*. Florida State University.

HAYS, J. D., LOZANO, J. A., SHACKLETON, N. J. & IRVING, G. 1976. Reconstruction of the Atlantic and western Indian Ocean sectors of the 18,000B.P. Antarctic Ocean. *In*: Investigation of Late Quaternary Paleoceanography and Paleoclimatology. *Geological Society of America, Memoir* **145**, 337–372.

HEEZEN, B. C. & HOLLISTER, C. D. 1971. *The Face of the Deep*. Oxford University Press, Oxford.

HOLLISTER, C. D. & MCCAVE, I. N. 1984. Sedimentation under deep-sea storms. *Nature*, **309**, 220–225.

HOWE, J. A. & PUDSEY, C. J. 1999. Antarctic Circumpolar Deepwater: a Quaternary paleoflow record from the northern Scotia Sea, South Atlantic Ocean. *Journal of Sedimentary Research*, **69**, 847–861.

JORDAN, R. W. & PUDSEY, C. J. 1992. High-resolution diatom stratigraphy of Quaternary sediments from the Scotia Sea. *Marine Micropalaeontology*, **19**, 201–237.

LOCARNINI, R. A., WHITWORTH, T., III & NOWLIN, W. D. JR. 1993. The importance of the Scotia Sea on the outflow of Weddell Sea Deep Water. *Journal of Marine Research*, **51**, 135–153.

MCCAVE, I. N. 1984. Erosion, transport and deposition of fine-grained marine sediments. *In*: D. A. V. STOW & D. J. W. PIPER (eds) *Fine-grained sediments: deep-water processes and facies*. Special Publication, Geological Society, **15**, 35–69.

MCCAVE, I. N. 1985. Properties of suspended sediment over the HEBBLE area on the Nova Scotian Rise. *Marine Geology*, **66**, 169–188.

MCCAVE, I. N., MANIGHETTI, B. & ROBINSON, S. G. 1995. Sortable silt and fine sediment size/composition slicing: parameters for paleocurrent speed and palaeoceanography. *Paleoceanography*, **10**, 593–610.

MORETON, S. G. & SMELLIE, J. L. 1998. Identification and correlation of distal tephra layers in deep sea sediment cores, Scotia Sea, Antarctica. *Annals of Glaciology*, **27**, 285–289.

MORLEY, J. J. & HAYS, J. D. 1979. *Cycladophora davisiana*: a stratigraphic tool for Pleistocene North Atlantic and interhemispheric correlation. *Earth and Planetary Science Letters*, **44**, 383–389.

NOWLIN, W. D. JR. & CLIFFORD, M. 1982. The kinematic and thermohaline zonation of the Antarctic Circumpolar Current at Drake Passage. *Journal of Marine Research*, **40**, 481–507.

Ó COFAIGH, C., DOWDESWELL, J. A. & PUDSEY, C. J. 2001. Late Quaternary iceberg rafting along the Antarctic Peninsula continental rise and in the Weddell and Scotia seas. *Quaternary Research*, **56**, 308–321.

ORSI, A. H., WHITWORTH, T. III & NOWLIN, W. D. JR. 1995. On the meridional extent and fronts of the Antarctic Circumpolar Current. *Deep-Sea Research*, **42**, 641–673.

PUDSEY, C. J. 1992. Late Quaternary changes in Antarctic Bottom Water velocity inferred from sediment grain size in the northern Weddell Sea. *Marine Geology*, **107**, 9–33.

PUDSEY, C. J. & HOWE, J. A. 1998. Quaternary history of the Antarctic Circumpolar Current: evidence from the Scotia Sea. *Marine Geology*, **148**, 83–112.

SEA ICE CLIMATIC ATLAS (1985) Vol. 1, Antarctic. Naval Oceanography Command, Asheville, North Carolina.

SHIMMIELD, G. B., DERRICK, S., MACKENSEN, A., GROBE, H. & PUDSEY, C. J. 1994. The history of biogenic silica, organic carbon and barium accumulation in the Weddell Sea and Antarctic Ocean over the last 150,000 years. *In*: R. ZAHN *et al.* (eds) *Carbon Cycling in the Glacial Ocean: Constraints on the Ocean's Role in Global Change*. Proceedings of the NATO ARW, **I 17**, 555–574.

TECTONIC MAP OF THE SCOTIA ARC. 1985. Sheet BAS (Misc.) **3**. British Antarctic Survey, Cambridge.

WEATHERLY, G. L. 1993. On deep-current and hydrographic observations from a mudwave region and elsewhere in the Argentine Basin. *Deep-Sea Research*, **40**, 939–961.

WHITWORTH, T., III, NOWLIN, W. D. JR. & WORLEY, S. J. 1982. The net transport of the Antarctic Circumpolar Current through Drake Passage. *Journal of Physical Oceanography*, **12**, 960–971.

ZENK, W. O. 1981. Detection of overflow events in the Shag Rocks Passage, Scotia Ridge. *Science*, **213**, 1113–1114.

Contourite sedimentation in the Falkland Trough, western South Atlantic

ALEX P. CUNNINGHAM, JOHN A. HOWE[1] & PETER F. BARKER

British Antarctic Survey, High Cross, Madingley Road, Cambridge CB3 0ET, UK
[1]*Present address: Scottish Association for Marine Science, Dunstaffnage Marine Laboratory, P.O. Box 3, Oban, Argyll PA34 4AD, Scotland, UK*

Abstract: The Falkland Trough is a west–east bathymetric deep that separates the Falkland Plateau from the North Scotia Ridge in the western South Atlantic. It lies in the path of Circumpolar Deep Water flowing within the Antarctic Circumpolar Current (ACC), and Weddell Sea Deep Water flowing beneath the ACC east of Shag Rocks passage. Marine geophysical and sediment core data demonstrate the influence of ambient bottom currents on deposition in this area, and reveal two styles of contourite sedimentation: (1) deposition of glauconite-rich sandy contourites in exposed areas of the Falkland Plateau and Falkland Trough, where vigorous ACC bottom currents control sedimentation, and (2) deposition of biogenic sandy contourites, muddy contourites and hemipelagites (western Falkland Trough), and muddy diatom ooze (eastern Falkland Trough), in the form of two elongate sediment drifts, which have developed in the presence of more sluggish bottom currents. The drift sediments contain a depositional record of bottom current flow through the glacial cycle (southern-origin bottom water flow in the east, and probably ACC flow in the west); analyses of core data from the western Falkland Trough suggest a reduction in bottom current strength during the Last Glacial Maximum at present depths of > 2500 m below sea level.

Table 1. *Principal characteristics*

location	Falkland Trough, western South Atlantic
setting	Deposition in the presence of the Antarctic Circumpolar Current, and deeper, more sluggish Weddell Sea Deep Water
age	Neogene? to Recent
drift type	Plastered drifts on the exposed Falkland Plateau slope, elongate confined drifts in more sheltered parts of the Falkland Trough
dimensions	Plastered drifts: width 3–5 km, length < 15 km Confined drifts: west Falkland Trough sediment drift (WFTD), width ≤ 20 km, length > 130 km; east Falkland Trough sediment drift (EFTD), width ≤ 46 km, length = 302 km
seismic characteristics	Plastered drifts: (1) 3.5 kHz profiles show regular, fairly continuous reflections to > 50 m below sea floor; (2) airgun profiles show very shallow acoustic stratification (drifts imaged within the source wavelet reflected from the sea-floor); (3) GLORIA sonographs show textured sea floor with moderate backscatter. Confined drifts: (1) 3.5 kHz profiles show laterally persistent reflections to the limit of penetration beneath the apex of each mound, which converge as component layers pinch out toward flanking non-depositional depressions; (2) airgun profiles show sub-parallel, divergent, undulating, wavy and migrating-wave reflection configurations; (3) GLORIA sonographs crossing the WFTD show featureless sea floor with very low backscatter.
sediment characteristics	Falkland Plateau slope and central Falkland Trough sediments: sandy contourites including foraminiferal sands (bioturbated medium to fine sand, moderately sorted, 60% biogenic component), glauconite-rich sands (bioturbated fine to very fine sand, very well sorted), and muddy contourites (bioturbated sand-silt, moderately to poorly sorted). West Falkland Trough drift sediments: foraminiferal sands (as above), diatomaceous hemipelagic muds (bioturbated silt, moderately to poorly sorted), sandy-muddy contourites (bioturbated silt, moderately to poorly sorted).

Introduction

The Falkland Trough is a west–east bathymetric deep extending 1300 km from the South American continental shelf to the Malvinas Outer Basin in the western South Atlantic (Fig. 1). The trough axis lies 3000–4000 m below sea level from 56 to 41°W, and separates the Falkland Plateau to the north from the North Scotia Ridge (NSR) to the south. The NSR is a component of the Scotia arc, an eastward-closing loop of islands and submarine ridges which connects the Andean Cordillera of South America to the Antarctic Peninsula, and encloses the Scotia Sea (Barker & Dalziel 1983; Barker *et al.* 1991). Previous studies (e.g. Ludwig & Rabinowitz 1982; Barker *et al.* 1991; Cunningham *et al.* 1998) have shown that the northern flank of the ridge constitutes a large and continuous accretionary prism, which formed in connection with presumed mid–late Cenozoic N–S tectonic convergence between the South American and Scotia plates. Platt & Philip (1995) suggested active convergent deformation of Falkland Trough sediments in a foreland tectonic setting at Burdwood Bank, but convergence elsewhere within the trough has ceased.

Methods

We present bathymetric data (Tectonic Map 1985, shown in Fig. 2), seismic reflection profiles, 3.5 kHz echograms, side-scan sonographs and sediment cores collected across the Falkland Trough by the University of Birmingham and British Antarctic Survey (BAS).

Four seismic reflection profiles (located in Fig. 2, and shown in Figs 3a–d) are reproduced in this study: profile SHA734-6 (Fig. 3a) obtained during RRS *Shackleton* cruise 734 (1974) using a single Bolt 1500C airgun source (chamber capacity = 4.8 L) and a six-channel hydrophone streamer (50 m group interval); profiles BAS878-03 and BAS878-04 (Figs 3b,c) obtained during RRS *Discovery* cruise 172 (1987) using a source of 3 Bolt 1500C airguns (combined chamber capacity = 15.1 L) and a 48-channel hydrophone streamer (50 m group interval); and profile BAS923-S26 (Fig. 3d) acquired during RRS *James Clark Ross* cruise JR04 (1993) using a single Bolt 1500C airgun source (chamber capacity = 4.8 L), and a four-channel hydrophone streamer (50 m group interval).

3.5 kHz sub-bottom profiles (track coverage shown in Fig. 2) were collected on BAS cruises between 1984 and 1995. The 3.5

From: STOW, D. A. V., PUDSEY, C. J., HOWE, J. A., FAUGÈRES, J.-C. & VIANA, A. R. (eds)
Deep-Water Contourite Systems: Modern Drifts and Ancient Series, Seismic and Sedimentary Characteristics.
Geological Society, London, Memoirs, **22**, 337–352. 0435-4052/02/$15.00 © The Geological Society of London 2002.

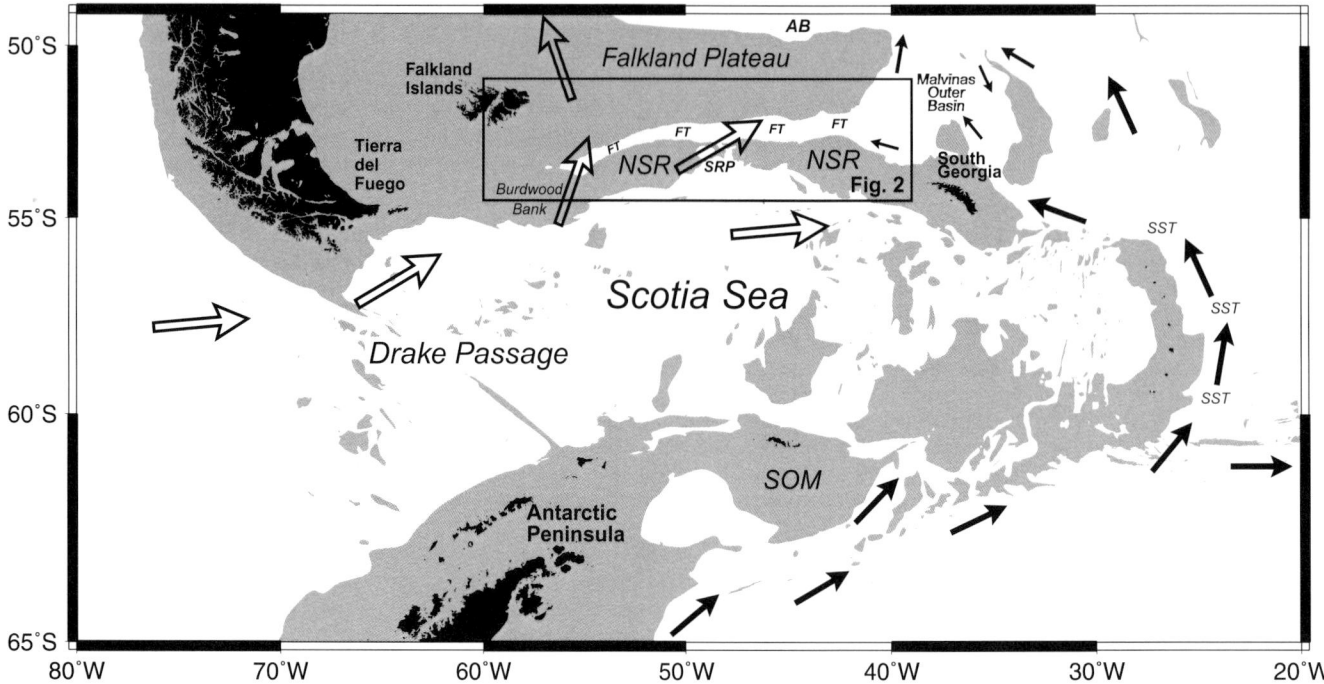

Fig. 1. The western South Atlantic Ocean with areas shallower than 3000 m below sea level shaded. Bottom flow directions described by Gordon (1966), Hollister & Elder (1969), Georgi (1981), Nowlin & Zenk (1988) and Locarnini *et al.* (1993) are shown by open arrows (ACC) and closed arrows (Weddell Sea Deep and Bottom Water). The area shown in Figure 2 is outlined. NSR, North Scotia Ridge; SOM, South Orkney microcontinent; FT, Falkland Trough; SRP, Shag Rocks passage; SST, South Sandwich Trench; AB, Argentine Basin.

kHz data are presented in the form of an interpreted echo character map (Fig. 4) and selected profiles (Figs 5 and 6).

During RRS *Charles Darwin* cruise 37 (1989), BAS conducted the first long-range, side-scan sonar survey of the NSR region using the GLORIA *Mk II* side-scan sonar (Cunningham *et al.* 1998). Side-scan sonographs reproduced here (Fig. 7) have a swath width of 45 km (30 s recording), and survey speeds of 15–18 km h^{-1} resulted in cross-line spatial sampling resolutions of c. 45 m, and in-line resolutions of c. 120 m, degrading to c. 900 m at far range (Searle *et al.* 1990).

BAS has obtained sediment cores from the Falkland Trough in water depths of 1200–3800 m (Howe *et al.* 1997). Five cores (located in Fig. 2, and shown in Figs 8, 9 and 10) are described in this study: KC096 and KC100 (Figs 8 and 10a) from the exposed Falkland Plateau slope and central Falkland Trough, and GC062, KC097 and KC098 (Figs 9 and 10b) from the western Falkland Trough. Sedimentological analyses included particle size analysis, measurement of magnetic susceptibility and smear slide analysis for terrigenous-biogenic ratios.

Objectives

This paper summarises results of recent investigations by Cunningham & Barker (1996), Howe *et al.* (1997), Cunningham *et al.* (1998) and Cunningham (1999). Our principal objectives are: (1) to understand better the effect of vigorous Antarctic Circumpolar Current (ACC) bottom currents and more sluggish Weddell Sea Deep Water (WSDW) flow on sedimentation in the Falkland Trough region, and (2) to describe the sedimentary record of bottom current flow within contourite drift sediments deposited in the Falkland Trough.

Oceanographic setting

The Falkland Trough lies in the path of Circumpolar Deep Water within the ACC, which flows eastward through Drake Passage, but then diverges as sea-floor topography permits: shallow Circumpolar Deep Water flows northeast over the western NSR and Falkland Plateau as the Falkland Current (Peterson & Whitworth 1989; Peterson 1992), but deeper Circumpolar Deep Water is retained within the Scotia Sea as far as Shag Rocks passage, a gap in the NSR at 48°W (Fig. 1). Thereafter, Circumpolar Deep Water flows north through this gap, and then across the eastern Falkland Plateau. A more southerly component of the ACC also flows eastward within the Scotia Sea (Peterson & Whitworth 1989; Grose *et al.* 1995).

Farther south, cold, fresh Weddell Sea Deep Water (potential temperature, –0.7°C to +0.2°C) flows north through gaps in the South Scotia Ridge into the Scotia Sea, where it underlies the ACC at a sub-horizontal interface (Locarnini *et al.* 1993). Weddell Sea Deep Water also flows northward along the South Sandwich Trench, through the Malvinas Outer Basin, and eventually into the Argentine Basin and beyond (Georgi 1981; Mantyla & Reid 1983, and shown in Fig. 1).

Direct oceanographic measurements are sparse in the vicinity of the Falkland Trough. Zenk (1981) and Wittstock & Zenk (1983) reported results of a year-long current meter mooring *M259* located directly on the 3000 m sill of Shag Rocks passage (Fig. 2). Bottom potential temperatures varied between –0.2°C and +0.7°C, and current speeds reached 60 cm s^{-1}; there was a strong correlation between low potential temperature and strong west–northwest bottom-current flow at speeds above 15 cm s^{-1}. On these grounds, Zenk (1981) argued for strong northwestward overflow events involving 'Scotia Sea bottom water', derived from Weddell Sea Deep Water, after mixing with Circumpolar Deep Water within the Scotia Sea. More recently, Locarnini *et al.* (1993) described bottom potential temperature in the Scotia Sea region, and suggested that Weddell Sea Deep Water flowing northward in the Scotia Sea probably fails to reach Shag Rocks passage, since the +0.2°C bottom potential temperature contour which defines its northern limit meets the NSR c. 110 km farther east. Also, re-examination of the data of Zenk (1981) suggested that the cold overflow events detected in the gap were more closely correlated with westward flow than northward flow. On these grounds,

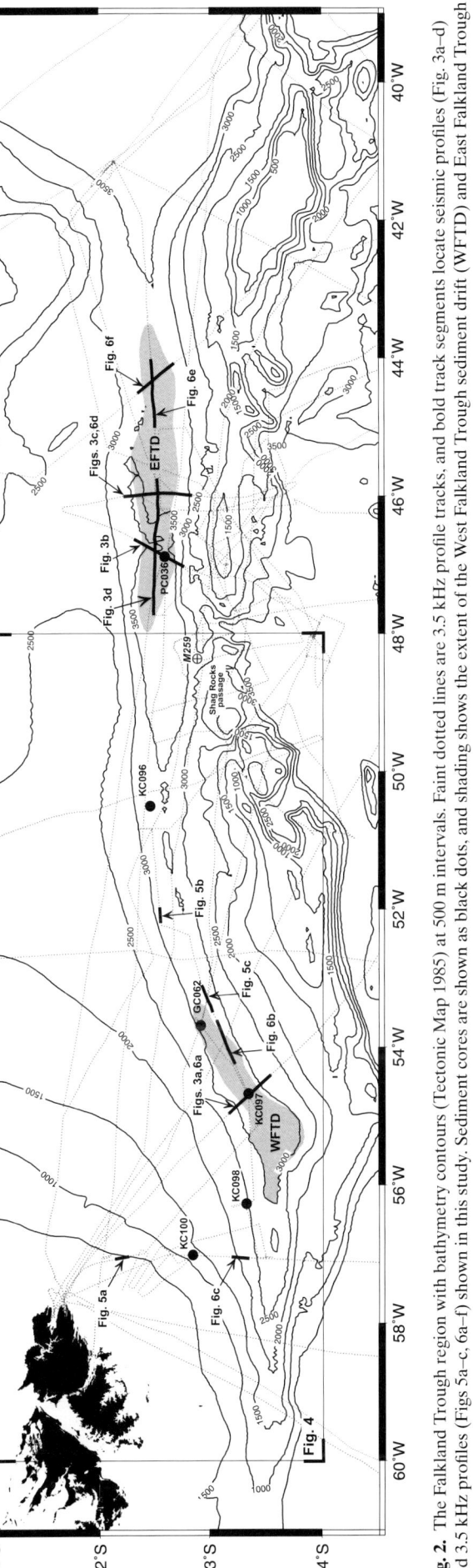

Fig. 2. The Falkland Trough region with bathymetry contours (Tectonic Map 1985) at 500 m intervals. Faint dotted lines are 3.5 kHz profile tracks, and bold track segments locate seismic profiles (Fig. 3a–d) and 3.5 kHz profiles (Figs 5a–c, 6a–f) shown in this study. Sediment cores are shown as black dots, and shading shows the extent of the West Falkland Trough sediment drift (WFTD) and East Falkland Trough sediment drift (EFTD) inferred from 3.5 kHz data. The area of the 3.5 kHz echo character map (Fig. 4) is outlined, and location of the M 259 current meter mooring of Zenk (1981) is marked.

Locarnini *et al.* (1993) inferred westward flow of Weddell Sea Deep Water along the Falkland Trough from the Malvinas Outer Basin, and cast doubt on the existence of northwestward 'Scotia Sea bottom water' flow through Shag Rocks passage. However, there was significant scatter in the aggregated data, and the bottom-current regime in the Falkland Trough remained unresolved.

Bathymetry

The bathymetry of the Falkland Trough (Fig. 2) has been described in previous regional studies (e.g. Lonardi & Ewing 1971; Rabinowitz *et al.* 1978; Tectonic Map 1985). West of 56°W, the Falkland Trough forms a bathymetric depression in the South American continental shelf between Burdwood Bank and the Falkland Islands, which deepens steadily eastward from shelf depth to *c.* 3000 m below sea level (Fig. 2). From 56 to 41°W, the trough reaches depths of 3000–4000 m below sea level, and lies between the Falkland Plateau and the NSR. At its eastern limit, the Falkland Trough opens into the Malvinas Outer Basin, an oceanic basin extending > 4000 m below sea level east of the Falkland Plateau (Fig. 1). Near 48°W, the southern slope of the trough is breached by Shag Rocks passage (Fig. 2), a gap in the NSR with a sill depth of 3000 m.

Seismic characteristics: seismic reflection profiles

Falkland Trough

Sediment drifts. Seismic reflection profiles (Fig. 3) show two mounded sediment bodies in the Falkland Trough, described here as the west Falkland Trough and east Falkland Trough sediment drifts (WFTD and EFTD, located in Fig. 2). The WFTD and EFTD are both isolated, elongate sediment mounds, which are mainly limited to depths of > 3000 m below sea level within the Falkland Trough (mapped extent, Figs 2 & 4). Both sediment bodies lie along the trough axis and show: (1) an elongate, mounded external form, (2) strong acoustic stratification, and (3) marginal bathymetric depressions (e.g. SHA734–6 at 0135 and 0355 h, and BAS878–03 at SPs 2050 and 2750, Fig. 3a, b). The EFTD is the larger of the two mounds, and shows a maximum vertical relief of *c.* 250 m (measured between its apex and the bordering marginal depressions), and a maximum time-thickness on BAS878–03 of *c.* 1.5 s two-way-time (STWT). To the north, the mounds overlie acoustically-stratified Falkland Plateau sediments, and farther south, deformed, acoustically-opaque sediments of the NSR accretionary prism (described below). EFTD and WFTD sediments show diverse internal reflection patterns including subparallel, divergent, undulating and wavy reflections, and migrating-wave acoustic facies (Fig. 3a–d). Deeper EFTD sediments show draped, wavy or undulating reflections which mimic the folds of the underlying NSR accretionary prism (BAS878–04 at 5.4 STWT near SP 1700, and BAS923–S26 at 5.2 STWT near SP 450, Fig. 3c, d). Undulating reflection patterns are also seen in older WFTD sediments (SHA734–6 at *c.* 4.6 STWT near 0245, Fig. 3a). Farther north, EFTD reflections locally onlap older southward dipping Falkland Plateau sediments (BAS878–03 at 5.4 STWT near SP 2680, and BAS878–04 at 5.3 STWT near SP 1240, Fig. 3b, c), although these discordant surfaces cannot be traced far, and some younger EFTD sediments may be coeval with Falkland Plateau strata. Reflection patterns within the mounds also record their growth: faint reflections record the northward migration of the northern flank of the WFTD (SHA734–6 at *c.* 4.3 STWT near 0335 hrs, Fig. 3a), and small but consistent variations in thickness of EFTD units suggest that its axial depocentre has migrated northward with time. Shallow EFTD strata show regular, coherent, laterally continuous

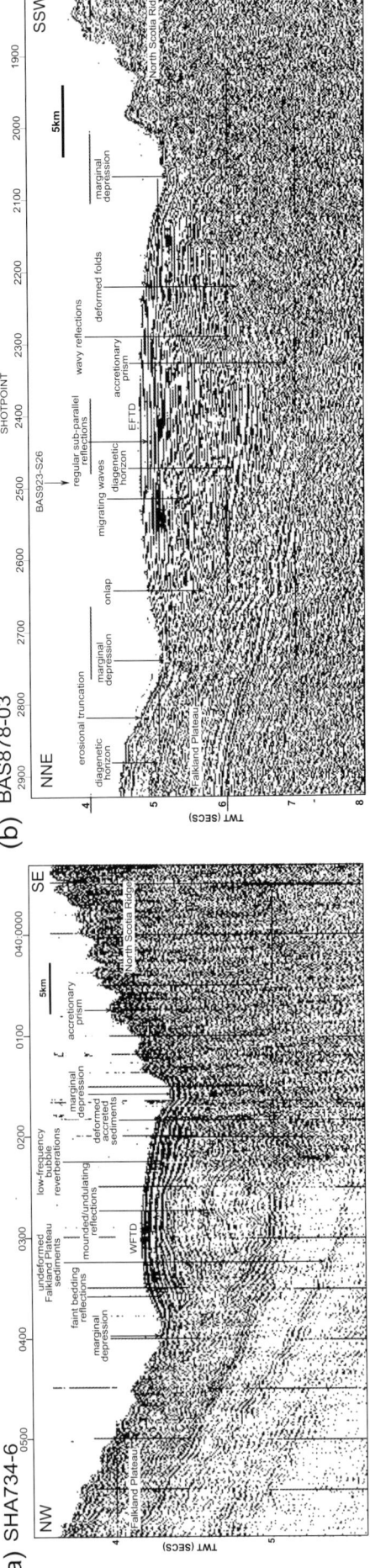
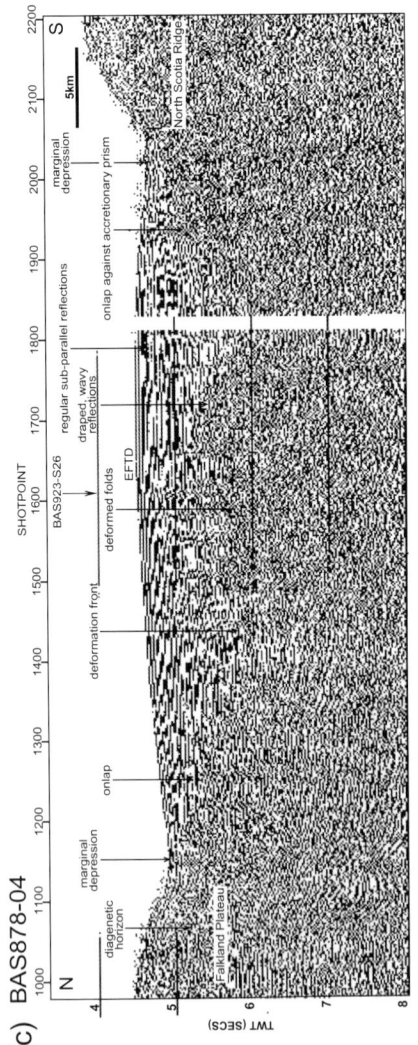

Fig. 3. (a) Single-channel seismic reflection profile SHA734-6 crossing the WFTD (located in Figure 2, vertical exaggeration (VE) = 18:1 at the sea floor). SHA734-6 lies along course 318°T. Profile shown is a bandpass filtered shipboard monitor record. Broad, shallow, high-amplitude reflections which parallel the sea floor beneath the surface of the WFTD are airgun source bubble pulse reverberations; primary geological reflections appear as higher frequency arrivals which cut across these events. These faint reflections record the northward migration of the northern flank of the WFTD. (b) Multi-channel seismic reflection profile BAS878-03 crossing the EFTD (located in Fig. 2, VE = 6:1 at the sea floor). BAS878-03 lies along 020°T and intersects profile BAS923-S26 at SP 2490 (at SP 1658 on BAS923-S26, Fig. 3d). These data were processed to 24-fold common-mid-point stack using standard procedures, and imaged using a Stolt F-K time migration algorithm. (c) Multi-channel seismic reflection profile BAS878-04 crossing the EFTD (located in Fig. 2, VE = 6:1 at the sea floor). BAS878-04 lies along 180°T and intersects 4-channel profile BAS923-S26 at SP 1610 (at SP 380 on BAS923-S26, Fig. 3d). This line crosses the EFTD c. 55 km east of BAS878-03, where the mound has a simpler internal acoustic architecture. Processing as for BAS878-03. (d) Four-channel seismic reflection profile BAS923-S26 obtained along the EFTD (located in Fig. 2, VE = 9:1 at the sea floor). The profile shown is an unmigrated 4-fold shot stack. BAS923-S26 lies along 270°T, close to the long axis of the EFTD (as shown in Fig. 2), and shows discontinuous acoustic facies of the Falkland Plateau and NSR overlain by the acoustically-stratified sediments of the drift. On this E-W strike profile, the base of the EFTD is not well defined, and corresponds to a broad upward increase in reflection continuity.

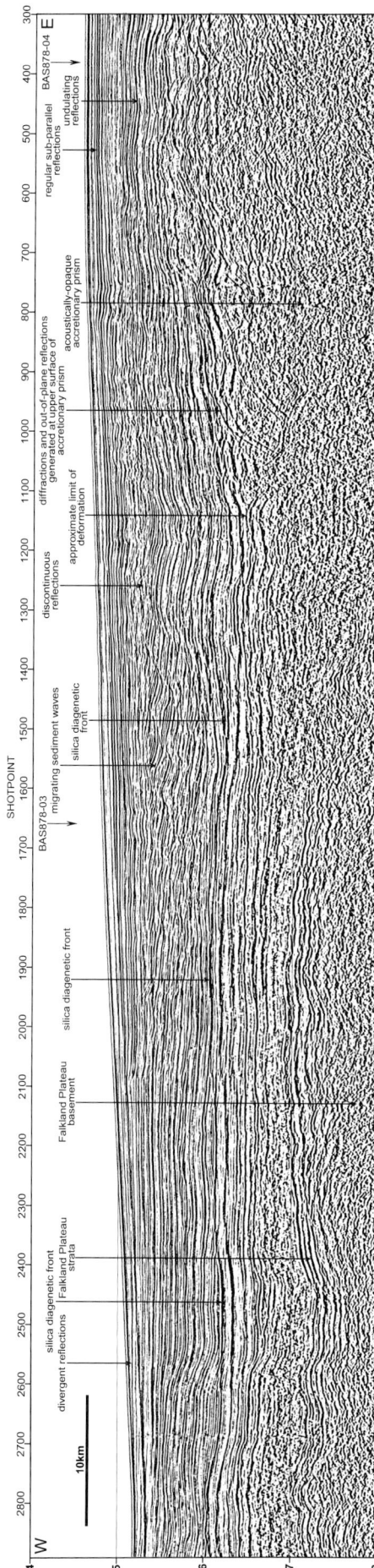

reflections (e.g. BAS923–S26 at 4.7 STWT near SP 530, Fig. 3d), which commonly converge as component layers pinch out near the edge of the mound (e.g. BAS878–03 at 5.1 STWT near SP 2690, and BAS923–S26 at 5.2 STWT near SP 2710, Fig. 3b, d).

Seismic reflection profiles show a sediment wave field buried within the EFTD at 5.0–6.0 STWT (BAS878–03 between SPs 2300 and 2680, and BAS923–S26 between SPs 800 and 2200, Fig. 3b, d). The sediment waves are more clearly imaged on east–west profile BAS923–S26, with apparent wavelengths of 3.8–4.6 km and amplitudes of up to c. 80 m (assuming an acoustic velocity of 1600 m s^{-1}). These buried bedforms show a consistent eastward component of migration.

Profile BAS878–03 also shows a moderately continuous reflection within Falkland Plateau and EFTD sediments (e.g. at 6.1 STWT near SP 2470, Fig. 3b). This often diffuse reflection is time-transgressive, and cuts across primary bedding plane reflections. The origin of this prominent reflection is uncertain, although it has been interpreted as a silica diagenetic front (Cunningham 1999).

Falkland Plateau and North Scotia Ridge

To the north, WFTD and EFTD sediments overlie southward-dipping Falkland Plateau strata. Older Falkland Plateau strata ('Falkland Plateau Lower Sequence' of Cunningham 1999) show high-amplitude, laterally continuous reflections, whereas younger Falkland Plateau strata ('Falkland Plateau Upper Sequence' of Cunningham 1999) show diverse reflection patterns, often with poor lateral reflection continuity. Profile BAS878–03 shows erosional truncation of these younger strata at the sea floor (near SP 2817, Fig. 3b). To the south, WFTD and EFTD sediments overlie deformed sediments of the NSR accretionary prism. These sediments are acoustically opaque, with few laterally persistent reflections.

Seismic characteristics: 3.5 kHz high-resolution acoustic profiles

3.5 kHz sub-bottom profiles (track coverage, Fig. 2) have been used to map sea-floor echo character across the western Falkland Trough (Howe *et al.* 1997, and Fig. 4). Profiles reproduced here (located in Fig. 2, shown in Figs 5 and 6) show geological features which reflect strong bottom current control of deposition including current-eroded depressions, areas of hard, sediment-free and thinly sedimented sea floor, sediment drifts and migrating sediment waves.

Falkland Trough

Sediment drifts. 3.5 kHz profiles crossing the WFTD and EFTD show laterally persistent reflections to the limit of acoustic penetration beneath the apex of each mound, which converge as component layers pinch out toward the flanking depressions (Fig. 6a, d). WFTD and EFTD strata also pinch out toward Shag Rocks passage (e.g. Fig. 6b), where we observe a hard sea floor (Fig. 4), with only a thin and patchy cover of recent sediment. Existing 3.5 kHz profiles show that the WFTD is up to 20 km wide and > 130 km long, and that the EFTD is up to 46 km wide and 302 km long (Fig. 2).

Migrating sediment waves. 3.5 kHz profiles (Figs 2, 6e and f) crossing the Falkland Trough between 44 and 45°W show migrating sediment waves at the surface of the EFTD. Correlation of five prominent sediment wave crests between the profiles (M1, M2, M4, M5 and M6, Fig. 6e, f) reveals consistent NE–SW crestal alignments between 045°T and 051°T, true wavelengths of 2.0–5.0 km and peak-to-trough heights of 10–95 m (Cunningham

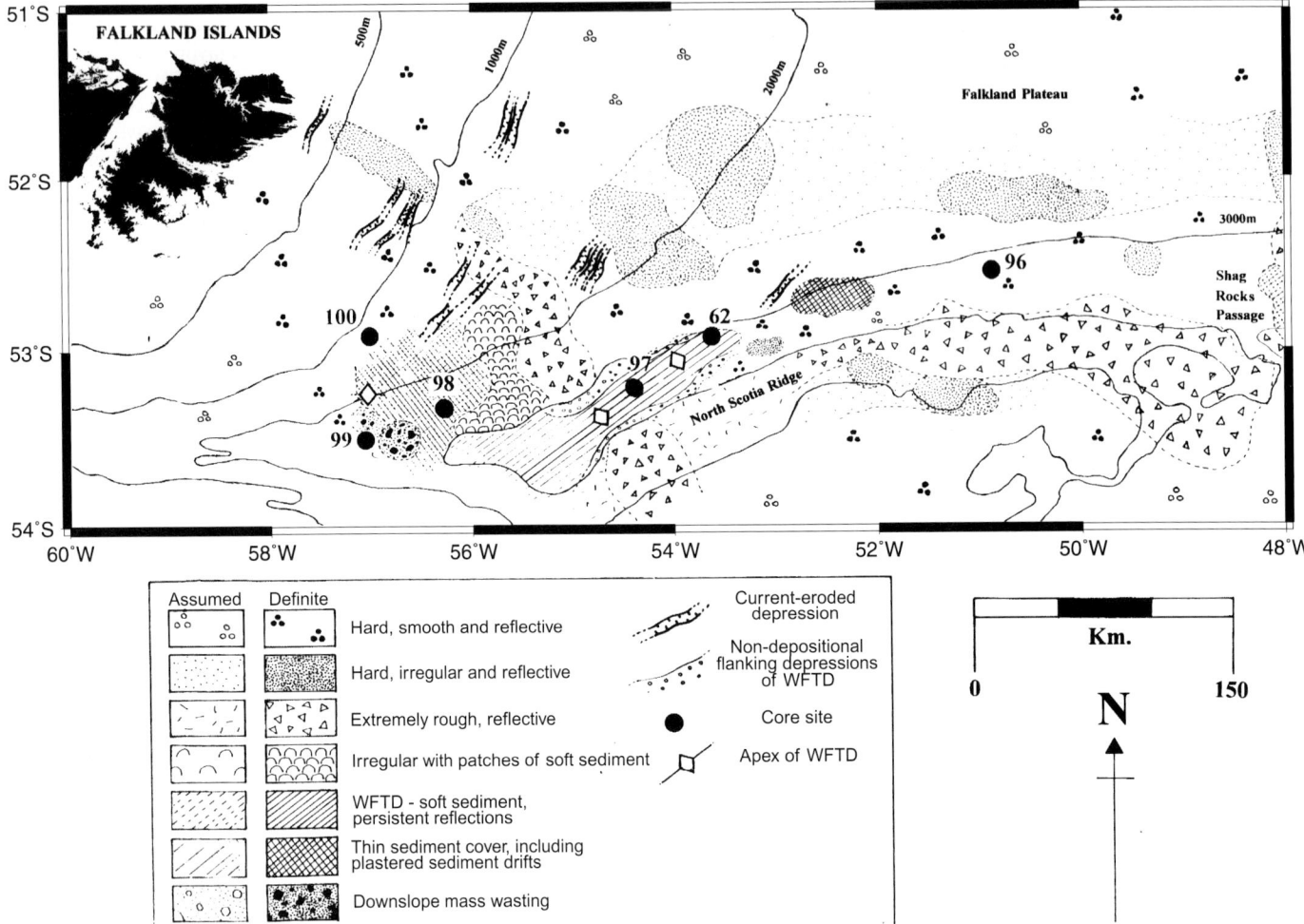

Fig. 4. Echo character map of the western Falkland Trough, compiled from 3.5 kHz sub-bottom profiles (Howe *et al.* 1997). 3.5 kHz track distribution is shown in Figure 2.

& Barker 1996). On both profiles, the sediment waves show varying degrees of asymmetry, but bedform migration directions remain consistently toward the south and east. There is no evidence of ponding within the sediment wave troughs or recent erosion of higher accumulation sediment wave flanks, which suggests that these bedforms are probably active.

Falkland Plateau and North Scotia Ridge

Current-eroded depressions. 3.5 kHz profiles crossing the Falkland Plateau slope show occasional large, asymmetric depressions. These features (located in Fig. 4, example shown in Fig. 5a) are commonly 35–150 m deep, up to 1.5 km wide, and can be traced up to 40 km alongslope. 3.5 kHz mapping (Fig. 4) shows that they are commonly aligned with the regional bathymetric contours. SCS profiles (e.g. fig. 6 of Cunningham *et al.* 1998) show no evidence of glide planes beneath the depressions, and they are unlikely therefore to result from slope failure. Hence, we attribute these features to erosion by bottom currents flowing within the ACC.

Hard, sediment-free and thinly sedimented sea floor. 3.5 kHz profiles also show areas of thin recent sediment cover (or entirely lacking recent sediment) on the southern Falkland Plateau slope (hard sea-floor echo character, Fig. 4), where bottom currents preserve fault scarps at the sea floor (Fig. 6b), and on the central Falkland Trough floor (Fig. 5c). Farther south, bottom currents also preserve deformational fabric of the NSR accretionary prism, which is commonly represented by hyperbolic echoes (extremely rough, reflective sea floor, Fig. 4), with little evidence of recent sedimentation (e.g. southernmost part of profile shown in Fig. 6a).

Plastered drifts. 3.5 kHz profiles show current-influenced sediments deposited on the lower Falkland Plateau slope (thin sediment cover, Fig. 4). These small, asymmetric sediment bodies have the form of a plastered sheet, and show regular, fairly continuous reflections extending > 50 m below the sea floor (Fig. 6c). We have traced these features 3 to 5 km downslope, and up to 15 km alongslope.

Seismic characteristics: side-scan sonographs

GLORIA long-range side-scan sonographs record acoustic energy back-scattered from the seabed. This may be influenced by both the large-scale topography of the sea-floor, and by its texture or roughness. Long, narrow, strongly reflecting lineaments in sonographs commonly represent fault scarps or folds. Rough, unsedimented sea floor yields strong backscatter, whilst heavily sedimented areas appear much darker. Figure 7 shows a digital mosaic of GLORIA sonographs crossing the western Falkland Trough, plotted so that areas of low backscatter appear dark, and areas of high backscatter appear bright.

Falkland Trough

Sediment drifts. The WFTD is represented in GLORIA sonographs by a featureless area of extremely low backscatter which

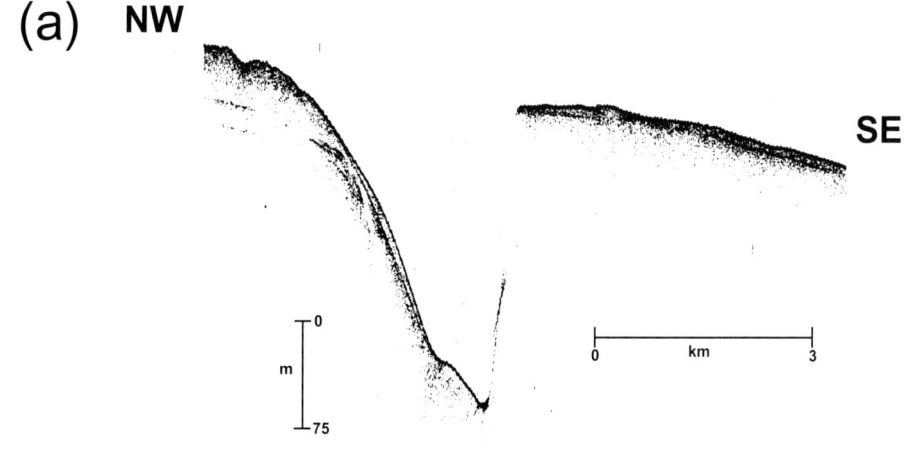
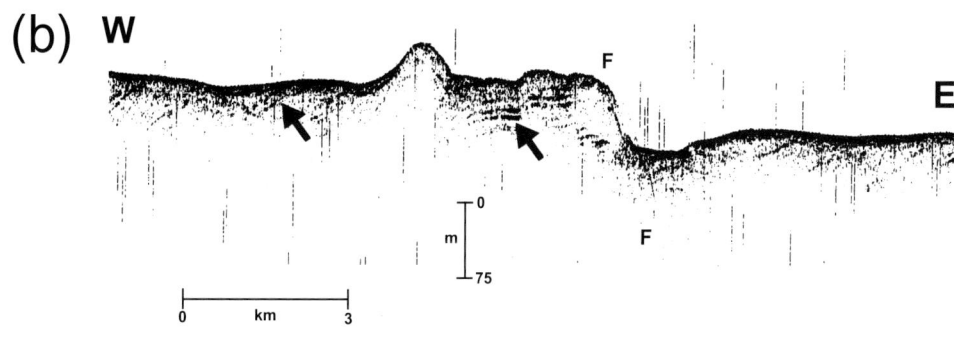
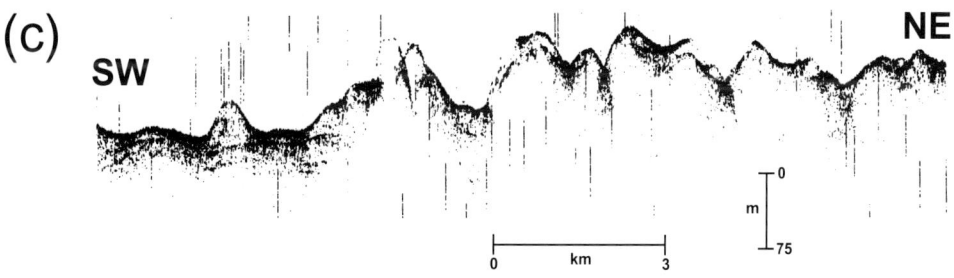

Fig. 5. 3.5 kHz profiles (located in Fig. 2) showing sediment-free and thinly sedimented sea floor in areas of strong current control (Howe *et al.* 1997). (**a**) Eroded asymmetric depression on the Falkland Plateau slope. (**b**) Rough sea-floor topography on the southern Falkland Plateau slope with exposed, contorted bedrock strata (labelled with arrows). F–F, fault scarp. (**c**) Rough sea-floor topography on the Falkland Trough floor, with some slight sediment cover in the lee of ridges.

contrasts strongly with the bright, textured sea floor of the adjacent Falkland Plateau slope and NSR (Fig. 7). The very low backscatter is consistent with the smooth, well-sedimented sea floor apparent in 3.5 kHz profiles (e.g. Fig. 6a). At its eastern limit, the low backscatter representing the WFTD fades steadily in sonographs, where 3.5 kHz profiles (Fig. 5b) show thinning and pinch out of the shallowest drift sediments toward Shag Rocks passage. Here also, sonographs show that the WFTD does not occupy the deepest part of the Falkland Trough, but lies instead on the southern Falkland Plateau slope (Fig. 7).

Falkland Plateau and North Scotia Ridge

Sonographs crossing the Falkland Plateau slope show a strongly textured sea floor, with broad areas of moderate backscatter, adjacent shadow and prominent lineaments. Between 800 and 1400 m below sea level, sonographs show very bright lineaments (Fig. 7) which correspond to current-eroded depressions observed on 3.5 kHz profiles (e.g. Fig. 5a). At greater depth, sonographs show broad, diffuse areas of moderate backscatter and adjacent shadow (Fig. 7); although the bright areas of sea-floor which form

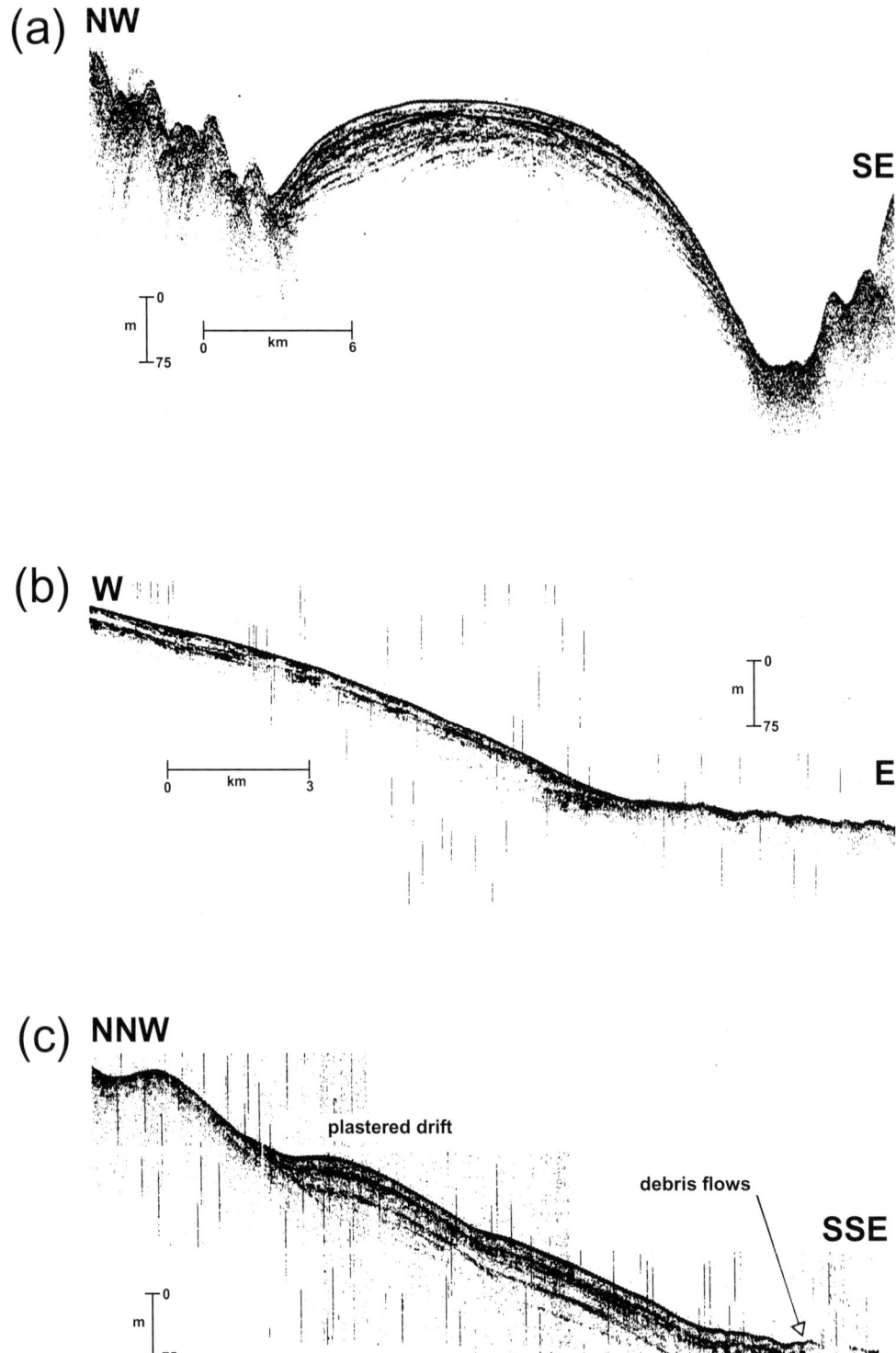

Fig. 6. 3.5 kHz profiles (located in Fig. 2) showing drift sedimentation and migrating sediment waves in the Falkland Trough (Cunningham & Barker 1996; Howe *et al.*, 1997; Cunningham 1999). (**a**) Profile acquired with SHA734–6 (Fig. 3a) across the West Falkland Trough sediment drift. (**b**) Profile showing the relationship between the eastern (distal) edge of the West Falkland Trough sediment drift, and the current-swept floor of the Falkland Trough in the vicinity of Shag Rocks passage. (**c**) Plastered drift sediments on the southern Falkland Plateau slope with laterally-equivalent down-slope debris flows encroaching on the floor of the Falkland Trough. (**d**) Profile acquired with BAS878–04 (Fig. 3c) across the East Falkland Trough sediment drift. (**e**) East–west profile showing migrating sediment waves at the surface of the East Falkland Trough sediment drift. M1–M6 locate bedforms described in the text. The point of intersection with the profile shown in (f) is marked with a vertical arrow. (**f**) Northwest–southeast profile showing migrating sediment waves at the surface of the East Falkland Trough sediment drift. The point of intersection with the profile shown in (e) is marked with a vertical arrow.

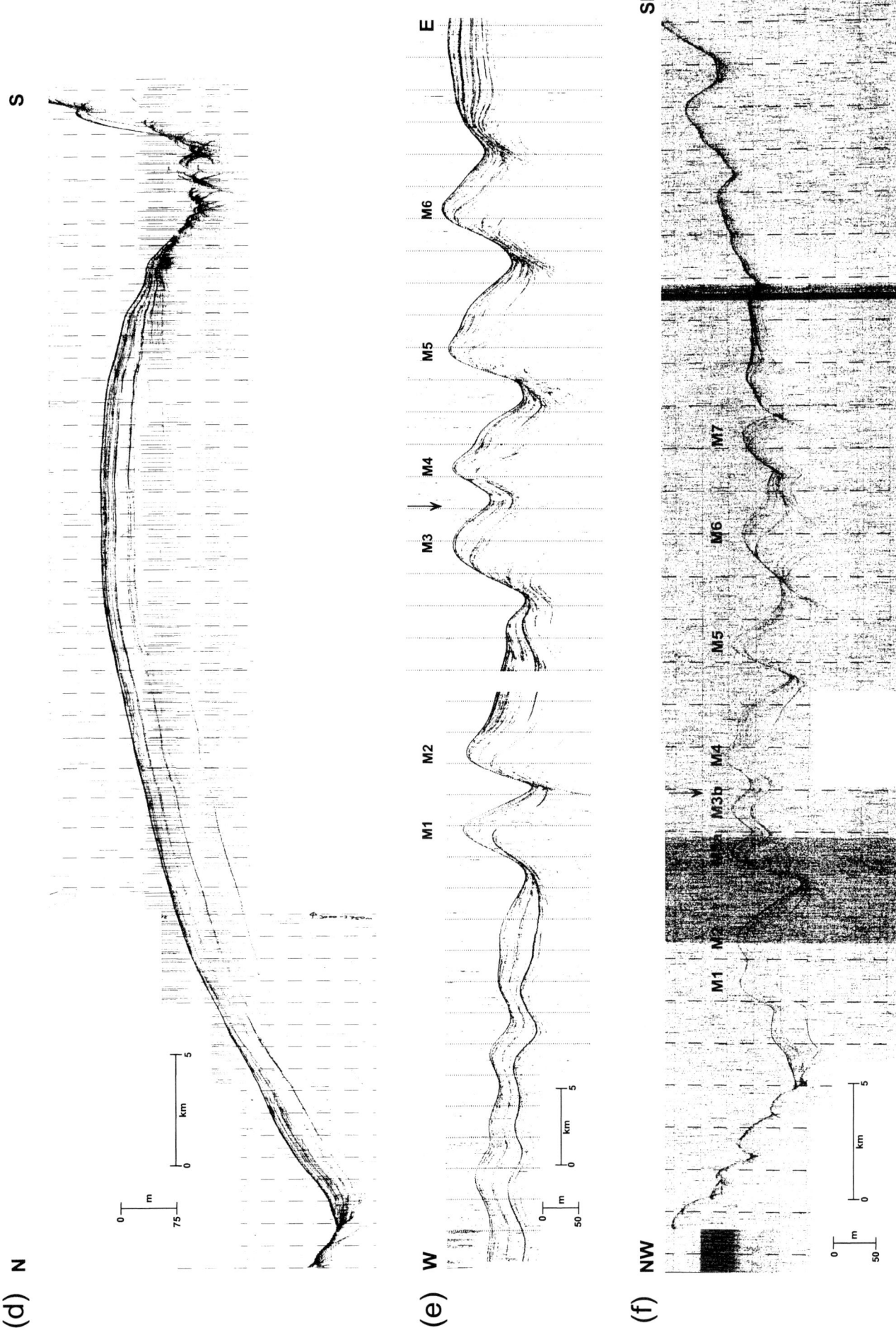

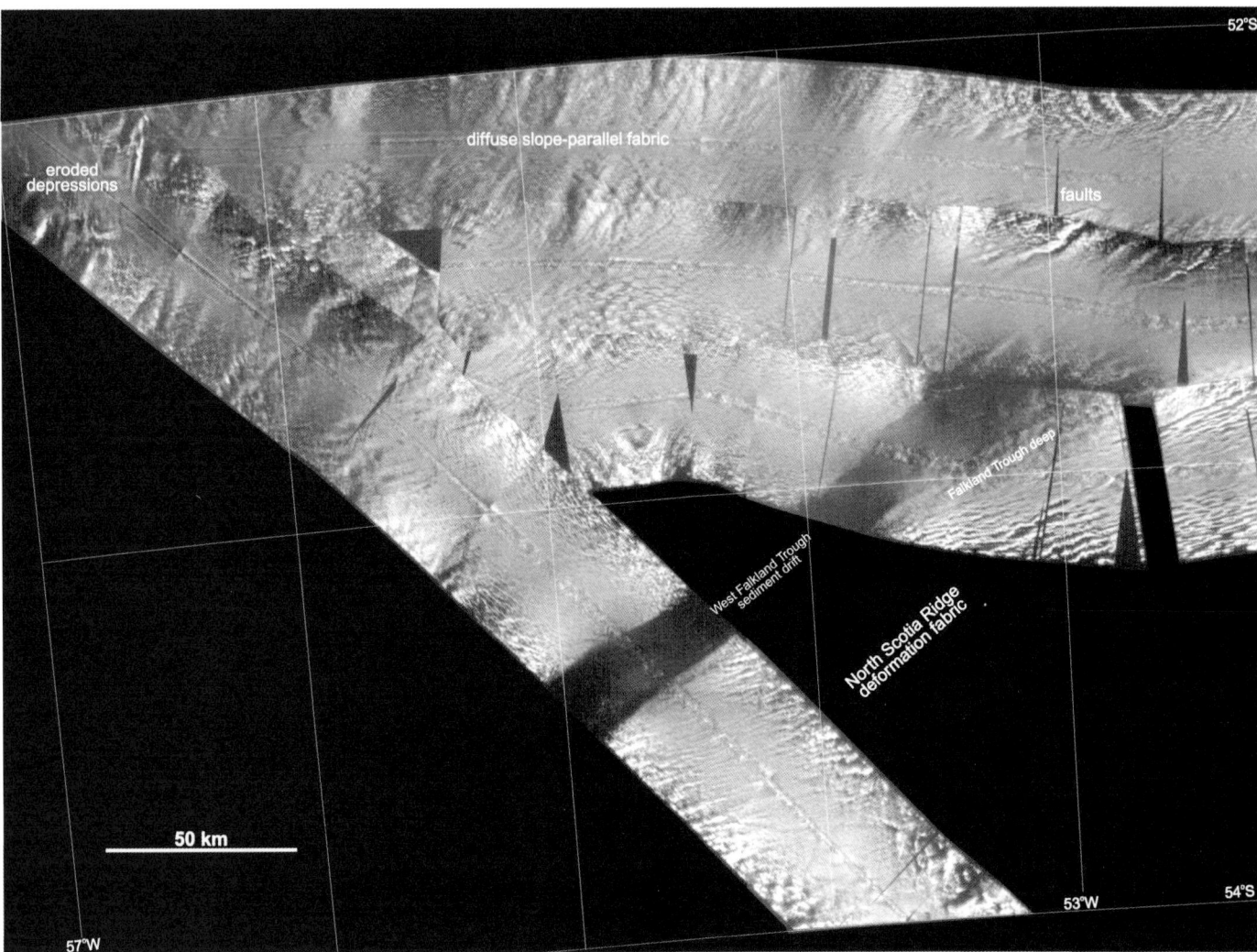

Fig. 7. Digital mosaic of GLORIA sonographs obtained across the western Falkland Trough (Howe *et al.* 1997; Cunningham *et al.* 1998). These data have been corrected for ray path geometry (slant-range correction), variations in ship speed (anamorphic correction), and beam and refraction effects (shading correction: Searle *et al.* 1990). Coincident sonographs were digitally edited and combined using Mini Image Processing Software (Chavez 1986), and a linear contrast stretch was applied before display.

this fabric have poorly defined, gradational edges, the fabric still appears to be aligned with the bathymetric contours. This diffuse fabric is represented on seismic and 3.5 kHz profiles only by variations in surface roughness; if it is influenced by sea-floor geology, it is on a scale smaller than seismic profiles display.

Farther east, prominent E–NE striking lineaments can be traced over 25 km along the Falkland Plateau slope at 2500 to 3500 m depth (Fig. 7). 3.5 kHz data (Fig. 5b) show that one of these lineaments corresponds to a fault scarp (*c.* 75 m height).

Sonographs crossing the northern flank of the NSR show a prominent striped acoustic fabric at depths of > 2000 m below sea level, characterized by very bright, sub-parallel lineaments which are closely aligned with the bathymetric contours (Fig. 7). This fabric represents backscatter from open, symmetric to gently asymmetric folds at the surface of the NSR accretionary prism. Fold wavelengths are 1 to 4 km, and individual folds can be traced in sonographs over 20 km.

Core stratigraphy and sedimentology

In this study, we describe five sediment cores collected in water depths of between 1500 and 3500 m: KC096, KC097, KC098, KC100 and GC062 (Howe *et al.* 1998, located in Fig. 2). Cores collected from the elevated Falkland Plateau slope (KC100, Figs 8 & 10a) and central Falkland Trough (KC096, Fig. 8) contain comparatively high proportions of sand, whereas cores from more sheltered parts of the western Falkland Trough (GC062, KC097 and KC098, Figs 9 & 10b) consist mainly of finer grained sediments. We first describe cores from the elevated Falkland Plateau slope and central Falkland Trough, followed by those from the western Falkland Trough.

Sediment cores from the elevated Falkland Plateau slope and central Falkland Trough

KC096 and KC100 (Figs 8 and 10a) core tops are foraminiferal sands which overlie glauconite-rich sands, interbedded with muds and sandy muds. The glauconite-rich sands contain dropstones in the more easterly core KC096. Relative age dating was undertaken using Last Appearance Datum (LAD) of radiolaria and diatoms (Fig. 8). For core KC096, two LADs indicate ages of 0.2–0.4 Ma at 1.20–1.75 m core depth. For core KC100, four LADs provide ages of 0.8–1.9 Ma from 1.6–2.4 m core depth. These data suggest sedimentation rates of < 1 cm ka^{-1}.

Foraminiferal sands. All core tops described in this study are pale cream-coloured, foraminifera-rich sands (additional thinner foraminifera-rich horizons are also present within core KC096,

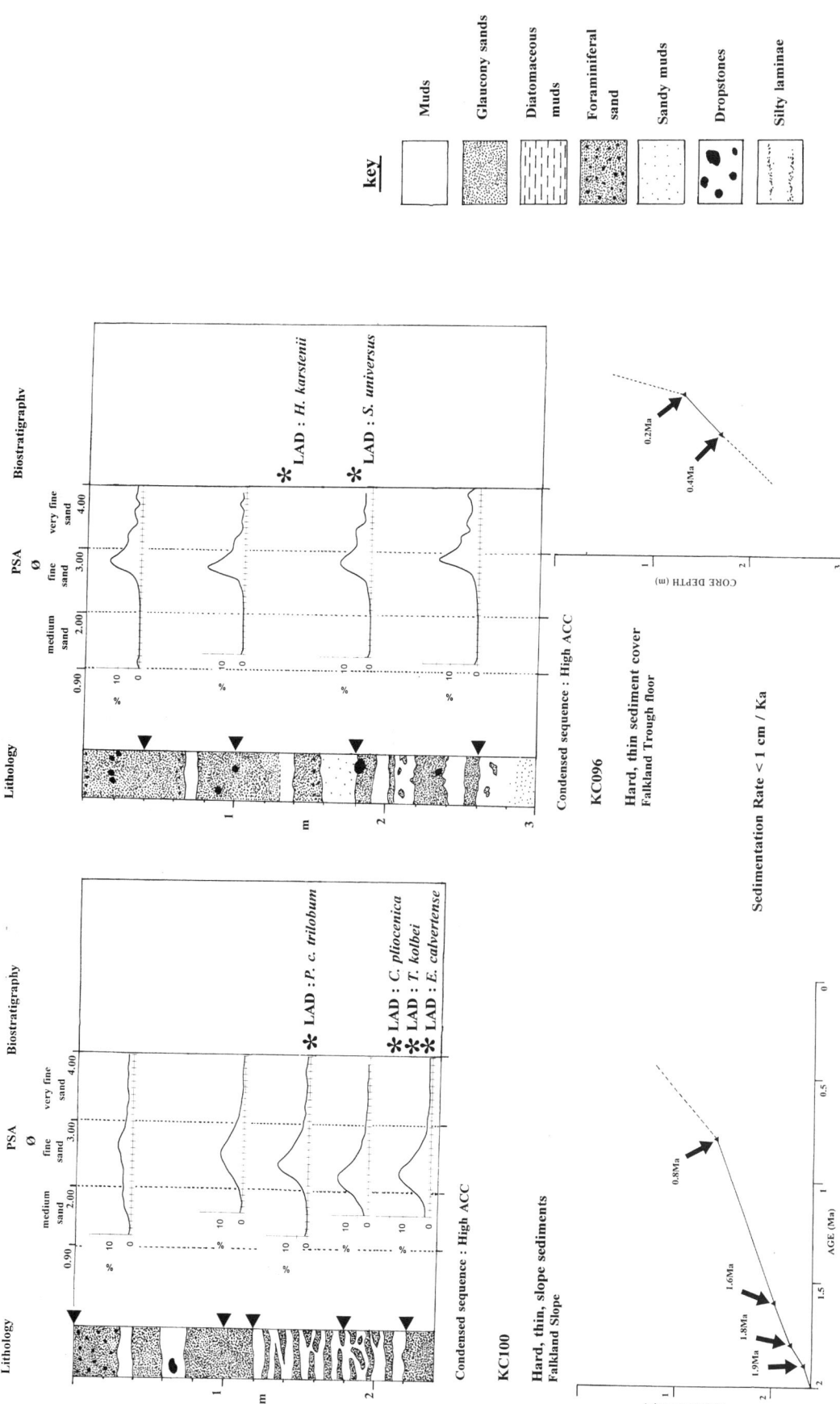

Fig. 8. Selected particle size frequency curves for cores KC096 and KC100 from the elevated Falkland Plateau slope and central Falkland Trough (located in Fig. 2). Biostratigraphic dating indicates sediments extending down to the Pliocene in KC100, and to the Mid-Pleistocene in KC096. Last Appearance Datum (LAD) positions of radiolaria are labelled. For core KC096: *Hemidiscus karstenii* LAD at 0.2 Ma, and *Stylatractus universus* at 0.4 Ma. For core KC100: *Pterocanium charybdeum trilobum* LAD at 0.8 Ma, *Cycladophora pliocenica* LAD at 1.6 Ma, *Thalassiosira kolbei* LAD at 1.8 Ma and *Eucyrtidium calvertense* LAD at 1.9 Ma.

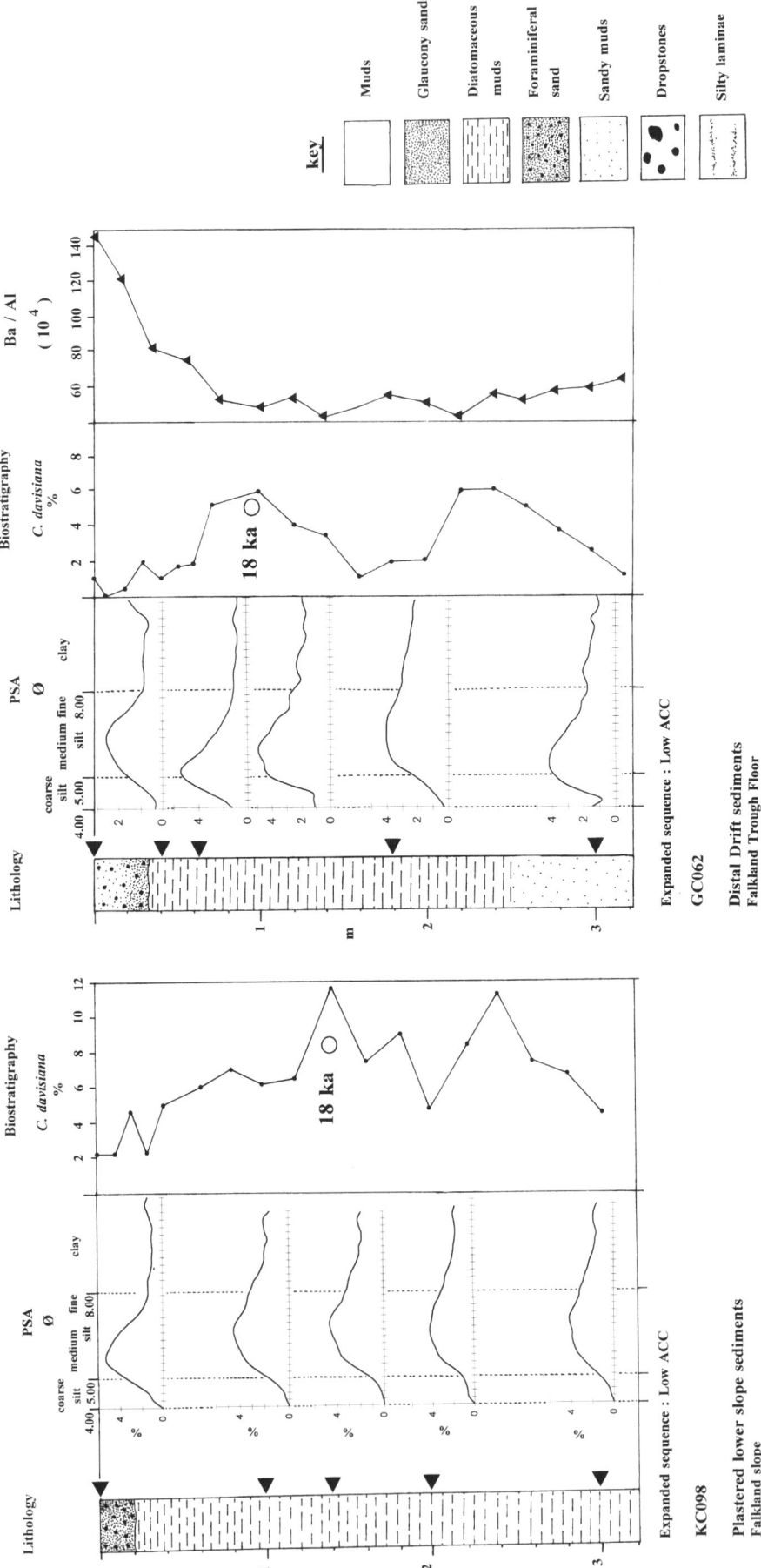

Fig. 9. Particle size frequency curves and core logs of cores KC098 and GC062 from the western Falkland Trough (located in Fig. 2). Also shown are *Cycladophora davisiana* curves (2–12%) for both cores and the Ba/Al ratio for core GC062. Ba/Al decreases downcore at 0.4 m; this decrease in palaeoproductivity is interpreted as the base of the Holocene. The radiolarian *Hemidiscus karstenii* LAD (present in the more condensed cores shown in Fig. 8) is absent here, suggesting sediments younger than 0.2 Ma.

Fig. 8). This lithofacies ranges in thickness from 10 to 40 cm and is totally homogeneous, a result of intense bioturbation. Lower contacts with the muds and diatomaceous muds are bioturbated and irregular. In KC096 and KC100, the sands are medium to fine, and moderately sorted with a mode between 2 and 3 Ø. Carbonate content varies within these units from 42 to 84%. The sands contain 60% biogenic material (mainly foraminifera, with some 10% diatoms, and radiolarians) and 40% terrigenous material. Howe *et al.* (1997) interpreted these sediments as sandy contourites.

Glauconite-rich sands. The glauconite-rich sands are greenish-black in colour and range in thickness from 10 to 50 cm. These sediments are heavily bioturbated with large *Planolites* and *Chondrites* burrow systems (Fig. 10a). Contacts with sediments above and below are bioturbated or sharp. The sands are fine to very fine grained, and very well sorted, with a mode at 2.5 Ø (Fig. 8). No internal structures are visible, apart from rare muddier lamination, with thin horizons of dropstones. The glauconite is very well-rounded and may be produced partly by contemporary authigenic formation on the slope and partly through erosion of outcropping Cretaceous and Tertiary glauconite-rich strata by strong bottom currents. Howe *et al.* (1997) interpreted these sediments as sandy contourites.

Muds. The greenish-grey homogeneous muds range in thickness from 5 to 60 cm, and occur as bioturbated units within the sandy contourite sequences (Fig. 8). Large *Chondrites* burrow systems are again evident. The muds are moderately to poorly sorted with bimodal grain size distributions in the sand-silt range (2–4 Ø and 5–7.5 Ø). Some silty lamination is present, and dropstones, diatoms and radiolarians occur throughout. Howe *et al.* (1997) interpreted these sediments as muddy contourites.

Sediment cores from the western Falkland Trough

GC062, KC097 and KC098 (Figs 9 and 10b) core tops are foraminiferal sands (described above) which overlie diatomaceous muds. In KC098, the diatomaceous muds extend to the base of the core. In GC062, diatomaceous muds extend to about 2.5 m, and overlie sandy muds which extend to the base of the core. Diatom assemblages and Ba/Al ratios suggest that both cores are in the *Thalassiosira lentiginosa* zone (0–0.6 Ma). The abundance of *Cycladophora davisiana* increases downcore, with the first abundance peak in each core correlated to the Last Glacial Maximum at 18 ka (Fig. 9). Sedimentation rates average 3 to 4 cm ka^{-1} on the WFTD (Howe *et al.* 1997).

Diatomaceous hemipelagic muds. These bioturbated, laminated to homogeneous, diatomaceous, green silty muds occur below the foraminiferal sands. Diatom content varies from 27 to 52% in core GC062, but is lower in more westerly cores from the WFTD (8 to 10% in core KC097: Howe *et al.* 1997). Diatom assemblages are dominated by *Fragilariopsis kerguelensis* and resting spores of *Chaetoceros* spp. Particle size curves display moderately to poorly sorted sediment with a silt mode between 6 and 7 Ø (Fig. 9). Monosulphidic patches are common throughout. The poor sorting, fine grain size, and high diatom content suggests deposition within a low energy, weak current influence, hemipelagic environment.

Sandy muds. These sediments occur as bioturbated units at the base of some drift cores (e.g. core GC062, Fig. 9). Grain size distributions are similar to the diatomaceous muds, but with a larger component (up to 10%) of fine terrigenous sand. The distributions are bimodal, with a silt peak between 5 and 6 Ø. Howe *et al.* (1997) suggested that these sediments were deposited in the presence of episodic ACC bottom currents.

Discussion

Although the ACC is principally a wind-driven current, previous studies demonstrate the dominant influence of ACC bottom flow on sedimentation. Seismic reflection profiles crossing Drake Passage (Barker & Burrell 1977) and the northern Scotia Sea (Pudsey & Howe 1998; 2002) show sediment accumulation in isolated banks, under strong bottom current control, and piston cores (Burckle & Hays 1974; Saito *et al.* 1974; Ciesielski *et al.* 1982) show pre-Quaternary sediments widely exposed at the sea bed on the Falkland Plateau. Similarly, DSDP cores provide evidence of prolonged periods of non-deposition and erosion on the Falkland Plateau since the inception of the ACC in the early Miocene (Barker *et al.* 1976). Marine geophysical and sediment core data presented in this study show widespread current control of deposition in this region, and reveal two styles of contourite sedimentation: (1) deposition of glauconite-rich sandy contourites on the Falkland Plateau slope and central Falkland Trough floor where vigorous ACC bottom currents control sedimentation; and (2) deposition of biogenic sandy contourites, muddy contourites and finer-grained hemipelagites (western Falkland Trough), and muddy diatom ooze (eastern Falkland Trough), where thick drift deposits have accumulated in the presence of more sluggish southern-origin bottom water, or deepest ACC flow.

Current-influenced sedimentation and erosion on the Falkland Plateau slope and central Falkland Trough floor

Sediment cores KC096 and KC100 (Figs 8 and 10a) sampled highly condensed, glauconitic sandy contourites. Howe *et al.* (1997) suggested that these sediments correspond to the coarse components of the contourite facies model of Stow & Piper (1984), and that deposition occurred in the presence of high velocity, sustained bottom currents which winnowed away the fines. ACC bottom flow in this region may be accelerated by interaction with the sea-floor topography of the Falkland Plateau and NSR.

Side-scan sonographs of the Falkland Plateau southern slope show a textured sea-floor, with prominent lineaments which correspond to large, steep-sided eroded depressions. One prominent lineament can be traced for 17 km in sonographs, and corresponds to an asymmetric depression, apparently created by differential erosion at the thinning up-slope edge of a sediment sequence that is widespread over the Falkland Plateau (Cunningham *et al.* 1998). The agent of erosion is the ACC, although seismic profiles (fig. 6 of Cunningham *et al.* 1998) suggest that the locus of the depression is controlled in part by underlying geology. We suspect that finer, diffuse slope-parallel fabric on the Falkland Plateau is also current-controlled. Sonographs of the Falkland Trough near Shag Rocks passage also show moderate to high backscatter, and such lineaments as are seen are characteristic of the Falkland Plateau. By inhibiting sedimentation, the ACC also preserves tectonic fabric across the elevated NSR.

Drift sedimentation in the Falkland Trough

Thick accumulations of fine-grained sediment are best preserved in sheltered parts of the Falkland Trough where bottom currents have remained sufficiently sluggish to permit prolonged pelagic-hemipelagic sedimentation. In particular, we describe two large elongate sediment mounds in the trough, the WFTD and EFTD, and smaller isolated drifts on the Falkland Plateau southern slope.

West Falkland Trough sediment drift. The WFTD lies within the Falkland Trough west of 53°W, at depths of > 2900 m below sea level. Sediment cores from the drift (e.g. GC062, Fig. 9) contain a 0.1–0.4 m layer of foraminiferal sand, underlain by fine-grained diatomaceous hemipelagites. These incorporate a significant (up to 55%) terrigenous sediment component, probably reworked and transported down-slope from Burdwood Bank, the Falkland Islands and southern South America, with a small proportion of ice-rafted material.

Although WFTD sediments incorporate a significant terrigenous component, the depositional geometry of the mound suggests that ambient bottom currents, rather than down-slope flows, have maintained a controlling influence on its formation. In particular, its internal acoustic architecture and mounded, gently asymmetric external form resembles that of contourite sediment drifts reported in studies of the Mediterranean and Sunda forearc (Reed et al. 1987; Marani et al. 1993). Near 54°30'W, the drift flanks are symmetrically disposed within the trough, and reflections showing thinning and pinch out suggest intensified bottom current flow at each drift margin. These characteristics may be explained by sedimentation within a cyclonic loop of Circumpolar Deep Water, with westward flow at the southern margin of the drift, and eastward flow at its northern margin; similar flow has been inferred at shallower depths by Piola & Gordon (1989). Alternatively, WFTD sediments may have been deposited within a loop of Weddell Sea Deep Water, as proposed for the EFTD farther east (Cunningham & Barker 1996, and below). However, this is considered less likely, since thick drift sediments do not extend continuously along the Falkland Trough, and Weddell Sea Deep Water appears confined within the eastern Falkland Trough by the ACC (Locarnini et al. 1993).

Near its eastern limit, the drift lies not within the deepest part of the trough, but instead extends along the base of the Falkland Plateau slope. Also, it progressively thins and widens, and adopts the form of a thin plastered sheet. We attribute this change in depositional style to an eastward increase in bottom current activity, acting on a relatively small volume of terrigenous sediment transported downslope.

East Falkland Trough sediment drift. The EFTD lies in the Falkland Trough between 43 and 48°W, at depths of > 3000 m below sea level. Piston core PC 036 (Jordan & Pudsey 1992, located in Fig. 2) shows that youngest EFTD drift sediments are predominantly biogenic, with an average diatom content of 70 to 80%, 10% carbonate near the top and base of the core, and a small (partly ice-rafted) terrigenous component. This contrasts with the substantial terrigenous component sampled in the WFTD, and reflects increased biogenic productivity in the vicinity of the Antarctic Polar Front, and a virtual absence of local terrigenous sources. EFTD sediments have a likely sedimentation rate of 5 cm ka^{-1} (Shimmield et al. 1993).

The EFTD shows characteristics which indicate deposition in the presence of ambient bottom currents. In particular, the internal acoustic architecture and mounded external geometry of the drift reflect current control of sedimentation, with intensified flow at its margins. Although the EFTD lies beneath the path of the ACC, oceanographic data show that the eastern Falkland Trough is occupied by Weddell Sea Deep Water, which has been sampled along the trough as far west as 47°56'W (R. Locarnini, pers comm. 1993). Hence, the EFTD is presently maintained by bottom currents associated generally with the northward transit of southern-origin bottom water. We suggest that non-deposition at the southern drift margin is consistent with the westward flow of Weddell Sea Deep Water along the trough from the Malvinas Outer Basin, as suggested by Locarnini et al. (1993). However, the fate of the west-flowing Weddell Sea Deep Water remains uncertain, and non-deposition at the northern drift margin could be caused by the eastward flow of returning Weddell Sea Deep Water, or by the eastward flow of Circumpolar Deep Water within the ACC.

Migrating sediment waves. Acoustic profiles reveal two separate sediment wave fields within the EFTD: active sediment waves at

Fig. 10. (a) Photograph of core KC100 showing bioturbation. (b) X-radiographs of core KC097 showing lamination partly obliterated by bioturbation.

the drift surface near 44°30'W (Figs 6e & f), and sediment waves buried at intermediate depth within the drift near 47°W (Figs 3b & d). Sediment waves have been associated with ambient thermohaline flow across sediment drifts (Flood et al. 1993) and turbidity current flow across deep-sea channel levees (e.g. Normark et al. 1980; Carter et al. 1990). At present, no cores have been obtained from the sediment waves in the eastern Falkland Trough, although on the basis of their regional setting, we believe they formed in connection with ambient bottom current flow. Cunningham & Barker (1996) related the active sediment waves at the drift surface to westward flow of Weddell Sea Deep Water along the Falkland Trough.

History of Antarctic Circumpolar Current flow

The Falkland Trough sediment drifts contain a depositional record which is sensitive to variations in ACC and southern-origin bottom water flow through the glacial cycle. In particular, cores KC098 and GC062 (Fig. 9) from the western Falkland Trough record a reduction in ACC bottom current flow during the Last Glacial Maximum at depths of > 2500 m below sea level.

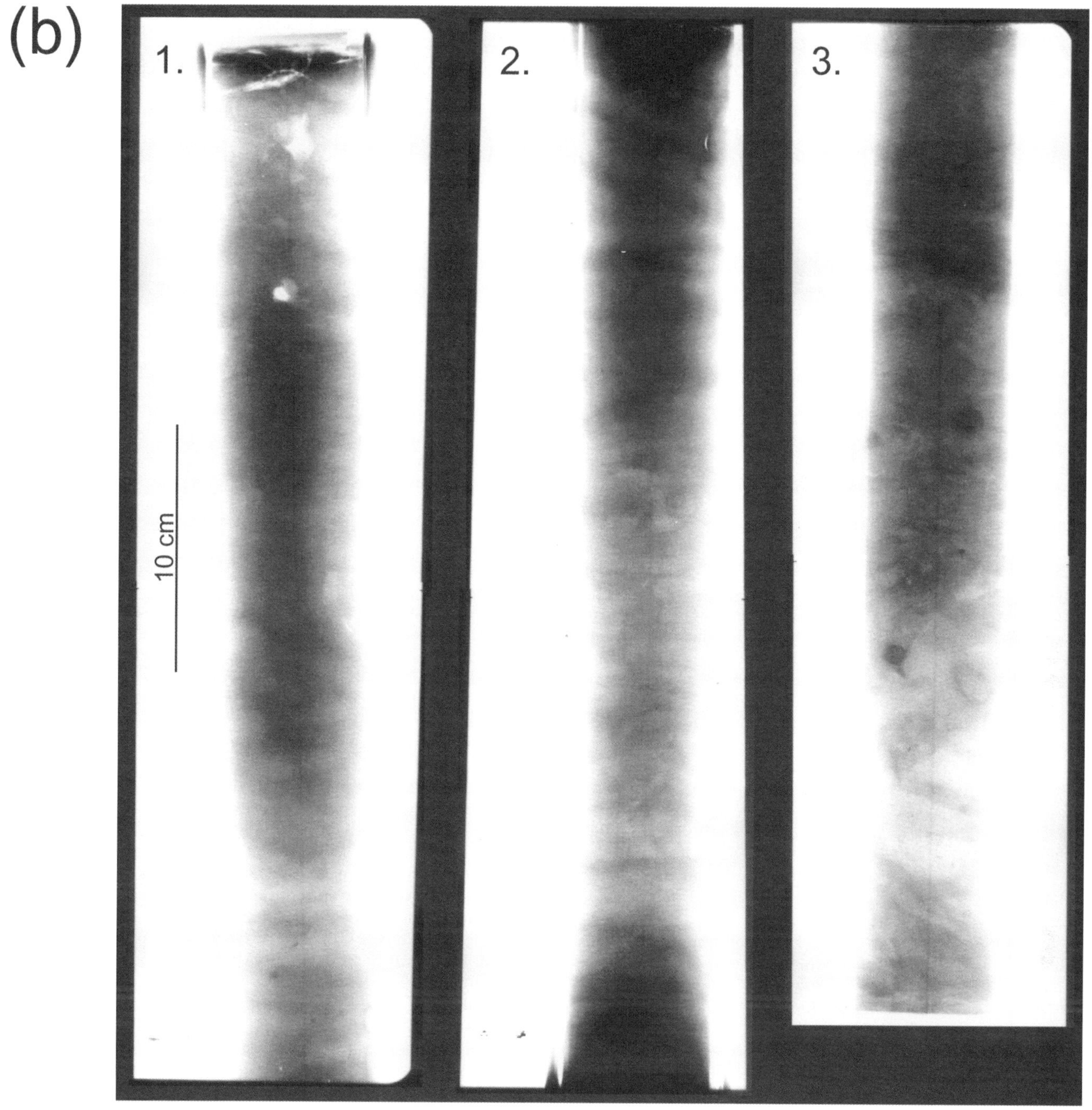

Foraminifera-rich sandy contourites in core tops suggest a relatively uniform style of sedimentation across the western Falkland Trough and Falkland Plateau slope during the Holocene. The thickest capping of foraminifera-rich sand occurs in the western Falkland Trough cores (KC098 and GC062, Fig. 9); cores from more exposed areas (e.g. KC096, Fig. 8) usually have a much thinner veneer of biogenic sand. These sediments record open marine conditions, with stronger, sustained ACC flow across the WFTD.

During the Late Pleistocene (late oxygen isotope stage 2), fine-grained, diatomaceous sediments were deposited across the WFTD, which suggests that the drift was bathed by more sluggish ACC currents during glaciation. During the Last Glacial Maximum, the fall in sea level would have reduced the ocean section accommodating northward ACC flow above the western NSR, and some parts of the ridge would have been subaerial. This may have reduced the flow of Circumpolar Deep Water into the western Falkland Trough at that time.

We thank D. Stow and G. Ercilla for their constructive reviews, and C. Pudsey and C. Ó Cofaigh for assistance with interpretation and presentation of sedimentological data. Figures 4, 5, 6a–c, 8 and 9 are reprinted from Marine Geology, 138, Howe, J. A., Pudsey, C. J. & Cunningham, A. P., Pliocene–Holocene contourite deposition under the Antarctic Circumpolar Current, Western Falkland Trough, South Atlantic Ocean, 27–50, Copyright 1997, with permission from Elsevier Science.

References

BARKER, P. F. & BURRELL, J. 1977. The opening of Drake Passage. *Marine Geology*, **25**, 15–34.

BARKER, P. F. & DALZIEL, I. W. D. 1983. Progress in geodynamics in the Scotia Arc region. *In*: CABRE, R. (ed.) *Geodynamics of the Eastern*

Pacific Region, Caribbean and Scotia Arcs. American Geophysical Union, Washington, D. C., Geodynamics Series, **9**, 137–170.

BARKER, P. F., DALZIEL, I. W. D. & SHIPBOARD SCIENTIFIC PARTY 1976. Evolution of the southwestern Atlantic Ocean basin: results of Leg 36, Deep Sea Drilling Project. *In*: BARKER, P. F., DALZIEL, I. W. D. *ET AL*. (eds) *Initial Reports of the Deep Sea Drilling Project*. US Government Printing Office, Washington, **36**, 993–1014.

BARKER, P. F., DALZIEL, I. W. D. & STOREY, B. C. 1991. Tectonic development of the Scotia arc region. In: Tingey, R. J. (ed.) *The geology of Antarctica*. Oxford University Press, New York, monographs on geology and geophysics, **17**, 215–248.

BURCKLE, L. H. & HAYS, J. D. 1974. Pre-Pleistocene sediment distribution and evolution of the Argentine continental margin and Falkland Plateau (abs.) Geological Society of America, Abstracts with Programs, **6**, 673–674.

CARTER, L., CARTER, R. M., NELSON, C. S., FULTHORPE, C. S. & NEIL, H. L. 1990. Evolution of Pliocene to Recent abyssal sediment waves on Bounty Channel levees, New Zealand. *Marine Geology*, **95**, 97–109.

CHAVEZ, P. S. 1986. Processing techniques for digital sonar images from GLORIA. *Photogrammetric Engineering and Remote Sensing*, **52**, 1133–1145.

CIESIELSKI, P. F., LEDBETTER, M. T. & ELLWOOD, B. B. 1982. The development of Antarctic glaciation and the Neogene palaeoenvironment of the Maurice Ewing Bank. *Marine Geology*, **46**, 1–51.

CUNNINGHAM, A. P. 1999. *Geophysical Investigations of the North Scotia Ridge*. Unpublished PhD thesis, University of London, London, England.

CUNNINGHAM, A. P. & BARKER, P. F. 1996. Evidence for westward-flowing Weddell Sea Deep Water in the Falkland Trough, western South Atlantic. *Deep-Sea Research*, **43**, 643–654.

CUNNINGHAM, A. P., BARKER, P. F. & TOMLINSON, J. S. 1998. Tectonics and sedimentary environment of the North Scotia Ridge region revealed by side-scan sonar. *Journal of the Geological Society, London*, **155**, 941–956.

FLOOD, R. D., SHOR, A. N. & MANLEY, P. D. 1993. Morphology of abyssal mudwaves at Project MUDWAVES sites in the Argentine Basin. *Deep-Sea Research II*, **40**, 859–888.

GEORGI, D. T. 1981. Circulation of bottom waters in the southwestern South Atlantic. *Deep-Sea Research*, **28A**, 959–979.

GORDON, A. L. 1966. Potential temperature, oxygen and circulation of bottom water in the Southern Ocean. *Deep-Sea Research*, **13**, 1125–1138.

GROSE, T. J., JOHNSON, J. A. & BIGG, G. R. 1995. A comparison between the FRAM (Fine Resolution Antarctic Model) results and observations in the Drake Passage. *Deep-Sea Research I*, **42**, 365–388.

HOLLISTER, C. D. & ELDER, R. B. 1969. Contour currents in the Weddell Sea. *Deep-Sea Research*, **16**, 99–101.

HOWE, J. A., PUDSEY, C. J. & CUNNINGHAM, A. P. 1997. Pliocene–Pleistocene contourite deposition under the Antarctic Circumpolar Current, western Falkland Trough, South Atlantic Ocean. *Marine Geology*, **138**, 27–50.

JORDAN, R. W. & PUDSEY, C. J. 1992. High-resolution diatom stratigraphy of Quaternary sediments from the Scotia Sea. *Marine Micropaleontology*, **19**, 201–237.

LOCARNINI, R. A., WHITWORTH III, T. & NOWLIN, W. D. 1993. The importance of the Scotia Sea on the outflow of Weddell Sea Deep Water. *Journal of Marine Research*, **51**, 135–153.

LONARDI, A. G. & EWING, M. 1971. Sediment transport and distribution in the Argentine Basin. 4. Bathymetry of the continental margin, Argentine Basin and other related provinces. Canyons and sources of sediments. *In*: AHRENS, L. H., RUNCORN, S. K. & UREY, H. C. (eds) *Physics and Chemistry of the Earth*. Pergamon Press, Oxford, **8**, 18–121.

LUDWIG, W. J. & RABINOWITZ, P. D. 1982. The collision complex of the North Scotia Ridge. *Journal of Geophysical Research*, **87**, 3731–3740.

MANTYLA, A. W. & REID, J. L. 1983. Abyssal characteristics of the world ocean waters. *Deep-Sea Research*, **30**, 805–833.

MARANI, M., ARGNANI, A., ROVERI, M. & TRINCARDI, F. 1993. Sediment drifts and erosional surfaces in the central Mediterranean: seismic evidence of bottom-current activity. *Sedimentary Geology*, **82**, 207–220.

NORMARK, W. R., HESS, G. R., STOW, D. A. V. & BOWEN, A. J. 1980. Sediment waves on the Monterey Fan levee: a preliminary physical interpretation. *Marine Geology*, **37**, 1–18.

NOWLIN, W. D. & ZENK W. 1988. Westward bottom currents along the margin of the South Shetland Island Arc. *Deep-Sea Research*, **35**, 269–301.

PETERSON, R. G. 1992. The boundary currents in the western Argentine Basin. *Deep-Sea Research*, **39**, 623–644.

PETERSON, R. G. & WHITWORTH III, T. 1989. The Subantarctic and Polar Fronts in relation to deep water masses through the southwestern Atlantic. *Journal of Geophysical Research*, **94**, 10817–10838.

PIOLA, A. R. & GORDON, A. L. 1989. Intermediate waters in the southwest South Atlantic. *Deep-Sea Research*, **36**, 1–16.

PLATT, N. H. & PHILIP, P. R. 1995. Structure of the southern Falkland Islands continental shelf: initial results from new seismic data. *Marine and Petroleum Geology*, **12**, 759–771.

PUDSEY, C. J. & HOWE, J. A. 1998. Quaternary history of the Antarctic Circumpolar Current: evidence from the Scotia Sea. *Marine Geology*, **148**, 83–112.

PUDSEY, C. J. & HOWE, J. A. 2002. Mixed biosiliceous–terrigenous sedimentation under the Antarctic Circumpolar Current, Scotia Sea. *In*: STOW, D. A. V., PUDSEY, C. J., HOWE, J. A., FAUGÈRES, J.-C. & VIANA, A. R. (eds) *Deep-Water Contourite Systems: Modern Drifts and Ancient Series, Seismic and Sedimentary Characteristics*. Geological Society, London, Memoirs, **22**, 325–336.

RABINOWITZ, P. D., DELACH, M., TRUCHAN, M. & LONARDI, A. 1978. *Bathymetry of the Argentine continental margin and adjacent areas*. American Association of Petroleum Geologists, Tulsa, Oklahoma, Argentine Map Series.

REED, D. L., MEYER, A. W., SILVER, E. A. & PRASETYO, H. 1987. Contourite sedimentation in an intraoceanic forearc system: eastern Sunda arc, Indonesia. *Marine Geology*, **76**, 223–241.

SAITO, T., BURCKLE, L. H. & HAYS, J. D. 1974. Implications of some pre-Quaternary sediment cores and dredgings. In: HAY, W. W. (ed.) *Studies in Palaeo-Oceanography*. Society of Economic Palaeontologists and Mineralogists, Tulsa, Special Publications, **20**, 6–35.

SEARLE, R. C., LEBAS, T. P., MITCHELL, N. C., SOMERS, M. L., PARSON, L. M. & PATRIAT, P. H. 1990. GLORIA Image Processing: The State of the Art. *Marine Geophysical Researches*, **12**, 21–39.

SHIMMIELD, G., DERRICK, S., PUDSEY, C., BARKER, P., MACKENSEN, A., GROBE, H. 1993. The use of inorganic chemistry in studying palaeoceanography of the Weddell Sea. *In*: HEYWOOD, R. B. (ed.) *University Research in Antarctica, 1989–1992. Proceedings of the British Antarctic Survey Antarctic Special Award Scheme Round 2*. British Antarctic Survey, Cambridge, 99–108.

STOW, D. A. V. & PIPER, D. J. W. 1984. Deep-water fine-grained sediments: facies models. *In*: STOW, D. A. V. & PIPER, D. J. W. (eds) *Fine-grained Sediments: Deep-water Processes and Facies*. Geological Society, London, Special Publications, **15**, 611–646.

TECTONIC MAP OF THE SCOTIA ARC. 1985. 1:3 000 000, BAS(Misc.) 3. British Antarctic Survey, Cambridge.

WITTSTOCK, R.-R. & ZENK, W. 1983. Some current observations and surface T/S distribution from the Scotia Sea and the Bransfield Strait during early austral summer 1980/1981. *Meteor Forschungs-Ergebnisse, Reihe A/B*, **24**, 77–86.

ZENK, W. 1981. Detection of overflow events in the Shag Rocks Passage, Scotia Ridge. *Science*, **213**, 1113–1114.

Sediment drifts and deep-sea channel systems, Antarctic Peninsula Pacific Margin

M. REBESCO[1], C. J. PUDSEY[2], M. CANALS[3], A. CAMERLENGHI[1], P. F. BARKER[2], F. ESTRADA[3] & A. GIORGETTI[1]

[1]*Istituto Nazionale di Oceanografia e di Geofisica Sperimentale (OGS), Borgo Grotta Gigante 42/c, 34010 Sgonico (TS), Italy (e-mail: mrebesco@ogs.trieste.it)*
[2]*British Antarctic Survey, Madingley Road, Cambridge CB3 0ET, UK*
[3]*Universitat de Barcelona, Campus de Pedralbes, Barcelona E-08028, Spain*

Abstract: Twelve sedimentary mounds are identified on the upper continental rise of the Pacific Margin of the Antarctic Peninsula. All these mounds are produced by a varying degree of interaction of along-slope bottom water flow with downslope turbidity currents. These mounds provide a complete range of intermediates between two end members: the sediment drift and the channel levee. Surface sediments on drift 7 suggest that the mechanisms for the supply and transport of sediment include entrainment of material from turbidity currents within ambient bottom currents, and pelagic settling from the sea surface, including biogenic and glacially derived material. The long-lasting activity of these mechanisms is documented by the data provided by four DSDP and ODP drill sites. Bathymetric and seismic data, both at a large, comprehensive scale and at a small, detailed scale, show the geometry of the sedimentary mounds and their relationships with the adjacent turbidity current channel systems. These data allow the determination of some diagnostic criteria to identify the sediment drifts.

The series of twelve large sedimentary mounds, elongated orthogonal to the Antarctic Peninsula Pacific continental margin, are examples of mixed drift systems produced by the interaction of downslope and alongslope processes. The margin has been extensively studied over the past decade by a series of Italian, German, British and joint research programmes, and has also been the focus of DSDP and ODP drilling. This paper focuses on Drift 7, an elongate body some 200 km long, 70 km wide and up to 1 km thick (Table 1, Fig. 1). It synthesises the principal results derived from study of a variety of data types, including detailed bathymetric and swath bathymetric data, seismic reflection profiles and sub-bottom (TOPAS) profiles, seabed photos, and sediment core samples.

Geological and oceanographic setting

The Pacific Margin of the Antarctic Peninsula has been characterised by rapid terrigenous sedimentation since the late Miocene (Tucholke *et al.* 1976; Larter & Barker 1989). Glacially-derived sediment has been redistributed downslope and alongslope by turbidity currents and bottom currents. West of Drake Passage, the axis of the eastward-flowing Antarctic Circumpolar Current (ACC) lies about 60°S (Nowlin & Klinck 1986). A narrow counter current flows south-westward close to the margin (Gordon 1966; Nowlin & Zenk 1988; Camerlenghi *et al.* 1997*a*).

On the continental rise, in water depths of 2700–3700 m, there are 12 large sediment mounds elongated orthogonally to the margin (Fig. 1). They are thought to be formed of material originally supplied from turbidity currents flowing in deep-sea channels extending from the margin to the abyssal plain, and redistributed by south-westerly-flowing bottom currents (Rebesco *et al.* 1994, 1996; McGinnis & Hayes 1995).

Present day bottom current flow in the area of drift 7, one of the largest drifts (Fig. 2) was reconstructed (Camerlenghi *et al.* 1997*a*; A. Crise, pers. comm.) from data recorded in three deep moorings deployed by the R/V OGS-Explora during two cruises of the Progetto Nazionale Ricerche in Antartide (PNRA). Mooring ST-01 was deployed at 3475 m depth on the south-west, steeper side of drift 7, Mooring ST-02 at 3338 m depth on the gentler north-eastern side, and Mooring ST-03 (equipped with two current meters at 8 m and 60 m above the sea bed respectively) at 3580 m on the distal north-western side.

Table 1. *Principal characteristics*

Location:	**Antarctic Peninsula Pacific Margin**
Setting:	Upper continental rise
Age:	Mid Miocene-Present
Drift type:	Detached drift (with influence of turbidity currents)
Dimensions:	Up to about $200 \times 70 \times 1$ km
Seismic facies and attributes:	Asymmetric, with a gentler side underlain by conformable reflectors and a steeper side underlain by more discontinuous reflectors with frequent unconformities and erosional truncations.
Sediment facies and attributes:	Very fine-grained, showing glacial-interglacial cycles in composition, texture and sedimentary structures (interglacials are bioturbated, contain ice-rafted debris and biogenic silica and carbonate; glacials are laminated clays).

The direction of the bottom water flow is controlled by drift topography, as shown by the mean current direction. The general SW flowing circulation follows the isobaths and appears to be geostrophically adjusted at least for a large part of the year (Fig. 2). The mean current velocity 8 m above the seabed is between 4 cm s^{-1} at ST-03 and 6.2 cm s^{-1} at ST-01 ($\pm$ 2.8 cm s^{-1}), and speed never exceeded 20 cm s^{-1}. This flow is capable of transporting fine sediment particles, but not of eroding the sediment. The potential temperature is remarkably stable between 0.11 ± 0.01°C and 0.13 ± 0.02°C. These values, even in the absence of salinity records, suggest that the bottom layer consists of modified Circumpolar Deep Water (CDW), which constitutes the largest volume of water in the Southern Ocean (Carmack 1990).

The observed bottom water flow is consistent with deposition of Holocene hemipelagic sediments during a 'drift maintenance' stage. Indicators of palaeoceanographic conditions during glacial periods are at present too scarce to fully understand how the past oceanographic conditions influenced the evolution of the drifts.

From: STOW, D. A. V., PUDSEY, C. J., HOWE, J. A., FAUGÈRES, J.-C. & VIANA, A. R. (eds)
Deep-Water Contourite Systems: Modern Drifts and Ancient Series, Seismic and Sedimentary Characteristics.

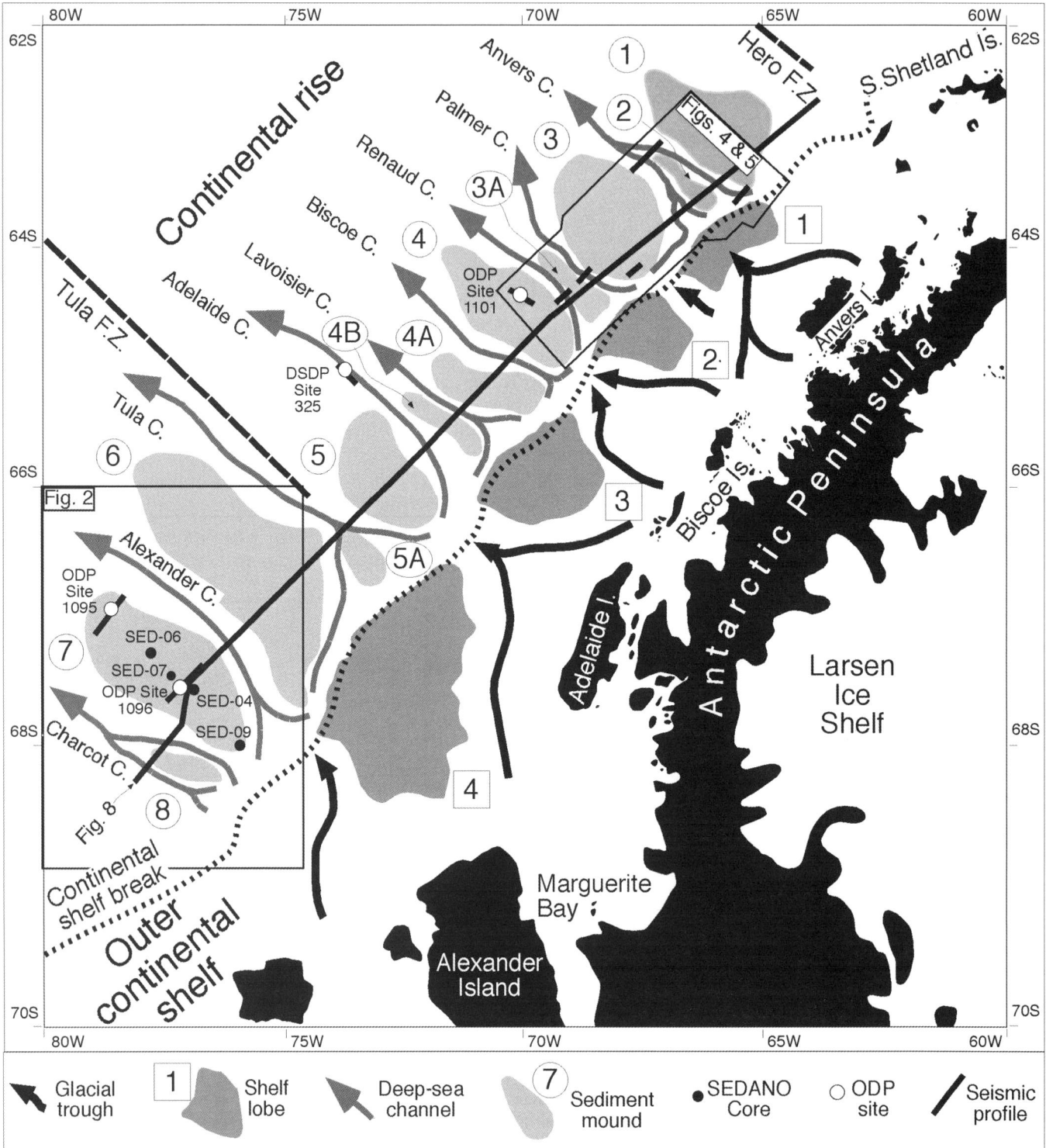

Fig. 1. Location map of the study area. Sediment drifts and mounds of the continental rise are shaded in light grey and progressively numbered from north to south according to the system used by Rebesco *et al.* (1996). Deep sea channels, indicated by dark grey arrows, are newly named in this paper. Prograding lobes of the continental shelf break are shaded in dark grey and numbered from north to south according to Larter *et al.* (1997). Glacial troughs are indicated by black arrows. Main fracture zones are indicated by dashed lines. Seismic profiles are shown as black lines. Location of DSDP and ODP sites and cores are also shown.

Bathymetry

General bathymetry

The bathymetric map of Figure 3, which covers the entire continental margin from shelf to abyssal plain, shows the main physiographic characteristics of this glacial system. Like other parts of the Antarctic continental margin (Cooper *et al.* 1991) the continental shelf is over-deepened with a seafloor that generally dips landward. Shelf bathymetric relief is high (up to 1000 m) due to the presence of overdeepened basins (Domack *et al.* 1994; Rebesco *et al.* 1998*a*) and glacial troughs carved by ice streams (Pope & Anderson 1992; Pudsey *et al.* 1994; Canals *et al.* 2000). Four main lobes, defined by oceanward-convex trends of the

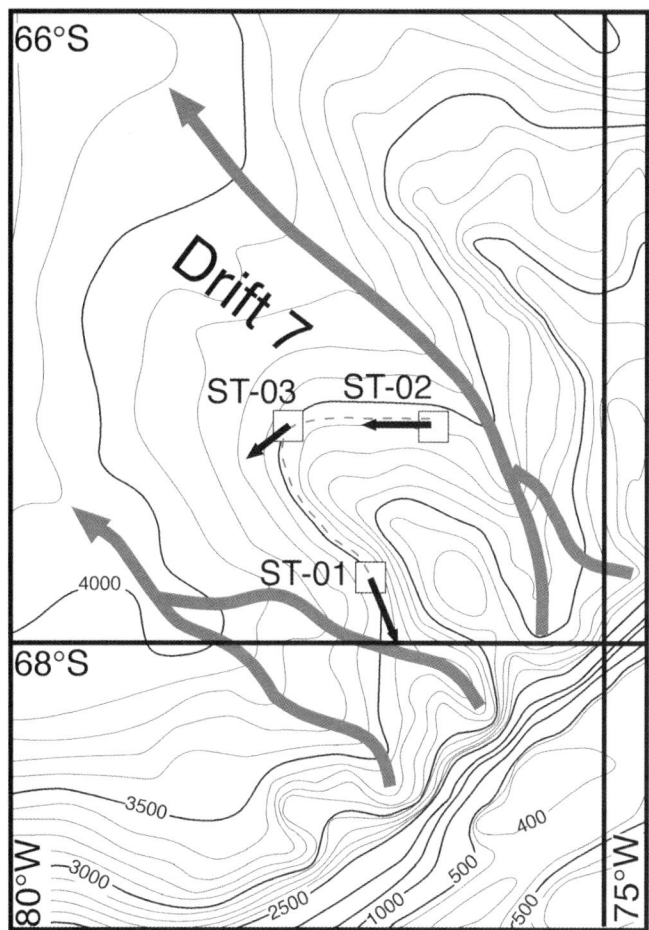

Fig. 2. Location map of current-moorings on drift 7. See location in Figure 1. The mean direction of the current measured 8 m above the seabed is indicated with arrows at the location of the three moorings. The inferred bottom water flow path along the isobath between the moorings is shown by a dashed line.

continental shelf break, can be correlated with prograding wedges consisting of coalescing banks of the outer shelf (Larter & Cunningham 1993; Larter et al. 1997; Rebesco et al. 1998b). These four lobes are separated by large symmetric, U-shaped troughs with gentle sides (Vanney & Johnson 1976a, b; Pope & Anderson 1992; Rebesco et al. 1998b). The continental slope (about 500 to 3000 m water depth) is very steep. The gradient of the four lobes (>13°) is generally steeper than that of the slope at the mouth of the troughs. Despite such steepness, there is no evidence of present-day major slope failure, nor of major canyons cutting into the slope.

The upper limit of the continental rise south of 67°S is deeper (almost 3500 m) than in the northeast part (around 3000 m) as a consequence of a major step (in correspondence of the Tula Fracture Zone) within the stepwise younging of basement age northeastward along the margin (Tucholke & Houtz 1976). The Bellingshausen Abyssal Plain lies below 4800 m (Vanney & Johnson 1976a, b). The irregular relief of the upper continental rise results from the 12 sediment mounds separated by deep-sea channel systems, originating at the base of the slope between the lobes. The channels are named for the first time in this paper (Fig. 1).

Since some of the channels run parallel to the margin for tens of kilometres before turning seaward towards the outer rise, the sediment mounds are commonly separated from the continental slope by a broad 10–20 km wide erosional depression. All the 12 mounds have a distinct bathymetric expression. Nine of them were numbered progressively from north to south and interpreted as sediment drifts, with varying degree of interaction of bottom currents with downslope turbidity current processes (Rebesco et al. 1994, 1996, 1998b; McGinnis & Hayes 1995; McGinnis et al. 1997). Three additional mounds (mounds 3A, 4B, and 5A) are newly mapped here, mainly on the basis of swath bathymetry and seismic data. All the sediment mounds share many common features, though each one is different from the others. They are preferentially located between the shelf lobes (i.e. in front of the shelf troughs). They are elongated in a direction approximately orthogonal to the margin, with a wider, thicker central body and two narrow ends. The largest mounds have their summits at an average depth of 2700 m, are up to 200 km long by 70 km wide, and attain an elevation of nearly 1 km above the adjacent channels. The majority of the mounds are asymmetric with a steep side (typically sloping 2°) and a gently-dipping side (typically 0.8°) that meet to form a long and narrow crest. Most mounds (drift 1, 2, 3, 4A, 6 and 7) have their steep side oriented toward the southwest.

The mounds merge into the lower continental rise, at a depth of about 3500 m northeast of the Tula Fracture Zone, and about 4000 m farther southwest. The lower continental rise north of the Tula Fracture Zone is a gently sloping region where the 3600 to 4000 m isobaths show an outward-convex curvature (Palmer deep sea fan of Vanney & Johnson 1976a, b). The 4100 to 4600 m isobaths are roughly SW–NE striking and correspond to a more uneven area where numerous seamounts protrude through a thin drape of sedimentary cover (Tucholke & Houtz 1976; Vanney & Johnson 1976a, b; Larter & Barker 1991a). The 4700 m deep South Shetland Trench, limited to the southwest by the Hero Fracture Zone, belongs to the active part of the margin and is not discussed here.

Swath bathymetry

The detailed swath bathymetric map of Figure 4, collected during the GEBRAP'96 cruise onboard BIO Hesperides covers the margin west of Palmer Archipelago from the outer shelf (325 m water depth) to the upper continental rise (3800 m).

The outer continental shelf is a relatively flat, landward-dipping area crossed by glacial troughs, marked by iceberg scours and including some shallow banks that correspond to till deltas (Canals et al. 1998). Water depth ranges from 325 m to 450 m, with the north-eastern sector being particularly uneven. The shelf edge is sinuous with two main seaward convex lobes. These lobes correspond to the prograding lobes 1 and 2 of Larter et al. (1997). The swath bathymetry map reveals that, in detail, the northeasternmost lobe (Lobe 1) is actually made up of two minor lobes. Farther to the north of Lobe 1 there is another minor shelf lobe, indicated by a change in the steepness of the continental slope. The northeast trending continental slope of Lobe 1 is narrow (10 km) and very steep (up to 22°), following the sinuous outline of the shelf edge. The slope is incised by small straight gullies that do not extend up to the shelf edge. The depth of the base of the continental slope ranges from approximately 2500 m in the south, down to 3000 m in the north.

The upper continental rise is characterised by five elongate sediment mounds whose NW-trending crests are roughly orthogonal to the shelf edge. Four of these mounds correspond to drifts 1, 2, 3, and 4 of Rebesco et al. (1996). A new mound (mound 3A) is here identified between drifts 3 and 4. The largest imaged mound is drift 3, some 125 km wide between the Palmer and South Anvers channels to either side (Fig. 1). The gentler sides of the mounds are mostly undisturbed, as exemplified by the level surface north of the crest of drift 3 (Fig. 5). Conversely, the crests and steep sides of the mounds are commonly affected by small curved scarps. Drifts 1, 2, and 3 have a steeper SW side than NE side. Drift 4, which apparently displays a steep NE side, is only

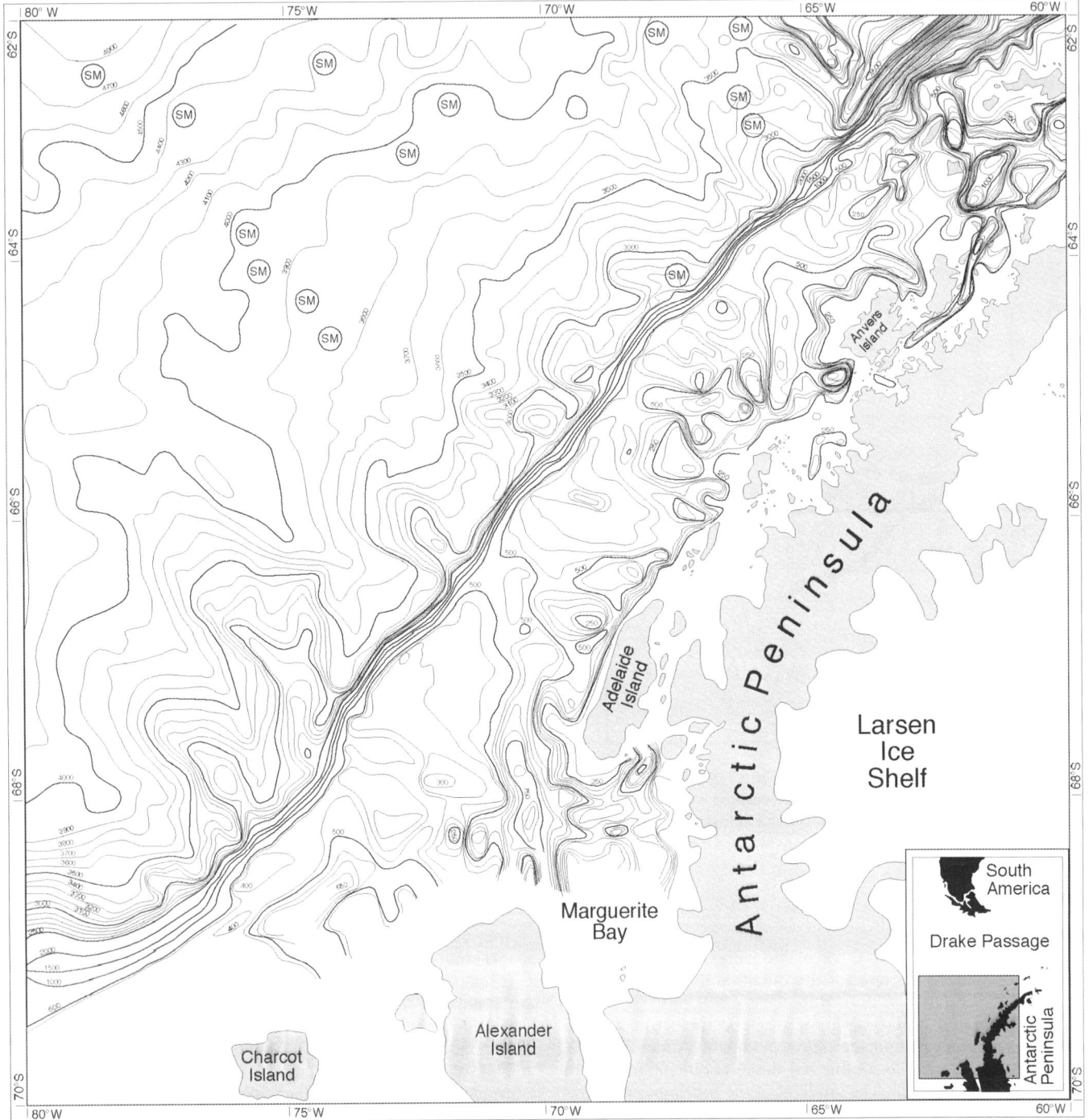

Fig. 3. General bathymetric map of the western margin of the Antarctic Peninsula between 62–70°S and 60–80°W, from Rebesco *et al.* (1998*b*). A coloured version of this figure, including both contours and data locations, is available at the web site of Terra Antartica (http://www.mna.unisi.it/TAP/mapcol.eps). Contours in corrected metres. Contour interval: 50 m on the continental shelf (shallower than 500 m); 500 m on the continental slope (between 500 and 2500 m); 100 m on the continental rise (deeper than 2500 m). SM, seamount.

partially imaged by the swath bathymetric survey. Mound 3A is a relatively more symmetrical and subdued feature. These mounds are separated by depressions, mostly parallel to the mounds' crests, which contain erosional deep-sea channels with poorly developed levees (Tomlinson *et al.* 1992). Slope gradients can reach up to 10° locally in the channel margins. Three seamounts of probable volcanic origin have been identified in the deepest region of the northernmost mound.

Channel systems vary greatly both in shape and morphological complexity. The southern systems (Renaud Channel between drift 4 and mound 3A, and Palmer Channel between mound 3A and drift 3) consist of two slightly sinuous channels that originate at the base of the continental slope. The Palmer Channel is wide and only slightly incised, in contrast to the straight and strongly incised Renaud Channel between drift 4 and mound 3A. The northern channel system has a complex morphology, displaying a dendritic pattern with many tributaries that extend back to the base of the slope. Three main catchments, collecting many gullies and smaller channels, feed the system, and evolve downslope into two main channels: North Anvers Channel that runs between drifts 1 and 2, and South Anvers Channel between drifts 2 and 3. These channels finally converge to form the biggest channel in the area.

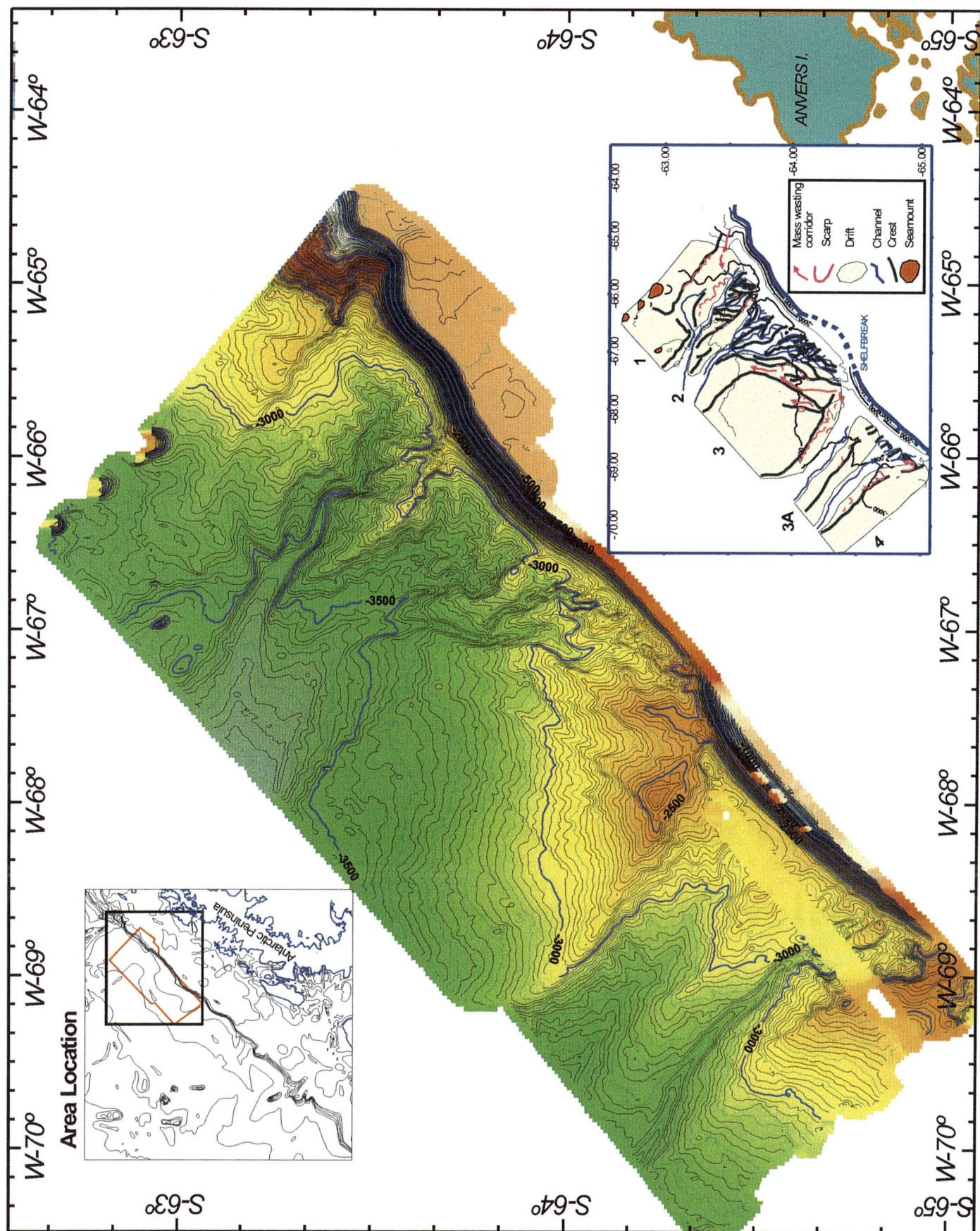

Fig. 4. Multibeam coloured bathymetric map of the five northern mounds (contour interval 25 m). See location in Figure 1. The upper inset is a location map superposed on the GEBCO bathymetry. The lower inset is an interpretation of the main morphosedimentary features.

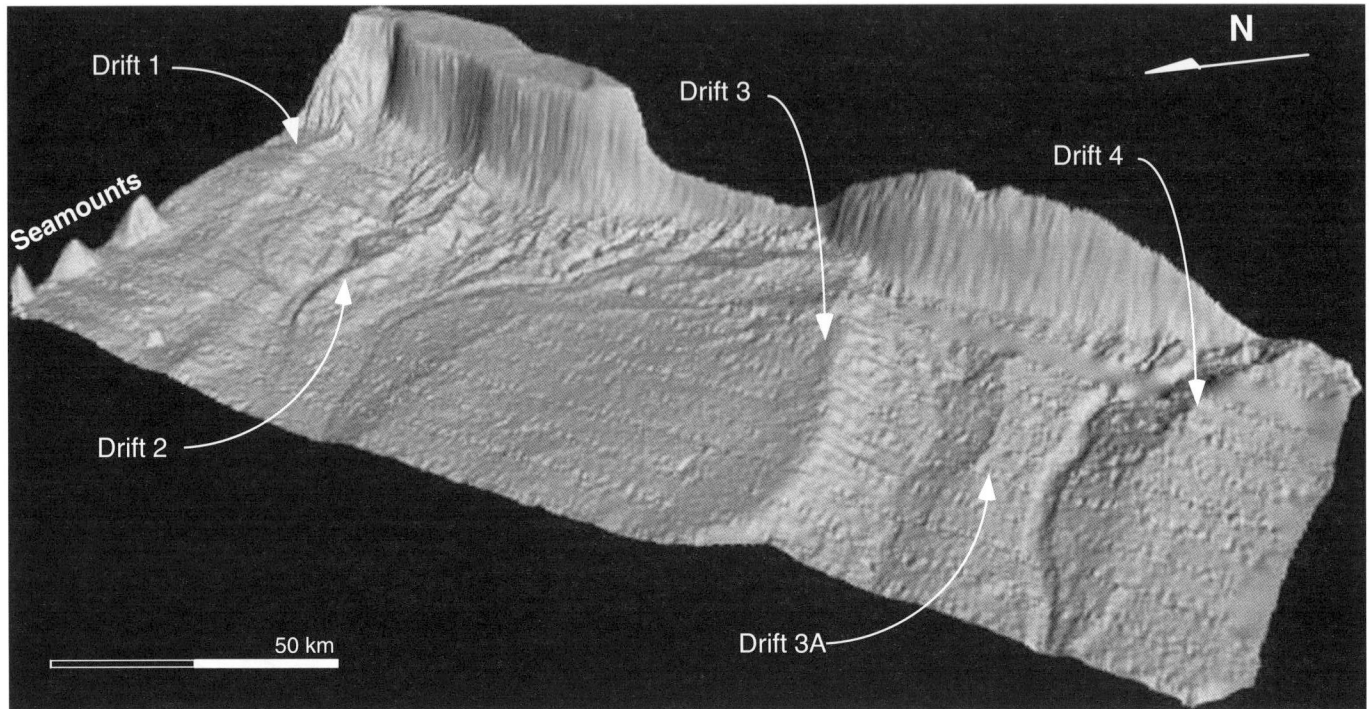

Fig. 5. 3D digital terrain model of the continental slope and rise area of the five northern mounds, same area as Figure 4. View from west. Note the variations in size of the various mounds, the distinct morphologies of the channel systems, the gullies that cut the continental slope, and the three volcanic seamounts near the northeastern corner.

Stratigraphic context

In piston and gravity cores, late Quaternary lithological cyclicity has been related to glacial-interglacial cycles using diatom and radiolarian biostratigraphy and barium stratigraphy (Pudsey & Camerlenghi 1998; Barker et al. 1999; Pudsey 2000). Intervals of diatom-bearing foraminifer-bearing mud alternate with intervals of almost barren mud with poorly-preserved diatoms. Radiolarians are common in core-top samples and are of omega zone age (0–0.43 Ma; Hays 1965). The diatoms are also of late Quaternary age (*Thalassiosira lentiginosa* zone, 0–0.6 Ma; Gersonde & Burckle 1990), with *Hemidiscus karstenii* identified in isotope stage 7. Foraminifera are also consistent with a Late Quaternary age (Silvia Spezzaferri, pers. comm). Barium data show 900–1600 ppm biogenic Ba in the interglacial facies (see Section 6b) and 0–150 ppm in the glacial facies (Pudsey & Camerlenghi 1998; Pudsey 2000).

The continental rise of the Pacific margin of the Antarctic Peninsula was drilled at four sites during one DSDP leg and one ODP leg (see Fig. 1), providing a Neogene stratigraphy.

Site 325 of DSDP Leg 35 (Hollister et al. 1976) was located in 3748 m water depth near the transition between the upper and lower continental rise according to the recent definition of Rebesco et al. (1998b). The main objective of drilling was to establish the age of the oceanic basement, with palaeoceanography as a side objective. Site 325 is located in the distal part of the wide zone of channels that separate drift 5 and mound 4B. This site penetrated 718 m of a sedimentary succession composed of terrigenous turbidites and ice-rafted debris (Fig. 6a). The hole was spot cored and recovered only 35 m of sediment from ten cores. Lithologies were mainly clay/claystone to sand/sandstones with ice rafted debris, with poorly preserved diatoms in places, rare nannofossils in the lower cores and conglomerates at the base of the hole. The upper lithologic unit (0–528 mbsf) is generally finer grained than the lower one. The oldest sediment recovered was claystone of Oligocene to Early Miocene age range. A major hiatus was identified between the two lithologic units, from approximately 15 to 8 Ma. This mid-Miocene hiatus has been attributed to the reduction in terrigenous supply as a result of margin uplift following ridge-crest subduction (Larter & Barker 1991b), and is therefore diachronous along the margin, but it can be used to place limits on the timing of transition from 'pre-drift' deposition to 'drift growth' on the continental rise (Rebesco et al. 1997).

Sites 1095, 1096, and 1101 of ODP Leg 178 (Barker et al. 1999) were located on the upper continental rise. The objective of drilling was to recover a continuous high resolution record of glacial processes occurring on the continental margin over the last 10 million years. Sites were located on drifts 7 and 4 because the elevation of these sites above the channels was believed to shield these sediments from high-energy turbidity current flows.

Sites 1095 and 1096 (Fig. 6b, c) were drilled on drift 7. The entire sequence from the Mid-Miocene to Present was obtained from two sites. Site 1096 recovered the more expanded upper part of the section, down to the Early Pliocene (4.7 Ma) by penetrating 608 m in 3152 m water depth. The more distal Site 1095 was located in 3842 m water depth at the transition between sediment drift and turbiditic lobe. It sampled 570 m of a more condensed succession at least down to the Late Miocene (10 Ma and possibly older). Continuous coring, multiple-hole drilling, and excellent magnetostratigraphic control provided a composite record of sedimentation over the last ten million years.

Site 1101 (Fig. 6d) was drilled in 3280 m water depth on drift 4 in order to obtain a comparative lithostratigraphic record from a more northerly position than the other two sites. 218 m of continuously deposited fine grained sediment extending to the mid-Pliocene (3.1 Ma) was sampled.

The uppermost few tens of metres of sediment (down to about 50 m) at the three sites, are composed of alternating grey terrigenous laminated and brown massive, bioturbated, foraminifer- and diatom-bearing silty clays. These lithologies are very similar to those recovered in piston and gravity cores from all other sediment drifts of this margin (Pudsey 2000), and indicate the glacial–interglacial alternation of sedimentation. No evidence of

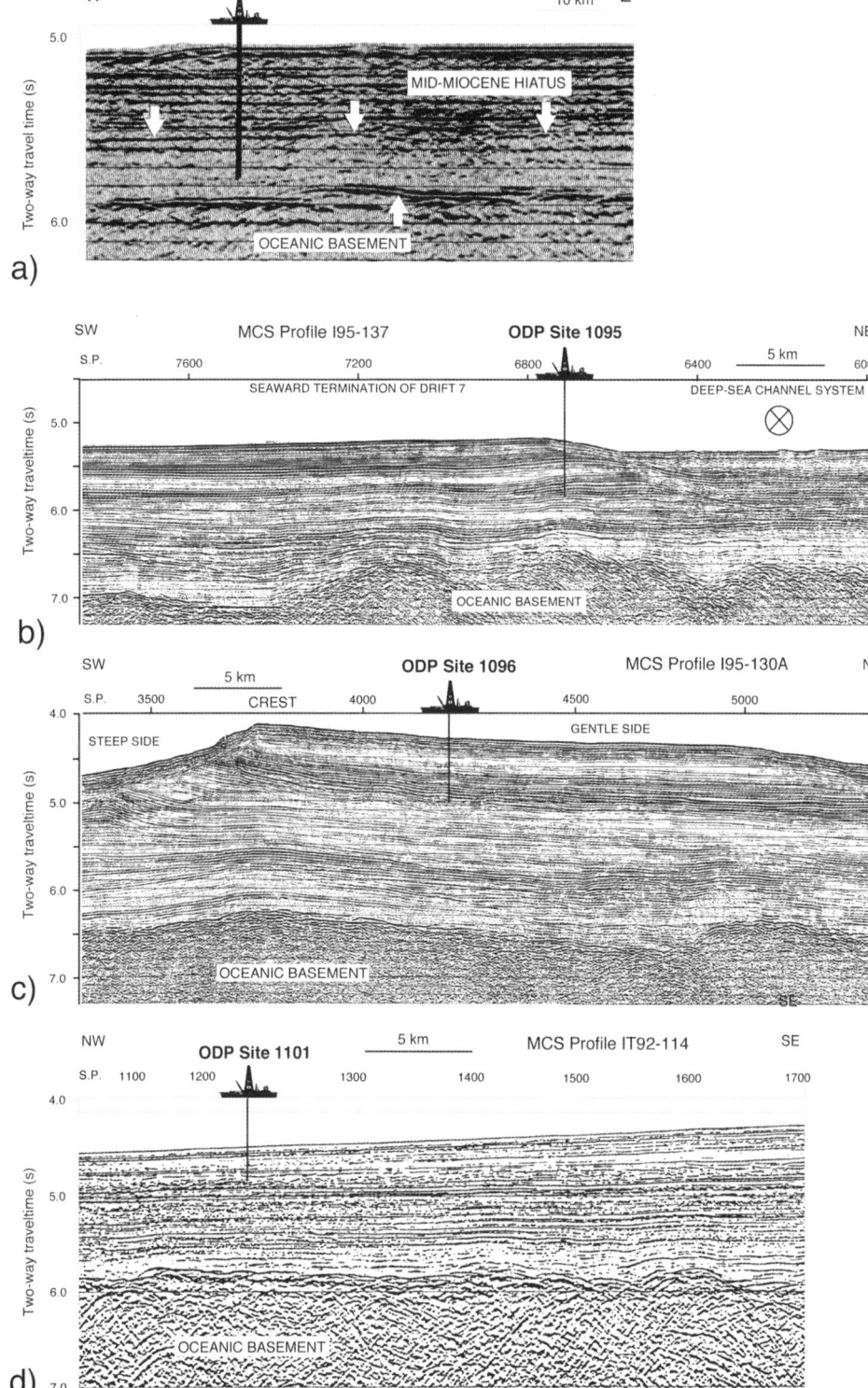

Fig. 6. Location of DSDP and ODP sites on multichannel seismic profiles, with estimated penetration on a two-way travel time scale. See location in Figure 1. (**a**) DSDP Site 325 re-positioned on a modern seismic profile (modified after Larter & Barker 1991*b*); (**b**), (**c**), and (**d**) ODP sites located on Leg 178 site survey profiles (Barker, Camerlenghi, Acton *et al.* 1999).

turbiditic deposition can be found in these sediments, and deposition appears to be dominated by redeposition of suspended fines from terrigenous turbidity currents in deep sea channels during glacials, and by hemipelagic/pelagic deposition during interglacials.

Deeper in the sections colour changes are less obvious, and parallel-laminated silt and mud turbidites become common. At Site 1096, a mixed contourite-turbidite succession occurs from 32.8 to 173.0 mbsf, with generally low biogenic content. Turbidite silts are thin, and subordinate to muds of likely contouritic origin. From 143–608 mbsf, very thinly laminated and generally non-bioturbated clays deposited from dilute turbidity currents alternate with intensely bioturbated homogenous silty clays. At Site 1095, thick and repetitive sequences of green laminated silt and mud were recovered, becoming dark greenish grey laminated claystones towards the base of the recovered section. Sharp-based, graded, variably laminated fine sands and silts and laminated silty clays, interbedded with more massive facies, represent a largely turbidite succession. At Site 1101, foraminifer-bearing layers alternate with barren laminated or massive intervals between 53.3 and 142.7 mbsf. Deeper in the section, the biogenic component is low, and massive clayey silt and diamict occur. Ice-rafted debris is scattered throughout the facies drilled at the three sites, and appears concentrated within bioturbated intervals.

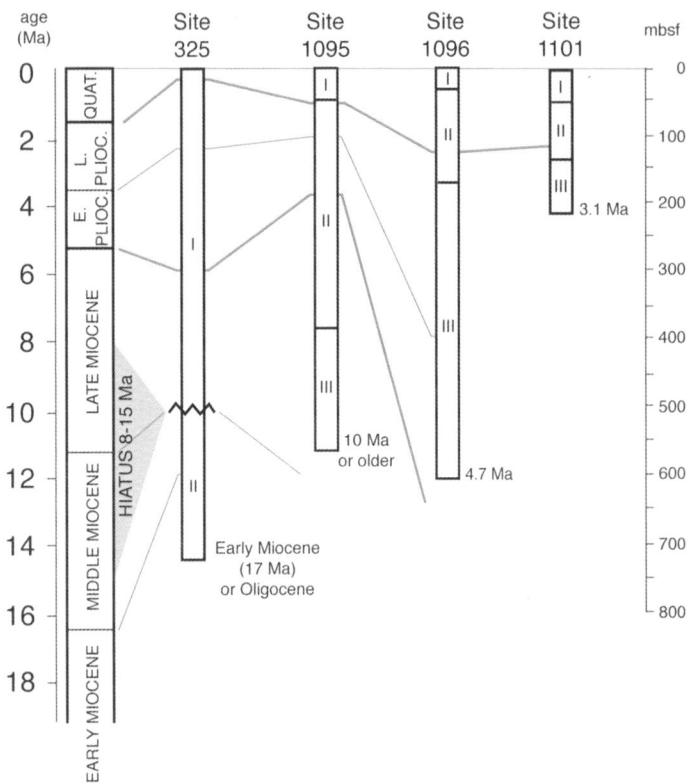

LITHOSTRATIGRAPHIC UNITS

Site 325
Unit I: Silty clay, silty claystone and claystone
Unit II: Sandstone, siltstone, and claystone

Site 1095
Unit I: Laminated and massive brown and dartk grey diatom-bearing silty clay and clay
Unit II: Sharp-based graded variably laminated silty clays, silts, and fine sands interbedded with massive bioturbated muds.
Unit III: Parallel laminated claystone and siltstone, with minor bioturbation and lower Ice Rafted Debris content than the units above.

Site 1096
Unit I: Laminated and massive, bioturbated clay and diatom-bearing silty clay as in Unit I of Site 1095.
Unit II: Parallel laminated silt and mud couplets.
Unit III: Thinly laminated silty clays alternating with bioturbated homogeneous silty clay.

Site 1101
Unit I: Alternation of massive and laminated clayey silt and diatom-bearing silty clay.
Unit II: Alternation of foraminifer-bearing silty clays and laminated silty clays.
Unit III: Primarily massive clayey silt and diatom-bearing silty clay with thin diamict levels.

Fig. 7. Simplified lithostratigraphic logs of DSDP and ODP sites on the continental rise west of the Antarctic Peninsula, with main chronostratigraphic ties. Note that stratigraphy at site 325 is poorly constrained. Age of the base of each site is also indicated.

In summary, deposition at these three sites was different, ranging from dominantly hemipelagic on the drift crest and centre to dominantly turbiditic at the distal site. All sites revealed a more or less pronounced cyclicity in turbidite abundance, bioturbation and ice-rafted debris, reflected in cyclicity in colour, magnetic susceptibility, and bulk density. This is considered to reflect the cyclic provision of glacial sediments to the uppermost continental slope. Sedimentation rates (see Fig. 7) were highest on the drift crest (18 cm ka^{-1} in Unit III, Site 1096) and lowest on the distal flank (5 cm ka^{-1} in the time equivalent units of Site 1095). At all three sites, the rates decreased through the Pliocene and into the Pleistocene (as low as 2.5 cm ka^{-1} in the Late Pleistocene of Site 1095). The gradual decrease in the rate of sedimentation observed at all rise sites is believed to reflect an overall trend of decreasing input of glacial sediment from the continental shelf, rather than changing palaeoceanographic conditions (i.e. bottom current direction and intensity) in the deep sea (cf. Barker 1995).

Seismic characteristics

Reflection profiles

The 900 km long composite multichannel seismic profile of Figure 8, striking parallel to the margin between 62°45′S and 68°15′S, crosses the 12 sedimentary mounds, with the largest (drifts 6 and 7 in the southwest) attaining a relief of about 1 km. In addition to the data in Figure 8, a considerable multichannel seismic

Fig. 8. Composite multichannel seismic profile striking parallel to the margin. See location in Figure 1. This 900 km long profile is composed of six different profiles acquired during four cruises of R/V OGS-Explora.

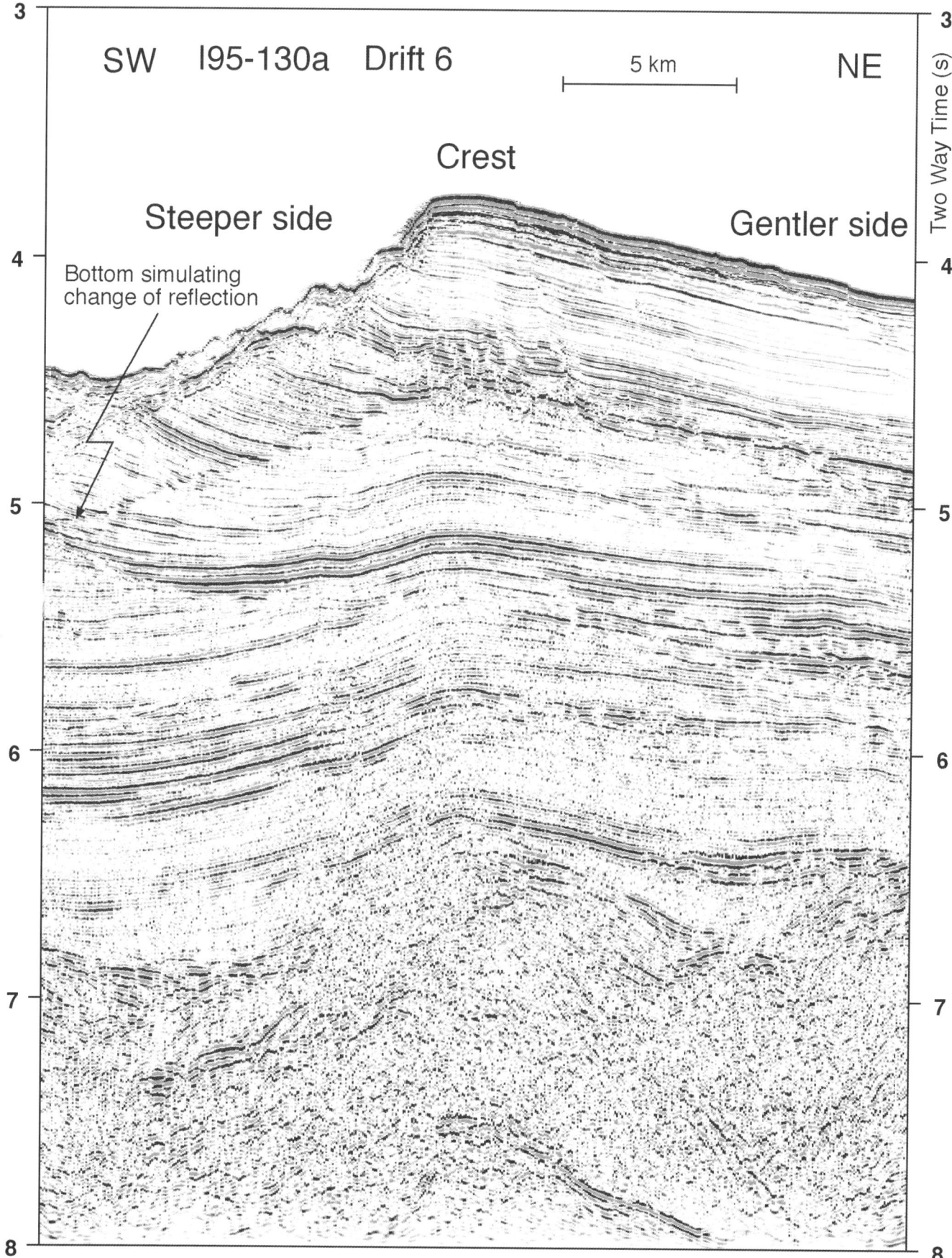

Fig. 9. Seismic profile I95-130a showing the crest of drift 6. The internal structure of the drift is different on either side. The steeper side (SW) is characterized by high reflectivity, abundant terminations and undulations of reflectors. A prominent change in reflectivity parallel to the sea bottom is particularly evident on this side of the drift. In contrast, the gentler side (NE) is relatively smooth, and underlain by continuous, highly reflective units, with linear, parallel or sub-parallel internal reflectors conformable with the sea floor.

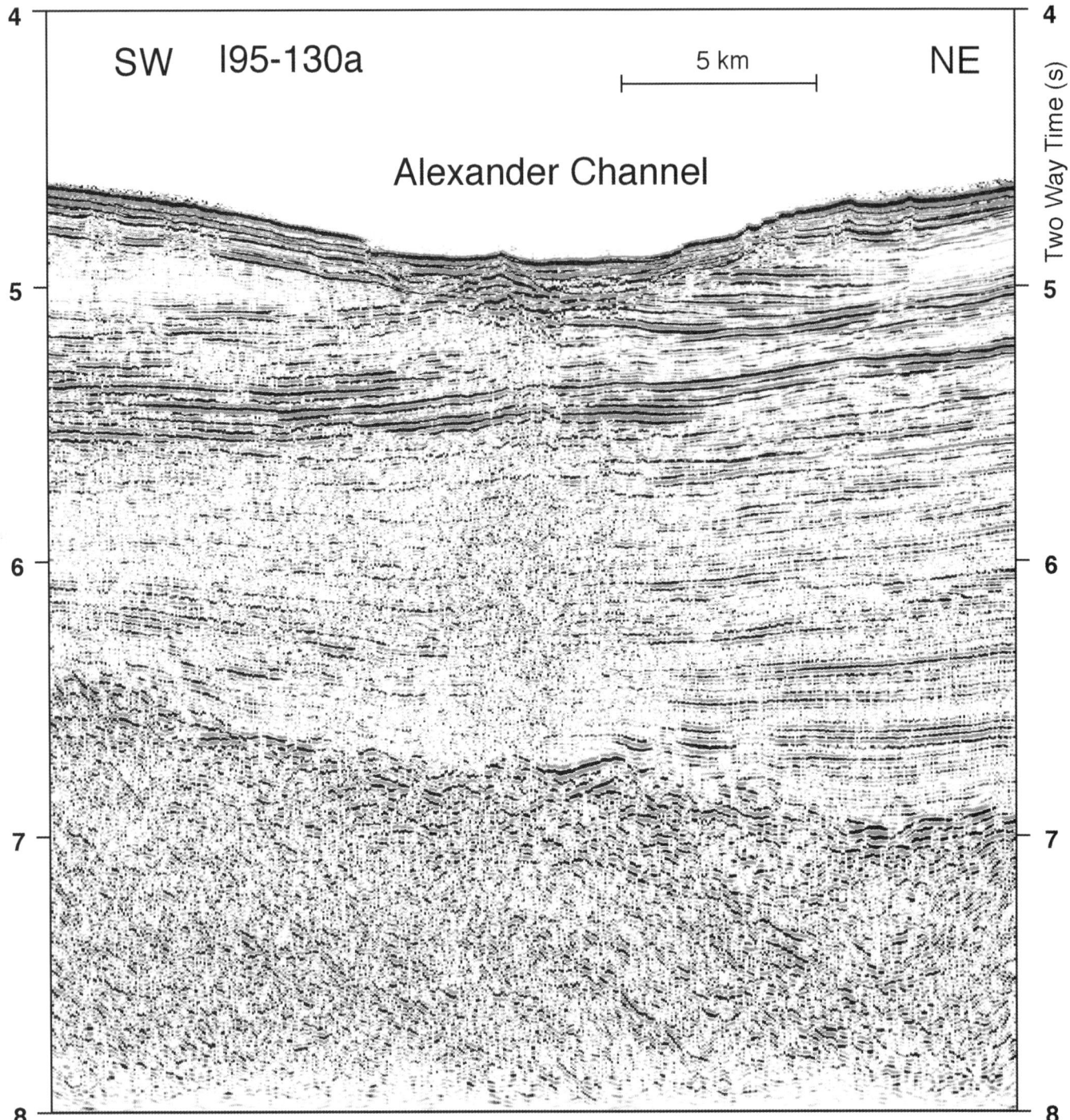

Fig. 10. Seismic profile I95-130a showing Alexander Channel that separates drift 6 and 7. The acoustic facies of the channel levee systems is very different from that of the drift. The channel system, composed by two branches, is evidenced by high amplitude, discontinuous, reflectors at its floor, surrounded by transparent facies. A levee, evident on the NE of the channel, is characterized by relatively well stratified and rapidly wedging-out deposits. Older, buried channels with dimensions comparable to the modern one are possibly detectable within the inter-drift area.

reflection coverage exists for this margin, acquired mainly on board R/V OGS Explora in 1990, 1992, 1995, and 1997 but with minor contributions also from UK, USA and Spanish cruises. The northeastern half is covered also by GLORIA sidescan and single-channel (watergun) seismic survey reported by Tomlinson *et al.* (1992) and Rebesco *et al.* (1996).

The majority of the mounds (drifts 1, 2, 3, 4A, and 7) have an asymmetric external shape with a steeper, rougher SW side and a gentler, smoother NE side. Two mounds (drifts 5 and 8) have the opposite geometry (NE sides are steeper). Drift 6 and possibly drift 4 both consist of a concave-up plateau formed by the gentler sides of two coalesced drifts. The remaining three mounds (mounds 3A, 4B and 5A) are smaller and more symmetrical. Dip seismic profiles show that the mounds generally have gently-dipping NW sides merging with the lower continental rise, and steeply-dipping SE sides facing the continental slope.

The internal structure of these mounds is different beneath the gentle and the steeper sides (Fig. 9). The steeper sides are characterized by high reflectivity, abundant lateral terminations and undulations, and sea-floor terminations of reflectors, by either erosion or non-deposition. Prominent changes in reflectivity parallel to the seafloor, probably produced by a diagenetic change

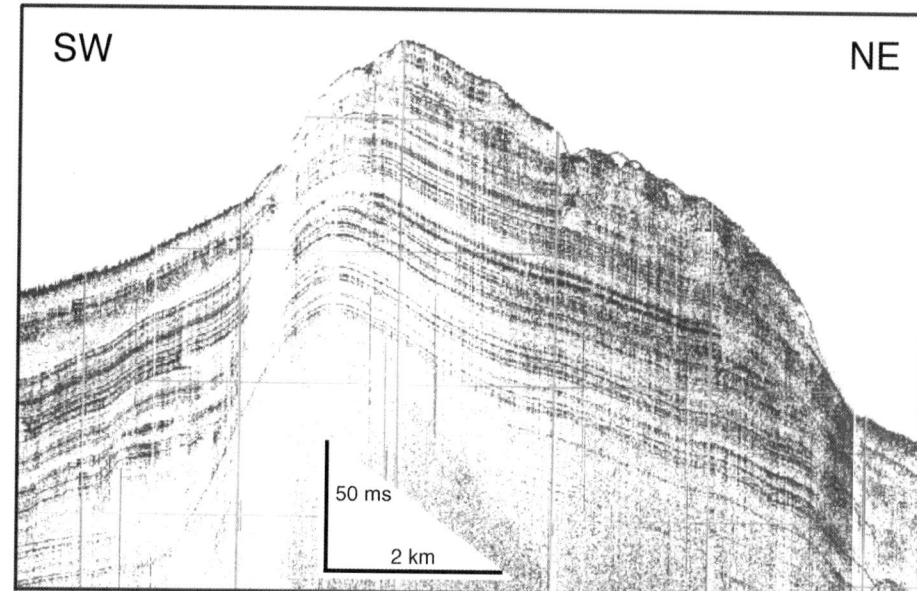

Fig. 11. Parametric source (TOPAS) profile across the crest of drift 3. Note the dominant parallel stratified facies, and the chaotic and transparent facies near the surface on the NE side, related to local lateral instability. Vertical exaggeration x 33. See location in Figure 1 (see also Figures 4 and 12).

deep in the sediments, are particularly evident on the steep sides of the mounds. In contrast, the gentler sides are relatively smooth, and are underlain by continuous, highly reflective units, with linear, parallel or sub-parallel internal reflectors conformable with the sea floor.

The mounds are separated by deep-sea channels traversing the deep areas between mounds. Channels are only evident between the mounds and not on the mounds themselves. The acoustic facies of these channel levee systems is very different from that of the mounds. Channel floors are characterised by high amplitude, discontinuous, reflectors, surrounded by transparent facies (Fig. 10). Levees, where present, are shown by relatively well stratified deposits abruptly wedging out at the side of the channel. Buried channels with dimensions comparable to the modern ones are detectable in seismic profiles within the three upper sequences. They show a limited lateral shift and are confined to the intermound areas. In the three deeper seismic sequences, the channels are replaced by enigmatic northeast dipping seismic reflectors that appear to cut across other horizons. These are interpreted to represent the traces of local, diachronous hiatuses of limited temporal extent enclosed within continuous depositional areas (Rebesco *et al.* 1997).

The history of sedimentation on the continental rise was reconstructed by Rebesco *et al.* (1997), who identified six major seismic units. Units were dated (very approximately in some cases) by correlation with DSDP Site 325 and known tectonic events, and by regional climatic events. This history is summarized in a three-stage scheme, as follows: (1) a 'Drift-maintenance Stage' (5 Ma to the present) characterized by preservation and enhancement of the elevation of the drifts; (2) a 'Drift-growth Stage' (15–5 Ma) showing substantial variations in thickness as a consequence of increasing bottom current activity and larger glacial sediment supply from the margin; (3) a 'Pre-drift Stage' (36–15 Ma) characterized mainly by subparallel reflectors representing a dominantly turbiditic sequence.

Sub-bottom profiles

The forty parametric source (TOPAS) profiles acquired during the GEBRAP'96 cruise onboard BIO Hesperides show that the outer continental shelf and slope are characterized by opaque and hyperbolic acoustic facies, while the continental rise displays mainly parallel stratified and chaotic acoustic facies, supplemented by opaque and hyperbolic facies. The mounds are mostly represented by parallel stratified facies. The generally high lateral continuity of this facies is locally interrupted by sheet and lens shaped bodies characterized by chaotic to transparent facies (Fig. 11). The crests of the mounds commonly have small scarps on both sides. Channel heads are mainly occupied by opaque and hyperbolic acoustic facies (Fig. 12a). Channels become highly erosive downslope (Fig. 12b), although they can also be filled and smoothed by massive deposits showing chaotic and transparent facies (Fig. 12c). Opaque acoustic facies prevail in the deepest, widest channel reaches (Fig. 12d).

Sediments

Sediments: seabed photos

Studies of bottom photographs from the area by Hollister & Heezen (1967), Sullivan *et al.* (1973) and Dangeard *et al.* (1977) showed evidence for strong bottom currents (in the form of aligned motile organisms, partly obliterated bioturbation structures, and scours round small obstacles) on the continental slope near Adelaide Island and in the vicinity of the Polar Front. Turbid bottom water was observed on the lower continental rise west of 90°W. Elsewhere on the continental rise, bottom photographs showed abundant tracks, trails and faecal structures, with scattered ice-rafted pebbles, suggesting weak bottom currents (Fig. 13).

Sediments: core description and facies

A total of 17 gravity cores have been collected from drift 7 and 15 from the other drifts (Camerlenghi *et al.* 1997b; Pudsey & Camerlenghi 1998; Pudsey 2000; Lucchi *et al.* in press). The sediments cored on drift 7 are predominantly terrigenous in composition and very fine-grained; sediment facies confirm the extent of the drift inferred from seismic data. Cores show a cyclicity between brown, bioturbated, diatom-bearing mud with foraminifera (core tops and additional thin units downcore) and grey, laminated, barren mud (thicker units between the brown layers). Contacts between brown and grey units are gradational and bioturbated. Ice-rafted debris is present but generally sparse. Cores on the steep sides of the drift recovered a condensed section with thinner cycles and probable hiatuses.

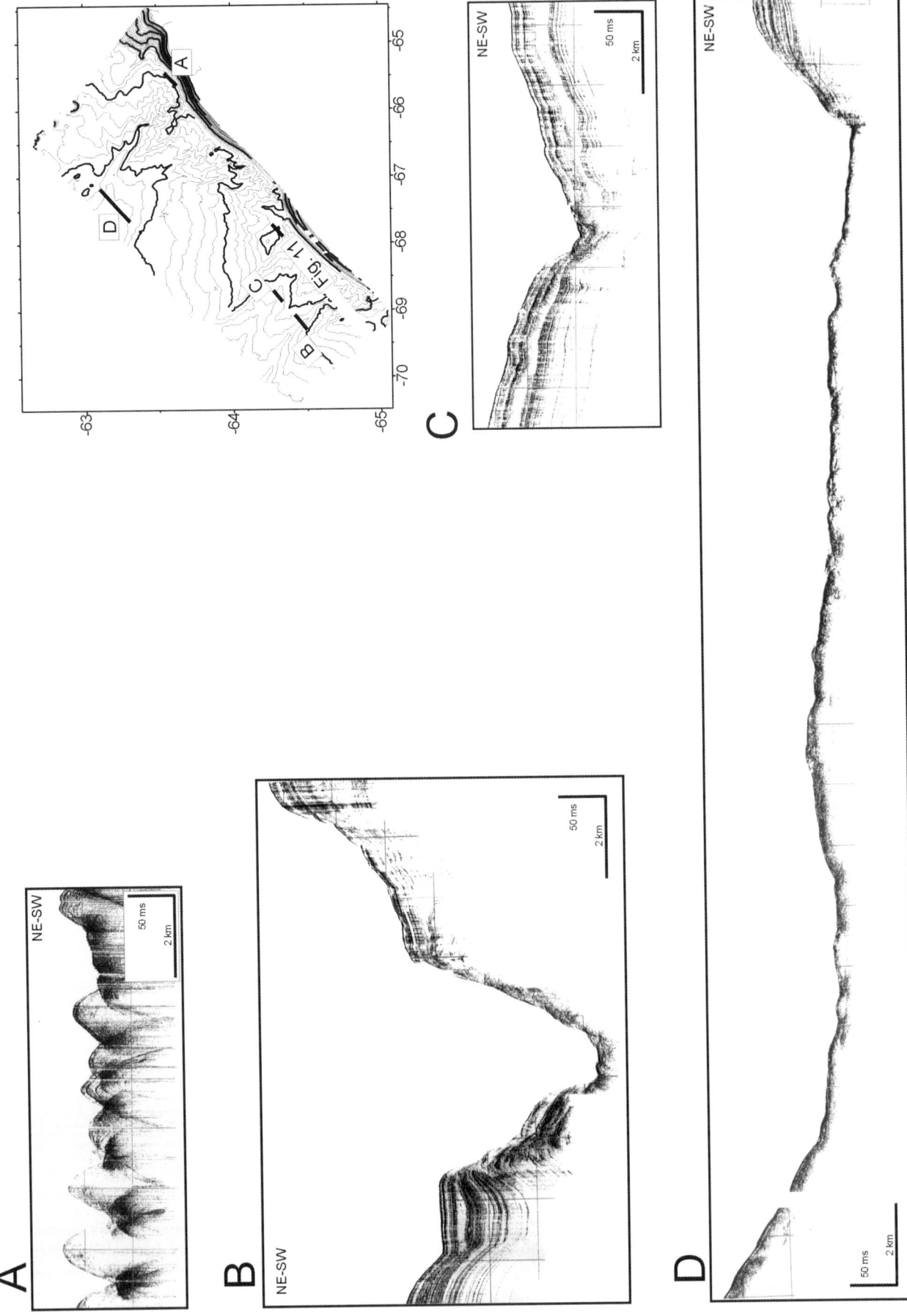

Fig. 12. TOPAS profiles illustrating the inter-drift channelled drainage systems on the continental rise, from upper to lower channel reaches. (a) Hyperbolic facies typical of the uppermost reaches of small erosive channels. (b) Erosive channel showing chaotic and hyperbolic facies in the thalweg and southwest wall; the east wall is slumped. (c) Mud-flow filled channel reach; post-flow erosion in the thalweg is also apparent. (d) Distal channel 12 km wide, characterized by chaotic, opaque and hyperbolic facies; inner small channels eroding the main channel floor can be also observed. Vertical exaggeration × 33. See location in Figure 1, see also Figure 4.

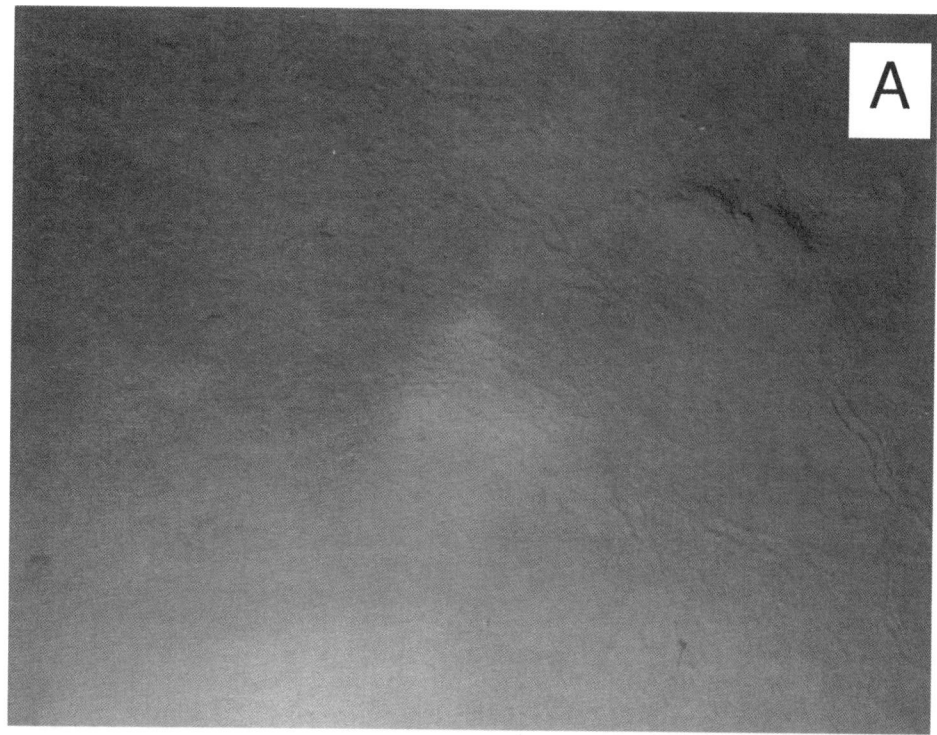

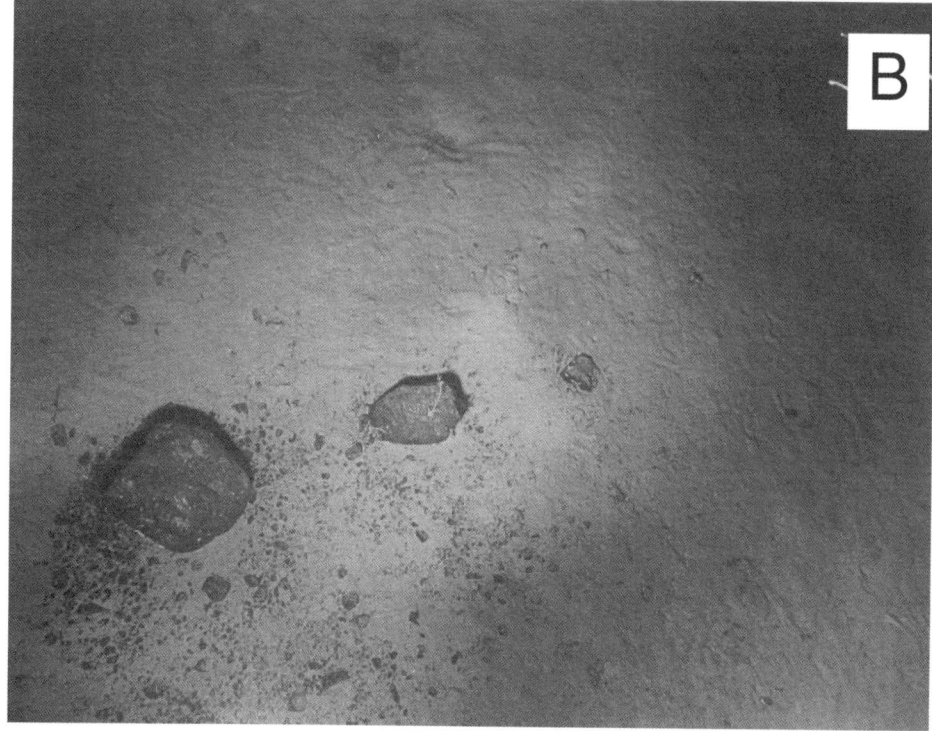

Fig. 13. Seabed photos taken in 1963–1964 from USNS Eltanin (Goodell 1964, 1965). (**a**) Eltanin 5-16 (63° 58'S, 67° 56'W, 2950 m water depth) Pale brown clayey silt (4.3% sand, 71.1% silt, 24.6% clay). Many echinoid trails and worm tracks can be seen. Sparse pebbles, no evidence for current activity; (**b**) Eltanin 5-22 (65° 06'S, 70° 41'W, 3109 m) Dark yellowish brown clayey silt (7.0% sand and gravel, 60.7% silt, 32.3% clay). Echinoid trails and worm tracks similar to above; note also patchy distribution of coarse angular ice-rafted debris.

The top of each core (Unit A of Pudsey & Camerlenghi 1998) consists of 0.1–0.2 m of structureless olive brown diatom-bearing mud with common dispersed IRD, overlying 0.2–0.4 m of olive brown to grey or dark grey bioturbated, faintly laminated mud. Carbonate content (foraminifer fragments) is 0.4–0.8% and organic carbon up to 0.4%, both decreasing downwards in the unit. The burrows are mainly of *Planolites* type and parallel to bedding.

Below Unit A, grey and dark grey silty clay forms a thick unit (2.2–5.5 m thick) in each core, designated Unit B. Parallel to lenticular lamination and thin bedding are outlined by slight contrasts in colour and in X-ray absorption. The lamination is rather indistinct and irregular; there are neither sharp-based graded units nor any signs of erosion. Bioturbation is rare and the burrows are very small (1–2 mm). Some distinctive reddish marker layers 1–5 cm thick can be correlated from core to core; in X-radiographs, these layers are seen to be laminated on a sub-millimetre scale. Wavy lamination is common in the upper part of Unit B in most cores, while the lower part tends to be more homogeneous. IRD is less common than in Unit A and occurs in thin layers (3–15 mm) rather than being dispersed. Unit B is thickest near the centre and north-east side of the drift, and thins markedly to the southeast, southwest and northwest.

The next brown unit, Unit C, attains a thickness of 1.0 m in cores 6 and 7; in core 9 its apparent base may be a hiatus (Fig. 14). It consists of light olive brown mud (8–12% diatoms, 4–6% foraminiferal carbonate) overlying olive grey to greyish brown mud (2–5% diatoms, 1% carbonate) then olive brown to light

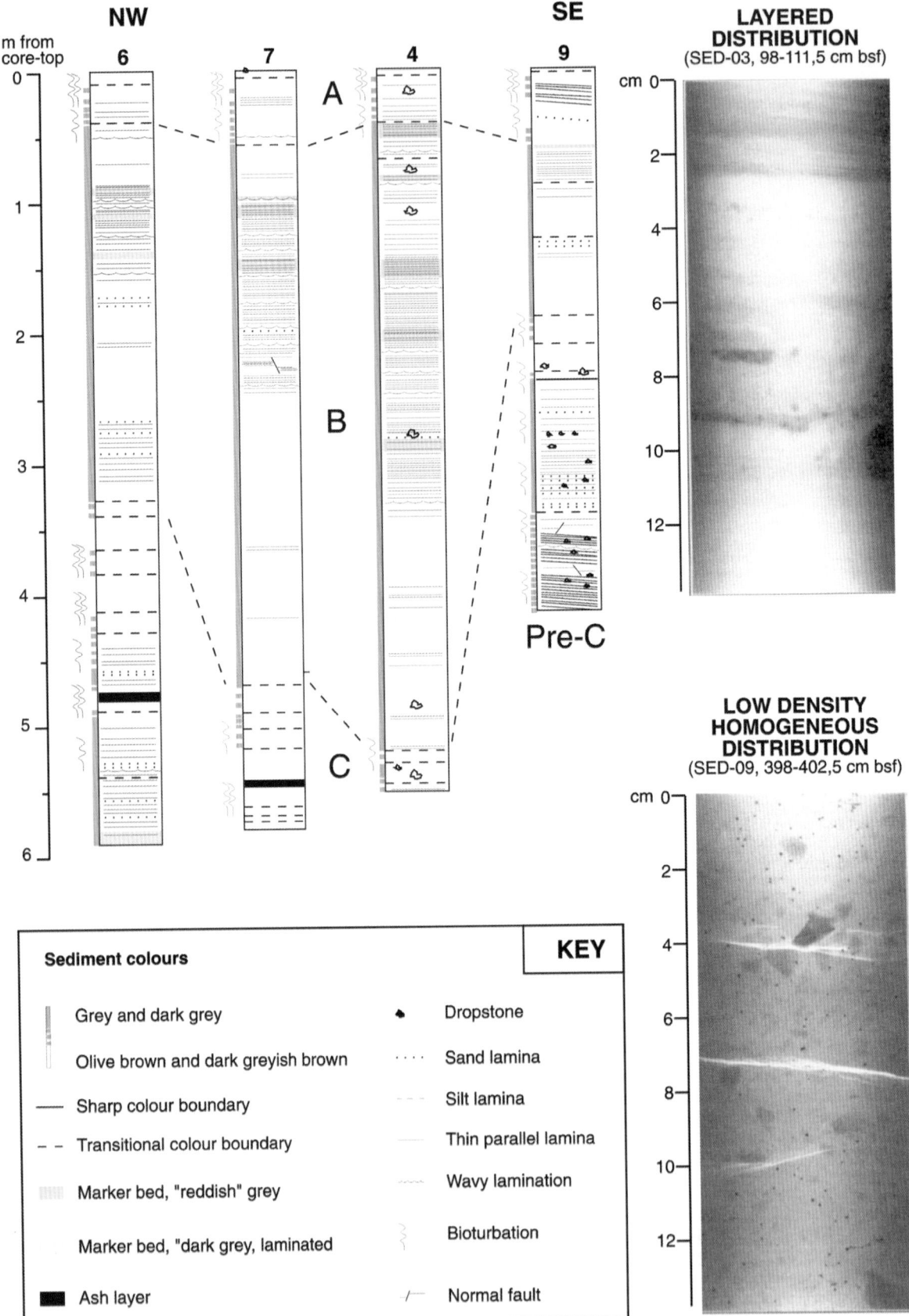

Fig. 14. Detailed core logs from visual descriptions, dip transect on drift 7 (Fig. 1). Only the larger ice-rafted pebbles are shown. Units A–D were defined using sediment colour changes combined with magnetic susceptibility data. 'Pre-C' denotes older sediments below a sharp contact at the base of unit C Also: X-radiographs of (**a**) homogeneous and (**b**) layered IRD occurrence in core 9. The photographs illustrate the central 6 cm of the 9 cm core diameter. The small black dots in (**a**) are manganese micronodules; the white lines are coring-induced fractures. From Pudsey & Camerlenghi (1998).

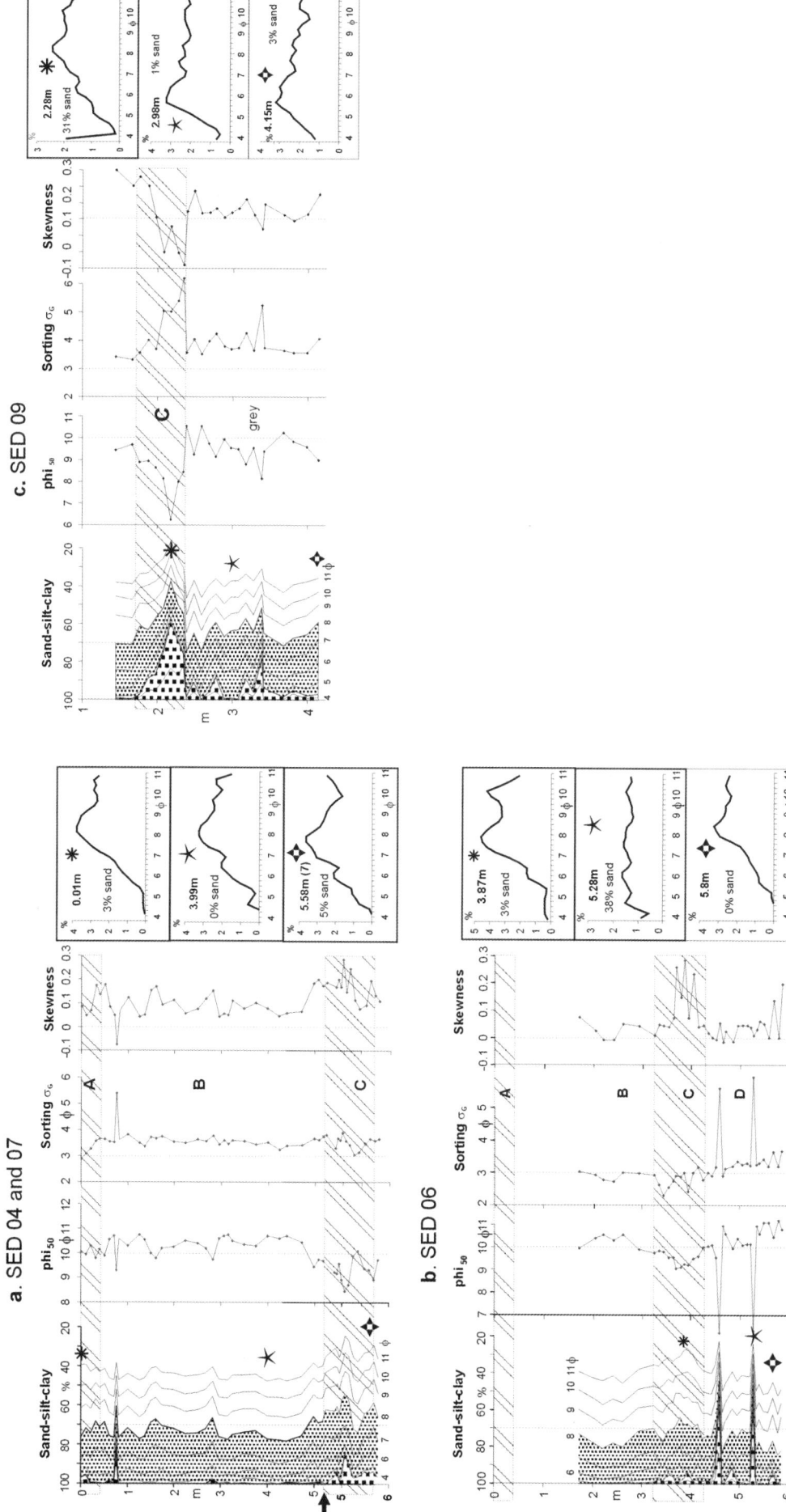

Fig. 15. Grain-size data, measured by sieving (sand %) and Sedigraph (silt and clay). (**a**) Cores 4 and 7, spliced together at top of Unit C (arrow). Size frequency histograms of the fine fraction (100% = total sediment) at 0.25Φ class interval at core top (0.01 m), 3.99 m and 5.58 m correspond to the star symbols on the downcore plot. Note that there is 30–45% of material finer than the Sedigraph measurement limit of 11Φ. There is commonly a mode near the silt-clay boundary. (**b**) Core 6, including size frequency histograms for units C and D. The sample at 5.28 m contains 38% sand. (**c**) Core 9, most of Unit C shows negative skewness. The very firm sediments underlying Unit C are all very poorly sorted, but have variable sand:silt:clay ratios and median diameters. They have more medium and coarse silt (10–15% in the range 4–6Φ) than the other cores, and most samples have a mode in the range 5.5–6Φ. From Pudsey & Camerlenghi (1998).

olive brown mud (20–27% diatoms, 4–16% carbonate). The planktonic foraminifer *Neogloboquadrina pachyderma* (sinistral) is the most common species. Colour transitions are gradational, and any original sedimentary structures have been obliterated by bioturbation. Dispersed IRD is common (Fig. 14c).

Core 6 (lowest 1.6 m) recovered another unit of grey laminated mud, Unit D, also seen in the lowest 0.1 m of core 7. Unit D is faintly laminated and bioturbated and the contact with Unit C is gradational and burrowed. IRD is common and occurs mainly dispersed in the upper half of the unit, and in thin layers in the lower half. Additional cores from the distal part of drift 7 and in the adjacent channels contain coarse-grained turbidite beds.

Sediments: summary analytical results

Particle size analyses on selected cores from drift 7 show the sediments are fine-grained and very poorly sorted. We present the data as downcore plots of sand-silt-clay, median diameter, sorting and skewness, with representative size frequency histograms of the fine fraction, in Figure 15.

Unit A was measured in all core tops except core 9, and downcore in core 4 (Fig. 15a). It contains 2–4% sand, 21–29% silt and 67–77% clay. There is no evident relationship between grain size and position of the core on the drift. Median grain size is about 10 Φ, there is a mode in the range 8–9 Φ and sorting is very poor (σ_G about 3).

Unit B generally has less than 1% sand, 21–28% silt, and >70% clay (Fig. 15a, b). Median grain size is 10–10.5 Φ and the mode is in the range 7.3–8.4 Φ with very poor sorting (σ_G about 3.5). Some samples have up to 32% silt with a weak secondary mode at 6–6.5 Φ. Throughout Units A and B, the proportion of medium and coarse silt (4–6 Φ) is very low, generally <8% of total sediment. Unit C contains 2–6% sand (locally up to 30–40%), 30–37% silt and 60–68% clay. Up to half of the sand is coarse (>0.5 mm), angular and interpreted as ice-rafted. Core 6 has somewhat less silt and more clay. Median grain size is 9–9.5 Φ and the mode is at 7.5–8 Φ, commonly with a secondary mode at 5.5–6 Φ. Sorting is very poor with σ_G of 2.5–3.5.

Unit D in core 6 has generally less than 1% sand except in two thin layers at 4.58 and 5.28 m, where both the sand and the fine fraction are extremely poorly sorted (Fig. 15b). Elsewhere silt forms 18–28% and clay 72–82%. Median diameter is 10–11.2 Φ, mode 8–8.5 Φ and sorting is very poor (σ_G = 3–3.5).

Discussion

Supply and transport of sediment in the late Quaternary

As shown by core data, drift 7 sediments are predominantly terrigenous, very fine-grained, very to extremely poorly sorted, and generally lack a mode in the silt size range. These features point to deposition from suspension with negligible current winnowing. In particular, the positive (fine) skewness throughout most cores attests to the absence of any process that removed fine material. The indistinct, parallel to lenticular lamination in Units B and D suggests small and irregular fluctuations in the supply and transport of suspended mud. The style of lamination is very similar to 'plumites' on the Labrador continental margin (Hesse *et al.* 1997). Such lamination may also have been present in Units A and C, prior to thorough bioturbation. These sediments are transitional between hemipelagites and muddy contourites. The absence of graded laminated units (see Stow & Bowen 1980), or indeed of any sharp-based silt-sand beds except in condensed sections from the steep sides of the drift, indicates a turbidite origin is unlikely.

The similarity of fine-fraction size distributions downcore suggests there was little variation in bottom current strength between glacial and interglacial parts of cycles. In each core, unit C has the coarsest median grain size and least poor sorting, which is evidence for marginally stronger currents in Unit C time. Grain-size variations that are more clearly attributable to bottom-current activity occur only in cores on the steep sides of the drift, where sandy, negatively skewed samples may have had fine material winnowed out by currents. Their texture contrasts with the unsorted sandy muds in Unit D in core 6, which have near-zero skewness and are thought to result from ice-rafting without current sorting.

The benthic nepheloid layer can be supplied with suspended sediment by a number of mechanisms, including: (1) current erosion farther upstream; (2) entrainment of material from turbidity currents; and (3) pelagic settling from the sea surface, including biogenic and glacially derived material (McCave 1986). On the Antarctic Peninsula Pacific margin, bottom currents are weak and rather steady, so mechanism (1) is probably insignificant. Mechanisms (2) and (3) are both important, and the amount and type of sediment supply are likely to have varied over glacial-interglacial cycles (Pudsey & Camerlenghi 1998).

Glacial Unit B is thickest near the centre and northeast side of drift 7, and thins to the southeast, southwest and northwest. This suggests that most fine sediment was supplied to the contour-following current and nepheloid layer near the slope, or from the channel on the north-east side of the drift (consistent with the entrainment of the fine fraction of turbidity currents; Fig. 16). The relative importance of supply to the nepheloid layer by turbidity currents and by meltwater plumes is, as yet, unknown.

Large scale geometry and depositional model

As pointed out by Rebesco *et al.* (1996), the largest mounds are most plausibly interpreted as sediment drifts. They cannot be explained as large levees on the northeastern side of the channels, because the Coriolis effect would cause overbank deposition southwest of the channels, with a short, steep northeast slope, which is the opposite of what we observe.

A generic model (Fig. 17), applicable to the (Plio-Pleistocene) drift maintenance stage, and most probably also to the preceding stage of rapid drift growth (late Miocene), was proposed by Rebesco *et al.* (1997). It takes into account the entrainment of material from turbidity currents as the major source of supply to the benthic nepheloid layer. The model shows a section through the axis of a progradational lobe and the adjacent drift during glacial maximum when a grounded ice stream transported unsorted basal till to the continental shelf edge. Small-scale slumps on the uppermost slope undergo downslope transition into debris flows, then turbidity currents feeding the main channel via tributaries on the uppermost rise. Suspended fines are entrained in SW-flowing bottom currents and deposited down-current to develop and maintain the drifts. However, subsequent turbidity current flow in the channels and slope instability on the steeper drift slopes tend to remove sediment from those areas, leaving a permanent sediment increment only on the gentle sides of the drifts.

The 12 mounds are hence generally interpreted as sediment drifts controlled by along slope bottom-water flow with varying degree of interaction with downslope turbidity current processes. These mounds provide a complete range of intermediary steps between two end members: the sediment drift (best represented by drift 7) and the channel levee (best represented by mound 5A, the SW levee of the north Tula channel).

The criteria that we consider as diagnostic to distinguish sediment drifts from channel-levees are the following:

(a) asymmetry: drifts have one side distinctly steeper than the other;
(b) orientation of the steep side: in the drifts, the steep side is the

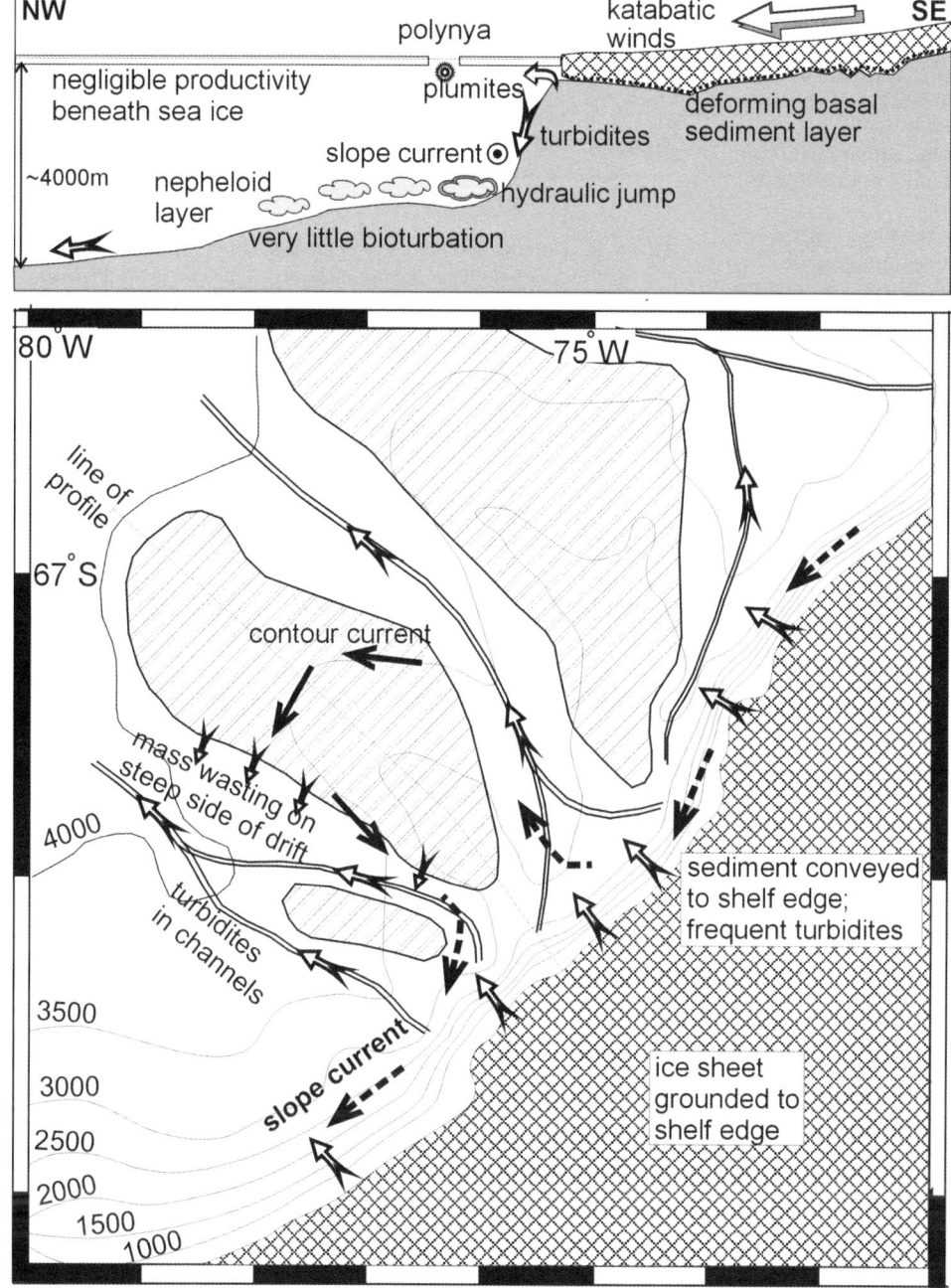

Fig. 16. Inferred sediment transport processes during glacial periods in the area of drift 7. See location in Figure 1. Downslope flow shown as open arrows, alongslope flow shown as black arrows where measured, dotted where inferred (after Pudsey & Camerlenghi 1998).

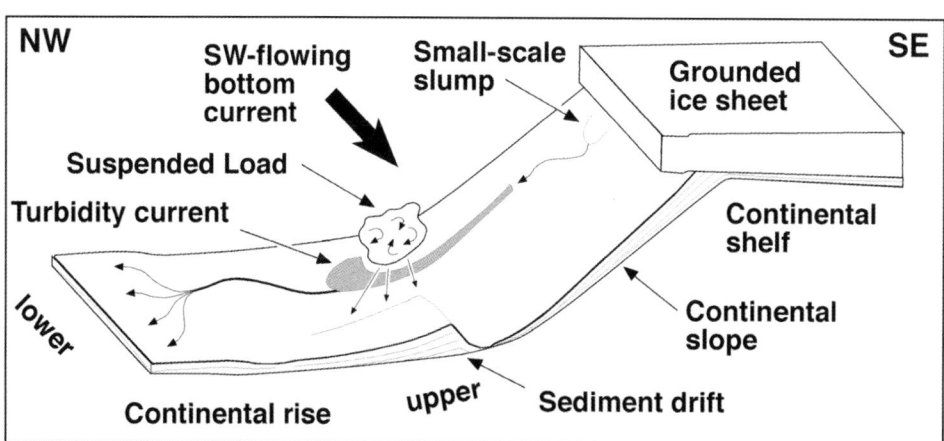

Fig. 17. Synthesis cartoon model of the depositional and oceanographic processes inferred to occur along the Antarctic Peninsula Pacific margin during a glacial maximum (after Rebesco et al. 1997).

SW one, hence not facing the NE upstream (feeding) channel;
(c) large dimensions: drifts are generally wider than 40 km;
(d) large elevation above the adjacent channels: the crest of the drifts rises at least 400 m above the adjacent thalweg;
(e) internal geometry: drifts have continuous, subparallel, conformable reflectors beneath the gentler side, and truncated, chaotic reflectors beneath the steeper side;
(f) relationship with the adjacent continental slope: drifts are mostly separated from the slope by erosive depressions;
(g) location: drifts are preferentially located in correspondence to the present-day shelf troughs, in between the shelf lobes.

This work has been funded by the Progetto Nazionale Ricerche in Antartide (PNRA) through the SEDANO (Sediment Drifts of the Antarctic Offshore) Project, the British Natural Environment Research Council (NERC), and the Spanish cooperative project 99120 from 'Comisión de Intercambio Cultural, Educativo y Científico entre España y los Estados Unidos de América'. Kristeen Roessig at the Antarctic Research Facility, Florida State University, kindly supplied the photographs in Figure 13.

References

BARKER, P. F. 1995. The proximal marine sediment record of Antarctic climate since the late Miocene. *In*: COOPER, A. K., BARKER, P. F. & BRANCOLINI, G. (eds) *Geology and Seismic Stratigraphy of the Antarctic Margin*. American Geophysical Union, Antarctic Research Series, **68**, 25–58.

BARKER, P. F., CAMERLENGHI, A., ACTON, G. D., ET AL. 1999. Proc. ODP, Init. Repts., 178 [CD-ROM]. Available from: Ocean Drilling Program, Texas A&M University, College Station, TX 77845-9547, U.S.A.

CAMERLENGHI, A., CRISE, A., ACCERBONI, E., LATERZA, R., PUDSEY, C. J. & REBESCO, M. 1997a. Ten-month observation of the bottom current regime across a sediment drift of the Pacific margin of the Antarctic Peninsula. *Antarctic Science*, **9**, 424–431.

CAMERLENGHI, A., REBESCO, M. & PUDSEY, C. J. 1997b. High resolution terrigenous sedimentary record of a sediment drift on the continental rise of the Antarctic Peninsula pacific margin (initial results of the 'SEDANO' Program). *In*: RICCI, C. A. (ed.) *The Antarctic Region: Geological Evolution and Processes*. Terra Antarctica Publication, Siena, 705–710.

CANALS, M., ESTRADA, F., URGELES, R. & GEBRAP 96/97 TEAM 1998. Very high-resolution seismic definition of glacial and postglacial sediment bodies in the continental shelves of the northern Trinity Peninsula region, Antarctica. *Annals of Glaciology*, **27**, 260–264.

CANALS, M., URGELES, R. & CALAFAT, A. M. 2000. Deep sea-floor evidence of past ice streams off the Antarctic Peninsula. *Geology*, **28**, 31–34.

CARMACK E. C. 1990. Large-scale physical oceanography of the polar ocean. *In*: SMITH, W. O. JR (ed.) *Polar Oceanography*. Academic Press, London.

COOPER, A. K., BARRETT, P. J., HINZ, K., TRAUBE, V., LEITCHENKOV, G. & STAGG, H. M. J. 1991. Cenozoic prograding sequences of the Antarctic continental margin: a record of glacio-eustatic and tectonic events, *Marine Geology*, **102**, 175–213.

DANGEARD, L., VANNEY, J. R. & JOHNSON, G. L. 1977. Affleurements, courants et facies dans la Zone Antarctique du Pacifique orientale (Mers de Bellingshausen et d'Amundsen). *Annales de l'Institut océanographique*, Paris, **53**, 105–124.

DOMACK, E. W., MCCLENNEN, C., MANLEY, P. & ISHMAN, S. 1994. Very high resolution stratigraphy of the Late Quaternary glacial marine sediments in fjords and offshore basins, Antarctic Peninsula. *Terra Antartica*, **1**, 269–270.

GERSONDE, R. & BURCKLE, L. H. 1990. Neogene diatom biostratigraphy of ODP Leg 113, Weddell Sea. *In*: BARKER, P. F, KENNETT, J. P. ET AL. (eds) *Proceedings of the Ocean Drilling Program, Scientific Results*, **113**. Ocean Drilling Program, College Station, Texas, 761–789.

GOODELL, H. G. 1964. *Marine Geology of the Drake Passage, Scotia Sea and South Sandwich Trench (USNS Eltanin cruises 1–8)*. PhD thesis, Florida State University, Tallahassee, Florida.

GOODELL, H. G. 1965. *Marine Geology: USNS Eltanin cruises 9–15*. Florida State University, Tallahassee, Florida.

GORDON, A. L. 1966. Potential temperature, oxygen and circulation of bottom water in the Southern Ocean. *Deep-Sea Research*, **13**, 1125–1138.

HAYS, J. D. 1965. Radiolaria and late Tertiary and Quaternary history of Antarctic Seas. *In*: LLANO, G. A. (ed.) *Biology of the Antarctic Seas II*. American Geophysical Union, Antarctic Research Series, **5**, 125–184.

HESSE, R., KHODABAKHSH, S., KLAUCKE, I. & RYAN, W. B. F. 1997. Asymmetrical turbid-plume deposition near ice-outlets of the Pleistocene Laurentide ice sheet in the Labrador Sea. *Geo-Marine Letters*, **17**, 179–187.

HOLLISTER, C. D. & HEEZEN, B. C. 1967. The floor of the Bellingshausen Sea. *In*: HERSEY, J. B. (ed.) *Deep-Sea Photography*. Johns Hopkins Press, Baltimore, 177–189.

HOLLISTER, C. D. CRADDOCK, C., ET AL. 1976. *Initial Reports of the Deep Sea Drilling Project*, **35**, Washington (U.S. Govt Printing Office), 930 pp.

LARTER, R. D. & BARKER, P. F. 1989. Seismic stratigraphy of the Antarctic Peninsula Pacific margin: A record of Pliocene-Pleistocene ice volume and paleoclimate. *Geology*, **17**, 731–734.

LARTER, R. D. & BARKER, P. F. 1991a. Effects of ridge crest-trench interaction on Antarctic-Phoenix spreading: forces on a young subducting plate. *Journal of Geophysical Research*, **96**, 19583–19607.

LARTER, R. D & BARKER, P. F. 1991b. *Neogene interaction of tectonic and glacial processes at the Pacific margin of the Antarctic Peninsula*. International Association of Sedimenologists, Special Publications, **12**, 165–186.

LARTER, R. D. & CUNNINGHAM, A. P. 1993. The depositional pattern and distribution of glacial- interglacial sequences on the Antarctic Peninsula Pacific Margin. *Marine Geology*, **109**, 203–219.

LARTER, R. D., REBESCO, M., VANNESTE, L. E., GAMBOA, L. A. P. & BARKER, P. F. 1997. Cenozoic Tectonic, Sedimentary and Glacial History of the Continental Shelf West of Graham Land, Antarctic Peninsula. *In*: COOPER, A. K. & BARKER, P. F. (eds) *Geology and Seismic Stratigraphy of the Antarctic Margin, Part 2*. American Geophysical Union, Antarctic Research Series, **71**, 1–27.

LUCCHI, R. G., REBESCO, M., BUSETTI, M., CABURLOTTO, A., COLIZZA, E. & FONTOLAN, G. in press. Sedimentary Processes and Glacial Cycles on the Sediment Drifts of the Antarctic Peninsula Pacific Margin: Preliminary Results of SEDANO-II Project. *In*: GAMBLE, J., SKINNER, D. & HENRYS, S. (eds) *Proceedings of the VIII° International Symposium on Antarctic Earth Sciences*. New Zealand Journal of Geology and Geophysics, Royal Society of New Zealand.

MCCAVE, I. N. 1986. Local and global aspects of the bottom nepheloid layer in the world ocean. *Netherlands Journal of Sea Research*, **20**, 167–181.

MCGINNIS, J. P. & HAYES, D. E. 1995. The roles of down-slope and along-slope depositional processes: southern Antarctic Peninsula margin. *In*: COOPER, A. K., BARKER, P. F. & BRANCOLINI, G. (eds) *Geology and Seismic Stratigraphy of the Antarctic Margin*. American Geophysical Union, Antarctic Research Series, **68**, 141–156.

MCGINNIS, J. P., HAYES, D. E. & DRISCOLL, N. W. 1997. Sedimentary processes across the continental rise of the southern Antarctic Peninsula. *Marine Geology*, **141**, 91–109.

NOWLIN, W. D. JR. & KLINCK, J. M. 1986. The physics of the Antarctic Circumpolar Current, *Reviews of Geophysics*, **24**, 469–491.

NOWLIN, W. D. & ZENK, W. 1988. Westward bottom currents along the margin of the South Shetland Island Arc. *Deep Sea Research*, **35**, 269–301.

POPE, P. G. & ANDERSON, J. B. 1992. Late Quaternary glacial history of the northern Antarctic Peninsula's western continental shelf: evidence from the marine record. *In*: ELLIOT, D. H. (ed.) *Contributions to Antarctic Research III*. American Geophysical Union, Antarctic Research Series, **57**, 63–91.

PUDSEY, C. J. 2000. Sedimentation on the continental rise west of the Antarctic Peninsula over the last three glacial cycles. *Marine Geology*, **167**, 313–338.

PUDSEY, C. J., BARKER, P. F. & LARTER, R. D. 1994. Ice sheet retreat from the Antarctic Peninsula shelf. *Continental Shelf Research*, **14**, 1647–1675.

PUDSEY, C. J. & CAMERLENGHI, A. 1998. Glacial-interglacial deposition on a sediment drift on the Pacific margin of the Antarctic Peninsula. *Antarctic Science*, **10**, 286–308.

REBESCO, M., LARTER, R. D., BARKER, P. F., CAMERLENGHI, A. & VANNESTE, L. E. 1994. The history of sedimentation on the continental rise west of the Antarctic Peninsula. *Terra Antartica,* **1**, 277–279.

REBESCO M., LARTER R. D., CAMERLENGHI, A. & BARKER, P. F. 1996. Giant sediment drifts on the continental rise west of the Antarctic Peninsula. *Geo-Marine Letters*, **16**, 65–75.

REBESCO, M., LARTER, R. D., BARKER, P. F., CAMERLENGHI, A. & VANNESTE, L. E. 1997. History of Sedimentation on the Continental Rise West of the Antarctic Peninsula. *In*: COOPER, A. K. & BARKER, P. F. (eds) *Geology and Seismic Stratigraphy of the Antarctic Margin (Part 2)*. American Geophysical Union. Antarctic Research Series, **71**, 29–49.

REBESCO, M., CAMERLENGHI, A., DE SANTIS, L., DOMACK, E. & KIRBY, M. 1998a. Seismic stratigraphy of Palmer Deep: a fault bounded Late Quaternary sediment trap on the inner continental shelf, Antarctic Peninsula Pacific margin. *Marine Geology*, **15**, 89–110.

REBESCO, M., CAMERLENGHI, A. & ZANOLLA, C. 1998b. Bathymetry and morphogenesis of the continental margin west of the Antarctic Peninsula. *Terra Antartica*, **5**, 715–725.

STOW, D. A. V. & BOWEN, A. J. 1980. A physical model for the transport and sorting of fine-grained sediment by turbidity currents. *Sedimentology*, **27**, 31–46.

SULLIVAN, L., THORNDIKE, E., EWING, M. & EITTREIM, S. 1973. *Nephelometer measurements, Hach turbidity measurements and bottom photographs from Conrad Cruise 15*. Lamont-Doherty Geological Observatory Technical Report **8-CU-8-73**.

TOMLINSON, J. S., PUDSEY, C. J., LIVERMORE, R. A., LARTER, R. D. & BARKER, P. F. 1992. Long-range sidescan sonar (GLORIA) survey of the Antarctic Peninsula Pacific Margin. *In*: YOSHIDA, Y., KAMINUMA, K. & SHIRAISHI, K. (eds) *Recent Progress in Antarctic Earth Science*. Terra Sci Publications, Tokyo, 423–429.

TUCHOLKE, B. E. & HOUTZ, R. E. 1976. Sedimentary record of the Bellingshausen Basin from seismic profile data. *In*: HOLLISTER, C. D., CRADDOCK, C. ET AL. (eds) *Initial Reports of the Deep Sea Drilling Project*, **35**, Washington (U.S. Govt Printing Office), 197–227.

TUCHOLKE, B. E., HOLLISTER, C. D., WEAVER, F. M. & VENNUM, W. R. 1976. Continental rise and abyssal plain sedimentation in the Southeast Pacific Basin – Leg 35 DSDP. *In*: HOLLISTER, C. D., CRADDOCK, C. ET AL. (eds) *Initial Reports of the Deep Sea Drilling Project*, **35**, Washington (U.S. Govt Printing Office), 359–400.

VANNEY, J. R. & JOHNSON, G. L. 1976a. Geomorphology of the Pacific continental margin of the Antarctic Peninsula. *In*: HOLLISTER, C. D., CRADDOCK, C. ET AL. (eds) *Initial Reports of the Deep Sea Drilling Project*, **35**, Washington (U.S. Govt Printing Office), 279–289.

VANNEY, J. R. & JOHNSON, G. L. 1976b. The Bellingshausen-Amundsen basins (Southeastern Pacific): Major sea-floor units and problems. *Marine Geology*, **22**, 71–101.

Current controlled deposition on the Wilkes Land continental rise, Antarctica

C. ESCUTIA[1], C. H. NELSON[2], G. D. ACTON[1], S. L. EITTREIM[3], A. K. COOPER[3], D. A. WARNKE[4] & J. M. JARAMILLO[1]

[1]*Texas A&M University, Ocean Drilling Program, 100 Discovery Drive, College Station, Texas 77845, USA*
(e-mail: Escutia@odpemail.tamu.edu)
[2]*Texas A&M University, Oceanography Department, College Station, Texas 77845, USA*
[3]*US Geological Survey, 345 Middlefield Road, Menlo Park, California 94025, USA*
[4]*California State University, Hayward, California 94542-3088, USA*

Abstract: Turbidite, contourite and hemipelagic deposition are the main components of Wilkes Land continental rise sedimentation above the regional unconformity WL2. On the continental shelf, unconformity WL2 marks the start of shelf progradation, which is interpreted to correspond with the onset of glacial conditions in this segment of the east Antarctic margin. Unusually large (i.e. up to 900 m relief and 18 km between levee crests) channel-levee deposits, and high relief (up to 490 m) mounded contourite-style deposits develop above unconformity WL1b. Unconformity WL1b overlies unconformity WL2 and is interpreted to have formed under a fully continental glacial regime where ice streams reached the palaeo-continental shelf edge. Based on an analysis of multichannel seismic profiles and sediment cores, we differentiate three phases in the development of the sedimentary unit between WL1b and the present seafloor. From older to younger these are: Phase 1, dominated by turbidite deposition; Phase 2, dominated by turbidite and contourite deposition with significant mound building; and Phase 3, dominated by turbidite and contourite deposition without active mound building. We hypothesize that building of the mounds during Phase 2 corresponded with times of expansion of the Antarctic ice-sheet when vast amounts of sediment were eroded from the continent and continental shelf. The large amount of unsorted glacial sediment supplied to the outer shelf apparently travelled down the slope canyons and rise channels as turbidity current flows to feed the usually large continental rise channel-levee complexes. The suspended fines of the turbidity flows were then entrained in a palaeo-nepheloid layer and carried by the westward flowing palaeo-contour currents until their deposition in the mounds. During Phase 3, sediment supply to the continental rise, although important in volume and capable of turbidite and contour-current deposition, was insufficient to support further building of the mounds. We believe the decrease in sediment supply to the continental rise from Phase 2 to Phase 3 could be the result of a change on sediment depocentres, with most of the sediment supplied to the margin during Phase 3 being trapped on the continental shelf. We believe that ultimately these changes are related to the stage of glacial evolution of the continent.

Wilkes Land is the section of the East Antarctic coast from 100–142°E, and this study focuses on the continental margin off the eastern half of this coastline. Preliminary studies of sediment cores collected during the USNS-Eltanin 1968–1971 and 1979 (Payne & Conolly 1972), US Geological Survey 1984 (Hampton *et al.* 1987) and Deep Sea Drilling Project (DSDP) Sites 268 and 269 (Piper & Brisco 1975) cruises, showed that turbidity currents, contour currents, and hemipelagic settling are the main modes of sedimentation on the Wilkes Land continental rise. Cenozoic channel and overbank deposition on the Wilkes Land continental rise were also reported from multichannel seismic stratigraphic analysis by Eittreim & Smith (1987), Hampton *et al.* (1987), Tanahashi *et al.* (1987), and Wannesson (1991). Escutia *et al.* (1997, 2000) noted the importance of the interplay between turbidity currents and bottom currents in the present Wilkes Land continental rise morphology and stratigraphy. This paper illustrates the morphologic, seismic, and sedimentological character of the Antarctic Wilkes Land continental rise bottom current deposits. The paper also discusses the influence of the evolution of Antarctic glaciation in the development and evolution of these deposits.

Figure 1 shows the location of sediment cores and the grid of multichannel seismic profiles collected from the Wilkes Land margin that were available for this study. Multichannel seismic profiles were collected during the 1981–1982 austral summer by the Institute Français du Pétrole (IFP); in 1982 and 1983 by the Japan National Oil Corporation (JNOC); and in 1984 by the United States Geological Survey (USGS). The IFP seismic profiles used for this study (Wannesson *et al.* 1985) were obtained with a 2000 m, 48 channel streamer and 36 l tuned airgun array. The JNOC seismic profiles (Tsumuraya *et al.* 1985) were collected with a 600 m, 24 channel streamer and 9.2 l airgun. The seismic profiles collected by the USGS (Eittreim & Smith 1987) were collected with a 2400 m, 24 channel streamer and 21 l airgun array. The shot interval was 50 m in each case.

Geological and oceanographic background

The Wilkes Land margin is believed to have formed in the Cretaceous as a consequence of an extensional tectonic episode that separated the Antarctic and the Australian continents (Cande & Mutter 1982; Veevers 1987). The basement across the margin, as inferred from multichannel seismic data, consists of block-faulted continental crust and oceanic crust (Eittreim & Smith 1987; Eittreim 1994). The transition zone from continental to oceanic crust is characterized by deep marginal rift basins with a maximum 8 km sediment thickness (Eittreim 1994). A prominent erosional unconformity that terminates at the southern edge of the oceanic crust has been interpreted by Eittreim & Smith (1987) as the breakup unconformity separating 3 km thick syn-rift sediment below from up to 5 km thick post-rift sediment above.

Syn-rift strata are not well layered, and have discontinuous, weak reflections that dip to the north and downlap onto the base of the sequence (Eittreim & Smith 1987; Eittreim 1994). Post-rift sediments are well stratified and continuous over the oceanic crust and become less stratified and less continuous in a landward direction. Post-rift evolution on the Wilkes Land margin is characterized by deposition of thick sedimentary sequences on the continental shelf, slope, and rise that are divided by two major unconformities named WL2 and WL1 (Tanahashi *et al.* 1994). On the continental shelf, unconformity WL2 is an erosional surface

From: STOW, D. A. V., PUDSEY, C. J., HOWE, J. A., FAUGÈRES, J.-C. & VIANA, A. R. (eds)
Deep-Water Contourite Systems: Modern Drifts and Ancient Series, Seismic and Sedimentary Characteristics.
Geological Society, London, Memoirs, **22**, 373–384. 0435-4052/02/$15.00 © The Geological Society of London 2002.

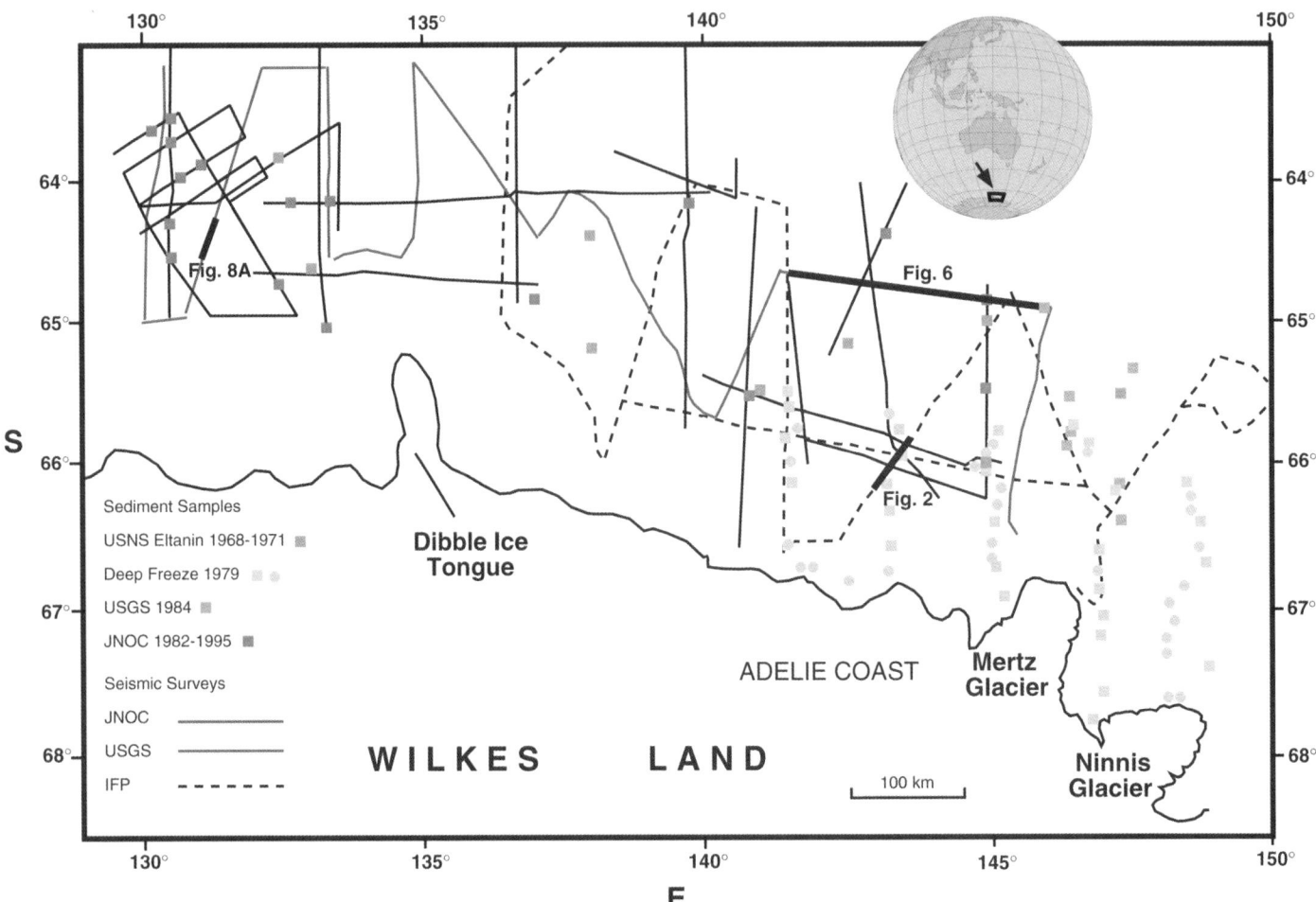

Fig. 1. Map showing: 1) the location of multichannel seismic data recorded by the Institut Français du Petrole (IFP), the Japan National Oil Corporation (JNOC) and the United States Geological Survey (USGS) on the Wilkes Land continental margin; 2) the location of surface sediment cores available for this study; and 3) the location of seismic profiles in Figures 2, 6 & 8A are also shown.

that truncates 300–600 m of middle-shelf and outer-shelf strata and is interpreted to mark the onset of glacial conditions on this margin (Eittreim et al. 1995). Strata above unconformity WL2 mostly prograde across the shelf and downlap onto WL2. Unconformity WL2 can be traced seaward into the continental rise deposits, where it correlates with an up-section increase in turbidite deposition and development of sediment mounds (Escutia et al. 1997, 2000) (Fig. 2). The inferred time for the formation of unconformity WL2 is Eocene, on the basis of indirect correlation with DSDP 269 (Wannesson 1991; Eittreim et al. 1995) as well as a nearby piston core sample from the continental rise dated as Cretaceous to early Eocene (Tanahashi et al. 1994). Unconformity WL1 has eroded and truncated about 350 to 700 m of the sedimentary unit above WL2, and forms the lower boundary onto which strata of the outer shelf downlap (Fig. 2). Unconformity WL1 marks a change in the geometry of the progradational wedge, from lower angle prograding strata below to steeply-dipping prograding strata above the unconformity (foreset slopes up to about 10°) (Fig. 2).

Current measurements, that were taken for short time periods (one hour) during the USNS Eltanin cruises 32–35, 37 and 39 in the South Indian Basin, show a great irregularity of current directions (Eittreim et al. 1971). When current directions from the individual current meters are averaged however, the mean flow pattern is a clockwise gyre (Fig. 3). These measurements show that the present day circulation of the Antarctic Bottom Water (AABW) along the Wilkes Land continental slope and rise is from east to west (Eittreim et al. 1971). Salinity distribution indicates that dense Ross Sea water spills into the South Indian Basin along the continental margin at about 140°E, where it flows westward below 2000 m depth (Eittreim et al. 1971; Gordon & Tchernia 1972). The highest velocities (> 15 cm s^{-1}) in the South Indian Ocean occur along the Antarctic margin (Table 1).

High-velocities and irregular flow lead to current shear turbulence in the bottom waters which, in turn, lead to the development of a nepheloid layer at about 2500 m (Eittreim et al. 1972) (Fig. 4). The nepheloid layer extends about 100 km across the Wilkes Land continental slope and rise, and has a thickness of about 600 m. It is believed that the newly formed Antarctic Bottom Water (AABW) entrains particles from the Antarctic shelf and from erosion of the continental slope and deep basin floor.

Bathymetry

The Wilkes Land continental shelf has an average width of 125 km and an average water depth of 450–500 m (Fig. 5). It exhibits an overdeepened and landward-sloping bathymetric profile that is caused principally by glacial erosion and flexural loading by grounded ice (Ten Brink & Cooper 1992). The topography is irregular, with depths ranging from 200 m on outer-shelf banks to 1000 m in shelf troughs. The troughs, which shoal from inner shelf (> 1000 m) to outer shelf (500 m), are the paths of ice streams that extended across the shelf during times of glacial maxima. The adjacent outer shelf banks are areas that have been bypassed by the most recent ice streams and where grounded ice has been relatively immobile or stagnant (Fig. 5).

The continental slope is steep (gradient 1:9–1:30) and narrow

IFP 107

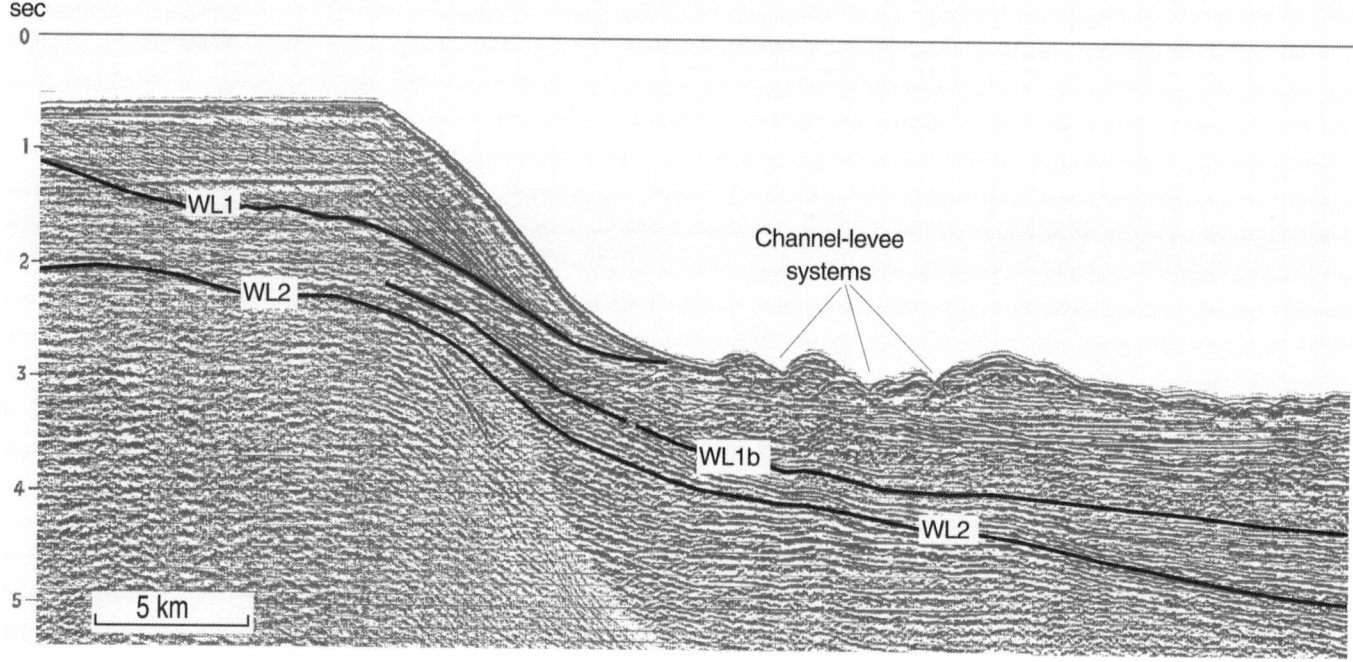

Fig. 2. Seismic profile across the Wilkes Land continental margin showing the unconformities that separate the three main sedimentary units recognized on the continental rise above unconformity WL2.

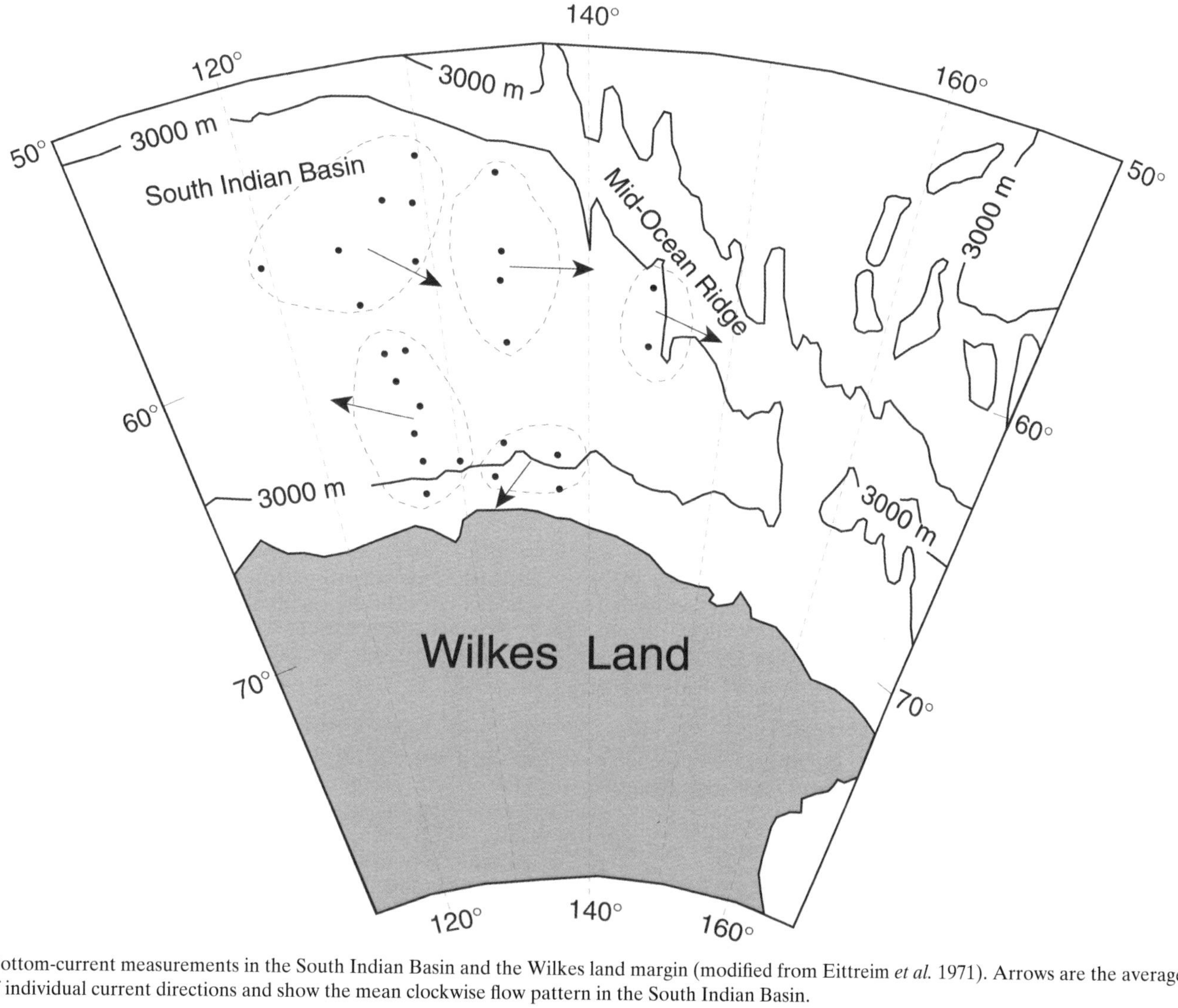

Fig. 3. Bottom-current measurements in the South Indian Basin and the Wilkes land margin (modified from Eittreim *et al.* 1971). Arrows are the average vector of individual current directions and show the mean clockwise flow pattern in the South Indian Basin.

Table 1. *Location of Eltanin 33 and 37 current meter stations and current meter results*

Cruise	Station	Latitude (S)	Longitude (E)	Corrected Depth (m)	Measured Velocities (cm s^{-1})	Mean Velocity (cm s^{-1})	Direction	Difference between compasses	Duration (minutes)
Eltanin 33	5	65°4′	139°48′	4432	3, 3, 4	3.2	160	30	32
Eltanin 33	6	67°03′	136°51′	4510	3, 2, 3	2.5	015	34	34
Eltanin 37	1	65°55′	144°57′	2013	>15	>15	191	–*	5
Eltanin 37	4	65°17′	140°50′	2196	11, 7, 8	8.8	153	–*	15
Eltanin 37	5	65°54′	139°01′	622	7	7.1	106	–*	5
Eltanin 37	6	65°55′	138°54′	684	>15, 11, >15, >15	>15	170	–*	23
Eltanin 37	7	65°14′	137°51′	2238	6, 4, 5, 6	5.2	305	27	25
Eltanin 37	8	64°32′	138°03′	3056	3, 5, 6, 6, 8	5.8	353	97**	34
Eltanin 37	9	64°04′	138°03′	3440	>15, >15, >15, >15, >15, >15	>15	185	38	34

* No camera compass reading.
** Large difference between two compasses; camera compass used.

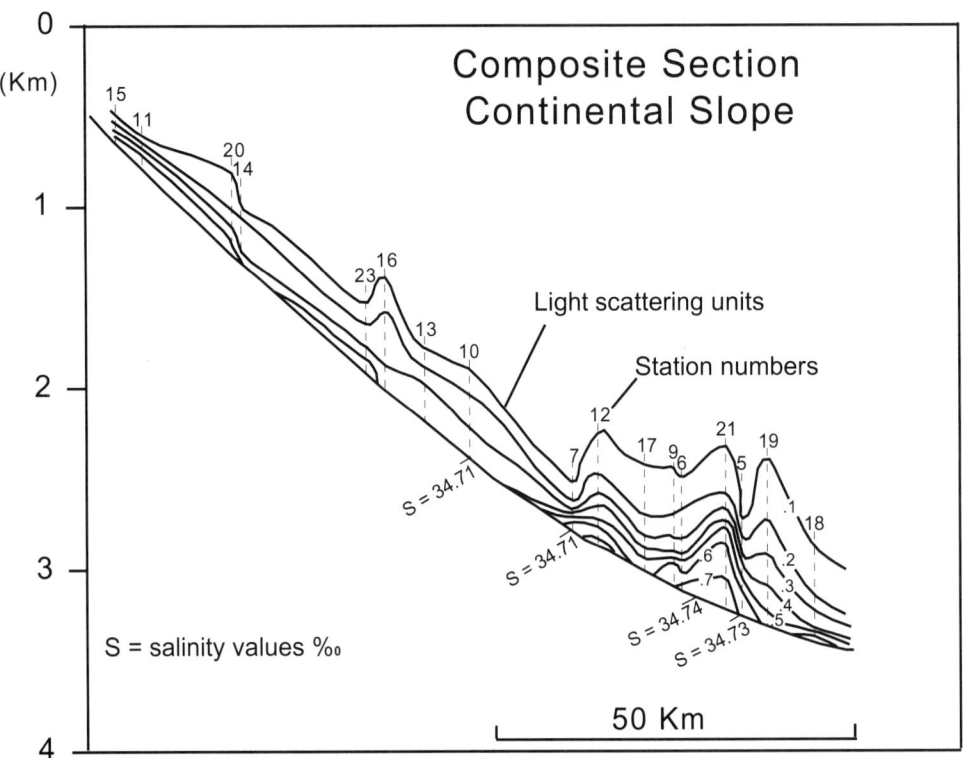

Fig. 4. Composite section of near-bottom nepheloid layer modified from Eittreim *et al.* (1972). Light scattering units are the log ratio of film exposure E to the exposure in clearest water E_0.

(15 km average, but up to 25 km), and extends from the shelf break to about 2000–2500 m water depth. The upper continental rise from approximately 2000 m to 3000 m water depth is also relatively steep (average gradients greater than 1:100), whereas the lower rise from about 3000 m to 4000 m water depth is less steep (average gradients less than 1:150) (Chase *et al.* 1987; Hampton *et al.* 1987). The morphology of the Wilkes Land lower slope and upper continental rise is also rugged because of a complex network of tributary slope canyons and rise channels (Fig. 5).

Seismic stratigraphic framework

The present model of glacial stratigraphy for the Wilkes Land margin (Eittreim *et al.* 1995; Escutia *et al.* 1997, 2000; Barker *et al.* 1998) is based on the analysis of the existing grid of multichannel and high-resolution seismic profiles, aided by the available information from sediment cores collected from this margin (Figs 1 and 2). The model indicates that supply of sediment to the shelf, slope and rise is strongly cyclic. Episodes of ice advance result in erosional surfaces, overcompacted sediment, and steeply prograded wedges on the outer continental shelf troughs and on the upper slope. The large volume of unconsolidated and unsorted material supplied by the ice streams causes slope instability and the generation of turbidity currents that, in the upper rise, develop channel-levee complexes characteristic of deep-sea fans. Fine-grained overbank sediment is likely entrained in ambient bottom-currents to form the high-relief mounded deposits in some of the fan interchannel areas. During interglacial periods, the outer shelf, slope, rise, deep-sea fans and sediment mounds receive mainly biogenic sediment and minor amounts of ice-rafted debris, and are subjected to reworking by slumping and/or oceanic currents.

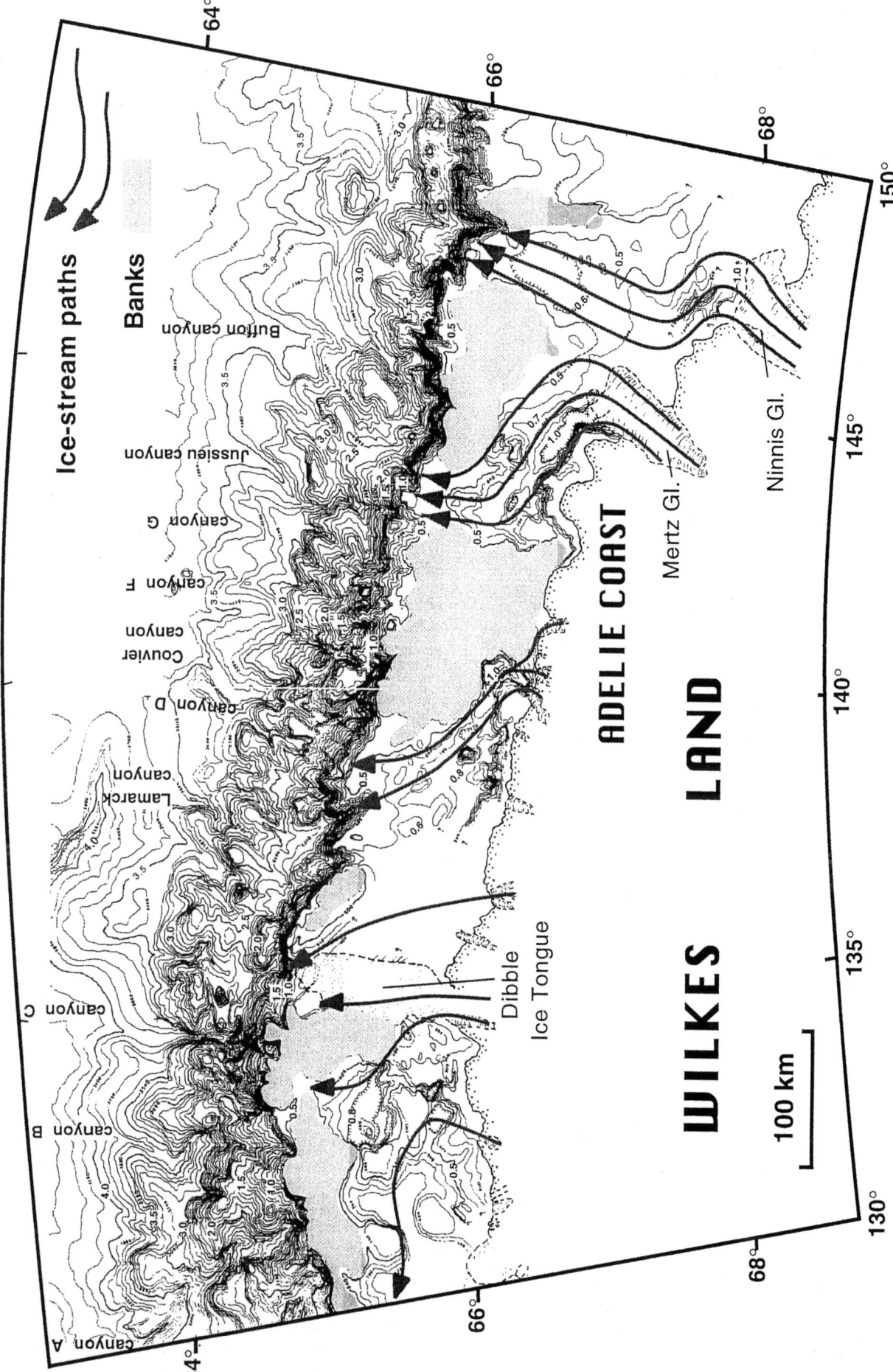

Fig. 5. Wilkes Land bathymetric map modified from Chase *et al.* (1987) and Eittreim *et al.* (1995). Shaded areas are outer-shelf banks less than 400 m. Cross-shelf troughs lie between these banks. Closely spaced soundings and seismic profiles do not exist on the Wilkes land margin therefore, this bathymetric map does not represent the final configuration of the Wilkes Land margin seafloor.

Within the Wilkes Land margin succession the principal marker is unconformity WL2, which can be traced from the shelf, where it marks the onset of progradation on the Wilkes Land margin (Eittreim & Smith 1987), to the continental rise where it correlates with an up-section increase in turbidite and contourite deposition (Escutia et al. 1997, 2000) (Fig. 2). Three acoustic units can be differentiated on the continental rise above unconformity WL2, bounded by two regional unconformities named, from older to younger, WL1c and WL1b (Figs 2 and 6). The acoustic unit directly above unconformity WL2 is characterized by horizontally stratified and continuous reflectors of varying amplitude, that pinchout and onlap unconformity WL2 in a landward direction. Above unconformity WL1c, stratified and continuous reflectors correlate laterally with irregular and discontinuous reflectors of varying amplitude that are indicative of channels and channel-fill deposits. Parallel–subparallel reflectors thin away from these channels and indicate levee deposition. The acoustic unit above unconformity WL1b has a more disrupted acoustic character than acoustic units below and is characterized by stratified and continuous reflectors, irregularly stratified and contorted reflectors of varying continuity and amplitude, and undulating reflectors.

Seismic characteristics

The formation of the Wilkes Land continental rise sediment mounds is the result of the complex interaction of turbidite, countour-current and hemipelagic sedimentation. Large channel-levee and sediment mounds have developed above the WL1b unconformity. Three phases are recognized in the development of the upper sedimentary unit above unconformity WL1b (Fig. 6). Phase 1 is dominated by channel-levee deposition, Phase 2 is dominated by channel-levee and contourite deposition with significant mound building, and Phase 3 is dominated by turbidite deposition with decreased contourite deposition and without active mound building. Phase 2 thus, corresponds with a period of large sediment mound building. During Phase 3, although contour-current deposition is still and active process, mound growth ceases and overbank deposits from the turbidity flows traveling through the channels that bound the mounds are observed onlapping the gentler flanks of the mounds (Fig. 6).

Evidence for Turbidite Sedimentation

Turbidity-current flows are responsible for the formation of the channel-levee complexes that develop in the Wilkes Land continental rise at the mouths of the canyons that incise the continental slope. Escutia et al. (2000) discussed in detail the morphologic and seismic character, and the environmental significance of the Wilkes Land deep-sea fan deposits, and concluded that: (1) the canyon-channel drainage patterns are most similar to other high-latitude and elongate, muddy fans (Stow 1981; Hesse 1995) (Fig. 7); (2) the unusually high relief (up to 900 m) of the leveed channels is consistent with other fans fed by ice-sheets at the palaeo-shelf edge (Fig. 8B); (3) the surface gradients are generally steeper than those observed on most other submarine fans (Nelson & Nilsen 1984); and (4) the overall morphology of the overbank deposits and interchannel areas have been strongly modified from that of typical submarine fans by the action of contour currents.

Evidence for Contourite Sedimentation

Sediment mounds on the Wilkes Land margin are identified in seismic reflection profiles as depositional features with positive relief that develop on the continental rise between 2500 m and 3000–3500 m water depth (Fig. 6). They typically form between some of the deep-sea fan channels that drain the adjacent continental margin. They differ morphologically from levee deposits however, in that the highest relief of the mounds is in the interchannel area and they thin away towards the channels. The sediment mounds are elongated in outline and trend N–S to NNE–SSW. In along-margin sections, the sediment mounds crossed by our seismic profiles have widths between 25 and 50 km and are typically asymmetric. They exhibit a sharp crest, with a relief that varies between 275 and 490 m, that separates a shorter and steeper erosional flank to the west from a longer and gentler flank to the east. The thickness of the seismic units within Phase 2 deposits compared to thickness of the same units in the surrounding areas suggest higher-rates of deposition within the sediment mounds during sedimentation of Phase 2.

Reflection configurations indicative of contour-current sedimentation (i.e. undulating, parallel to subparallel and generally continuous reflectors and convex upward reflectors that both, onlap and downlap onto the underlying unconformity) are clearly evident within the sediment mounds (Fig. 6). The wavy and mounded reflectors are consistent with apparent strong paleo-currents along the Wilkes Land continental rise during Phase 2 of continental rise sedimentation above the WL1b unconformity. The asymmetry of the mounds suggests that these paleo-currents moved in a general E–W direction, similar to the direction of the AABW currents that sweep the Wilkes Land continental rise at present. Building of the mounds during Phase 2 was not a continuous process, because the mounds were scoured and modified by cut and fill sequences during their construction. On the west flank of the mounds, acoustic reflector configurations characteristic of channel cutting, levee deposition and sediment waves are common, whereas on the gentler flank (east flank) the reflectors are more stratified and continuous. The internal acoustic characteristics of the sediment mounds suggest an interplay between turbidity currents and bottom-currents in their formation.

The spatial distribution, morphology and internal seismic characteristics described for the Wilkes Land sediment mounds are similar to other mounded deposits around Antarctica such as the ones on the Antarctic Peninsula (McGinnis & Hayes 1994; Rebesco et al. 1996; McGinnis et al. 1997; Barker et al. 1999), and in Prydz Bay (Kuvaas & Leitchenkov 1992; O'Brien et al. 2001).

Sediment characteristics

Interaction between turbidite and contourite deposition is also observed in sediment cores from the Wilkes Land continental rise. DSDP Site 268 was drilled to a depth of 474.5 m on the Wilkes Land lower continental rise 680 km west of our study area (Hayes et al. 1975). The oldest sedimentary deposits (160–474.5 m) include silty clay, laminated silty clay, clayey silt and chert with calcareous nannofossils dated as mid-Oligocene or older to early Miocene. Differences between the silt laminae in this unit and silt laminae found in demonstrable turbidite-dominated sequences led Piper & Brisco (1975) to interpret these sediments as contourites. In contrast, the younger unit (0–160 m) dated Pliocene and Pleistocene, dominated by clay, silty clay with common silt laminae, fine sand beds 2 to 20 cm thick with a high *Neogloboquadrina pachyderma* content and diatom ooze, was interpreted to be mainly of turbidite origin. A seismic profile obtained by the Japanese National Oil Corporation (JNOC) in 1982 shows that Site 268 sampled one of the continental rise interchannel deposits, but the quality and length of the seismic line does not allow us to differentiate if it sampled a levee or a mounded deposit. Sedimentation rates of 4.1 cm ka^{-1} and 3.1 cm ka^{-1} were calculated for the lower and the upper sediment units from Site 268 (Hayes et al. 1975). These sedimentation rates were calculated in spot (non-continuous) cores and without taking into consideration a known unconformity in the lower unit and possibly other non-recovered unconformities, and should therefore be considered as minimum values.

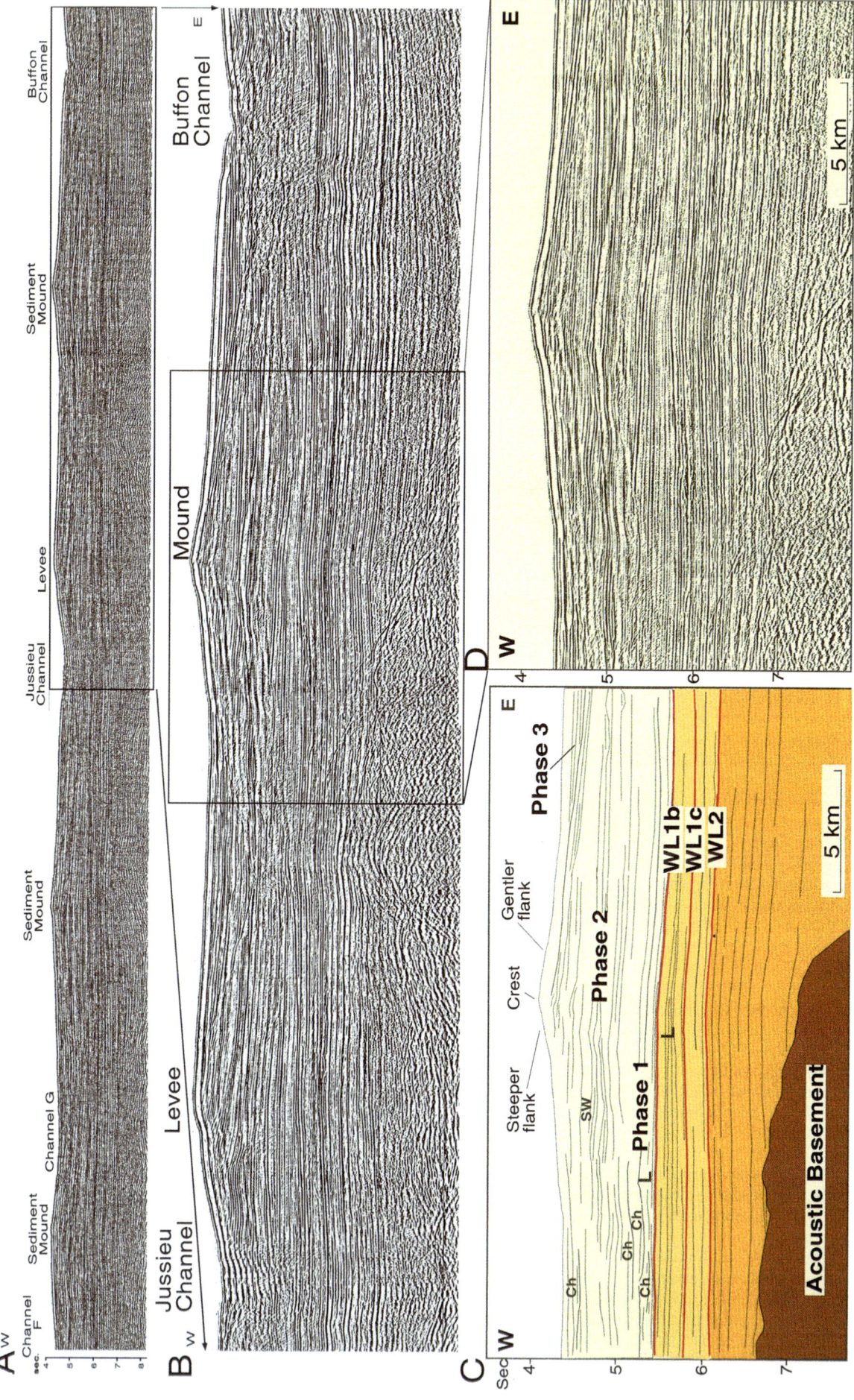

Fig. 6. (A) Seismic profile along the continental rise with annotated location of channels, levee and sediment mounded deposits. (B) Detailed section of the seismic profile above including two channels, the east levee of the Jussieu Channel, and one of the sediment mounded deposits. (C & D) interpreted and uninterpreted detail of the sediment mounded deposit. Interpretation: Ch, channels; L, levees and SW, sediment waves. See Figure 1 for location of profile.

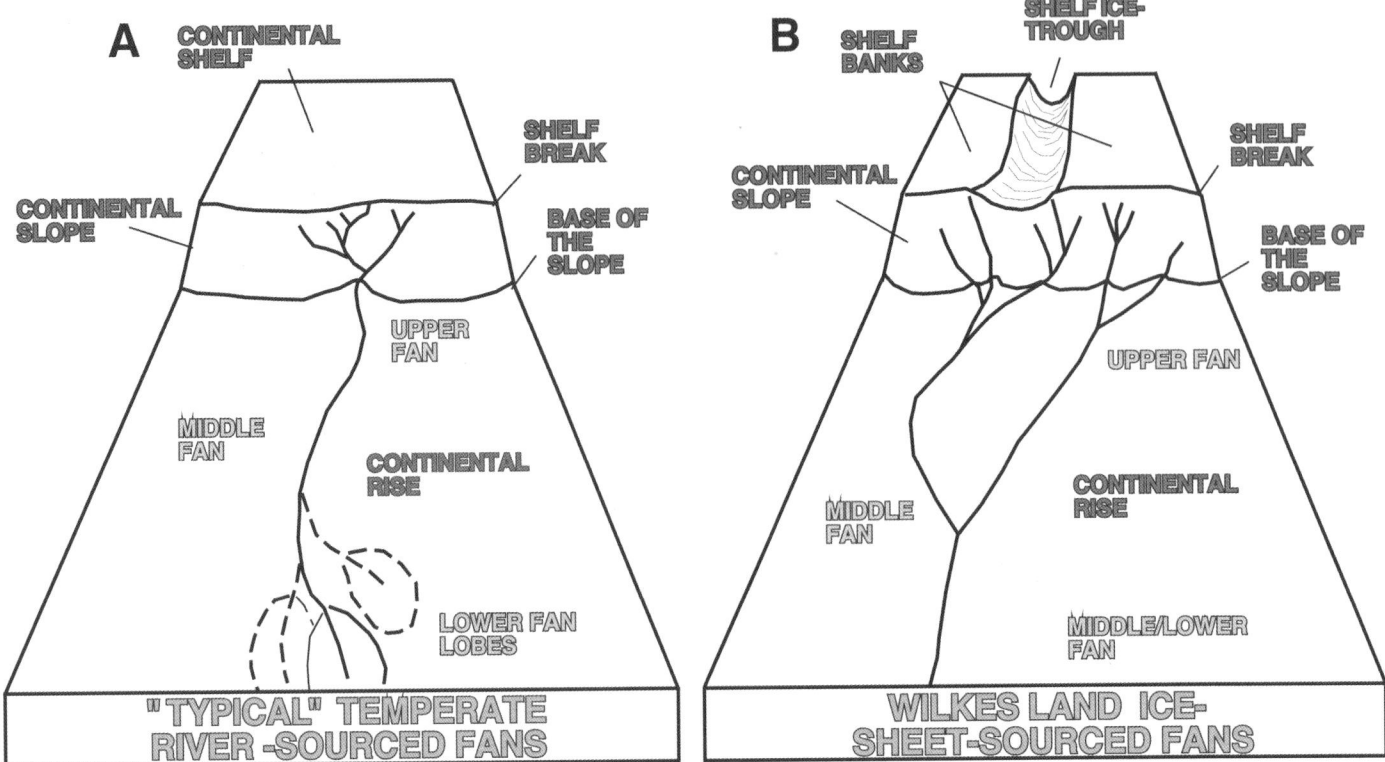

Fig. 7. Cartoon comparing: (A) canyon-channel network patterns of 'typical' fluvial-marine sediment sourced fans deep-sea fans based on Nelson & Nilsen (1984), and (B) canyon-channel network patterns of the Wilkes Land deep-sea fans from this paper.

Surface cores collected from the Wikes Land continental rise during the USNS Eltanin (Payne & Conolly 1972), Deep Freeze 79 (Domack 1982), and USGS 1984 (Hampton *et al.* 1987) cruises, have been analysed previously in some detail (i.e. grain size, x-radiographs, physical properties). We have reanalysed these cores for: (1) consistency in the descriptions, (2) biostratigraphic analyses in the light of more recent diatom biostratigraphy for the Southern Ocean (Gersonde *et al.* 1990; Barron *et al.* 1991) as well as radiolarian and foraminiferal biostratigraphic zonations; and (3) magnetostratigraphic studies, which had previously been limited to low alternating field (AF) demagnetization of discrete samples from the Eltanin cores (Watkins & Kennett 1972).

Three of the USGS gravity cores, Cores A8G1, A9G1 and A10G1, were collected from the Wilkes Land continental rise interchannel areas at 2635 m, 3037 m and 3379 m water depth, respectively. These cores are 368 to 375 cm long and record sedimentation during the upper part of Phase 3. They consist of alternating massive mud and laminated units (Fig. 9). Massive mud intervals contain common Pleistocene diatom assemblages, whereas the laminated units contain few to abundant fragmented and mostly reworked diatoms. Individual laminae are thin (*c*. 1 mm thick) and can be continuous across the core or terminate laterally. Laminae can have a wavy or straight geometry (Fig. 9A). Internal structures such as graded bedding or cross bedding are not evident in visual description of the cores or in X-radiographs. Laminae can be distinguished by changes in grain size between clay and silt. Sediment within the laminae is generally well sorted silt.

Sedimentation rates were estimated from paleomagnetic measurements made on U-channel samples (up to 1.5 m long samples with a 2 cm by 2 cm cross section) and discrete samples from selected cores. Samples were progressively AF demagnetized to peak fields of 60 mT or higher, and generally gave linear demagnetization paths indicating stable single-component magnetizations. Several cores only penetrated Brunhes age sediments and so yielded only minimum sedimentation rates. However, we identified the Brunhes/Matuyama magnetic polarity reversal in two cores, DF79-19 and DF79-02, which were collected on the continental rise during the Deep Freeze 79 cruise. Core DF79-19 comes from off the Mertz Bank just east of the shelf trough and Core DF79-02 comes from off the Adelie Bank just west of the shelf trough. The Brunhes/Matuyama occurs at a depth of 256 cm in Core DF79-19 and at about 320 cm in Core DF79-02, giving relatively slow sedimentation rates of 0.33 cm ka^{-1} and 0.41 cm ka^{-1}, respectively.

Discussion

On the Wilkes Land continental shelf, unconformity WL2 marks the change from predominantly aggradational sequences below to dominantly progradational sequences above. Eittreim *et al.* (1995) interpreted this change in geometry to mark the first arrival of grounded ice sheets to the Wilkes Land continental shelf. The scale and geometry of this unit is similar to the glacial outwash deposits of late Eocene to early Oligocene age recovered during ODP Leg 119 in Prydz Bay (Hambrey *et al.* 1991). In the Ross Sea, the change from aggradational to progradational sequences has also been interpreted as indicating a pronounced decrease in meltwater reflecting the transition from sub-polar to polar climate (Anderson & Bartek 1994; De Santis *et al.* 1995).

On the continental rise, the change up-section in seismic facies from unconformity WL2 to the present seafloor (i.e. from stratified and continuous to irregularly stratified, contorted and discontinuous reflectors) is interpreted to reflect the response of sedimentation to the progradation of the margin as a result of general cooling and increased glaciation on this segment of the East Antarctic margin (Escutia *et al.* 1997). We believe the unusually large channel-levee complexes and contourite sediment mounds above unconformity WL1b develop under a fully continental glacial regime where ice-sheet sources fed glacial ice streams that reached the outer palaeo-continental shelf. Although

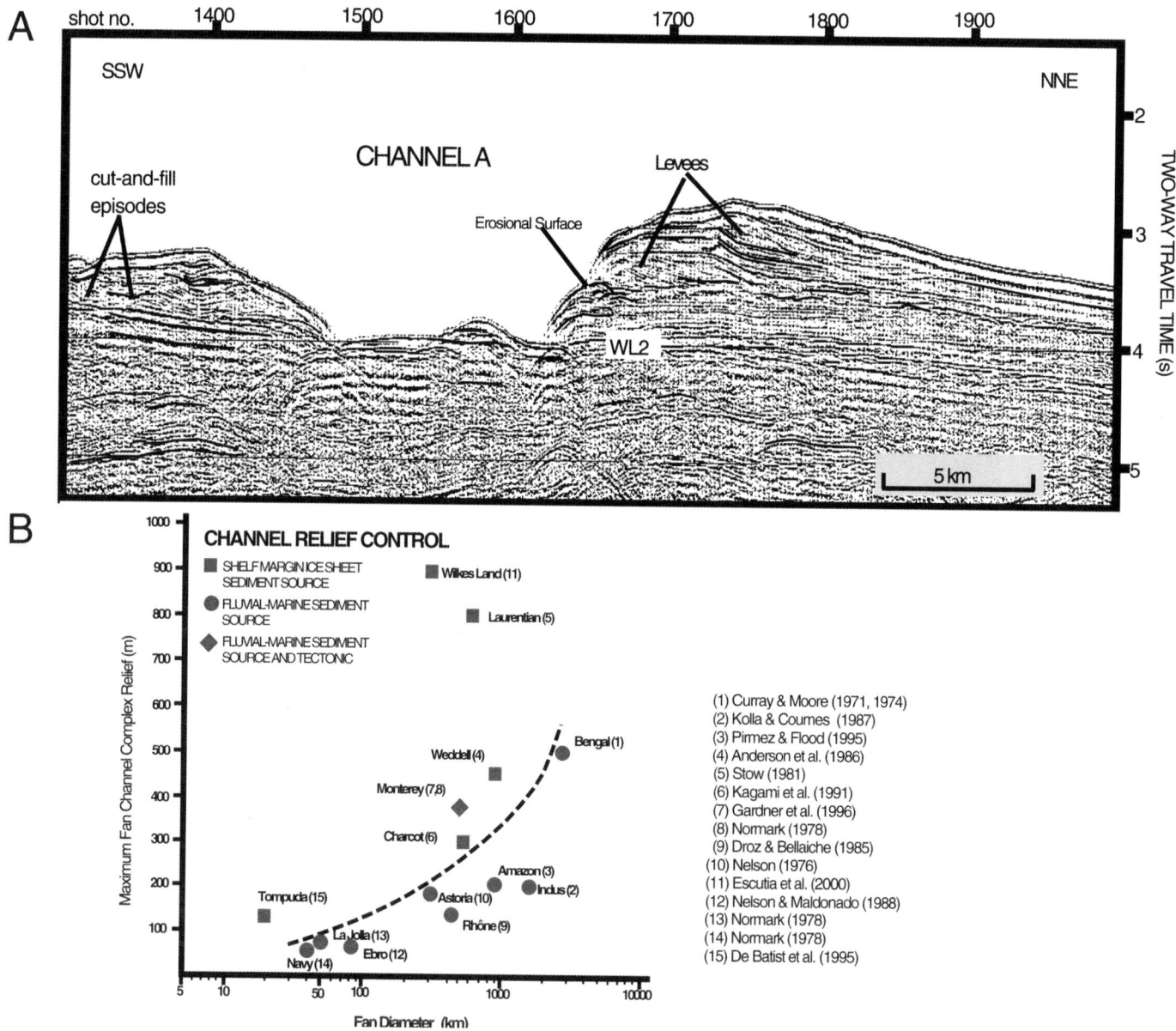

Fig. 8. (A) Seismic profile with interpretation across Channel A on the Wilkes Land continental rise. Channel A is the largest mapped channel of the Wilkes Land fans. See Figure 1 for location of profile. (B) Graph of upper-channel relief *vs.* fan length. The dashed line separates most fluvial-marine sediment sourced fans (below the line) from fans that are sourced by shelf margin ice sheets (above the line). Upper-fan channel relief in fluvial-marine sourced fans generally increases with fan length. Unusually high upper-fan channel relief for the size fan is observed in most shelf margin ice-sheet sourced fans. Numbers in parenthesis key reference sources used to determine channel relief and fan length.

direct seismic ties with the shelf cannot be established (i.e. WL1b unconformity can be traced only from the rise to the lower slope), this period appears to correlate with a time when sedimentation in front of the continental shelf troughs is characterized by very steep (i.e. *c.* 10°) prograding foresets (Fig. 2). Ice-stream delivery of a high volume of unconsolidated and sediment of all grain sizes to the outer shelf and upper continental slope during times of glacial maxima resulted in sediment failures that led to the development of the present-day network of canyons across the margin. The wide range of grain sizes in the source area favoured the development of deep channels and high levee-interchannel areas (e.g. Stow 1981).

We postulate that suspended fines from the turbidity-currents that originate during times of glacial maxima, are entrained in a nepheloid layer associated with westward flowing palaeo-bottom-currents like those of the AABW that presently sweep the Wilkes Land continental rise below 2500 m water depth. Current meter data (Figs 3 & 4), cores from DSDP Site 268, surface sediment cores (Fig. 9), and seismic reflection profiles (Fig. 6), provide evidence that contour-currents and contourite deposition have been active in the Wilkes Land continental rise from the time of sedimentation of Phase 2 until the late Pleistocene. Contourite deposition is a major component of the lower unit recovered at DSDP Site 268 and an important component of the Pleistocene laminated intervals in surface sediments such as USGS cores A8G1 (Fig. 9), A9G1 and A10G1, and other cores from the Wilkes Land continental rise. Although contour-current deposition seems to be an active process in areas of the Wilkes Land continental rise where sediment is supplied by a nepheloid layer, it is only during Phase 2 of sedimentation when there is active mound building.

We hypothesize that the transition from the mound building Phase 2 to the non-mound building Phase 3, signals a decrease in sediment input to the continental rise. Building of the sediment mounds during Phase 2 appears to take place during a time when a larger volume of sediment was supplied from the continental

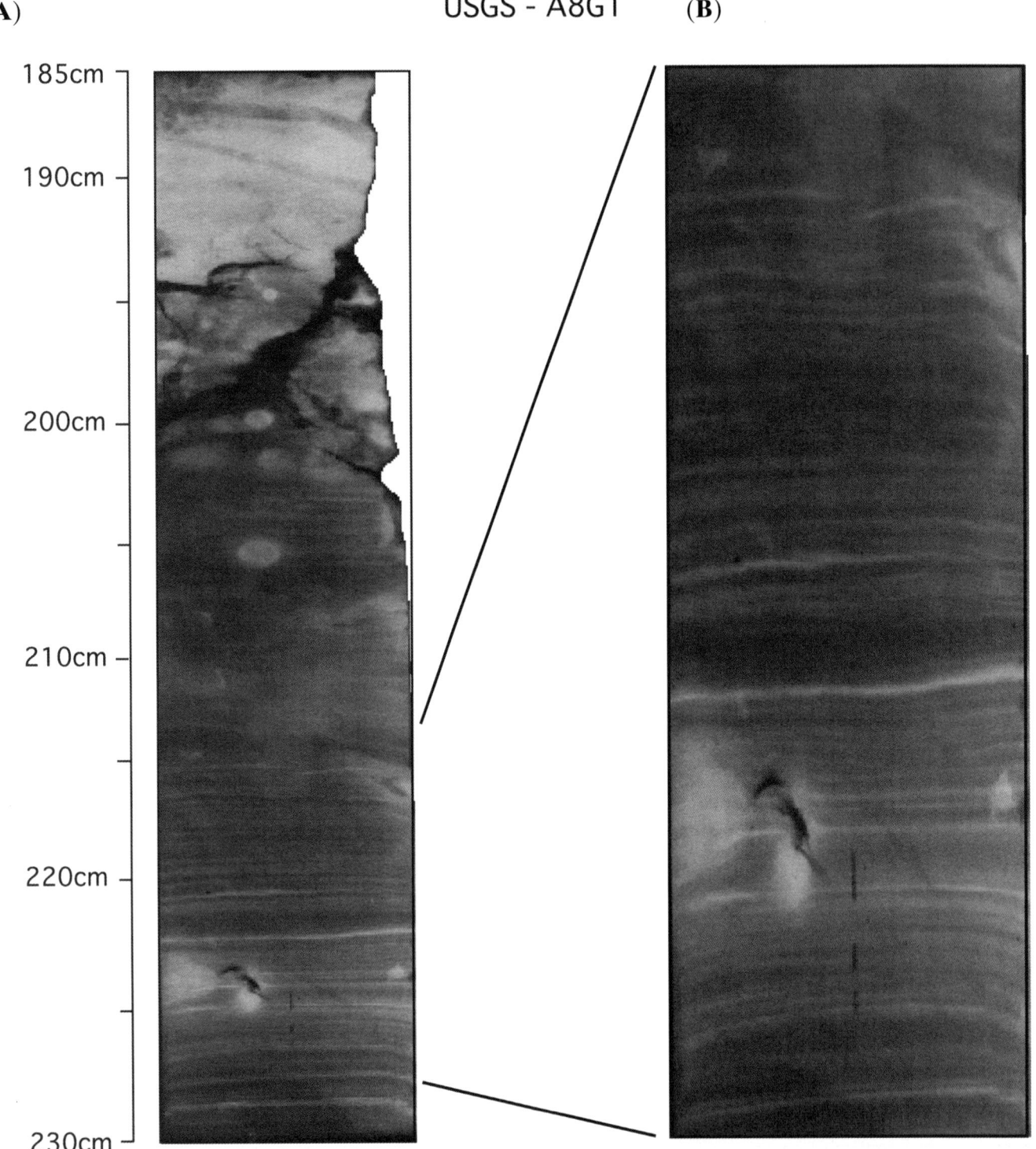

Fig. 9. (**A**) X-radiograph of US Geological Survey Core A8G1 (185–230 cm), which consists of Pleistocene massive mud (185–200 cm) and laminated mud and silty-mud (200–230 cm). (**B**) Detail of the laminae in the laminated interval.

shelf to the continental rise and abyssal plain, most likely the result of expansion of the Antarctic ice-sheet. Larger sediment input to the mounds during Phase 2 is suggested in seismic profiles by the increase in thickness of Phase 2 seismic units within the mounds when compared to the same units in surrounding areas (Fig. 6). Additionally, sedimentation of Phase 2 appears to correlate with the deposition of very steep foresets by grounded ice-streams as the margin prograded rapidly (i.e. just below and above unconformity WL1). During Phase 3, sediment supply to the continental rise, although important in volume and capable of turbidite and contourite deposition, appears not have been sufficient to support further building of the mounds. Lower sediment input to the continental rise during the upper part of Phase 3 is suggested by the geometry and thickness of the Phase 3 sediment package in seismic profiles (Fig. 6).

A decrease in the sediment input to the continental rise from Phase 2 to present is also indicated by variation in sedimentation rates. Minimum rates measured in sediment cores from Site 268 show a decrease in sedimentation from the lower unit (> 4.1 cm ka^{-1}) to the upper unit (3.1 cm ka^{-1}). In surface cores of middle Pleistocene to present age, sedimentation rates are 0.33 cm ka^{-1} to 0.41 cm ka^{-1}. A decrease in sedimentation rates at about 2.5 Ma was also reported from Sites 1095, 1096 and 1101 drilled during Leg 178 (Barker *et al.* 1999), where sedimentation rates decreased from 5–18 cm ka^{-1} in the upper Miocene and lower Pliocene to 2.5–8 cm ka^{-1} during the upper Pliocene and Pleistocene. We hypothesize that the decrease in sediment input to the Wilkes Land continental rise, from Phase 2 to recent, could be the result of a change on sediment depocentres with most of the sediment supplied to the margin during Phase 3 being trapped on the continental shelf. A similar pattern of decreasing sedimentation rates was observed at Site 1165 drilled during Leg 188 in Prydz Bay (Shipboard Scientific Party 2000), where sedimentation rates decreased from *c.* 1.5 cm ka^{-1} during the upper Miocene to *c.* 0.7 cm ka^{-1} during the middle Pleistocene. We believe the decrease in sediment input to the continental rise from Phase 2 to recent is ultimately related to the stage of glacial evolution of the Antarctic continent.

New research on existing surface sediment cores from the Wilkes Land continental rise included in this paper was possible with the support by National Science Foundation grant #OPP-9815085.

References

ANDERSON, J. B. & BARTEK, L. R. 1992. Cenozoic glacial history of the Ross Sea revealed by intermediate resolution seismic reflection data combined with drill site information. *In*: KENNETT, J. P. & WARNKE, D. A. (eds) *The Antarctic Palaeoenvironment: a Perspective on Global Change.* AGU Antarctic Research Series, American Geophysical Union, Washington DC, **56**, 231–263.

ANDERSON, J. B., WRIGHT, R. & ANDREWS, B. 1986. Weddell Fan and associated abyssal plain, Antarctica; Morphology, sediment processes, and factors influencing sediment supply. *Geo-Marine Letters*, **6**, 121–129.

BARKER, P. F., BARRETT, P. J., ET AL. 1998. Ice sheet history from Antarctic Continental margin sediments: the ANTOSTRAT approach. *Terra Antartica*, **5**, 737–760.

BARKER, P. F., CAMERLENGHI, A., ACTON, G. D., ET AL. 1999. Proceedings of the Ocean Drilling Program, Initial Reports, **178**, CD-ROM, ODP, College Station, Texas.

BARRON, J., LARSEN, B. & SHIPBOARD SCIENTIFIC PARTY 1991. Proceedings of the Ocean Drilling Program, Scientific Results, **119**.

CANDE, S. C. & MUTTER, J. C. 1982. A revised identification of the oldest sea-floor spreading anomalies between Australia and Antarctica. *Earth and Planetary Science Letters*, **58**, 151–160.

CHASE, T. E., SEEKINS, B. A., YOUNG, J. D. & EITTREIM, S. L. 1987. Marine topography of offshore Antarctica. *In*: EITTREIM, S. L. & HAMPTON, M. A. (eds) *The Antarctic Continental Margin. Geology and Geophysics of Offshore Wilkes Land.* Circum-Pacific Council for Energy and Mineral Resources Earth Sciences Services, Houston, **5A**, 147–150.

CURRAY, J. R. & MOORE, D. G. 1971. Growth of the Bengal deep-sea fan and denudation in the Himalayas. *Geological Society of America Bulletin*, **82**, 563–572.

CURRAY, J. R. & MOORE, D. G. 1974. Sedimentary and tectonic processes in the Bengal deep-sea fan and geosyncline. *In*: BURK, C. A. & DRAKE, C. L. (eds) *The Geology of Continental Margins.* Springer-Verlag, New York, 617–627.

DEBATISTE, M., VAN CAUWENBERGHE, T. ET AL. 1995. First results of an integrated geophysican survey in northern Lake Baikal, Russia. *EOS, Transactions, American Geophysical Union, Abstracts*, **76**, F551.

DESANTIS, L., ANDERSON, J. B., BRANCOLINI, G. & ZAYATZ, I. 1995. Seismic record of late Oligocene through Miocene glaciation on the central and eastern continental shelf of the Ross Sea. *In*: COOPER, A. K., BARKER, P. F. & BRANCOLINI, G. (eds) *Geology and Seismic Stratigraphy of the Antarctic Margin.* American Geophysical Union, Antarctic Research Series, **68**, 235–260.

DOMACK, E. W. 1982. Sedimentology of glacial and glacial marine deposits on the George V-Adelie continental shelf, East Antarctica. *Boreas*, **11**, 79–97.

DROZ, L. & BELLAICHE, G. 1985. The Rhône deep-sea fan: morphostructure and main growth pattern. *American Association of Petroleum Geologist Bulletin*, **69**(1), 460–479.

EITTREIM, S. L. 1994. Transition from continental to oceanic crust on the Wilkes-Adelie margin of Antarctica. *Journal of Geophysical Research*, **99**, 24 189–24 205.

EITTREIM, S. L. & SMITH, G. L. 1987. Seismic sequences and their distribution on the Wilkes Land margin. *In*: EITTREIM, S. L. & HAMPTON, M. A. (eds) *The Antarctic Continental Margin. Geology and Geophysics of Offshore Wilkes Land.* Circum-Pacific Council for Energy and Mineral Resources Earth Sciences Series, Houston, **5A**, 15–43.

EITTREIM, S. L., BRUCHHAUSEN, P. M. & EWING, M. 1972. Vertical distribution of turbidity in the south Indian and south Australian basins. *In*: HAYES, D. E. (ed.) *Antarctic Oceanology II. The Australian-New Zealand Sector.* Antarctic Research Series, **19**, 51–58.

EITTREIM, S. L., COOPER, A. K. & WANNESSON, J. 1995. Seismic stratigraphic evidence of ice-sheet advances on the Wilkes Land margin of Antarctica. *Sedimentary Geology*, **96**, 131–156.

EITTREIM, S. L., GORDON, A. L., EWING, M., THORNDIKE, E. M. & BRUCHHAUSEN, P. M. 1971. The nepheloid layer and observed bottom currents in the Indian-Pacific Antarctic Sea. *In*: GORDON, A. L. (ed.) *Studies in Physical Oceanography – A tribute to Georg Wüst on his 80th birthday.* Gordon and Breach, New York, 19–35.

ESCUTIA, C., EITTREIM, S. L. & COOPER, A. K. 1997. Cenozoic glaciomarine sequences on the Wilkes Land continental rise, Antarctica. *Proceedings Volume-VII International Sysposium on Antarctic Earth Sciences*.

ESCUTIA, C., EITTREIM, S. L., COOPER, A. K. & NELSON, C. H. 2000. Morphology and acoustic character of the Antarctic Wilkes Land turbidite systems: ice-sheet sourced versus river-sourced fans. *Journal of Sedimentary Research*, **70**, 84–93.

GARDNER, J. V., BOHANNON, R. G., FIELD, M. E. & MASSON, D. 1996. The morphology, processes, and evolution of the Monterey Fan: A revisit. *In*: GARDNER, J. V., FIELD, M. E. & TWICHELL, D. C. (eds) *Geology of the United States Seafloor: The View from GLORIA.* Cambridge University Press, Cambridge, 193–220.

GERSONDE, R., ABELMANN, A. ET AL. 1990. Biostratigraphic synthesis of Neogene siliceous microfossils from the Antarctic Ocean, ODP Leg 113 (Weddell Sea). *Proceedings of Ocean Drilling Program, Scientific Results.* Ocean Drilling Program, College Station, TX, **113**, 915–936.

GORDON, A. L. & TCHERNIA, P. 1972. Waters of the continental margin off the Adelie Coast, Antarctica. *In*: HAYES, D. E. (ed.) *Antarctic Oceanology II. The Australian-New Zealand Sector.* Antarctic Research Series, **19**, 59–70.

HAMBREY, M. J., EHRMANN, W. U. & LARSEN, B. 1991. Cenozoic glacial record of the Prydz Bay continental shelf, East Antarctica. *Proceedings of the Ocean Drilling Program, Initial Results.* Ocean Drilling Program, College Station, TX, **119**, 77–132.

HAMPTON, M. A., EITTREIM, S. L. & RICHMOND, B. M. 1987. Post-breakup sedimentation on the Wilkes Land Margin, Antarctica. *In*: EITTREIM, S. L. & HAMPTON, M. A. (eds) *The Antarctic Continental Margin: Geology and Geophysics of Offshore Wilkes Land.* Circum-Pacific Council for Energy and Mineral Resources Earth Sciences Series, Houston, **5A**, 75–88.

HAYES, D. E., FRAKES, L. A. ET AL. 1975. Initial Reports of the Deep Sea Drilling Project, **28**. US Government Printing Office, Washington.

HESSE, R. 1995. Continental slope and basin sedimentation adjacent to an ice margin: a continuous sleeve-gun profile across the Labrador Slope, Rise and Basin. *In*: PICKERING, K. T., HISCOTT, R. N., KENYON, N. H., RICCI LUCCHI, F. & SMITH, R. D. A. (eds) *Atlas of Deep Water Environments: Architectural Style in Turbidite Systems.* Chapman & Hall, London, 14–17.

KAGAMI, H., KURAMOCHI, H. & SHIMA, Y. 1991. Submarine canyons in the Bellingshausen and Riiser-Larsen Seas around Antarctica. *Proceedings NIPR Sysmposium Antarctic Geosciences*, **5**, 84–98.

KOLLA, V. & COUMES, F. 1987. Morphology, internal structure, seismic stratigraphy, and sedimentation of Indus Fan. *The American Association of Petroleum Geologists Bulletin*, **71**, 650–677.

KUVAAS, B. & LEITCHENKOV, G. 1992. Glaciomarine turbidite and current controlled deposits in Prydz Bay, Antarctica. *Marine Geology*, **108**, 365–381.

MCGINNIS, J. P. & HAYES, D. E. 1994. Sediment drift formation along the Antarctic Peninsula. *Terra Antartica*, **1**, 275–276.

MCGINNIS, J. P., HAYES, D. E. & DRISCOLL, N. W. 1997. Sedimentary processes across the continental rise of the southern Antarctic Peninsula. *Marine Geology*, **141**, 91–109.

NELSON, C. H. 1976. Late Pleistocene and Holocene depositional trends, processes, and history of Astoria deep-sea fan, northeast Pacific. *Marine Geology*, **20**, 129–173.

NELSON, C. H. & NILSEN, T. H. 1984. Modern fan morphology and stratigraphy. *In*: NELSON, C. H. & NILSEN, T. H. (eds) *Modern and Ancient Deep-Sea Fan Sedimentation*. Society of Economic Palaeontologist and Mineralogist, Lecture Notes, **14**, 38–108.

NELSON, C. H. & MALDONADO, A. 1988. Factors controlling depositional patterns of Ebro turbidite systems, Mediterranean Sea. *American Association of Petroleum Geologists Bulletin*, **72**(6), 698–716.

NORMARK, W. R. 1978. Fan valleys, channels and depositional lobes on modern submarine fans: Characters for recognition of sandy turbidite environment. *The American Association of Petroleum Geologists Bulletin*, **62**, 912–931.

NORMARK, W. R., PIPER, D. J. W. & STOW, D. A. 1983. Quaternary development of channels, levees, and lobes on middle Laurentian Fan. *The American Association of Petroleum Geologists*, **67**, 1400–1409.

O'BRIEN, P. E., COOPER, A. K., RICHTER, C., *ET AL.* 2001. Proceedings of the Ocean Drilling Program, Initial Reports, **178**, CD-ROM, ODP, College Station, Texas.

PAYNE, R. R. & CONOLLY, J. R. 1972. Turbidite sedimentation off the Antarctic continent. *Antarctic Research Series*, **19**, 349–364.

PIPER, D. J. W. & BRISCO, C. D. 1975. Deep-water continental margin sedimentation, DSDP LEG 28, Antarctica. *In*: HAYES, D. E., FRAKES, L. A. *ET AL.* (eds) *Initial Reports of the Deep Sea Drilling Project*, **28**. US Government Printing Office, Washington, 727–755.

PRIMEZ, C. & FLOOD, R. D. 1995. Morphology and structure of Amazon Channel. *In*: FLOOD, R. D., PIPER, D. J. W., KLAUS, A. *ET AL.* (eds) *Proceedings of the Ocean Drilling Program, Initial Reports*. Ocean Drilling Program, College Station, TX, **155**, 23–45.

REBESCO, M., LARTER, R. D., CAMERLENGUI, A. & BARKER, P. F. 1996. Giant sediment drifts on the continental rise west of the Antarctic Peninsula. *Geo-Marine Letters*, **16**, 65–75.

SHIPBOARD SCIENTIFIC PARTY 2000. *Leg 188 Preliminary Report: Prydz Bay-Cooperation Sea, Antarctica: glacial history and paleoceanography. ODP Prelim. Rpt., 88 [Online]*. World Wide Web Address: http://www-odp.tamu.edu/publications/prelim/188_prel/188prel.pdf.

STOW, D. A. V. 1981. Laurentian Fan: Morphology, sediments, processes, and growth pattern. *The American Association of Petroleum Geologists*, **65**, 375–393.

TANAHASHI, M., SAKI, T., OIKAWA, N. & SATO, S. 1987. An interpretation of the multichannel seismic reflection profiles across the continental margin of the Dumont d'Urville Sea, off Wilkes Land, East Antarctica. *In*: EITTREIM, S. L. & HAMPTON, M. A. (eds) *The Antarctic Continental Margin: Geology and Geophysics of Offshore Wilkes Land*. Circum-Pacific Council for Energy and Mineral Resources Earth Sciences Series, Houston, **5A**, 1–14.

TANAHASHI, M., EITTREIM, S. & WANNESON, J. 1994. Seismic stratigraphic sequences of the Wilkes Land margin. *Terra Antarctica*, **1**(2), 391–393.

TEN BRINK, U. S. & COOPER, A. K. 1992. Modeling the bathymetry of Antarctic continental margins. *In*: YOSHIDA, Y., KAMINUMA, K. & SHIRAISHI, K. (eds) *Recent Progress in Antarctic Earth Science*. Terra Scientific Publishing, Tokyo, 763–772.

TSUMURAYA, Y., TANAHASHI, M., SAKI, T., MACHIHARA, T. & ASAKURA, N. 1985. Preliminary report of the marine geophysical and geological surveys off Wilkes Land, Antarctica in 1983–1984. *Memoirs of National Institute of Polar Research* Special Issue, **37**, 48–62.

VEEVERS, J. J. 1987. The conjugate continental margin of Antarctica and Australia. *In*: EITTREIM S. L. & HAMPTON M. A. (eds) *The Antarctic Continental Margin: Geology and Geophysics of Offshore Wilkes Land*. Circum-Pacific Council for Energy and Mineral Resources Earth Sciences Series, Houston, **5A**, 45–74.

WANNESSON, J. 1991. Geology and petroleum potential of the Adelie Coast margin, East Antarctica. *In*: St. John, B. (ed.) *Antarctica as an Exploration Frontier*. American Association of Petroleum Geologists, Studies in Geology, **31**, 77–87.

WANNESSON, J., PERLAS, M., PETTITPERRIN, B., PERRET, M. & SEGOUFIN, J. 1985. A geophysical survey of the Adelie margin, East Antarctica. *Marine Petrology and Geology*, **2**, 192–201.

WATKINS, N. D. & KENNETT, J. P. 1972. Regional sedimentary disconformities and upper Cenozoic changes in bottom water velocities between Australia and Antarctica. *In*: HAYES, D. E. (ed.) *Antarctic Oceanology II. The Australian-New Zealand Sector*. Antarctic Research Series, **19**, 273–294.

Eastern New Zealand Drifts, Miocene–Recent

LIONEL CARTER[1] & I. NICHOLAS McCAVE[2]

[1]*National Institute of Water and Atmosphere, P.O. Box 14 901, Kilbirnie, Wellington, New Zealand*
(e-mail: l.carter@niwa.cri.nz)
[2]*Department of Earth Sciences, University of Cambridge, Downing Street, Cambridge CB2 3EQ, UK*

Abstract: Sediment from New Zealand passes into the north-flowing SW Pacific deep western boundary current (DWBC) to form widespread drifts. South of 49°S, where the DWBC is reinforced by the Antarctic Circumpolar Current, drifts are small and, in the case of *Campbell 'skin' Drift*, reduced to a veneer over oceanic sediments. North of 49°S deposition prevails with (1) *Chatham Terrace Drifts* – 350 km long and 320 m thick deposits around eastern Chatham Rise at 3000 m depth, (2) *Chatham Deep Drift* – 400 m thick, 300 km long ridge around the 4500–5000 m deep rise base, (3) *Louisville Moat Drift* – 400 m of sediment within an isostatic depression along Louisville Seamount Chain and (4) *Rekohu Drift* – a 250 km long, 480 m thick ridge of Oligocene to Recent sediment. These drifts are sparsely surveyed, but two others have sufficient detail to be case studies. *North Chatham Drift* plasters northern Chatham Rise at 2000–4500 m depth. Drilling recovered 587 m of nannofossil ooze/chalk with cyclical amounts of terrigenous detritus and reworked subantarctic microfossils. Deposition has been almost continuous since the early Miocene. *Hikurangi Fan-drift* is a 300 km-long lobe, developed down-current of Hikurangi Channel/Fan through redirection of turbidity current plumes by the DWBC at 4850–>5500 m depth. Seismically, the drift core comprises 200–360 m of probable Miocene palaeo-drift formed prior to the Pleistocene inception of Hikurangi Channel when 400 m of fan-drift accumulated.

Geological and oceanographic setting

As the deep western boundary current (DWBC) enters the SW Pacific Ocean, it passes along the 4500 km long margin of the New Zealand microcontinent (Fig. 1). This complex of submarine rises and plateaux sits astride the collisional boundary between the Australian and Pacific plates (Fig. 1 inset). The boundary also transects New Zealand where rapid tectonic uplift of readily erodible rocks, volcanism, and a vigorous climate generate large volumes of sediment that approach c. 1.7% of the annual suspended load delivered to the world ocean (data of Griffiths & Glasby 1985; Milliman & Syvitski 1992). Much of this load is captured by three submarine channel systems: Solander, Bounty and Hikurangi, which guide turbidity currents up to 1400 km before disgorging them directly into the DWBC (Carter *et al.* 1996). As a result, a suite of drifts have formed along the microcontinental margin (Fig. 2).

The abyssal circulation comprises the thermohaline DWBC which, below 49°S, flows in consort with the Antarctic Circumpolar Current (ACC). Although mainly wind-driven, the ACC has deep-reaching effects as manifested by its high abyssal eddy kinetic energy (Hollister & McCave 1984; Daniault & Ménard 1985), and extensive erosion of the ocean floor (Gordon 1975). In fact, the ACC dominates the abyssal flow around southernmost Campbell Plateau (Carter & Wilkin 1999).

Initially, the deep currents intercept the margin at Macquarie Ridge (Gordon 1972). Most of the flow is forced around the ridge to recirculate into nearby Emerald Basin as a series of eddies and meanders (Boyer & Guala 1972). However, gaps in Macquarie Ridge allow current filaments to pass directly into Emerald Basin and adjacent Solander Trough housing the Solander Channel and Fan (Schuur *et al.* 1998). The perturbed flow exits Emerald Basin and continues northeast along the 3000–3500 m high flanks of Campbell Plateau.

En route, the ACC diverts to the east (Bryden & Heath 1985; Orsi *et al.* 1995) and from 49°S, the DWBC travels north beneath several flows related to the Subtropical Front (Carter & McCave 1994). Immediately north of 49°S, the steep western boundary of Campbell Plateau is replaced by the gently sloping floor of Bounty Trough where the DWBC broadens and decelerates. Further north, however, the flow re-intensifies, this time against Chatham Rise (Warren 1973). Upon rounding the eastern end of the rise, the DWBC strikes northwest with the main flow directed along subdued topography that steepens markedly along the flank of Hikurangi Plateau. A small branch of the DWBC is also guided along Louisville Seamount Chain (McCave & Carter 1997). Eventually, both branches of the DWBC intercept Kermadec Ridge and Trench where the flow is steered north towards the central Pacific (Whitworth *et al.* 1999). Again, a pronounced western boundary, this time with >4000 m relief, causes the flow to strengthen, and nearly all the volume transport of the 700 km wide current occurs within 150 km of Kermadec Ridge.

The DWBC typically occupies depths >2000 m with most of the transport at depths >3250 m. Not unexpectedly, for a flow that supplies most of the water filling the Southwest and North Pacific basins, its volume transport is high. It is also highly variable with a temporal average and variance of $16.0 \times 10^6 m^3 s^{-1} \pm 11.9 \times 10^6 m^3 s^{-1}$ (Whitworth *et al.* 1999). Most of this transport involves Circumpolar Deep Water (CDW) which McCave & Carter (1997) subdivided into (i) upper CDW at c. 1400–2900 m depth, and (ii) lower CDW occupying the rest of the water column.

Compared to Atlantic drifts (e.g. Faugères *et al.* 1993), the SW Pacific drifts are not well known. They have been identified only in the last eight years from regional seismic and 3.5 kHz data plus a few cores (Carter & McCave 1994, 1997; McCave & Carter 1997). Only recently, have these deposits been targets of the Ocean Drilling Program (Carter *et al.* 1999). This state of knowledge has necessitated a dual approach to this chapter. First, we present a regional perspective of the drifts, followed by two case histories of the larger and better known deposits.

Eastern New Zealand Drifts

Emerald Basin Drifts

Schuur *et al.* (1998) documented several drifts in Emerald Basin including small mounded deposits within faulted depressions, and a prominent sheet drift on the central basin floor (Fig. 3). Seismically, the drifts have numerous, sub-parallel wavy reflections that Schuur *et al.* (1998) grouped into a lower, structurally disrupted unit overlain by a less deformed unit. Separating those units is an unconformity of possible Pliocene age. Data from DSDP Site 278 in the central Emerald Basin (Kennett *et al.* 1975) suggest that

From: STOW, D. A. V., PUDSEY, C. J., HOWE, J. A., FAUGÈRES, J.-C. & VIANA, A. R. (eds)
Deep-Water Contourite Systems: Modern Drifts and Ancient Series, Seismic and Sedimentary Characteristics.
Geological Society, London, Memoirs, 22, 385–407. 0435-4052/02/$15.00 © The Geological Society of London 2002.

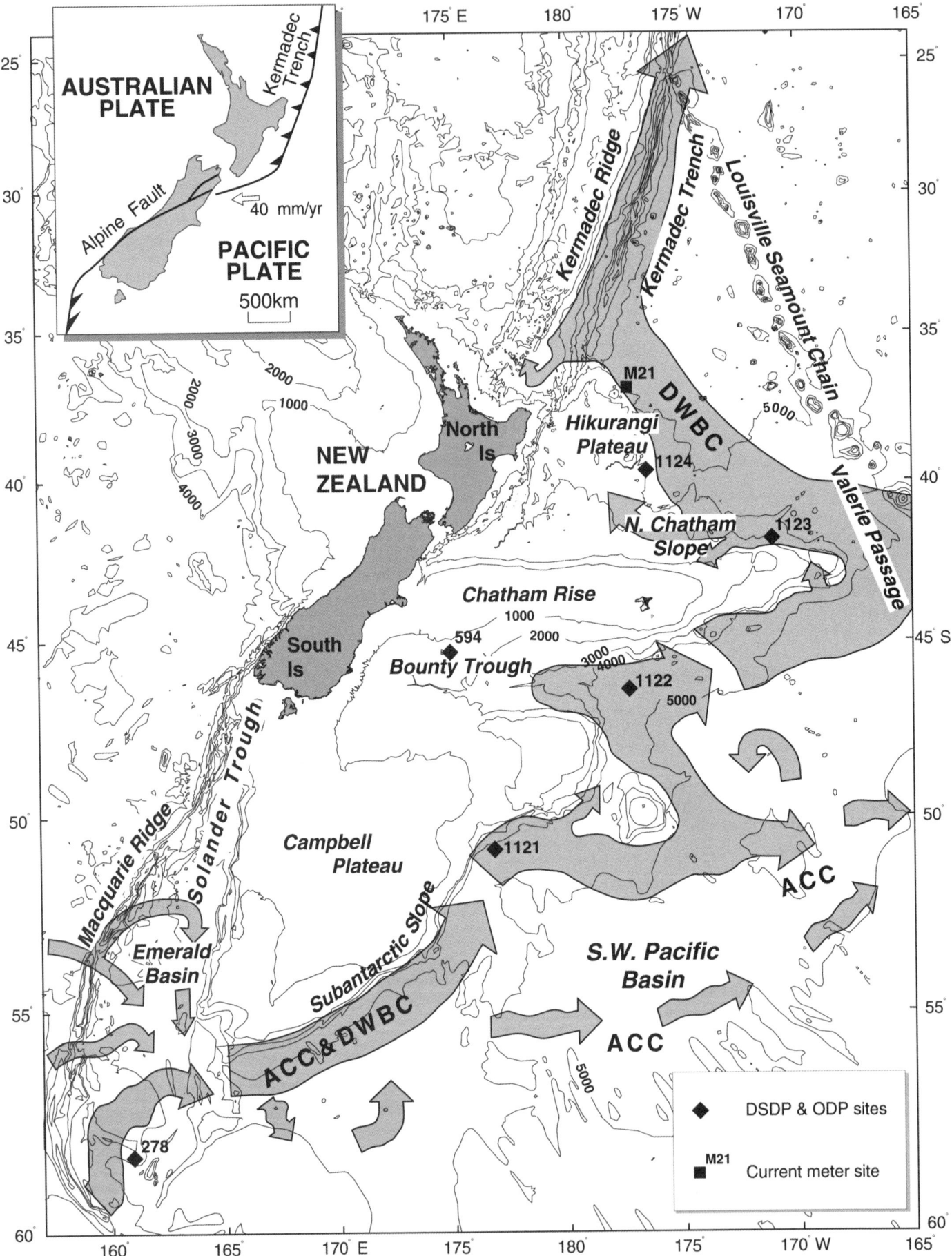

Fig. 1. Bathymetry and generalized pathway of the deep western boundary current (DWBC) and Antarctic Circumpolar Current (ACC) (from Carter *et al.* 1998). Inset outlines the Australian/Pacific plate boundary (after Lewis & Pettinga 1993).

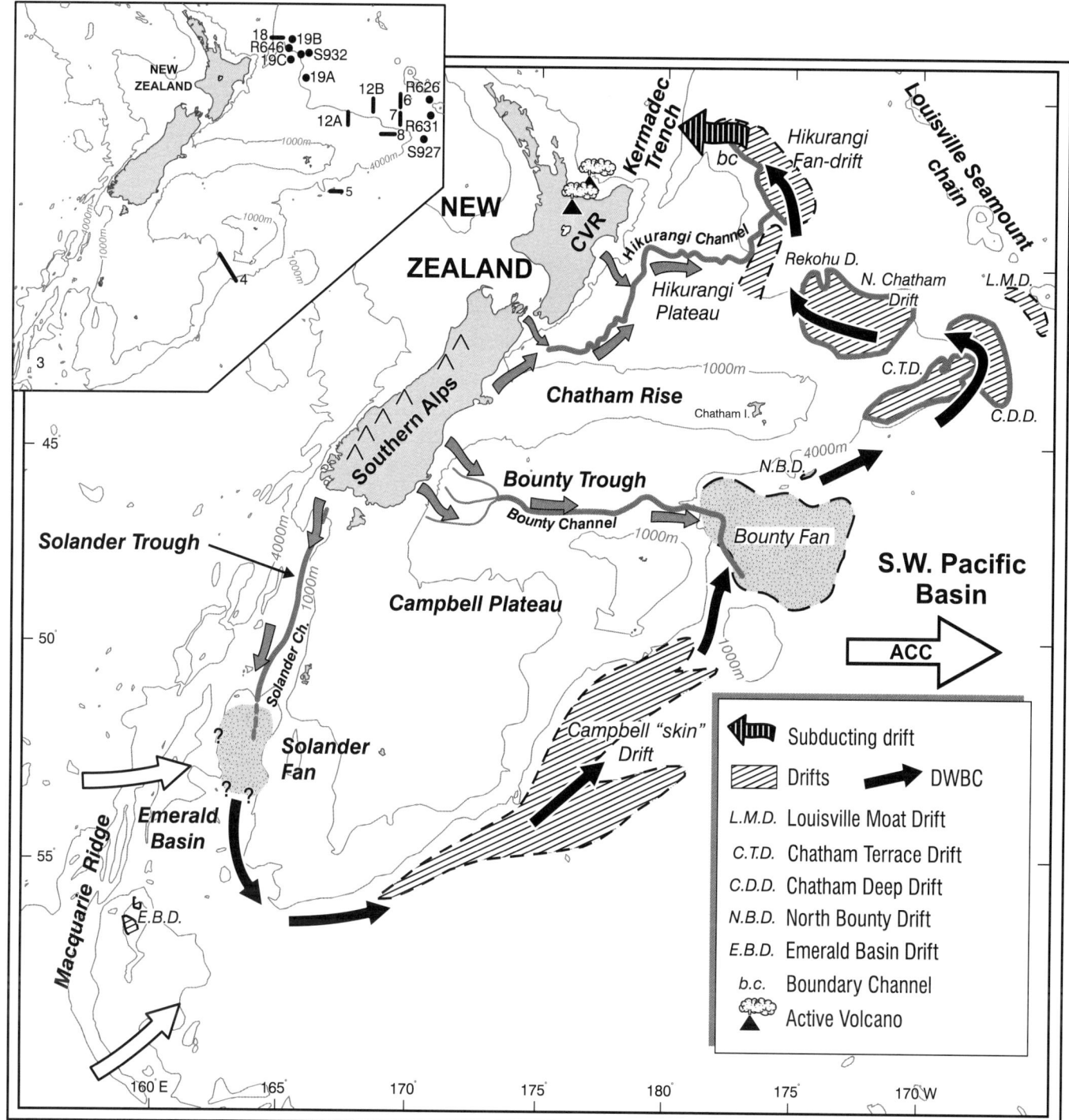

Fig. 2. Location of main drifts formed beneath the DWBC/ACC. Central Volcanic Region (CVR) generates a significant proportion of the terrigenous flux to the deep ocean. Inset has positions of the illustrated seismic profiles designated by figure number, and locations of cores mentioned in the text.

drift sediments are mainly calcareous and siliceous oozes with a small terrigenous component. In northern Emerald Basin, the terrigenous input to the drifts may be more substantial as suggested by the presence of closely spaced, parallel and continuous seismic reflectors which Schuur et al. (1998) interpret as Pliocene or younger turbidites from the Solander Fan.

Fed by the Solander Channel, the fan represents the first significant terrigenous input into the ACC/DWBC (Carter & McCave 1997). The channel transferred sediment from southern New Zealand along the floor of Solander Trough, especially during sea-level lowstands. Fan growth probably began in the Miocene with the generation of terrigenous sediment from the developing Australian/Pacific plate boundary (e.g. Nelson 1985). Such growth presumably accelerated around 6.4 Ma when the dominant plate motion changed from strike slip to compression (Walcott 1998), increasing uplift of the Southern Alps and thereby enhancing the terrigenous supply.

Despite the injection from Solander Channel much of the terrigenous load appears to have bypassed the pelagic-dominant drifts of central and southwestern Emerald Basin. Carter & McCave (1997) suggest that the ACC/DWBC transports terrigenous detritus out of eastern Emerald Basin and around the southern Campbell Plateau (Fig. 1).

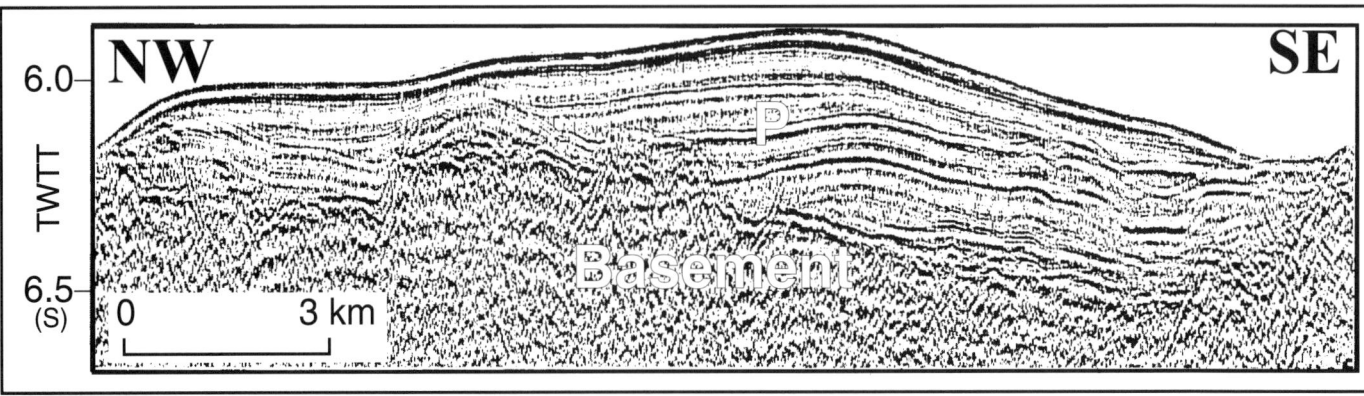

Fig. 3. Eight channel reflection seismic section (RV *Rig Seismic*) across an Emerald Basin drift with a major unconformity of suggested Pliocene (P) age. Modified from Schuur *et al.* (1998).

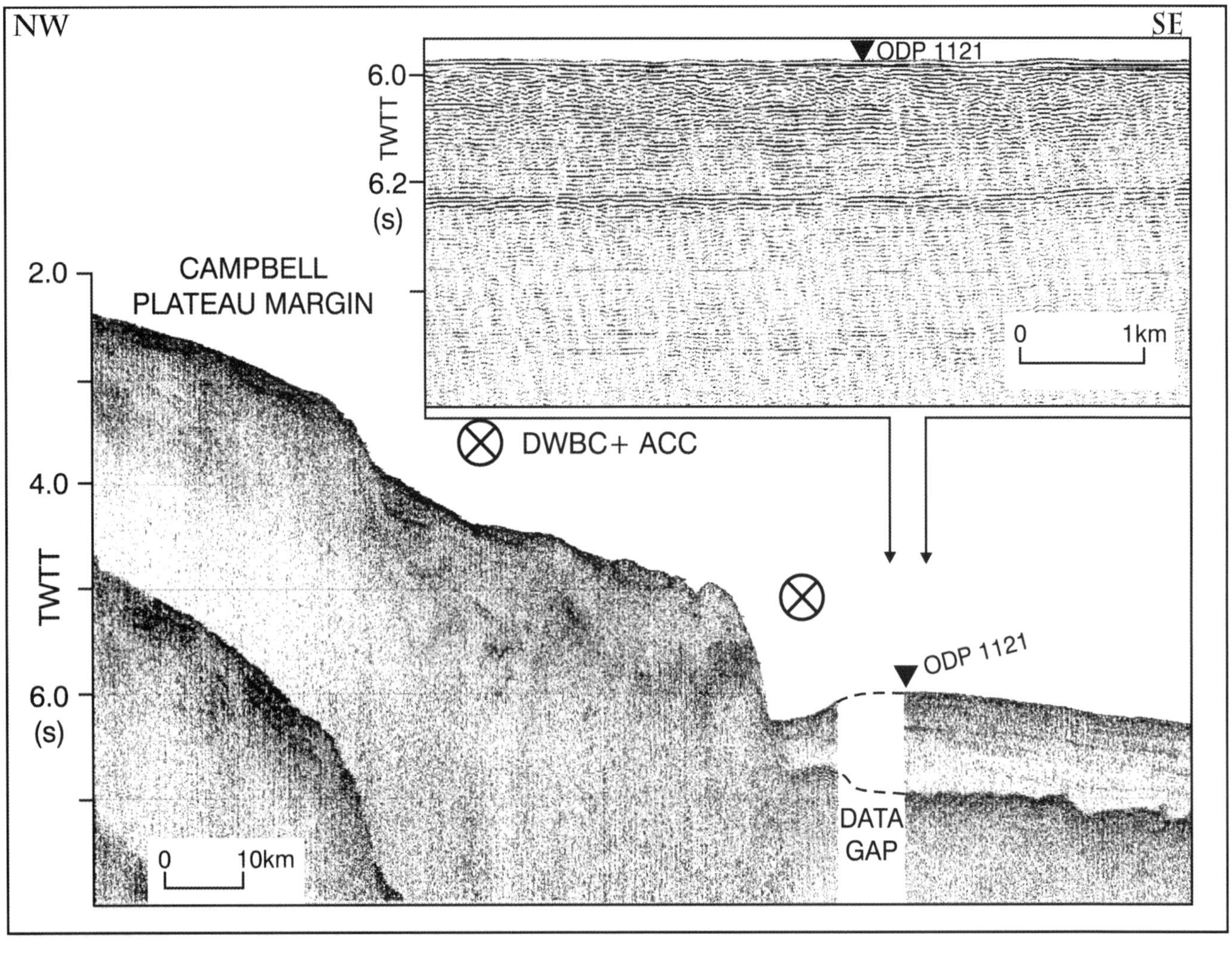

Fig. 4. General seismic outline of the Campbell 'skin' drift, present as a 15–32 m thick layer on a ridge of oceanic sediment that has been sculpted by the ACC and DWBC. Inset is high resolution multichannel profile across the skin drift at ODP Site 1121 (SSP 1999a). The irregular wavy reflectors just below 6 s TWTT are a response to chertified zones. Located on Figure 2 inset.

Campbell 'Skin' Drift

From Emerald Basin, the sediment-bearing ACC/DWBC passes northeast along Campbell Plateau. *En route* some sediment may be diverted into the SW Pacific Basin by an eastward branch of the ACC at 56°S (Orsi *et al.* 1995). However, sediment also continues northeast as demonstrated by active bedforms and a strong benthic nepheloid layer (Carter & McCave 1997). Using unprocessed, single channel seismic profiles from USNS *Eltanin* (e.g. Hayes *et al.* 1976), Carter & McCave (1997) suggested that sediment entrained by the ACC/DWBC was deposited to form *Campbell drift* along the foot of Campbell Plateau at 4000 to 5100 m depth (Fig. 4). This wedge to ridge-like deposit was estimated to have up to 160 m of possible Neogene drift. Indeed, drift sediment, consisting of current-reworked terrigenous and foraminiferal sands interbedded with mud, was subsequently recovered in a 3 m long kasten core (NIWA number Y11; Carter *et al.* 1997). However, long cores recovered from ODP Site 1121

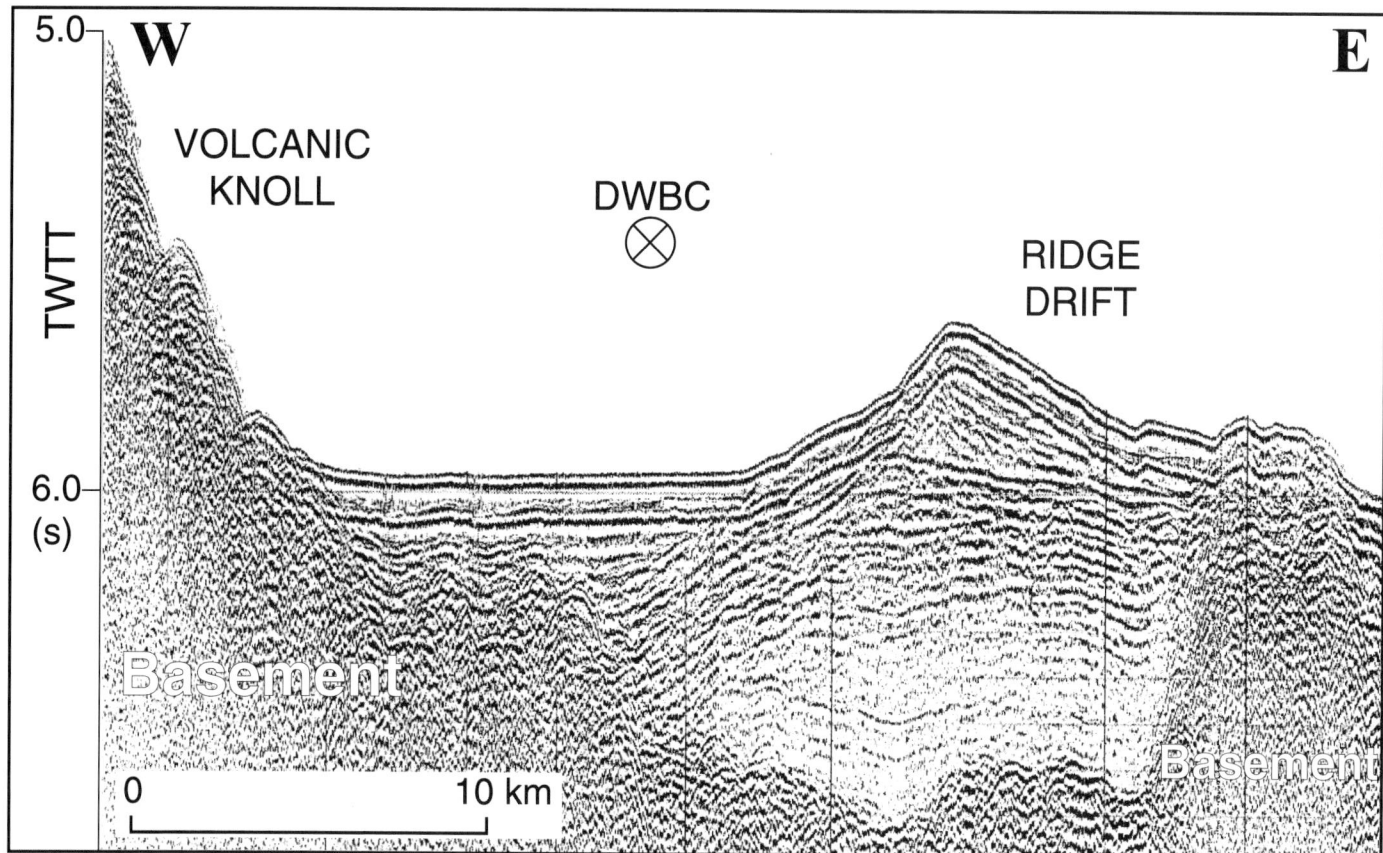

Fig. 5. Single channel seismic profile across a north Bounty drift with the DWBC directed into plane of the page. Located on Figure 2 inset.

show the drift component to be only 15–32 m thick, and is comprised of manganese-nodule bearing, reworked sands and silts of probable late Neogene age (Shipboard Scientific Party [henceforth 'SSP'] 1999a). Between 15–32 metres below sea floor, Palaeocene, deep oceanic calcareous and siliceous biogenic oozes were recovered. To describe such thin deposits of current reworked sediment we propose the term *skin drift*.

North Bounty Drifts

Deceleration of the DWBC across Bounty Trough has allowed the Bounty Fan to form right across the current's path. Nevertheless, direct sediment injection *via* Bounty Channel, coupled with current winnowing of the middle to lower fan, have provided material to the DWBC for one or more ridge drifts to form north of the fan (Carter & Carter 1996). One ridge is 200 m high and *c*. 70 km long with the crest sub-parallel to the current (Fig. 5). Its profile is asymmetric with the steeper flank facing the western boundary formed by the steep sides of Chatham Rise. Although this material is directly downstream of the fan, there is no evidence that it is dominantly sandy. Indeed samples in the general area suggest it is probably a slightly sandy mud. Thus, the deposits of a major fan, reworked by the biggest boundary current in the world, have not yielded significant contourite sands.

Louisville Moat Drift

As the DWBC passes through Valerie Passage, between the Louisville Seamount Chain and Chatham Rise, it bathes a moat along the southwestern side of the chain (Fig. 2; Watts *et al.* 1988).

Between 240 and 400 m of sediment have accumulated in the depression to form the *Louisville Moat Drift* (Fig. 6; Carter & McCave 1994). Despite these thicknesses, the drift has little relief above the 4700–4900 m deep ocean floor surrounding the moat. In seismic section, the drift is a body of near-transparent sediment with faint internal reflectors. The surface is irregular and partly inherited from the underlying basement. Additional to the inherited topography, 3.5 kHz profiles exhibit small scale, regular hyperbolae that possibly represent erosional furrows or some transport bedform (Fig. 6 inset; McCave & Carter 1997). Core R626 retrieved mud with scattered tephra layers. The median grain size of the core top was 8.68 Φ (2.4 µm) reflecting high amounts of clay and fine silt.

Chatham Deep Drift

An irregular ridge of sediment, up to *c*. 400 m thick, extends *c*. 300 km around the base of the Chatham Rise at 4500–5000 m depth (Figs 2 and 7). Termed *Chatham Deep Drift* by Carter & McCave (1994), it has a semi-transparent acoustic response with faint discontinuous but ordered reflections. The impact of the abyssal flow is shown by prominent moats, well developed non-migratory sediment waves, and small regular hyperbolic reflections which may come from furrows or depositional bedforms. Reoccurrence of hyperbolic reflectors through 3.5 kHz sections suggests fluctuations in drift sedimentation (Fig. 7b). Kasten core S927 penetrated mud which, at the surface, has a median grain size of 8.08 Φ (3.7 µm) and a dominant 7.4 Φ (5.9 µm) silt mode. This sediment is slightly coarser and better sorted than that of the moat drift (R626), and probably reflects the location of the deep drift beneath the main flow of the DWBC. The sedimentation rate over the past 150 ka is *c*. 1.98 cm ka^{-1}.

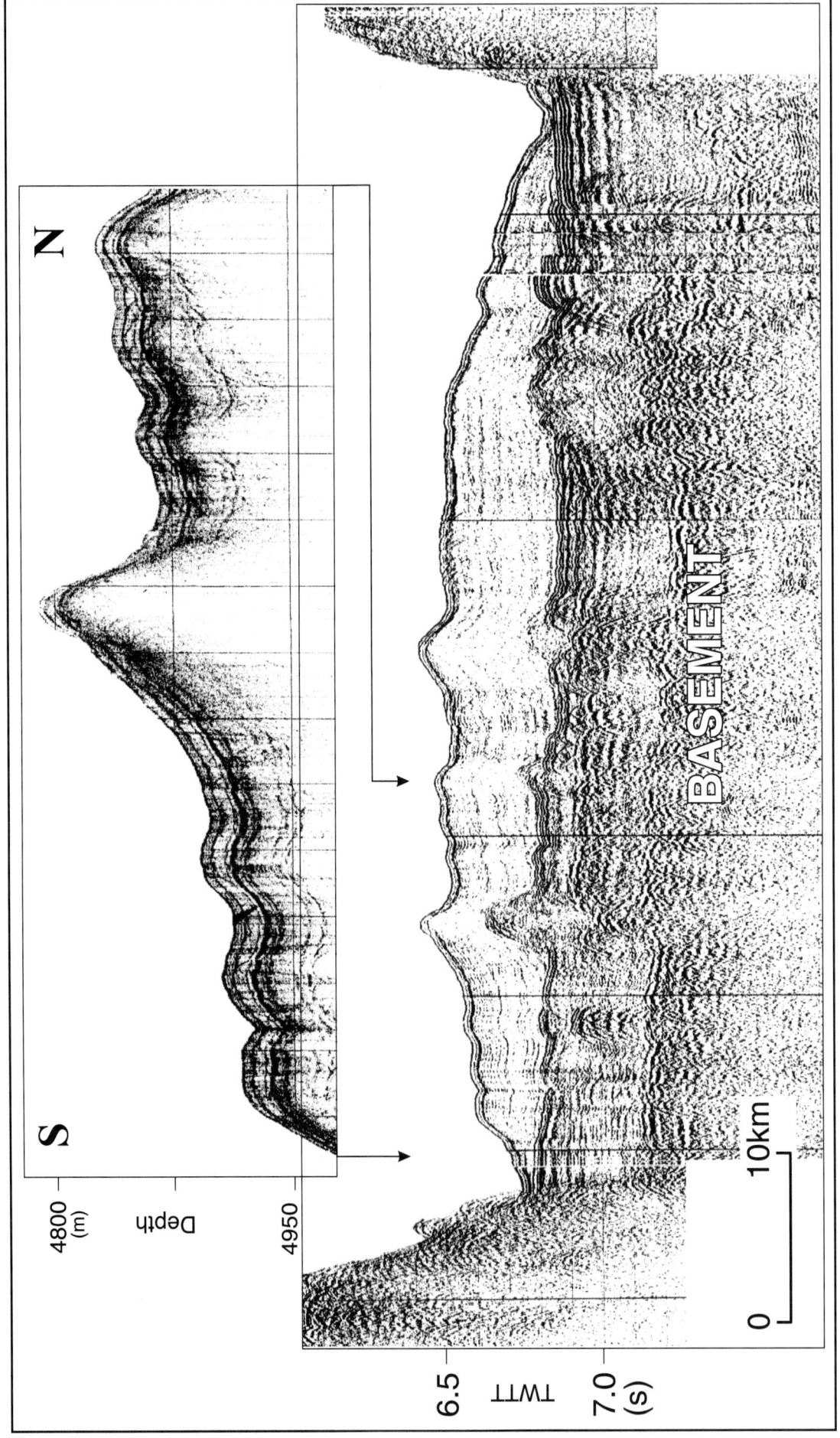

Fig. 6. Louisville Moat Drift outlined in a single channel seismic section showing reduced deposition/erosion at drift edges where DWBC is intensified against moat walls. Inset is 3.5 kHz profile with undulations that appear to be inherited from the basement. Located on Figure 2 inset.

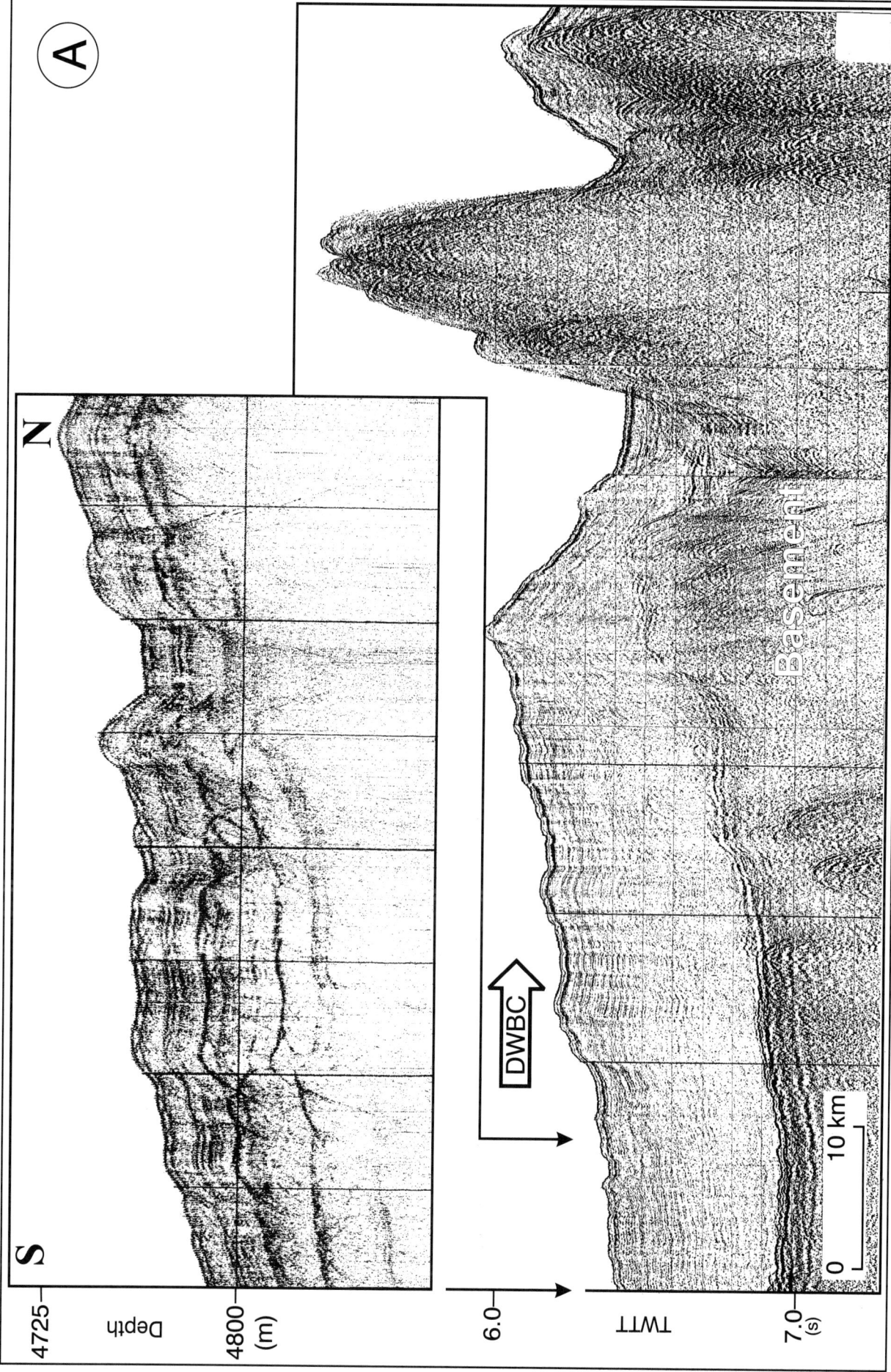

Fig. 7. (**A**) Single channel seismic section run obliquely across Chatham Deep Drift high-lighting its acoustically weak internal structure, moating against volcanic knoll, with irregular sediment waves and small hyperbolae evident in the 3.5 kHz profile (inset). (**B**) 3.5 kHz profile across the deep drift (see Fig. 8) with hyperbolae on the western face, and truncated reflectors on the eastern face which forms a local boundary for the DWBC. Located on Figure 2 inset.

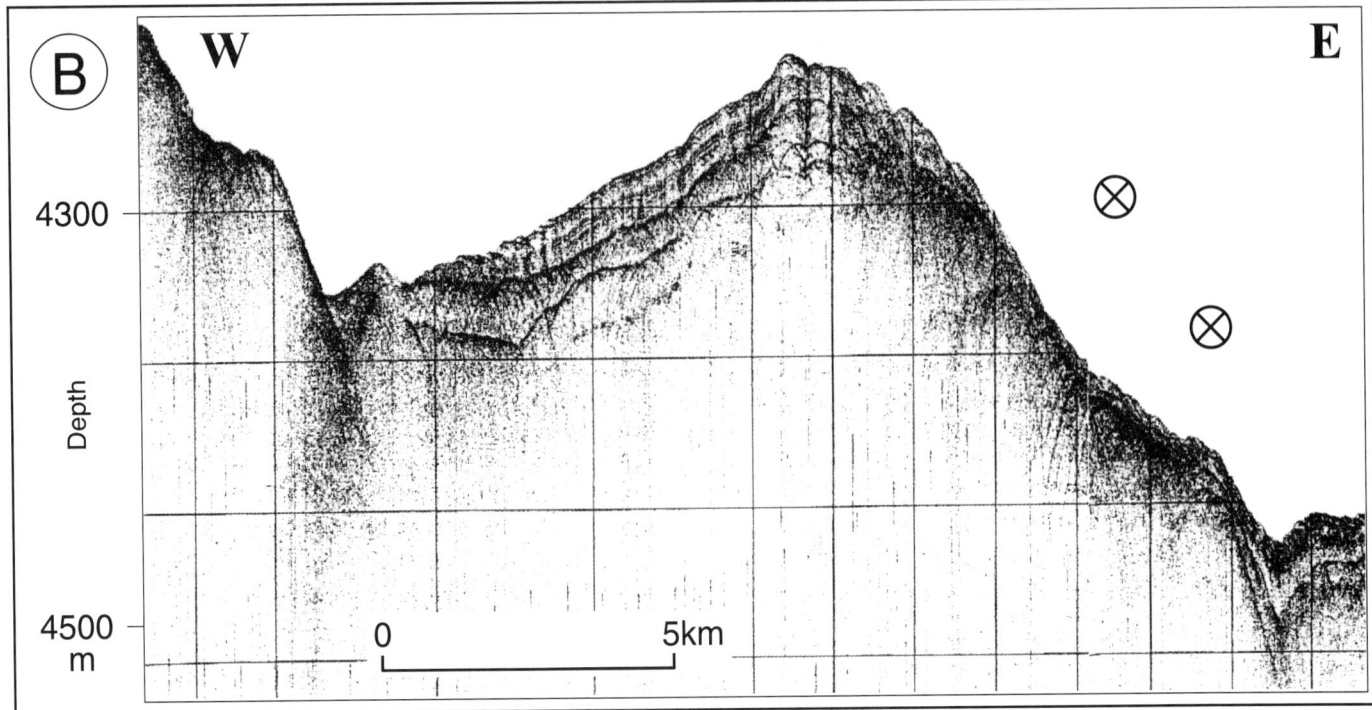

Fig. 7(B).

Chatham Terrace Drifts

At the top of the slope, above the *Chatham Deep Drift*, sinuous sediment bodies wrap around the end of Chatham Rise to form an irregular terrace at about 3000 m depth (Carter & McCave 1994). Designated the *Chatham Terrace Drifts*, these deposits are up to 350 km long and are usually < 320 m thick (Fig. 8). Irregular, discontinuous internal reflectors with unconformities point to bouts of deposition punctuated by intermittent erosion, although results from ODP Leg 181 indicate that such 'unconformities' can be seismic responses from diagenetic fronts (Carter *et al.* 1999). Terrace drift surfaces are rough, in part due to relief inherited from older topography, but also due to current effects such as scour moats and non-migratory sediment waves.

Rekohu Drift

The next drift down-current of Chatham Rise is *North Chatham Drift* (see below) and beyond that lies *Rekohu Drift* (Carter & McCave 1994). This deposit is a 250 km long ridge whose crest deepens northwards from 3600 m to 4190 m (Fig. 9). Thus, it serves to direct the lower DWBC to the north, whereas the flow above the ridge crest probably continues northwest across Hikurangi Plateau. The drift crest is irregular with some relief inherited from the volcanic basement, but other features are caused by the flow, e.g. sediment waves and occasional small erosional/depositional bedforms manifested as small regular hyperbolae. The drift terminus is further modified by slumping.

Rekohu Drift reaches 480 m thickness and comprises (1) a basal seismic Unit B (*c.* 400 m max.) of acoustically indistinct sediment passing upsection into faintly layered material for which Carter & McCave (1994) suggested a Miocene age, and (2) a thin cover (*c.* 80 m thick) with closely spaced, parallel and continuous reflections belonging to seismic Unit A of probable Plio–Pleistocene age (SSP 1999*c*). Testing and confirmation of the Rekohu seismic stratigraphy was provided by ODP Site 1124 (Fig. 10). There, the drift rests on the regional Marshall Paraconformity which probably marks the first inflow of the DWBC into the SW Pacific Ocean at about 32 Ma (Kennett 1977). Apart from hiatuses of 6 Ma in the early Miocene, and *c.* 3Ma in the mid-Miocene, both of which may relate to increased strength of the DWBC, the overlying drift provides a record of sedimentation spanning the last 27 Ma.

Nannofossil chalk, containing varying amounts of terrigenous detritus, prevailed from the late Oligocene to late Miocene. Significantly, middle Miocene terrigenous detritus contains mica which is interpreted as coming from the New Zealand Southern Alps via Bounty Channel (Carter & Mitchell 1987). Alternations of nannofossil ooze and silty clay overlie the chalk at a boundary approximating the onset of more distinct bottom reflectors in seismic Unit B (Figs 9 and 10). Better acoustic contrast from the late Miocene to Recent may reflect cyclic changes in lithology. Glacial phases have elevated levels of terrigenous sediment caused by increased supply, concomitant with dissolution of pelagic carbonate especially during the late Pliocene–Pleistocene, as suggested by piston and kasten core data of Weaver *et al.* (1998). The continued presence of mica, coupled with concentrations of subantarctic diatoms, confirms contributions from southern sources carried by the DWBC. Additional terrigenous sediment may have come from the nearby Hikurangi Channel which diverted towards the central part of Rekohu Drift sometime in the Pleistocene, probably around 1.2 Ma when Site 1124 witnessed an abrupt increase in sedimentation rate (Lewis *et al.* 1998; SSP 1999*c*). Finally, Rekohu Drift contains airfall tephra derived from explosive volcanism on the North Island (e.g. Carter *et al.* 1995). At least 134 macroscopic tephra are present, making up to 5% of the thickness of the Pliocene section and 13% of the Pleistocene section (SSP 1999*c*).

Case Study – North Chatham Drift

Geological and oceanographic setting

North Chatham Drift lies on the side of Chatham Rise which extends 1550 km due east from the South Island, emerging at the Chatham Islands and descending to the 5000 m deep Valerie Passage at its eastern extremity (Figs 1 and 2). It is underlain by

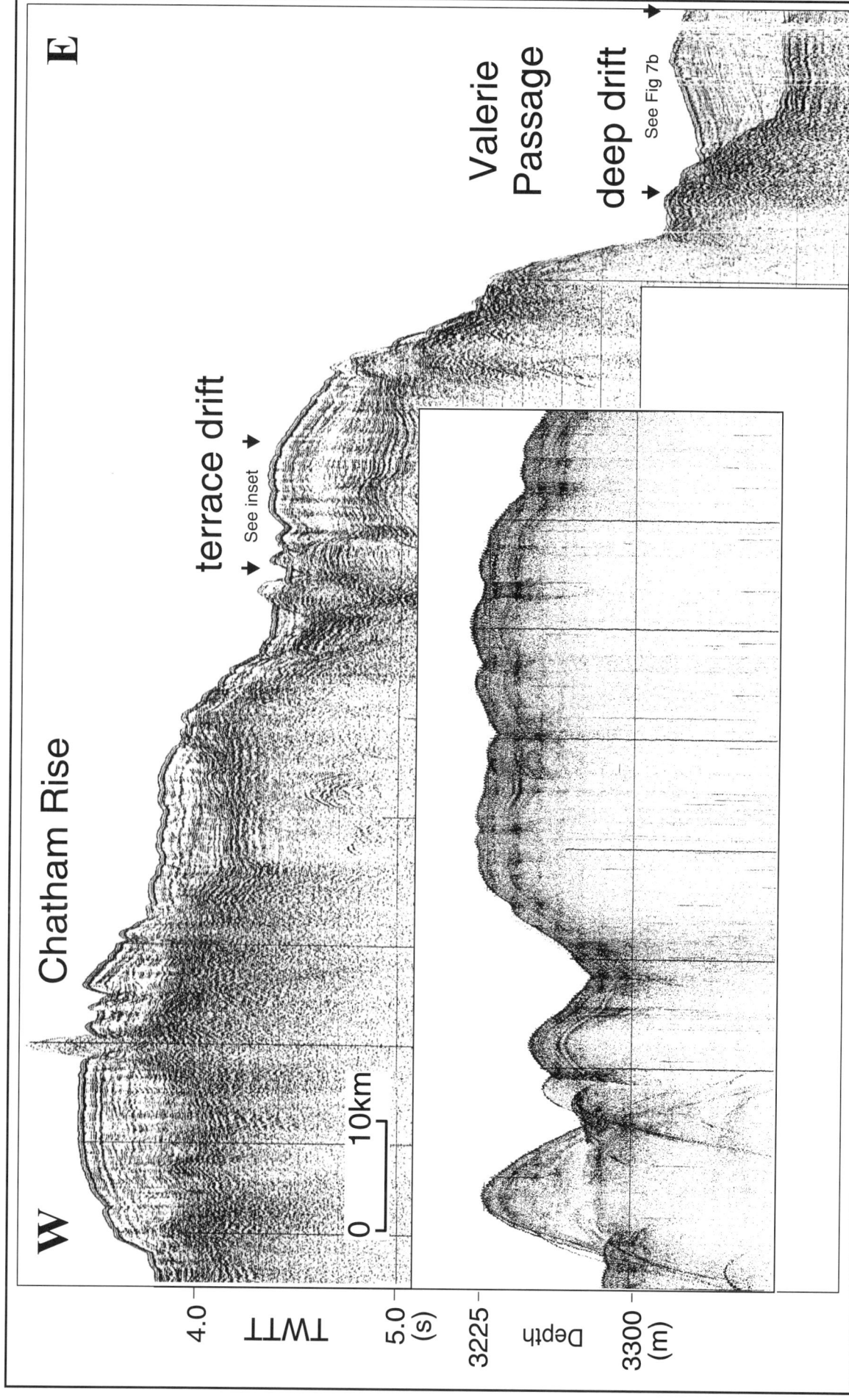

Fig. 8. Chatham Terrace Drift and Deep Drift on single channel seismic profile together with sediment waves and scour zone high-lighted on the 3.5kHz profile (inset). Located on Figure 2 inset.

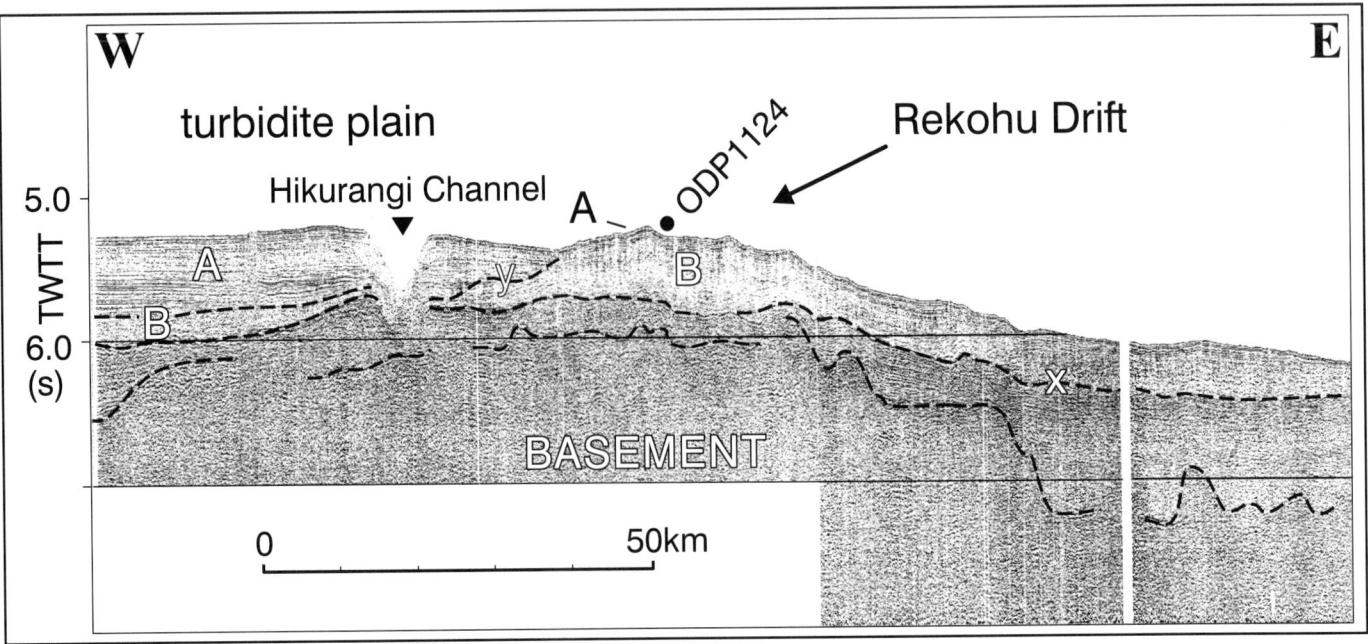

Fig. 9. Rekohu Drift on a single channel seismic profile showing its damming of the turbidite plain associated Hikurangi Channel. Seismic units A and B are delimited by reflectors Y and X respectively. Located on Figure 16 inset.

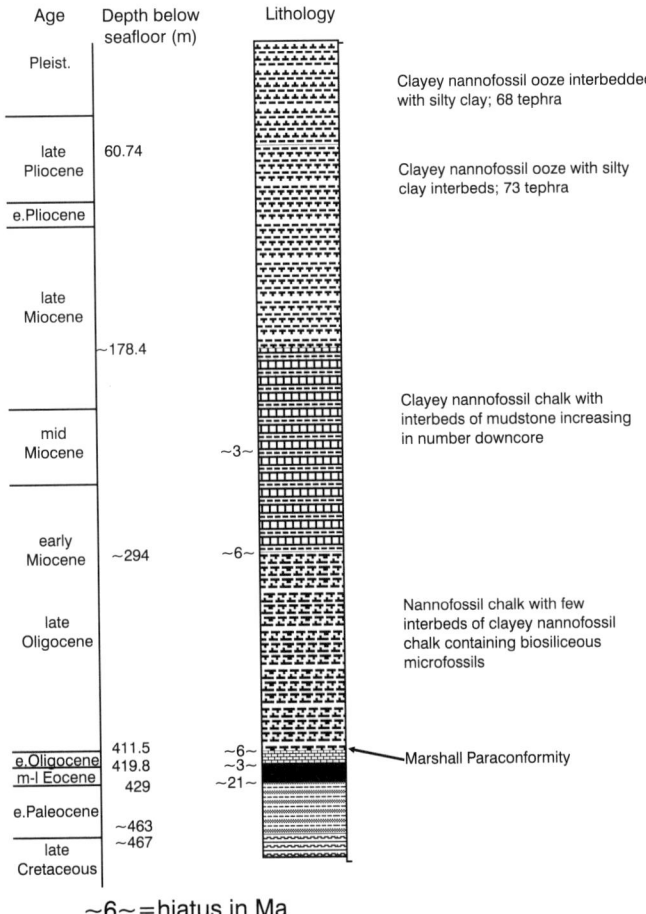

Fig. 10. Lithostratigraphy through Rekohu Drift which rests on the regional Marshall Paraconformity. Data from ODP Site 1124 (SSP 1999c). Located on Figures 1 and 9.

Mesozoic metamorphic basement (Torlesse terrane) in the west and by Mesozoic sedimentary rocks east of about 175°W (Wood et al. 1989). The drift is of the plastered type (McCave & Tucholke 1986), forming an extensive apron on the north side of the rise and extending from c. 2000 m to below 4500 m depth.

There are four basic water masses in the region, one of which can be refined into smaller units on the basis of the detail portrayed in CTD profiles (Fig. 11). The uppermost water column comprises subtropical and Australasian Subantarctic waters, located respectively north and south of Chatham Rise. These waters meet at the east-west trending Subtropical Front, which is confined along the Rise crest. To the east, beyond Chatham Rise, the convergence zone becomes diffuse and loops south (Heath 1985). The depth range of c. 600–1400 m is occupied by Antarctic Intermediate Water (AAIW), formed by sinking at the Antarctic Polar Front, and characterized by temperatures of 3–7°C and a salinity minimum of 34.36–34.50 (Fig. 11). Chatham Rise partially intercepts a region of northward transport of this watermass.

Water between c. 1400 and 2900 m depth is tentatively correlated with upper Circumpolar Deep Water (UCDW), described by Gordon (1975) from the region of the DWBC further south off Campbell Plateau. North of Chatham Rise, close to the western boundary, UCDW is distinguished mainly by an oxygen minimum (Fig. 11) and relatively low dissolved silica content. The remainder of the water column, from 2900 m to the ocean floor, is lower Circumpolar Deep Water (LCDW) comprising a high salinity zone (c. 2900–3800 m depth) and a colder, lower salinity zone (>3800 m) (e.g. Gordon 1975; Warren 1973). The upper layer has a salinity maximum of 34.72–34.73 which is the signature of North Atlantic Deep Water (NADW) derived from the South Atlantic and transported to the South Pacific by way of the Antarctic Circumpolar Current and DWBC (Warren 1981; Webb et al. 1991). The salinity maximum deepens northwards along the main western boundary; from 2900 m off Campbell Plateau, 3050 m in Valerie Passage, 3320 m off Hikurangi Plateau and 3500 m at Kermadec Ridge. The basal part of the LCDW owes its density to an infusion of cold (–0.7 to 0°C) Weddell Sea Deep Water (WSDW) and probably Ross Sea Deep Water together with water formed off the Adelie Land coast (Rintoul 1998), which is mixed with NADW in the Antarctic Circumpolar Current (Gordon 1975).

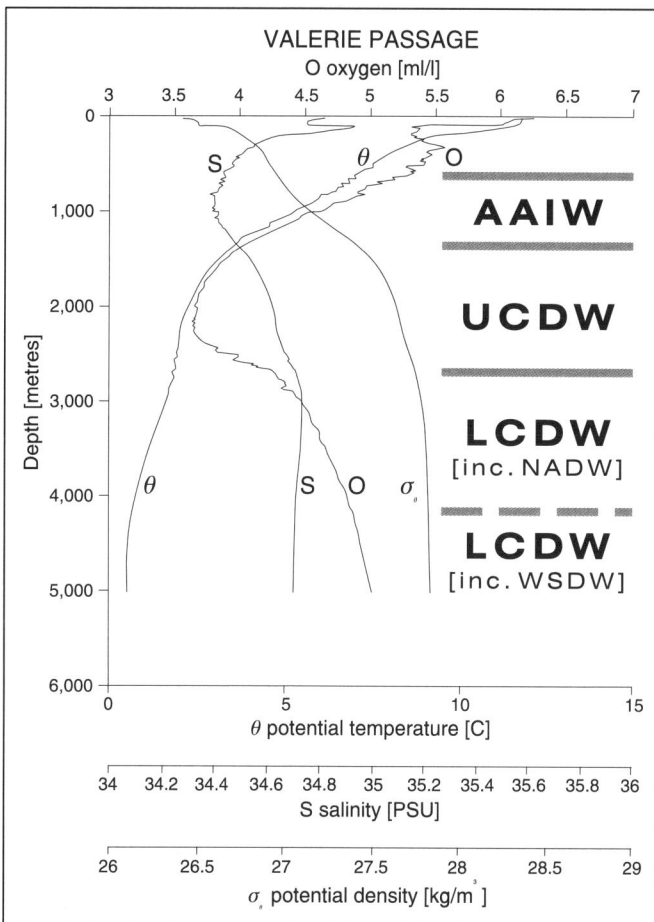

Fig. 11. Hydrographic (CTD) profile from Valerie Passage (Station R631, Fig. 2 inset) with indicative features of the main water masses including Antarctic Intermediate Water (AAIW), Upper Circumpolar Deep Water (UCDW), Lower Circumpolar Deep Water (LCDW) with North Atlantic Deep Water (NADW) and Weddell Sea Deep Water (WSDW) and possibly other Antarctic sources.

Geostrophic velocities, derived by Warren (1973), Whitworth et al. (1999) and McCave & Carter (1997), are relative to a 'zone of no motion' estimated to be around 2000 m depth. This choice of reference level is supported by evidence from the eastern flank of Chatham Rise where the upper limit of moat-scouring currents is about 2150 m (Carter & McCave 1994). Off southern Chatham Rise, geostrophic speeds near the bottom are 3–4 cm s^{-1}, but these increase to 5 cm s^{-1} as the DWBC passes through the steep-walled confines of Valerie Passage (Fig. 1). The current slows to c. 2 cm s^{-1} over the gently sloping western boundary of North Chatham Slope, but off Rapuhia Scarp and Kermadec Trench the steep flanks cause the DWBC to narrow and accelerate the geostrophic speed to a maximum of 18 cm s^{-1}. There are no direct current meter measurements in the vicinity of northern Chatham Rise. Elsewhere in the region, measurements are broadly consistent with geostrophic estimates (Whitworth et al. 1999). Thus, although the flow is volumetrically large and locally displays fast flow against steep boundaries, McCave & Carter (1997) concluded it was relatively sluggish at present, a view corroborated by the presence of relatively dilute bottom nepheloid layers.

Bathymetry

The DWBC approaches the drift near the eastern end of Chatham Rise, where the current swings from a northeast to northwest course (Fig.1). In the vicinity of the DWBC, the rise crest descends eastward and is dissected by several current-scoured, north–south channels separating terraces in depths of 2000–3500 m. Beyond that an abrupt scarp drops to the floor of Valerie Passage. This corridor, separating Chatham Rise from the Louisville Seamount Chain, is the major gateway for the DWBC to the SW Pacific Ocean. Northwest of Valerie Passage the topography becomes subdued. The northern flank of eastern Chatham Rise is a gently inclined slope that descends gradually from the rise crest to the 4500 m isobath over a distance of more than 150 km. North Chatham Rise gives way to Rekohu Embayment, a broad re-entrant at 4200–4600 m depth.

Seismic stratigraphy

The north slope of eastern Chatham Rise is blanketed by a major drift lying between 169°W and 175°W in depths of 2050–4500 m (Figs 12 and 13). It gradually increases in thickness upslope, reaching a maximum of over 1.0 s TWTT at 2500 m depth. Drift > 0.6 s thick lies above 3500 m depth and is located on the left-hand side of the principal zone of fast flow. The thicker part of the drift is dissected into two parts by a scour zone downstream of Broughton Gap, one of three scoured passages across the top of eastern Chatham Rise (Fig. 13).

Two principal seismic reflectors are recognized in the drift.

(1) Reflector X and its associated sequence of distinctive parallel reflectors. The shallow part of North Chatham Drift lies on the continental crust of the rise (Fig. 12a) whereas its toe is in deep water over a disconformity on older pelagic sediments and oceanic basement (Fig. 12b). Reflector X is traceable throughout. Within this basal part of the drift, another reflector, designated X′, defines a thin wedge with a similar shape to the present drift, thickening upslope to a maximum of about 0.3s (Fig. 12a). The X′ reflector is caused by a sharp increase in density and sonic velocity due to incipient silica cementation (SSP 1999b). At ODP Site 1123 this occurs in sediment of middle Miocene age (c. 13 Ma) at c. 450 metres below seafloor. ODP Site 1123 did not penetrate reflector X. However, X does *not* appear to be the Marshall Paraconformity of early Oligocene (c. 32 Ma) age because drilling revealed > 45 m of Eocene–Oligocene micritic limestone below the paraconformity. It is possible that, by comparison with ODP Site 1124 (SSP, 1999c), this limestone overlies more siliceous sediment and the contact between them causes the reflector. The onset of the modern deep ocean circulation is believed to date from the Marshall Paraconformity, but the seismically contoured drift (Fig.13a) includes more than that thickness of sediment.

(2) Reflector Y, though marked in places, is not consistently recognizable to allow separate isopachs for seismic Units A and B, so they are compounded in Figure 13a. In general, however, the thickness of Unit A increases upslope together with the increase in Unit B. Reflector Y is primarily a change of acoustic character, occurring at c. 140 mbsf (c. 4 Ma, early Pliocene) at ODP Site 1123 (SSP 1999b). There is no obvious lithological change at this level, though tephra layers are more frequent above this depth. (The strictly sub-bottom-parallel signature of the top 0.17s is most probably due to bubble–pulse reverberation)

Chatham Rise crest is marked by outcrops of Palaeogene or older basement (Wood et al. 1989). The junction between drift and basement outcrop is marked by a moat at 2250 m depth, but at 2100 m and shallower there is no moat (Fig. 14d, e). We suggest that the long-term level of no motion (velocities below those required for scouring even when the current is concentrated along a hard rock surface) is around 2150 m, in good agreement with the hydrography (see Warren 1973).

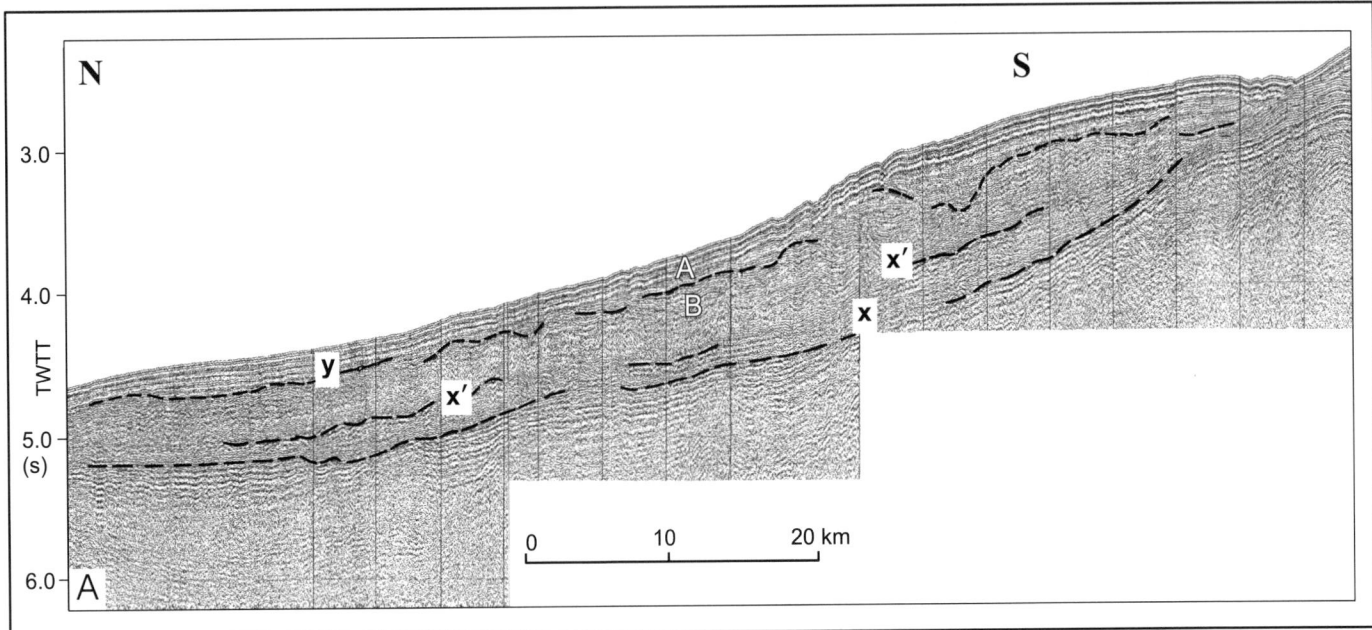

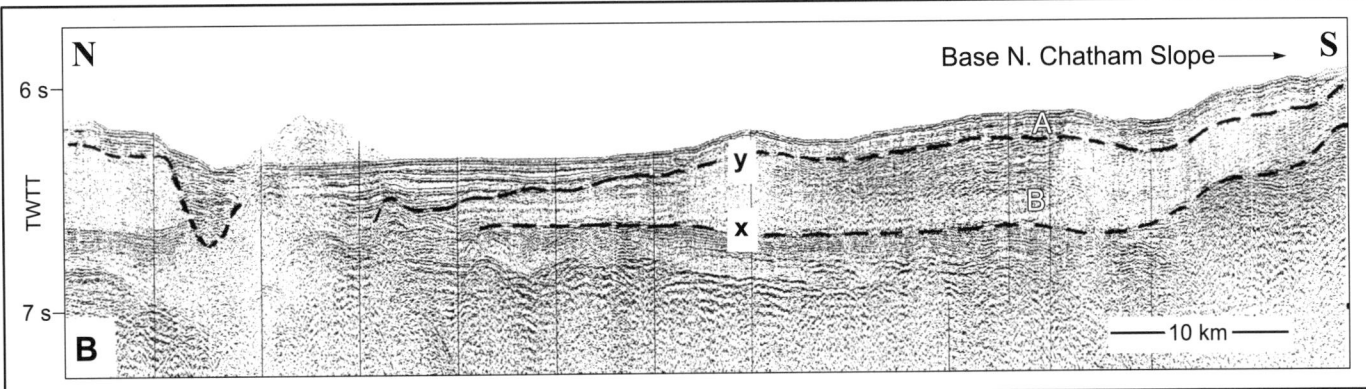

Fig. 12. Airgun seismic profiles highlighting the structure and stratigraphy across the North Chatham Drift. Reflector X' is due to a sharp change in density and velocity in sediments of middle Miocene age (*c.* 13 Ma) (SSP 1999*b*).

Acoustic character: 3.5 kHz profiles

The surface of the drift is dominated by two 3.5 kHz signatures: uniform pelagic/hemipelagic drape (Fig. 14a, b), and irregular undulations climbing vertically (Fig. 14c). In both of these reflector β can be clearly seen 10–17 ms below the surface. A core from a region where β is at 10 ms has an average sedimentation rate of 3.5 cm ka^{-1} from 0–100 ka. This suggests that the top of β is perhaps related to a change in sediment character at the marine isotope stage 8/7 transition dated at 245 ka. However, at ODP Site 1124 reflector β may be caused by the +0.8 m thick Matahina tephra dated to 0.34 Ma. But, on North Chatham Drift (ODP Site 1123) the tephra is not particularly thick, reflector β is pronounced and corresponds to an age close to the 245 ka previously deduced (SSP 1999*b*).

Echo-types identified on 3.5 kHz profiles are classified according to a scheme based on that of Jacobi & Hayes (1982). A total of 17 echo-types are recognized in this area from ocean-floor reflectivity and micromorphology, and the distribution of these types is portrayed as an echo character chart (Fig. 13b). In areas where 3.5 kHz data are sparse, interpolation between survey lines has been guided by bathymetry.

The crest of eastern Chatham Rise is mantled mainly by pelagic sediments characterized by strong, continuous surface and subsurface reflections that can be traced to 30–40 m below the sea-bed (echo-type IB). On Chatham Rise, three prominent gaps have been eroded by the DWBC: (from east to west) East Chatham, Broughton and Lavrentyev gaps (Fig. 1; Carter & McCave 1994; Fig. 12). These appear to begin 20–40 km south of Chatham Rise crest and extend north and then northwest down the northern flank of the rise for an additional 70–90 km. Near East Chatham Gap, the hemipelagic/pelagic drape displays non-migratory sediment waves of variable amplitude and wavelength (B-2A). Within the gaps themselves echo-types imply current action with zones of strong surface reflection and little penetration (IIA). Downstream of the gaps, the echo-types indicate decreasing current speeds across the drift. Down from Lavrentyev Gap, for example, hard reflectors and hyperbolae (IIA and IIIC-3) are replaced by pelagic drape covered with small hyperbolae (IIID-3) that are probably current induced (Jacobi & Hayes 1982).

The gently sloping surface of North Chatham Drift is mantled by a large field of irregular, non-migratory sediment waves (B2-A) as well as normal hemipelagic/pelagic drape (IB). To the north, flat-lying to gently undulating pelagic layers (IB) dominate the abyssal floor (> 4500 m deep). Northwest of the drift is a small turbidite pond containing material that has flowed off the side of North Chatham Rise.

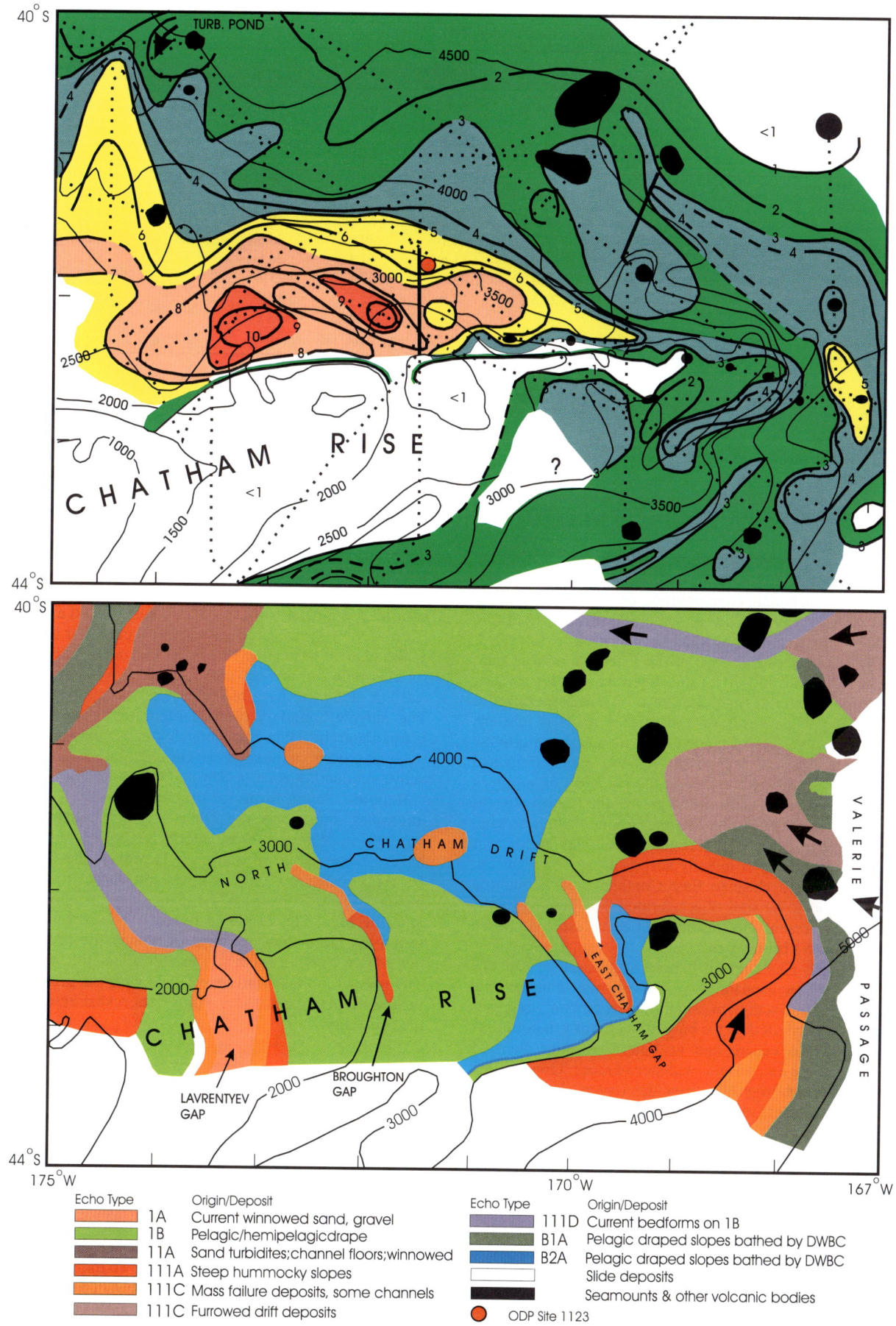

Fig. 13. (**A**) Isopachs of North Chatham Drift sediment thickness above reflector X (bold contours with values in tenths of a second TWTT, thin lines are depths in metres). The location of ODP Site 1123 is shown by a red dot, the solid line adjacent to that is the track of seismic shown in Figure 4A, and the solid line at c. 169° 30′W is the track of Figure 4B. (**B**) Echo character chart showing the distribution of the main echo-types and microphysiographic features with zones of inferred weak and strong currents marked by arrows.

Table 1. *Calcium carbonate and percent sand, silt and clay of the total surficial sediment, together with selected terrigenous components including d_m 10–63 µm which is the mean size of the 'sortable silt' component of McCave et al. (1995)*

Sediment body	Depth (m)	CaCO$_3$ %	Total sediment				Terrigenous fraction	
			Sand %	Silt %	Clay %	Silt/clay	d_m 10–63 µm	Silt/clay
Hikurangi Fan-drift – distal	5002	4.2	29.2	38.1	32.7	1.16	14.5	1.01
Hikurangi Fan-drift – mid	5235	3.6	13.5	52.2	34.3	1.52	13.4	1.24
Hikurangi Fan-drift	4851	5.3	0.5	76.1	23.5	3.24	15.9	1.37
Hikurangi Fan channel	4761	8.3	0.2	61.5	38.1	1.61	13.1	1.13
Hikurangi Plateau (turbidite plain)	2890– 3785	17.8– 22.7	4.1– 11.6	59.1 64.3	36.8 24.1	1.61 2.67	16.2– 19.2	1.22– 1.57
Rekohu Drift	4099	21.4	13.8	62.4	23.7	2.63	15.5	1.36
North Chatham Drift	4240	22.3	4.0	53.4	42.5	1.26	–	0.91
Louisville Moat Drift	4908	4.5	1.5	55.4	43.1	1.29	15.8	1.1
Chatham Deep Drift	4802	7.9	0.9	67.5	31.6	2.14	–	–

Sediments: sample data

There are no bottom photographs from North Chatham Drift, so our knowledge is based purely on bottom sample calibration of 3.5 kHz profiles.

Drifts and other deposits in depths > 4700 m have carbonate contents < 10%. A carbonate v. depth plot for all samples indicates the local carbonate compensation depth is about 4750 m (McCave & Carter 1997). This is somewhat deeper than the regional 4000–4500 m depth proposed for this area by Berger & Winterer (1974). At depths < 3500 m the Chatham area is relatively rich in carbonate (50–70%) due to lesser dilution by terrigenous turbidite material than elsewhere on the New Zealand margin. The IB pelagic cover on the terrace drifts of eastern Chatham Rise has a surface of sandy ooze with a carbonate range of 52–67%. The coarsest sediments (c. 39% sand composed mainly of foraminifera) occur either on the rise crest or southern flank facing upcurrent. Sediments from the down-current northern flank of the drift proper are finer grained with less sand (2–22%), lower carbonate and terrigenous silt-clay ratios < 1 (Table 1). This difference is initially surprising as down-current samples were collected close to East Chatham and Broughton Gaps, which appear to be sites of accelerated flow judging by the 3.5 kHz data. The present currents may not be as fast as supposed from interpreted 3.5 kHz records.

Data from North Chatham Drift present a paradox. On one hand, the eastern New Zealand margin has a vast sediment and current system with sediment supplied by major deep-sea channels to be redistributed by the biggest deep western boundary current in the world ocean, and in Valerie Passage there is abundant topographic and seismic evidence of scour. On the other hand, most bottom photographs from the region show tranquil conditions, current meters record generally slow flows, mud waves are mainly irregular non-migrating types, nepheloid layers have low suspensate concentrations, bottom sediments contain little current-sorted silt, and sedimentation rates on contourite drifts are not high. We suspect that modern currents are slower than their glacial period counterparts (Hall et al. 2001), and that most of the sediment dispersal and erosional topography occurs at those times.

Case Study – Hikurangi Fan-drift

Geological and oceanographic setting

Hikurangi Fan-drift lies along the abyssal SW Pacific Ocean floor marginal to Rapuhia Scarp forming the eastern side of Hikurangi Plateau (Figs 2 & 15). Feeding the fan-drift is Hikurangi Channel. This active conduit transfers sediment from the Australian/Pacific plate boundary, 1400 km across the 3500–3750 m deep Hikurangi Plateau, to the plateau margin (Lewis 1994; Lewis et al. 1998). There, the channel splits with a (1) short branch merging northwestward into a boundary channel scoured along the base of Hikurangi Plateau by the DWBC and (2) a branch swinging southeast (Fig. 2). A recent survey shows this southeast branch continues a further 600 km, curving east to feed a small distal fan (Lewis 1999). However, the main fan occurs near the channel split at the plateau base (Fig. 15). From this point, sediments destined for the main fan are entrained by the DWBC and deposited for 300 km along the base of the plateau (McCave & Carter 1997). The narrow boundary channel separates the fan-drift from the plateau margin. Thus, this elongate deposit extends from the relatively passive zone of south Hikurangi Plateau to the tectonically active subduction zone of Kermadec Trench (Fig. 1).

Rapuhia Scarp forms a prominent western boundary that intensifies and steers the DWBC to the northwest in depths > c. 3500 m. Waters at those depths are dominated by LCDW and have a mean speed of 5 cm s^{-1}. However, the flow is variable. Currents measured 1000 m above the fan-drift had at least 25 bursts > 15 cm s^{-1} over a 674 day long deployment (current meter mooring M21, Fig. 1). Some bursts were short-lived, but others lasted for more than ten days. Although the NW flow prevails, periods of weak (≤ 5 cm s^{-1}) flow to the south were also noted. The measured maximum speed of 17 cm s^{-1} is probably exceeded near Rapuhia Scarp where other data reveal active erosion (see next sections).

Bathymetry

The main depocentre of Hikurangi Fan is at 4600–4850 m depth with the shallower part associated with a well developed left bank levee of the SE channel branch (facing down-channel) that formed preferentially under Coriolis and DWBC forces (Carter & McCave 1994). The drift component has grown NW of the fan to form a broad ridge that is separated from the 900–1200 m high Rapuhia Scarp by a scoured boundary channel (Fig. 16). A change in scarp alignment at 37.5°S separates two morphological zones of the drift/ boundary channel couple. South of 37.5°S, the drift ridge is aligned 330° and has a simple rounded profile that becomes asymmetric down-current, the steeper side facing Rapuhia Scarp (Fig. 16). Boundary channel relief increases down-current (92–170 m) as does the depth of the channel floor (4600–5322 m). By comparison, the fan-drift and channel, north of 37.5°S, are aligned 315° and exhibit variable morphologies associated with

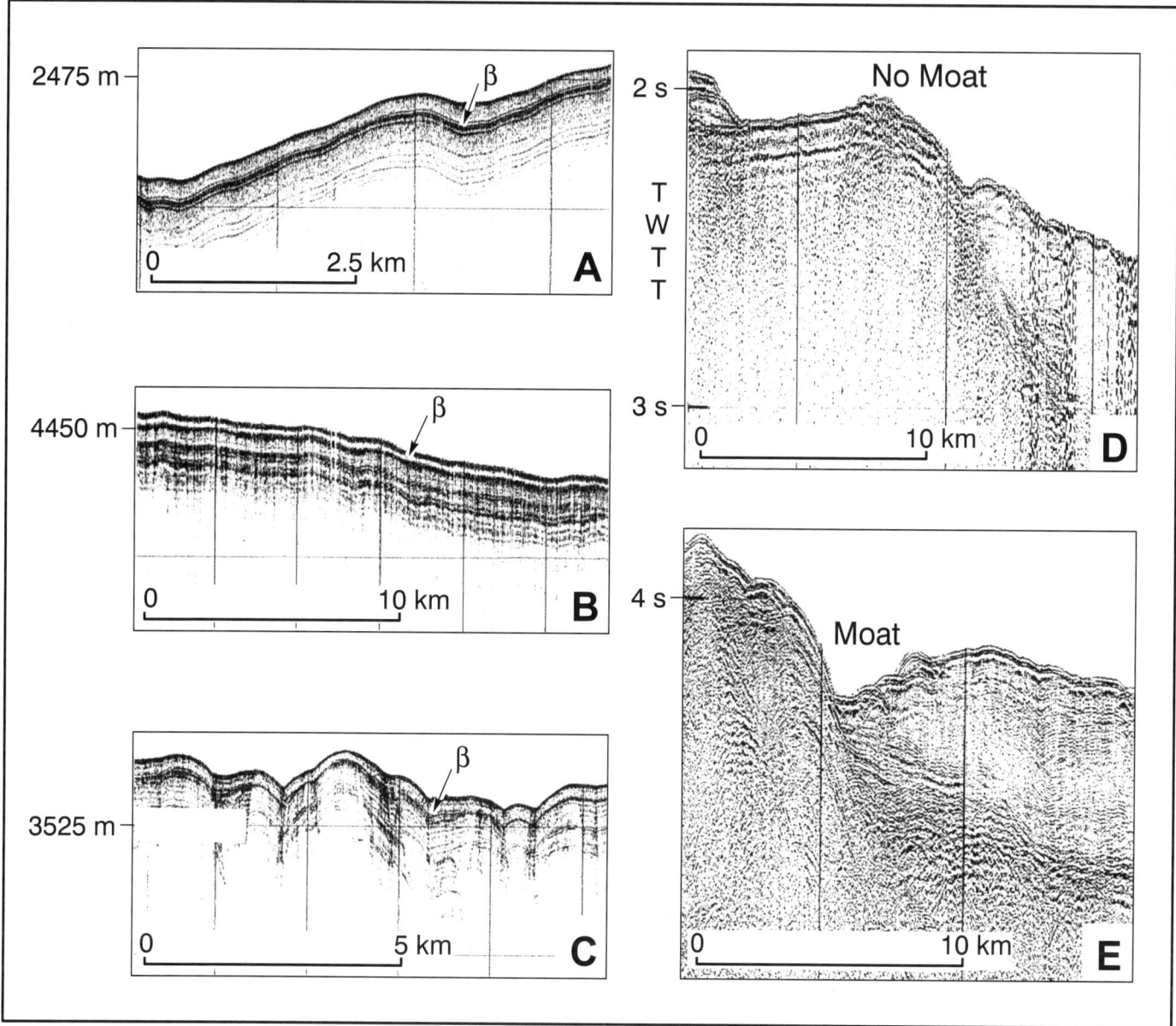

Fig. 14. 3.5 kHz profiles from North Chatham Drift with its (**a, b**) hemipelagic/drape, (**c**) irregular, vertically climbing waves. Airgun profiles record (**d**) a lack of moating near the 'level of no motion' and (**e**) well-developed moats within the DWBC.

localized mass failure, frequent volcanic knolls and scour moats (Fig. 17a, b). Eventually the channel merges with the regional ocean floor at c. 6700 m (McCave & Carter 1997). At that depth, the fan-drift descends into Kermadec Trench on the subducting Pacific Plate (Fig. 18).

Seismic characteristics: reflection profiles

Between 200–360 m of mainly weak, discontinuous reflectors (Unit B) rest on a well defined regional unconformity. The top of Unit B is also an erosional unconformity. On the main Hikurangi Fan, this upper unconformity is essentially a series of palaeo-channels and levees that migrate upsection to the west (Fig. 15). Palaeo-channels are absent within the fan-drift where the top of Unit B has a more subdued palaeo-relief.

Overlying Unit B is a sequence of distinct, closely spaced, continuous reflections (Unit A) representing the main body of the fan-drift. On the fan proper, Unit A is up to 400 m thick, but it thins to 200 m mid-way along the fan-drift, and to < c. 50 m where the deposit descends into Kermadec Trench (Fig. 18). For most of its length, the eastern side of the fan-drift displays sediment waves throughout much of Unit A (see next section).

Stratigraphic context

A tentative stratigraphy for Hikurangi Fan-drift is based on a comparison with nearby ODP Site 1124 (Fig. 10; SSP 1999c). The regionally extensive unconformity at a maximum sub-seabed depth of c. 800 m, is interpreted to be the Oligocene Marshall Paraconformity which transects nannofossil chalk and mudstone of early Oligocene to Palaeocene age (SSP 1999c). Drift deposits of probable Miocene age (Unit B) rest above the paraconformity. The youngest sequence (Unit A) of well ordered reflections has probably accumulated since the Pliocene or later (Carter &

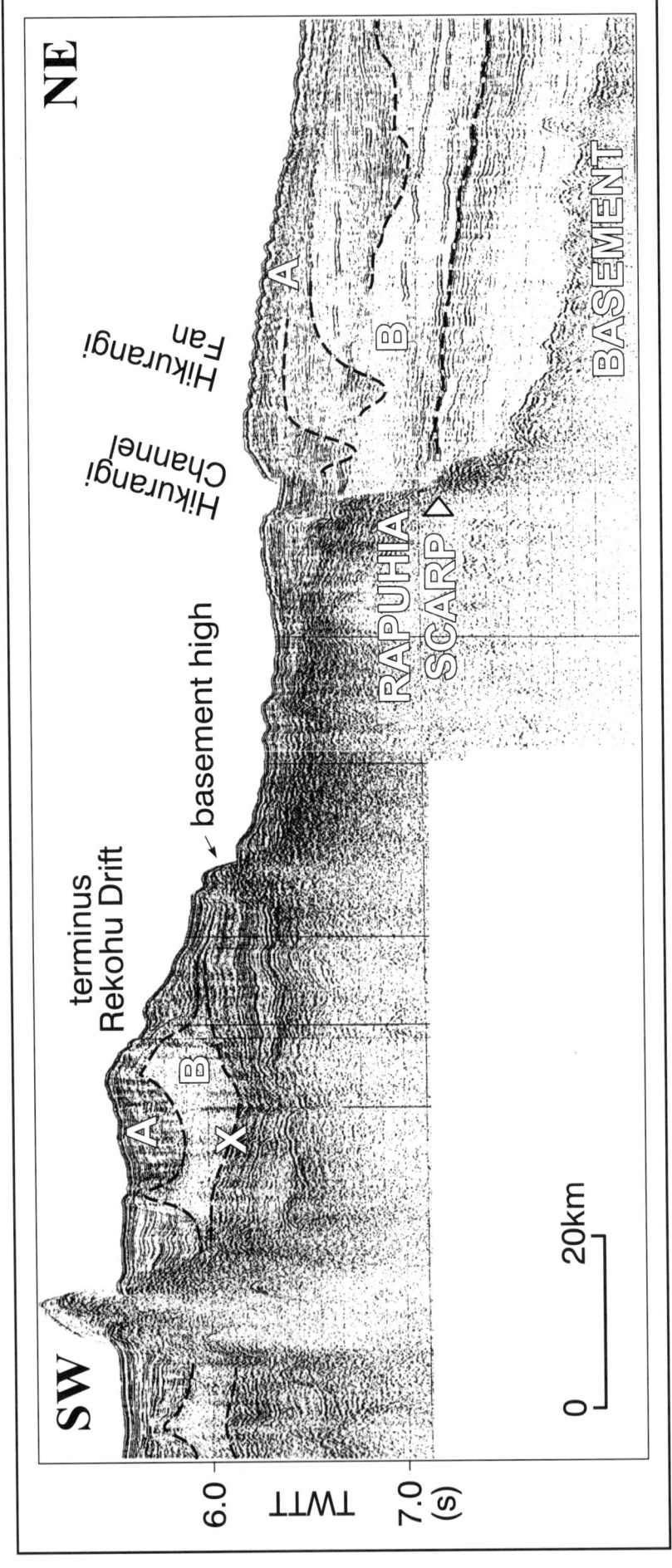

Fig. 15. Single channel seismic profile through Rekohu Drift terminus and nearby Hikurangi Fan with acoustically indistinct seismic Unit B (Miocene) unconformably overlain by well stratified seismic Unit A (Plio-Pleistocene) which is locally thrown into sediment waves on the left bank levee. Located on Figure 16 inset.

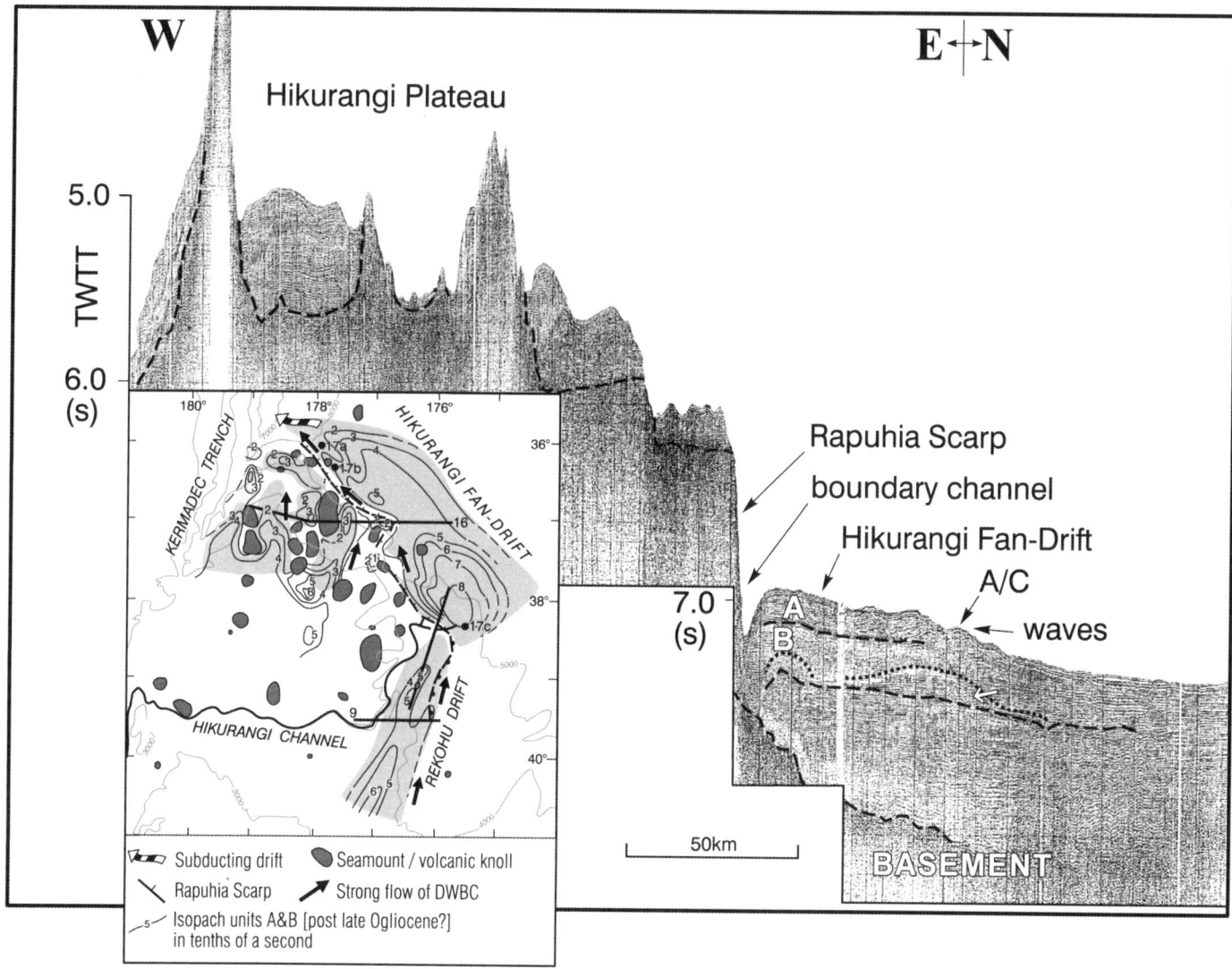

Fig. 16. Single channel seismic section across the Hikurangi Fan-drift with prominent boundary channel scoured along base of Rapuhia Scarp. Known extent and thickness of fan-drift (in inset) along with positions of illustrated seismic profiles designated by figure number. (A/C, alteration of course.)

McCave 1994; Lewis *et al.* 1998). The laminated character of these reflections is probably a response to their turbidity current origin and to the presence of numerous tephra layers, e.g. ODP Site 1124 recorded 134 macroscopic tephra.

Seismic characteristics: 3.5 kHz

Most of the Hikurangi Fan levees and Fan-drift have an echo-character dominated by closely spaced, continuous to discontinuous, sub-parallel reflections. This signature has the hallmark of turbidites but is also probably influenced by tephra layers (Fig. 17; McCave & Carter 1997). The exception is those fan levees facing into the DWBC. In such settings, levees exhibit strongly reflective surfaces suggesting the presence of winnowed coarse sediment.

Much of the fan-drift is mantled by sediment waves that are commonly 3–15 m high and 0.5–1.5 km in wavelength. Wave fields are interrupted by zones of regular hyperbolae presumably caused by small bedforms (e.g. Damuth 1980). As the drift approaches Kermadec Trench, it becomes increasingly prone to mass failure as shown by zones of irregular surface hyperbolae underlain by layers with an incoherent internal fabric (Fig. 18). Near the trench, the 3.5 kHz character in undeformed parts of the drift takes on a more pelagic/hemipelagic appearance with reflections that are better organised and more continuous than the turbidite-dominant facies up-current.

Compared to the layered sediments of the fan-drift, the adjacent boundary channel is highly reflective (Fig. 17). This response is consistent with current winnowing and the concentration of coarse, reflective sediment; a contention that is confirmed by bottom photographs.

Sediments: bottom photographs

Both the fan and its drift extension are shown to be hydraulically tranquil settings as attested by muddy substrates with a myriad of well preserved biological trace structures that include delicate tracks and trails, pits, burrows, symmetric mounds, feeding marks, and fecal strings (Fig. 19). Such tranquility contrasts with the current-swept boundary channel. A discontinuous layer of rippled, gravelly sand appears to be moving across a current-scoured mud floor. Localized undermining of this floor has resulted in substrate collapse. Ripple crests are aligned perpendicular to the channel axis, but their direction of migration is unclear. However, the pronounced deflection of epibenthic organisms such as stalked crinoids indicate a northwesterly flow as confirmed by current meter data from station M21 (Figs 1 & 19).

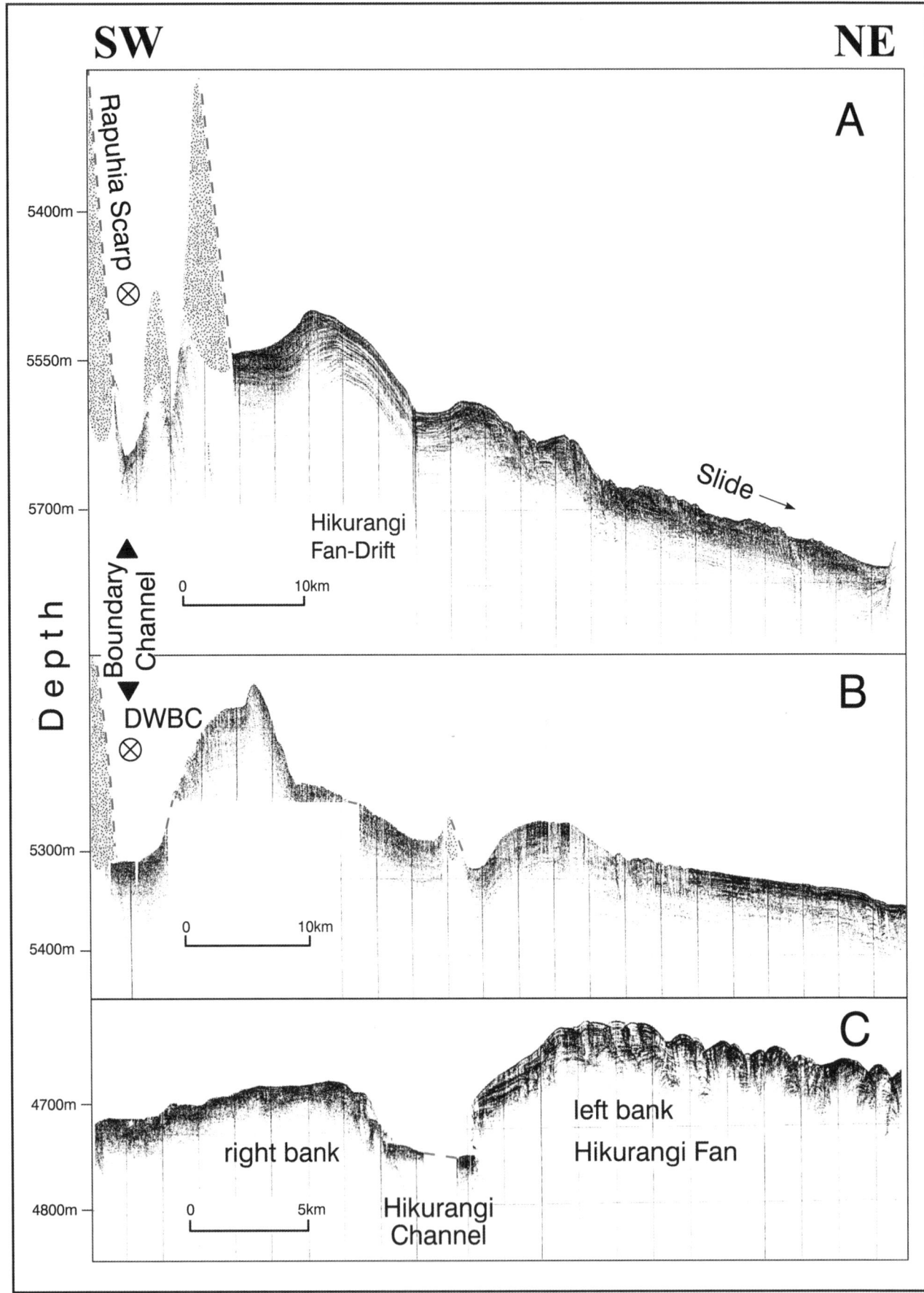

Fig. 17. 3.5kHz profiles along Hikurangi Fan-drift showing changes in morphology (volcanic knolls, slide) as the deposit approaches the Australian/Pacific plate boundary. Located on Figure 2 inset.

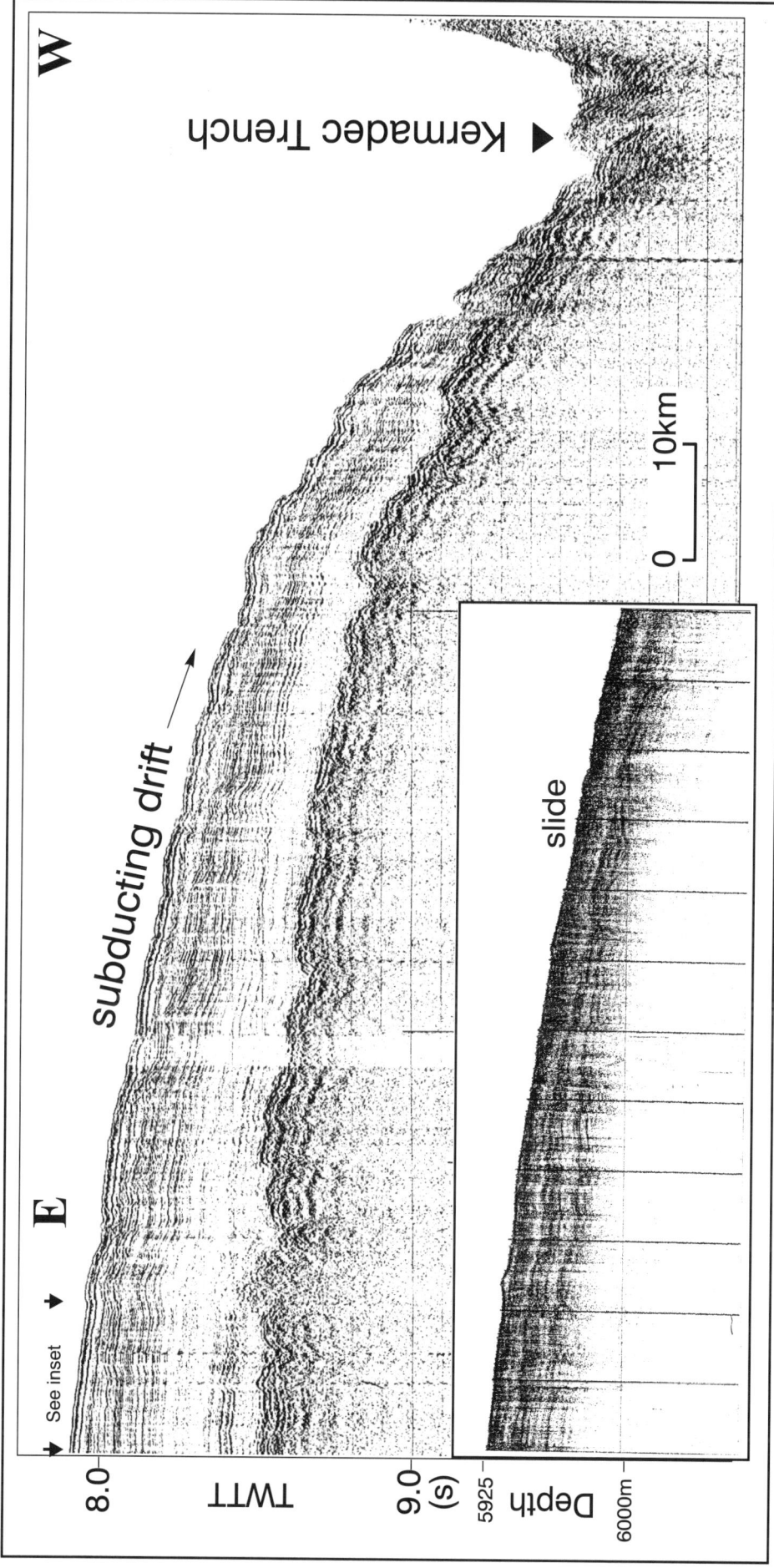

Fig. 18. Single channel seismic section of Hikurangi Fan-drift on the subducting Pacific plate. Crumpled, acoustically massive layers at surface (see also 3.5 kHz inset) and listric faulting suggest gravitational instability on the increased slope. Located on Figure 2 inset.

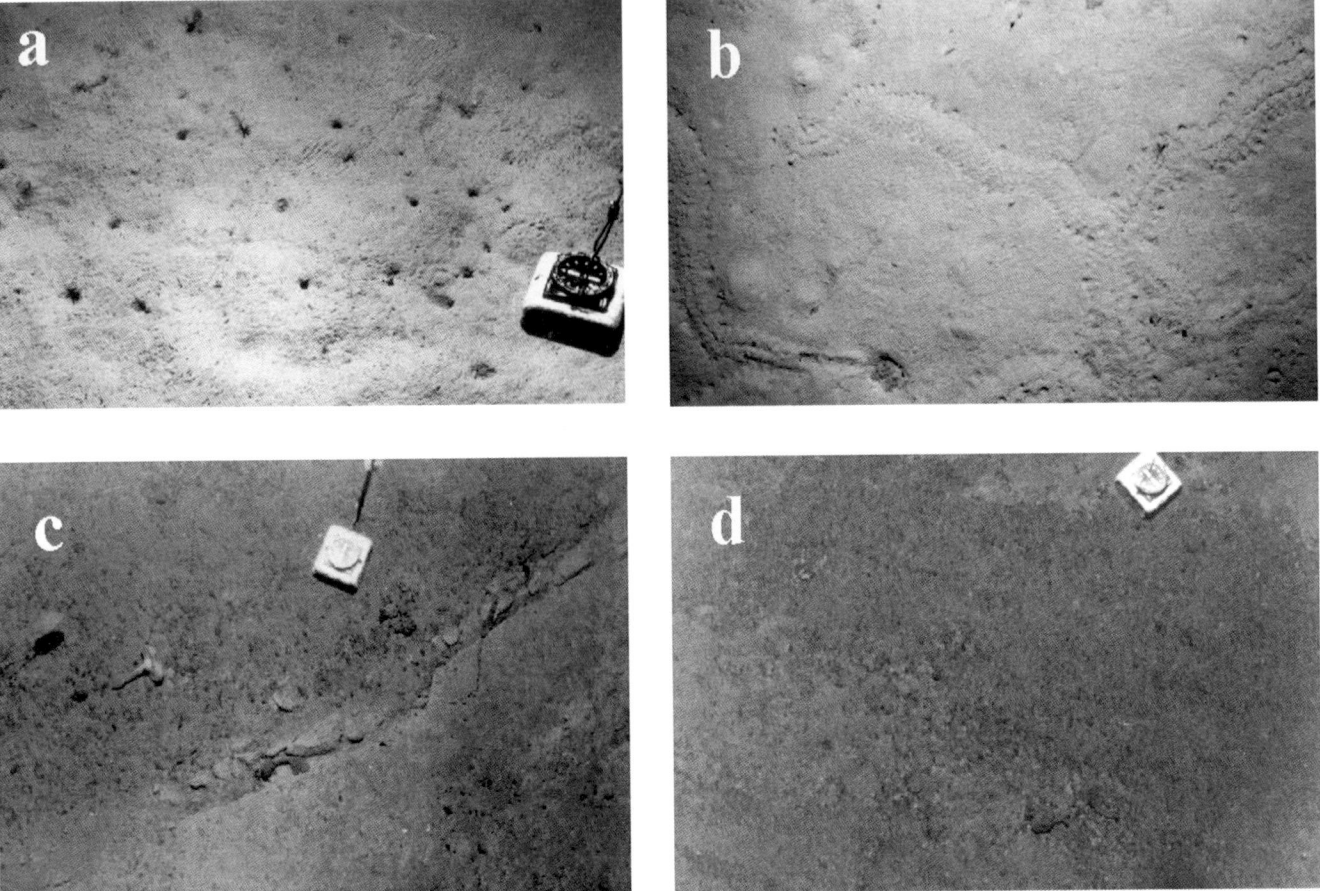

Fig. 19. The tranquil depocentre on Hikurangi Fan (a, Q857) and Fan-drift (b, R643) compared to the current-swept boundary channel (c, d, S933). Stations located on Figure 2 inset.

Sediments

As Hikurangi Fan-drift is near to or below the local carbonate compensation depth of 4750 m, sediments are principally terrigenous muds with <10% $CaCO_3$ (Table 1; McCave & Carter 1997). On the fan-drift surface, sediments have a variable sand content that reflects mainly different contributions of airfall tephra rather than hydraulic processes. By comparison, surficial mud becomes finer grained down-current of Hikurangi Fan as evinced by decreases in sortable silt (10–63 μm) and the silt/clay ratio (Fig. 20). Underlying these surficial muds is a suite of mud turbidites. Core S932 (CHAT 8K), from mid-way along the fan-drift at 5235 m, penetrated a succession of mud turbidites interspersed with bioturbated layers (Figs 21 and 22). Core R646 (CHAT 14K), located further down-current at 5321 m, also recovered turbidites but they were usually but not invariably thinner and visually less distinct than their counterparts at S932. They are strikingly seen in X-radiographs (Fig. 22). The topography around a seamount protruding through the fan-drift is typical of the pattern of erosion and deposition expected under a Coriolis-controlled flow around an obstacle when the flow is to the northwest (McCave & Carter 1997). From these observations it appears that the fan-drift has formed from turbidity current plumes directed along the margin by the DWBC. Plumes may have resulted from simple overspill of Hikurangi Channel or from avulsion at the prominent bend where the channel swings from northeast to southeast, or both these processes. The presence of a thick turbidite at distal site R646 but not a proximal site S932, suggests some turbidites may be locally generated through mass failure of sediments on Rapuhia Scarp.

Discussion – Eastern New Zealand Oceanic Sedimentary System (ENZOSS)

Carter *et al.* (1996) proposed the ENZOSS model as a means of unifying provenance, transport and depositional aspects of the eastern New Zealand drifts. ENZOSS demonstrates that sediment generated at the Australian/Pacific plate boundary is captured by three major channel systems (Solander, Bounty and Hikurangi) and delivered to the DWBC flowing along the New Zealand continental margin. A series of sediment drifts has developed northwards along the margin in accord with the abyssal flow. Presently, the northernmost deposit, Hikurangi Fan-drift, is subducting into the mantle to undergo melting and contribute to the arc volcanism of the plate boundary (Gamble *et al.* 1996). Thus, one ENZOSS cycle is completed and another begins.

Since 1995, new data from the Solander Channel/Fan complex (Schuur *et al.* 1998), Hikurangi Fan-drift (Lewis 1999) and ODP Leg 181 (Carter *et al.* 1999) allow refinement of the ENZOSS model and an insight into its evolution.

Inception of an abyssal circulation around New Zealand is unclear. ODP Site 1124 revealed the presence of a major hiatus extending from 37 to 49 Ma which is consistent with widespread erosion recorded by Kennett & von der Borch (1985). The palaeoceanographic circumstances behind the event are unclear, but it may have resulted from the onset of cold bottom water production off Antarctica (e.g. Barrett 1996) concomitant with the formation of a proto-ACC following the shallow breaching of the Tasman Rise. A second major hiatus, marked by the Marshall Paraconformity, is attributed to the full opening of the South

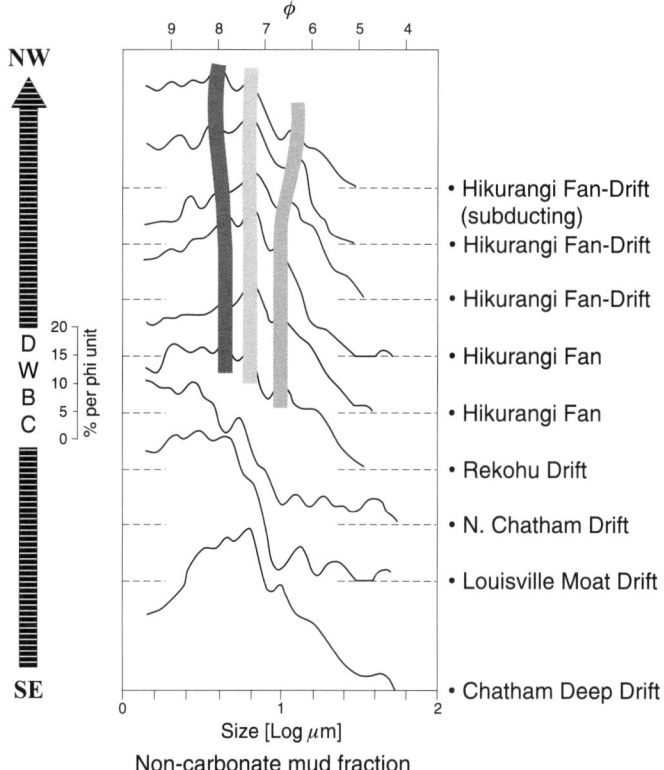

Fig. 20. Grain-size frequency curves for surface sediments along the path of the main deep drifts revealing (1) coarsening at Rekohu Drift and Hikurangi Fan-drift suggesting modern Rekohu is receiving sediment from Hikurangi Channel and (2) down-current fining of the Fan-drift concomitant with a variable increase in >10 μm silt due to volcanic ash or variable current or both these factors.

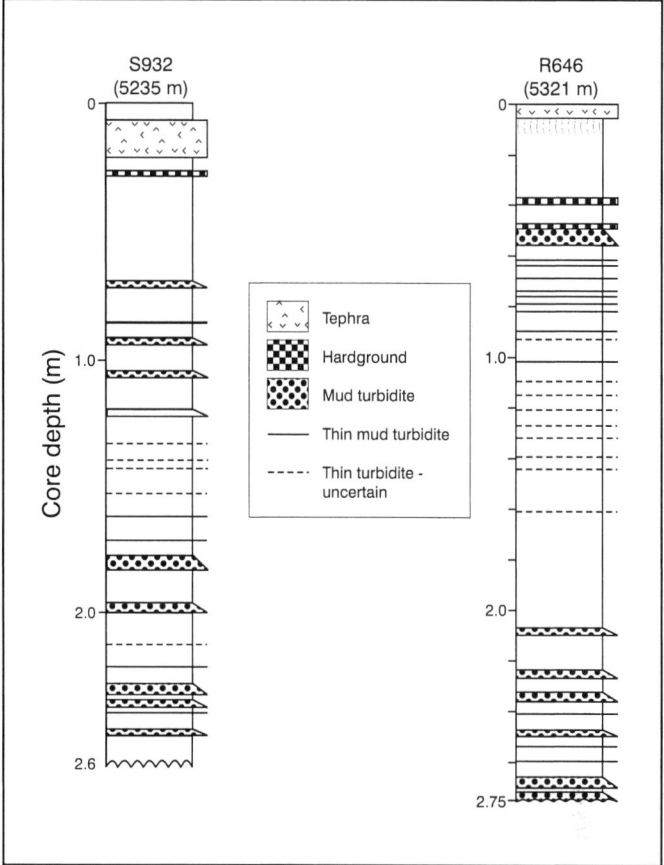

Fig. 21. Turbidites, redirected by the DWBC, play a major role in Hikurangi Fan-drift formation as attested by frequency in cores which are located on Figure 2 inset. This underestimates the frequency of turbidites as many silt-clay units shown in X-radiographs (Fig. 22) are not recorded in visual core descriptions.

Tasman gateway about the early Oligocene (Carter & Landis 1972; Kennett & von der Borch 1985; Kennett 1977). A somewhat later Oligocene opening of the Drake Passage is indicated by some authors (e.g. Barker & Burrell 1977) resulting in a continuous circum-Antarctic seaway. An unimpeded ACC that was possibly reinforced by bottom water generated at the ancestral East Antarctic ice sheet, provided the ingredients for a vigorous abyssal circulation. Consequently, oceanic sediments were eroded during and after the Marshall event and transported north to form the nucleii of North Chatham (ODP Site 1123), Rekohu (ODP Site 1124) and Hikurangi drifts (Carter & McCave 1994). Thus, the proto-ENZOSS was born. However, the system received little terrigenous detritus judging by the prevalence of pelagites even near Solander and Bounty channels (e.g. Kennett et al. 1975; Nelson 1985).

Uplift along the Southern Alps began about 8 Ma or slightly earlier (e.g. Tippett & Kamp 1993). This event presumably marked the inception of the Solander Fan (Schuur et al. 1998), and the introduction of terrigenous sediment into Bounty Trough (Nelson 1985). That the Solander and Bounty conduits guided sediment into the DWBC, is attested by an increase in the terrigenous flux about 9.5–8 Ma at Rekohu Drift (Carter et al. 1999). Such an increase was probably unrelated to Hikurangi Channel. In late Miocene times, the channel appears to have passed along the base of the North Island slope to feed a turbidite plain near the present head of Kermadec Trench (Lewis et al. 1998).

Abyssal erosion renewed in the Pliocene, possibly in response to a strengthened ACC, is indicated by the restriction of eroded sediments to south of Chatham Rise. In Solander Trough the fan was eroded, possibly by a powerful jet of the ACC/DWBC which entered the region through a newly formed gap in Macquarie Ridge at 53.5°S (Schuur et al. 1998). Sediment entrained within the jet probably moved out of Emerald Basin and along the margin of Campbell Plateau, but little of this material accumulated *en route* judging by the dearth of Pliocene beds in the Campbell 'skin drift' (ODP Site 1121). Down-stream at Bounty Fan (ODP Site 1122) and Rekohu Drift (ODP Site 1124) there are increases in Pliocene sedimentation rates but it is unclear if the increases result from erosion in the deep south, increased discharge from Bounty Channel, or both these factors.

ENZOSS was fully established in the Pleistocene. Increased uplift along the New Zealand mountain ranges and enhanced erosion under palaeoclimatic extremes produced a flood of terrigenous detritus. Eustatic lowerings of sea level allowed rivers to discharge directly into canyon/channel systems which steered turbidity currents directly into the path of the boundary flow. The Pleistocene also may have witnessed the diversion of the Hikurangi Channel to its present location (Lewis et al. 1998) resulting in a major phase of fan-drift growth. This event may be marked by a sharp increase in the terrigenous flux rate at nearby Rekohu Drift about 1.6 Ma (Hall et al. in press). Seismic profiles suggest Hikurangi Fan-drift developed over older drift deposits which, judging by sediment at ODP Site 1124, were partly derived from Bounty Trough and further south (SSP 1999c).

The Pleistocene thus became the epoch of maximum drift growth, although the effect of individual glacial and interglacial cycles on this growth has yet to be resolved.

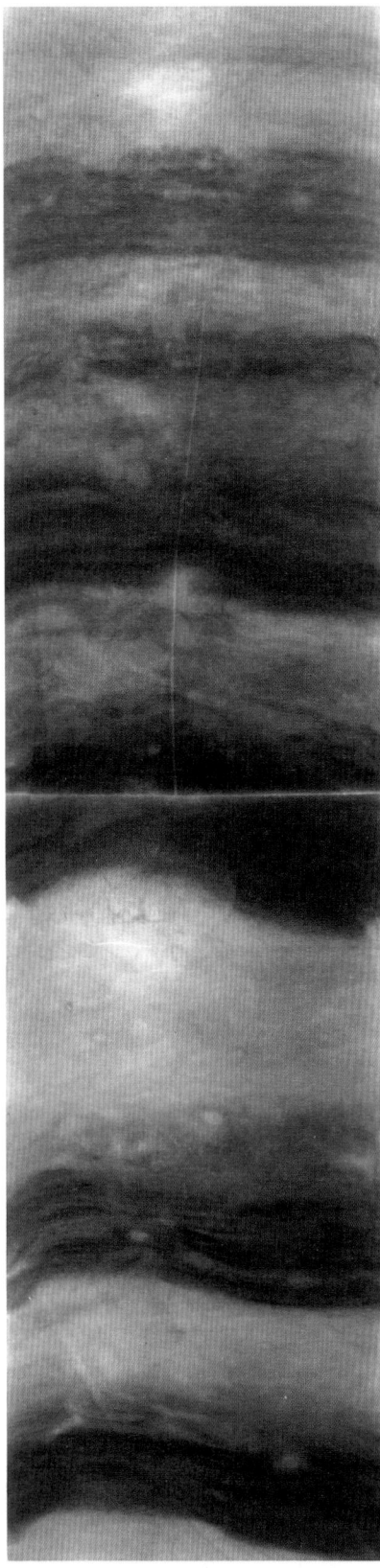

Fig. 22. X-radiograph of 35 to 96 cm in Kasten core CHAT 8K (NIWA Site S932) (cf. Fig. 21) from Hikurangi Fan-drift. This shows fine grained turbidites which 3.5 kHz profiles indicate to be forming mud waves with an antidune behaviour (migrating probably upstream towards the boundary channel along Rapuhia Scarp). Darker material is more silt-rich, lighter tones are from finer material with higher water content. Turbidites from 35–64 cm (top half of the figure) were not apparent visually on the fresh core, whereas those in the lower half are recorded in the visual core log on Figure 21. Their average thickness in this core is c. 10 cm.

References

BARKER, P. F. & BURRELL, J. 1977. The opening of Drake Passage. *Marine Geology*, **25**, 15–34.

BARRETT, P. J. 1996. Antarctic palaeoenvironment through Cenozoic times – a review. *Terra Antarctica*, **3**, 103–119.

BERGER, W. H. & WINTERER, E. L. 1974. Plate stratigraphy and the fluctuating carbonate line. *International Association of Sedimentologists Special Publication*, **1**, 11–48.

BOYER, D. L. & GUALA, J. R. 1972. Model of the Antarctic Circumpolar Current in the vicinity of the Macquarie Ridge. *In*: HAYES, D. E. (ed.) *Antarctic Oceanology II – The Australian-New Zealand Sector*: Antarctic Research Series, **19**, 79–94.

BRYDEN, H. L. & HEATH, R. A. 1985. Energetic eddies at the northern edge of the Antarctic Circumpolar Current in the Southwest Pacific. *Progress in Oceanography*, **14**, 65–87.

CARTER, R. M. & CARTER, L. 1996. The abyssal Bounty Fan and lower Bounty Channel; evolution of a rifted margin sedimentary system. *Marine Geology*, **130**, 181–202.

CARTER, R. M. & LANDIS, C. A. 1972. Correlative Oligocene unconformities in Southern Australasia. *Nature*, **237**, 12–13.

CARTER, L. & MCCAVE, I. N. 1994. Development of sediment drifts approaching an active plate margin under the SW Pacific Deep Western Boundary Current. *Paleoceanography*, **9**, 1061–1085.

CARTER, L. & MCCAVE, I. N. 1997. The sedimentary regime beneath the deep western boundary current inflow to the Southwest Pacific Ocean. *Journal of Sedimentary Research*, **67**, 1005–1017.

CARTER, L. & MITCHELL, J. S. 1987. Late Quaternary sediment pathways through the deep ocean, east of New Zealand. *Paleoceanography*, **2**, 409–422.

CARTER, L. & WILKIN, J. 1999. Abyssal circulation around New Zealand – a comparison between observations and a global circulation model. *Marine Geology*, **159**, 221–239.

CARTER, L., CARTER, R. M., LEWIS, K. B., MCCAVE, I. N., NELSON, C. S., WEAVER, P. P. E. & ALLOWAY, B. 1997. SW Pacific Gateway: palaeohydrography of the deep Pacific source – site survey data. *Report* **441** lodged ODP databank, Lamont Doherty Earth Observatory, New York.

CARTER, L., CARTER, R. M., MCCAVE, I. N. & GAMBLE. J. 1996. Regional sediment recycling in the abyssal Southwest Pacific Ocean. *Geology*, **24**, 735–738.

CARTER, L., GARLICK, R. D., SUTTON, P., CHISWELL, S., OIEN, N. A. & STANTON, B. R. 1998. Ocean Circulation New Zealand. *National Institute of Water and Atmosphere Chart Miscellaneous Series 76*.

CARTER, L., NELSON, C. S., NEIL, H. L. & FROGGATT, P. C. 1995. Correlation, dispersal, and preservation of the Kawakawa Tephra and other late Quaternary tephra layers in the Southwest Pacific Ocean. *New Zealand Journal of Geology and Geophysics*, **38**, 29–46.

CARTER, R. M., MCCAVE, I. N. *ET AL*. 1999. Proceedings of the Ocean Drilling Program: *Initial Reports*, **181**: College Station, Texas Ocean Drilling Program. CD-ROM and World Wide Web Address: http:\\www.odp.tamu.edu/publications/181IR.

DAMUTH, J. E. 1980. Use of high-frequency (3.5–12 kHz) echograms in the study of near-bottom processes in the deep sea: a review. *Marine Geology*, **38**, 51–75.

DANIAULT, N. & MÉNARD, Y. 1985. Eddy kinetic energy distribution in the Southern Ocean from altimetry and FGGE drifting buoys. *Journal of Geophysical Research*, **90**, 11 877–11 889.

FAUGÈRES, J.-C., MEZERAIS, M-L. & STOW, D. A. V. 1993. Contourite drift types and their distribution in the North and South Atlantic Ocean basins. *Sedimentary Geology*, **82**, 189–206.

GAMBLE, J., WOODHEAD, J., WRIGHT, I. & SMITH, I. 1996. Basalt and sediment geochemistry and magma petrogenesis in a transect from oceanic island arc to rifted continental margin arc: The Kermadec Hikurangi Margin, SW Pacific. *Journal of Petrology*, **37**, 1523–1546.

GORDON, A. L. 1972. On the interaction of the Antarctic Circumpolar Current and the Macquarie Ridge. *In*: HAYES, D. E. (ed.) Antarctic Oceanology II – The Australian-New Zealand Sector: *Antarctic Research Series*, **19**, 71–78.

GORDON, A. L. 1975. An Antarctic oceanographic section along 170°E. *Deep-Sea Research*, **22**, 357–377.

GRIFFITHS, G. A. & GLASBY, G. P. 1985. Input of river-derived sediment to the New Zealand continental shelf: I. Mass. Estuarine, *Coastal and Shelf Science*, **21**, 773–787.

HALL, I. R., CARTER, L., HARRIS, S. E. in press. Major depositional events under the deep Pacific inflow. *Geology*, in press.

HALL, I. R., MCCAVE, I. N., SHACKLETON, N. J., WEEDON, G. P. & HARRIS, S. E. 2001. Intensified deep Pacific inflow and ventilation in Pleistocene glacial times. *Nature*, **412**, 809–811.

HAYES, D. E., HOUTZ, R., TALWANI, M., WATTS, A. B., WEISSEL, J. & AITKEN, T. 1976. U.S.N.S. *Eltanin*. Cruises 39–45: *Preliminary Report* **24**, Lamont-Doherty Geological Observatory, Palisades, New York.

HEATH, R. A. 1985. A review of the physical oceanography of the seas around New Zealand-1982. *New Zealand Journal of Marine and Freshwater Research*, **19**, 79–124.

HOLLISTER, C. D. & MCCAVE, I. N. 1984. Sedimentation under deep sea storms. *Nature*, **309**, 220–225.

JACOBI, R. D. & HAYES, D. E. 1982. Bathymetry, microphysiography and reflectivity characteristics of the West African Margin between Sierra Leone and Mauretania. *In*: VON RAD, U. ET AL. (eds) *Geology of the Northwest African Continental Margin*. Springer, New York, 182–212.

KENNETT, J. P. 1977. Cenozoic evolution of Antarctic glaciation, the Circum-Antarctic Ocean, and their impact on global paleoceanography. *Journal of Geophysical Research*, **82**, 3843–3860.

KENNETT, J. P. & VON DER BORCH, C. C. 1985. Southwest Pacific Cenozoic palaeoceanography. *In*: KENNETT, J. P., VON DER BORCH ET AL. (eds) *Initial Reports of the Deep Sea Drilling Project*. Ocean Drilling Program, College Station, TX, **90**, 1493–1517.

KENNETT, J. P., VON DER BORCH, C. C. ET AL. 1975. *Initial Reports of the Deep Sea Drilling Project*, **Leg 29**. U. S. Government Printing Office, Washington, DC.

LEWIS, K. B. 1994. The 1500 km long Hikurangi Channel: trench axis channel that escapes its trench, crosses a plateau, and feeds a fan drift. *Geo-Marine Letters*, **14**, 19–28.

LEWIS, K. B. 1999. Distal turbidity currents that flow through powerful abyssal current: the Hikurangi Channel 2000 km from Kaikoura. *Geological Society of New Zealand, conference abstracts*, **107A**, 87.

LEWIS, K. B. & PETTINGA, J. R. 1993. The emerging, imbricate frontal wedge of the Hikurangi Margin. *In*: BALLANCE, P. F. (ed.) *South Pacific Sedimentary Basins: Sedimentary Basins of the World*. Elsevier Science Publishers B. V., Amsterdam, 225–250.

LEWIS, K. B., COLLOT, J-Y. & LALLEMAND, S. E. 1998. The dammed Hikurangi Trough: a channel fed trench blocked by subducting seamounts and their wake avalanches (New Zealand-France GeodyNZ Project). *Basin Research*, **10**, 441–468.

MCCAVE, I. N. & CARTER, L. 1997. Sedimentation beneath the Deep Western Boundary Current off northern New Zealand. *Deep-Sea Research*, **44**, 1203–1237.

MCCAVE, I. N. & TUCHOLKE, B. E. 1986. Deep current-controlled sedimentation in the western North Atlantic. *In*: VOGT, P. R. & TUCHOLKE, B. E. (eds) *The Geology of North America. (The Western North Atlantic Region)*. Geological Society of America, Boulder. **Vol. M**, 451–468.

MCCAVE, I. N., MANIGHETTI, B. & ROBINSON, S. G. 1995. Sortable silt and fine sediment size/composition slicing: parameters for palaeocurrent speed and palaeoceanography. *Paleoceanography*, **10**, 593–610.

MILLIMAN, J. D. & SYVITSKI, J. P. M. 1992. Geomorphic tectonic control of sediment discharge to the ocean: the importance of small, mountainous rivers. *Journal of Geology*, **100**, 525–544.

NELSON, C. S. 1985. Lithostratigraphy of Deep Sea Drilling Project Leg 90 drill sites in the Southwest Pacific: an overview. *In*: KENNETT, J. P., VON DER BORCH, C. C. ET AL. *Initial Reports Deep Sea Drilling Project*. Ocean Drilling Program, College Station, TX, **90**, 1471–1491.

ORSI, A. H., WHITWORTH III, T. & NOWLIN JR, W. D. 1995. On the meridional extent and fronts of the Antarctic Circumpolar Current. *Deep-Sea Research*, **42**, 641–673.

RINTOUL, S. R. 1998. On the origin and influence of Adelie Land Bottom Water. In: Ocean, Ice, and Atmosphere – Interactions at the Antarctic Continental Margin. *Antarctic Research Series*, **75**, 151–171.

SCHUUR, C. L., COFFIN, M. F., FROHLICH, C., MASSELL, C. G., KARNER, G. D., RAMSEY, D. & CARESS, D. W. 1998. Sedimentary regimes at the Macquarie Ridge Complex: interaction of Southern Ocean circulation and plate boundary bathymetry. *Paleoceanography*, **13**, 646–670.

SHIPBOARD SCIENTIFIC PARTY 1999*a*. 5. Site 1121: The Campbell 'Drift'. *In*: CARTER, R. M., MCCAVE, I. N., RICHTER, C. & CARTER, L. (eds) *Proceedings ODP, Initial Reports*. Ocean Drilling Program, College Station, TX, **181**, 1–62. (CD-ROM and World Wide Web Address: http://www.odp.tamu.edu/publications/181IR/chap 05/chap 05.htm).

SHIPBOARD SCIENTIFIC PARTY 1999*b*. 7. Site 1123: North Chatham Drift – a 20 Ma record of the Pacific Deep Western Boundary Current. *In*: CARTER, R. M., MCCAVE, I. N., RICHTER, C., CARTER, L. ET AL., *Proceedings ODP, Initial Reports*. Ocean Drilling Program, College Station, TX, **181**, 1–184. (CD-ROM and World Wide Web Address: http://www.odp.tamu.edu/publications/181IR/chap 07/chap 07.htm).

SHIPBOARD SCIENTIFIC PARTY 1999*c*. 8. Site 1124: Rekohu Drift – from the K/T Boundary to the Deep Western Boundary Current. *In*: CARTER, R. M., MCCAVE, I. N., RICHTER, C., CARTER, L. ET AL., *Proceedings ODP, Initial Reports*. Ocean Drilling Program, College Station, TX, **181**, 1–137 (CD-ROM and World Wide Web Address: http://www.odp.tamu.edu/publications/181IR/chap 08/chap 08.htm).

TIPPETT, J. M. & KAMP, P. J. J. 1993. Fission track analysis of the late Cenozoic vertical kinematics of continental Pacific crust, South Island, New Zealand. *Journal of Geophysical Research*, **98**, 16119–16148.

WALCOTT, R. I. 1998. Modes of oblique compression: Late Cenozoic tectonics of the South Island of New Zealand. *Reviews of Geophysics*, **36**, 1–26.

WARREN, B. A. 1973. TransPacific hydrographic sections at latitudes 43°S and 28°S; the SCORPIO Expedition – deep water. *Deep-Sea Research*, **20**, 9–38.

WARREN, B. A. 1981. Deep circulation of the world ocean. *In*: WARREN, B. A. & WUNSCH, C. (eds) *Evolution of Physical Oceanography*. MIT Press, Cambridge, MA. 6–41.

WATTS, A. B., WEISSEL, J. K., DUNCAN, R. A. & LARSON R. 1988. Origin of the Louisville Ridge and its relationship to the Eltanin Fracture Zone System. *Journal of Geophysical Research*, **93**, 3051–3077.

WEAVER, P. P. E., CARTER, L. & NEIL, H. L. 1998. Response of surface water masses and circulation to late Quaternary climate change, east of New Zealand. *Paleoceanography*, **13**, 70–83.

WEBB, D. J., KILWORTH, P. D., COWARD, A. C. & THOMPSON, S. R. 1991. *The FRAM Atlas of the Southern Ocean*. Natural Environment Research Council, Swindon, U.K.

WHITWORTH, T., WARREN, B. A., NOWLIN, W. D., PILLSBURY, R. D. & MOORE, M. I. 1999. On the deep western-boundary current in the Southwest Pacific Basin. *Progress in Oceanography*, **43**, 1–54.

WOOD, R. A., ANDREWS, P. B. & HERZER, R. H. 1989. Cretaceous and Cenozoic geology of the Chatham Rise region, South Island, New Zealand. *New Zealand Geological Survey, Basin Studies*, **3**.

Neogene contourites, Miura–Boso forearc basin, SE Japan

DORRIK A. V. STOW[1], YUJIRO OGAWA[2], IN TAE LEE[3], KYOHIKO MITSUZAWA[4]

[1]SOES-SOC, University of Southampton, Southampton SO14 3ZH, UK (e-mail: davs@soc.soton.ac.uk)
[2]Institute of Geoscience, University of Tsukuba, Tsukuba 305-8571, Japan
[3]Oceanography Dept., Chonam National University, Kwangju 500-757, Korea
[4]JAMSTEC, Yokosuka 237-0061, Japan

Abstract: The mid to late Miocene Misaki Formation sediments of the Miura and Boso peninsulas, south central Japan, were deposited in the Pacific-facing forearc region of the proto Izu-Bonin arc at bathyal depths. The hemipelagic background facies, composed mainly of calcareous biogenic and pumiceous volcaniclastic material, are interbedded with thin to thick scoriaceous beds of turbiditic and pyroclastic fall origin. Careful study in the field and in the laboratory of these fine-grained background sediments has revealed the marked influence of bottom currents at certain horizons in producing intervals with distinct muddy contourite characteristics. These include: a general absence of primary structures due to intense bioturbation, some diffuse layering, irregular concentrations of coarser-grained material, sharp and erosive top and bottom lamina contacts, rare micro-cross-lamination disrupted by bioturbation that was continuous with deposition, and a mixed pelagic biogenic (commonly fragmented), volcaniclastic and terrigenous composition. Small-scale cyclicity of variations in grain-size and structural features can be related in part to episodic volcaniclastic input and in part to fluctuation in bottom current strength. Evidence for bottom current reworking of the tops of thin-bedded sandy turbidites is equivocal, and further work is required to resolve this debate. The recognition of Miocene-age contourites from the NW Pacific provides further evidence for the existence of active deep-ocean circulation in the Pacific at this time. However, it is not possible to determine which current system, Antarctic Bottom Water or deep Kuroshio Current, was responsible for these outcrop examples of fossil contourites.

In the search for reliable outcrop analogues of contourites, the Mio-Pliocene Miura Group of SE Japan provides a good example of fine-grained calcareous volcaniclastic hemipelagites that have, in part, been influenced by bottom currents. These were initially recognized as probable contourites during a joint British–Japanese research programme in the 1980s, and first described at the 13th International Geological Congress in Nottingham, UK (Stow & Faugères 1990). They have been referred to in subsequent publications on the area (Lee & Ogawa 1998; Stow et al. 1998a), as well as in a general review of fossil contourites (Stow et al. 1998b).

However, the original muddy-silty contourites have not yet been fully described and illustrated, whereas the sandy, cross-laminated facies documented by Lee & Ogawa (1998) are considered by the senior author (Stow et al. 1998a, b) to be thin-bedded turbidites. The aim of this joint paper, therefore, is to present the data pertaining to a muddy contourite interpretation, to propose the section as a type example of fossil contourites that are closely interbedded with other deep-water volcaniclastic facies, and to further discuss the possibility of bottom current reworked sandy turbidites.

More recently, Ito (1996, 1997, 2002) has interpreted parts of the overlying succession in the Boso peninsula, the Plio-Pleistocene Kasuza Group, as the result of bottom current reworking and winnowing of turbidite tops. This interpretation is also questioned by the present senior author and joint work is currently in progress to attempt to resolve this issue.

A wealth of stratigraphic and structural data already exists from previous work in the region. This has been used to support detailed fieldwork on the well-exposed sedimentary succession of the Miura Group throughout the Miura and Boso peninsulas.

Geological and oceanographic setting

Geological setting

The Miura and Boso peninsulas are located just south of Tokyo, flanking the entrance to Tokyo Bay from Okinoyama Bank (Figs 1 and 2). They lie on the NE side of the Izu collision zone between the Izu-Bonin and Honshu island arc systems, and adjacent to Sagami Trough (a short segment of oceanic trench) immediately north from the Boso trench–trench–trench triple junction (Ogawa & Taniguchi 1988, 1989). The Neogene age Miura Group is one of the chief components of the onland geology in this area. It has

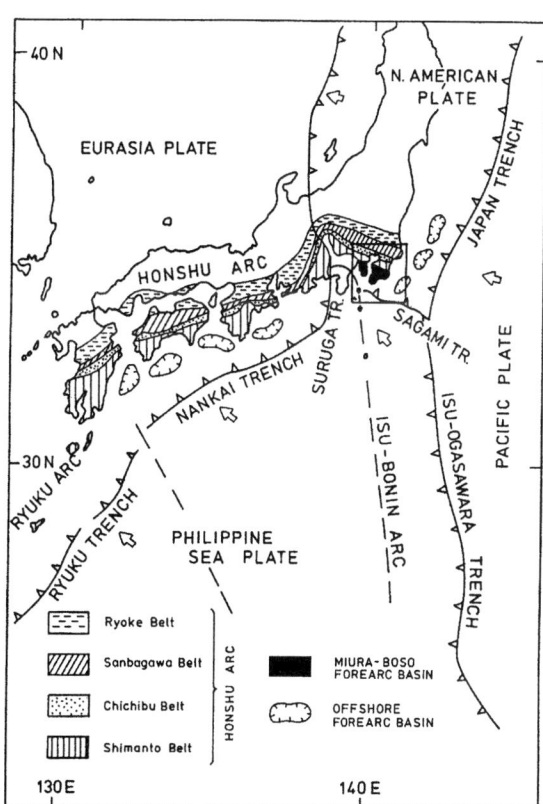

Fig. 1. Regional plate tectonic setting of the Miura-Boso basin. Study area (boxed) is shown in Figure 2 (From Stow et al. 1998a).

From: STOW, D. A. V., PUDSEY, C. J., HOWE, J. A., FAUGÈRES, J.-C. & VIANA, A. R. (eds)
Deep-Water Contourite Systems: Modern Drifts and Ancient Series, Seismic and Sedimentary Characteristics.
Geological Society, London, Memoirs, **22**, 409–419. 0435-4052/02/$15.00 © The Geological Society of London 2002.

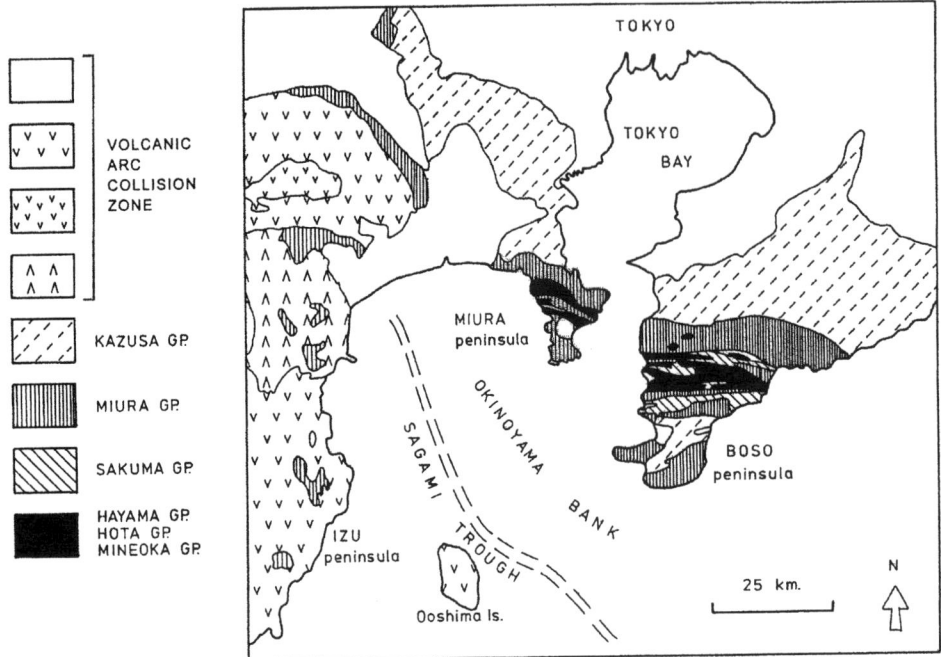

Fig. 2. Simplified geological map of the Miura-Boso peninsulas and adjacent areas. (From Stow et al. 1998a).

been interpreted as a shallowing-upward succession that accumulated over a 10.5 Ma period from about 13 Ma to 2.5 Ma in a forearc slope-basin setting, as the Philippine Sea plate moved northwards at a rate of 2.5 cm a^{-1} (Soh et al. 1991; Taira & Ogawa 1991). Continued collision between the two arc systems resulted in the accretion of the Miura Group from the Philippine forearc region onto the Honshu arc at around 2 Ma.

This accretion involved considerable dextral oblique-slip motion and associated structural deformation (Taira et al. 1982; Pickering et al. 1990). The latter resulted in a complex series of folds and minor faults, wet-sediment deformation, injection and veining, and bedding-parallel to sub-parallel thrusts. Nonetheless, apart from significant section duplication, the sedimentary succession shows very good preservation and is well exposed, especially in coastal exposures around both peninsulas (Soh et al. 1989; Stow et al. 1998a).

Oceanographic setting

The present day pattern of bottom currents in the NW Pacific has been summarized by Lee & Ogawa (1998) from various sources (Fig. 3, Table 1). It is dominated by south and SW-directed North Pacific Deep Water (NPDW) and generally north-flowing Antarctic Bottom Water (AABW). NPDW lies at depths in excess of 2000 m and down to at least 5800 m in the Japan Trench, where long-term current measurements have recorded steady SW-flow over a period of one year with a maximum velocity of 15–30 cm s^{-1}. AABW, having found its way along the Mariana and Izu-Bonin trenches or through the Yap gateway and north across the Philippine Sea, is then forced to turn either back towards the south or to continue northeast around the Pacific rim. Long-term bottom current measurements on the Pacific flank of the Izu-Bonin Arc show two directions. Deeper than 4000 m, there is a southerly-directed current interpreted as the NPDW, whereas above 2000 m AABW flow is steadily towards the north. In both cases maximum velocities of around 15–20 cm s^{-1} have been recorded.

Of the surficial currents around the Japanese mainland, only the warm-water Kuroshio current is of sufficient magnitude to have any influence at depth. Whereas surface velocities can be in excess of 180 cm s^{-1}, speeds as high as 40 cm s^{-1} have been measured at 2000 m water depth (Taft 1978; Fukazawa et al. 1985). Regular periodicity in both deep and shallow currents typically show a tidal, seasonal and/or longer term component. Measurements of internal tides with velocities in excess of 50 cm s^{-1} have been recorded from 2000 m water depth in both the Suruga and Sagami Troughs (Okada & Ohta 1993).

It seems most likely that a very close precursor of the present day circulation system in this part of the NW Pacific Ocean would have been in existence from at least early to mid-Miocene time. Development or intensification of AABW and its penetration into the North Pacific occurred in response to Antarctic cooling in the mid-Miocene (Wright & Miller 1993) and the development of the NPDW would have occurred at a similar time. Both the warm-water Kuroshio and cold-water Oyashio currents were in existence at around 17–16 Ma (Tsuchi & Ingle 1992). Although Stow & Faugères (1991) propose an AABW influence on the Miura Group, and Ito (1997, 2002) suggests a deep Kuroshio current influence for the Kazusa Group, we cannot be certain which deep water system was present during the time period considered here.

Palaeobathymetry

The overall sediment facies and ichnofacies characteristics of the Miura Group clearly show a shallowing upward succession from deep water (undisturbed turbidites and bioturbated hemipelagites of the Misaki Formation) to shallow water (high-energy sandy tidal and channel-fill facies of the Hasse Formation) (Soh et al. 1989, 1991; Stow et al. 1998a). There is also a trend towards a deeper eastern part of the inferred forearc basin, now represented by Boso peninsula sediments. For the older parts of the Misaki Formation, i.e. mid-Miocene time period, palaeodepth estimates based on benthic foraminiferal assemblages range from 2000 to 3000 m for the Miura peninsula to over 4000 m for the Boso peninsula (Ando et al. 1989; Akimoto et al. 1991). Rapid shallowing then took place through the mid to late Miocene, with eventual uplift and accretion during the Pliocene. Palaeodepth estimates from part of the Hasse Formation are around 100–200 m.

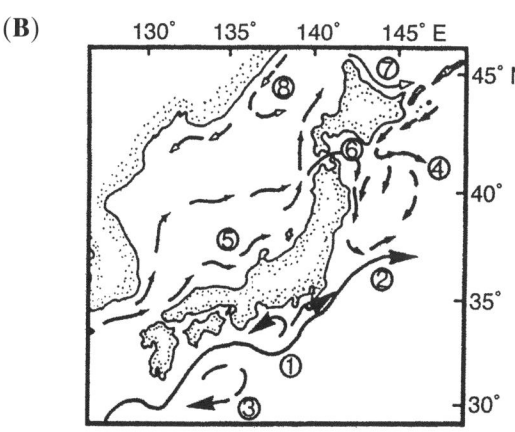

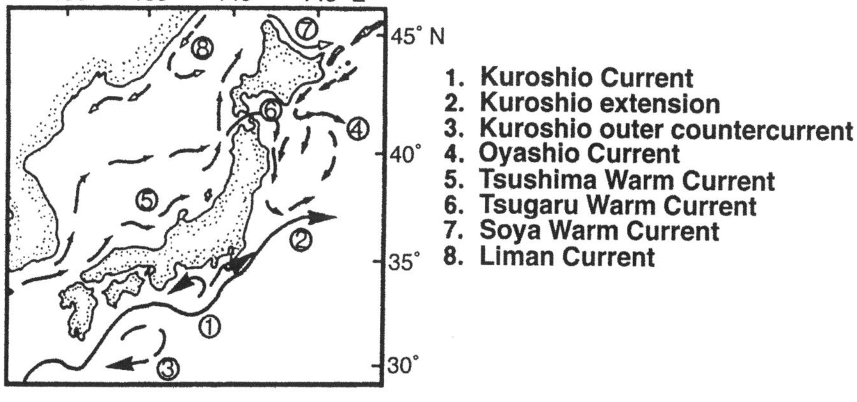

Fig. 3. (**A**) Present-day deep-water circulation in the NW Pacific region south of Japan, summarized from various sources (see Lee & Ogawa 1998). Numbers refer to location of current measurements in Table 1. (**B**) Present-day surface circulation around the Japanese islands. (From Ito 1997).

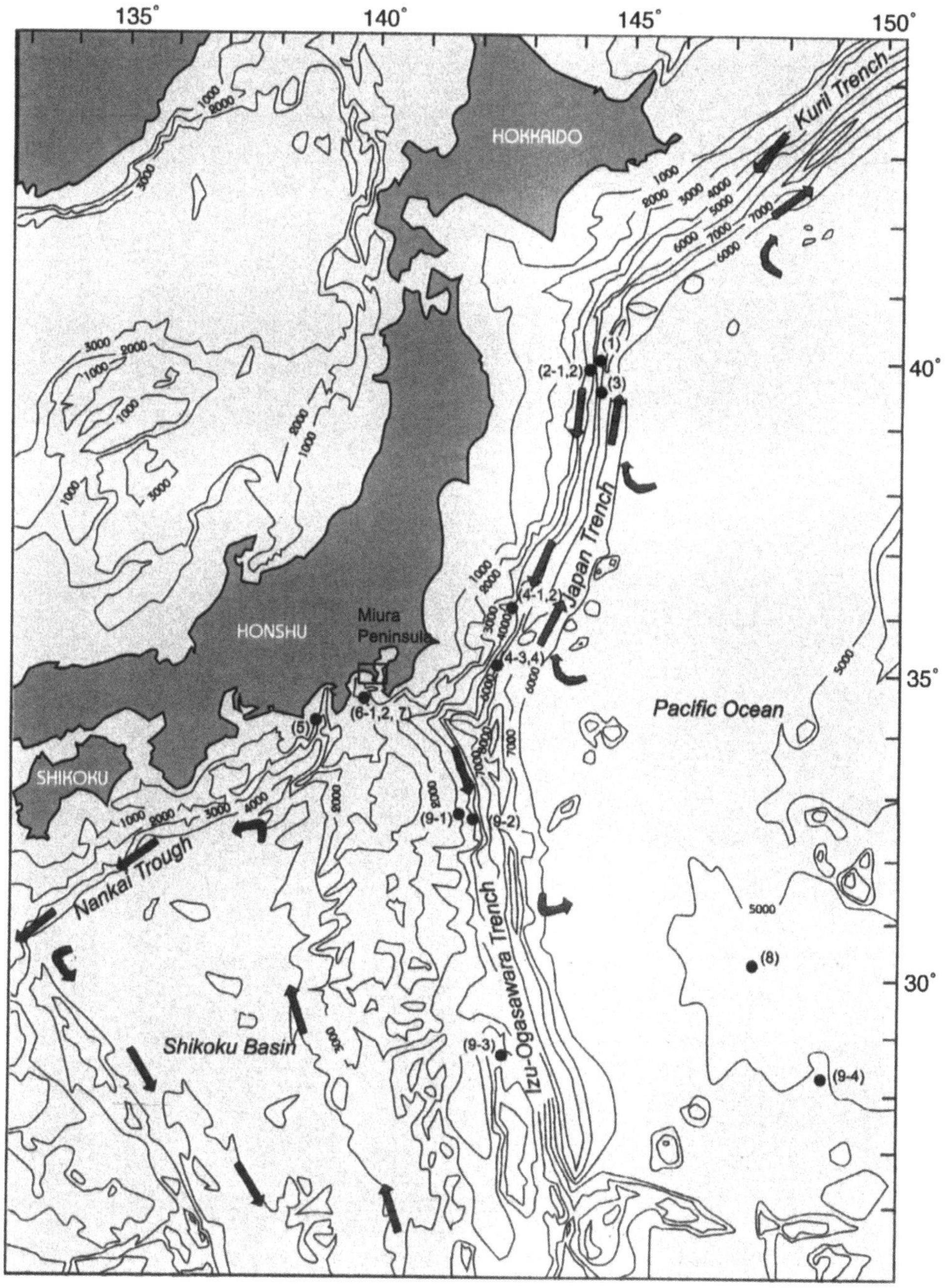

Table 1. *Summary data on present-day bottom currents from the NW Pacific region, south and east of Japan. Mostly based on long-term current-metre moorings as well as some direct seafloor observations*

Location	Depth (m)	Depth of current metre	Velocity (cm s^{-1}) mean-max	Dominant direction	Origin	Reference	Map key
Japan Trench	6400		30	SW	NPDW	Ogawa et al. 1996	1
Japan Trench	5805	5755	10 20	SSW	NPDW (+tide)	Mitsuzawa & Holloway 1998	2–1
Japan Trench	4220	4185	10 20	SSW	NPDW (+tide)	Mitsuzawa & Holloway 1998	2–2
Japan Trench	5000–6000			SSW to NNE	(video)	Horiuchi et al. 1993	3
Japan Trench	3100	2043	1	SSW	NPDW	Hollock & Teague 1996	4–1
		3043	1	SSW	NPDW		
Japan Trench	4400	2160	1	SSW	NPDW	Hollock & Teague 1996	4–2
		3145	2	SSW	NPDW		
		4169	2	SSW	NPDW		
Japan Trench	6500	409	30		Kuroshio	Hollock & Teague 1996	4–3
		1909	10				
		2824	5				
		4824	2	NNE			
Japan Trench	5400	517	15		Kuroshio	Hollock & Teague 1996	4–4
		1984	4				
		2984	3				
		4984	4	NNE			
Suruga Bay	1980	850	25	NE or SW	Tidal	Mitsuzawa et al. 1991	5
		1160	30	NE or SW	Tidal		
		1970	70	NE or SW	Tidal		
Sagami Trough	Approx 2000	2063	2	S	NPDW	Terramoto & Taira 1985	6–1
		2042	4	S	NPDW		
		1979	2	S	NPDW		
Sagami Trough	1597	1590	4	S	NPDW	Terramoto & Taira 1985	6–2
Sagami Trough	1174	1173	5 20	SW		Momma et al. 1998	7
Western Pacific	6200	5000	10	All directions	Eddy currents	Imawaki & Takano 1982	8
Izu-Bonin Ridge	3220	1820	20	NE	AABW	Chaen 1998	9–1
		2820	15	N or S			
Izu-Bonin Ridge	5260	3860	15	N or S		Chaen 1998	9–2
		4860	15	SE			
Izu-Bonin Ridge	4300	3900	20	SE	NPDW	Chaen 1998	9–3
			10	NW or N			
Izu-Bonin Ridge	5550	4100	15	SE	NPDW	Chaen 1998	9–4

Stratigraphic context

The general age and stratigraphy of the Miura Group are well known from palaeontological and palaeomagnetic studies. These data have been summarised in several recent papers (e.g. Kanie *et al.* 1991; Soh *et al.* 1991; Lee & Ogawa 1998; Stow *et al.* 1998*a*) and are illustrated in Figure 4. The basic subdivision is into the lower Misaki Formation, which is conformably to unconformably overlain by the upper Hasse Formation (also known as Hatsuse). However, the base of the Misaki formation is either unconformable or poorly exposed and the top is marked by a significant unconformity, so that a precise age range is less easily established. Estimates of the basal age range from 13 Ma to 10 Ma, while the top is generally taken as 2.5 Ma. Transition from the Misaki to Hasse formations appears to be diachronous from late Miocene in the Boso peninsula to early Pliocene in the Miura peninsula.

Recently much effort has been made to refine dating and attempt basinwide correlation using ash marker beds with some degree of success, particularly for the Miura Peninsula (Fig. 5) (Horiuchi & Taniguchi 1985; Kanie & Hattori 1991; Lee & Ogawa 1998).

Estimates of the total thickness of each formation and of the Miura Group as a whole are also subject to some uncertainty as a result of section duplication and thickening along bedding-parallel thrust faults. The figures given in Figure 4, from over 900 m to over 2000 m for the Misaki Formation are now considered to be

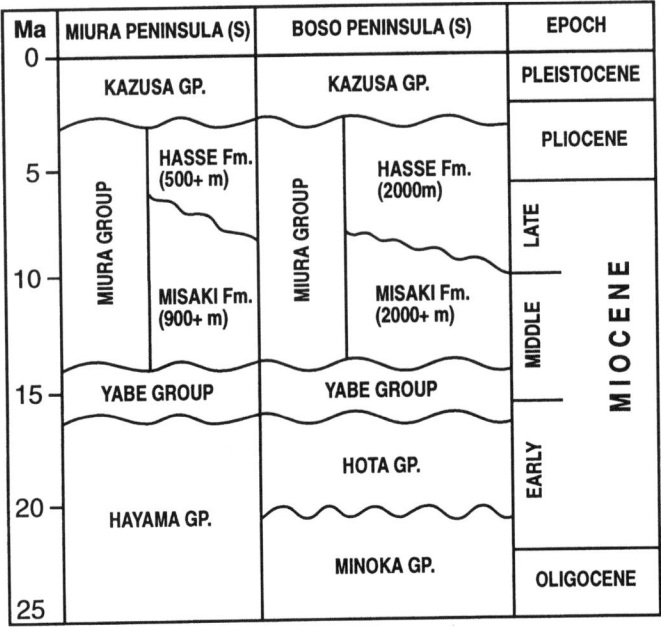

Fig. 4. Summary stratigraphy for the Miura–Boso area. (From Stow *et al.* 1998*a*). There is still uncertainty regarding the basal age of the Miura Group.

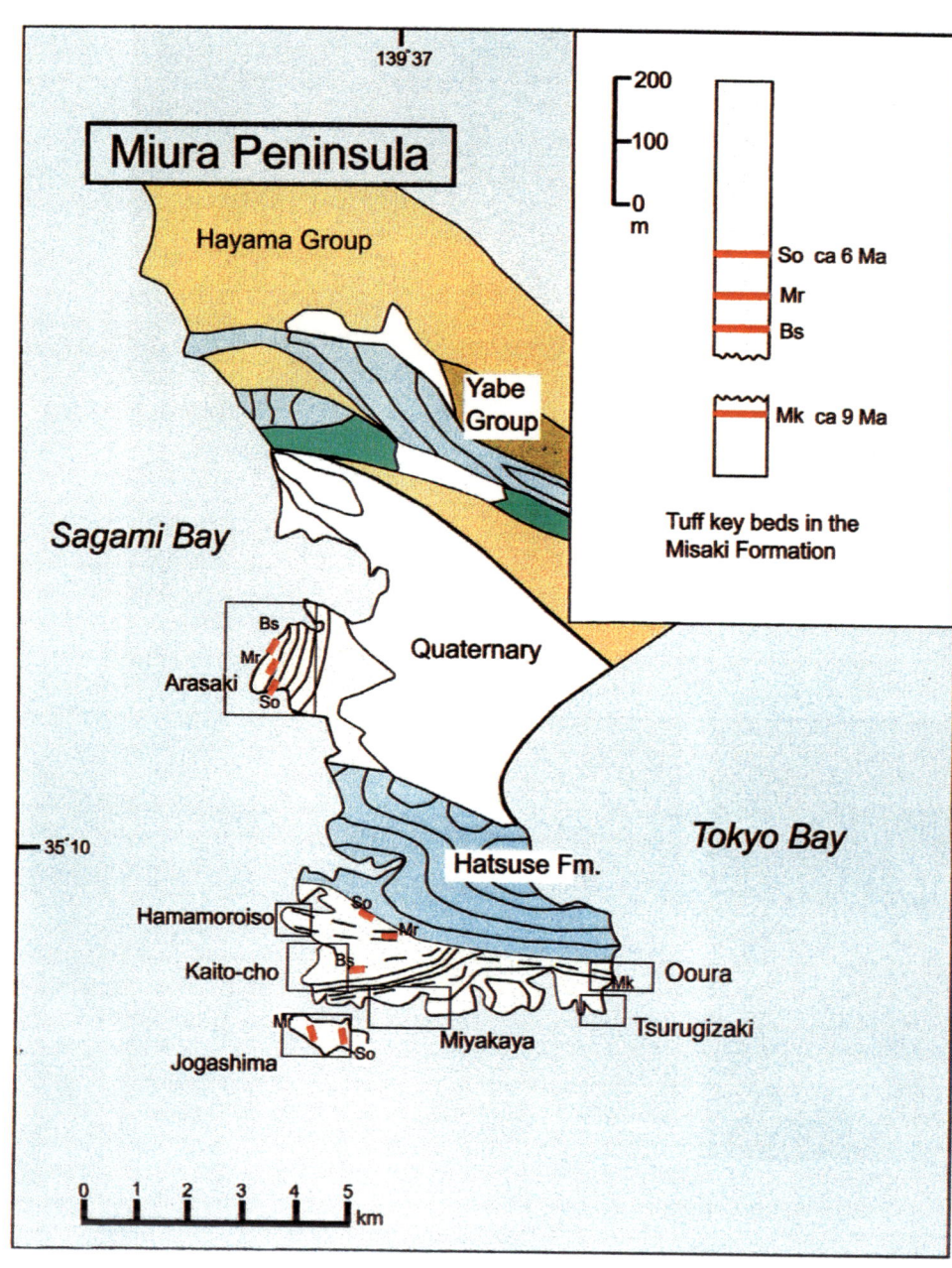

Fig. 5. Schematic geological map of the Miura Peninsula showing distribution of the Misaki Formation, representative key tuff horizons (with radiometric ages), and principal localities referred to in text. (Modified from Lee & Ogawa 1998).

overestimates, perhaps by about 50%. If we take a thickness range of 700–1400 m for the Misaki Formation deposited in a time period of 7 Ma, then we obtain an average sedimentation rate of 100–200 m Ma^{-1} (10–20 cm ka^{-1}). The Hasse Formation probably accumulated more rapidly than this. Note that these figures are for compacted, lithified sediment and include both rapidly deposited event beds as well as more slowly deposited background hemipelagites and contourites.

Sediment facies

Misaki formation facies

Based on previous detailed description and interpretation of the Misaki formation (Stow *et al.* 1998*a*), we recognize six separate facies groups comprising 22 individual facies (Fig. 6). These include coarse-grained scoriaceous beds (Facies Groups A, B and C), fine-grained pumiceous beds (Facies Group D), tuffaceous beds (Facies Group E) and disturbed or chaotic strata (Facies Group F).

The scoriaceous beds are distinctive dark-coloured layers making up 5–45% (average 10–20%) of the succession. They are composed mainly of basic to intermediate composition lava clasts and grains, with rare pumice fragments, set in a matrix of altered glass and vitric ash. The scoria are typically low-alkali tholeiite series basaltic andesite, similar in composition to the present day Izu-Oshima Island volcanics. They are event beds, mainly deposited by turbidity currents, debris flows and direct pyroclastic fall processes.

The dominant (background) facies making up 55–95% of the Misaki formation are the light-coloured, fine-grained (mud-silt grade), pumiceous sediments. They are composed of variable admixtures of pumice, clays, biogenic material and scattered scoriaceous debris. Bioturbation is dominant throughout. These are interpreted as slowly deposited hemipelagites and contourites.

At certain horizons through the succession are distinctive, light-coloured, thin-bedded, acidic composition tuffaceous beds. These are the ash layers that can be correlated across the basin on the basis of their characteristic geochemical signatures, and some of which have been radiometrically dated. Their sedimentary structures indicate that most have been deposited by low-concentration turbidity currents.

Fig. 6. Photographs of representative facies from the Misaki Formation. (**A**) Pumiceous hemipelagite facies dominant, with thin dark scoriaceous horizons and brown sideritic nodules. (**B**) Scoriaceous turbidite facies dominant, with thin pumiceous horizons. (**C**) Scoriaceous subaqueous pyroclastic-fall facies. (**D**) Mixed composition volcaniclastic turbidite, showing Bouma A–E divisions, sharp erosive base and large pumiceous clasts.

A variety of disturbed and chaotic strata are recognized including clear slump and debris flow units on the one hand, and sediment injection bodies on the other. The latter are most likely related to the period of post-depositional disturbance caused by compression, eduction and accretion of the Miura forearc basin succession.

Contourite characteristics

Within the background hemipelagic facies, there are certain intervals that show a greater influence of bottom current activity during sediment accumulation (Fig. 7). The evidence for this includes:

- A general absence of primary sedimentary structures through much of the facies, but intervals with clear but diffuse layering (indistinct parallel lamination).
- Intervals with remnant parallel and rare cross-lamination of dark scoriaceous material within the light-coloured pumiceous mudstone.
- Irregular concentrations and elongate lenses of coarse scoriaceous and bioclastic material.
- Sharp and/or erosive contacts associated with zones of diffuse lamination and slightly coarser grain sizes.
- An ichnofacies assemblage dominated by *Chondrites*, *Helminthoides*, *Planolites* and *Zoophycos* together with intense bioturbation throughout, but an apparent absence of *Zoophycos* and larger trace fossils in the more laminated zones.
- These diffusely laminated intervals show bioturbation continuous with deposition, together with localised sharp horizons of non-deposition or minor erosion below which a slightly different ichnofacies is apparent, including short vertical burrow systems.
- A grain size that is typically poorly-sorted mudstone, but varies from sand-rich muddy silt to silty clay, and includes localised lenses or layers of slightly better sorted coarse silt and sand-size material.
- Cyclic grain size variation between more and less silty (sandy) mudstones over intervals of about 10 cm to 100 cm, in some cases probably due to increased volcanic input of sandy scoriaceous material, but in other cases varying in phase with the diffusely laminated intervals and so interpreted as fluctuation in the strength of bottom current activity.
- A mixed pelagic/biogenic and volcanogenic/terrigenous

Fig. 7. Selected photographs of contourite facies from the Misaki Formation. (**A**) Uniform silty mud with indistinct parallel lamination, vague bioturbational mottling and rare burrows. Ooura, Miura Peninsula – width of view approx. 15 cm. (**B**) Uniform muddy silt with indistinct parallel lamination, intense bioturbational mottling, common burrows. Note minor erosive/non-deposition horizons (or omission surfaces) with more vertical burrows evident. Pale buff-coloured silty-sandy layer near base. Ooura, Miura Peninsula – width of view approx. 15 cm. (**C**) As for **B**, with buff-coloured sandy contourite towards top. Ooura, Miura Peninsula – width of view approx. 15 cm. (**D**) Typical pumiceous silty mud contourite facies from near Mera on Boso Peninsula. Note silty lamination, somewhat diffuse in parts and disturbed by burrowing. Width of view approx. 15 cm. (**E**)Muddy to silty hemipelagite-contourite facies with bioturbation, burrows, scattered nodules and dispersed scoriaceous grains. Tsurugusaki, Miura Peninsula – width of view approx. 15 cm. (**F**) As for **E**, with clearer but still diffuse layering of silty material within muddy background. Tsurugusaki, Miura Peninsula – width of view approx. 15 cm.

composition, including foraminifers, nannofossils, diatoms and radiolarians, dominant pumiceous glass with intermediate to high acidic composition and minor basaltic scoriaceous grains, and clays from both detrital and volcanic alteration sources.
- Much of the coarser biogenic material is fragmeneted and, in some cases, iron stained.

Most of these features are consistent with deposition of fine muddy and silty contourites under the influence of weak to moderate, fluctuating bottom currents (Fig. 8). One distinctive facies is a better sorted pumiceous siltstone (to very fine silty sandstone) with plane parallel lamination, without scattered scoria and with minor bioturbation throughout, continuous with

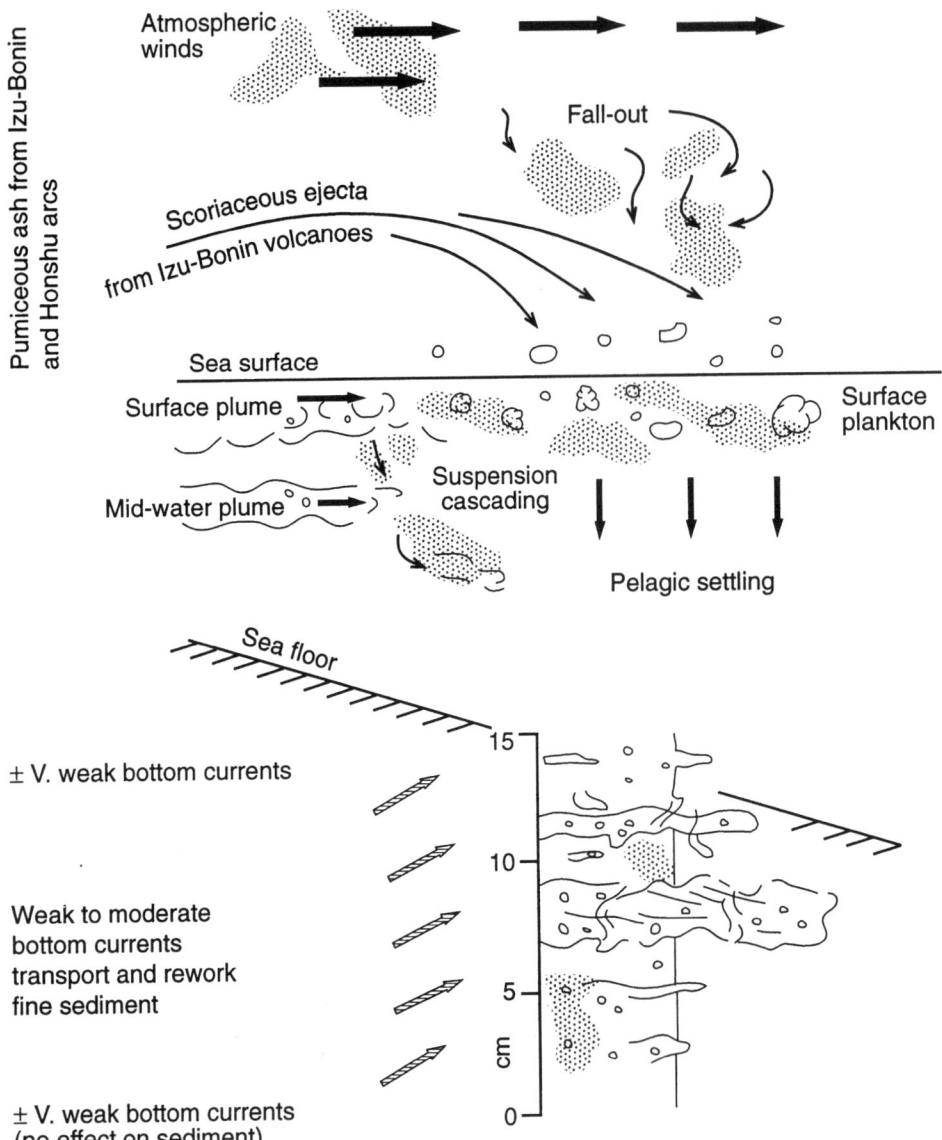

Fig. 8. Combination of hemipelagic and bottom current processes in the deposition of Misaki Formation fine-grained sediments. (From Stow et al. 1998a).

deposition. This might be interpreted as a higher-energy laminated contourite.

Distribution of fine-grained contourites

Due to the problems of structural complexity and sequence repetition, it has proved difficult to reconstruct the true vertical and lateral facies relationships. It is especially difficult to correlate between the rather different successions on Miura and Boso peninsulas.

A first order distribution of seven principal facies associations was presented by Stow et al. (1998a), and is reproduced here as Figure 9. This serves to illustrate as much the differences between sections as their correlation. Overall, there is a coarsening-upward trend from the Misaki formation, through a transitional facies to the Hasse formation, coincident with basin shallowing. Work on the geochemical characterisation of specific tuff beds has allowed for some detailed correlation between sections on the Miura peninsula (Fig. 5) (Horiuchi & Taniguchi 1985; Lee & Ogawa 1998).

At present we can only note that features of contourites within the fine-grained pumiceous mudstones are recognized in many parts of Miura peninsula, including Arasaki, Hamamoroiso, Jogashima, Miyakawa, Tsurugisaki and Ooura. We have also noted similar features at Mera on the Boso peninsula. They appear to range in age from near the base of the Misaki formation exposed to near the top. Towards the top, the facies indicate progressively shallower water and hence a greater influence of current activity. These currents, however, are interpreted as more likely of tidal or shelf origin, rather than deeper water bottom currents.

Discussion

Are they contourites?

Knowing how difficult it is to positively identify contourites in ancient successions on land, and being fully aware that the literature abounds with false and misleading identifications (Lovell & Stow 1981; Pickering et al. 1989; Stow et al. 1998b), it is important to critically examine our evidence and claim as presented here. We will attempt to follow the three-stage procedure and criteria laid down by Stow et al. (1998b), but must note that these are rather different from the criteria recently advocated by Shanmugam (2000).

(a) *Large-scale:* For rocks of Neogene age the palaeoceanographic reconstruction is generally very good, so that we are

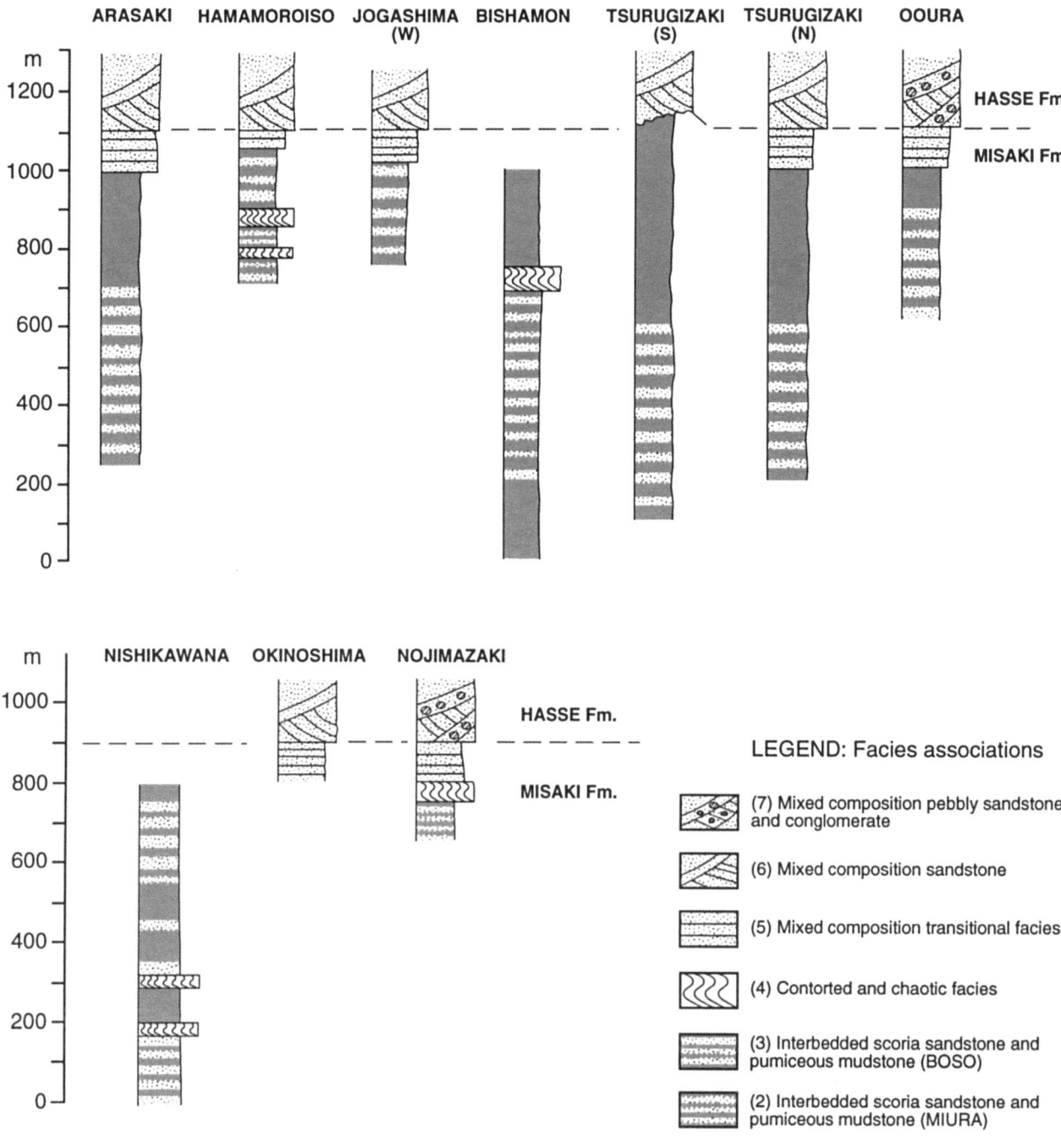

Fig. 9. Facies associations and their distribution within the Miura Group, Miura and Boso Peninsulas. (From Stow *et al.*, 1998a).

confident of the original depositional setting in a slope to forearc basin of the Izu-Bonin Arc. The estimates of deep-water palaeobathymetry are also reasonably accurate. From modern oceanographic data, there is clear evidence for the action of two or three geostrophic current systems capable of moving and depositing at least fine-grained sediments in the Izu-Bonin deep water slope region. These current systems were almost certainly present from early to mid-Miocene time. The data therefore support the potential for depositing contourites in rocks of this age and area.

(b) *Medium-scale:* The contourite facies are closely interbedded with turbidites, hemipelagites and other deep-water facies as might be expected in a slope-basin setting. They do not, therefore, exhibit any distinctive drift geometry other than presumed intercalated sheets. Palaeocurrent measurements have mostly been taken from the coarser grained facies, which were originally deposited by turbidity currents or pyroclastic fall. These data are discussed below in terms of the possible reworking by bottom currents. Kanamatsu (1995) has investigated the magnetic fabric (anisotropy of magnetic susceptibility) of both the coarse and fine-grained facies. For the latter, his results show mixed current directions of flow to the SSE, NNW and NE, when corrected for inferred plate rotation during emplacement. These directions are not incompatible with an element of alongslope current influence. Systematic regional trends in mineralogy, geochemistry or textural attributes have not been observed. However, the mixed composition of volcaniclastic ash in the pumiceous facies, together with bioclastic debris and terrigenous clays, would also support the supply of far-travelled material by bottom currents into an otherwise local source. In summary, there is little at this scale of observation to either positively support or refute a contourite interpretation, but several indicators that do favour at least a weak bottom current influence.

(c) *Small-scale:* There is abundant evidence at the scale of the sediment facies – in terms of structures, ichnofacies characteristics, textures and composition – that support the influence of weak to moderate bottom currents on the otherwise hemipelagic background sediment. This evidence is outlined above and we would therefore interpret the system as a contourite-hemipelagite continuum. The cyclic grain-size variation through the section is probably in part due to episodicity in the primary supply of volcaniclastic material. However, close inspection reveals the coincidence of change in other sedimentary features with grain size in some cases – for example, an increase in primary diffuse lamination and bioturbational hiatuses with coarser grain size. This suggests that fluctuation in long-term bottom-current velocity has also played a part.

Reworked turbidites? – re-examining the evidence

Most of the coarse-grained scoriaceous beds as well as the fine-grained tuffaceous layers were clearly deposited by turbidity currents and direct pyroclastic fall. All recent and detailed sedimentological studies in the area support this conclusion (Ogawa & Taniguchi 1988, 1989; Soh *et al.* 1989; Pickering *et al.* 1990; Stow *et al.* 1998a). However, if bottom currents were active during deposition of the background hemipelagite-contourite continuum as argued above (see also Stow & Faugères 1991; Lee & Ogawa 1998; Stow *et al.* 1998a, b), then it is possible that they were sufficiently strong to winnow and rework the tops of some turbidites. Lee & Ogawa (1998) have argued that this is indeed the case, especially for pale-coloured pumiceous beds showing parallel or cross lamination, whereas Stow *et al.* (1998a) note nothing unusual about the Misaki formation turbidites that would suggest any bottom current influence.

The evidence presented centres largely on measurements of palaeocurrent indicators, dominantly cross-bedding, but supported also by magnetic fabric data from Kanamatsu (1995). All authors agree that there is a dominant SE (corrected) palaeocurrent trend for Misaki formation facies on the Miura peninsula, which veers to more easterly on Boso. This fully supports mainly turbidity current supply from the Izu-Bonin arc located somewhere west of the slope-forearc basin system. All further note some local variation, with minor ENE and NE directions observed. In general, we do not see this variability as incompatible with a turbidite system; in fact, it is the norm for such deep-water systems. However, Lee & Ogawa (1998) have also documented minor NW directed trends in some localities, which they interpret as indicative of bottom currents generated by internal tides reworking the tops of turbidites. This they note is most common in the middle to upper Misaki formation, but absent from older parts of the succession.

In writing this joint paper, we agree to differ in our interpretation of whether or not bottom currents (internal tides or alongslope currents) have had any significant influence on the tops of turbidites. Further work is clearly required to help resolve this debate, and the apparent reverse-flow units need particular scrutiny.

The senior author (DAVS) acknowledges support from a number of sources while carrying out the field and laboratory work for this study, including an NERC research grant, British Council collaborative funds, and Royal Society and IGCP432 travel monies. He would also like to thank the many colleagues who have helped in the field and in discussion, including A. Taira, W. Soh, H. Taniguchi, M. Ito, J.-C. Faugères and K. T. Pickering, as well as the reviewers of an earlier version of this manuscript. The associate authors (YO, ITL, KM) further acknowledge laboratory assistance provided by Hyun Sang Min, K. Suzuki and H. Isobe.

References

AKIMOTO, K., UCHIDA, E. & ODA, M. 1991. Palaeoenvironmental reconstruction by benthic foraminifers from middle to late Miocene in the Misaki Formation, southern Miura Peninsula. *Chikyu Earth monthly*, **13**, 24–30.

ANDO, J., TANAKA, Y. & HASEGAWA, S. 1989. Sedimentary environment of the Miura Group in the southern Miura Peninsula. Abstracts Volume, 96th Annual Meeting. *Geological Society of Japan*, 216.

CHAEN, M. 1988. Report on study of deep-water circulation processes. *In*: TERAMOTO, T. (ed.) *Monbusho Grant Report.* Tokyo, 73–84.

FUKAZAWA, O. *ET AL.* 1985. The Kuroshio Current in Shikoku Basin. *In*: KAJIURA, K. (ed.) *Oceanic Characteristics and their Changes.* Kouseisha-Kouseikaku, Tokyo, 89–120.

HANAMURA, Y. & OGAWA, Y. 1993. Layer-parallel faults, duplexes, imbricate thrusts and vein structures of the Miura Group: keys to understanding the Izu forearc sediment accretion to the Honshu forearc. *The Island Arc*, **3**, 126–41.

HORIUCHI, K. & TANIGUCHI, H. 1985. Study on the correlation of tuff key beds in the Miura Group, Southern Miura Peninsula, Central Japan. *Science Report of Nihon University*, **20**, 11–31.

HORIUCHI, K., MOMMA, H., MITSUZAWA, K. & FUJIOKA, K. 1993. Holothurians and bottomcurrents on the oceanward slope of the Japan trench. *Proceedings of JAMSTEC Symposium on Deep Sea Research*, 41–48.

IMAWAKI, S. & TAKANO, K. 1982. Low-frequency eddy kinetic energy spectrum in the deep Western North Pacific. *Science*, **216**, 1407–1408.

ITO, M. 1996. Sandy contourites of the lower Kazusa Group, a middle Pleistocene forearc basin fill in the Boso Peninsula, Japan. *Sedimentary Geology*, **88**, 219–230.

ITO, M. 1997. Spatial variations in turbidite-to-contourite continuums of the Kiwada and Otadai formations in the Boso Peninsula, Japan: an unstable bottom-current system in a Plio-Pleistocene forearc basin. *Journal of Sedimentary Research*, **67**, 571–582.

ITO, M. 2002. Kurosiho Current-influenced sandy contourites from the Plio-Pleistocene Kazusa forearc basin, Boso Peninsula, Japan. *In*: STOW, D. A. V., PUDSEY, C. J., HOWE, J. A., FAUGÈRES, J.-C. & VIANA, A. R. (eds) *Deep-Water Contourite Systems: Modern Drifts and Ancient Series, Seismic and Sedimentary Characteristics.* Geological Society, London, Memoirs, **22**, 421–432.

KANAMATSU, T. 1995. *A study of the magnetic fabric of the sediment in the accretionary complex, Boso and Miura Peninsulas, Central Japan.* PhD Thesis, Tokyo University, Japan.

KANIE, Y. & HATTORI, M. 1991. Report of the symposium on *Chronology and palaeoenvironmental aspects of the Miura Group, central Japan* held at the 97th annual meeting of the Geological Society of Japan in 1990. *Journal Geological Society of Japan*, **97**, 849–864.

KANIE, Y., OKADA, H., SASAHARA, Y. & TANAKA, H. 1991. Calcareous nannoplankton age and correlation of the Neogene Miura Group between the Miura and Boso Peninsulas, Southern-Central Japan. *Journal Geological Society of Japan*, **97**, 135–155.

KITAZATO, H. 1997. Palaeogeograpic change in central Honshu, Japan, during the Cenozoic, in relation to the collision of the Izu-Ogasawara Arc with the Honshu Arc. *The Island Arc*, **6**, 144–157.

KODAMA, K., OKA, S. & MITSUNHI, T. 1980. *Geology of the Misaki District.* Geological Survey of Japan, Quadrangle Series, Tokyo, **8**.

KOTAKE, N. 1989. Taphonomy of the trace fossils in bathyal deposits of the Boso Peninsula, Central Japan. *Benthos Society of Japan*, **35**, 53–60.

MITSUZAWA, K. 1992. Measurement of currents and temperature inside the submarine caldera in Kaikata Seamount. *Proceedings JAMSTEC Symposium on Deep Sea Research*, **8**, 163–168.

MITSUZAWA, K., KUMAI, R., NAKAO, A., YAMAGUCHI, N. & FUKASAWA, M. 1991. *Observation of bottom currents in the Suruga Trough.* JAMSTEC Technical Report, 45–50.

LEE, IN TAE & OGAWA, Y. 1998. Bottom-current deposits in the Miocene–Pliocene Misaki Formation, Izu forearc area, Japan. *The Island Arc*, **7**, 315–329.

LOVELL, J. P. B. & STOW, D. A. V. 1981. Identification of ancient sandy contourites. *Geology*, **9**, 347–349.

OGAWA, Y. 1985. Variety of subduction and accretion processes in Cretaceous to recent plate boundaries around southeast and central Japan. *Tectonophysics*, **112**, 493–518.

OGAWA, Y. & TANIGUCHI, H. 1988. Geology and tectonics of the Miura-Boso Peninsulas and the adjacent area. *Modern Geology*, **12**, 147–168.

OGAWA, Y. & TANIGUCHI, H. 1989. Origin and emplacement process of the basaltic rocks in the accretionary complexes and structural belts in Japan in view of minor element analysis and mode of occurrence. *Journal of Geological Society of Japan*, **98**, 118–132.

OKADA, H. & OHTA, S. 1993. Photographic evidence of variable bottom-current activity in the Suruga and Sagami Bays, Central Japan. *Sedimentary Geology*, **82**, 221–237.

PICKERING, K. T., AGAR, S. M. & PRIOR, D. J. 1990. Vein structure and the role of pore fluids in early wet-sediment deformation, late Miocene volcaniclastic rocks, Miura group, SE Japan. *In*: KNIPE, R. J. & RUTTER, E. H. (eds) *Deformation Mechanisms, Rheology and Tectonics*. Geological Society, London Special Publications, **54**, 417–430.

PICKERING, K. T., HISCOTT, R. N. & HEIN, F. J. 1989. *Deep-Marine Environments: Clastic Sedimentation and Tectonics*. Unwin Hyman, London.

SHANMUGAM, G. 2000. 50 years of the turbidite paradigm (1950s–1990s): deep-water processes and facies models – a critical perspective. *Marine & Petroleum Geology*, **17**, 285–342.

SOH, W., PICKERING, K. T., TAIRA, A., & TOKUYAMA, H. 1991. Basin evolution in the arc-arc Izu collision zone, Mio-Pliocene Miura Group, central Japan. *Journal of the Geological Society, London*, **148**, 317–330.

SOH, W., TAIRA, A., OGAWA, Y., TANIGUCHI, H., PICKERING, K. T. & STOW, D. A. V. 1989. Submarine depositional processes for volcaniclastic sediments in the Mio-Pliocene Misaki formation, Miura Group, central Japan. *In*: TAIRA, A. & MASUDA, F. (eds) *Sedimentary Facies in the Active Plate Margin*. Terra Scientific Publications, Tokyo, 619–630.

STOW, D. A. V. & FAUGÈRES, J.-C. 1991. *Miocene contourites from the proto Izu-Bonin forearc region, southern Japan*. Abstract Volume, 13th International Sedimentological Congress, Nortingham, UK.

STOW, D. A. V., FAUGÈRES, J.-C., VIANA, A. & GONTHIER, E. 1998b. Fossil contourites: a critical review. *Sedimentary Geology*, **115**, 3–32.

STOW, D. A. V., TAIRA, A., OGAWA, Y., SOH, W., TANIGUCHI, H. & PICKERING, K. T. 1998a. Volcaniclastic sediments, process interaction and depositional setting of the Miocene–Pliocene Miura Group, SE Japan. *Sedimentary Geology*, **115**, 351–382.

TAFT, B. A. 1978. Structure of the Kuroshio south of Japan. *Journal of Marine Research*, **36**, 77–117.

TAIRA, A. & OGAWA, Y. 1991. Cretaceous to Holocene forearc evolution in Japan and its implication to crustal dynamics. *Episodes*, **14**, 205–212.

TAIRA, A., OKADA, H., WHITAKER, J. H. McD. & SMITH, A. J. 1982. *The Shimanto Belt of Japan: Cretaceous–Lower Miocene Active Margin Sedimentation*. Geological Society, London, Special Publications, **10**, 5–26.

TANIGUCHI, H., OGAWA, Y. & SOH, W. 1991. *Tectonic development of the Izu Arc and Proto-Izu Arc*. Journal of Geography, Tokyo Geographical Society, **100**, 514–29.

TSUCHI, R. & INGLE, J. C., EDS. 1992. *Pacific Neogene: Environment, Evolution and Events*. University of Tokyo Press, Tokyo.

WRIGHT, J. D. & MILLER, K. G. 1993. Southern Ocean influences on late Eocene to Miocene deep-water circulation. *Antarctic Research Series*, **60**, 1–25.

Kuroshio Current-influenced sandy contourites from the Plio–Pleistocene Kazusa forearc basin, Boso Peninsula, Japan

MAKOTO ITO

Department of Earth Sciences, Chiba University, Chiba 263-8522, Japan

Abstract: Lithofacies features of bottom current influenced deep-sea sandy deposits (sandy contourites) have been studied by detailed outcrop analyses of the Plio–Pleistocene infill of the Kazusa forearc basin on Boso Peninsula, Japan. Sandy contourites are characterized by traction structures, such as parallel lamination and ripple cross-lamination, together with minor inverse grading and wave ripple-lamination. Ripple cross-lamination is commonly associated with mud drapes and flaser and lenticular bedding. These sedimentary structures do not display any regular vertical sequences. Sandy contourites are commonly associated with turbidites and comprise turbidite-to-contourite continuums. In general, sandy contourites are better sorted than associated turbidites and have sharp or gradational basal contacts with underlying turbidites or hemipelagites, and sharp upper contacts with overlying hemipelagites. The framework composition of sandy contourites, in contrast, is largely equivalent to that of the associated turbidites. Palaeocurrent directions of sandy contourites are variable in time and space and alongslope- and upslope-directed palaeocurrents are dominant. Turbidite-to-contourite continuums show spatial and temporal variation in bed thickness and lithofacies features. These variations suggest varying intensity of deep-sea bottom currents in the Kazusa forearc basin. Analogous flow conditions have been documented from the modern, deep-water mass of the Kuroshio Current. Thus, the observed variations in lithofacies features and palaeocurrents of turbidite-to-contourite continuums can best be interpreted in terms of fluctuating strength of the deep-water palaeo-Kuroshio bottom current during the Pliocene through Pleistocene.

The interpretation of thin-bedded, well-sorted sandstones in deep-water successions as either turbidites or contourites remains controversial. This is particularly true of ancient successions now exposed on land (Stow & Lovell 1979; Stow *et al.* 1998). Whereas, some authors describe sandy contourites as poorly-sorted, mainly structureless and bioturbated (Stow *et al.* 1996, 1998, 2002), others believe them to be clean, well-sorted and with abundant evidence of traction structures such as parallel and cross-lamination (Shanmugam *et al.* 1993; Shanmugam 2000). While there is general agreement that a strong, semi-permanent, bottom current system flowing along a continental margin will rework turbidites, both during and after deposition, the particular features that characterize this turbidite to contourite continuum are still a matter of debate (Stanley 1993; Stow *et al.* 1998).

Having previously interpreted the Plio–Pleistocene Kazusa Group on Boso Peninsula as a dominantly turbidite fill of a forearc basin, that includes bottom current reworking and sandy contourites (Ito 1996, 1997), this contribution further explores the case for sandy contourites. The criteria for interpretation closely follow those of Shanmugam *et al.* (1993), and are clearly different from those used by Stow *et al.* (2002) in their recognition of Mio–Pliocene contourites in the closely related Miura Basin.

Geological and paleoceanographic setting

The Kazusa forearc basin developed in response to the west-northwestward subduction of the Pacific plate beneath the Eurasia plate at the Izu-Bonin Trench (Katsura 1984; Ito & Masuda 1988) (Fig. 1A). The sedimentary infill of the basin is defined as the Kazusa Group and is exposed on the Boso Peninsula of Japan (Fig. 1C). The Kazusa Group is as thick as 3000 m and developed from about 2.4 to 0.45 Ma. The group records a series of marine environments ranging from deep-sea basin-plain, through submarine-fan and slope, to shallow sea (Ito & Katsura 1992). In general, the lower part of the Kazusa Group is characterized by deep-sea basin-plain, submarine-fan, and lower slope sediments in the northeastern area, and upper slope to shelf sediments in the southwestern area (Ito 1998*a*). In the upper part, the group is represented by upper slope and shelf sediments with minor nearshore sediments.

The Boso Peninsula is situated in the central part of Japan, and the Kuroshio Current flows along the southeastern offshore of the peninsula (Fig. 1B). Based upon palaeoecological analyses of marine faunas of the Kazusa Group (e.g. Aoki 1968; Ujihara 1986; Baba 1990; Igarashi 1994), this area is interpreted to have been affected during the Pliocene and Pleistocene not only by the warm-water Kuroshio Current but also by the cold-water Oyashio Current in response to glacial and interglacial environmental changes. These analyses, however, indicate that deposition of the lower Kazusa Group discussed in this paper was influenced mainly by the warm-water Kuroshio Current and that the deep-sea environment of the Kazusa forearc basin was dominated mainly by the deep-water mass of that current. Fluctuation in speed and direction of the palaeo-Kuroshio Current appears to have controlled spatial and temporal variation in strength of a bottom-current system in and around the Boso Peninsula area during the time of deposition of the lower Kazusa Group.

Depositional and sequence-stratigraphic setting

The sandy contourites discussed in this paper are mainly from the lower part of the Kazusa Group in the northeastern area of the Boso Peninsula. The lower Kazusa Group is represented by interbedded sandstones and siltstones intercalated with volcanic ash beds (Ito & Katsura 1992). In general, sandstones and siltstones are characterized by vertical sequences of sedimentary structures that are diagnostic of turbidites (e.g. Bouma 1962; Lowe 1982; Pickering *et al.* 1989). The sedimentary environment is interpreted to have been one in which turbidity currents alternated with hemipelagic fallout in supplying sediments to submarine fans and associated deep-sea environments (Hirayama & Nakajima 1977; Katsura 1984; Ito 1998*a*). Northeastward-flowing palaeocurrents are dominant (Katsura 1984; Ito 1998*a, b*) and benthic marine faunas in siltstones indicate 400–1500 m palaeowater depth (Kitazato 1986; Baba 1990). Interbedded siltstones are moderately and locally intensely burrowed with ichnofaunas such as *Chondrites*, *Zoophycos*, and *Thalassinoides*.

The deep-sea deposits of the lower Kazusa Group are characterized by repetition of sandstone-dominated and siltstone-dominated intervals. Each pair of one sandstone-dominated and

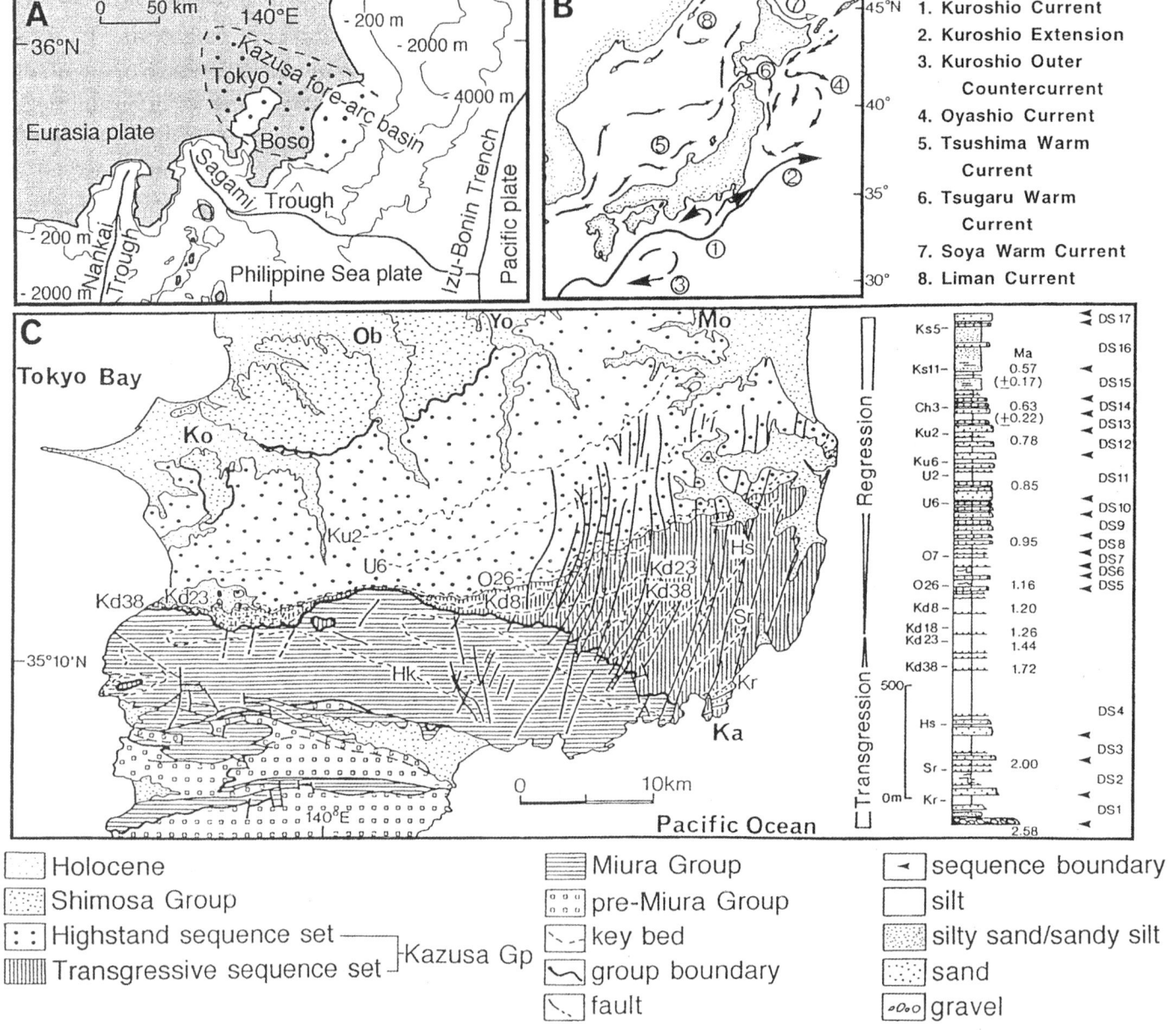

Fig. 1. **A**, Plate-tectonic framework of the Kazusa forearc basin (Simplified from Yamada *et al.* 1990). **B**, Modern surface current systems around the Japanese Islands (Simplified from Nishiyama 1990). **C**, Geological sketch map of the central part of the Boso Peninsula (Modified from Ito 1996). Lefthand side letters with numerals of the composite section of the Kazusa Group indicate stratigraphic horizons and the coding of some volcanic ash beds. Distribution of some volcanic ash beds is also indicated by dashed lines in the geological sketch map. Age data and sequence-stratigraphic classification (DS1-17; DS, depositional sequence) of the Kazusa Group are given on the right side of the composite section. Localities: Ko, Koito River; Ob, Obitsu River; Yo, Yoro River; Mo, Mobara; Ka, Katsuura.

one siltstone-dominated interval (cycle thickness is as much as 270 m thick) is interpreted to be a depositional sequence of about 0.2–0.05 million years duration (Ito & Katsura 1992; Ito 1998*a*, *b*). Each depositional sequence is further divided into lowstand, transgressive, and highstand systems tract deposits on the basis of geometry, lithofacies features, and bounding-surface characteristics of the deposits (Ito 1998*a*). The distinctive associations of lowstand, transgressive, and highstand systems tracts in deep-sea deposits are interpreted to reflect a narrow shelf, steep slope, and high sediment supply in the Kazusa forearc basin (Ito 1998*a*, *b*). The development of high-frequency depositional sequences of the lower Kazusa Group can best be interpreted in terms of glacial and interglacial eustatic sea-level cycles (Ito & Katsura 1992).

Sediment facies

Three different types of lithofacies organization were identified in couplets of sandstones and siltstones of the lower Kazusa Group (Ito 1997) (Fig. 2). These three types account for the majority of beds in the lower Kazusa Group and indicate the turbidite-to-contourite continuum.

Type A – Sandstone beds of this type (Beds 1–100 cm thick) show vertical sequences of sedimentary structures that are typical of the Bouma (1962) model and fine upward to overlying graded and/or ungraded siltstones with gradational contacts (Figs 2, 3, and 4A). Graded siltstones commonly contain plant fragments.

TURBIDITE-to-CONTOURITE CONTINUUMS

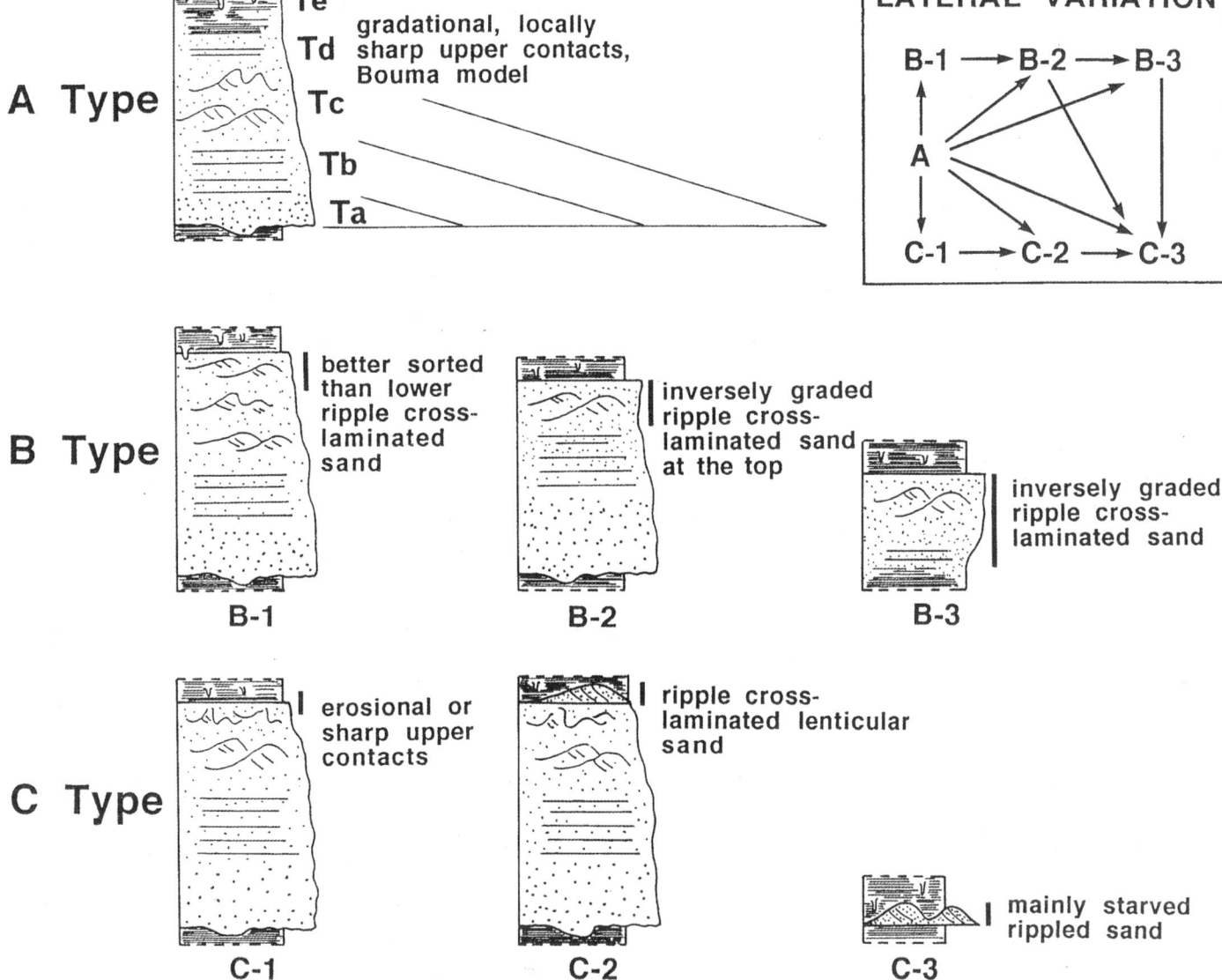

Fig. 2. Three types of turbidite-to-contourite continuums of the lower Kazusa Group. Positions of bottom current reworked deposits are indicated by vertical bars on the right hand side of each section. Right top box illustrates lateral relationships among the seven subtypes (A–C-3). Vertical scale is variable from 1 to 10 cm.

Couplets of sandstones and siltstones of Type A are interpreted to be turbidites and hemipelagites. Graded siltstones are interpreted as turbidite mudstones, and ungraded siltstones are interpreted as mainly hemipelagites rather than ungraded turbidite mudstones (cf. Stow & Shanmugam 1980).

Type B – Sandy deposits of Type B are characterized by sharp upper contacts beneath ungraded siltstones and a general lack of the upper parallel-laminated and graded siltstone divisions (Figs 2 and 3). Upper parts of ripple cross-laminated sandstones show better-sorting and palaeocurrents different from lower parts (Type B-1) (Figs 3, 4B and 5A). Locally, better-sorted ripple cross-laminated sandstones show inverse grading and rest on parallel laminated sandstones (Type B-2) (Figs 5B and 6A, B). Locally, the top of rippled sandstones are defined by erosional surfaces and are overlain directly by ungraded siltstones (Fig. 6C). Ungraded siltstones, associated with Type B, as well as Types A and C, locally exhibit inverse grading with very coarse silt and/or very fine sand grains (Fig. 6D). Some inversely-graded ripple cross-laminated sandstones with gradational basal contacts and sharp upper contacts are encased in ungraded siltstones (Type B-3) (Fig. 5–C).

The lack of either an upper parallel-laminated division or a graded siltstone division can be interpreted in two ways. One possibility is bypassing of the grain-size population in which upper parallel-laminated and graded siltstone divisions appear to have developed during a waning flow stage. The other possible mechanism is the winnowing and resedimentation of sand and silt either from the evolving load of turbidity currents flowing beneath bottom currents, or from the top of freshly deposited turbidites.

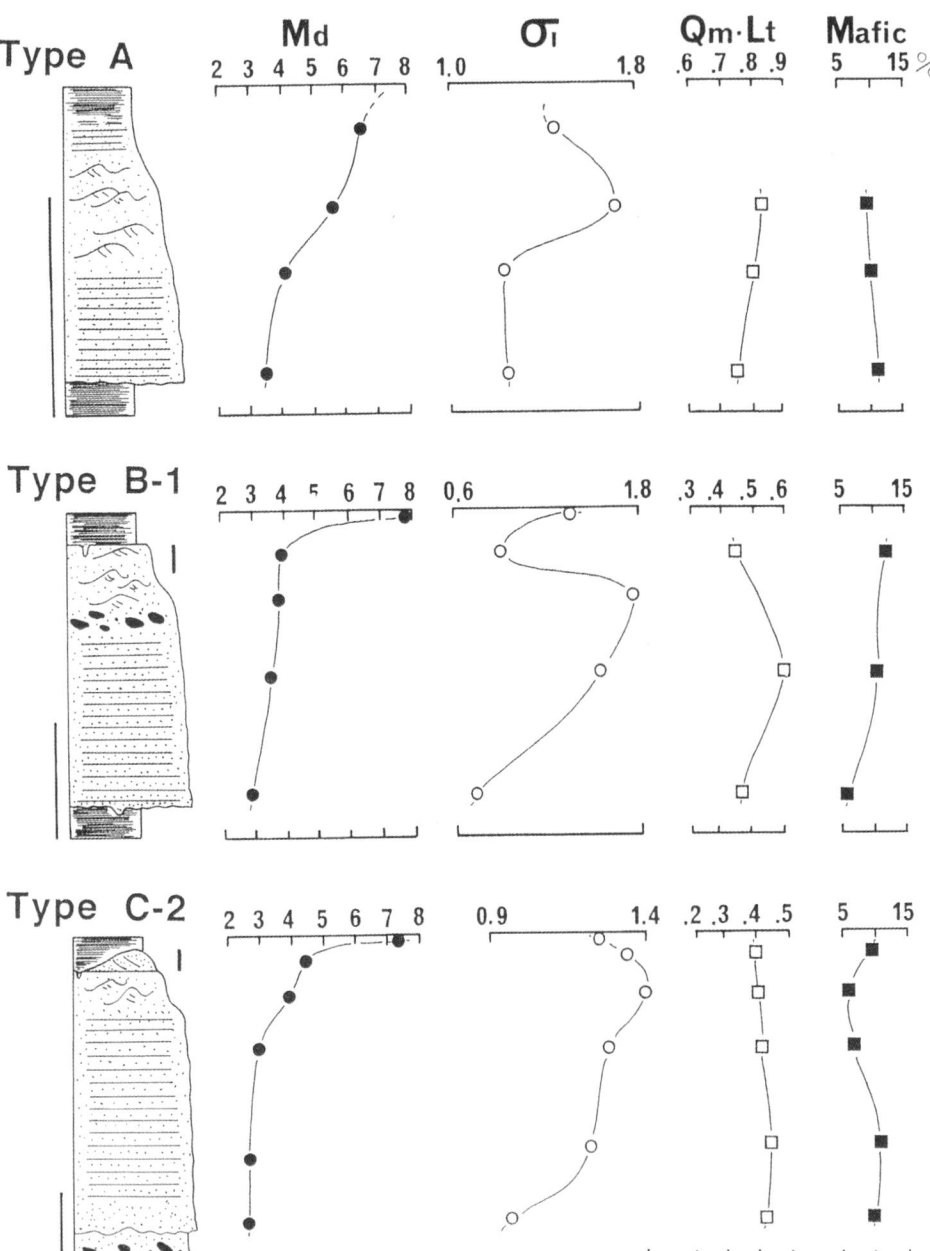

Fig. 3. Examples of vertical sequences of sedimentary structures, textural features and framework composition of three different types of turbidite-to-contourite continuums of the lower Kazusa Group. Md, median grain size; σ_I, inclusive graphic standard deviation; Qm-Lt, relative abundance of monocrystalline quartz grains (Qm) compared to total lithic fragments (Lt) (Qm/(Qm + Lt)); Mafic, relative abundance of mafic mineral grains. Grain size was analyzed using standard sieves and laser diffraction grain size analyzer (Shimazu SALD-1100). Methods of sample preparation and point-counting for framework composition used in this study were the same in those of Ito (1994). Vertical bars on the left side of each log indicate 10 cm.

Locally developed erosional upper surfaces also indicate post-depositional erosion by bottom currents. Better sorting and different palaeocurrents of upper parts of ripple cross-laminated divisions support the second alternative.

Type C – Type C beds are also characterized by a general lack of either an upper parallel-laminated division or a graded siltstone division. Ripple cross-laminated and/or convoluted sandstone divisions have sharp upper contacts with ungraded siltstones (Type C-1) (Figs 2 and 4C). Laterally, this sharp contact separates the sandstone bed from a better sorted, ripple cross-laminated, very fine sandstone or very coarse siltstone about 1–5 cm thick (Type C-2) (Figs 2, 3 and 4D). Internally, ripple cross-laminae are associated with mud drapes. Locally, lenticular beds of ripple cross-laminated very fine sandstone or very coarse siltstone 1–5 cm thick with sharp basal and upper contacts are encased in ungraded siltstones (Type C-3) (Fig. 2). Palaeocurrents of such starved rippled deposits are commonly different from those of associated Type A beds by more than 90°.

Sharp upper contacts of ripple cross-laminated and/or convoluted sandstone divisions are probably the result of erosion of turbidites by deep-sea bottom currents. Internal mud drapes in ripple cross-laminated sandstones and siltstones indicate dynamic conditions oscillating between traction and suspension deposition that are believed by some workers to characterize bottom current reworked sands and silts (Mutti 1992; Shanmugam *et al.* 1993). Better sorted, ripple cross-laminated sandstones indicate more thorough erosion and redeposition of turbidites.

Vertical sequence of sedimentary structures

Sandy contourites associated with turbidites are commonly characterized by several different types of sedimentary structures (Fig. 7). These sedimentary structures are characterized by traction structures, such as parallel lamination and current-ripple lamination, together with minor inverse grading. Furthermore, internal erosion surfaces overlain by blankets of mudstones are

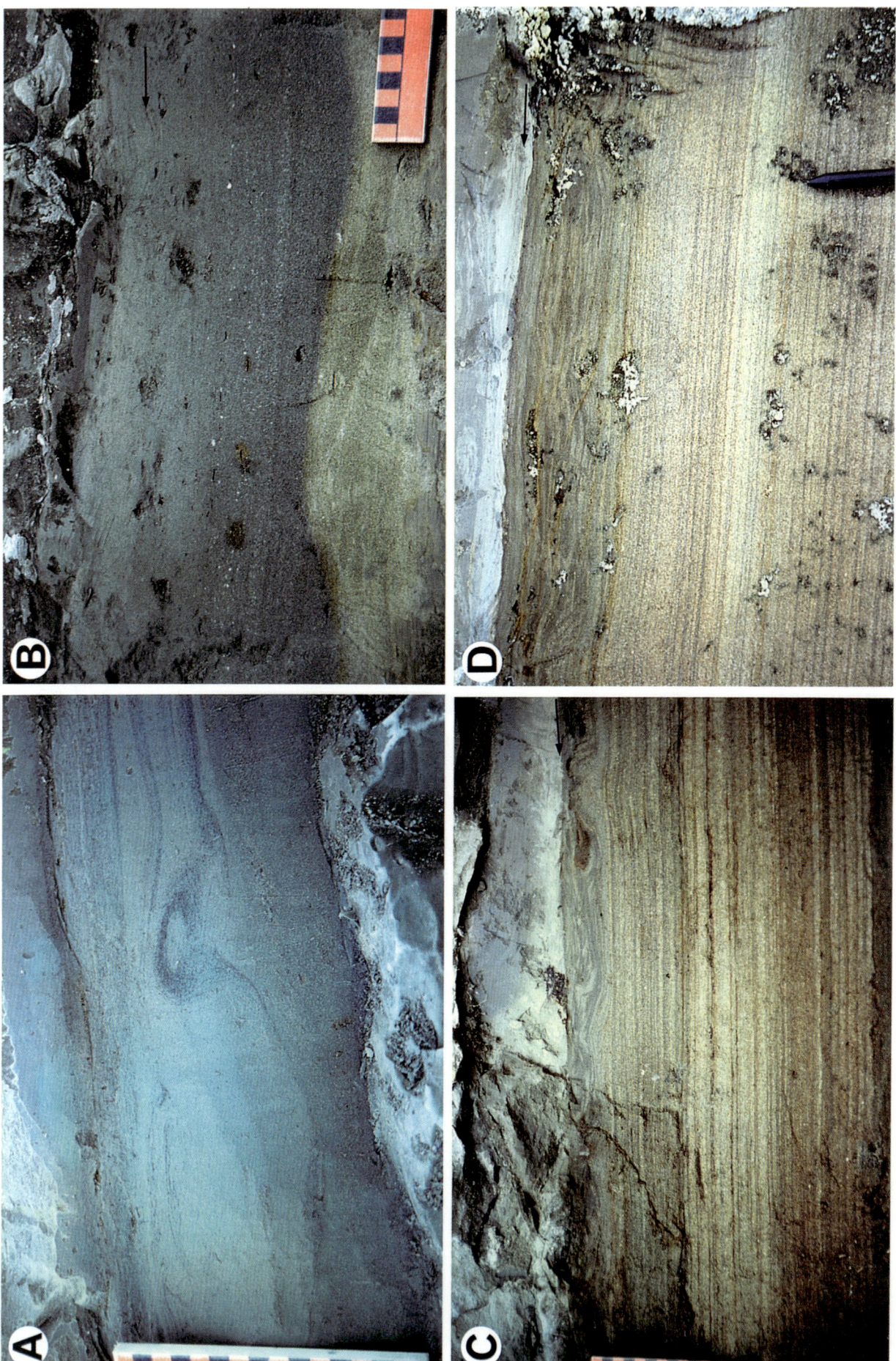

Fig. 4. (A) An outcrop example of turbidite sandstones and siltstones and overlying ungraded bioturbated hemipelagic siltstones of Type A lithofacies indicating Tcde turbidite at Yamase, Isumi Town. Scale on the lefthand side shows 1 cm interval. Vertical changes in textural features are given in Figure 3. (B) An outcrop example of Type B-1 lithofacies characterized by turbidite sandstones and overlying ripple cross-laminated sandy contourite (arrowed) with gradational basal contacts and upper sharp contacts beneath hemipelagites at Kodaki, Misaki Town. Scale on the left bottom shows 1 cm interval. Vertical changes in textural features are given in Figure 3. (C) An outcrop example of Type C-1 lithofacies characterized by sharp upper contact of a turbidite (arrowed) at Taito, Misaki Town. Scale shows 1 cm interval. (D) An outcrop example of Type C-2 lithofacies characterized by a turbidite sandstone bed and an overlying ripple cross-laminated sandy contourite bed (arrowed) at Taito, Misaki Town. Ball-point pen gives scale. Vertical changes in textural features are given in Figure 3.

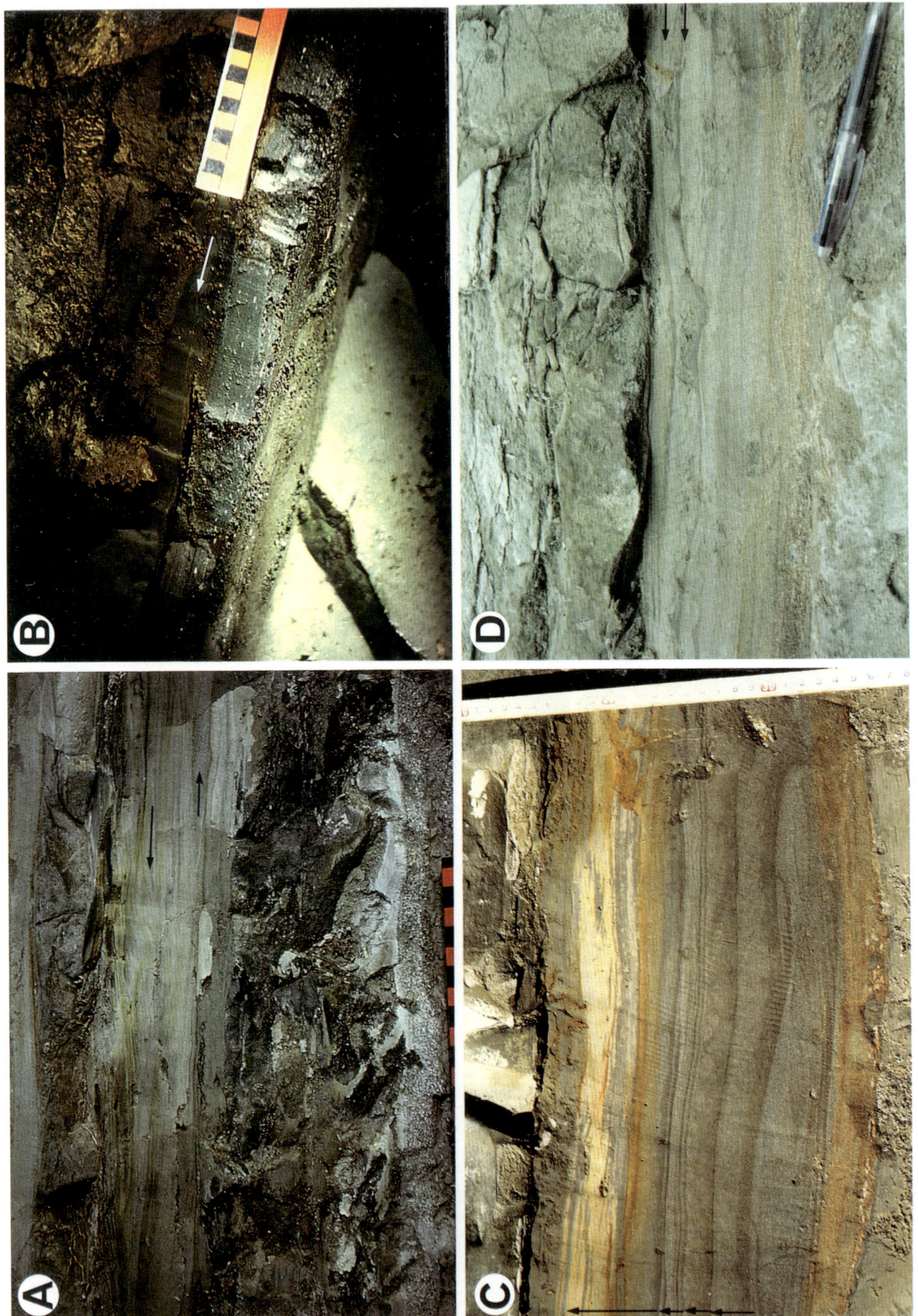

Fig. 5. (**A**) An outcrop example of a ripple cross-laminated turbidite and overlying ripple cross-laminated sandy contourite with reverse palaeocurrent direction at Kuniyoshi, Isumi Town. Arrows indicate paleocurrent directions. Scale on the bottom shows 1 cm interval. (**B**) An outcrop example of an inversely graded, ripple cross-laminated sandy contourite (arrowed) at Orikisawa, Kimitsu City. Scale shows 1 cm interval. (**C**) An outcrop example of inversely graded units (arrowed) and associated ripple cross-laminated and parallel-laminated units of a sandy contourite bed at Shimonohara, Misaki Town. Scale on the right-side show 1 cm interval. Each inversely graded unit (arrowed) has a gradational base and a sharp upper contact. (**D**) An outcrop example of internal erosion surfaces (arrowed) overlain by mudstone blankets in a sandy contourite at Oi, Ohara Town.

Fig. 6. (A) A ripple cross-laminated (c) and parallel-laminated (d) turbidite is overlain by mudstones (m) with a sharp contact (white arrowed) that coarsen upward to a ripple cross-laminated (r) sandy contourite at Kuniyoshi, Isumi Town. Scale is 1 cm interval. (B) A ripple cross-laminated (c) and parallel-laminated turbidite coarsens upward to a ripple cross-laminated (r) sandy contourite with a gradational contact at Kodaki, Misaki Town. Scale is 1 cm interval. (C) A ripple cross-laminated (r) sandy contourite gradationally rests on a ripple cross-laminated (c) and parallel-laminated (d) turbidite and is characterized by an erosional upper surface (arrowed) that is directly overlain by ungraded siltstones at Taito, Misaki Town. Lens cap is 6 cm in diameter. (D) Ungraded siltstones locally exhibit inverse grading (white arrowed) with very coarse silt and very fine sand grains at Yugura, Otaki Town. Scale is 1 cm interval.

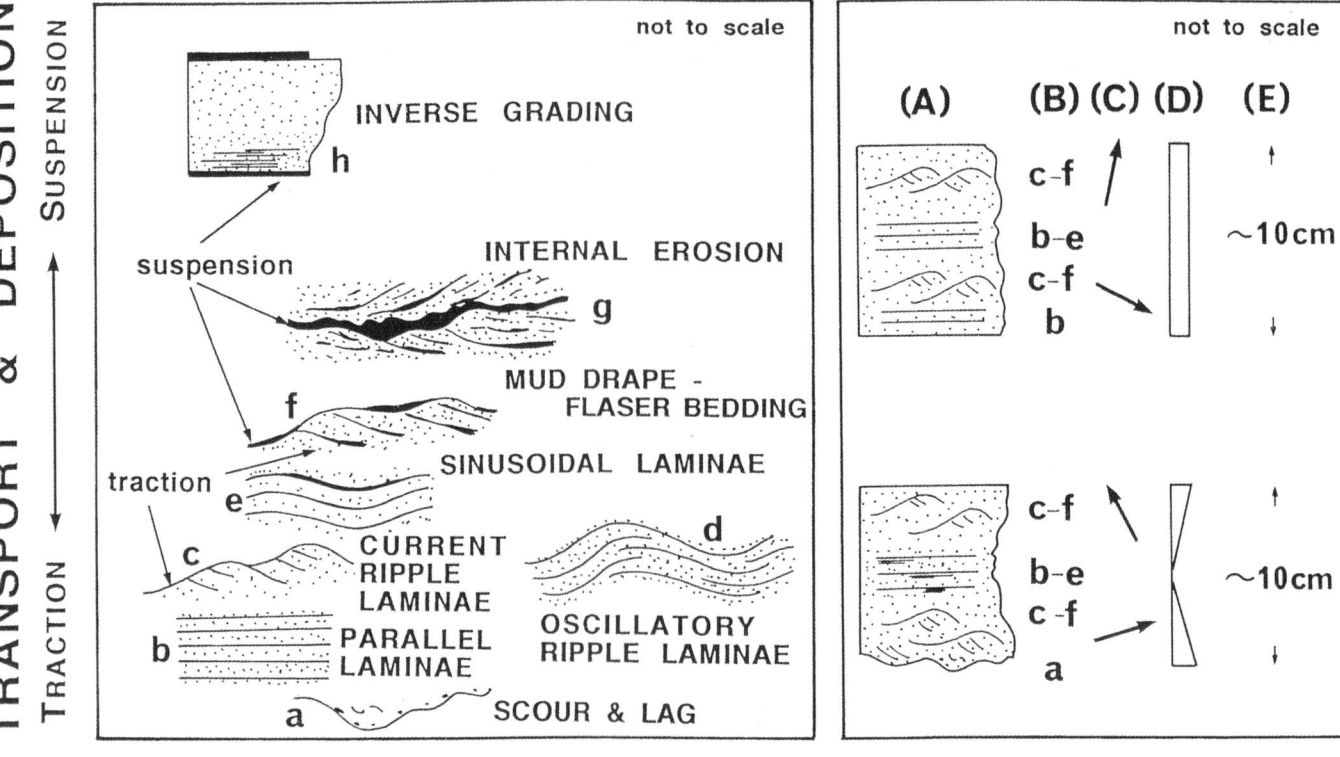

Fig. 7. Left: Major sedimentary structures characteristics of sandy contourites of the lower Kazusa Group in terms of their depositional processes and variability of palaeocurrent directions. Right: Two examples of vertical sequences of sedimentary structures of sandy contourites of the lower Kazusa Group.

also observed (Figs 5D and 7). In particular, ripple cross-lamination is commonly associated with mud drapes and flaser and lenticular bedding. In general, these sedimentary structures of sandy contourites do not display any regular vertical sequences (Ito 1996) (Fig. 7). Such features indicate oscillating traction and suspension deposition in a deep-sea environment and are interpreted to indicate fluctuation in bottom currents.

Palaeocurrents

Palaeocurrent directions were measured based mainly on dip directions of lamina planes of rippled sandstone beds. Palaeocurrent directions of sandy contourites are variable compared to those of associated turbidites (Fig. 8). In general, turbidites indicate northeastward to southeastward downslope transport in the Kazusa forearc basin. In contrast, some palaeocurrents of sandy contourites fall in a range of directions from north to north west, and other palaeocurrents fall in a range of directions from south to southwest (Fig. 8). These palaeocurrents indicate largely alongslope bottom currents, associated with some upslope- and downslope-directed palaeocurrents. In a sequence-stratigraphic framework of the lower Kazusa Group, no distinct relationship between palaeocurrent directions of sandy contourites and associated systems tracts was evident (Fig. 8).

Composition

Concentration of heavy minerals along laminae has been interpreted to be one of diagnostic features of sandy contourites as a result of multistage erosion and reworking of turbidites. However, sandy contourites of the lower Kazusa Group do not show any distinct concentration of heavy minerals compared to associated turbidite sandstones (Figs 3 and 9). Furthermore, sandy contourites are characterized by a framework composition similar to that of associated turbidites (Figs 9 and 10). These petrographic features of sandy contourites are interpreted to indicate that erosion and reworking of turbidites were not necessarily multistage events and winnowing and redistribution of sands and silts were not necessarily affected by bottom currents that completely modify petrographic features.

Lateral variation

Lateral variation in lithofacies features and palaeocurrent directions documents spatial variation in intensity and direction of deep-sea bottom currents in the Kazusa forearc basin. Lateral variation can be mapped within the framework of precise bed-by-bed correlation of sandstone beds. Figure 11 is an example of correlation of a sandstone bed of middle-fan and outer-fan

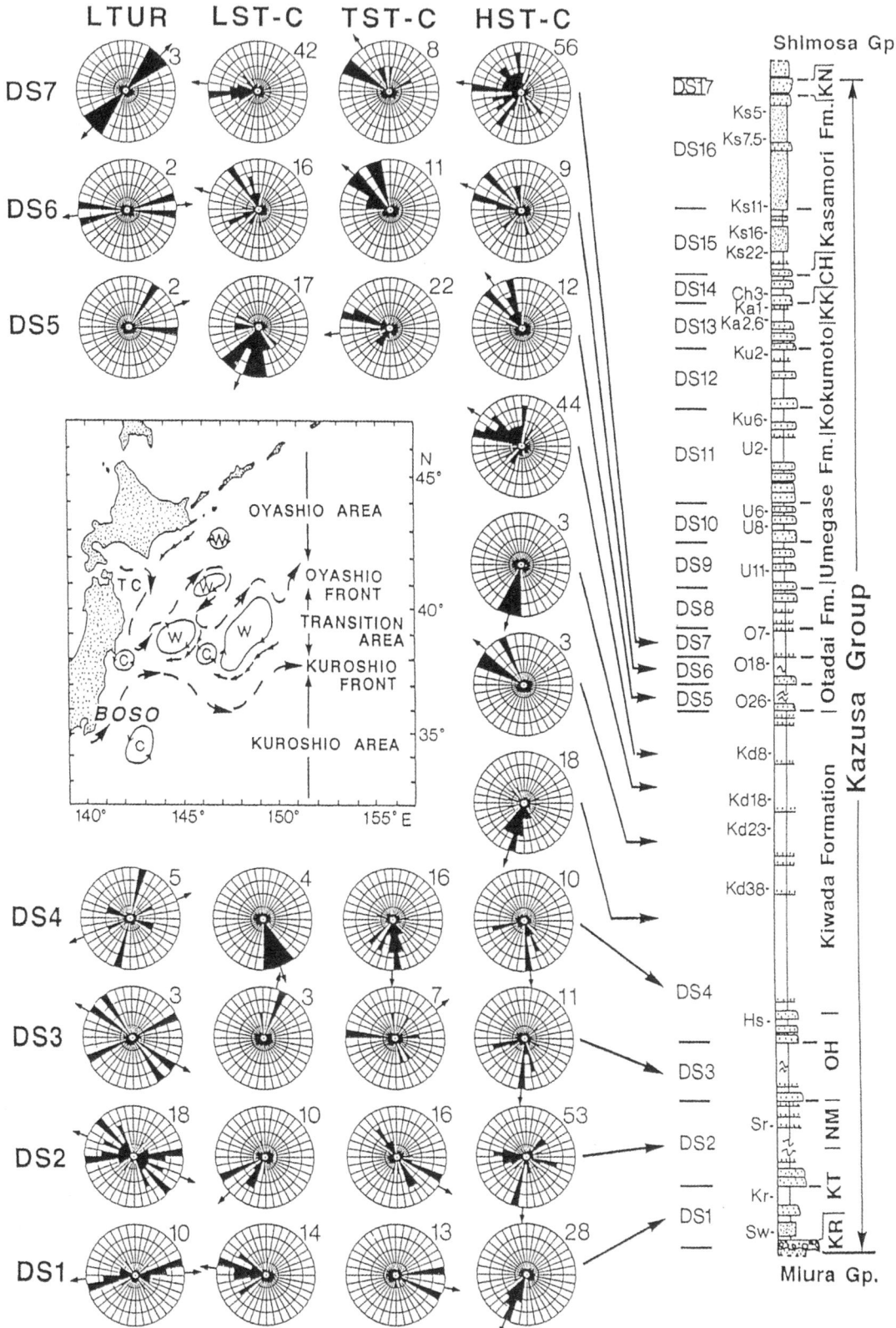

Fig. 8. Temporal variation in palaeocurrent directions of lowstand turbidites (LTUR) and sandy contourites (LST-C, lowstand; TST-C, transgressive; HST-C, highstand) (Modified from Ito 1996). Sequence-stratigraphic classification of the Kazusa Group is from Ito and Katsura (1992) and Ito (1998a). Rose diagrams of LTUR indicate flow lineation except for that of DS5. The number of readings is given in the upper right of each rose diagram. Arrows indicate vector means. North is upper for palaeocurrents. Inset map indicates modern surface current systems around the northern part of the Japanese Islands (TS, Tsugaru Current; C, cold eddy; W, warm eddy) (Simplified from Kawai 1972). KR, Kurotaki Formation; KT, Katsuura Formation; OH, Ohara Formation; KK, Kakinokidai Formation; CH, Chonan Formation; KN, Kongochi Formation. See Figure 1 for other explanations.

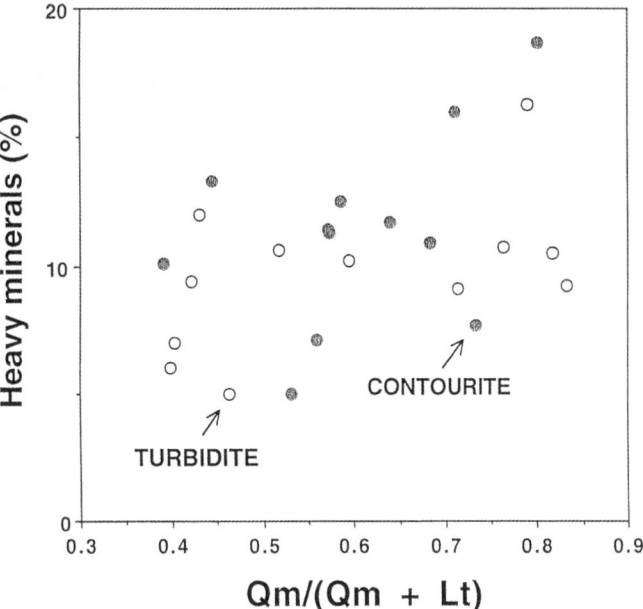

Fig. 9. Heavy mineral contents against the relative abundance of monocrystalline quartz grains (Qm) to total lithic fragments (Lt) of sandy contourites and associated turbidites of the lower Kazusa Group. Methods of sample preparation and point-counting for framework composition used in this study were the same as those of Ito (1994).

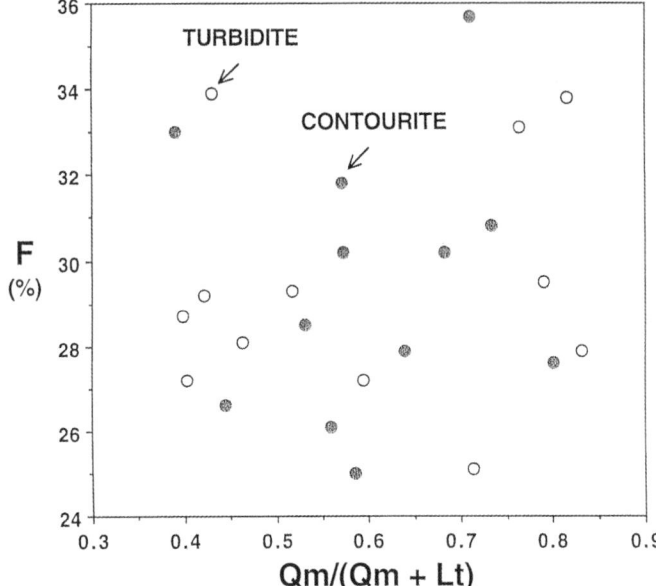

Fig. 10. Largely equivalent framework composition of sandy contourites and associated turbidites of the lower Kazusa Group. F, total feldspar grains; Qm, monocrystalline quartz grains; Lt, total lithic fragments.

deposits (Otadai Formation), illustrating lateral changes in bed thickness and the imprint of bottom current modification. A Type A bed in the middle-fan environment changes laterally into first a Type B-1 bed and then a Type C-3 bed, through a Type A bed again, toward the outer-fan environment. However, some Type A beds are persistent and do not show any modification, indicating that bottom currents were sometimes absent or very weak. Palaeocurrent measurements from the contourite part of beds of Types B and C are different from those measured from Type A turbidites and are laterally variable.

Discussion and conclusions

Better sorting and dominant alongslope palaeocurrent directions of sandy deposits associated with turbidite sandstones are interpreted to be a result of winnowing and redeposition of deep-sea beds by bottom currents (Stanley 1993). Ripple cross-lamination and parallel lamination contained in such reworked sandy deposits (sandy contourites) do not show any distinct vertical sequences such as the Bouma model and are interpreted to have been formed by tractional bottom currents. Reflected turbidites also have different types of vertical sequences of sedimentary structures (Pickering *et al.* 1989). These sedimentary structures in reflected turbidites, however, indicate reversed palaeocurrent directions and fine upward to hemipelagic mudstones. In contrast, the sedimentary structures described in this paper do not show such reversed palaeocurrents and have sharp upper contacts with overlying hemipelagites. Temporal and spatial fluctuation in speed of a turbidity current is also interpreted to develop different types of vertical sequences of texture and sedimentary structures in a deep-sea environment (Kneller 1995). According to this model, bypassing of finer grain populations may occur in a waning current that experiences either constriction of a flow as it pass around obstacles or acceleration due to an increase in slope of the sea floor. Large obstacles that divide the Kazusa forearc basin into sub-basin is not probable based upon the interpretation of depositional environments and geometry of the basin. Internal erosion surfaces and overlying blankets of mudstones, together with internal mud drapes, also are interpreted to indicate fluctuation in flow speeds for erosion and deposition in a deep-sea environment. Fluctuation in sedimentary processes in a deep-sea environment can be attributed to varying currents such as a long-term bottom current system (Stow 1994). Furthermore, variation in palaeocurrent directions across internal erosion surfaces also are observed and indicates varying flow conditions of a long-term bottom current system rather than a short-term turbidity current system.

The Kuroshio Current is one of the strongest surface currents flowing at velocities up to and at times greater than 180 cm s^{-1} (Taft 1978). Measurements of current speed in the deep-water mass of the Kuroshio Current off the Japanese Islands recorded speeds as high as 40 cm s^{-1} in 2000 m water depth (Fukazawa *et al.* 1985). Current speeds and directions in the deep-water mass of the Kuroshio Current are unstable, and the deep currents are not necessarily identical to those of the surface (Taira & Teramoto 1981). These variations in speed and direction of the deeper part of the Kuroshio Current are considered analogous to those responsible for the deposition of sandy contourites in the Kazusa forearc basin. Although thermohaline-driven deep-sea currents off the Japanese Islands such as the Western Boundary Undercurrent have potential for developing sandy contourites in the Kazusa forearc basin, this current, in general is located farther southeast and offshore of the Boso Peninsula, that is, along the Japan and Izu-Bonin trenches (Hallock & Teague 1996). A deep-sea tidal current is another alternative process for developing sandy contourites in the Kazusa forearc basin. Such deep-sea tidal currents as high as 40 cm s^{-1} in more than 2000 m water depth has been measured in the Sagami and Suruga Troughs (Taira & Teramoto, 1985). However, palaeocurrent directions of sandy contourites of the lower Kazusa Group are not bi-directional, but instead indicate both nearly alongslope and upslope or downslope-directed migration of small bedforms. Consequently, unstable speeds and directions of the deep-water part of palaeo-Kuroshio Current are the most likely causal mechanism of such variable lithofacies and scattered flow directions in the Kazusa forearc basin.

I thank S. Li, J-C. Faugères and D. A. V. Stow for their valuable comments on early versions of the manuscript. This research has been supported in

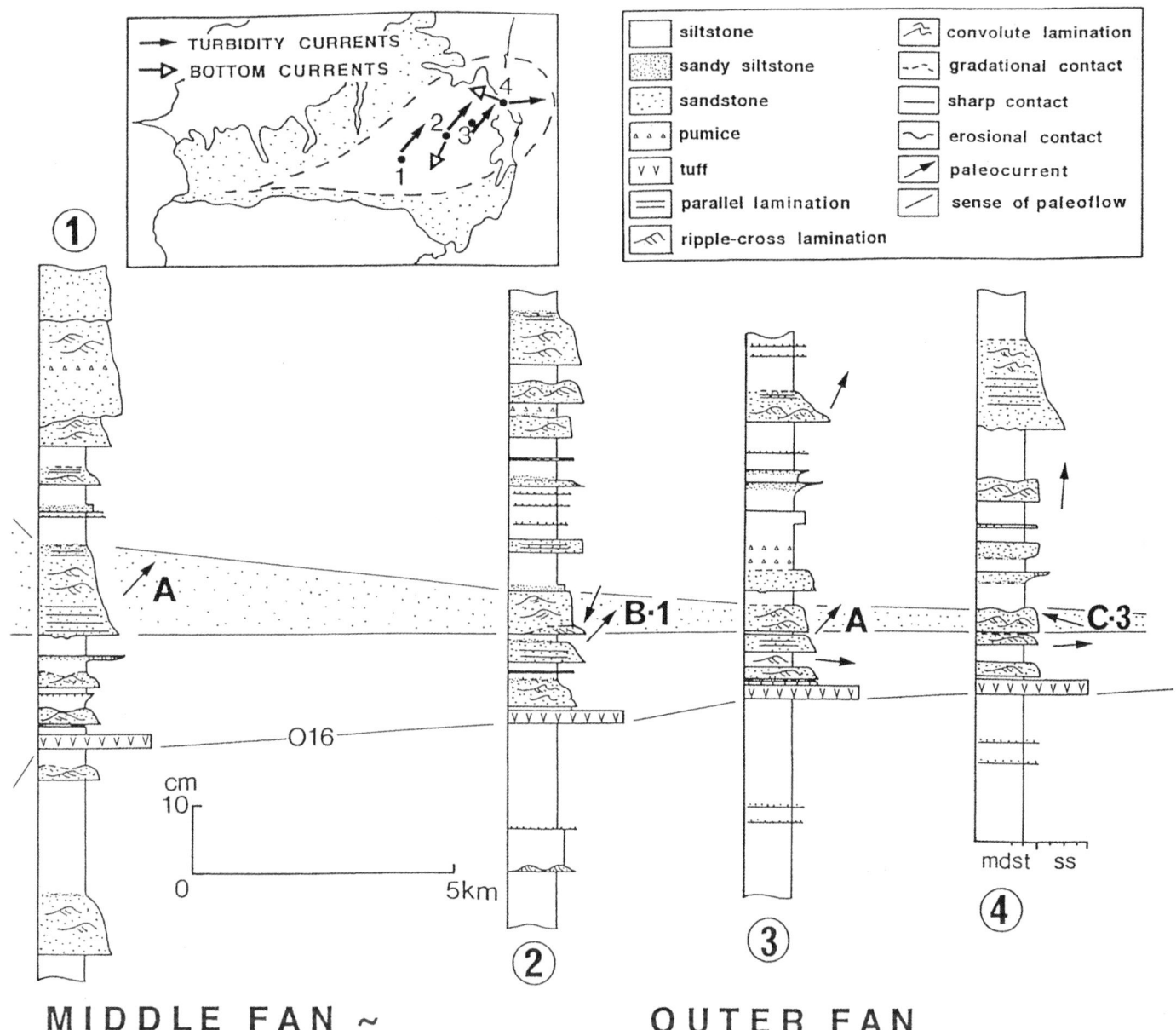

Fig. 11. An example of lateral variation in bed thickness and lithofacies of turbidite-to-contourite continuums of a sandstone bed of the lower Kazusa Group (Ito 1997). A, B-1, and C-3 are lithofacies types discussed in the text. Inset map shows locations of measured sections. O16 is the coding of a volcanic ash bed.

part by Grant-in-Aid for Scientific Research from the Ministry of Education, Science and Culture of Japan (grant 07640594) and from the Japan Society for the Promotion of Science (grant 12640437).

References

AOKI, N. 1968. Benthonic foraminiferal zonation of the Kazusa Group, Boso Peninsula. *Transactions and Proceedings of the Palaeontological Society of Japan*, New Series, **70**, 238–266.

BABA, K. 1990. *Molluscan Fossil Assemblages of the Kazusa Group, South Kwanto, Central Japan*. Tokyo, Keiyo Yochisha (in Japanese).

BOUMA, A. H. 1962. *Sedimentology of Some Flysch Deposits*. Elsevier, Amsterdam.

FUKAZAWA, O., TERAMOTO, T. ET AL. 1985. The Kuroshio Current in the Shikoku Basin. *In*: KAJIURA, K. (ed.) *Oceanic Characteristics and Their Changes*. Kouseisha-Kouseikaku, Tokyo, 89–120 (in Japanese).

HALLOCK, Z. R. & TEAGUE, W. J. 1996. Evidence for a North Pacific Deep Western Boundary Current. *Journal of Geophysical Research*, **101**, 6617–6624.

HIRAYAMA, J. & NAKAJIMA, T. 1977. Analytical study of turbidites, Otadai Formation, Boso Peninsula, Japan. *Sedimentology*, **24**, 747–779.

IGARASHI, A. 1994. Paleoceanographic changes during the deposition of the middle Pleistocene Kazusa Group, central Japan: Estimation based on the principal components analysis of planktonic foraminifera. *Journal of The Geological Society of Japan*, **100**, 348–359 (in Japanese with English abstract).

ITO, M. 1994. Compositional variation in depositional sequences of the upper part of the Kazusa Group, a middle Pleistocene forearc basin fill in the Boso Peninsula, Japan. *Sedimentary Geology*, **88**, 219–230.

ITO, M. 1996. Sandy contourites of the lower Kazusa Group in the Boso Peninsula, Japan: Kuroshio-Current-influenced deep-sea sedimentation in a Plio–Pleistocene forearc basin. *Journal of Sedimentary Research*, **A66**, 587–598.

ITO, M. 1997. Spatial variation in turbidite-to-contourite continuums of the Kiwada and Otadai Formations in the Boso Peninsula, Japan: A unstable bottom current system in a Plio–Pleistocene forea*Journal of Sedimentary Research*, **67**, 571–582.

ITO, M. 1998a. Submarine fan sequences of the lo Plio–Pleistocene forearc basin fill in *Sedimentary Geology*, **122**, 69–93.

ITO, M. 1998b. Contemporaneity of component units of the lowstand systems tract: An example from the Pleistocene Kazusa forearc basin, Boso Peninsula, Japan. *Geology*, **26**, 939–942.

ITO, M. & KATSURA, Y. 1992. Inferred glacio-eustatic control for high-frequency depositional sequences of the Plio–Pleistocene Kazusa Group, a forearc basin fill in Boso Peninsula, Japan. *Sedimentary Geology*, **80**, 67–75.

ITO, M. & MASUDA, F. 1988. Late Cenozoic deep-sea to fan-delta sedimentation in an arc-arc collision zone, central Honshu, Japan: sedimentary response to varying plate-tectonic regime. *In*: NEMEC, W. & STEEL, R. J. (eds) *Fan Deltas: Sedimentology and Tectonic Settings*. Glasgow, Blackie, 400–418.

KATSURA, Y. 1984. Depositional environments of the Plio–Pleistocene Kazusa Group, Boso Peninsula, Japan. *Science Report, Institute of Geoscience, University of Tsukuba, Section C, Geological Sciences*, **5**, 69–104.

KAWAI, H. 1972. Hydrography of the Kuroshio extension. *In*: STOMMEL, H. & YOSHIDA, K. (eds) *Kuroshio: Its Physical Aspects*. Tokyo University Press, Tokyo, 235–352.

KITAZATO, H. 1986. Paleogeographic changes in the South Fossa Magna region. *Earth Monthly*, **8**, 605–611 (in Japanese).

KNELLER, B. 1995. Beyond the turbidite paradigm: physical models for deposition of turbidites and their implication for reservoir prediction. *In*: HARTLEY, A. J. & PROSSER, D. J. (eds) *Characterization of Deep Marine Clastic Systems*. Geological Society of London, Special Publication, **94**, 31–49.

LOWE, D. R. 1982. Sediment gravity flows: II. Depositional models with special reference to the deposits of high-density turbidity currents. *Journal of Sedimentary Petrology*, **52**, 279–294.

MUTTI, E. 1992. *Turbidite Sandstones*. Milan, Agip S.p.A.

NISHIYAMA, H. 1990. Oceanic current systems around the Japanese Islands. *In*: The Oceanographical Society, Coastal Oceanography Research Committee (ed.) *Coastal Oceanography of Japanese Islands*. Supplementary Volume: Tokai University Press, Tokyo, 121–142.

PICKERING, K. T., HISCOTT, R. N. & HEIN, F. J. 1989. *Deep Marine Environments Clastic Sedimentation and Tectonics*. London, Unwin Hyman.

SHANMUGAM, G. 2000. Fifty years of the turbidite paradigm (1950s–1990s): deep-water processes and facies models – a critical perspective. *Marine & Petroleum Geology*, **17**, 285–342.

SHANMUGAM, G., SPALDING, T. D. & ROFHEART, D. H. 1993. Process sedimentology and reservoir quality of deep-marine bottom current reworked sands (sandy contourites): an example from the Gulf of Mexico. *American Association of Petroleum Geologists, Bulletin*, **77**, 1241–1259.

STANLEY, D. J. 1993. Model for turbidite-to-contourite continuum and multiple process transport in deep marine settings: examples in the rock record. *Sedimentary Geology*, **82**, 241–255.

STOW, D. A. V. 1994. Deep sea processes of sediment transport and deposition. *In*: PYE, K. (ed.) *Sediment Transport and Depositional Processes*. Oxford, Blackwell, 257–291.

STOW, D. A. V. & LOVELL, J. P. B. 1979. Contourites: their recognition in modern and ancient sediments. *Earth-Science Reviews*, **14**, 251–291.

STOW, D. A. V. & SHANMUGAM, G. 1980. Sequence of structures in fine-grained turbidites: comparison of recent deep-sea and ancient flysch sediments. *Sedimentary Geology*, **25**, 23–42.

STOW, D. A. V., FAUGÈRES, J.-C., VIANA, A. & GONTHIER, E. 1998. Fossil contourites: a critical review. *Sedimentary Geology*, **115**, 3–32.

STOW, D. A. V., READING, H. G. & COLLINSON, J. 1996. Deep Seas. *In*: READING, H. G. (ed.) *Sedimentary Environments and Facies*. Blackwell Scientific Publications, 380–442.

STOW, D. A. V., OGAWA, Y., LEE, I. T. & MITSUZAWA, K. 2002. Neogene contourites, Miura–Boso forearc basin, SE Japan. *In*: STOW, D. A. V., PUDSEY, C. J., HOWE, J., FAUGÈRES, J.-C. & VIANA, A. (eds) *Deep-Water Contourite Systems: Modern Drifts and Ancient Series, Seismic and Sedimentary Characteristics*. Geological Society, London, Memoirs, **22**, 409–419.

TAFT, B. A. 1978. Structure of Kuroshio south of Japan. *Journal of Marine Research*, **36**, 77–117.

TAIRA, K. & TERAMOTO, T. 1981. Velocity fluctuations of the Kuroshio near the Izu Ridge and their relationship to current path. *Deep-Sea Research*, **28A**, 1187–1197.

TAIRA, K. & TERAMOTO, T. 1985. Bottom currents in Nankai Trough and Sagami Trough. *Journal of Oceanographical Society of Japan*, **41**, 388–398.

UJIHARA, A. 1986. Pelagic gastropod assemblages from the Kazusa Group of the Boso Peninsula, Japan and Plio–Pleistocene climatic changes. *Journal of The Geological Society of Japan*, **92**, 639–65 (in Japanese with English abstract).

YAMADA, N., SAITO, E. & MURATA, Y. 1990. *Computer-generated geologic map of Japan*. Geological Survey of Japan, **Map Series 2**, Scale 1:2000 000.

Ordovician carbonate contourite drifts in Hunan and Gansu Provinces, China

SHUNSHE LUO[1], ZHENZHONG GAO[1], YOUBIN HE[1] & DORRIK A. V. STOW[2]

[1]*Department of Geology, Jianghan Petroleum University, Hubei, China*
[2]*Southampton Oceanography Centre, University of Southampton, Southampton SO14 3ZH, UK*
(e-mail: davs@soc.soton.ac.uk)

Abstract: The Early Ordovician continental margin at the southern edge of the Yangtze Platform, China, is represented by a succession of deep-water carbonate sediments near Jiuxi. A distinctive elongate mound-like form, some 120 km long, 25 km wide and 350–450 m thick, has been identified between shallow-water platform carbonates and deeper-water mudstones, and interpreted as the *Jiuxi* drift. The principal facies include calcilutites, calcisiltites, calcarenites, calcirudites, and bioclastic contourites, interpreted as contourites on the basis of their slope location, alongslope palaeocurrent indicators, features of traction flow processes coupled with intense bioturbation, and distinctive contourite sequences. During the Middle Ordovician, along the western slope of the Ordos Platform, Gansu Province, a narrow deep gateway (here called the *Pingliang gateway*) opened up between two deep-water basins. An elongate, mounded, carbonate body (70 km × 15 km × 150 m thick), developed along the eastern slope of this gateway, has been interpreted as the *Pingliang* contourite drift. A similar suite of contourite facies and sequences is found as at Jiuxi, although there is a greater abundance of calcarenitic contourites in the Pingliang drift, as well as other evidence that suggests a more active bottom current regime in this Ordovician gateway region.

Two examples of fossil carbonate contourite drifts have been described from Lower and Middle Ordovician sedimentary rocks in China. These are located, respectively, on an inferred palaeocontinental margin of the middle Yangtze Terrane, near Jiuxi in the Hunan Province of southern China, and on the western margin of Ordos Platform, in the Gansu Province of NW China (Fig. 1). They have been documented separately as the *Jiuxi* and *Pingliang* drifts in previous papers by the current authors and their associates (Duan *et al.* 1990, 1993; Gao *et al.* 1995, 1998).

The main objective of this paper, therefore, is to summarise the principal sedimentary characteristics of each drift, emphasising their broad similarity as well as certain differences that exist, in order to provide two further case studies of fossil contourite systems for this compilation (Tables 1 & 2). We then briefly discuss the evidence for their contourite interpretation, the nature of the contourite sequence, and the likely controls that influenced their deposition during the Ordovician period.

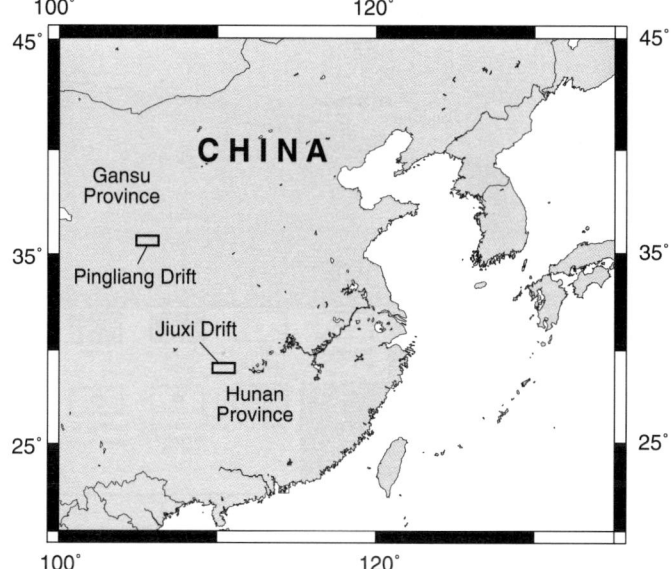

Fig. 1. China location map showing study areas in Hunan Province (Jiuxi Drift) and Gansu Province (Pingliang Drift).

Table 1. *Principal characteristics of Jiuxi drift*

Location:	Jiuxi, Taoyuan county, northern Hunan, China.
Setting:	Mid-slope of palaeocontinental margin of the S Yangtze Platform.
Age:	Early Ordovician (Tremadocian)
Drift type:	An elongate mounded drift extending along paleoslope
Dimensions:	Some 120 km long, 25 km wide, 350–450 m thick
Sediments:	Contourite facies – calcilutites, calcisiltites, calcarenites, calcirudites, and bioclastic contourites; associated facies – hemipelagites, pelagites and turbidites.

Table 2. *Principal characteristics of Pingliang drift*

Location:	Guanzhuang, Pingliang, Gansu province, China.
Setting:	The carbonate slope on the west margin of Ordos Platform.
Age:	Middle Ordovician (Llandeilo)
Drift type:	An elongate mounded drift extending along the palaeoslope.
Dimension:	Some 70 km long, 15 km wide, 150 m thick.
Sediments:	Contourite facies – calcilutites, calcisiltites and calcarenites; associated facies – hemipelagites, pelagites and turbidites.

Geological and stratigraphic setting

In the Early Ordovician, Jiuxi was part of the southern continental margin of the Yangtze microplate (Duan *et al.* 1988). It was a relatively stable passive margin, represented by the Yangtze shallow-water carbonate platform in the north, a broad slope region on which the drift developed, and a deep-water basin to the

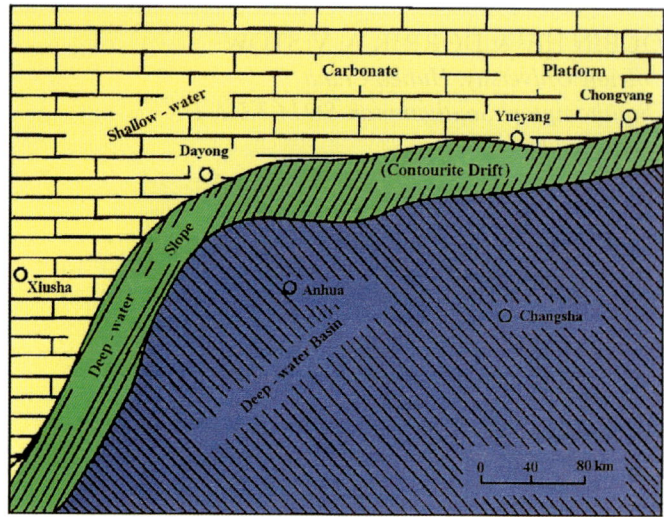

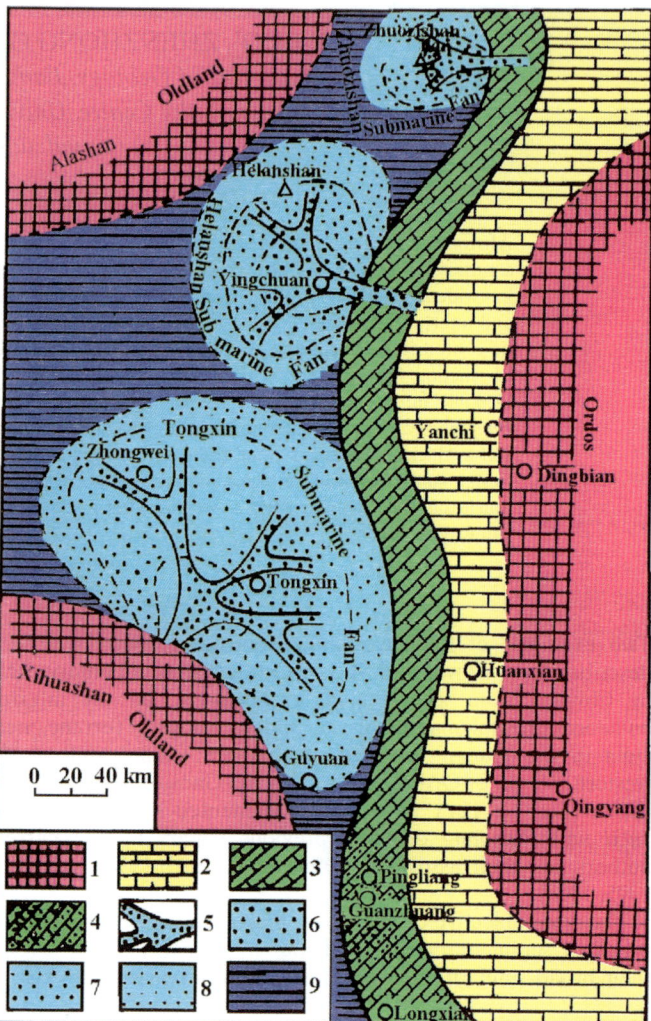

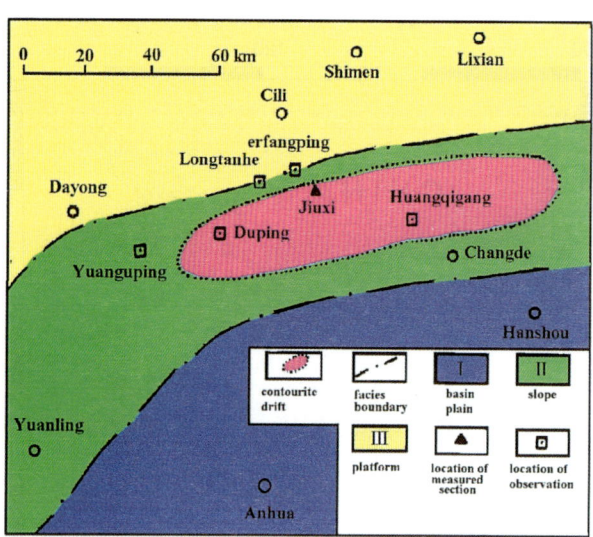

Fig. 2. (A) Early Ordovician palaeogeography showing regional facies distribution for the northern Hunan Province, China. (B) Detail for Jiuxi drift region.

south (Fig. 2). Previous work has shown that the area of Jiuxi contourite drift had been a deep-water environment dominated by gravity flow sedimentation in the Middle and Upper Cambrian, and remained so during the early Ordovician global sea level rise.

By contrast, the Pingliang area, located on the southwest slope of the Ordos Platform, was more active tectonically during deposition of the drift (Sheng 1986). In the Early Ordovician, the area was a shallow platform region, but this foundered during the early Middle Ordovician and rapidly developed into an expanding deep-water trough (Fig. 3). The basal Pingliang formation is dominated by gravity flow deposits, which pass upwards into quieter conditions with more pelagites, hemipelagites and accumulation of the contourite drift unit. There are no Late Ordovician deposits preserved, at least in part because the trough was in the process of closing and being uplifted.

Palaeotopography and palaeobathymetry

Sea-floor topography can exert an important influence on the velocity and flow pathway of bottom currents. The regional facies distribution and inferred palaeogeography of the early Ordovician

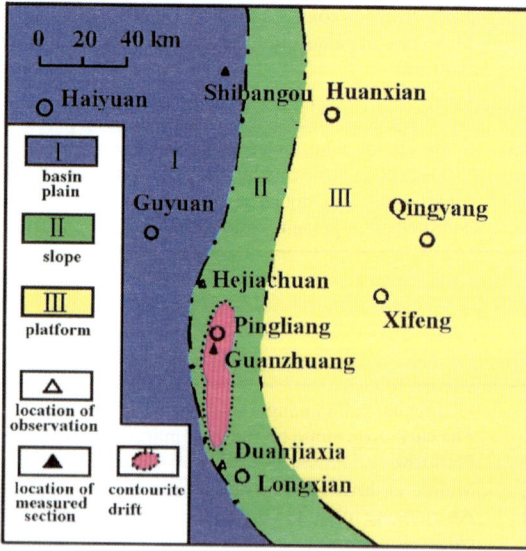

Fig. 3. (A) Middle Ordovician palaeogeography of the western margin of Ordos platform showing regional facies distribution: 1, Oldland; 2, Carbonate platform; 3, Carbonate slope; 4, Contourite drift; 5, Submarine channel; 6, Inner fan; 7, Middle fan; 8, Outer fan; 9, Margin of fan and basin plain; (from Gao *et al.* 1995). (B) Detail for Pingliang drift region.

northern Hunan Province (including the Jiuxi drift) is shown in Figure 2. From this, we may deduce that the palaeotopographic pattern was a relatively simple platform-slope-basin system, with a change in strike orientation of the slope from SW–NE west of Duping to more or less E–W in the area of drift development, as supported by limited current direction data (grain alignment, cross-lamination) from the contourite facies. There were no evident topographic barriers, although the change in strike of the slope may have resulted in the change of direction and velocity of contour current and hence led to drift accumulation.

The regional facies distribution and palaeogeographical map for the western margin of Ordos platform (including the Pingliang drift) in the middle Ordovician is shown in Figure 3. A shallow-water carbonate platform surrounded Ordos Oldland in this part, and was flanked by a carbonate slope leading down into a relatively narrow deep-water basin, the Helanshan–Qilanshan trough. Further platform areas, Xihuashan Oldland and Alashan Oldland formed partial barriers to the southwest and northwest, respectively. Further to the south of Ordos platform was Qinling trough, which was connected to the Helanshan–Qilanshan trough through a narrow gateway. Pingliang contourite drift was developed on the eastern slope of this gateway, perhaps under the influence of enhanced bottom flow through this topographic restriction. Limited current direction data from the contourite deposits suggest flow towards the north–northwest.

The successions in both Jiuxi and Pingliang contourite drifts, in addition to the contourites, contain a range of gravity flow deposits as well as pelagic and hemipelagic deposits. In Jiuxi the turbidites are mostly fine-grained, base-cut-out, distal deposits. The fossil assemblage within Jiuxi drift includes graptolites, trilobites, a few small brachiopods and cephalopods, together with a range of conodonts. These area represented by enconodonts and vacuole paraconodonts, such as *Cordylodus*, *Eoconodontus* and *Belodella*, which are typical deep-water conodont biofacies. Ichnofossils are common but monotonous, including *Glockeria* and a scribble grazing trace, which are also typical of deep-water. Deposits of the deeper-water basin are typified by dark grey shales with small micritic limestone lenses, and dominantly graptolite fossils. The palaeo-water depth of the basin has been inferred to be around 1200 m (Gao *et al.* 1985).

In the succession of Pingliang, the pelagites and hemipelagites are siliceous rocks, siliceous shales, shales and micritic limestone. Fossils include abundant graptolites as well as some radiolarians, sponge spicules and thin-shelled ostracods, all typical of deep-water environments. Gravity flow deposits include a range of turbidites and debrites, which suggest a gradually deepening basin towards the middle Ordovician. The palaeowater depth inferred by Mei *et al.* (1982) was about 1000 m. Both drifts are therefore considered as mid-slope drifts, following the terminology of Stow *et al.* (1998), and the Pingliang drift to be closely related to a palaeo-gateway.

Sediment facies

As outlined above, the range of facies recovered in both areas are typical of deep-water successions the world over (Pickering *et al.* 1989; Stow 1994; Stow *et al.* 1996). We focus here only on those facies interpreted as contourites. For both drifts, the principal contourite facies (Fig. 4) are all carbonate-rich and include:

Calcilutites: homogeneous, bioturbated, silty muds, with minor silt laminae.

Calcisiltites: mottled, bioturbated, alternation of silty and muddy grade irregular laminae.

Calcarenites: bioturbated, muddy to silty sand grade, irregular laminae and cross-laminae.

More minor facies types encountered in the Jiuxi drift only are:

Calcirudites: coarse-grained, mud-free facies, lenticular beds, with sharp, irregular upper contacts.

Bioclastic contourites: coarse-grained (mainly sandy), dominant bioclasts, irregular lenticular beds. The main sedimentary characteristics of these facies are summarised in Tables 3 and 4. Some of the features that are interepreted as most diagnostic of contourites are discussed further below.

Sedimentary structures (Fig. 4)

Jiuxi and Pingliang drifts have closely similar sedimentary structures in the dominant fined-grained contourite facies, calcilutites and calcisiltites, but show certain differences in the calcarenitic facies. Bioturbation and burrowing is the most common feature of the calcilutites. Most of the burrows are horizontal, some oblique or vertical, and are filled with calcisilts, micrite or sparite cements; their diameters are 1–5 mm and their observed lenghs from several mm to more than 10 mm. Bioturbation creates a generally chaotic or mottled aspect, although some of the irregularity of beds and abrupt contacts of laminae may be primary rather than secondary features. Rare calcisilts tend to be concentrated into discontinuous thin laminae (1–3 mm thick), with sharp or gradational upper and lower contacts. Bedding is not sharply defined and there are no regular sequences of structures nor clear grading.

The calcisiltitic contourite facies also shows common bioturbation throughout, and an irregular mottled appearance in some cases. Where primary structures are preserved, they consist of alternating laminae and thin beds of calcisiltite and micritic limestone on a cm-scale. Boundaries between lithologies range from sharp to gradational. Some small-scale ripples can be found in calcisiltitic contourite facies, and the dip direction of laminae is parallel to the strike of the palaeoslope.

The calcarenitic contourites of the Jiuxi drift typically exhibit an irregular, subparallel stratification on a mm-cm scale, and flat to irregular boundaries. Contacts are both sharp and gradational, with the top contacts more often the abrupt ones. There are commonly small lenses, irregular streaks and some thin irregular beds, which are locally coarser-grained and well-sorted. Fabric analysis shows very significant current features, the long-axes of grains are coincident with the strike of the regional palaeoslope as determined from regional palaeogeographic analysis. Vertically, there are marked changes of grain size, and both normal and reverse grading are very common. Bioturbation and distinct burrow traces are common throughout.

In the Pingliang drift, the calcarenitic contourites tend to be more abundant, and with more common intervals of relatively well-preserved cross-lamination. The dip direction of cross-laminae are coincident with the strike of the regional palaeoslope. Lensoid dissolution surfaces and stylotitic structures are also common in some thick-bedded contourite units, and these tend to lack any primary structures. Thick-bedded contourites commonly developed stylotitic structures, cross-lamination is not obvious. Vertically, there are changes of grain size, and both normal and reverse grading are common.

Texture and composition (Fig. 4)

The principal facies in both drifts are the finer grained carbonate-rich contourites, calcilutites and calcisiltites. The former are generally poorly sorted, whereas the latter are from poorly to moderately well sorted. The calcarenite facies are also poorly to moderately well sorted in Jiuxi, but better sorted in the Pingliang drift, suggesting a greater degree of winnowing and removal of fines. General grain-size parameters for the different facies as well

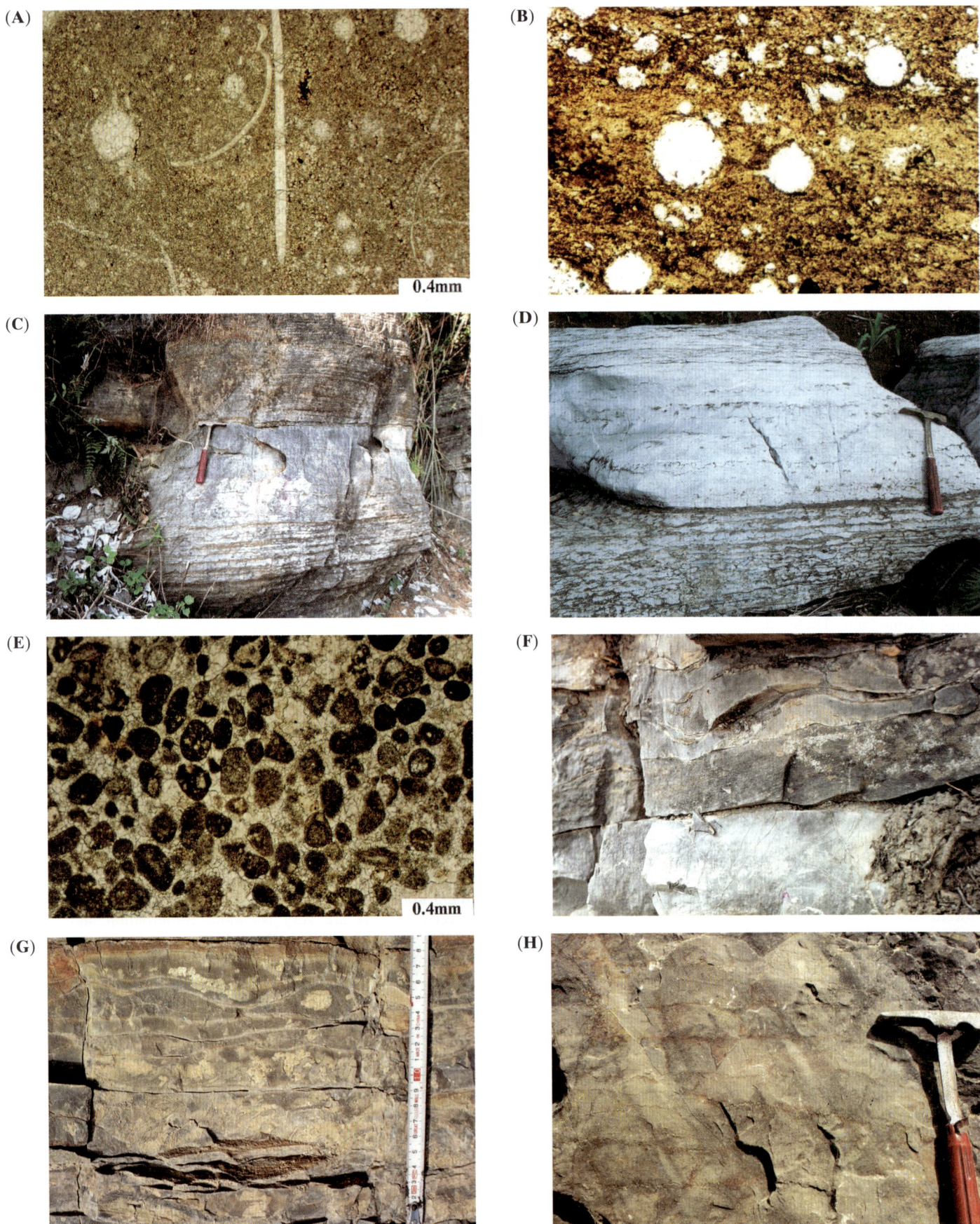

Fig. 4. Typical contourite facies and microfacies from the Jiuxi and Pingliang drifts. (**A**) Calcilutite microfacies, showing calcispheres, sponge spicules and thin-shelled ostracods (Jiuxi). (**B**) Siliceous calcilutites with radioarians (Jiuxi). (**C**) Part of standard contourite sequence showing mottled calcisilts grading up into calcarenites, and then a sharp upper contact back into mottled calcisilts and calcilutites at the very top (Jiuxi). (**D**) Part of contourite sequence with mottled calcisilts and indistinctly layered calcarenites (Jiuxi). (**E**) Calcarenite microfacies (Pingliang). (**F**) Thin-bedded calcarenite with irregular layering, lenticular bedding and dissolution horizons (Pingliang). (**G**) Thin-bedded calcarenite with irregular layering, lenticular bedding and poorly preserved cross-lamination (Pingliang). (**H**) Thick-bedded calcarenite with sylolites (Pingliang).

Table 3. *Main characteristics of contourite facies in Jiuxi drift*

	Calcilutitic contourite	**Calcisiltitic contourite**	**Calcarenitic contourite**	**Calciruditic contourite**	**Bioclastic contourite**
Structure	homogeneous, bioturbated, + distinct burrows; thin silt laminae – discontinuous; rare cross-stratification	irregular thin bedding/lamination Sharp-gradational contacts, bioturbation and distinct burrows, rare cross-strat	irregular to lenticular thin beds and laminae, erosive surfaces very common, normal-reverse grading common, sharp-gradational contacts, bioturbation and burrows common	beds with erosive lower and abrupt upper contacts	lenticular bedding, some cross-laminae, basal contacts commonly erosive and top surfaces flat or wavy
Texture	silty mud, poorly sorted, ungraded	muddy silt, moderate-well sorted	silt to sand-sized, poor to well sorted, grains subrounded and bioclasts broken	sand-gravel-sized moderate-well sorted, no clay fraction, fair to good roundness	sand and silt sized; sorting moderate, roundness very variable
Fabric	unclear	unclear	grain orientation parallel to the bottom current	grain orientation parallel to the bottom current, some imbrication	
Composition	mixed biogenic and terrigenous material, quartz and deep-water bioclastics	biogenic material from pelagic and resedimented sources	mixed biogenic and terrigenous quartz sands and pelagic-benthic biogenic material	biogenic material from resedimented sources	mixed biogenic and terrigenous biogenics from pelagic, benthic resedimented sources
Sequence	decimetric cycles of grain-size with calcisiltitic and calcarenitic contourite	decimetric cycles of grain-size with calcilutitic and calcarenititic contourites	decimetric cycles of grain-size with calcisiltitic and calcilutitic contourites	does not occur within standard cyclic sequence	non-standard sequence, interbedded with calcilutite

Table 4. *Main characteristics of contourite facies in Pingliang drift*

	Calcilutitic contourite	**Calcisiltitic contourite**	**Calcarenitic contourite**
Structure	Dominantly homogeneous, bioturbational mottling common, irregular silt laminae	Generally bioturbated, some ripples with a set height of several cm	Thin-irregular bedding, some parallel and cross-lamination preserved; stylotitic structures well developed in thicker beds normal and inverse grading common
Texture	Dominantly silty mud; calcisilt 5–40%, sorting poor to medium	Dominantly silt sized; sorting medium to good; mostly matrix is micritic, rarely sparitic	Sand-sized, may be relatively free of mud and well sorted; tendency to low or negative skewness
Fabric	Unclear	Indication of long grain orientation parallel to the bottom current	Indication of long grain orientation parallel to the bottom current
Composition	Dominantly biogenic material from pelagic and resedimented sources	Dominantly biogenic composition	Dominantly biogenic composition, more rarely with quartz sands, biogenic material from benthic and resedimented sources
Sequence	Typically arranged in decimetric cycles of grain-size with calcisiltitic and calcarenitic contourites	Typically arranged in decimetric cycles of grain-size with calcisiltitic and calcarenitic contourites	Typically arranged in decimetric cycles of grain-size with calcisiltitic and calcarenitic contourites

Table 5. *Textural and compositional attributes of contourites from Jiuxi (J) and Pingliang (P) drifts*

FACIES		Clay-size %	Silt-size %	Sand-size %	Gravel-size %	Bioclasts %	Clastics %
Calcilutite	J	45–65	2–20			2–20	3–30 (clays)
	P	50–70	5–40			2–10	0–10
Calcisiltite	J	30–55	40–60	0–5		2–10	Quartz
	P	30–55	40–60	0–5		2–10	Quartz
Calcarenite	J	10–40	10–30	40–60 +spar			8–20 (qtz)
	P	0–5	10–40	40–60 +spar			
Calcirudite	J	0–3	5–10	15–20	Approx. 75		0–5 (qtz)
Bioclastic contourite	J	5–10	0–10	10–20	> 70		0–5 (qtz)

Table 6. *Grain-size parameters for Piangling drift calcarenitic contourites*

Sample number	Mean grain-size (M_Z)	Standard deviation (σ_1)	Skewness (Sk_1)	Kurtosis (K_G)
91-B_{70}(1)	1.55ϕ	0.596	0	0.982
105-B_{79}	2.17ϕ	0.777	0.007	0.942
107-B_{81}	1.75ϕ	0.65	0.014	0.967
110-B_{83}	1.61ϕ	0.683	0.008	0.972
110-B_{84}	1.55ϕ	0.549	0.003	0.976
115-B_{85}	2.09ϕ	0.565	0.006	0.983
115-B_{87}	1.71ϕ	0.640	0.019	0.981
115-B_{90}	1.74ϕ	0.724	–0.002	0.952
115-B_{91}	1.55ϕ	0.797	0.005	0.976
127-B_{95}	1.47ϕ	0.726	–0.029	0.971
128-B_{96}	1.34ϕ	0.790	0.002	0.972
129-B_{97}	2.49ϕ	0.486	0.012	0.998
Mean value	1.74ϕ	0.665	0.004	0.973
	Mainly medium sand-size	Good sorting	Symmetrical	Normal distribution

Note: These parameters have been calculated according to the formula of Folk & Ward (1957)

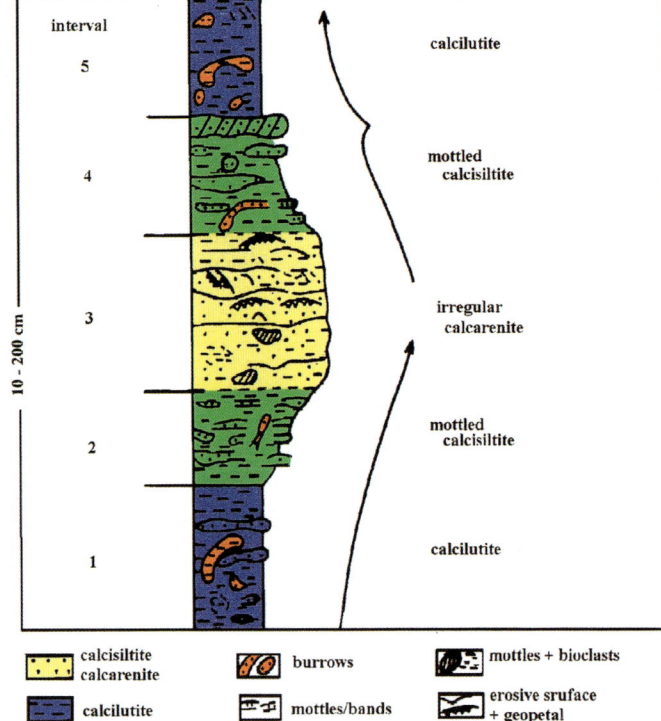

Fig. 5. Typical contourite sequence showing coupled negative to positive grading from the Jiuxi contourite drift (from Duan *et al.* 1993).

as for the winnowed calcarenitic contourites are given in Tables 5 and 6. The calcirudite and bioclastic facies, restricted to Jiuxi drift, also appear to be have had the fine fraction largely or completely removed, but both show a bimodal rather than well-sorted grain size distribution. The coarser fraction comprises micrite intraclasts and bioclastic debris respectively, and the finer fraction carbonate silt to fine sand. Where quartz sand is present in any of the coarse-grained facies, it is mostly well-rounded, suggesting either long distance transport and/or a long period of reworking.

The compositional attributes of the different facies for each drift system are summarized in Tables 3 and 4. Clearly, carbonate is the dominant component of all facies, and includes both biogenic debris (planktonic, benthic and resedimented) and a non-biogenic or indeterminate fraction. Fine micritic cement dominates, but sparite may be present in the calcarenites and coarser facies. Recognizable fossils include trilobites, graptolites, brachiopods, crinoids and cephalopods, many of which are fragmented and probably reworked, as well as some radiolarians, ostracods and sponge spicules. The terrigenous fraction is variable up to about 30% in some samples, mainly comprising clays and quartz.

Facies distribution

Contourite sequences

Based on detailed observation of outcrops, the most common small-scale facies arrangement in both Jiuxi and Pingliang drifts comprises the five standard contourite divisions in a coupled negative to positive, symmetrically graded sequence (Fig. 5). These divisions are, from base to top:

(1) calcilutite – bioturbated, burrowed, homogeneous to indistinctly laminated;
(2) calcisiltite – bioturbated, burrowed, mottled, irregularly banded, rare cross-laminae;
(3) calcarenite – bioturbated, mottled, irregular and discontinuous laminae/lenses, some cross-lamination preserved;
(4) calcisiltite – bioturbated, burrowed, mottled, irregularly banded, rare cross-laminae;
(5) calcilutite – bioturbated, burrowed, homogeneous to indistinctly laminated.

The contacts between divisions varies from gradational to sharp or erosive, and the thickness of the complete sequence is typically 30–80 cm, locally up to 2 m.

Incomplete sequences are also common in both drifts, the most typical of these being an absence or marked reduction in thickness of the calcisiltite divisions 2 and 4 (Fig. 6). In the Pingliang drift, there are further intervals that consist only or very largely of

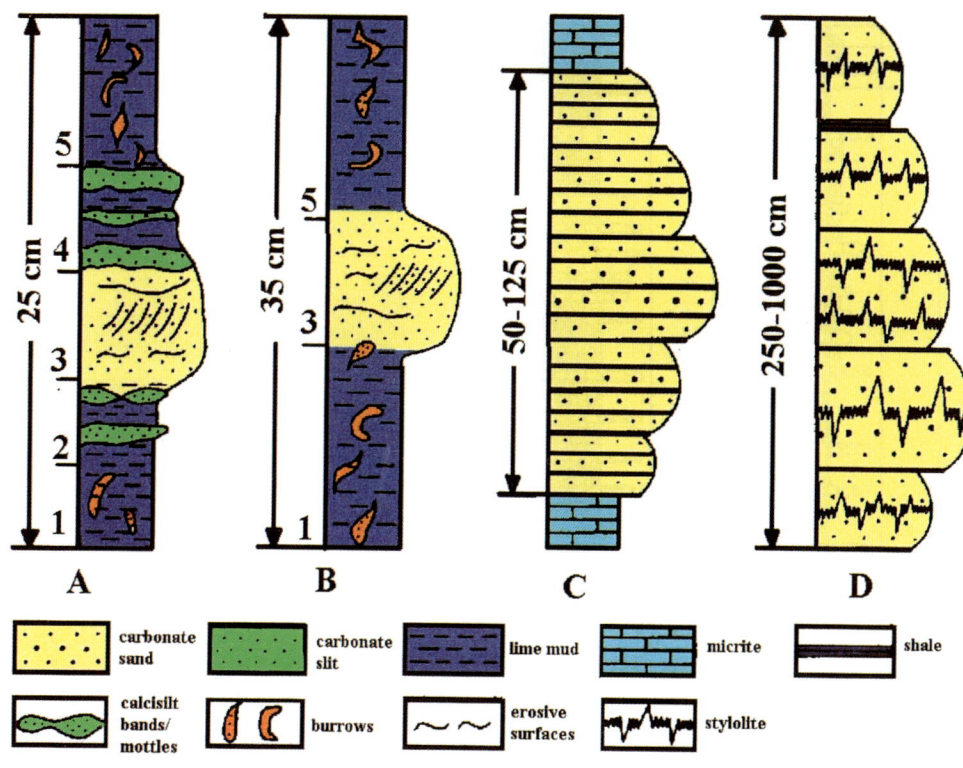

Fig. 6. Contourite sequences from the Pingliang contourite drift. (**A**) complete contourite sequence, but with attenuated calcisilt divisions; (**B**) incomplete contourite sequence without calcisilt divisions; (**C**) sequence dominated by medium-bedded calcarenitc contourites; (**D**) sequence consisting only of thick-bedded calcarenitc contourites with abundant stylolites; (from Gao *et al.* 1995).

calcarenitic contourites. Based on the thickness of individual calcarenite layers and their internal characteristics, two sequence types can be identified. The first consists mainly of medium-bedded calcarentic contourites. Individual beds are 10–25 cm thick. A somewhat irregular parallel layering is well developed and displayed by an alternation of algal clast-rich and agal clast-poor calcisiltitic–calcarenitic beds. In vertical succession, each calcarenite bed shows symmetrical grading from fine-coarse-fine. These beds are themselves associated in 50 to 125 cm thick vertical successions which also display a fine-coarse-fine trend. The second sequence type consists mainly of thick-bedded calcarenitic contourites. Individual beds of sparry calcarenite are 0.5–2 m thick and display a fine-coarse-fine grain-size trend, but with the positively graded interval usually thicker than the negatively graded interval. Styolites are common within beds. In some cases, thin micritic limestone or shale interbeds are present between calcarenite beds, probably as dissolution horizons, but mostly calcarenite beds are in sharp contact with each other. Individual beds are stacked into fine-coarse-fine successions.

Distribution of sequences

Jiuxi and Pingliang drifts are both elongate mounded bodies extending along the palaeocontinental slope for 120 km and 70 km respectively. However, no significant differences in contourite facies or sequence type have been noted regionally, and no marked proximal to distal trends have yet been observed. The drift thicknesses range up to about 450 m and 150 m respectively. In the Jiuxi drift, the contourite sequences are almost uniformly distributed throughout the succession at approximately 3–4 m intervals, with a greater proportion of the finer-grained facies throughout (Fig. 7). Whereas in the Pingliang drift, the distribution is less regular and shows an asymmetric coarsening-upward to fining-upward sequence (Fig. 8). The lower part of the succession comprises mainly pelagites, hemipelagites, fine-grained turbidites and rare contourites. Upwards, the proportion of calcisiltite and calcarenites increases, somewhat patchily distributed within a dominant background of calcilutite contourites. Towards the upper part of the drift succession, calcarenitic contourites become dominant, forming between 80% and 90% of the sequence. This proportion decreases slightly to between 50% and 60% near the very top of the drift.

Sedimentation rates and cyclicity

Based on the maximum thickness of each drift and the estimated duration of the time interval over which construction continued, it is possible to calculate average rates of sedimentation for the Jiuxi drift as 3.5–4.0 cm ka^{-1}, and for the Pingliang drift as 1.1–1.5 cm ka^{-1}. Although these rates must be taken as very approximate, they do illustrate a threefold difference between the two drifts. The regular distribution of sequences throughout the Jiuxi drift further allows an estimate of the average periodicity of one contourite sequence at around 200 000 years.

Discussion

Evidence for contourites

Recognizing the controversy that exists concerning the recognition of fossil contourites, it is important first to justify our interpretations by following the criteria and three-stage approach set out by Stow *et al.* (1998).

(1) At the *large scale,* palaeogeographical reconstructions suggest a continental margin setting for the Jiuxi drift and a gateway setting between two larger marine basins for the Pingliang drift. Both are therefore compatible with drift development. We cannot comment on the likely palaeocirculation patterns for either period or region, but note that the Ordovician period has been implicated before for fossil contourites (Faugères & Stow 1993). In addition, the late Ordovician has been recognized as a likely time for either global or partial glaciation, so that during the period of drift growth in the middle and early Ordovician, there would have been the potential for marked climate differences between polar and

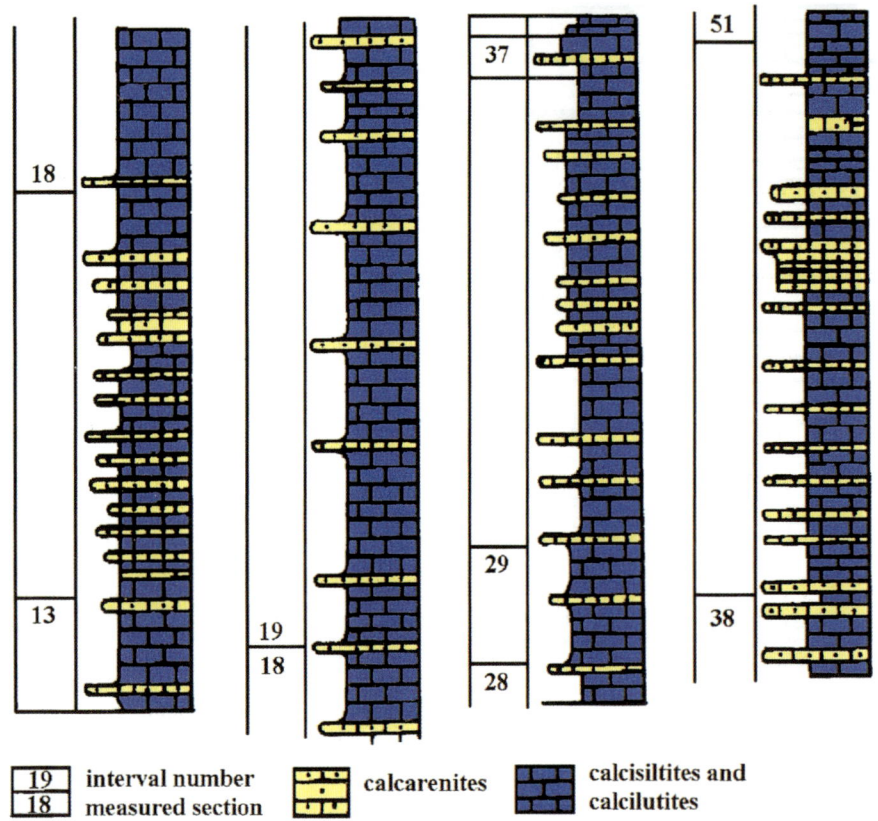

Fig. 7. Measured section through part of the Jiuxi contourite drift, showing vertical distribution of contourite sequences (from Duan et al. 1993).

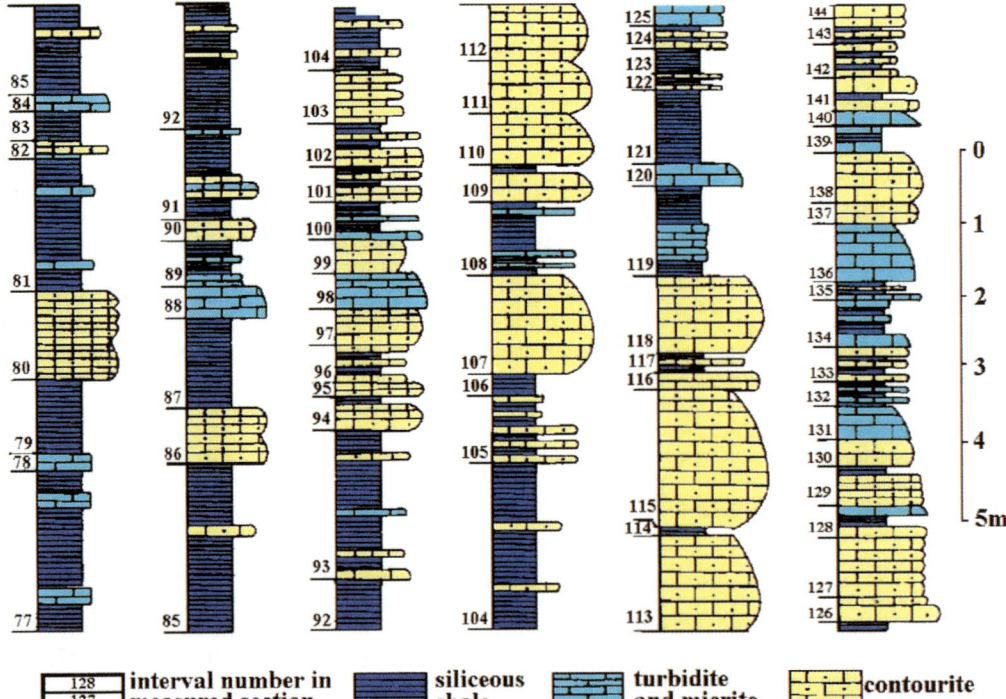

Fig. 8. Measured section through part of the Pingliang contourite drift, showing vertical distribution of contourite sequences (from Gao et al. 1995).

equatorial regions, and hence for the initiation of thermohaline bottom-water circulation.

(2) At the *medium scale*, both the associated facies and fossil evidence support relatively deep-water environments (i.e. mid-slope depths around 1000 m). The overall geometries for the inferred contourites in each region show regional thickening of the contourite succession that demonstrates an along-slope, elongate, mounded form. These elongate bodies are contained within slope to basinal facies or, in the case of the Pingliang drift, marked by an upper unconformity. This unconformity overlies an apparently condensed, calcarenite-dominated, succession, and may therefore have included a regional hiatus related to bottom current erosion. However, the duration of the unconformity and its relationship to the longer-term tectonic evolution of the region, makes it impossible now to recognise any original hiatus.

(3) At the *small-scale*, the general sedimentary characteristics are typical of those described from many modern drift systems. Burrowing and bioturbation are very common throughout indicating relatively slow and continuous sedimentation rather than episodic events. Some evidence of irregular parallel and local cross-lamination is preserved, as well as irregular concentrations and lenses of coarser material. Any current directional data obtained indicate alongslope currents. The composition is dominantly carbonate-rich with a varied assortment of bioclastic grains of pelagic, benthic and resedimented origin. Some admixture of terrigenous clays and rounded quartz grains occurs. Typical contourite sequences occur over 10–200 cm of section (Stow *et al.* 1986).

In summary, we can be confident in concluding that these sediments are true examples of fossil contourites from the Ordovician period. They are quite typical of other carbonate-rich fossil contourites described to date (Bein & Weiler 1976; Pickering *et al.* 1989; Kahler & Stow 1998; Stow *et al.* 1998).

Sequence interpretation

The characteristics of the vertical sequences described above are completely different from those of turbidites, tempestites or any other event deposit (Aigner 1985; Pickering *et al.* 1989; Stow *et al.* 1996). The standard, symmetrical contourite sequence may reflect regular periodicity in mean bottom current velocity, in turn related to climatic variation or another external factor. This periodicty averages 200 000 years for the Jiuxi drift, and the relative proportion of calcarenite facies is around 10%. However, variation in the nature and rate of sediment input cannot be wholly excluded as a contributary factor, although the mixed composition of the calcarenites and common micritic clasts, do not favour a mainly upslope or shelfal source of material for the drift construction.

Within the Pingliang drift, the successions consisting mainly or entirely of calcarenites are interpreted as representing prolonged periods of high-energy bottom currents. Grain-size variation also indicates probable fluctuation in current intensity, but the general absence of micrite or lutite components suggests that significant winnowing and erosion were persistent. Overall, the calcarenite facies makes up around 36% of the section, and more towards the upper parts. Clearly bottom current intensity was greater through the Pingliang gateway than over the Jiuxi slope region. This would also explain the relatively lower average rates of sedimentation for the Pingliang drift.

Factors influencing drift growth

Considering the nature of the two Ordovician drift systems described, we can recognize the influence on their development of several of the key factors noted as significant for modern drifts (e.g. Faugères *et al.* 1993; Stow *et al.* 1996). (1) *Sea-floor topography*. The mid-slope location and variation of slope trend in the palaeo-Hunan region probably led to initial development of the Jiuxi drift in that area. Flow constriction through the Pingliang gateway would have intensified bottom currents leading to deposition of the Pingliang system as a channel-related lateral drift, subject to periodic current intensification, winnowing and erosion. (2) *Sea-level highstand*. Eustatic sea-level was relatively high through much of the Ordovician, partly as a result of rapid sea-floor spreading (Worsly *et al.* 1984). This probably favoured drift development rather than downslope processes for the Jiuxi region. (3) *Tectonic activity*. For the Pingliang drift, by contrast, it was clearly tectonic development of the region rather than eustatic sea-level that created the necessary deep-water gateway setting for drift formation. (4) *Sediment supply*. For both drifts, there was an adequate supply of mainly carbonate material throughout the period of drift growth. The source of this material included pelagic supply, benthic and resedimented bioclastics, as well as alongslope winnowing and reworking. Terrigenous supply was limited but present.

Many colleagues are thanked for their contribution to field and laboratory investigations, as well as to the discussion and development of our ideas. The support of technical staff at our respective institutes is also recognized. Earlier work on the Jiuxi drift by T. Duan is fully acknowledged, as well as the work of regional geological teams in both areas. The senior author (SL) acknowledges support of a Jianghan Petroleum University fellowship, which allowed him to work for a six-month period with DAVS at Southampton University during completion of this manuscript. J. Howe and M. Ito provided very helpful reviews of an earlier version of this manuscript.

References

AIGNER, T. 1985. *Storm Depositional Systems*. Lecture Notes in Earth Sciences, Springer Verlag, Berlin.

BEIN, A. & WEILER, Y. 1976. The Cretaceous Talme Yafe Formation, a contour current shaped sedimentary prism of calcareous detritus at the continental margin of the Arabian Craton. *Sedimentology*, **23**, 511–532.

DUAN, T. Z., GAO Z. Z., ZENG, Y. F. & STOW, D. A. V. 1993. A fossil carbonate contourite drift on the Lower Ordovician palaeocontinental margin of the middle Yangtze Terrane, Jiuxi, northern Hunan, southern China. *Sedimentary Geology*, **82**, 271–284.

DUAN, T. Z., GUO, J. H., GAO, Z. Z., LI, Z. H. & ZENG, Y. F. 1990. A Lower Ordovician carbonate contourite drift on the margin of the South China palaeocontinent in Jiuxi, northern Hunan. *Acta Geologica Sinica*, **2**, 133–143. (in Chinese with English abstract).

DUAN, T. Z., ZENG, Y. F. & GAO, Z. Z. 1988. Analysis of the tectonic evolution of a paleocontinental margin in South China based on sedimentary history. *Oil and Gas Geology*, **9**(4), 410–420. (in Chinese with English abstract).

FAUGÈRES, J.-C. & STOW, D. A. V. 1993. Bottom-current-controlled sedimentation: a synthesis of the contourite problem. *Sedimentary Geology*, **82**, 287–297.

GAO, Z. Z. & DUAN, T. Z. 1985. Gravity-displaced deposits of deep-sea carbonates in West Hunan and East Guizhou. *Acta Sedimentologica Sinica*, **3**(3), 7–22. (in Chinese with English abstract).

GAO, Z. Z., ERIKSSON, K. A., HE, Y. B., LUO, S. S. & GUO, J. H. 1998. *Deep-Water Traction Current Deposits*. Science Press, Beijing, New York, 57–105.

GAO, Z. Z., LUO, S. S., HE, Y. B., ZHANG, J. S. & TANG, Z. J. 1995. The Middle Ordovician contourite on the west margin of Ordos. *Acta Sedimentologica Sinica*, **13**(4), 16–26. (in Chinese with English abstract).

GONTHIER, E., FAUGÈRES, J.-C. & STOW, D. A. V. 1984. Contourite facies of the Faro Drift, Gulf of Cadiz. *In*: STOW, D. A. V. & PIPER, D. J. W. (eds) *Fine-Grained Sediments: Deep-Water Processes and Facies*. Geological Society, London, Special Publications, **15**, 275–292.

KAHLER, G. & STOW, D. A. V. 1998. Turbidites and contourites of the Palaeogene Lefkara Formation, Southern Cyprus. *Sedimentary Geology*, **115**, 215–231.

MEI, Z. C., CHEN, J. W., LU, H. Y. & LI, W. H. 1982. Deep water carbonate debris flow in the Middle Ordovician Pingliang Formation of Fuping, Shaanxi. *Oil and Gas Geology*, **3**(1), 49–56. (in Chinese with English abstract).

PICKERING, K. T., HISCOTT, R. N. & HEIN, F. J. 1989. Contourite drifts. In: *Deep-Marine Environments: Clastic Sedimentation and Tectonics*. Unwin Hyman, London, 219–245.

SHENG, X. F. 1986. Problems on the classification of Ordovician System of China. *Bulletin of the Institute of Geology*, CAGS, **15**, 26–36. (in Chinese).

STOW, D. A. V. 1994. Deep-sea processes of sediment transport and deposition. *In*: PYE K. (ed.) *Sediment Transport and Depositional Processes*. Blackwell, Oxford, 257–291.

STOW, D. A. V. & PIPER, D. J. W. 1984. Deep-water fine-grained sediments: Facies Models. *In*: STOW, D. A. V. & PIPER, D. J. W. (eds) *Fine-Grained Sediments: Deep-Water Processes and Facies*. Geological Society, London, Special Publications, **15**, 611–645.

STOW, D. A. V., FAUGÈRES, J.-C. & GONTHIER, E. 1986. Facies distribution and textural variation in Faro Drift contourites; velocity fluctuation and drift growth. *Marine Geology*, **72**, 71–100.

STOW, D. A. V., FAUGÈRES, J.-C., VIANA, A. R. & GONTHIER, E. 1998. Fossil contourites: a critical review. *Sedimentary Geology*, **115**, 3–31.

STOW, D. A. V., READING, H. G. & COLLINSON, J. C. 1996. Deep seas. *In*: READING, H. G. (ed.) *Sedimentary Environments: Processes, Facies and Stratigraphy*. Blackwell, Oxford, 395–453.

WORSLY, T. R., NANCE, D. & MOODY, J. B. 1984. Sea Level and Plate Dynamics. *In*: HAQ, B. U. & MILLIMAN, J. D. (eds) *Marine Geology and Oceanography of Arabian Sea and Coastal Pakistan.* VNR and SAE, 233–251.

Fossil contourites: type example from an Oligocene palaeoslope system, Cyprus

DORRIK A. V. STOW[1], GISELA KAHLER[2] & MIKE REEDER[3]

[1]*SOES-SOC, Southampton University, Waterfront Campus, Southampton SO14 3ZH, UK (e-mail: davs@soc.soton.ac.uk)*
[2]*20 Malmesbury Road, Shirley, Southampton, UK*
[3]*Gaffney, Cline and Associates, Bentley Hall, Blacknest, Alton, Hampshire GU34 4PU, UK*

Abstract: As part of a wider programme on the recognition and decoding of contourite sediments, we propose the adoption of part of the Lefkara Formation on Cyprus as the first in a series of type examples of fossil contourites in ancient series exposed on land. A clear case can be made for interpreting these mainly Oligocene age sediments as carbonate-rich contourites. They combine faint structures due to current action with pervasive bioturbation, typically show cyclic grain size alternation, and have compositional differences from the more easily identifiable calciturbidites. One particularly diagnostic contourite facies is a lenticular, thin-bedded calcarenite–cacisiltite from which the finer bioclastic material and terrigenous clays have been winnowed away. This facies is a well developed condensed unit, which occurs at the same stratigraphic interval throughout Cyprus. Variations in thickness and rates of sedimentation of the contourite units, as well as the presence of widespread hiatuses and alongslope palaeocurrent trends are consistent with drift development on a lower slope apron in the closing Tethys ocean.

Much as geologists would like there to be a simple way to recognize contourites in the field, this is not the case. The processes of contourite accumulation are complex and the resultant features are a subtle combination of different depositional influences. Whereas this is also true of many deep-water facies, turbidites are more commonly easily recognized. It is partly the frustration of being unable to easily identify contourites, coupled with early errors in identification of contourites on land, that has led to 'great controversy over the recognition and interpretation of fossil contourites exposed in ancient series on land' (Stow *et al.* 1998).

In this recent critical review of fossil contourites (Stow *et al.* 1998), only very few ancient examples are accepted as being closely comparable to modern contourites, for which our knowledge is much more extensive. One of those referred to is part of the Palaeogene Lefkara Formation on Cyprus, first outlined by Kahler & Stow (1998). A similar conclusion was reached by Pickering *et al.* (1989), who listed only one example as acceptable – the Cretaceous Talme Yafe Formation in Israel (Bein & Weiler 1976).

The aim of this paper is to document clearly the case for interpreting the mainly Oligocene age carbonate sediments of parts of southern Cyprus as contourites, dominantly calcareous and some siliceous biogenic contourites. We therefore propose that this can be used as a type example of fossil contourites exposed on land.

In presenting the case, we follow the three stage approach to identification as first proposed by Lovell & Stow (1981) and refined by Stow *et al.* (1998). Simply stated, the following criteria must be met:

(1) *Small-scale (outcrop or core):* do the sediments in question have the full range of facies characteristics typical of known modern contourites?
(2) *Medium-scale (formation, region):* do regional trends, sediment distribution and geometry, presence of hiatuses or condensed sequences, and other features, support a bottom current influence?
(3) *Large-scale (system or continent):* is the palaeoceanographic setting as deduced from independent evidence compatible with bottom current activity?

Geological setting

Between Maastrichtian and early Miocene, deep sea sediments of the Lefkara Formation were deposited in the area that is now Cyprus over newly formed (late Cretaceous) ocean crust (Moores & Vine 1971; Robertson 1990). After an initial period dominated by slow pelagic sedimentation, there was a gradual increase in the influx of turbidites derived from the north during the early and middle Eocene. This was followed by a return to slow pelagic deposition and then, during the latest Eocene to early Miocene, by the influence of bottom currents (Kahler 1994; Kahler & Stow 1998). The depositional setting evolved through this period from a relatively deep oceanic basin to a carbonate slope apron system.

Sedimentation was strongly influenced by tectonic activity, including uplift of the Kyrenia Range in northern Cyprus and continued uplift of the Troodos ophiolite core in central Cyprus, as well as the onset of subduction to the south of the island. The palaeoceanographic setting in a closing Tethys seaway is believed to have been a further influential factor, particularly with regard to intensification of bottom current flow.

The Lefkara Formation can be divided into four lithological units (Fig. 1): the Lower Marl, the Chalk and Chert, the Chalk, and the Upper Marl units. Contourites are recognised in the upper parts of the Chalk Unit and throughout the Upper Marls.

Evidence at outcrop (small-scale)

Of the six localities studied in detail by Kahler (1994), the Lymbia Motorway section presented the best evidence for fossil contourites, as described by Kahler & Stow (1998) (Fig. 2). However, the state of this exposure has deteriorated with time. Further detailed investigation has revealed an even better type locality in the Upper Marls (or Upper Marls equivalent) on the western margin of Pissouri Basin near Petra Tou Romiou, in the section referred to as lenticular/fissile micrites by Stow *et al.* (1995). These two sections are documented here as good candidates for type contourite localities (Figs 3 and 4).

Sediment facies and structures

The first appearance of contourites in the Lymbia section is in the upper part of the Chalk Unit. This mainly comprises massive chalks of presumed pelagic origin, but with parts showing faint laminae together with micro-burrowing. Laminae are often isolated and indistinct and no clear structural sequences exist. Higher up, some of the bedded chalks are calcarenitic in grade due to a greater proportion of foraminifers. The overlying Upper

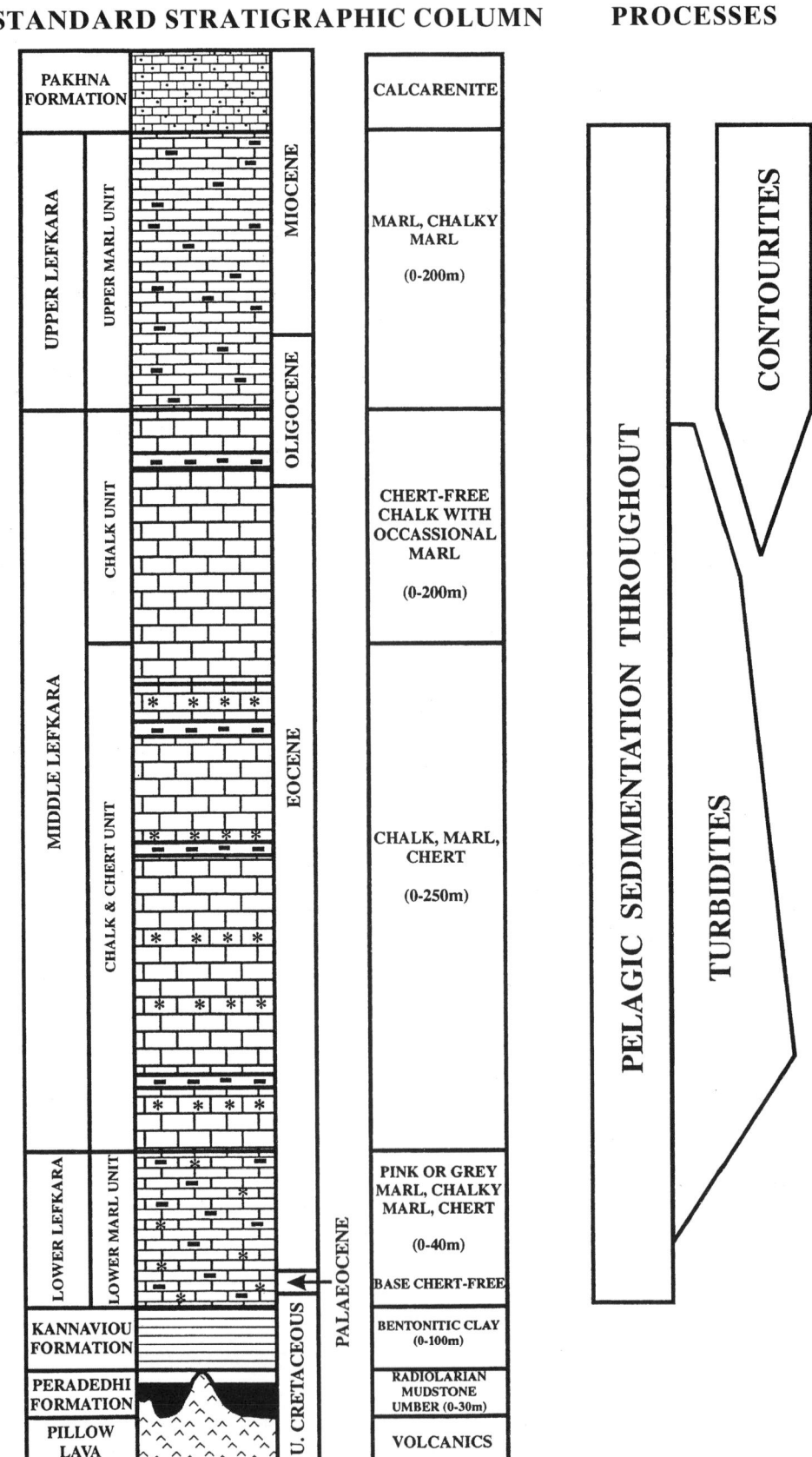

Fig. 1. Lithostratigraphic column for the Lefkara Formation in Cyprus, showing age, lithological units, thickness variation and principal depositional processes.

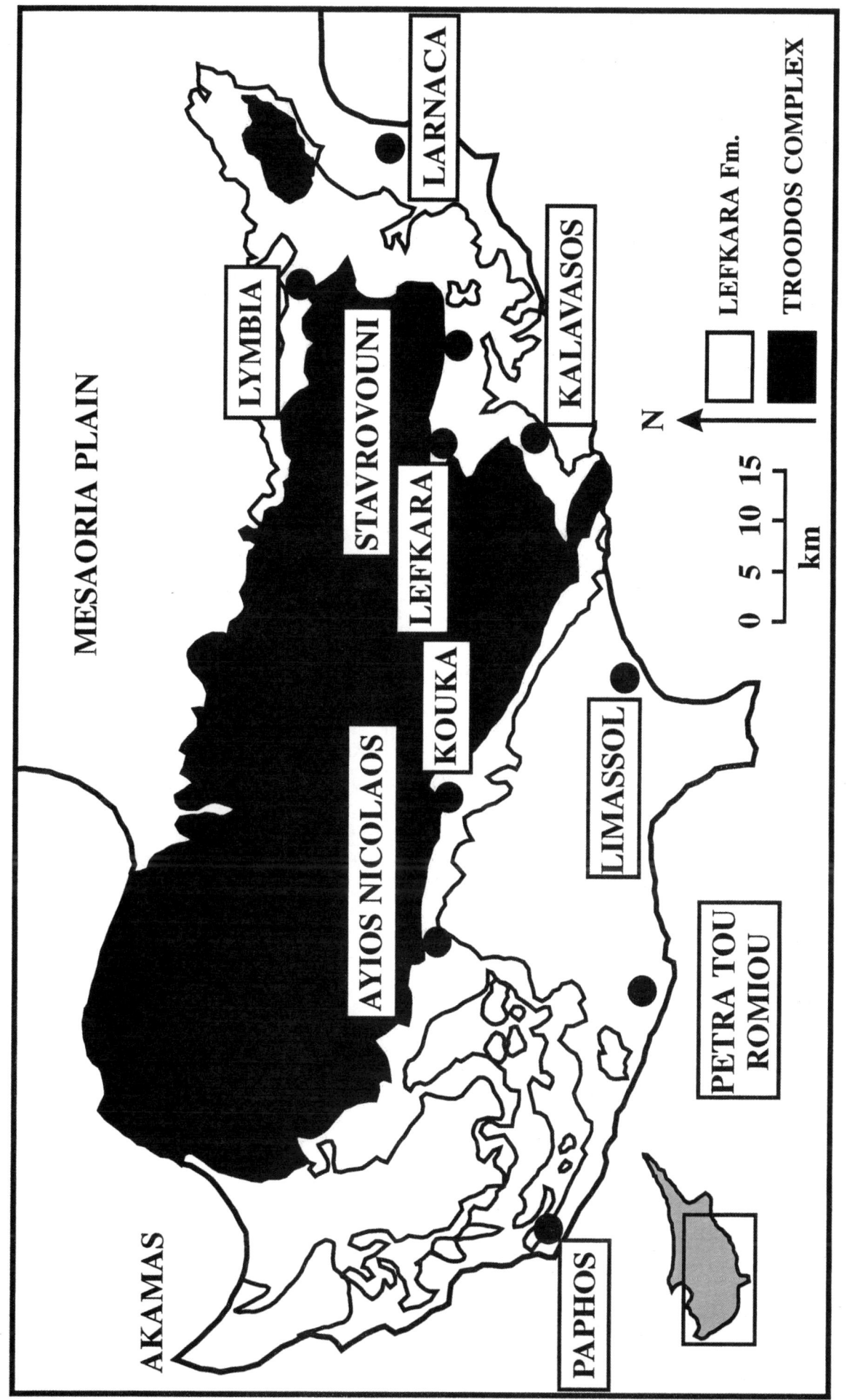

Fig. 2. Simplified geological map of southern Cyprus showing the location of sections studied in detail, as well as principal towns along the south coast.

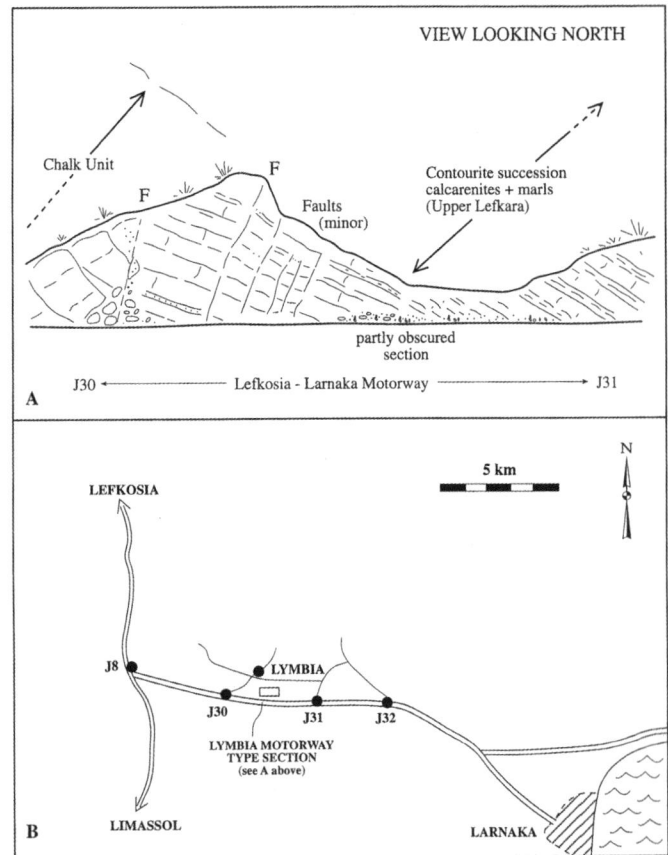

Fig. 3. Locality map and outcrop sketch for Lymbia fossil contourite type section.

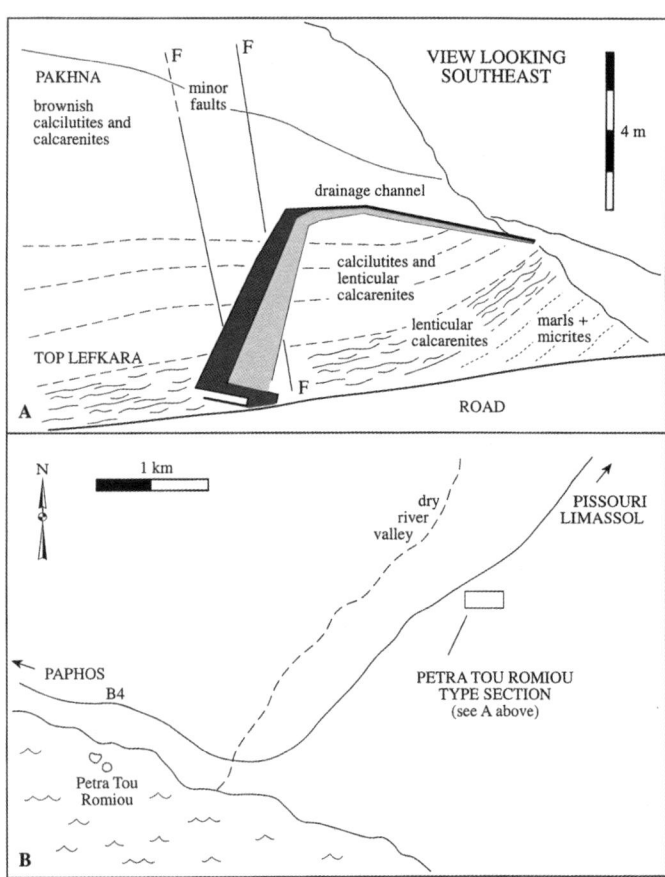

Fig. 4. Locality map and outcrop sketch for Petra Tou Romiou fossil contourite type section.

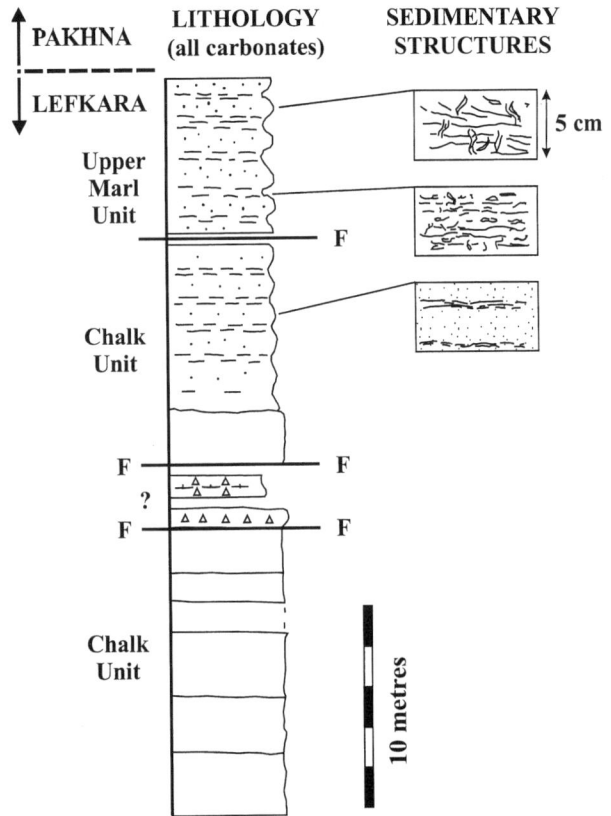

Fig. 5. Logged section of upper Lefkara Formation at Lymbia. The upper part of the Chalk Unit and the Upper Marls Unit show regular calcarenite–calcilutite grain size cyclicity, together with faint indications of lamination and intense bioturbation. Lenticular, thin-bedded calcarenites become more common upwards. This section is now badly weathered.

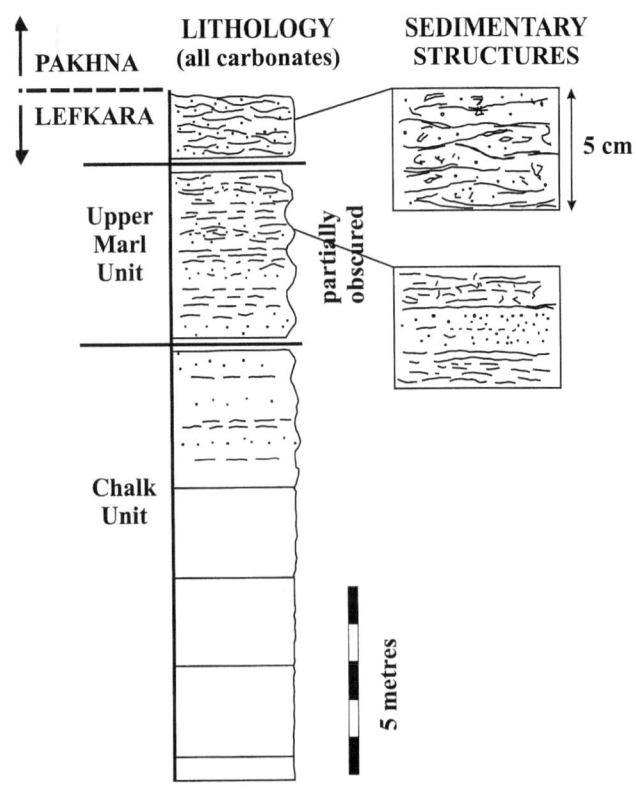

Fig. 6. Logged section of upper Lefkara Formation at Petra Tou Romiou. The upper part of the Chalk Unit and the Upper Marls Unit show regular calcarenite–calcilutite grain size cyclicity, together with faint indications of lamination and intense bioturbation. Lenticular, thin-bedded calcarenites are more common than in the Lymbia section, particularly towards the top of the Upper Marls unit.

Marl equivalent comprises interbedded marly units (a few dm thick) and chalk or calcarenite units, which contain parallel laminae, cross-laminae and strong bioturbation throughout. No grading or turbidite-like sequences are present, but the cyclic alternation of coarser-grained calcarenites and finer marly calcilutites could be interpreted in terms of the standard contourite model (Stow *et al.* 1986). These features are illustrated in Figures 5 and 7.

The contourite influence at Petra Tou Romiou also begins towards the top of the Chalk Unit with faint lamination and burrowing that passes upwards into a more lenticular-bedded calcarenitic chalk. This is overlain by a poorly exposed mud-rich unit of the Upper Marls followed by a relatively thin interval (< 5 m) of lenticular to fissile bedded micrites and calcarenites. These show faint parallel and cross-lamination in parts as well as bioturbation throughout. They are interpreted as calcilutitic to calcarenitic contourites formed beneath relatively strong bottom currents that winnowed away any finer grained material (Figs 6 and 7). This facies is particularly diagnostic at all the sections examined.

Microfacies

Microfacies analysis of samples from the Lymbia and Petra Tou Romiou sections reveal the sediments to be mainly packed biomicrites (40–60% bioclasts), with the bioclasts dominated by planktonic foraminifers, many of which are fragmented. Associated pelagic facies are mostly sparse biomicrites. The other areas with contouritic sediments identified by Kahler & Stow (1998) show that the contourite microfacies in fact ranges from sparse to packed biomicrites, presumably depending on the strength of the bottom currents involved. In the section at Lefkara, some of the contourites show silicification due to dissolution of radiolarians, together with micro-cross lamination and burrowing.

Fig. 7. Outcrop photographs of fossil contourites from Lymbia (**A–C**) and Petra Tou Romiou (**D–F**) type sections. (**A**) Alternating calcilutite-calcarenite layers reflect cyclic grain-size changes in contourites. Width of section 1.5 m. (**B**) Detail from A showing lenticular, thin-bedded calcisiltites and more weathered calcilutites. Width of section 25 cm. (**C**) Detail from A showing calcilutites with bioturbation and indistinct lamination. Width of section 15 cm. (**D**) Part of the upper lenticular, thin-bedded contourite unit, showing some grain-size cyclicity. Width of section 1.0 m. (**E**) Detail from D showing lenticular, thin-bedded, calcarenite contourites. Width of section 20 cm. (**F**) Detail from D showing lenticular, thin-bedded, calcarenite contourites. Width of section 15 cm.

Fig. 7(B).

Fig. 7(C).

Fig. 7(D).

Textural attributes

Grain size is a less significant parameter in ancient carbonate-rich contourites, in particular, as the high degree of diagenetic alteration can mask any original textural distribution. In the case of the Lefkara contourites, compaction has reduced the sediment thickness by up to 2/3, porosity has markedly decreased, some weaker grains have fragmented and pressure dissolution has led to the partial disappearence of calcareous bioclasts. In contrast, neomorphic processes, mainly inside bioclast voids but partly also within the matrix, have increased grain sizes. SEM studies have shown the matrix to be dominantly 2–3 μm sized rhombohedra, with strongly altered coccolith relics occasionally still visible. Larger bioclasts are planktonic foraminifers and radiolarians 50–500 μm in size.

Certain gross characteristics do probably reflect the original grain size. The beds of lenticular and partly laminated calcarenites are most likely the result of bottom current winnowing, cleaning and sorting during deposition, as well as subsequent enhancement of coarse carbonate spar by diagenetic processes. The cylcic alternation of these beds or units with finer grained marly intervals may also reflect an original grain size oscillation due to fluctuation in current strength.

Composition

The sand fraction of the chalks and marls throughout the Lefkara Formation comprises planktonic foraminifers and radiolarians, with only rare benthics and lithoclasts. In the Chalk and Chert Unit, the once high proportion of radiolarians has been removed by post-depositional remobilization associated with chert diagenesis. In the Chalk Unit, radiolarians tend to dominate over foraminifers, whereas they are almost absent in all material from the Upper Marls. It is possible that these hydrodynamically light microfossils were selectively removed by more active bottom

Fig. 7(E).

Fig. 7(F).

Ma	Epoch		Ayios Nicolaos	Kouka	Kalavasos	Lefkara	Stavrovouni	Lymbia
16.6	MIOCENE	E					?	
17.6			H		?	H		? *
23.4	OLIGOCENE	L	H	H		H		H
28.2			H	H	H	H		H
31.6		M	H	H	H	H		H
34.0		E	H	H	*	H		
36.3	EOCENE			H		H	H	
37.3		L		H		H	H	
38.1				H		H	H	
40.2				H				
42.6				H				
43.0		M					fault	
46.0					* ?		?	?
49.0					*	?		
52.0								
53.4		E						
55.2								
56.1								
58.2	PALEO-CENE	L	?					
58.8								

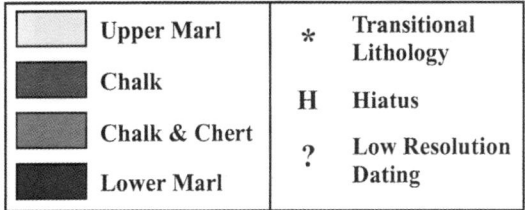

- Upper Marl
- Chalk
- Chalk & Chert
- Lower Marl
- * Transitional Lithology
- H Hiatus
- ? Low Resolution Dating

Fig. 8. Stratigraphic correlation of the four lithological units within the Lefkara Formation based on careful biostratigraphic study by Kahler (1994). Note the marked diachroneity of units and the timing and extent of hiatuses.

currents at this time, or that the bottom circulation affected water chemistry promoting their dissolution. Comparatively heavy benthic foraminifers are slightly enriched in the Upper Marls, perhaps as a winnowed residue, together with rather more transported allochems in the Lymbia section especially. There is also notable occurrence of bioclasts reworked from older strata.

In all the contourites studied, as well as in most of the rest of the Lefkara Formation, carbonate content is above 70%, with the exception of some of the diagenetic cherts. The increased clay content of the Upper Marls could be the result of three main factors: (a) an increased flux of terrigenous material in the bottom currents due to increased input from turbidity currents derived from the Kyrenia Range in the north; (b) selective removal of hydrodynamically light microfossils by moderately strong currents; and (c) preferential dissolution due to changes in water chemistry brought about by increased bottom currents. However, particularly strong currents seem to have removed most of the clays leaving only the heavier microfossils to form the calcarenitic contourites. Consistent with an increase in clay is an increase in detrital quartz content in the Upper Marls from the Lymbia section in particular.

Regional evidence (medium-scale)

Thickness variation

Due to lack of exposure and to local faulting, it is not everywhere possible to find a complete section of the Upper Marl unit. We have not yet, therefore, obtained a complete picture of thickness variation of this unit across Cyprus but note, from previous work as well as our own studies, that marked differences do occur, from 0–200 m. Assuming that at least some of this variation reflects original depositional thickness, it further supports the influence of bottom currents in shaping contourite drift sedimentation during this interval, notwithstanding the diachronous nature of the unit (Kahler & Stow 1998).

Maximum thicknesses of the Upper Marls occur on the southern flank of the Troodos Massif, for example in the Mandria–Kouka area, with lesser thickness at Kalavasos (20 m) and the thinnest development at Lymbia and Petra Tou Romiou (< about 10 m). These last are the two sites at which the bottom currents appear to have been strongest.

Sedimentation rates and hiatuses

Based on the detailed dating carried out by Kahler (1994), we can estimate the approximate rates of sedimentation for the contourite intervals of the Lefkara Formation. These vary from < 1 m Ma^{-1} to 17 m Ma^{-1} for the Chalk unit and from 1m Ma^{-1} to 13 m Ma^{-1} for the Upper Marls, all of which indicate slow to extremely slow oceanic accumulation. Clearly, the lower rates apply where the thicknesses are least and bottom current intensity was presumably stronger and hence non-deposition or erosion most likely.

Indeed, hiatuses are evident in all sections where closely spaced biostratigraphic sampling has been carried out (Fig. 8). These are most intense and widespread in the Oligocene, but also occur in the late Eocene and early Miocene, exactly the time period during which contourite accumulation occurs. They are not present during any of the rest of the Lefkara deposition.

Fig. 9. Examples of thin-bedded, lenticular calcarenite contourites from the topmost part of the Upper Marls lithostratigraphic unit within the Stavrovouni (**A**), Lefkara (**B**), and Kalvasos (**C**) sections.

Fig. 9(B).

Fig. 9(C).

Facies variation and paleocurrent direction

The contourite facies are clearly calcicontourites, ranging from calcilutites to calcarenites. A distinctive unit (5–15 m thick) of lenticular-bedded and laminated/bioturbated calacarenites occurs at most locations in a stratigraphically equivalent interval, immediately overlying the finer-grained, marl-dominated part of the Upper Marls unit (Fig. 9). We include this facies within the Upper Marls and suggest that it reflects a period of time when bottom currents were stronger over a broad area of deposition, leading to winnowing, non-deposition and development of a condensed sedimentary unit. Calcilutites dominate elsewhere and in other parts of the Upper Marls lithostratigraphic unit.

In general, the contourite influence has occurred subsequent to the main turbidite input. In the western sections examined, turbidites are overlain by pelagic chalks and then, near the top of the Chalk unit, bedded cherty intervals and chalk-marl cycles are interpreted as being influenced by bottom currents. In some of the eastern sections (e.g. at Lymbia and Lefkara), however, an interplay of both downslope and alongslope processes seems likely, with top-cut thin turbidites resulting from bottom current winnowing and erosion.

Whereas palaeocurrent directions for most of the turbidites are mainly NE–SW, consistent with derivation from the Kyrenia Range, Robertson (1976) noted atypical and highly variable current directions towards the top of the succession at Lefkara, some of which were aligned perpendicular to palaeoslope. This favours alongslope reworking of turbidite tops by bottom currents.

Palaeoceanographic setting (large-scale)

The Lefkara Formation as a whole was deposited in a closing small ocean basin, probably in a basin plain to distal slope-apron setting (Robertson *et al.* 1991; Kahler 1994; Kahler & Stow 1998). Palaeowater depth, based on sediment facies, microfossil assemblages and preservation, was in the range of 2000–3000 m. The nearest land lay to the north, now represented by the Kyrenia Range some 25–50 km distant, but which may have been at least twice as far away during the Oligocene. A schematic depositional setting for the late Oligocene to early Miocene is shown in Figure 10.

The nature of bottom circulation in the region during the Oligocene is not well known. However, this period witnessed both constriction of the main Tethyan surface flow from the Indian to the Atlantic Ocean and initiation of circum-Antarctic flow (Haq 1984) (Fig. 11). Whereas earlier in the Palaeogene, bottom waters were made up of warm high salinity water masses generated at low latitudes, by Oligocene time, deep cold water was also being generated at high northern and southern latitudes. For the Cyprus region, it seems most likely that continued constriction of the Tethys seaway led to an intensification of the westward directed Tethys current, which was therefore capable of influencing sedimentation at bathyal depths. Alternatively, there may have been a deep counterflow to the east. Widespread hiatuses have been noted in oceanic sediments worldwide, with maxima in the late Eocene, Oligocene and early Miocene. These are generally linked to climatic cooling, eustatic sealevel lowstand and enhanced bottom circulation (Miller *et al.* 1987).

Tectonic activity in the region, including submergence of the

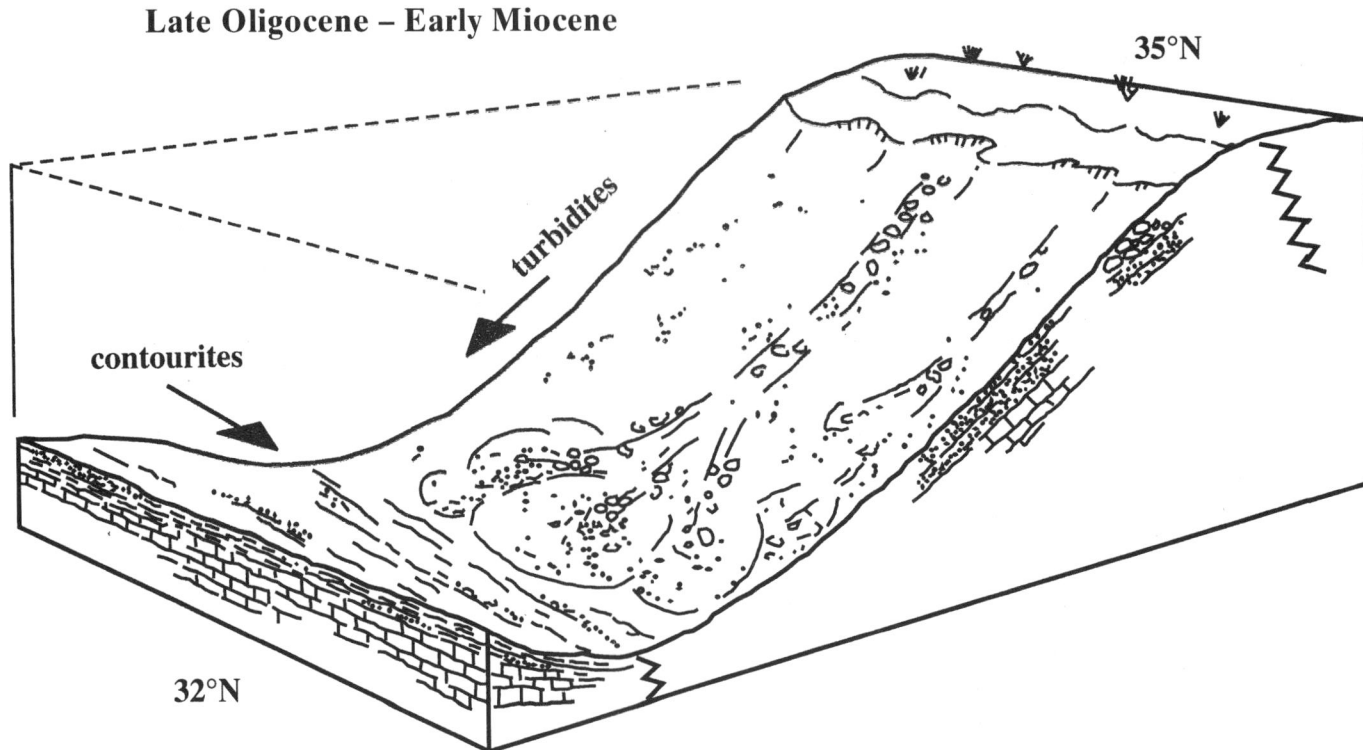

Fig. 10. Inferred depositional setting for the Lefkara Formation during the Oligocene acme of contourite sedimentation. The influence of turbidity current input was generally less at this time, although some distal turbidites are interbedded with the mainly contourite/pelagite section.

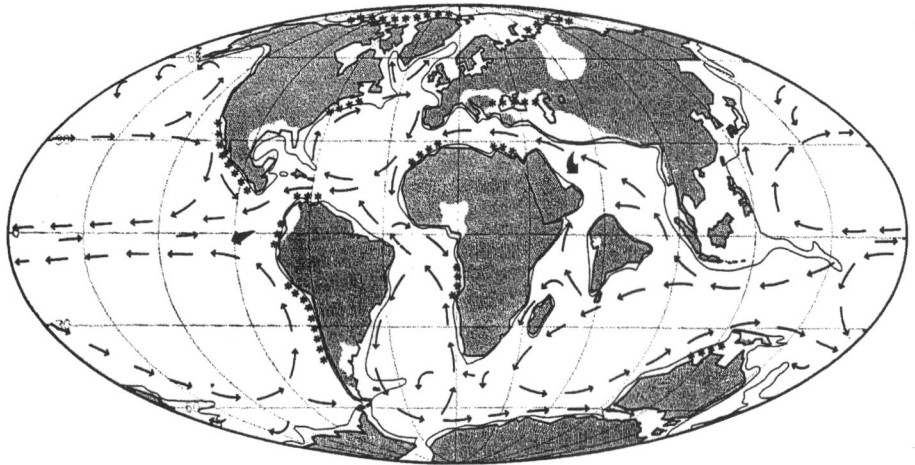

A. 60-58 Ma

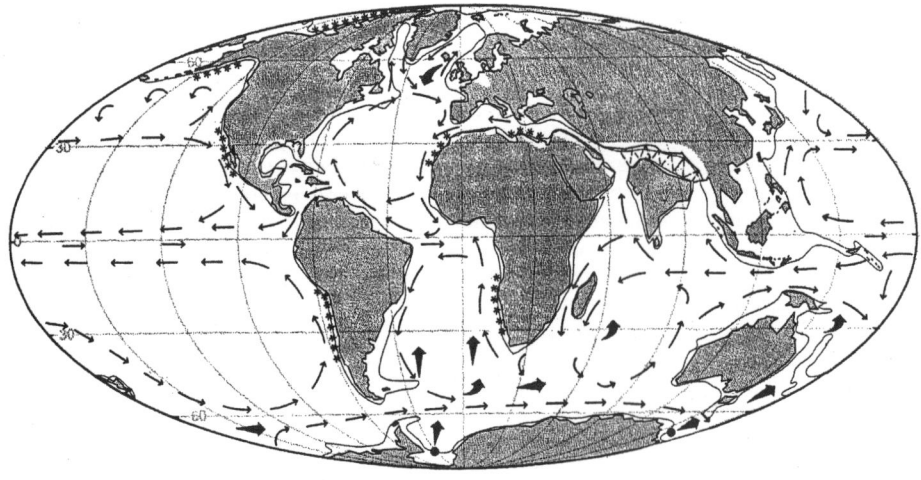

B. 31-30 Ma

Fig. 11. Palaeogeography, palaeoceans and inferred global palaeocirculation in: (**A**) the early late Paleocene (*c.* 60–58 Ma); and (**B**) the mid-Oligocene period (*c.* 31–30 Ma) (from Haq 1984).

Kyrenia Range, uplift of the Troodos terrane and subduction to the south of Cyprus may also have led to the local intensification of bottom currents due to increased slope gradients (Robertson *et al.* 1991).

Discussion and conclusions

There is considerable evidence that has allowed us to infer bottom current influence on sedimentation during accumulation of the upper parts of the Lefkara Formation in Cyprus. The principal lithostratigraphic units affected are the upper parts of the Chalk and the whole of the Upper Marls, ranging in age from late Eocene to early Miocene with an acme of influence during the Oligocene. A three stage approach to contourite identification has been followed.

(1) The sediments are carbonate-rich contourites that show generally faint structures (in parts only) indicative of current deposition, together with bioturbation throughout. Coarser grained calcarenites and finer grained calcilutites both occur, in some cases in cyclic alternation. Grain fragmentation and reworking of older bioclasts are evident. These features, as well as sediment composition, are notably different from those of Lefkara Formation turbidites, which mainly occur lower in the succession. They show close similarity with calcicontourites described from other 'confirmed' ancient series (Bein & Weiler 1976; Duan *et al.* 1993) as well as with the standard contourite facies model (Stow *et al.* 1986, 1996).

(2) There are marked variations in thickness, accumulation rates and facies of the principal contourite units across the region. Overall, very low sedimentation rates prevail (< 10 m Ma^{-1}), together with distinctive condensed intervals and widespread hiatuses. Locally, palaeocurrent measurements show alongslope reworking of the tops of turbidites. These features are compatible with contourite drift construction (Faugères *et al.* 1993) along the eastern and southern margins of the rising Troodos terrane, as well as with moat areas that experienced higher velocity bottom currents and more marked erosion.

(3) Deposition occurred on a carbonate slope-apron system in the closing Tethys seaway. The depositional setting and likelihood of bottom currents both independently support our case for contourites.

The contourite facies are very easy to overlook in the field. Firm identification necessitates a combination of evidence gained from further laboratory work and literature review, as presented in this paper. There is a clear need for type examples of fossil contourites exposed in ancient series on land, that are well documented and readily accessible for examination by geologists and oceanographers. We provisionally propose the adoption of these Oligocene examples from Cyprus, especially the sections at Lymbia and Petra Tou Romiou, as the first of a series of type samples to be established worldwide. At the same time, we recognize that further work is required on these examples as well as on comparable Oligocene successions elsewhere in the Tethyan region.

This work was initiated by GK as part of her doctoral thesis at Southampton University. Thanks are due to the many people involved in that effort, both at Southampton and in Cyprus. Dr C. Xenophontos and Y. Panayides are thanked for their invaluable assistance in the field. This paper forms part of the IGCP Project 432 (Contourite Watch) programme.

References

BEIN, A. & WEILER, Y. 1976. The Cretaceous Talme Yafe Formation: a contour-current shaped sedimentary prism of calcareous detritus at the continental margin of the Arabian Craton. *Sedimentology*, **23**, 511–532.

DUAN, T., GAO, Z., ZENG, Y. & STOW, D. A. V. 1993. A fossil carbonate contourite drift on the Lower Ordovician palaeocontinental margin of the middle Yangtze Terrane, Jiuxi, Northern Huan, Southern China. *Sedimentary Geology*, **82**, 271–284.

FAUGÈRES, J.-C., MEZERAIS, M.-L. & STOW, D. A. V. 1993. Contourite drift types and their distribution in the North and South Atlantic ocean basins. *Sedimentary Geology*, **82**, 189–206.

HAQ, B. U. 1984. Palaeoceanography: a synoptic view of 200 million years of ocean history. *In*: HAQ, B. U. & MILLIMAN, J. D. (eds) *Marine Geology and Oceanography of Arabian Sea and Coastal Pakistan*. Van Nostrand Reinhold Co., New York, 201–232.

KAHLER, G. 1994. *Stratigraphy and sedimentology of the Palaeogene Lefkara Formation, Cyprus*. Unpublished PhD thesis, University of Southampton, UK.

KAHLER, G. & STOW, D. A. V. 1998. Turbidites and contourites of the Paleogene Lefkara Formation, southern Cyprus. *Sedimentary Geology*, **115**, 215–231.

LOVELL, J. P. B. & STOW, D. A. V. 1981. Identification of ancient sandy contourites. *Geology*, **9**, 347–349.

MILLER, K. G., FAIRBANKS, R. G. & MOUNTAIN, G. S. 1987. Tertiary oxygen isotope synthesis, sea level history and continental margin erosion. *Paleoceanography*, **2**(1), 1–19.

MOORES, E. M. & VINE, F. J. 1971. The Troodos Massif, Cyprus and other ophiolites as oceanic crust: evolution and implications. *Philosophical Transactions of the Royal Society of London*, **268**, 443–466.

PICKERING, K. T., HISCOTT, R. N. & HINE, F. J. 1989. *Deep Marine Environments*. Unwin Hyman, London.

ROBERTSON, A. H. F. 1976. Pelagic chalks and calciturbidites from the Lower Tertiary of the Troodos Massif, Cyprus. *Journal of Sedimentary Petrology*, **46**, 1007–1016.

ROBERTSON, A. H. F. 1990. Tectonic evolution of Cyprus. *In*: MALPAS, J., MOORES, E. M., PANAYIOTOU, A. & XENOPHONTOS, C. (eds) *Ophiolites Oceanic Crust Analogues*. Proceedings of the Symposium 'Troodos 87', Geological Survey Department, Nicosia, Cyprus, 235–250.

ROBERTSON, A. H. F., EATON, S., FOLLOWS, E. J. & MCCALLUM, J. E. 1991. *The role of local tectonics versus global sea-level changes in the Neogene evolution of the Cyprus active margin*. International Association of Sedimentologists, Special Publications, **12**, 331–369.

STOW, D. A. V., BRAAKENBURG, N. E. & XENOPHONTOS, C. 1995. The Pissouri fan-delta complex, southwestern Cyprus. *Sedimentary Geology*, **98**, 245–262.

STOW, D. A. V., FAUGÈRES, J.-C. & GONTHIER, E. 1986. Facies distribution and drift growth during the late Quaternary, Faro Drift, Gulf of Cadiz. *Marine Geology*, **72**, 71–100.

STOW, D. A. V., FAUGÈRES, J.-C., VIANA, A. R. & GONTHIER, E. 1998. Fossil contourites: a critical review. *Sedimentary Geology*, **115**, 3–31.

STOW, D. A. V., READING, H. G. & COLLINSON, J. C. 1996. Deep Seas. *In*: READING, H. G. (ed.) *Sedimentary Environments*. Blackwell Scientific Publications, Oxford, 395–453.

Index

Page numbers in *italics* refer to Figures and page numbers in **bold** refer to Tables

AABW *see* Antarctic Bottom Water
AAIW *see* Antarctic Intermediate Water
abyssal currents 24, 26, 39, 46–47, 273
    zonation 36
abyssal plain 1, 11
abyssal sheet *11*, 11
ABW *see* Arctic Bottom Water
ACC *see* Antarctic Circumpolar Current
Adelaide Island 361
Adelie Coast *374*, *377*
Adventure Bank *172*, *174*, *178*, *181*, *183*, *187*
Agulhas Current 271, *273*, 281
Agulhas Plateau 271–288
    bathymetry *272*
    bedforms 273, *277*, 281, 285
    geological setting 271, *272*
    oceanographic setting 271, *272–273*
    seismic characteristics 273, *274–287*, 281, 285
    seismostratigraphic framework 272–273, *272*
Agulhas Return Current 271
Alagoas Basin *245*
Albacora Terrace 261
Alboran Sea 138, 142, 155–170
Almirante Camara canyon 249, *250*, 253, 259, 261, *261*
Almirante Passage 11
Alvarez Cabral channel 140, *143*, 144, 145, *147–148*, 150
Antarctic Bottom Water (AABW) 8, 11, 39–55, 209–225, 239–288, 305
    velocity 237, 374, 410
Antarctic Circumpolar Current (ACC) 9, 251, 273, 281, 293, 325–353, 385–407
Antarctic Coastal Current 305, 315, 317
Antarctic Intermediate Water (AAIW) 251–271, *273*, 287, *394*, *395*
Antarctic Peninsula Pacific Margin 135, 317, 353–371
    bathymetry 353–356, *356–357*
    geological setting 353, *354*
    oceanographic setting 353, *358*, 360
        current measurements *355*
    sediments
        core description and facies 361, 363, *367–368*
        cyclicity 361
        depositional model 364, 366, 370
        grain size analysis 363–364, *368*
        sedimentation rates 360
        supply and transport 364
    seismic characteristics
        reflection profiles *354*, *355*, *359–363*
        sub-bottom profiles 361, *364–365*
    stratigraphic context 358–360, *360*
        barium trace concentrations 358
        biostratigraphy 358
        magnetostratigraphy 358, 360, *367*
Antarctic Surface Water 290
Anvers Channel *354*, 355–356
Arctic Bottom Water (ABW) 9
Arctic Intermediate Water 57, 61
Argentinian Basin *11*, 11, 210, 335
ASW *see* Atlantic surface water
Atlantic Surface Water (ASW) 57, 59, 61, *142*, *156*
    velocity 155
Atlantic water (AW/AI) 138, *174*, *186–187*, 188, 192
    velocity 175
AW/AI *see* Atlantic water

Bahama Banks 39, 44, *45*, 50–54
Bahia Seamounts 239–241, *241*, *243–245*, 246
Balearic Abyssal Plain 175, *187*
Baltic Ice Lake (BIL) 121–122

Baltic Sea 121–136
    bathymetry *122*
    contourites *130–131*
    geological setting 121–122
    oceanographic setting 122, *123*, 132
        gateway 121–122, 129, *130*, 132
        hydrographic measurements 121–122, *124*, *126–9*
        nepheloid layers 122, 129
    sediment characteristics 132
        grain size analysis *125*
        transport rates 129
    seismic characteristics 129, *130–132*
Baltic Sea Research (BALTICA) 121
baroclinic current 271
Barra Fan 12, 65, 85–119
    bathymetry *99*, 101, *118*
    contourite sand sheet *116*, 116–118, *118*
    geological setting 85, 101
    oceanographic setting 85, *86*, 101, *102*
        bottom current influence 89
        current velocity 85, 101
        current winnowing 87, 89
        seabed topography *118*
    seafloor polishing and sand spillover 114–116, *117*
    sediment
        bedforms 93–96, 103–104, 109, *110–111*
        climate-controlled deposition 94
        depositional patterns 89
        facies 87, *87–90*, 89–90, 104, 109–111, *110–114*
        glaciomarine sediments *88*, 89
        grain size analyses 87, 89–90, *90*, *92*, 101, 109, *116*
        magnetic susceptibility 87, *89*, 89, *92*
        mudline v. sandline *116*, 118–119
    seismic characteristics 85, *86*, 90–96, 103–108
        3.5 kHz echocharacter 104, *105–108*
        seafloor morphology 103–104
    stratigraphy 101–102, *101–102*
        biostratigraphy 101
        carbon-14 90, 94, 96, 101
        ice-rafted debris, relative abundance 102
        lead-210 101
Bartolomeu Dias Planalto (sheeted drift) 137, *138*, 140, *143*, 144, 150
base-only contourites 83
Bellingshausen Abyssal Plain 355
Benthic Boundary Layer 2–3 *see also* HEBBLE
benthic storms 2, 9, 18–24, 237, *291*, *327*, 335
Bermuda Rise 44, 47, 50, *54*
Betic-Rif orogenic belt 137–138, *141*
Blake Outer Ridge 1, 50
Bode Verde fracture zone 239–240
bottom current 2, 7–20, 249–259
    bedforms 26–29, *29–32*, 35
    intensity 26, *29*, *214*, 273, 360, 428, 441
        depth dependency 36
    interaction with topography 201, 325
    flow lofting 188
    processes 39, 51
    velocity 18, 26 ,121
    winnowing 72, 188, 449 *see also* current winnowing
    see also abyssal currents; contour currents
bottom-current reworked turbidites 8, 16–17, 19, 236, 409, 418–424, 453
bottom water anoxia 188
Bounty Fan 12, 389
Bounty Trough (submarine channel system) 385, *386–387*, 389, 392, 404–405
Brazil Current 237, 249–270
Brazil Intermediate Counter Current *251–2*, 251, *257–258*, 263, *264*, *270*

Brazilian continental margin 8, 209–270
  bathymetry *241*
  geological setting 239–240
  oceanographic setting 240
  seismic characteristics *242*
    abyssal erosion 241
    sediment drifts 242, 246
  stratigraphic context 240
    biostratigraphy 240
    well-log control 240
  see also Columbia channel-levee system; Vema contourite fan
Broughton Gap 395–398
Bruce Bank 326
Buffon Channel *377, 379*
Burwood Bank 337, *338*, 339

Caicos Outer Ridge 39, *40*, *43*, *46*, *49*, 50, *54*
Campbell Plateau 385–388, 394, 405
Campos Basin 12, 239, 249–270
  bathymetry *251*, *264*
  geological setting 249, *256*
  oceanographic setting 249–251, *257*, *263–264*, 263
  sediments 253, *254*, 256–258, 263, *265–266*
    bedforms *255*, *262*, 265, *267–270*, 270
    sedimentation rate 259
  seismic characteristics 251, *253–254*, *264*, *266–269*
  seismic stratigraphy *269*
  stratigraphic setting 251, 253, *254*, 256
    biostratigraphy 258
    carbon-14 258
    highstand 259, *269*, 270
    isotope analyses 258
    lowstand 259, 270
canyon-channel drainage 378, *380*
Cape Hatteras 51
Capraia sill 191, *192–193*, 201
carbonate compensation depth 212, 224, 404
carbonate contourites 14, 16, 433–455
CDW *see* Circumpolar Deep Water
Ceuta Canyon 155–157, *156–157*, *160*, *162*, 165
Ceuta Drift 155–170
  bathymetry 155–156, *156–157*
  contourite recognition 165
  morphology 155–156, *158*
  sea-level changes 165
  sediment characteristics 158–159, *165–167*
  seismic characteristics 157–158, *159–164*, 168
  stratigraphic framework 156–157, *166*
    lowstand sequences 165, 168
channel-levee systems *329*, 353–384 *see also* Columbia channel-levee system; Vema contourite fan; Weddell Sea
channel-related drifts 11–12, *13*, 73–84, 171–189, 209–238, *243–245*, 246, 353–371, 385–407
Chatham-Kermadec Margin 12, *386*
Chatham Rise 385, *386–387*, 389, 392, 394–396, 405
Chesapeake Drift 50
Circumpolar Antarctic Current *see* Antarctic Circumpolar Current
Circumpolar Deep Water (CDW) 290, 325, 338, 350–353, 385, 394–395
  velocity 398
Coats Land 289, *289*, 305, *306–308*, 315
Columbia channel-levee system (Columbia fan-drift system) 223–238
  bathymetry 223–224, *224*
  geological setting 223
  oceanographic setting 223
    contour processes 234–235
    contourite-turbidite interactions 236–237
    turbidite processes 235–236
  sediments *228*, *230*, 230, 232–237
    bedforms 228
    grain size *231–232*, *234–235*
  seismic characteristics
    echofacies mapping *224*, 225, 228–230
    seismic reflection profiles 225, *226–227*
  stratigraphy 224
    biostratigraphy 224
    carbonate curves 224
    excess thorium-230 224
companion drift-fan 12
composite slope-front fan 12
condensed section 259, *348*, 440
confined drifts *11*, 12, 171–189, *244*, 246, 337–352
contour currents (contour following bottom currents) 1, 7–8, 228, *242*, 252, 271, *295*, 312, *369*, 373, 378, 381
contourite
  definition 1, 8
  examples 7
  facies model 14–18, 81, 85, 235, 435–439, 447–449, 455
  originator 1, 7
  recognition of 18–19, *19*
  sequences 16–18
    link with bottom current velocity 19
    link with palaeoclimate 18, 83–84
    link with palaeocirculation 18
    link with periodicity 19
  size *11*
  see also base-only contourites; contourite drift; fossil contourites; glacigenic contourites; lag contourites; laminated contourites; mid-only contourites; shallow-water contourites; top-only contourites; turbidite-contourite channel-levee system; turbidite-contourite channel system; turbidite-contourite continuum; turbidite-contourite sheets; volcaniclastic contourites
contourite drifts 8, 10–12, *11 see also* channel-related drifts; confined drifts; elongate mounded drifts; infill drifts; modified drift-turbidite systems; patch drifts; plastered drifts; separated drifts; sheeted drifts
contourite facies *see* carbonate contourites; manganiferous contourites; mud-clast contourites; muddy contourites; sandy contourites; sandy muddy contourites; siliciclastic contourites; silty contourites; silty muddy contourites; silty sandy contourites;
contourite fan *11*, 209–222
contourite-turbidite process interaction 236–237, 239, *243*, 246, 373
Coriolis Force 9, 11, 122, 138, 150, 175, 192, 219, 230–232, 240, 245–246, 252–253, 273, *282*, 317, 321, 364, 398, 404
Corsica Basin 191
Corsica Channel 191–208
  bathymetry *192–193*, 195
  drift
    history and palaeoceanographic changes 207
    morphology and topography *193*, *198–200*, 201, 207
  geological setting 191–192
  oceanographic setting 192, 195
  sediments 201, *204*
    bedforms 191
  seismic characteristics 191, 194–206
    cyclicity 197
  stratigraphic context 194, *198*, *200*, *202–203*
    biostratigraphy 195, *196*
    carbon–14 195, *196*
    highstand system tracts 191
    lowstand system tracts 191, 198, *203*, 207
    magnetic-susceptibility patterns 195, *196*
crag and tails features *30–31*, 109, *111*
Crary Fan 306–310, 312, 315, 317, 321
Crary Trough 305, *307*
current-controlled deposition 39, 44, 49
current direction indicators 29
current reworking 1, 26
current shear turbulence 374
current winnowing 401

Danish Straits 121–122
debris flows 24–26, 73–86, 94, 103, 249, 308–322, *329*, 413
debrites 12, 83, 90–96, 104, 171, 177, 180
Deep Atlantic Surficial Water *142*
deep Mediterranean Sea Water 11

deep-sea channels 11, 353–371
Deep Sea Drilling Program (DSDP) 2, 8, 14 *see also* Ocean Drilling Program
    Leg 1 2
    Leg 28 *46*, 48, *51*
    Leg 35 2
    Leg 35 Site 325 *354*, *359–360*, 361
    Leg 72 Site 515 210, 212, *216*, 225
    Site 23 240, 242
    Site 24 240, 242
    Site 268 373, 378, 381–382
    Site 269 373
    Site 278 385, *386*
deep-sea fans 378
Deep Western Boundary Current (DWBC) 21–23, 36, 385–407
definitions 7–8
detached drift *11*, 325–336, 353–371
Diego Cao channel 140, 144–145, 150
discontinuities 157, *161*, *211*, 220, 225, 281
Donegal fan 85, *86*, 99, 101
down-slope processes *25*, 39, 51
Drake Passage 251, 325–353, 405
drifts (drift deposits) 8, 19, *45*, 50 *see also* contourite drifts; detached drifts; moat drifts; mixed drift system; sediment drifts; shallow-water drifts; skin drifts
Dronning Maud Land 289, *289*, 305–306, *306–308*, 312–5, 317
DSDP *see* Deep Sea drilling Programme
DWBC *see* Deep Western Boundary Current

East Chatham Gap 396, *397*, 398
East Falkland Trough sediment drift 339–342, *345*, 350
Eastern New Zealand Oceanic Sedimentary System (ENZOSS) 404–405
echo-character mapping 2, 104–108, 212, *215–217*, *224*, 225, 228–230
echosounder 13, 129
    2–7 kHz CHIRP SONAR 195, *198*, *200*, 202
    3.5 kHz 2, 25–6, *100*, 101, 195, *198*, *224*
    4 kHz 2
    7.5 kHz 101
    12 kHz 1–2, *264*
    20–30 kHz ELAC 121, *130–131*
    300 j UNIBOOM 195
eddies 263, *264 see also* intrahalocline eddies
eddy kinetic energy 9, *10*, 21, *23*, 36, 335, 385
Elba Canyon *192–193*, 195, 201
Elba Channel *192–193*, 195
Elba Ridge 191–195, 198, 200, *202–203*, 206
elongate mounded drifts 11–13, 39–72, 137–170, 191–208, 239–248, 271–288, 305–336, 433–442
Emerald Basin 385, *386*, 405
ENAM2 research program 85
Enderby Basin 306, 322
Ewing, John 1–2
Ewing, Maurice 'Doc' 1–2
expansion zone 265, *266–8*, *270*, 270
Explora Escarpment 306
extended turbidite bodies *11*

Faroe Bank 67
Faroe Bank Channel 73, *73*, 75
Faroe Shelf 65, *66*, 73, *76*
Faroe–Shetland Basin 73
Faroe–Shetland Channel 11, 18, 65–85
    bathymetry 66, *73*, 75–76
    deep-water
        circulation 65, *66*
        sedimentation 75
    drift
        contourite deposition 71–72
        style 71
        geological setting 74–75
            glacigenic mass flow slope-apron deposits *73*
        oceanographic setting 65–66, *66*, 75
            bottom currents, influence on sedimentation 83
            current velocity 65–66, 75
            linear gullies and fans *73*, 83
            palaeotopographic high 67, 71
        sediment characteristics 68–71, 78, 80–83
            cyclicity 81, 83–84
            facies 68–69, 71, 78, *80–82*
            sedimentation rates 83
        seismic characteristics 66–67, *67–68*, 76, *78–79*
        stratigraphy *69*, 76, *77*, *81*
            biostratigraphy 76, 83
            carbon-14 76
            oxygen isotopes 76, 83
Falkland Current 338
Falkland Plateau 271, 337–352
Falkland Trough 12, *326*, 335, 337–352
    bathymetry 337, 339, *339*
    oceanographic setting *338*, 338–339
    sediment characteristics
        bedforms 341, 350
        biogenic-terrigenous ratios 338
        core sedimentology 338, 346, *347–348*, 349
        grain size analyses 346
        sedimentation rates *347*
    seismic characteristics
        3.5 kHz acoustic profiles 337, *337*, 339, 341–342, *342–345*
        seismic reflection profiles 337, *337*, 339, *339–341*, 341
        side-scan sonar *337*, 338, 342, *346*, 349
    stratigraphy 346, *347–348*
        barium/aluminium ratio *348*, 349
        biostratigraphy *347–348*, 349
    see also East Falkland Trough sediment drift; West Falkland Trough sediment drift
fan-drift deposit 12, *223*, 241
Faro-Albufereira drift complex 12, 137–154
    bathymetry *138*, 140, *141*, *143*
    climate changes 153
    Corvina borehole *140*, 141
    drift origin and development 150, 152
    geological setting 137–138, *139–142*
    oceanographic setting
        bottom current velocity 153
        circulation pattern 138
        palaeoceanography 140
    sediment characteristics
        bedforms 145, *149–50*
        contourite facies cyclicity 153
        facies 145–147, *149–151*
        sedimentation rates and hiatuses 147, 148, 150
    seismic characteristics *139–140*, 141, 144–145, 150
        facies cyclicity 146–147, *152*, 152–153
    stratigraphic context *140–141*, 141, *144–145*
        biostratigraphy 141, *145*
        oxygen isotopes 141, *145*, *151*, 153
Faro-Cadiz Planalto (sheeted drift) 137, *138*, 140, *143*
Faro Canyon 140, *143*, 144, 152
Feni Drift 96
Filchner Ice Shelf 289, 305–306, *306*
fossil contourites 8, 18–19, 239, 409–455 *see also* Lefkara, Miura, Jiuxi, Kazusa, Pingliang fossil drift systems
    identification criteria 443
Fram Strait 58, 61, 74–75
funnelling zone (acceleration zone) 265, *267–268*, *270*, 270

GAOR *see* Greater Antilles Outer Ridge
gateways 8–11, 74–75, 121–132, 171–189, 271, 287, 405, 435, 439–441 *see also* Baltic Sea gateway; Shetland-Rockall gateway; Straits of Gibraltar; Yap gateway
Geike Bulge *99*, 101
Gela Basin *172*, *174*, 175, 176, *178*, *184*, 187
geostrophic currents 1, 85, 116–117, 191, 230, *264*, 271, 321, 353, *395*
Gibralter Canyon 155
glacial-interglacial cycles 12, 290, 317, 331–336, 353–371, 405, 421–422

Glacial North Atlantic Deep Water *258*, 258–259
Glacial North Atlantic Intermediate Water 259
glacigenic contourites 73–84
glacio-eustatic loading 306
glacio-eustatic sea-level changes 157, 207
glaciomarine dumpstone 109, *112*, *114–115*
glaciomarine hemipelagites 12, 81, 83, 90, 94, 109, *112*
global cycles *242*, 246
GLORIA side-scan sonar 26, *337*, 342, *346*, 361
Golo Fan 191, *192*
Gotland Basin Experiment (GOBEX) 121
Gotland Deep 127, 132
gravity flows 39, 44, 53, 94–96, 187–192, 207, *242*, 258–270, 434–435
Greater Antilles Outer Ridge (GAOR) 39–55
    bathymetry *40*, *46*
    development 50–54
    drift position and growth, controls on 53
    oceanographic setting
        current measurements 47, *49*, 51, 53
        hydrography and abyssal currents 44, 47, *49*, *51*, 53
        layered valleys *44–45*, 53
        Outer Ridge sediments 39, 44
    sediment
        accumulation rates 44
        bedforms 53
        chlorite abundance *47*, 47, 51
        current-controlled deposition 49–50, 54, *54*
        properties 48–49, 51, 53
        suspended sediment 47–48, 53
        transport 51
    seismic character 51
    seismic stratigraphy *41–46*, 48–50
Greenland-Scotland Ridge 74–75
growth faults *309*, 310
Grussai Canyon 249
Guadilquivir channel 140
Gulf of Cadiz 11–12, 137–154
Gulf Stream 9, 21, *22*, 36
Gweilo Sequence 90–91

halocline 122, *125*, 129
halothermal circulation 246
Hatteras Abyssal Plain 39, *44*, 47, 49–50, *53*
HEBBLE (High Energy Benthic Boundary Layer Experiment) 2–3, 8, 21–38
Hebridean Margin (Hebrides Shelf) 12, 65–67, *99*, 117
Hebrides Slope *100*, 104
    current velocity 85
Hebrides Terrace Seamount *86*, *99*, 101
Heezen, Bruce 1, 7
hemipelagites 8, 26, *28*, 69–72, 87–90, 94, 104, 109, *112–115*, 165, 171–188, 225, 256–258, 289–303, *308*, 317, 321, *328*, *330*, 349, 353, 359–360, 364, 373, 378, 397–401, 410–418, 421–427, 434–435
    bottom-current affected 8, 409
    horseshoe patterns *28*
    see also glaciomarine hemipelagites; volcaniclastic hemipelagites
Hero Fracture Zone 355
hiatuses 11, 18–19, 132, 141, 152, 224, 271–273, 358–361, 392, 440, 451–453
High Energy Benthic Boundary Layer Experiment *see* HEBBLE
Hikurangi fan-drift system 223
Hikurangi Plateau 385, 392, 394, *398*, 398, *401*
Hikurangi Trough (submarine channel system) 12, 385, 392, *394*, 398, *400–402*, 404–405
Hollister, Charles 1–5, 7
Horseshoe Abyssal Plain *141–142*
Hunan Province, China 433–442
hybrid turbidite-contourite facies 16
hydrocarbon exploration and production 7–8, 65, 73, 129, 137, 220, 246, 249, 270

ice-rafted material 16, 78–83, 293, 317–321, 333–336, 350, 358–367, 376
Ice Shelf Water (ISW) 305, 321–322
ichnofacies 19, *266*

infill drifts *11*, 12, 85–119, 137–154, 249–270
intercalated turbidite-contourite bodies *11*, 305–323
International Geological Correlation Programme Project 432 7, 19
intrahalocline eddies 129
Ionian Abyssal Plain *172*, *174*, 175, 180, *182*, *186–187*
Ionian Sea *173*
ISW *see* Ice Shelf Water
Itapemirim Canyon 249, *250*, 261, *261*, *264–265*, 267
Izu-Bonin Arc and Trench system 409–410, *412*, *416*, 418, 421, *422*, 430

Jane Bank *292*
Jane Basin *289–290*, 290, *292*, 293, *296*, *298*
Jean Charcot Seamount *209*, *214*, 217
Jiuxi Drift fossil contourite, China 18, 433–442
    drift growth 441
    evidence for contourites 439
    facies 435, *436–437*
        contourite sequences 438, *438–439*
        cyclicity 439, 441
        distribution 439, *440*
        sedimentation rates 439
    geological setting 433–434, *433–434*
    palaeobathymetry 435
    palaeotopography 434–435
    sediment
        composition 438, *437–438*
        structures 435, *437*
        texture 435, *437–438*
    sequence interpretation *437*, 441
    stratigraphic setting 433–434
Judd Deeps *73*, *75*, 76
Jussieu Channel *377*, *379*

Kane Gap 11
Kazusa forearc basin fossil contourite, SE Japan 19, 421–432
    geological setting 421, *422*
    palaeoceanographic setting 421
    sediments
        bedforms 428, 430
        depositional setting 421
        facies 422–425, *422–423*, *425–431*, 430
        heavy minerals 428, *430*
        Kazusa Group 421
        palaeocurrents 428, *429*, 430
    sequence-stratigraphic setting 421–422, 428
        highstand sequences 422, *422*, *429*
        lowstand sequences 422, *429*
        transgressive sequences 422, *422*, *429*
Kermadec Ridge 385, *386*, 394
Kermadec Trench 385, *386–387*, 395, 398–405
Kuroshio Current, Japan 9, 410, *411*, 421–432
Kyrenia Range, Cyprus 443, 451, 453, 455

Labrador Continental Margin 364
Labrador Sea Water (LSW) 65, 85, *86*
lag contourites 11
laminated contourite, high-energy 416
Large Igneous Provinces 271
Laurentian Channel 21
Laurentian Fan 12, *22*, 25, 32
Lavrentyev Gap 396, *397*
Lefkara Formation fossil contourite, Cyprus 18, 443–455
    geological setting 443, *444–445*, 454
    Lymbia fossil contourite 443, *446–448*, 451, 453, 455
    palaeoceanographic setting 453–455
    palaeocurrent direction 453
    Petra Tou Romiou fossil contourite 443, *446*, *449–450*, 451, 455
    sediment
        composition 449, 451
        facies 443, *446*, 447, *452–453*, 453
        sedimentation rates 451, 455
        structures *446*, 447

textural attributes 449
thickness variation 451
stratigraphy *451*
hiatuses 451, 453
lowstand 453
Levantine Basin 175
Levantine Sea 9
Levantine Sea Intermediate Water (LIW) 174–207
Ligurian Basin 191–192, *192*, 207
Linosa Trough (Linosa Graben) *172*, *174*, 175, *178*, *187*
listric faulting *309*, 310, *402*
LIW *see* Levantine Sea Intermediate Water
Lofoten Basin 57
Lofoten Drift 57–64
bathymetry *57–58*
drift
axis 59
relief 59
glaciomarine environment 61
oceanographic setting 57–59
continental slope current velocity 57
sea-floor gradient 59
slope gradient 57
surface-water winnowing 61
palaeoenvironmental model *63*
sediment
facies 57, 60–61
flux 59, 61
properties 61
source area 62
seismic facies 57, 59–60
stratigraphic context 59, 61, *61*
Louisville Seamount Chain 385, *386–387*, 389, 395
LSW *see* Labrador Sea Water

Macquarie Ridge 385, *386–387*, 405
Maghrebian Thrust (Maghrebian Arc) *173*, 175
Malta Escarpment (Sicily-Malta Escarpment) *172*, *174*, 175, 177, *182*, *185*, *187*
Malta-Medina Channel (Medina Graben) *172*, *174*, 175, *182*
Malta Sill *172*, 175, 188
Malta Trough (Malta Graben) 171–189
Malvinas Outer Basin 337–339, *338*, 350
manganiferous contourites 15–16, 219
Marshall Paraconformity 392, *394*, 395, 399, 404–405
mass-movement deposits 25, 85, *187*, 239, *261*, *269*, 312, 322, 326, *342*, *357*, *369*, *376*, *397*, 401
Medina Escarpment 175, *182*
Medina Wrench Zone (Straits of Sicily Rift Zone) *173*, 175
Mediterranean Deep Water 65, *142*, 159, 192
velocity 138, 165
Mediterranean Intermediate Water (MIW) *142*
velocity 138
Mediterranean Outflow Water (MOW) 137–138, *142*, 145, 150, 152
Mediterranean Sea 9, 101, 155–208
Mediterranean Undercurrent *156*
velocity 155
megaturbidite *187*, 188
Meiji drift 12
meltwater overflow plumes 89, 364
Messina Plate *173*, 175, *187*
Messinian salinity crisis 140, 157, 188, 195
mid-only contourites 71
Milankovitch cyclicity 18
Misaki Formation fossil contourite, Japan 18
Miura-Boso forearc basin, SE Japan 409–419
geological setting 409–410, *409–410*, *413*
oceanographic setting 410, *411*
current measurements *411–412*
palaeobathymetry 410, 418
sediments
bedforms 418
contourite characteristics 413–414, 416–418
facies 410, 413, *414–415*, *417*
Hasse (Hatsuse) Formation 410, *412*, 417
grain size cyclicity 416, 418
Kazusa Group 410
Misaki Formation facies 410, *412–413*, 413–414, *416*, 417–418
Miura Group 409–410, *417*, *422*
turbidites 418
stratigraphic context 412–413, *412*
biostratigraphy 410, 412
palaeomagnetism 412, 418
radiometric age *413*
MIW *see* Mediterranean Intermediate Water
mixed drift system 8, 188, 223–238
moat drifts *269*, *387*, 389, *390*, *398*
moats 12, 15–16, 58, 66–68, 71, 177, 193–206, 246, 271, *280*, *328*, 389–392, 399
Modified Atlantic water (MAW) *see* Atlantic Water
modified drift-turbidite systems *11*, 12, 19, 21–38, 85–119, 239–248, 353–371, 385–407 *see also* companion drift-fan; composite slope-front fan; fan-drift deposit
Moroccan Slope 155–170
morphological features 13–14
mound and tail features 29, *30–31*
MOW *see* Mediterranean Outflow Water
mud-clast contourites 11, 14, 16, 81, 214–216, 219, 235
mud diapirism 155, 159, *168–169*
muddy contourites 21–38, 68–69, 81, 87–115, 121–136, 145–153, 207–238, 290, *294*, *337*, 349, 364
mudline *116*, 118–119
multibeam bathymetry 155, *157*

NADW *see* North Atlantic Deep Water
Nares Abyssal Plain 39–53
NASW *see* North Atlantic Surficial Water
Navidad sill 39, *46*, 47, *49*
NEAW *see* Northeast Atlantic Water
nepheloid layer 3, 9, 47–50, 89, 207, 364–376, 381, 388, 398
nephelometer profiles *46*, *47*, *50*, 121
New England Seamounts 24
New Zealand Drifts 11, 385–407
bedforms *393*, *396*, 398–401, *406*
Campbell 'Skin' Drift 387–389, 405
Chatham Deep Drift *387*, 389, 392, *392–393*, *398*
sedimentation rate 389
Chatham Terrace Drifts *387*, 392, *393*
Emerald Basin Drifts 385, *387–388*, 387
lowstands 387
geological setting 385, *398*
Hikurangi Fan-Drift
bathymetry 398–389
geological setting 398, 404, 405
oceanographic setting *387*, 398, 401, 405
sediments *398*, 401, 404–406
seismic characteristics 399–403
Louisville Moat Drift *387*, 389, *390*, *398*
North Bounty Drifts *387*, 389, *389*
North Chatham Drift 392
bathymetry 395
geological setting 392, 394, *397*, 405
oceanographic setting *386–387*, 394–395, 405
sediments 396, 398, *398*, 405
seismic stratigraphy 395–396, *396*, 399
oceanographic setting 385, *386–387*
Rekohu Drift 392, *394*, *398*, *400–401*, 405
sedimentation rate 405
North Atlantic Deep Water (NADW) 44, 47, *50*, 65, 85–102, 251–264, 271, 394–395
velocity 21
North Atlantic Surficial Water (NASW) *142*
North Pacific Deep Water (NPDW) 410
North Scotia Ridge 325–326, *328*, 336, 337–352
North Sea 121, *122*, 122, *123*
North Sea Bottom Water *125*

North Tula Channel *365*
Northeast Atlantic Water (NEAW) 65
Northwest Atlantic Mid-Ocean Channel 241
Norwegian Current 57–58
Norwegian-Greenland Sea 57–59, 61–62
Norwegian Sea Deep Water (NSDW) 71, 85, *86*, *102*
  velocity 65
Norwegian Sea Overflow Water (NSOW) 21, *22*, 36, 44, *76*
  velocity 75
Nova Scotian Rise 8, 21–38, 44
  contourite fan associations *27*, *28*
  oceanographic setting 26
    current speeds 36
    down-slope processes 26, *27*
    relationship to glacial events 26, 35, 37
  sediment
    accumulation 26
    bedforms 26
    flux 26
    properties 29, 32, *33*, 36–37
    sedimentation rate 34
  stratigraphy *34*
    dating evidence 26
  turbidite *34*
NPDW *see* North Pacific Deep Water
NSDW *see* Norwegian Sea Deep Water
NSOW *see* Norwegian Sea Overflow Water

ocean-climate link 7
Ocean Drilling Program (ODP) *see also* Deep Sea Drilling Program
  Leg 119 380
  Leg 178 Site 1095 *354*, 358, *359–360*, 382
  Leg 178 Site 1096 *354*, 358, *359–360*, 382
  Leg 178 Site 1101 *354*, 358, *359–360*, 382
  Leg 181 392, 404
  Leg 188 Site 1165 382
  Site 693 307–308
  Site 697 290, *292*, 293
  Site 1121 *386*, *388*, 388, 405
  Site 1122 *386*, 405
  Site 1123 *386*, 395–6, *397*
  Site 1124 *386*, 392, *394*, 395–396, 399, 401, 404–405
Okinoyama Bank 409, *409*
olistostromes 25–26
Ordos Platform, China 434–435, *434*
Oyashio Current 410, *411*, 421, *422*

palaeocanyon 160, *162*, 165
palaeoceanography 404, 418, 443, *454*
palaeocirculation 2, 8, 18–19, 258, 285, 439, *454*
palaeogeography 2, 7–8, 353, 358, 422, *434*, *454*
palaeoproductivity 331, *348*
Palmer Archipeligo 355
Palmer Channel *354*, 355–356
Palmer Deep Sea Fan 355
Pantellaria Trough (Pantelleria Graben) *172*, *174*, 175, 177, *178*, 180, *183–187*
patch-drifts *11*, 65–72, 104, 129, 177, *181 see also* slope (patch) sheets
Peach Slide *86*, 90, *96*, 101, *103*, 109
pelagic sedimentation *444*, 447, 453, *454*
Pernambuco Abyssal Plain 240
Pernambuco Contourite Fan 241, *244*, 246
Pernambuco Seachannel 239–241, *243–245*, 246
photographs, seabed 2, 13, 26, 28–32, 36, 47, 53, 70–72, 104, 109–117, 121, 132–135, 137, *166–167*, 293, 298, 315, 332, 353, *404*
Pingliang Drift fossil contourite, China 18, 433–442
  contourites, evidence for 439
  drift growth 441
  facies 435, *436–437*, 438
    contourite sequences 438–439, *439*
    cyclicity 439
    distribution of sequences 439, *440*

    sedimentation rates 439
  geological setting 434, *433–434*
  palaeobathymetry 435
  palaeotopography 435
  sediment
    composition 438, *437–438*
    structures 435, *437*
    texture 435, *437–438*
  sequence interpretation *437*, 441
    highstand 441
  stratigraphic setting 434
plastered drifts 201, *290*, 290, *294*, 325–352, 394 *see also* slope (plastered) sheets
plume, turbid *264*
plumites 364, *369*
Portimao Canyon 140, *143*, 144, 152
Powell Basin *289–290*, *292*, 293, *298*
Prydz Bay 378, 380, 382
Puerto Rico Trench 39, *40*, 44, 47, 49–54

Rangitata (Mount Curl) tephra 395
Rapuhia Scarp 395, 398, *401–402*, 404, *406*
Renaud Channel *354*, 356
reversed lee-wave model 96
Rio Grande Rise 209, *209*, *214*
Rockall Basin (Trough) 9, 11, 18, 65–72, 85–97, 99–119
  bathymetry 66
  drift
    contourite deposition 71–72
    style 71
  geological setting
    glacigenic slope apron 67
    glaciomarine origin 68
  oceanographic setting 65–66, *66*
  sediment characteristics 68–71
    bedforms 67, 71
    facies 68–69, 71
  seismic characteristics 66–67, *67–68*
  stratigraphy *69*
Rockall Plateau 65, *66*, *86*
Ronne Ice Shelf 289, *289*, 305–306, *306*
Rosemary Bank *99*
Ross Sea water 374, 380, 394

SACW *see* South Atlantic Central Water
Sagami Trench 409–410, *412*, 430
salt diapirism 21, 24–26, 138, *142*, 144, 249
sandline *116*, 118–119
sandy contourites 68–119, 153, 220, 259, *268*, 290–293, *337*, 349–351, *414–415*, 421–430
sandy muddy contourites *337*
Sao Francisco Deep Sea Fan 239, 246
Sao Paulo Abyssal Gap (Sao Paulo Channel) 209–212, *214*, *217*, 219
Sao Paulo Plateau 249, *251*, 251, *261*
Sao Tome Canyon 250, 259, 261, *261–262*, 264–265, *267*, *270*
sapropels, cyclic deposition 207
Scotia Sea 289, 325–337, *338*
  geological setting 325
    channel levee system *329*
  oceanographic setting 325, *326*
    bottom-current flow *327*, 334–335
    glacial-interglacial ACC flow 335–336
  sediment characteristics
    bedforms *332*, 335
    contourite identification *328–330*, 334
    core description and facies 332–333
    cyclicity, biogenic-terrigenous 332–333, *333*
    grain-size data 334, *334–335*, 336
    sedimentation rate 331
  seismic characteristics
    3.5 kHz profiles 325, *329–330*
    reflection profiles 325, *328*

stratigraphic context 326
  barium stratigraphy 331, *331*
  biostratigraphy 331, *331*
  oxygen isotope stages 331, *331*
  thorium dating 331, *331*
Scotia Sea Bottom Water 338
Scotland–Iceland–Greenland topographic barrier 9
sea-level falls 246, 251
seafloor polishing *105*, 114–115, *117*, 263
seamount *45*, *209*, 223, 239, *356–358*, 385–404 *see also* Bahia Seamounts; Hebrides Terrace Seamount; Jean Charcot Seamount; Louisville Seamount Chain; New England Seamounts; Vitoria Trindade ridge
seaward-dipping reflectors *243, 245*
sediment drifts 7–20 *see also* drifts, contourite drifts, and the various other drift types indexed
  distal 39–55
sediment facies 14–16, *16, 18*
sediment transport 1, 9
seismic attribute analysis 13
seismic features 8, *13, 25*
  depositional seismic units scale 12–13
  drift scale 12–13
  seismic facies 13, *14*, 18–19
seismic processing 239
seismic reflection profiling 1, 10, 24–25, 41–51, 157, *178*, 191, *328*, 353
  3.5 kHz acoustic profiles 57–58, *60*, 137, 140–141, 144, 171, 176–178, *181–184, 187*, 209, 223, 249, *250, 255–256*, 258, *265–268*, 290, *294, 297, 314–316, 329–330*, 337, *339*, 341–345, 389–393, *399, 402*
  airgun profiles 76, *78*, 141, *159–164*, 239, 290, 337, 373, *396, 399*
  analogue sparker 57–60, 76, *78*, 141, *146*, 176, 178, 194–195, *198, 200, 202–206*
  boomer survey 67, *68*, 194
  deep tow 13, 25
  deep tow boomer 85, *91–94*
  deep tow sparker *78*
  multichannel profiles 48, 57–58, *60*, *145*, 249, *252–253*, 272, *297*, 308–309, *311–313*, 337, 339–341, 359–361, 373, *379, 381, 388*
  Parasound profiles 310, 313–315
  single-channel seismic profile *296, 389–391, 393–394, 400–401, 403*
  sleeve gun profile 85, *95*, 157
  water-gun 209, 223, 225, 361–363
  see also echocharacter mapping; echosounder; sidescan sonar; sub-bottom profiler; swath bathymetric mapping
seismic stratigraphy *24*, 24–25, 36, 157, 224, 373
separated drifts *11*, 57–64, 198–203, 251, 259, *290*
Sergipe Basin 240, *243*, 246
Sergipe Fracture Zone 239–240, *244*
Shag Rocks passage 337–339, *338*, 341, *342*, 343, *344*, 349
shallow-water contourites 8, 16, 115, 121–136
shallow-water drifts 10, 19, 121–136
sheeted drifts 10–13, 19–38, 65–119, 137–154, 171–208, *242*, 246, 385 *see also* Bartolomeu Dias Planalto; Faro-Cadiz Planalto
Shetland-Rockall gateway 65
Sicily Gateway 9, 171–189, 191
  bathymetry *172*, 175, *176, 181–184, 187*
  geological framework 171, *173*, 175, *187*
  oceanographic setting *174*, 175
    bottom currents 171, *174*
  sediment characteristics
    facies associations and depositional processes 180, 185, *187*
    physiographic regions and facies *172, 176*, 177, 179–180, *181–184*
    sedimentation rates and hiatuses *181–184*, 185, 187, 188
  seismic characteristics 176–177, *181–184*
  sequence stratigraphy
    highstand system tracts 188
    lowstand system tracts 185
  stratigraphic context 175–176, *177*
    biostratigraphy 176, *187*
    carbon-14 176
    oxygen isotope176, *177*
    sapropel chronology176, *177, 187*
    tephra chronology176, *177, 187*

  see also Straits of Sicily
Sicily Sill *172*, 188
Sicily-Tunisian Platform 171–189
side-scan sonar 2, 13, 25–26, 32, 72, *103*, 137, 150, 249–259, *262*, 337–349, 361
silica diagenesis front 341
siliciclastic contourites 14–16, *17*, 19
sill 122, *182–183*, 188–192, 207, 305–306, 338 *see also* Capraia sill; Malta Sill; Navidad Sill; Sicily Sill
silty contourites 15, *87*
silty muddy contourites 68–69, 81, 87–94, 145, *210*, 233, 263, *292*, 409
silty sandy contourites 94, 96, 109–115, 145, *290, 292*
Silver Abyssal Plain 39–54
Skerki Channel *172, 181, 186*
skin drift 389
slope-apron system 85, 104
slope (patch) sheets *11*
slope (plastered) sheets 11, *11*
slump 12, 308–309, 322, 326, *329*
Sohm Abyssal Plain 24, 26
Solander Fan 385, *387*, 387
Solander Trough (submarine channel system) 385, *386–387*, 404–405
South Agulhas Basin *272*
South Atlantic Central Water (SACW) 251, *252*, *257–258*, 263 *see also* Brazil Intermediate Counter Current
South Brazilian Basin 209–239
South Sandwich Trench *338*, 338
South Scotia Ridge 289, *289*, 305, *306*, 325
South Shetland Trench 355
Southern Ocean bottom circulation 271
Southern Ocean Water (SOW) *258*, 258–259
spillover *105*, 115–117, 249, 263, 265, 270, 404
stacked channel deposits *24*
Straits of Gibraltar 10, 138, *142*, 153, 155–159, *162*
Straits of Sicily 171
sub-bottom profiler
  2–8 kHz Geo-Chirp 121, 129
  3.5 kHz 26, *27, 28*, 337, *337*
  TOPAS profiles 353, 361, *364–365*
submarine canyons 249
Sula Sgeir Fan 65, 85, *86*, 99
Sumba drift 12
Suruga Trench 410, *412*, 430
swath bathymetric mapping 13, 101, 137, 293, *295*, 353, 355

Tabajara Canyon 249
Talme Yafe Formation fossil contourite, Israel 18, 443
Tasman Rise 404–405
tempestites 441
tephra 182–188, 392–404, 412, *422 see also* Rangitata (Mount Curl) tephra
terminology 8
Tethys Seaway 443, 453, 455
thermohaline circulation 7–9, 12, 175, 239–240, 246, 249, 290, 321, 350, 385, 430, 440
top-only contourites 70–71, *81*, 82, 84
top-truncated turbidites *223, 232*, 236–237
Triffid Channel, E. Canada 32, *33*, 36
Troodos ophiolite 443, *445*, 451, 455
trough-mouth fan 99
Tuff 412, *413*
Tula Fracture Zone 355
turbidite channel system 353
turbidite-contourite channel-levee system 223–238, 359
turbidite-contourite channel system 137–154
turbidite-contourite continuum 421, *422–423*, 431
turbidite-contourite sheets 225
turbidite pond 396
turbidites 1, 7–8, 12, 14, 16, 19, 25–26, 73–85, 87–96, 146–153, 171–192, 223–238, 290–293, 305–322, 353–369, 385–430, *440–444 see also* bottom-current reworked turbidites; megaturbidite; top-truncated turbidites
Tyrrhenian Sea *173*, 175, 191–208

unconformities 24, 141, *164*, *168*, 195, 212, 220, 240–242, 251, 272, *313*, 374, 376, 378, 380, *388*, 392, *400* *see also* disconformities; hiatuses
Upper Circumpolar Deep Water *see* Antarctic Circumpolar Current

Valerie Passage *386*, 389, 392–395, *397*, 398
Vema Channel 11, 209–219, 230, 233, 241
Vema contourite fan 209–222, 224, 239
   bathymetry *209–210*, 210
   climate 216
   drift morphology *217*
   oceanographic setting 209–210, *214*
   sediments *219–221*
     bedforms 212
     current control on 219
     facies 212, *214*, 214–215
     palaeocurrent variation 215
     sedimentation rate 216
   seismic characteristics
     echofacies mapping 212, *215–217*
     seismic reflection profiles 210–213
   stratigraphy 210
     biostratigraphy 210
     carbonate curves 210, *219–221*
     oxygen isotopes 216, *219–220*
     thorium-230 210, *219–221*
Vema Gap 39–54
Vestkapp, Antarctica 306, *307–308*,
Vitoria Trindade ridge (Vitoria Trindade lineament or Seamounts) 209, 223–225, 228, 236–237
volcaniclastic contourites 14, 409
volcaniclastic hemipelagites 409
volcanism 271, 305–404, 412–414, 421–422, *444 see* tuffs

watermass exchange 271
WBUC *see* Western Boundary Undercurrent
Weddell Abyssal Plain 306
Weddell Gyre 289–323, 325
   velocity 290, *291*, 306, 336
Weddell Sea 8, 11, 289–323, 331
   bathymetry 306–307
   geological setting 289–290, 305
     channel-levee system 307–312, 315, 321
   oceanographic setting 290–291, *291*, 305–306, *306*
     bottom-water flow history 293, 302
     current measurements 306, *308*
     hydrodynamic interpretation 321–322
   sediment characteristics
     bedforms *294–297*
     contourite-turbidite facies *305*, 317–321
     core description and facies 293, *299–302*, 315, 317
     cyclicity, glacial-interglacial 293
     debris-flow facies *312*, 317, *319–320*
     hemipelagic facies *308*, 317, *318–321*, 321
     grain size data 293, *299*, *301–302*, 317, *318–321*, 321
     sedimentation rates 290, 293, 307–315, 317, 321–322
     shelf facies 317, *318–319*
   seismic characteristics
     3.5kHz/Parasound mapping 308, *310*, 313, 315, *314–316*
     reflection profiles 307–313
     seismic data 293, *294–297*
   stratigraphic context 290, 293, 307–308
     biogenic barium 290, 293
     biostratigraphy 290
     carbon-14 *318*, 320
     excess thorium-230 290, 293
     magnetostratigraphy 290
     oxygen isotopes 307, *319*
Weddell Sea Bottom Water 290, 305, 338, *338*
Weddell Sea Deep Water 290, 325, 338–339, 350, 394, *395*
Wegener Canyon 306, *307*, *313*
West Agulhas Basin *272*, 285
West Antarctic Peninsula Margin 12, 16
West Falkland Trough sediment drift 339–350
West Shetland Shelf 65, *66*, 73, *73*, *76*, 76
West Shetland Slope *73*, 75, 83
Western Alboran Basin 155–157, 159, *163–164*, 165, *169*
western boundary Brazil Current 249
Western Boundary Undercurrent (WBUC) 1–2, *22*, 36, 39–57, 430
Wilkes land, Antarctica 373–384
   bathymetry 374, *374*, *377*
   geological setting 373–374
   oceanographic setting 374
     current measurements 374, *375–376*
   sediment characteristics 378, 380
     cyclic supply 376
     sedimentation rates 378, 380, 382
   seismic characteristics *375*, *379*, *381*, 381
     contourite sedimentation evidence 378
     turbidite sedimentation evidence 378
   seismic stratigraphic framework 376, 378
   stratigraphy
     biostratigraphy 378
     magnetostratigraphy 380
wind-driven circulation 8–9, *9*, 349, 385
Woods Hole Oceanographic Institution 2–3
Wust, George 1, 7
Wyville-Thomson Ridge 65–85, *86*, *99*, *102*

Xanuen Bank 155
X-radiographs *35*, *61–62*, *69*, *166–167*, *218*, *233*, *317–321*, *333*, *351*, 361, *367*, 378, 380, *382*, 404, *406*

Yangtze Terrane 433
Yap gateway 410